GENERA OF THE POTTIACEAE: MOSSES OF HARSH ENVIRONMENTS

GENERA OF THE POTTIACEAE: MOSSES OF HARSH ENVIRONMENTS

By Richard H. Zander

Illustrated by Patricia M. Eckel

BULLETIN OF THE BUFFALO SOCIETY OF NATURAL SCIENCES
VOL. 32 BUFFALO, N.Y. 1993

The Buffalo Society of Natural Sciences
Buffalo Museum of Science
1020 Humboldt Parkway
Buffalo, New York 14211-1293 USA

Bulletin of the Buffalo Society of Natural Sciences
Volume 32

Published December 1993
(Volume 33 was published in 1988.)

Cover illustration: see page 128.

Library of Congress Cataloging-in-Publication Data

Zander, Richard Henry. 1941–
Genera of the Pottiaceae: mosses of harsh environments / by Richard H. Zander; illustrated by Patricia M. Eckel.
p. cm. — (Bulletin of the Buffalo Society of Natural Sciences. ISSN 0096-4131; v. 32)
Includes bibliographical references and index.
ISBN 0-944032-51-6 (hardcover)
1. Pottiaceae—Classification. 2. Pottiaceae—Identification. I. Title. II. Series.
OK539.P8Z36 1993
588'.2—dc20 93-37139 CIP

ISSN 0096-4131
ISBN 0-944032-51-6

The Bulletin of the Buffalo Society of Natural Sciences first appeared in 1873. Please contact the Publications Sales Division, Buffalo Museum of Science, 1020 Humboldt Parkway, Buffalo, NY 14211 USA, for back issues and price lists.

TABLE OF CONTENTS

PREFACE

This study is a taxonomic treatment of the moss family Pottiaceae at the genus level, and is summarized in the Table of Contents. It was begun in 1984 but involves more than two decades of specialization on the part of the author. The genera are viewed in the traditional sense as groups of species sharing either a unique combination of several character states (or a majority of such states) or standing out in sharing (or having, in the case of the several monotypic genera) one or more unusual features. Subfamilies and tribes, however, are distinguished with the aid of cladistic analysis, answering the question: "What would be the best suprageneric classification assuming the family as presently conceived was monophyletic and its genera as now circumscribed were each monophyletic?" Certain generic concepts have been considerably emended to increase the chances that future study of the individual genera will find them at least in large part monophyletic.

As critics have noted, the original Hennigian cladistic method of evolutionary analysis is overly simplistic in that evolution as we understand it is incompletely modeled, e.g. there are no provisions for dealing with introgression, convergence, fossils or anagenetic change. Thus, no matter how excellent the data and "rigorous" the analysis, the simple model produces a simplified and surely at least partially incorrect evolutionary hypothesis. On the other hand, the present popularity of phylogenetic cladistics among taxonomists reflects a general apprehension that, for all its faults, it remains an excellent method of analyzing large taxonomic data sets, and it is possible to modify the method to some extent when we actually have information about the aspects unmodeled. It succeeds in using large data sets to parsimoniously group taxa, however simplistically, on the basis of shared, recently evolved traits, a valuable first step in evaluating evolutionary relationships. That the large-scale results are commonly similar to evolutionary relationships suggested by past systematic study, as is the case in this work, is further support for an assumption that the higher resolution possible in cladistic studies also reflects valid hypotheses of relationships. The details of standard cladistics-generated evolutionary hypotheses, however, should be examined with other methods that reflect better theory.

In practice, the present suprageneric classification is based totally on a cladistic analysis. That convergence in sporophyte features is a reality in many genera of the Pottiaceae is obvious to the specialist, and is well illustrated in this text. In view of the above reservations, I have had no qualms about differentially weighting the characters used in the analysis against possible convergence through reduction. Also, when discussing genera, I point out those relationships between certain taxa that seem plausible, probable and due to convergence, even if such relationships are contrary to the results of the cladistic analysis. Nor do I spurn the past work of non-cladists. In pursuit of a phylogenetic classification, a revisionist can with traditional methods build on what is already known, or start afresh with a new, much heralded but immature analytic technique, or, as here, in the face of a taxonomic group impossible to comprehend in only one lifetime, necessarily attempt both.

Considerable new nomenclature is presented in this treatment: four new subfamilies, four new tribes, two new genera, two new subgenera, and two new sections, and one new variety. There are also 37 new names for taxa and 317 new combinations. The total number of recognized taxa in the Pottiaceae now includes seven subfamilies, six tribes, 76 genera, 1457 species, 31 subspecies, 536 varieties, 339 formae and seven subformae.

Part of the cost of research and publication has been underwritten by the National Science Foundation, Grant BSR-8314843. I thank H. Bischler and R. A. Pursell for help in obtaining specimens from the Bizot herbarium at PC, William D. Reese for comments on portions of the text draft, and J. Cargill, R. Halling, W. Margadant and G. Zijlstra for correspondence regarding nomenclature and associated problems. J.-P. Frahm and P. J. Lightowlers made valuable comments on slide mounting media. W. R. Buck, R. Magill and W. Reese gave helpful observations on various taxonomic and nomenclatural topics. D. Wagner discussed terminology for directions of helical twist in plant parts. W. Gall, R. Vineyard, E. DeLuna, B. Mishler and B. Tan provided valuable comments on the phylogenetic analysis, but the execution, results and interpretation are entirely my own. C. Delgadillo, D. Griffin, B. Murray, D. Norris, P. Sollman, and B. Tan tried out the keys and provided many worthwhile observations. M. R. Crosby commented on details of the myriad nomenclatural changes. Correspondence and discussion with more than a few bryologists over the years helped firm the taxonomy developed here, and I thank these colleagues. Many workers have openly shared with me their opinions on aspects of taxonomy of the Pottiaceae; I have, to the best of my recollection, given their ideas, when reflected here, proper attribution. I also appreciate the diligence and courtesy extended to me by the curators at the institutions from which I borrowed specimens; this work could not have been done without their cooperation and patience. I greatly appreciate the help of Allen Press, especially G. Dresser and K. Blair, in expediting the printing of this volume from our camera-ready copy. The Buffalo Museum of Science and especially its Director, Ernst Both, provided significant research support and an atmosphere conducive to uninterrupted, long-term study.

This book is dedicated to Patricia M. Eckel, colleague and spouse, who provided unflagging support and sensitive suggestions throughout the long evolution of this work. Her fine illustrations furnish an essential reinforcement for the generic concepts proposed here and round out the descriptions in an elegant manner.

INTRODUCTION

The taxonomy of the Pottiaceae is considered by many to be less than easy and largely the province of adepts. This is due to intrinsic problems such as polymorphy, taxonomic importance of anatomical characters, reduced size of the plants, obscure areolation, and sterility of many specimens, and to extrinsic factors such as lack of regional identification manuals (especially for tropical areas), few revisions, many poorly conceived "geographic" species, and the heterogeneity and large size of some genera.

The latest intensive treatment of the family at the generic level was published by Brotherus in 1924, being a series of short descriptions with keys to genera and lists of species (some species lists presented with keys). Of the 85 genera recognized in the family (of 795 in the mosses *fide* Crosby & Magill 1981) just previous to the present study, Brotherus described only 60 and illustrated only 41. His keys generally require a specimen that has a sporophyte. Many new genera and generic synonymies have been published since 1924 in various journals that may be hard to obtain by botanists without easy access to the large libraries of developed countries. A major problem in identification of mosses of tropical and other non-Western areas is the lack of a modern key to genera that reflects recent (past 50 years) work on taxonomy. The twenty-five accepted genera of Pottiaceae not found in Brotherus' (1924–25) work must be tracked down through a large and scattered literature.

This study is an overview of the family providing a taxonomic introduction and means of identification for the genera of the Pottiaceae. The family is the largest in the Bryopsida in number of genera, 76 being recognized here, and consists of an agglomeration of many small, fairly well-defined genera and a few large, poorly defined genera much in need of revision. Because the author believes that important morphological transformation series may cut across presently recognized generic boundaries, no concerted attempt was made at extensive revision (that is, reapportionment of groups of species) at the generic level except in the *Phascum-Pottia-Desmatodon-Tortula* complex and in the *Weissia-Trichostomum* relationship. Instead, a key to recognized genera, a cladistic evaluation, comprehensive descriptions, discussion of problems, a bibliography, detailed illustrations, and a list of the correct names of the taxa are provided to stimulate further study, floristic, taxonomic and phylogenetic. This is not a "type revision"; the descriptions are based on the study of a selection of specimens of representative species, names of which are listed with citations of herbaria consulted. The discussions are the result of the author's 25 years of specialization in the group and of new observations made during this study.

PREVIOUS WORK

The more than 1400 species of the Pottiaceae are characteristic of variable or harsh environments, and may form a conspicuous portion of the vegetation of ruderal, arid land, alpine or arctic areas. They exhibit a great variety of apparent morphological, physiological and genecological adaptations to their particular environments, mostly poorly or not at all understood.

The Pottiaceae is here reviewed with certain modifications at the generic level reflecting modern recognition that morphological reduction series are characteristic of many genera of the Pottiaceae, *cf.* discussions below of *Barbula*, *Bryoerythrophyllum*, *Didymodon*, *Hennediella*, *Tortula*, *Trichostomum* and *Weissia*, among others.

Several common species of Pottiaceae in England were described and illustrated by J. Dillenius (1741) in his early bryological manual. J. Hedwig (1801) described and illustrated many pottiaceous species from regions worldwide in his "Species Muscorum," from which moss nomenclature (except that of *Sphagnum*) begins. The genera of Hedwig, however, were usually taxonomically heterogeneous.

The name Pottiaceae was first used by Bruch et al. (1843) in the "Bryologia Europaea" (1836–1855). The authors included only three genera, while other genera now placed in the Pottiaceae were recognized in segregate families (Phascaceae, Weissiaceae or Trichostomaceae), these based largely on sporophytic characters. Most other authors in the 19th Century followed the practice of dividing the Pottiaceae into smaller families. Mitten (1859) recognized only one family, however, the Trichostomaceae, which he later called the Tortulaceae (Mitten 1869). Brotherus (1902–09) also recognized only one family, the Pottiaceae, with 46 genera in four subfamilies. Later, Brotherus (1924–25), in a treatment for Engler and Prantl's "Die Natürlichen Pflanzenfamilien," recognized 71 genera in five subfamilies in the Pottiaceae. Brotherus' treatment remained the standard compiliative work on the family to this date. Recent authors generally follow the single family concept of the Pottiaceae, but *Cinclidotus* is variously treated in the Pottiaceae or in its own family, the Cinclidotaceae. Saito (1975a) gave a detailed discussion of the historical development of family and suprageneric taxonomic concepts for the Pottiaceae.

Three papers published since Brotherus' (1924–25) treatment contributed significantly to generic and suprageneric classification.

Hilpert's (1933) "Studien zur Systematik der Trichostomaceen" was a treatment at the generic level of the "Trichostomaceae," which was composed of what have come to be recognized (*sensu* Saito 1975a) as the tribes Trichostomeae, Barbuleae and Pleuroweisieae (= Hyophileae here) of the Pottiaceae. Hilpert recognized three subfamilies and three tribes in the Trichostomaceae while the rest of the genera that Brotherus had recognized in the Pottiaceae *sensu lato* were placed by Hilpert in the Pottiaceae and Cinclidotaceae, and not studied. Hilpert discussed at length morphology and taxonomic importance of characters of the areolation, costa, curvature of the leaf margins, differentiation of areolation at the leaf margins, papillae and mamillae, propagula, perichaetial leaves, annulus and peristome. He pointed out important differences between the Trichostomaceae and the Pottiaceae in such morphological characters as leaf shape, areolation, costal anatomy and paraphysis shape. The morphology of each genus was discussed in relationship to related genera. Many species names were transferred from one genus to another. He described as new the genera *Semibarbula, Prionidium* and *Macroglossum.* He suggested that *Sarconeurum* was probably near to or a synonym of *Tortula*, that *Leptodontiopsis* belonged in the family Orthotrichaceae, that *Chrysoblastella* and *Rhamphidium* (see Excluded Taxa below) belong in the Ditrichaceae, and that *Weisiopsis* belonged in the Pottiaceae *sensu stricto* (= Pottioideae *sensu* Saito 1975a). The chief value of Hilpert's (1933) work is that it provided the only discussion of the many tropical genera published since Brotherus' (1924–25) treatment, and presented modern charts showing purported supraspecific phylogenetic relationships.

Chen's (1941) "Studien über die ostasiatischen Arten der Pottiaceae" is a landmark treatment of the Pottiaceae. Primarily a floristic work concerning only those species in eastern Asia, this study included detailed descriptions and extraordinary illustrations of species in 32 genera, discussions of morphological characters at the genus level, floristic relationships of eastern Asian species, and a lengthy evaluation of intra-familial phylogeny with a detailed chart of presumed relationships.

Chen recognized six subfamilies, the Cinclidotoideae, Pottioideae, Trichostomoideae, Eucladioideae, Leptodontioideae and Barbuloideae. These were distinguished by such characters as capsule and costal anatomy, leaf shape and margin recurvature, length of the operculum relative to that of the theca, differentiation of the leaf base and morphology of the laminal papillae. Characters that distinguished the genera were variations and combinations of these plus such characters as plant size, shape of leaf apex, length of the costa relative to that of the leaf, etc. Chen made many new combinations at the species level, and described a few new species and the genera *Reimersia* and *Bellibarbula*. He provided *Bryoerythrophyllum* as a *nomen novum* for the illegitimate homonym *Erythrophyllum* (Lindb. in Braithw.) Loeske, and was the first worker to recognize *Bryoerythrophyllum* as a rather large genus. Of the generic complexes most juggled about in identification manuals today, Chen lumped *Hymenostomum* and *Astomum* under *Weissia*, *Didymodon* under *Barbula*, and *Hymenostylium* under *Gymnostomum*, but recognized as separate *Tortula* and *Syntrichia* (*sensu* Chen, not Zander).

Saito's (1975a) "Monograph of Japanese Pottiaceae," gave extensive discussion of the generic and suprageneric classification of the Japanese taxa, and presented a nicely explained classification system based in part on several new characters and thorough morphological and anatomical study. The most conservative characters in the family were considered to be curvature of the leaf margins, shape of the area of differentiated basal leaf cells, the occurrence of gemmae, and characters of the sporophyte. Saito found that variations in the peristome (reduction series and "complication" series) within certain generic complexes were correlated with variation in other characters of the sporophyte, such trends varying in parallel. He made new statuses of names at the intrageneric level and described a new subgenus (*Barbula* subg. *Odontophylla* Saito). He merged *Astomum* and *Hymenostomum* with *Weissia*, *Hymenostylium* with *Gymnostomum*, and *Syntrichia* (*sensu* Chen) p.p. with *Tortula* (in an earlier paper, Saito 1973a), but kept as separate *Barbula, Bryoerythrophyllum* and *Didymodon.* Saito, like Chen (1941), provided excellent illustrations of all species including such important details as papillae morphology and transverse sections of the leaves. Saito's major contributions to supraspecific classification are the well-characterized and well-justified subfamilies, tribes, genera, subgenera and sections of Japanese Pottiaceae, often based in part on new anatomical and morphological characters.

Ninety genera were recognized in the Pottiaceae by Crosby and Magill (1981) in their "Dictionary of Mosses." Since the publication of this list, revisionary and floristic work has resulted in the synonymizing of some generic names with others and in the description of new genera, reducing the total to 85 at the beginning of the present study in 1985. There are, however, only 76 recognized genera at its completion, reflecting considerable synonymy and removal of genera to other families.

My recent publications on the family, too, have been generally limited to one floristic region, the New World or parts of it. In this study, however, several genera were deemed insubstantial and synonymized with other genera with previously published names. For example, *Trichostomopsis* (Zander 1978e) and *Husnotiella* (Zander 1981c) were made synonyms of *Didymodon*, and *Barnesia* a synonym of *Streptocalypta. Leptodontiella* was erected as a segregate of *Leptodontium* (Zander & Hegewald 1976). I agreed for the most part with Saito (1975a) in distinctions between *Barbula* and *Didymodon* (Zander 1978e, 1979f, 1981c). I used color reactions of leaf laminae to various acidic and basic reagents (Zander 1980a) as characters valuable in distinguishing *Barbula* and *Bryoerythrophyllum*, showing that twisted peristomes are not restricted to the former genus. I used the phenological feature of date of sporophyte maturation to characterize some genera of the Pottiaceae and to place certain species in their proper genus (Zander 1979d). My taxonomic methods and concepts are more extensively summarized elsewhere (Zander 1981c, 1982e, 1982g, 1985b).

I have specialized in the taxonomy of the Pottiaceae since 1967. My larger papers include treatments of *Leptodontium* (1972), *Tuerckheimia* (1978f); and *Streptocalypta* (1983a) in the New World; the tribe Pleuroweisieae in Middle America (1977c); *Barbula* and *Pseudocrossidium* (1979f) and *Didymodon* (1978e) in North America north of Mexico; *Bryoerythrophyllum* and *Morinia* [= *Mironia*] in the New World (1978g); *Barbula, Pseudocrossidium* and *Bryoerythrophyllum* p.p. in Mexico (1981a); and *Didymodon* in Mexico and California (1981c). Treatments of 33 of the genera of Mexican Pottiaceae for a moss flora of Mexico (A. Sharp et al. 1993), and of the Trichostomoideae for a moss flora of the North American Arctic and Greenland (G. Mogensen, ed.) have been completed.

Significant recent taxonomic studies in the Pottiaceae by other authors include treatments of *Aloina, Aloinella* and *Crossidium* by Delgadillo (1975a and in Sharp et al. 1993); of *Weissia* subg. *Astomum* in Africa by Crundwell and Nyholm (1972a, 1974); of *Tortula* sect. *Rurales* by Kramer (1978, 1980, 1988); of *Globulinella* by Magill (1977a); of *Pottia* in Great Britain by Chamberlain in A. Smith (1978); of *Hymenostyliella* and *Trichostomopsis* by Robinson (1970, 1971a); of *Tortula* by Mishler (in Sharp et al. 1993) of *Triquetrella* by Casas de Puig et al. (1993); of *Phascum* by Guerra et al. (1991); and of *Weissia* by Stoneburner (1985)—see also the section on Practical Identification below and the Bibliography of works on the Pottiaceae. A relatively few identification manuals, such as those of Gangulee (1969–80) for eastern India, Crum and Anderson (1981) for eastern North America, Catcheside (1980) for South Australia, Magill (1981) for South Africa, Norris and Koponen (1989) for Papua New Guinea, Eddy (1991) for Malesia (Malaysia and the East Indies), and the multiple author treatment edited by Sharp et al. (1993) for Mexico, provide exacting taxonomic treatments and considerable new information on the Pottiaceae.

Recent developments in taxonomic techniques have resulted in the use of new taxonomic characters in the Pottiaceae. Most of these are simply the results of more detailed observation of morphology and anatomy. Some examples may be mentioned. Saito (1974a, 1975a) asserted that the subfamily Trichostomoideae may be distinguished from other subfamilies of the Pottiaceae by the five, rather than four, amphithecial layers of the capsule. Insertion patterns of rhizoids are distinctive features distinguishing the Pottiaceae from certain other families with similar gametophytes according to Norris (in Norris & Zander, 1981). Saito (1975a) made great use of the morphology of axillary hairs in distinguishing genera of the Pottiaceae. He did not consider the hymenium of *Hymenostomum* to be an organ distinct from the columella, and, for this reason, recognized *Hymenostomum* only at subgeneric rank under *Weissia*. The scanning electron microscope was used in studies of spore and lamina ornamentation for taxonomic evaluations of some genera of Pottiaceae by Lewinsky (1974), Saito and Hirohama (1974a,b), Zander (1972), Zander and Vitt (1979), and others.

The present study reflects an emphasis on leaf characteristics in defining higher categories begun by C. Müller with his "Synopsis Muscorum" (1849) and continued by Braithwaite (1887) and Mitten (1869) in their works. Most recent authors, including myself, have emphasized at least some of a number of unobtrusive anatomical or morphological features as being of considerable importance at the generic and specific levels, e.g. presence or absence of a stem central strand, or of a stem hyalodermis, or of a costal epidermis or hydroid strand (Begleiter cells); the morphology of the laminal papillae; characters of the laminal cell walls and areolation patterns; position and size of propagula, and so forth. These relatively new characters, however, have been used in revisionary or floristic studies of only a small proportion of the species and, until the present work, of the genera of the family. A large number of species remain unstudied or uncritically examined since their original description during the main exploratory phase of bryological floristic development in the 1800's and early 1900's.

The genera studied are illustrated with one or more species, these selected to demonstrate the range of variation in morphological and anatomical features considered of taxonomic importance. Usually pictured are at minimum a habit, leaf, leaf apex, leaf basal cells, sections of stem and leaves; also commonly shown are upper laminal papillae, propagula, peristome details, operculum, and calyptra. Since size as a taxonomic character is usually given as a range, magnifications are not provided for the illustrations; however, the reader may assume that for all but very small and very large taxa, figures of comparable features are enlarged to a similar extent. Actual measurements for plants and their morphological features should be obtained from the descriptions.

TAXONOMIC CHARACTERS

Comments on specific morphological and anatomical features of taxonomic importance are given in the following paragraphs, while evaluations of phylogenetic polarizations of characters are related in the section on cladistics. A number of treatments describe the morphology of the mosses in detail; those of Flowers (1973a) and of Saito (1975a) include much up-to-date discussion and illustration of the morphology of the Pottiaceae.

Color

Emphasis is placed on natural coloration (see discussion of Mårtensson & Nilsson 1975) as a character only for those taxa in which color correlates with other, taxonomically important characters, e.g. *Erythrophyllopsis* (red leaves) or *Scopelophila* (blackish leaves). Even such genera with characteristically red or black leaves can be mimicked by placing a related but less colorful genus in KOH solution to develop leaves of a red coloration or in ferric chloride solution to create blackish leaves (*cf.* Field 1977). Thus, substrates that are highly alkaline or are associated with iron deposits may produce false color characters in certain taxa, just as characteristic coloration of other taxa may be suppressed in acid or iron-deficient habitats. As a rule, highly colored plant parts are yellow in acid solution and red in alkali, but this often varies per genus. A basic chemistry seems to permeate collections growing on calcareous rock, judging, for instance, from the strong bubbling of gas commonly produced from capsules of calciphiles placed in dilute HCl solution. One of the long-used characters of *Barbula convoluta* is the yellow seta. This seems to be a stable character, as the color becomes merely light orange in dilute KOH solution, quite different from the deep red evoked in, say, the setae of *B. unguiculata*. The peristomes of both species, however, show the same reactions to dilute HCl and KOH solutions, being light yellow in the first and deep red in the second. Changes in color can be expected, too, for observations of peristomes in lactophenol gel or Hoyer's solution mounting media (both acidic) or the highly alkaline KOH solution that might be used to help remove the operculum. For accuracy, color of plant parts should be used as a taxonomic character only when such parts are immersed in the acid or basic media standard for such observations.

The present taxonomic treatment emphasizes color responses of leaf laminal walls to two-percent KOH solution. These color reactions are usually either brick red or a bright yellow, occasionally orange, seldom negative (colorless, no change, or with the color of chlorophyll). Apparently (R. Mues, pers. comm.), the yellow and red reactions are characteristic of phenolic compounds—those compounds with only one or no hydroxyl groups giving a yellow color in KOH, while those with two or more closely associated hydroxyls will yield red. All other descriptions of coloration are given here as the natural state, and the general taxonomic value of these is not emphasized. Often, the yellowish green of chlorophyll masks somewhat the reaction of the cell walls to KOH (e.g. in the case of *Chenia leptophylla*, which at first glance appears light yellowish green in KOH under the dissecting microscope); when in doubt, the taxonomist should look at the lamina under high magnification to determine the actual color of the cell walls (which are actually bright purplish red in *C. leptophylla*).

The color reaction of the laminae to KOH solution is apparently unaffected by at least some chemical treatments involved in collection of mosses. Plants of *Bryoerythrophyllum ferruginascens* were boiled in 38% formaldehyde solution and also in 95% ethyl alcohol and, after rinsing, the laminae retained their ability to switch from red to yellow in dilute hydrochloric acid and back to red in two-percent KOH.

Stems

The stem of the Pottiaceae when highly developed (e.g. *Timmiella* spp.) has four morphologically distinctive layers, the central hydroid strand (hereafter usually termed simply the central strand) of thin-walled, often partially collapsed cells; the central cylinder of parenchyma; the cortex, often differentiated as a sclerodermis of stereid or substereid cells; and the hyalodermis of enlarged epidermal cells, usually thin-walled and often collapsed in mature parts of the stem (e.g. Pl. 1, f. 7). There is much variation in the presence or absence of these features, and in the degree of expression, but these are often taxonomically important and examining stem sections must become standard practice.

Central strand. Hilpert (1933) found that the central strand was sometimes absent in species in which it was otherwise usually present, *Trichostomum tenuirostre* being a case in point. Saito (1975a) indicated that *Didymodon asperifolius* always lacked the central strand, but Zander (1978e) questioned this. In *Hymenostylium recurvirostrum*, one variety occasionally has a central strand, but it is otherwise absent (Pl. 32, f. 1) in the typical variety. In spite of some variation, however, presence or absence of the central strand is generally a good character at the generic level in the Pottiaceae, being, for instance, always absent in *Leptodontium* (Pl. 36, f. 2–3; 37, f. 10) and *Scopelophila* (Pl. 47, f. 2). Occasionally, the stem is hollow (Pl. 3, f. 2), in which case a central strand was probably present (section near the stem apex to check this). "Satellite" strands visible in sections in the cortex of the stem occur in *Aloinella* and *Calymperastrum*, and are connected with the leaf strand, but not with the central strand of the stem.

Central cylinder. The central cylinder of cells is usually of medium-sized, thin-walled parenchymatous cells, but may occasionally be anatomically distinctive. Its cells may be uncommonly large, as in species of *Barbula* (Pl. 43, f. 2), or thick-walled, as in many species of *Trichostomum* (Pl. 11, f. 7), *Tortella* (Pl. 18, f. 16) and related taxa.

Cortex. The cortex of the stem of Pottiaceae usually consists of cells with smaller lumens than those of the central cylinder. It is sometimes differentiated as sclerodermis, which is well developed in many genera of Pottiaceae and consists of small, thick-walled, longitudinally fusiform cells, the "stereids." In most cases, the presence of a sclerodermis is easily determined. In *Barbula*, especially, the sclerodermis is usually strongly differentiated from the enlarged parenchyma of the central cylinder, but occasionally, the outer cortex consists of "substereid" (Pl. 31, f. 2; 39, f. 3) cells difficult to interpret.

Hyalodermis. This consists of usually one layer of enlarged, thin-walled cells on the surface of the stem, commonly in combination with a sclerodermis or at least an outer cortex of small-lumened cells. Like the central strand, this character is variable in expression in some taxa, both as to presence and degree of differentiation, yet it is often of value taxonomically. In *Leptodontium* (Pl. 36–37), the hyalodermis is characteristic of certain sections of the genus, but not of others (Zander 1972). A collapse of the thin outer walls of the hyalodermis in mature stem parts of many but not all taxa may give a "fluted" appearance to the stem section.

Axillary hairs. The filaments (Pl. 44, f. 21; 59, f. 15–16) borne in the axils of the leaves of Pottiaceae are generally from 2–15 cells in length, consisting of hyaline, uniseriate, cylindrical cells or rarely these bulging and thus producing a moniliform filament (*cf. Molendoa* and *Gymnostomiella*). The basal 1–2 cells are either similar to the rest or are more thick-walled and brownish or yellowish. The terminal cell is apparently never terminally pored as in the axillary mucilage hairs of species of certain other families (e.g. *Brachymitrion jamesonii* Tayl. of the Splachnaceae), but occasionally may be similarly swollen (e.g. *Globulinella globifera*). Ligrone (1986) has described the ultrastructure and cytochemistry of the mucilage-secreting axillary hairs of *Timmiella*.

Leaves

The cauline leaves of the Pottiaceae are typically lanceolate to spathulate. Although the basal cells are differentiated in most genera, leaves with more or less sheathing bases are less common, and are characteristic of the *Leptodontium* (Pl. 36–37) group, *Pleurochaete* (Pl. 16), and certain other genera—these taxa usually of relatively large physical size. Such leaves may be squarrose-recurved, reflexed at the top of the basal sheath (Pl. 3–4) or rarely wasp-waisted as in *Timmiella* (Pl. 1). Leaf shape may be variable in a characteristic fashion such that, within the same species, there appears to be a standard differential development of character states. In species with much variation in leaf shape, leaf width commonly varies far less than leaf length, perhaps because stem width also varies proportionately less than leaf length. Thus, in some one species or genus, shorter leaves are ovate, and are characteristically stiff and erect in stance when dry, and longer leaves are generally scarcely wider to sometimes twice as wide, but may be three to four times as long as are the short leaves, and then mostly lanceolate in shape, more easily twisting or curling when dry simply because they have a high length to width ratio. So what may be perceived as a number of characters (including leaf length and stance), correlated with plant size may all be based on a single, simple developmental feature found in many species with variable leaf shapes. Leaves that twist, flex or curl or become tubulose when dry may in that way lessen heat stress. In vascular plants, research on *Spartina pectinata* Link (Gramineae) by S. Heckathorn and E. DeLucia (Anon. 1990b) demonstrated that leaf rolling reduced surface exposure to sunlight, lowering leaf temperature by five degrees Cent.; J. Lebkuecher and W. Eickmeier (Anon. 1990b) found that leaf curling in *Selaginella lepidophylla* (Hook. & Grev.) Spring. (Selaginellaceae) reduced high radiation damage that occurred during desiccation. See also discussion of surface wax by Proctor (1979b).

There are at least six distinct trends in the family towards elaboration of specialized photosynthetic leaf tissue:

1. Corrugated or rasp-like leaves, such as are developed in *Anoectangium aestivum* variants (see Zander 1976).
2. Costal lamellae on the ventral laminal surface, as in *Pterygoneurum* (Pl. 72, f. 9) or species of *Acaulon* (Pl. 102, f. 12, 24–25), or the dorsal surface as reported in species of *Hymenostyliella*.
3. Ventral costal filaments, as in *Aloina* (Pl. 75, f. 8, 14), *Aloinella*, *Crossidium* and *Pseudocrossidium*.
4. Bulging ventral costal surface, as in the ventral surface of certain species of *Tortula* sect. *Tortula*, especially *T. atrovirens* (Pl. 85, f. 8) and *T. revolvens* (Pl. 89, f. 4), with the epidermal cells generally higher than wide in transverse section.
5. Protected tissue inside leaf margin revolutions consisting of swollen, thin-walled and coarsely hollow-papillose tissue in species of the genus *Pseudocrossidium* (Pl. 27, f. 13, 17), and in *Hilpertia* (Pl. 112, f. 4–5) forming a densely chlorophyllose cylinder along each lateral margin.
6. Ventrally bulging and dorsally nearly flat upper laminal cells that may enhance photosynthesis in periods of dim light and dew in arid habitats by focusing light on the chloroplasts. Papillae may obscure the asymmetric bulge of such cells, and sections are often necessary to demonstrate these. Leaves with this distinctive areolation are found, for example, in genera of the *Hyophila* complex (Hyophileae *sensu* this treatment; Pl. 54, f. 7, 16) and in *Gertrudiella* and *Timmiella*. Upper laminal cells that are strongly bulging on both sides, as in some species of *Syntrichia* (Pl. 105, f. 7), may be an equivalent adaptation to dim light. Delgadillo (1984) listed several taxa characteristic of the Yucatan Peninsula of Mexico with a similar phenotype including ventrally bulging upper laminal cells.

Costa

Wyatt (1985) has reviewed the complex and various terminology that has been applied to moss costal anatomy.

Genera that consistently lack a ventral epidermal layer are *Aschisma* (Pl. 81, f. 7–8, 17), *Gymnostomiella*, *Leptodontiella*, *Leptodontium* (Pl. 36, f. 8), *Reimersia*, *Streptotrichum*, *Trachyodontium*, and *Triquetrella*. This is also characteristic of *Hymenostylium* (Pl. 31, f. 7, 16) except in one variety of *H. recurvirostrum* (Pl. 32, f. 6).

The costal hydroid strand (the Begleiter cells of many earlier authors, and hereafter referred to in context simply as the hydroid strand) when present is found between the guide cells and the dorsal stereid band, or rarely also between the guide cells and the ventral stereid band (e.g. *Calymperastrum*, Pl. 9, f. 9; *Barbula riograndensis*, Pl. 45, f. 12). The hydroid strand is usually single, but in some taxa (e.g. *Pseudocrossidium*, Pl. 27, f. 8, 13, *Timmiella, Erythrophyllopsis*), there may be two or more distinct strands all dorsal to the guide cells. In very young leaves, the several individual, thin-walled cells of the strand may be seen in section (Pl. 9, f. 9; 69, f. 10–11), but in mature leaves, the strand is usually represented by an angular, three-armed or sometimes star-shaped (with concave sides) opening between the central guide cells and the stereid band (Pl. 3, f. 5; 23, f. 9). Taxa that consistently lack a stem central strand apparently never have a leaf hydroid strand. In taxa with both stem and leaf strands, there is no evidence of a hydroid "leaf trace" connecting the leaf and the stem strands. In some taxa with a stem central strand, the leaf strand may be seen (in stem sections) penetrating the outer cortex for a short distance (Pl. 77, f. 2).

The location of the hydroid strand in the center of the single stereid band of *Hennediella stanfordensis* (Pl. 98, f. 11), *Phascopsis rubicunda* (Pl. 100, f. 8–10), *Acaulon robustum* and other taxa, and the central location of the single stereid band in the costa of *Scopelophila ligulata* (Pl. 47, f. 6) and *Streptopogon* spp., may both be explained as the guide cells being absent in essentially double stereid-banded species, but correlation of other characters indicates that this is probably not the case, except possibly the latter in *Scopelophila*. Also, developmental series studied by Saito (1975a, p. 388) indicate that the ventral stereid band of Barbuleae is elaborated from the same tissue that forms the parenchymatous layer between the guide cells and ventral epidermis in Pottieae. Additionally, certain taxa with variable development of the ventral stereid band (e.g. species of *Streptocalypta*, *Pseudocrossidium* and *Tortula*) show what is apparently a thickening of the walls of pre-existing cells ventral to the guide cells, giving the appearance of what seems to be a second stereid band.

The shape of the dorsal stereid band is considered important here in taxonomically distinguishing between advanced genera of the Pottieae. It must be noted, generally, that the dorsal stereid band in species with much reduced leaves and apparently narrowed costa will be reduced from a crescent-shaped state to a rounded state (e.g. *Gyroweisia*). Taxa with rounded dorsal stereid bands in the upper part of the leaf may show crescent-shaped bands in the lower part of the leaf (e.g. *Leptobarbula*, Pl. 78, f. 9–10, and species of *Tortula*). This character is, as are many in the Pottiaceae, of greatest value with plants of comparatively large stature and, in any case, sections should be taken at midleaf, most definitely above a sheathing leaf base.

Upper Laminal Cells

Cell wall thickness may have a significant effect on survival in some species. Gerson (1987) found that thickened cell walls in mosses prevent feeding by mites with short mouth parts.

For consistency, measurements of laminal cells in this paper are of the width (perpendicular to the length of the leaf) with indication of length to width (in that order) ratio following. Norris and Koponen (1989) indicated that laminal cells arranged in rows was characteristic of the Pottiaceae, and could be used to distinguish such genera as *Streptopogon* from, for instance, those of the Splachnaceae.

Trigones. Cell wall thickenings where three walls are contiguous (the so-called "corners" of cells, Pl. 52, f. 4) are characteristic of species of several taxa, including *Calyptopogon*, *Hymenostylium*, *Hymenostyliella*, *Leptodontium*, and *Reimersia*, correlated with hygric habitat. *Leptodontium*, *Hymenostylium* and *Reimersia* are apparently closely related (see phylogenetic analysis).

Papillae and mamillae. The laminal cells of most taxa may be described as papillose, but this is an overly general term for a rather complex superficial ornamentation. Ignoring papillae *sensu stricto* for the moment, examination of transverse sections of laminae demonstrates that species generally have characteristically superficially bulging or flat or slightly convex superficial cell walls. This character has been little used taxonomically (the exceptions are treatments of *Tortula* by Kramer 1980 and Mishler 1986a) firstly because it is masked by papillae when they are present and, secondly, because secondary thickening of the cell walls can modify the degree of convexity. Some taxa (e.g. species of *Tortula*) that are quite mamillose, with deep grooves around each laminal cell such that the cells meet only in a narrow band around their periphery, are commonly not described as being intensely bulging because of obscuring high papillae.

Taxa with ventrally mamillose and dorsally flat or weakly convex laminal cells are distinctive in appearance (Pl. 1, f. 2; 52, f. 6; 55, f. 11; 60, f. 5; 64, f. 8–9; 65, f. 8), and the lens-like (commonly non-papillose) nature of the ventral bulge may play some part in focusing light during photosynthesis. Recently, R. Donahue (Anon. 1990a) used fiber-optic light detectors, SEM techniques, and a gas-exchange system to trace light paths within leaves of *Thermopsis montana* Nutt. (Leguminosae). Shade plants had cells with significantly more convex surfaces than sun plants, capturing diffuse light at greater angles and focusing it on a particularly sensitive photosynthetic region of the cell. In the same report (Anon. 1990a), G. Martin used finely detailed molds of leaf surfaces to make replicas in agarose gel. He found that shade leaves can intensify captured light up to 26 times better than sun leaves, while in intense light potentially damaging radiation may be focused on a protective, absorbing layer. As for papillae, Simon (1987) has pointed out that the papillae of *Tortula ruralis* increase its leaf surface area by a factor of 30 to 40. Proctor (1979a) discussed papillae as efficient capillary systems in the Pottiaceae.

How these observations might apply to the majority of species of Pottiaceae, which are well known to survive in harsh environments, would be worthwhile to investigate. The character of ventrally convex and dorsally flattened upper laminal cells is often recognized as contributing to distinctiveness at the generic level (e.g. *Gertrudiella*, *Luisierella*, and other genera and species). *Weissia condensa*, however, exhibits variation in degree of ventral mamillosity, this additionally obscured by its papillae. *Weissia breutelii*, *W. veviridis* and other species, likewise, have ventrally mamillose leaf cells, yet are not otherwise distinguished as a group from other members of the genus. *Hyophila* is a genus with ventrally mamillose and dorsally flat cells characteristic of some species, but not others. There is a strong possibility that eperistomate taxa of *Trichostomum* are masquerading as *Hyophila* species, and, especially because of the large size of the genus *Trichostomum* and its evident heterogeneity, *Hyophila* will probably be significantly reduced in size by some future monographer.

The papillae are often taxonomically important features at the infrageneric level, but, like cell wall convexity, their appearance may be modified by secondary thickening of the cell walls. Hollow papillae, e.g. those characteristic of species of *Leptodontium* sect. *Verecunda*, may be "filled in" with secondary wall thickening in some specimens. Likewise, papillae that are characteristically bifid may, in some thick-walled specimens of the same species, appear to be fused into massive "multiplex" papillae (Pl. 38, f. 13).

Hagen (1929, p. 14) described three distinctive types of laminal papillae in the Pottiaceae: (1) large, rounded or platelike, scattered over the lumens, (2) cylindrical and centered over the lumens or otherwise occurring in groups, and (3) ring or horseshoe-shaped papillae. In the literature, many species are commonly described as having "cee-shaped" papillae. In most cases, these are actually hollow, simple papillae (like blisters). Because they are arranged on a convex cell wall surface, they are viewed from above at an angle, and are seen in optical section because of the shallow depth of field at high magnification. The optical section of the walls of these blisters takes the form of the letter "c" or "o". Andrews (1945) correctly called "ring-shaped or horseshoe-shaped" papillae an "optical illusion," urging more careful focusing on the part of the microscopist.

Simple papillae in the Pottiaceae exclusive of Leptodontioideae occur usually only singly or in mixed ones and twos over each upper cell lumen (Pl. 10, f. 8); plants morphologically similar to Pottiaceae and with many simple papillae over each lumen may be looked for in the Orthotrichaceae or Rhachitheciaceae. Bifid papillae (Pl. 22, f. 8; 107, f. 17) are common in the Pottiaceae, occurring usually two or more over each upper laminal cell lumen. There may be a high saddle of tissue between the two salients, and in this case the pairs form bifid papillae that are genuinely cee-shaped in both morphology and gross appearance, i.e. the cee shape is not an optical section of a thin hemisphere. Some species, e.g. *Leptodontium flexifolium*, may have optically cee-shaped papillae in thin-walled plants and bifid papillae in thick-walled plants, so there may be a developmental relationship between the two types. Scanning electron micrographs of the papillae of the Pottiaceae have been published by Mishler (1985b, 1986a, 1987a), Proctor (1979a), Robinson (1974) and Zander (1972), among others.

Laminal thickness. The laminae of some taxa are typically multistratose (Pl. 1, f.2; 3, f. 5), yet this character is variable at the specific level in others. Some genera may have some species with bistratose laminae (Pl. 108, f. 15) and some species with unistratose laminae, but most genera have unistratose laminae. Considerable variation is evident in the genus *Didymodon* (Zander 1982a), especially the species related to *D. rigidulus* and *D. vinealis*. Sayre (1952) briefly described similar variation in *Grimmia* (Grimmiaceae). When laminal cells are bistratose in patches medially, each of the cells is usually only half as thick through (measured from ventral surface of the leaf to the dorsal) as the cells of the unistratose portion, giving the appearance of a single laminal cell with an interior periclinal wall (Pl. 4, f. 8; 13, f. 15). When leaf margins are bistratose, however, the cells are generally each as thick through as the medial cells, inflating the border whether the marginal cell walls are thickened (Pl. 17, f. 8; 33, f. 8) or similar to the medial cells in thickness (Pl. 8, f. 8; 9, f. 9; 23, f. 9).

Basal Laminal Cells

The basal laminal cells are sometimes called hyalocysts (*cf.* Edwards 1980a) when much enlarged and lacking chlorophyll. They are usually also epapillose or weakly papillose and sharply differentiated from the smaller, green, papillose upper cells. Hyalocysts are characteristic of Calymperaceae, but are also found occasionally in the Pottiaceae in various species of such genera as *Leptodontium*, *Bryoerythrophyllum* and *Tortula*. In most species of Pottiaceae, however, the basal laminal cells more or less gradually intergrade in size and degree of papillosity with the upper cells. The basal cells may be wide or little wider than the upper cells, and are generally smooth and elongated. Well-differentiated basal cells are commonly bordered along the leaf margins by narrow, rectangular cells that may serve to inhibit breakage of the leaf, the basal cells apparently being weak structurally. Occasionally a distinct area of unusually thickened rectangular cells is discernible in the leaf just distal to the area of hyaline basal cells, as in species of *Trichostomum* subg. *Oxystegus*.

Edwards (1980a) noted that resorption pores are found in the hyalocysts of all three major traditional groups of the Pottiales, in the leucobryaceous leaf, and in *Sphagnum*. He reviewed past references to such pores by other authors, and was surprised that they have not been mentioned more commonly. He found, in Calymperaceae of west tropical Africa, variation in number per cell, position and shape, such characters being of taxonomic value at the family and infrageneric levels. Zander and Cleef (1982) described such pores in *Kingiobryum* Robins., now recognized in the Dicranaceae but which may belong to the Pottiaceae.

Propagula

Propagula in the Pottiaceae are generally borne on stalks arising from the axils of cauline leaves (e.g. some species of *Barbula* (Pl. 43, f. 20; 44, f. 10), *Didymodon*, and *Hyophila*), but species with propagula borne on long rhizoids buried in the soil are also common (e.g. species of *Barbula*, *Chenia* and *Tortula*). Less frequently, propagula may be born on the tips or margins of leaves (e.g. *Streptopogon* spp., Pl. 41, f. 12–14; 42, f. 17), on the ventral surface of the costa (e.g. *Didymodon* spp., *Syntrichia rigescens* (Pl. 110, f. 19), *Trichostomum tenuirostre*), and on a naked branch (e.g. *Leptodontium stoloniferum*, Pl. 37, f. 7). My report of propagula in a cup at the apex of a leaf (*Leptodontium stellaticuspis*, Zander 1972) is incorrect; the cup apparently produces only rhizoid initials. Rhizoids are also commonly produced at unmodified leaf tips of *Leptodontium viticulosoides* and *Didymodon occidentalis*. Long-elliptical propagula of species of *Streptopogon* have longitudinal internal walls (Pl. 41, f. 16); similarly shaped propagula of species of Calymperaceae lack longitudinal internal walls.

Unicellular propagula are rather rare in the Pottiaceae, to date having been found in *Bryoerythrophyllum inaequalifolium* (Pl. 24, f. 8; Zander 1968), *Didymodon revolutus* (Zander 1981c), and *Didymodon perobtusus* (Zander 1978c). The three species are curiously similar in their small stature and broadly obtuse, elliptical to ovate leaves, among other characters, and may be products of convergent evolution. The fine nature of the diaspores is a possible adaptation for non-sexual long-distance dispersal or local saturation of a large and homogeneous habitat.

Whitehouse (1980) discussed both the literature and provided his own observations on gemmae produced directly on the protonema of many species of Pottiaceae. To date, such protonemal propagula are known for species of the pottiaceous genera *Barbula*, *Desmatodon*, *Didymodon*, *Eucladium*, *Gyroweisia* and *Tortula*, and may well be a common occurrence in hygric habitats.

Sexuality

Terminology for sexuality in the mosses is discussed at length by Lewis (1961), Zander (1984) and Wyatt (1985), among many others. Wyatt (1985) proposed a terminology for bryophytes reflecting the great variation in sexual condition in this group. Unfortunately, Wyatt indicated that, although distinguishing between haploid and diploid sexuality was important, this could ascertained by "context." This is surely insufficient, being a source of confusion when discussing sexuality in general. Allen and Magill (1987) have published a good response to Wyatt's article, emphasizing the need for a separate terminology for bryophytes. The Wyatt system is recommended here as it nicely parallels the terminology used for angiosperms but with the use of "-oicous" endings (to indicate gametophyte sexuality) (see Zander 1984 for a review of the history of "-oecious" and "-oicous" endings in bryophyte terminology).

Perichaetia

The perichaetial leaves of the Pottiaceae may be similar to those of the remainder of the plant, or differentiated, and sheathing the inner perichaetium (and ultimately the basal portion of the seta) to various degrees. The basal cells of differentiated perichaetial leaves of Pottiaceae taxa are generally elongate and rectangular with a small percentage of cells having tapering, blunt ends. They are seldom prosenchymatous (i.e., with tapering, pointed proximal and distal ends). They may have thickened, porose walls (e.g. *Trachyodontium*) or thin walls (e.g. *Bryoerythrophyllum* spp.).

Perigonia

Perigonia (Pl. 1, f. 8; 15, f. 18; 16, f. 9, 16; 18, f. 9; 38, f. 9) are generally bud-shaped in the Pottiaceae, occasionally flattened. Some species have swollen, nearly spherical perigonia with tightly appressed bracts, while some others have perigonia that are small and loosely foliate, thus often difficult to find on gross examination. The perigonia occur terminally or laterally on entirely perigoniate plants, or laterally in the axils of the leaves of sporophyte-bearing, monoicous plants, in which case they are sometimes flattened.

Theca

The thickness of exothecial cell walls in the descriptions of the genera presented here refers to that of the anticlinal walls. The superficial walls may be, and usually are, much thicker than the anticlinal walls (Pl. 59, f. 11; 79, f. 10). The stomates (Pl. 93, f. 8) are usually rather transparent compared to the other exothecial cells, but this is obvious only in split-open capsules. Stomates in the Pottiaceae are confined to the base of the capsule above or on the short neck, and their numbers are generally correlated with capsule size. They are seldom entirely absent, and variation among the genera is discussed by Paton (1957). The exothecial walls of *Byroceuthospora*, *Eucladium*, *Trachycarpidium* (Pl. 80, f. 9), *Uleobryum* (Pl. 89, f. 9) and species of *Weissia* are bulging, to a greater or lesser degree, while capsules of *Ganguleea* (Pl. 65, f. 10–11) and species of *Weisiopsis* (Pl. 67, f. 20) are distinctly plicate, or, in the case of *Tetrapterum* (Pl. 79, f. 10), are weakly winged.

Peristome

The peristomes of most taxa are either straight or nearly so, or are twisted to various degrees counterclockwise (as viewed from above with direction of twist coming upward); peristomes are seldom twisted clockwise. The counterclockwise direction of torsion is known as sinistrorse, and may also be described as the appearance of the threads of a standard, right-handed screw (right-handed because it is screwed in with an over-the-top right twist, not because the threads twist to the right). Of all genera studied, only *Timmiella* has, in some species, a peristome distinctly twisted clockwise; the peristome of *Leptodontiella* is either straight or very weakly twisted clockwise. Most genera have spiculose peristome teeth, while a few (e.g. *Leptodontium*) are characterized by smooth or striate peristome teeth. A basal membrane (Pl. 84, f. 12–13; 97, f. 16; 106, f. 9), may be differentiated in various genera.

Operculum

The operculum is usually rostrate and straight or weakly curved or inclined. When immature, rostrate opercula are often merely long-conic, e.g. as in *Trichostomum brachydontium*.

Calyptra

The calyptrae of Pottiaceae species are usually cucullate (Pl. 51, f. 11), but are sometimes mitrate (Pl. 42, f. 8) or conic-mitrate (Pl. 17, f. 17; 41, f. 9). Species of *Microbryum*, *Hypodontium* and *Streptopogon* may have papillose calyptrae.

Spores

A list of published electron micrographs of spores of Pottiaceae was given by Ireland (1987). In the present treatment the main character used taxonomically is relative size of the spores. Anisospory in the Pottiaceae is discussed in the treatment of *Leptodontium*.

TECHNIQUE

Study of the Pottiaceae entails techniques common to the study of most mosses with a few modifications that address the need to examine features of internal anatomy, use of color reactions to KOH solution, or to make permanent microscopic mounts of delicate tissue. Zander (1979g, 1980a, 1983b) and others (e.g. Lightowlers 1981; Long 1982b; Frahm 1981, 1990a) have detailed aspects of this, but a review is appropriate here.

Packets are folded from 16-pound substance 100 percent rag or buffered paper to 10 cm in height and 14 cm in width using an easily constructed device consisting of a square board to which is affixed a low ledge below with a metal template the bottom of which is screwed to the ledge, held parallel and about 3 mm above the board. The template is slightly less than 10 by 14 cm in dimension. Packets can be folded quickly and accurately about this flange. Subpackets for fragments are made from thin rag or buffered paper (we use a rag tracing paper) and may be cut to various sizes and folded into square or triangular packets. Glue for labels is made from polyvinyl alcohol dissolved in water and used as refill for commercially available "roll-on" glues. Commercially available polyvinyl acetate glue is also adequate and perhaps less liable to degradation through crosslinking. (A planchet of polyvinyl acetate glue was dried in 1970 and to date retains its original elasticity after about 20 years exposed to fluorescent lighting. Evidently any plasticizer included in the formulation is not volatile or degraded in this time frame.) Water-based glues are, of course, subject to weakening in very humid environments.

In annotating specimens, care must be taken to use inks that do not fade with time. This is also true with making slide labels, which may fade surprisingly quickly when exposed to sunlight. India ink is, of course, ideal. Attaching annotation labels to packets may be done with glue or with plastic or plastic-coated paper clips (to avoid rust marks). It is helpful to keep some insect-mounting pins available for attaching annotation labels to specimens borrowed from herbaria that require such pinning.

Certain standard and some modified microdissecting tools are used in the study and processing of specimens. Fine watchmaker's forceps may be further sharpened with a file. Dissecting probes are re-constructed with sewing needles of medium size and little flexibility inserted into a standard wooden or composition probe handle to replace the original comparatively coarse and bendable needle. Single-edge razor blades for sectioning should be discarded after five to ten uses since they dull fast. One holds a leaf or stem crosswise with a stiff dissecting needle, then slices the material with a razor blade held longitudinally against the far side of the needle, meanwhile rolling the needle slowly towards oneself to gradually expose uncut portions of the material. Practice (and a relatively fresh blade) makes this technique quite effective, even for very small leaves. Remember to scrape off sections (especially stem sections) adhering to the razor blade with a dissecting needle after cutting. The usual pair of compound and dissecting microscopes are needed, but using an additional illuminator with the dissecting microscope for fine dissections rather than just a single lamp will prove surprisingly advantageous for observation of fine features.

Round cover slips of 18-mm size and the thickest weight (number 2) are used for routine examinations in water or KOH solution because of ease of handling and cleaning for reuse. Square 18-mm cover slips of medium thickness (number 1) are used for making permanent mounts because they are less expensive and can be luted (i.e., sealed to prevent evaporation) more easily. To avoid annoying spills, a box of cover slips may be glued to the microscope base or to the bottom of a low, flat tray.

Stains are not standardly used, although they might be worthwhile in examining the pores in the basal laminal cells of some taxa, notably species of the genus *Tortula*. Color reactions to alkali provide, however, important data, and a two percent potassium hydroxide solution is kept at hand in a plastic squeeze bottle. Potassium hydroxide solution helps hydrate dry plants (*cf.* G. Smith 1971, p. 2). Hagen (1929, p. 14) suggested the use of a 10 percent KOH solution soak for the leaves of Pottiaceae species to enlarge papillae and make them more easily visible; this effect has not been evident to me, at least at the two percent concentration recommended. Glass containers should not be used because KOH reacts with the walls to form a precipitate. Dilute hydrochloric acid is a valuable reagent to help determine, from observation of gas bubbles, whether or not a collection was made from a calcareous habitat. It will also help clean plants of limy incrustations.

Pohlstoffe (Wagner 1981; Christy 1987) may be used to

hydrate specimens quickly for examination. The formula for the stock solution is one part di-octyl sodium sulfosuccinate, 24 parts methanol and 75 parts water. A few drops of the stock solution are added to a dropping bottle of distilled water to make a fine wetting agent. Mounts on microscope slides for examination at high magnification should, however, always be made in one or two drops of two percent potassium hydroxide solution. This brings out characteristic color reactions in leaves and other plant parts. This also helps remove the operculum (Lauridsen 1972) to reveal the peristome. Heating the slide with a butane cigarette lighter (if the flame does not touch the glass slide, there is no soot) aids in loosening the operculum, and placing a cover slip on the preparation before heating lessens evaporation. If the peristome persists in being broken off at the base when removal of the peristome is attempted, allow an intact capsule to soak in a mixture of KOH and Pohlstoffe 15 to 20 minutes or longer. Since Pohlstoffe will eventually precipitate from basic solutions, one or two drops of a concentrated (4 g in 20 cc water) stock solution of sodium N-lauroylsarcosine (trade name "Gardol") may be added to the bottle of KOH solution as an effective surfactant, although small amounts of any commercial detergent may substitute.

Taking time to make a good microscopic preparation will save time later during identification. My standard method is to place one or two drops of two-percent KOH solution on a microscope slide, and add a stem of the plant to the slide. The material is cleaned of debris (at this time one can search for rhizoid-borne propagula) and the laminal color reaction to the KOH solution is noted. If the material is to be transferred to an acidic mountant like lactophenol gel (see below), a drop or two of dilute hydrochloric acid solution is now added. The best preparation consists of one slide with free leaves that have been removed from the stem, especially near the apex to reveal the axillary hairs, and sections of the stem and leaf, and a second slide with the capsule showing the peristome, spores and perichaetial leaves. Taking care to flip some of the leaves so that the ventral side is uppermost will save time later. When making permanent slides, two cover slips may be mounted on each slide with a slide label on the left for consistency.

Permanent slides may be made with Hoyer's solution (Anderson 1954) or polyvinyl lactophenol (Frahm 1990a), these being very convenient mountants that need not be luted, especially if sufficient glycerine is used in formulating Hoyer's solution to counteract dry air in the storage area.

Many species of Pottiaceae, however, have large, delicate laminal cells that collapse in Hoyer's solution or polyvinyl lactophenol. Lactophenol gel (Zander 1983b), which does not collapse the cells of most species, may be used instead for sensitive species or as a standard mountant for all species. It has a high index of refraction. The formula for lactophenol gel is:

30 cc lactic acid (= 2–hydroxypropanoic acid)
15 g phenol, crystal (= carbolic acid)
15 cc distilled water
6 g methyl cellulose, powder (= cellulose methyl ether, of viscosity 25 cP in 2% solution or lowest viscosity available)
35 cc ethylene glycol (= 1,2–ethanediol).

Mix the phenol with the lactic acid, dissolving it with gentle heat. Add the water and stir. Heat to just boiling (use a fume hood). Add the methyl cellulose powder and stir vigorously into the hot solution to dissolve (reheat if necessary). Add the ethylene glycol last. Pour into a glass cylinder and let stand to allow bubbles and undissolved material to rise. After a day or two, remove any floating particles and pour the clear liquid into a storage bottle. A small bottle with an applicator wand built into the lid or a plastic squeeze bottle with a fairly wide aperture (4 mm) is used to place a drop or two on a slide. Specimens incrusted with carbonate deposits must first be soaked in a drop or two of dilute hydrochloric acid to prevent bubble formation in the lactophenol gel.

Specimens soaked in KOH should also be neutralized with a drop or two of dilute HCl before mounting in lactophenol gel. A precipitate may appear when overly moist plants are placed in lactophenol gel but this is redissolved on stirring. Plants may be manipulated, and leaves and stems sectioned, while in the gel.

The slides prepared as noted above, after placement of the cover slip, are adequate as semi-permanent mounts for one or two months. Permanent mounts may be made by sealing the cover slip to the slide. So as to further solidify the gel and help prevent migration of the plant material, water portion of the mounting fluid may be allowed to evaporate somewhat by leaving the freshly mounted slide exposed in a dry place, such as a fume hood, overnight, but sealing the slide immediately upon preparation is generally best. To seal the slide, clean cover slips and slides must be used to insure a complete bond. Clear fingernail polish, although commonly used as a lutant, is generally inadequate since pyroxylin (nitrocellulose) does not adhere well to glass.

A good lutant is poly (ethyl methacrylate) with butyl benzyl phthalate as plasticizer (as Krystalon™, Harleco, Gibbstown, NJ 08027 USA), an artificial balsam with good long-term adherence to glass. Another, more easily obtainable sealant with excellent adhesion to glass, and good flexibility when dry (but somewhat reactive, so do not store slides in strong light) is the commercial liquid "polyurethane" gloss finish for wooden floors, of various formulations but based mainly on diisocyanates (such as tolylene diisocyanate). Isocyanates adhere to glass extremely well apparently because they react with an always present thin film of water strongly adsorbed to glass (Skeist 1962). Keep the polyurethane container well sealed and, if necessary, add a few drops of artist's drying retardant to the commercial product to slow the gelling during polymerization caused by oxidation of linseed oil, a common additive. Any lutant will stay liquid longer and be more easy to use if kept in a "balsam bottle" or, better, in a small, disposable applicator bottle. Keep the bottle more than half full to help exclude air. Also acceptable as glass sealants include various commercially available silicone rubber glues or caulks, and "hot-melt glue" applied with a glue gun, but these are difficult to work with and leave a ridge that may obstruct the high-dry objective lens. Apply lutants liberally to ensure a good seal; most seals can be scraped off in patches if a morphological detail is obscured.

Disposable serviettes or wipers are important to have on hand to clean tools and slides of reagents. A small fan is valuable at times to disperse any acrid fumes.

Sexing specimens is often required for identification, but this can be as difficult as sectioning. Poor lighting makes this more difficult, and one should not be satisfied with only a single standard illuminator for the dissecting microscope. Dry plants may show bulbiform perigonia better than moist ones, but plants should be moistened before removal or manipulation of leaves. Sexuality cannot be determined accurately without location of antheridia (in perigonia, in antheridiate buds or naked on the stem, and preferably as observed on several plants in the collection to weigh degree of variation). Perichaetiate collections entirely lacking antheridia are often assumed, with some degree of confidence, to be dioicous, but note Steere's (1940) discussion of dichogamy in *Syntrichia princeps* (as *Tortula princeps*). Perigoniate plants are best looked for near the periphery of a clump of sporangiate plants. Most genera show a distinct apical thickening of the plant apex in perigoniate plants, but this is not always the case. Autoicous buds can be searched for by carefully stripping leaves of archegoniate plants. Paroicy and synoicy is determined by attentive dissection of perichaetia with fine forceps under high magnification under the dissecting microscope with bright illumination. Antheridiate buds or short perigoniate plants located on soil at the very base of archegoniate plants are a sign of possible rhizautoicy, but this cannot be definitely determined without dissection of the rhizoid system or cultivation of single spores.

Measurements are made with a transparent plastic metric ruler or a more accurate "Minitool" 5-mm ruler marked in 0.1 mm increments. An ocular micrometer for microscopic use is of course essential.

Specialists in the taxonomy of particular groups are often asked by other bryologists to render identifications of unusual specimens or other puzzling collections. The accepted way to have this done is to send a "duplicate for identification," which is a labeled portion of a collection that the specialist can retain for his/her herbarium. A name for each specimen sent is then returned to the sender by the specialist. If duplicates cannot be made, each specimen to be returned should be clearly marked "unicate." The cover letter should point out which specimens may be retained by the specialist and which returned. Generally, a quick response may be had if only one to three specimens are sent at a time; a package of several specimens may be placed aside by a specialist pending availability of a larger block of time for work on them. Identifications of large sets may not be completed for many months, and, for these, preliminary inquiries of the specialist should be made. Botanists preparing grant proposals that necessarily involve more than a little identification work by others should consider inclusion of support through subcontracting.

PRACTICAL IDENTIFICATION

Recommended identification guides for Pottiaceae include Brotherus' (1924–25) world treatment, and bryological manuals for particular floristic regions (see Lane 1978b and Schofield 1985, p. 412 ff). Two family monographs stand out: the treatment for China by Chen (1941) and that for Japan by Saito (1975a).

Among generic monographs, revisions or keys, sometimes only regional, that are valuable for identification are those of Bell (1974) for *Willia*; Delgadillo (1975a) for *Aloina*, *Aloinella* and *Crossidium*; Crundwell and Nyholm (1972a, 1974) for *Weissia* and (1962) *Tortella*; Kramer (1978, 1980, 1988) for *Tortula* sect. *Rurales* (here treated as (*Syntrichia*); Lightowlers (1986a,b,c) for *Tortula s. lat.*; Magill (1977a) for *Globulinella*; Robinson (1970, 1971a) for *Trichostomopsis* (here treated as *Didymodon* sect. *Asteriscium*) and *Hymenostyliella*; Saito (1973a) for Pottioideae; Salmon (1903) for *Streptopogon*; Sloover (1977) for *Gymnostomiella*; Stoneburner (1985) for *Weissia*; Warnstorf (1916) for *Pottia* (here split between various genera of Pottioideae); Zander for Pleuroweisieae (the species here placed in other sections, 1977c), *Barbula*, *Bryoerythrophyllum* and *Pseudocrossidium* (1979f, 1981a), *Didymodon* (1981c), *Leptodontium* (1972), *Neohyophila* (= *Plaubelia*, 1983c), *Scopelophila* (1967), *Streptocalypta* (1983a), and *Tuerckheimia* (1978f).

Floristic treatments with especially interesting or useful treatments of the Pottiaceae or portions thereof, in addition to being thorough and well illustrated, are those of Agnew & Vondráček (1975) for Iraq, Catcheside (1980) for South Australia, Crum and Anderson (1981) for eastern North America, Dixon (1913–1929) and Sainsbury (1955) for New Zealand, Noguchi (1988) for Japan, Eddy (1991) for Malesia (Malaysia and the East Indies), Flowers (1973a) for the North American West, Sharp et al. (1993) for Mexico, Gangulee (1969–80) for eastern India, Magill (1981) for South Africa, Norris and Koponen (1989) for Papua New Guinea, Nyholm (1989) for Finland and Scandinavia, Savicz-Ljubitzkaja and Smirnova (1970) for northern Asia, A. Smith (1978a) for the British Isles, and Steere (1938a,b, 1939a) for North America. There are, of course, a number of other fine treatments of smaller areas, or which examine only relatively common species, or which are not or are little illustrated, but the above are those estimated to prove most valuable for practical identification.

Many taxa of moss families other than the Pottiaceae approach the Pottiaceae in gametophytic features and may be confounded with them when sterile. It will be to the advantage of bryological taxonomists, both beginners and adepts, to keep in mind the following list of Pottiaceae "look alikes": Andreaeaceae: *Andreaea*; Andreaeobryaceae: *Andreaeobryum*; Aulacomniaceae: *Aulacomnium*; Bartramiaceae: *Breutelia*; Diphysciaceae: *Diphyscium*; Cinclidotaceae: *Cinclidotus*; Dicranaceae: *Aongstroemiopsis*, *Cnestrum*, *Dichodontium*, *Dicranoweisia*, *Holomitrium*, *Kingiobryum*, *Oreoweisia*, *Pseudohyophila*, *Rhabdoweisia*, *Wilsoniella*; Ditrichaceae: *Astomiopsis*, *Ceratodon*, *Cheilothela*, *Rhamphidium*; Encalyptaceae: *Bryobartramia*, *Bryobrittonia*; Grimmiaceae: *Jaffueliobryum*, *Tridontium*; Orthotrichaceae: *Amphidium*, *Kleioweisiopsis*, *Leptodontiopsis*, *Uleastrum*, *Zygodon*; Ptychomitriaceae: *Campylostylium*, *Ptychomitrium*; Rhachitheciaceae: *Rhachithecium*, *Tisserantiella*; Serpotortellaceae: *Serpotortella*; Splachnaceae: *Brachymitrion*; Splachnobryaceae: *Splachnobryum*; Timmiaceae: *Timmia*; Viridivelleraceae: *Viridivellus*.

MOST RECENT SUPRAGENERIC CLASSIFICATION

The latest arrangement of genera in suprageneric categories for the Pottiaceae at the world level, which may serve for purposes of comparison, is that of Walther (1983):

Trichostomoideae
- Trichostomeae: *Timmiella, Pseudotimmiella, Trichostomum, Rhamphidium, Hymenostyliella, Weissia* (including *Hymenostomum* and *Astomum*), *Chionoloma, Phasconica, Aschisma, Trachycarpidium, Kleioweisiopsis.*
- Tortelleae: *Tortella, Pleurochaete, Oxystegus, Pseudosymblepharis, Stephanodictyon, Serpotortella, Barnesia, Streptocalyptra.*
- Barbuleae: *Barbula* (including *Didymodon* and *Trichostomopsis*), *Bryoerythrophyllum, Streblotrichum, Semibarbula, Husnotiella, Pseudocrossidium, Erythrophyllopsis, Morinia* [= *Mironia*], *Gertrudiella, Sebillea, Prionidium, Bellibarbula, Geheebia, Hydrogonium, Leptobarbula, Tetrapterum.*
- Hyophileae: *Hyophila, Gymnostomiella, Teniolophora, Uleobryum.*
- Eucladieae: *Eucladium, Gyroweisia, Gymnostomum, Hymenostomum, Tridontium, Reimersia.*
- Pleuroweisieae: *Anoectangium, Molendoa, Pleuroweisia.*

Pottioideae
- Pottieae: *Tortula* (including *Syntrichia*), *Desmatodon, Stegonia, Neohyophila* (= *Plaubelia*), *Crumia, Globulinella, Uleopsis, Willia, Sarconeurum, Phascopsis, Aloina, Crossidium, Pterygoneurum, Aloinella, Ulea* (= *Uleopsis*), *Streptopogon, Calyptopogon, Tisserantiella, Pottia, Acaulon, Phascum, Byroceuthospora, Hyophilopsis.*
- Scopelophileae: *Scopelophila, Weisiopsis.*

Cinclidotoideae
- Cinclidoteae: *Dialytrichia, Cinclidotus, Pachyneuropsis.*

Leptodontioideae
- Leptodontieae: *Leptodontium, Tuerckheimia, Triquetrella, Rhexophyllum, Luisierella, Leptodontiella, Streptotrichum.*

Walther's compiliation is clearly a synthesis of the major elements of Brotherus' (1924–25), Chen's (1941) and Saito's (1975a) treatments, and several recent papers of various authors in addition to his own ideas (see also discussion of Cladograms 17–19 under Phylogenetic Analysis). In the main, it serves well as a summary of modern thought for comparison with the taxonomic concepts and nomenclature recognized in the present study. Although there is little basis for comparison since Walther's arrangement is not based on rigorous analysis of all characters, the present study supports certain of the groupings above (see Table of Contents for an overview) but proposes many new relationships.

Study with electron (TEM) and light microscopy of the spore wall ornamentation of several genera of European Pottiaceae led Rejment-Grochowska (1978) to divide the Pottiaceae into informal "sections" grouping the following taxa: *Cinclidotus*; *Anoectangium*; *Eucladium, Gymnostomum, Gyroweisia, Hymenostylium, Hymenostomum, Weissia, Trichostomum, Tortella*; *Barbula, Bryoerythrophyllum*; *Pottia, Acaulon, Phascum, Pterygoneurum, Desmatodon*; *Tortula*; *Aloina.* Saito and Hirohama's (1974a) similar study (but using a scanning rather than a transmission electron microscope) of 19 species of 14 genera of Japanese Pottiaceae (one collection examined for each species) found only one apparent difference at the suprageneric level: the Trichostomoideae differed from the other taxa studied by the presence of "multistalked verrucae" on most but not all of its species studied. They pointed out that major differences between spore ornamentation may exist within the same genus (e.g. between *Trichostomum crispulum* and *T. platyphyllum*), while spores of different genera may be quite alike.

A major new character (Zander 1980a, 1989) that greatly affects the classification of the taxa recognized here is the upper laminal color reaction to two percent potassium hydroxide solution. The genera are for the most part well and consistently characterized intragenerically by KOH laminal color reaction, and suprageneric classification (see below) is in part based on this character. The species of *Tortula*, when split into KOH yellow (*Tortula s. str.*) and KOH red (largely *Syntrichia* and *Hennediella*) sets, are more similar to each other than to groups outside the Pottioideae, and share such apparently advanced characters as the broad leaf shape, lack of a ventral stereid band even in well-developed plants, and rather large upper laminal cells.

EVOLUTIONARY RELATIONSHIPS

Taxonomically important character states are generally not morphologically complex in the Pottiaceae, e.g. leaf margin recurvature, presence or absence of a stem central strand or a hyalodermis, red color reaction to KOH, etc., see also discussion of Rohrer (1985, p. 230), for a similar problem with characters in the Hylocomiaceae. Character states are apparently easily re-evolved (or perhaps de-suppressed). This contributes much homoplasy or taxonomic "noise." For many of these character states, functions, if any, are not clear, but their appearance may be pleiotropic to complex physiological functions.

The occasional presence of a twisted peristome in genus *Ditrichum* (Ditrichaceae; *cf.* discussion by Robinson 1968) is difficult to interpret as indicative of either primitive or advanced status for this character in respect to the base of the Pottiaceae tree. There is, apparently, evidence for reticulate evolution in at least some mosses (Wyatt et al. 1992). On the other hand, the Pottiaceae has many morphological and anatomical characters, and an analysis based on these many characters ought to provide a testable evolutionary hypothesis or at least point to an acceptable classification.

The Pottiales

Fleischer (1920) defined the Pottiales in the commonly accepted modern sense as: small, mostly turf-forming, upright plants; leaves radially arranged on the stem, always with a costa, seldom much-elongate or setaceous, of one layer of cells; upper laminal cells parenchymatous; sporophyte acrocarpous, seldom cladocarpous (i.e. borne on a short side branch); capsule mostly upright and smooth, commonly cleistocarpous; peristome single or absent, of 16 teeth, the dorsal layer thicker than the ventral, usually papillose, sometimes with a basal membrane. Families included were Calymperaceae, Syrrhopodontaceae, Encalyptaceae, and Pottiaceae. Robinson (1971b) on the other hand placed the genera of the Pottiales in a much larger Dicranales, a judgment the present study tends to support.

Shaw et al. (1987, 1989) have indicated that *Buxbaumia* (Buxbaumiaceae) and *Diphyscium* have a haplolepideous peristome. The relationship of *Buxbaumia* to *Diphyscium* (Diphysciaceae) is clear in its capsule shape and peristome (*cf.* Taylor 1962, p. 205; Crum & Anderson 1981). *Diphyscium* is similar to families of the Pottiales (see also Shaw et al. 1987, p. 68) in its gametophyte morphology, without unique characters (the bistratose laminal cells are staggered, overlapping each other as in *Timmiella*). The basal membrane of the pleated peristome does not have the tesselations of that of the high basal-membraned species of Pottieae, but the ornamentation is similarly branching-spiculose (note that Shaw et al. 1989 showed evidence that the haplolepideous peristome is homologous to the cilia of *Bryum*-type peristomes, not the endostome segments). There seems to be no problem with the inclusion of both the Buxbaumiaceae and the Diphysciaceae (including *Diphyscium*, *Theriotia* and *Muscoflorschuetzia*) in the Pottiales along with Calymperaceae and Encalyptaceae (the latter includes *Bryobartramia*, but see cogent remarks on the non-haplolepideous character of the peristome of *Encalypta* by Edwards 1979, p. 342). The leaves of *Buxbaumia aphylla* Hedw. are deep red in KOH. The three species of *Diphyscium* seen in this study had laminae reddish orange in KOH, although the cell walls of the laminae of young leaves were often uncolored or merely vaguely pinkish. The cladograms of Mishler (1986b) and Mishler and Churchill (1984) show the Buxbaumiales as the closest sister group for the Bryales.

Another family worthy of classification near the Pottiales (but see Vitt 1984) is the Timmiaceae, of which the single genus, *Timmia*, has an inner peristome strikingly like the high basal-membraned, fenestrate peristome typical of many *Tortula* (*s. lat.*) species, except that the inner peristome of *Timmia* is more coarsely granulose and does not show the 2:3 primary peristomial layer to inner peristomial layer ratio pattern characteristic of the Pottiaceae and the Haplolepideae in general as discussed by Edwards (1979). The gametophyte has many features otherwise characteristic of the Pottiaceae: ligulate leaves with ventrally mamillose and dorsally flat cell walls, sheathing leaf base; upper laminal cells isodiametric and bulging ventrally but only weakly convex dorsally; and the costal transverse section with two stereid bands, hydroid strand, guide cells, and ventral and dorsal epidermises well differentiated. In addition, the laminal KOH reaction is uncolored in young leaves (laminae clear green *in toto*, the walls are uncolored) with reddish orange blotches in patches, especially on the leaf base (*T. megapolitana* Hedw.), but pink in mature leaves. Akiyama and Nishimura (1993, p. 191), on the other hand, felt that the Timmiaceae was only remotely related to the Bryineae on the basis of branch development analysis.

Encalypta species examined (*E. ciliata* Hedw., *E. mutica* Hagen, *E. procera* Bruch, *E. rhaptocarpa* Schwaegr., *E. spathulata* C. Müll., *E. streptocarpa* Hedw. and *E. vulgaris* Hedw.) in the course of this study have KOH color reactions characteristically bright yellow (becoming red-orange in older leaves) in the upper lamina and deep red in the basal cells, and species of Encalyptaceae have inner peristomes with high basal membranes (Horton 1982). Edwards (1979) indicated, however, that *Encalypta*, because it did not have the characteristic 2:3 pattern of peristomial development, was probably not a member of the Haplolepideae, while Shaw et al. (1987, p. 68) indicated a close phylogenetic relationship between the Encalyptaceae and Buxbaumiaceae based on their own work and that of others. In any case, a high basal membrane is probably not an advanced character in the Pottiaceae, since four acrocarpous families with papillose or ventrally mamillose, isodiametric upper laminal cells (Buxbaumiaceae, Diphysciaceae, Encalyptaceae and Timmiaceae) have high basal-membraned inner peristomes. Of these families, all have either reddish orange or deep red laminal KOH color reactions (in *Encalypta* restricted to the basal cells), which indicates that a yellow KOH reaction is an advanced character.

Edwards' (1979) indication that the Ptychomitriaceae (two stereid bands, yellow in KOH) belong with the Haplolepideae is agreeable on general morphological grounds: none of the major gametophyte features would be out of place in the Pottiales, while the sporophyte is also similar to those of the Pottiales, excepting only the large, plicate, mitrate calyptra of *Ptychomitrium*.

Species of Calymperaceae examined reacted to KOH with a

yellow color. Edwards (1980a) characterized the Calymperaceae as having hyalocysts abruptly differentiated from the green upper laminal cells and having resorption pores, teniolae (a strong, mostly intramarginal leaf border) present, and propagula borne on the lamina. Only the presence of resorption pores is constant, according to Edwards. All the characters noted by Edwards (1980a) are present in the Pottiaceae. In combination, however, these characters are rather different than those of the Pottiaceae. Reese (1987b) listed several advanced characters of the Calymperaceae (Pottiales), including tropical environment, corticolous habitat, well-defined cancellinae (basal "windows" of hyaline, inflated laminal cells) with superficial pores, reduced or absent peristomes, broad and sheathing leaf bases, and propagula usually present. The differences between Calymperaceae and Pottiaceae are discussed at length by Reese (1987b) and Reese and Zander (1988). *Hypodontium* (q.v.) is here included in the Pottiaceae although previously long accepted as Calymperaceae (see also discussion of *Calymperastrum*); Reese (1987b) felt that *Hypodontium* is intermediate between Calymperaceae and Pottiaceae.

Regarding evidence of gametophytic characters, certain families with somewhat similar gametophytes and haplolepideous peristomes (Calymperaceae, Encalyptaceae, Dicranaceae, Ditrichaceae and Seligeriaceae) are shown by Kawai (1968) to have both one and two stereid bands, and costal morphology with clear differentiation into dorsal and ventral epidermises and medial guide cells. Other families gametophytically similar to the Pottiaceae (Timmiaceae, Ptychomitriaceae) have two stereid bands. Other, less closely related families (Mniaceae, Polytrichaceae) have one or two stereid bands. The Orthotrichaceae, which often have the general appearance of Pottiaceae, have mostly a single stereid or substereid band and poor costal differentiation into cellular layers. The Bartramiaceae, on the other hand, have a single band and good cellular differentiation, as have certain genera of Grimmiaceae. Pleurocarpous families appear to have internally fairly undifferentiated, thin costae. Given the often rather high degree of costal differentiation of closely related families and of such relatively primitive families as Polytrichaceae (see cladogram of Mishler 1986b), one might expect that the Pottiaceae and related families were derived from ancestors in which two stereid bands and possibly a high basal membrane, too, had already been evolved. Additional evidence against convergence is that certain pottiaceous taxa normally having a single stereid band (species of *Pseudocrossidium*, *Tortula s. lat.*, *Streptocalypta*) may develop a few stereid cells between the ventral epidermis and the guide cells, an exact positioning more indicative of partial desuppression (*cf.* Basile & Basile 1984) of a ventral stereid band than of *de novo* elaboration of one. Thus, the line of costal evolution in Pottiaceae probably began with taxa with two stereid bands and progressed through loss of the ventral band to those with a single strong stereid band. The fact that some taxa that usually have two stereid bands may have only one band in occasional collections of small stature is probably not apropos here, since genera with only one band may be quite large in stature (e.g. *Crumia*, *Hennediella*, *Syntrichia* and *Tortula*) in some species and have their own reduction series involving sporophyte characters and stem length but without much variation in costal morphology. Reduced sporophytes are correlated with small gametophyte stature (but *cf. Willia*).

Cladogram 1, discussed at length below, of both the genera of the Pottiaceae and haplolepideous and diplolepideous genera selected for having gametophytes similar to those of the Pottiaceae, with scored character states restricted to those found in the Pottiaceae and using *Polytrichum* (Polytrichaceae, which has many of the anatomical and morphological characters of the Pottiaceae) as outgroup, shows most non-pottiaceous genera derived from lines of the Pottiaceae. This may not show the relationships of non-pottiaceous genera to each other, since non-pottiaceous characters are not used, but it at least indicates likely relationships based on shared apomorphies of these genera with the Pottiaceae. The distance in steps of non-pottiaceous genera from the root of the tree indicates the degree of weighting of particular pottiaceous or non-pottiaceous (and not used) characters that would be necessary to bring these genera down to the base of the tree; it also indicates that certain genera (e.g. *Ceratodon* and *Encalypta*) high in the tree have a large number of advanced characters that can only doubtfully be accommodated by weighting. It will be important to study details of the peristomes of all taxa if sufficient additional characters are to be found to justify viewing the Pottiaceae as presently constituted as monophyletic.

Intrafamilial relationships of the genera

It will be apparent from the discussions presented here that there are probably far more genera worthy of recognition than are presently accepted; however, much more work will be necessary to adequately identify and describe these additional genera. In the absence of a massive cladistic evaluation at the species level, there are two obvious places to look for unrecognized genera; first, as transformation series within large genera, such as *Trichostomum* (q.v.), and, second, as rather different species brought together in small "wastebasket" genera, such as *Gyroweisia*. If clusters of closely related species were recognized as genera or sections, there would be at least double the number of supraspecific taxa presently accepted. On the other hand, segregation of such clusters would leave an overabundance of species not assignable to any segregate and remaining in the original genera by default.

"Good" characters may be variable. Many species in the Barbuleae vary in production of a second stereid band (usually in well-developed plants or larger leaves). Thus, it is the *potential* for development of a second stereid band that provides this traditionally major character. A complication is the occasional production of a second stereid band in some species otherwise clearly belonging to the Pottioideae. One of the aims of this study was to examine the possibility that separating genera at the suprageneric level on the basis of presence or absence of a second stereid band may mask relationships of closely related species losing a second stereid band in a reduction series. Certainly small plants of *Didymodon* and *Gymnostomum*, among other genera, have only one stereid band. It was found, however, that there were few (e.g. *Calyptopogon* and *Streptopogon*) clear-cut overall morphological similarities between species traditionally distinguished by numbers of costal stereid bands that might be inferred as being due to a close phylogenetic relationship marred only by lack of a second stereid band, i.e. the number of stereid bands is a valuable character.

A "separate evolution of gametophytic and sporophytic characters" (*cf.* discussion by Rohrer 1988) is often invoked to explain incongruent character assemblages. Although sporophyte characters have much utility at the suprafamilial level in the mosses, only a few such characters in the Pottiaceae were found important at the generic level (*cf.* treatments of *Aschisma*, *Trachycarpidium* and related genera) in the Pottiaceae. Many character states once widespread among ancestors of present-day mosses may now be suppressed or lost. In the Pottiaceae, few uncommon characters, if any, are found in one genus and not also in another, apparently distantly related, genus. Striking convergent evolution is apparent in a large number of cases. This may even include the one most obvious "unique" character in the family, the twisted peristome of many of the genera. Although Mishler (1985b, p. 389) could identify no apomorphies for the Pottiaceae, the twisted peristome may be considered a synapomorphy for the family if genera of apparently reduced and simplified morphology, including shortening (and subsequent straightening) or loss of the peristome are considered derived, or if such genera are considered relicts of larger lineages with members having twisted peristomes now extinct. A twisted peristome, in any case, is present in widely divergent pottiaceous genera and not in others, and it is present in two species of *Ditrichum* (Ditrichaceae), namely *D. tortipes* (Mitt.) Par. and *D. ambiguum* Best (*cf.* Crum & Anderson 1981, Grout 1927 and Robinson 1968). These two species of Ditrichaceae (discussed further under treatment of *Barbula*) have a distinctly twisted peristome much like that of the Pottiaceae, and the laminae of these species have yellow KOH reactions. These two species or the genus *Ditrichum* itself may actually belong near *Barbula* sect. *Hydrogonium*. One might note, in this respect, that *Distichium* (Ditrichaceae) has a pottiaceous peristome, albeit mostly untwisted.

Apparent convergences may be genetically complex: one might cite the production of a propaguliferous leaf apex in *Leptodontium proliferum*, *Tortella fragilis*, and *Syntrichia angustifolia* with associated elongation of the upper marginal laminal cells, and the fleshy capsules of *Tridontium* (here placed with the Scoulerioideae of the Grimmiaceae) and *Dialytrichia*; but more usually such convergences involve apparently simple traits, such as basal margins of lamina serrulate in *Eucladium verticillatum*, *Molendoa hornschuchiana* and occasional collections of *Gymnostomum aeruginosum*; the lack of a ventral stereid band in *Calyptopogon, Pseudocrossidium, Streptopogon* and *Tortula*; systylious capsules in *Hymenostylium* and species of *Hennediella* (taxa previously recognized under *Desmatodon*).

Although certain families of mosses (e.g. Archidiaceae, Ephemeraceae) are rightly characterized by relatively small sporophytes with short setae and cleistocarpous capsules, the assumption that these characteristics are conservative in the Pottiaceae has proven to be unacceptable to certain modern authors, e.g. those dealing with the *Weissia-Hymenostomum-Astomum* complex. In the Pottieae, *Pterygoneurum* is an example of a genus with so distinctive a gametophytic character (the ventral costal lamellae) that species with sporophytes with peristomes and species without peristomes have long been acceptable within that genus. Elsewhere in the Pottieae, species with eperistomate capsules have occasionally been thrown by previous authors into genera (e.g. *Tortula*) that are generally peristomate but with gametophytes that are fairly heterogeneous, and the complexity of these groups of disparate species has discouraged further work by traditional methods. Such reduction is evident in other acrocarpous families; for instance, Fife (1985), in his treatment of the genera of the Funariaceae, stated that "The dominant evolutionary trend in the Funariaceae has been the progressive simplification of the sporophyte." He associated large stature, small spores and "elaborate morphological structures for the regulation of spore release" in patchy environments through wind dispersal, being most suitable for opportunistic species, and at the other end of the spectrum small stature, large spores and lack of peristomes or even opercula is associated with growth in habitats widely and continuously or at least periodically available, often characterized by disturbance or flooding. A parallel is evident in the Pottiaceae, although the gametophytes of the Pottiaceae are not nearly as uniform in morphology as those of the Funariaceae. This last supports my apprehension of several independent reduction series. On the other hand, Mishler and Churchill (1987) emphasized that "postulating reduction series...depends on the use of outgroup comparison, not appeal to an *a priori* evolutionary scenario...," a caveat addressed below in the phylogenetic analysis.

Longton (1988) reviewed the literature on life history strategies in desert-dwelling bryophytes and noted that "most bryophytes are poikilohydric: they have little control over rates of water uptake or loss, but in compensation the gametophytes of many species can tolerate severe and prolonged cytoplasmic desiccation." He found several trends, among them: "...towards reduced gametophyte life-span, early and prolific production of spores and/or asexual propagula, and monecism in sexually reproducing species: shifts from K- to r-selection and from tolerance to evasion of environmental stress" and "...towards reduced duration of habitat availability but with habitats tending to recur predictably within a community. Parallel trends towards increase in spore size and loss of feature[s] promoting spore dispersal in regularly sexually-reproducing species." The apparent morphological reduction series seen in some genera of Pottiaceae (notably *Hennediella*, *Tortula* and *Weissia*) is possibly associated with increased fitness in the environments of arid lands.

Many nomenclatural novelties are presented here, most being just transfers for long-unstudied exotic species necessary to reflect modern generic constructs, but many others involve recognition of probable evolutionary tracks often signalled by unique character states or character state combinations.

A beginning at a new synthesis based on recognition of apparent transformation series that may cut across traditional generic limits is indicated in the discussion of *Trichostomum*, *Weissia* and related genera, and the *Phascum-Pottia-Desmatodon-Tortula* complex. An alternative view emphasizing the taxonomic utility of peristome morphology in the Pottiaceae is presented at length by Hagen (1929, p. 8–13), probably as a reaction to Dixon's (1924) extensive generic lumping in the family. Reduction of the sporophyte in Pottiaceae (and other acrocarpous families colonizing dry areas) may be assessed as an evolutionary adaptation correlated with two environmental influences: (1) capsules may be exerted to their detriment above the boundary layer of stagnant humid air for considerable periods of time (sporophytes generally take one year to mature, *cf.* discussion by Zander 1979d), and (2) fitness of mosses with massive diaspores, such as large spores and cleistocarpous capsules, or with diaspores developed close to the substrate ("atelochory" in the terminology of van der Pijl 1972, and "precinctiveness" of Carlquist 1966, 1974) may be improved in patchy environments (*cf.* discussion by Zander 1979f).

Phylogenetic evaluations: discussion

Phylogenetic opinions of past authors were apparently based on one or more of a combinations of "marker characters," being unique or rare shared characters, of presumed great phylogenetic weight, to distinguish genera and higher taxa. The mass of non-weighted characters were apparently assumed to be considerably less important and prone to abundant convergence and parallelism. One must evaluate, however, the assumption that morphological convergence between taxa in a particular (especially stressful) environment may involve the evolution quite naturally of a combination of many characters the existence of which in clearly unrelated taxa in the same environment must otherwise be imagined as improbably fortuitous. (Thus the particular low weighting of 22 reduction-related characters in the cladistic analysis below.)

Taxonomic evaluation of intrageneric relationships is here mostly based on general phenetic similarity; when transformation series are clear, however, the generic limits as given presumably reflect probable intrageneric phylogenetic relationships, i.e. grouping those taxa most readily derived from common ancestors through evolution of traits that are unique or rare in the immediate group. In the case of what appear to be extant little-different descendants of ancestral species, the decision whether to keep closely related species representing a branching series together in one genus or assign them (perhaps arbitrarily) to their respective, different phyletic lines is here usually made in favor of the latter. Thus, *Pseudocrossidium crinitum* is quite closely related to the genus *Barbula* (differing mainly in the shape of the section of its dorsal stereid band and in the sheathing perichaetial leaves), yet because it may be construed as a stem species for the *Pseudocrossidium* transformation series, it is placed with that genus.

It may also prove fruitful to examine the Barbuleae (as was begun by Saito in his lucid treatment of the distinguishing characteristics of *Barbula* and *Didymodon*) and identify transformation series that begin in genera with well developed peristomes and end in those with reduced or absent peristomes. This may result in some splitting of the well-peristomed genera and possible disappearance of some eperistomate genera, e.g. *Gymnostomum* and *Gyroweisia*. A typical transformation series is described in a synopsis of *Pseudocrossidium* (Zander 1979f) in which a south to north transformation series from the Andes to the Arctic showed that *P. revolutum* and *P. hornschuchianum* belonged in the genus even though lacking many of the more salient characters (but possessed of the distinctive costal section anatomy).

Unfortunately, in this light, revisions of single genera of the Pottiaceae will commonly not reveal complete transformation series. If such series are found to be worthwhile as contributing to the clarification of generic limits, future revisionists must resign themselves to careful examination of several genera at once to distinguish the full range of potential series. At present, the genera of the Pottiaceae are in some cases neither "natural" nor instructive of possible phylogenetic relationships, and, in fact, many present generic definitions seem to cut across natural lines. Exceptions are probably those genera with sporophytes that have unusual (rare or autapomorphic) features or modifications (mainly of the peristome) different than what may be interpreted as mere reduction. These include *Aschisma* (exothecial cells in palisade-like rows and stomates absent), *Leptodontium* (and other taxa with unvaryingly smooth teeth), *Steptotrichum* (elaboration of internal lateral cell walls interior to the teeth), *Leptodontiella* and *Trachyodontium* (teeth cleft into several rami), *Uleobryum* (and other genera with pustulate exothecial cells), and *Weisiopsis* (16 spaced teeth).

In the absence of a species-based cladistic study, genera are here characterized either by an observed decided lack of species of transitional morphology between two groups (a phenetic gap) or by some obvious division of evolutionary directions (non-rigorous evaluation of salient advanced characters). It is, of course, most satisfactory when both obtain, such as is the case with some of the smaller, isolated genera like *Tetrapterum*. Transitional species between apparent major evolutionary thrusts, however, are disconcertingly common. It is clear that cladistic evaluations are capable of generating more rigorous (based on defined methodologies) and more reasonable classifications than an "omnispection" technique grouping taxa with general overall resemblance. This study attempts a classification based on cladistically derived lineages of genera as these are presently delimited.

An emphasis on using only presumed derived characters (Hennig 1966) as a basis for phylogenetic analysis has been extensively rationalized recently, and has been used in studies of the mosses (e.g. Bremer, 1981; Cao & Vitt 1986; Churchill 1981, 1985; Frahm 1991b; Koponen 1968, 1973, 1982; Mishler 1985b,c, 1986a,b; Mishler & Churchill 1984, 1985b; Rohrer 1985, 1988; Waters et al. 1992, and others). On the other hand, cladistics has become a new orthodoxy (see also Heywood 1983 and Robinson 1986a) that, in sweeping clean, has discounted the value of past classifications; even taxonomy is eliminated from one recent definition our field: "...phylogenetics and biogeography, together referred to as systematics" (Erwin 1991)! Judging from the good match of the three traditional Pottiaceae subfamilies (Trichostomoideae, Barbuloideae and Pottioideae) with the results of the present study, formal parsimony is reflected to a significant extent in past attempts at the creation of a simplest classification that takes into account the available data with emphasis on shared characters or character state combinations, even if the effort does not involve a large data set, base relationships only on derived characters, and use an exact algorithm. It is expected that the generic assignments of infrageneric taxa will be supported to a considerable extent by future species-level cladistic studies, except in genera pointed in the taxonomic treatments as clearly heterogeneous (e.g. *Barbula*, *Didymodon*, *Hyophila*, and *Trichostomum*).

Plausible phylogenetic relationships are satisfactorily inferable within identified transformation series, but the direction of evolution is sometimes far from obvious. A series might start in the middle, at either end, or in two or more places. Assuming, however, that similar complexes of many apomorphic characters are more likely to be derived from shared ancestors than evolved anew for each set of individuals supports an interpretation of most transformations as reduction series. This does not imply that reduction is a process that occurs more easily than elaboration, but that an explanation of a transformation series as reduction requires the assumption of fewer parallel evolutionary events than does elaboration.

There are two ways to view "splitting" and "lumping." The first is to see these as two different ways of deviating from good taxonomy. The second, and better approach, is to see these as analytic and synthetic methods of dealing with information, splitting thus being appropriate at exploratory stages and lumping at revisionary stages.

Many of my publications attribute rather wide variation to certain species and species complexes (e.g. *Didymodon fallax* complex, *Didymodon vinealis* complex, *Hymenostylium recurvirostrum* complex, *Molendoa sendtneriana s. lat.*). For example, see charts of variation in *Hymenostylium* (Zander 1977c, Zander & Eckel 1982), which demonstrate independently varying character states and are surely evidence of considerable genetic variation. It should not be surprising that such intraspecific phenetically distinctive swarms exist and can be expected to exemplify complex intraspecific genetics inasmuch as genetic expression in mosses is expected to be essentially genotypic in the gametophytic generation. Additionally, the considerable variation within species or species complexes confounds easy identification. Often, characteristics that may not always be present, but are nevertheless distinctive, may be helpful. For instance, peculiar to the *Didymodon vinealis* group is the usual absence of the quadrate ventral costal cells at the extreme leaf apex, resulting in a short, boat-shaped groove bottomed by epapillose elongate cells. *Hymenostylium* species commonly have longitudinally elongate medial laminal cells, but this is not always the case.

Species are here seen as inferential populational units composed of individuals sufficiently similar to form a probable biologically predictable unit, and when this is not clear (such as when species are known from single gatherings or when taxonomy is doubtful), a species is put forward as representing a presumed "basic taxonomic unit." Genera are more or less clear-cut apparent phyletic series of species or groups of such phyletic series, or simply groups of similar species when no evolutionary series are evident. Unusual characters are viewed here with some diffidence: the lens-like central thickening of the exothecial cells of *Byroceuthospora*, *Uleobryum* and *Trachycarpidium* is a clearly advanced feature, yet this is not a complicated character and may well have been developed through evolutionary convergence. If these three genera were seen, however, as a single suprageneric taxon, the quite different morphology of the gametophytes would require that they be considered remnants of a once larger, complex group with taxa of intermediate morphology now extinct. Since these are arid-land genera with austral affinities, such a view is an acceptable possibility and the cladistic analysis below supports it.

PHYLOGENETIC ANALYSIS

The object of the study was to obtain a most-parsimonious (minimized homoplasy) hypothesis of evolutionary relationships of the genera of the Pottiaceae as presently conceived, and to use this as a basis for a suprageneric classification. This evaluation is complex and a précis, as overview, follows.

SUMMARY OF THE ANALYSIS

Timmia (Timmiaceae, diplolepideous) and *Polytrichum* (Polytrichaceae, nematodontous) have gametophytes that are quite similar to certain genera of Pottiaceae, but are considered distant from the Pottiaceae and from each other because of their much different peristomes, which they share with other genera not in the Pottiaceae. Because these two genera are both distant and share many characteristics with the Pottiaceae, they are necessarily expected to appear at the base of any cladogram including both them and the Pottiaceae. The Polytrichales is shown to be more primitive than the Bryales in cladistic evaluations by Mishler (1986b) and Mishler and Churchill (1984) although the Buxbaumiales is the nearest sister group (but which is not used here as an outgroup because of a lack of characters that is probably due to reduction).

Of the various haplolepideous genera in families evaluated as potential outgroups (Cladograms 1–6), *Ptychomitrium* (Ptychomitriaceae) was the outgroup that best kept *Timmia* and *Polytrichum* low in the tree when ten non-pottiaceous genera were added to the Pottiaceae data set. Analysis was then made with only a single non-pottiaceous genus as outgroup. With *Ptychomitrium* as outgroup (Cladogram 10), *Timmiella* was found to be the most primitive pottiaceous genus, and this genus was also at the base of the tree with *Polytrichum* (Cladogram 7) and *Timmia* (Cladogram 9) as outgroups. *Timmiella* (Pottiaceae) was then used (Cladograms 13–16) as functional outgroup to avoid the possibility that *Ptychomitrium* was less distant from the Pottiaceae than *Aloina* (*cf.* Cladograms 1, 4 and 8) or *Gertrudiella*, and to avoid problems introduced by reduction in *Ptychomitrium*.

Cladograms 1–4 and 7–10 support the use of *Timmiella* as functional outgroup in Cladograms 13–16. They likewise show that at least some genera of Pottiaceae retain characteristics of the more primitive mosses *Polytrichum* and *Timmia*. Also, putative sister group genera (*Bryobartramia, Ceratodon, Diphyscium, Ditrichum, Encalypta, Grimmia, Syrrhopodon*) in the Pottiales, Dicranales, and Grimmiales are mostly too modified in morphology to serve as outgroups in that too many characters are lacking or fixed in states deemed here to be the result of evolutionary reduction. These cladograms demonstrate that many genera are found in similar or even identical lineages in cladograms generated with different outgroups, implying evolutionary development of quite distinctive synapomorphies and supporting more firmly the present hypothesis of relationship.

Cladograms 11 and 12 are from data sets restricted to terminal taxa of two major branches of Cladograms 9 (outgroup *Timmia*) and 10 (outgroup *Ptychomitrium*) that are identical between cladograms. These consensus trees are based on a known number of equally parsimonious trees and generally support details of branching patterns in the other cladograms, for which only a portion of all trees could be kept in computer memory and thus contribute to the consensus tree.

The cladogram that presents the hypothesis used here for suprageneric classification is the single tree shown in Cladogram 16. Character state changes for this tree are detailed in Cladogram 15. This single tree was chosen from more than 1250 equally most-parsimonious trees as summarized in the strict consensus tree in Cladogram 14. The major subclades of Cladogram 16 are identified as the subfamilies and tribes recognized in this treatment.

INTRODUCTION:

Outgroup Selection

The argument for past emphasis on sporophyte characters in distinguishing genera of the Pottiaceae is (1) such characters work well in distinguishing taxa in other groups, and (2) taxa sharing similar sporophytes in the Pottiaceae also share similar gametophytes. The present study of the genera of the Pottiaceae has shown, however, that although the general morphology of the well developed capsule and its peristome is similar and typically pottiaceous (with notable exceptions, e.g. *Leptodontiella*, *Streptotrichum*, *Trachycarpidium* and relatives), there is considerable variation in expression of individual sporophyte features, i.e. in degree rather than in kind. Also, detailed morphological analyses demonstrate that this variation is often considerable among taxa with gametophytes that share many morphological and anatomical features not previously evaluated across the family (e.g. the genera *Hennediella*, *Tortula* and *Weissia* as emended here). Arguments for special weighting of sporophytic characters (*cf.* Crosby 1974, Dixon 1932, Miller 1979) in the Musci generally are not supported at the generic level in the Pottiaceae.

The rationale used here for selecting an outgroup with shared presumed primitive characters is that haplolepideous sister groups to the Pottiaceae (e.g. Calymperaceae, Ditrichaceae, Encalyptaceae, Grimmiaceae) are evolutionarily much reduced and simplified in both gametophytic and sporophytic morphology. This is true when such groups are compared to the range of morphotypes seen in the Pottiaceae. It is necessary to examine the other haplolepideous and even diplolepideous and nematodontous taxa for genera retaining characters of the ancestral morphology. The "generalized" ancestor is not necessarily a gametophytically relatively characterless taxon (e.g. *Grimmia*), but may be character-rich.

Ptychomitrium (Ptychomitriaceae, haplolepideous) was found to be the closest sister group to the Pottiaceae that was not so morphologically reduced as to have important plesiomorphic character states (also present in more distant ancestors *Timmia* and *Polytrichum* as will be demonstrated below) much modified or completely eliminated. This genus is considered the primary outgroup, and was used to establish a functional outgroup among the pottiaceous genera; cladograms were generated with additional alternate non-pottiaceous outgroups in an attempt to evaluate character state polarizations more globally.

Although its peristome is nematodontous, *Polytrichum* is considered here an acceptable outgroup because it shares many characters with the Pottiaceae that are not associated with apparent reduction. *Polytrichum* is evidently a moss with very primitive characters, sharing with ferns, for instance, a leaf trace connecting the leaf hydroid strand and the stem central strand, and it has been demonstrated cladistically to be of a lineage more primitive than the Bryales (discussion of Mishler & Churchill 1984). *Polytrichum* is also distant from the Pottiaceae because of (1) the several unique characters of its peristome, which presuppose a considerable lineage involved in their development, and (2) the number of the species in its genus (and family), representing multiple evolutionary events. The similarity of the gametophytes of several genera of the Pottiaceae with *Polytrichum* indicates that the Pottiaceae, unlike other families of the Pottiales, retains members with primitive morphologies. Characters of *Polytrichum* that are, on the other hand, typical of reduced members of the Pottiaceae include epapillose upper laminal cells, lamellae present on the ventral surface of the costa, and annulus vesiculose.

The heterolepideous genus *Timmia* (Timmiaceae, diplolepideous) is also a major source of information on plesiomorphic features of the Pottiaceae. This genus of the monotypic family Timmiaceae is extraordinarily similar in gametophytic characters to *Timmiella*, *Gertrudiella* and other large and presumably non-reduced genera of the Pottiaceae. Like *Polytrichum*, *Timmia* has a wealth of characters, all of which are found in the Pottiaceae except the following: inclined capsule position, stomates occurring in several rows in the lower half of the capsule, exothecial cells with sinuous walls, operculum short-mammillate, outer peristome is present, and inner peristome, although similar to that of the Pottiaceae in being filamentous, with 64 segments, and the Primary Peristomial Layer to Inner Peristomial Layer cell ratio (*cf.* Edwards 1979, Shaw et al. 1989) is apparently that of the Diplolepideae rather than the Haplolepideae. *Timmia* shares the diplolepideous type of peristome with other families of non-pottiaceous taxa, and would not be expected to appear high in the Pottiaceae tree. Vitt (1984) treats *Timmia* as quite a derived taxon in the Bryales.

Timmia, *Timmiella* and *Gertrudiella* are also quite like *Polytrichum* in range and development of gametophytic characters; *Polytrichum* is possibly farther removed from the Pottiaceae than *Timmia* because of its nematodontous peristome, but it must be recognized that the sharing of numerous gametophytic characters generally not found in similar combination elsewhere in the mosses is evidence of a close phylogenetic relationship between these haplolepideous, diplolepideous and nematodontous genera. The weighting of the three peristome types depends (or should depend) on the number of characters involved in each and is not pursued to any extent in this work. Consideration should be given to the possibility that haplolepideous families may have evolved more than once, and that past weighting of certain characters of the peristome (e.g. presence or absence of the outer peristome) should be eliminated. Shaw et al. (1989) found that although peristome developmental data unite haplolepideous mosses, there is as yet no information as to whether or not this is a synapomorphic condition or not.

Timmia also has characters associated with reduced members of the Pottiaceae, including stem sclerodermis absent, leaves tubulose above, costa elliptical in transverse section, sometimes monoicous, and spores rather large. It is here considered that the gametophytes of *Polytrichum*, *Ptychomitrium* and *Timmia* represent the character states of a non-reduced ancestor of the Pottiaceae better than do other Pottiales. The possibility that the Pottiaceae may be been derived from a reduced immediate ancestor (e.g. other Pottiales) is less probable because the character state combinations of the many Pottiaceous taxa of large stature, which share so many characters with *Polytrichum* and *Timmia*, must then have been derived independently. The nematodontous peristome of *Polytrichum* and its relatives in the Polytrichaceae may well have been derived from arthrodontous and secondarily eperistomate ancestors with gametophytes similar to that of *Timmia*.

Again, outgroups for the phylogenetic analysis were not selected from only apparently close sister groups, e.g. other Pottiales such as Calymperaceae or Encalyptaceae, or from the Grimmiaceae (see Churchill 1981), because of the potential masking effects of extensive reduction in these groups, at least compared to taxa of the Pottiaceae. This is clearly demonstrated in cladograms including genera of non-pottiaceous families (Cladograms 1–4) in which some non-pottiaceous genera appear, not at the bottom of the tree, but at or near the ends of branches, among what are here considered advanced pottiaceous taxa. Cladograms 5 and 6, with *Encalypta* and *Grimmia* respectively as outgroups, place *Timmia* and *Polytrichum* together at the end of an apparently highly evolved subclade comprising the here-accepted (Cladograms 14–16) basal stem of the Pottiaceae. This would require re-evolution of a large number of character states that in combination phenocopy the gametophytes of two unrelated non-pottiaceous taxa.

The essential characters of many non-pottiaceous sister genera are those of reduction, and are therefore likely to result in convergence. If this is the case, the primitive members of such groups are unknown, and may well not be as similar to the stem genera of Pottiaceae as are *Ptychomitrium*, *Polytrichum* and *Timmia*, which appear near the base of the trees.

Thus, if *Polytrichum* and *Timmia* are both distant from the Pottiaceae and from each other because (1) of the number of unique characters in their peristomes, (2) the number of species (major evolutionary events) in their genera, (3) the fact that other genera share the non-pottiaceous characteristics, and (4) they share more traits with the Pottiaceae than other mosses, then they should appear at the base of the Pottiaceae cladogram. This assumes that the Pottiaceae is monophyletic, of which the twisted peristome found in various subclades is evidence. Of the eight non-pottiaceous genera (other than *Polytrichum* and *Timmia*) that were used as outgroups, only *Ptychomitrium* forced *Polytrichum* and *Timmia* low in the tree. The other seven non-pottiaceous genera are considered, therefore, too modified (probably by morphological reduction and simplification) to act as outgroups in calculating a hypothetical phylogenetic tree.

Other Comments

Although two species of *Ditrichum* (Ditrichaceae), *D. tortipes* (Mitt.) Par. and *D. ambiguum* Best, have the twisted peristomes otherwise unique to the Pottiaceae, the gametophytes of that genus are much reduced, and a case might be made for deriving the genus from ancestors of *Barbula* sect. *Hydrogonium*. The cladograms, however, do not support for this (but see Cladogram 4).

Encalypta (Encalyptaceae), is variously placed in the Haplolepideae or Diplolepideae (see Edwards 1979 and Vitt 1984). *Encalypta* is apparently a closer sister group than *Timmia* because of the quasi-haplolepideous peristome, but there is considerable modification of characters, including monoicy, loss (compared to *Timmia*) of stem central strand (in most species), gain of yellow KOH reaction of upper laminal cells, loss of costal ventral stereid band, and elaboration of additional layers of costal guide cells. *Bryobartramia* (Encalyptaceae, see treatment of excluded taxa) is a taxon probably reduced from *Encalypta*-like ancestors (note that this study will suggest that monotypic genera of reduced morphology are probably relicts of once larger and more complex genera now all but extinct), demonstrating convergence in traits associated with reduction in the Pottiaceae: protonema persistent, small size of gametophyte; very short seta; spherical and cleistocarpous capsule; large spores; and papillose calyptra (that of *Encalypta* is occasionally papillose).

Autapomorphic characters (e.g. the hyaline exothecial cells of *Uleobryum*) were not included in the data set in that they are of no value in determining relationship.

METHODS

The phylogenetic inference method used was that of parsimony; this method has been successful in reconstructing a complex, known, real phylogeny, and predicts ancestral character states (Atchley & Fitch 1991, Hillis et al. 1992). This obtains apparently in spite of problems of present computerized techniques of phylogenetic inference via parsimony, for example, in dealing with treatment of missing entries representing "unknown data, inapplicable data, and polymorphic taxa. Each of those potential sources of ambiguity is logically (if not computationally) different..." (Platnick et al. 1991; also see Doyle 1993, Faith 1991 and Robinson 1986a). Although, ideally, is it best to analyze the limits of genera from a cladogram in which species are the terminal taxa (Funk 1985), this was unachievable with the large number of species involved. The relationships of the genera reflected in the cladograms produced in this study deal with traditional generic concepts derived from overall similarity of the species (see above) and apprehension of morphological "gaps" between genera. Inasmuch as only 75 morphological characters were used, a data set with more than a thousand species would produce a very poorly resolved cladogram. It may be hoped that the present study has resolved the family into smaller groups related by shared apomorphic characters; these groups may be studied separately in the future.

Maximally parsimonious trees were generated using the program Hennig86 (Farris 1988, 1989) with an extended branch-swapping algorithm, that is, the preliminary multiple tree-generating command "mhennig*" followed by an extended branch-breaking command, "bb*". The full data set consists of 75 characters and 86 taxa including ten genera not in the Pottiaceae; the "Pottiaceae" data set consists of a subset of 76 taxa. All multistate characters were treated as additive (ordered) in that all such characters could be viewed as having transitional character states; character states that were not viewed as transitional (e.g. shape of propagula) and which were therefore probably governed by different genetic systems were treated as states of different characters. A character that is variable in state (polymorphic in the sense of Mishler 1990) is scored with a dash ("–")as "unknown, undefined, or missing" (Farris 1988). The particular algorithm used in Hennig86, because of the large data set, apparently produces only a "heuristic approximation," not a guaranteed minimal tree or trees; exact algorithms are, apparently, impossibly time-consuming for data sets larger than 15 to 25 taxa (Sanderson 1990). On the other hand, phylogenetic analysis of this large data set offers a window on evolution in a large and complex taxonomic group.

The order in which data is presented to the computer program affects, according to Maddison (1991), the length of the tree when heuristic tree-searching algorithms are used and retention index is less than .67 and number of terminal taxa is greater than 20. So, to find a short tree for each different outgroup used, the data set was subjected to a minimum of 30 computer runs based on random orderings of the rows of taxon data. A relentless search for the "shortest tree" was deemed useless, however, because of abundant homoplasy, and because the general concurrence of critical groupings between the short trees obtained with different outgroups was considered contributing at least sufficient resolution of patterns for purposes of suprageneric classification.

The number of equally parsimonious trees represented in most strict consensus cladograms is only a fraction of the number possible because of limitations in computer memory; however, the loss of resolution in the consensus trees based on the first 1000 or more trees is very small compared to consensus trees based on only the first 100 trees, and Cladograms 11 and 12 support the fine structure of the rest of the trees, being consensus trees based on smaller data subsets and a definite, much smaller number of equally parsimonious trees.

An Homoplasy-Excess-Ratio analysis (Archie 1989 a,b), which demonstrates the difference between the length of trees generated by the data set and that of those created by randomized data, was not performed. This is because the data set is certainly non-random since it reflects considerable sorting in past studies. A data randomization study is appropriate for studies that require identification of possibly random initial data sets (as in gene studies, *cf.* Waters et al. 1992), or to compare two or more data sets to judge which is more non-random.

The particular polarization of character states of the outgroups used here supports in large part Miller's (1979) list of commonly accepted polarizations or "principles for moss systematics" (Miller warned of occasional reversals in these generalizations in particular families). These polarizations were presumably derived from non-rigorous evaluations but akin to the outgroup criterion. Some of the 26 polarizations listed by Miller are relevant to the present study of the Pottiaceae. Presumed *primitive* traits are summarized as follows: large size, well-developed stem central strand, distinct stem sclerodermis (vs. an undifferentiated cortex), strong costa, epapillose laminal cells; while *advanced* traits include excurrent costa, very thin- or very thick-walled laminal cells, presence of propagula ("specialized diaspores"), axillary hairs with brownish basal cells (vs. all hyaline hairs), monoicy, sporophyte with short seta and immersed capsule, stomata absent, cleistocarpy, and peristome reduced or absent. These intuitive ideas, when compared (1) with the polarizations indicated in the list of characters below and (2) with the character states of the various outgroups contributing to the classification used in this study (*Polytrichum, Ptychomitrium, Timmia* and *Timmiella* in the Data Set) are largely acceptable at least as they apply to the present study.

Advanced traits may be convergent across family lines. For instance, Frahm (1991b) found the following traits to be apomorphic in the Campylopodioideae of the Dicranaceae, many of which are likewise apomorphic in the Pottiaceae: "presence of alar cells, leaf-borne rhizoids, incrassate laminal cells, hyaline leaf tips and strongly differentiated perichaetial leaves, hyalocysts in transverse section of costa...in the gametophyte and presence of an annulus, fringed calyptra, filiform peristome teeth, long lid, stomata and large spore size..." in the sporophyte. Vitt (1984) described an hypothetical ancestor to the Bryales as intolerant of desiccation; perennial; acrocarpous; essentially branchless; stems with hydroids and leptoids; leaves entire, strongly costate, spiralled, little spreading, unistratose; laminal cells thin-walled, rectangular, mostly homogeneous, alar cells not different; laminal papillae absent; dioicous; perigonial paraphyses numerous, thin-walled, apical cell enlarged; seta elongate, with hydroids and leptoids; capsule cutinized, with superficial stomates, photosynthetic, annulus and operculum present, exostome inwardly thickened, endostome segments oppositely placed and partly fused to a membrane, cilia absent; spores numerous and homogeneous in size; calyptra mitrate, hairless and epapillose. Eddy (1991) postulated as an "archetype" for the Pottiaceae "an erect, tufted or gregarious moss, 0.5–2 cm tall, with lanceolate, rather opaque leaves that lack either a border, teeth or specialized basal cells"; the present study has managed to clothe this simple pro-pottiaceous structure with many additional plesiomorphic character states.

CHARACTER STATES OF THE DATA SET

The character states scored for the data set are listed below; (R) = reduction-related character, of which there are 22 and weighted lower than other characters in most cladograms, (0×) = character not used at all in Cladogram 15 (i.e., not phylogenetically informative), (1×) = character occurred once in Cladogram 15 (i.e. phylogenetically very informative), and * = the hypothesized plesiotypic character state for the Pottiaceae. Character numbers correspond to the those in the data matrix used with Hennig86, and begin with zero.

0. (R, 0×) plant conflux
   0. gregarious or scattered
   1. *caespitose, usually in a mat or turf
1. (R) length of stem
   0. short, less than 1.0 cm, usually less than 0.6 cm
   1. *relatively long, usually 1.0 cm or greater
2. (1×) stem transverse section shape
   0. *rounded-pentagonal
   1. triangular
3. stem central strand presence
   0. absent
   1. *present
4. stem sclerodermis presence
   0. *not or little differentiated from central cylinder cells
   1. clearly differentiated
5. stem hyalodermis
   0. *absent
   1. present (sometimes small)
6. (1×) axillary hair basal cell walls
   0. hyaline and all cells of hair similar
   1. *of 1 or more cells with thicker or darker colored walls
7. leaves when dry
   0. tubulose
   1. *occasionally channeled but not distinctly tubulose
8. leaves when wet
   0. appressed to weakly spreading
   1. *widely spreading to squarrose
9. (0×) leaves, size gradation
   0. *gradually becoming larger distally on stem
   1. rosulate (abruptly larger distally)
10. leaf shape
    0. long-triangular to linear-lanceolate
    1. *lanceolate
    2. broadly ligulate-elliptical to spathulate
11. (R) leaf length
    0. less than 1.5 mm
    1. 1.5 mm to 3.0 mm
    2. *3.0 mm or more
12. leaf ventral surface above midleaf

0. *nearly plane to broadly channeled
1. keeled
13. rather deep, narrow groove along costa
0. *absent
1. present
14. margins
0. sharply incurved
1. *incurved or involute
2. plane
3. recurved to revolute
15. marginal toothing
0. entire or minutely and evenly crenulate
1. denticulate or serrulate
2. *strongly toothed
16. margins specially denticulate
0. *not denticulate only below
1. denticulate only near leaf base or at top of leaf sheath
17. (1×) upper marginal cells differentiation
0. *same number of layers as medial cells
1. always differentiated as a bistratose (or more) border
18. upper marginal cells, elongation
0. *not longer than medial cells (sometimes larger)
1. rectangular and clearly longer than medial cells
19. (1×) leaf apex
0. rounded
1. *acute to acuminate
20. leaf apex ventral surface
0. *flattened, channeled or keeled
1. cucullate
21. (R) leaf base
0. little differentiated in shape
1. ovate to elliptical
2. *sheathing the stem and commonly oblong and with "shoulders"
22. (1×) costa
0. ending before the leaf apex
1. *percurrent or apiculate or short-excurrent as a mucro
2. excurrent as an awn
23. costal ventral cells viewed from above
0. *quadrate to very short-rectangular, occasionally in many layers
1. longitudinally elongate 3:1 or more
24. number of rows of cells across ventral surface of costa
0. usually 2 but up to 4
1. 4–6
2. 6–10
3. *usually 10 or more
25. shape of transverse section of costa
0. *round to semicircular
1. distinctly flattened, usually reniform
26. ventral stereid band
0. absent
1. *present
27. (1×) size of ventral stereid band
0. *smaller than the dorsal or of nearly equal size
1. larger than the dorsal
28. transverse section of dorsal stereid band
0. round or semicircular
1. *clearly flattened or ventrally indented, reniform or crescent-shaped
29. ventral costal epidermis
0. absent
1. *present
30. dorsal costal epidermis
0. absent
1. *present
31. guide cells
0. usually 0
1. 2–6
2. *commonly more than 6
32. (0×) number of guide cell layers
0. *1–2
1. 3 or more
33. hydroid strand
0. absent
1. *present
34. (R, 0×) ventral costal outgrowths
0. *absent or not a bulging pad of cells
1. a bulging pad of cells
35. (R, 1×) ventral costal filamentous outgrowths (note: exclusive of lamellae)
0. *absent or not of filaments
1. of separate filaments three or more cells in length
36. width of medial upper laminal cells
0. *small to medium sized, 7–14 µm
1. rather large, 15–17 µm
2. very large, 18–25 µm
3. extremely large, 25 µm or more
37. (0×) length to width ratio of medial upper laminal cells
0. *1–2:1
1. 3–4:1 or more
38. (0×) upper laminal cell walls
0. *thin to evenly thickened, lumens quadrate to rounded
1. trigonous or porose, lumens irregularly angular or stellate
39. superficial walls of upper laminal cells
0. flat or very weakly convex on both sides
1. strongly convex to bulging on both sides
2. *ventrally bulging-mamillose, weakly convex dorsally
40. layering of upper laminal cells
0. *medially unistratose
1. medially bistratose, at least in patches
41. (R, 0×) dorsal superficial laminal cell walls
0. *about same thickness as the ventral or weakly thicker near costa
1. throughout distinctly thicker than the ventral
42. (1×) upper laminal papillae
0. absent
1. *present
43. kind of papillae when present
0. simple, hemispherical
1. *bifid to multifid to columniform to capitulate
44. number of papillae per lumen when present
0. one or occasionally two
1. *2–6, usually bifid or multifid
2. many, usually 6 or more, simple or bifid
45. basal cell group
0. not or little differentiated from upper medial cells
1. *clearly differentiated approximately straight across leaf or rising higher medially
2. differentiated as a vee, with at least laterally

differentiated cells rising higher marginally as a border tapering distally

46. propagula presence
    0. *always absent or extremely rare and associated with unusually moist conditions
    1. often present and characteristic
47. (1×) position of propagula when present
    0. not borne on rhizoids as brood bodies, or absent
    1. borne on rhizoids as brood bodies
48. position of propagula when present
    0. not born in leaf axils, or absent
    1. borne in axils
49. (1×) position of propagula when present
    0. not borne on leaf costa or lamina, or absent
    1. born on leaf costa or lamina
50. shape of axillary propagula when present
    0. not clavate or filamentous
    1. clavate or filamentous
51. shape of axillary propagula when present
    0. not branching or stellate
    1. branching or stellate
52. (R) sexual condition
    0. *dioicous
    1. monoicous
53. (0×) perichaetium
    0. *terminal on main axis
    1. lateral on main axis at ends of very short branches
54. perichaetial leaves
    0. *similar to cauline leaves or occasionally smaller or somewhat enlarged
    1. distinctly different in size or morphology, sometimes strongly sheathing
55. (R) length of seta
    0. nearly absent to short, less than 1 cm
    1. *elongate, 1 cm or longer
56. (R) seta twist
    0. *usually twisted clockwise
    1. never twisted
    2. usually twisted counterclockwise
57. (R) theca length
    0. short, 1.5 mm in length or less
    1. *usually between 1.5 and 3.5 mm in length
    2. 3.5 mm in length or more
58. (R, 1×) shape of theca
    0. spherical and non-apiculate
    1. spherical and apiculate
    2. *ovoid to cylindrical
59. (R, 1×) capsule dehiscence
    0. cleistocarpous
    1. *stegocarpous
60. (R, 1×) capsule ornamentation
    0. *surface nearly smooth
    1. surface evenly mamillose or with distinct protuberances of strongly bulging cells basally or throughout
61. (R, 1×) exothecial cells
    0. with walls superficially thickened centrally and lens-like
    1. *with evenly thickened superficial walls
62. (R, 0×) capsule
    0. cleistocarpous and rupturing mainly along weak transverse walls at butt ends of long-rectangular exothecial cells
    1. *rupturing irregularly if cleistocarpous, or stegocarpous
63. (1×) stomates
    0. absent
    1. *present
64. annulus
    0. of weakly differentiated cells
    1. *of vesiculose cells, often in two or more rows
65. (R) peristome teeth
    0. absent
    1. *present
66. (R) peristome teeth when present
    0. rudimentary, a low plate or short-elliptical in shape
    1. 16, often cleft to near base into two or more rami, or of 16 pairs
    2. *32 similar rami
67. (0×) ornamentation of peristome teeth
    0. smooth
    1. striate or ridged
    2. *spiculose or papillose
68. (R) peristome teeth twist
    0. untwisted or very weakly so
    1. *distinctly twisted
69. (0×) peristome teeth if twisted
    0. clockwise
    1. *counterclockwise
70. (0×) calyptra ornamentation
    0. *smooth or nearly so
    1. papillose, distinctly roughened or strongly mamillose
71. (R) length of calyptra
    0. short, less than 1 mm
    1. 1–3 mm
    2. *more than 3 mm
72. (R) spore diameter
    0. *8–15 μm
    1. more than 15 μm
73. KOH color reaction of upper laminal cell walls
    0. essentially yellow
    1. essentially orange
    2. *essentially red, usually a definite brick red
74. (R, 1×) calyptra shape
    0. *cucullate or at least long-conic and cleft, usually more than 1 mm in length
    1. mitrate or short-conic and uncleft, commonly lobed, usually less than 1 mm in length
    2. campanulate-inflated

Discussion of Character States and Polarity

Outgroups appropriate for analysis need to possess most of the characters of the Pottiaceae, but be sufficiently distant phylogenetically as to share a common ancestor with all genera of the Pottiaceae. Inasmuch as closely related families are much reduced, choosing the appropriate outgroup required considerable analysis, described below. Basically, the rationale was that *Timmia* and *Polytrichum* were similar in morphology gametophytically to certain genera of the Pottiaceae but had peristomes of apparently considerable evolutionary distance. The particular haplolepideous genus used here as outgroup that placed these two distant but related genera low in the tree when both pottiaceous and non-Pottiaceae genera were included in the cladogram would be the appropriate outgroup. Ten non-pottiaceous genera were used as outgroups to construct a series of cladograms, which were then evaluated individually and compared. Three genera were similar in gametophyte characters to those of particular apparently non-reduced Pottiaceae, but which have quite different peristomes: *Polytrichum* (nematodontous), *Timmia* (diplolepideous) and *Ptychomitrium* (haplolepideous, like the Pottiaceae). These genera grouped near the base of the tree whenever one or the other of the three were used as outgroup. The characters of these genera are taken as plesiomorphic.

It is more probable that the many character states (75 characters are scored in this study) for each character of large-statured genera of Pottiaceae being in large part identical to that of *Polytrichum*, *Timmia* and *Ptychomitrium* is due to sharing those of a common ancestor than to multiple development of these many identical traits. There is no evidence that such traits are not homologous. On the other hand, the probability that *Polytrichum* is derived from a diplolepideous ancestor is a position not supported by Edwards' (1979, 1984) extensive evaluations of peristome structure.

In a study of this large number of terminal taxa it is doubtful that many of the characters are globally homologous because of their apparent lack of genetic complexity. These 75 characters were selected for analysis because homology of the character states is supported by similar morphological positions of the states, the fact that the characters remain similar through transformation series intragenerically, and because other characters correlate with them intragenerically. Additional work, however, along the lines reviewed by Wiley (1981, p. 130–158), might reduce homoplasy in the resultant analysis.

The many quantitative characters used in this analysis have assigned states that are not arbitrarily circumscribed, but correspond to perceived discontinuities in ranges of values, meaning that the genera as a whole can usually be easily assigned a particular range of values (e.g. lengths, diameters) for each state of each character. No statistical studies, however, were done beyond an informal evaluation of the actual data for the genera as shown in the data file compiled for the genera during the study, in which discontinuities in measured states for the 76 genera were usually obvious.

The list below evaluates the polarity of each character of the Pottiaceae used in this analysis. There was no one outgroup taxon found that had all states plesiotypic as per this analysis. The plesiotypic state is simply stated as such when candidate outgroups *Polytrichum*, *Timmia* and *Ptychomitrium* have a similar, presumed homologous feature, otherwise exceptions are discussed. Characters for *Polytrichum* are taken in part from G. Smith (1971); *Encalypta* in part from Horton (1982 & 1983); *Timmia* in part from Brassard (1979, 1980 & 1984). Discussions of taxa other than these three are provided to better describe particular shared characters or variations that are probably apotypic and commonly associated with reduction. A hypothetical sister group based on this evaluation was, however, not used as outgroup in the actual analysis.

Polarity of Character States

0. Plants growing caespitose is a plesiomorphic condition; the gregarious or scattered state is associated with reduction.
1. A relatively long (more than 1 cm) stem is plesiomorphic, although *Ptychomitrium* has a rather short stem.
2. Rounded-pentagonal stem sections are plesiomorphic, being found in *Ceratodon*, *Encalypta*, *Polytrichum* and *Timmia*; triangular stem sections are rare in the Pottiaceae.
3. Presence of a stem central strand is plesiomorphic, found in *Ceratodon*, *Polytrichum*, *Timmia* and *Ptychomitrium* but usually absent in *Encalypta*; absence of a central strand is, however, not clearly associated with reduction in the Pottiaceae.
4. A stem sclerodermis is clearly differentiated in *Ceratodon* and *Polytrichum* but not is so in *Encalypta* or *Timmia* (outer cortical cells are smaller and only weakly thickened); one might assume the condition of *Polytrichum* is plesiomorphic in association with the large size and complexity of the plant. The situation is variable in *Ptychomitrium*.
5. Lack of a stem hyalodermis is plesiomorphic, being absent in *Ceratodon*, *Encalypta* (occasionally present in patches), *Polytrichum* and *Timmia*, but present in *Ptychomitrium*.
6. Basal cells of the axillary hair being colored rather than hyaline (as is the remainder of the hair) is advanced; this is true for *Ceratodon* and *Timmia*, but not *Polytrichum*. In *Encalypta* (several species examined), the basal cells are brownish, but entirely hyaline hairs have been reported, Horton 1982). *Ptychomitrium* is variable in respect to this character.
7. Leaves variously broadly channeled or tubulose when dry is advanced, this being variable in *Encalypta* and in *Timmia* and not tubulose in *Ceratodon*, *Polytrichum* or *Ptychomitrium*. Again, the condition of the genus with largest plant size is taken as plesiomorphic.
8. The leaves of *Polytrichum* and *Timmia* are strongly spreading, and nearly squarrose when wet; the condition is plesiomorphic. Those of *Ptychomitrium* are less strongly differentiated in this respect.
9. Leaves becoming gradually larger on the stem is the plesiomorphic condition.
10. *Polytrichum*, *Timmia*, *Ptychomitrium* and *Ceratodon* have lanceolate leaves; the condition is plesiomorphic. The spathulate to ovate-lanceolate leaves of *Encalypta* are taken as apomorphic, and, as in the *Tortula* group of Pottiaceae, associated convergently with loss of the ventral stereid band and enlargement of upper laminal cells.
11. Leaves usually relatively long, 3.0 mm or more, is the condition with *Polytrichum* and *Timmia*, and larger species of *Ptychomitrium*, and is plesiomorphic. *Encalypta* leaves are likewise rather large.
12. Although *Ceratodon* has keeled leaves, *Polytrichum*, *Timmia*, *Ptychomitrium* and *Encalypta* have broadly channeled leaves, and the latter is the plesiomorphic condition.
13. Although *Ceratodon* (and many robust genera of Pottiaceae) have leaves narrowly grooved along the costa, *Polytrichum*, *Timmia*, *Ptychomitrium* and *Encalypta* have a ventrally rather flat upper leaf surface, which is the plesiomorphic state.
14. Type of marginal flexion is a difficult character to polarize satisfactorily but is taxonomically critical. The upper leaf margins of *Polytrichum*, *Timmia*, *Ptychomitrium*, *Diphyscium* and the Dicranaceae in general are plane to incurved above (like those of *Trichostomum* in the Pottiaceae), those of *Ceratodon* are recurved, those of Grimmiaceae plane to recurved. *Encalypta* has both plane and recurved margins. Both incurved and recurved leaf margins are found in the

Pottiaceae. If such flexion were equivalent evolutionarily (e.g. both providing marginal stiffening), then both directions of marginal recurvature would be apomorphic. Infolded margins, however, provide additional protection in arid environments, and this indicates an advanced condition. The state for *Polytrichum*, however, is taken as plesiomorphic with respect to the Pottiaceae.

15. The upper margins of *Polytrichum*, *Timmia*, *Ptychomitrium* and some species of *Diphyscium* are commonly strongly and distantly toothed, and those of *Ceratodon* are denticulate near the apex; *Encalypta* has entire margins. Robust species in the Pottiaceae commonly have dentate margins, and, given that robustness is here taken as plesiomorphic, the clearly dentate condition is considered likewise plesiomorphic.
16. The marginal ornamentation of *Polytrichum*, *Timmia*, *Ptychomitrium*, *Encalypta*, *Bryobartramia* and *Ceratodon* is never of denticulation restricted to the upper portion of the leaf base; leaf margins not denticulate only below is plesiomorphic. *Molendoa* is occasionally denticulate at the top of the leaf sheath margin while *Eucladium* is nearly always so.
17. *Polytrichum*, *Timmia*, *Encalypta* and *Ceratodon* lack bistratose leaf borders, but these are common in *Ptychomitrium* and *Diphyscium*; the unistratose margin is considered here plesiomorphic.
18. Cells of leaf margins similar to those of the medial leaf is the plesiomorphic condition.
19. Acute leaf apices are plesiomorphic; the apices are obtusely acute to rounded in *Encalypta*, but this is correlated with an apomorphic broadening of leaves and loss of ventral stereid band.
20. The cucullate leaf apex is an apomorphic character.
21. While *Ptychomitrium* and *Ceratodon* have a merely ovate leaf base, the leaf base of *Polytrichum* and *Timmia* sheathes the stem and has "shoulders," the latter being the plesiomorphic state. The leaf base of *Encalypta* is occasionally differentiated in shape, but this is usually masked by the broadened distal portion of the leaf.
22. A percurrent costa is plesiomorphic.
23. *Polytrichum*, *Timmia*, *Ptychomitrium* and *Encalypta* have quadrate to very short-rectangular ventral costal cells, while *Ceratodon* has elongate ventral costal cells; the former is the plesiomorphic state.
24. *Polytrichum* and *Timmia* have a broad costa commonly more than 10 ventral cells across, which is the plesiomorphic state. The costa of *Encalypta* is 6–8 ventral cells in breadth, while *Diphyscium* commonly is 15 or more costal ventral cells across. *Ptychomitrium*, however, has a narrower costa.
25. The costal section of *Timmia* is elliptical, with the ventral surface usually distinctly convex, similar to that of *Encalypta*; *Ceratodon* and *Polytrichum* have a reniform costal section, the latter shape is probably plesiomorphic. The costal section of *Ptychomitrium* is semicircular and is probably derived.
26. Two stereid bands are present in *Polytrichum*, *Timmia*, *Ptychomitrium* and *Ceratodon*, which is the plesiomorphic character. *Diphyscium* commonly has two stereid bands while *Encalypta* has only one.
27. The ventral stereid band being smaller than the dorsal is the plesiomorphic state.
28. *Polytrichum*, *Timmia*, *Ptychomitrium* and *Ceratodon* have reniform, strongly flattened ventral stereid bands. The same is the case with *Encalypta* and *Diphyscium*.
29. Ventral costal epidermis being present is the plesiomorphic state. The lack of this is also rare in the Pottiaceae.
30. Dorsal costal epidermis being present is the plesiomorphic state; it is commonly absent in *Encalypta*, and is rare in Pottiaceae. Perhaps the similar *Syntrichia*, for which the lack of a dorsal costal epidermis is diagnostic, is an example of convergent evolution with *Encalypta*.
31. *Polytrichum* and *Timmia* usually have more than six guide cells (Brassard 1979 illustrates only six for *Timmia*); *Ptychomitrium* and *Ceratodon* have 2–6 guide cells, and *Encalypta* 6–8; a large number of guide cells is the plesiomorphic state.
32. A single layer of guide cells is the plesiomorphic state; *Encalypta* commonly has more than one; this is rare in Pottiaceae.
33. A hydroid strand (occasionally multiple) is present in *Polytrichum*, *Timmia* and *Ceratodon* and is the plesiomorphic state; a hydroid strand is variably present in *Encalypta* and absent in *Ptychomitrium*.
34. Although the ventral surface of the costa of both *Polytrichum* and *Timmia* is distinctly convex, it does not constitute a pad of cells (the longitudinal lamellae of *Polytrichum* are not considered a pad). Lack of a bulging pad of cells is the plesiomorphic state.
35. Ventral outgrowths absent or at least not of filaments is plesiomorphic; the presence of lamellae in *Polytrichum* is apparently advanced and is only matched in the Pottiaceae by *Pterygoneurum*.
36. The upper laminal cells of *Polytrichum*, *Timmia* and *Ptychomitrium* are rather small, ca. 9–14 µm in width, while those of *Ceratodon* are somewhat wider, and those of *Encalypta* 7–18 µm in width; the first state is plesiomorphic.
37. A short length to width ratio is plesiomorphic.
38. Upper laminal cells with thin to evenly thickened walls is the plesiomorphic state.
39. The upper laminal cells of *Ceratodon*, *Polytrichum* and *Ptychomitrium* are weakly convex and those of *Encalypta* strongly convex on both superficial surfaces (occasionally somewhat more bulging ventrally than dorsally in *Encalypta*), and those of *Timmia* are ventrally strongly mamillose and dorsally nearly flat (except *T. sibirica*, in which this difference is both poorly developed and masked by dense papillae). An areolation similar to that of *Timmia* is present in a far larger number of genera of the Pottiaceae than previously recognized (though often masked by papillae). The Calymperaceae (Pottiales) commonly has ventrally bulging upper laminal cells, which is probably the plesiomorphic condition for the Pottiaceae.
40. The unistratose leaf condition is plesiomorphic, though variable in *Polytrichum* and *Ptychomitrium*.
41. The superficial cell walls of the upper leaf being of equal thickness on both sides of the leaf is plesiomorphic. Although the dorsal cell walls of the leaf are commonly thicker than the ventral in the lower portion of the leaves of *Timmia*, the ventral and dorsal walls are of similar thickness in the majority of the limb.
42. The leaves of *Ceratodon*, *Ptychomitrium* and *Polytrichum* lack papillae, those of *Encalypta* are papillose, and those of *Timmia* are variously papillose or smooth. Inasmuch as

most taxa in the Pottiaceae, including all robust, morphologically complex taxa of Pottiaceae (excepting *Streptopogon*), have papillose upper leaf cells, the papillose condition is presumed plesiomorphic and that of *Polytrichum* is advanced.

43. The papillae of *Timmia* are coarse and bifid, and those of *Encalypta* are generally well developed and bi- or trifid. The bifid to multifid state is plesiomorphic. None of the taxa was always capitulate or columniform.

44. Both *Timmia* and *Encalypta* have 2–4 papillae per lumen, the plesiomorphic condition. *Polytrichum* is scored, however, as characteristic not present.

45. The basal cells of *Polytrichum*, *Ptychomitrium* and *Timmia* are differentiated straight across the leaf base, thus being the plesiomorphic condition. They are poorly differentiated in *Ceratodon* and differentiated straight across or rising higher medially in *Encalypta*.

46. Propagula are unknown in *Polytrichum*, *Ptychomitrium* and *Timmia*, and are axillary and filamentous in both *Ceratodon* and *Encalypta*. The former state is the plesiomorphic condition.

47–51. Not polarized.

52. *Ceratodon* and *Polytrichum* are dioicous, *Encalypta* is largely autoicous but occasionally dioicous, and *Timmia* is largely dioicous but also monoicous. *Ptychomitrium* is monoicous. Monoicy, being associated with reduction in the Pottiaceae, is here regarded as apotypic (but *cf.* discussion of Stark and Castetter 1987, p. 195, who find no support for the hypothesis that monoicy is generally favored in arid lands).

53. The terminal perichaetium is the plesiotypic condition.

54. Although the perichaetial leaves are somewhat differentiated in *Ceratodon*, in *Polytrichum*, *Ptychomitrium*, *Timmia* and *Encalypta* they are little differentiated, which is considered here the plesiomorphic state.

55. An elongate seta is the plesiomorphic state.

56. The seta is twisted clockwise in *Polytrichum*, *Ptychomitrium*, *Timmia* and *Ceratodon*. In *Encalypta* it is twisted counterclockwise. The former is taken as plesiomorphic.

57. The thecae of both *Polytrichum*, *Timmia* and *Ceratodon* range from 2.5 to 3.0 mm in length, which is taken to be the plesiomorphic condition. The measurements for *Encalypta* are 1–3 mm, probably associated with reduction.

58. An ovoid to cylindrical capsule is the plesiomorphic state.

59. The stegocarpous condition is plesiomorphic.

60. Smooth capsule walls is the plesiomorphic state.

61. Evenly thickened superficial exothecial cell walls is the plesiomorphic condition.

62. Capsules stegocarpous is the plesiomorphic condition.

63. Stomates present is the plesiomorphic state.

64. The annulus is of vesiculose cells in *Ptychomitrium*, *Timmia*, *Encalypta* and *Ceratodon*, but is weakly developed in *Polytrichaceae*; the latter is probably the plesiomorphic state for the Pottiaceae.

65. Peristome teeth present is the plesiomorphic state for the Pottiaceae.

66. There are 64 inner peristome rami in *Timmia*, 16 of the homologous equivalent in *Encalypta* and *Ptychomitrium*, and 16 pairs of teeth fused at the joints in *Ceratodon*. Inasmuch as there are clear reduction series in certain genera of the Pottiaceae beginning with species with 32 similar rami, this latter condition is taken as plesiomorphic. Peristome teeth characters are scored as character not present in *Polytrichum*.

67. The ornamentation of the teeth of *Timmia* and *Encalypta* (endostomes) and of *Ptychomitrium* and *Ceratodon* is papillose to weakly spiculose in the first two and strongly spiculose in the last two. Ridged, striate or smooth are derived states.

68. Although the peristome teeth of *Ptychomitrium*, *Timmia*, *Encalypta* and *Ceratodon* are straight, having twisted peristome teeth is a major, unique character state of the Pottiaceae, and occurs in morphologically disparate intrafamilial groups; it is assumed to have been a feature of the immediate ancestry of the Pottiaceae and is here taken as plesiomorphic.

69. Peristome teeth, when twisted, wind mostly counterclockwise in the Pottiaceae, rarely clockwise. *Ptychomitrium*, *Timmia*, *Encalypta* and *Ceratodon* have straight teeth, while *Ditrichum tortipes* (Ditrichaceae) has its peristome and cells of the operculum clearly twisted counterclockwise. Since the last taxon may not represent a sister group but may instead be derived from *Barbula* sect. *Hydrogonium*, there is no good outgroup indication of the state of the hypothetical ancestor of the Pottiaceae. A counterclockwise twist is considered plesiomorphic based on ingroup comparisons: twisted peristomes in various major groups of the Pottiaceae are nearly all wound counterclockwise, and parallel derivation from a clockwise ancestor is far less probable than the occasional reversal.

70. Calyptrae are smooth in *Ptychomitrium*, *Timmia* and *Ceratodon*, which is the plesiomorphic state, but variously smooth, papillose or prorulose in *Encalypta*. This character is scored as "not smooth" for *Polytrichum*.

71. The calyptrae of *Polytrichum* and *Timmia* are generally 5–6 mm in length while those of *Ptychomitrium* and *Ceratodon* are reduced in size; the former is the plesiomorphic condition. Those of *Encalypta* are 2–10 mm in length.

72. The spores of *Timmia* are commonly 12–20 µm in diameter, those of *Ceratodon* are mostly 11–15 µm in diameter, and those of *Encalypta* 7–80 µm in diameter. Because larger spore size is generally associated with gametophytic reduction in the Pottiaceae, the smallest spore size state, also typical of *Polytrichum*, is taken as plesiomorphic.

73. *Ceratodon* is yellow in KOH; *Encalypta* is KOH yellow in the upper leaf but the basal cells react red, and mature leaves become KOH red or orange above; *Buxbaumia*, *Diphyscium* and *Polytrichum* leaves are red in KOH. *Ptychomitrium* is variable in KOH reactions. The upper lamina or upper (distal on the stem) leaves of *Timmia* in KOH are either yellow, orange, colorless, or faintly pink above (in *T. austriaca* Hedw.—walls were examined at high magnification under microscope to distinguish the bright yellow-green of the cell contents), occasionally with medial patches of brick-red blush, occasionally intensely brick-red at damaged portions of the leaves; the leaf bases and costae are generally a deep, clear yellow with large brick-red blotches. The botchy condition is similar to that of certain *Tortula* and *Crossidium* species, which are here considered essentially yellow in KOH because most species have strongly KOH yellow upper laminae of both young and mature leaves. *Tortula* (as emended here) and *Crossidium* species do not have leaves entirely with walls pink or red.

The upper laminal cells of the more mature leaves (lower on the stem) of *Timmia* species, generally, have distinctly pink walls throughout; the mature leaf has much the color response as have those of *Bryoerythrophyllum*, here considered essentially red in KOH. Some species of *Bryoerythrophyllum* (*cf.* treatment of that genus) have KOH yellow young leaves but a typically red reaction of more mature leaves. *Timmia* has an essentially red color response of the upper laminae to KOH, which is here considered the plesiomorphic character state for the Pottiaceae.

74. The mitrate calyptra is characteristic only of reduced taxa in the Pottiaceae (excepting *Streptopogon*, which because of its evident lack of characters may be derived from highly reduced ancestors, *cf.* Cladogram 14, with little subsequent elaboration except in stature), and is here considered apomorphic. *Ptychomitrium* has a mitrate calyptra. The inflated campanulate calyptra typical of the Encalyptaceae is probably an elaboration of the mitrate type.

THE CLADOGRAMS

Sixteen cladograms are presented here to summarize the cladistic analyses. Cladograms 17 through 19 interpret diagrammatically trees of other workers at the suprageneric level. In the cladograms given here, dichotomous branching at a node is shown as "—|" while multiple branching at a node is symbolized by a longer vertical line with several horizontal lines appended to the right. Thus all terminal taxa and subclades attached to the right of a single vertical line are branches, e.g. "G" is a multiple-branched node in the consensus trees of Cladograms 3 and 4, and so on. The outgroup is also shown as coming from a multiple-branched node in all cladograms because the algorithm, apparently, cannot evaluate the relative phylogenetic position of the outgroup and the first branch; for the purposes of this study, this can be ignored.

Cladograms 1 through 6 are based on the full data set, which includes information on ten genera not in the Pottiaceae: *Polytrichum* (Polytrichaceae); *Bryobartramia* and *Encalypta* (Encalyptaceae); *Ceratodon* and *Ditrichum* (Ditrichaceae); *Diphyscium* (Diphysciaceae); *Grimmia* (Grimmiaceae); *Ptychomitrium* (Ptychomitriaceae); *Syrrhopodon* (Calymperaceae); and *Timmia* (Timmiaceae). Cladograms 2–4 do the following: (1) compare the cladograms of genera relatively distant from the Pottiaceae, *Polytrichum* and *Timmia*, to those of putative sister groups of the Pottiaceae, and (2) identify the closest appropriate haplolepideous sister group to the Pottiaceae as those two (*Ditrichum and Ptychomitrium*) placing the relatively distant genera *Polytrichum* and *Timmia* closest to the base of the tree. Because *Ditrichum* is potentially a derivative of the *Barbula* lineage or at least may belong to the Merceyoideae (Cladogram 4), *Ptychomitrium* is probably the ideal sister for use as outgroup in analysis of the Pottiaceae.

All the listed non-pottiaceous genera were examined individually as outgroups; Cladograms 5 and 6, with *Encalypta* and *Grimmia* respectively as outgroups demonstrate that, like cladograms with *Bryobartramia*, *Ceratodon*, *Ditrichum*, *Diphyscium*, and *Syrrhopodon* (these not shown), sister groups that are apparently reduced and simplified in morphology create cladograms that place *Polytrichum* and *Timmia* high in the tree. Because the data set used is the same in Cladograms 1–6, the branching structure of Cladograms 2–6 would be identical if they had been able to be calculated exactly. To the extent that the actual heuristically derived trees of Cladograms 2–6 agree, this ideal branching pattern is approximated. Branching patterns of Cladograms 2–6 (as well as the other cladograms) that are well resolved support the classification indicated by lettered lineages in Cladogram 16.

Cladograms 7 through 10 include only pottiaceous genera in addition to a non-pottiaceous outgroup. Subclades that are similar throughout the 16 cladograms and which are also formally named in Cladogram 16 are marked with letters at ancestral nodes. Cladograms 11 and 12 analyze character state changes in two highly evolved subclades that comprise the same species and the same structure in both Cladograms 9 (*Timmia* as outgroup) and 10 (*Ptychomitrium* as outgroup); these are approximately the same as the Merceyoideae and Pottioideae of Cladogram 16 with the major exception that the Hyophileae is not a distinct subclade. Cladograms 13 through 16 involve only genera in the Pottiaceae, use *Timmiella* as functional outgroup, and summarize the best hypothesis for phylogenetic relationships and of a projected suprageneric classification.

Cladograms 1, 2, 7, and 8 have *Polytrichum* as outgroup, *Timmia* is outgroup in numbers 3 and 9, *Ptychomitrium* is outgroup in numbers 4 and 10, *Encalypta* is outgroup in number 5, and *Grimmia* is outgroup in number 6.

Cladograms 1–10 and 13–14 are strict consensus trees, summarizing the information from many equally parsimonious trees. Other cladograms are of single trees selected from sets of equally parsimonious trees (one set for each outgroup), usually through the use of the functional ingroup and functional outgroup (FIG/FOG) method of Watrous and Wheeler (1981) on various subclades that have multiple-branched nodes in the consensus tree.

Cladograms 1, 7 and 13 weight all characters alike (at weight 1). All other cladograms in this analysis have 22 reduction-related characters weighted at 1:15 (i.e. reduction-related characters are weighted at 1, non-reduction-related at 15). The reason for this particular level of weighting is that cladograms generated at successively higher weightings but below 1:15 are different from each other, but at 1:15 and higher weighting ratios, they are the same. Because reduction series are observed internal to each of many genera in the Pottiaceae, reduction series may be expected to be several and parallel within the family. These reduction series should be distinguished from each other on the basis of characters other than those generally associated with reduction intragenerically. The 1:15 weighting provides for this. Reduction-related characters cannot be eliminated because, although they may not be expected to be homologous globally, they may be important clues to relationships in particular lineages.

The first cladogram (Cladogram 1) clearly shows what are comparatively reduced taxa, such as *Crossidium* and *Aschisma*, clustered in the same lineage. To see if such taxa were associated by characters other than those of reduction, twenty-two characters were selected for the weighting above as either directly associated with reduction in size and complexity, or which are elaborations (e.g. costal ornamentation, capsule wall ornamentation) associated with reduction in the Pottiaceae. The characters each commonly contribute at least two additive steps in the clade. The apotypic states of these reduction-related characters, numbered according to the main character list above, are: 0 gregarious (as opposed to caespitose); 1 stem short; 11 leaf short; 21 leaf base not differentiated; 34 bulging pad of cells present on costa; 35 filamentous costal outgrowths present; 41 dorsal superficial laminal cells walls clearly thicker than the ventral; 52 monoicous; 55 seta short; 56 seta not twisted; 57 theca short; 58 theca spherical; 59 cleistocarpous; 60 exothecial cells bulging; 61 exothecial cells with lens-like medial thickenings; 62 capsule cleistocarpous and rupturing along weak transverse walls; 65 peristome absent; 66 teeth short when present; 68 teeth untwisted when present; 71 calyptra short; 72 spores large; and 74 calyptra mitrate.

RESULTS (see also Summary above):

Phylogenetic Position of Non-Pottiaceous Genera

In Cladograms 1–4, *Ceratodon*, *Grimmia* and *Syrrhopodon*, representing three different families, are associated with what are here termed (in Cladogram 16) the Merceyoideae (equivalent to a combination of the traditional Barbuloideae and Leptodontioideae). Also, *Bryobartramia*, *Diphyscium* and *Encalypta*, representing two different families, are associated with the Pottioideae group. This may be evidence either of derivation of these families from Pottiaceae stock (thus a primitive member of the Haplolepideae) or of polyphyly, or simple "long branch attraction" of very distant relatives (*cf.* Albert & Mishler 1992); distinguishing the possibilities requires extensive study at a broader level than is appropriate in this study. The presence of the unusual twisted peristome in three different lineages of the Pottiaceae indicates monophyly. But note support for Robinson's (1971b) classification which groups all the above under the Dicranales, eliminating the Pottiales as an order.

In the sense of Farris (1974; see Wiley 1981, p. 92), the Pottiaceae would be paraphyletic if at least one other family not sharing advanced distinctive pottiaceous characters with other genera is included in the tree; this is a problem since the twisted peristome is commonly reduced to conditions (straight, rudimentary or absent) indistinguishable from similar non-homologous conditions. The Pottiaceae would be also be polyphyletic if the Ditrichaceae were recognized as a separate family possessing a twisted peristome in some members, requiring the Pottiaceae to be distinguished solely as a polythetic group. Work needs to be done to determine the phylogenetic importance of peristome characters through actual analysis across families.

Major Pottiaceous Groups

The various cladograms with outgroups of one of ten different non-pottiaceous genera with three very different peristome types (diplolepideous, nematodontous, haplolepideous) produced cladograms with much in common. It is quite possible that the three outgroup taxa with the most similar cladograms (*Polytrichum*, *Timmia* and *Ptychomitrium*) are phylogenetically closely related. Weighting the non-reduction related characters at 15 gave greater resolution to the cladograms, and was designed to limit artificial pairings on the cladogram due to convergence of non-homologous traits.

Cladograms 5 and 6 with *Encalypta* and *Grimmia* as outgroups demonstrate a similarity in many subclades with subclades found in cladograms created with other outgroups. These cladograms indicate that the classification arrived at here reflects to a great degree three major groups or branches that are discernible in relatively well resolved cladograms, such as Cladograms 3, 4, 5 and possibly 2 (among those including non-pottiaceous taxa). In Cladogram 5, the traditional Trichostomoideae (dealt with in Cladogram 16 as a basal stem requiring names for otherwise minor branches) can be seen as a major branch.

Thus the three major branches of the network shared by the cladograms (i.e. similar Steiner minimal trees) are approximately the traditional Trichostomoideae, Barbuloideae and Pottioideae. If *Polytrichum* and *Timmia* are not closely related to the Pottiaceae, and these two genera are simply evolutionarily convergent to the traditional Trichostomoideae and to each other, then the correct outgroup for the Pottiaceae might better be a taxon similar to the pottiaceous genera near the joining of the three major branches, which are certainly more similar to other families in the Pottiales. No non-pottiaceous genus analyzed is close to the joining except *Ditrichum* in Cladogram 4. Possible ingroups basal to such a pottiaceous tree include *Leptobarbula, Hymenostyliella, Tetracoscinodon* and *Tetrapterum.*

This line of attack might be pursued but, in this study, evolutionary convergence of *Polytrichum, Timmia* and the traditional Trichostomoideae (as exemplified by *Timmiella*) is clearly less probable than derivation of the Pottiaceae from a shared ancestor of all three. If the immediate ancestor of the Pottiaceae were similar (in being relatively reduced) to other families in the Pottiales, then even a fourth group, the Ptychomitriaceae (haplolepideous), would join the taxa required to independently evolve the large complex of morphological and anatomical characters characteristic of robust taxa of the traditional Trichostomoideae. Cladograms 5 and 6 are here considered improbable as hypotheses because the two outgroups for these do not restrict *Polytrichum* and *Timmia* to the base of the tree. If convergence of several lines of taxa with *Timmiella*-like character states of the gametophyte were likely, as might be caused by simple, single-gene desuppression of character states, then there would probably be *Polytrichum*- or *Timmia*-like gametophytes having sporophytes characteristic of the Dicranaceae or Grimmiaceae or Orthotrichaceae; but there are not. Thus, convergence of complex gametophytes is unlikely while reduction as advanced traits, common in many families, is apparently the best justification for placing the traditional Trichostomoideae low in the tree.

Taxa that are apparently morphologically reduced occurred in Cladogram 1 (all characters at weight 1) either alone (e.g. *Anoectangium, Leptobarbula, Tetracoscinodon*) or were clustered, i.e. *Aloinella*, *Crossidium* and *Globulinella* with *Tortula*, and *Acaulon*, *Aschisma*, *Byroceuthospora*, *Microbryum*, *Phascopsis*, *Pterygoneurum*, *Stegonia*, *Stonea* and *Trachycarpidium* with *Tetrapterum* in the Pottioideae subclade. These taxa are also either alone or clustered, respectively, in Cladogram 13, all characters also at weight 1, but which is less resolved. The reduced taxa remain clustered when non-reduced characters are weighted at 15:1 in generally 1–3 lineages in Cladograms 2–4 and 8–10 and 14, which implies that these taxa are closely related. The grouped reduced taxa continue to be grouped even when the importance of reduction-related characters is minimalized.

Two major traditional suprageneric taxa, the Pottioideae (with spathulate leaves with a single stereid band in the costa and margins commonly recurved) and the Barbuloideae *cum* Leptodontoideae (here treated as the Merceyoideae—with lanceolate leaves and recurved margins, the basal cells largely reaching higher in the leaf medially), are well distinguished in most cladograms as two large subclades terminating the tree.

The genera traditionally comprising the Leptodontioideae clearly belong with the Merceyoideae, as do most genera of the traditional Pleuroweisieae; the former group is well distinguished as a subclade of the Merceyoideae, the latter genera are rather scattered within the Merceyoideae, which reflects well on my previous comment (Zander 1977c, p. 240) that the Pleuroweisieae probably has little phylogenetic coherence. Inasmuch as the two groups (approximately the Merceyoideae and Pottioideae of Cladogram 16) comprise identical taxa and identical structures in Cladograms 9 and 10 (subclades "F" and "G" of both trees), an analysis of character state changes was made, which consisted of analysis of the subclades alone and selection of single trees from each of the resulting two consensus trees (Cladograms 11 and 12).

Cladogram 11 includes only those terminal taxa distal to node F in Cladograms 9 and 10, approximately the Merceyoideae recognized in Cladogram 16. *Tetracoscinodon* is the outgroup in this analysis. Four subclades are distinguishable, *Tetracoscinodon* (sole member), the *Barbula* group, the *Leptodontium* group, and the *Bryoerythrophyllum* group. This is support for these groups to be treated as tribes (see Cladogram 16) in that they are fairly stable among the various cladograms with different outgroups shown here. Polytomies in the consensus tree occurred distal to the immediate ancestor of *Barbula*. A functional ingroup and functional outgroup (FIG/FOG) analysis (Watrous and Wheeler 1981) was done with *Anoectangium* as FIG and including all taxa distal to *Anoectangium* as FOG. Tree 24 of the 127 trees contributing to the consensus tree (Cladogram 11) was selected because its Barbuleae branch matched that of the FIG/FOG result. The characters incompletely map to tree 24 (Cladogram 11) in that no character state changes were found for node "X" of Cladogram 11, the immediate ancestor of *Barbula*; node "X" may be taken as an artifact signifying a multiple-branched node in its own immediate ancestor, leading directly to *Barbula*, to the *Scopelophila*, *Gymnostomiella* and *Didymodon* lineage, and to the *Gymnostomum* and *Molendoa* lineage. Cladogram 11, now tempered with information from other cladograms, shows the general evolutionary structure of the Merceyoideae. Note that Cladogram 11 has the *Leptodontium* subclade derived from a more primitive ancestor than is given in Cladogram 10. The consensus tree of Cladogram 14 shows a similar number of polytomies in the Barbuleae; the structure of the Barbuleae of the single tree reflected in Cladograms 15 and 16 was selected from equally parsimonious trees based on the above FIG/FOG analysis.

Cladogram 12 summarizes an analysis of the Pottioideae as shown in Cladograms 9 and 10 with *Leptobarbula* as outgroup. As FIG/FOG analysis was to no avail in further resolving this part of Cladogram 10, but tree 5 of 8 equally parsimonious trees was chosen, being the first tree to place *Weissia* and *Weissiodicranum* together on a single branch, as is suggested by sufficiently resolved Cladograms 1, 3, 7, 13 and 14 (but not Cladogram 8). In spite of the polytomies low in the strict consensus tree, this configuration results in a map of character state changes that retains much the same groupings for this subclade as seen in other cladograms. The structure of the Pottioideae in Cladogram 12 is much different than that shown in Cladogram 14 although the taxa are similar (*Aschisma*, *Byroceuthospora* and *Uleobryum*, among others are included in the Pottioideae in Cladogram 14). In both cladograms, the Pottioideae subclade is not easily describable as a morphologically integral unit; "broad leaves" is about the only readily discernible character state applicable to most taxa. Cladogram 12 reflects apparent evolution along a continuous lineage, while Cladogram 14 shows two distinct lineages (also *cf.* Cladogram 8).

I agree with Mishler (1985) that the genus *Tortula* as traditionally viewed is paraphyletic, in that the genera into which many of the traditional *Tortula* species are segregated here, *Hennediella*, *Syntrichia* and *Tortula s. str.*, are in three separate subclades in the Pottioideae in most of the cladograms. Cladograms of Mishler's (1985b, 1986a, 1990) previous work with a number of species of *Tortula s. lat.* support to a considerable extent the above three groups recognized here as genera. The Pottioideae, as here emended with the addition of several taxa related to *Weissia* and *Hyophila*, is a monophyletic group, the basal synapomorphies given in Cladogram 15 being: stem sclerodermis little differentiated, leaf base little differentiated, and seta less than 1 cm in length.

The traditional Trichostomoideae (taxa with largely lanceolate, plane leaves, often with a narrow marginal band of hyaline cells running up from the base) is here dissected into different lineages in many cladograms; these appear in most cases at the base of the tree. The traditional Trichostomoideae is here split into five subfamilies in Cladogram 16 (four are evident in Cladogram 14 with all characters weighted equally), each of which is also rather distinctive morphologically as a taxonomic grade. Norris and Koponen (1989) in a recent description of the subfamilies of Pottiaceae, found the traditional Trichostomoideae to share the following characters: most abundant at tropical latitudes, leaf margins plane, areolation of leaf and epidermis continuous (no leaf butress of shortened cells formed), hyalodermis commonly differentiated, basal laminal cells often running up the margins in a vee, and a persistent felt of protonema commonly developed. Most of these traits apply to the subclades at the base of the Pottiaceae tree, but, apparently significant dichotomies at the base of the tree require splitting (and recognition at the subfamily level), which, by the way, is of a higher order than the dichotomy between the Merceyoideae and Pottioideae since such splitting is nearer the base of the tree.

As *Ptychomitrium* is apparently the closest of the candidate sister groups to the Pottiaceae that keep the distant taxa *Polytrichum* and *Timmia* near the base of the tree (trees with non-pottiaceous taxa in the data set other than *Encalypta* and *Grimmia* not shown), *Timmiella* is found to be the most primitive extant genus of Pottiaceae (Cladogram 10); this is also supported in Cladogram 3, while *Timmiella* is near the base of Cladograms 1–2, 4, and 7–9. Although *Aloina* and *Gertrudiella* are also found near or at the base of the last mentioned cladograms (see especially Cladogram 4), an analysis with *Timmiella* as functional outgroup was performed and taken to be the best phylogenetic hypothesis for the family (Cladograms 13–16).

Cladogram 13 is the strict consensus tree with *Timmiella* (Pottiaceae) as functional outgroup and all characters weighted the same. Three lineages are identical with the named lineages in Cladogram 16, while many small branches are identical or similar in structure and terminal taxa. Presumed reduced taxa (e.g. four small branches including *Aschisma*, *Phascopsis*, *Luisierella*, and *Crossidium*, respectively) are grouped to some extent. As in Cladogram 14, the Pottioideae is split into (at least) two lineages

Cladogram 14 is the strict consensus tree with *Timmiella* as functional outgroup and with non-reduction characters weighted at 15:1; it can be viewed as the "raw results" of this phylogenetic study. Cladogram 14 is more resolved than Cladogram 13. A striking difference between this cladogram and Cladograms 9 and 10, which agree with each other in many respects, is that the Pottioideae (1) is split into two lineages, (2) includes the *Aschisma*, *Byroceuthospora* and *Trachycarpidium* group, otherwise placed with the traditional Trichostomoideae, and (3) derives the *Aloina, Aloinella, Crossidium, Globulinella, Pterygoneurum and Stegonia* group from the *Weissia* branch of the *Hyophila* lineage (as in Cladogram 13).

The *Weissia-Hyophila* group is characterized by largely ligulate to spathulate leaves with incurved margins, upper laminal cells commonly strongly bulging ventrally and nearly flat dorsally. This group is distinguished in some cladograms (see especially Cladogram 8 and Cladograms 14–16), and appears at the base of the Pottioideae lineage in others. It is here recognized at the tribal level (as Hyophileae, see Cladogram 16) because (1) it is a morphologically distinctive group, and (2) it is surprising that ancestors of the end members of such a highly structurally modified group with such an unusual (tropics, especially the West Indies) geographic distribution (see Cladogram 14) should be the source group for the traditional Pottieae (as in Cladograms 9 and 10).

The presumed reduced taxa that are grouped in Cladogram 13 continue to be associated with each other in Cladogram 14, indicating that their relationships are not due to convergence; on the other hand, other reduced taxa, such as *Leptobarbula*, *Tetrapterum*, *Stegonia*, *Pterygoneurum*, *Microbryum*, and *Trachycarpidium*, that are poorly resolved in Cladogram 13 are now well resolved with higher weighting of non-reduced characters.

The Phylogenetic Hypothesis

A single tree (shown in both Cladograms 15–16) was selected from the more than 1250 equally parsimonious trees contributing to the consensus tree of Cladogram 14. Multiple-branching nodes in the Merceyoideae were resolved by selecting from the 1250 trees those several trees with the structure of the Merceyoideae of the single tree of Cladogram 11. Of these fewer trees, those that (1) placed *Aschisma, Byroceuthospora, Trachycarpidium* and *Uleobryum* in the same subclade and (2) which also derived *Saitoella* from an earlier ancestor than the immediate ancestor of *Microbrym* were selected because elsewhere in the cladogram generally relict and southern taxa have more primitive traits than larger and more widespread genera (this particular structure is an example of the cladist's "reciprocal illumination" and must not be used as an example supporting any generalization about relict, southern taxa). Of these yet fewer trees, one tree was selected that (1) put all remaining traditional Trichostomoideae that were at multiple branches in the consensus tree into one subclade because it allowed a classification that was both simple and did not imply a degree of resolution greater than was warranted, (2) put *Calymperastrum* near the base of this subclade because Gondwana taxa are commonly at the base of other subclades (reciprocal illumination—do not use this particular structure to support the generalization), (3) placed *Eucladium* next in evolutionary order since this monotypic genus is isolated morphologically and may be a relict (reciprocal illumination again), and (4) placed *Trichostomum* and *Streptocalypta* as most highly evolved because of the comparatively larger size of the genera in numbers of species, implying recent evolution. The single tree (tree number 1126) that met all these above criteria is the basis for Cladograms 15 and 16 and for the present suprageneric classification.

The traits of hypothesized ancestors of named subclades in Cladogram 15 are detailed in the following chart. These states characterize a particular shared ancestor in a most parsimonious tree, but may change later in each of the named lineages (subfamily names capitalized entirely).

STATES AT MAJOR ANCESTRAL NODES

- **TIMMIELLOIDEAE** (functional outgroup)
- Guide cells fewer, 2–6: **ERYTHROPHYLLOPSOIDEAE**
  - **GERTRUDIELLOIDEAE**
  - Leaf base not sheathing; lamina unistratose
    - Leaves long-linear; ventral stereid band larger than dorsal: **CHIONOLOMOIDEAE**
    - Leaf hydroid strand absent
      - Axillary hairs completely hyaline; dorsal costal epidermis absent: **TRICHOSTOMOIDEAE**
      - Leaves shorter, 1.5–3.0 mm; rows of cells on ventral surface of costa fewer, 4–6; costal guide cells fewer, 2–6; theca shorter, less than 1.5 mm
        - **Tetracoscinodontieae**
        - Leaf keeled above; costa grooved ventrally: **MERCEYOIDEAE**
          - Costa flattened in section; leaf hydroid strand present; theca longer, 1.5–3.5 mm; peristome teeth of 32 similar rami: **Bryoerythrophylleae**
          - Leaf margins recurved to revolute; costal ventral cells elongate; rows of cells across ventral surface of costa 2(–4)
            - Stem central strand absent; ventral costal epidermis absent; upper laminal cells superficially flat or weakly convex: **Leptodontieae**
            - Laminal papillae simple; seta nearly absent to short, less than 1 cm in length
              - Dorsal stereid band section round or semicircular; peristome teeth absent: **Barbuleae**
        - Stem hyalodermis absent; perichaetial leaves sheathing
          - Leaves tubulose when dry; leaf margins incurved or involute; upper laminal cell walls ventrally bulging and dorsally nearly flat: **Hyophileae**
          - Stem sclerodermis not or little differentiated; leaf base little differentiated in shape; seta nearly absent to short, less than 1 cm in length: **POTTIOIDEAE**
            - Stem short, less than 1 cm in length; peristome of 32 similar rami; peristome distinctly twisted: **Pottieae**

The assumptions are several but they are reasonable and it is felt that Cladogram 16 is the best hypothesis on which to base a classification at this time. Again, reciprocal illumination here helps select that single cladogram (of many of equal length) that is (1) most internally consistent in all subclades in geographic distribution of the terminal taxa, and (2) most consistent with other cladograms that do resolve polytomies. Because Cladogram 14 summarizes the results of the study without modification based on geographical ranges, the geographic data is summarized with it (q.v.) rather than with Cladogram 16.

Cladogram 15 summarizes the character state changes of this study's projected hypothesis of the phylogenetic relationships of the Pottiaceae. Cladogram 16 gives a better view of the structure of Cladogram 15, and letters the basal ancestral nodes of the named subclades. The lettered subclades are also noted in other cladograms when similar (or identical in the case of Cladograms 13 and 14), indicating considerable support for these relationships. Cladogram 16 is a version of Cladogram 15 that clearly shows the basis for the classification in the taxonomic section of this work. Note that Cladogram 16: (1) is supported in large part by cladograms based on the non-pottiaceous outgroups *Polytrichum*, *Timmia* and *Ptychomitrium*, (2) makes few major changes in traditional classification, and (3) introduces hypotheses that are of considerable interest and that have a logical, rigorous basis.

There are 72 ancestral nodes in Cladogram 15 (on which Cladogram 16 is based) that have synapomorphies, plus one node to which no data on character state changes map. Of the 75 total scored characters, 14 were unused, these being characters numbered 0, 9, 16, 32, 34, 37, 38, 41, 51, 53, 62, 67, 69, and 70. These characters evidently have little phylogenetic importance in the Pottiaceae, at least taking into account the assumptions (including FIG/FOG and reciprocal illumination) on which this single tree is based.

The 62 scored characters used in Cladogram 15 supported 97 different character state changes. Of these, 59 *state changes* (61%) were true synapomorphies, i.e. only occurred once in the cladogram. Of the 97 state changes in Cladogram 15, 38 (39%) were homoplasous synapomorphies. Of these 38, there were 17 character state changes occurring twice in the cladogram, 14 thrice, 5 four times, 1 five times (character 57:0–1R), and 1 six times (character 46:0–1). (Note that this refers to state changes in the cladogram not the number of times the character itself occurs in the cladogram.) The characters contributing to the many homoplasous synapomorphies are not unimportant, and indicate possible modes of convergence of lineages in addition to the characters selected prior to analysis as indicative of reduction.

Eighteen *characters* only occurred once in Cladogram 15. These 18 are considered here the characters most important to the taxonomy of the Pottiaceae, and are (as numbered in the list above): 2 shape of stem section; 6 color of axillary hair basal cells; 17 number of layers of marginal laminal cells; 19 leaf apex rounded or acute; 20 leaf apex cucullate or otherwise; 22 costa ending before or at the leaf apex or excurrent; 27 ventral stereid band larger or smaller than the dorsal; 35 ventral costal outgrowths present as filaments or otherwise; 40 upper laminal cells medially unistratose or bistratose; 42 upper laminal papillose or smooth; 47 propagula when present borne on rhizoids as brood bodies or otherwise; 49 propagula when present borne on leaf costa or lamina or otherwise; 58 shape of theca; 59 capsule cleistocarpous or stegocarpous; 60 capsule smooth or mamillose; 61 exothecial cell superficial walls with central lens-like thickenings or evenly thickened; 63 stomates present or absent; 74 calyptra cucullate, mitrate or campanulate. That some of these characters are among the reduction-related characters weighted low in the analysis indicates that they are nevertheless important in at least some lineages, and must not be eliminated from analysis.

By way of summary, traditional groupings are supported in large part and grossly dissimilar genera are not associated, thus the subclades make "sense" to the specialist. Such support for classical groupings indicates that traditional methods of "omnispection" and the like are, to a certain point, effective. Inasmuch as the branches named here at the subfamily and tribal level are reflected in most of the cladograms generated in the course of this study, they may be expected to appear in at least approximate form in future studies.

The major subclades are formally named at the subfamily and tribal level in the taxonomic section. For these, the "sequencing convention" of Nelson (1972, 1974) is used because of the considerable asymmetric depth of the tree; that is, the taxonomic levels recognized here—subfamilies and tribes—are mostly not coordinate with each other according to depth in the tree. The nine subfamilies are the Timmielloideae, Gertrudielloideae, Chionolomoideae, Erythrophylloideae, Trichostomoideae, Merceyoideae (including tribes Tetracoscinodontieae, Barbuleae, Leptodontieae and Bryoerythrophylleae), and Pottioideae (with two tribes, the Hyophileae and Pottieae).

Minor Groups

There are lesser trends in evolution in the Pottiaceae that are shown in hypotheses presented by the cladograms. Subclades that are similar (the same or nearly the same taxa and structure) to those in Cladogram 16 are marked with letters on the basal nodes in Cladograms 1–10 and 13–14. These generally support the structure of Cladogram 16.

Certain genera always appear at the base of the tree: *Gertrudiella*, *Timmiella*, *Erythrophyllastrum* and *Erythrophyllopsis*. The several lineages into which the traditional Trichostomoideae are broken in the various cladograms include these "strong" (because they appear in several of the different cladograms) generic associations: *Calyptopogon*, *Pleurochaete* and *Tortella*; *Chionoloma*, *Pachyneuropsis* and *Pseudosymblepharis*; and *Aschisma*, *Byroceuthospora*, *Trachycarpidium* and *Uleobryum* (these latter are found in the Hyophileae of the Pottioideae in Cladograms 14–16). Taxa related to *Leptodontium* are presented as a well distinguished subclade: *Leptodontiella*, *Leptodontium*, *Streptotrichum* and *Trachyodontium*, while *Hymenostylium*, *Reimersia* and *Triquetrella* are closely related. *Anoectangium* is associated with the *Barbula* group, while Merceyoideae that are KOH red are placed in a separate branch: *Bryoerythrophyllum*, *Mironia*, *Pseudocrossidium* and *Rhexophyllum*, with *Dialytrichia* (KOH yellow) at the base of the branch.

The fact that most reduced taxa, these with no confounding mismatches in geographical ranges (see Cladogram 14), are grouped together (in at least two groups) indicates what are probably monophyletic lineages, without masking convergence. The most clearly grouped of the taxa of small stature are: (1) *Aschisma*, *Byroceuthospora*, *Trachycarpidium* and *Uleobryum*; and (2) *Stegonia*, *Pterygoneurum*, *Aloinella*, *Crossidium* and *Globulinella*, probably including *Aloina*. The second group, although of small plants, is not of highly reduced morphology in comparison with terminal taxa with immediate ancestors inserted deeper in the subclade. These, consisting of the *Weissia-Hyo-*

phila group at the base of the Pottioideae lineage in Cladograms 9 and 10 and recognized as the Hyophileae in Cladogram 16, consist of plants of small stature, less simplified than the first group of taxa above, but commonly with reduced or absent peristomes and thus more reduced than the second group; these, too, have a certain geographic integrity (largely tropical). Of course, if there were more than the 22 identified reduction-related characters acting to produce false associations, these groups might be split further. This should be pursued.

Within the Pottioideae, the KOH red genera are generally segregated as a group (with *Willia* or sometimes *Microbryum* at a basal branch), except that highly reduced taxa (e.g. *Crossidium* and *Globulinella*) of the subfamily are generally clustered together without regard for KOH reaction.

The position of *Aloina* is problematic. It is placed near *Polytrichum* in Cladograms 1, 4, 7 and 8, and, more traditionally, near *Aloinella* in Cladograms 2, 3, 9, 10 and 14–16. One might make a case that the absence of *Polytrichum* in a cladogram introduces a false relationship of *Aloina* with *Aloinella* since the new outgroup (*Timmia* or *Ptychomitrium*) is not sufficiently distant from the Pottiaceae, but this is not true for Cladogram 2 (outgroup *Timmia*), in which *Aloina* is associated with *Aloinella* not *Polytrichum*. Pending further study, *Aloina* is placed with *Aloinella*, following Cladograms 14–16, and the association with *Polytrichum* is considered anomalous, perhaps an example of convergence together with lack of sufficient data to address the problem.

Position of Reduced Taxa

If one assumes, for the purposes of argument, that all genera develop from a single species to a series of morphologically diverse species in a genus and finally senesce towards extinction by species being eliminated from the genus in response to environmental pressures, certain hypotheses can be developed from the cladograms, notably Cladogram 14. Although one might assume that highly morphologically reduced and simplified genera may have been derived from the immediate ancestor of extant similar genera of greater stature and complexity, the present data do not support this. Somewhat reduced genera appear near the base of most subclades (*Calymperastrum* in the *Tortella* lineage of Cladograms 2 and 8, *Tetracoscinodon* in the Merceyoideae (Cladogram 14 and elsewhere), *Anoectangium* at the base of the Barbuleae (Cladogram 14 and elsewhere), *Weissia* in the *Weissia-Hyophila* group of various cladograms and Cladogram 14, and *Leptobarbula* at the base of the Pottioideae (Cladogram 14 and elsewhere.

This may indicate that relatively reduced taxa were ancestors of these lineages. If such is the case, the twisted peristome of the Pottiaceae may have appeared several times, possibly as a result of genetic desuppression (*cf.* Basile & Basile 1984). On the other hand, simplified taxa at the base of subclades may be remnants of ancient lineages now largely extinct. The latter seems a more likely scenario because the possibility that more highly evolved (morphologically reduced) taxa are more likely to survive recent environmental change than coeval taxa retaining ancestral (non-reduced) features is more probable than multiple evolution or even desuppression of the characteristic twisted Pottiaceae peristome. The presence of the character state change of loss of peristome in the shared ancestor of *Anoectangium* and the rest of the Barbuleae (in both Cladograms 11 and 15) must be dealt with, however, as it does not seem to be an artifact of the poor resolution of the remainder of the Barbuleae (*cf.* also Cladograms 2, 3, 4, 8, 9, 10).

Also, genera with complex morphology and no or few reduced species (e.g. *Tortella* and *Syntrichia*) commonly occur at the ends of lineages. These may be viewed as very recent taxa, not having yet accumulated genetic diversity; apparently large-statured genera of few or one species, such as *Dolotortula*, *Calyptopogon*, *Sagenotortula* and *Teniolophora*, are extremely recently evolved and are just beginning evolutionary development. Some extant taxa at the middle or ends of evolutionary lines commonly include many species and consist of plants of both large stature and complex morphology and of smaller stature with simplified, reduced morphology, e.g. *Bryoerythrophyllum*, *Hennediella*, *Pseudocrossidium*, *Tortella*, *Tortula* and *Trichostomum* (all as emended here). These genera are apparently evolutionarily more mature, having spread into habitats that provide evolutionary pressures toward reduction and having been in existence long enough to respond to such pressures during speciation. Genera consisting only or largely of reduced species, e.g. *Anoectangium*, *Leptobarbula*, *Tetracoscinodon*, *Tetrapterum*, *Trachycarpidium*, *Triquetrella* and *Weissia*, are found largely near the base of subclades or near the base of the tree, and may be taken to be relicts of genera whose large-statured members of complex morphology have become extinct with changing environments. Thus, among primitive genera, presumed non-reduced species have been extinguished by evolutionary pressures before the reduced species of such genera, leaving reduced relicts.

There are some clear exceptions to this that require explanation. For instance, large-statured taxa near the base of the tree, e.g. *Erythrophyllastrum*, *Erythrophyllopsis*, and *Gertrudiella*, may be taken to be large-statured relicts surviving in relictual environments (i.e. Andean microhabitats). The position of, for example, the *Acaulon*, *Sarconeurum* and *Stonea* subclade seems anomalous, but, perhaps, there the ancestral lineage consisting of the austral genera *Tetrapterum*, *Willia*, *Phascopsis* and *Stonea* is a line of relictual taxa that is actually separate from the remainder of the Pottieae.

The data in general, however, is suggestive and future work should examine whether or not the number of species and the morphology of extant genera correlates with how deep in the tree their immediate ancestors are, and on how the relative harshness of relictual habitats affects phylogenetic position, the less severe habitats preserving ancestral characters.

Geographic Trends

There appears to be considerable support for the evolutionary hypothesis of Cladogram 16 from the geographic distributions of the taxa given for the consensus tree of Cladogram 14, of which there are no major mismatches. Although many of the genera must be simply described as wide-ranging, certain distributional trends are correlated with phylogenetic groupings. The basal-most branches are either of widespread or Andean distribution. The Merceyoideae and Pottioideae include a large number of cosmopolitan taxa, with the remainder largely restricted to the Andes.

The members of the Chionolomoideae (see Cladogram 16) are characteristic of lowland tropical areas, especially those of eastern Asia. The Hyophileae (see Cladograms 8 and 14–16) encompasses a majority of members with ranges in the lowland tropics, especially the West Indies. Certain highly evolved genera (*Aloinella*, *Crossidium*, *Globulinella*, *Pterygoneurum* and *Stegonia*), considerably reduced in stature and generally grouped together in the Hyophileae, are apparently monophyletic and adapted for warm, highly arid habitats. Frey and Kürschner (1983, 1988a) assigned a number of circum-Mediterranean, central Asian and western USA Pottiaceae species to a "Circum-Tethyan" floristic zone, and indicated that these taxa may be Triassic relicts of ancestral populations of areas around a mid-continental warm sea formed during the breakup of Pangaea. The Hyophileae includes several genera with species characteristic of the Circum-Tethyan flora, and this tribe may have originated there.

The nine mainly Gondwanaland (austral) genera (*Calymperastrum*, *Calyptopogon*, *Hypodontium*, *Phascopsis*, *Sarconeurum*, *Stonea*, *Tetracoscinodon*, *Tetrapterum*, *Willia*) are not grouped in any of the subclades, excepting *Phascopsis*, *Sarconeurum*, *Stonea* and *Willia*, which are clustered in the Pottioideae. This indicates that the Gondwanaland ancestral group contributed to all traditional taxonomic groups and many of the presently recognized subclades. Because Gondwana taxa occur at the base of the Merceyoideae (see Cladogram 14 and 16), the Pottieae, and some of the subclades making up the traditional Trichostomoideae, it is probable that differentiation of these lineages occurred in austral regions.

The Pottioideae, on the other hand, may have been derived from a pantropical *Hyophila*-like ancestor, with austral genera derived secondarily (*cf.* especially Cladograms 2, 9 and 10) but do include some austral taxa. On the other hand, the ancestral stock contributing to the five austral genera (see Cladogram 16) of the Pottieae may simply be extinct (certainly the nine austral genera are reduced genera each of few species and mostly occur in marginal habitats).

The difficulty in resolving the Barbuleae may be due to continued survival of populations representative of ancestral stock, which would add morphological variation that does not contribute to analysis. It may be possible to better resolve the subclade by eliminating from analysis all taxa of intermediate morphology apparently surviving (on a geological time scale) in refugial areas. This would have to be investigated at the species level, with due attention to distinguishing if possible recent divergent evolution in isolated habitats. Identification of the extant species with the most primitive character states in each genus would be very helpful.

The most ancient stock of the Pottiaceae could be considered Andean, as judged from the distribution of the taxa of the basal branches, but the Andes are geologically rather recent. Possibly the Andes have served as refugia for basal taxa (these usually well distinguished morphologically) evolved elsewhere, probably Gondwanaland.

Classification

Traditional classification systems presuppose that taxa can be described as coherent morphological units that have some phylogenetic significance, most taxa sharing either at least one distinctive character state (monothetic) or possess most of a distinctive set of character states (polythetic). The taxonomic requirement is that the taxa share at least to some degree these significant character states. If major subclades are considered suprageneric taxa, however (as in Cladogram 16), such subclades are best defined by the character state changes at the basal ancestral node. Any sharing of traits in the subsequently derived taxa of the subclade is dependent on the degree of morphological divergence among those taxa. In the present treatment, subclades named as suprageneric taxa will be identified by the character state changes at the immediate ancestral node and will also be described morphologically to the best extent possible as though they were traditional grades. Taxonomists with some reservations about using cladistics as a basis for classification can be reassured that the subclades here formally recognized as subfamilies and tribes are in most cases rather well distinguished morphologically, and some largely match traditional groupings.

An evaluation of the cladograms indicates that, terminating the tree, there are two highly evolved lineages of many taxa, corresponding to two traditional subfamilies—the Merceyoideae (previously known as Barbuloideae) and the Pottioideae. There are also several single or few-taxa lineages at the base of the cladograms that are each given subfamily status. The Trichostomoideae, as recognized in Cladogram 16, is an arbitrary construct from the many equally parsimonious trees contributing to the strict consensus tree of Cladogram 14, and its structure is dependent on "reciprocal illumination."

Cladogram 16 is the basis for the classification since (1) it provides a single tree as a hypothesis for evolutionary relationships, (2) it summarizes best the structure of Cladogram 14 while reflecting the structures of the majority of the rest of the cladograms, and (3) because its functional outgroup, *Timmiella*, is selected largely based on Cladogram 10 with the outgroup *Ptychomitrium*, a taxon that may be taken as the nearest little-modified sister group of the Pottiaceae.

Additional Discussion

At the species level, lines of evolution involving reduction of sporophyte and gametophyte characteristics in some cases can be visualized as an exclamation mark (!), in which a line of species exhibiting features intermediate in a probable reduction series terminates in a distinctive, highly reduced species or group of species. The line may represent extant flotsam of evolution ranging from the peristomate to the eperistomate: why the terminal group, often cleistocarpous, is commonly distinguishable at the supraspecific but seldom suprageneric level is not presently understood but may be simply a function of general strong modification of otherwise easily recognizable features (e.g. of the leaf adapting to clasp the capsule). Further study is necessary.

Examples of this reduction pattern include *Weissia* and its well-characterized subgenus *Phasconica*; and, at the species level, *Syntrichia bartramii*, a dwarf variant of *S. ruralis* from the southwest of the United States, apparently trailing populations identifiable as *S. intermedia* in its evolutionary wake; and *Trichostomum caespitosum*, distinguished from *Trichostomum crispulum* by its small size and differentiated perichaetial leaves.

Trichostomum sections may be examples of unconnected evolutionary patterns. A cladistic evaluation of the relationships would probably be more instructive in this case than sorting into taxonomic categories on the basis of perceived gaps, as is done here in the taxonomic section.

At the generic level, one might expect small-statured genera to be at the end points of lineages of larger-statured genera, but upon cladistic analysis (see above) this does not seem to be the case. Lineages often have apparently reduced genera at the base of branches. Also, generalizing from the appearance of the taxa, one might expect the following reduction pairs (the second genus sharing ancestors with reduced species of the first genus): *Tortella* and *Calymperastrum*; *Quaesticula* and *Plaubelia*; *Barbula* and *Scopelophila*; *Tortula* and *Stegonia*; *Syntrichia* and *Stonea*; *Willia* and *Phascopsis*; *Microbryum* and *Saitoella*; but in fact, of these, only *Willia* and *Phascopsis* (see Cladogram 14) show what might be interpreted as reduction in series at the generic level. Clearly what have been considered previously as minor characters are of considerable importance as evolutionary markers. One can see, however, a few pairs of reduced and non-reduced genera at the end of lineages that might qualify as reduction series resulting in a distinct genus: *Bellibarbula* and *Gyroweisia* (Cladogram 11), *Didymodon* and *Gymnostomiella*, and *Aloina* and *Aloinella*, but most pairs at the end of lineages are of nearly equal plant stature.

Directions of evolution may be inferred from the advanced characters of end members of the subclades in Cladograms 11, 12 and 15. It is doubtful that genera evolve as a unit (unless one postulates sufficiently common reticulate evolution, c.f. Wyatt et al. 1992). On the other hand, present morphology restricts and thus may "guide" future evolutionary possibilities, i.e. evolution may take place in species of similar morphology in similar habitats undergoing similar environmental changes, resulting in a group of species within the genus that has evolved similarly. Assuming that a newly evolved genus has as its ancestor one species (or, better, one mutated individual), not a whole genus, one might expect that more than one extant species showing similar evolutionary directions within a genus might telegraph the next phase in evolution *at the generic level*. Such is not the case.

Actually, it is not surprising that highly evolved genera are difficult to imagine as simple extensions of the evolutionary directions of other extant genera on the basis of easily observed characteristics. This is because many of the most significant character state changes are anatomical as opposed to gross morphological traits.

Character weighting was employed to allow for the fact that characters associated with possible reduction (or elaboration) involving clearly related intrageneric taxa in transformation series are almost certainly not globally homologous in a data set of this size. The series are usually ascribed to reduction here, because parallel loss or reduction is explainable with fewer assumptions than parallel gain or elaboration.

Concentrated study of both poorly and densely speciated relatives was important to clearly distinguish states that represent evolutionary steps. Use of unordered, multistate characters and poorly studied characters may introduce false weighting if (1) truly intermediate evolutionary steps are unobserved and thus unscored or (2) certain state changes are not recognized as representing multiple synapomorphies. Character states in this study are represented by those seen in densely speciated taxa and are used globally on the assumption that homoplastic mutations recur as similar states (this was observed in related densely speciated taxa). This allows automatic weighting through use of multiple ordered, additive states for groups in which much extinction has taken place and intermediate states are absent. Ordered, additive states are recognized as such by correlation with other states in transformation series. Features with similar functions that could be treated as non-ordered states (e.g. propagula types) are here considered non-homologous and are used as distinct characters even when the presence of one feature appears to exclude the others.

COMPARISON WITH THE CLASSIFICATIONS OF PREVIOUS AUTHORS

Three authors have provided phylogenetic trees of the Pottiaceae or portions of the Pottiaceae. These may be compared with the various cladograms developed in this study, especially Cladograms 2, 8 and 10 as well as Cladogram 16. The similar subclade structure of many of the cladograms developed in the present study is given in Cladogram 16, which is a single tree selected from many equally parsimonious trees of Cladogram 14 by FIG/FOG analysis and "reciprocal illumination."

Saito's (1975a) phylogeny (Cladogram 17) is the closest to that of the cladograms generated in this study (assuming the classical Trichostomoideae is a paraphyletic group). Chen (1941; Cladogram 18) recognized the *Weissia-Hyophila* group as a distinct suprageneric group but included only *Hyophila* and *Uleobryum* in it. The *Tortella* group is separated by Chen from *Trichostomum* at the tribal level; it is part of a lineage separate from *Trichostomum* in several cladograms in the present study. Hilpert (1933) (Cladogram 19) clearly places the *Leptodontium* group near the *Barbula* lineage. Although there are some significant comparisons, previous authors' family trees are relatively simple at the suprageneric level, and mostly different in structure from that recognized here.

Some authors have presented classifications without actually describing a tree. Brotherus (1924–25) divided the Pottiaceae into the Pleuroweisioideae, Cinclidotoideae, Trichostomoideae, Merceyoideae, and Pottioideae. In his treatment of the geography of moss distributions, Herzog (1926, p. 95) recognized the Trichostomaceae and Pottiaceae as separate families. He split the Trichostomaceae into the Pleuroweisioideae and the Trichostomeae, the latter included the Barbuleae, Hyophileae (*Hyophila*, *Weisiopsis*, and *Dialytrichia*), and Leptodontieae. Walther's (1983) classification recognized the Trichostomeae (Tortelleae, Barbuleae, Hyophileae, Eucladieae, Pleuroweisieae), Pottioideae (Pottieae, Scopelophileae), Cinclidotoideae (Clinclidoteae), and Leptodontioideae (Leptodontieae), a classification similar to that of Saito (1975a) but the suprageneric taxa were constituted of genera in considerably different combinations.

EVALUATION

The present phylogenetic hypothesis is in many respects at odds with past evaluations of phylogenetic relationship and systematic position at the suprageneric level. Because of a lack of resolution and possible inaccuracies caused by abundant homoplasy, inherent problems with the branch-swapping algorithm, and the immaturity of cladistic methodology, the systematic classification used in the present treatment of the Pottiaceae must be seen as reflecting only the congruent results of the several analyses of the data set in light of these conditions.

It is here considered that cladistic analysis is the best method at this time for organizing the information from this large data set into the basis for a supraspecific classification. Many lineages are identifiable with considerable certainty in the present work (discussion of Cladogram 14), especially involving genera with one or few species.

Cladogram 1

Data set includes 9 non-pottiaceous genera in addition to the outgroup, *Polytrichum*. This is a strict consensus tree. All characters are weighted the same. Tree length 557, consistency index .17, retention index .56, number of equally parsimonious trees is more than 1115. Lettered ancestral nodes denote lineages that are similar to the named lineages of Cladogram 16.

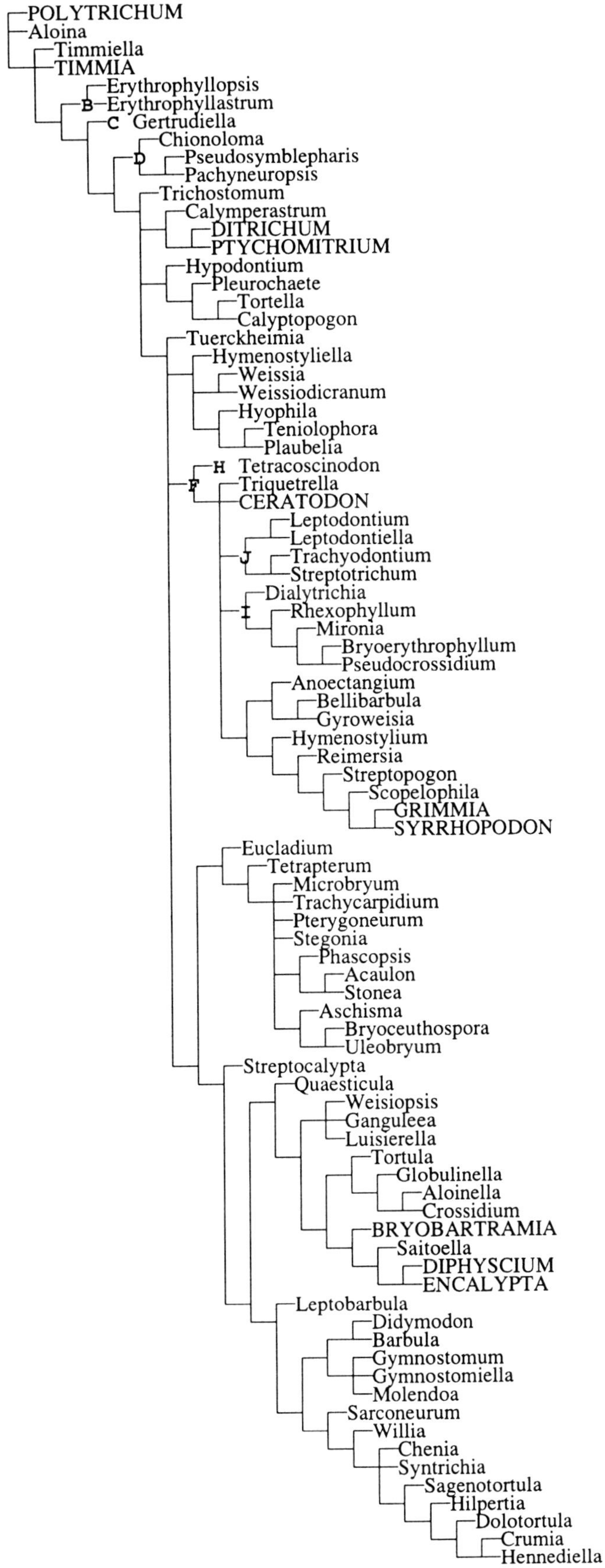

Cladogram 2

Data set includes 9 non-pottiaceous genera in addition to the outgroup, *Polytrichum*. This is a strict consensus tree; 22 reduction-related characters are weighted at 1:15. Tree length 6024, consistency index .17, retention index .57, number of equally parsimonious trees is more than 1115. Lettered ancestral nodes denote lineages that are similar to the named lineages of Cladogram 16.

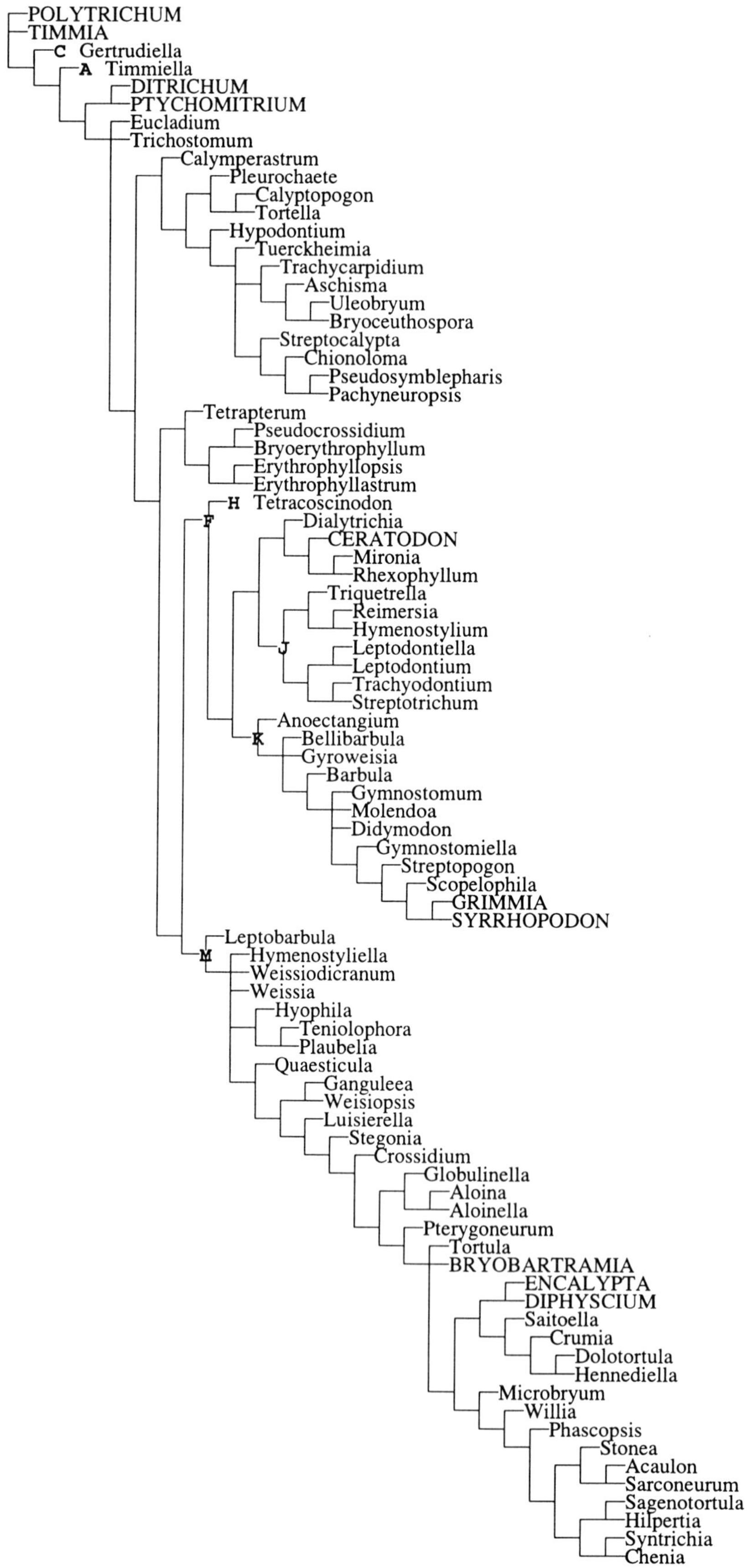

Cladogram 3

Data set includes 9 non-pottiaceous genera in addition to the outgroup, *Timmia*. This is a strict consensus tree; 22 reduction-related characters are weighted at 1:15. Tree length 6034, consistency index .17, retention index .57, number of equally parsimonious trees is more than 1115. Lettered ancestral nodes denote lineages that are similar to the named lineages of Cladogram 16.

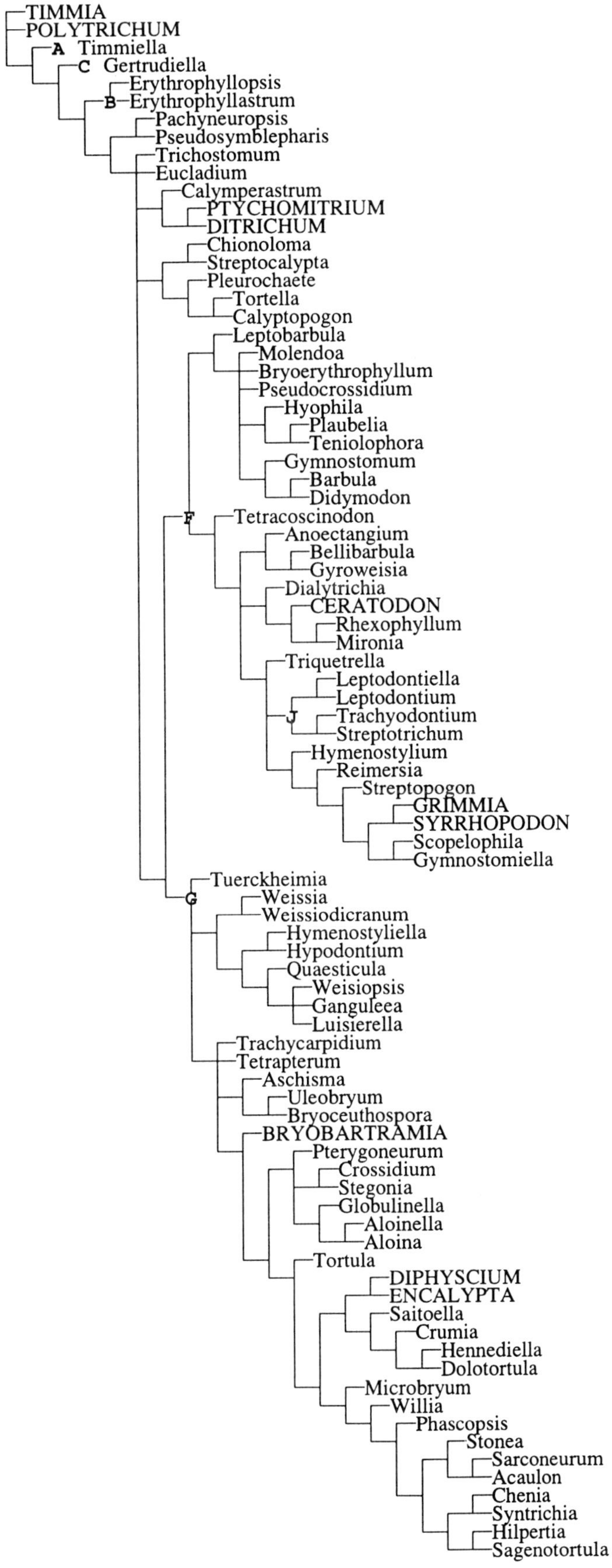

Cladogram 4

Data set includes 9 non-pottiaceous genera in addition to the outgroup, *Ptychomitrium*. This is a strict consensus tree; 22 reduction-related characters are weighted at 1:15. Tree length 6037, consistency index .17, retention index .57, number of equally parsimonious trees is more than 1115. Lettered ancestral nodes denote lineages that are similar to the named lineages of Cladogram 16.

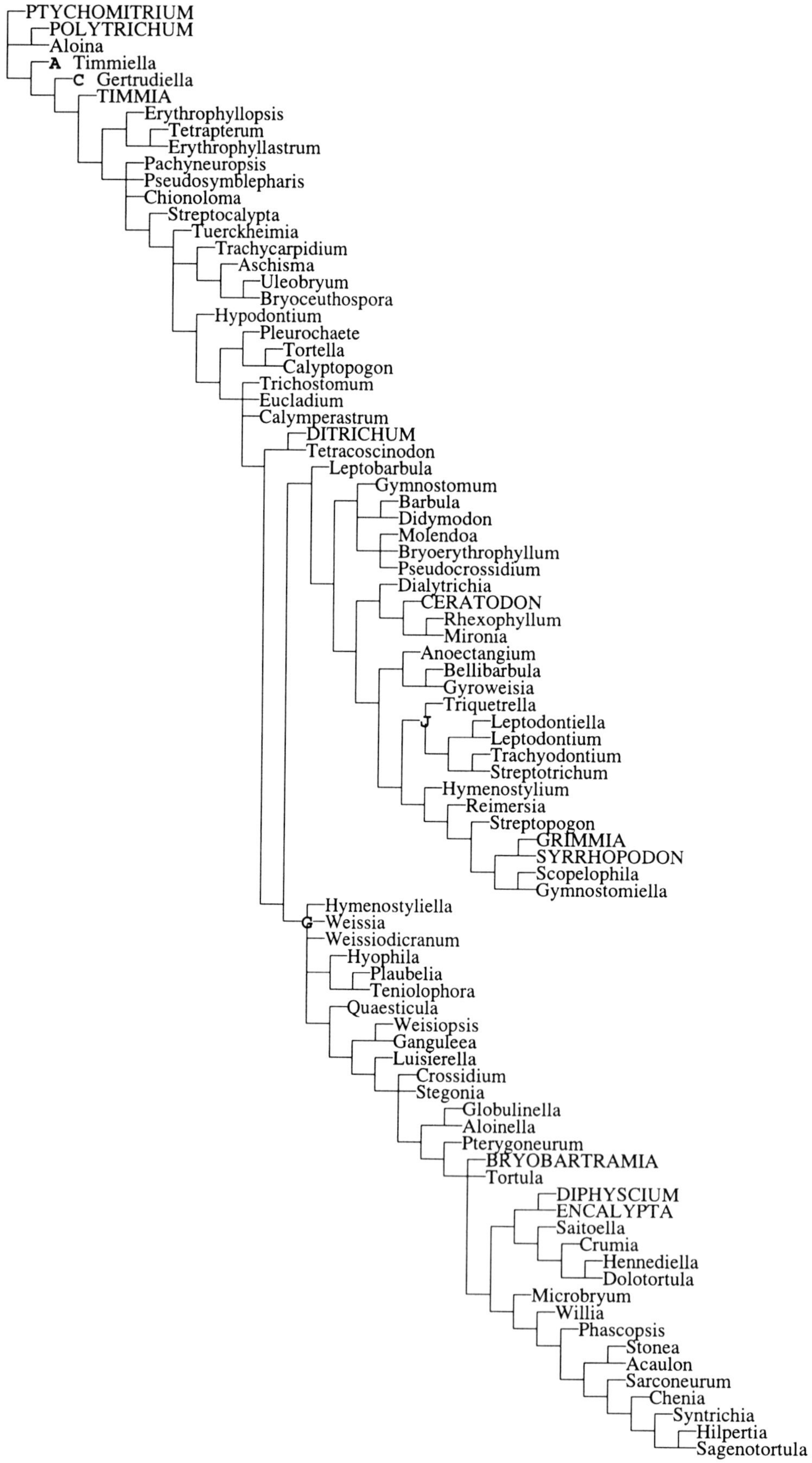

Cladogram 5

Data set includes 9 non-pottiaceous genera in addition to the outgroup, *Encalypta*. This is a strict consensus tree; 22 reduction-related characters are weighted at 1:15. Tree length 6033, consistency index .17, retention index .57, number of equally parsimonious trees is more than 1115. Lettered ancestral nodes denote lineages that are similar to the named lineages of Cladogram 16.

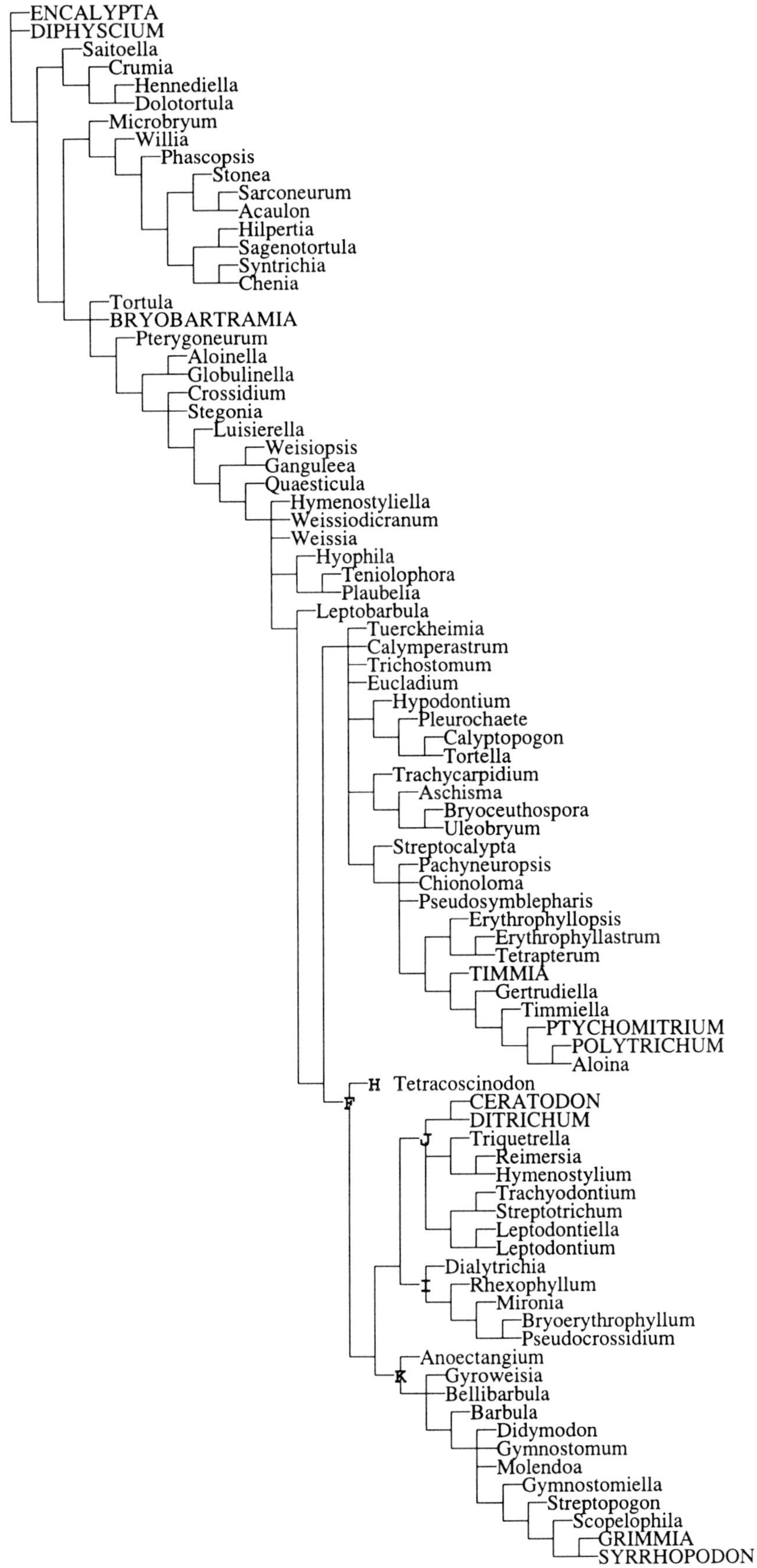

Cladogram 6

Data set includes 9 non-pottiaceous genera in addition to the outgroup, *Grimmia*. This is a strict consensus tree; 22 reduction-related characters are weighted at 1:15. Tree length 6034, consistency index .17, retention index .57, number of equally parsimonious trees is more than 1115. The lettered ancestral node is of a lineage similar to a named lineage of Cladogram 16.

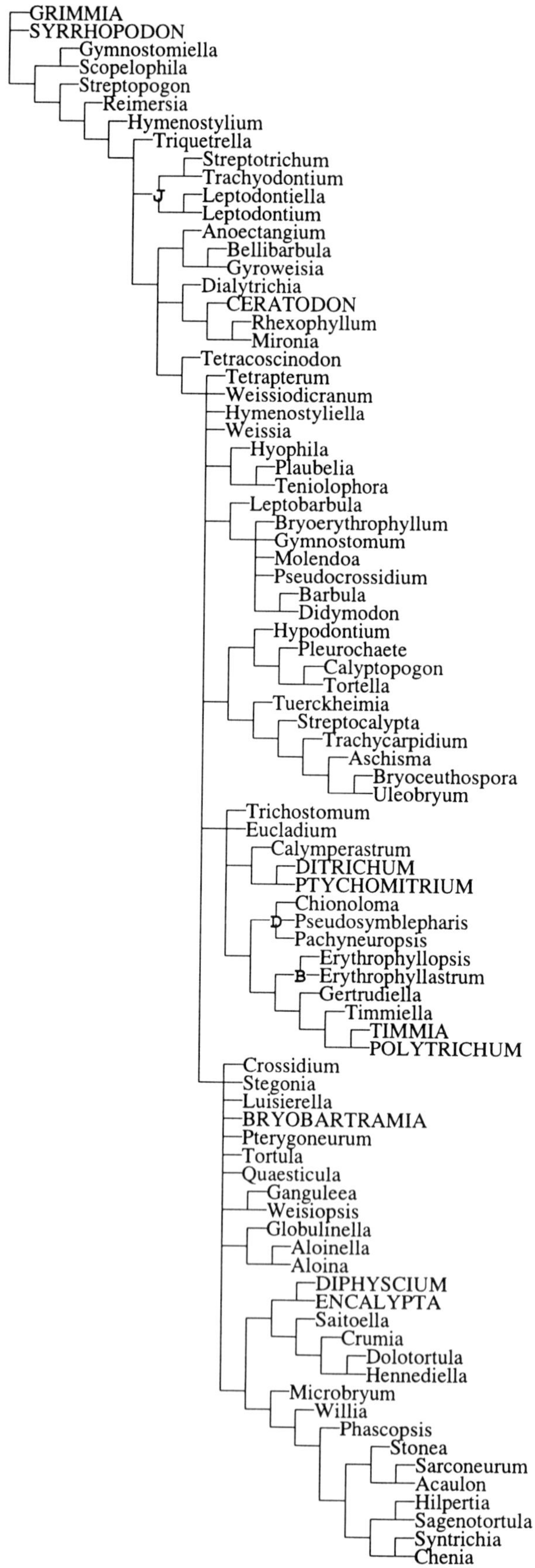

Cladogram 7

Data set restricted to pottiaceous genera except for the outgroup, *Polytrichum*. This is a strict consensus tree; all characters are weighted the same. Tree length 491, consistency index .19, retention index .57, number of equally parsimonious trees is more than 1241. Lettered ancestral nodes denote lineages that are similar to the named lineages of Cladogram 16.

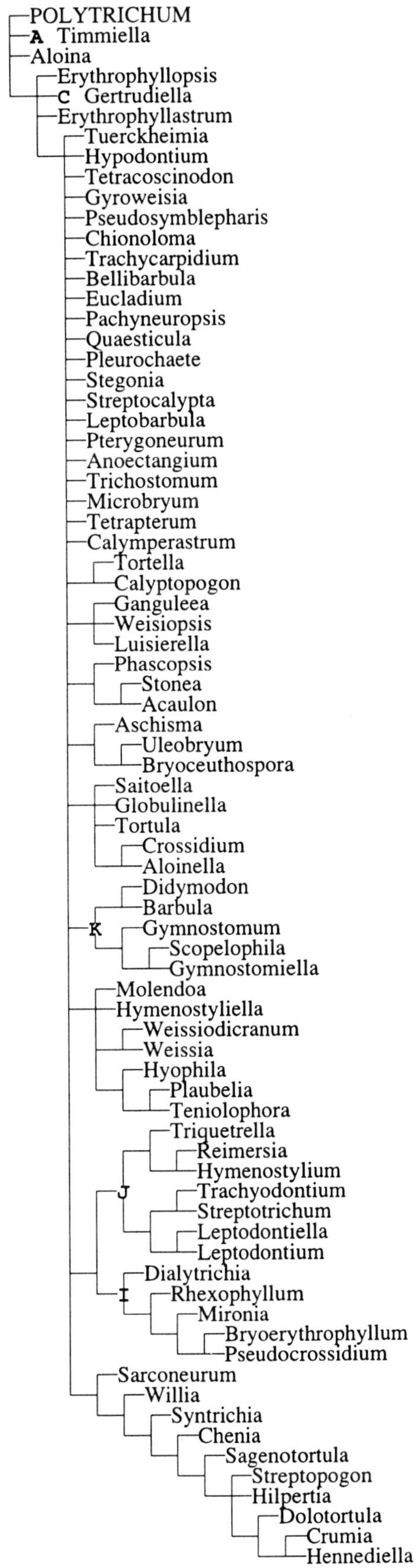

Cladogram 8

Data set restricted to pottiaceous genera except for the outgroup, *Polytrichum*. This is a strict consensus tree; 22 reduction-related characters are weighted at 1:15. Tree length 5375, consistency index .19, retention index .57, number of equally parsimonious trees is more than 1241. Lettered ancestral nodes denote lineages that are similar to the named lineages of Cladogram 16.

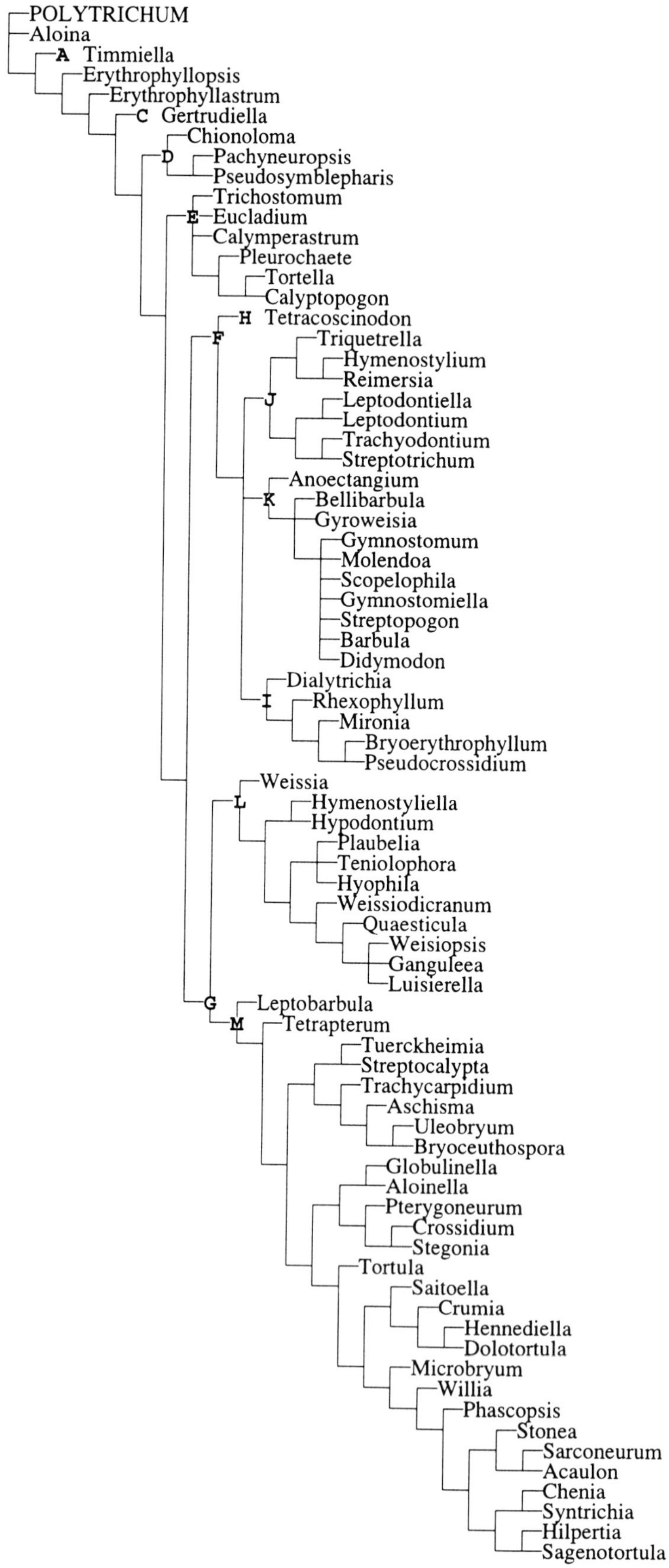

Cladogram 9

Data set restricted to pottiaceous genera except for the outgroup, *Timmia*. This is a strict consensus tree; 22 reduction-related characters are weighted at 1:15. Tree length 5290, consistency index .19, retention index .58, number of equally parsimonious trees is more than 1241. Lettered ancestral nodes denote lineages that are similar to the named lineages of Cladogram 16.

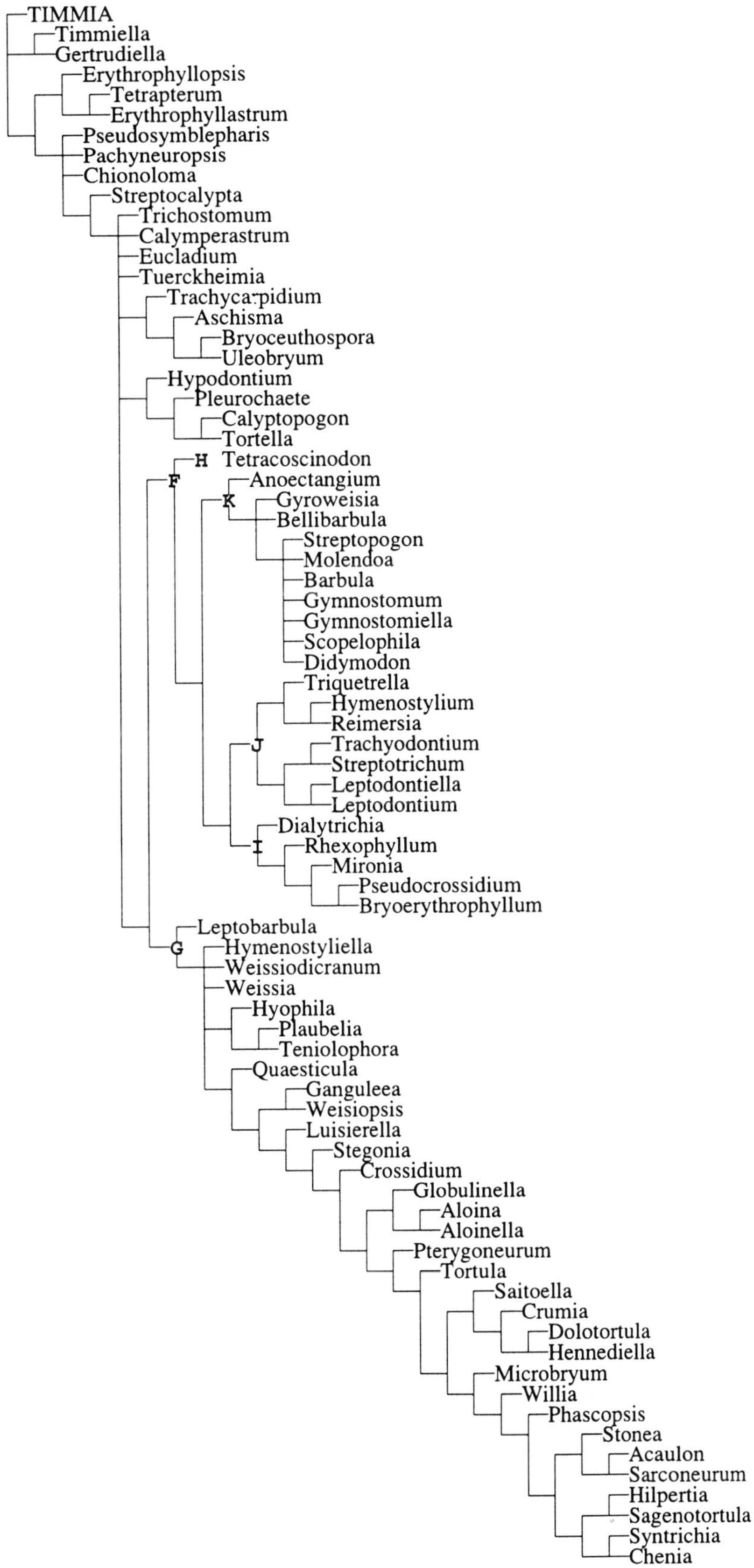

Cladogram 10

Data set restricted to pottiaceous genera except for the outgroup, *Ptychomitrium*. This is a strict consensus tree; 22 reduction-related characters are weighted at 1:15. Tree length 5341, consistency index 9, retention index .57, number of equally parsimonious trees is more than 1241. Lettered ancestral nodes denote lineages that are similar to the named lineages of Cladogram 16.

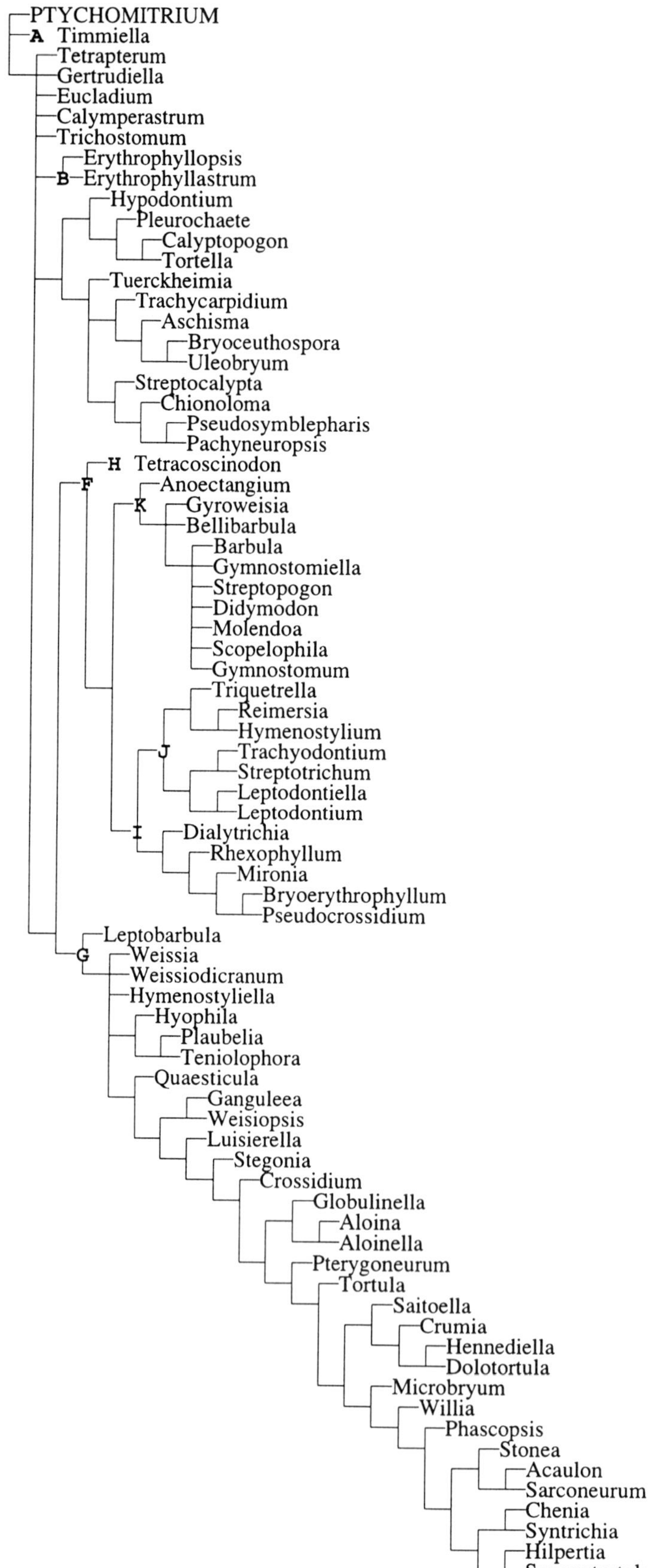

Cladogram 11

Data set restricted to genera in Merceyoideae "F" subclade of Cladograms 9 and 10; outgroup is *Tetracoscinodon*. Tree 24 of 127 equally parsimonious trees, 22 reduction-related characters are weighted at 1:15. Tree length 1494, consistency index .48, retention index .62. Character state changes shown for ancestral nodes. "X" denotes a node for which character state changes do not map, implying an ancestral multiple-branched node.

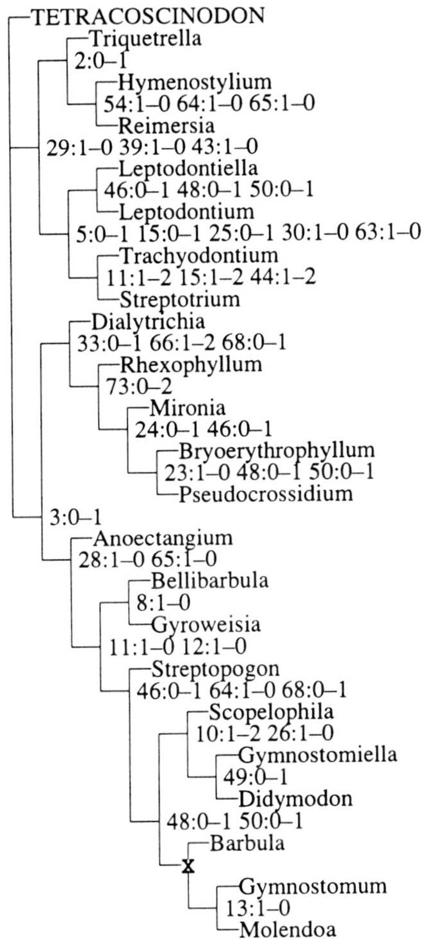

Cladogram 12

Data set restricted to genera in Pottioideae "G" subclade of Cladograms 9 and 10, outgroup is *Leptobarbula*. Tree 5 of 8 equally parsimonious trees, 22 reduction-related characters are weighted at 1:15. Tree length 2247, consistency index .37, retention index .59. Character state changes shown for ancestral nodes.

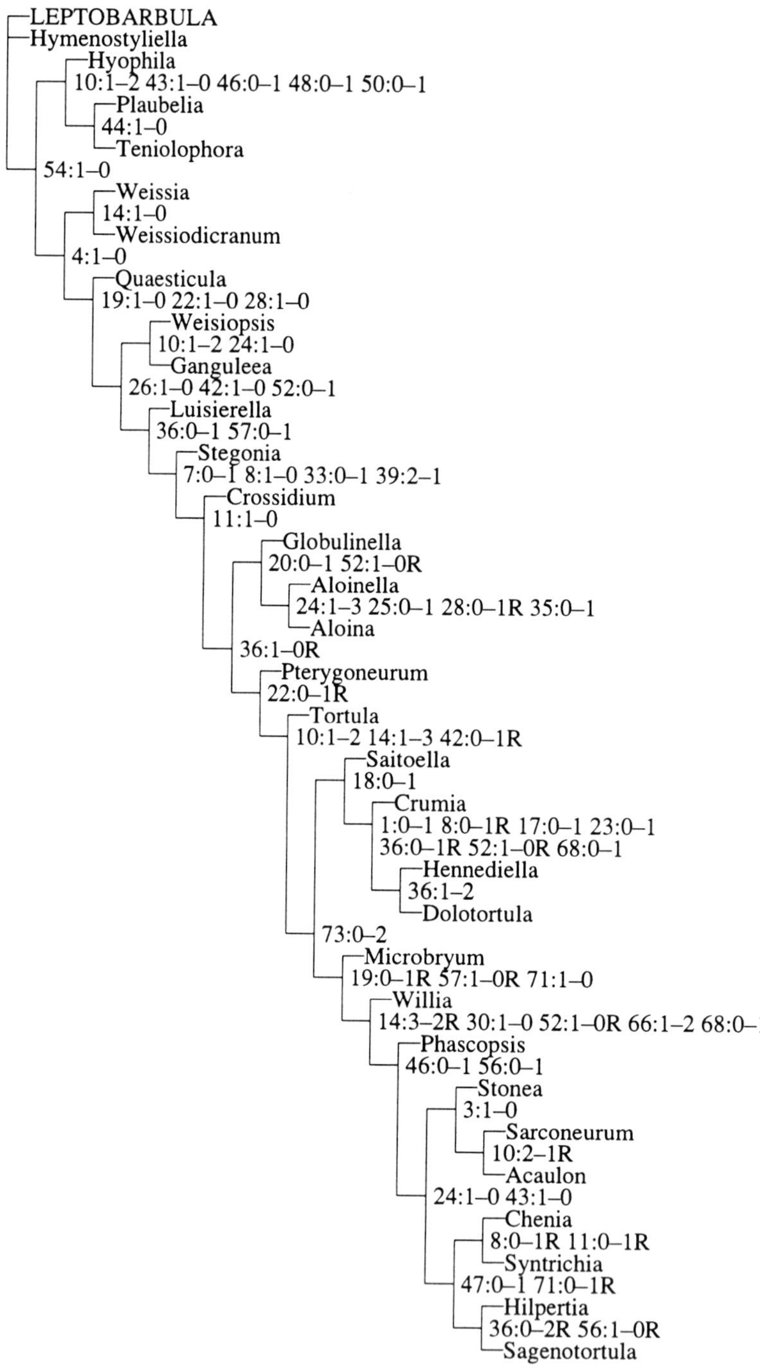

Cladogram 13

Data set restricted to pottiaceous genera. The functional outgroup is *Timmiella*. This is a strict consensus tree; all characters are weighted the same. Tree length 485, consistency index .20, retention index .57, number of equally parsimonious trees is more than 1259. The lettered ancestral nodes are of lineages identical to the named lineages of the single tree of Cladogram 16, on which the present classification is based.

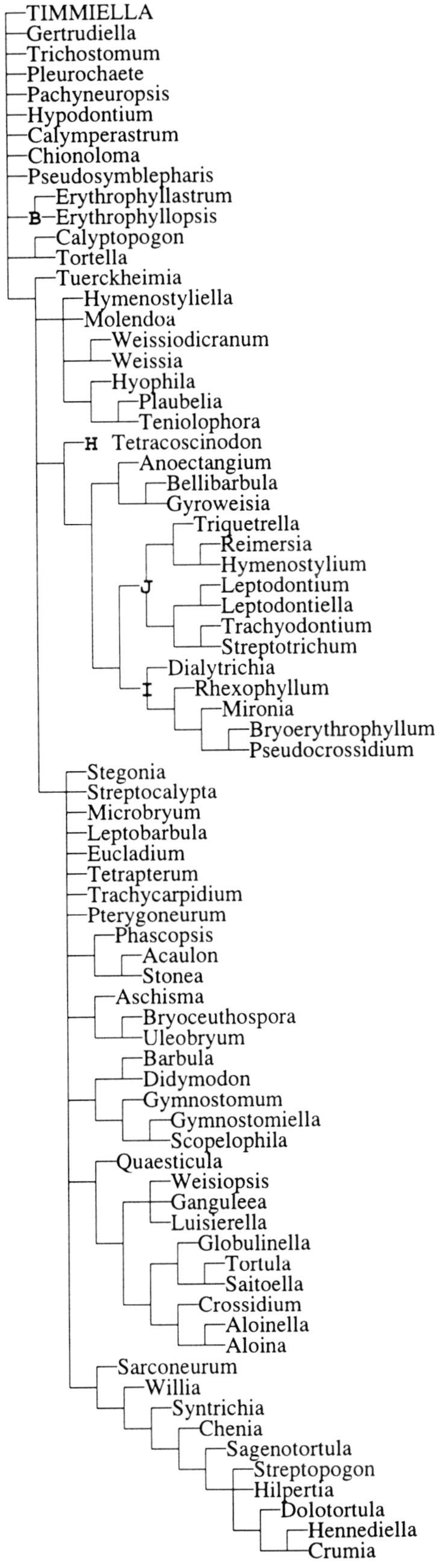

Cladogram 14

Data set restricted to pottiaceous genera. The functional outgroup is *Timmiella*. This is a strict consensus tree; 22 reduction-related characters are weighted at 1:15. Tree length is 5270, consistency index .19, retention index .57, number of equally parsimonious trees is more than 1250. Lettered ancestral nodes denote lineages that are identical to the named lineages of the single tree of Cladogram 16, on which the present classification is based. Geographic ranges ("*" denotes austral taxa) are given.

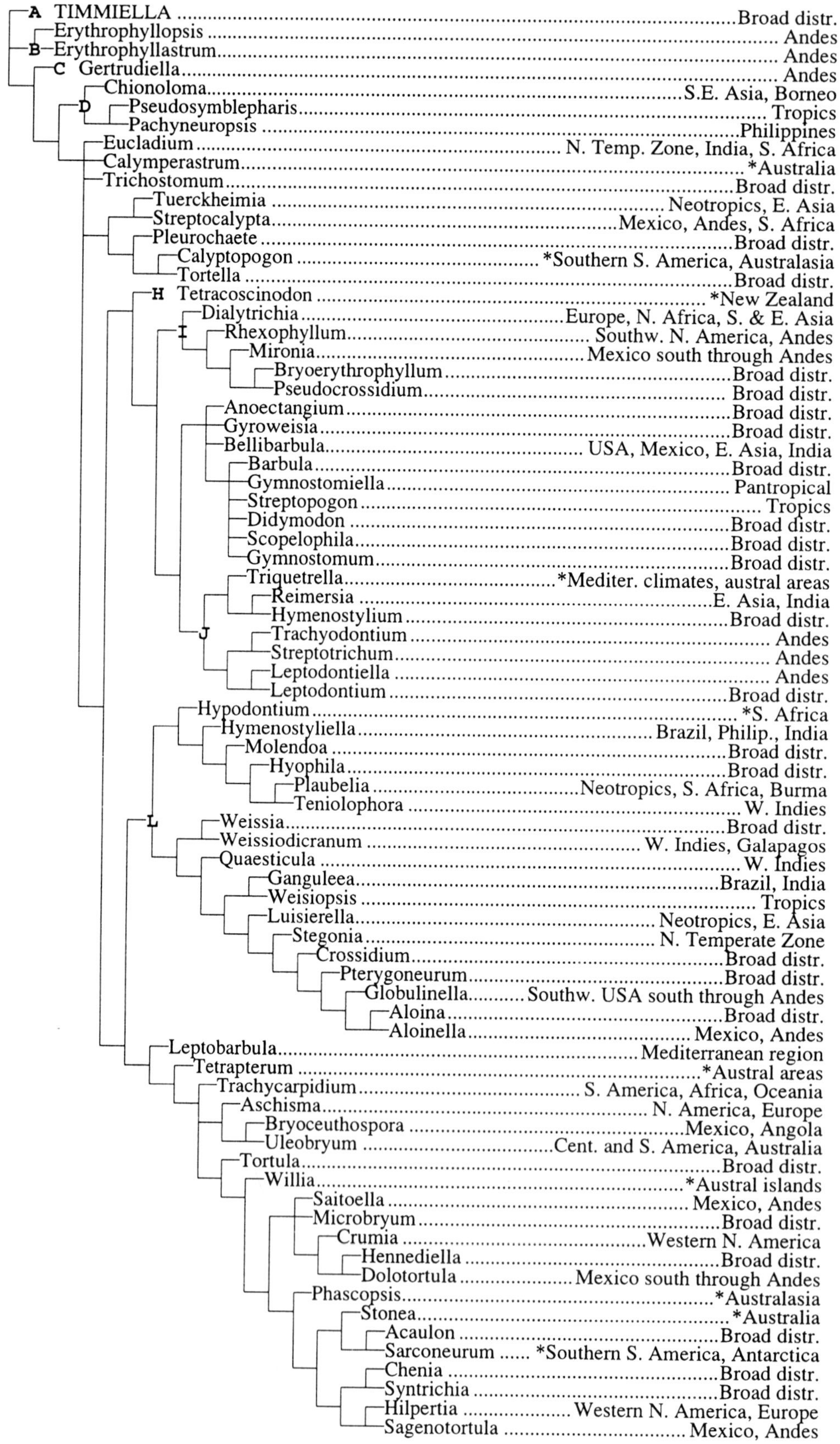

Cladogram 15

Data set restricted to pottiaceous genera. The functional outgroup is *Timmiella*. This is tree 1126 of more than 1250 equally parsimonious trees, 22 reduction-related characters are weighted at 1:15. Tree length is 5270, consistency index .19, retention index .57. This trees shows ancestral character state changes on which Cladogram 16 is based. "X" denotes a node for which character state changes do not map, implying an ancestral multiple-branched node.

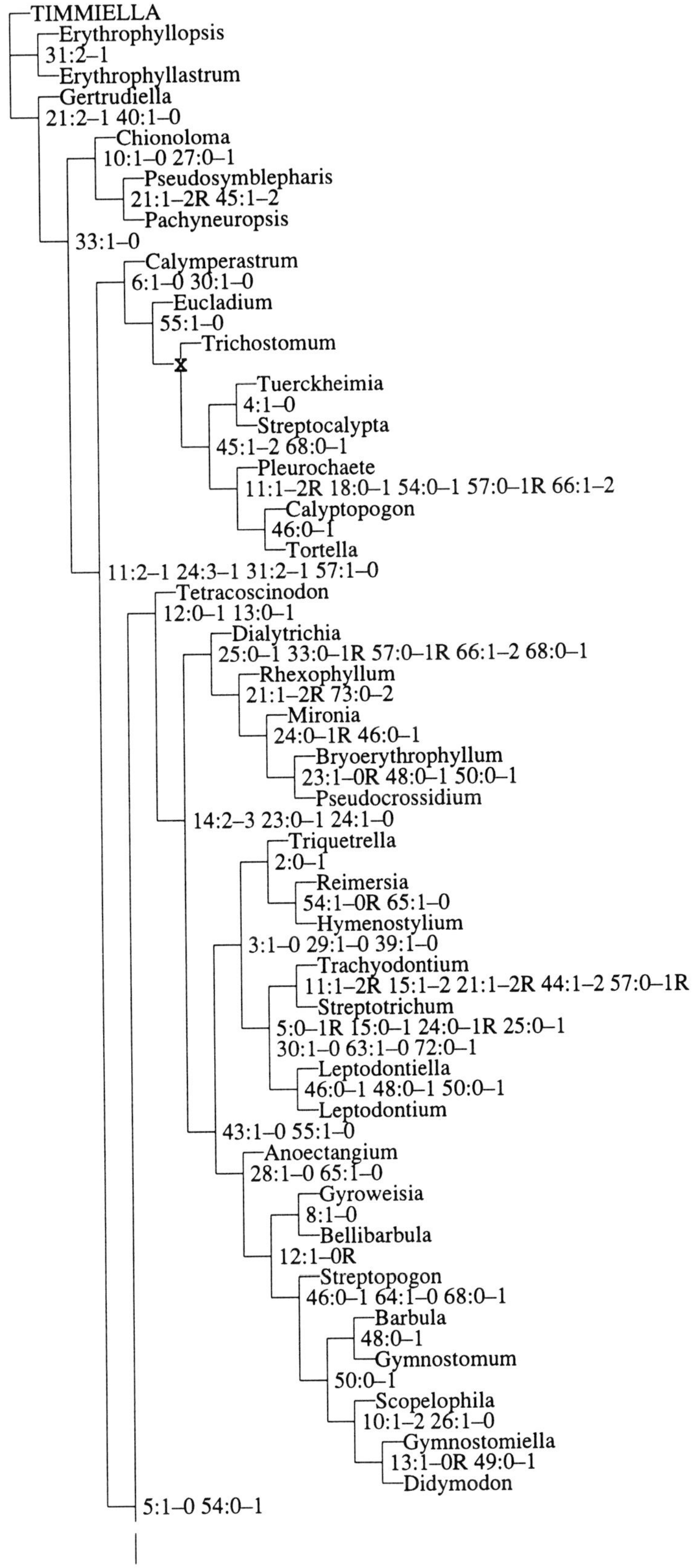

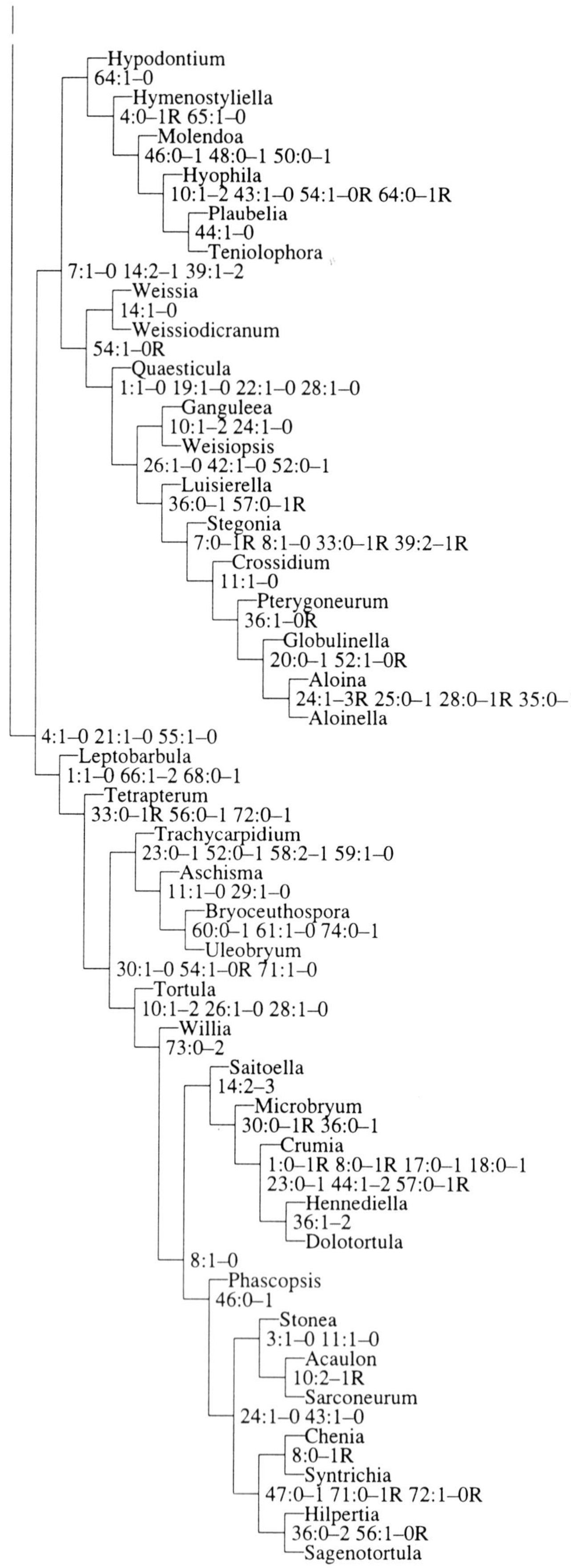
Hypodontium
64:1–0
Hymenostyliella
4:0–1R 65:1–0
Molendoa
46:0–1 48:0–1 50:0–1
Hyophila
10:1–2 43:1–0 54:1–0R 64:0–1R
Plaubelia
44:1–0
Teniolophora
7:1–0 14:2–1 39:1–2
Weissia
14:1–0
Weissiodicranum
54:1–0R
Quaesticula
1:1–0 19:1–0 22:1–0 28:1–0
Ganguleea
10:1–2 24:1–0
Weisiopsis
26:1–0 42:1–0 52:0–1
Luisierella
36:0–1 57:0–1R
Stegonia
7:0–1R 8:1–0 33:0–1R 39:2–1R
Crossidium
11:1–0
Pterygoneurum
36:1–0R
Globulinella
20:0–1 52:1–0R
Aloina
24:1–3R 25:0–1 28:0–1R 35:0–1
Aloinella
4:1–0 21:1–0 55:1–0
Leptobarbula
1:1–0 66:1–2 68:0–1
Tetrapterum
33:0–1R 56:0–1 72:0–1
Trachycarpidium
23:0–1 52:0–1 58:2–1 59:1–0
Aschisma
11:1–0 29:1–0
Bryoceuthospora
60:0–1 61:1–0 74:0–1
Uleobryum
30:1–0 54:1–0R 71:1–0
Tortula
10:1–2 26:1–0 28:1–0
Willia
73:0–2
Saitoella
14:2–3
Microbryum
30:0–1R 36:0–1
Crumia
1:0–1R 8:0–1R 17:0–1 18:0–1
23:0–1 44:1–2 57:0–1R
Hennediella
36:1–2
Dolotortula
8:1–0
Phascopsis
46:0–1
Stonea
3:1–0 11:1–0
Acaulon
10:2–1R
Sarconeurum
24:1–0 43:1–0
Chenia
8:0–1R
Syntrichia
47:0–1 71:0–1R 72:1–0R
Hilpertia
36:0–2 56:1–0R
Sagenotortula

Cladogram 16

Phylogenetic arrangement of the Pottiaceae in the present study. Same as Cladogram 15 but without character state changes indicated. Subclades recognized as suprageneric taxa are indicated by letters at ancestral nodes: "A" Timmielloideae, "B" Erythrophyllopsoideae, "C" Gertrudielloideae, "D" Chionolomoideae, "E" Trichostomoideae, "F" Merceyoideae, "G" (and all its ancestors in the family) Pottioideae, "H" Tetracoscinodontieae, "I" Bryoerythrophylleae, "J" Leptodontieae, "K" Barbuleae, "L" Hyophileae, "M" Pottieae.

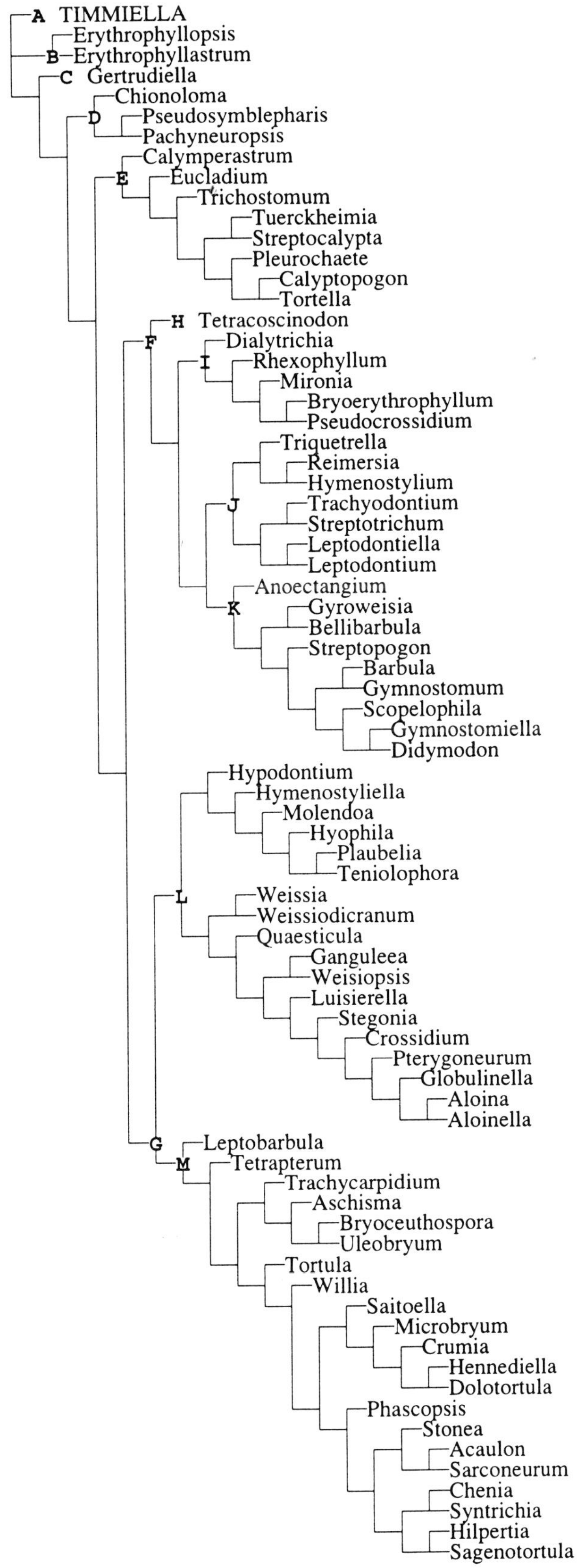

Cladogram 17

Phylogenetic arrangement of the Pottiaceae by Saito (1975).

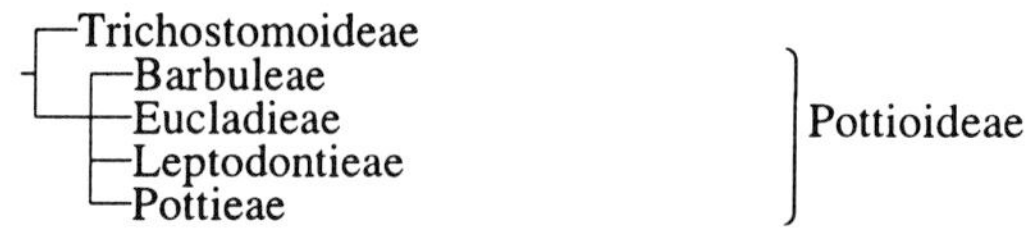

Cladogram 18

Phylogenetic arrangement of the Pottiaceae by Chen (1941).

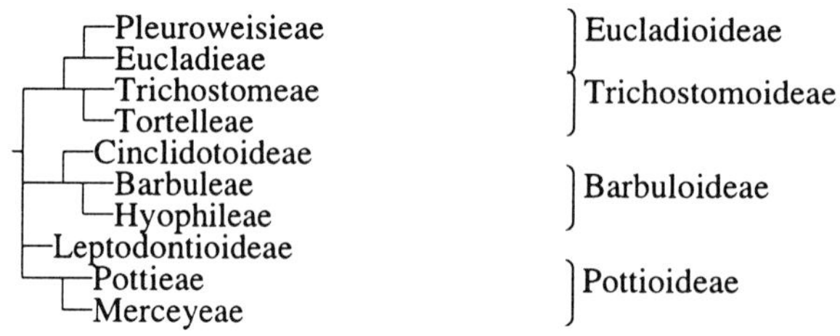

Cladogram 19

Phylogenetic arrangement of the Pottiaceae by Hilpert (1933).

DATA SET

This data set contains morphological information on the 76 genera of the Pottiaceae, and on ten (fully capitalized) non-pottiaceous genera used as outgroups in analysis. A dash ("-") indicates a character that is variable in state or otherwise "unknown, undefined, or missing" (Farris 1988). There are 86 taxa and 75 characters; the first character is numbered as character "0".

```
                     0         1         2         3         4         5         6         7
                     012345678901234567890123456789012345678901234567890123456789012345678901234
POLYTRICHUM          1101100010120012000102103110111201000000-00--10-----0001012101110-----12020
BRYOBARTRAMIA        100-000100-0002000000110100-01-10-0000000011110-----101000100111-0----01102
CERATODON            11010011-0111131000101111110111101 00-00-000--00-----001-0121011-11110-0--00
DIPHYSCIUM           10000001002-002-0-0001-020101112000000 0--0---10------0101121011-11-20-01021
DITRICHUM            1-01100-000-0-2-0-010--1-110111100000100000--00------0-1012101111112-101000
ENCALYPTA            0-0100--002-00-0000000-0200-11-20-00-001001111-01010-00-2-210111-1120-----2
GRIMMIA              110-00010011-1-00-0101-1000-00010-000000-00----01110-0--1-210111---20-0100-
PTYCHOMITIUM         0101-1-00012002201011110101011110000000 0100--10-----10000121011111120-01--1
SYRRHOPODON          1100000-001---2-0--10-1-1010---10000000-00--11100110100-1121011101020--1000
TIMMIA               1101001-101200-2000102103010111201000002 00-1110------0010121011111-20-02-20
Acaulon              000000-1001-002-000100-1000-01110-100-010--0-0000000-00010000111------00121
Aloina               -00-000-00-2001-00-012-0310-11120101200-110--1000000-00-0121011111121101020
Aloinella            1101001000-0001-00-01000210-11-1010100010--111000000001-0-21011111-20-01100
Anoectangium         11-110-1101-11-000010-110-0-0-110000000--0--110000000111002101111 0----01000
Aschisma             -00100100010002-00-100100010-0010-00000-001112000000100010100100------10100
Barbula              1-011--1101--1--000-0-110010-1110-00000100--1111101100--00210111-1221101000
Bellibarbula         11011011001-0130000101--0010011100000 0--001-10000000001000210111 10----01120
Bryoceuthospora      000-0001101-0022000100011010000101000001 00100-000000-01010101011-------0101
Bryoerythrophyllum   1-0111-1101-----00010--0101011-10-000000-01111101010-0--0-2101111--2-101020
Calymperastrum       10-11101102-0020010102101010110101000001001121000000---------------------0-
Calyptopogon         110011-11012--2-00110-11000-1101010000 11001-12100101-010012101111122111 10-0
Chenia               100110111-210-3-000-0---000-011101000001000--1110000000-----01111-221101-2-
Chionoloma           110101--100200-100110110301111020000000-001112000000---------------------0-
Crossidium           -00100-100-000--000--1-0100-01-1010110--0--111000000-0--0-21011111121101000
Crumia               110-001110220030011-00-1100-0111000010 01001-210000000000-012101111112 1101-10
Dialytrichia         11011011101-1130010-0-11011011-10100000100111100000000 0-012101110122 1101000
Didymodon            11-1101110---03-0-0-0---0----111000000-1-01-1-11111000--0-210111-1---1010-0
Dolotortula          11010001112-0030011000-1000-011101002001000--0000000000101210111112211--02-
Erythrophyllastrum   1101-1111011-0200-010210301001110100 0001-0111100000000110121011111020-01020
Erythrophyllopsis    11-101111012-0200001021021101111010000 0110111100000000 0-0121011111-20---02-
Eucladium            1-00-10110-1002110010-1-101011-1000000-1001-110000000000102101110 1120-01000
Ganguleea            10-0-00011210010000000-0000-011100000002000--10000001110002101111 0------00-
Gertrudiella         110111011012003-000101-0300-11121100000200-0110000000--------------------1-
```

Globulinella	-0010--100-000-00000100010--0111011000--000--110000000000-21011111120-01000
Gymnostomiella	1001-0111020002-000-00-1000-00110000100100101110111000100021011100----010-0
Gymnostomum	11-10-11101000-000010-0-0010-1-10000000-001-11100000000000210111-0----0-000
Gyroweisia	100-101100-00--0000-0--110--01-10000000--01-11110000-0-00-2101111-020-0-0-0
Hennediella	1-01---110--002-01110-1-100-01-101002000001-21000000-00-2-2101111--2110--20
Hilpertia	1-01-0-1001-003-01010021000-010101002--101-01011000010100-21011111221101-20
Hymenostyliella	11011010001200l-0001001-10101111000000012000--100000001100-21011100----01000
Hymenostylium	11-0101110--11-000010-11001000-100000--00010110000000000021011100----00000
Hyophila	1--11-001-2100--000-00-0101011110-00000-00-011101011-0000-21011110----01-00
Hypodontium	11010110101-00---01-0210201-1102000000002001102000000000102121011101100-11100
Leptobarbula	1001-01110100020000100-00110111000000000001111000000000100-2101111221101000
Leptodontiella	1100111110111131000101-1111010010000000000001-1-1010100010002101101111000110-
Leptodontium	11001--1101-113-00010--1111010010000000--001-11101110-01-0121011-111-0-0-100
Luisierella	000000101---00-0000-01001-0-1001000000002000--2000000100001210111l--20-010-0
Microbryum	000--0-1002-003-000100--100-011101001101001-1-000000100020--01110--20--0120
Mironia	11011011101-113-0101021-11101111010000000001111100000000110-2101111221101020
Molendoa	11-11-0-10--00----010---1-1011110000000-00111110101001100021011100----01000
Pachyneuropsis	110100111002002001010210301-1112000000001001112000000---------------------0-
Phascopsis	10010-1100-00020000-001-100-01010-000001001-1110000000000010-00111------00122
Plaubelia	10011010112-001-000-00-010-00111010000000200-00-10100000-0012101111l120-0-000
Pleurochaete	110111011012002100110-102-1011-10000000-00111200000000111012101111112211020-0
Pseudocrossidium	1-01--01101---30000-0--011-011110101000100111110111000--0121011111-211020-0
Pseudosymblepharis	1101-1-110--00-00001021030111l-20000000-00111200000000--012101111--20-01000
Pterygoneurum	-00100-1001-001-000---20100-01110100000-00-01-00000010002-2101110-020-0-10-
Quaesticula	100100001011001000000-0001010011100000000-001001000000000-0021011111221101000
Reimersia	011010-1101-1120000102110010101100000-10000--100000000000021011100----01000
Rhexophyllum	110111-11011113200010211011010110-000001101111000000000110121011110----01020
Sagenotortula	11010001002-0020000100--000-110101003001000--111000000010-2101111221102020
Saitoella	1001000100-000300110100-100-01-0-0100001001111000000---------------------2-
Sarconeurum	1-0-0-11101-002000010-1000-0010101001000001-11100000---------------------2-
Scopelophila	110000-11021-1--00010---000---110000000-000--110101000000-210111-0----00000
Stegonia	00010011001100--000-00-1100-01-101101101010--1000000100-0---01111--20-0-100
Stonea	000000110020002-000110-0000-01-1001000010010111000000--------------------2-
Streptocalypta	10010--110--002-00-----010-01102-000000-001112000000-000002101111-221101000
Streptopogon	1-0010-1001--03-0------1000-0--1000021-1000--0100100-0-0012101110--211-1--1
Streptotrichum	1100111110121132000102-111101001000000000001021000000000101121011011120-01100
Syntrichia	110-0--1102-----0-0----0000--101--0000-1-01-11110100-0012-2101111122110-020
Teniolophora	11011010102100--011-0010201011110000000200-001101010000------------------0-
Tetracoscinodon	110010-10001112000010-1-1010111100000001001111000000000110021011111l-0-01100
Tetrapterum	1001001110110020000101102-1011110100000-001111000000101011200101------011-0
Timmiella	1-0111-011120022000102103-10111201000002100--1000000-0010-210111-1---002100
Tortella	110-0101101-00200001011-101011-10000000-001112100000-01-012-0111--22110--00
Tortula	1-01001110--0-3-0---0---100-01-101100001001-11000000-00-2---01111--2110--00
Trachycarpidium	-001001110110-20000100-1101011-10-00100-0011120000000100010201111------00101
Trachyodontium	110011111012113201110211111010010000000-001021000000000110221011011120----0-
Trichostomum	11---1--101-002-0-01-1101-10-1-100000001001111000000-00-0-2-01111---0-01--0
Triquetrella	111-10111011113-000101110010-011000000--001-00000000001101210111111-0-00000
Tuerckheimia	11-100--10-1002100010-10101011-1000000-1001-2-0000000000000210111--0-----00-
Uleobryum	-001000110-0012000--001110100-0101000001001-11000000100010101011------10101
Weisiopsis	1001-01010-10-20000-01--0-0-0-11000000-2000--1000000100000210111--1-0--1000
Weissia	100101101011000000010-10101011-10-00000-00111-000000-0--0---01111---0-0-100
Weissiodicranum	110100-000110000000010-10101011110000000020011-1000000-00------------------0-
Willia	11010011102-012-00010---100--1010100000100111100000000100-2101111-22110-02-

TAXONOMIC SECTION

Family POTTIACEAE

Pottiaceae Schimp., Coroll. Bryol. Eur. 24, 1855 [1856], *nom. fam. cons.*

The family name is based on *Pottia*, which was named by Friedrich Ehrhart in his Beiträge zur Naturkunde (1787) for "Cel. Joh. Frid. Pott," a physician and professor of botany at Brunswick (mistakenly Braunschweig according to Müller 1853), author of an unpublished "Flora Brunswickensis."

Plants usually turf-forming or loosely caespitose, green above and brown below, irregularly branching. Stems short or to several cm in length, mostly pentagonal in transverse section, central strand usually present, hyalodermis usually absent; axillary hairs several cells in length, sometimes the basal 1–3 cells brownish. *Leaves* usually appressed and often contorted when dry, spreading when wet, *ovoid to lanceolate or lingulate*, ca. 1.5 to 3.5 mm in length, margins usually recurved below, occasionally plane, entire or sometimes dentate above, occasionally bordered by thick-walled or elongate cells or cells in layers; apex rounded-obtuse to more commonly narrowly acute; base usually ovate to oblong, occasionally sheathing the stem; *costa ending a few cells below the apex to short-excurrent or awned, in medial transverse section usually* with a differentiated epidermis ventrally or on both sides, *one or two stereid bands, guide cells in one or seldom more than one layer, hydroid strand occasionally present (sometimes multiple); upper laminal cells usually subquadrate*, occasionally hexagonal or rarely short-rectangular, mostly ca. 9–16 µm wide, 1(–2):1, usually in one layer, *walls* mostly evenly thickened, superficially flat to bulging, *sometimes bulging only ventrally; papillae usually present over the upper laminal cells*, solid or occasionally hollow, usually bifid but occasionally simple, sometimes flattened or compound; *basal cells usually differentiated*, usually clear, smooth or lightly papillose above, rectangular, generally filling the base medially, sometimes rising marginally in a vee-shape, occasionally bulging. Asexual reproduction not uncommon, by multicellular (rarely unicellular) propagula borne on stalks in the leaf axils or more seldom on leaves, or obovoid brood bodies borne on rhizoids in the soil. Perichaetia and perigonia terminal or occasionally lateral on short branchlets. Dioicous or monoicous, occasionally apparently rhizautoicous. Perichaetial leaves often sheathing in the lower part and then with elongate-rhomboidal cells in lower portion, usually larger than the stem leaves, long-oval to long-lanceolate. Perigoniate plants occasionally smaller than the perichaetiate, seldom nearly stemless and bud-like. *Sporophyte often in transformation series of apparent peristome reduction and seta shortening. Seta usually elongate*, often twisted; *capsules ovoid to cylindric*, mostly stegocarpous, occasionally spherical and then cleistocarpous and rhexolytic; *neck usually small or nearly absent*; annulus mostly of 1–2 rows of vesiculose cells, occasionally revoluble or deciduous in pieces; operculum short-conic to short-rostrate, cells in straight or oblique rows; usually elongate, *peristome teeth occasionally absent, more usually erect or twisted usually clockwise*, yellow, orange or red, rudimentary or *consisting of 16 mostly twice cleft, spiculose, striate or papillose, lanceolate teeth, or 32 linear, usually densely spiculose filiform divisions*, the basal membrane usually low or absent, *occasionally very high and trabeculate. Calyptra cucullate*, smooth, occasionally mitrate. Spores ca. 10–15 µm in diameter. Laminal color reaction yellow to orange-red or red in two percent KOH solution. Reported chromosome number generally x = 13.

The conservation of the name Pottiaceae (Magill 1977b; Greuter 1988) apparently was unnecessary. The original discussion of Hyophilaceae Hampe, Linnaea 4: 68, 1847, is actually given as a footnote in which was said: "Ich habe diesen Namen den Früheren: *Calympereae* oder *Syrrhopodonteae* vorgezogen, weil sich diese Namen auf die Form der Haube, oder auf das Peristome beziehen; die Beschaffenheit des Peristoms kann dem Familiennamen nicht zum Grunde gelegt werden, wo das Peristom oft felt." Argument may be made that there is no description here ("wo das Peristom oft felt" being merely rhetorical and applying to Hyophilaceae, Calympereae and Syrrhopodonteae equally), hence Hyophilaceae was not validly published prior to Pottiaceae; also, according to G. Zjilstra (pers. comm.), this is not a diagnosis since the phrase including "oft" does not cover the whole group. The name Tortulaceae, used for the Pottiaceae in some older literature, is a synonym of Ephemeraceae (Crosby & Magill 1981). Recently, Saito (1975a) agreed with Hilpert (1933) and Podpěra (1954) in excluding *Cinclidotus* from the Pottiaceae and placing it into a separate family, the Cinclidotaceae Schimp. (= Ripariaceae Schimp. see Crosby & Magill 1981), a disposition followed here. For additional information on excluded genera and species, see the discussion of Excluded Taxa.

The authorities for the names mentioned in the text are those of the names recognized as correct in the updated list of taxa (see Table of Contents). The specific and infraspecific nomenclature is mainly that of the Index Muscorum (van der Wijk et al. 1959–69), with additions and emendations based on research since the publication of that work as summarized in the updated list of taxa. Some of the names of subgenera and sections acknowledged as correct here may not, however, represent taxonomically valid or properly placed groups, but are merely starting points for revisionary study. Thorough revisionary work is required for most genera; for this reason, no emphasis is placed on citation of putative type specimens in that lectotypification, inappropriate here, is required for a majority of them; the specimens examined in the course of this study were annotated and their herbarium designations (Holmgren et al. 1990) are given here.

All new combinations, new names and new synonymy are based on examination of material obtained on loan from the herbaria cited, or in relatively few cases, are based on my judgement that the novelty is clearly warranted. The vast majority of specimens seen during this study are "authentic" or syntype material as far as determinable with the literature immediately available, but their exact identity as types, although probable, remains to be more accurately ascertained through revisionary treatments and associated lectotypification not appropriate here. This study could not be done adequately, however, without a great deal of name shuffling.

In the nomenclature list at the beginning of each generic treatment, names following the genus name are not indented if they are correct (e.g. accepted infrageneric names) while synonyms (of the genus or of infrageneric names) are indented. Synonyms that cannot be referred to a particular infrageneric taxon are placed immediately below the generic name. Infrageneric taxa are grouped by category, being arranged alphabetically (after the typical name), and include both those recognized by the present author and those unstudied supraspecific names given as correct names by van der Wijk et al. (1959–69). Synonyms are grouped by taxonomic category (subgenus, section, subsection), then arranged alphabetically.

The citation of a species name under the heading "species examined" does not mean that the species has been extensively studied by me. It merely indicates that well-identified or syntype or type or otherwise authentic specimens of the species were examined. These fit well within the genus concept presented in the description, or at least cannot be easily referred to a different genus.

KEY TO SUPRAGENERIC TAXA

The subfamilies and tribes recognized here are subclades defined by character state changes at ancestral nodes, but these groups are fairly coherent morphologically and may be described in the fashion of a standard key. The couplets of this key reflect the coordination of the subclades in Cladogram 16. See the section on morphology above or the glossary for explanation of specialized terms.

1. Upper lamina bistratose medially and the cells not vertically aligned (i.e. not directly over each other) near the costa but grading to vertically evenly stacked towards the leaf margin, leaves broadly to linearly lanceolate .... Subfamily **Timmielloideae**
1. Upper lamina unistratose or if bistratose then cells situated directly over one another throughout 2
    2. Upper laminal cells ventrally mamillose medially but several rows of cells bulging on both sides marginally, costal guide cells forming a thick-walled, multilayered cylinder .... Subfamily **Gertrudielloideae**
    2. Upper laminal cells similarly bulging or not throughout leaf, guide cells either not multilayered or if so then thin-walled 3
        3. Leaves lanceolate, margins plane to weakly incurved, apex acute, base sheathing, upper lamina KOH red, stereid bands two, guide cells 4–6, rows of cells across ventral surface of costa 10(–16) .... Subfamily **Erythrophyllopsoideae**
        3. Not this combination of characters 4
            4. Leaves long-linear, margins plane, ventral stereid band larger than the dorsal .... Subfamily **Chionolomoideae**
            4. Leaves lanceolate to spathulate, ventral stereid band absent or generally smaller than the dorsal 5
                5. Sclerodermis commonly poorly differentiated, hyalodermis commonly present, leaves lanceolate, margins plane to weakly incurved, upper laminal cells KOH yellow, costa lacking a differentiated dorsal epidermis, clavate axillary propagula rare .... Subfamily **Trichostomoideae**
                5. Not this combination of characters 6
                    6. Stem sclerodermis commonly well differentiated from cells of central cylinder, which have abruptly larger lumens, leaves usually broadly lanceolate to narrowly elliptical, usually with two costal stereid bands, leaf base commonly differentiated in shape and ovate or rectangular, upper laminal cells equally convex on both free surfaces, clavate axillary propagula commonly present in some genera .... Subfamily **Merceyoideae** 7
                        7. Stem black, leaves long-triangular, capsule with a circumstomal ring .... Tribe **Tetracoscinodontieae**
                        7. Not this combination of characters 8
                            8. Upper lamina usually KOH red, dorsal stereid band usually reniform .... Tribe **Bryoerythrophylleae**
                            8. Not this combination of characters 9
                                9. Stem central strand absent .... Tribe **Leptodontieae**
                                9. Stem central strand usually present, or if absent then costa with one stereid band .... Tribe **Barbuleae**
                    6. Stem sclerodermis commonly not or poorly differentiated from cells of central cylinder, which generally grade in size into the cortical cells, leaves usually broadly ligulate to spathulate, usually with one stereid band in the costa, leaf base usually little differentiated in shape, sometimes upper laminal cell free surfaces ventrally bulging and dorsally weakly convex, clavate axillary propagula rare .... Subfamily **Pottioideae** 10
                        10. Upper laminal cells usually bulging ventrally and weakly convex dorsally .... Tribe **Hyophileae**
                        10. Upper laminal cells equally convex on both free surfaces .... Tribe **Pottieae**

KEY TO THE GENERA

This key was developed in part using the DELTA (DEscription Language for TAxonomy) system (Dallwitz 1974, 1980; Dallwitz & Paine 1986; Partridge et al. 1988). Data matrices were created using the descriptions in the taxonomic section. In order to make the key work, total variation was not scored for certain of the larger genera in that rare and unusual character states were sometimes not included. This was done to account for problems in key creation that are due to the presence in a genus of (1) species probably incorrectly assigned to a particular genus and requiring further study for correct disposition, (2) species possessing several of the reliable characteristics of that genus but with one or more additional characters considered reliable for a different genus, and (3) much reduced species with occasional secondary lack of features (e.g. ventral stereid band) whose absence was a key characteristic of other genera (e.g. genera of Pottieae). Gametophytic characters are emphasized in the key, reflecting taxonomic importance in the actual treatments, but closeness of taxa in the key does not imply a close phylogenetic relationship. Because artificial distinctions of sporophyte characteristics previously used to distinguish genera are not recognized, many genera necessarily key out in more than one place. To keep the number of couplets to a manageable minimum, reliable technical characters involving sectioning and color reactions to two percent potassium hydroxide solution are necessary at the beginning of the key. Emphasis on more immediately observable characters are practicable only in regional treatments. The key characters below are those of leaf morphology and anatomy unless otherwise noted.

The number of times each genus appears in this key is dependent on the variation within the genus of the characters considered reliable for identification. This is a measure of the internal complexity of the genus, whether due to taxonomic heterogeneity or simple breadth of variation of a monophyletic taxon. An index of such complexity for each suprageneric group is the number of times of their genera occur in the key divided by the number of genera in the group, as follows: Timmielloideae, 1.0; Erythrophyllopsoideae, 1.0; Gertrudielloideae, 1.0; Chionolomoideae, 1.7; Trichostomoideae, 3.9 (*Trichostomum* occurs 11 times!); Merceyoideae, 2.8; tribe Tetracoscinodontieae, 1.0; tribe Bryoerythrophylleae, 2.2 (*Pseudocrossidium* occurs 4 times); tribe Leptodontieae, 1.5 (*Triquetrella* occurs 4 times); tribe Barbuleae, 4.2 (*Didymodon* and *Gyroweisia* each occur 9 times, *Gymnostomum* 7 times); Pottioideae, 2.3; tribe Hyophileae, 2.5 (*Hyophila* and *Weissia* each occur 9 times); tribe Pottieae, 2.2 (*Aschisma* and *Trachycarpidium* each occur 4 times). It is probable that, after revision, the Trichostomoideae and Barbuleae will prove less of a source of variation in taxonomic characters; taxa with low ratios are apparently presently well understood or at least easily characterized.

1(0). Ventral costal stereid band absent (costa with a single stereid band) 2
Ventral costal stereid band present (costa with two stereid bands) 51

2(1). Ventral costal outgrowths absent 3
Ventral costal outgrowths present as a pad of cells, lamellae or filaments 41

3(2). Stem central strand absent 4
Stem central strand present 16

4–39 Costa with one stereid band, costal outgrowths absent, central strand absent

4(3). KOH color reaction of upper laminal cell walls essentially yellow or orange 5
KOH color reaction of upper laminal cell walls essentially red, usually a definite brick red 11

5(4). Superficial walls of upper laminal cells similarly shaped on both sides of lamina; leaves when dry occasionally channeled but not distinctly tubulose 6
Superficial walls of upper laminal cells ventrally bulging-mamillose, weakly convex dorsally; leaves tubulose when dry 10

6(5). Medial upper laminal cells small to medium sized, 7–14 µm in width 7
Medial upper laminal cells large, commonly more than 14 µm in width 8

7(6). Stem sclerodermis not or little differentiated; upper laminal papillae absent; basal cell group differentiated and rising higher medially; length of stem usually 1.0 cm or more; basal cells usually with straight walls, little wider than upper medial cells; wide distribution .... *Scopelophila*
Stem sclerodermis clearly differentiated; upper laminal papillae present; basal cell group differentiated approximately straight across leaf; length of stem less than 1.0 cm, usually less than 0.6 cm; basal cells inflated and bulging, considerably wider than upper medial cells; wide distribution .... *Gyroweisia*

8(6). Margins plane to incurved or involute; transverse section of dorsal costal stereid band clearly flattened or ventrally indented, reniform or crescent-shaped; hydroid strand present; stem hyalodermis present (sometimes indistinct); southern South America, Australia .... *Calyptopogon*

Margins recurved to revolute; transverse section of dorsal costal stereid band round to semicircular; hydroid strand absent; stem hyalodermis absent 9

9(8). Transverse section of costa round or ovate or elliptical; basal cell group not or little differentiated from upper medial cells; leaves widest below midleaf; number of rows of cells across ventral surface of costa usually 2 but up to 4; stem sclerodermis clearly differentiated; tropics .... *Streptopogon*
Transverse section of costa semicircular; basal cell group clearly differentiated, usually larger, less papillose, walls thinner; leaves widest at or above midleaf; number of rows of cells across ventral surface of costa commonly 4 or more; stem sclerodermis not or little differentiated; western North America .... *Crumia*

10(5). Basal cell group differentiated as a vee, with at least laterally differentiated cells rising higher marginally as a tapering border; transverse section of dorsal costal stereid band clearly flattened or ventrally indented, reniform or crescent-shaped; ventral costal epidermis absent; dorsal costal epidermis absent; leaf base clearly differentiated in shape; American tropics, eastern Asia .... *Luisierella*
Basal cell group differentiated straight across leaf or rising higher medially; transverse section of dorsal costal stereid band round to semicircular; ventral costal epidermis present; dorsal costal epidermis present; leaf base little differentiated in shape; Brazil, India .... *Ganguleea*

11(4). Medial upper laminal cells small to medium sized, 7–14 µm in width 12
Medial upper laminal cells large, commonly more than 14 µm in width 13

12(11). Dorsal costal epidermis absent; basal cell group clearly differentiated, usually larger, less papillose, walls thinner; theca ovoid to cylindrical; capsule stegocarpous; leaves usually widest at or above midleaf; wide distribution .... *Syntrichia*
Dorsal costal epidermis present; basal cell group not or little differentiated from upper medial cells; theca spherical; capsule cleistocarpous; leaves widest below midleaf; wide distribution .... *Acaulon*

13(11). Margins plane to incurved or involute; length to width ratio of medial upper laminal cells 1–2:1 14
Margins recurved to revolute; length to width ratio of medial upper laminal cells 2–4:1 or more 15

14(13). Basal cell group differentiated as a vee, with at least laterally differentiated cells rising higher marginally as a tapering border; transverse section of dorsal costal stereid band clearly flattened or ventrally indented, reniform or crescent-shaped; costal ventral cells longitudinally elongate 3:1 or more; stem sclerodermis clearly differentiated; superficial walls of upper laminal cells strongly convex to bulging on both sides of lamina; southern South America, Australasia .... *Calyptopogon*
Basal cell group differentiated straight across leaf or rising higher medially; transverse section of dorsal costal stereid band round to semicircular; costal ventral cells quadrate to very short-rectangular; stem sclerodermis not or little differentiated; superficial walls of upper laminal cells flat or very weakly convex on both sides of lamina; southern South America, Antarctica .... *Sarconeurum*

15(13). Hydroid strand absent; upper laminal papillae absent; theca longer, usually more than 1.5 mm in length; calyptra longer, 1.0 mm or more in length; plants caespitose, usually in a mat or turf; tropics .... *Streptopogon*
Hydroid strand present; upper laminal papillae present; theca short, less than 1.5 mm in length; calyptra short, less than 1.0 mm; plants gregarious or scattered; wide distribution .... *Microbryum*

16–40 Costa with one stereid band, costal outgrowths absent, central strand present

16(3). Medial upper laminal cells small to medium sized, 7–14 µm in width 17
Medial upper laminal cells large, commonly more than 14 µm in width 31

17(16). Dorsal costal epidermis absent 18
Dorsal costal epidermis present 23

18(17). KOH color reaction of upper laminal cell walls essentially yellow or orange 19
KOH color reaction of upper laminal cell walls essentially red, usually a definite brick red 20

19(18). Basal cell group differentiated as a vee, with at least laterally differentiated cells rising higher marginally as a tapering border; transverse section of dorsal costal stereid band clearly flattened or ventrally indented, reniform or crescent-shaped; stem sclerodermis not or little differentiated; costal ventral cells quadrate to very short-rectangular; guide cells commonly more than 6; Mexico, Andes, South Africa .... *Streptocalypta*
Basal cell group differentiated straight across leaf or rising higher medially; transverse section of dorsal costal stereid band round to semicircular; stem sclerodermis clearly differentiated; costal ventral cells longitudinally elongate 3:1 or more; guide cells 2–6; wide distribution .... *Gyroweisia*

20(18). Capsule cleistocarpous; calyptra short, less than 1.0 mm; Australasia .... *Phascopsis*
Capsule stegocarpous; calyptra longer, 1.0 mm or more in length 21

21(20). Dorsal superficial laminal cell walls about same thickness as the ventral or weakly thicker near costa; basal cell group clearly differentiated, usually larger, less papillose, walls thinner; leaves widest at or above midleaf 22
Dorsal superficial laminal cell walls throughout distinctly thicker than the ventral; basal cell group not or little differentiated from upper medial cells; leaves widest below midleaf; western North America, central Europe .... *Hilpertia*

22(21). Seta nearly absent to short, less than 1 cm; perichaetial leaves distinctly different in size or morphology, strongly sheathing the seta; austral oceanic islands .... *Willia*
Seta elongate, 1 cm or longer; perichaetial leaves similar to cauline leaves or occasionally smaller or somewhat enlarged; wide distribution .... *Syntrichia*

23(17). Superficial walls of upper laminal cells similarly shaped on both sides of lamina 24
Superficial walls of upper laminal cells ventrally bulging-mamillose, weakly convex dorsally 29

24(23). Ventral costal epidermis absent 25
Ventral costal epidermis present 26

25(24). Leaf ventral surface above midleaf nearly plane to broadly channeled; leaves widest at or above midleaf; rather deep, narrow groove along costa absent; perichaetium terminal on main axis; seta nearly absent to short, less than 1 cm; pantropical .... *Gymnostomiella*
Leaf ventral surface above midleaf keeled; leaves widest below midleaf; rather deep, narrow groove along costa present; perichaetium lateral on main axis at ends of very short branches; seta elongate, 1 cm or longer; wide distribution .... *Anoectangium*

26(24). Hydroid strand absent 27
Hydroid strand present; wide distribution .... *Pseudocrossidium*

27(26). Perichaetium terminal on main axis, leaves usually broadly channeled across the ventral surface 28
Perichaetium lateral on main axis at ends of very short branches, leaves usually strongly keeled; wide distribution .... *Anoectangium*

28(27). Length of stem less than 1.0 cm, usually less than 0.6 cm; basal cells inflated and bulging, considerably wider than upper medial cells; wide distribution .... *Gyroweisia*
Length of stem usually 1.0 cm or more; basal cells usually with straight walls, little wider than upper medial cells; wide distribution .... *Didymodon*

29(23). KOH color reaction of upper laminal cell walls essentially yellow; margins plane to incurved or involute; transverse section of dorsal costal stereid band round to semicircular; guide cells 2–6; stem hyalodermis absent 30
KOH color reaction of upper laminal cell walls essentially orange; margins revolute to near apex; transverse section of dorsal costal stereid band clearly flattened or ventrally indented, reniform or crescent-shaped; guide cells commonly more than 6; stem hyalodermis present (sometimes indistinct); Andes .... *Gertrudiella*

30(29). Margins plane or very weakly incurved; transverse section of costa round or ovate or elliptical; hydroid strand absent; leaf base clearly differentiated in shape; number of rows of cells across ventral surface of costa usually 2 but up to 4; tropics .... *Weisiopsis*
Margins clearly incurved to involute; transverse section of costa semicircular; hydroid strand present; leaf base little differentiated in shape; number of rows of cells across ventral surface of costa commonly 4 or more; American tropics, South Africa, Burma .... *Plaubelia*

31(16). Margins plane to incurved or involute 32
Margins recurved to revolute 35

32(31). Transverse section of dorsal costal stereid band round to semicircular; upper laminal papillae present 33
Transverse section of dorsal costal stereid band clearly flattened or ventrally indented, reniform or crescent-shaped; upper laminal papillae absent; Mexico, Andes .... *Sagenotortula*

33(32). Ventral costal epidermis absent; hydroid strand absent; superficial walls of upper laminal cells strongly convex to bulging on both sides of lamina; pantropical .... *Gymnostomiella*

Ventral costal epidermis present; hydroid strand present; superficial walls of upper laminal cells flat or very weakly convex on both sides 34

34(33). Number of rows of cells across ventral surface of costa usually 2 but up to 4; upper marginal cells in same number of layers as medial cells; upper marginal cells not longer than medial cells (sometimes larger); papillae per lumen 2–6, usually bifid or multifid; basal cell group differentiated and rising higher medially; southern South America, Antarctica...... *Sarconeurum*
Number of rows of cells across ventral surface of costa commonly 4 or more; upper marginal cells differentiated as a bistratose (or more) border; upper marginal cells rectangular and clearly longer than medial cells; papillae per lumen many, usually 6 or more, simple or bifid; basal cell group differentiated approximately straight across leaf; wide distribution .... *Hennediella*

35(31). KOH color reaction of upper laminal cell walls essentially yellow or orange....36
KOH color reaction of upper laminal cell walls essentially red, usually a definite brick red....37

36(35). KOH color reaction of upper laminal cell walls essentially yellow; hydroid strand present; medial upper laminal cells small, 7–10 μm in width; papillae per lumen 2–6, usually bifid or multifid; wide distribution .... *Tortula*
KOH color reaction of upper laminal cell walls essentially orange; hydroid strand absent; medial upper laminal cells medium sized, 11–14 μm in width; papillae per lumen many, usually 6 or more, simple or bifid; western North America....... *Crumia*

37(35). Dorsal costal epidermis absent; dorsal superficial laminal cell walls throughout distinctly thicker than the ventral; margins revolute; perichaetial leaves distinctly larger in size; western North America, central Europe .... *Hilpertia*
Dorsal costal epidermis present; dorsal superficial laminal cell walls about same thickness as the ventral or weakly thicker near costa; margins recurved; perichaetial leaves similar to cauline leaves or occasionally smaller or somewhat enlarged .38

38(37). Basal cell group not or little differentiated from upper medial cells....39
Basal cell group clearly differentiated, usually larger, less papillose, walls thinner 40

39(38). Medial upper laminal cells 15–17 μm in width; leaf apex obtuse or acute to acuminate; upper laminal papillae present; sexual condition monoicous; upper marginal cells in same number of layers as medial cells; wide distribution.............*Microbryum*
Medial upper laminal cells commonly 18 μm in width or more; leaf apex broadly rounded; upper laminal papillae absent; sexual condition dioicous; upper marginal cells differentiated as a bistratose (or more) border; Mexico, Central America, Andes .... *Dolotortula*

40(38). Upper laminal papillae absent (except sometimes along extreme margins of leaf); sexual condition dioicous; plants caespitose, usually in a mat or turf; length to width ratio of medial upper laminal cells 1–2:1; calyptra longer, 1.0 mm or more in length; wide distribution .... *Chenia*
Upper laminal papillae present; sexual condition monoicous; plants gregarious or scattered; length to width ratio of medial upper laminal cells 2–4:1 or more; calyptra short, less than 1.0 mm; wide distribution ....*Microbryum*

41–50 Costa with one stereid band, costal outgrowths present

41(2). KOH color reaction of upper laminal cell walls essentially yellow or orange....42
KOH color reaction of upper laminal cell walls essentially red, usually a definite brick red....48

42(41). Ventral costal outgrowths differentiated as filaments or lamellae 43
Ventral costal outgrowths differentiated as a bulging pad of cells 46

43(42). Transverse section of dorsal costal stereid band round to semicircular; transverse section of costa round to semicircular......44
Transverse section of dorsal costal stereid band clearly flattened or ventrally indented, reniform or crescent-shaped; transverse section of costa distinctly flattened, usually reniform....45

44(43). Ventral costal outgrowths differentiated as longitudinally elongated lamellae; spore diameter more than 15 μm; annulus of weakly differentiated cells; wide distribution....*Pterygoneurum*
Ventral costal outgrowths differentiated as separate filaments three or more cells in length; spore diameter 8–15 μm; annulus of vesiculose cells, often in two or more rows; wide distribution .... *Crossidium*

45(43). Margins plane to incurved or involute; axillary hairs with 1 or more basal cells with thicker or darker-colored walls; leaf apex cucullate; leaves tubulose when dry; spore diameter more than 15 μm; Mexico, Andes ....*Aloinella*
Margins recurved to revolute; axillary hair basal cell walls hyaline and all cells of hair similar; leaf apex flattened, channeled or keeled; leaves when dry occasionally channeled but not distinctly tubulose; spore diameter 8–15 μm; wide distribution .... *Pseudocrossidium*

46(42). Dorsal superficial laminal cell walls about same thickness as the ventral or weakly thicker near costa; length to width ratio of medial upper laminal cells 1–2:1 47
Dorsal superficial laminal cell walls throughout distinctly thicker than the ventral; length to width ratio of medial upper laminal cells 2–4:1 or more; North Temperate Zone .... *Stegonia*

47(46). Margins plane to incurved or involute; medial upper laminal cells 7–14 µm in width; leaf apex cucullate; upper laminal papillae absent; southwestern USA south through Andes .... *Globulinella*
Margins recurved to revolute; medial upper laminal cells commonly larger than 14 µm in width; leaf apex flattened, channeled or keeled; upper laminal papillae present; wide distribution .... *Tortula*

48(41). Ventral costal outgrowths differentiated as filaments or lamellae; medial upper laminal cells commonly larger than 14 µm in width; transverse section of dorsal costal stereid band clearly flattened or ventrally indented, reniform or crescent-shaped; leaf base clearly differentiated in shape; transverse section of costa distinctly flattened, usually reniform; wide distribution .... *Aloina*
Ventral costal outgrowths differentiated as a bulging pad of cells; medial upper laminal cells small to medium sized, 7–14 µm in width; transverse section of dorsal costal stereid band round to semicircular; leaf base little differentiated in shape; transverse section of costa round to semicircular 49

49(48). Stem central strand absent; margins plane to incurved or involute; number of rows of cells across ventral surface of costa usually 2 but up to 4; medial upper laminal cells medium sized, 11–14 µm in width; leaf apex obtuse or acute to acuminate 50
Stem central strand present; margins recurved; number of rows of cells across ventral surface of costa commonly 4 or more; medial upper laminal cells small, 7–10 µm in width; leaf apex broadly rounded; Mexico, Andes .... *Saitoella*

50(49). Basal cell group not or little differentiated from upper medial cells; leaf apex flattened, channeled or keeled; costal ventral cells longitudinally elongate 3:1 or more; leaves widest below midleaf; sporophyte commonly present, spherical, seta short; wide distribution .... *Acaulon*
Basal cell group clearly differentiated, usually larger, less papillose, walls thinner; leaf apex cucullate; costal ventral cells quadrate to very short-rectangular; leaves widest at or above midleaf; sporophyte unknown; Australia .... *Stonea*

51–67 Costa with two stereid bands, upper laminal cells KOH yellow or orange, central strand absent

51(1). KOH color reaction of upper laminal cell walls essentially yellow or orange....52
KOH color reaction of upper laminal cell walls essentially red, usually a definite brick red....171

52(51). Stem central strand absent....53
Stem central strand present....68

53(52). Margins entire or minutely and evenly crenulate....54
Margins denticulate or serrulate to toothed....60

54(53). Ventral costal epidermis absent....55
Ventral costal epidermis present....57

55(54). Basal cell group not or little differentiated from upper medial cells; peristome teeth present; seta elongate, 1 cm or longer; theca longer, usually more than 1.5 mm in length; annulus of vesiculose cells, often in two or more rows; austral areas and Mediterranean climates in the North Temperate Zone .... *Triquetrella*
Basal cell group clearly differentiated, usually larger, less papillose, walls thinner; peristome teeth absent; seta nearly absent to short, less than 1 cm; theca short, less than 1.5 mm in length; annulus of weakly differentiated cells....56

56(55). Transverse section of dorsal costal stereid band round to semicircular; medial upper laminal cells small, 7–10 µm in width; upper laminal papillae present; calyptra short, less than 1.0 mm; wide distribution.... *Hymenostylium*
Transverse section of dorsal costal stereid band clearly flattened or ventrally indented, reniform or crescent-shaped; medial upper laminal cells medium sized, 11–14 µm in width; upper laminal papillae absent; calyptra longer, 1.0 mm or more in length; eastern Asia, India.... *Reimersia*

57(54). Basal cell group differentiated as a vee, with at least laterally differentiated cells rising higher marginally as a tapering border; wide distribution .... *Tortella*
Basal cell group differentiated straight across leaf or rising higher medially....58

58(57). Leaf ventral surface above midleaf nearly plane to broadly channeled 59
Leaf ventral surface above midleaf keeled; New Zealand .... *Tetracoscinodon*

59(58). Costal ventral cells quadrate to very short-rectangular; stem hyalodermis present (sometimes indistinct); length of stem usually 1.0 cm or more; basal cells usually with straight walls, little wider than upper medial cells; wide distribution .... *Trichostomum*
Costal ventral cells longitudinally elongate 3:1 or more; stem hyalodermis absent; length of stem less than 1.0 cm, usually less than 0.6 cm; basal cells inflated and bulging, considerably wider than upper medial cells; wide distribution. *Gyroweisia*

60(53). Margins denticulate only near leaf base or at top of leaf sheath; North Temperate Zone, India, South Africa........ *Eucladium*
Margins denticulate or serrulate to toothed near apex or throughout........ 61

61(60). Margins plane to incurved or involute; leaf ventral surface above midleaf nearly plane to broadly channeled; rather deep, narrow groove along costa absent........ 62
Margins recurved; leaf ventral surface above midleaf keeled; rather deep, narrow groove along costa present........ 63

62(61). Ventral costal epidermis absent; hydroid strand present; theca spherical; superficial exothecial cell walls with a central and lens-like superficial thickening; stem hyalodermis absent; Mexico, Angola .... *Bryoceuthospora*
Ventral costal epidermis present; hydroid strand absent; theca ovoid to cylindrical; superficial exothecial cell walls evenly thickened; stem hyalodermis present (sometimes indistinct); wide distribution........ *Trichostomum*

63(61). Basal cell group not or little differentiated from upper medial cells........ 64
Basal cell group clearly differentiated, usually larger, less papillose, walls thinner 65

64(63). Dorsal costal epidermis absent; stem hyalodermis present (sometimes indistinct); number of rows of cells across ventral surface of costa commonly 4 or more; transverse section of costa distinctly flattened, usually reniform; stomates absent; Andes .... *Leptodontiella*
Dorsal costal epidermis present; stem hyalodermis absent; number of rows of cells across ventral surface of costa usually 2 but up to 4; transverse section of costa round to semicircular; stomates present; austral areas and Mediterranean climates of the North Temperate Zone .... *Triquetrella*

65(63). Papillae per lumen 2–6, usually bifid or multifid........ 66
Papillae per lumen many, usually 6 or more, simple or bifid 67

66(65). Theca short, less than 1.5 mm in length; Andes........ *Leptodontiella*
Theca longer, usually more than 1.5 mm in length; wide distribution........ *Leptodontium*

67(65). Upper marginal cells in same number of layers as medial cells, not longer than medial cells (sometimes larger); seta nearly absent to short, less than 1 cm; theca usually 1.5–3.5 mm in length; marginal cell walls not thicker than those of medial cells; Andes........ *Streptotrichum*
Upper marginal cells differentiated as a bistratose border, rectangular and clearly longer than medial cells; seta elongate, 1 cm or longer; theca more than 3.5 mm in length; marginal cell walls thicker than those of medial cells; Andes .... *Trachyodontium*

68–170 Costa with two stereid bands, upper laminal cells KOH yellow or orange, central strand present

68(52). Margins entire or minutely and evenly crenulate........ 69
Margins denticulate or serrulate to toothed........ 148

69(68). Hydroid strand absent........ 70
Hydroid strand present 130

70(69). Inflated, banana-shaped alar cells absent or alar cells merely swollen and rounded 71
Inflated, banana-shaped alar cells present and decurrent as a pad on the stem; Mexico, West Indies .... *Weissiodicranum*

71(70). Superficial walls of upper laminal cells similarly shaped on both sides of lamina........ 72
Superficial walls of upper laminal cells ventrally bulging-mamillose, weakly convex dorsally........ 119

72(71). Ventral costal stereid band smaller than the dorsal or of nearly equal size 73
Ventral costal stereid band distinctly larger than the dorsal 116

73(72). Margins plane to incurved or involute 74
Margins recurved to revolute 107

74(73). Margins plane or very weakly incurved 75
Margins clearly incurved to involute 105

75(74). Dorsal costal epidermis absent 76
Dorsal costal epidermis present 89

76(75). Capsule cleistocarpous, i.e., rupturing (usually) irregularly 77
Capsule stegocarpous, i.e., dehiscing by an operculum 80

77(76). Basal cell group differentiated as a vee, with at least laterally differentiated cells rising higher marginally as a tapering border 78
Basal cell group differentiated straight across leaf or rising higher medially (occasionally rising weakly along extreme basal margin as a line of elongate cells); wide distribution .... *Trichostomum*

78(77). Ventral costal epidermis absent; theca spherical; stomates absent; calyptra papillose, distinctly roughened or strongly mamillose; central North America, Europe .... *Aschisma*
Ventral costal epidermis present; theca ovoid to cylindrical; stomates present; calyptra smooth or nearly so 79

79(78). Leaf base little differentiated in shape; axillary hairs with 1 or more basal cells with thicker or darker colored walls; capsule surface evenly mamillose or with distinct protuberances of strongly bulging cells basally or throughout; stem hyalodermis absent; theca short, less than 1.5 mm in length; South America, Africa, Oceania .... *Trachycarpidium*
Leaf base clearly differentiated in shape; axillary hair cell walls hyaline and all cells of hair similar; capsule surface nearly smooth; stem hyalodermis present (sometimes indistinct); theca longer, usually more than 1.5 mm in length; wide distribution .... *Tortella*

80(76). Transverse section of dorsal costal stereid band round to semicircular 81
Transverse section of dorsal costal stereid band clearly flattened or ventrally indented, reniform or crescent-shaped 84

81(80). Costal ventral cells quadrate to very short-rectangular 82
Costal ventral cells longitudinally elongate 3:1 or more 83

82(81). Costa ending before the leaf apex; wide distribution .... *Gymnostomum*
Costa percurrent to excurrent; wide distribution .... *Trichostomum*

83(81). Stem sclerodermis not or little differentiated; length of stem usually 1.0 cm or more; basal cells usually with straight walls, little wider than upper medial cells; wide distribution .... *Gymnostomum*
Stem sclerodermis clearly differentiated; length of stem less than 1.0 cm, usually less than 0.6 cm; basal cells inflated and bulging, considerably wider than upper medial cells; wide distribution .... *Gyroweisia*

84(80). Guide cells 2–6; length of stem usually 1.0 cm or more 85
Guide cells commonly more than 6; length of stem less than 1.0 cm, usually less than 0.6 cm; Mexico, Andes, South Africa .... *Streptocalypta*

85(84). Stem hyalodermis absent 86
Stem hyalodermis present (sometimes indistinct) 87

86(85). Medial upper laminal cells small, 7–10 μm in width; leaf apex obtuse to broadly acute; costa ending before the leaf apex; wide distribution .... *Gymnostomum*
Medial upper laminal cells medium sized, 11–14 μm in width; leaf apex narrowly acute to acuminate; costa percurrent to excurrent; American tropics, eastern Asia .... *Tuerckheimia*

87(85). Basal cell group differentiated as a vee, with at least laterally differentiated cells rising higher marginally as a tapering border; perichaetial leaves distinctly different in size or morphology, sometimes strongly sheathing; wide distribution .... *Tortella*
Basal cell group differentiated straight across leaf or rising higher medially; perichaetial leaves similar to cauline leaves or occasionally smaller or somewhat enlarged 88

88(87). Costa ending before the leaf apex; wide distribution .... *Gymnostomum*
Costa percurrent to excurrent; wide distribution .... *Trichostomum*

89(75). Transverse section of costa round to semicircular 90
Transverse section of costa distinctly flattened, usually reniform 103

90(89). Guide cells 2–6 91
Guide cells commonly more than 6; Philippines .... *Pachyneuropsis*

91(90). Capsule cleistocarpous, i.e., rupturing (usually) irregularly 92
Capsule stegocarpous, i.e., dehiscing by an operculum 94

92(91). Basal cell group differentiated as a vee, with at least laterally differentiated cells rising higher marginally as a tapering border 93
Basal cell group differentiated straight across leaf or rising higher medially; wide distribution .... *Trichostomum*

93(92). Axillary hairs with 1 or more basal cells with thicker or darker colored walls; stem hyalodermis absent; capsule surface evenly mamillose or with distinct protuberances of strongly bulging cells basally or throughout; theca short, less than 1.5 mm in length; South America, Africa, Oceania .... *Trachycarpidium*
Axillary hair basal cell walls hyaline and all cells of hair similar; stem hyalodermis present (sometimes indistinct); capsule surface nearly smooth; theca longer, usually more than 1.5 mm in length; wide distribution .... *Tortella*

94(91). Stem sclerodermis not or little differentiated 95
Stem sclerodermis clearly differentiated 99

95(94). Stem hyalodermis absent 96
Stem hyalodermis present (sometimes indistinct) 97

96(95). Medial upper laminal cells small, 7–10 µm in width; leaf apex obtuse to broadly acute; costa ending before the leaf apex; wide distribution .... *Gymnostomum*
Medial upper laminal cells medium sized, 11–14 µm in width; leaf apex narrowly acute to acuminate; costa percurrent to excurrent; American tropics, eastern Asia .... *Tuerckheimia*

97(95). Basal cell group differentiated as a vee, with at least laterally differentiated cells rising higher marginally as a tapering border; perichaetial leaves distinctly different in size or morphology, sometimes strongly sheathing; wide distribution .. *Tortella*
Basal cell group differentiated straight across leaf or rising higher medially; perichaetial leaves similar to cauline leaves or occasionally smaller or somewhat enlarged 98

98(97). Costa ending before the leaf apex; wide distribution .... *Gymnostomum*
Costa percurrent to excurrent; wide distribution .... *Trichostomum*

99(94). Perichaetium terminal on main axis 100
Perichaetium lateral on main axis at ends of very short branches; wide distribution .... *Molendoa*

100(99). Costal ventral cells quadrate to very short-rectangular 101
Costal ventral cells longitudinally elongate 3:1 or more 102

101(100). Leaf base little differentiated in shape; leaves widest at or above midleaf; wide distribution .... *Hyophila*
Leaf base clearly differentiated in shape; leaves widest below midleaf; wide distribution .... *Trichostomum*

102(100). Basal cells usually with straight walls, little wider than upper medial cells; wide distribution .... *Barbula*
Basal cells inflated and bulging, considerably wider than upper medial cells; wide distribution .... *Gyroweisia*

103(89). Perichaetium terminal on main axis 104
Perichaetium lateral on main axis at ends of very short branches; wide distribution .... *Molendoa*

104(103). Peristome twisted; leaf base (of sterile leaves) little differentiated in shape; superficial walls of upper laminal cells flat or very weakly convex on both sides of lamina; stem hyalodermis absent; basal cell group differentiated and rising higher medially; length of stem less than 1.0 cm, usually less than 0.6 cm; Mediterranean region .... *Leptobarbula*

Peristome straight or absent; leaf base commonly clearly differentiated in shape; superficial walls of upper laminal cells strongly convex to bulging on both sides of lamina; stem hyalodermis present (sometimes indistinct); basal cell group differentiated approximately straight across leaf; length of stem usually 1.0 cm or more; wide distribution......*Trichostomum*

105(74). Transverse section of dorsal costal stereid band round to semicircular .... *Quaesticula*
Transverse section of dorsal costal stereid band clearly flattened or ventrally indented, reniform or crescent-shaped 106

106(105). Upper laminal margins not sharply and narrowly infolded; leaf base little differentiated in shape; leaves widest at or above midleaf; stem sclerodermis clearly differentiated; axillary hair basal cell walls hyaline and all cells of hair similar; wide distribution .... *Hyophila*
Upper laminal margins sharply and narrowly infolded; leaf base clearly differentiated in shape; leaves widest below midleaf; stem sclerodermis not or little differentiated; axillary hairs with 1 or more basal cells with thicker or darker colored walls; wide distribution .... *Weissia*

107(73). Stem sclerodermis not or little differentiated; wide distribution .... *Gymnostomum*
Stem sclerodermis clearly differentiated 108

108(107). Ventral costal epidermis absent; austral regions and Mediterranean climates of the North Temperate Zone .... *Triquetrella*
Ventral costal epidermis present 109

109(108). Perichaetium terminal on main axis 110
Perichaetium lateral on main axis at ends of very short branches; wide distribution .... *Molendoa*

110(109). Transverse section of dorsal costal stereid band round to semicircular 111
Transverse section of dorsal costal stereid band clearly flattened or ventrally indented, reniform or crescent-shaped 114

111(110). Leaf plane to broadly channeled ventrally, costa lacking a ventral groove 112
Leaf channeled or keeled, costa with a rather deep, narrow groove ventrally 113

112(111). Length of stem less than 1.0 cm, usually less than 0.6 cm; basal cells inflated and bulging, considerably wider than upper medial cells; wide distribution .... *Gyroweisia*
Length of stem usually 1.0 cm or more; basal cells usually with straight walls, little wider than upper medial cells; wide distribution .... *Didymodon*

113(111). Basal cells usually with straight walls, little wider than upper medial cells; wide distribution .... *Barbula*
Basal cells inflated and bulging, considerably wider than upper medial cells; wide distribution .... *Gyroweisia*

114(110). Leaf plane to broadly channeled ventrally, costa lacking a ventral groove 115
Leaf channeled or keeled, costa with a rather deep, narrow groove ventrally .... *Barbula*

115(114). Leaves widest at or above midleaf; axillary hair basal cell walls hyaline and all cells of hair similar; peristome teeth absent; leaves tubulose when dry; wide distribution .... *Hyophila*
Leaves widest below midleaf; axillary hairs with 1 or more basal cells with thicker or darker colored walls; peristome teeth present; leaves occasionally channeled but not distinctly tubulose when dry; wide distribution .... *Didymodon*

116(72). Stem hyalodermis absent 117
Stem hyalodermis present (sometimes indistinct); tropics .... *Pseudosymblepharis*

117(116). Stem sclerodermis not or little differentiated; Philippines .... *Pachyneuropsis*
Stem sclerodermis clearly differentiated 118

118(117). Length of stem less than 1.0 cm, usually less than 0.6 cm; basal cells inflated and bulging, considerably wider than upper medial cells; wide distribution .... *Gyroweisia*
Length of stem usually 1.0 cm or more; basal cells usually with straight walls, little wider than upper medial cells; wide distribution .... *Didymodon*

119(71). Basal cell group differentiated as a vee, with at least laterally differentiated cells rising higher marginally as a tapering border 120
Basal cell group differentiated straight across leaf or rising higher medially 125

120(119). Ventral costal stereid band smaller than the dorsal or of nearly equal size 121
Ventral costal stereid band distinctly larger than the dorsal 124

121(120). Ventral costal epidermis absent; stomates absent; capsule cleistocarpous and rupturing mainly along weak transverse walls at butt ends of long-rectangular exothecial cells; central North America, Europe .... *Aschisma*
Ventral costal epidermis present; stomates present; capsule rupturing irregularly if cleistocarpous, or stegocarpous 122

122(121). Upper laminal margins not sharply and narrowly infolded; guide cells commonly more than 6 123
Upper laminal margins sharply and narrowly infolded; guide cells 2–6; wide distribution .... *Weissia*

123(122). Medial upper laminal cells small, 7–10 µm in width; papillae per lumen one or occasionally two; calyptra papillose, distinctly roughened or strongly mamillose; number of rows of cells across ventral surface of costa usually 6 or more; South Africa .... *Hypodontium*
Medial upper laminal cells medium sized, 11–14 µm in width; papillae per lumen usually two or more; calyptra smooth or nearly so; number of rows of cells across ventral surface of costa 4–6; Mexico, Andes, South Africa .... *Streptocalypta*

124(120). Calyptra smooth or nearly so; papillae per lumen usually two or more; spore diameter 8–15 µm; upper marginal cells not longer than medial cells (sometimes larger); tropics .... *Pseudosymblepharis*
Calyptra papillose, distinctly roughened or strongly mamillose; papillae per lumen one or occasionally two; spore diameter more than 15 µm; upper marginal cells rectangular and clearly longer than medial cells; South Africa .... *Hypodontium*

125(119). Transverse section of dorsal costal stereid band round to semicircular; Mexico, West Indies .... *Quaesticula*
Transverse section of dorsal costal stereid band clearly flattened or ventrally indented, reniform or crescent-shaped 126

126(125). Transverse section of costa round or ovate or elliptical 127
Transverse section of costa semicircular 128

127(126). Upper laminal margins not sharply and narrowly infolded, with an intramarginal border of elongate cells; leaf base little differentiated in shape; stem sclerodermis clearly differentiated; leaves widest at or above midleaf; stem hyalodermis absent; West Indies .... *Teniolophora*
Upper laminal margins sharply and narrowly infolded, leaf not bordered intramarginally by differentiated cells; leaf base clearly differentiated in shape; stem sclerodermis not or little differentiated; leaves widest below midleaf; stem hyalodermis present (sometimes indistinct); wide distribution .... *Weissia*

128(126). Perichaetium terminal on main axis; upper laminal cell walls thin to evenly thickened, lumens quadrate to rounded 129
Perichaetium lateral on main axis at ends of very short branches; upper laminal cell walls clearly trigonous or porose, lumens irregularly angular or stellate; Brazil, India, Philippines .... *Hymenostyliella*

129(128). Upper laminal margins not sharply and narrowly infolded; leaf base little differentiated in shape; stem sclerodermis clearly differentiated; axillary hair basal cell walls hyaline and all cells of hair similar; leaves widest at or above midleaf; wide distribution .... *Hyophila*
Upper laminal margins sharply and narrowly infolded; leaf base clearly differentiated in shape; stem sclerodermis not or little differentiated; axillary hairs with 1 or more basal cells with thicker or darker colored walls; leaves widest below midleaf; wide distribution .... *Weissia*

130(69). Ventral costal outgrowths absent 131
Ventral costal outgrowths present as a pad of cells, lamellae or filaments 147

131(130). Dorsal costal epidermis absent 132
Dorsal costal epidermis present 138

132(131). Basal cell group differentiated as a vee, with at least laterally differentiated cells rising higher marginally as a tapering border 133
Basal cell group differentiated straight across leaf or rising higher medially 135

133(132). Ventral costal epidermis absent; stomates absent; calyptra papillose, distinctly roughened or strongly mamillose; leaf apex obtuse to broadly acute; capsule cleistocarpous and rupturing mainly along weak transverse walls at butt ends of long-rectangular exothecial cells; central North America, Europe .... *Aschisma*
Ventral costal epidermis present; stomates present; calyptra smooth or nearly so; leaf apex narrowly acute to acuminate; capsule rupturing irregularly if cleistocarpous, or stegocarpous 134

134(133). Margins plane or very weakly incurved; upper laminal margins not sharply and narrowly infolded; leaf base little differentiated in shape; costal ventral cells longitudinally elongate 3:1 or more; stem hyalodermis absent; South America, Africa, Oceania .... *Trachycarpidium*
Margins clearly incurved to involute; upper laminal margins sharply and narrowly infolded; leaf base clearly differentiated in shape; costal ventral cells quadrate to very short-rectangular; stem hyalodermis present (sometimes indistinct); wide distribution .... *Weissia*

135(132). Transverse section of dorsal costal stereid band round to semicircular; Central and South America, Australia .... *Uleobryum*
Transverse section of dorsal costal stereid band clearly flattened or ventrally indented, reniform or crescent-shaped 136

136(135). Leaf ventral surface above midleaf nearly plane to broadly channeled; margins plane to incurved or involute; narrow groove along ventral surface of costa absent; costal ventral cells quadrate to very short-rectangular; transverse section of costa round to semicircular 137
Leaf ventral surface above midleaf keeled; margins recurved; rather deep, narrow groove along ventral surface of costa present; costal ventral cells longitudinally elongate 3:1 or more; transverse section of costa distinctly flattened, usually reniform; Europe, North Africa, southern and eastern Asia .... *Dialytrichia*

137(136). Margins plane or very weakly incurved; upper laminal margins not sharply and narrowly infolded; leaves widest at or above midleaf; axillary hair basal cell walls hyaline and all cells of hair similar; stem sclerodermis clearly differentiated; Australia .... *Calymperastrum*
Margins clearly incurved to involute; upper laminal margins sharply and narrowly infolded; leaves widest below midleaf; axillary hairs with 1 or more basal cells with thicker or darker colored walls; stem sclerodermis not or little differentiated; wide distribution .... *Weissia*

138(131). Capsule cleistocarpous 139
Capsule stegocarpous 142

139(138). Basal cell group differentiated as a vee, with at least laterally differentiated cells rising higher marginally as a tapering border 140
Basal cell group differentiated straight across leaf or rising higher medially 141

140(139). Margins plane or very weakly incurved; upper laminal margins not sharply and narrowly infolded; leaf base little differentiated in shape; costal ventral cells longitudinally elongate 3:1 or more; stem hyalodermis absent; South America, Africa, Oceania .... *Trachycarpidium*
Margins clearly incurved to involute; upper laminal margins sharply and narrowly infolded; leaf base clearly differentiated in shape; costal ventral cells quadrate to very short-rectangular; stem hyalodermis present (sometimes indistinct); wide distribution .... *Weissia*

141(139). Margins plane or very weakly incurved, not sharply and narrowly infolded above; stem hyalodermis absent; capsule cleistocarpous and rupturing mainly along weak transverse walls at butt ends of long-rectangular exothecial cells; austral areas .... *Tetrapterum*
Margins clearly incurved to involute, sharply and narrowly infolded above; stem hyalodermis present (sometimes indistinct); capsule rupturing irregularly if cleistocarpous, or stegocarpous; wide distribution .... *Weissia*

142(138). Costal ventral cells quadrate to very short-rectangular 143
Costal ventral cells longitudinally elongate 3:1 or more 146

143(142). Transverse section of dorsal costal stereid band round to semicircular; American tropics, South Africa, Burma .... *Plaubelia*
Transverse section of dorsal costal stereid band clearly flattened or ventrally indented, reniform or crescent-shaped 144

144(143). Upper laminal margins not sharply and narrowly infolded; axillary hair basal cell walls hyaline and all cells of hair similar 145
Upper laminal margins sharply and narrowly infolded; axillary hairs with 1 or more basal cells thicker or darker colored; wide distribution .... *Weissia*

145(144). Transverse section of costa round to semicircular; peristome teeth absent; leaves widest at or above midleaf; leaves tubulose when dry; calyptra 1.0–3.0 mm; wide distribution .... *Hyophila*
Transverse section of costa distinctly flattened, usually reniform; peristome teeth present; leaves widest below midleaf; leaves when dry occasionally channeled but not distinctly tubulose; calyptra more than 3.0 mm; wide distribution .... *Pseudocrossidium*

146(142). Transverse section of costa round to semicircular; theca short, less than 1.5 mm in length; upper marginal cells in same number of layers as medial cells; wide distribution .... *Barbula*
Transverse section of costa distinctly flattened, usually reniform; theca longer, usually more than 1.5 mm in length; upper marginal cells differentiated as a bistratose (or more) border; Europe, North Africa southern and eastern Asia ... *Dialytrichia*

147(130). Ventral costal outgrowths differentiated as filaments or lamellae; margins recurved to revolute; transverse section of dorsal costal stereid band clearly flattened or ventrally indented, reniform or crescent-shaped; leaf apex flattened, channeled or keeled; transverse section of costa distinctly flattened, usually reniform; wide distribution .... *Pseudocrossidium*
Ventral costal outgrowths differentiated as a bulging pad of cells; margins plane to incurved or involute; transverse section of dorsal costal stereid band round to semicircular; leaf apex cucullate; transverse section of costa round to semicircular; southwestern USA south through Andes .... *Globulinella*

148(68). Margins denticulate only near leaf base or at top of leaf sheath 149
Margins denticulate or serrulate to toothed near apex or throughout 150

149(148). Basal cell group differentiated as a vee, with at least laterally differentiated cells rising higher marginally as a tapering border; dorsal costal epidermis absent; superficial walls of upper laminal cells ventrally bulging-mamillose, weakly convex dorsally; stem sclerodermis not or little differentiated; peristome teeth present; South Africa .... *Hypodontium*
Basal cell group differentiated straight across leaf or rising higher medially; dorsal costal epidermis present; superficial walls of upper laminal cells similarly shaped on both sides of lamina; stem sclerodermis clearly differentiated; peristome teeth absent; wide distribution .... *Molendoa*

150(148). Dorsal costal epidermis absent 151
Dorsal costal epidermis present 157

151(150). Ventral costal stereid band smaller than the dorsal or of nearly equal size; upper laminal margins not sharply and narrowly infolded 152
Ventral costal stereid band distinctly larger than the dorsal; upper laminal margins sharply and narrowly infolded near apex; Southeast Asia, Borneo .... *Chionoloma*

152(151). Ventral costal epidermis absent; theca spherical 153
Ventral costal epidermis present; theca ovoid to cylindrical 154

153(152). Superficial exothecial cell walls with a central and lens-like thickening; costal ventral cells longitudinally elongate 3:1 or more; number of rows of cells across ventral surface of costa commonly 4 or more; axillary hair basal cell walls hyaline and all cells of hair similar; papillae per lumen one or occasionally two .... *Bryoceuthospora*
Superficial exothecial cell walls thickened; costal ventral cells quadrate to very short-rectangular; number of rows of cells across ventral surface of costa usually 2 but up to 4; axillary hairs with 1 or more basal cells with thicker or darker colored walls; papillae per lumen usually two or more; central North America, Europe .... *Aschisma*

154(152). Perichaetium terminal on main axis; perichaetial leaves similar to cauline leaves or occasionally smaller or somewhat enlarged 155
Perichaetium lateral on main axis at ends of very short branches; perichaetial leaves distinctly different in size or morphology, strongly sheathing the seta; wide distribution .... *Pleurochaete*

155(154). Guide cells 2–6; length of stem usually 1.0 cm or more 156
Guide cells commonly more than 6; length of stem less than 1.0 cm, usually less than 0.6 cm; Mexico, Andes, South Africa .... *Streptocalypta*

156(155). Stem hyalodermis absent; American tropics, eastern Asia .... *Tuerckheimia*
Stem hyalodermis present (sometimes indistinct); wide distribution .... *Trichostomum*

157(150). Superficial walls of upper laminal cells similarly shaped on both sides of lamina 158
Superficial walls of upper laminal cells ventrally bulging-mamillose, weakly convex dorsally 165

158(157). Rather deep, narrow groove along costa absent 159
Rather deep, narrow groove along costa present 164

159(158). Margins plane to incurved or involute 160
Margins recurved to revolute 163

160(159). Perichaetium terminal on main axis; upper marginal cells not longer than medial cells (sometimes larger); perichaetial leaves similar to cauline leaves or occasionally smaller or somewhat enlarged 161
Perichaetium lateral on main axis at ends of very short branches; upper marginal cells rectangular and clearly longer than medial cells; perichaetial leaves distinctly different in size or morphology, strongly sheathing the seta; wide distribution .... *Pleurochaete*

161(160). Leaves widest at or above midleaf; wide distribution .... *Hyophila*
Leaves widest below midleaf 162

162(161). Stem hyalodermis absent; American tropics, eastern Asia .... *Tuerckheimia*
Stem hyalodermis present (sometimes indistinct); wide distribution .... *Trichostomum*

163(159). Leaves widest at or above midleaf; axillary hair basal cell walls hyaline and all cells of hair similar; peristome teeth absent; leaves tubulose when dry; wide distribution .... *Hyophila*
Leaves widest below midleaf; axillary hairs with 1 or more basal cells with thicker or darker colored walls; peristome teeth present; leaves when dry occasionally channeled but not distinctly tubulose; wide distribution .... *Didymodon*

164(158). Ventral costal epidermis absent; basal cell group not or little differentiated from upper medial cells; theca longer, usually more than 1.5 mm in length; calyptra short, less than 1.0 mm; stem transverse section triangular; austral areas and Mediterranean climates of the North Temperate Zone .... *Triquetrella*
Ventral costal epidermis present; basal cell group clearly differentiated, usually larger, less papillose, walls thinner; theca short, less than 1.5 mm in length; calyptra longer, 1.0 mm or more in length; stem transverse section rounded-pentagonal; wide distribution .... *Barbula*

165(157). Hydroid strand absent 166
Hydroid strand present 169

166(165). Basal cell group differentiated as a vee, with at least laterally differentiated cells rising higher marginally as a tapering border; leaf base clearly differentiated in shape; wide distribution .... *Pleurochaete*
Basal cell group differentiated straight across leaf or rising higher medially; leaf base little differentiated in shape 167

167(166). Transverse section of costa round or ovate or elliptical; upper marginal cells differentiated intramarginally as a bistratose border; upper marginal cells rectangular and clearly longer than medial cells; marginal cell walls thicker than those of medial cells; West Indies .... *Teniolophora*
Transverse section of costa semicircular; upper marginal cells in same number of layers as medial cells; upper marginal cells not longer than medial cells (sometimes larger); marginal cell walls not thicker than those of medial cells 168

168(167). Leaves widest at or above midleaf; upper laminal cell walls thin to evenly thickened, lumens quadrate to rounded; axillary hair basal cell walls hyaline and all cells of hair similar; perichaetium terminal on main axis; annulus of vesiculose cells, often in two or more rows; wide distribution .... *Hyophila*
Leaves widest below midleaf; upper laminal cell walls clearly trigonous or porose, lumens irregularly angular or stellate; axillary hairs with 1 or more basal cells with thicker or darker colored walls; perichaetium lateral on main axis at ends of very short branches; annulus of weakly differentiated cells; Brazil, India, Philippines .... *Hymenostyliella*

169(165). Leaf base little differentiated in shape; guide cells 2–6; seta nearly absent to short, less than 1 cm; upper laminal cells medially unistratose 170
Leaf base clearly differentiated in shape; guide cells commonly more than 6; seta elongate, 1 cm or longer; upper laminal cells medially bistratose, the cells staggered, not one directly above the other in section; wide distribution .... *Timmiella*

170(169). Transverse section of dorsal costal stereid band round to semicircular; axillary hairs with 1 or more basal cells with thicker or darker colored walls; peristome teeth present; American tropics, South Africa, Burma .... *Plaubelia*
Transverse section of dorsal costal stereid band clearly flattened or ventrally indented, reniform or crescent-shaped; axillary hair basal cell walls hyaline and all cells of hair similar; peristome teeth absent; wide distribution .... *Hyophila*

171–181 Costa with two stereid bands, upper laminal cells KOH red

171(51). Margins entire or minutely and evenly crenulate 172
Margins denticulate or serrulate to toothed 179

172(171). Transverse section of dorsal costal stereid band round to semicircular 173
Transverse section of dorsal costal stereid band clearly flattened or ventrally indented, reniform or crescent-shaped 176

173(172). Margins plane to incurved or involute; costal hydroid strand present 174
Margins recurved to revolute; costal hydroid strand absent 175

174(173). Dorsal costal epidermis absent; medial upper laminal cells commonly larger than 14 µm; transverse section of costa round or ovate or elliptical; number of rows of cells across ventral surface of costa usually 2 but up to 4; superficial walls of upper laminal cells flat or very weakly convex on both sides of lamina; southern South America, Antarctica .... *Sarconeurum*
Dorsal costal epidermis present; medial upper laminal cells 7–14 µm in width; transverse section of costa semicircular; number of rows of cells across ventral surface of costa commonly 4 or more; superficial walls of upper laminal cells strongly convex to bulging on both sides of lamina; Andes .... *Erythrophyllastrum*

175(173). Rather deep, narrow groove along costa absent; peristome teeth present; spore diameter 8–15 µm; wide distribution .... *Didymodon*
Rather deep, narrow groove along costa present; peristome teeth absent; spore diameter more than 15 µm; USA, Mexico, India, eastern Asia .... *Bellibarbula*

176(172). Stem sclerodermis not or little differentiated; Andes .... *Erythrophyllopsis*
Stem sclerodermis clearly differentiated 177

177(176). Stem hyalodermis absent 178
Stem hyalodermis present (sometimes indistinct); wide distribution .... *Bryoerythrophyllum*

178(177). Hydroid strand absent; rather deep, narrow groove along costa absent; number of rows of cells across ventral surface of costa usually 2 but up to 4; superficial walls of upper laminal cells strongly convex to bulging on both sides of lamina; wide distribution .... *Didymodon*
Hydroid strand present; rather deep, narrow groove along costa present; number of rows of cells across ventral surface of costa commonly 4 or more; superficial walls of upper laminal cells flat or very weakly convex on both sides of lamina; Mexico south through Andes .... *Mironia*

179(171). Ventral costal epidermis absent; southwestern USA, Mexico, Andes .... *Rhexophyllum*
Ventral costal epidermis present 180

180(179). Stem hyalodermis absent 181
Stem hyalodermis present (sometimes indistinct); wide distribution .... *Bryoerythrophyllum*

181(180). Hydroid strand absent; rather deep, narrow groove along costa absent; number of rows of cells across ventral surface of costa usually 2 but up to 4; superficial walls of upper laminal cells strongly convex to bulging on both sides of lamina; wide distribution .... *Didymodon*
Hydroid strand present; rather deep, narrow groove along costa present; number of rows of cells across ventral surface of costa commonly 4 or more; superficial walls of upper laminal cells flat or very weakly convex on both sides of lamina; Mexico south through Andes .... *Mironia*

Subfamily TIMMIELLOIDEAE

Timmielloideae Zand., *subfam. nov.* Type: *Timmiella* (De Not.) Limpr.

Plantae rosulatae; filum centrale robustum hyalodermidem evolvens; folia in statu sicco tubulosa marginibus planis, serrulatis praedita; costa percurrens, lata, stratis stereidarum duobus et filo hydroideo praedita; cellulae supernae laminales in regione folii mediana bistratosae, juxta costam altero strato cellularum supra alterum excentrico praeditae, ventraliter tumescentes, dorsaliter subconvexae; dentes peristomii stricti vel sinistrorsum extus visi subtorti.

Plants rosulate; central strand strong, hyalodermis present; leaves tubulose when dry, margins plane, serrulate; costa percurrent, broad, with two stereid bands, hydroid strand present; upper laminal cells bistratose in the middle portion of the leaf and not directly stacked over each other near the costa, superficially bulging ventrally and weakly convex dorsally; peristome teeth straight or twisted weakly clockwise.

Of all terminal taxa, *Timmiella*, the sole member of this subclade, shares the most plesiomorphic characters with the outgroups *Polytrichum*, *Timmia* and *Ptychomitrium* (see cladograms and discussion above). The distinguishing autapomorphy is laminal cells medially bistratose but not situated directly over one another near the costa (as seen either from above or in transverse section). This unusual feature is matched in *Diphyscium* (Diphysciaceae), a genus with many of the characteristics of the Pottiaceae (see discussion above). The following major traits in combination aid in distinguishing the genus: costa very wide, with multiple hydroid strands; leaf cells epapillose, ventrally bulging and dorsally nearly flat; peristome twisted clockwise (rare in the Pottiaceae) or straight. The distribution of the subfamily is rather general, being most common in temperate areas.

1. TIMMIELLA — Plates 1–2

Timmiella (De Not.) Limpr., Laubm. Deutchl. 1: 590, 1888. Type: *Timmiella anomala* (BSG) Limpr.

Trichstomum subg. *Eutrichum* Schimp., Corol. 28, 1856.

Barbula subg. *Timmiella* (De Not.) Kindb., Eur. N. Amer. Bryin. 2: 245, 1897.

Barbula sect. *Anomalae* BSG, Bryol. Eur. 2: 75, 107 (fasc. 13–15 Mon. 13. 45), 1846.

Trichostomum sect. *Timmiella* De Not., Cronac. Briol. Ital. 1: 14, 1866.

From *Timmia*, a genus + *-ella*, diminutive; resembling the genus *Timmia*.

Plants loosely caespitose to cushion-forming, *often rosulate*, green or often dark green above, brown below. *Stems* branching irregularly or simple, to 1.5 cm in length, transverse section rounded-pentagonal, often flattened, *central strand very strong*, sclerodermis of 1–3 layers (occasionally substereid), *hyalodermis present, weak*; axillary hairs of ca. 7 cells, all hyaline or basal cell firm-walled; sparsely radiculose. *Leaves incurved and tubulose when dry*, spreading when moist, *long-elliptical to ligulate or broadly lanceolate*, *often wasp-waisted*, 3.5–5.0 mm in length, upper lamina broadly channeled across leaf, *margins plane to weakly incurved, distantly denticulate or serrulate to near base, usually strongly serrate near apex, lamina bistratose except along margins*; *apex acute*; base often broadened, elliptical, somewhat sheathing, occasionally with distinct shoulders; *costa percurrent, tapering to apex and much broadened below midleaf, superficial cells ventrally quadrate and bulging*, dorsally elongate and smooth, 6–19 rows of cells across costa ventrally at midleaf, costal transverse section flattened, reniform or elliptical, *two stereid bands present, ventral epidermis present, this bistratose or occasionally tristratose*, dorsal epidermis present, unistratose, guide cells 6–16 in 1 layer, *hydroid strand present*, often multiple, often on both ventral and dorsal sides of guide cells; *upper laminal cells* quadrate to rounded-hexagonal, 9–12 µm in width, 1:1, walls mostly evenly thickened, lumens occasionally rounded, *superficially bulging ventrally and nearly flat dorsally, cells bistratose medially but near the costa not situated directly over each other*; *papillae absent*; *basal cells differentiated straight across leaf*, bulging-rectangular, 10–18 µm in width, mostly 3–4:1, walls thin, hyaline to yellowish. Dioicous or monoicous (autoicous, synoicous or apparently rhizautoicous). *Perichaetia terminal*, inner leaves scarcely different from cauline leaves, base often shortly and broadly clasping. Perigonia lateral on archegoniophore stem, as somewhat flattened buds, or very small and terminal on perigoniate plant. Seta ca. 1.0–2.0 cm in length, 1 per perichaetium, yellowish to reddish brown, twisted clockwise; theca 3.0–5.0 mm in length, brown, long-elliptical to long-cylindrical, occasionally somewhat ventricose, exothecial cells 13–20 µm in width, ca. 4–6:1, evenly thick-walled, stomates phaneropore, on neck of theca, annulus of weakly differentiated cells or of up to several layers of highly vesiculose cells, revoluble or deciduous in pieces; *peristome teeth of 16 paired teeth or 32 evenly spaced rami*, linear-lanceolate to filamentous, papillose to closely branching-spiculose, occasionally also striate below, rudimentary or to 300–700 µm, with several articulations, *straight or twisted clockwise weakly once or less*, basal membrane absent or very low, essentially smooth. Operculum long-conic to rostrate, 0.6–1.7 mm in length, cells in straight rows or twisted clockwise. Calyptra cucullate, smooth, ca. 3.5 mm in length. Spores 10–13 µm in diameter, yellowish brown, weakly papillose. *Laminal color reaction to KOH yellow.* Reported chromosome number n = 13, 13+m, 14, 14+m, 14+m+1acc, 15.

Found on rock and soil; in arid lands and mountainous areas of North and South America, Europe, Asia and Africa.

Timmiella is characterized by the following combination of distinguishing features: very strong stem central strand (Pl. 1, f. 7); laminal margins plane to weakly incurved, denticulate to dentate; costa very wide, with multiple hydroid strands; laminal cells medially bistratose but not vertically aligned near the costa, epapillose, ventrally bulging and dorsally nearly flat (Pl. 1, f. 2); peristome straight (Pl. 2, f. 6) or twisted clockwise (Pl. 1, f. 5).

Timmiella is one of only two genera in the Pottiaceae with

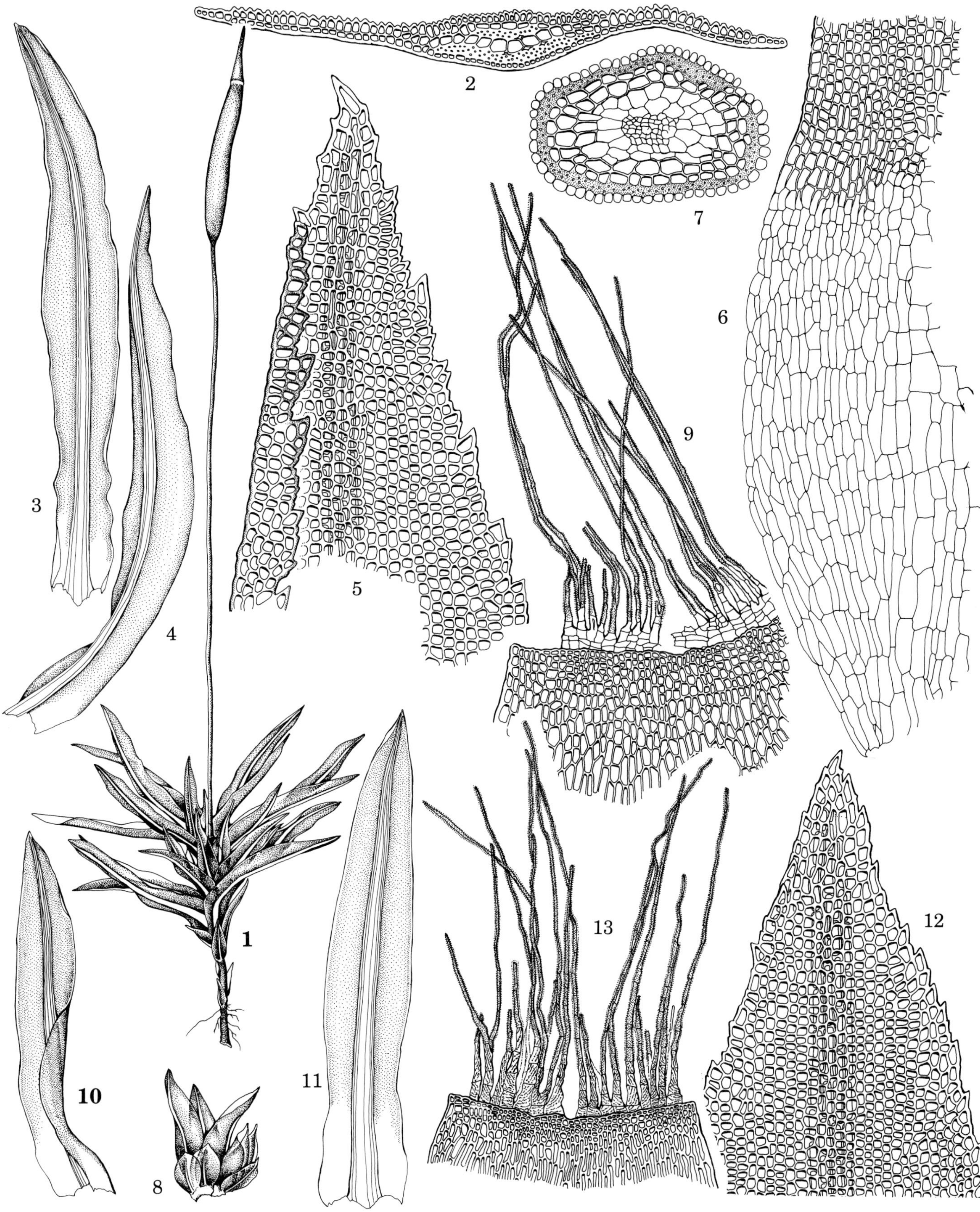

Plate 1. ***Timmiella.* 1–9. *T. anomala.*** 1. Habit. 2. Transverse section of stem. 3–4. Two leaves. 5. Leaf apex. 6. Basal cells. 7. Transverse section at midleaf. 8. Clustered lateral perigoniate buds. 9. Peristome. **10–13. *T. barbuloides.*** 10–11. Two leaves. 12. Leaf apex. 13. Peristome.

peristome teeth (in at least some species) twisted clockwise (*Leptodontiella*, with one species, has teeth that are straight or twisted weakly clockwise). Characters important in *Timmiella* are variations in sexual condition, annulus development, and peristome size and degree of torsion. *Timmiella corniculata* is dioicous, contrary to Brotherus' (1924–25) key. Propagula are only doubtfully present in the genus; small, clavate, multicellular propagula, ca. 50–70 μm in length, were found loosely associated with gametophores in the type of *T. brevidens*, but may be due to admixture of other species.

Additional literature: Castaldo-Cobianchi et al. (1982), Chopra and Kapur (1989), Gambardella et al. (1980), Kapur (1989, 1991), Ligrone (1986), Ligrone et al. (1980a,b).

Number of accepted species: 13.

Species examined: *T. anomala*, *T. barbuloides*, *T. brevidens* (BM), *T. corniculata* (NY), *T. diminuta* (NY), *T. flexiseta* (NY), *T. pelindaba* (NY), *T. subintegra* (BM), *T. umbrosa* (NY).

New heterotypic synonymy: *Timmiella argentinica* Broth. = *Timmiella acaulon* (C. Müll.) Zand.

New combinations: *Timmiella acaulon* (C. Müll.) Zand., *comb. nov.* (*Trichostomum acaulon* C. Müll., Linnaea 42: 320, 1879; *Tortella acaulon* (C. Müll.) Broth.).

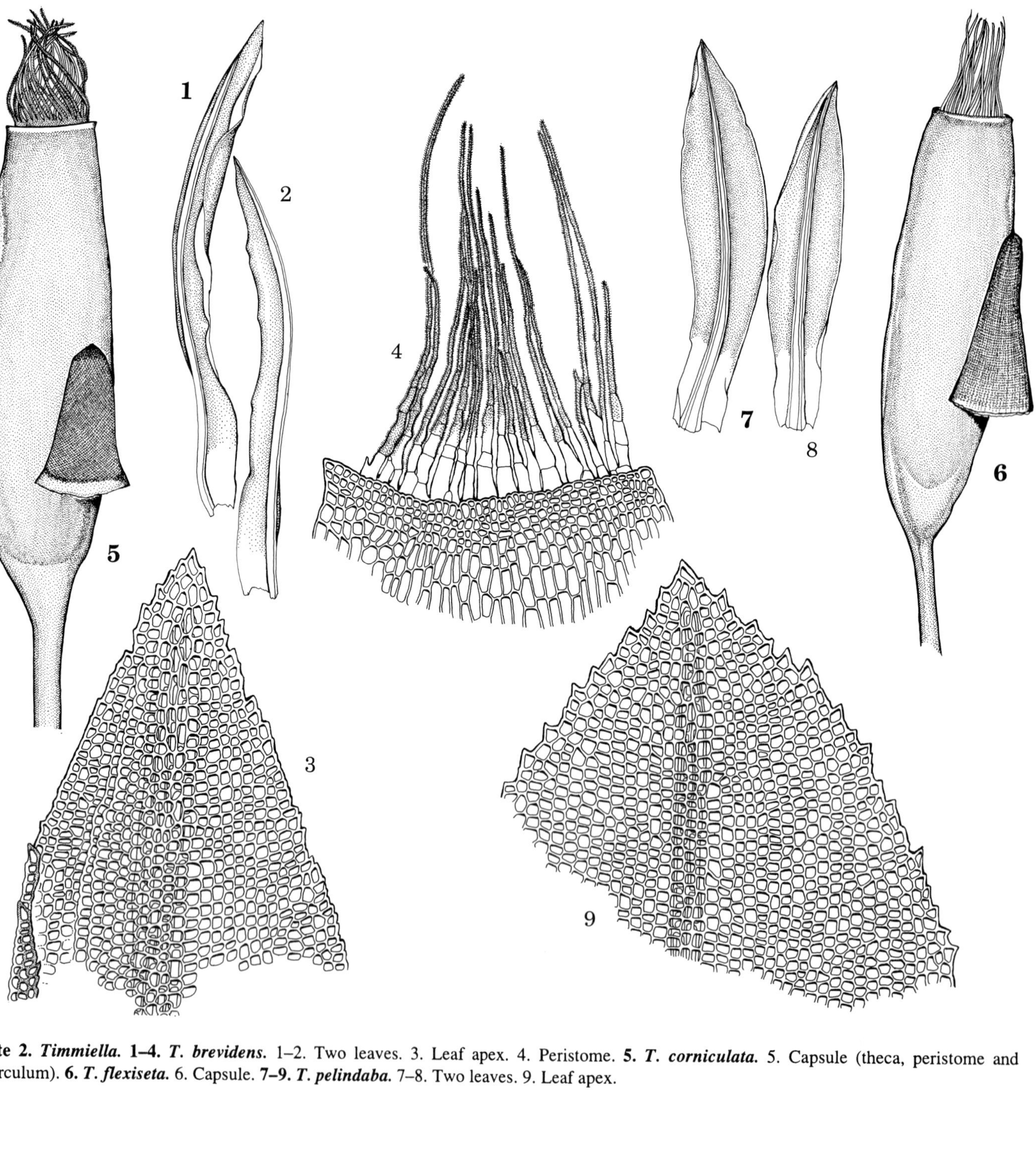

Plate 2. ***Timmiella.*** **1–4.** ***T. brevidens.*** 1–2. Two leaves. 3. Leaf apex. 4. Peristome. **5.** ***T. corniculata.*** 5. Capsule (theca, peristome and operculum). **6.** ***T. flexiseta.*** 6. Capsule. **7–9.** ***T. pelindaba.*** 7–8. Two leaves. 9. Leaf apex.

Subfamily ERYTHROPHYLLOPSOIDEAE

Erythrophyllopsoideae Zand., *subfam. nov.* Type: *Erythrophyllopsis* Broth. in Herz.

Plantae magniusculae, altitudine 2.0–3.5 cm; folia lanceolata marginibus planis, apicibus acutis, basibus vaginantibus, laminis in solutione KOH rubrescentibus, costae cellulas ducium 4–6 et trans superficiem ventralem cellulas 10–16 evolventi praedita.

Plants rather large, 2.0–3.5 cm in height, leaves lanceolate, with plane margins, apices acute, bases sheathing; upper lamina KOH red, costa with 4–6 guide cells and 10–16 cells across the ventral surface.

With *Timmiella* as outgroup, this subclade is distinguished from lower branches on the tree by the advanced character of costal guide cells fewer, 2–6. The general appearance of the two included genera, *Erythrophyllastrum* and *Erythrophyllopsis*, is intermediate between that of the traditional Trichostomoideae and the Barbuloideae, especially in the lanceolate leaves with plane margins. Other distinguishing traits are the acute leaf apex, sheathing leaf base, KOH red upper lamina, two costal stereid bands, 4–6 guide cells, and many (10–16) rows of cells across ventral surface of costa. The species of this small subfamily are rare and restricted to the Andes.

2. ERYTHROPHYLLOPSIS — Plate 3.

Erythrophyllopsis Broth. in Herz., Biblioth. Bot. 87: 41, 1916.
Type: *Erythrophyllopsis boliviana* Broth., Bolivia, Cerros de Malaga, Herzog 4371, syntype, M, also Tunarisee, Herzog 4765, syntype, NY.

From *Erythrophyllum*, a genus + ὄψσις, -εως, appearance; resembling the genus *Erythrophyllum*.

Plants forming cushions, dark green to reddish brown above, reddish brown below. Stems branching occasionally, to 2–3 cm in length, transverse section rounded-triangular to pentagonal, central strand strong, sclerodermis weakly developed, substereid, *hyalodermis of 1 layer of usually collapsed cells*; axillary hairs ca. 10 cells in length, the basal 1–2 cells more firm walled or brown, otherwise hyaline; sparsely radiculose. *Leaves* appressed below and incurled above when dry, *squarrose* when moist, *long-lanceolate,* 3.5–4.5 mm in length, upper lamina broadly channeled to keeled, *margins plane,* entire, occasionally denticulate at leaf apex; apex narrowly acute; *base strongly sheathing, with distinct shoulders, bordered by 4–6 rows of narrower, often thicker-walled cells; costa* percurrent or ending in an apiculus, *rather broad* below midleaf, lamina inserted about 45°, costal superficial cells quadrate and papillose, but becoming smooth and elongate in distal 1/4 of leaf ventrally, dorsally elongate, ca. 10 rows of cells across costa ventrally at midleaf, *costal transverse section reniform,* stereid bands ventrally distinct, the dorsal crescent-shaped, epidermis present ventrally and dorsally, guide cells 4–6 in 1 layer, 1–3 hydroid strands usually present, weakly distinguishable; *upper laminal cells bistratose* across the leaf, quadrate, 7–12 µm in width, 1(–2):1(–2), walls evenly thickened, superficially weakly convex on both sides of lamina; *papillae bifid, 2–6 per lumen, low, scab-like, flattened, crowded,* solid or hollow; *basal cells sharply differentiated nearly across the leaf base but rising highest medially,* rectangular, ca. 9–12 µm (to 16 µm above) in width, mostly 4–5:1, walls thin. Dioicous. Perichaetia terminal, inner leaves little different from the cauline, sheathing the seta. Perigonia terminal, weakly gemmate, inner leaves shortly lanceolate. Seta 0.5–1.5 cm in length, 1 per perichaetium, reddish brown, twisted weakly clockwise; theca ca. 3.3 mm in length, reddish brown, cylindrical, exothecial cells rectangular, 3–5:1, ca. 13–18 µm in width, thin-walled, stomates at base of capsule, phaneropore, annulus of ca. 3–4 rows of vesiculose cells, persistent; peristome teeth much reduced or 16, truncate to shortly lanceolate, papillose, 50–130 µm in length, with up to 5 articulations, straight, basal membrane low or absent in height, papillose. Operculum conic-rostrate, ca. 1 mm in length, cells straight. Calyptra not seen. Spores 13–15 µm in diameter, light brown, essentially smooth. *Laminal KOH color reaction red.*

A monotypic genus of the Andes of Bolivia and Argentina, found on moist rocks at high elevations.

Erythrophyllopsis, known from one species, *E. fuscula*, is characterized by a combination of characters: red coloration in KOH, leaves bistratose above (Pl. 3, f. 5), with plane margins and a strongly differentiated sheathing leaf base having "shoulders" (Pl. 3, f. 3–4) at the top of the base, and peristome rudimentary or short (illustrated by Herzog 1916). It is similar to *Bryoerythrophyllum* by the red coloration, leaves with differentiated leaf base and upper laminal cell walls evenly thickened, with papillae crowded and bifid to multiplex. My previous treatment of this genus (Zander 1977a) included *Trichostomum andinum*, a morphologically similar taxon here placed in a separate genus, *Erythrophyllastrum*, (*q.v.* The upper lamina of *Erythrophyllopsis*, which has been figured by previous authors (Herzog 1916; Brotherus 1924–25; Hilpert 1933), is entirely bistratose. *Mironia* is similar but has leaves bistratose only along the upper margins, and has a well developed peristome. *Leptodontium* has a similarly lanceolate leaf with a broad, sheathing base, but is yellow in KOH and lacks a stem central strand, among other differences. There may be a confusion with *Didymodon* sect. *Vineales* because of the lanceolate, KOH-red leaves with dorsally quadrate costal cells; the leaf apex has ventrally elongate costal cells in a boat-shaped groove and is occasionally abruptly constricted, *cf. D. occidentalis.* This last taxon, however, has a quite different costa, it being dorsally rounded and protuberant, and the guide cells commonly in two layers. Hilpert (1933) placed *Erythrophyllopsis* in close relationship with what are here dealt with as *Didymodon* sect. *Asteriscium*, *Gertrudiella* and *Bryoerythrophyllum.*

Erythrophyllopsis challaensis is here transferred to *Didymodon* ((*q.v.*). It differs from *E. fuscula* in slightly larger upper laminal cells, weakly serrulate basal marginal cells, shorter and broader basal cells which are transversely slit through cell wall resorption of the medial portion of the longitudinal cell walls, and a yellow color reaction to KOH.

Number of accepted species: 1.

Species examined: *E. fuscula* (M, NY).

New heterotypic synonymy: *Didymodon semivaginatus* (Britt.) Broth. = *Erythrophyllopsis fuscula* (C. Müll.) Hilp.

3. ERYTHROPHYLLASTRUM Plate 4.

Erythrophyllastrum Zand., *gen. nov.* Type: *Erythrophyllastrum andinum* (Sull.) Zand. Holotype: Peru, Andes, U.S. Expl. Exp. Wilkes, 1838–1842, FH.

Plantae magnae, usque altitudine 3.5 cm; caules in sectione interdum triangulares, filo centrali robusto praediti, hyalodermidem e stratis 1–3 texturae laxae compositam in maturitate subcollapsam evolventes; folia in statu madido late patentia, late lanceolata, longitudine 3.5–4.5 mm, marginibus planis, integris, uni- vel bistratosis, apice acuto, base subvaginanti, lamina superna in solutione KOH rubrescenti, costa in sectione semicirculari strata stereidarum duo evolventi praedita.

From *Erythrophyllum*, a genus + *-astrum*, an ending implying deception; mimicking *Erythrophyllum*.

Plants forming cushions, *green to reddish brown above, reddish brown below. Stems* branching irregularly, to 3.5 cm in

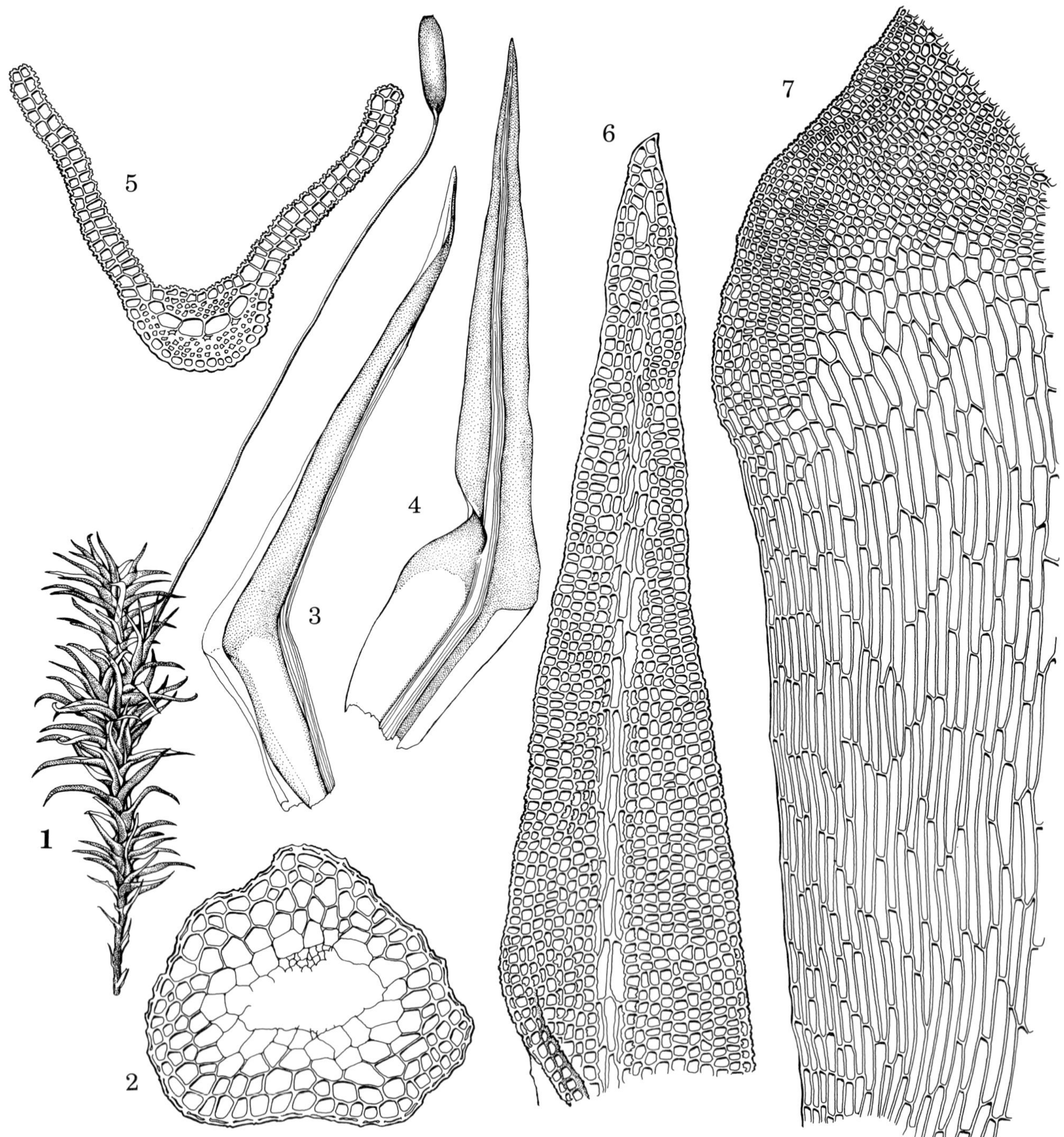

Plate 3. *Erythrophyllopsis*. 1–7. *E. fuscula*. 1. Habit. 2. Transverse section of stem. 3–4. Two leaves. 5. Transverse section at midleaf. 6. Leaf apex. 7. Basal cells.

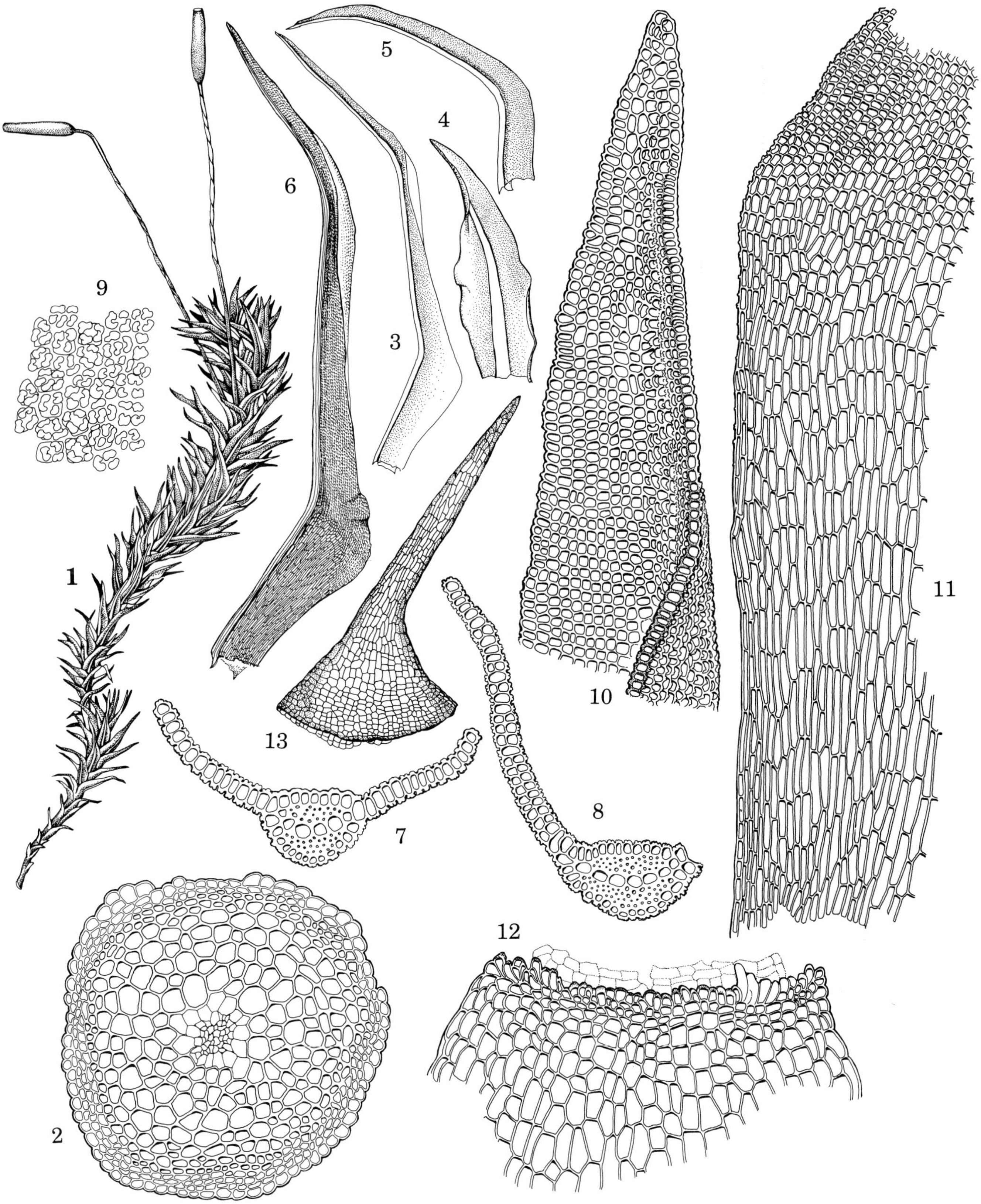

Plate 4. ***Erythrophyllastrum.*** **1–13.** ***E. andinum.*** 1. Habit. 2. Transverse section of stem. 3–6. Four leaves, one enlarged. 7–8. Two transverse sections, midleaf. 9. Upper laminal papillae. 10. Leaf apex. 11. Basal cells. 12. Rudimentary peristome. 13. Operculum.

length, transverse section rounded-pentagonal, *central strand strong, sclerodermis weakly developed, substereid, hyalodermis of 1(–3) cell layers of lax cells, often not collapsed*; axillary hairs long, to 20 cells in length, basal 1–4 cells yellowish; radiculose below. *Leaves* appressed-incurved and curled when dry, *widely spreading when moist, lanceolate,* 2–3 mm in length, upper lamina broadly channeled to keeled, broadly grooved along costa, *margins plane, entire, upper lamina unistratose or variably bistratose in large patches or throughout*; apex acute; *base broadly sheathing below, with weak shoulders*; *costa* percurrent, with lamina inserted laterally, superficial cells quadrate, papillose, often short-rectangular near apex ventrally, dorsally usually quadrate near apex, rectangular below, *10–16 rows of cells across costa ventrally at midleaf, costal transverse section semicircular,* stereid bands two and strong, the dorsal semi-circular in shape, ventral and dorsal epidermises present, the dorsal medially weak, *guide cells (4–)6 in 1 layer,* hydroid strand(s) very weak, often multiple; *upper laminal cells small, subquadrate,* 7–10 µm in width, 1:1, walls thin to evenly thickened, superficially weakly convex on both sides of lamina; papillae bifid, 2–3 per lumen to multiplex, generally hollow, crowded; *basal cells differentiated across the sheathing base,* rectangular, ca. 13 µm in width, mostly 3–4:1, walls thin. Propagula not seen. Dioicous. Perichaetia terminal, inner leaves lanceolate, somewhat larger than the cauline, to 4.5 mm in length, sheathing the seta in lower 1/3, lower cells rectangular, walls weakly porose. Perigonia terminal, weakly gemmate. Seta ca. 1 cm in length, 1 per perichaetium, reddish brown, twisted clockwise; theca 1.5–2.0 mm in length, reddish brown, cylindrical, exothecial cells short-rectangular, ca. 2:1, thin-walled, stomates at base of capsule, phaneropore, annulus of 1–2 rows of vesiculose cells; *peristome teeth rudimentary,* consisting of a few low plates, to 30–45 µm in height. Operculum long-rostrate, oblique, ca. 1.2 mm in length, cells straight. Calyptra conic-cucullate, smooth, ca. 1.5 mm in length. Spores ca. 13 µm in diameter, light brown, weakly papillose. *Laminal KOH color reaction red.*

A monotypic genus found in the Andes of Colombia and Peru, on moist rock at high elevations (ca. 3700–4000 msm).

This taxon was previously placed with *Erythrophyllopsis* (Zander 1977a), which is superficially similar by its red coloration; stem with large central strand; leaves with plane margins, broad, flat costae and sheathing leaf bases; and upper lamina bistratose. *Erythrophyllastrum* differs from *Erythrophyllopsis* by the former's 1(–3) layers of stem hyaloderm cells (Pl. 4, f. 2), comprising loose tissue little collapsed with maturity; stems sometimes triangular in section; the leaves commonly shorter and more broadly lanceolate, sometimes reflexed above the base and widely spreading to squarrose; the leaf base less strongly sheathing, sometimes without highly differentiated "shoulders" (Pl. 4, f. 4); upper laminal cells variably entirely bistratose or bistratose in patches or entirely unistratose (Pl. 4, f. 7–8); both costal section and dorsal stereid band semicircular or nearly so (Pl. 4, f. 7–8, both being reniform in *Erythrophyllopsis*); dorsal costal cells usually quadrate near the leaf apex; and peristome even more reduced, to a few small plates (Pl. 4, f. 12).

Through its plane, reflexed, broadly keeled leaves with sometimes weakly sheathing base, small, dense upper laminal cells, and basal cells usually differentiated completely across the leaf, *Erythrophyllastrum* may have much the general aspect of species of *Barbula* sect. *Convolutae* (e.g. *B. amplexifolia*).

Number of accepted species: 1.

Species examined: *E. andinum* (BUF).

New combination: *Erythrophyllastrum andinum* (Sull.) Zand., *comb. nov.* (*Trichostomum andinum* Sull., U.S. Expl. Exp. Wilkes Musci 5, 1859; *Erythrophyllopsis andina* (Sull.) Zand.).

Subfamily GERTRUDIELLOIDEAE

Gertrudielloideae Zand., *subfam. nov.* Type: *Gertrudiella* Broth.

Plantae filum centralem et hyalodermidem caulis evolventes; folia in statu madido squarrosa, magniuscula, longitudine 3.5–4.5 mm, apice acuto, base vaginanti, costa e strato stereidarum unico composita et cylindro multistratoso e cellulis ducium parietibus crassis praeditis composito, cellulis laminae supernae solum in margine folii papillosis, in regione mediano ventraliter tumescentibus dorsaliter planis sed in margine folii in aliquot seriebus utrinsecus tumescentibus praedita.

Plants with strong stem central strand, hyalodermis present; leaves squarrose when wet, rather large, 3.5–4.5 mm in length, apex acute, base sheathing, costa with one stereid band and a multilayered cylinder of thick-walled guide cells; upper laminal cells papillose only marginally, medially bulging ventrally and flat dorsally but marginally bulging on both sides in several rows.

This tribe is more highly evolved than the Timmielloideae by the advanced traits of leaf base not sheathing and upper lamina unistratose. There are two autapomorphies: the upper laminal cells medially ventrally bulging and dorsally flat but the marginal cells bulging on both sides in several rows, and the costal guide cells forming a thick-walled, multilayered cylinder. This is a rare, monotypic, Andean subfamily.

4. GERTRUDIELLA **Plate 5.**

Gertrudiella Broth., Nat. Pfl. ed. 2, 11: 528, 1925. Type: *Gertrudiella validinervis* (Herz.) Broth.

Gertrudia Herz., Biblioth. Bot. 87: 44, 1916, *hom. illeg. non* K. Schumann, 1900. Type: *Gertrudia validinervis* Herz.

From Gertrude, a name + i + *-ella*, diminutive; T. Herzog's (1916) original name for this genus was a dedication to his wife.

Plants in cushions, green above, brown to red-brown or blackish-brown below. Stems branching occasionally, to 2.0 cm in length, transverse section rounded-pentagonal to elliptical, *central strand very strong* and well differentiated, sclerodermis present, hyalodermis present but firm-walled; axillary hairs of 5–7 clear cells. *Leaves crowded*, weakly spreading, weakly to strongly contorted or curled when dry, *squarrose from a short, sheathing base* when moist, *lanceolate*, 3.5–4.5 mm in length, upper lamina broadly channeled across leaf, *margins strongly revolute to near apex, entire or occasionally weakly dentate near apex*; apex narrowly acute, occasionally narrowly obtuse; base long-elliptical; *costa percurrent to excurrent as a short, denticulate awn*, superficial cells quadrate ventrally, elongate dorsally, *10–15 rows of cells across costa ventrally at midleaf*, costal transverse section semicircular, *stereid band single*, strong and reniform, *epidermis of 1 layer of strongly hollow-papillose cells ventrally*, epidermal cells weakly developed dorsally, *guide cells 15–20 in 3–4 layers, hydroid strand present*; *upper laminal cells* subquadrate to irregularly hexagonal, *small*, 8–10 µm in width, 1:1, walls evenly thickened, *superficially bulging-mamillose ventrally, flat dorsally in medial portion of lamina, bulging on both sides in 10–20 rows marginally*; *papillae present on marginal cells, low, simple, hollow*; basal cells differentiated medially, *rectangular, little wider than the upper,*

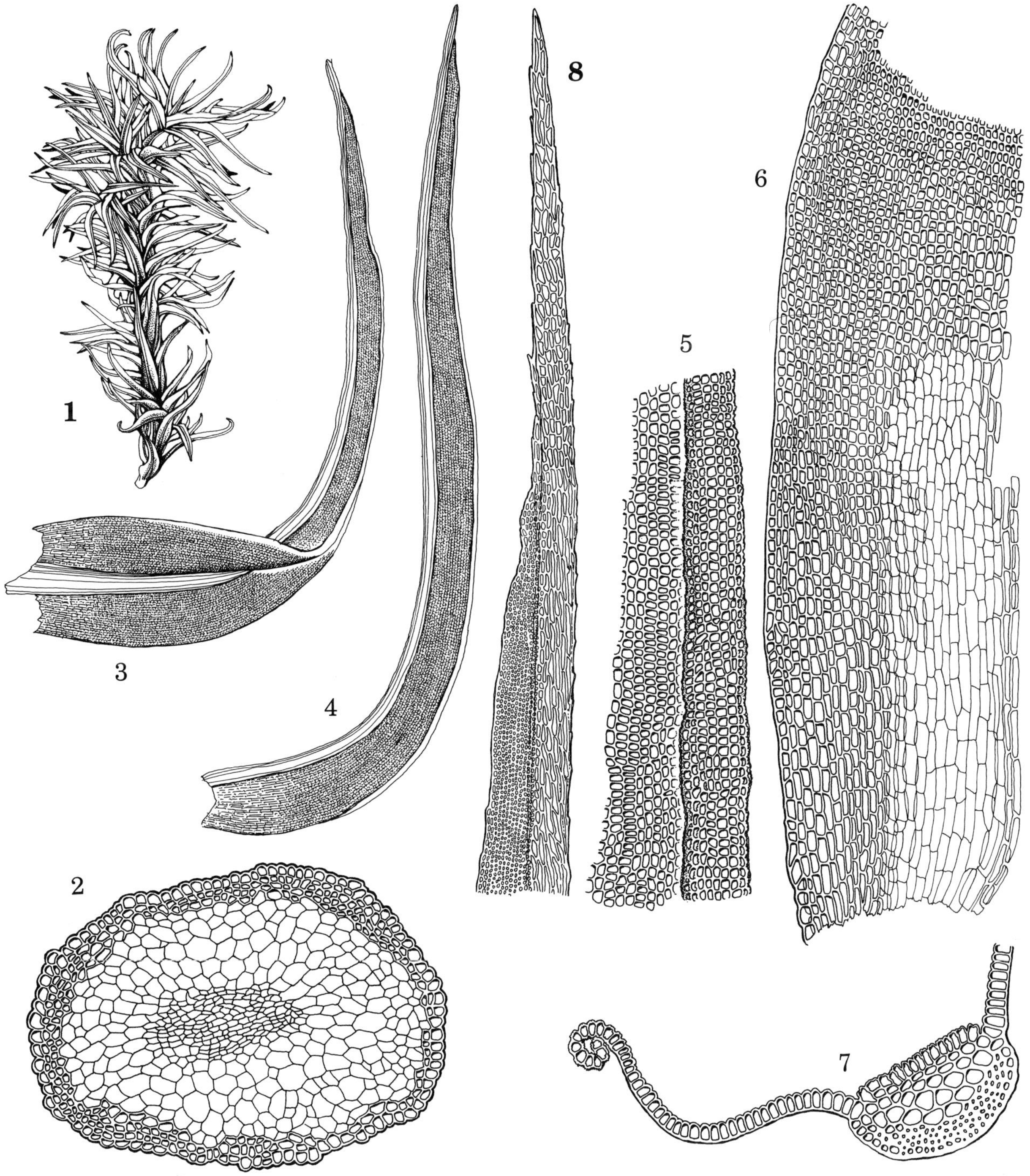

Plate 5. ***Gertrudiella*. 1–6. *G. validinervis*.** 1. Habit. 2. Transverse section of stem. 3–4. Two leaves. 5. Upper marginal cells. 6. Basal cells. 7. Transverse section at midleaf. **8. *G. validinervis*** var. ***serratopungens***. 8. Leaf apex.

3–4:1, walls thin, occasionally transversely slit across longitudinal walls. Apparently dioicous. Perichaetia unknown. Perigonia appearing as axillary buds near stem apex, inner leaves deltoid, outer leaves elongate. Sporophyte unknown. Laminal KOH color reaction orange.

Rare, restricted to Bolivia where it has been collected on stone in rather dry areas, at low to high (300–3200 m) elevations.

This monotypic genus is well characterized, mainly by the ventrally mamillose medial cells of the lamina, several rows of marginal cells bulging on both sides and low-papillose, epidermal cells on the ventral surface of the costa numerous and strongly hollow-papillose, and guide cells rather thick-walled and forming a multilayered cylinder (Pl. 5, f. 7). Hilpert (1933) included *Trichostomum ferrugineum* Herz. in *Gertrudiella* on account of the several layers of guide cells. This species is actually a *Didymodon* (with much the same general appearance of *D. occidentalis* of western North America) by the orange reaction to KOH, lanceolate leaves deeply and widely channeled ventrally along the stout costae, and upper laminal margins bistratose, and is here treated as *Didymodon herzogii* (*nom. nov.* in treatment of *Didymodon*), differing from related taxa of sect. *Vineales* by the microstomous, eperistomate capsule and medial basal cells transversely resorbed across longitudinal walls to form slits. The last character has also been described for *Kingiobryum* Robins. (see Zander & Cleef 1982) of the Dicranaceae. Although *G. validinervis* shows some slitting of basal laminal cells, it and *D. herzogii* are not closely related. *Kingiobryum*, which is eperistomate, may well belong to the Pottiaceae; further study is needed.

Gertrudiella shares several characters with various species of *Pseudocrossidium*, among them revolute margins, differentiation of medial upper laminal cells (especially *P. leucocalyx*) and presence of ventral costal epidermal cells, a single stereid band, guide cells in more than one layer, and small upper laminal cells. It is, however, apparently not closely related. *Streptocalypta* has a similar arrangement of thick-walled guide cells but has quite different areolation. The guide cell cylinder of *Gertrudiella* is probably mainly of structural importance; sections reveal that there is no direct connection with the equally well-differentiated stem central strand (Pl. 5, f. 2).

Biblography: Herzog (1916), Hilpert (1933).

Number of accepted species: 1.

Taxa examined: *G. validinervis* (BUF and isotypes at JE, L, W), *G. validinervis* var. *serratopungens* (JE, L).

New combination: *Gertrudiella validinervis* (Herz.) Broth. var. *serratopungens* (Herz.) Zand., *comb. nov.* (*Gertrudia validinervis* Herz. var. *serratopungens* Herz., Biblioth. Bot. 87: 45, 1916).

Subfamily CHIONOLOMOIDEAE

Chionolomoideae Zand., *subfam. nov.* Type: *Chionoloma* Dix.

Plantae magniusculae; caules plerumque usque altitudine 4 cm; folia longilinearia, flagelliformia, apice anguste acuto, base plerumque vaginanti, marginibus planis, costa stratum ventrale stereidarum maius quam stratum dorsale evolventi et filum hydroideum effacienti praedita.

Plants relatively large; stems often to 4 cm in length; leaves long-linear, whip-like, apex narrowly actute, base often sheathing, margins plane; costa with ventral stereid band larger than the dorsal, and hydroid strand absent.

This subclade is distinguished from the Timmielloideae, Erythrophyllopsoideae and Gertrudielloideae by the advanced character of hydroid strand absent. The genera share two advanced character states: leaves long-linear and ventral stereid band larger than the dorsal. The three included genera have long, whip-like leaves with plane margins, and are characteristic of tropical areas of low or medium elevation.

5. CHIONOLOMA Plate 6.

Chionoloma Dix., J. Bot. 60: 102, 1922. Type: *Chionoloma induratum* Dix.

From χιων, χιόνος, snow + o + λῶμα, -ατις, fringe, hem, border.

Plants growing in a dense turf, yellowish green above, *light reddish brown below. Stems* branching occasionally, ca. 4.0 cm in length, transverse section rounded-pentagonal, central strand distinct, often strong, *sclerodermis not differentiated from central cylinder, which is of incrassate, large-lumened cells, hyalodermis present*, usually collapsed in mature parts of stem; axillary hairs of several hyaline cells, basal 1–2 also hyaline, but with somewhat thicker walls; radiculose and also *closely invested with a verrucose or papillose red tomentum. Leaves* incurved, curled and often tubulose when dry, spreading when moist, *long-linear-lanceolate*, ca. 6–7(–8) mm in length, upper lamina broadly channeled to shallow-grooved along costa near apex, *margins plane to weakly incurved, often sharply incurved near apex, distantly weakly denticulate along border, bordered in lower 1/2–3/4 of leaf by 2–7 rows of unistratose, rectangular, epapillose, thick-walled, porose cells*; apex very narrowly acute; base long-elliptical, weakly sheathing; *costa excurrent as a cylindrical smooth mucro*, superficial cells quadrate and papillose ventrally, elongate dorsally, *10–12 rows of cells across costa ventrally at midleaf*, costal transverse section elliptical to rounded-triangular, *stereid bands two, the ventral* dorsal band stronger than the reniform, epidermis present ventrally, entirely absent dorsally, *guide cells 8–10* in 1 layer, hydroid strand absent; *upper laminal cells small, rounded-quadrate, often short-rectangular*, 6–9 µm in width, 1–2:1, walls thick, obscured by papillae, *medially free walls bulging ventrally and weakly convex dorsally*, bulging marginally on both sides of lamina; *papillae massive, multiplex-capituliform, crowded, with many small salients*; *basal cells differentiated* (but not sharply so) *across leaf, not or weakly grading into marginal border*, rectangular,

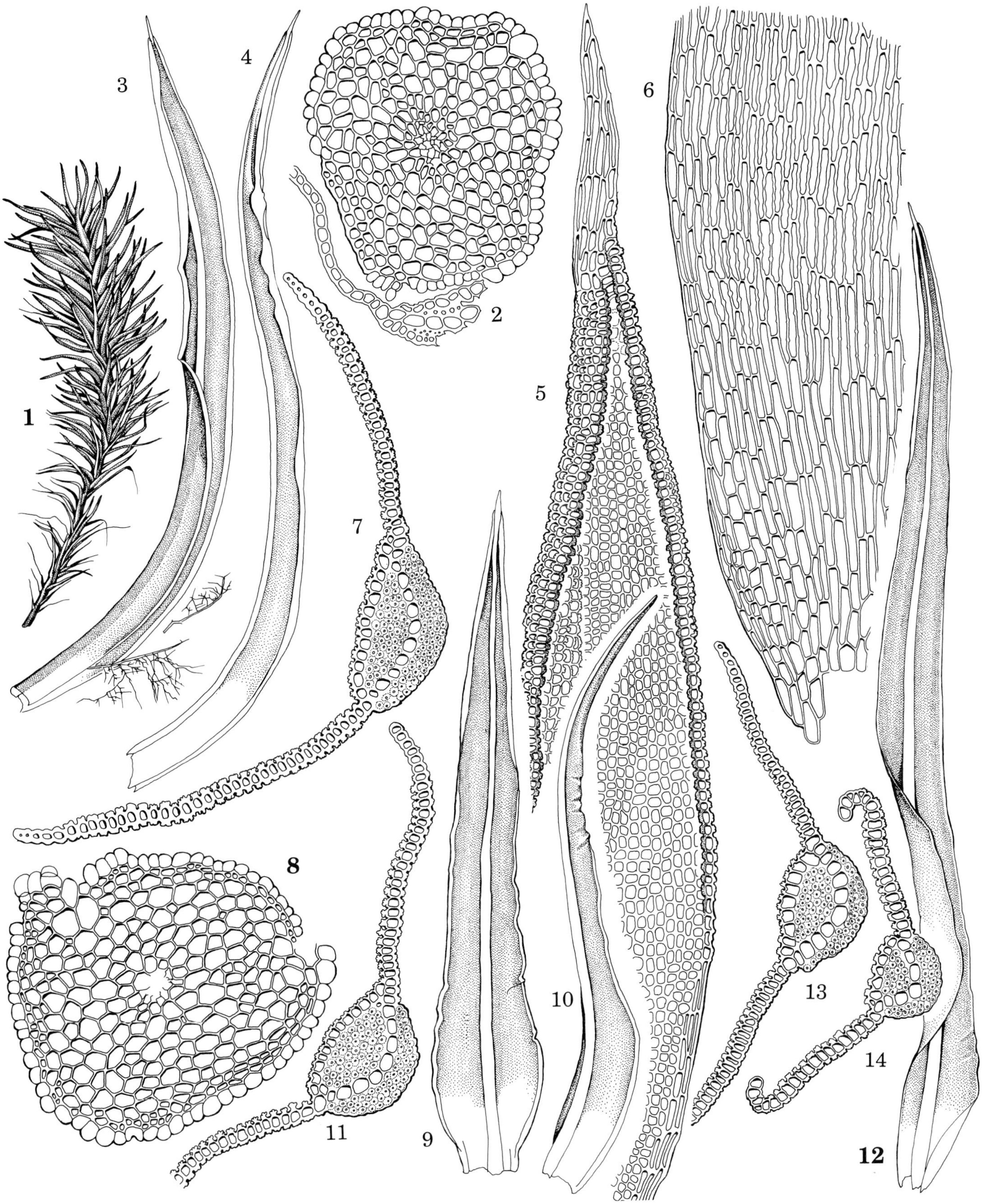

Plate 6. ***Chionoloma*****. 1–7.** ***C. induratum*****.** 1. Habit. 2. Transverse section of stem. 3–4. Two leaves and tomentum. 5. Leaf apex. 6. Basal cells. 7. Transverse section at midleaf. **8–11.** ***C. latifolium*****.** 8. Transverse section of stem. 9–10. Two leaves. 11. Transverse section at midleaf. **12–14.** ***C. longifolium*****.** 12. Leaf. 13. Transverse section at midleaf. 14. Transverse section near leaf apex.

Plate 7. ***Pseudosymblepharis*****. 1–7.** ***P. schimperiana*****.** 1. Habit. 2. Transverse section of stem. 3. Leaf. 4. Leaf apex. 5. Basal cells. 6. Transverse section at midleaf. 7. Peristome teeth. **8–10.** ***P. angustata*****.** 8–9. Two leaves. 10. Leaf apex. **11–13.** ***P. duriuscula*****.** 11–12. Two leaves. 13. Leaf apex. **14–16.** ***P. mauiensis*****.** 14–15. Two leaves. 16. Leaf apex. **17–20.** ***P. subduriuscula*****.** 17–18. Two leaves. 19. Leaf apex. 20. Transverse section at midleaf. **21–23.** ***P. syrrhopodontoides*****.** 21–22. Two leaves. 23. Leaf apex.

mostly 10–13 µm in width, 3–5:1, *thin-walled and hyaline near insertion grading to incrassate and porose in upper part of leaf base*. Perichaetia, perigonia and sporophyte unknown. Laminal KOH color reaction deep yellow or yellowish orange above, reddish brown below.

Found on limestone in Burma, Thailand, Malaysia and Borneo.

This genus has much the appearance of *Tortella*, but differs in having four areas of very distinct laminal areolation: quadrate upper laminal cells; a marginal border of long-rectangular, porose, incrassate cells (Pl. 6, f. 5); short-rectangular, incrassate and porose medial upper leaf base cells (similar to those characteristic of *Trichostomum* subg. *Oxystegus*), and hyaline, thin-walled rectangular cells of the lower leaf base (Pl. 6, f. 6). The border of *Chionoloma* is similar to that of *Pleurochaete* in that it does not clearly meld with the basal cells to reach the costa below, and the upper medial laminal cells are also often bulging strongly ventrally but weakly so dorsally; but *Pleurochaete* differs from *Chionoloma* by the costal section reniform, the ventral stereid band smaller than the dorsal, the leaf margin plane or broadly channeled near the apex, and the marginal leaf border of relatively thin-walled cells. *Pleurochaete* is unusual in being pleurocarpous, and the sexual position in *Chionoloma* is unknown. The incurved upper margins (Pl. 6, f. 14), the rather thick papillae (best seen in section), and the leaf border of thick-walled cells of *Chionoloma* are similar to those of *Hypodontium*.

Only the quadrate upper leaf cells, the upper ventral costal cells, and cells of the uppermost portion of the leaf base are papillose. The stem section (Pl. 6, f. 2) is distinctive and similar to that of many *Trichostomum* species, with all cells incrassate except those of the central strand and the hyalodermis. The leaf section shows an unusually large number of ventral epidermal cells (in costal sections) and of guide cells (but both are each in one layer), and the section is otherwise similar to that of *Pseudosymblepharis* in the ventral stereid band being stronger than the dorsal and the costal section often rounded-triangular. The red tomentum (Pl. 6, f. 3) is also a distinctive feature, reminiscent of that of some species of *Leptodontium*, a genus which, however, differs in the lack of a central strand and of a differentiated ventral costal epidermis. The extreme upper laminal margins are sometimes sharply and narrowly incurved like those of *Weissia*, but that genus does not have the marginal border or characteristic costal section; *W. jamaicensis* is similar in size, the distinct rectangular leaf base, and its ventral stereid band is commonly larger than the dorsal, but the costal section is round and the upper laminal cells are mainly isodiametric, with walls not so highly thickened. The three species of *Chionoloma* are not especially different from one another. The genus was synonymized with *Pseudosymblepharis* by Eddy (1990, *cf.* Menzel 1992) as merely "exceptionally robust plants in which the pellucid leaf border ascends high into the leaf limb, but the latter character is extremely variable, even within a single species." The denticulate border of *Chionoloma* will immediately distinguish this genus from *Pseudosymblepharis*.

Additional literature: Dixon (1922a).

Number of accepted species: 3.

Species examined: *C. induratum* (BM), *C. latifolium* (BM), *C. longifolium* (BM).

6. PSEUDOSYMBLEPHARIS **Plate 7.**

Pseudosymblepharis Broth., Nat. Pfl. ed. 2, 10: 261, 1924.
Lectotype: *Pseudosymblepharis papillosula* Card. & Thér. *fide* Saito, J. Hattori Bot. Lab. 39: 439, 1975.

From ψευδής, falsehood + o + *Symblepharis*, a genus; resembling the genus *Symblepharis*.

Plants growing in clumps or turf, green above, brown or reddish brown below. *Stems* branching often and irregularly, often long, to 2–4(–6) cm in length, transverse section rounded-pentagonal, central strand usually present, occasionally small, *cells of central cylinder often thick-walled, sclerodermis usually weakly developed and composed of substereid cells, hyalodermis present, sturdy*, collapsed only in very mature parts of stem; axillary hairs to 16 cells in length, all hyaline or basal 1–3 cells yellow; rhizoids sparse or stem occasionally weakly red-tomentose. *Leaves* incurved, contorted to spiralled when dry, *spreading to squarrose from a usually sheathing base when moist, lanceolate to linear*, 1–7 mm in length, upper lamina often narrow, flat to weakly tubulose, *margins plane, entire* or occasionally weakly dentate at apex, margins often eroded by fragmentation, occasionally decurrent; *apex* subulate, sharp, occasionally narrowly obtuse, *often fragile and broken*; *base usually strongly sheathing, ovate to rectangular, with "shoulders"*; costa excurrent as a sharp, cylindrical mucro or occasionally percurrent, superficial cells quadrate to short-rectangular, occasionally elongate near apex ventrally, elongate dorsally, *(4–)8–15 rows of cells across costa ventrally at midleaf*, costal transverse section semicircular to ovate, *stereid bands strong, the dorsal usually larger than the ventral*, epidermis ventrally present, dorsally present, weak or absent, guide cells (4–)6–9 in 1 (occasionally partially bistratose) layer, *hydroid strand absent*; upper laminal cells subquadrate to short-rectangular, 8–10 µm in width, 1(–2):1, walls evenly thickened, superficially weakly convex on both sides, occasionally strongly bulging ventrally and weakly convex dorsally; papillae bifid, 3–5 crowded over each lumen or fused into one capitulate papilla, or cells occasionally nearly smooth; *differentiated basal cells filling leaf base, often reaching up margins in a vee*, rectangular to bulging-rectangular, 13–25 µm in width, 3–5:1, walls thin, becoming thicker above, occasionally porose medially. Dioicous. Perichaetia terminal (these occasionally appear lateral because of subperichaetial elongate, overtopping branches), inner leaves little different from cauline leaves or high-sheathing. Perigonia terminal, inner leaves little different from the cauline leaves. Seta 0.7–2.0 cm in length, 1 per perichaetium, brown to reddish brown, twisted clockwise below, counterclockwise above; theca (0.5–)2.0–2.5 mm in length, reddish brown, cylindrical, exothecial cells rectangular, 20–30 µm in width, 2–4:1, walls thin, stomates present at base of theca, phaneropore, annulus usually of 2–4 rows of vesiculose cells, deciduous in pieces; *peristome teeth 16, rather short, occasionally absent, yellow to whitish*, triangular to linear, cleft and perforated to near base, spiculose, 125–300 µm in length, with many articulations, straight, basal membrane when present 35–50 µm in height, spiculose. Operculum rostrate, (0.4–)1.2 mm in length, cells straight. Calyptra cucullate, smooth, ca. 2 mm in length. Spores 11–15 µm in diameter, yellow-brown, papillose. Laminal KOH color reaction golden yellow-orange, occasionally deep yellow. Reported chromosome number n = 13, 14.

A genus found in most tropical and warm-temperate areas, occurring on soil, rock (mostly calcareous) and bark.

This genus is like both *Tortella* and *Trichostomum* subg. *Oxystegus* in the plane or broadly incurved leaf margins (Pl. 7, f. 20) and well differentiated, often inflated basal cells that usually extend up the margins in a more or less distinct vee (Pl. 7, f. 9, 18). The broadly sheathing leaf base with distinct "shoulders" has been used in the past as a major character, but is actually found to some extent in other genera, and is poorly or not at all developed in some species of *Pseudosymblepharis*. An additional and possibly better character is the relative size of the ventral stereid band (Pl. 7, f. 6, 20). In *Pseudosymblepharis* the ventral stereid band is almost always distinctly larger than the dorsal, and is often strongly bulging ventrally. *Tortella* may be additionally distinguished by the usually better developed peristome, while *Trichostomum* subg. *Oxystegus* differs, gametophytically, by the vee of basal cells poorly developed or absent. There may well prove to be no acceptable distinction between *Pseudosymblepharis* and *Trichostomum* subg. *Oxystegus*, excepting the color of the peristomes, yellow to white in the former and red in the latter (also see discussion of Norris and Koponen 1989). In leaf shape, *P. indica* might be recognized in *Trichostomum* subg. *Oxystegus* (*cf.* Hilpert 1933) but the costa is quite thick, with the ventral stereid band larger than the dorsal (at least in the larger, mature leaves). Problematically, *T.* (*Oxystegus*) *hybernicus* has the sheathing leaf base of *Pseudosymblepharis* and a ventral stereid band equal to or greater than the dorsal, but the rectangular upper laminal cells and thin-walled cells of the central cylinder associate it clearly with *Trichostomum tenuirostris*. The genus *Pseudosymblepharis* has considerable resemblance to *Symblepharis* Mont. (Dicranaceae), but gametophytes of the latter genus may be distinguished by their smooth leaf cells. The generitype of *Trichostomum*, *T. brachydontium*, has the ventral stereid band much larger than the dorsal in most very robust specimens, much as is the case in *Pseudosymblepharis*. *Weissia jamaicensis* has a leaf shape and costal structure similar to that of *Pseudosymblepharis* but the sharply and narrowly incurved upper laminal margins are characteristic only of *Weissia s. lat.* and *Chionoloma*. This entire relationship needs careful revision.

Tortella has 32 filamentous teeth, longer than 500 µm, these strongly twisted. *Trichostomum* (including subgenus *Oxystegus*) has 16 teeth cleft variously to near that base, shorter than 400 µm, and straight to very weakly twisted. Future evaluation may find that *Tortella* and *Trichostomum* (*s. lat.*) may better be divided by characters other than those of the sporophyte and basal cell area shape of the leaves. A similar situation has occurred in the *Barbula-Didymodon* group, which was in the past distinguished along similar lines of peristome morphology, but which is now (Saito 1975a) separated into three genera (*Bryoerythrophyllum* being segregated from *Didymodon*) by a combination of several gametophytic characters (see treatments below).

There may be quite a difference in stature between various collections of the same species, with certain variation correlated with plant size. In *P. schimperiana*, for instance, small plants may be only weakly sheathing at the leaf base, the hyaline basal cells of such collections may reach up the margins to half the length of the leaf, and the leaves are more commonly fragile. Examined collections of the widespread species Old World species *P. angustata* (holotype, NY!) are nearly identical to the New World *P. schimperiana*, but differ in lacking a stem central strand (weak in *P. schimperiana*). A revision of this genus, however, may yet demonstrate synonymy and a pan-tropical distribution of *P. angustata*.

Number of accepted species: 11.

Additional literature: Crum (1952a).

Species examined: *P. angustifolia* (MICH), *P. angustata* (BUF, NY), *P. cavernarum* (H), *P. circinnatula* (H), *P. duriuscula* (BUF, NY), *P. indica* (BM, NY), *P. khasiana* (NY), *P. mauiensis* (DUKE, NY), *P. perlongifolia* (NY), *P. schimperiana* (BUF, CANM, DUKE, NY, SPA, TENN), *P. subduriuscula* (BUF, NY), *P. syrrhopodontoides* (BM), *P. verrucosa* (H).

New heterotypic synonymy: *Tortella grossiretis* Bartr. = *Pseudosymblelpharis schimperiana* (Par.) Crum. *Pseudosymblepharis bartramii* Thér. ex Bartr. = *Pseudosymblepharis schimperiana* (Par.) Crum.

New combinations:

Pseudosymblepharis cavernarum (Broth.) Zand., *comb. nov.* (*Trichostomum cavernarum* Broth., Denkschr. Ak. Wiss. Wien Math. Nat. Kl. 83: 283, 1926).

Pseudosymblepharis circinnatula (Broth. in Völtzk) Zand., *comb. nov.* (*Trichostomum circinnatulum* Broth. in Völtzk, Reise Ostafr. 3: 54, 1908).

Pseudosymblepharis khasiana (Mitt.) Zand., *comb. nov.* (*Tortula khasiana* Mitt., J. Linn. Soc. Bot. Suppl. 1: 29, 1859; *Barbula khasiana* (Mitt.) Jaeg.; *Trichostomum khasianum* (Mitt.) Broth.; *Oxystegus khasianus* (Mitt.) Gangulee).

Pseudosymblepharis perlongifolia (Fröhl.) Zand., *comb. nov.* (*Trichostomum perlongifolium* Fröhl., Rev. Bryol. Lichénol. 31: 92, 1962).

Pseudosymblepharis syrrhopodontoides (Dix.) Zand., *comb. nov.* (*Tortella syrrhopodontoides* Dix., J. Bot. 76: 228, 1938.).

Pseudosymblepharis verrucosa (Broth. & Par.) Zand., *comb. nov.* (*Trichostomum verrucosum* Broth. & Par., Öfv. Finsk. Vet. Soc. Förh. 51A(17): 12, 1909; *Oxystegus verrucosus* (Broth. & Par.) Hilp.).

7. PACHYNEUROPSIS **Plate 8.**

Pachyneuropsis H. Mill., Taxon 19: 822, 1970, *nom. nov.* for *Pachyneurum* Bartr. Type: *Pachyneuropsis bartlettii* (Bartr.) H. Mill.

Pachyneurum Bartr., Philippine J. Sci. 68: 100, 1939, *hom. illeg. non* Amann, 1912.

From *Pachyneurum*, a genus + νευρον, nerve, sinew, tendon + ὄψσις, -εως, appearance; resembling the genus *Pachyneurum*.

Plants in cushions, *light green above*, tan below. *Stems* branching often, ca. 1.5 cm in length, transverse section rounded-pentagonal, *central strand present, sclerodermis present but weakly developed, hyalodermis present*; *axillary hairs 6–15 cells in length, becoming swollen and firm-walled throughout*; rhizoids sparse. *Leaves* incurved, somewhat twisted when dry, spreading when moist, long-linear, ca. 5 mm in

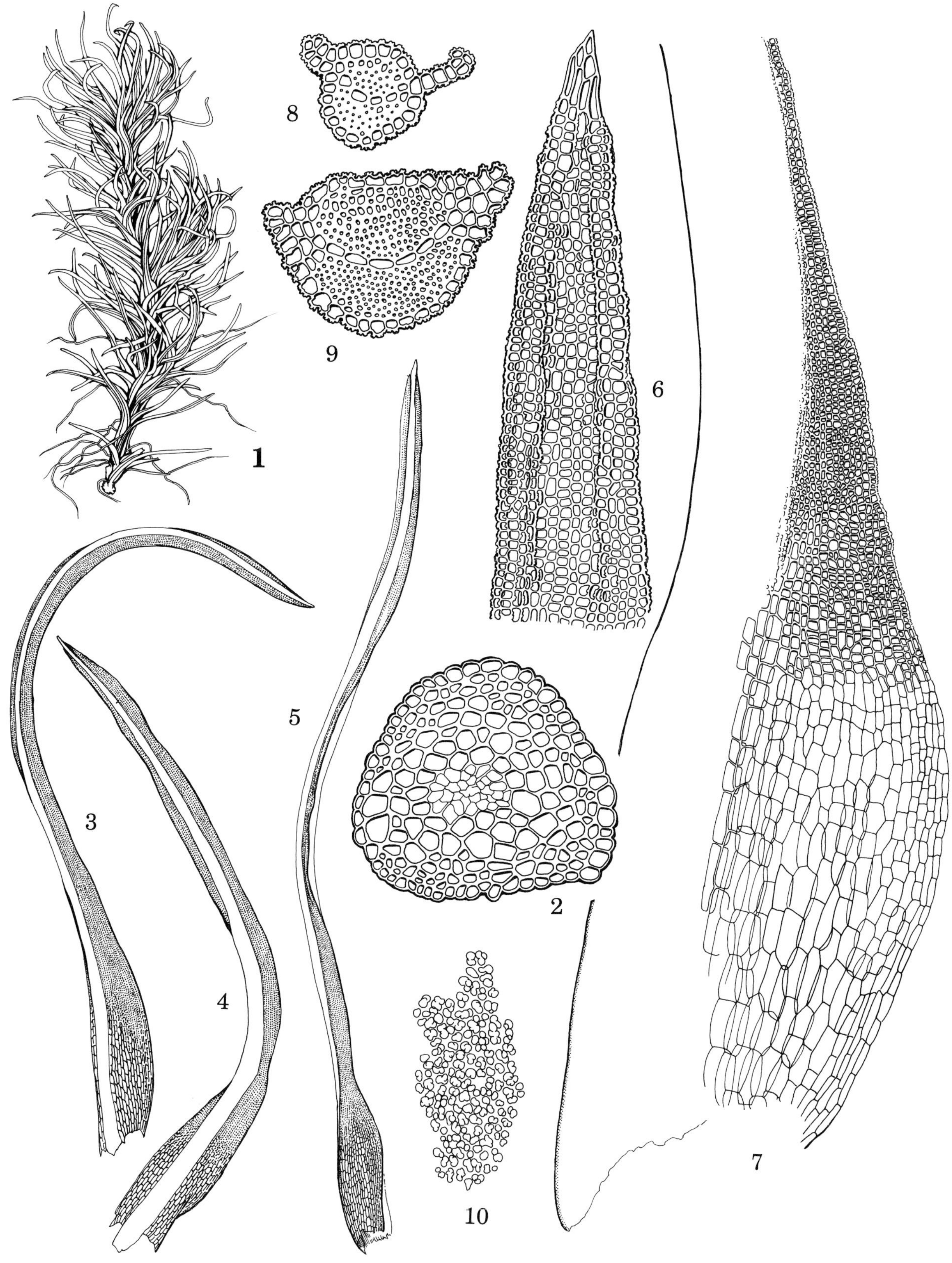

Plate 8. ***Pachyneuropsis*.1–10.** ***P. bartlettii*.** 1. Habit. 2. Transverse section of stem. 3–5. Three leaves. 6. Leaf apex. 7. Basal cells. 8–9. Two transverse sections near midleaf. 10. Upper laminal papillae.

length, *lamina usually very narrow*, the blade on each side seldom as wide as the costa and *occasionally absent above midleaf*, often narrower on one side than the other, upper lamina broadly channeled but costa bulging ventrally, *margins plane, entire, bi(tri)stratose* from just above leaf base to near apex; apex narrowly acute; *base abruptly ovate, with distinct shoulders*; *costa very stout*, short-excurrent as a sharp, smooth mucro or smooth hyaline apiculus, superficial cells quadrate and papillose on both sides, *ca. 10–13 rows of cells across costa ventrally at midleaf*, costal transverse section circular, *ventral stereid band stronger than dorsal*, dorsal band reniform, epidermis present on both sides, *guide cells ca. 8 in 1 layer*, hydroid strand absent; upper laminal cells rounded-quadrate, 8–10 µm in width, 1(–2):1, walls evenly thickened, lumens somewhat angular, superficially bulging on both sides; papillae bifid, 3–4 per lumen, to multiplex, crowded, solid; *basal cells* differentiated across leaf, reaching somewhat higher along margins, *inflated rectangular, bulging*, to 25 µm, much wider than upper cells, 2–4:1, walls thin-walled, hyaline. Gametoecia and sporophyte unknown. Laminal KOH color reaction yellow except leaf insertion red.

Endemic to the Philippines. It is known from a single collection: Luzon, Rizal Province, Mantalban, Bartlett 14392, holotype, FH, without indication of substrate.

Pachyneuropsis is morphologically close to *Pseudosymblepharis* in the linear leaf shape with abruptly broadened base (Pl. 8: 3–5) and a tendency to extreme length, the thick costa with ventral stereid band larger than the dorsal (Pl. 8, f. 8–9), among other characters. It differs in the very narrow or absent upper lamina with bi- to tristratose margins, and the costa even thicker than is usual in the latter genus. Bartram (1939) suggested a passing similarity to *Barbula pachyloma* Broth., which he had not seen, but this latter species (transferred to *Cinclidotus* as *C. involutus* Hilp., *nom. nov.*, by Hilpert 1933) is actually a good species of *Barbula*.

Number of accepted species: 1.

Species examined: *P. bartlettii* (FH).

Subfamily TRICHOSTOMOIDEAE

Trichostomoideae (BSG) Limpr. in Broth., Nat. Pfl. 1(3): 381, 1902 "Trichostomeae" used as a subfamily name.

Trichostomaceae Schimp., Syn. 141, 1860 "Trichostomeae." Type: *Trichostomum* Bruch, *nom. cons.*

Trichostomeae BSG in Schimp., Syn. 1: 168, 1872, apparently used as a family name, see Saito, J. Hattori Bot. Lab. 39: 416, 1975.

Trichostomataceae Crosby & Magill, Dict. Mosses 42, 1977, *orth. err pro* Trichostomaceae Schimp., 1860.

Trichostomeae Limpr., Laubm. Deutsch. 1: 519, 1888, rank not given.

Eucladioideae Chen, Hedwigia 80: 40, 1941.

Trichostomeae Dix., Stud. Handb. Brit. Moss. 205, 1924.

Tortelleae Chen, Hedwigia 80: 142, 1941.

Eucladieae Chen, Hedwigia 80: 55, 1941.

This subclade is distinguished from lower branches on the tree by the advanced characters of leaves shorter, 1.5 to 3.0 mm in length; rows of cells across the ventral surface of the costa at midleaf fewer, 4–6; costal guide cells fewer, 2–6; and theca shorter, less than 1.5 mm in length. The traits at the immediate ancestral node are axillary hairs completely hyaline and dorsal costal epidermis absent. The eight included genera may be characterized generally by at least combinations of the most of the following characters: stem sclerodermis poorly differentiated from the central cylinder, hyalodermis commonly present; leaves lanceolate; margins plane or weakly recurved below and plane or erect, occasionally incurved above; upper laminal papillae crowded; basal cells occasionally reaching much higher along the margins to form a differentiated vee; upper lamina KOH reaction yellow (seldom yellowish orange or reddish brown); costa lacking a differentiated dorsal epidermis; clavate propagula absent. This subfamily is widely distributed; the basal taxon, *Calymperastrum*, is austral.

Saito (1975a) cited the number of amphithecial cell layers at the mouth of the capsule as one of the major characters distinguishing the Trichostomoideae (including *Timmiella*, *Tortella* and *Weissia* according to Saito) from the Pottioideae (including Barbuleae, Eucladieae, Leptodontieae and Pottieae all *sensu* Saito), the former with five layers and the latter with only four. He illustrated these conditions for eleven species in eight (*Barbula*, *Bryoerythrophyllum*, *Desmatodon*, *Didymodon*, *Tortella*, *Tortula s. str.*, *Trichostomum* and *Weissia*) different peristomate genera of the Japanese moss flora. This anatomical difference certainly appears promising but should be reevaluated in the light of cladistic study of all characters. Although Catcheside (1980) has echoed, with reservations, the use of Saito's amphithecial layer distinction for the traditional Trichostomoideae, a thorough study of taxa worldwide is needed using techniques of the plant anatomist. Sectioning capsules lengthwise freehand is a considerable challenge and wasteful of capsules if not immediately successful. It has been impossible to use this character in practice. Saito and Hirohama (1974a) pointed out that "multistalked verrucae" as spore ornamentation are unique to but not always present (as seen with the SEM) in the Trichostomoideae (*sensu* Saito) for the 19 species in 14 genera of Japanese Pottiaceae that they studied.

8. CALYMPERASTRUM Plate 9.

Calymperastrum Stone, J. Bryol. 14: 315–318, 1986. Type: *Calymperastrum latifolium* (Hampe) Stone.

From *Calymperes*, a genus + *-astrum*, an ending implying deception; mimicking the genus *Calymperes*.

Plants in a low turf, *yellowish green above, yellowish brown below*. Stems not branching, ca. 6 mm in length, transverse section rounded-pentagonal to triangular, central strand distinct, generally with an *satellite strand just below the sclerodermis*, sclerodermis of 1–2 layers of thick-walled cells, hyalodermis weakly developed in patches; axillary hairs ca. 6 cells in length, all hyaline; *papillose, reddish brown tomentum on lower part of stem. Leaves* spreading-incurved and conduplicate from an appressed base when dry, spreading when moist, *long-spathulate*, 2.0–3.1 mm in length, upper lamina broadly concave, *margins plane to weakly incurved above, entire, bistratose in 1(–3) rows from shoulder to 3/4 length of leaf*, cell walls of border slightly thickened; apex rounded-acute and bluntly apiculate; *base strongly differentiated* in shape in lower 1/3 of leaf, *cuneate, with distinct shoulders*, basal cells narrower along margins; *costa* strong and tapering above, *percurrent*, superficial cells quadrate to short-rectangular and papillose ventrally, elongate and weakly papillose dorsally, 4–6 rows of cells across costa ventrally at midleaf, *costal transverse section nearly round*, stereid bands two, usually strong ventrally (the cell walls often thicker than those of dorsal band, especially in sections made through leaf base) and generally larger than the dorsal, distinct to nearly absent dorsally, epidermis well differentiated ventrally but absent dorsally, guide cells small and 4 in 1 layer, *hydroid strand strong*, in a central position just dorsal to the guide cells and often also ventrally; *upper laminal cells subquadrate to 5–*

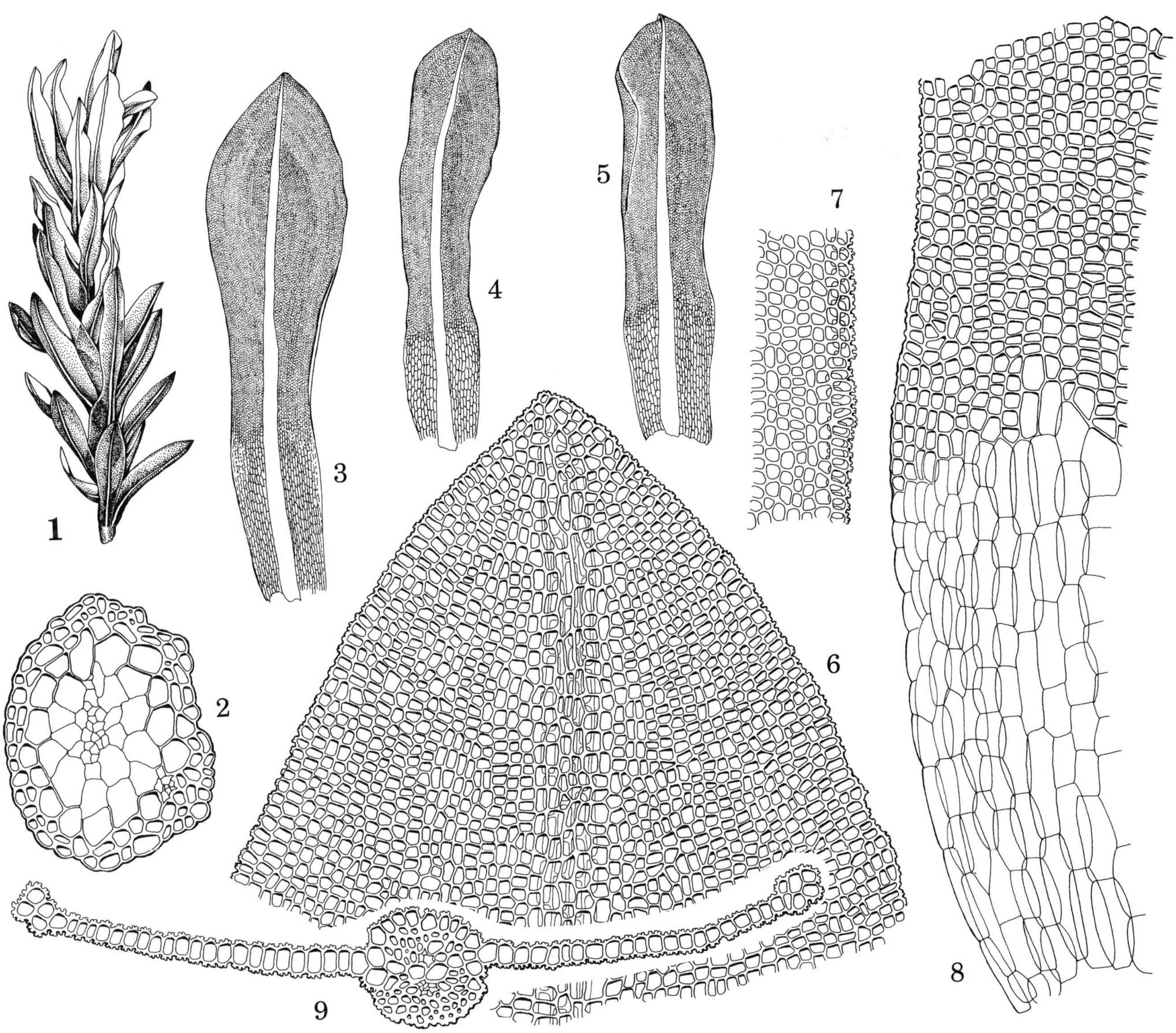

Plate 9. *Calymperastrum*. 1–9. *C. latifolium*. 1. Habit. 2. Transverse section of stem showing a satellite strand. 3–5. Three leaves. 6. Leaf apex. 7. Marginal laminal cells showing bistratose border. 8. Basal cells. 9. Transverse section at midleaf.

or 6–sided, often slightly longer than wide, 9–13 μm in width, 1(–2):1, *walls thin to* weakly evenly thickened, *superficially strongly bulging ventrally, less strongly bulging dorsally; papillae small, crowded, bifid, solid, ca. 8 per lumen*; basal cells strongly differentiated from upper cells, rising higher and larger medially, weakly inflated, rectangular, ca. 20 μm in width, 2–4:1, walls thin, hyaline, lacking internal or superficial pores. Sexual structures and sporophyte unknown. *Laminal KOH color reaction yellow.*

A monotypic genus endemic to western Australia, growing on cycads.

This genus can be excluded from the Calymperaceae mainly by the hydroid strand (Pl. 9, f. 9) in the leaf. It appears to have many of the characteristics of *Bryoerythrophyllum* and *Mironia* (especially those of the areolation), but it differs markedly in being yellow in KOH. From *Dialytrichia*, which also has a bistratose leaf margin (Pl. 9, f. 7), it differs in its broadly concave ventral laminal surface (not deeply grooved along the costa), its plane leaf margins (not recurved below midleaf), its round (not semicircular) costal section and semicircular (not crescent-shaped) section of dorsal stereid band. *Calymperastrum* is also somewhat like *Trichostomum* especially in the plane to incurved margins but a hydroid strand is rare in that genus, and the only *Trichostomum* species seen with a bistratose leaf margin, *T. marginatum*, only doubtfully belongs to *Trichostomum*.

Calymperastrum is also like certain species of *Leptodontium* (e.g. *L. stoloniferum*) in the leaf shape and strongly differentiated basal cells, but the presence of a central strand, a hydroid strand and a differentiated costal epidermis are not characters of *Leptodontium*. Transverse sections of the stem show satellite strands just interior to the sclerodermis, usually one per section and correlated with a broad stem ridge decurrent from each costa. The satellite strand appears to be a leaf hydroid strand (these being unusually strong in *Calymperastrum*) extending into the central cortex but not reaching the stem central strand (Pl. 9, f. 2). *Aloinella*, an unrelated taxon, has a similar stem morphology. A more thorough description of the genus is given by Stone (1986), who saw additional material.

This monotypic genus appears to be, like *Hypodontium*, transitional in morphology between the Pottiaceae and the Calymperaceae, approaching the latter through the following combination of characters: arboreal habitat, leaves with rounded apices and plane, thickened margins, costa percurrent, ending in an apiculus, the ventral stereid band of cells with walls more strongly thickened that those of the dorsal (*cf. Syrrhopodon richardsii* Dix.), upper laminal cells more strongly bulging ventrally than dorsally, basal cells strongly differentiated from the upper cells (but not as sharply demarcated as is usual in *Calymperes* and *Syrrhopodon*), and laminal KOH reaction yellow. The ventral stereid cells being more strongly thickened than those of the dorsal band is a character not found elsewhere in the Pottiaceae, but is apparently occasional in the Calymperaceae. The only unique character to my knowledge distinguishing *Calymperastrum* from genera of the Calymperaceae is the presence of a leaf hydroid strand. Other characters cited in the original description (Stone 1986) as indicative of Pottiaceae are also found in various genera of both the Calymperaceae and the Pottiaceae, including the stem central strand, papillae on the upper cells of the sheathing leaf base, bifid papillae on both surfaces of laminal cells, and absence of foliar gemmae. The significance of the presence of seriate papillae on the dorsal costal surface cannot be evaluated here, and information on the sporophyte is lacking.

Number of accepted species: 1.

Species examined: *C. latifolium* (MELU).

9. EUCLADIUM — Plate 10.

Eucladium B.&S. in BSG, Bryol. Eur. 1: 93, 1846 (fasc. 33–36 Mon. 1). Type: *Eucladium verticillatum* (Brid.) B&S.

Mollia subg. *Eucladium* (B.&S.) Lindb., Musci Scand. 21, 1879.

Weissia subg. *Eucladium* (B.&S.) Kindb., Eur. N. Amer. Bryin. 2: 283, 1897.

Weissia sect. *Saxicolae* Nees & Hornsch., Bryol. Germn. 2(2):P 26, 97, 1831, *p.p.*

Mollia sect. *Eucladium* (B.&S.) Braithw., Brit. Moss Fl. 1: 230, 439, 1885.

Weissia sect. *Eucladium* (B.&S.) Dix., Stud. Handb. Brit. Moss. 210, 1896.

From εὖ, well, very, true, good + κλάδος, branch, twig, stem, repeatedly branched or forked stem + *-ium*, characteristic of.

Plants growing in turfs or cushions, *bright to dark green above*, pale green to yellowish brown below. Stems branching irregularly, 0.5–2.0 cm in length, transverse section elliptical, *central strand absent, sclerodermis usually absent, hyalodermis present, inflated; axillary hairs of 5–10 clear cells*; occasionally weakly radiculose. *Leaves* appessed to erect-spreading from base, incurved above when dry, weakly spreading to spreading-recurved when moist, *oblong- to linear-lanceolate*, 1.5–2.5 mm in length, *upper lamina broadly channeled, margins plane, denticulate on lower margins*, rarely entire; apex broadly to narrowly acute or subulate; base scarcely differentiated in shape to ovate; *costa strong, usually excurrent as a stout mucro*, decurrent at base, superficial cells quadrate to elongate ventrally, elongate dorsally, ca. 6 rows of cells across costa ventrally at midleaf, costal transverse section semicircular to elliptical, 2 stereid bands present, epidermis present ventrally, usually present dorsally, guide cells 4–7 in 1(–2) layers, hydroid strand absent; *upper laminal cells* subquadrate (occasionally rectangular medially), smaller at leaf margins, 8–10 μm in width, 1:1(–2), walls moderately thick-walled, often irregularly thickened, somewhat bulging superficially; *papillae low, indistinct, simple* to occasionally multifid, scattered to centered, 2–5 per cell; *basal cells usually strongly differentiated* across leaf or medially, *bulging-rectangular to rhomboidal*, 12–15 μm in width, 2–5:1, walls hyaline, thin. Dioicous. Perichaetia terminal, inner leaves ovate-lanceolate, to 2.5 mm in length, sheathing, lower cells rhomboidal in lower half. *Perigonia* terminal, *inner leaves long-lanceolate, similar to cauline*. Seta ca. 5–6 mm in length, 1 per perichaetium, yellow, twisted little or not at all; theca ca. 1 mm in length, yellow-brown, ovoid to cylindrical, *exothecial cells short-rectangular, thin-walled, bulging*, stomates phaneropore, present at base of theca, annulus of ca. 2 rows of weakly vesiculose cells; *peristome teeth 16, lanceolate, entire to variously cleft*, yellow, papillose, rudimentary or to 300 μm, with up to several articulations, straight, basal membrane low, papillose. Operculum conic-rostrate, ca. 0.5–0.8 mm in length, cells in straight rows. Calyptra cucullate, smooth, ca. 2.5 mm in length. Spores 9–14 μm in diameter, pale, essentially smooth. Laminal KOH color reaction yellow. Reported chromosome number n = 13.

Found on calcareous rock in wet places, especially in spray of waterfalls, across Europe and Asia, northern and southern Africa, and North America (including Mexico).

Brotherus (1924–25) accepted the combination *Eucladium irroratum* (Mitt. in Hook.) Jaeg., as did Dixon (1923). The rigid, erect leaves and the wet, calcareous habitat of this New Zealand species certainly suggest this genus, but technical characters indicate that that species belongs to a monotypic genus, *Tetracoscinodon*. *Eucladium* may now be considered a monotypic genus.

The most distinctive characters of *Eucladium* are the bright green leaves, lack of a stem central strand or sclerodermis and presence of a somewhat inflated hyalodermis (Pl. 10, f. 2), the broad costa, plane leaf margins, which are denticulate below (Pl. 10, f. 6) except in certain Bermuda populations, laminal cells generally larger medially than at the margins above midleaf (Pl. 10, f. 5) and bulging-hyaline at the leaf base (Pl. 10, f. 6), laminal papillae simple (Pl. 10, f. 8), perigonial leaves little different from the cauline, and exothecial cells thin-walled and bulging. *Eucladium* has several morphological features that are similar to those of *Leptodontium* and *Hymenostylium*: lack of a stem central strand, lanceolate leaves, trigonous upper laminal cells, inflated basal laminal cells, peristome similar to that of *L. viticulosoides*. The denticulate lower leaf margins are matched in *Molendoa hornschuchiana*. The narrowly lanceolate leaf shape, broad costa and large laminal cells ornamented with rather large, simple papillae are reminiscent of *Tuerckheimia*. The generic relationships are, on analysis, however, with *Trichostomum*.

Additional literature: Dalby (1966), Dixon (1912), Dunk and Dunk (1973), Głowacki (1909), Nagano (1959), Osada (1958), Saito (1972a).

Number of accepted species: 1.

Species examined: *E. verticillatum*.

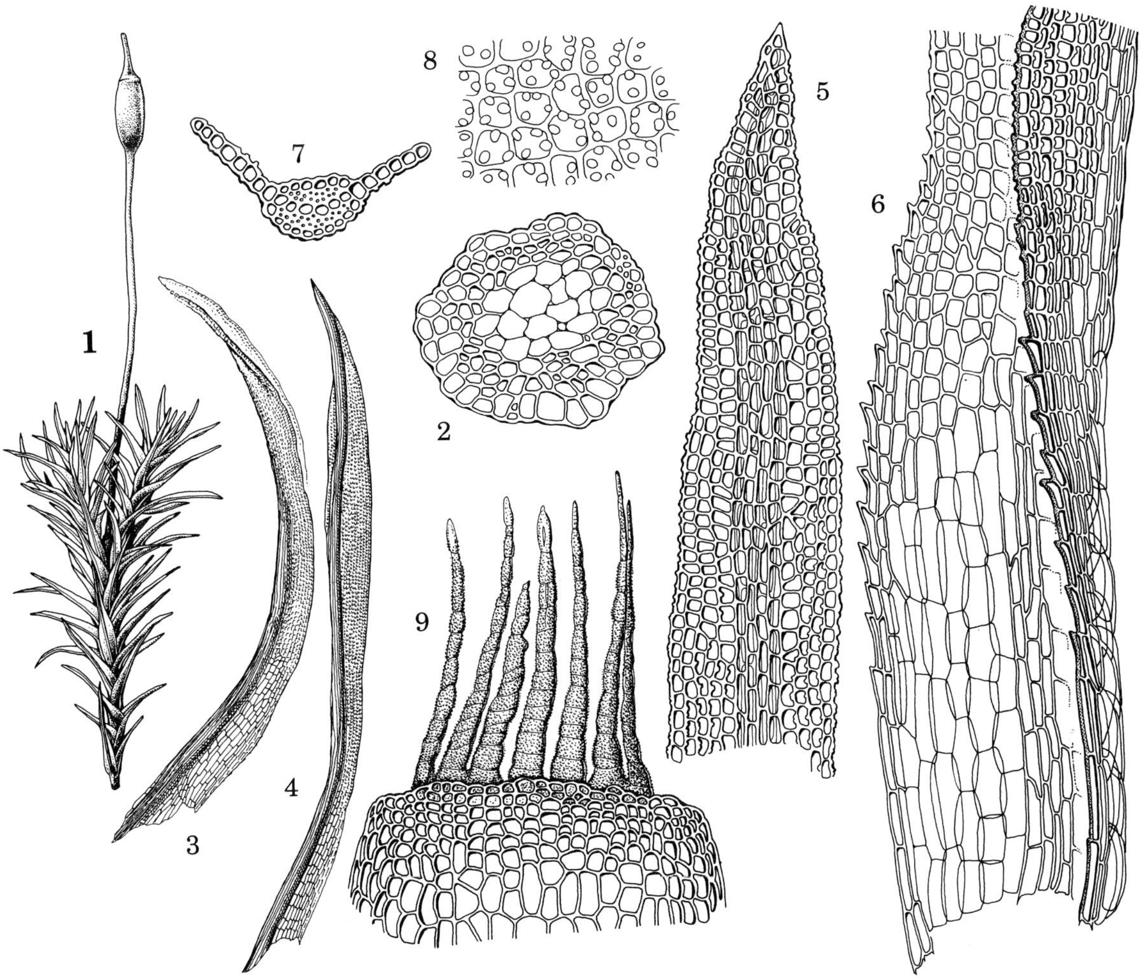

Plate 10. ***Eucladium*. 1–9. *E. verticillatum*.** 1. Habit. 2. Transverse section of stem. 3–4. Two leaves. 5. Leaf apex. 6. Basal cells. 7. Transverse section at midleaf. 8. Papillae. 9. Peristome.

10. TRICHOSTOMUM **Plates 11–13.**

Trichostomum Bruch, Flora 12: 396, 1829, *nom. cons. non* Hedw., 1801. Lectotype: *Trichostomum brachydontium* Bruch.

Subg. *Trichostomum* (Hedw.) Turn., Musc. Hib. Spic. 35, 1804 (as autonym).

Trichostomum subg. *Trichostomum* Lor., Bryol. Notizb. 20: 1865, *nom. illeg.*

Didymodon subg. *Trichostomum* Kindb., Eur. N. Amer. Bryin. 2: 272, 1897. Type: *Trichostomum* Hedw., *nom. rej., p.p.*

Bryum sect. *Trichostomum* (Hedw.) Relh., Fl. Cantabr. ed 2: 422, 1802, *nom. illeg.*

Trichostomum sect. *Lancifolia* B.&S. in BSG, Bryol. Eur. 2: 119 [fasc. 18–20 (Trichost.): 5], 1843.

Trichostomum sect. *Pycnophyllum* C. Müll., Syn. 1: 567, 1849.

Tortula sect. *Trichostomum* Mitt., J. Linn. Soc. Bot. 12: 142, 146, 1869, *nom. illeg.*

Subg. *Crispuliformes* (Kindb.) Zander, *comb. et stat. nov.* see below.

Didymodon sect. *Crispuliformes* Kindb., Eur. N. Amer. Bryin. 2: 272, 1897. Type: *Didymodon crispulus* (Bruch) Wils.

Subg. *Laminanchium* Zander, *subg. nov.* see below. Type: *Trichostomum tortelloides* (Broth. & Dix.) Zand.

Subg. *Oxystegus* Limpr., Laubm. Deutschl. 1: 569, 1888. Type: *Trichostomum cylindricum* (Bruch ex Brid.) C. Müll. *fide* Saito, J. Hattori Bot. Lab. 39: 436, 1975, *hom. illeg.*

Oxystegus (Limpr.) Hilp., Beih. Bot. Centralbl. 50: 666, 1933.

Stephanodictyon Dix., J. Linn. Soc. Bot. 50: 86, 1935. Type: *Stephanodictyon borneense* Dix.

Paraleptodontium Long, J. Bryol. 12: 181, 1982. Type: *Paraleptodontium recurvifolium* (Tayl.) Long.

Didymodon subg. *Oxystegus* (Limpr.) Roth, Eur. Laubm. 1: 304, 1904.

Sect. *Campylopus* Arnott, Mém. Soc. Linn. Paris 5: 244, 1827.

Sect. *Leptomitrium* Wallr., Fl. Crypt. Germ. 1: 170, 1831.

From θρίχ, τρῖχός, hair + o + στόμα, -ατος, mouth; a peristome of filiform teeth.

Plants turf-forming, yellowish green above, medium brown to light brown below. Stems branching irregularly, to ca. 3 cm in length, transverse section rounded-pentagonal, occasionally rounded-triangular, central strand present or absent, cells of central cylinder thin- or thick-walled, *sclerodermis variously developed but often weak,* of substereid cells or of one or two layers of stereid cells, *hyalodermis present or seldom absent*; axillary hairs ca. 10 cells in length, basal 1–3 cells thicker walled or all hyaline; sparsely radiculose, or occasionally with a thin red tomentum. *Leaves* incurved, occasionally tubulose and often catenulate or reflexed at top of appressed base when dry, spreading to squarrose when moist, *oblong, elliptical or ligulate to long-lanceolate,* 1.5–2.5(–5.0) mm in length, *upper lamina flat or broadly channeled,* leaves seldom keeled, occasionally grooved along costa, *margins usually plane,* seldom broadly incurved to tubulose, *entire to crenulose-notched or occasionally dentate* in the upper 1/2–3/4, occasionally with a narrow, less papillose border above or throughout, rarely 2–3 rows of marginal cells bistratose; apex narrowly to broadly acute or rounded, occasionally fragile and broken, occasionally sharply reflexed or constricted, seldom cucullate; base elliptical to rectangular, occasionally sheathing or not differentiated in shape; *costa usually excurrent as a smooth, sharp mucro,* occasionally ending 2–3 cells below apex, seldom short-awned, superficial cells quadrate or occasionally elongate (especially near apex) ventrally, elongate or seldom short-rectangular dorsally, 2–6(–8) rows of cells across costa ventrally at midleaf, costal transverse section semicircular, reniform or ovate, two stereid bands present, occasionally very weak ventrally, *ventral stereid band smaller or about same size as dorsal stereid band,* epidermis present ventrally, weak or absent dorsally, guide cells 2–4(–6) in 1 layer or seldom 1–3 scattered bistratose pairs, *hydroid strand absent or very seldom present*; upper laminal cells rounded-quadrate, occasionally rectangular or hexagonal, seldom transversely elliptical along margin, 6–12(–18) µm in width, 1(–2):1, seldom bistratose along margins or in small medial patches, *walls usually evenly thickened, superficially usually strongly convex on both sides* or seldom somewhat more bulging ventrally than dorsally; *papillae usually bifid, crowded, 2–6 per lumen, occasionally single, multiplex and completely covering lumen*; *basal cells differentiated across leaf or occasionally only medially or rising weakly along margins,* rectangular, *seldom bulging superficially,* 7–15 µm in width, 3–5:1, walls thin to somewhat thickened, occasionally porose. *Propagula rare,* on rhizoids or ventral surface of costa, of several cells, vermiform to irregular in shape, occasionally branching. *Dioicous or occasionally autoicous. Perichaetia terminal, inner leaves usually little different from cauline leaves,* rarely oblong-sheathing and constricted at the apex, lower basal cells inflated rectangular to rhomboid, hyaline. Perigonia terminal, outer leaves often similar to the cauline, inner leaves elongate-triangular, or occasionally present in leaf axils of archegoniate plants as flattened buds. Seta 0.4–1.5 cm in length, 1 or seldom 2–3 per perichaetium, yellowish brown, occasionally reddish, twisted clockwise, in cleistocarpous species with an abscission layer just below capsule; theca ca. 1–3 mm in length, yellowish brown, occasionally orange, cylindrical or elliptical, occasionally slightly curved, exothecial cells thin-walled, rectangular, stomates phaneropore at base of theca, *stegocarpous or seldom cleistocarpous, annulus of 1–4 rows of vesiculose cells, persistent or seldom revoluble*; *peristome teeth 16, usually rather short, occasionally rudimentary or absent, ligulate to filamentous, entire or occasionally irregularly cleft 2–3-fid or perforate, papillose, striate or spiculose, occasionally smooth,* occasionally rather distant, to 400 µm, with several articulations, *straight or seldom weakly twisted counterclockwise,* basal membrane absent or low but distinct, papillose to spiculose, mouth of capsule occasionally closed by a hymenium. Operculum long-conic to rostrate, ca. 0.4–0.7 mm in length, cells straight, seldom weakly twisted counterclockwise or not differentiated. Calyptra cucullate, smooth, ca. 2.0–2.5 mm in length. Spores 8–20 µm in diameter, yellow to brown, essentially smooth to strongly papillose. Laminal KOH color reaction medium orange to yellowish orange, occasionally yellow. Reported chromosome number n = 12, 13, 13+m.

A large genus found on soil, rock (often calcareous) or organic material on all continents except Antarctica.

Trichostomum has been a large, "wastebasket" genus of species that cannot be easily assigned to other genera. It differs from similar *Barbula* (Merceyoideae) species with plane or erect upper leaf margins (*B.* sect. *Convolutae*) by the cells of the outer layer of the stem usually with large lumens (Pl. 11, f. 14;

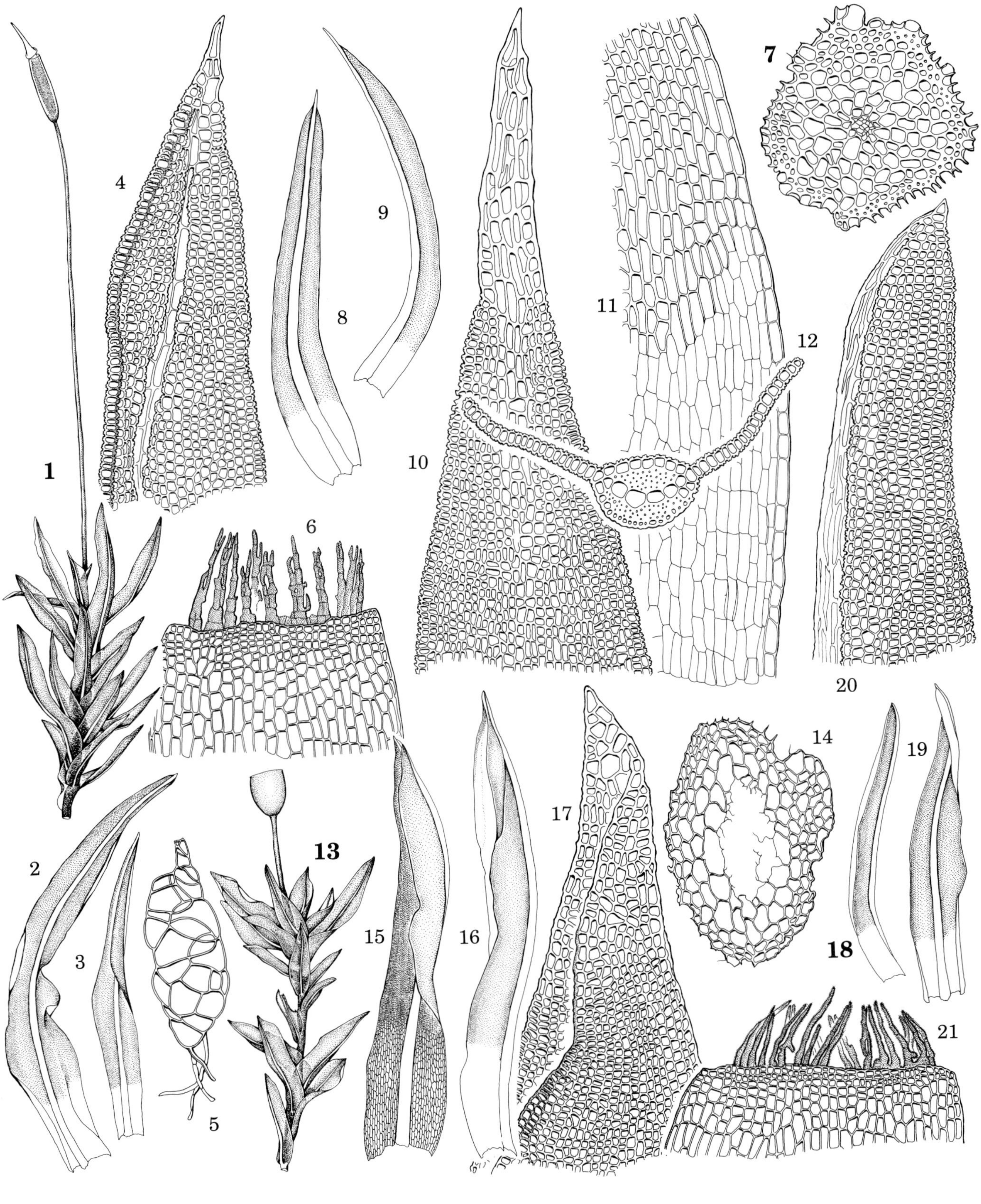

Plate 11. ***Trichostomum.*** **1–6.** ***T. tenuirostre.*** 1. Habit. 2–3. Two leaves. 4. Leaf apex. 5. Propagulum. 6. Peristome. **7–12.** ***T. aequitoriale.*** 7. Transverse section of stem. 8–9. Two leaves. 10. Leaf apex. 11. Basal cells. 12. Transverse section at midleaf. **13–16.** ***T. contractum.*** 13. Habit. 14. Transverse section of stem. 15–16. Two leaves. 17. Leaf apex. **18–21.** ***T. crispulum.*** 18–19. Two leaves. 20. Leaf apex. 21. Peristome.

13, f. 11), lower leaf margins plane, upper leaf margins seldom dentate (but commonly minutely serrulate by projecting cell walls), costal hydroid strand seldom present (Pl. 11, f. 12), dorsal costal epidermal cells sometimes not differentiated (Pl. 12, f. 21), ventral stereid band generally large, almost or about the thickness of the dorsal band, perichaetial leaves generally not differentiated from the cauline, and propagula (Pl. 11, f. 5) very rarely present. The genus differs from *Hyophila* by the mostly narrower leaves that are broadest at the base, laminal cells usually somewhat bulging on both exposed surfaces (rather than just one surface though this is variable in *Hyophila*), propagula usually lacking, and peristome usually present (among other characters, see *Hyophila*); however, the thick-walled exothecial cells of *Hyophila* (forming a series of concatenated semicircles as seen in transverse section of the theca) are matched in some eperistomate species of *Trichostomum*.

I have found no good distinctions between *Trichostomum* and *Oxystegus* at the generic level. There are no characters of the sporophyte that acceptably distinguish the two; for instance, *Trichostomum* (subg. *Trichostomum*) *brachydontium* may have the spiral peristome tooth ornamentation typical of *T.* (subg. *Oxystegus*) *tenuirostre*). Gametophytically, subg. *Oxystegus* differs only weakly from subg. *Trichostomum* in (1) generally plane (occasionally tubulose when dry) upper leaf laminae, (2) a tendency toward broadly sheathing leaf bases with the differentiated marginal cells somewhat running up the margins as in *Tortella*, (3) medial upper basal cells thick-walled and rectangular (forming a group sensibly different from the upper cells and lower basal cells), and (4) the upper laminal cells often distinctly enlarged and rounded rectangular, 1–2:1. The first two of these characters are, however, also present to some degree in *Trichostomum* sect. *Trichostomum*, and the last is neither constant nor easily gauged. Crum and Anderson (1981) likewise proposed that *Trichostomum* include *Oxystegus*. Saito (1975a) recognized *Oxystegus* but did not discuss it as a genus. His key to genera of the Trichostomoideae separated *Oxystegus* and *Pseudosymblepharis* from *Trichostomum* by the presence of a central strand in the last; Eddy (1990) used the same character to separate the two. This distinction does not, however, hold outside of Japan or, apparently, Malesia. Stoneburner (1985) expressed the opinion that differences between *Trichostomum* and *Oxystegus* were probably artificial. Norris and Koponen (1989) distinguished between *Oxystegus* and *Trichostomum* by "a suite of differing, although somewhat overlapping characters": *Trichostomum* having leaves with lengths less than 6:1, blunt papillae grouped over the lumens, and laminal marginal cells thick-walled near the shoulders; *Oxystegus* having leaves longer than 6:1, sharp papillae grouped around the lumens, and cells of the shoulders thin-walled. These differences are recognized here at the subgeneric level. They also indicated that although *Oxystegus* and *Pseudosymblepharis* are morphologically close gametophytically, the red peristome of the former contrasts with the whitish peristome of the latter.

There are trends in subg. *Oxystegus* towards three morphotypes of robust stature: a lingulate leaf as in *Trichostomum recurvifolium* (= *Paraleptodontium* Long 1982b, a synonym), a lanceolate leaf (e.g. *Trichostomum hibernicum*), and a linear leaf (e.g. *Trichostomum borneense*, originally published as the type of *Stephanodictyon*). Such tendencies are duplicated in infraspecific variation in the central species of the *Oxystegus* complex—*T. tenuirostre*. Eddy (1990) made *Stephanodictyon* a synonym of *Pseudosymblepharis* (along with *Chionoloma*, here recognized as a good genus).

Of species common in North Temperate Zone climates, *Trichostomum (*subg. *Oxystegus) tenuirostre* is rather different from *T. crispulum* and *T. brachydontium* (both of subg. *Trichostomum*) in its plane margins, vaguely vee-shaped area of differentiated basal cells, and less well-developed peristome; as a genus, however, *Oxystegus* cannot be maintained at the world level because of considerable variation in expression and combination of characters among the species. For instance, the difference between the peristomes of *T. crispulum* (often 32 filamentous rami like of those of *Tortella*, but sometimes reduced) and *T. brachydontium* (16 variously cleft or perforated lanceolate teeth) is greater than that between the peristomes of *T. brachydontium* (generitype of *Trichostomum*) and *T.* (subg. *Oxystegus*) *tenuirostre*.

Tortella can be separated from *Trichostomum* on the basis of a single relatively constant character, the vee-shaped basal cell region, with the usually long and twisted peristome and generally larger upper laminal cells being associated trends; see Cladogram 15 for analysis of phylogenetically important character state changes. Species of *Trichostomum* may also have basal cells differentiated in a (less distinctive) vee-shape, and those with flattened autoicous buds (e.g. *Trichostomum fragilifolium*, Pl. 12, f. 14) differ gametophytically from certain *Tortella* species (like *Tortella humilis*) with similar buds only in the basal cells being differentiated evenly across the leaf rather than running higher up the margins. Quite probably, *Trichostomum*, *Pseudosymblepharis* and possibly also *Tortella* represent a complex of several transformation series. This needs to be studied. Future revision of the *Trichostomum* complex may turn up a satisfactory generic classification recognizing several separate phyletic series, with perhaps *Trichostomum brachydontium*, *T. crispulum*, *Weissia jamaicensis* or even species of *Tortella* variously at the unreduced end of each.

Species of *Weissia* (and of *Tetrapterum*) with clearly plane or erect-incurved upper laminal margins (including subg. *Hymenostomum* and subg. *Astomum*) examined during this study are here transferred to *Trichostomum s. lat.* Groups of *Trichostomum* species very similar to each other in areolation and certain other gametophytic characters may be construed (until analysis demonstrates otherwise) as separate reduction series toward smaller stature of gametophyte, shorter seta or capsule (the latter occasionally almost spherical), short or absent peristome, and monoicy, and are recognized below as subgenera.

Trichostomum subg. *Trichostomum*

Leaves ovate to long-lanceolate, apex usually rounded, plane; upper margins plane; costa thick and excurrent in a stout, sharp mucro; basal cells moderately differentiated across leaf base.

Occasionally the upper margins in some leaves are rather narrowly incurved, like *Weissia* species. A reduction series (large to small stature with concomitant reduction of morphological details) might include: *Trichostomum brachydontium*, *T. planifolium*, *T. urceolare*, *T. sinaloense* (Pl. 13, f. 16–19), *T. termitarum*, *T. williamsii*, *T. unguiculatum* and *T. austrocrispum*. Also belonging here is *T. incertum*.

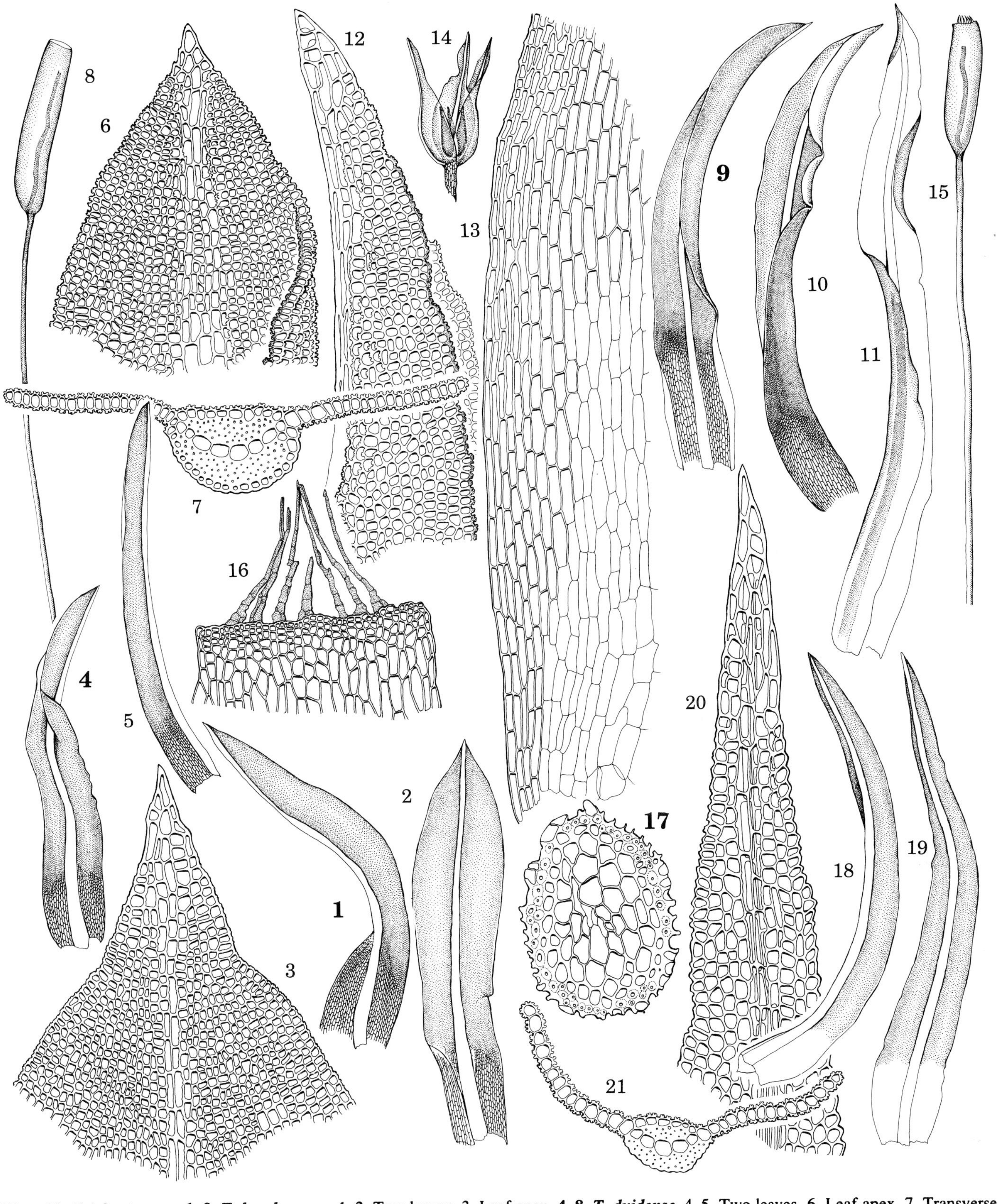

Plate 12. ***Trichostomum.*** **1–3.** ***T. bombayense.*** 1–2. Two leaves. 3. Leaf apex. **4–8.** ***T. duidense.*** 4–5. Two leaves. 6. Leaf apex. 7. Transverse section at midleaf. 8. Capsule. **9–16.** ***T. fragilifolium.*** 9–10. Two leaves. 11. Perichaetial leaf. 12. Leaf apex. 13. Basal cells. 14. Autoicous bud. 15. Capsule. 16. Peristome. **17–21.** ***T. pallidens.*** 17. Transverse section of stem. 18–19. Two leaves. 20. Leaf apex. 21. Transverse section at midleaf.

Trichostomum subg. *Crispuliformes* (Kindb.) Zander, *comb. and stat. nov.* Basionym: *Didymodon* sect. *Crispuliformes* Kindb., Eur. N. Amer. Bryin. 2: 272, 1897. Type: *Trichostomum crispulum* Bruch ex F. A. Müller.

Leaves elliptical to long-lanceolate; apex rounded-acute, often somewhat cucullate; upper margins broadly incurved; costa evenly tapering and generally ending before the apex or percurrent; basal cells moderately differentiated across leaf base.

This taxon includes the series: *Tortella inflexa*, *Trichostomum crispulum* (Pl. 11, f. 18–21), *T. castaneum*, *T. caespitosum*, *T. brittonianum*, *T. subangustifolium*, *T. atrocaule*, *T. perligulatum*, and possibly *Weissia crispa*. Also belonging here are *T. connivens* and *T. pulicare*.

Trichostomum subg. *Laminanchium* Zander, *subg. nov.* Type: *Trichostomum tortelloides* (Broth. & Dix.) Zand.

Folia longiligulata, et plerumque in regione supra basem subvaginantem angustata et regione sub apice, apice acuto, plano, marginibus supernis et interdum infernis planis, costa decrescenti excurrentique, mucronem formanti, cellulis basalibus valde in regione mediana distinctis et in marginali cellulis subangustatis praedita.

Leaves long-ligulate, often pinched just above the somewhat sheathing base and also just below the apex; apex acute, plane; upper margins and sometimes the lower margins plane; costa tapering and excurrent as a mucro; basal cells strongly differentiated medially with narrower cells marginally.

The upper medial superficial cell walls of most species of this subgenus are ventrally bulging and dorsally nearly flat, and are also commonly papillose. Few sporophytes have been seen, but these appear to uniformly lack peristomes and sometimes have rather short setae. Species included here are similar to *Barbula* sect. *Convolutae* by the mainly medially differentiated basal cells, but the basal cells of species of sect. *Convolutae* are rather thick-walled and not strongly differentiated. Species of subg. *Laminachium* are: *Trichostomum bombayense* (Pl. 12, f. 1–3, and see discussion of Townsend 1983), *T. contractum* (Pl. 11, f. 13–16), *T. criotum* (autoicous, with flattened axillary perigonia), and *T. tortelloides* (Pl. 13, f. 20–23). See also discussion of *Calymperastrum*.

Trichostomum subg. *Oxystegus* Limpr. Type: *Trichostomum tenuirostre* (Hook. & Tayl.) Lindb.

Leaves lanceolate to linear-lanceolate, occasionally with a sharply dilated base; apex rounded to narrowly acute, plane; upper margins plane; costa tapering and excurrent as a short mucro; basal cells moderately differentiated, often running up the margins somewhat as in *Tortella*, and bordered distally by a region of thick-walled rectangular cells.

Species of this subgenus could include the following series: *Trichostomum tenuirostre* (Pl. 11, f. 1–6), *T. duidense* (Pl. 12, f. 4–8), *T. melanostomum*, *T. spirale* and *T. abyssinicum*. There are certainly many other names that might go here.

Trichostomum caespitosum, having leaves with small upper laminal cells and broadly incurved margins, and being autoicous, was recognized in *Pottia* recently by both Smith (1978) and Corley et al. (1981), but clearly fits in a reduction series involving *T. crispulum*. Its distinctly differentiated perichaetial leaves are, however, unusual in the genus.

The idea that *Trichostomum* might include taxa with much reduced or absent peristomes has been entertained recently by several bryologists (pers. comm.) with interest in the group. Magill (1981) pointed out that *Trichostomum brachydontium* has rudimentary peristomes or gymnostomous capsules both in South Africa and South America. He also noted that the gametophyte of the gymnostomous and rhexolytically operculate species *Phasconica tisserantii* (= *T. unguiculatum cf.* Crundwell & Nyholm 1974) resembles that of *T. brachydontium*. Brotherus (1924–25) avoided the problem of the relationship of species with *Trichostomum*-like gametophytes and cleistocarpous capsules by referring all these to *Tetrapterum*.

Tuerckheimia is quite like *Trichostomum* in general appearance but the leaf bases of the former genus are not differentiated in shape, the basal cells are merely short-rectangular in a small area near the insertion, the upper laminal cells have a massive, multiplex papilla over the center of each lumen (but not covering the lumen as in *Oxystegus*), and the leaves (often widest near the middle) are pale yellow in KOH solution.

Non-type specimens from India collected by Ramskuhl (BM) and identified as *Desmatodon longirostris* (Griff.) Mitt. (= *Merceyopsis longirostris* (Griff.) Broth. & Dix.) proved to be the superficially similar *Scopelophila cataractae*; the correct name for *D. longirostris* is now *Trichostomum contractum*, *nom. nov.* (see below).

Additional literature: Bartram (1924c), Blumrich (1916), Brown (1897), Crum and Anderson (1958a), Dixon (1911b), Frisvoll (1978), Hammerschmid (1915), Herzog (1907), Lett (1901), Redfearn (1976), Zander (1978d,h, 1982a,c,i, 1985a).

Number of accepted species: 130, plus 2 combinations remaining in *Oxystegus*.

Species examined: *T. abyssinicum* (PC), *T. acutiusculum* (H), *T. aequitoriale* (BM), *T. arcticum* (BUF, C, CANM, NY), *T. atrocaule* (TNS), *T. austrocrispum* (NY), *T. bombayense* (NY—typified by Townsend 1983), *T. borneense* (BM), *T. brittonianum* (NY), *T. castaneum* (NY), , *T. connivens* (NY), *T. criotum* (BUF), *T. cylindrotheca* (NY), *T. distans* (NY), *T. duidense* (BUF, NY), *T. eckelianum* (NY), *T. exulatum* (NY), *T. fallax* (L), *T. fragilifolium* (NY), *T. hyalinoblastum* (NY), *T. imshaugii* (NY), *T. incertum* (NY), *T. laticostatum* (PC), *T. lindigii* (NY), *T. melanostomum* (NY), *T. mitteneanum* (NY), *T. orthodontum* (NY), *T. ovatifolium* (H), *T. pallidens* (BM), *T. perannulatum* (BM), *T. perligulatum* (COLO), *T. planifolium* (MICH, NY), *T. platyphyllum* (NY), *T. portoricense* (BUF, NY), *T. pulicare* (PC), *T. recurvifolium* (BUF), *T. sinaloense* (FH, MICH, NY, TENN), *T. spirale* (BUF, PC), *T. subangustifolium* (PC), *T. subintegrum* (H), *T. tenuirostre*, *T. termitarum* (NY), *T. tortelloides* (FH), *T. unguiculatum* (PRE), *T. urceolare* (BM), *T. wagneri* (H), *T. williamsii* (NY).

New heterotypic synonymy: *Barbula wollei* Aust. (*Didymodon wollei* (Aust.) Aust.) = *Trichostomum tenuirostre* var. *holtii* (Braithw.) Dix. *Hyophila usambarica* Broth. = *Trichostomum brachydontium* Bruch. *Phasconica tisserantii* P. de la Varde = *Trichostomum unguiculatum* (Mitt.) Zand. *Stephanodictyon obscurirete* Dix. = *Trichostomum brachydontium* Bruch. *Trichostomum molariforme* Zand. = *Trichostomum portoricense* Crum & Steere. *Trichostomum schlimii* C. Müll. = *Trichostomum tenuirostre* (Hook. & Tayl.) Lindb. *Weissia tortivelata* Williams = *Trichostomum tenuirostre* var. *gemmiparum*

(Schimp.) Zand.

New names, combinations and statuses (see also discussion section above):

Trichostomum abyssinicum (Thér.) Zand., *comb. nov.* (*Weissia abyssinica* Thér., Bull. Mus. Hist. Nat. Paris 34: 116, 1928).

Trichostomum acutiusculum (Broth.) Zand., *comb. nov.* (*Hyophila acutiuscula* Broth., Bot. Jahrb. 20: 183, 1894).

Trichostomum arboreum (Mitt.) Zand., *comb. nov.* (*Weissia*

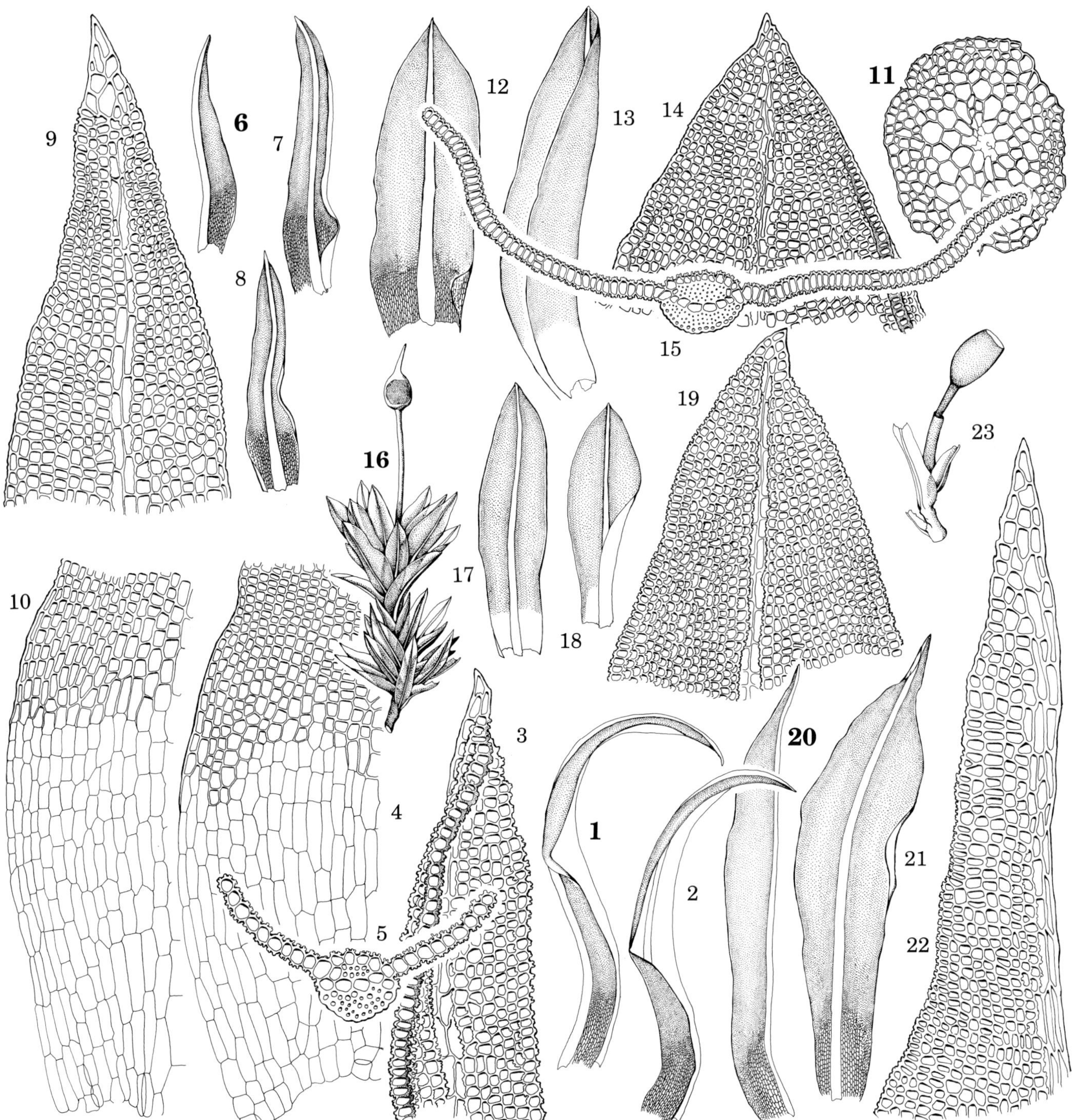

Plate 13. *Trichostomum.* 1–5. *T. hyalinoblastum.* 1–2. Two leaves. 3. Leaf apex. 4. Basal cells. 5. Transverse section at midleaf. **6–10. *T. perannulatum.*** 6–8. Three leaves. 9. Leaf apex. 10. Basal cells. **11–15. *T. platyphyllum.*** 11. Transverse section of stem. 12–13. Two leaves. 14. Leaf apex. 15. Transverse section at midleaf. **16–19. *T. sinaloense.*** 16. Habit. 17–18. Two leaves. 19. Leaf apex. **20–23. *T. tortelloides.*** 20–21. Two leaves. 22. Leaf apex. 23. Sporophyte.

arborea Mitt., J. Linn. Soc. Bot. 12: 138, 1869; *Hyophila arborea* (Mitt.) Jaeg.).

Trichostomum atrocaule (Saito) Zand., *comb. nov.* (*Weissia atrocaulis* Saito, J. Hattori Bot. Lab. 39: 425, 1975).

Trichostomum austrocrispum (Beck.) Zand., *comb. nov.* (*Phascum austrocrispum* Beck., Trans. New Zealand Inst. 26: 274, 1894; *Astomum austrocrispum* (Beck.) Broth.).

Trichostomum austrocrispum var. *longifolium* (R. Br. ter) Zand., *comb. et stat. nov.* (*Phascum longifolium* R. Br. ter, Trans. New Zealand Inst. 26: 308, 1894).

Trichostomum borneense (Dix.) Zand., *comb. nov.* (*Stephanodictyon borneense* Dix., J. Linn. Soc. Bot. 50: 86, 1935).

Trichostomum brittonianum Zand., *nom. nov.* (*Hymenostomum flavescens* E. Britt. in N. Britt. & Millsp., Bahama Fl. 485, 1920; *Weissia flavescens* (Britt. in N. Britt. & Millsp.) Reese, Bryologist 94: 54, 1991; *non Trichostomum flavescens* Dix.).

Trichostomum castaneum (Crum & Steere) Zand., *comb. nov.* (*Hymenostomum castaneum* Crum & Steere, Amer. Midland Nat. 60: 12, 1959).

Trichostomum criotum Zand., *nom. nov.* (*Hyophila perannulata* Ren. & Card., Bull. Soc. R. Bot. Belg. 34(2): 60, 1896; *non Trichostomum perannulatum* Dix. et P. Vard.).

Trichostomum contractum Zand., *nom. nov.* (*Gymnostomum longirostre* Griff., Calcutta J. Nat. Hist. 2: 480, 1842; *Merceyopsis longirostris* (Griff.) Broth. & Dix.; *Merceya longirostris* (Griff.) Wijk & Marg. Type: India, Khasia Hills, "123–a 1," lectotype, BM; *non Trichostomum longirostre* (Web. & Mohr) Hartm.).

Trichostomum deciduaefolium (Saito) Zand., *comb. nov.* (*Weissia deciduaefolia* Saito, J. Hattori Bot. Lab. 39: 429, 1975).

Trichostomum eckelianum Zand., *nom. nov.* (*Trichostomum cirrhatum* Hampe, Icon. Musc. 28, 1844, *hom. illeg.*; *Tortella cirrhata* Broth., Nat. Pfl. 1(3): 397, 1902).

Trichostomum exulatum Zand., *nom. nov.* (*Phascum vernicosum* C. Müll. ex Roth, Aussereur. Laubm. 212, 1911; *Tetrapterum vernicosum* (Roth) Broth. Type: Brazil, Santa Catharina, Ule 8, isotype, NY; *non Trichostomum vernicosum* Ren. & Card.).

Trichostomum finukamactum Zand., *nom. nov.* (*Stephanodictyon angustinerve* Fröhl., Rev. Bryol. Lich. 31: 92, 1962; *non Trichostomum angustinerve* Card.), not seen.

Trichostomum imshaugii (Vitt) Zand., *comb. nov.* (*Barbula imshaugii* Vitt, Bryologist 74: 464, 1971 [1972]); the leaf shape, the variably but often strongly excurrent costa, and a red KOH reaction make this species curiously reminiscent of the genus *Willia*, but it is placed here by the plane leaf margins and the two stereid bands in some leaves.

Trichostomum incertum (Mitt.) Zand., *comb. nov.* (*Weissia incerta* Mitt., Phil. Trans. Roy. Soc. London 168: 389, 1879; *Tortula incerta* (Mitt.) Broth.).

Trichostomum (subg. *Oxystegus*) *ligulaefolium* (Broth. & Par.) Zand., *comb. nov.* (*Hyophila ligulaefolia* Broth. & Par., Rev. Bryol. 31: 44, 1904).

Trichostomum lindigii (Hampe) Zand., *comb. nov.* (*Systegium lindigii* Hampe, Ann. Sc. Nat. Bot. ser. 5, 3: 337, 1865; *Astomum lindigii* (Hampe) Jaeg.).

Trichostomum melanostomum (Mitt.) Zand., *comb. nov.* (*Weissia melanostoma* Mitt., J. Linn. Soc. Bot. 12: 138, 1869; *Hyophila melanostoma* (Mitt.) Jaeg.).

Trichostomum mitteneanum Zand., *nom. nov.* (*Weissia umbrosa* Mitt., J. Linn. Soc. Bot. 12: 133, 1869; *Hymenostomum umbrosum* (Mitt.) Kindb.; *non Trichostomum umbrosum* C. Müll.).

Trichostomum ovatifolium Zand., *nom. nov.* (*Hymenostomum anomalum* Broth. in Herz., Biblioth. Bot. 87: 29, 1916; *non Trichostomum anomalum* (BSG) Schimp.).

Trichostomum pallidens (Dix.) Zand., *comb. nov.* (*Pseudosymblepharis pallidens* Dix., J. Bombay Nat. Hist. Soc. 39: 776, 1937).

Trichostomum perligulatum (Flow. ex Crum) Zand., *comb. nov.* (*Weissia perligulata* Flow. ex Crum, Bryologist 76: 291, 1973); this species is monoicous, and is not a synonym of *T. crispulum* as per Stoneburner and Wyatt (1985).

Trichostomum planifolium (Dix.) Zand., *comb. nov.* (*Weissia planifolia* Dix., Rev. Bryol. Lichénol. 1: 179, 1928).

Trichostomum pulicare (Besch.) Zand., *comb. nov.* (*Hymenostomum pulicare* Besch., Ann. Sc. Nat. Bot. ser. 6, 9: 299, 1880).

Trichostomum recurvifolium (Tayl.) Zand., *comb. nov.* (*Bryum recurvifolium* Tayl., Ann. Mag. Nat. Hist. 11: 208, 1843; *Oxystegus recurvifolius* (Tayl.) Zand., *comb. nov.*; *Paraleptodontium recurvifolium* (Tayl.) Long).

Trichostomum sinaloense (Bartr.) Zand., *comb. nov.* (*Weissia sinaloensis* Bartr., Bryologist 28: 64, 1925; *Hyophila sinaloensis* (Bartr.) Bartr.).

Trichostomum soulae (C. Müll. in Ren. & Card.) Zand., *comb. nov.* (*Ptychomitrium soulae* C. Müll. in Ren. & Card., Bull. Soc. R. Bot. Belg. 33(2): 118, 1895; *Oxystegus soulae* (C. Müll. in Ren. & Card.) Wijk & Marg.).

Trichostomum soulae var. *corticicola* (Ren. & Card.) Zand., *comb. et stat. nov.* (*Barbula corticicola* Ren. & Card., Bull. Soc. R. Bot. Belg. 35(1): 309, 1897; *Oxystegus soulae* var. *corticicola* (Ren. & Card.) Wijk & Marg.).

Trichostomum subangustifolium (Thér.) Zand., *comb. nov.* (*Hyophila subangustifolia* Thér., Smiths. Misc. Coll. 85(4): 14, 1931; *Weissia subangustifolia* (Thér.) Zand.).

Trichostomum tenuirostre var. *gemmiparum* (Schimp.) Zand., *comb. nov.* (*Didymodon cylindricus* var. *gemmiparus* Schimp., Syn. Musc. ed. 2 165, 1876; *Oxystegus tenuirostris* var. *gemmiparus* (Schimp.) Zand.).

Trichostomum termitarum (C. Müll.) Zand., *comb. nov.* (*Weissia termitarum* C. Müll., Hedwigia 39: 267, 1900).

Trichostomum tisserantii (P. de la Varde) Zand., *comb. nov.* (*Phasconica tisserantii* P. de la Varde, Rev. Bryol. Lichénol. 7: 231, 1934).

Trichostomum tortelloides (Broth. & Dix.) Zand., *comb. nov.* (*Calymperes tortelloides* Broth. & Dix., J. Bot. 48: 306, 1910; suggested as being Pottiaceae by Reese, Bryologist 88: 106, 1985)

Trichostomum unguiculatum (Mitt.) Zand., *comb. nov.* (*Systegium unguiculatum* Mitt., J. Linn. Soc. Bot. 22: 304, 1886; *Astomum unguiculatum* (Mitt.) Broth.; *Weissia unguiculata* (Mitt.) Crundw. & Nyh.).

Trichostomum urceolare (Hampe) Zand., *comb. nov.* (*Hyophila urceolaris* Hampe, Vid. Medd. Naturh. For. Kjoebenh. ser. 3,2: 269, 1870; *Hymenostomum urceolare* (Hampe) Hampe).

Trichostomum williamsii Zand., *nom. nov.* (*Astomum chilense* Williams, Bull. Torrey Bot. Club 42: 393, 1915; *non Trichostomum chilense* Mont.).

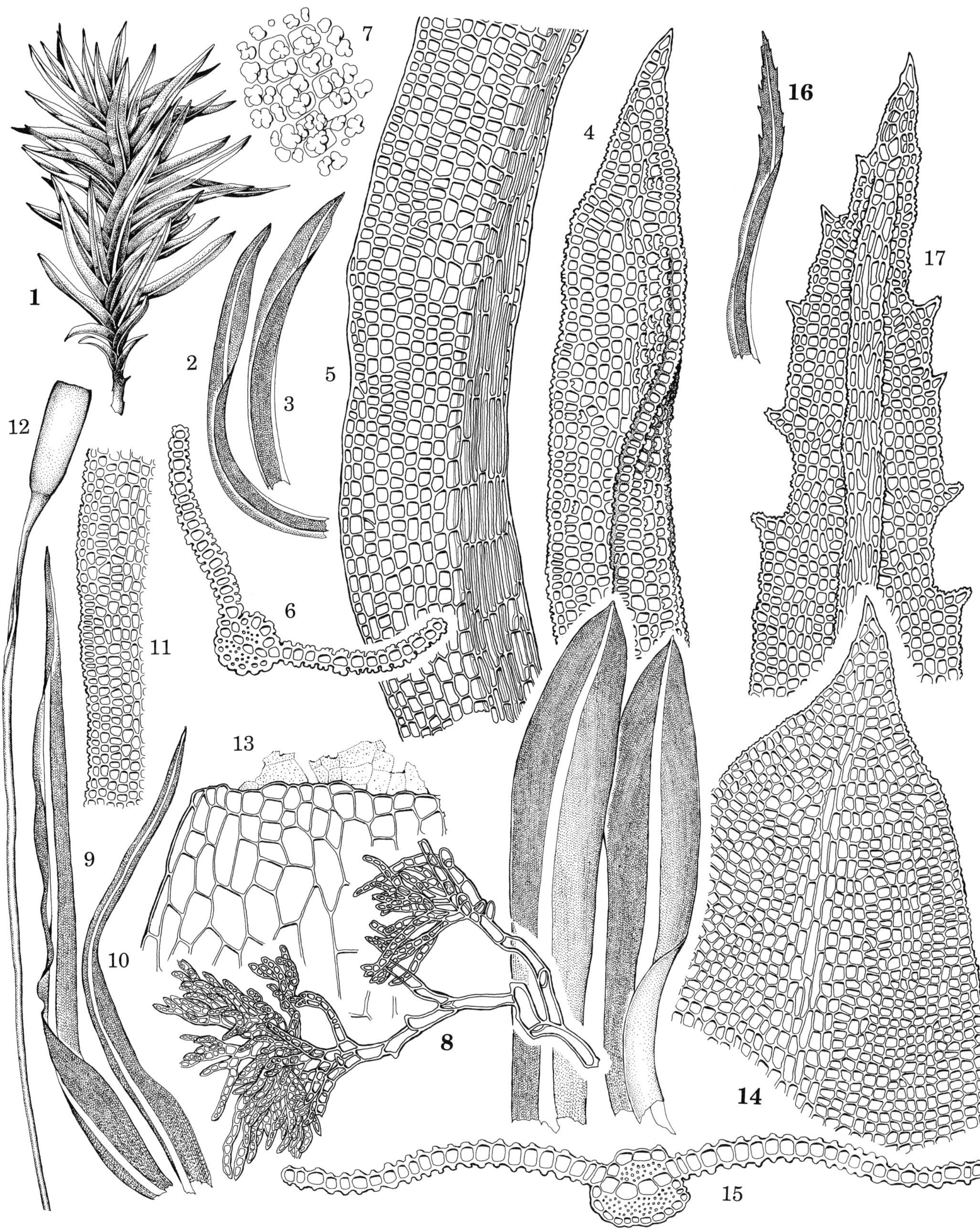

Plate 14. ***Tuerckheimia.*** **1–7.** ***T. svihlae.*** 1. Habit. 2–3. Two leaves. 4. Leaf apex. 5. Basal cells. 6. Transverse section at midleaf. 7. Papillae. **8–13.** ***T. guatemalensis.*** 8. Heterotrichous protonema. 9–10. Two leaves. 11. Upper laminal cells. 12. Theca. 13. Remnants of peristome. **14–15.** ***T. robusta.*** 14. Leaf apex. 15. Transverse section at midleaf. **16–17.** ***T. valeriana.*** 16. Leaf. 17. Upper marginal cells.

11. TUERCKHEIMIA **Plate 14.**

Tuerckheimia Broth., Öfv. F. Vet.-Soc. Förh. 52A (7): 2, 1910. Type: *Tuerckheimia guatemalensis* Broth.

Named for H. von Türckheim, who collected plants in Guatemala from 1877 to 1908 and in the West Indies from 1909 to 1910.

Plants growing in a turf, *light to dark green and somewhat glaucous throughout*, seldom to commonly branching. Stem to 2.5 cm in length, in transverse section rounded-pentagonal or triangular, *central strand present*, sclerodermis absent or weakly developed, *hyalodermis rarely differentiated*; *axillary hairs up to 16 cells in length*, the basal cell occasionally brown; red tomentum sometimes present; heterotrichous, persistent protonema occasionally present, with green, much branched, aerial chloronemata and green to red-brown caulonemata. *Leaves about equal-sized to near base, spreading-incurved from the insertion and subtubulose, weakly twisted to crisped or catenulate when dry*, widely spreading when moist, *oblong- to linear-lanceolate*, to 3.2 mm in length; *margins plane to weakly incurved, entire or sometimes deeply dentate*, occasionally bistratose in patches; apex narrowly acute; *base not or weakly differentiated in shape to short-ovate*; *costa percurrent and ending in an apiculus to stoutly excurrent as a short mucro, ventral superficial cells quadrate to short-rectangular, papillose*, 4–6 cells across costa at midleaf, costa in transverse section circular or ovate, ventrally commonly bulging, *with two stereid bands*, both usually strong, the dorsal crescent-shaped, two or more guide cells in one layer, *a ventral and sometimes a dorsal epidermis present, hydroid strand absent*; *upper laminal cells subquadrate to hexagonal or elliptical, ca. 10–14 µm in width, 1:1, somewhat bulliform, walls evenly thickened* or occasionally trigonous or irregularly thickened, lumens angular or rounded; *papillae usually massive, simple, bifid or multifid, usually centered over the lumens*, (1–)2–4(–6) salients per lumen; *basal cells differentiated in a small group* across the leaf base, smooth, rectangular, yellowish to hyaline. Dioicous. Perichaetia terminal, leaves abruptly sheathing below, otherwise little different from cauline leaves, somewhat larger or smaller. Perigonia terminal, gemmate. Seta ca. 4–8 mm in length, reddish to yellowish brown, twisted clockwise above; theca ellipsoidal to short-cylindrical, brown, ca. 1.0–1.5 mm in length, stomates phaneropore, at base of theca, annulus not vesiculose to strongly so but of persistent cells; peristome absent or present but eroded (or possibly rudimentary), consisting of a basal membrane 20–25 µm in height, papillose, retaining the extreme basal portions of one or two narrow papillose teeth. Operculum long-rostrate, ca. 1 mm in length, cells in straight rows. Calyptra not seen. Spores ca. 8–13 µm in diameter, essentially smooth. *Laminal KOH color reaction yellow*.

Found on mostly calcareous substrates in moist areas in U.S.A. (southeastern states and Alaska), Mexico, Central America, and Eastern Asia.

The major characteristics of *Tuerckheimia* are the narrow, stoutly mucronate leaves with broadly channeled ventral surfaces, acute apices, plane margins, and generally massive papillae centered over the lumens of the often rather large and bulliform upper laminal cells (Pl. 14, f. 6, 7, 15). The basal laminal cells are poorly differentiated (Pl. 14, f. 5) and do not run up the margins as a vee, as in *Tortella*. Fruiting plants are rare and reproduction in some species may take place largely through fragile leaf apices. The deeply dentate leaves of *Tuerckheimia valeriana* are suggestive of *Leptodontium* but the central strand and short cells of the ventral surface of the costa are features distinguishing *Tuerckheimia*; *T. valeriana*, for which sporophytes are unknown, may actually be a *Ptychomitrium* with anomalous papillae.

There is a resemblance to *Eucladium*, with its long-lanceolate leaves, large upper laminal cells generally larger medially and with simple papillae, and the generally stoutly excurrent mucro. The upper laminal cells of *Tuerckheimia robusta* have the angular lumens of *Hymenostylium*. The narrow, plane leaves of *Tuerckheimia* are like of those of *Trichostomum*, but species of that genus generally have small cells and crowded, multiplex, flattened papillae covering most of the cell. K. Saito annotated the type specimen of *T. guatemalensis* (Pl. 14, f. 8–13) with a new combination (unpublished) in *Trichostomopsis* (a genus here referred to *Didymodon* sect. *Asteriscium*), apparently based on bistratose upper margins. These last were made much of by Brotherus (1910) and Bartram (1949), but are merely patches of bistratose cells not comparable to the completely bistratose margins of, e.g. *Didymodon umbrosus*, which has similar long-lanceolate leaves. *Didymodon* sect. *Asteriscium* is easily distinguished by the much differentiated leaf base and general lack of a ventral stereid band. *Quaesticula* is similar in areolation but has infolded upper laminal margins, and a rounded-acute, often cucullate leaf apex.

Remnants of a papillose basal membrane that were found at a capsule mouth (Pl. 14, f. 13) in the single known specimen (Guatemala, Livingston, Tuerckheim, 1908, holotype, H) of *T. guatemalensis* indicate the presence of a peristome in this one species. Additional material is necessary to ascertain the exact morphology of this feature.

Crosby et al. (1992) found *Gymnostomum angustifolium* Saito and thus its combination *Tuerckheimia angustifolia* (Saito) Zand. to be invalid names because a single element was not cited as holotype. The correct name for that taxon is now *T. svihlae* (Bartr.) Zand.

Additional literature: Iwatsuki and Sharp (1958), Zander (1978f).

Species recognized: 4.

Species examined: *T. guatemalensis* (H, NY), *T. robusta* (BM), *T. svihlae* (BUF, DUKE, MEXU, TENN), *T. valerianum* (DUKE, FH, MICH).

New heterotypic synonymy: *Tuerckheimia angustifolia* (Saito) Zand. = *Tuerckheimia svihlae* (Bartr.) Zand.

New combinations: *Tuerckheimia robusta* (Dix.) Zand., *comb. nov.* (*Merceyopsis robusta* Dix., Ann. Bryol. 3: 59, 1930). *Tuerckheimia svihlae* (Bartr.) Zand., *comb. nov.* (*Trichostomum svihlae* Bartr., Rev. Bryol. Lichénol. 23: 245, 1954; *Oxystegus svihlae* (Bartr.) Gangulee).

12. STREPTOCALYPTA **Plate 15.**

Streptocalypta C. Müll., Linnaea 42: 353, 1879. Type: *Streptocalypta lorentziana* C. Müll.

Barnesia Card., Rev. Bryol. 37: 122, 1910. Type: *Barnesia tortelloides* Card.

From στρεπτός, twisted + o + κάλυπτός, covered, a calyptra, a veil, cover, lid.

Plants loosely caespitose, green above, light brown below.

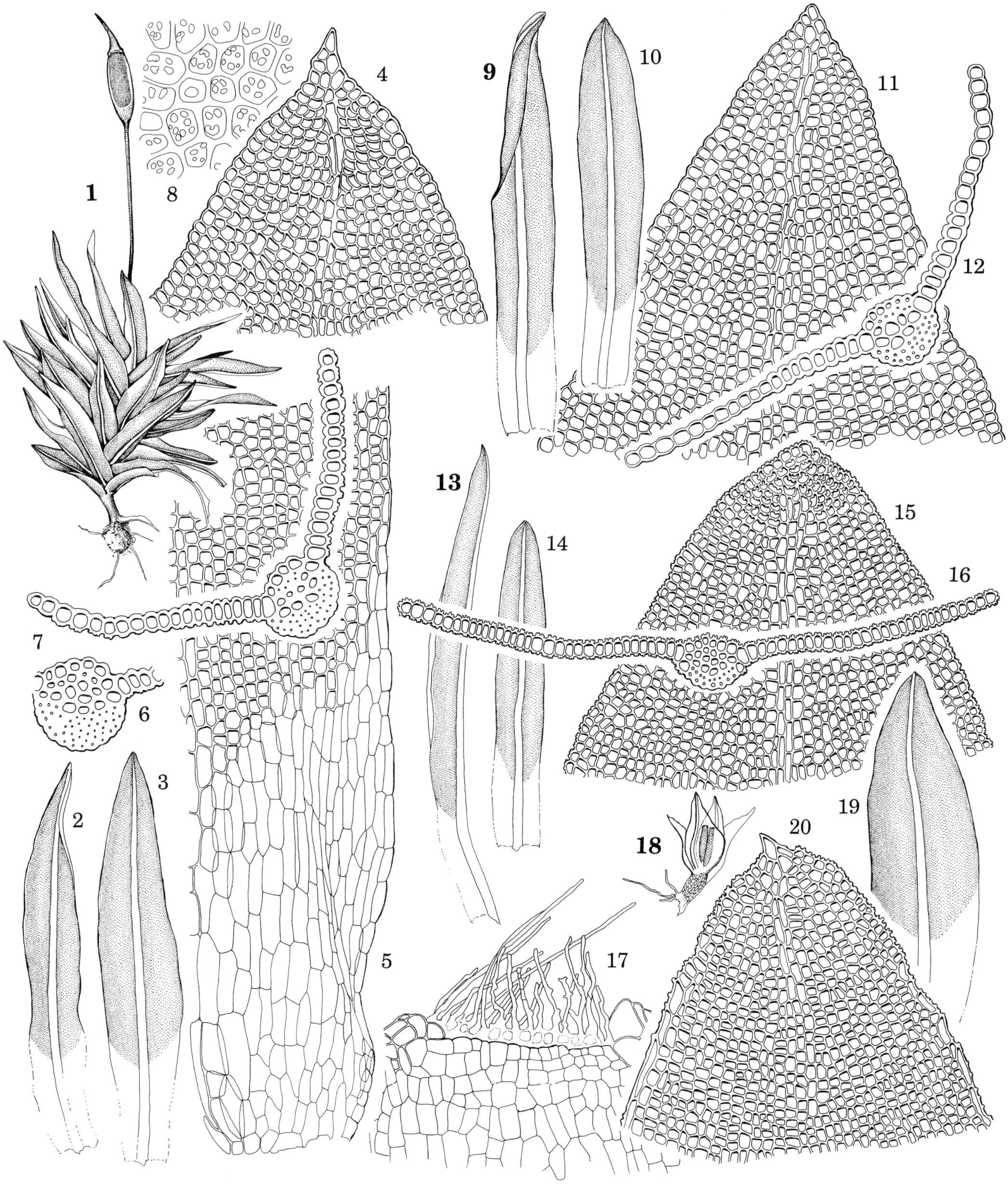

Plate 15. ***Streptocalypta*. 1–8. *S. lorentziana*.** 1. Habit. 2–3. Two leaves. 4. Leaf apex. 5. Basal cells. 6–7. Transverse sections from midleaf. 8. Upper laminal papillae. **9–12. *S. pulchriretis*.** 9–10. Two leaves. 11. Leaf apex. 12. Transverse section at midleaf. **13–17. *S. santosii*.** 13–14. Two leaves. 15. Leaf apex. 16. Transverse section at midleaf. 17. Peristome. **18–20. *S. tortelloides*.** 18. Perigoniate branch (autoicous bud). 19. Leaf. 20. Leaf apex.

Stems to 0.6 cm in length, rounded pentagonal in transverse section, central strand present or occasionally absent, sclerodermis not or little differentiated, hyalodermis occasionally differentiated, axillary hairs of up to 10 hyaline cells or the basal cells brownish; weakly radiculose. *Leaves incurved and contorted when dry*, spreading when wet, *ligulate-lanceolate, long-elliptical or spathulate*, to ca. 3.0 mm in length, upper lamina flat to broadly channeled across leaf; *margins plane or weakly incurved, entire or distantly weakly denticulate above, sometimes bordered by elongate or less papillose cells*; *apex acute to rounded, usually apiculate*, occasionally somewhat cucullate; base little differentiated in shape to ovate; costa ending up to 4 cells below the apex or percurrent to short-excurrent, superficial cells ventrally quadrate and dorsally elongate, ca. 5–7 cells across costa ventrally at midleaf, *costal transverse section with ventral stereid band absent or present but weak, 1–4 layers of many (totalling ca. 7–20) guide cells (these essentially of equal size and often with thickened walls)*, dorsal stereid band strong, reniform, dorsal epidermis not or little differentiated, *hydroid strand absent*; upper laminal cells quadrate to hexagonal, ca. 10–13 µm in width, 1:1, superficially bulging on both sides or more strongly ventrally than dorsally; papillae bifid to sometimes multiplex, often rather small, generally centered and 2–4 over each lumen; *basal cells sharply differentiated, extending up the leaf margins in a vee*, bulging-rectangular, 10–35 µm in width, 3–4:1, walls thin, smooth. Dioicous or monoicous (synoicous, paroicous, autoicous or rhizautoicous). Perichaetia terminal, inner leaves little differentiated or narrower. Seta elongate, to 1 cm in length, brown, twisted clockwise, 1 per perichaetium; theca long-elliptical to cylindrical, to 1.5 mm in length, yellowish brown, exothecial cells rectangular to rhomboidal, walls thin to evenly thickened, stomates phaneropore, at base of theca; annulus of vesiculose cells; *peristome absent or of 32 straight or weakly twisted counterclockwise* or rarely clockwise at the base, *filamentous, densely spiculose teeth* up to 500 µm in length, basal membrane low, spiculose or weakly papillose. Operculum long-conic to rostrate, cells straight or twisted weakly clockwise, 0.5–1.2 mm in length. Spores (8–)10–15 µm in diameter, yellow to brown, weakly papillose. Calyptra cucullate, smooth, 1.5–2.0 mm in length. *Laminal KOH color reaction yellow or light olive.*

Found on soil and rock in mountainous regions of Latin America and South Africa. Quite rare; known only from two states in Mexico and single localities in Bolivia, Argentina and South Africa (Natal).

This genus is similar to *Tortella* in the plane leaf margins, basal laminal cells strongly differentiated in a vee-shape (Pl. 15, f. 5), small upper laminal cells, and spiculose, generally twisted peristome (slightly and probably secondarily twisted clockwise at the base—Pl. 15, f. 17—in *S. santosii* as its opercular cells are in straight rows), but *Streptocalypta* is distinguished by its internal costal structure of several layers of guide cells, general lack of a ventral stereid band (Pl. 15, f. 6, 7, 12, 16), and absence of a peristome in at least one species. *Gertrudiella* likewise has a cylinder of mostly equal-sized guide cells but has strongly recurved upper leaf margins and lacks the basal vee. Species of certain taxa in other tribes (e.g. *Tortula* and *Didymodon* sect. *Asteriscium*) do have more than one layer of guide cells, but are easily distinguished by other characters. The character of multistratose guide cells is distinctive for the genus *Streptocalypta*, but is less pronounced in *S. pulchriretis* than in other species. The peristome of *S. pulchriretis* is extremely fragmentary in the single known collection; this species is very similar to *S. lorentziana*.

Although a few stereid cells are commonly present ventrally in *S. lorentziana* (Pl. 15, f. 7) and usually several ventral stereids in *S. pulchreretis* (Pl. 15, f. 12), the usual absence of a ventral stereid band in the genus may be viewed as a derived character, and is duplicated in *Pseudocrossidium* of the Barbuleae. Species of *Tortula*, of the Pottieae, may occasionally exhibit a few ventral stereid cells.

A misprint in my previous treatment of the genus (Zander 1983a) gives lengths of the setae of *S. santosii* and *S. tortelloides* in millimeters; it should have read "cm".

Additional literature: Magill (1981, discussion as *Weisiopsis pulchriretis*).

Number of accepted species: 3.

Species examined: *S. lorentziana* (BUF, F, H), *S. pulchriretis* (BM), *S. santosii* (FH, MICH), *S. tortelloides* (NY, PC, TENN).

New combinations: *Streptocalypta pulchriretis* (Dix.) Zander (*Weisiopsis pulchriretis* Dix., Trans. Roy. Soc. S. Afr. 18: 252, 1930).

13. PLEUROCHAETE — Plate 16.

Pleurochaete Lindb., Oefv. K. Vet. Ak. Foerh. 21: 253, 1864. Type: *Pleurochaete squarrosa* (Brid.) Lindb.

Barbula subg. *Pleurochaete* (Lindb.) Schimp., Syn. ed. 2: 220, 1876.

Tortella subg. *Pleurochaete* (Lindb.) Limpr., Laubm. Deutschl. 1: 607, 1890.

Barbula sect. *Pleurochaete* (Lindb.) C. Müll., Linnaea 39: 400, 1875.

Barbula sect. *Squarrosae* Lesq. & James, Man. N. Am. Moss. 130, 1884, *nom. illeg. incl. sect. prior.* Type: *Barbula squarrosa* Brid.

Barbula sect. *Squarrosa* Lázaro é Ibiza, Bot. Descr. Comp. Fl. Esp. 1: 586, 1896, *nom. illeg. incl. Barbula* sect. *Pleurochaete* (Lindb.) C. Müll. Type: *Barbula squarrosa* Brid.

Barbula subsect. *Squarrosae* C. Müll., Linnaea 39: 402, 1875.

From πλευρά, side, or πλευρόν, rib + o + χαίτη, long hair, mane; referring to the laterally borne sporophytes.

Plants forming a deep or sprawling turf, green above, brown below. *Stems* branching irregularly, *to 4.0 cm in length*, transverse section rounded-pentagonal, central strand small, *walls of central cylinder cells thin to weakly thickened*, sclerodermis of 2–4 rows thick-walled cells, hyalodermis present; *axillary hairs to 15 cells in length*, cells all hyaline; indumentum weakly radiculose. *Leaves spreading and strongly contorted when dry, squarrose-recurved above a sheathing base when moist, oblong-lanceolate*, 3–5 mm in length, upper lamina broadly channeled, *margins plane* but occasionally recurved to revolute along leaf base, *denticulate in upper 1/3 of leaf to throughout*, 1–2 marginal rows often weakly papillose distally beyond a *border of hyaline, thin-walled rhomboidal cells not merging below with inner basal cells* but instead extending as a distinct hyaline strip to leaf insertion, 4–7 cells in width below, narrowing upwards, reaching barely higher than shoulder of leaf base to 3/4 leaf length; apex usually sharply acute, occasionally broadly acute; base ovate to rectangular, often broadly

Plate 16. ***Pleurochaete.*** **1–10.** ***P. squarrosa*** **var.** ***squarrosa*** 1. Habit. 2–3. Two leaves. 4. Leaf apex. 5. Upper marginal cells. 6. Upper laminal papillae. 7. Laminal cells at leaf shoulder. 8. Transverse section of upper part of leaf. 9. Perigoniate bud. 10. Peristome. **11–16.** ***P. squarrosa*** **var.** ***luteola.*** 11. Transverse section of stem. 12. Leaf. 13. Midleaf marginal cells. 14. Basal cells. 15. Perichaetiate branch. 16. Perigoniate bud. **17–20.** ***P. malacophylla.*** 17–18. Two leaves. 19. Leaf apex. 20. Transverse section at midleaf.

sheathing and with distinct shoulders; costa short-excurrent as a sharp mucro, superficial cells quadrate to short-rectangular and papillose from top of sheathing leaf base to near apex ventrally, elongate and smooth dorsally, 6–8 rows of cells across costa ventrally at midleaf, *costal transverse section semicircular or reniform, stereid bands strong ventrally and dorsally, larger dorsally*, epidermis differentiated in one layer ventrally, present or absent dorsally, guide cells 4–6 in 1 layer, hydroid strand absent; upper laminal cells subquadrate, 8–11 µm in width, 1(–2):1, walls thin or evenly weakly thickened, superficially bulging on both exposed sides or only ventrally and then weakly convex dorsally; papillae bifid, 1–4 per lumen; basal cells differentiated and rising higher medially, distinct from the 4–6 rows of border cells, rectangular to rhomboidal, 10–20 µm in width, 2–5:1, walls thin to evenly thickened or porose. Dioicous. *Perichaetia on very short lateral branches*, inner leaves long-lanceolate, narrower than cauline leaves, to 5 mm in length, sheathing the seta, lower cells rhomboidal to rectangular and porose, reaching to 3/4 length of inner leaves. Perigonia lateral on the stem, small. Seta 1.3–1.7 cm in length, 1 per perichaetium, orangish to reddish brown, twisted clockwise; theca 2.0–2.8 mm in length, light yellowish or reddish brown, cylindrical, exothecial cells rectangular to rhomboidal, 20–25 µm in width, 3–4:1, walls thin, stomates phaneropore, at base of theca, annulus of 3–5 rows of vesiculose cells, persistent; *peristome teeth 32, filamentous*, branched spiculose, 300–1000 µm in length, often broken, with several articulations, *twisted once counterclockwise*, basal membrane apparently absent or to 35 µm high in height, low-spiculose. Operculum long-conic, 0.9–1.7 mm in length, cells twisted counterclockwise. Calyptra cucullate, smooth, ca. 4 mm in length. Spores 10–13 µm in diameter, yellowish brown, papillose. Laminal KOH color reaction deep yellow to orange. Reported chromosome number n = 13.

Found in dry areas on soil and rock (generally calcareous), occasionally on tree roots; North, Central and South America, Europe, North and Central Africa, the Middle East and Asia.

Crum and Anderson (1981) pointed out that *Pleurochaete* is quite like *Tortella* in several respects but differs in that the differentiated thin-walled marginal cells, which extend up from the insertion often to above midleaf, do not form a coherent basal vee extending medially to the costa (Pl. 16, f. 7, 14). The inner basal cells, instead, form a distinct region, much as in the case with some but not all *Pseudosymblepharis* species. *Pleurochaete* is much like the Asian *Chionoloma* in leaf shape, marginal strip of elongate cells (Pl. 16, f. 13), and upper medial cells often bulging much greater ventrally than dorsally, but the latter has thick-walled border cells and the upper laminal margins are sharply incurved as in *Weissia*.

Variation in the degree of extension of the basal marginal cells up the leaf margins and in the length of the inner basal cells is not quite continuous, but is correlated with plant stature. *Pleurochaete squarrosa* has a largely Temperate Zone and paleotropical facies (*P. squarrosa s. str.*, found in Europe, Ethiopia, the Congo, southern U.S.A., Mexico) and a neotropical facies (*P. luteola*, found in southeastern U.S.A., e.g. Tennessee, Zander 4333, BUF, and Arkansas, Crum & Anderson's exsiccat Mosses of North America 951 as *P. squarrosa*; and Latin America). These morphotypes are distinguishable in most specimens examined. Crum and Anderson (1981), however, felt that the neotropical variant is only a "robust expression," not worthy of a separate name; Crum (1951) indicated that, where growing sympatrically, these may sometimes be difficult to name as one or the other. In any case, a varietal name is provided here for workers like myself who wish to distinguish among the two. The var. *luteola* (Pl. 16, f. 11–16) differs in being more robust, with a more highly sheathing leaf base, and the border of thin-walled marginal cells being strongly denticulate and extending farther up the leaf margins. This parallels the distinction between *Molendoa hornschuchiana* and *M. sendtneriana*, the former differing in stature (largest in the genus) and denticulate lower leaf margins; however, the larger species of *Molendoa* is distributed primarily across the Alps and Himalayas (one station is in Alaska), while the latter ranges from the Andes and North American Cordillera across the Arctic and throughout the Eurasian continent in arctic-montane situations. Thus, if the each of these two genera developed intraspecifically through reduction, then *Pleurochaete* probably originated in Gondwanaland and *Molendoa* is Laurasian, but vice versa if large stature and marginal denticulation are elaborations.

African material identified as *Pleurochaete beccarii* at BM and NY is *P. squarrosa* var. *squarrosa*; type material of *P. beccarii* at TR could not be obtained on loan but is probably also synonymous. *Pleurochaete malacophylla* is doubtfully distinct from *P. squarrosa* var. *squarrosa*.

Additional literature: Nebel (1990), Quarterman (1956), Wyatt and Stoneburner (1982).

Number of accepted species: 4.

Species examined: *P. malacophylla* (BM), *P. squarrosa*.

New combinations and statuses: *Pleurochaete squarrosa* (Brid.) Lindb. var. *luteola* (Besch.) Zand., *comb. et stat. nov.* (*Trichostomum luteolum* Besch., Mém. Soc. Sc. Nat. Cherbourg 16: 178, 1872; *Pleurochaete luteola* (Besch.) Thér.).

14. CALYPTOPOGON **Plate 17.**

Calyptopogon (Mitt.) Broth., Nat. Pfl. 1(3): 419, 1902.

Streptopogon sect. *Calyptopogon* Mitt., Phil. Trans. R. Soc. London 168: 33, 1879. Type: *Calyptopogon mnioides* (Schwaegr.) Broth.

From κάλυπτός, covered + o + πώγων, -ωνος, beard; referring to the ragged margin of the calyptra.

Plants growing in tufts or mats, yellowish green above, brown below. *Stems* branching occasionally, 2–4 cm in length, transverse section rounded-pentagonal, *central strand absent*, sclerodermis present, *hyalodermis present*, occasionally only in patches; axillary hairs to 7 cells in length, all hyaline or 1–2 basal cells thick-walled; sparsely to densely radiculose. *Leaves* incurved or catenulate, *margins extremely undulate when dry*, widely spreading and reflexed when moist, *long-ovate to ovate-lanceolate*, 3.5–4.5 mm in length, *upper lamina flat to somewhat keeled, margins plane*, nearly entire, minutely serrulate with projecting transverse cell walls and papillae, *bordered intramarginally throughout length of leaf by 3–4 rows of rectangular epapillose porose cells* as an extension of the region of differentiated basal cells; apex acute, often narrowly channeled, extreme cells bordering the mucro rhomboid; base little differentiated in shape but tubulose-clasping; *costa* evenly tapering, *excurrent as a papillose mucro*, superficial cells short-rectangular and distantly papillose ventrally, elongate and distantly papillose with forward-pointing cell ends dorsally, ca. 4

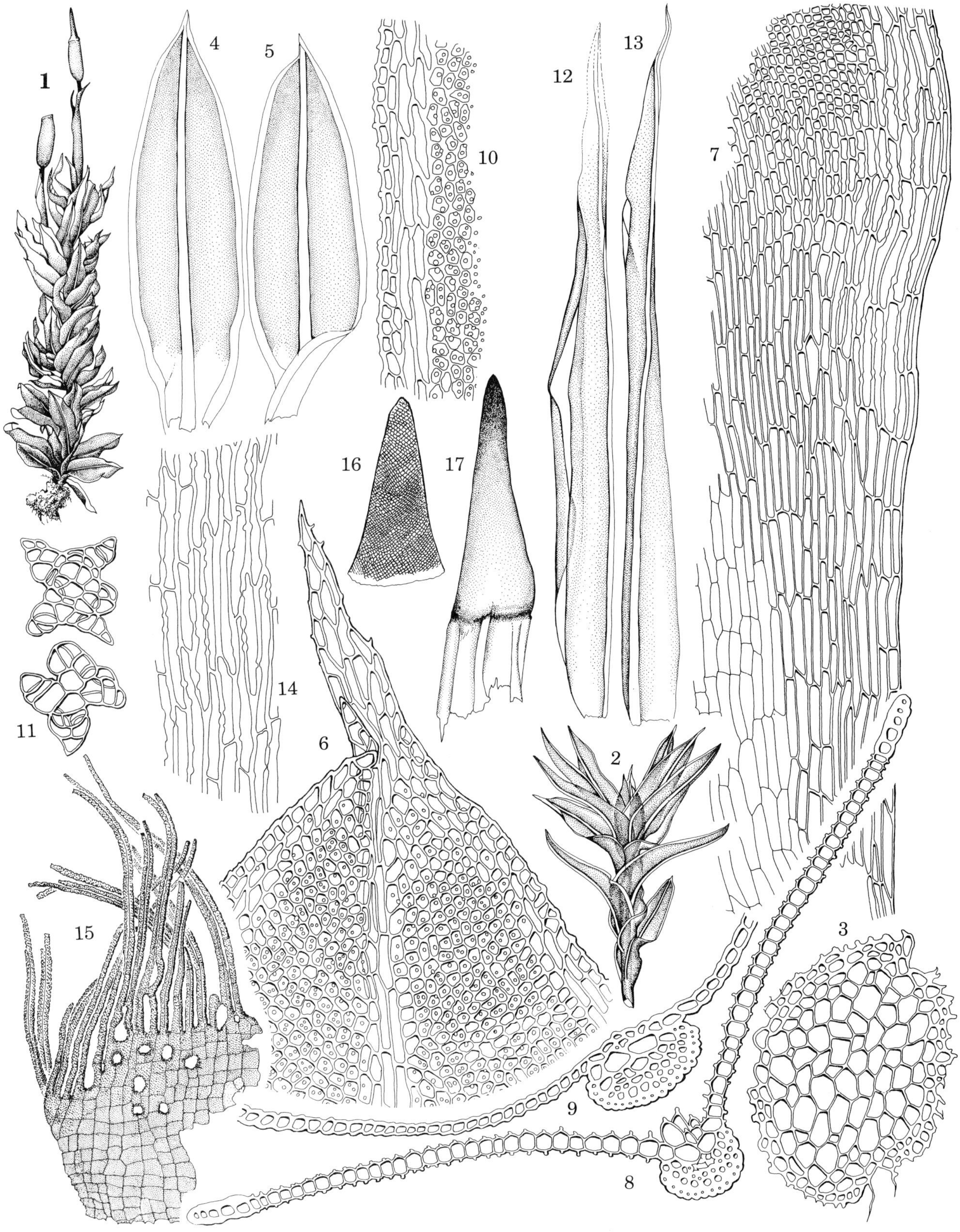

Plate 17. *Calyptopogon*. 1–17. *C. mnioides*. 1. Perichaetiate plant. 2. Perigoniate plant. 3. Transverse section of stem. 4–5. Two leaves. 6. Leaf apex showing papillae. 7. Basal cells. 8. Transverse section at midleaf. 9. Transverse section near leaf base. 10. Upper leaf margin. 11. Propagula. 12–13. Two perichaetial leaves. 14. Cells of perichaetial leaves. 15. Peristome teeth. 16. Operculum. 17. Calyptra.

rows of cells across costa ventrally at midleaf, *costal transverse section* ovate, *stereid band absent ventrally*, *strong dorsally* and crescent-shaped, lumens of stereid cells smaller towards dorsal surface, substereid near guide cells, ventral epidermis of bulging or conic cells, dorsal epidermis absent, guide cells 2–4(–6) in 1(–2) layers, *hydroid strand very large*, ventral costal cells often protuberant as scattered, ovate, thick-walled cells; *upper laminal cells irregularly hexagonal*, occasionally longitudinally or transversely elliptic, ca. 12–16 µm in width, ca. 1:1, *walls trigonous*, porose, superficially weakly convex on both sides; *papillae small, simple to bifid, conic-spiculose, solid, 3–4 per lumen*, evenly scattered over the lamina; *basal cells* differentiated across leaf, *rising higher marginally and fusing with the intramarginal border*, rectangular, little wider than upper cells, to 5–6:1, walls evenly thickened to porose. *Propagula multicellular, caltrop-shaped*, 20–30 µm in diameter, *borne on the ventral surface of the costa*. Dioicous. *Perichaetia* terminal, *inner leaves strongly differentiated, linear-lanceolate*, to 12–13 mm in length, *strongly sheathing, convolute, cells entirely rectangular to long-rhomboidal, thick-walled and porose*. Perigonia born terminally on an equal-sized perigoniate plant as gemmate buds, often several per plant, each ending a branch. *Seta short*, ca. 0.5 cm in length, 1 per perichaetium, brown, twisted clockwise; theca 1.8–2.2 mm in length, brown, short-rectangular, occasionally curved, exothecial cells rectangular, evenly thickened, ca. 20 µm in width, 2–4:1, stomates phaneropore, at base of capsule, annulus of 3–4 rows of somewhat vesiculose, persistent cells; *peristome* often coming off with the operculum, *teeth 32, filamentous*, densely low-spiculose, to 500 µm, with many articulations, twisted once counterclockwise, *basal membrane high*, to 300 µm in height, with round, thin-walled windows, densely low-spiculose. *Operculum conic*, ca. 1.5 mm in length, cells twisted counterclockwise. Calyptra cucullate, somewhat rough above, ragged and longitudinally undulate below, smooth, ca. 3 mm in length. Spores ca. 13–15 µm in diameter, brown, essentially smooth. Laminal KOH color reaction red, often blotchy, or yellow with red blotches, or yellow-orange.

This monotypic rupestral genus is restricted to southern South America, New Zealand, Tasmania and Australia.

Sainsbury (1955) described the calyptra as mitriform, but it is actually long-cucullate, weakly pleated and strongly frayed at the base (Pl. 17, f. 17). The genus is distinctive in the combination of the trigonous upper laminal cells with small, simple to bifid, solid papillae (Pl. 17, f. 6); intramarginal border of porose, rectangular, epapillose cells (Pl. 17, f. 10); single costal stereid band (Pl. 17, f. 8–9); nerve-borne, short-branching propagula (Pl. 17, f. 11); convolute-sheathing perichaetial leaves of thick-walled porose cells (Pl. 17, f. 12–13); and 32 filamentous peristome teeth (Pl. 17, f. 15). Sections of the costa near the base show more than one hydroid strand and a clear differentiation of tissues into a crescent of 2 or 3 superficial rows of stereids dorsally, and substereid cells just ventral to that crescent, the significance of which is presently unclear. Older axillary hairs apparently become thick-walled basally; hairs near the apex are entirely hyaline.

The ventral stereid band is absent, as is characteristic of highly evolved taxa of the Pottioideae. This genus is strongly reminiscent of *Syntrichia papillosa* (Pottieae), a species of broad distribution fruiting, however, only in Australia (*cf.* Catcheside 1980 and Dixon 1923). *Syntrichia papillosa* shares the following features with *Calyptopogon*: large, collenchymatous upper laminal cells; papillae simple (occasionally bifid); stereid band crescent-shaped; no dorsal epidermis; hydroid strand strong; propagula clavate with bulging cells, borne on ventral surface of costa; perichaetial leaves strongly sheathing the costa, laminal cells rhomboidal and hyaline throughout; calyptra large; KOH red. *Syntrichia papillosa* was thought by Kramer (1980) to possibly form its own section of *Tortula*. It differs from *Calyptopogon* by its stem having a (weak) central strand; awned, concave leaves lacking a leaf border and not strongly undulate marginally when dry; upper laminal cells smooth or only singly papillose; apically dorsally spinose costa; and perichaetial leaf cell walls not thickened. *Syntrichia pagorum* has, like *C. mnioides*, a semicircular costal section, which lacks dorsal epidermal cells and bulges laterally over the dorsal surface of the leaf, and the species likewise fruits only in austral regions (*cf.* Catcheside 1980; Stone 1971); *S. pagorum* differs, however, in the only weakly differentiated perichaetial leaves and the relatively small, multiplex-papillose upper laminal cells characteristic of *Syntrichia*. The cladistic study, however, places *Calyptopogon* in the Trichostomoideae.

Number of accepted species: 1.

Species examined: *C. mnioides* (NY, US).

15. TORTELLA — Plates 18–19.

Tortella (Lindb.) Limpr., Laubm. Deutschl. 1: 599, 1888, *nom. cons.* Lectotype: *Tortella caespitosa* (Schwaegr.) Limpr.

Mollia subg. *Tortella* Lindb., Musci Scand. 21, 1879.

Tortella subg. *Tortella* Limpr., Laubm. Deutschl. 1: 600, 1888, *nom. illeg.*

Barbula subg. *Tortella* (Lindb.) Kindb., Eur. N. Amer. Bryin. 2: 245, 1897.

Tortella subg. *Eutortella* Roth, Eur. Laubm. 1: 344, 1903, *nom. illeg.*

Trichostomum subg. *Tortelloidea* Roth, Eur. Laubm. 1: 315, 1903.

Tortula sect. *Tortuosae* De Not., Mem. R. Acc. Sc. Torino 40: 288, 1838. Type: *Tortula tortuosa* Hedw.

Tortula sect. *Caespitosae* De Not., Mem. R. Acc. Sc. Torino 40: 287, 1838. Type: *Tortula caespitosa* Schwaegr.

Barbula sect. *Tortuosae* BSG, Bryol. Eur. 2: 86, 1842 (fasc. 13–15 Mon. 24). Type: *Barbula tortuosa* (Hedw.) Web. & Mohr.

Barbula sect. *Tortella* C. Müll., Syn. 1: 599, 1849, *nom. illeg. incl. sect. prior.*

Barbula sect. *Fragiles* Schimp., Syn. 181, 1860. Type: *Barbula fragilis* (Hook. & Wils.) BSG.

Tortula sect. *Fragiles* (Schimp.) Lindb., Oefv. K. Svensk. Vet. Ak. Foerh. 21: 214, 1864. Type: *Tortella fragilis* (Hook. & Wils.) C. Hartm.

Mollia sect. *Tortella* (Lindb.) Braithw., Brit. Moss Fl. 1: 230, 247, 1885.

From *tortus* (a participle), twisted + *-ella*, diminutive.

Plants loosely caespitose or forming turfs or cushions, green to dark green above, brown to tan below. *Stems* branching occasionally or often, usually short but up to 4 cm in length, in transverse section rounded-pentagonal, central strand absent, weak or strong, *cells of central cylinder often thick-walled, sclerodermis present but usually weak, hyalodermis present*, occasionally weak or only in patches, *composed of cells that are not or little collapsed when mature*; *axillary hairs long*, ca. 10–20 cells in length, *all hyaline*; weakly radiculose, occasionally matted with reddish brown rhizoids. *Leaves* often crowded, when dry appressed, erect or incurved, often twisted or contorted, occasionally spiralled about stem, spreading when moist, *ligulate to long-lanceolate*, rarely spathulate, 1.5–7.0 mm in length, upper lamina flat, broadly channeled across leaf or weakly channeled along costa, *margins plane, occasionally erect or narrowly incurved or tubulose, entire*, weakly crenulate with projecting papillae or seldom weakly dentate near apex or below midleaf, rarely bordered with elongate cells near apex; apex subulate or narrowly to broadly acute, occasionally rounded or cucullate; *base oblong, elliptical or short-ovate*, occasionally long-sheathing or not differentiated in shape; costa ending 1–4 cells below apex, percurrent or more usually short-excurrent as a mucro, superficial cells quadrate or elongate, occasionally only quadrate at midleaf ventrally, elongate dorsally, 2–6(–10) rows of cells across costa ventrally at midleaf, costal transverse section semicircular or ovate, *stereid bands two*, weak to strong ventrally, strong dorsally, epidermis present ventrally and present, weak or absent dorsally, guide cells 4–6(–8) in 1 layer, *hydroid strand absent*; upper laminal cells quadrate to hexagonal, occasionally bistratose, 7–15 µm in width, 1(–2):1, walls thin to evenly thickened, superficially convex on both sides; papillae spiculose, bifid or occasionally simple, 3–6 per lumen, seldom capituliform and centered; *basal cells differentiated across leaf in a vee-shape*, usually sharply differentiated, occasionally forming only a low vee or weakly differentiated from the upper cells, rectangular, occasionally bulging, ca. 8–30 µm in width, 4–6:1, walls thin or evenly thickened to porose, usually smooth. Asexual reproduction occasional, apparently by fragile leaf apices or portions of lamina. Dioicous or occasionally autoicous. *Perichaetia terminal*, inner leaves usually long-lanceolate, to 6 mm in length, usually sheathing the seta, lower cells long-rhomboidal in lower 1/2. Perigonia terminal, weakly gemmate, inner leaves smaller, or occurring as flattened, stalked, buds in leaf axils of perichaetiate plants. Seta ca. 0.7–3.0 cm in length, 1 per perichaetium, yellowish, reddish or light brown, twisted clockwise below, occasionally counterclockwise above, seldom straight; theca ca. (1.0–)1.5–3.0 mm in length, yellowish to reddish brown, cylindrical or occasionally elliptical, exothecial cells rectangular, 20–45 µm in width, 3–5:1, seldom hexagonal, walls thin, stomates at base of theca, phaneropore, annulus of 1–4 rows of weakly or strongly vesiculose cells, persistent; *peristome teeth 32, linear*, spiculose, occasionally branching-spiculose, ca. 550–1400 µm, with many articulations, *twisted counterclockwise*, usually two or three times, basal membrane absent or low, papillose to spiculose, or rarely absent, or capsule rarely cleistocarpous and then elliptical and long-apiculate. Operculum long-conic to rostrate, ca. 1.0–2.5 mm in length, cells twisted counterclockwise, seldom undifferentiated. Calyptra cucullate, smooth, 2.5–3.5 mm in length. Spores 8–20 µm in diameter, yellowish brown, essentially smooth or papillose. Laminal KOH color reaction usually yellow, sometimes yellowish orange or reddish brown, leaves often yellow when immature but orange lower on the stem. Reported chromosome number n = 7, 13, 13+m, 14, 15, 26, 30, 52.

Found on soil, rock or organic substrates on all continents.

Tortella and *Trichostomum* (including subg. *Oxystegus*) are distinguished at present by characters that are somewhat variable. A thorough reevaluation is needed through revision of both genera together. *Tortella simplex*, for instance, has much the appearance of *Trichostomum brachydontium*, especially in the characteristic weakly reflexed apex ending in a stout mucro, and its basal cells are only weakly differentiated as a vee; this species may be more closely related to *Trichostomum* than to *Tortella* in spite of its somewhat twisted peristome.

The vee-shaped area of basal cells is generally easily distinguished in most species of *Tortella*, yet, in some, the differentiated basal cells do not rise very high along the margins (Pl. 18, f. 6, forming a rather low vee) or they are not sharply different in size and shape or thickness of cell walls from the upper laminal cells. Some species with a vee-shaped basal cell region (e.g. *Tortella flavovirens*) have untwisted, often short peristomes, similar to those of *Trichostomum s. lat.*

Two autoicous species, *Tortella lilliputanum* (Pl. 19, f. 1–3) and *T. fruchartii*, have cleistocarpous capsules and a distinctly vee-shaped basal laminal cell area, plus a stem section with a characteristic well developed tortellaceous hyalodermis. The gametophytes are clearly *Tortella*; the two taxa are much alike and may be conspecific. Species of *Tortella* with reduced sporophytes (*T. eckendorffii* lacks a peristome) are presently few, but more should be added to *Tortella* when *Hyophila* and other eperistomate genera are revised, e.g. *Tortella walkeri* is eperistomate and stegocarpous.

Pseudosymblepharis is usually distinguishable by the narrow upper laminae and sharply broadened and sheathing leaf bases, but specimens of *Tortella tortuosa* with cirrhate leaves have much the appearance of *Pseudosymblepharis*. Some species of *Pseudosymblepharis*, also, are much like *Tortella* in the broad upper laminae. *Trichostomum hibernicus* has these characters of *Pseudosymblepharis* but is obviously closely related to *Trichostomum* (subg. *Oxystegus*) *tenuirostre* by the upper laminal areolation of evenly thickened, rectangular cells, although the well developed sclerodermis is a character more common in subg. *Trichostomum*. Collections of *Pseudosymblepharis schimperianum* of small stature have been in the past assigned to *Tortella* (e.g. the synonyms *Tortella mollissima*, *Tortella richardsii* and *Tortella subfragilis*) because of a tendency for such small plants to have less well-marked sheathing leaf bases. The names *Tortella*, *Trichostomum*, *Pseudosymblepharis* and *Oxystegus* may actually represent only one genus, characterized by a usually vee-shaped basal cell arrangement, a vee-shaped transverse section of the dorsal stereid band (in most species), an often thick-walled central cylinder, and sturdy, non-collapsing hyaloderm cells, but further research is necessary. *Barbula*, *Didymodon*, *Bryoerythrophyllum*, and even *Pseudocrossidium* have occasionally been treated even in recent times (e.g. Nyholm 1989) as a single genus under the name *Barbula s. lat.*, and it has been a temptation to similarly unify *Tortella* and *Trichostomum s. lat.* here pending further study. As a possible parallel to be pursued in future studies of *Tortella* and *Trichostomum*, the above four groups of Merceyoideae have been successfully separated in recent studies (Saito 1975a; Zander 1978e, 1978g, 1979f, 1981a, 1981c) as genera each with considerable variation in the development of the peristome, but taxonomically distinguishable by gametophytic features. Because it is thought, however, that good gametophytic characteristics will eventually be used *to place at least the type species of* Tortella and *Trichostomum* in separate genera, these two generic names

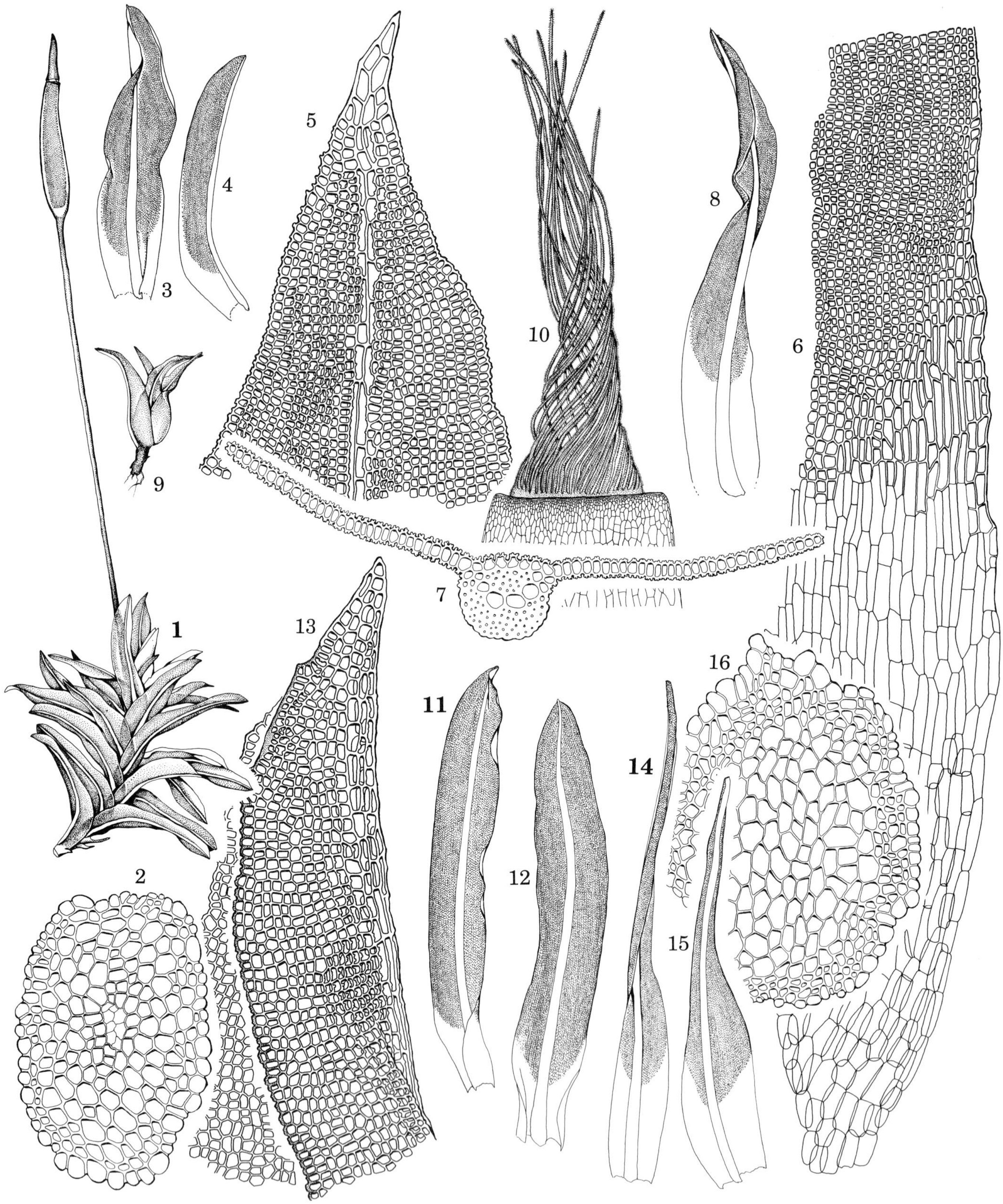

Plate 18. ***Tortella.*** **1–10.** ***T. humilis.*** 1. Habit. 2. Transverse section of stem. 3–4. Two leaves. 5. Leaf apex. 6. Basal cells 7. Transverse section at midleaf. 8. Perichaetial leaf. 9. Perigonial bud. 10. Peristome. **11–13.** ***T. bryotropica.*** 11–12. Two leaves. 13. Leaf apex. **14–16.** ***T. fragilis.*** 14–15. Two leaves. 16. Transverse section of stem.

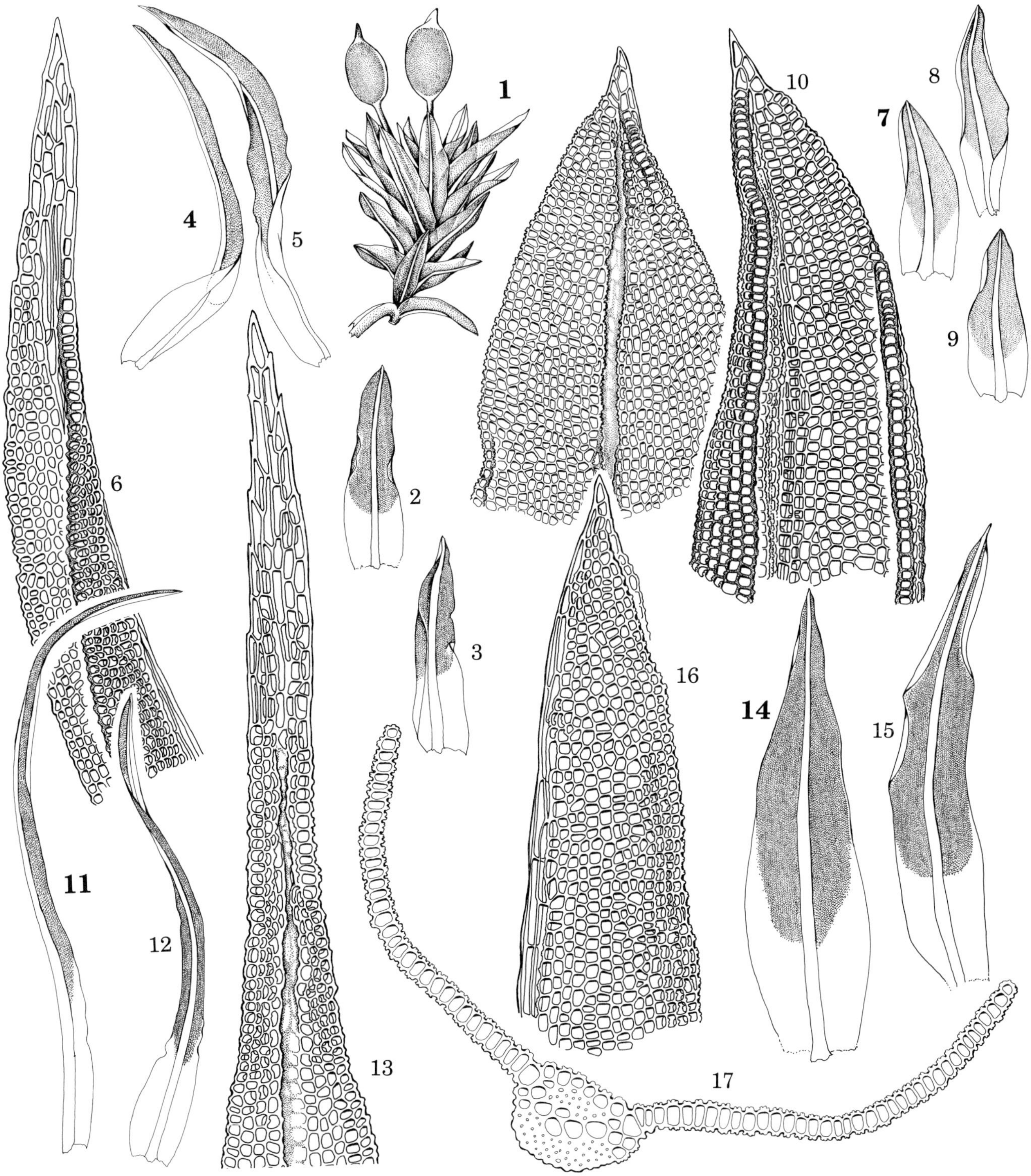

Plate 19. ***Tortella.*** **1–3** ***T. lilliputana.*** 1. Habit. 2–3. Two leaves. **4–6.** ***T. japonica.*** 4–5. Two leaves. 6. Leaf apex. **7–10.** ***T. rubripes.*** 7–9. Three leaves. 10. Leaf apex. **11–13.** ***T. tortuosa.*** 11–12. Three leaves. 13. Leaf apex. **14–17.** ***T. xanthocarpa.*** 14–15. Two leaves. 16. Leaf apex. 17. Transverse section at midleaf.

are accepted here as representing different taxa, albeit with the counsel that the present descriptions are based on species many of which will probably be redistributed at some future date among two or more generic names. It is emphasized here that *Tortella* cannot be adequately revised taxonomically unless *Trichostomum s. lat.* and *Hyophila* are reviewed at the same time.

Species of *Tortella* and some related genera may usually be distinguished from many other plane-margined taxa, mainly those of the Barbuleae, by the stem section. Although presence of a central strand is variable, as a rule the hyalodermis is present, composed of rectangular (in transverse section) cells that are seldom collapsed in mature parts of the stem, and the sclerodermis is generally only weakly developed, often of substereid cells only slightly smaller than those of the central cylinder or of a few scattered stereid cells. Often, the cells of the central cylinder are thick-walled. Taxa of Merceyoideae with a hyalodermis have it composed of rounded cells, these usually collapsed except in the region of the extreme stem apex, and have a distinct sclerodermis of one or more layers of stereid cells.

Pleurochaete has a similar vee of basal cells, but in that genus the thin-walled basal marginal cells form a group easily distinguished from the thicker-walled inner basal cells.

Selected bibliography: Crundwell and Nyholm (1962, 1963), Dixon (1900), Gorton and Eakin (1958), E. J. Hill (1913), Nicholson (1910), Zander and Hoe (1979).

Number of accepted species: 53.

Species examined: *T. acaulon* (NY), *T. alpicola* (BUF, CANM, NY), *T. bryotropica* (BUF), *T. cirrifolia* (NY), *T. cryptocarpa* (NY), *T. cyrtobasis* (BM), *T. densa* (DUKE, NY), *T. eckendorffii* (PC), *T. flavovirens* (BUF), *T. fragilis*, *T. fruchartii* (H, NY), *T. germainii* (NY), *T. goniospora* (NY), *T. hildebrandtii* (NY), *T. humilis*, *T. inclinata* (BUF), *T. inflexa* (BUF), T. *japonica* (BUF), *T. knightii* (MO, NY), *T. lilliputana* (NY, S), *T. linearis* (NY), *T. nitida* (BUF), *T. novae-valesiae* (H), *T. pseudocaespitosa* (NY), *T. rigens* (DUKE, NY), *T. rubripes* (NY), *T. simplex* (US), *T. somaliae* (NY), *T. tortuosa*, *T. xanthocarpa* (H, NY).

New heterotypic synonymy: *Astomum latifolium* Broth. in Roth = *Tortella fruchartii* (C. Müll.) Zand. *Tortella tortelloides* (Greene) Robins. in Llano. = *Tortella alpicola* Dix. *Trichostomum sitkanum* Card. & Thér. = *Tortella tortuosa* var. *arctica* (Arnell) Broth. in Fedch.

New combinations: *Tortella cryptocarpa* (Broth.) Zand., *comb. nov.* (*Astomum cryptocarpum* Broth., Bih. K. Svensk. Vet. Ak. Handl. 26 Afd. 3(7): 19, 1900).

Tortella eckendorffii (P. Varde) Zand., *comb. nov.* (*Hymenostomum eckendorffii* P. Varde, Rev. Bryol. Lichénol. 11: 170, 1939).

Tortella fruchartii (C. Müll.) Zand., *comb. nov.* (*Phascum fruchartii* C. Müll., Flora 71: 4, 1888; *Astomum fruchartii* (C. Müll.) Broth.).

Tortella goniospora (C. Müll.) Zand., *comb. nov.* (*Barbula goniospora* C. Müll., Hedwigia 37: 131, 1898).

Tortella lilliputana (C. Müll. ex Roth) Zand., *comb. nov.* (*Phascum lilliputanum* C. Müll. ex Roth., Aussereur. Laubm. 212, 1911 "*liliputanum*"; *Tetrapterum lilliputanum* (C. Müll. ex Roth) Broth.).

Tortella linearis (Web. & Mohr.) Zand., *comb. nov.* (*Barbula linearis* Web. & Mohr, Ind. Mus. Pl. Crypt. 2, 1803; *Tuerckheimia linearis* (Web. & Mohr) Britt.; *Oxystegus linearis* (Web. & Mohr.) Hilp.).

Tortella somaliae (C. Müll.) Zand., *comb. nov.* (*Hyophila somaliae* C. Müll., Linnaea 40: 293, 1876).

Tortella walkeri (Broth.) Zand., *comb. nov.* (*Hyophila walkeri* Broth., Rec. Bot. Surv. India 1: 317, 1899).

Subfamily MERCEYOIDEAE

Merceyoideae Broth., Nat. Pfl. ed. 2, 10: 246, 1924.

The Merceyoideae subclade (as constituted in Cladograms 14–16) is readily discernible in general form and composition of terminal taxa in Cladograms 2–5 and 8–10; while *Tetracoscinodon* is the basal taxon of the Merceyoideae in Cladograms 2, 5 and 8–10. The traits of the ancestral node distinguishing the Merceyoideae (and the Pottioideae) from branches lower in the tree (i.e. Timmielloideae, Erythrophyllopsoideae, Gertrudielloideae, Chionolomoideae and Timmielloideae) are stem hyalodermis absent and perichaetial leaves sheathing. The character states at the immediate ancestral node of the Merceyoideae lineage are leaves keeled above midleaf and costa grooved ventrally. The Merceyoideae may be generally described as taxa with stem sclerodermis usually differentiated from the inner cylinder of cells, leaves lanceolate, basal portion of leaf commonly differentiated in shape and ovate to elliptical, laminal margins usually recurved below, upper laminal cells comparatively small, costa usually with two stereid bands and often grooved along the ventral surface which in most taxa is only 2(–4) rows of cells across, and clavate propagula commonly present in certain genera (rarely present in other subfamilies).

Tribe TETRACOSCINODONTIEAE

Tetracoscinodontieae Zand., *tribus nov.* Type: *Tetracoscinodon* R. Br.

Plantae elongatae, longitudine 1–3 cm, tenues; caules nigri in sectione sclerodermide praediti, cellulis amplificatis cylindri centralis et filo centrali carentes; folia longitriangularia vel linearia, apicem angustatum, setaceum evolventia, costa stratis stereidarum duobus praedita; theca annulum crassum, circumstomalem evolvens.

Plants elongate, 1–3 cm in length, thin; stem black, in section with sclerodermis, enlarged cells of the central cylinder, and central strand absent; leaves long-triangular to linear, with a narrow, setaceous apex; costa with two stereid bands; theca with a thick circumstomal ring.

The tribe Tetracoscinodontieae is monotypic; it is apparently the most primitive member of the Merceyoideae. It has the characters of the single species, *Tetracoscinodon irroratus*, notably the black stem with hyalodermis absent, comparatively large central cylinder cells below a distinct sclerodermis, and central strand absent; a long-triangular or linear leaf shape and setaceous apex; superficially bulging upper laminal cells with crowded, bifid papillae; mouth of capsule with a thickened circumstomal ring, and laminae with a yellow color in KOH. Known only for New Zealand from very wet habitats.

16. TETRACOSCINODON — Plate 20.

Tetracoscinodon R. Br. ter, Trans. & Proc. New Zealand Inst. 29: 532, 1897. Type: *Tetracoscinodon hectorii* R. Br. ter.

From τετρα-, four + κόσκινον, sieve + ὀδούς (ὀδών), ὀδόντος, tooth.

Plants growing in clumps or turf, green above, tan below. *Stems blackish brown*, branching occasionally, 1–3 cm in length, transverse section rounded-pentagonal, *central strand absent*, *cells of the central cylinder large*, 30–45 µm in diam., sclerodermis present, of 1–2 layers of stereids, hyalodermis absent; axillary hairs ca. 10 cells in length, all hyaline or basal cell thicker walled; rhizoids sparse. *Leaves* when dry appressed-incurved, *when moist rigid*, *weakly spreading from the insertion*, *long-triangular to linear*, 2–3 mm in length, *upper lamina broadly and deeply channeled across leaf*, margins plane or seldom weakly recurved, entire (minutely crenulate by projecting cell walls and papillae); *apex setaceous and very narrowly obtuse*; *base triangular, broadest at insertion*; *costa percurrent*, superficial cells on both sides of costa variously papillose and quadrate or elongate and smooth, 4–6 rows of cells across costa ventrally at midleaf, costal transverse section semicircular, two stereid bands present, epidermis present on both sides of costa, guide cells 4–6 in 1 layer, *hydroid strand absent*; upper laminal cells quadrate to short-rectangular, 8–12 µm in width, 1–2:1, walls evenly thickened, superficially bulging on both sides; *papillae bifid, crowded, 3–6 per lumen*; differentiated basal cells filling the triangular leaf base, rectangular, little wider than upper cells, 3–5:1, walls evenly thickened, smooth, bulging more strongly dorsally than ventrally, somewhat longer on the margins. Dioicous. *Perichaetia* terminal, *inner leaves lanceolate*, little longer than the cauline, *convolute-sheathing in lower half*, lower cells long rhomboidal in sheathing portion. Perigonia terminal, not gemmate, inner leaves sheathing below, paraphyses linear. Seta ca. 1 cm in length, 1 per perichaetium, brown, twisted clockwise; *theca* 0.6–1.5 mm in length, *dark brown*, *short-cylindrical*, exothecial cells rectangular, thin-walled, stomates phaneropore, at base of theca, mouth of capsule rather thick-walled with a somewhat swollen *circumstomal ring several layers of cells in thickness bearing above it 4–6 rows of dark brown, vesiculose annular cells*; *peristome teeth ca. 16, flat, short-triangular, strongly incurved, much perforated and often cleft into 2–4 rami, smooth*, covered externally by a hyaline membrane, 150–200 µm in length, with ca. 6–10 articulations, straight, basal membrane absent or present but low, smooth. Operculum long-conic, usually inclined, 0.5–1.0 mm in length, cells straight. Calyptra cucullate, smooth, ca. 2 mm in length. *Spores large*, 25–35 µm in diameter, brown, papillose.

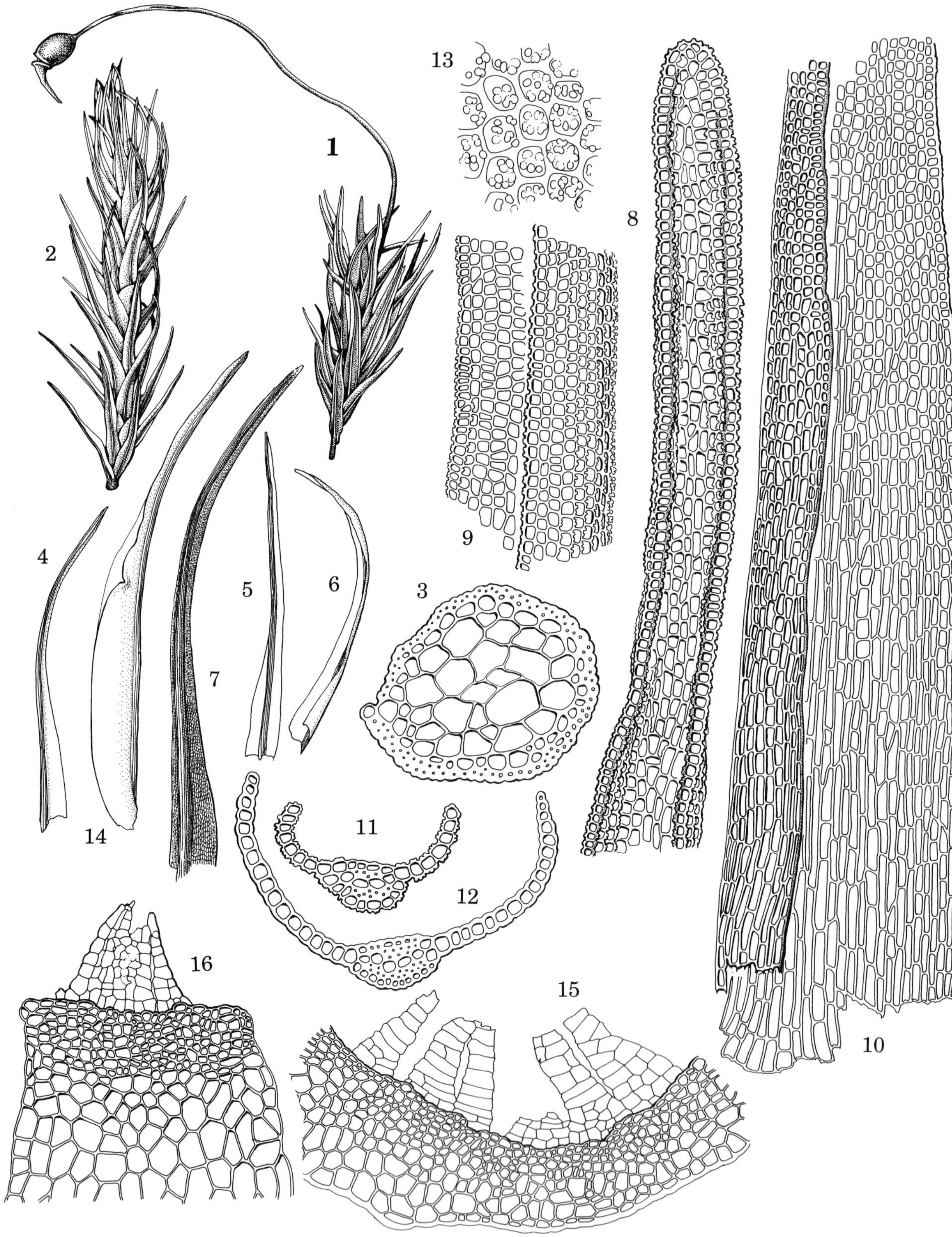

Plate 20. ***Tetracoscinodon.*** **1–16.** ***T. irroratus.*** 1. Plant with sporophyte. 2. Habit of sterile plant. 3. Transverse section of stem. 4–7. Four leaves. 8. Leaf apex. 9. Upper marginal cells. 10. Basal cells. 11. Transverse section through upper leaf. 12. Transverse section through leaf base. 13. Upper laminal papillae. 14. Perichaetial leaf. 15. Thecal mouth (vertical view) showing portion of annulus, inner cells of circumstomal ring and peristome teeth. 16. Peristome teeth (lateral view).

Laminal KOH color reaction yellow.

This is an infrequent taxon restricted to New Zealand where it is found usually encrusted with limestone deposits on dripping rocks.

Dixon (1923) was correct in synonymizing Brown's *Tetracoscinodon hectorii* with *Eucladium irroratum* as evidenced by authentic material of both names at NY, but this species differs widely from *Eucladium* ((*q.v.*) in important features that require recognition of a separate genus. *Tetracoscinodon* shares many apparently significant characters with *Leptobarbula* and *Barbula*, such as the stem section showing rather large central cylinder cells and a strong sclerodermis (Pl. 20, f. 3), a long-triangular leaf shape (Pl. 20, f. 4–7), superficially bulging upper laminal cells with crowded bifid papillae (Pl. 20, f. 13), and yellow color in KOH. It shares with *Tridontium* (Grimmiaceae, Scoulerioideae) and *Dialytrichia* a hygric habitat and relatively pachydermous capsule. Like *Tridontium* (see Excluded Taxa), the peristome (Pl. 20, f. 15–16) of *Tetracoscinodon* has an adherent hyaline membrane, but in the former genus the teeth are closely spiculose. The peristome of *Tetracoscinodon*, in being smooth, extremely thin periclinally, and in having transverse joints that often do not go all the way across the tooth but instead angle down to intersect the next upper or lower joint, has the appearance of derivation from the outer rather then the inner peristomial layer. This needs further study. The most salient characters of *Tetracoscinodon* are its black stems with large cells in the central cylinder and lacking a central strand; the rigid, long-triangular to linear leaves with narrowly obtuse apices and papillose, quadrate to short-rectangular superficial cells dorsally; the rather large and crowded bifid upper laminal papillae; the rather thick-walled basal cells (Pl. 20, f. 10); the circumstomal ring of the capsule; and the calcareous incrustation generally found on the lower parts of the plants.

Additional literature: Brotherus (1924–25), Sainsbury (1955).

Number of accepted species: 1.

Species examined: *T. irroratus* (NY).

New combination: *Tetracoscinodon irroratus* (Mitt. in Hook. f.) Zand., *comb. nov.* (*Weissia irrorata* Mitt. in Hook. f., Handb. New Zealand Fl. 404, 1867).

Tribe BRYOERYTHROPHYLLEAE

Bryoerythrophylleae Zand., *tribus nov.* Type: *Bryoerythrophyllum* Chen.

Plantae plerumque magnae; folia plerumque lanceolata, lamina superna plerumque in solutione KOH rubra, marginibus recurvis vel revolutis, saepe denticulatis vel dentatis, costa ventraliter e 2(–4) seriebus cellularum composita, in sectione complanata, strata stereidarum duo evolventi, strato dorsali saepe in sectione reniformi, filo hydroideo praedita; peristomium e 32 dentibus similaribus compositum.

Plants often large, leaves usually lanceolate, upper lamina usually red in KOH, margins recurved or revolute and often denticulate or toothed, costa with 2(–4) rows of cells ventrally, flattened in section, with two stereid bands, dorsal stereid band often reniform in section, hydroid strand present, and peristome teeth of 32 similar rami.

This subclade of the Merceyoideae is distinguished from the Tetracoscinodontieae by the leaf margins recurved to revolute below midleaf, costal ventral cells elongate, and 2(–4) rows of cells across ventral surface of costa. The traits of the immediate ancestral node are: costa flattened in section and possessing a hydroid strand, theca 1.5–3.5 mm in length, and peristome teeth of 32 similar rami. The Bryoerythrophylleae consists largely of taxa that have laminae red in KOH, upper margins commonly denticulate or toothed, the dorsal costal epidermis often absent, and a dorsal stereid band that is reniform in section. Ancestors of this tribe were probably adapted to hygric environments. The tribe's distribution is nearly worldwide, generally in mountainous areas.

17. DIALYTRICHIA Plate 21.

Dialytrichia (Schimp.) Limpr., Laubm. Deutschl. 1: 691, 1888. Type: *Dialytrichia brebissonii* (Brid.) Limpr.

Barbula subg. *Dialytrichia* Schimp., Flora 47: 211, 1865.

Barbula sect. *Dialytrichia* (Schimp.) Schimp., Syn. ed. 2: 222, 1876.

From διαλύ-, separated, disbanded (διαλύειν, to part asunder) + θρίχ, τρῖχός, hair; referring to the fine peristome teeth

Plants hydrophilic, forming a turf, yellowish green above, light brown below. Stems branching occasionally, to 2–3 cm in length, transverse section pentagonal, central strand strong, sclerodermis weak, hyalodermis absent; axillary hairs of 6–15 cells, basal 1–4 cells yellow and thicker walled; radiculose below. *Leaves appressed to contorted, conduplicate when dry*, weakly spreading when moist, *oblong to oblong-lanceolate*, 2.5–3.5 mm in length, *upper lamina narrowly grooved along costa, margins recurved below midleaf, entire, thickened as a 2–4-stratose border 2–3 cells in width, border cells quadrate* and walls weakly thickened; *apex broadly acute to rounded*; base oblong, scarcely differentiated in shape, weakly sheathing; *costa* tapering, *excurrent as a short, stout mucro, superficial cells elongate*, weakly papillose ventrally, with small, simple scattered papillae dorsally, 2–4 rows of cells across costa ventrally at midleaf, *costal transverse section reniform*, two stereid bands present (of substereid cells), dorsal band lunate, ventral epidermis present but weakly differentiated, dorsal absent or very weakly differentiated, guide cells 4 in 1 layer, *hydroid strand present*, often more than one; *upper laminal cells subquadrate, small*, 8–10 µm in width, 1:1, walls thin to weakly evenly thickened, superficially strongly bulging on both sides of lamina; *papillae small, bifid, solid, 3–4 per lumen*; basal cells differentiated across leaf, rising higher medially, rectangular, little wider than upper cells, 2–4:1, walls thin to evenly weakly thickened. Dioicous. Perichaetia terminal, inner leaves oblong-lanceolate, to 3 mm in length, sheathing in lower 1/2–2/3, cells rhomboidal and thin-walled in sheathing portion. Perigonia terminal as a single, small, inconspicuous gemmate bud. Seta ca. 0.8–1.0 cm

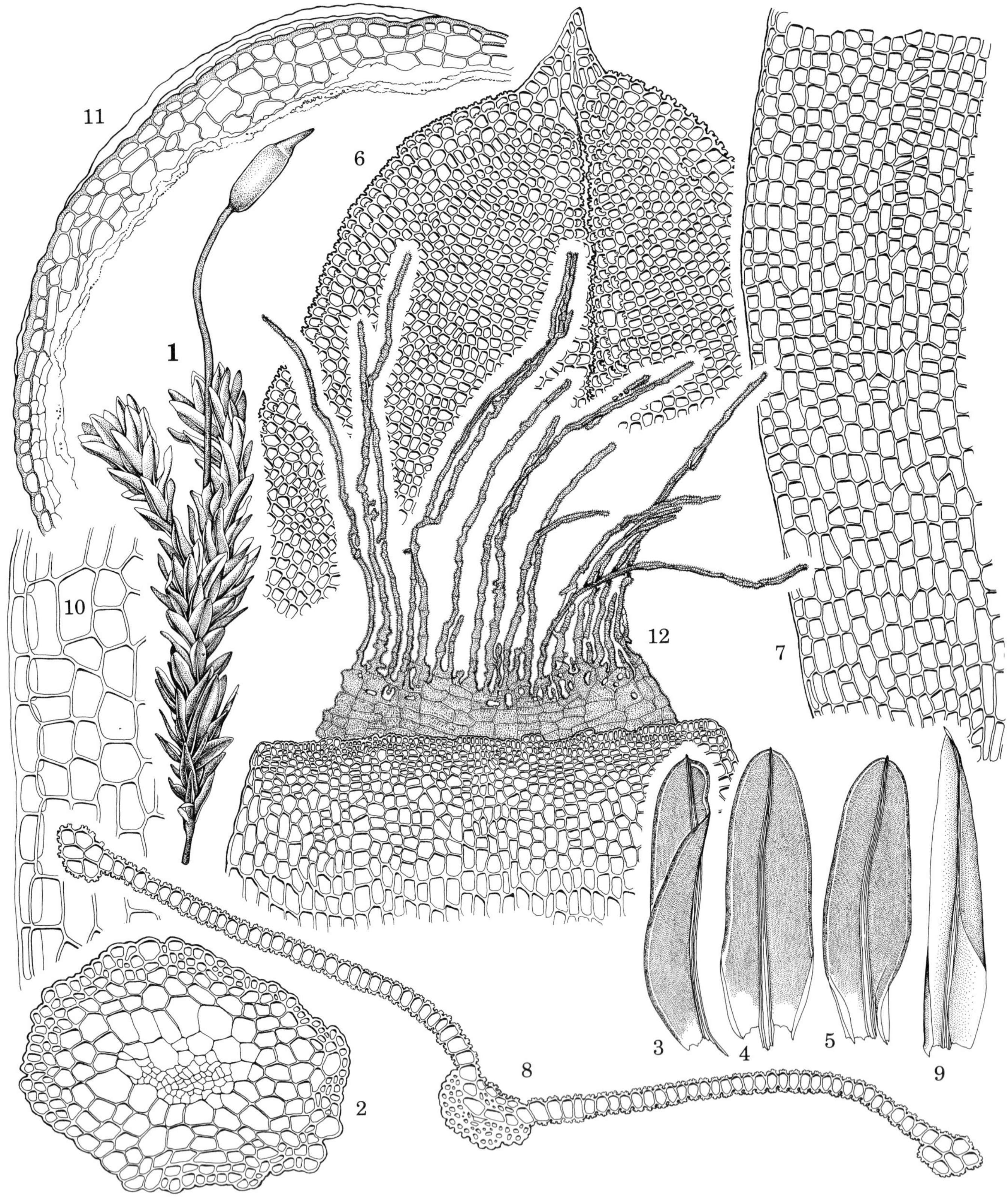

Plate 21. ***Dialytrichia.*** **1–12.** ***D. mucronata.*** 1. Habit. 2. Transverse section of stem. 3–5. Three leaves. 6. Leaf apex. 7. Basal cells. 8. Transverse section at midleaf. 9. Inner perichaetial leaf. 10. Exothecial cells. 11. Transverse section of thecal wall. 12. Peristome teeth.

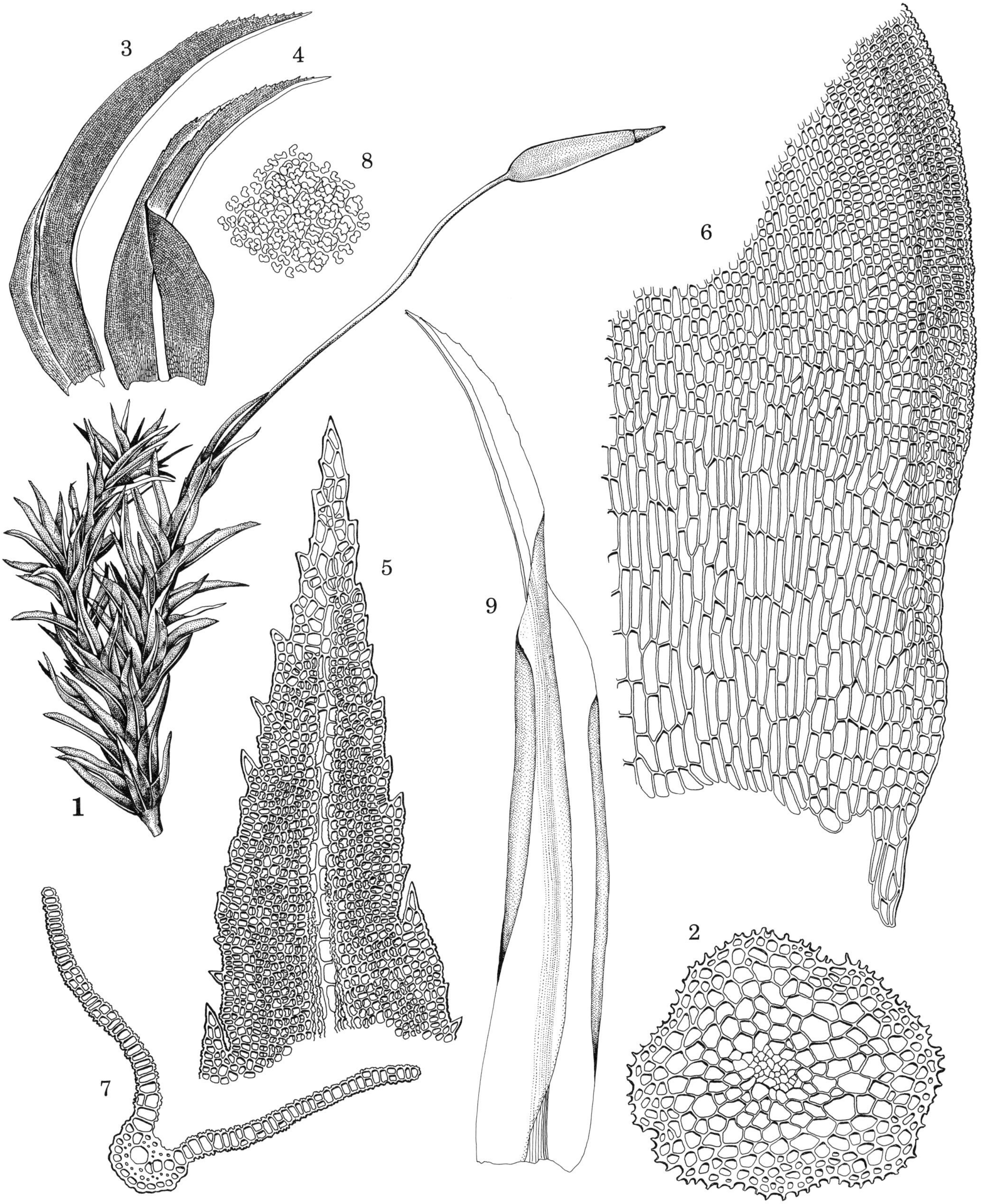

Plate 22. ***Rhexophyllum.*** **1–9.** ***R. subnigrum.*** 1. Habit. 2. Transverse section of stem. 3–4. Two leaves. 5. Leaf apex. 6. Basal cells. 7. Transverse section at midleaf. 8. Upper laminal papillae. 9. Perichaetial leaf.

in length, 1 per perichaetium, reddish brown, twisted clockwise; theca 2.5–3.0 mm in length, yellowish to orangish brown, cylindrical, occasionally curved, *thecal walls fleshy*, 3–4 cells in thickness, exothecial cell walls evenly thickened, 20–28 µm in width, 2:1, stomates phaneropore, on the fleshy neck, annulus of 2–4 rows of small, weakly vesiculose cells; *peristome teeth 32*, somewhat paired, filamentous, finely spiculose, ca. 700 µm, with many articulations, *weakly twisted counterclockwise*, basal membrane 75–90 µm in height, finely spiculose. Operculum conic, ca. 1.3 mm in length, cells twisted counterclockwise. Calyptra cucullate, smooth, ca. 3 mm in length. Spores 13–15 µm in diameter, light brown, finely papillose. *Laminal KOH color reaction yellow to yellowish orange.*

A monotypic genus found on wet calcareous rock (occasionally submerged); Europe, northern Africa, and eastern and southern Asia.

Corley et al. (1981) regard the single species of this genus as belonging to *Cinclidotus*, which Saito (1975a) regards as comprising a separate family in its own right, the Cinclidotaceae; the peristome (Pl. 21, f. 12) and laminal areolation (Pl. 21, f. 6) of *Dialytrichia*, however, are those of the Pottiaceae. The yellow KOH reaction; oblong leaves (conduplicate when dry) narrowly grooved along the costa ventrally; fine, solid, crowded, bifid papillae; mucronate apex; presence of two stereid bands in the costa (Pl. 21, f. 8); and a striking superficial resemblance to *Barbula unguiculata* are characters indicating that *Dialytrichia* is correctly placed with the Merceyoideae; cladistic study puts it with the Bryoerythrophylleae, however, rather than the Barbuleae.

It is probable that *Dialytrichia* represents a largely palearctic representative of a now largely extinct, possibly Andean assemblage (*Mironia*, for instance, has three species, two Andean, one widely distributed along the Latin American Cordillera), paralleling the situation with *Pseudocrossidium*, a large, essentially Andean genus that likewise has a few northern and in its case much reduced outlier species in the Northern Hemisphere, including Europe. *Dialytrichia* has the KOH yellow color reaction that is plesiotypic in its lineage (see Cladogram 15), and the one extant species may have survived through adaptation to a specialized habitat.

Additional literature: Bizot and Roux (1968), Györffy (1911), Sérgio and Sim-Sim (1984).

Number of accepted species: 1.

Species examined: *D. mucronata* (BUF, DUKE, NY).

18. RHEXOPHYLLUM — Plate 22.

Rhexophyllum Herz., Biblioth. Bot. 87: 38, 1916. Type: *Rhexophyllum laciniatum* Herz.

Neocardotia Thér. & Bartr. in Thér., Smiths. Misc. Coll. 85(4): 12, 1931. Type: *Neocardotia subnigra* (Mitt.) Thér. & Bartr. in Thér.

From ῥῆξις, -εως, a breaking + ο + φύλλον, leaf; the upper leaf margins of this genus split easily.

Plants forming a turf, *green to blackish green above, reddish brown below*. Stems often branching, ca. 2–4 cm in length, transverse section rounded-pentagonal, *central strand present*, sclerodermis strong, hyalodermis present; axillary hairs of ca. 10 cells, all hyaline or basal 1–3 cells brownish; rhizoids often present, reddish brown. *Leaves* conduplicate, twisted, appressed-incurved when dry, *widely spreading to squarrose when moist, lanceolate*, mostly 2.0–3.0 mm in length, *keeled above, margins recurved in lower 1/4–1/2, sharply dentate to erose-dentate* in upper 1/4–1/2; apex acute, fragile, often absent; base rectangular, somewhat sheathing, basal margins short-decurrent; *costa* excurrent as a sharp, smooth mucro, *superficial cells elongate ventrally*, 3–4 across costa at midleaf, short-rectangular to quadrate above, *serrulate dorsally* by projecting papillae at distal ends of cells, *costal transverse section reniform*, two stereid bands present, the dorsal crescent-shaped, *epidermis absent or weakly differentiated ventrally*, present dorsally as cells with enlarged lumens, guide cells 2–4 in 1 layer, *hydroid strand sometimes present*; *upper laminal cells* subquadrate to rounded-hexagonal, mostly 7–10 µm in width, 1:1, walls thin, superficially bulging, *bistratose in patches across leaf*; papillae crowded, multifid or 1–4 irregular plates per lumen, solid or hollow; basal cells differentiated across leaf base, rectangular, 8–15 µm in width 3–5:1, walls thin. Dioicous. *Perichaetia* terminal, *inner leaves highly differentiated*, long-lanceolate, 4–6 mm in length, convolute-sheathing, lower cells long-rhomboidal in lower 1/2–7/8. Perigonia terminal, leaves little differentiated from the cauline. Seta ca. 1 cm in length, 1 per perichaetium, brown, twisted clockwise above; theca 2.5–3.5 mm in length, brown, ellipsoidal to cylindrical, exothecial cells short-rectangular, 1–3:1, stomates present at base of theca; annulus of ca. 2 rows of strongly vesiculose cells, deciduous in pieces; *peristome absent*, operculum short-conic to conic-rostrate, 0.5–1.0 mm in length, cells straight. Calyptra cucullate, smooth, ca. 3.0 mm in length. Spores 13–15 µm in diameter, weakly papillose, brown. *Laminal KOH color reaction red.*

Known from southwestern U.S.A., Mexico, and the Andes of Bolivia and Peru, where it is found on soil, rock and trees in montane situations.

The genus is readily recognized by the plant coloration: blackish green above and reddish brown below; leaves lanceolate with an oblong base (Pl. 22, f. 3–4), squarrose when moist, upper margins sharply and deeply dentate (Pl. 22, f. 5); upper laminal cells bistratose in patches across the leaf (best seen in section, Pl. 22, f. 7); dorsally serrulate costa; and eperistomate capsule. It is similar to *Leptodontium* in the leaf shape and stance, elongate cells on the ventral costal surface (Pl. 22, f. 5), reniform costal section, and much differentiated perichaetial leaves, but differs mainly in the red color reaction with KOH, bistratose laminae and a hydroid strand (at least occasionally) in the costa.

Thériot and Bartram (Thériot 1931) felt that the genus was related to *Leptodontium* because of the habit of the plant, the dentate cauline leaves, and the strongly sheathing perichaetial leaves. Hilpert (1933) placed *Rhexophyllum* near *Leptodontium* and *Triquetrella*, considering the eperistomate condition to be a reduction from a leptodontioid peristome structure. He suggested that the small leaf cell size, the dense papillae, and the patchy bistratose upper lamina are xeromorphic adaptations. Cladistic analysis, however, places it with the Bryoerythrophylleae.

Additional literature: Zander (1976).

Number of accepted species: 1.

Species examined: *R. subnigrum* (BUF, FH, JE, MEXU, NY, TENN, US).

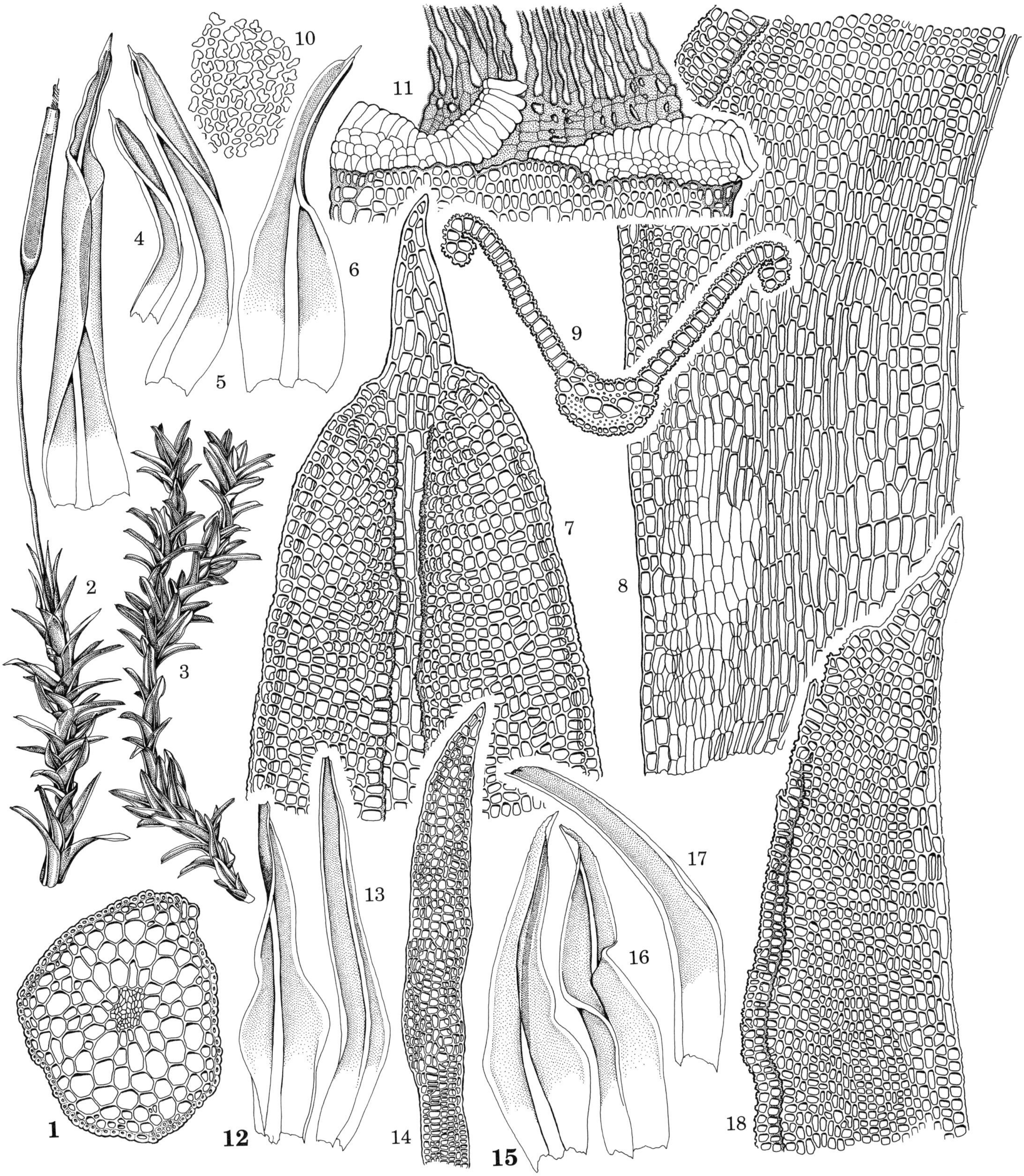

Plate 23. ***Mironia*. 1–11. *M. stenotheca*.** 1. Transverse section of stem. 2. Habit of perichaetiate plant. 3. Habit of perigoniate plant. 4–6. Three leaves. 7. Leaf apex. 8. Basal cells. 9. Transverse section at midleaf. 10. Upper laminal papillae. 11. Peristome at capsule mouth. **12–14. *M. crassicuspis*.** 12–13. Two leaves. 14. Propaguloid leaf apex. **15–18. *M. ehrenbergiana*.** 15–17. Three leaves. 18. Leaf apex, lateral view.

19. MIRONIA Plate 23.

Mironia Zander, *nom. nov.*

Morinia Card., Rev. Bryol. 37: 124, 1910. Type: *Morinia trichostomoides* (Besch.) Card., Mexico, D.F., Desierta Vieja, Bourgeau 1335, holotype, PC; isotype, NY. (*Non Morinia* Berlese & Bresadola 1889 ≡ *Rinomia* Nieuwl., *nec Morinia* Linn. 1753.)

Named for L'Abbé F. Morin, author of "Anatomie comparée et expérimentale de la feuille des Muscinées. Anatomie de la nervure appliquée à la classification," 1893, Thèse, Rennes, France. Like the fungus name *Rinomia*, the new name is a simple anagram.

Plants in cushions, yellowish or reddish green above, *reddish brown below*. Stems occasionally branching, to 4.0 cm in length, rounded-pentagonal in transverse section, central strand strong, sclerodermis present, hyalodermis absent; axillary hairs of ca. 6–10 cells, the basal 1–2 yellowish or brownish. *Leaves* twisted, appressed to erect when dry, *erect-spreading to squarrose* from top of sheathing base when wet, *ovate- to oblong-lanceolate*, 1.5–3.5 mm in length; *margins* narrowly recurved in lower 3/4, entire or dentate, *narrowly bistratose in upper 1/3*; apex broadly to narrowly acute, *occasionally thickened as a fragile or deciduous multistratose propagulum*; *base oblong and sheathing*; costa percurrent or excurrent as a short, sharp mucro, superficial cells ventrally quadrate or elongate, dorsally elongate, ca. 4–6 cells across ventral surface of costa, *costal transverse section reniform*, with two stereid bands (the ventral often weak), epidermis present on both sides (dorsally weak), guide cells 2–5 in 1 layer, *hydroid strand present*; upper laminal cells subquadrate, occasionally short-rectangular, ca. 10 µm in width, 1(–2):1, walls thin to evenly thickened, superficially flat; *papillae solid, low, flattened, granular to multifid*, 2–6 per lumen; *basal cells differentiated across leaf, filling the sheathing leaf base* and reaching somewhat higher medially, rectangular, ca. 10 µm in width, 3–5:1, walls moderately thickened to porose. Dioicous. *Perichaetia* terminal, *inner leaves oblong-lanceolate to long-oval, sheathing*, often to 6 mm in length, basal cells long rhomboidal in lower half to most of leaf. Perigonia terminal, gemmate, inner leaves ovate-triangular. Seta ca. 1–2 cm in length, yellow- to red-brown, twisted counterclockwise above, clockwise below; theca (1–)3–6 mm in length, red-brown, long-cylindrical; exothecial cells rectangular, walls thin; stomates present at base of theca, phaneropore; annulus of 1–3 rows of strongly vesiculose cells, deciduous in strips; *peristome of 32 filamentous, red, densely spiculose teeth*, up to 1.2 mm in length, of many articulations, *twisted counterclockwise about 2.5 times*, basal membrane low but distinct, spiculose. Operculum long-conic, ca. 1.3 mm in length, cells in counterclockwise twisted rows. Calyptra cucullate, smooth, ca. 4 mm in length. Spores ca. 8–11 µm in diameter, light brown to yellowish, weakly papillose. *Laminal KOH color reaction red*, seldom olive or yellow.

The genus is restricted to Mexico, Central America and the northern Andes (Venezuela, Ecuador, Colombia), generally found at high elevations on a variety of substrates, including soil, rock and bark.

Mironia is similar to *Bryoerythrophyllum* in the red coloration of the plants (especially in reaction to KOH), the well differentiated leaf base; the upper leaf cells with characteristically thin to evenly but rather weakly thickened walls, these often somewhat sinuose; the low, generally solid (in mature leaves) and bifid laminal papillae, 4–6 scattered over each lumen (Pl. 23, f. 10); and the often reniform costal transverse section generally with one or more hydroid strands (Pl. 23, f. 9). It differs in the leaves commonly keeled, with narrowly bistratose margins, and leaf apices sometimes thickened, fragile or deciduous (Pl. 23, f. 14). The peristome is very long and twisted; likewise, some species recently transferred to *Bryoerythrophyllum* from *Barbula* have well-developed peristomes (Zander 1980a).

The comparatively large, lanceolate, occasionally dentate, keeled, often squarrose leaves with highly differentiated leaf bases and the much modified perichaetial leaves are also characters of *Leptodontium*, but *Mironia* differs in the presence of an epidermis and hydroid strand(s) in the costa and of a central strand in the stem (Pl. 23, f. 1). The general leaf shape is similar to that of *Barbula*, but this last genus has unistratose leaf margins and, like *Leptodontium*, a yellow response to KOH solution.

Additional literature: Hilpert (1933), Zander (1978g, 1983d), Zander et al. (1980).

Number of accepted species: 3.

Species examined: *M. crassicuspis* (BUF, FH, MICH, TENN), *M. ehrenbergiana* (BM, BUF, DUKE, FH, NY, PC, TENN, US), *M. stenotheca* (BUF, MEXU, TENN).

New heterotypic synonymy: *Didymodon killipii* Williams = *Mironia ehrenbergiana* (C. Müll.) Zand.

New combinations: *Mironia crassicuspis* (Robins.) Zand., *comb. nov.* (*Barbula crassicuspis* Robins., Bryologist 67: 446, 1964; *Morinia crassicuspis* (Robins.) Zand.). *Mironia ehrenbergiana* (C. Müll.) Zand., *comb. nov.* (*Barbula ehrenbergiana* C. Müll., Synop. Musc. 1: 636, 1849; *Morinia ehrenbergiana* (C. Müll.) Thér.). *Mironia ehrenbergiana* var. *elongata* (Wils. in Mitt.) Zand., *comb. nov.* (*Barbula elongata* Wils. in Mitt., Kew J. Bot. 3: 51, 1851; *Morinia ehrenbergiana* var. *elongata* (Wils. in Mitt.) Zand.). *Mironia stenotheca* (Thér.) Zand., *comb. nov.* (*Barbula stenotheca* Thér., Smiths. Misc. Coll. 85(4): 21, 1931; *Morinia stenotheca* (Thér.) Zand.).

20. BRYOERYTHROPHYLLUM Plates 24–25.

Bryoerythrophyllum Chen, Hedwigia 80: 4, 1941. Type: *Bryoerythrophyllum recurvirostrum* (Hedw.) Chen.

Globulina C. Müll., Nuov. Giorn. Bot. Ital., N.S. 4: 39, 1897, *hom. illeg. non* Link in Nees 1820, *nec* Turpin 1827, *nec* Velen. 1834, *nec* Spegaz. in Sacc. 1891. Type: *Globulina boliviana* C. Müll.

Erythrophyllum (Lindb. in Braithw.) Loeske, Hedwigia 47: 175, 1908, *hom. illeg. non* J. Ag., 1872.

Erythrobarbula Steere, Bryologist 54: 199, 1951, *nom illeg. incl. gen. prior.*

Didymodon subg. *Erythrophyllum* Limpr., Laubm. Deutschl. 1: 405, 1887. Type: *Didymodon rubellus* BSG.

Barbula subg. *Erythrophyllum* (Lindb. in Braithw.) C. Jens., Skand. Bladmfl. 243, 1939.

Barbula subg. *Erythrobarbula* Szafr., Fl. Polska Mchy 1: 213, 1957 [1958], *nom. inval. descr. polon.*

Barbula sect. *Erythrophyllum* Lindb. in Braithw., Brit. Moss Fl. 1: 260, 1887. Type: *Barbula rubella* Mitt. in Lindb.

Didymodon sect. *Orthocarpae* Broth., Nat. Pfl. 1(3): 405, 1902.

Didymodon sect. *Amblystegioideae* Broth., Nat. Pfl. 1(3): 405, 1902.

From βρύον, a moss + o + ἐρυθρός, red + o + φύλλον, leaf; referring to the red cast of older leaves.

Plants turf-forming or loosely caespitose, usually green above and *red-brown below*. Stems occasionally branching, short or to several cm in length, rounded-pentagonal in transverse section, *usually with central strand*, sclerodermis usually present, hyalodermis usually absent; axillary hairs several cells in length, sometimes with 1–3 brownish basal cells. *Leaves* appressed when dry, spreading when wet, ovate, elliptical or lanceolate,

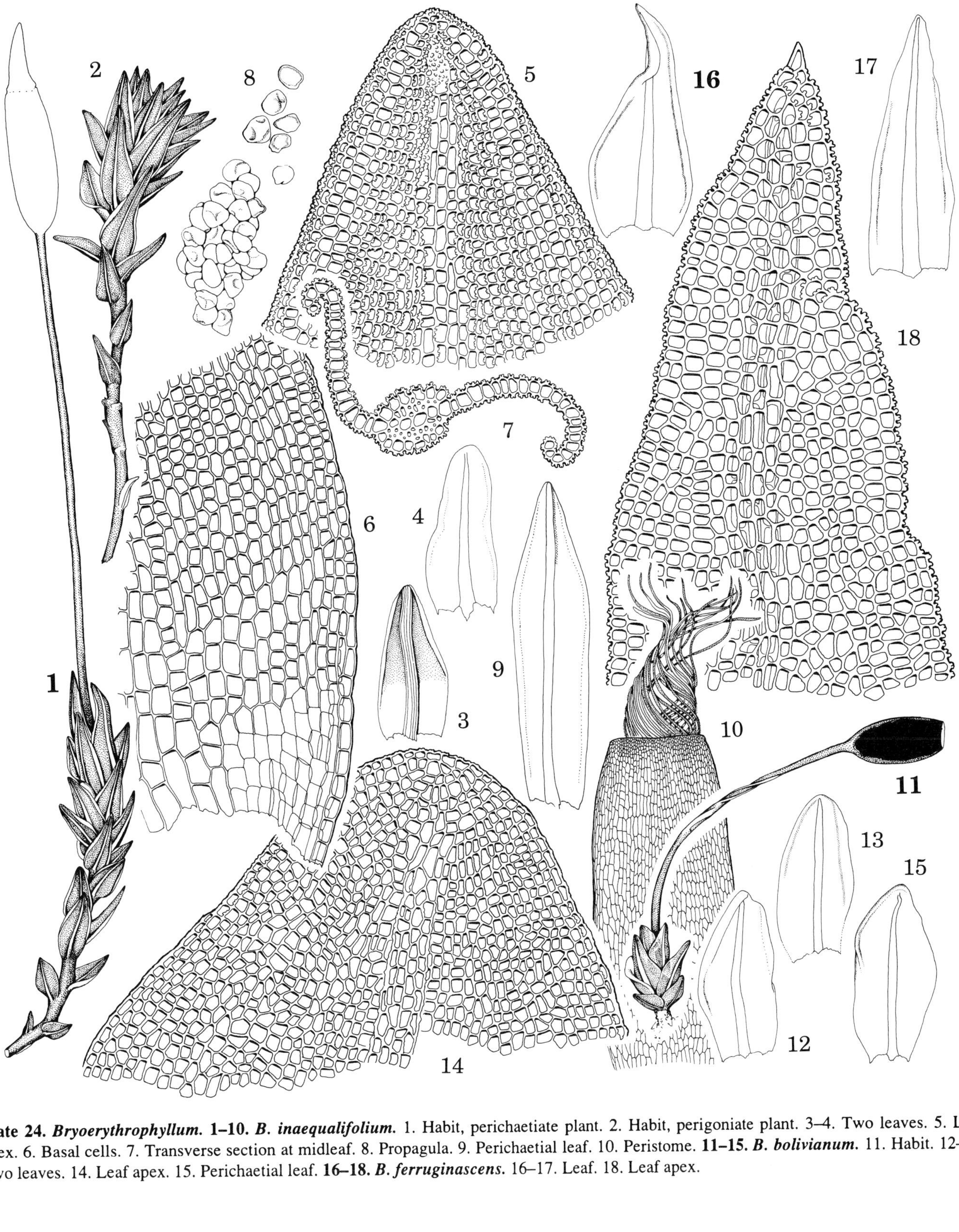

Plate 24. ***Bryoerythrophyllum.*** **1–10.** ***B. inaequalifolium.*** 1. Habit, perichaetiate plant. 2. Habit, perigoniate plant. 3–4. Two leaves. 5. Leaf apex. 6. Basal cells. 7. Transverse section at midleaf. 8. Propagula. 9. Perichaetial leaf. 10. Peristome. **11–15.** ***B. bolivianum.*** 11. Habit. 12–13. Two leaves. 14. Leaf apex. 15. Perichaetial leaf. **16–18.** ***B. ferruginascens.*** 16–17. Leaf. 18. Leaf apex.

1–5 mm in length, often grooved along the costa, marins usually recurved below or to near apex, seldom plane, *entire or more often dentate above*, often *with a border of 3–4 rows of thick-walled cells*; apex rounded-obtuse to acute; *base usually ovate to oblong and sheathing*; *costa* ending a few cells below the apex to short-excurrent, *with quadrate, papillose cells ventrally*, cells elongate dorsally, in transverse section showing a differentiated epidermis ventrally or on both sides, ca. 2–6 rows of cells across costa ventrally at midleaf, two stereid bands present, guide cells 2–4 in 1(–2) layers, one or more hydroid

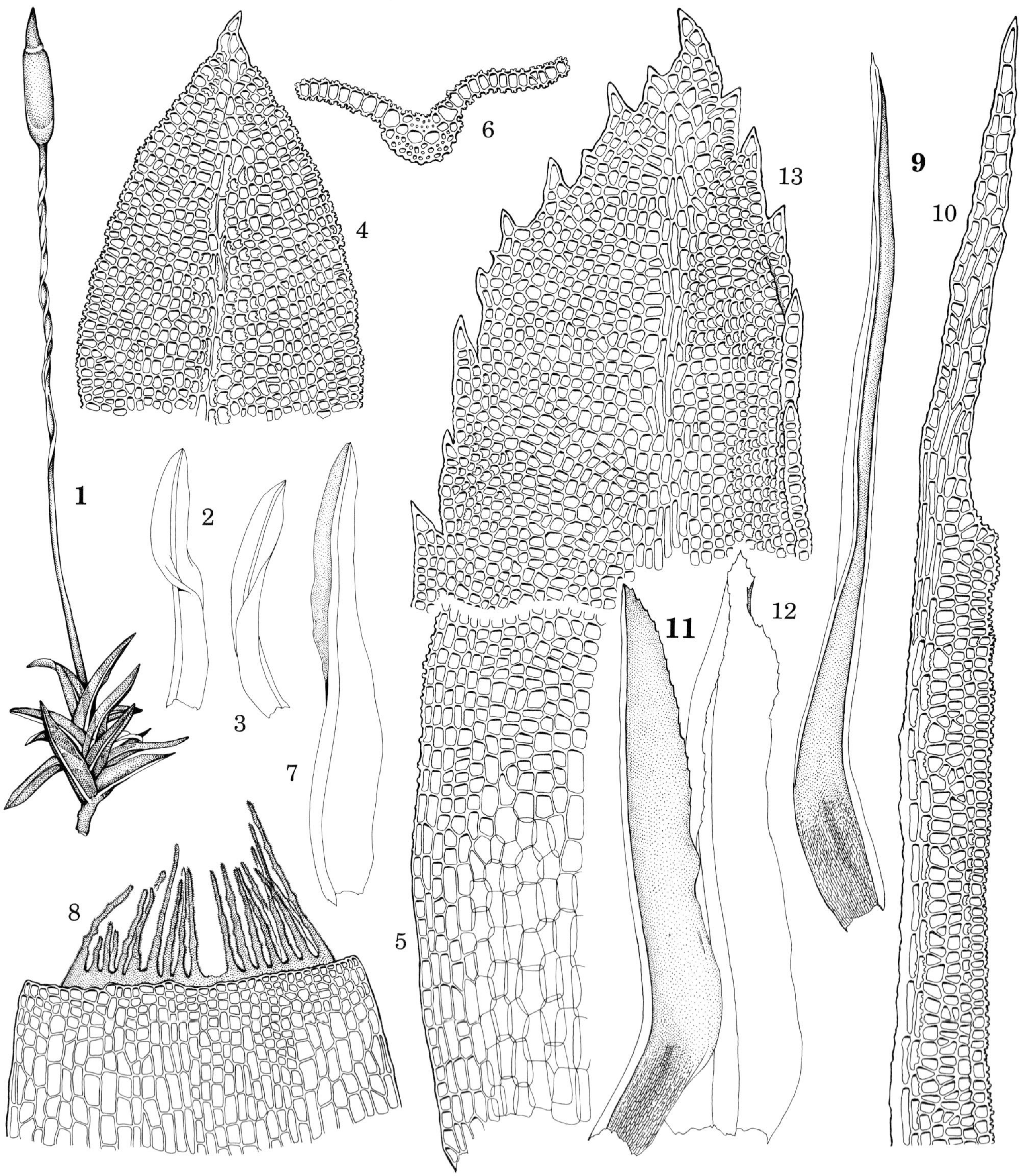

Plate 25. ***Bryoerythrophyllum.*** **1–8.** ***B. binsii.*** 1. Habit. 2–3. Two leaves. 4. Leaf apex. 5. Basal cells. 6. Transverse section at midleaf. 7. Perichaetial leaf. 8. Peristome. **9–10.** ***B. rubrum.*** 9. Leaf. 10. Leaf apex. **11–13.** ***B. wallichii.*** 11–12. Two leaves. 13. Leaf apex.

strands occasionally present; upper laminal cells subquadrate to short-rectangular, ca. 10–15 µm in width, 1(–2):1, *walls evenly thickened,* weakly sinuose, *superficially nearly flat,* occasionally bulging; *papillae solid, or occasionally hollow, bifid, or occasionally flattened or compound, mostly 4–6 per lumen, obscuring the lumens; basal cells usually hyaline and rectangular, filling most of the base medially,* little wider than upper cells, *occasionally bulging,* 2–5:1. *Asexual reproduction occasional, by unicellular propagula in masses in the leaf axils or obovoid brood bodies borne on rhizoids in the soil.* Dioicous or occasionally monoicous. Perichaetia and perigonia terminal. Perichaetial leaves usually sheathing in the lower half, larger than the stem leaves, long-oval to long-lanceolate, lower cells long rhomboidal in lower half. Perigoniate plants occasionally small, bud-like. Seta elongate, twisted clockwise below, often counterclockwise above, usually 1 per perichaetium; theca ellipsoidal to cylindric, usually red-brown, 0.7–2.5(–4.0) mm in length, sometimes curved; stomates phaneropore, at base of theca, *annulus of 1–2 rows of vesiculose cells,* often revoluble or deciduous in pieces; operculum short-conic to short-rostrate, cells in straight or oblique rows; *peristome none, rudimentary, or consisting of 16 or 32 linear rami,* erect to twisted to 2–4 times counterclockwise, usually densely spiculose, basal membrane low or absent. Operculum short-conic to short-rostrate, 0.2–1.2 in length. Calyptra cucullate, smooth, ca. 2 mm in length. Spores ca. 8–15 µm in diameter, light brown, lightly papillose. *Laminal KOH color reaction red to orange-red.* Reported chromosome number n = 13, 13+m, 14, 26+2m.

An essentially cosmopolitan genus. More extensive treatments of this taxon for the New World have been done by Zander (1978g, 1980a, 1981a).

Bryoerythrophyllum has much the appearance of *Barbula* or *Didymodon* species, but is easily distinguished by a combination of features, most of which are usually present: red coloration (at least in older parts of stems—*Bryoerythrophyllum fuscinervium,* for instance, generally having immature leaves at the stem apex with a yellow KOH reaction), the red color usually evident even without KOH; bifid, crowded laminal papillae obscuring the upper laminal cell lumens; usually flat superficial laminal cell walls (easily seen in section, Pl. 24, f. 7; 25, f. 6); and well-differentiated leaf basal cells (Pl. 24, f. 6; 25, f. 5). Like *Barbula* and *Didymodon, Bryoerythrophyllum* exhibits a full range of peristome development, some species eperistomate (e.g. *B. boliviana,* Pl. 24, f. 11), some with rudimentary peristomes (e.g. *B. rotundifolia*), others with straight, well-developed teeth (e.g. *B. binsii,* Pl. 25, f. 8, and *B. recurvirostrum*), and finally those with long, twisted peristomes (e.g. *B. inaequalifolium,* Pl. 24, f. 10). Only recently has it been recognized (Saito 1975a; Zander 1978g, 1980a) that this genus is rather large and includes many species previously placed in *Barbula* and *Didymodon. Bryoerythrophyllum fuscinervium* and the similar *B. columbianum* have a rather distinctive broad and bulging ventral costal epidermis of a single layer of cells and may be confused with *Pseudocrossidium* species (such as *P. chilense*) but the red KOH reaction and distinct ventral stereid band are diagnostic. *Bryoerythrophyllum* differs from *Syntrichia,* which is also KOH red, in having two stereid bands, a usually narrower leaf shape and dorsal costal epidermis commonly differentiated. The upper areolation of rather large cells with evenly thickened walls is similar to that of *Trichostomum* subg. *Oxystegus* but the red color in KOH, presence of a stem central strand and costal hydroid strand(s), and the basal cells extending farthest up the leaf medially distinguish *Bryoerythrophyllum* (e.g. *cf. B. chimborazense*). The genus *Mironia* is very similar, but differs in the keeled leaves with bistratose upper margins (the upper lamina of *Bryoerythrophyllum,* in one species, *B. sharpii,* may be bistratose throughout) and, in two species of *Mironia,* the presence of propaguloid leaf apices.

The genus *Globulina* was published validly by C. Müller (1897a) in a combined genus and species description. Although Müller indicated that *Seligeria globifera* Hampe (≡ *Globulinella globifera* (Hampe) Steere) was related to and was probably also a member of the genus ("*Globulina* mihi"), he left the genus monotypic with *G. boliviana.* Later, Müller (1901) established *Seligeria* subg. *Globulina* C. Müll., with *S. globifera* the only species, being thus the type of that, different, taxon. Steere (Steere & Chapman 1946) proposed *Globulinella* as a new genus, with *Globulina* C. Müll. *non* Link in Nees as a synonym, but excluded the type of *Globulina, G. boliviana,* later in the publication. Steere apparently did not intend *Globulinella* as a *nom. nov.* for *Globulina* as was suggested by Magill (1977a). This confusion is reflected in the Index Nominum Genericorum (Farr et al. 1979). Actually, *Globulina* C. Müll., *hom. illeg.,* is not a synonym of *Globulinella* Steere, but is instead a synonym of *Bryoerythrophyllum* Chen. *Seligeria* subg. *Globulina* C. Müll. is a synonym of the valid genus *Globulinella* Steere, both taxa having the same type species, *G. globifera* (lectotype *fide* van der Wijk et al., 1959–69).

The type of *Husnotiella glossophylla* Herz. is not at L, JE or W and may have been destroyed during the Second World War (pers. comm. F. K. Meyer at JE). Specimens collected by Hosseus in Argentina (FH, W) identified by Bartram as this or as *H. baueri* are eperistomate collections of the rather variable *Didymodon tophaceus.* The description and illustrations of *H. glossophylla* itself indicate that it is probably the same as *Bryoerythrophyllum calcareum.*

Additional literature: Chen (1941), Crum (1957a), Long (1982a, 1982b), Ratkowsky (1980).

Number of accepted species: 27.

Species examined: *B. andersonianum* (BUF, TENN), *B. binnsii* (BUF), *B. bolivianum* (BUF, JE, TENN), *B. byrdii* (Bartr.) Zand. (NY), *B. calcareum* (BUF, FH, MICH, TENN, US), *B. campylocarpum, B. chimborazense* (NY), *B. columbianum* (UBC, US), *B. ferruginascens, B. fuscinervium* (BUF, NY), *B. hostile* (JE), *B. inaequalifolium* (BUF, DUKE, NY, TENN), *B. jamesonii, B. ligulare* (NY), *B. recurvirostrum, B. rotundatum* (H), *B. rubrum* (BUF, NY), *B. sharpii* (BUF), *B. wallichii* (NY, US).

New heterotypic synonymy: *Bryoerythrophyllum columbianum* var. *atacamense* Zand. & Lewis in Lewis = *Bryoerythrophyllum fuscinervium* (Mitt.) Zand. *Didymodon integrifolius* Broth. in Mildbr. var. *paucidentatus* Thér. = *Bryoerythrophyllum campylocarpum* (C. Müll.) Crum. *Didymodon luzonensis* Bartr. = *Bryoerythrophyllum ferruginascens* (Stirt.) Giac. (the type of the former at FH is of fruiting material and includes the characteristic propagula). *Didymodon merceyoides* Broth. in Herz. = *Bryoerythrophyllum campylocarpum* (C. Müll.) Crum. *Didymodon pelichucensis* Williams = *Bryoerythrophyllum campylocarpum* (C. Müll.) Crum. *Hyophila calymperoides* Thér. & Nav. = *Bryoerythrophyllum campylocarpum* (C. Müll.) Crum.

New combinations: *Bryoerythrophyllum byrdii* (Bartr.)

Zand., *comb. nov.* (*Barbula byrdii* Bartr., Ann. Missouri Bot. Gard. 25: 720, 1938). *Bryoerythrophyllum chimborazense* (Mitt.) Zand., *comb. nov.* (*Tortula chimborazensis* Mitt., J. Linn. Soc. Bot. 12: 163, 1869; *Didymodon chimborazensis* (Mitt.) Broth. in Par.). *Bryoerythrophyllum fuscinervium* (Mitt.) Zand., *comb. nov.* (*Tortula fuscinervia* Mitt., J. Linn. Soc. Bot. 12: 154, 1869; *Barbula fuscinervia* (Mitt.) Jaeg. *Bryoerythrophyllum ligulare* (Mitt.) Zand., *comb. nov.* (*Tortula ligularis* Mitt., J. Linn. Soc. Bot. 12: 156, 1869; *Barbula ligularis* (Mitt.) Jaeg.).

21. PSEUDOCROSSIDIUM Plate 26–27.

Pseudocrossidium Williams, Bull. Torrey Bot. Club 42: 396, 1915. Type: *Pseudocrossidium chilense* Williams.

Barbula sect. *Revolutae* BSG, Bryol. Eur. 2: 89, 1842 (Fasc. 13–15 Mon. 27). Type: *Barbula revoluta* Brid. in Schrad.

Tortula sect. *Revolutae* (BSG) Spruce, Ann. Mag. Nat. Hist. ser. 2, 3: 377, 1849.

Barbula sect. *Platyneuron* Kindb., Eur. N. Amer. Bryin. 2: 246, 1897. Type: *Barbula platyneura* C. Müll. & Kindb.

Barbula sect. *Pseudocrossidium* (Williams) Nyholm, Ill. Fl. Nordic Mo. 2: 102, 1989.

Barbula subsect. *Revolutae* (BSG) Chen, Hedwigia 80: 209, 1941.

From ψευδής, false + o + *Crossidium*, a genus; a false *Crossidium*.

Plants growing in cushions or turf, yellowish green to brown above, brown to reddish brown below. Stems branching often, ca. 0.3–2.0 cm in length, transverse section rounded-pentagonal, central strand usually present, often strong, sclerodermis weakly differentiated, hyalodermis absent or weakly differentiated; axillary hairs of 5–8 cells, all hyaline or occasionally basal 1–3 cells thicker walled; weakly radiculose. *Leaves* appressed and often spiralled when dry, weakly or widely spreading when moist, *ovate or ligulate to lanceolate*, 0.5–3.0 mm in length, *upper lamina channeled or grooved along costa, margins recurved to broadly revolute or spiralled*, entire or occasionally weakly denticulate near apex, *the rolled margins occasionally differentiated as cylindrical photosynthetic organs of thin-walled, hollow-papillose cells*; apex acute to rounded; base scarcely differentiated in shape to oblong; *costa* often broad and flat, *often swollen medially, excurrent as a mucro or short, smooth awn, occasionally long-awned, superficial cells* quadrate to short-rectangular, papillose ventrally, *often differentiated as a pad of papillose, thin-walled photosynthetic filaments*, elongate, smooth or papillose to rough or weakly toothed dorsally, 2–5 rows of cells across costa ventrally at midleaf, *costal transverse section reniform to circular, stereid bands present or absent ventrally, present and usually strong and flattened crescent-shaped* (occasionally semicircular) *dorsally*, ventral and dorsal epidermis present, the latter often weak, *guide cells 2–4(–9) in 1–2 layers, hydroid strand present*, often multiple; *upper laminal cells subquadrate to hexagonal, often transversely elongated*, 8–16(–18) µm in width, 1:1(–3), walls evenly thickened, occasionally weakly trigonous, superficially weakly convex to bulging on both sides; papillae rarely absent, crowded, usually hollow, occasionally plate-like or bifid to multiplex, usually crowded, occasionally capitulate and solid; basal cells differentiated medially, occasionally across leaf, rectangular, 11–13(–23) µm in width, 2–6:1, walls thin to evenly thickened, occasionally porose, hyaline or occasionally orange. Propagula occasionally present, borne on ventral surface of costa or in leaf axils, clavate or spherical, 40–50 µm in length. Dioicous. *Perichaetia* terminal, *inner leaves little different from the cauline leaves or more commonly highly differentiated, enlarged, often awned, often convolute-sheathing*, lower cells not differentiated or rectangular to rhomboidal throughout. Perigonia gemmate. Seta 1.0–1.7 in length, 1 per perichaetium, yellowish to reddish brown, twisted clockwise; theca 1.6–3.0(–3.6) in length, yellowish to reddish brown, elliptical to cylindrical, occasionally curved, exothecial cells short-rectangular, 16–20 µm in width, 2–3:1, thin-walled to evenly thickened, stomates phaneropore, at base of theca, annulus of 2–4 rows of vesiculose cells, persistent; *peristome teeth 16, cleft to base or 32, linear, densely spiculose*, 350–1000 µm in length, with many articulations, twisted to once twisted counterclockwise, occasionally straight, basal membrane low or absent, weakly spiculose. Operculum short- to long-conic or conic-rostrate, 0.6–2.1 mm in length, cells counterclockwise. Calyptra cucullate, smooth, 3.2–3.5 mm in length. Spores 8–15 µm in diameter, yellow to light brown, essentially smooth to weakly papillose. Laminal KOH color reaction yellow to orange, occasionally with red blotches. Reported chromosome number n = 13.

A large genus largely growing on soil and rock at high elevations; present in mountainous regions of the New World, Europe, the Middle East, Africa, and Australasia.

Important characters for this genus are the usual small size or absence of the ventral stereid band and the broadly crescent-shaped dorsal stereid band with a clearly differentiated dorsal epidermis of cells often with semicircular lumens through differential thickening of the walls (Pl. 26, f. 7, 21; 27, f. 8, 13, 17). Other important characters are not consistent in appearance through the genus, but are often striking when found: leaves often ending in a short or long awn (Pl. 26, f. 5); differentiation of photosynthetic tissue either as a ventral pad of costal filaments (Pl. 26, f. 20, 21) or within rolled margins (Pl. 27, f. 17) or both (Pl. 27, f. 13); medial cells more papillose and thicker through (the distance between the two superficial walls) than the marginal cells (Pl. 26, f. 7; 27, f. 8); and perichaetial leaves abruptly enlarged and sheathing the seta (Pl. 26, f. 8).

A few stereid cells of a second stereid band may be found in occasional specimens of what has been called *P. aureum* in Mexico and southwestern U.S.A. This is synonymized (independently by Sollman 1990) with *P. crinitum*, which usually has two stereid bands. The North American material is quite like the type of the South American *Barbula arenicola*, specimens of which have a single stereid band but which is also here placed in the synonymy of *P. crinitum*. *Pseudocrossidium crinitum* (Pl. 26, f. 1–8) is an essentially Gondwanaland taxon, being found in southern parts of Africa, South America and Australasia, which may indicate a southern origin for the genus. The fact that *P. crinitum* also occurs in Mexico and southwestern U.S.A., albeit as a sterile, depauperate (shorter awned and smaller stature) population, shows that there can be considerable northward extension of the ranges of species of ultimately southern derivation. A simple explanation of the evolutionary history of the genus, following the thread of a previous discussion (Zander 1979f) and discounting long-distance dispersal as a factor, would be that ancestors essentially identical to modern *P.*

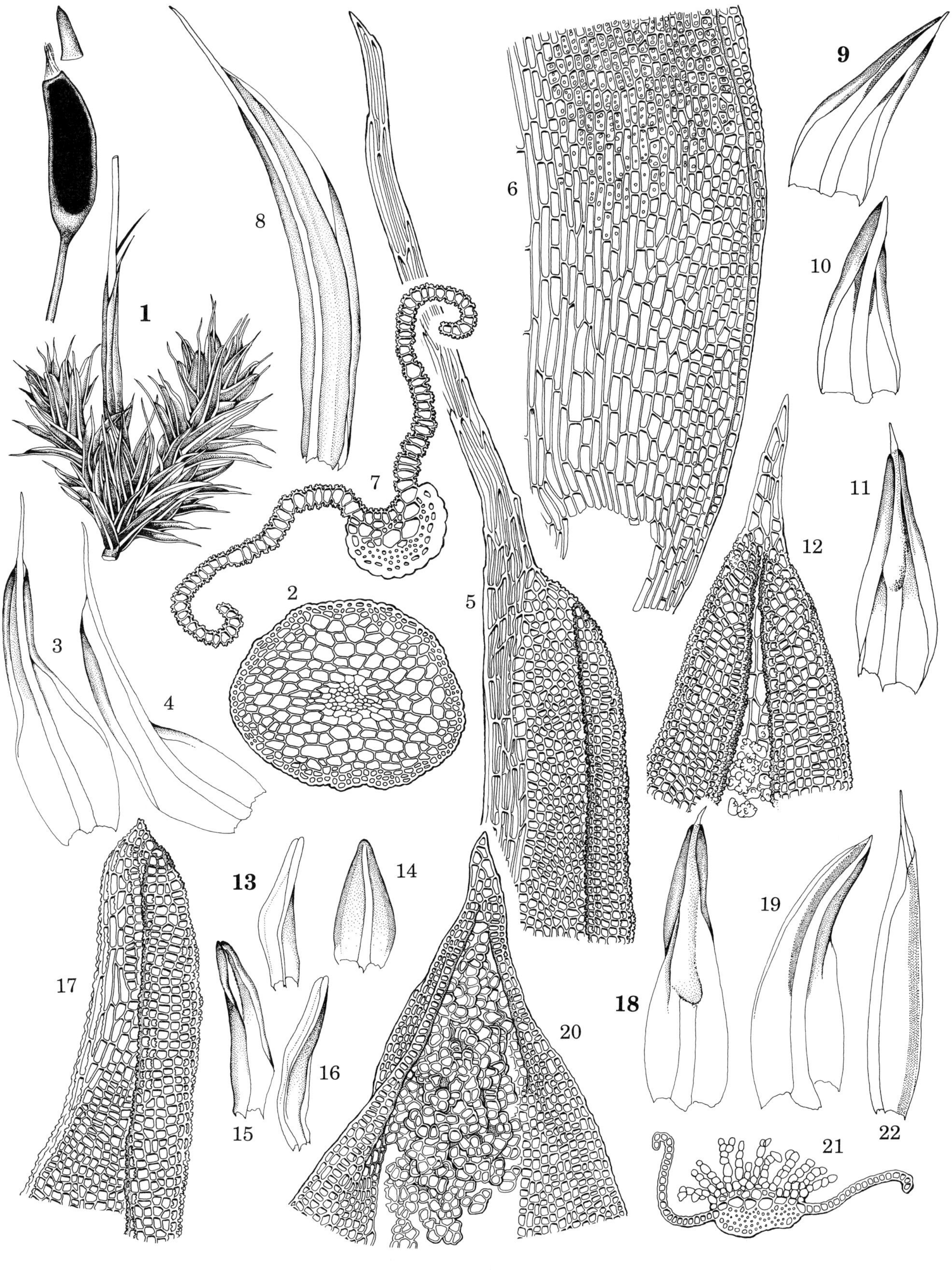

Plate 26. ***Pseudocrossidium.*** **1–8.** ***P. crinitum.*** 1. Habit. 2. Transverse section of stem. 3–4. Two leaves. 5. Leaf apex. 6. Basal cells. 7. Transverse section at midleaf. 8. Perichaetial leaf. **9–12.** ***P. apiculatum.*** 9–11. Three leaves. 12. Leaf apex. **13–17.** ***P. chilense.*** 13–16. Four leaves. 17. Leaf apex. **18–22.** ***P. elatum.*** 18–19. Two leaves. 20. Leaf apex. 21. Transverse section at midleaf. 22. Perichaetial leaf.

crinitum spread across Gondwanaland to be later isolated in austral areas through tectonic plate separation. In migrating northward along the Andes, ancestors of *P. crinitum* developed into the several species now there, these characterized by loss of the ventral stereid band and elaboration of photosynthetic tissue along the leaf margins or ventral costal surface. In North America, descendants of the derived species *P. replicatum* lost many of the characters of *Pseudocrossidium* while developing into the essentially high-northern latitude taxa *P. revolutum* and *P. hornschuchianum* (Pl. 27, f. 1–5), which may be referred to the genus by the general lack of a ventral stereid band, the highly revolute leaf margins with walls somewhat thinner and more papillose than the medial portion of the leaf, and the presence of propagula on the ventral surface of the costa (possibly a vestige of the ventral costal elaboration). Thus, *P. crinitum* and *P. revolutum*, though once both regarded as *Barbula* species, are actually at opposite ends of a complex north-south evolutionary series through mid-Andean taxa of *Pseudocrossidium*. The northernmost taxa subsequently became established in southern Africa and Australia probably through human agency. Evidence for anthropogenic disjunction is the discovery of *P. hornschuchianum* in North America in Massachusetts in the U.S.A. (Mishler & Miller 1983) and in British Columbia, Canada (Tan et al. 1981), associated with parks or gardens; Arts (1988) has reported rhizoidal tubers in the closely related *P. revolutum*. In the southern hemisphere, *P. crinitum* can be separated into two morphotypes, which may rate recognition as taxa (if so, types of synonyms, *cf.* Catcheside 1980, Magill 1981,

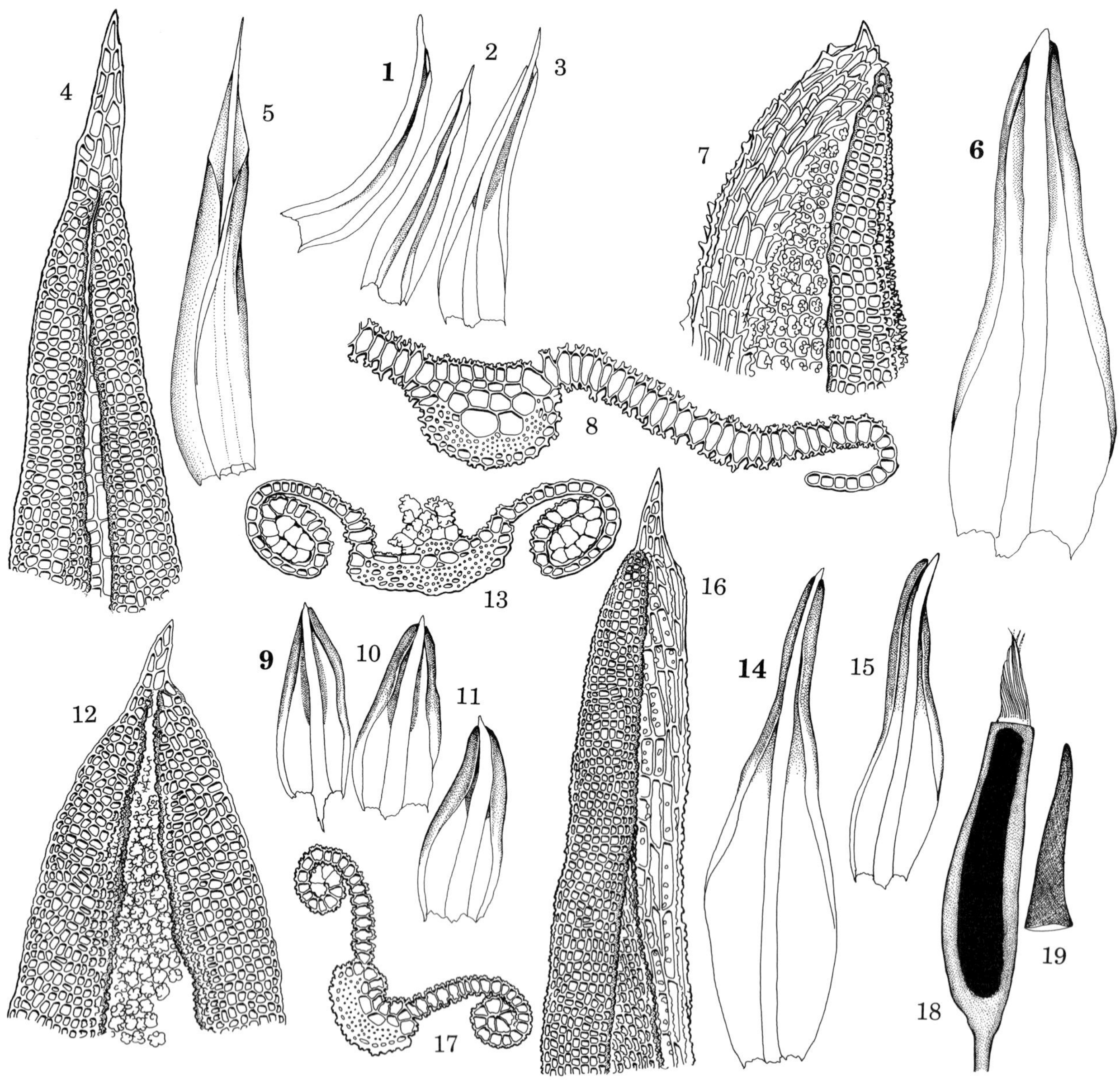

Plate 27. ***Pseudocrossidium.* 1–5. *P. hornschuchianum.*** 1–3. Three leaves. 4. Leaf apex. 5. Perichaetial leaf. **6–8. *P. leucocalyx.*** 6. Leaf. 7. Leaf apex, lateral view. 8. Transverse section at midleaf. **9–13. *P. pachygastrellum.*** 9–11. Three leaves. 12. Leaf apex. 13. Transverse section at midleaf. **14–19. *P. replicatum.*** 14–15. Two leaves. 16. Leaf apex, lateral view. 17. Transverse section at midleaf. 18. Capsule. 19. Operculum.

Weber 1972, should be consulted for the earliest name, and this would require a careful revision of the complex): one has a recurved leaf margin, coarse upper laminal papillae, dorsal costal epidermis absent or weakly developed, and ventral stereid band strong, while the second has revolute upper laminal margins, delicate upper laminal papillae, dorsal costal epidermis well developed, and ventral stereid band absent or weakly developed. Awned species of *Syntrichia* may have much the same appearance as *P. crinitum* (obtusely and broadly short-lanceolate leaves with densely papillose upper laminal cells) but differ in their distinctive red KOH color reaction, *P. crinitum* reacting deeply yellow.

Pseudocrossidium leucocalyx (Pl. 27, f. 6–8) is unusual for the genus in its costa ending in a distinctive conical, smooth cell and rough dorsally with both sharp prorulae and solid simple papillae; the upper laminal cells medially strongly bulging on both sides, with high, solid, capitulate and spiculose papillae, but marginal cells smaller in several rows, weakly bulging and smooth; and lamina bright yellow in KOH. It may belong elsewhere, possibly in a monotypic genus of its own. It is superficially similar to *Hypodontium* species, which differ significantly, however, in their strongly incurved upper laminal margins and two strong stereid bands in the costa. *Pseudocrossidium porphyreoneurum* is a South African species having a long-mucronate costa and strongly revolute leaf margins, but is unusual in its semicircular (not crescent-shaped) stereid band and basal laminal cells not differentiated from the upper cells; it is placed here only tentatively (*cf.* Magill 1981, p. 213). The genus requires a thorough revision for adequate evaluation.

Although *P. elatum* (Pl. 26, f. 18–22) is similar to *Crossidium* in its smooth upper laminal cells and rather rounded section of the costa, it is recognized here in *Pseudocrossidium* (following Delgadillo 1975a) because of its elongate stem, lanceolate leaves,a few stereid cells occasionally differentiated immediately below the ventral costal filaments, and the well-differentiated dorsal costal epidermis. The species remains, however, uncomfortably intermediate in morphology between *Pseudocrossidium* and *Crossidium* (especially *C. spiralifolium* of South Africa), while differing from both by the quite elongate patch of ventral costal filaments and poorly differentiated basal laminal cells. Eventual recognition of this species in a monotypic genus may be the best reflection of relational distances.

Additional literature: Churchill (1990), Delgadillo and Zander (1984), Frey and Kürschner (1988c).

Number of accepted species: 16.

Species examined: *P. apiculatum* (BUF, NY, US), *P. austrorevolutum* (NY), *P. carinatum* (NY), *P. chilense* (BUF, US), *P. crinitum* (NY, SPA), *P. elatum* (F, NY), *P. excavatum* (NY), *P. hornschuchianum*, *P. leucocalyx* (CU, FH, NY, US), *P. mendozense* (NY), *P. pachygastrellum* (L), *P. perrevolutum* (NY), *P. porphyreoneurum* (NY), *P. replicatum*, *P. revolutum*, *P. steerei* (BUF).

New heterotypic synonymy: *Barbula arenicola* Dus. = *Pseudocrossidium crinitum* (Schultz) Zand.

New homotypic synonymy: *Pseudocrossidium obtusulum* (Lindb.) Crum & Anderson = *Pseudocrossidium revolutum* var. *obtusulum* (Lindb.) Tan, Zand. & T. Tayl.

New combinations:

Pseudocrossidium austrorevolutum (Besch.) Zand., *comb. nov.* (*Barbula austrorevoluta* Besch. in Britt., Bull. Torr. Bot. Cl. 23: 480, 1896).

Pseudocrossidium carinatum (Gill. ex Grev.) Zand., *comb. nov.* (*Tortula carinata* Gill. ex Grev., Edinburgh J. Nat. Geogr. Sc. 2:2, 1830), near *P. crinitum.*

Pseudocrossidium crinitum (Schultz) Zand., *comb. nov.* (*Barbula crinita* Schultz, Nov. Act. Ac. Leop. Car. 11(1): 226, 1823).

Pseudocrossidium mendozense (Mitt.) Zand., *comb. nov.* (*Tortula mendozensis* Mitt., J. Linn. Soc. Bot. 12: 154, 1869; *Barbula mendozensis* (Mitt.) Jaeg.).

Pseudocrossidium perrevolutum (C. Müll.) Zand., *comb. nov.* (*Barbula perrevoluta* C. Müll., Linnaea 43: 486, 1882).

Pseudocrossidium perrevolutum var. *acutifolium* (C. Müll.) Zand., *comb. nov.* (*Barbula subrevoluta* var. *acutifolia* C. Müll., Linnaea 42: 335, 1879; *Barbula perrevoluta* var. *acutifolia* (C. Müll.) Par.), not seen.

Pseudocrossidium perrevolutum var. *linearifolium* (C. Müll.) Zand., *comb. nov.* (*Barbula subrevoluta* var. *linearifolia* C. Müll., Linnaea 42: 335, 1879; *Barbula perrevoluta* var. *linearifolia* (C. Müll.) Par.), the isotype at NY seems to be different at the species level and needs evaluation.

Pseudocrossidium porphyreoneurum (C. Müll. ex Vent.) Zand., *comb. nov.* (*Barbula porphyreoneura* C. Müll. ex Vent. in Nuovo Giorn. Bot. Ital. 4: 13, 1872; *Tortula porphyreoneura* (C. Müll.) Townsend).

Tribe LEPTODONTIEAE

Leptodontieae Herz., Geogr. d. Moose 101, 1926.
Leptodontioideae (Herz.) Hilp., Beih. Bot. Centralbl. 50: 679, 1933.

This subclade of the Merceyoideae, including *Leptodontium* and other genera once recognized as Leptodontioideae, is distinguished from the Tetracoscinodontieae by the following advanced character states of the node common to the Leptodontieae and Barbuleae: laminal papillae simple; seta nearly absent to short, less than 1 cm in length. The traits of the immediate ancestral node of the Leptodontieae are: stem central strand absent, ventral costal epidermis absent, and upper laminal cells superficially flat or weakly convex. The tribe may be further characterized generally by upper laminal margins commonly denticulate or toothed, costa flattened in section, and dorsal costal epidermis often absent. Analysis shows *Hymenostylium*, previously treated in the Pleuroweisieae (Zander 1977c), to be a member of this group. Although the geographic distribution of the tribe is nearly worldwide, certain genera are restricted to the Andes.

22. TRIQUETRELLA **Plates 28–29.**
Triquetrella C. Müll., Öst. Bot. Zeit. 47: 421, 1897. Lectotype: *Triquetrella tristicha* (C. Müll.) C. Müll. *fide* Grout, Moss Fl. N. Amer. 1: 170, 1938.

From *triquetrus*, three-edged, three-angled + *-ella*, diminutive; referring to the leaves commonly arranged in three rows.

Plants growing in mats, yellowish or blackish green above, brown or blackish brown below. *Stems* branching occasionally, to 7 cm in length, *transverse section rounded-triangular, central strand absent or present and small*, sclerodermis present, of 1–3 layers of stereid cells, hyalodermis absent; axillary hairs to 9 cells in length with cells weakly bulging, basal 1–2 cells thick-walled; weakly radiculose below. *Leaves in 3 distinct rows, the rows straight or weakly spiralling* in either direction, appressed- to spreading-incurved when dry, *widely spreading to squarrose and strongly reflexed when moist, triangular to ovate-triangular*, ca. 2 mm in length, upper lamina keeled, narrowly channeled along costa, margins recurved in middle 3/4–4/5 of leaf, *entire to weakly serrulate near apex*; apex narrowly acute to short-acuminate; base ovate, *basal leaf margins broadly long-decurrent*; *costa percurrent, superficial cells elongate and weakly papillose ventrally*, rhombic-quadrate near apex and papillose dorsally, 2(–4) rows of cells across costa ventrally at midleaf, costal transverse section ovate to semicircular, 2 stereid bands present, ventral epidermis absent, the dorsal weakly differentiated, guide cells 2–4 in 1 layer, *hydroid strand absent*; *upper laminal cells* rounded rhombic to quadrate, ca. 9–11 µm in width, 1:1, occasionally elongate transversely, *walls thickened, weakly trigonous*, superficially weakly convex to distinctly bulging on both sides; *papillae spiculose, mostly simple to bifid, 1 per* lumen; *basal cells differentiated weakly in a very small group medially*, occasionally across insertion, rectangular, little wider than upper cells, 2–4:1, walls thick-walled, somewhat porose. Dioicous. *Perichaetia both terminal and lateral in the same species and occasionally on the same plant* (acrocarpous and pleurocarpous), often in clusters, *inner leaves ovate*, rounded or short-acuminate to apiculate, to 2.5 mm in length, *strongly sheathing seta*, lower cells rhombic, thick-walled throughout. Perigonia terminal and lateral in clusters. Seta ca. 1.5–2.3 cm in length, 1 per perichaetium, yellow, twisted clockwise; theca 1.6–2.1 mm in length, yellowish brown, elliptical, exothecial cells short-rectangular, 20–25 µm in width, thin-walled, stomates phaneropore, at base of theca, annulus of ca. 3 rows of smaller, vesiculose cells; *peristome teeth 16, variously cleft* (usually to base) or occasionally perforate, *subulate, transparent, smooth or very weakly spirally striate*, 110–180 µm in length, with ca. 6 articulations, straight, basal membrane absent. Operculum conic, ca. 0.7–0.8 mm in length, cells straight. Calyptra cucullate, smooth, ca. 2.6–2.8 mm in length. Spores ca. 10–13 µm in diameter, light brown, papillose. Laminal KOH color reaction orange to yellowish orange.

A small genus found on soil or rock generally in or near dry, Mediterranean climates; Europe (Spain), South Africa, western North America (California), southern South America, Australia, Tasmania and New Zealand.

Triquetrella papillata (Pl. 28, f. 1–11, NY!) has perichaetia that in some plants are terminal and in others are distinctly lateral on the stem, often clustered below a terminal perichaetium (*cf. Ganguleea*). Perigonia, too, are clustered both terminally and laterally near the stem apex on separate, entirely perigoniate plants. The "acropleurocarpous" condition may be an anomaly, but whether this also occurs elsewhere in the genus cannot be immediately determined since perichaetia-bearing herbarium specimens are uncommon. Catcheside (1980) indicated that sporophytes are rare in South Australia, and Brotherus (1924–25) found the genus mostly sterile.

Triquetrella is similar to *Leptodontium* in the 16 smooth or nearly smooth peristome teeth, mostly cleft to the base (Pl. 28, f. 11; 29, f. 7); strongly differentiated perichaetial leaves (Pl. 28, f. 10); stem usually without a central strand (Pl. 28, f. 3); squarrose, strongly reflexed cauline leaves; and absence of a differentiated epidermal layer on the ventral surface of the costa (Pl. 28, f. 9). *Triquetrella* differs from *Leptodontium* by the occasional presence of a stem central strand (in *T. californica*, Pl. 28, f. 12–14, which also lacks the otherwise characteristic triangular section of the stem); the short, triangular cauline leaves forming three distinct rows (Pl. 28, f. 1–2); and sharp, spiculose upper laminal papillae (blunt and columnar in sect. *Leptodontium*). An intermediate taxon is *Leptodontium paradoxicum*, which shares with *Triquetrella* two characters rare in *Leptodontium*: entire leaves and dorsal costal epidermis differentiated. On the other hand, Eddy (1990) said about *Triquetrella* that "the sum of its gametophytic and sporophytic characters strongly suggests that the genus is misplaced in the Pottiaceae," without further explanation.

There are differences between the species of *Triquetrella* in length of the theca, leaf shape, acumination of the leaf apex, length of the decurrencies, and in presence or absence of the stem central strand, that might be used to distinguish taxa, but,

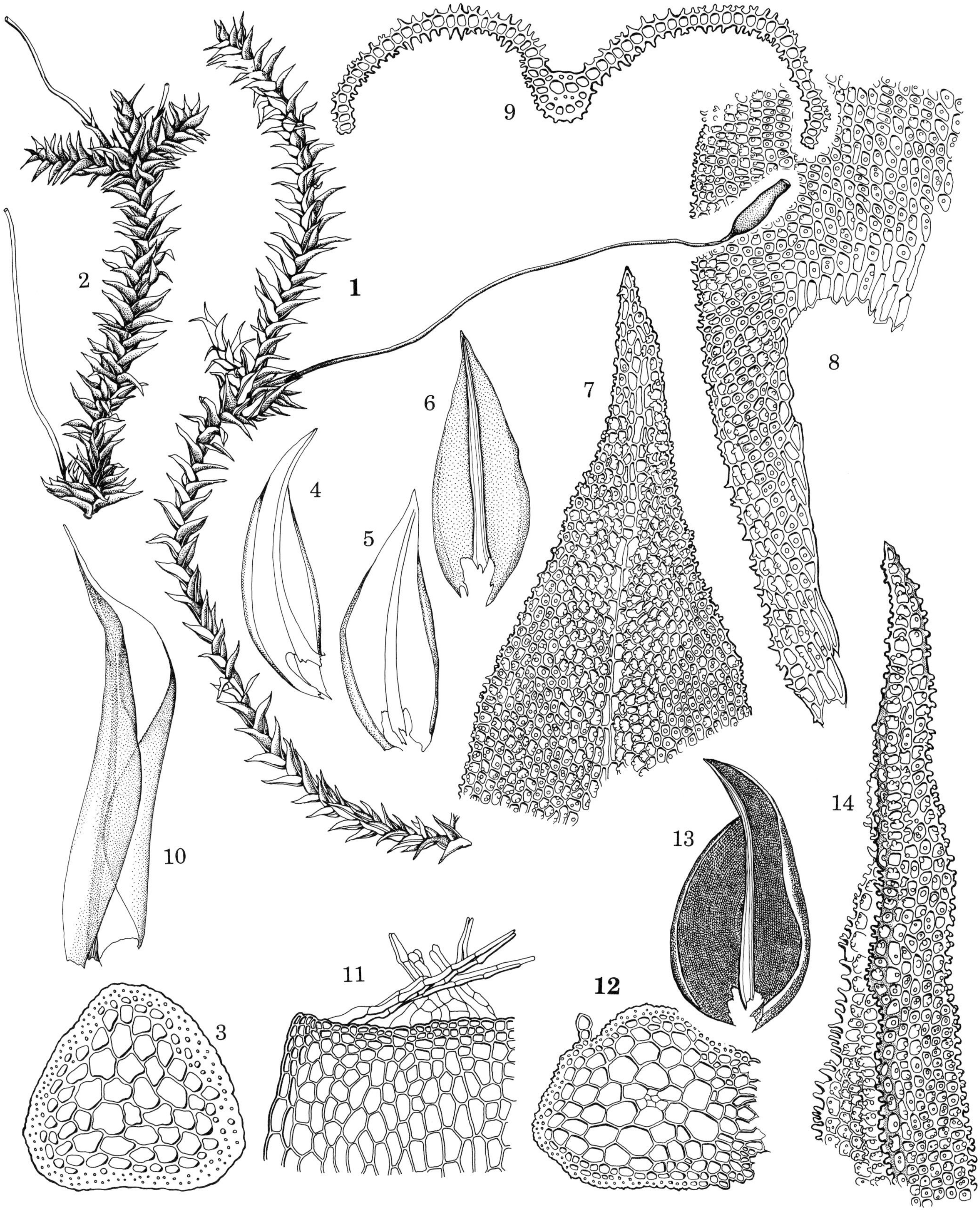

Plate 28. ***Triquetrella*. 1–11. *T. papillata*.** 1. Habit. 2. Stem with clustered terminal and lateral perichaetia. 3. Transverse section of stem. 4–6. Three leaves. 7. Leaf apex. 8. Leaf base and decurrency. 9. Transverse section at midleaf. 10. Perichaetial leaf. 11. Peristome. **12–14. *T. californica*.** 12. Transverse section of stem. 13. Leaf. 14. Leaf apex.

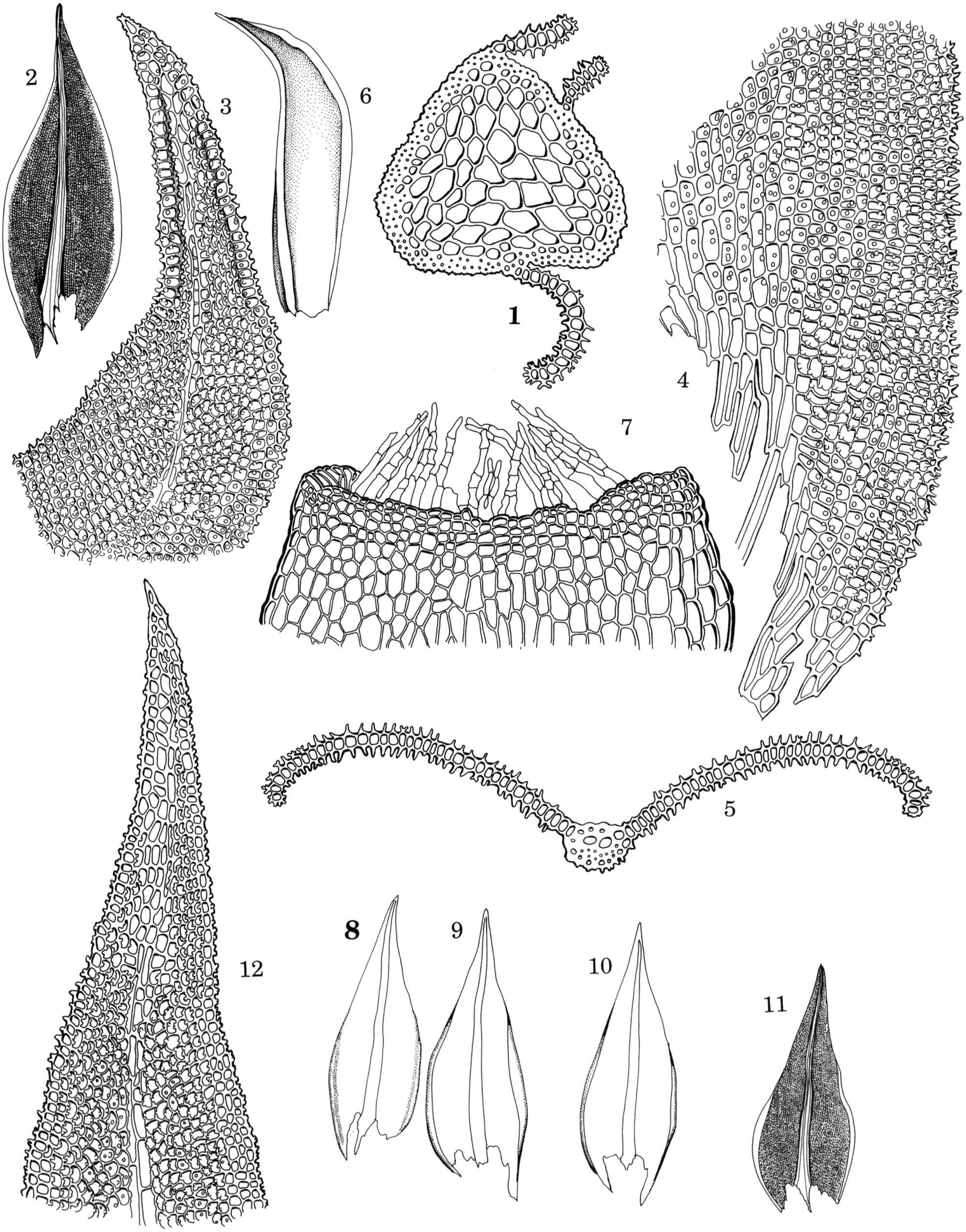

Plate 29. ***Triquetrella*. 1–7. *T. filicaulis*.** 1. Transverse section of stem (with attached leaf decurrencies). 2. Leaf. 3. Leaf apex. 4. Leaf base and decurrency. 5. Transverse section at midleaf. 6. Perichaetial leaf. 7. Peristome. **8–12. *T. tristicha*.** 8–11. Four leaves. 12. Leaf apex.

in general, the species are much alike. Casas et al. (1993) summarized the differences between most of the species in their study of *T. arapilensis*. *Triquetrella preissiana* of Australia has been synonymized with *T. papillata* by Watts and Whitelegge (1902) according to Stone and Scott (1976); the type of the former name (BM!) is indeed taxonomically *T. papillata*.

Additional literature: Cour (1955), Moore et al. (1982), Müller (1897b), Stark (1980), Zander (1980a).

Number of accepted species: 9.

Species examined: *T. arapilensis* (FH), *T. californica* (FH, NY), *T. filicaulis* (NY), *T. papillata* (BM, DUKE, FH, NY), *T. patagonica* (FH), *T. tristicha* (NY).

23. REIMERSIA Plate 30.

Reimersia Chen, Hedwigia 80: 62, 1941. Type: *Reimersia inconspicua* (Griff.) Chen.

Named for Hermann J. O. Reimers, 1893–1961, German bryologist and professorial advisor of P.-c. Chen.

Plants in cushions, light green above, light brown below. Stems branching irregularly, to 6 cm in length, *transverse section distinctly triangular, central strand absent, cells of central cylinder thick-walled*, sclerodermis present, *hyalodermis absent*; axillary hairs of ca. 10 cells, all clear or basal 1–3 cells yellow; weakly radiculose. *Leaves trifarious*, widely spreading when dry, *sharply squarrose-recurved when moist*, lanceolate, ca. 3.0 mm in length, *upper lamina strongly keeled above, margins plane*, entire; apex narrowly acute to acuminate; *base broadly ovate, sheathing the stem*, strongly convex in 2 small circular areas on both sides of costa at top of leaf base, *strongly decurrent at margins*; costa slender, excurrent as a sharp mucro, *superficial cells elongate ventrally and dorsally*, 2–4 rows of cells across costa ventrally at midleaf, costal transverse section semicircular, 2 stereid bands present (the ventral weak), *epidermis absent ventrally*, present dorsally, guide cells 2–4 in 1 layer, hydroid strand absent; *upper laminal cells quadrate to short-rectangular or rhomboidal*, 10–13 µm in width, 1–3:1, *thick-walled, trigonous or porose, superficially flat*, marginal cells narrower than medial; *papillae absent*; basal cells differentiated across the leaf, rectangular, scarcely wider than upper cells, to 4–5:1, walls thick-walled, porose. Dioicous. Perichaetia terminal, inner leaves similar to the cauline and sheathing the seta, laminal cells long rhomboidal to near apex. Perigonia terminal, gemmate. Seta 0.5–0.6 cm in length, 1 per perichaetium, yellowish brown, twisted clockwise; theca ca. 1.0 mm in length, yellowish brown, ovoid, macrostomous, neck differentiated, exothecial cell walls thick, stomates present on neck, annulus weakly differentiated, of 1–2 rows of weakly vesiculose cells; peristome absent. Operculum rostrate, erect or oblique, ca. 1.0–1.2 mm in length, cells in straight rows. Calyptra cucullate, smooth, ca. 2.0 mm in length. Spores ca. 10 µm in diameter, brown, essentially smooth. *Laminal KOH color reaction lemon yellow*. Reported chromosome number n = 13+"f".

Found on soil and calcareous rock in wet areas at moderately high elevations (2000 to 3000 m) in India, Nepal, China and the Philippines.

Reimersia is clearly like *Hymenostylium* in its hygric habitat; lack of a stem central strand (Pl. 30, f. 2) or of a ventral costal epidermis (Pl. 30, f. 7); trigonous or porose and generally rectangular upper laminal cells, which are larger medially (Pl. 30, f. 5); and eperistomate capsule with rostrate operculum. It resembles most closely *Hymenostylium recurvirostrum* var. *insigne* (Zander & Eckel 1982). The genus is easily recognized by the combination of the trifarious, epapillose leaves, which are sharply squarrose from a sheathing base; the triangular stem section; the elongate cells on the ventral surface of the costa; and the strongly porose areolation. The two concavities at the upper part of the sheathing base are distinctive, but not immediately evident on superficial examination. The capsule is not systylious, but this is not a character found in all taxa of *Hymenostylium* and is also not constant in the widespread *H. recurvirostrum*.

Chen (1941) provided a through review of the history (combinations in *Gymnostomum*, *Hymenostylium*, *Zygodon*, *Triquetrella*, *Pottia* and *Didymodon*) and morphological characters of the single species of this rare genus.

Number of accepted species: 1.

Species examined: *R. inconspicua* (BM, F, FH, NY, US).

24. HYMENOSTYLIUM Plates 31–32.

Hymenostylium Brid., Bryol. Univ. 2: 81, 1827. Type: *Hymenostylium xanthocarpum* (Hook.) Brid.

Gymnoweissia Mont. in Orbigny, Dict. Univ. Hist. Nat. 7: 402, 1849.

Hymenostelium Engl., Syllab. 48, 1892, *nom. inval. err. pro Hymenostylium* Brid.

Barbula subg. *Hymenostylium* (Brid.) Lindb., Musci Scand. 22, 1879.

Weissia subg. *Hymenostylium* (Brid.) Kindb., Eur. N. Amer. Bryin. 2: 283, 1897.

Gymnostomum sect. *Hymenostylium* (Brid.) Griff., Calcutta J. Nat. Hist. 2: 480, 1842.

Gymnostomum sect. *Vera* Griff., Calcutta J. Nat. Hist. 2: 478, 1842, *p.p.*, *nom. illeg.*

Pottia sect. *Hymenostylium* (Brid.) C. Müll., Syn. 1: 562, 1849.

Weissia sect. *Hymenostylium* (Brid.) Mitt., J. Linn. Soc. Bot. 12: 134, 1869.

Barbula sect. *Hymenostylium* (Brid.) Braithw., Brit. Moss Fl. 1: 258, 1887.

From ὑμήν, -ένος, membrane + o + στῦλος, pillar, column + *-ium*, characteristic of.

Plants growing in turfs or cushions, often flagellate, green, *often glossy*, occasionally glaucous above, *light brown below*. *Stems* often branching, to 3(–8) cm in length, occasionally papillose, *transverse section rounded-pentagonal to triangular, central strand usually absent*, sclerodermis present, *hyalodermis usually absent*; axillary hairs ca. 8 cells, basal 1–2 cells brownish or occasionally all hyaline; *often with a red tomentum*. *Leaves* often distant on stem, appressed-incurved, sometimes twisted or lax when dry, spreading, occasionally squarrose when moist, *ligulate to lanceolate or linear-lanceolate*, ca. 2.0(–3.5) mm in length, *upper lamina keeled, margins plane to broadly recurved* (occasionally revolute) *along 1 or both sides, entire* or rarely serrulate near apex by projecting cell walls, rarely bistratose in patches along margins or medially; *apex acute, occasionally obtuse or rounded*; *base scarcely differentiated in shape to oval or rectangular*, occasionally narrowly decurrent; *costa often stout*, sometimes ending 1–2 cells below apex or percurrent or more usually excurrent as a broad mucro, often "scalloped" along margins by projecting cell walls, *superficial cells usually elongate ventrally*, short- to long-rectangular dorsally, 2–4 or occasionally several rows of cells across costa ventrally at midleaf, costal transverse section semicircular to round, two stereid bands usually present, the dorsal crescent-shaped, *epidermis usually absent ventrally*, often absent dorsally, guide cells 2–4 in 1 layer, hydroid strand absent; *upper laminal cells usually heterogeneous in size and shape*, *quadrate to rectangular or rhomboidal*, ca. 8–10 µm in width, 1–3:1, *walls thin-walled to trigonous*, *often porose*, superficially flat to somewhat convex, seldom bistratose in

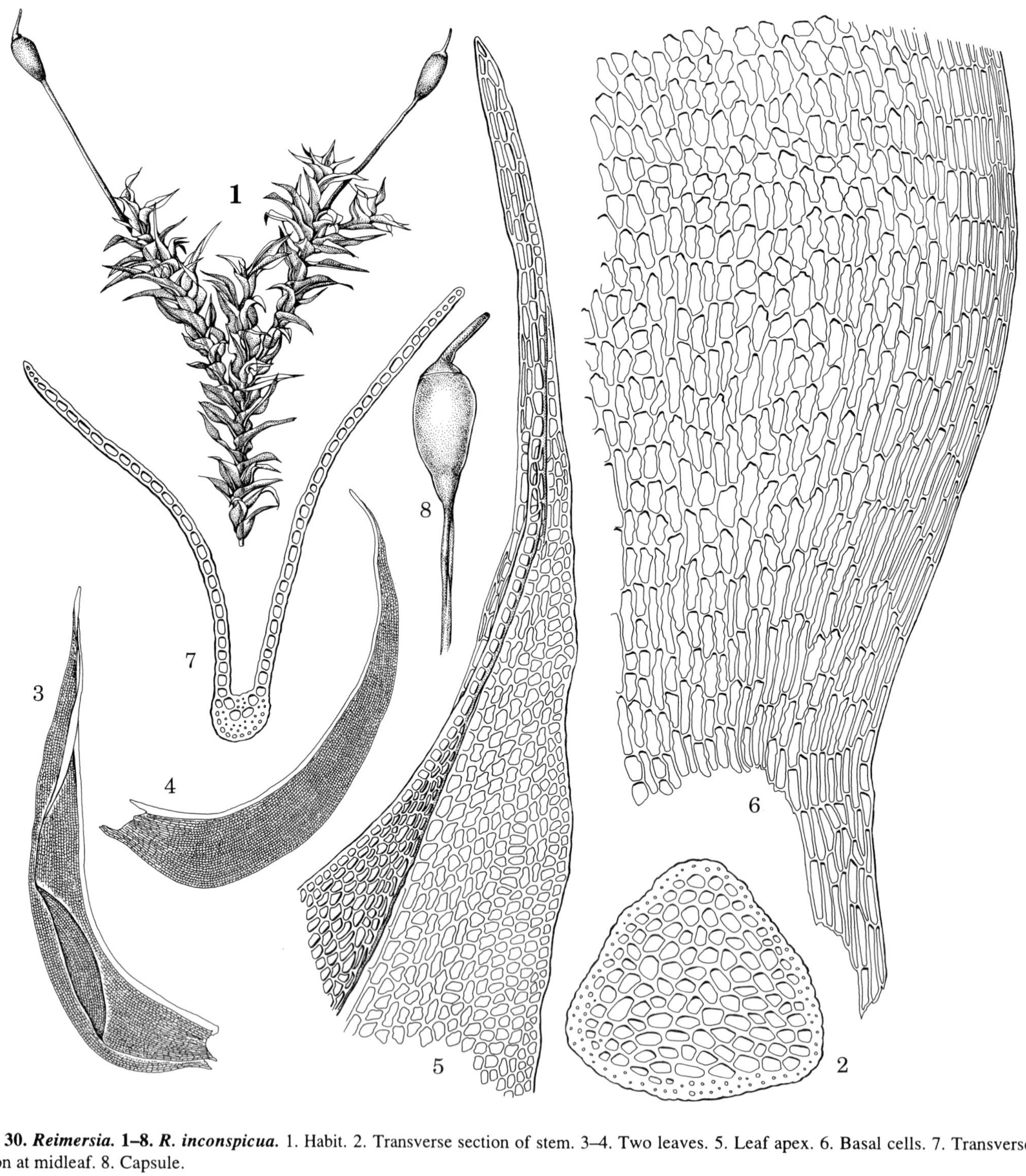

Plate 30. ***Reimersia.*** **1–8.** ***R. inconspicua.*** 1. Habit. 2. Transverse section of stem. 3–4. Two leaves. 5. Leaf apex. 6. Basal cells. 7. Transverse section at midleaf. 8. Capsule.

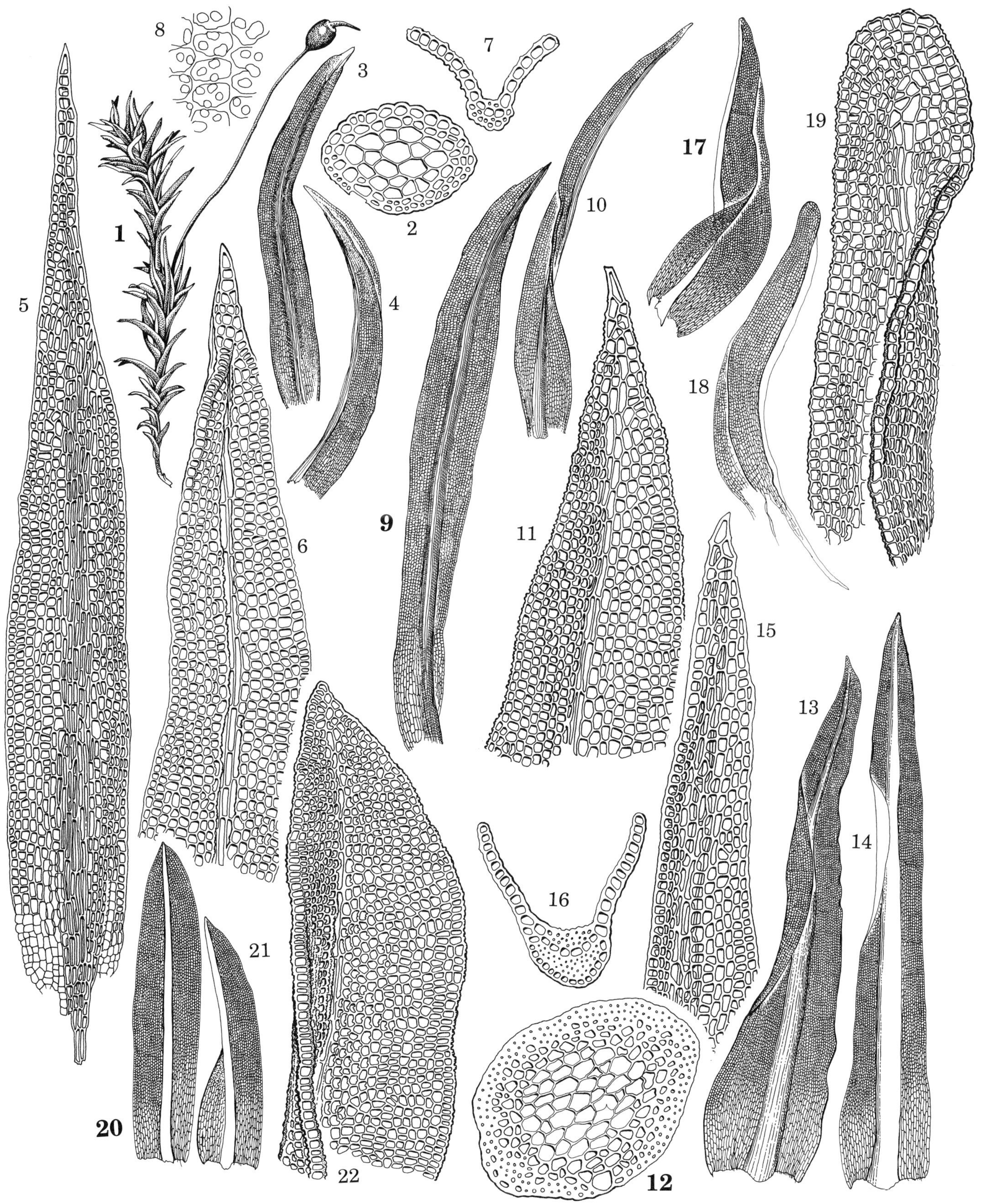

Plate 31. ***Hymenostylium*****. 1–8.** ***H. dicranelloides*****.** 1. Habit. 2. Transverse section of stem. 3–4. Two leaves. 5. Young leaf. 6. Leaf apex. 7. Transverse section at midleaf. 8. Upper laminal papillae. **9–11.** ***H. congoanum*****.** 9–10. Two leaves. 11. Leaf apex. **12–16.** ***H. crassinerve*****.** 12. Transverse section of stem. 13–14. Two leaves. 15. Leaf apex. 16. Transverse section at midleaf. **17–19.** ***H. filiforme*****.** 17–18. Two leaves. 19. Leaf apex. **20–22.** ***H. papillinerve*****.** 20–21. Two leaves. 22. Leaf apex.

patches; *papillae low, simple to granular, not obscuring lumens, centered to scattered,* rarely absent; basal cells differentiated across leaf, rectangular, little wider than upper cells, 2–4:1, walls thin to porose. Dioicous. Perichaetia terminal, inner leaves weakly differentiated, lanceolate, somewhat longer than the cauline, sometimes sheathing and lower cells inflated-rectangular in lower third. Perigonia terminal, gemmate. Seta to 1 cm in length, 1 per perichaetium, reddish or yellowish brown, twisted clockwise; capsule occasionally systylious, *theca* ca. 1 mm in length, yellowish or reddish brown, *ovoid to short-rectangular,* exothecial cells thin- to thick-walled, 1–4:1, stomates phaneropore, at base of capsule, annulus weakly vesiculose; *peristome absent.* Operculum narrowly rostrate, occasionally long-conic from a flaring base, oblique, ca. 0.5–1.0 mm in length, cells straight. Calyptra cucullate, smooth, ca. 1.0–1.5 mm in length. Spores ca. 13 µm in diameter, brownish, weakly papillose. Laminal KOH color reaction yellow. Reported chromosome number n = 12+m, 13.

Widespread in the world (North, Central and South America, Europe, Asia, Africa and Australasia), found generally on calcareous rock, rarely trees, in seepage, along streams and rivers, near waterfalls, at various elevations.

Hymenostylium is characterized by the usual lack of a stem central strand (Pl. 31, f. 2, 12; 32, f. 1, 10); mostly keeled, ligulate-lanceolate leaves with a tendency to trigonous cell walls and longitudinally elongate median leaf cells; costa with two stereid bands and the ventral epidermis usually lacking (Pl. 31, f. 7, 16; 32, f. 5, 13); upper laminal papillae mostly low, simple, not obscuring the lumens (Pl. 31, f. 8); capsules ovoid to short-rectangular, sometimes systylious; and peristome lacking. As with many hygrophiles, there is much morphological variation. In fact, in addition to presumed ecotypic differentiation, collections of this genus may be commonly found with leaves of rather different shape and areolation on the same stem. The amount of variation recognized for one species, *Hymenostylium recurvirostrum* (Pl. 32, all figures), in the New World (Zander 1977c; Zander & Eckel 1982) is inclusive of that seen in almost all species of the genus examined in the course of this study, indicating that the genus, on revision, should be found to include far fewer species than are presently accepted. The systylious capsule is not a constant feature in *H. recurvirostrum.*

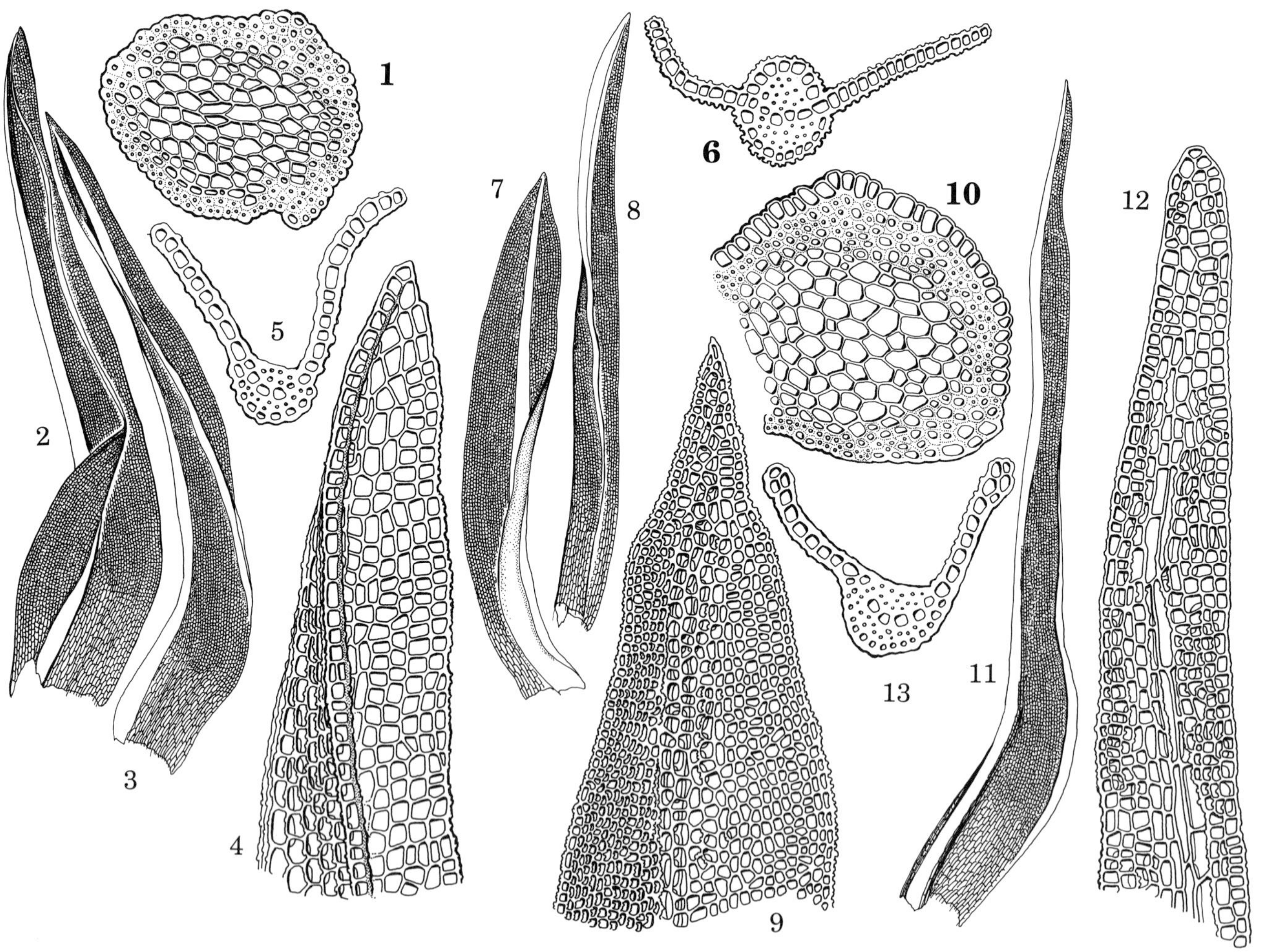

Plate 32. *Hymenostylium.* 1–5. *H. recurvirostrum* var. *recurvirostrum.* 1. Transverse section of stem. 2–3. Two leaves. 4. Leaf apex. 5. Transverse section at midleaf. ***6–9. H. recurvirostrum* var. *cylindricum.*** 6. Transverse section of stem. 7–8. Two leaves. 9. Leaf apex. ***10–13. H. recurvirostrum* var. *insigne.*** 10. Transverse section of stem. 11. Leaf. 12. Leaf apex. 13. Transverse section at midleaf.

Hymenostylium has many of the gametophyte characters of *Leptodontium*, including the usual absence of a stem central strand (always absent in *Leptodontium*); lanceolate, carinate leaves that are strongly recurved when moist; lower leaf margins often recurved; and ventral costal cells elongate (because the ventral epidermis is absent and the ventral stereid band is thus exposed). *Hymenostylium* is particularly similar to *L. viticulosoides* in the usual absence of a stem hyalodermis, often trigonous or porose laminal cell walls, upper medial laminal cells often longitudinally elongate, and laminal papillae simple. Some collections of *H. recurvirostrum* var. *cylindricum* from the West Indies with denticulate upper leaf margins bear a striking resemblance to *L. viticulosoides*. *Leptodontium* differs from *Hymenostylium*, however, in the dry to mesic habitat; broad, reniform costal section; convolute-sheathing perichaetial leaves; and long-cylindrical, peristomate capsule with well-developed annulus. Certain *Trichostomum* species (*T. tortelloides* and *T. contractum*) are also similar to *Hymenostylium* in the short, eperistomate capsules with long-conic opercula (these often falling off with the columella) and plane-margined leaves, but differ in the presence of a stem central strand and a ventral costal epidermis, broad acumination, and upper laminal cell walls thin and evenly bifid-papillose. There is, also, a similarity with *Didymodon* sect. *Fallaces*, which, in addition to having a similarly broad geographic range, also has simple laminal papillae and an exposed ventral stereid band; *D. waymouthii* and *D. brotheri*, which also grow in hygric habitats, have elongated, somewhat porose medial laminal cells, and lack a central strand, and are apparently the closest morphologically to *Hymenostylium* in *Didymodon*. Cladistic study shows a more distant relationship between *Hymenostylium* and *Didymodon* than between the former and *Leptodontium*.

Additional study is needed of West Indian populations of *H. recurvirostrum* of great morphological variability (Zander 1977c) with respect to similar variation often recognized at the specific level in Asia.

Additional literature: Andrews (1926, 1943), Azziz and Vohra (1988), Dixon (1927), Györffy (1905a), Khanna (1976).

Number of accepted species: 18.

Species examined: *H. congoanum* (BM), *H. contextum* (L), *H. crassinervium* (NY), *H. dicranelloides* (BM, NY), *H. filiforme* (BM), *H. hildebrandtii* (NY), *H. papillinerve* (BM), *H. recurvirostrum*, *H. rigescens* (H).

New heterotypic synonymy: *Barbula svihlae* Bartr. = *Hymenostylium recurvirostrum* (Hedw.) Dix. var. *recurvirostrum*. *Gyroweisia tophicola* (C. Müll.) Kindb. = *Hymenostylium recurvirostrum* (Hedw.) Dix. *Hymenostylium firmum* (C. Müll.) Broth. in Bartr. = *Hymenostylium recurvirostrum* (Hedw.) Dix. var. *recurvirostrum*. *Hymenostylium luzonense* Broth. (*H. recurvirostrum* var. *luzonense* (Broth.) Bartr.) = *Hymenostylium recurvirostrum* var. *cylindricum* (Bartr.) Zand. in Zand. & Eckel (based on *Hymenostylium glaucum* var. *cylindricum* Bartr., 1936). *Weissia venezuelensis* C. Müll. (*Gymnostomum venezuelense* (C. Müll.) Kindb.) = *Hymenostylium recurvirostrum* (Hedw.) Dix. *Molendoa burmensis* Bartr. (FH) = *Hymenostylium recurvirostrum* var. *cylindricum* (Bartr.) Zand. in Zand. & Eckel. *Molendoa sordida* (Mitt.) Steere (NY) = *Hymenostylium recurvirostrum* (Hedw.) Dix.

New combination: *Hymenostylium hildebrandtii* (C. Müll.) Zand., *comb. nov.* (*Weissia hildebrandtii* C. Müll., Linnaea 40: 298, 1876; *Gyroweisia hildebrandtii* (C. Müll.) Kindb.).

25. TRACHYODONTIUM **Plate 33.**

Trachyodontium Steere, Bryologist 89: 17, 1986. Type: *Trachyodontium zanderi* Steere.

From τρᾶχύς, rough + o + ὀδους (ὀδων), ὀδόντος, tooth + *-ium*, characteristic of; referring to the ornamentation of the peristome teeth.

Plants loosely caespitose to forming cushions, yellow-green above, light brown below. *Stems* often branching, to 3 cm in length, in transverse section rounded-pentagonal, *central strand absent*, sclerodermis differentiated but thin, hyalodermis strongly differentiated, of thin-walled, bulging cells, collapsed in older parts of stem; axillary hairs ca. 12 cells in length, basal 1–3 cells brownish; *occasionally with knotty tomentum below*. *Leaves* twisted-flexuose and spreading when dry, spreading-recurved to squarrose when moist, oblong-lanceolate, to 4 mm in length, keeled; *margins recurved in lower 1/2 to 1/3, dentate in upper 1/2 to 1/3, bordered by 1–3 rows of bistratose, elongate, thick-walled cells in upper 2/3*; apex acute; *base oblong, sheathing*; *costa* percurrent, *superficial cells elongate* and finely papillose on both surfaces, 4–6 rows of cells across ventral surface at midleaf, *costa in transverse section reniform*, with two stereid bands, *lacking differentiated epidermal cells*, hydroid strand absent, guide cells 4 in one layer; upper laminal cells quadrate to hexagonal or short-rectangular, ca. 12–14 µm in width, 1(–2):1, walls evenly thin, superficially convex but not strongly bulging; *papillae simple, small, scattered*, 6–8 per lumen; *basal cells strongly differentiated* across leaf base, reaching higher medially, rectangular, 10–14 µm in width, 4–7:1, thin-walled, mostly weakly papillose. Propagula absent. Dioicous. Perichaetia terminal, *inner leaves long-lanceolate*, to 10–11 mm in length, *sheathing the seta*, porose-rectangular, thick-walled cells reaching up 1/2 to 2/3 leaf length. Perigonia terminal, gemmate, inner leaves deltoid. Seta 1.2–1.4 cm in length, brown, twisted clockwise above; *theca* 5.5–6.0 mm in length, brown, smooth, *long-cylindrical*; exothecial cells rectangular, thick-walled; annulus of deciduous, vesiculose cells, *stomates absent*; *peristome of 64 linear teeth, arranged in 16 groups of 4*, to 600 µm in length, up to 13 articulations, straight, red-orange, *densely spiculose* grading below to low ridges, *basal membrane absent*. Operculum and calyptra not seen. Spores rather variable in size, 13–33 µm in diameter. *Laminal KOH color reaction yellow with occasional red patches*.

Found on large grass culms and small tree trunks, ca. 2650 m elevation, endemic to Ecuador.

The salient distinguishing features of this genus are the cartilaginous leaf border (Pl. 33, f. 8), simple laminal papillae (Pl. 33, f. 9), elongate cells on both surfaces of the costa (Pl. 33, f. 6), two stereid bands, and spiculose peristome teeth arranged in groups of four each (Pl. 33, f. 11). It has a superficial resemblance to *Calyptopogon* by the leaf shape and border, the simple papillae, and the lack of a stem central strand, but *Calyptopogon* is easily distinguished from *Trachyodontium* by its semicircular costal transverse section showing a single stereid band and a hydroid strand.

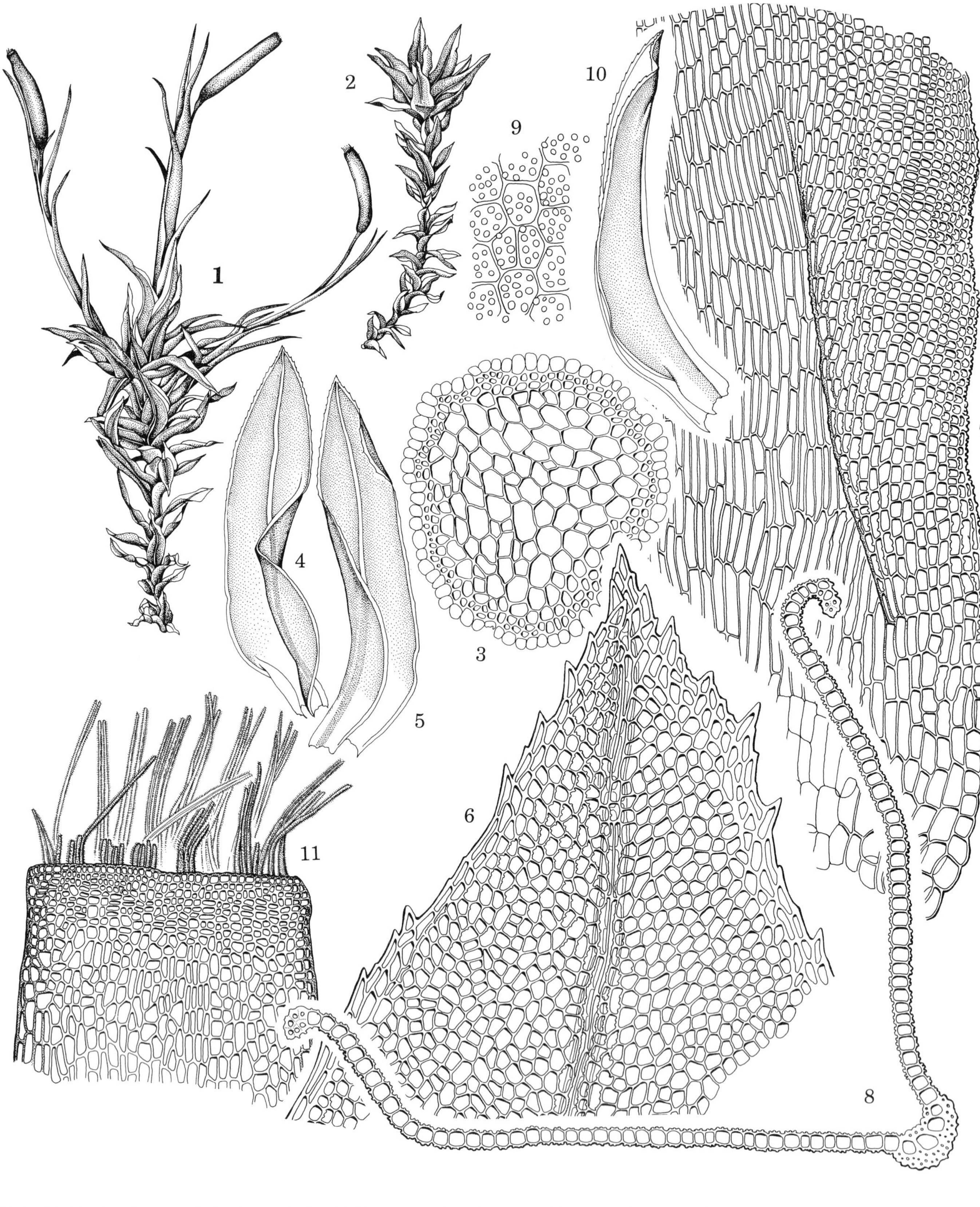

Plate 33. ***Trachyodontium.* 1–11. *T. zanderi.*** 1. Habit, perichaetiate plant. 2. Habit, perigoniate plant. 3. Transverse section of stem. 4–5. Two leaves. 6. Leaf apex. 7. Basal cells. 8. Transverse section at midleaf. 9. Upper laminal papillae. 10. Perichaetial leaf. 11. Peristome.

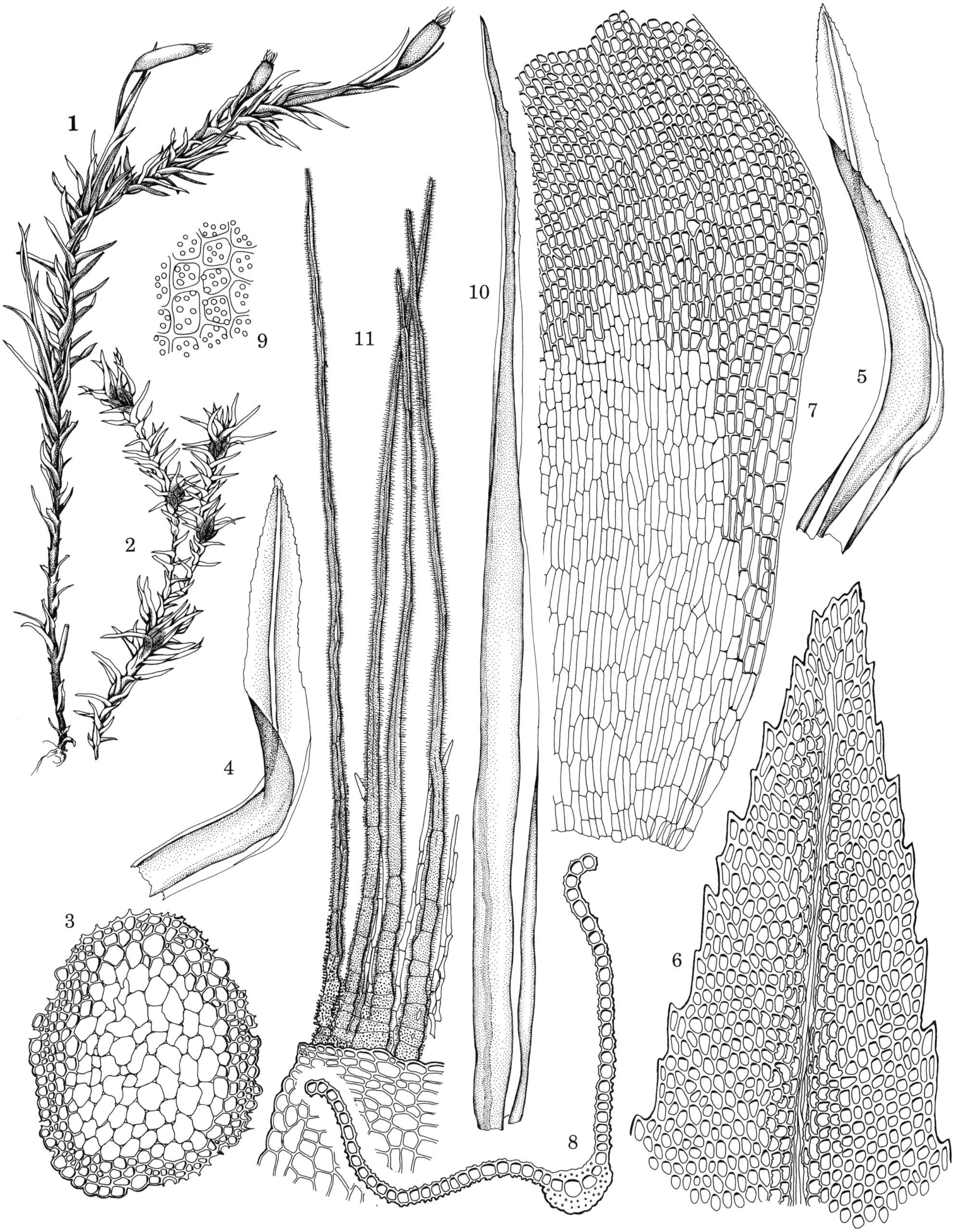

Plate 34. ***Streptotrichum.*** **1–11.** ***S. ramicola.*** 1. Habit of perichaetiate plant. 2. Perigoniate plant. 3. Transverse section of stem. 4–5. Two leaves. 6. Leaf apex. 7. Basal cells. 8. Transverse section at midleaf. 9. Upper laminal papillae. 10. Perichaetial leaf. 11. Portion of peristome (note perpendicular aspect of teeth formed from inner anticlinal walls).

Trachyodontium agrees with salient features of closely related members of the Leptodontieae in its large size, the convolute-sheathing perichaetial leaves (Pl. 33, f. 1), elongate-cylindrical capsule lacking stomates, and spreading-recurved to squarrose, dentate leaves with reniform costal transverse section showing two stereid bands (Pl. 33, f. 8). It is particularly close to *Leptodontium* through the characteristic laminal KOH color reaction and the lack of both a stem central strand and differentiated costal epidermal cells. *Leptodontium*, unlike *Trachyodontium*, has spirally grooved peristome teeth, but a closely related genus, *Streptotrichum*, has spiculose teeth. Another related genus, *Leptodontiella*, has 16 peristome teeth similarly divided into three or four rami (these not spiculose, however).

Spindle-shaped propagula are common in the material of *Trachyodontium* examined, but evidently belong to a *Zygodon* species occurring in mixture.

Literature: Steere (1986).

Number of accepted species: 1.

Species examined: *Trachyodontium zanderi* (BUF).

26. STREPTOTRICHUM — Plate 34.

Streptotrichum Herz., Biblioth. Bot. 87: 37, 1916. Type: *Streptotrichum ramicolum* Herz.

From στρεπτός, twisted + o + θρίχ, τρῖχός, hair; the peristome teeth are, however, not sensibly twisted.

Plants in loose mats, green above, tan below. *Stems* branching often, to 4 cm in length, transverse section rounded-pentagonal, *central strand absent, sclerodermis present, hyalodermis present;* axillary hairs of ca. 7 cells, the basal 2–3 brownish and firm-walled; *knotty tomentum present* but sparse. *Leaves* spreading-twisted when dry, *squarrose-recurved when moist, lanceolate*, ca. 3.0–3.3 mm in length, *upper lamina keeled, margins recurved below, dentate above midleaf*; apex acute; *base sheathing* in lower 1/4 to 1/3 of leaf; costa ending 2–4 cells below the apex, *superficial cells elongate on both sides of costa*, 4–7 rows of exposed stereid cells across costa ventrally at midleaf, *costal transverse section reniform*, two stereid bands present, *epidermis absent ventrally and dorsally*, guide cells 4 in 1 layer, hydroid strand absent; upper laminal cells quadrate to short-rectangular, ca. 9–12 µm in width, 1–2:1, walls evenly thickened, superficially flat to weakly convex; *papillae very small, simple, solid, ca. 6–7 per lumen*; basal cells differentiated across leaf base, rectangular, little wider than upper cells, 3–4:1, walls evenly thickened. Dioicous. *Perichaetia* terminal, or subterminal by innovations, *inner leaves long-lanceolate, to 8 mm in length, convolute-sheathing, hiding greater portion of seta, cells long-rhomboidal, thick-walled and porose.* Perigonia gemmate, terminal or subterminal by repeated innovations. *Seta rather short*, ca. 5–7 mm in length, 1(–2) per perichaetium, yellowish brown, not twisted; theca 2–3 mm in length, yellowish brown, cylindrical, occasionally somewhat ventricose, exothecial cells thin-walled, short-rectangular, *stomates absent,* annulus of 1–2 rows of vesiculose cells, persistent; *peristome teeth 16, bright orange-red, linear, cleft into two rami from near base, occasional interpolated rami arise from anticlinal walls at margins of regular teeth, densely branching-spiculose,* 500–1000 µm in length, with many articulations, straight or very weakly twisted counterclockwise, basal membrane low, ca. 70 µm high in height, sharply granulose to spiculose-papillose, occasionally striate in patches. (Operculum reportedly conic-rostrate, not seen.) Calyptra cucullate, smooth, ca. 3 mm in length. *Spores large*, ca. 20 µm in diameter, brown, low, spiculose-papillose. *Laminal KOH color reaction yellow, with red mottling.*

Endemic to Bolivia, where it is known from a single collection from a joint of a bamboo-like grass at 3400 m elevation.

Streptotrichum is close to *Leptodontium* in the lack of a stem central strand (Pl. 34, f. 3); the knotty tomentum; the squarrose-recurved, lanceolate leaves (Pl. 34, f. 4–5) with a reniform costal section lacking differentiated epidermal layers (Pl. 34, f. 8); and the highly differentiated perichaetial leaves (Pl. 34, f. 10). It differs from *Leptodontium*, as does the similar *Leptodontiella*, in having an unusual peristome, in this case of 16 deeply bifid, densely spiculose teeth, each of which occasionally bears an additional ramus or two formed from an interior anticlinal wall (Pl. 34, f. 11). These extra teeth are similar to the other teeth except they are shorter and appear edge on when viewed laterally from outside the capsule. *Streptotrichum* has much the same lax habit (Pl. 34, f. 1–2) and short capsules (though this is variable) as does *Leptodontiella*, but the peristome is quite different.

Literature: Herzog (1916).

Number of accepted species: 1.

Species examined: *S. ramicola* (Bolivia, Waldgrenze über Tablas, Herzog 2844, syntypes, JE, W).

27. LEPTODONTIELLA — Plate 35.

Leptodontiella Zand. & E. Hegew. Type: *Leptodontiella apiculata* (Zand.) Zand. & E. Hegew.

From *Leptodontium*, a genus + i + *-ella*, diminutive; resembling the genus *Leptodontium*.

Plants in loose mats, slender, often branching, greenish-brown. Stems to 3.0 cm in length, rounded-pentagonal in section, *central strand lacking*, sclerodermis strong, *hyalodermis distinct, collapsed*; axillary hairs of ca. 6 cells, basal 2 firm-walled, yellowish. *Leaves distant, appressed and flexuose when dry, spreading-recurved when moist, distinctly trifarious, oblong-lanceolate to lanceolate*, to 2.5 mm in length, carinate above; apex narrowly acute, *usually ending in a long, pellucid cell or fine apiculus*; margins weakly recurved below, decurrent, *distantly dentate in the upper 1/3*; base ovate in shape; upper laminal cells rounded-quadrate to short-rectangular, 9–12 µm in width, 1(–2):1; *costa* subpercurrent to percurrent, cells elongate and smooth ventrally and dorsally, ca. 4 cells across surface ventrally, with two stereid bands, the dorsal crescent-shaped, *lacking differentiated epidermal cells on both sides of the costa*, guide cells ca. 4(–6) in one row, *hydroid strand absent*; upper laminal papillae simple to bifid, spiculate or scablike, several scattered over each lumen; *walls* thin or collenchymatous, *superficially flat*; basal cells quadrate to short-rectangular and chlorophyllose to abruptly enlarged and hyaline medially, little wider than the upper cells, to 1–4:1, bordered marginally by several rows of quadrate to short-rectangular cells. *Propagula* occasionally present, borne on the stem, cylindrical, 70–120 µm in length, of ca. 3–5 uniseriate cells, occasionally with longitudinal internal walls, *dark* brownish green excepting a single transparent basal stalk cell.

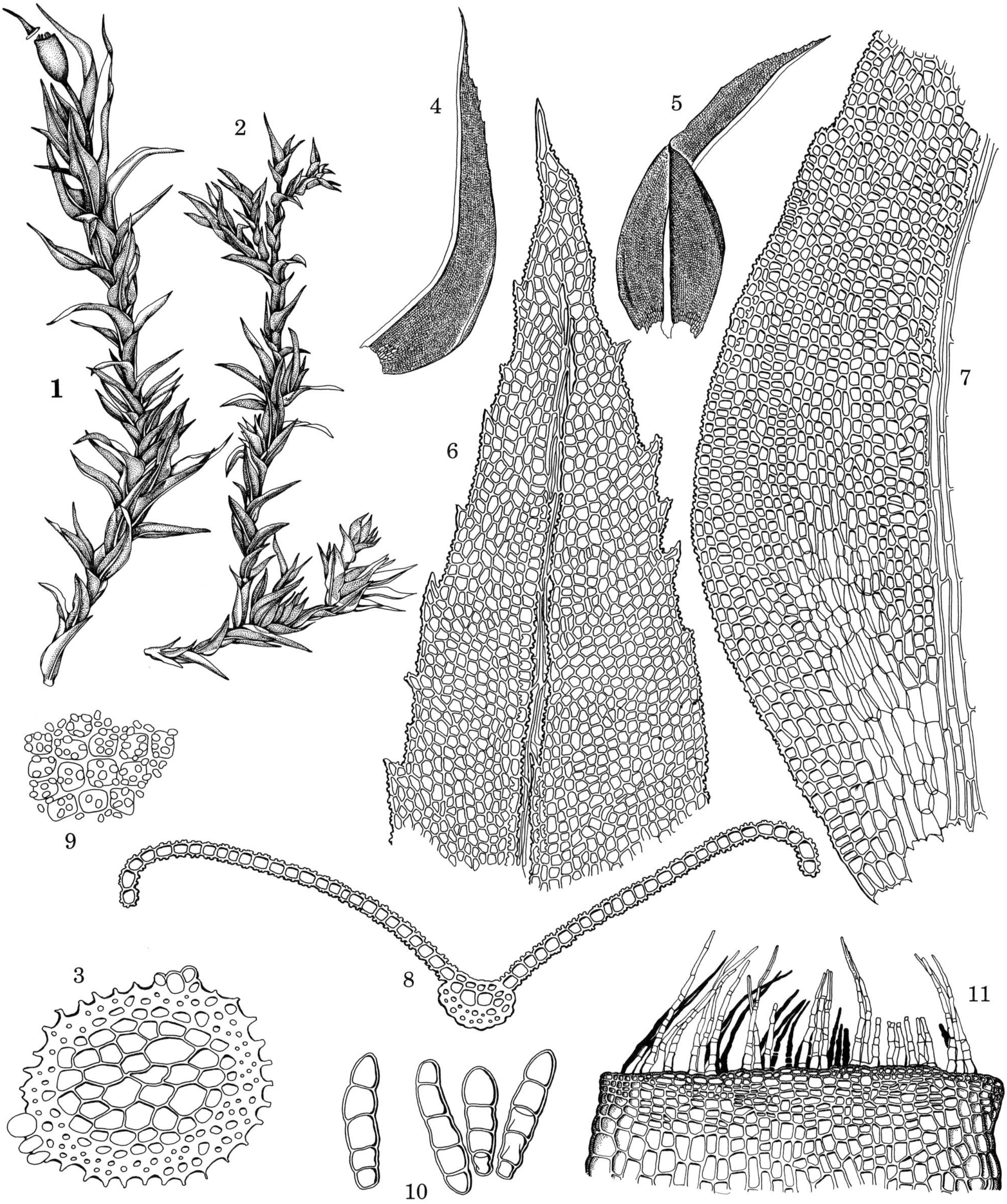

Plate 35. ***Leptodontiella.*** **1–11.** ***L. apiculata.*** 1. Habit of perichaetiate plant. 2. Perigoniate plant. 3. Transverse section of stem. 4–5. Two leaves. 6. Leaf apex. 7. Basal cells. 8. Transverse section at midleaf. 9. Papillae. 10. Propagula. 11. Peristome.

Dioicous. Perichaetia terminal, *perichaetial leaves* to 6 mm in length, *long-sheathing*, concealing most or all of the seta; *perigonia lateral on the stem*, gemmate. *Seta short*, to 2.0 mm in length, brown, twisted clockwise; theca short-elliptical to ovoid, brown, wide-mouthed, 1.0–1.2 mm in length, *without stomata*; *exothecial cells* 33–40 µm in width, *relatively short*, 2:1, walls thickened; annulus 2–4 cells in height, vesiculose, persistent; operculum conic-rostrate, ca. 0.5 mm in length, cells in straight rows; *peristome teeth* 16, inserted below the annulus, *straight or somewhat twisted clockwise*, criss-crossing perpendicularly when dry, forming an open cone or dome when moist, to 500 µm in length, yellow, *cleft to base in (2–)3–4 linear to filiform rami*, variously weakly fused below, *indistinctly spirally striate*, with 7–11 articulations. Calyptra cucullate, smooth, ca. 2 mm in length. Spores 15–20 µm in diameter, lightly papillose. *Laminal KOH color reaction yellow to yellow-orange.*

Found on rock, trees and shrubs at 600 to 4236 m elevation; endemic to Peru.

The genus is closely related to *Leptodontium* by the enlarged, sheathing perichaetial leaves (Pl. 35, f. 1); lateral perigonia (Pl. 35, f. 2) as in some species of *Leptodontium* (e.g. *Leptodontium wallisii*); the nearly straight, striate peristome teeth (Pl. 35, f. 11); the unistratose, carinate cauline leaves; and the costa with a ventrally exposed stereid band (Pl. 35, f. 8). *Leptodontiella* differs, however, in the very short seta (rare in *Leptodontium*), the short urn, the 16 peristome teeth each cleft into three or four rami (occasional in the polymorphic *L. viticulosoides*), and the propagula being cylindrical, uniseriate, and dark brownish green.

Sainsbury (1955) indicated that Brotherus' (1924–25) illustration of very regularly trifid peristome teeth for *Tridontium tasmanicum* Hook. f. (here placed in the Grimmiaceae, see Excluded Taxa) were somewhat misleading in that they are actually irregularly cleft bi-trifid, and are often not cleft to the base, although perforate below. The multi-fid peristome teeth of *Leptodontiella* are clearly cleft to the base, although there are marginal fusions.

Number of accepted species: 1.

Species examined: *Leptodontiella apiculata* (BUF, FH, H, US, herb. E. & P. Hegewald).

Additional literature: Zander (1972), Zander and Hegewald (1976).

28. LEPTODONTIUM — Plates 36–37.

Leptodontium (C. Müll.) Hampe ex Lindb., Öfv. K. Vet. Ak. Förh. 21: 227, 1864. Type: *Leptodontium squarrosum* (Hook.) Hampe in Lindb.

Sect. *Leptodontium*

Didymodon subg. *Leptodontium* (C. Müll.) Lor., Bryol. Notizb. 18, 1865.

Trichostomum sect. *Leptodontium* C. Müll., Syn. 1: 577, 1849.

Sect. *Verecunda* Zand., Bryologist 75: 230, 1972. Type: *Leptodontium flexifolium* (Dicks. ex With.) Hampe in Lindb.

Didymodon subg. *Leptodon* BSG, Bryol. Eur. 2: 2, 1851 (fasc. 46–47 Consp. 2: IV), *p.p.*

Sect. *Crassicostata* Zand., Bryologist 75: 256, 1972. Type: *Leptodontium pungens* (Mitt.) Kindb.

Sect. *Coronopapillata* Zand., Bryologist 75: 264, 1972. Type: *Leptodontium longicaule* Mitt.

Williamsia Broth., Nat. Pfl. 1(3): 1190, 1909, *hom. illeg. non* Merrill, 1908. Type: *Williamsia tricolor* Williams.

Williamsiella Britt., Bryologist 12: 62, 1909, *nom. nov.* for *Williamsia* Broth.

From λεπτός, peeled, fine, small, thin, delicate + ὀδούς (ὀδών), ὀδόντος, tooth + *-ium*, characteristic of.

Generally *robust plants*, in thick mats or short turf, greenish yellow- to orange-brown above, yellow- to red-brown below. Stems seldom to often branching, 1–5(–20) cm in length, transverse section rounded-pentagonal, *central strand absent*, outer cortex usually of thick-walled cells, hyalodermis often present, usually collapsed in mature parts of stem; axillary hairs of 6–16 cells, cells often bulging, with hyaline walls or basal 1–2 cells brownish; *tomentum* often present, red to brown, *knotty or kinky*. *Leaves* erect to spreading, twisted to contorted when dry, *spreading to squarrose-recurved when wet*, ovate- to long-lanceolate, occasionally lingulate or oblong, (1–)2–5(–8) mm in length, *keeled* above or channeled along costa, *margins recurved* in lower 1/3–3/4, rarely to near insertion, *usually dentate* in upper 1/3–1/2, rarely to near insertion, occasionally bordered above by 1–5 rows of less papillose, thick-walled cells; apex acute, occasionally rounded to narrowly obtuse; *base commonly rectangular, often sheathing*, basal margins sometimes long-decurrent; *costa* shortly excurrent, percurrent or ending 1–6(–15) cells below apex, *superficial cells elongate both ventrally and dorsally, costal transverse section reniform*, occasionally elliptical or semicircular, two stereid bands present, *epidermis absent both ventrally and dorsally*, guide cells 2–4 in 1 layer, *hydroid strand absent*; upper laminal cells subquadrate, mostly 11–15 µm in width, 1:1, walls thin or evenly to collenchymatously thickened, often trigonous, occasionally strongly bulging superficially; papillae variously simple, bifid, multifid, often hollow (oh- or cee-shaped in optical section), occasionally simple- or branching-columnar; *basal cells generally strongly differentiated* medially or across base, *often sharply demarcated and hyaline*, rectangular (occasionally somewhat inflated), slightly wider than upper cells, 2–5:1, walls thin to evenly or laterally thick-walled, occasionally porose. Propagula multicellular, clavate to obovate, borne on short stalks in leaf axils, occasionally on leaf apices or leafless branchlets. Flagellate branchlets occasionally present in axils. Usually dioicous, rarely autoicous or possibly rhizautoicous. *Perichaetia terminal, inner leaves usually long-lanceolate*, to 7–8 mm in length, *usually convolute-sheathing*, lower cells long-rhomboidal, porose or thin-walled. Perigonia terminal or lateral (as autoicous buds), or both lateral and terminal on antheridiate plants. Seta 0.3–3.0 cm in length, 1(–2) per perichaetium, twisted usually clockwise above; theca cylindrical, 2.0–3.5 mm in length, exothecial cells short-rectangular, moderately thick-walled; *stomates absent or present at base of theca*; *annulus of 2–7 rows of yellowish or reddish vesiculose cells, persistent or irregularly deciduous*; *peristome teeth 16, linear, occasionally rudimentary* (a preperistome rarely present), yellowish brown to reddish orange, *smooth to deeply striate*, usually 300–500 µm in length, of several articulations, *straight, basal membrane absent or very low*. Operculum conic to conic-rostrate, 0.5–1.0 mm in length, cells in straight rows. Calyptra cucullate, smooth, 2.5–4.0 mm in length. Spores homogeneous or occasionally heterogeneous, i.e., in one capsule half larger and chlorophyllose and half smaller and

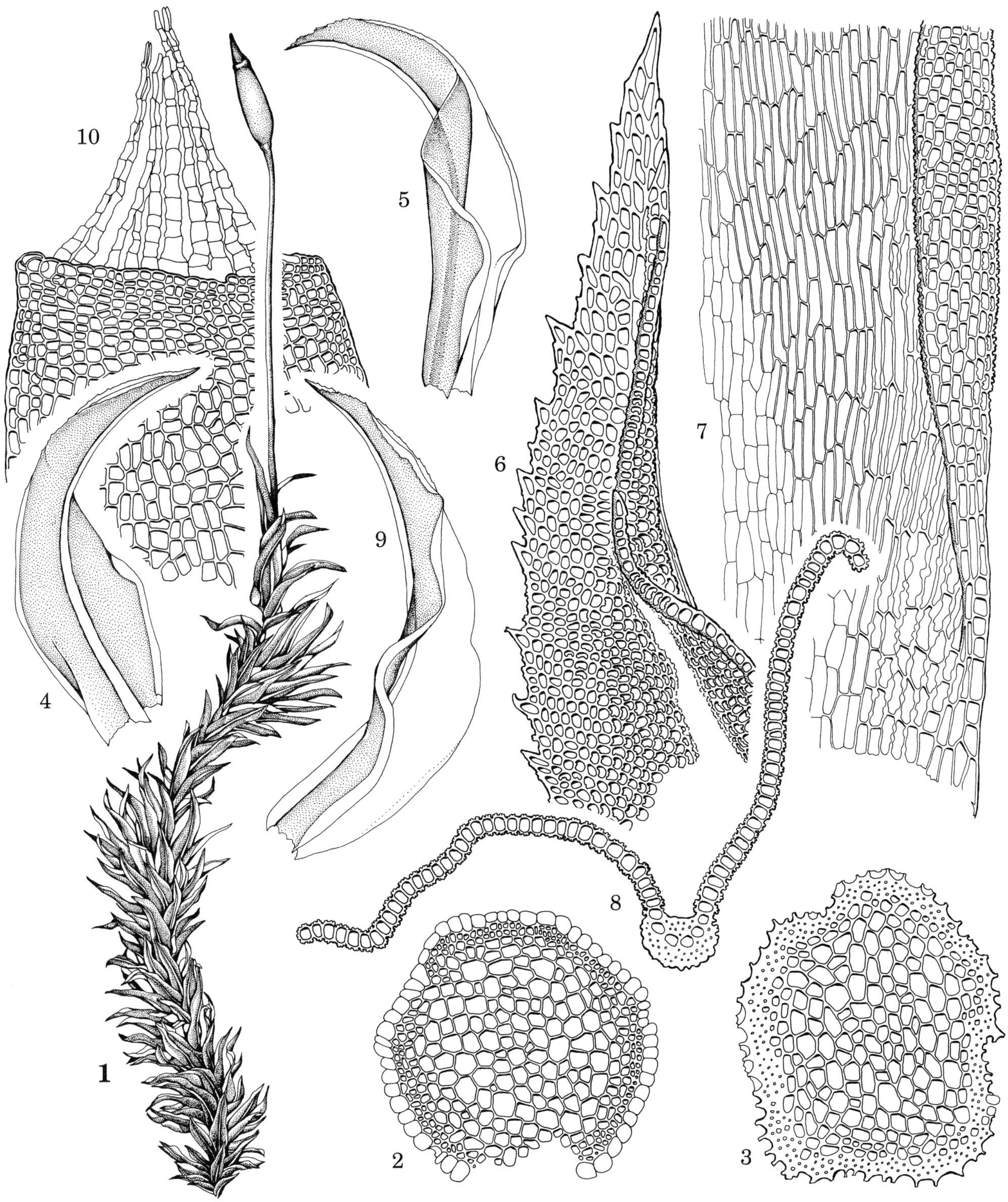

Plate 36. ***Leptodontium.*** **1–10.** ***L. pungens.*** 1. Habit. 2–3. Two transverse sections of stem; cells of hyalodermis bulge in immature stem and are collapsed in mature stem. 4–5. Two leaves. 6. Leaf apex. 7. Basal cells. 8. Transverse section at midleaf. 9. Perichaetial leaf. 10. Peristome.

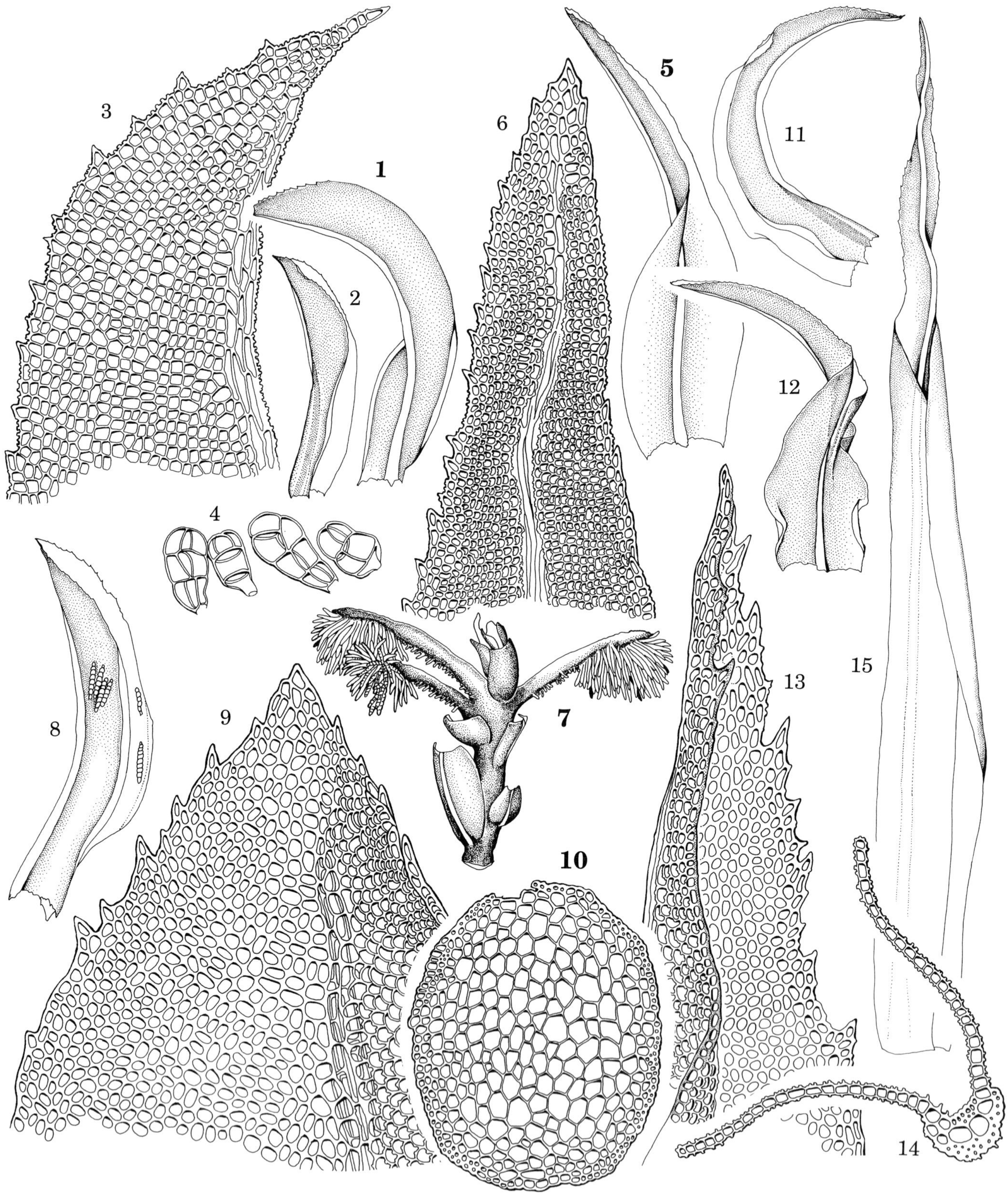

Plate 37. ***Leptodontium.*** **1–4.** ***L. flexifolium.*** 1–2. Two leaves. 3. Leaf apex. 4. Propagula. **5–6.** ***L. luteum.*** 5. Leaf. 6. Leaf apex. **7–9.** ***L. stoloniferum.*** 7. Stem apex with propaguliferous branchlets. 8. Leaf. 9. Leaf apex. **10–15.** ***L. viticulosoides* var. *viticulosoides.*** 10. Transverse section of stem. 11–12. Two leaves. 13. Leaf apex. 14. Transverse section at midleaf. 15. Perichaetial leaf.

brownish, weakly papillose, mostly ca. 17–20 µm in diameter. *Laminal KOH color reaction usually strong, yellow* or less often orange to yellowish orange, occasionally yellow with red blotches. Reported chromosome number n = 13.

This is a large genus found mainly in tropical, mountainous areas of the world, especially characteristic of high elevation fog forests, growing on soil, acidic rock, trees and shrubs.

The genus *Leptodontium* is superficially well distinguished by the robust size of the plants, squarrose-recurved and carinate leaves, sheathing leaf base, elongate and sheathing perichaetial leaves (Pl. 36, f. 9), and the straight peristome teeth (Pl. 36, f. 10), but there is significant variation in many of these characters among the species. More constant features include the absence of a stem central strand (Pl. 36, f. 2–3), absence of a differentiated epidermis over the costa (Pl. 36, f. 8), and smooth or striate peristome teeth (Pl. 36, f. 10). Although *Leptodontium* is, all in all, a surprisingly well-delimited genus (for the family), *Leptodontiella* and *Streptotrichum* are rather similar in morphology when sterile, and their descriptions and illustrations should be carefully studied to avoid confusion. These last two genera lack stomates in their capsules, as do many species of *Leptodontium*. Hollow papillae reported (Newton & Boyce 1987) as "mamillae" in British *L. flexifolium* are actually rather common in tropical collections of that species.

A revision for the New World (Zander 1972) reviewed the nomenclatural history, morphology and geographic distribution of the genus in detail, including much discussion of extra-American species. Sloover (1987) provided a key to the tropical African species of the genus.

Didymodon sect. *Fallaces* is rather similar in aspect to *Leptodontium*. *Didymodon erosodenticulatus* has many of the characteristics of *L. viticulosoides* (Pl. 37, f. 10–15), including reflexed, keeled leaves with serrate upper leaf margins and simple papillae, but differs significantly in the presence of a stem central strand, brownish basal cells of the axillary hairs, costal ventral epidermal layer present in at least some leaf sections, perichaetial leaves not strongly differentiated, and red-orange laminal KOH color reaction. *Hymenostylium* ((*q.v.*) has many of the characters of *Leptodontium*. The presence of upper laminal cell wall trigones in *Hymenostylium* and the related genus *Reimersia*, plus the keeled leaves, general absence of a dorsal costal epidermis and of a stem central strand are important characters indicating an ancestral link with *Leptodontium*, and this link is supported by the cladistic study. Trigones are variously present in *L. wallisii* (Peru, Hegewald 6948, BUF; Colombia, Van Cleef 249, BUF) and *L. viticulosoides*.

Leptodontium stellatifolium of Brazil has somewhat the appearance of *Barbula* sect. *Convolutae* and has an unusual bright orange coloration of its basal cells, but may be retained in *Leptodontium* by its striate peristome teeth.

Crum and Anderson (1981) recognized *L. excelsum* (Sull.) Brit. (given as a synonym of *L. viticulosoides* var. *panamense* [= var. *sulphureum*] by Zander 1972) as a good species endemic to the Appalachian Mountains of the southeastern United States, based largely on a uniformity of small size and flagellate appearance. Robust collections from Jackson Co., North Carolina (BUF), which they may not have seen, are quite like *L. viticulosoides* from Mexico, and their report of an autoicous sexual condition would require transference of this name to the synonymy of *L. viticulosoides* var. *viticulosoides*; the material available to me lacks perigonia. Actually, the var. *sulphureum* is commonly dimorphic (*cf.* Mexico, Oaxaca, Smith et al. 3028, TENN), the perigoniate plants with hair-like stems and distant leaves and the perichaetiate not flagellate.

Leptodontium stoloniferum (Pl. 37, f. 7–9) has the appearance of species of Calymperaceae, but it differs in the shape of the laminal papillae, in the stalk bearing the propagula being branch-like (not a modified leaf), and the propagula with internal longitudinal cross walls (Calymperaceae apparently have only uniseriate cells in their propagula). A specimen from Ecuador (Steere 27591, BUF, NY) has centered, capituliform papillae, which places the species in sect. *Coronopapillata*.

The anisosporous condition (heterogeneous spores, with two size classes) characteristic of the spores of *L. viticulosoides* var. *viticulosoides* (Anderson & Zander 1986) was also found, at least rarely, in *L. wallisii*; in one Bolivian collection (Cochabamba, Lewis 79–2217, F), about half the spores in a capsule were brown and collapsed, the other half green and turgid. Recognized in my (Zander 1972) revision, *L. viticulosoides* var. *exasperatum* is characterized by high, columnar upper laminal papillae and is variably isosporous or anisosporous in different collections. Two additional collections (Mexico, Bowers et al. 5252, 5264–f, both BUF and TENN) have capsules showing both conditions in the same collection; the latter specimen is clearly dioicous. Perhaps the lethal factors proposed by Mogensen (1978, 1981, *cf.* Andrews 1929, Wettstein 1928) in his discussion of false anisospory are the source of variation in spore size in *Leptodontium*; this should be investigated.

Additional literature: Frahm (1973, 1986), Herzog (1932), Janssens and Zander (1980), Long (1982b), Newton and Boyce (1987), Schumacker and de Zuttere (1981), Sloover (1987), Zander (1982b, 1983d), Zander and Hegewald (1976), Zander & Vitt (1979).

Number of accepted species: 39.

Species examined: see revision (Zander 1972) of species in North, Central and South America, also *L. aggregatum* (NY), *L. gemmascens* (NY), *L. hyalinum* (FH), *L. insolitum* (FH), *L. interruptum* (FH), *L. joannis-meyeri* (NY), *L. latifolium* (H), *L. paradoxicum* (BUF), *L. pumilum* (US), *L. styriacum* (NY), *L. taiwanense* (US).

New heterotypic synonymy: *Leptodontium ramosum* Crum & Richards = *Leptodontium stoloniferum* Zand.

Tribe BARBULEAE

Barbuleae Herz., Geogr. Moose 98, 1926.
Anoectangiaceae Schimp., Coroll. Bryol. Eur. 7, 1855 [1856].
Merceyaceae Casares-Gil, Fl. Ibérica, Musgos 247, 1932.
Gymnoweisieae Limpr., Laubm. Deutschl. 1: 222, 1888, rank not given. *Lectotyp. nov.*: *Gymnostomum* Hedw.
Barbuloideae (Herz.) Hilp., Beih. Bot. Centralbl. 20: 612, 1933.
Merceyeae (Broth.) Chen, Hedwigia 80: 265, 1941.

The Barbuleae is distinguished at the immediate ancestral node by two advanced features, the dorsal stereid band section round or semicircular in section and the peristome teeth absent. As per the discussion of Cladograms 11 and 15 in the section on phylogenetic analysis, the exact interrelationships of the genera of the Barbuleae are at best little more than guesswork. Both cladograms also indicate that loss of a peristome is an advanced character state in the immediate ancestor of the Barbuleae; it is difficult to suppose re-evolution of the peristome in this lineage unless it involved some simple suppression and desuppression mechanism.

Four members of the Barbuleae heretofore placed elsewhere—*Anoectangium* (Zander 1977c) with the traditional Pleuroweisieae and *Gymnostomiella*, *Scopelophila*, and *Streptopogon* with the Pottioideae (but *cf.* Walther 1983 who places *Gymnostomiella* with the Hyophileae)—are placed with the Merceyoideae in Cladograms 2–5, 8–10 and 14 in spite of the lack of a ventral stereid band in these genera. This last feature was previously considered a crucial character distinguishing the traditional Pottioideae (one stereid band) and Barbuloideae (= Merceyoideae here, two stereid bands). Apparently, like *Calyptopogon* in the Trichostomoideae, these four genera are examples of characteristically lanceolate-leaved lineages that have simply lost the ventral stereid band.

The distribution of the Barbuleae is nearly worldwide.

29. ANOECTANGIUM — Plate 38.

Anoectangium Schwaegr., Spec. Musc. Suppl. 1(1): 33, 1811, *nom. cons. non Anoectangium* Röhl, 1809, *nom. rejic.* Type: *Anoectangium compactum* Schwaegr.
Anictangium Hedw., Spec. Musc. 40, 1801, *nom. rejic.*
Anyctangium Hedw. in Lam. & Cand., Fl. Franc. 2: 444, 1805, orthogr. var.
Pleurozygodon Lindb., Utkast Nat. Grupp. Eur. Lavmoss. 35, 1878. Type: *Pleurozygodon aestivus* (Hedw.) Lindb.
Anoectangium subg. *Euanoectangium* Roth, Eur. Laubm. 1: 171, 1904, *nom. illeg.* Type: *Anoectangium compactum* Schwaegr.
Anoectangium subg. *Pleurozygodon* (Lindb.) Kindb., Eur. N. Amer. Bryin. 2: 317, 1897.
Gymnostomum sect. *Anictangium* (Hedw.) Leman, Dict. Sc. Nat. 20: 151, 1821.
Zygodon sect. *Anoectangium* C. Müll., Syn. 1: 683, 1849.

From ἀνοεχτός, opened + ἀγγεῖον, diminutive of ἄγγος, vessel or container, capsule; with a wide- or open-mouthed capsule.

Plants growing in turf or mats, green to yellow-brown above, light to dark brown below. Stems seldom branching, to 1(–3) cm in length, transverse section oval to rounded-triangular or pentagonal, *central strand present*, strong, outer cortex with smaller lumens, usually thick-walled, *hyalodermis absent*; axillary hairs of 3–10 cells, all hyaline or basal 1–2 brownish and with thicker walls; weakly radiculose or with red-brown tomentum. *Leaves* often distant or crowded, occasionally secund, appressed-incurved, often twisted when dry, spreading when moist, *ligulate to lanceolate*, occasionally triangular or acuminate, 1.0–1.5(–2.0) mm in length, *upper lamina strongly keeled, deeply grooved along costa, margins plane or weakly recurved in lower 1/2, entire*, occasionally finely crenulate or weakly denticulate above; apex broadly obtuse to sharply acute, usually apiculate, occasionally acuminate or somewhat cucullate; base scarcely differentiated in shape or ovate; *costa ending at the apiculus* or sometimes a few cells below apex or becoming short-excurrent as a mucro, seldom stoutly excurrent, *superficial cells elongate* or occasionally short-rectangular to quadrate near apex ventrally, elongate dorsally and generally papillose, 2–3 cells across costa ventrally at midleaf, costal transverse section oval to reniform, *stereid band single*, strong, semicircular to oval, epidermis absent ventrally (the guide cells superficially exposed) to weakly developed, epidermis usually distinct dorsally, guide cells 2–4 in 1 layer, *hydroid strand absent;* upper laminal cells subquadrate, occasionally hexagonal or elongate transversely or longitudinally in patches, (5–)7–9(–15) µm in width, 1(–2):1(–2), walls thin to greatly thickened, superficially flat to bulging, occasionally in extraplanar rows (very seldom bistratose); papillae multifid, often massively so, centered over the lumens or simple to bifid, scattered, rarely absent; basal cells differentiated in a small group at base of costa, short-rectangular, little wider than upper cells, 2–4:1, walls usually thick-walled. Dioicous. *Perichaetia terminal on short lateral branches, inner leaves* ovate-acuminate, 1.0–1.5 mm in length, *convolute-sheathing*, lower cells short-rhomboidal to near apex. Perigonia lateral. Seta 0.3–0.8 cm in length, 1 per perichaetium, yellow-brown, twisted clockwise below, occasionally counterclockwise above; theca 0.5–1.0(–1.5) mm in length, yellow-brown to brown, ovoid to elliptical, exothecial cells rectangular, walls thin, stomates phaneropore, at base of theca, annulus of two rows of weakly vesiculose cells; *peristome absent.* Operculum long-rostrate, 0.4–0.6(–1.8) mm in length, sometimes longer than the theca, inclined, cells in straight rows. Calyptra cucullate, smooth, 1.2–1.5(–2.0) mm in length. Spores 9–12(–19) µm in diameter, light brown, weakly to strongly papillose. *Laminal KOH color reaction yellow to yellow-orange.* Reported chromosome number n = 13.

A rather large but singularly homogeneous group distributed mainly in tropical, arctic or mountane areas of the world.

Anoectangium is unusual in that the distinctions between the many species are mainly in characters considered variable in other genera. A close study of the Middle American representation (Zander 1977c) resulted in extensive synonymy with only

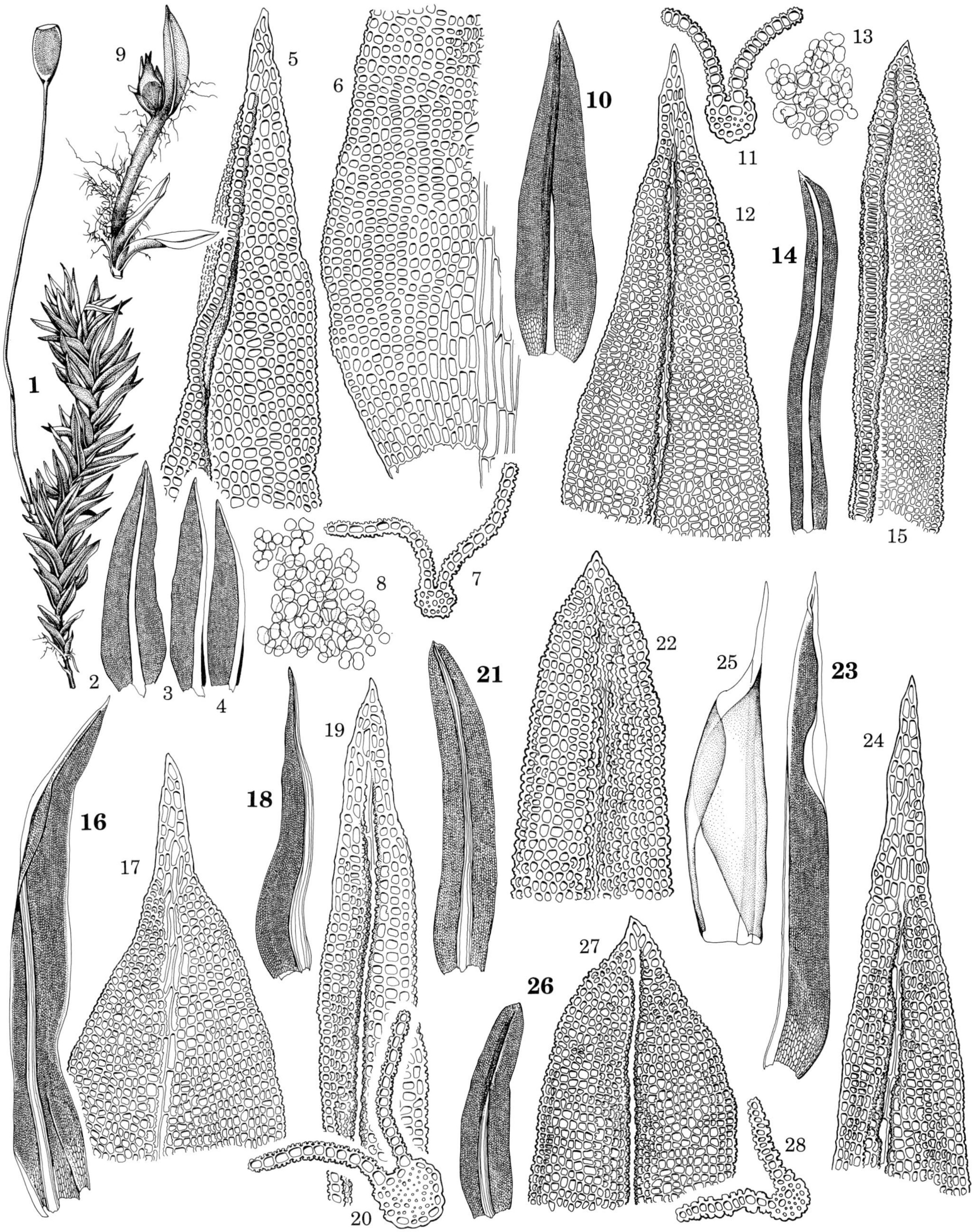

Plate 38 ***Anoectangium.*** **1–9.** ***A. afrocompactum.*** 1. Habit. 2–4. Three leaves. 5. Leaf apex. 6. Basal cells. 7. Leaf section. 8. Papillae. 9. Perigoniate branch. **10–13.** ***A. aestivum.*** 10. Leaf. 11. Leaf section. 12. Leaf apex. 13. Papillae. **14–15.** ***A. angustifolium.*** 14. Leaf. 15. Leaf apex. **16–17.** ***A. clarum.*** 16. Leaf. 17. Leaf apex. **18–20.** ***A. mafatense.*** 18. Leaf. 19. Leaf apex. 20. Leaf section. **21–22.** ***A. magnirete.*** 21. Leaf. 22. Leaf apex. **23–25.** ***A. strachyanum.*** 23. Leaf. 24. Leaf apex. 25. Perichaetial leaf. **26–28.** ***A. wilmsianum.*** 26. Leaf. 27. Leaf apex. 28. Leaf section.

one species, *A. aestivum* (Pl. 38, f. 10–13), recognized for the area. The most common expression of the species in the area was found to have small, superficially bulging laminal cells with multifid papillae centered over the lumens. Considered weakly distinguishable were certain uncommon, mostly local variants with various leaf shapes and usually superficially flat, larger upper laminal cells having thickened walls, often transversely or longitudinally elongate medially, the laminal papillae low, simple, seldom multifid. Some collections, however, showed independent segregation of these features.

Review of a series of exotic taxa for this study indicate that the Middle American variation is repeated on a larger and possibly more distinctly stepwise scale, with taxonomic distinctions dependent on combinations of leaf shape, and details of the areolation and papillae morphology. A Mexican morphological variant with a tendency for upper laminal cells to have massive, centered, capituliform papillae, and for the cells to protrude superficially on one side of the leaf or the other in spaced rows or patches (these often bistratose) was found to occur also in South Africa (as *A. wilmsianum*, Natal, Cathedral Peak Forest Station, Magill 5532, PRE). Whether this represents a vicariance event or polytopic differentiation is unknown; however, it may be pointed out that another South African taxon, *Tortula ammonsiana*, also occurs rarely in eastern North America, which is a disjunction of equal geographic magnitude.

Anoectangium is generally distinguishable from a similar genus with very short sporophyte-bearing branches arranged laterally on the axis, *Molendoa*, by its constant lack of a ventral costal stereid band (Pl. 38, f. 7, 11, 20, 28), but see the discussion of *Molendoa* for additional comments. Norris and Koponen (1989) suggested that "the apparent lateral perichaetia (Pl. 38, f. 1) may be interpreted as resulting from innovation below the still very immature perichaetia." The narrow leaves appear to distinguish *Anoectangium* from Pottioideae taxa with single stereid bands, and the rather strongly differentiated stem sclerodermis is clearly that of the Merceyoideae. The single stereid band, however, cannot be attributed to small plant size alone (as is the case in certain single-stereid banded collections of species of *Didymodon* and *Gymnostomum*, and other genera of Merceyoideae). *Syntrichia abruptinervis* is similar to *Anoectangium* species in the lanceolate leaves with a deep, narrow groove along the costa, but differs significantly in the inflated basal cells, costa arcuate in section, hydroid strand present, and the red KOH reaction. *Anoectangium* shares the distinctive deep groove running up the ventral side of the leaf at the costa with *Barbula* species, along with similar papillae and KOH color reaction, among other characters. *Molendoa*, on the other hand, is similar to *Didymodon* in many characters, though placed in the Hyophileae of Cladograms 14–16.

Additional literature: Geissler (1985), Malta (1921), Newton (1983), Rashid (1970), Saxena and Gill (1986), Saxena and Rashid (1980), Zander and Vitt (1979).

Number of accepted species: 47.

Species examined: *A. abyssinicum* (BM), *A. aestivum*, *A. afrocompactum* (NY), *A. angustifolium* (NY), *A. bicolor* (NY), *A. borbonense* (FH), *A. clarum* (BUF), *A. eukilimandscharicum* (BM), *A. hobsonii* (NY), *A. humblotii* (FH), *A. impressum* (BM, FH), *A. keniae* (PC), *A. lineare* (NY), *A. mafatense* (FH), *A. magnirete* (FH), *A. raphidostegium* (NY), *A. rufoviride* (DUKE), *A. shepherdae* (PC), *A. strachyanum* (BUF, NY), *A. thompsonii* (DUKE, NY), *A. walkeri* (DUKE), *A. wilmsianum* (NY).

New combinations: *Anoectangium keniae* (P. de la Varde) Zand., *comb. nov.* (*Gymnostomum keniae* P. de la Varde, Rev. Bryol. Lichénol. 22: 10, 1953). *Anoectangium shepherdae* (Card. & Dix.) Zand., *comb. nov.* (*Hymenostylium shepherdae* Card. & Dix., J. Bot. 48: 307, 1910).

New synonymy: *Anoectangium compactum* var. *alaskanum* Card. & Thér. Aml = *Anoectangium aestivum* (Hedw.) Mitt. *Anoectangium sordidum* Mitt. = *Hymenostylium recurvirostrum* (Hedw.) Dix.

30. GYROWEISIA — Plate 39.

Gyroweisia Schimp., Syn. Musc. Eur., ed. 2. 38, 1876, *nom. cons.* Lectotype: *Gyroweisia tenuis* (Hedw.) Schimp.

Weisiodon Schimp., Coroll. 9, 1856, *nom. rejic.* Type: *Weisiodon reflexus* (Brid.) Schimp.

Gyroveisia Schimp. ex Luis., Broteria ser. Bot. 8: 36, 1909, *orthogr. var.*

Gymnostomum subg. *Gymnoweisia* B.&S. in BSG, Bryol. Eur. 1: 78, 1846 (fasc. 33–36 Mon. 4). Lectotype: *Gyroweisia tenue* Hedw.

Weissia subg. *Weisiopsis* BSG, Bryol. Eur. 1: 5. 1851 (fasc. 46–47. Consp. 1: VII), *nom. illeg.* Type: *Weissia reflexa* Brid.

Trichostomum sect. *Weisiodon* (Schimp.) Lindb., Oefv. K. Vet. Ak. Foerh. 21: 213, 1864, as "*Weissiodon*." Type: *Trichostomum reflexum* (Brid.) Lindb.

Weissia sect. *Spathulidium* C. Müll., Linnaea 40: 298, 1876. Type: *Weissia tophicola* C. Müll.

From *gyrus*, circle + o + *Weis[s]ia*, a genus; a *Weissia*-like moss with a well-developed persistent annulus.

Plants low, gregarious or forming a thin turf, green above, tan below. *Stems short*, branching occasionally, to 0.4 cm in length, transverse section rounded-pentagonal, central strand present or absent, sclerodermis present (substereid), hyalodermis absent; axillary hairs of ca. 5(–10) cells, basal 1–3 brownish. *Leaves* appressed-incurved when dry, weakly spreading, strict to reflexed when moist, *narrowly ligulate to long-ovate*, ca. 0.7–1.4 mm in length, upper lamina shallow-grooved along costa or flat, *margins plane to weakly recurved*, *entire*, occasionally bistratose throughout; *apex rounded to rounded acute or obtuse*, sometimes acuminate or apiculate by a sharp cell; *base scarcely differentiated to ovate*, occasionally sheathing; costa percurrent or ending ca. 4–8 cells below apex, superficial cells elongate ventrally and dorsally, 2–8 rows of cells across costa ventrally at midleaf, costal transverse section semicircular to rounded, ventral stereid band absent or weak, often superficially exposed, dorsal band present and semicircular in section when well developed, epidermis usually present ventrally, often present dorsally, guide cells 2(–6) in 1 layer, *hydroid strand absent*; *upper laminal cells quadrate or short-rectangular*, ca. 8–11 µm in width, 1–2:1, walls thin to evenly thickened, superficially flat to convex; papillae hollow, simple to indistinctly bifid, scattered, ca. 4 per lumen, occasionally absent; basal cells differentiated across leaf base in lower 1/4 to 1/2 of leaf, rectangular, *commonly enlarged or inflated and hyaline*, usually little wider than upper cells, 3–5:1, walls thin to evenly thickened. *Propagula often present*, oval to spindle-shaped, of several cells, borne on basal rhizoids, brown.

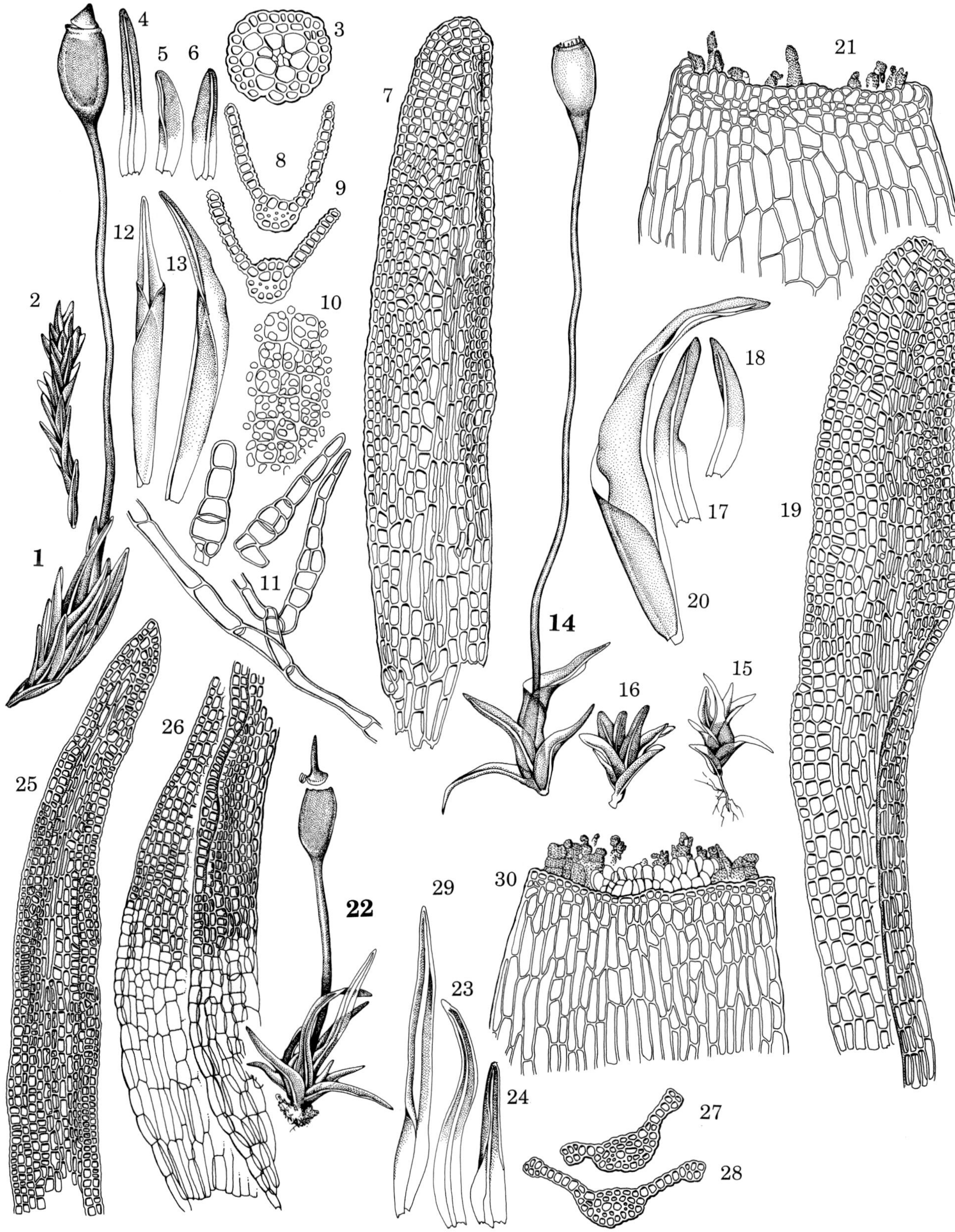

Plate 39. ***Gyroweisia.*** **1–13.** ***G. tenuis.*** 1. Sporangiate plant. 2. Sterile plant. 3. Section of stem. 4–6. Three leaves. 7. Areolation. 8–9. Two sections at midleaf. 10. Papillae. 11. Propagula. 12–13. Two perichaetial leaves. **14–21.** ***G. reflexa.*** 14. Sporangiate plant. 15. Perigoniate plant. 16. Sterile plant. 17–18. Two leaves. 19. Areolation. 20. Perichaetial leaf. 21. Peristome. **22–30.** ***G. yuenannensis.*** 22. Habit. 23–24. Two leaves. 25. Leaf apex. 26. Leaf base. 27–28. Two sections at midleaf. 29. Perichaetial leaf. 30. Peristome.

Dioicous or autoicous, occasionally heteroicous. Perichaetia terminal, inner leaves lanceolate, usually well differentiated, to 1.5 mm in length, often strongly sheathing the seta, lower cells rectangular to long rhomboidal. Perigonia terminal, gemmate, on somewhat smaller plants, or as buds at base of perichaetiate plant. Seta ca. 1.5–6.0 mm in length, 1 per perichaetium, yellowish brown, twisted clockwise; theca ca. 0.8–1.5 mm in length, yellowish brown, oval to short-cylindrical, neck sometimes well differentiated, exothecial cells short-rectangular to rhomboidal, thin-walled, stomates phaneropore, occasionally somewhat enlarged, at base of capsule, *annulus of 2–3 rows of highly vesiculose cells,* usually revoluble but often merely persistent; *peristome teeth absent or ca. 16, rudimentary,* short, ligulate or oblong and much perforate, lightly papillose to closely spiculose, ca. 30–80 µm in length, with ca. 3–4 articulations, straight, basal membrane low. Operculum short-conic to narrowly rostrate, ca. 0.2–0.6 mm in length, cells straight. Calyptra cucullate, smooth, ca. 0.7–1.4 mm in length. Spores ca. 8–14 µm in diameter, light brown, smooth to papillose. Laminal KOH color reaction yellow or orange. Reported chromosome number n = 13.

Found on thin soil over calcareous rock, in widely scattered localities across North America, Europe, the Middle East, Africa and China.

Gyroweisia has long been a "wastebasket" genus (Zander 1977c) wherein several small, feature-poor species have been set aside. Generally, these taxa have ligulate leaves with subpercurrent costae and enlarged, hyaline basal cells (Pl. 39, f. 7, 19, 25–26), a vesiculose annulus, and a rudimentary peristome or sometimes none at all (Pl. 39, f. 21, 30). Little attention has been given to characters of the areolation and anatomy, which may be used to better place the species. Past work (Zander 1977c; Hill 1981) and the present study has assigned many of these taxa to more appropriate genera, and even further reduction in the size of the genus is probable.

Gyroweisia may be viewed (see also Zander & Hermann 1986) as part of a complex evolutionary series, also including *Gymnostomum*, *Leptobarbula* and *Barbula* sect. *Convolutae*, involving morphological reduction of plant size and expression of the peristome. *Gyroweisia* differs from a morphologically similar genus, *Gymnostomum*, by the occasional presence of a peristome (albeit rudimentary), the large annulus, the sterile plants (Pl. 39, f. 2) distinctly smaller than the sporophyte-bearing gametophytes, basal leaf cells differentiated higher up the leaf (lower 1/4 to 1/2) (Pl. 39, f. 7, 19, 26), the more common presence of propagula (Pl. 39, f. 11—see also Sérgio 1984), and the perichaetial leaves much larger than the cauline (Pl. 39, f. 12, 13, 20, 29). The transformation series conceived above is, however, not supported by the cladistic analysis, probably because the analysis utilizes more than just the most obvious characteristics of the taxa involved.

Additional literature: Andrews (1922c), Conard (1945b), Crundwell (1981), Egunyomi and Olarinmoye (1978), Geheeb (1906b), Sérgio (1972b), Steere (1939c), Thériot (1923).

Number of accepted species: 6.

Species examined: *G. monterreia* (BUF), *G. reflexa* (BUF, DUKE, NY), *G. tenuis* (BUF, DUKE, MICH, NY), *G. yuenannensis* (H).

New homotypic synonymy: *Gyroweisia lindigii* (Hampe) Broth. = *Didymodon lindigii* (Hampe) Zand. mixed with *Didymodon tophaceus* (Brid.) Lisa and *Didymodon australasiae* (Hook. & Grev.) Zand., judging from the several apparent isotypes that are present at FH and NY.

31. BELLIBARBULA — Plate 40.

Bellibarbula Chen, Type: *Bellibarbula kurziana* Chen, Hedwigia 80: 223, 1941, India, "Sikkim, Phalloot Top," Kurz 2026, BM, *lectotyp. nov.*; NY, isotype.

From *bellus*, beautiful + i + *Barbula*, a genus.

Plants forming a low turf, *green to red-orange above, blackish red-brown below.* Stems branching occasionally, to 3.0 cm in length, transverse section rounded-pentagonal, central strand present, sclerodermis thick, hyalodermis absent; axillary hairs short, 3–4 cells in length, basal 1(–2) cells yellow; sparsely radiculose. *Leaves* appressed, incurved when dry, weakly spreading when moist, *short-lanceolate,* often concave at midleaf, ca. 0.7–1.5(–2.0) mm in length, upper lamina broadly and deeply channeled along costa, margins strongly recurved in lower 3/4 or to near apex, entire; apex rather narrowly blunt to acute and sometimes apiculate; *base ovate*; *costa* ending 2–4 cells below apex or percurrent, *sinuose above midleaf,* superficial cells elongate or quadrate near apex ventrally, elongate dorsally, ca. 3–4 rows of cells across costa ventrally at midleaf, costal transverse section semicircular or circular, *two stereid bands present,* epidermis present on both sides of costa, guide cells 4 in 1 layer, hydroid strand absent; *upper laminal cells rounded-quadrate to short-rectangular,* occasionally transversely elongated in patches, ca. 10–13 µm in width, 1–2:1(–2), *walls thin to incrassate, lumens angular to oval or elliptical,* superficially flat to weakly convex; *papillae small, simple to bifid, solid or hollow, 4–6 per lumen*; *basal cells not differentiated or weakly developed medially, quadrate to rectangular,* not wider than upper cells, 1(–4):1, walls thin- to thick-walled. Dioicous. *Perichaetia* terminal, *highly differentiated, inner leaves long-rectangular, apiculate,* to 2.0 mm in length, *convolute-sheathing, cells long-rhomboidal to near apex.* Perigonia terminal, gemmate, often in series from subterminal innovations. Seta ca. 4–6 mm in length, 1 per perichaetium, reddish brown, twisted clockwise; theca 0.8–1.4 mm in length, reddish brown, cylindrical, exothecial cells thin-walled, rhomboidal, stomates numerous at base of theca, *annulus of highly vesiculose cells, revoluble*; *peristome absent. Operculum short-conic,* ca. 0.4–0.5 mm in length, cells straight. Calyptra cucullate, smooth, ca. 2.0 mm in length. Spores 15–18 µm in diameter, brownish, lightly papillose. *Laminal KOH color reaction red.*

Found on rock in montane situations in the Himalayas of India, China (Yunnan), the Appalachians of the U.S.A., and mountain ranges in Mexico.

Bellibarbula was proposed by Chen (1941) for two species distinguished from other Barbuleae by strongly convolute-sheathing perichaetial leaves (Pl. 40, f. 10, 11, 16), peristome absent, ovate-lanceolate leaves, papillose upper laminal cells (Pl. 40, f. 9) and little-differentiated leaf bases (Pl. 40, f. 7). The genus proves to be similar to *Bryoerythrophyllum* in the red coloration and small, bifid laminal papillae. *Bryoerythrophyllum* shows a tendency towards *Bellibarbula* in that some of its species likewise have an eperistomate capsule with short-conic operculum. *Bellibarbula* is easily distinguished, however, by the costa usually sinuose above (Pl. 40, f. 4, 5, 13, 14), rather

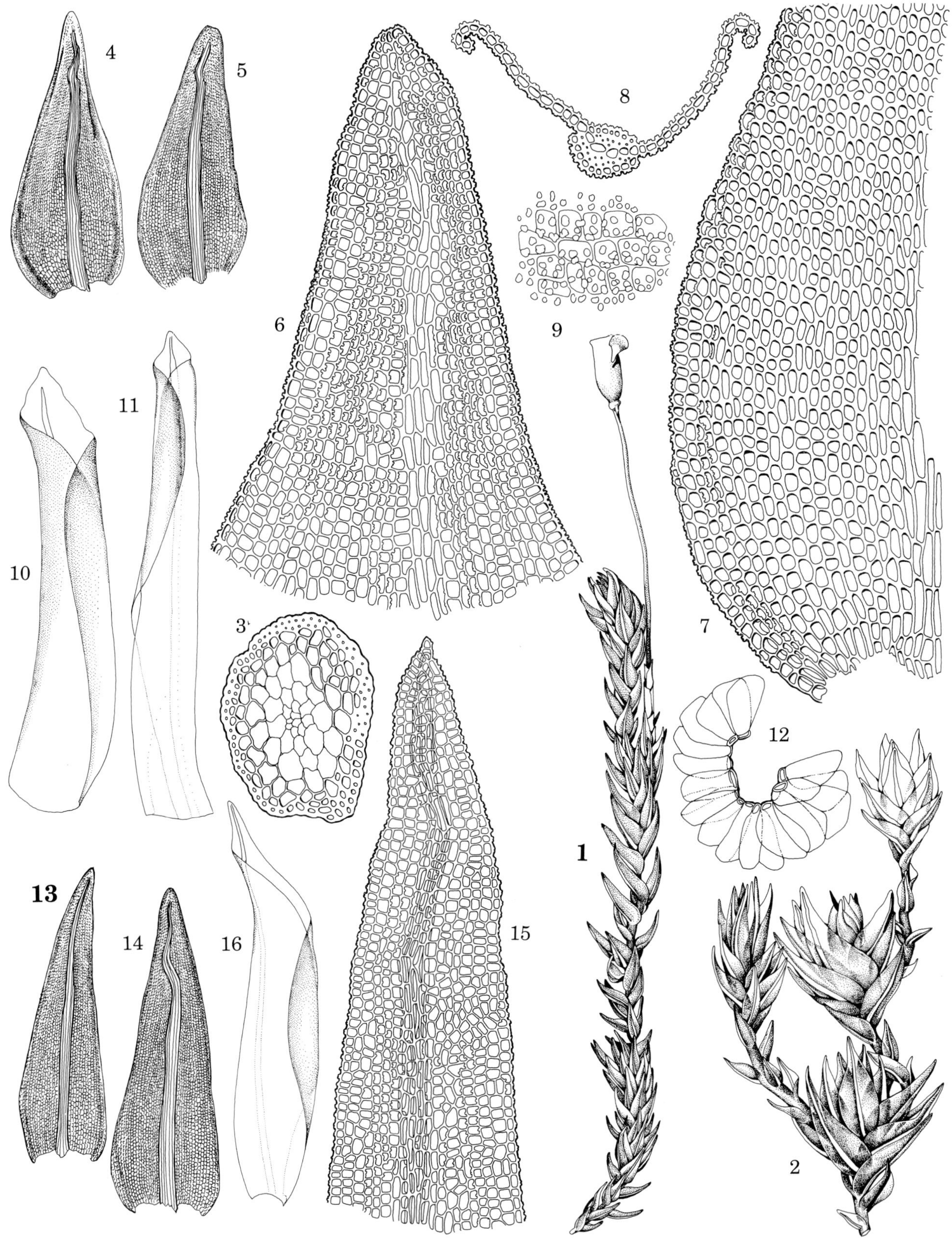

Plate 40. ***Bellibarbula.*** **1–12.** ***B. kurziana.*** 1. Sporangiate plant. 2. Perigoniate plant. 3. Section of stem. 4–5. Two leaves. 6. Leaf apex. 7. Basal cells. 8. Transverse section at midleaf. 9. Upper laminal papillae. 10–11. Two perichaetial leaves. 12. Fragment of revoluble annulus. **13–16.** ***B. recurva.*** 13–14. Two leaves. 15. Leaf apex. 16. Perichaetial leaf.

concave upper lamina (at least in collections with incrassate cell walls), and the highly differentiated, convolute-sheathing inner perichaetial leaves with long, rhomboidal cells extending to the apiculate apex. *Didymodon nigrescens* and related species are similar to *Bellibarbula* in the red coloration, incrassate upper laminal cells, generally elongate ventral costal cells and short basal cells, but differ in their very thin costae and lack of papillae or papillae low and massive.

Judging from study of a range of specimens and the types at BM and PC, the two species of *Bellibarbula* are only weakly distinguishable by the features mentioned in Chen's (1941) key and as discussed (in Chinese) by Chen (1963). The type of *B. kurziana* includes some plants with much the same morphology as the second species. *Bellibarbula obtusicuspis* of China (type at PC!) is clearly the same as what has been known as *Bryoerythrophyllum recurvum*, and a new combination is necessary along with extension of known morphological variation (description by Zander 1978g) and geographic range of the species (United States and Mexico).

Number of accepted species: 2.

Species examined: *B. kurziana* (BM, NY), *B. recurva* (BUF, DUKE, MICH, PC, TENN).

New heterotypic synonymy: *Bellibarbula obtusicuspis* (Besch.) Chen = *Bellibarbula recurva* (Griff.) Zand.

New combinations: *Bellibarbula recurva* (Griff.) Zand., *comb. nov.* (*Gymnostomum recurvum* Griff., Calcutta J. Nat. Hist. 2: 482, 1842, see discussion of nomenclature by Robinson 1968; *Didymodon recurvus* (Griff.) Broth; *Bryoerythrophyllum recurvum* (Griff.) Saito).

32. STREPTOPOGON Plates 41–42.

Streptopogon Wils. in Mitt., Kew J. Bot. 3: 51, 1851. Type: *Streptopogon erythrodontus* (Tayl.) Wils.

Striptopogon Hampe, Flora 45: 450, 1862, *orthogr. var.*

Sect. *Streptopogon*

Sect. *Streptopogon* (Wils.) Mitt., Phil. Trans. R. Soc. London 168: 33, 1879, *nom. superfl.*

Streptopogon sect. *Eustreptopogon* C. Müll., Gen. Musc. Fr. 423, 1900, *nom. illeg.*

Sect. *Calymperella* C. Müll., Hedwigia 33: 128, 1894.

Sect. *Streptopogonella* Demar. & P. Varde, Bull. Jard. Bot. Bruxelles 26: 270, 1956. Type: *Streptopogon calymperoides* Demar. & P. Varde.

From στρεπτός, twisted + o + πώγων, -ωνος, beard.

Plants growing in tufts or mats, green above, *reddish brown below*. Stems commonly branching, ca. 0.5–3.0 cm in length, transverse section rounded-pentagonal to elliptical, *central strand absent* (inner cells of central cylinder occasionally crushed), sclerodermis weakly developed to distinct, *hyalodermis absent*; axillary hairs of ca. 7–10 cells, basal 1–3 cells often brownish; often red-tomentose below. *Leaves* appressed-incurved to lax and weakly contorted when dry, widely spreading and recurved when moist, *ovate-lanceolate*, 2.0–6.5 mm in length, *upper lamina keeled to broadly channeled, margins recurved to revolute at leaf base or to near apex*, entire to denticulate or serrulate above, sometimes bordered with elongate cells below or throughout; apex acute to acuminate, occasionally rounded and cucullate; base scarcely differentiated to rectangular, somewhat decurrent; *costa* usually strongly bulging dorsally, percurrent to excurrent as a weakly denticulate awn, occasionally subpercurrent by several cells, *superficial cells on both sides elongate and smooth, 2 rows of cells across costa ventrally at midleaf, costal transverse section round, stereid band single*, usually strong and round to semicircular, ventral and dorsal epidermis present or occasionally absent, guide cells 2 in 1 layer, *hydroid strand absent*; *upper laminal cells rectangular to rhombic*, occasionally fusiform near apex leaf, 15–20 µm in width, *2(–3):1*, walls thin to thick, occasionally porose, superficially equally convex on both sides of lamina; *papillae absent*; *basal cells only weakly differentiated from the upper cells*, rectangular, to 25 µm in width, 3–5:1, walls thin to weakly thickened and porose. Uni- or multiseriate (with internal longitudinal walls) elliptical propagula borne on dorsal leaf apex, apex of costa or upper leaf margins, seldom remaining on mature leaves, clavate or cylindrical, of 8–16 cells. Dioicous, paroicous, synoicous or autoicous. *Perichaetia* terminal, *leaves somewhat enlarged*, inner leaves long-elliptical to oblong, to 9.0 mm in length, *not sheathing*, lower cells thin-walled and rectangular in lower 3/4. Perigonia in dioicous species terminal on branches of much-branching smaller plants. *Seta usually short*, 0.5–9.0 mm in length, 1(–2) per perichaetium, yellowish, occasionally twisted clockwise; theca 3.0–3.5 mm in length, yellowish brown, elliptical, often with a short, rugose neck, mouth of theca occasionally flaring or forming a narrow ring, columella may protrude past mouth of theca after removal of operculum, exothecial cells rectangular, walls thin to weakly thickened, stomates phaneropore, at base of theca, annulus of ca. 4 rows of weakly vesiculose cells, persistent; *peristome teeth 16–32* (reportedly absent in one species), *often coming off with operculum, often strongly spreading when wet*, twisted when dry, linear, filamentous, variously cleft, densely spiculose, ca. 1000–1400 µm, with several articulations, twisted counterclockwise ca. once, basal membrane 70–800 µm in height, densely spiculose. Operculum conic, ca. 0.9–2.0 mm in length, cells twisted counterclockwise ca. once. *Calyptra conic-mitrate, lobed below, strongly papillose* (*prorulose*) *with upward pointing simple papillae* (but reportedly smooth in three species), 2.0–3.0 mm in length. Spores ca. 13–15(–30) µm in diameter, light brown, densely papillose. *Laminal KOH color reaction usually red*, occasionally yellow to yellowish orange. Reported chromosome number n = 20.

Found on rocks and tree branches and trunks; Latin America, central and southern Africa, Madagascar and Hawaii.

Important characters for *Streptopogon* are the plant's reddish color below, lack of a central strand in the stem (Pl. 41, f. 2, 11), ovate-lanceolate leaves, costa with one stereid band and no hydroids (Pl. 41, f. 8; 42, f. 5, 14, 22), upper laminal cells smooth, basal cells little differentiated from the upper, clavate propagula often present on leaves or costa (Pl. 41, f. 14, 16, 19; 42, f. 17), perichaetial leaves not or little sheathing, seta often short (Pl. 41 1), calyptra usually scabrous-prorulose (Pl. 41, f. 9–10; 42, f. 9—as in *Hypodontium*), and laminal KOH reaction usually red. The peristome (Pl. 42, f. 7, 15) often is only weakly attached basally and may be removed with the operculum resulting in what appear to be gymnostomous capsules; however, Griffin (1979b) reported, from a collection with a single sporophyte with semidetached operculum and calyptra, what appears to be a true gymnostomous condition in *S. calymperes*.

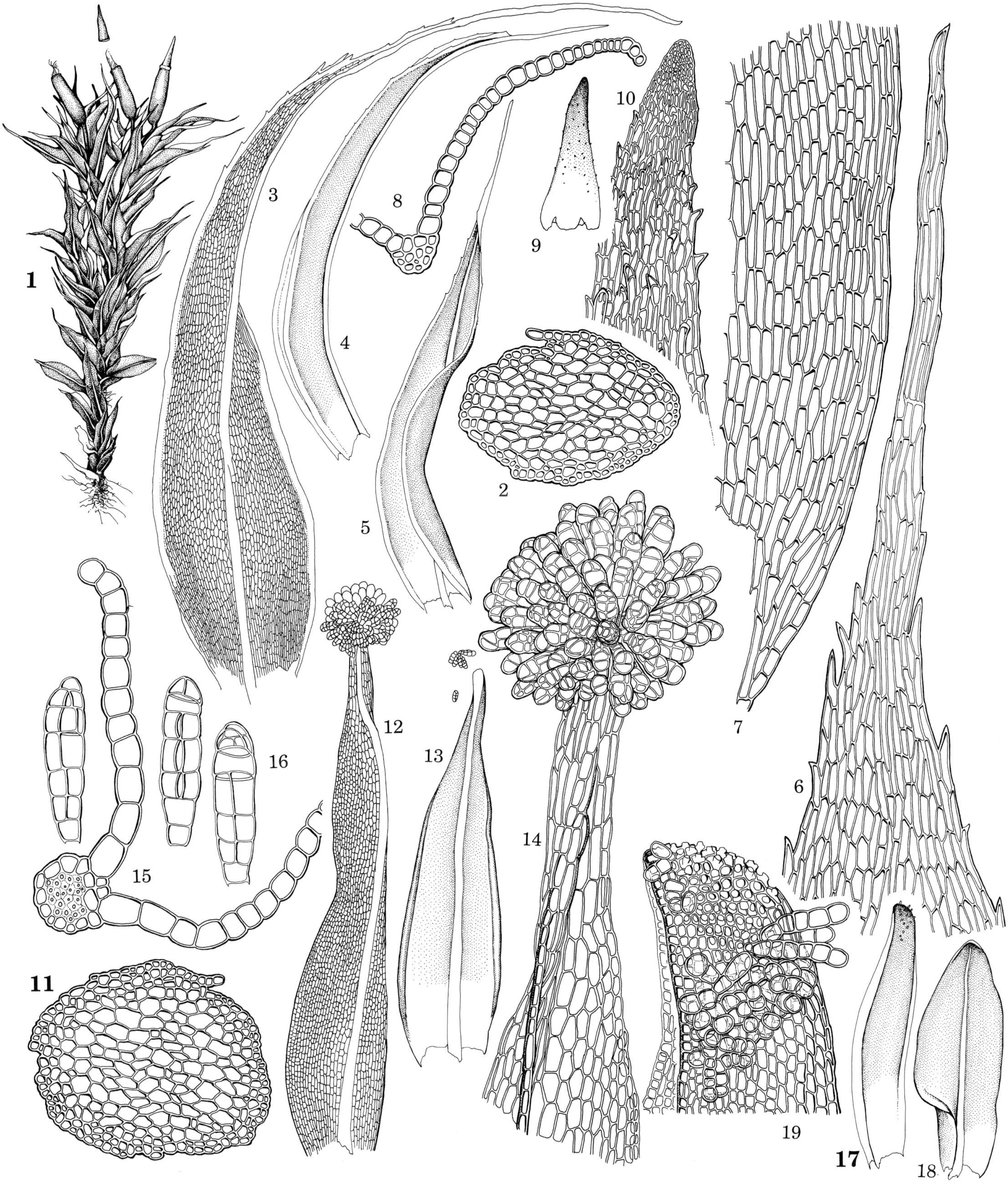

Plate 41. ***Streptopogon.*** **1–10.** ***S. erythrodontus.*** 1. Habit. 2. Transverse section of stem. 3–5. Three leaves. 6. Leaf apex. 7. Basal cells. 8. Transverse section at midleaf. 9. Calyptra. 10. Detail of calyptra. **11–16.** ***S. calymperes.*** 11. Transverse section of stem. 12–13. Two leaves. 14. Leaf apex. 15. Transverse section at midleaf. 16. Propagula. **17–19.** ***S. cavifolius.*** 17–18. Two leaves. 19. Leaf apex.

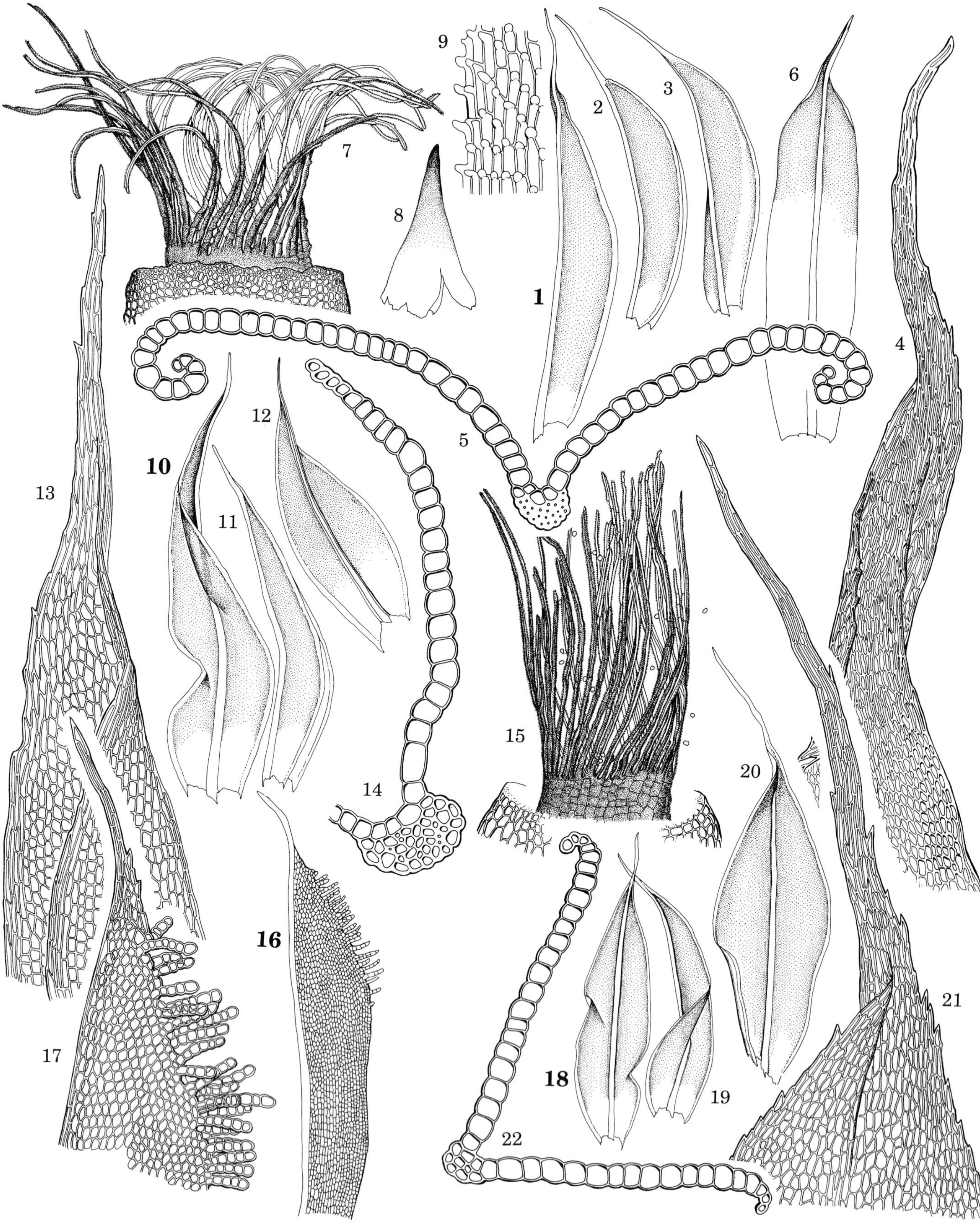

Plate 42. ***Streptopogon.*** **1–9.** ***S. clavipes.*** 1–3. Three leaves. 4. Leaf apex. 5. Section at midleaf. 6. Perichaetial leaf. 7. Peristome. 8. Calyptra. 9. Detail of calyptra. **10–15.** ***S. heterophyllus.*** 10–12. Three leaves. 13. Leaf apex. 14. Section at midleaf. 15. Peristome. **16–17.** ***S. matudianus.*** 16. Leaf. 17. Leaf apex. **18–22.** ***S. stenophyllus.*** 18–20. Three leaves. 21. Leaf apex. 22. Section at midleaf.

Crum (1952b) found both the synoicous and paroicous condition in a specimen of *S. heterophyllus* from southern Peru; the type of this species at W! is synoicous and autoicous. *Streptopogon stenophyllus* (Pl. 42, f. 18–22) and *S. calymperoides* are unique in their yellow to yellowish orange reaction to KOH; the former is, however, clearly a member of the genus, possessing characteristic weak costa, lack of a stem central strand, papillose long-mitrate calyptra, seta only slightly longer than the 2 mm-long capsule, and little-differentiated perichaetial leaves, while the latter, though known only from extremely fragmentary, sterile material, is much like *S. calymperes*. "*Streptopogon peruvianus*" (L!), although cited as a correct name in Index Muscorum (van der Wijk et al. 1959–1969) lacks a description, and is thus a *nomen nudum*.

Additional literature: Brown (1898a), Griffin (1986b), Salmon (1903), Sharp (1986).

Number of accepted species: 14.

Species examined: *S. calymperes* (NY), *S. calymperoides* (FH), *S. cavifolius* (TENN), *S. clavipes* (BUF), *S. erythrodontus* (NY), *S. heterophyllus* (L), *S. matudianus* (TENN), *S. stenophyllus* (PC).

New heterotypic synonymy: *Streptopogon australis* Mitt. = "*Daltonia angustifolia*? Dozy & Molk." *fide* H. Robinson's annotation of the type (NY); the type is definitely assignable to that widely distributed species; also note the strong yellow KOH reaction.

33. BARBULA — Plates 43–45.

Barbula Hedw., Spec. Musc. 115, 1801, *nom. cons. non* Loureiro, 1790. Lectotype: *Barbula unguiculata* Hedw.

Barbiferus Poir., Enc. Meth. Bot. Suppl. 1(2): 587, 1811, *nom. illeg.*

Mollia Schrank. ex Lindb., Utkast Nat. Grupp. Eur. Bladmoss. 38, 1878, *hom. illeg. incl. gen. prior. non Mollia* Gmell., 1791, *nec Mollia*, 1806, *nec Mollia* Mart., 1826, *nom. cons. (p.p. Barbula* Hedw. *et p.p. Tortula* Hedw.).

Sect. *Barbula*

Tortula subg. *Barbula* (Hedw.) De Not., Mem. R. Acc. Sc. Torino 40: 287, 1838.

Barbula subg. *Barbula* Schimp., Coroll. 31, 1856.

Barbula subg. *Helicopogon* (Mitt.) Lindb., Musci Scand. 22, 1879, *nom. illeg. incl. typ. gen.*

Barbula subg. *Eubarbula* (C. Müll.) Kindb., Eur. N. Amer. Bryin. 2: 246, 1897, *nom. illeg.*

Barbula subg. *Tortobarbula* Szafr., Fl. Polska Mchy 1: 213, 1957 [1958] *nom. inval. descr. polon.*

Bryum sect. *Barbula* (Hedw.) Relh., Fl. Cantabr. ed. 2: 426, 1802.

Barbula sect. *Barbula* Rebent., Prod. Fl. Neomarch. 257, 1804, *nom. illeg.*

Barbula sect. *Caulescentes* Hüb., Musc. Germ. 317, 1833, *nom. nud. incl. typ. gen.*

Tortula sect. *Unguiculatae* De Not., Mem. R. Acc. Sc. Torino 40: 287, 1838. Type: *Barbula unguiculata* Hedw.

Barbula sect. *Unguiculatae* BSG, Bryol. Eur. 2: 80, 1842 (fasc. 13–15 Mon. 18), *nom. illeg. incl. typ. gen.*

Barbula sect. *Senophyllum* C. Müll., Syn. 1: 606, 1849 [Not an error for "*steno-*", see protologue of *Pottia* sect. *Senophyllaria* C. Müll.]

Barbula sect. *Eubarbula* C. Müll., Syn. 1: 623, 1849, *nom. illeg. excl. typ. gen. cons.*

Tortula sect. *Barbula* (Hedw.) Mitt., J. Linn. Soc. Bot. 12: 144, 158, 1869.

Tortula sect. *Helicopogon* Mitt., J. Linn. Soc. Bot. 12: 142, 150, 1869.

Barbula sect. *Falax* Lázaro é Ibiza, Bot. Descr. Comp. Fl. Esp. 1: 586, 1869, *nom. illeg. incl. typ. gen. cons.*

Barbula sect. *Eubarbula* Lindb. ex Braithw., Brit. Moss Fl. 1: 261, 1887, *nom. illeg. excl. typ. gen. cons.*

Barbula sect. *Helicopogon* (Mitt.) Braithw., Brit. Moss. Fl. 1: 274, 1887.

Barbula sect. *Senophyllum* C. Müll. ex Podp., Consp. Musc. Eur. 200, 1954.

Sect. *Hyophiladelphus* C. Müll., Syn. 1: 604, 1849.

Barbula subg. *Hyophiladelphus* (C. Müll.) Zand., Phytologia 44: 201, 1979. Lectotype: *Barbula agraria* Hedw. *fide* Zander, Phytologia 44, 201, 1979.

Tortula sect. *Hyophiladelphus* (C. Müll.) Broth., Nat. Pfl. 1(3): 429, 1902.

Barbula sect. *Agrariae* Steere in Grout, Moss Fl. N. Amer. 1(3): 173, 1938, *nom. illeg.* Type: *Barbula agraria* Hedw.

Sect. *Bulbibarbula* C. Müll., Flora 62: 379, 1879.

Barbula sect. *Rhystobarbula* C. Müll., Gen. Musc. Fr. 463, 1900, *nom. illeg. incl. sect. prior.*

Tortula sect. *Rhystobarbula* Dix., J. Bot. 80: 41, 1941.

Tortula sect. *Bulbibarbula* (C. Müll.) Wijk & Marg., Taxon 7: 290, 1958. Type: *Tortula eubryum* C. Müll.

Sect. *Convolutae* B.&S. in BSG, Bryol. Eur. 2: 91, 1842 (fasc. 13–15 Mon. 29).

Streblotrichum P. Beauv., Mag. Enc. 5: 317, 1804. Lectotype: *Streblotrichum convolutum* (Hedw.) P. Beauv. *fide* Saito, J. Hattori Bot. Lab. 39: 499, 1975.

Tortula sect. *Convolutae* De Not., Mem. R. Acc. Sc. Torino 40: 287, 1838. Type: *Tortula convoluta* (Hedw.) Gaertn., Meyer & Scherb.

Tortula sect. *Leptopogon* Mitt., J. Linn. Soc. Bot. 12: 143, 156, 1869. Type: *Tortula calyculosa* Mitt., *lectotyp. nov.*

Tortula subg. *Streblotrichum* (P. Beauv.) Chev., Fl. Gen. Env. Paris 2: 51, 1827.

Barbula subg. *Odontophylla* Saito, J. Hattori Bot. Lab. 39: 499, 1975. Type: *Barbula hiroshii* Saito.

Barbula sect. *Leptopogon* (Mitt.) Lindb., Musc. Scand. 22, 1879.

Barbula subg. *Streblotrichum* (P. Beauv.) Limpr., Laubm. Deutschl. 1: 626, 1888.

Sect. *Hydrogonium* (C. Müll.) Saito, J. Hattori Bot. Lab. 39: 492, 1975.

Hydrogonium (C. Müll.) Jaeg., Ber. St. Gall. Naturw. Ges. 1877–78: 405, 1880 (Ad. 2: 669). Lectotype: *Hydrogonium ehrenbergii* (Lor.) Jaeg. *fide* Saito, J. Hattori Bot. Lab. 39: 492, 1975.

Didymodon subg. *Hydrogonium* (C. Müll.) Kindb., Eur. N. Amer. Bryin. 2: 273, 1897.

Barbula subg. *Hydrogonium* (C. Müll.) Fleisch., Musci Fl. Buitenzorg 1: 352, 1904.

Trichostomum sect. *Hydrogonium* C. Müll., Linnaea 40: 297, 1876.

Semibarbula Herz. ex Hilp., Beih. Bot. Zentralbl. 50: 626, 1933. Type: *Semibarbula indica* (Hook.) Hilp.

Hydrogonium sect. *Barbuliella* Chen, Hedwigia 80: 233, 1941.
Hydrogonium sect. *Euhydrogonium* Chen, Hedwigia 80: 233, 1941, *nom. illeg.*
Sect. *Pachynoma* (Mitt.) Par., Ind. Bryol. ed. 2, 3: 348, 1905.
Tortula sect. *Pachynoma* Mitt., J. Linn. Soc. Bot. 12: 143, 151, 1869.
Sect. *Pseudocrossidiella* Thér. in Felipp., Rev. Bryol. n. ser. 2: 216, 1930. Type: *Barbula subgrimmiacea* Thér.
Subsect. *Purpureaeformes* Kindb., Eur. N. Amer. Bryin. 2: 246, 1897. Type: *Barbula purpurea* C. Müll.

From *barba*, beard + *-ula*, diminutive; little beard, referring to the peristome of 32 hair-like divisions.

Plants loosely caespitose or forming cushions, *yellowish brown, brown or blackish above*, yellowish brown to reddish brown below. *Stems* branching irregularly, ca. 0.2–3.0 cm in length, transverse section rounded-pentagonal or irregular, central strand present or seldom absent, sclerodermis usually present, *hyalodermis occasionally present*; *axillary hairs 2–10 cells in length, usually all hyaline*, occasionally basal 1–2 cells firm-walled; rhizoids sparse to common. *Leaves* appressed-incurved to weakly spreading, *often contorted or twisted about stem* and occasionally catenulate when dry, spreading when moist, *spathulate, ligulate or more usually broadly lanceolate to long-triangular*, usually ca. 1–3 mm in length, *upper lamina usually deeply and narrowly grooved ventrally along costa*, occasionally only broadly concave, lamina unistratose or seldom bistratose in patches, *margins usually recurved in lower 1/2–2/3 of leaf*, occasionally plane, entire or occasionally denticulate near apex or above midleaf, seldom dentate, margins occasionally thick-walled, seldom bi- or multistratose; *apex rounded to obtusely acute, usually mucronate*, occasionally entire or short-apiculate; base usually weakly differentiated to ovate, sometimes broadened and sheathing, sometimes narrowly decurrent; *costa percurrent to short-excurrent as a sharp mucro, seldom short-awned*, occasionally ending a few cells below the apex, *superficial cells elongate or occasionally quadrate to short-rectangular ventrally,* usually elongate but sometimes quadrate to short-rectangular dorsally, 2–3(–5) rows of cells across costa ventrally at midleaf, costal transverse section oval to semicircular, *two stereid bands usually present*, usually small but distinct ventrally, usually strong dorsally and crescent-shaped but sometimes nearly semicircular, *epidermis usually differentiated ventrally, usually present but weakly differentiated dorsally*, guide cells usually 2–4 in 1 layer, hydroid strand occasionally present; *upper laminal cells quadrate to short-rectangular*, 6–13 µm in width, usually 1:1, *walls thin to evenly thickened*, usually superficially bulging on both sides, occasionally ventrally bulging and dorsally flat or nearly so; *papillae hollow or solid, multifid or bifid, 2–3 per lumen, seldom simple or absent, usually obscuring the lumens*, occasionally mamillose ventrally; *basal cells usually differentiated*, reaching across leaf or reaching higher medially or occasionally marginally, rectangular, *usually little wider than upper laminal cells*, 3–5:1, walls thin to evenly thickened. *Propagula when present borne on basal rhizoids or axillary on stalks, often rather large, 30–300 µm in length, of 1–50 cells, clavate to ovate, occasionally armed, multicellular, green to reddish. Dioicous*, possibly occasionally rhizautoicous. Perichaetia terminal, inner leaves little differentiated or ovate to long-lanceolate, sometimes strongly sheathing, lower cells long rhomboidal in lower 1/2 of leaf, occasionally to 3/4 or more. Perigonia gemmate, often prominent. Seta 0.5–1.5 cm in length, usually 1 per perichaetium, yellowish to reddish brown, twisted clockwise below, often counterclockwise above; *theca* usually 0.8–1.5 mm in length, yellowish to reddish brown, *ovate to long-cylindrical*, exothecial cells rectangular or occasionally rhomboidal, walls thin to evenly thickened, stomates phaneropore, at base of theca, annulus weakly differentiated to strong, of 1–3 rows of vesiculose cells, usually persistent, occasionally revoluble or deciduous in pieces; *peristome teeth of 32 narrow rami*, seldom short or rudimentary, filamentous to narrowly triangular, *usually densely spiculose*, with many articulations, *usually strongly twisted* 1.5 to 2 times counterclockwise, occasionally straight, to 1200 µm in length, *basal membrane low but distinct*, ca. 40 µm in height, granulate to spiculose. Operculum usually long-conic, ca. 0.8–1.7 mm in length, cells twisted counterclockwise. Calyptra cucullate, smooth, usually 1.5–3.0 mm in length. Spores usually 9–16 µm in diameter, light brown, weakly papillose. *Laminal KOH color reaction yellow*, occasionally yellowish orange. Reported chromosome number n = 10, 10+m, 11, 12+m, 13, 13+m, 13+2m, 14, 14+2m, 16, 24, 26.

A cosmopolitan genus found in a wide variety of habitats, mainly on soil and acid or calcareous rock.

Saito's (1975a) distinctions between the genera *Barbula* and *Didymodon* are to a large extent followed in this treatment. Axillary hair characters, however, are considered here to be more variable than indicated by Saito. In *Barbula*, the hairs are often but not always entirely much elongate and hyaline (Pl. 44, f. 21—as opposed to the short, basally brown-celled and otherwise hyaline hairs of *Didymodon*). Some species of *Barbula* have elongate, entirely hyaline hairs (e.g. *B. hiroshii* and *B. unguiculata*), some have firm-walled hair basal cells (e.g. *B. riograndensis*), and others may be variable within the same species (e.g. *B. indica*). (Completely hyaline hairs may be interpreted as all firm-walled rather than all thin-walled. Firm-walled basal cells are either hyaline or somewhat more yellowish than the distal cells. The cells of the completely hyaline hairs in *Barbula* are, in fact, rather firm-walled.) In *Didymodon*, however, the brownish, firm walls of the basal cells of the axillary hairs appear to develop earlier than they do in those species of *Barbula* that have them. Small, dark rectangles are generally visible under the microscope through the transparent bases of the leaves of the extreme stem apex of *Didymodon* species, while they are seldom so evident in *Barbula*. Careful stripping of the subapical leaves will uncover the young hairs.

Saito's (1975a) use of the shape of the superficial cells of the ventral surface of the costa as a distinction (*Barbula* with elongate cells—Pl. 44, f. 9, 14, 20, *Didymodon* sect. *Didymodon* with short or quadrate cells) is variable for *Barbula* world wide, though it apparently holds for the Japanese species he revised; Eddy (1990) also considered the elongate ventral costal cells to be taxonomically important in the Malesia area. The two genera, however, can usually be distinguished quickly by the morphology of the laminal papillae and the leaf apex. In *Barbula*, the upper laminal papillae are rough, knobby, obscuring the lumens, and protuberant along the upper laminal margins (Pl. 43, f. 5), while in *Didymodon* these are generally low, difficult to distinguish, and little evident along the upper margins. There are, of course, exceptions, notably *Barbula* sect. *Hydrogonium* and *Didymodon* sect. *Vineales*. *Bryoerythrophyllum* is quite like

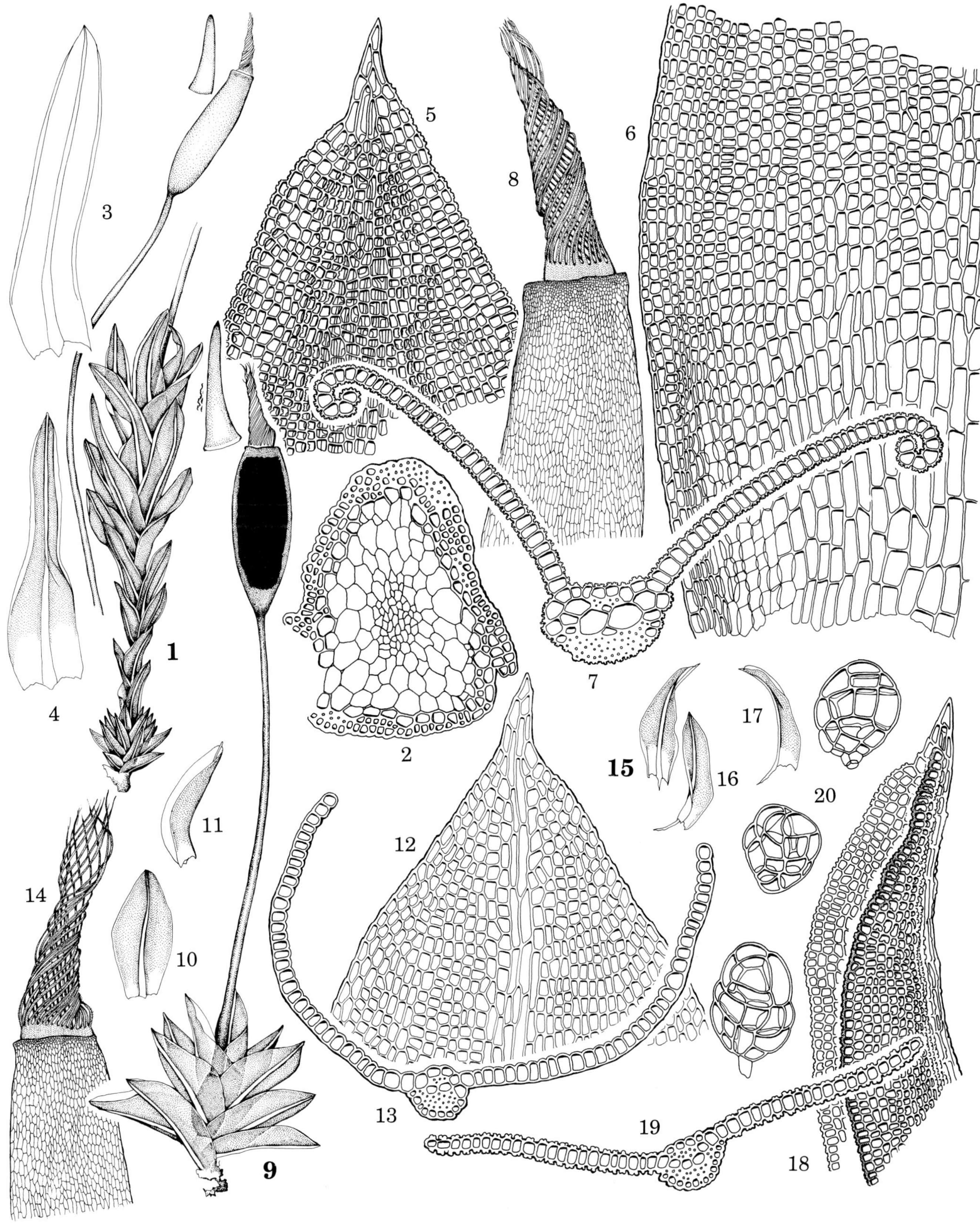

Plate 43. ***Barbula.*** **1–8.** ***B. costesii.*** 1. Habit. 2. Transverse section of stem. 3–4. Two leaves. 5. Leaf apex. 6. Basal cells. 7. Transverse section at midleaf. 8. Peristome. **9–14.** ***B. agraria.*** 9. Habit. 10–11. Two leaves. 12. Leaf apex. 13. Transverse section at midleaf. 14. Peristome. **15–20.** ***B. amplexifolia.*** 15–17. Three leaves. 18. Leaf apex. 19. Transverse section at midleaf. 20. Propagula.

Barbula in papillae morphology but reacts red, not yellow, to KOH. *Barbula* is unlike *Didymodon* in that its leaf apex usually ends in an apiculus of one or a few clear cells or in a rather strong, sharp mucro; in *Didymodon*, the costa ends before or in the apex, or, if excurrent, is rather blunt and opaque, seldom apiculate by one or more clear cells. As Saito (1975a) pointed out, in *Barbula* the propagula (Pl. 43, f. 20; 44, f. 5, 10) range from small and ovate to large and irregular in shape, while *Didymodon* has only small ovate propagula. Both *Barbula* and *Didymodon*, however, remain rather heterogeneous, and segregate genera will surely be described in the future to include, for instance, such oddities as *B. integrifolia* and *B. eubryum*. Compounding this are species of *Didymodon* still remaining as combinations in *Barbula*; transfer of some of these is done here, but much remains to be accomplished by revisionists. *Barbula* may be hypothesized as the start of a reduction series continuing through *Barbula* sect. *Convolutae*, *Leptobarbula*, *Gyroweisia* and *Gymnostomum*. *Gyroweisia* is distinct in the combination of enlarged basal cells, vesiculose and often revoluble annulus, and peristome absent or reduced. It may well be, however, that *Barbula* will be broken up in the future into segregate genera consisting of narrowly conceived reduction series of species crossing presently recognized limits of these similar genera.

Barbula sect. *Convolutae* is distinctive in the mostly plane leaf margins, usual presence of a stem hyalodermis, commonly strongly differentiated perichaetial leaves, and generally a yellow seta. It is problematically quite similar to *Trichostomum*, except for the presence of a twisted peristome, and may be related. However, in many species of sect. *Convolutae* the basal cells extend farthest up the leaf medially (in others merely straight across) or reaching highest at the midpoint between the costa and the margins on both sides). What is apparently the type of sect. *Convolutae*, *B. convoluta*, has a stem section like that of the Merceyoideae (central strand present, comparatively large inner cylinder cells, thin cortex of abruptly smaller stereid cells, hyalodermis little or not differentiated) rather than the Trichostomoideae (central strand variably present, inner cylinder cells often relatively small, cortex commonly of substereid cells, hyalodermis often well differentiated and sometimes in more than one layer). This is also true of *B. indica*, which, although having plane leaf margins, is clearly a *Barbula* by its stem anatomy, though probably best placed in *Hydrogonium* since the generitype of *Hydrogonium* is clearly a related but derived species. The type of *B. cancellata* (a name recently much in use in North America) at NY is *B. indica* var. *indica*; it bears small propagula. Section *Convolutae* characteristically has rather large propagula; axillary or rhizoidal propagula are typical of many genera of Barbuleae but are uncommon in genera of Pottieae or Trichostomoideae. It is quite possible, in any case, that upon revision at least certain species of sect. *Convolutae* (such as *B. amplexifolia*, Pl. 43, f. 15–20, which has the stem anatomy of the Trichostomoideae) will be seen to better belong in or near *Trichostomum*, or even *Tortella*. Norris and Koponen (1989), on the other hand, in their treatment of the bryophytes of an area in New Guinea, indicated that plane-margined *Barbula* species lack a hyalodermis and have little taper to the costae, while *Trichostomum* species with which they might be confused have a hyalodermis and tapering costae. Further investigation is necessary at the revision level. They also pointed out that sterile specimens of *Hydrogonium* can be distinguished from *Dicranella* (Dicranaceae) by the former's quadrate cells of the ventral surface of the costa and a regular arrangement of the laminal cells in rows.

The genus *Tetrapterum* is similar to sect. *Convolutae* in its gametophyte morphology, but differs significantly in the sporophyte. *Barbula calycina*, which may belong with *Tetrapterum*, is unusual in its basal cells differentiated into two groups of about equal size, of hyaline, thin-walled, smooth cells towards the margins, and yellow, thick-walled, papillose cells medially. This is reminiscent of a corresponding morphology in *Pseudosymblepharis* species. *Barbula subcalycina* has very lax basal cells reminiscent of those of *Tortella humilis*. Stone (1991) provided good distinctions between *B. calycina* and *B. subcalycina*. *Barbula calyculosa*, *B. fendleri* and *B. fidelis* are probably the same as *B. convoluta*.

Many species of sect. *Hydrogonium* (the type of which is *Barbula ehrenbergii*) may have evolved from ancestors quite like *B. indica* towards a hygric habitat and large size. *Barbula leucodontoides* (Pl. 44, f. 18–21—isotype, NY!) has a dorsally prorulose costa, and is surely very closely related to *B. indica*. Other species (e.g. *B. zambesiaca*) have rather flaccid leaves with reduced papillae and the dorsal surface of the costa is merely bumpy. *Barbula javanica*, with its ventrally bulging and dorsally smooth (or nearly so) upper lamina, may share ancestors with the Hyophileae. Note that cladistic analysis does not support a close relationship between *Hyophila* and *Barbula*, and perhaps the section *Hydrogonium* should be recognized at the generic level. *Barbula agraria* (Pl. 43, f. 9–14), with its broad leaves and ventrally bulging upper laminal cells may also belong with the Hyophileae. Both sect. *Hydrogonium* and sect. *Hyophiladelphus* have the essentially tropical lowland distribution characteristic of the Hyophileae. Cladistic analysis at the species level might detail support for this possible dismemberment of *Barbula* along these lines.

Barbula marginatula (Pl. 44, f. 22–25) is at an end point in the evolution of sect. *Hydrogonium* with the additional character of a cartilaginous, denticulate leaf border of elongate cells. Gangulee (1972) indicated that *B. marginatula* may represent a "new genus of barbuloid mosses" but if such were recognized the genus should include the types of *Hydrogonium* and *Semibarbula*, which clearly belong to the lineage. *Barbula pachyloma* (refered to *Cinclidotus* by Hilpert 1933 but recognized as a good species of *Barbula* by Eddy 1990 and Norris and Koponen 1989) may well belong here too, having multistratose upper laminal borders of stereid cells covered with parenchymatous cells (much like those of Calymperaceae species), but prorulae are not evident; it is much like *Dialytrichia* in appearance and moist habitat but differs in the stereid cells of the laminal margin, widely channeled (non-keeled) leaves, and quadrate ventral superficial cells of the costa.

Species previously referred to *Barbula* but which have long-awned, marginally highly revolute leaves with strongly flattened dorsal stereid band and the ventral band often absent, and the perichaetial leaves usually enlarged and convolute-sheathing are better recognized in *Pseudocrossidium* ((*q.v.*).

Certain species of *Ditrichum*, namely *D. tortipes* (Mitt.) Par. and *D. ambiguum* Best (*cf.* Crum & Anderson 1981, Grout 1927 and Robinson 1968—these two species probably conspecific), have once-twisted, densely spiculose peristome teeth and long-triangular, apically rough, setaceous leaves with rectangular, smooth upper laminal cells much like the leaves of forms of

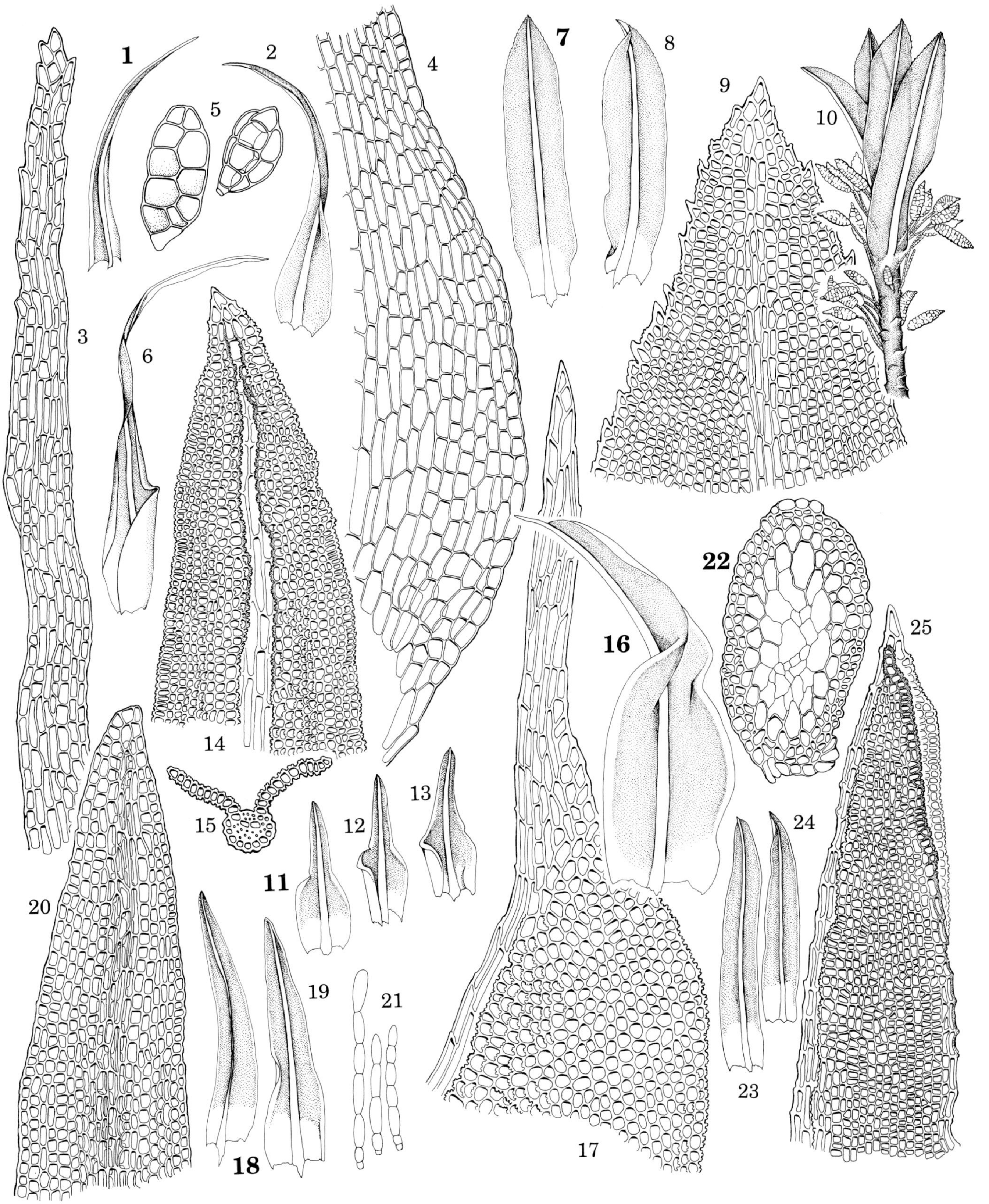

Plate 44. ***Barbula.*** **1–6.** ***B. comosa.*** 1–2. Two leaves. 3. Leaf apex. 4. Basal cells. 5. Propagula. 6. Perichaetial leaf. **7–10.** ***B. hiroshii.*** 7–8. Two leaves. 9. Leaf apex. 10. Propagula. **11–15.** ***B. hispaniolensis.*** 11–13. Three leaves. 14. Leaf apex. 15. Transverse section at midleaf. **16–17.** ***B. integrifolia.*** 16. Leaf. 17. Leaf apex. **18–21.** ***B. leucodontoides.*** 18–19. Two leaves. 20. Leaf apex. 21. Axillary hairs. **22–25.** ***B. marginatula.*** 22. Transverse section of stem. 23–24. Two leaves. 25. Leaf apex, lateral view.

Barbula arcuata. These *Ditrichum* species may actually belong in*Barbula*. Future investigation, at least of the *B. arcuata* complex, should deal with this question. *Barbula arcuata* differs significantly in the red rather than yellow, more strongly twisted peristome, and the leaves only weakly denticulate and margins unistratose. Its costal section, however, is strikingly like that of the *Ditrichum* species, especially in the strongly differentiated dorsal epidermal cells.

A quite unusual species is *Barbula hispaniolensis* (Pl. 44, f. 11–15), which is similar to *Bryoeyrthrophyllum ferruginascens*, except that the leaves are bright yellow in KOH solution and the costa has a narrow ventral groove (characteristic of *Barbula*). This species is retained here pending further study; one might hypothesize a chemical treatment used during collecting that might have changed the color reaction, but casual experiments show that neither boiling in formaldehyde solution or in ethyl alcohol causes irreversible changes to the red KOH reaction of true *B. ferruginascens*.

Additional literature: Abramova et al. (1967), Coats and Mahler (1985), Conard (1945a, 1951a), Crum (1956, 1965d, 1967a), Crundwell (1976), Dhingra-Babbar (1988), Dhingra-Babbar and Chopra (1985), Field (1990), Hoffmann (1957), Maheu (1908), Roorda van Eysinga (1972), Saito (1971a, Sharma and Chopra (1987), Sharma and Sharma (1980, 1986), 1975a), Steere (1938a, 1939b), Takio (1989), Takio et al. (1986), Weber (1972), Zander (1979f, 1981a). Most bibliographic references to *Barbula* also discuss species of *Didymodon sensu* Saito (1975a) and the present treatment.

Number of accepted species: 205, plus 8 remaining in *Hydrogonium*, 1 in *Semibarbula*, and 3 in *Streblotrichum*.

Species examined: *B. afrofontana* (NY), *B. agraria*, *B. arcuata* (BM, BUF, FH, NY, TENN), *B. amplexifolia* (BM, BUF, MICH, NY), *B. bicolor* (BUF), *B. calycina* (NY), *B. calyculosa* (NY), *B. clavicostata* (PC), *B. comosa* (NY), *B. convoluta*, *B. costesii* (NY), *B. crocea* (BUF), *B. ehrenbergii*, *B. enderesii* (BUF), *B. eubryum* (BUF, H, US), *B. eustegia* (NY), *B. fendleri* (NY), *B. fidelis* (NY), *B. hiroshii* (TNS), *B. hispaniolensis* (NY), *B. indica*, *B. integrifolia* (US), *B. isoindica* (NY), *B. javanica* (NY), *B. lambarenensis* (NY), *B. leucodontoides* (NY), *B. microcalycina* (NY), *B. munyensis*

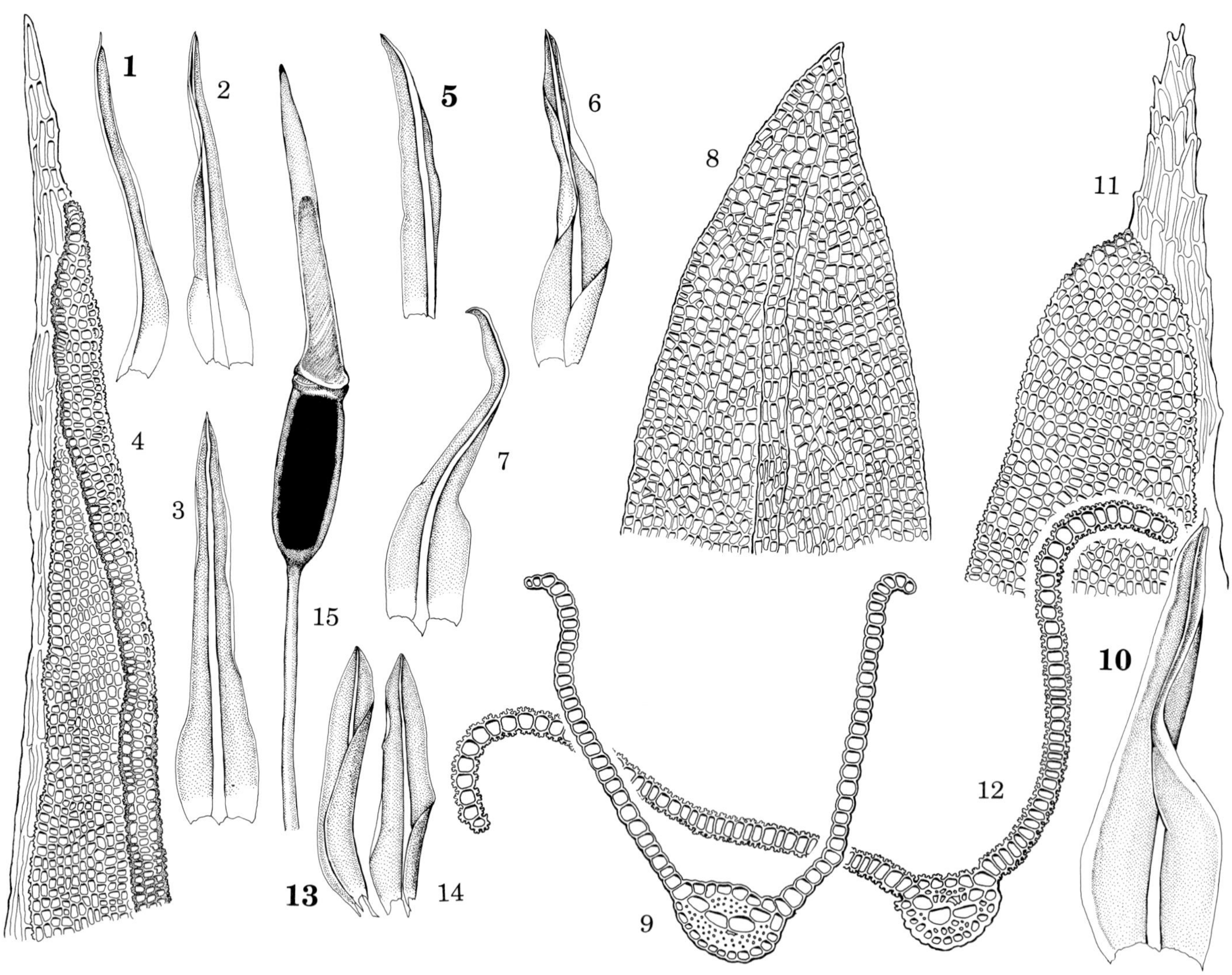

Plate 45. *Barbula*. 1–4. *B. munyensis*. 1–3. Three leaves. 4. Leaf apex. **5–9. *B. pseudoehrenbergii*.** 5–7. Three leaves. 8. Leaf apex, dorsal view. 9. Transverse section at midleaf. **10–12. *B. riograndensis*.** 10. Leaf. 11. Leaf apex. 12. Transverse section at midleaf. **13–15. *B. williamsii*.** 13–14. Two leaves. 15. Capsule with calyptra.

(NY), *B. occidentalis* (NY), *B. orizabensis* (BM, BUF, FH, TENN), *B. pachyloma* (BUF), *B. peruviana* (NY)—near *B. indica*, *B. pseudoehrenbergii* (NY), *B. rechingeri* (H), *B. riograndensis* (BUF, FH), *B. semirosulata* (PC), *B. spathulifolia* (BM), *B. subcalycina* (NY), *B. williamsii* (NY), *B. unguiculata*, *B. zambesiaca* (NY).

New heterotypic synonymy: *Barbula tonkinensis* (Besch.) Broth. = *Barbula indica* (Hook.) Spreng.

New combinations and new names: *Barbula clavicostata* (Ren. & Hilp.) Zand., *comb. nov.* (*Hyophila clavicostata* Ren. & Card., Act. Soc. Linn. Bordeaux 53: 21, 1898 and Prodr. Fl. Bryol. Madag. 123, 1898; *Hydrogonium clavicostatum* (Ren. & Card.) Hilp.). *Barbula isoindica* Zand., *nom. nov.* (*Pottia papillinervis* Lor., Moostud. 160, 1864; *Barbula papillinervis* (Lor.) Broth., *hom. illeg.*, Nat. Pfl. 1(3): 408, 1902). *Barbula semirosulata* Zand., *nom. nov.* (*Gymnostomiella rosulata* P. Varde, Rev. Bryol. Lichénol. 26: 1, 1957). *Barbula* (sect. *Hydrogonium*) *spathulifolia* (Dix. & P. Varde) Zand., *comb. nov.* (*Merceyopsis spathulifolia* Dix. & P. Varde, Arch. Bot. 1(8–9): 164, 1927).

34. GYMNOSTOMUM **Plate 46.**

Gymnostomum Nees & Hornsch., Bryol. Germ. 1: 112, 1823, *nom. cons., non Gymnostomum* Hedw., 1801, *nom. rejic.*
Lectotype: *Gymnostomum calcareum* Nees & Hornsch.

Sect. *Gymnostomum*

Gymnostomum subg. *Eugymnostomum* Schimp., Syn. 39, 1860, *nom. illeg.*

Gymnostomum subg. *Gymnostomum* Kindb., Eur. N. Amer. Bryin. 2: 284, 288, 1897, *nom. illeg.*

Bryum sect. *Gymnostomum* (Hedw.) Relh., Fl. Cantabr. ed. 2: 424, 1802.

Gymnostomum sect. *Holomitria* Wallr., Fl. Crypt. Germ. 1: 94, 1831, *p.p.*

Trichostomum sect. *Pycnocaulus* Lindb. ex Milde, Bryol. Siles. 106, 1869.

Sect. *Diastoma* Griff., Calcutta J. Nat. Hist. 2: 481, 1842.

From γυμνός, naked + ο +στόμα, -ατος, mouth; the peristome is absent.

Plants growing in turf or cushions, light to dark or olive green above, light to dark brown below. Stems branching often, to 2.7 cm in length, transverse section rounded-pentagonal to occasionally rounded-triangular, *central strand usually present*, weak, sclerodermis absent or occasionally present, hyalodermis present or absent; axillary hairs 3–10 cells in length, the basal 1–2 usually brownish; sparsely radiculose or occasionally red tomentose. *Leaves* appressed to appressed-incurved when dry, weakly spreading to spreading-recurved when moist, *usually ligulate, occasionally ovate to circular, short*, to 1.1(–1.8) mm in length, *upper lamina flat to broadly convex*, occasionally keeled, *margins plane or occasionally recurved below midleaf*, entire or minutely crenulate by projecting papillae, sometimes bistratose marginally above midleaf; *apex rounded obtuse to broadly acute, often apiculate*; base scarcely differentiated or seldom ovate-rectangular, sometimes denticulate marginally; *costa ending 2–5 cells below apex*, only occasionally percurrent, occasionally swollen at or above midleaf, *ventrally usually bulging, ventral superficial cells quadrate or short-rectangular, occasionally elongate*, dorsally elongate or occasionally short-rectangular to quadrate above midleaf, *both sides of costa papillose*, 2–4(–6) rows of cells across costa ventrally at midleaf, costal transverse section ovate to semicircular, stereid bands weak or lacking ventrally, present but often weak dorsally, *ventral epidermis present*, dorsal occasionally little differentiated, guide cells 2(–4) in 1 layer, hydroid strand absent; *upper laminal cells subquadrate*, 7–11 µm in width, 1:1, *walls thin to weakly evenly thickened*, seldom irregularly thickened and lumens angular, *homogeneous*, superficially flat to convex on both sides, often appearing wrinkled in section because of hollow papillae, cells of apex often in rows that "criss-cross" at right angles just below the apex; *papillae simple to bifid, low, small, scattered*, generally crowded, 3–5 per lumen, hollow or solid; basal cells differentiated across leaf or rising higher medially, rectangular, little wider than upper cells, 2–4:1, walls thin. Propagula occasionally present, spherical to obovoid or spindle-shaped, of 5–10 usually multiseriate cells, borne on branching stalks in leaf axils. Dioicous. *Perichaetia terminal*, inner leaves ovate-lanceolate, to 1.5 mm in length, sheathing below midleaf, sometimes marginally serrulate, cells rectangular and occasionally bulging below midleaf. Perigonia gemmate, terminal. Seta generally 0.3–0.6 cm in length, 1 per perichaetium, yellowish to reddish brown, twisted clockwise; theca 0.5–0.8 mm in length, yellowish to reddish brown, ovoid to elliptical, occasionally with a weak but high circumstomal ring, exothecial cells quadrate to rectangular, walls thin to somewhat thickened, stomates phaneropore, at base of theca, *annulus of 1–3 rows of smaller, transversely elongated, occasionally vesiculose cells; peristome teeth absent.* Operculum rostrate to conic-rostrate, ca. 0.4–0.5 mm in length, cells in straight rows. Calyptra cucullate, smooth, 0.5–1.2 mm in length. Spores 9–15 µm in diameter, brownish, essentially smooth to clearly papillose. Laminal KOH color reaction usually yellow to yellow-orange, occasionally red in patches. Reported chromosome number n = 13.

Found on rock (usually calcareous, occasionally acid) in very moist areas; widely distributed on most continents.

Gymnostomum might be viewed as a much reduced, hygrophilic segregate of *Barbula*, an end-member of a phyletic series beginning with *Barbula* sect. *Barbula*, and extending through *B.* sect. *Convolutae*, *Leptobarbula* and *Gyroweisia*. Cladistic analysis (Cladograms 11 and 14), however, does not support more than a close relationship with *Barbula*. The generally well developed papillae (Pl. 46, f. 9) of *Gymnostomum*, especially as they roughen the upper laminal margins, are like those of *Barbula*, but the firm-walled basal cells of the axillary hairs and the costa ending below the apex are more characteristic of *Didymodon*.

Large plants of *Gymnostomum aeruginosum* (e.g. U.S.A., Flowers, Utah, 331, US, and Czechoslovakia, Pilous 857, DUKE) have a tendency toward a leaf base differentiated in shape (becoming rectangular) but the apex of such large plants is more like that of *Hymenostylium*, the costa being percurrent or vaguely excurrent as a broadly triangular mucro, or percurrent and the apex rounded-acute, as is commonly the case in *Didymodon* sect. *Didymodon*, not excurrent and clearly distinct from the tissue of the lamina as is the case with *Barbula*. Hilpert's (1933) combination *Barbula mosis* (Lor.) Hilp. shows a sensitivity on his part to the *Barbula* relationship, as does K. Saito's annotation of a combination in *Barbula* on the type of

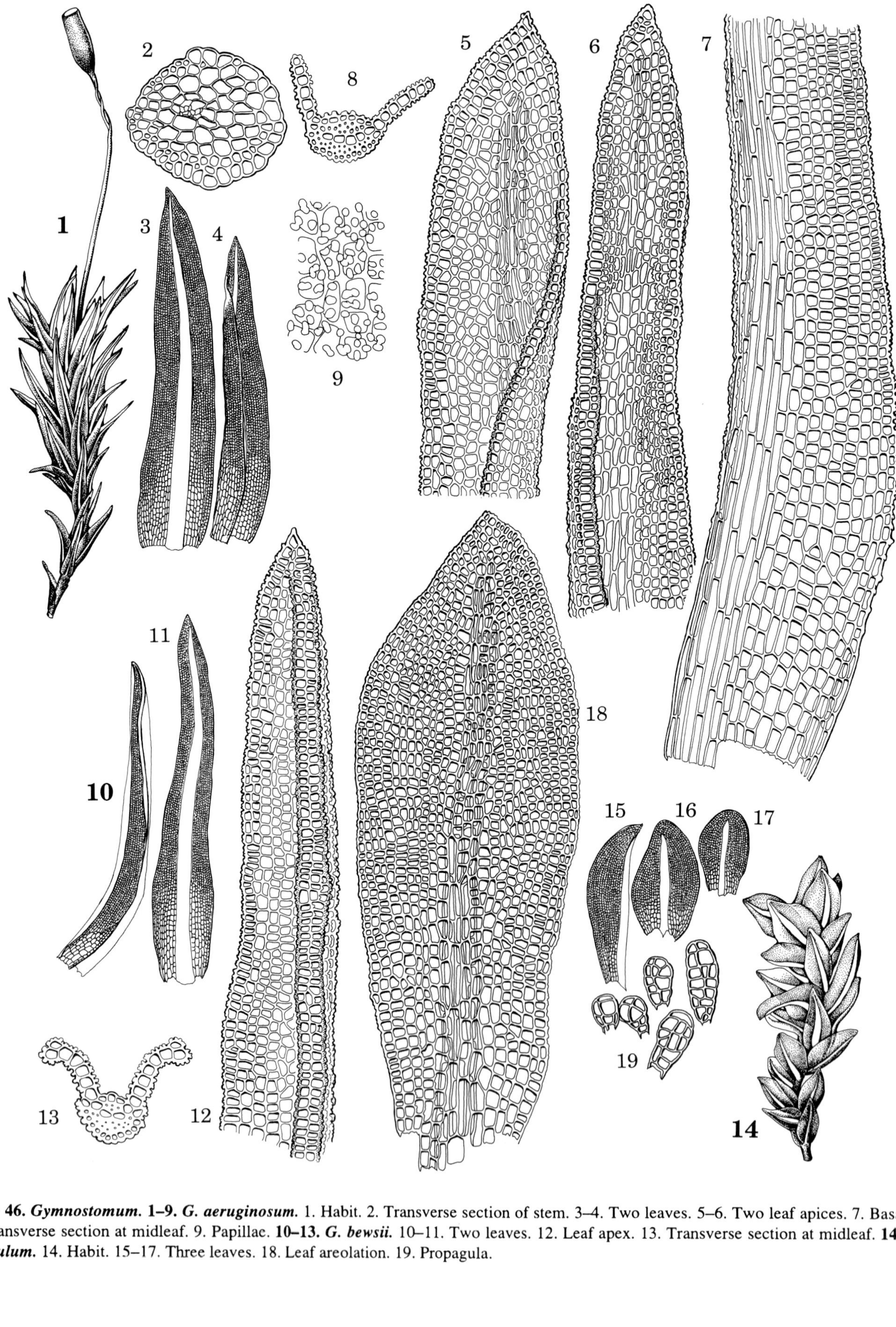

Plate 46. ***Gymnostomum.*** **1–9.** ***G. aeruginosum.*** 1. Habit. 2. Transverse section of stem. 3–4. Two leaves. 5–6. Two leaf apices. 7. Basal cells. 8. Transverse section at midleaf. 9. Papillae. **10–13.** ***G. bewsii.*** 10–11. Two leaves. 12. Leaf apex. 13. Transverse section at midleaf. **14–19.** ***G. viridulum.*** 14. Habit. 15–17. Three leaves. 18. Leaf areolation. 19. Propagula.

G. hymenostylioides (as *Merceyopsis*) at H, with the comment: "Though the species lacks peristome teeth, it seems to be on the same evolutionary (or reductive) line with *Barbula indica*. I think it may be more natural to treat it [as] a member of the genus *Barbula*."

Large specimens of *Gymnostomum aeruginosum* may also have distinctly serrulate lower leaf margins, much as is the case in *Molendoa hornschuchiana*, a large end-member of a stature gradient extending through *M. sendtneriana*, and the serrulate perichaetial leaves of some large specimens of *G. aeruginosum* and those of *G. bewsii* are quite similar to those of *M. sendtneriana*. Evidence of apparent phenocopy phenomena (including presence of propagula in small forms of *M. sendtneriana*) between the two genera at the small end of both of their stature gradients (Zander 1977c, p. 261) supports this relationship. *Eucladium*, however, also has denticulate lower laminal margins and is not closely related to *Gymnostomum* or *Molendoa*, and it is possible that this characteristic is merely typical of hygrophilic species. Although *Molendoa* when sterile can generally be distinguished from *Gymnostomum* by the percurrent or excurrent costa and more massive, scab-like laminal papillae, this is not always the case. Additional discussion on the morphological similarities of *Gymnostomum* and *Molendoa* has been given by Newton (1983), who favored a more analytic approach to species distinctions. Cladistic analysis gives differing results as to the relationship of *Gymnostomum* and *Molendoa*, but see discussion of Cladograms 14–16.

Within *Gymnostomum* itself, there is a clear reduction series based on leaf shape (ligulate to ovate) from *G. aeruginosum* (Pl. 46, f. 3–4) to *G. mosis*, with elaborations such as propagula in *G. virdulum* (Pl. 46, f. 19, and rarely in *G. aeruginosum*), thickened costae in *G. hymenostylioides*, and bistratose upper laminal margins in *G. bewsii* (Pl. 46, f. 13—often present in *G. aeruginosum* from the southeastern U.S.A.). *Gymnostomum viridulum* (Pl. 46, f. 14–19; *cf.* Sérgio 1984, also Whitehouse & Crundwell 1991, 1992) is doubtfully different from *G. mosis*, while *G. hymenostylioides* is very similar to both taxa in its leaf being widest at the middle but the latter has a longer leaf and very stout costa. Although I have synonymized *G. calcareum* with *G. aeruginosum* (Zander 1977c), because these appear to intergrade in the New World, the two are apparently distinct in Europe (see e.g. Whitehouse & Crundwell 1991, 1992) and, for this treatment to be of maximum service, both names are recognized here. Khanna (1976) has pointed out that "*G. recurvirostrum* [= *Hymenostylium recurvirostrum*]..., *G. calcareum*...and *G. aeruginosum* cannot be called clearly marked species because of the presence of intergrading forms among them, at least in the area under investigation [the Himalayas]. However, they are morphologically distinct from each other and statistical distances among them can be defined."

Eventually, certain of the names presently accepted as combinations in *Gymnostomum* will probably be assigned to other genera, such as *Hymenostylium* or *Anoectangium*. For instance, *G. chenii* is surely the same as *H. recurvirostrum* var. *cylindricum*, if one may judge from Saito's (1973c) detailed illustrations.

Additional literature: Andrews (1922b), Brown (1894a), Crum & Anderson (1956), Crundwell (1981), Geheeb (1906a), Khanna (1976), Nyholm and Hedenäs (1986), Pierrot (1973, 1989), Stirling (1968).

Number of accepted species: 24.

Species examined: *G. aeruginosum*, *G. bewsii* (PRE), *G. calcareum* (BUF), *G. hymenostylioides* (BM, H), *G. luisieri* (LISU), *G. mosis* (BUF, MO).

New combination: *Gymnostomum hymenostylioides* (Broth. & Dix.) Zand., *comb. nov.* (*Merceyopsis hymenostylioides* Broth. & Dix., J. Bot. 48: 302, 1910).

35. SCOPELOPHILA — Plate 47.

Scopelophila (Mitt.) Lindb., Acta Soc. Sci. Fenn. 10: 269, 1872. Type: *Scopelophila ligulata* (Spruce) Spruce.

Merceya Schimp., Syn. Musc. ed. 2: 852, 1876. Type: *Merceya ligulata* (Spruce) Schimp.

Merceyopsis Broth. & Dix. ex Dix., J. Bot. 48: 301, 1910. Type: *Merceyopsis pellucida* Broth. & Dix., J. Bot. 48: 301, 1910, India, W. Ghats, Panchgani, Sedgwick 35, H, *lectotyp. nov.*

Weissia sect. *Scopelophila* Mitt., J. Linn. Soc. Bot. 12: 134, 1869.

Pottia sect. *Orthotrichella* C. Müll., Gen. Musc. Fr. 392, 1900. Type: *Pottia gedeana* Lac.

From σκόπελος, lookout place, headland, a crag, high rock + o + φίλος, love, dear, beloved; referring to the characteristic rocky habitat.

Plants forming a thin or thick turf, greenish yellow to brown above, *blackish green or weakly iridescent metallic tan or yellowish brown* below. *Stems* seldom branching and then irregularly, to 4 cm in length, transverse section rounded-pentagonal, *central strand absent*, *sclerodermis absent*, *hyalodermis absent*; axillary hairs 3–5 cells in length, basal cell usually brownish; sparsely radiculose or occasionally with a thick, reddish brown tomentum. *Leaves* often crowded, incurved to spreading, contorted, usually carinate, occasionally with undulate upper margins when dry, spreading when moist, *lingulate to ligulate or oblanceolate*, *widest at or above midleaf*, to 2.5 mm in length, upper lamina narrowly grooved along costa or sometimes broadly channeled, *margins plane or somewhat recurved below*, entire to minutely crenulate or denticulate above, *often bordered by a few rows of thicker walled cells above*; *apex broadly acute or obtuse*, *often with a broad apiculus*, occasionally rounded; base scarcely differentiated in shape to long-elliptical, occasionally wasp-waisted, occasionally slightly decurrent; *costa slender*, percurrent or ending up to 2–8 cells below the apex, occasionally excurrent as a short mucro, superficial cells quadrate to rectangular ventrally, short-rectangular to elongate dorsally, 2–4 rows of cells across costa ventrally at midleaf, costal transverse section semicircular to round, *one stereid band present*, generally well distinguished from the superficial parenchymatous layers, epidermis absent or present ventrally, present dorsally, guide cells 2(–4) in 1 layer, *hydroid strand absent*; *upper laminal cells rounded-quadrate to hexagonal or short-rectangular*, ca. 8–14 µm, *often heterogeneous in size and shape*, 1(–2):1, *walls* thin to evenly thickened or weakly collenchymatous, *thicker near margins*, *superficially flat or somewhat bulging on ventral surface*; *papillae lacking*, occasionally low-verrucose; basal cells differentiated across leaf, extending higher medially, rectangular, occasionally inflated, scarcely wider than the upper cells to inflated, 2–3:1, walls hyaline or deep brown, bordered on margins by one or more rows of narrow rectangular cells. Propagula rare, on stalks from

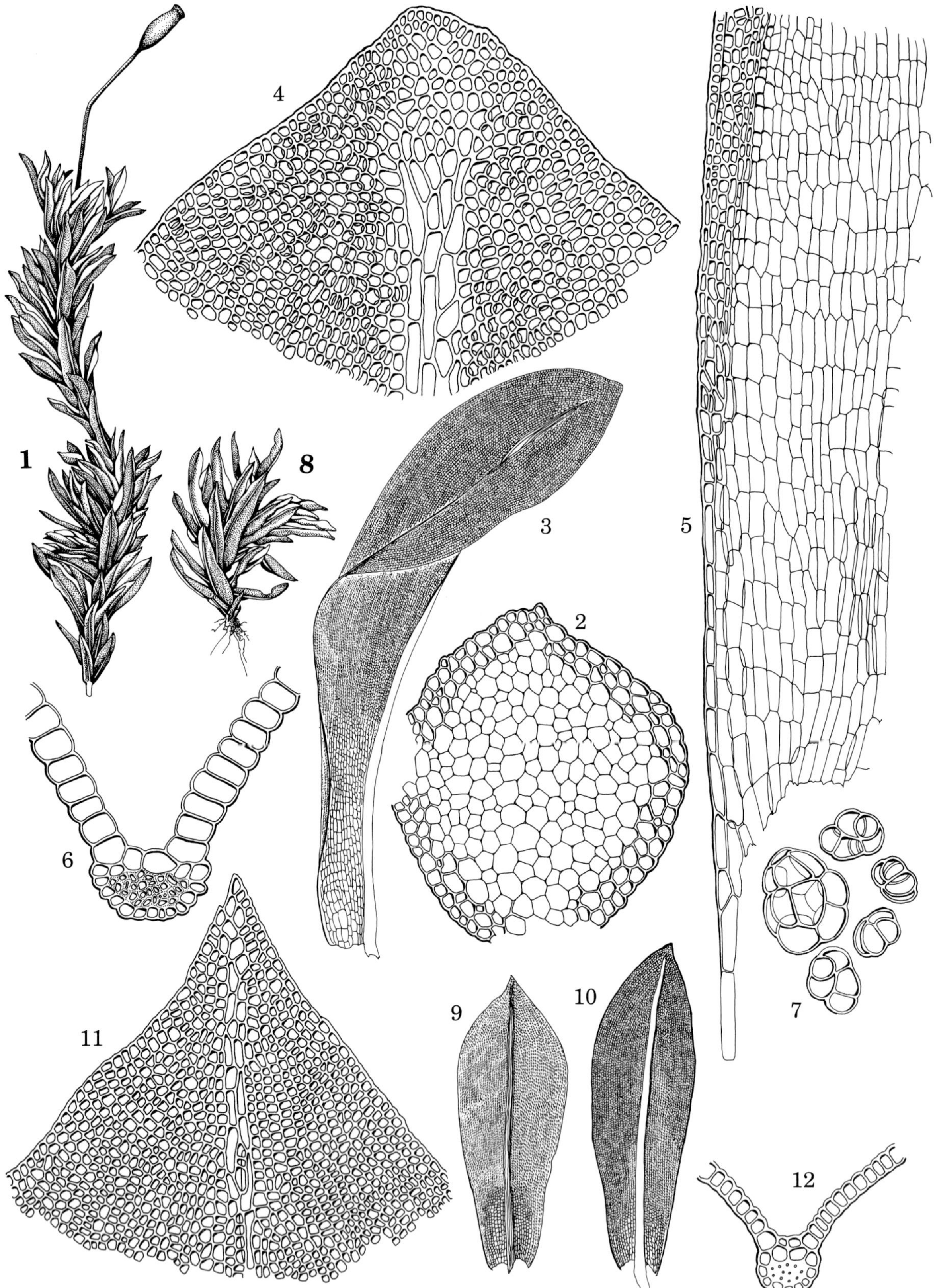

Plate 47. ***Scopelophila.*** **1–7.** ***S. ligulata.*** 1. Habit. 2. Transverse section of stem. 3. Leaf. 4. Leaf apex. 5. Basal cells. 6. Transverse section at midleaf. 7. Propagula. **8–12.** ***S. cataractae.*** 8. Habit. 9–10. Two leaves. 11. Leaf apex. 12. Transverse section at midleaf.

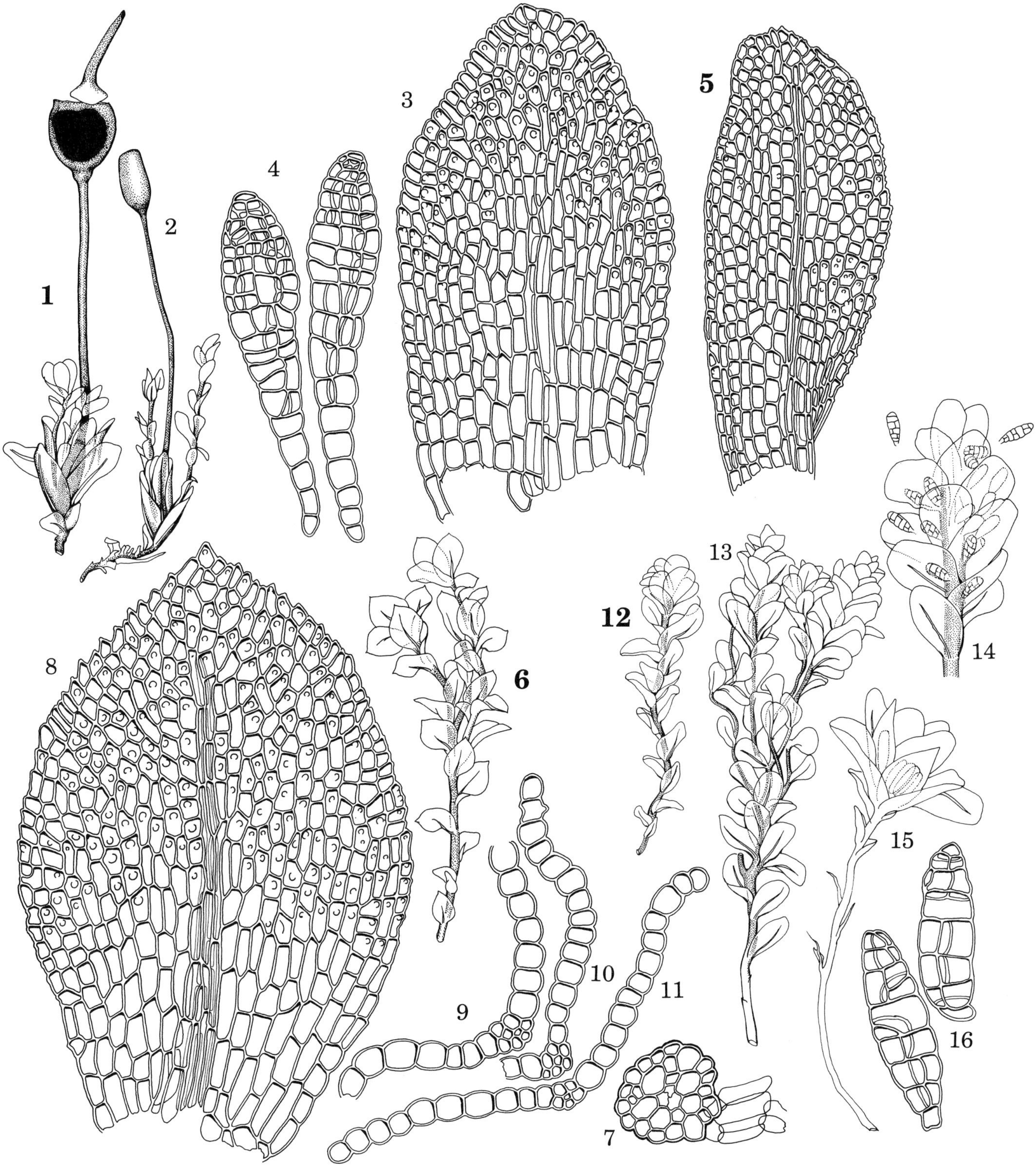

Plate 48. ***Gymnostomiella.*** **1–4.** ***G. vernicosa.*** 1–2. Two habits. 3. Leaf. 4. Propagula. **5.** ***G. longinervis.*** 5. Leaf. **6–11.** ***G. monodii.*** 6. Habit. 7. Transverse section of stem. 8. Leaf. 9–11. Transverse sections near midleaf. **12–16.** ***G. orcuttii.*** 12–14. Habits of sterile and propaguliferous plants. 15. Perigoniate plant. 16. Propagula.

the stem, greenish brown, clavate to ellipsoidal or filamentous and branching, ca. 12 µm in diameter, of 2 or more rounded cells. Dioicous. *Perichaetia terminal, inner leaves little differentiated from the cauline.* Perigonia terminal, outer leaves loosely sheathing, inner deltoid. Seta 2–6 mm in length, 1(–2) per perichaetium, brown to yellowish brown, twisted clockwise below; theca 0.6–2.2 mm in length, brown to yellowish brown or blackish, short-elliptical to short-cylindrical, exothecial cells quadrate-hexagonal to rectangular, occasionally bulging, 2–3:1, walls thin, stomates present and commonly difficult to distinguish or often apparently absent, phaneropore, at base of theca; annulus weakly differentiated or of 1–4 rows of vesiculose cells, deciduous in fragments; *peristome absent.* Operculum conic-rostrate, erect or oblique, ca. 0.7–1.5 mm in length, cells in straight rows. Calyptra cucullate, smooth, ca. 0.8 mm in length. Spores 8–13 µm in diameter, light brown, finely papillose or smooth. *Laminal KOH color reaction yellow to yellowish orange.* Reported chromosome number n = 13.

Scopelophila is a small genus usually associated with highly mineralized soils, found in mountainous areas of North, Central and South America, Europe, central Africa, Asia and Oceania (Hawaii).

Distinguishing characters of *Scopelophila* include the stem section of nearly homogenous, wide-lumened cells (Pl. 47, f. 2), lacking sclerodermis, hyalodermis, or central strand, the usually spathulate, oblanceolate leaves often with a distinctive broad apiculus (Pl. 47, f. 4), upper laminal cells lacking papillae and often heterogeneous in size and shape, costal section with a single stereid band imbedded in parenchymatous tissue (Pl. 47, f. 6, 12), and the capsule lacking a peristome. The leaves are often blackish in the lower parts of the plant, and this may be associated with iron ions in the substrate as the leaves of many mosses react with a black coloration to ferric chloride solution.

There is an interesting body of literature (reviewed by Persson 1948, among others, and most recently by Shaw and Anderson 1988) on *Scopelophila* as a "copper moss," meaning that it is one of a small number of taxa characteristically found growing in association with copper, zinc and iron ores.

The genus *Merceyopsis* was created (Dixon 1910) with seven species, apparently as an attempt to group certain eperistomate species (except in the case of *M. spathulifolia*, the type of which had only fragmentary sporophytes) that are seen here (see also Noguchi 1956) as members of *Barbula*, *Gymnostomum*, *Hymenostylium*, *Scopelophila*, *Trichostomum* or *Tuerckheimia* (see *Merceyopsis* in the list of recognized taxa for dispositions). Dixon's (1910, p. 298) preliminary discussion of his treatment indicates that the lectotype of *Merceyopsis* should be *M. pellucida* Broth. & Dix. ex Dix. (= *Scopelophila cataractae* (Mitt.) Broth. *cf.* Noguchi 1956; Zander 1967).

Additional literature: Arts (1988), Bartram (1924b), Corley and Perry (1985), Crundwell (1986), Frahm (1990b), Hoe (1973), Jones (1961), Lampton (1966), Lecointe and Schumacker (1988), Mårtensson and Berggren (1954), Melick (1975), Nagano and Schimizu (1973), Nagano et al. (1969), Noguchi and Furuta (1956), Noguchi and Ochi (1956), Reese (1989), Rumsey and Newton (1989), Satake (1990, 1991), Satake et al. (1988, 1990), Schatz (1955), Schumacker and Brugués (1991), Schumacker et al. (1989), Shaw (1987a,b), Shaw and Anderson (1989), Shaw and Beer (1989), Sotiaux et al. (1987), Takenaka and Satake (1991), Thyssen and Poelt (1958), Zander (1967, 1986b).

Number of accepted species: 3 plus 1 remaining in *Merceya.*

Species examined: *S. cataractae* (BUF, DUKE, FH, H, L, MEX, TENN, US), *S. infericola* (BUF), *S. ligulata.*

New heterotypic synonymy: *Desmatodon africanus* P. Varde (PC) = *Scopelophila cataractae* (Mitt.) Broth.

36. GYMNOSTOMIELLA — Plate 48.

Gymnostomiella Fleisch., Musci Fl. Buitenzorg 1: 309, 1904. Type: *Gymnostomiella vernicosa* (Hook.) Fleisch.

Gymnomitriella Sak., Bot. Mag. Tokyo 56: 221, 1942, *err. pro. Gymnostomiella* Fleisch.

Pottia sect. *Splachnobryella* C. Müll., Gen. Musc. Fr. 389, 1900. Type: *Pottia vernicosa* (Hook.) Hampe.

From *Gymnostomum*, a genus + i + *-ella*, diminutive; resembling the genus *Gymnostomum.*

Delicate plants in dense tufts or mats, light green to blackish green above, brown below. *Stems short*, often branching, 1.5–6.0 mm in length, transverse section rounded-pentagonal, *central strand present, comparatively large*, sclerodermis absent or weak, hyalodermis absent; *axillary hairs* ca. 3 cells in length, *moniliform*, basal cell with thicker walls than the distal cells; rhizoids present below. *Leaves* erect or laxly spreading when dry, erect to spreading, lax when moist, *obovate to oblong-obovate*, larger above, 0.3–0.4 mm in length, upper lamina flat, *margins plane, crenulate above by bulging cell walls*, occasionally serrulate, occasionally with 1 or 2 additional large teeth above; apex rounded or very broadly acute, occasionally apiculate by a sharp, conical cell; base not differentiated in shape; *costa* slender and ending at midleaf or ending near apex, seldom percurrent, superficial cells on both sides elongate, ca. 2 rows of cells across costa ventrally at midleaf, costal transverse section elliptical, *stereid band single, small, present centrally*, often absent and costal section appearing homogeneous, epidermis absent ventrally and present dorsally, guide cells 2 in 1 layer, *hydroid strand absent*; *upper laminal cells quadrate to hexagonal*, occasionally longer than broad, 12–18 µm in width, 1(–2):1, walls thin, delicate, superficially convex on both sides; *1–3 small, simple, hollow, conical papillae per lumen*, scattered over lamina; basal cells weakly differentiated across lamina, rectangular, little wider than upper cells, 2–3:1, walls thin. *Propagula elliptical to clavate, to 280 µm in length*, with ca. 6 transverse and 6 inner longitudinal walls, consisting of about 14 cells, *borne in leaf axils or on ventral surface of leaves*, sometimes 1–3 in a cup-like terminal rosette of leaves. Dioicous. Perichaetia terminal, inner leaves ovate, to 0.7 mm in length, sheathing, smooth, lower cells laxly rectangular, paraphyses moniliform. Perigonia terminal, paraphyses absent, plants often small, nearly stemless. Seta 3–6 mm in length, 1 per perichaetium, yellowish brown, twisted clockwise; *theca* 0.6–0.8 mm in length, black or brown, often shiny as if varnished, *ovate to short-elliptical, exothecial cells large*, rounded-hexagonal, 35–55 µm in diameter, mostly thin-walled, stomates phaneropore, at base of theca but often absent, annulus of about 2 rows of little-differentiated cells; peristome teeth absent. *Operculum relatively large*, obliquely rostrate from a low-conic base, 0.8–1.0 mm in length, cells weakly twisted clockwise. Calyptra cucullate, smooth, ca. 1.3 mm in length. Spores 11–15 μm in diameter, light brown, finely papillose. Laminal KOH color reaction yellow, or negative to black, or pink to deep purple. Reported chromosome number n = 13.

This is a small genus of southern and eastern Asia, Australia, northern and central Africa, subtropical and tropical North America, Central America, the West Indies and Brazil; found on limy rock usually in association with cyanobacteria (as is the case with *Luisierella barbula*).

In the absence of a peristome, this genus is placed in the Pottiaceae largely because of its papillose, obovate or spathulate leaves (Pl. 48, f. 3, 5, 8). Although without obvious close relatives, it has some similarity to the genus *Chenia* and *Hennediella* (Pottioideae) through the large hyaline laminal cells, simple papillae, and serrulate upper margins. Cladistic analysis indicates that a better phylogenetic hypothesis would be placement in the Barbuleae. Unusual characters are the central location of the single (sub)stereid band (Pl. 48, f. 9–11), the moniliform axillary hairs (approached in shape by the weakly bulging cells of the hairs of *Molendoa* species), the often nearly spherical capsule with unusually large exothecial cells and stomates often absent, and odd variation in KOH color reaction of the lamina, including black and purple. The possible relationships of *Gymnostomiella* with the Splachnobryaceae are discussed by Andrews (1949) and Crum (1949) and in an overview of the Splachnobryaceae by A. Koponen (1981). A convenient key to the species of *Gymnostomiella* was constructed from descriptions by Sloover (1977).

The propagula of *G. vernicosa* (Pl. 48, f. 4) are borne in leaf axils and in clusters of 1–3 in a somewhat swollen cup-like terminal rosette of ovate leaves with the general appearance of a perigonium. The propagula, with their clavate shape, internal transverse and longitudinal cross walls, are somewhat similar to antheridia. Thus, the propaguliferous sterile plants may easily be mistaken for perigoniate plants. Also, Fleischer (1902–22, Vol. 1, p. 310) correctly indicated that the perigonia of actual perigoniate plants of *G. vernicosa*, which differ from the propaguliferous sterile plants by being nearly stemless, lack the paraphyses usually expected in perigonia.

Of some significance is the relatively tiny size of the leaves in respect to that of the stem, sporophyte and propagula, in conjunction with the leaves' simple anatomy and serrulate upper margins. The leaves are apparently heterochronically paedomorphic (*cf.* discussion of Mishler 1986a of this phenonemon in *Tortula*) in that they may not have developed much beyond the "scale leaf" stage of very young shoots. It would be interesting to attempt to "force" the mature stage by modifying developmental processes (*cf.* Basile & Basile 1984).

Redfearn (1991) recently synonymized the American *G. orcuttii* (Pl. 48, f. 12–16) with the Asian *G. vernicosa*. These two taxa are here conservatively retained as separate species as the former often has multipapillose upper laminal cells and the latter unipapillose cells. *Gymnostomiella orcuttii* is, however, much like the Asian *G. burmensis*, and the variation thus may not be geographical. In any case, a revision of all species of the genus is sorely needed, and considerable synonymy might be expected, in all probability supporting Redfearn's evaluation.

Additional literature: Eckel (1985b), Potier de la Varde (1953), Schornherst (1944), Seki and Miyagi (1980), Stone (1985), Vital (1984).

Number of accepted species: 6.

Species examined: *B. burmensis* (FH), *G. longinervis* (NY), *G. monodii* (PC), *G. orcuttii*, *G. vernicosa* (BM).

37. DIDYMODON **Plates 49–50.**

Didymodon Hedw., Sp. Musc. 104, 1801. Lectotype: *Didymodon rigidulus* Hedw., *fide* Grout, Moss Fl. N. Amer. 1: 186, 1939.

Pottia sect. *Gomphoneuron* C. Müll., Linnaea 42: 310, 1879. Type: *Pottia lorentzii* C. Müll. (= *Didymodon lorentzianus* (C. Müll.) Broth. *fide* van der Wijk et al., Ind. Musc. 4: 540, 1967).

Pottia sect. *Senophyllaria* C. Müll., Linnaea 42: 311, 1879.

Sect. *Didymodon*

Didymodum Hedw. ex P. Beauv., Mag. Enc. 5: 309, 1804, *nom. illeg. incl. gen. prior.*

Didimodon P. Beauv., Mém. Soc. Linn. Paris (fasc. planch.): 3 f. 5, 1822, *nom. illeg. orthogr. var.*

Dydimodon Hedw. ex Arnott, Mém. Soc. Linn. Paris 5: 263, 1827, *nom. illeg. orthogr. var.*

Trichostomum subg. *Didymodon* (Hedw.) Turn., Musc. Hib. Spic. 34, 1804.

Didymodon subg. *Didymodon* (Hedw.) Boul., Fl. Crypt. Est Muscin. 504, 1872.

Didymodon subg. *Eudidymodon* Kindb., Eur. N. Amer. Bryin. 2: 273, 1897, *nom. illeg.*

Barbula sect. *Luridae* Moenk., Laubm. Eur. 281, 1927. Lectotype nov.: *Didymodon rigidulus* Hedw.

Barbula sect. *Acutae* Steere in Grout, Moss Fl. N. Amer. 1(3): 174, 1938.

Barbula sect. *Didymodon* (Hedw.) Giac., Atti Ist. Bot. Univ. Lab. Critt. Pavia ser. 5, 4: 208, 1947.

Barbula subsect. *Acutiformes* Kindb., Eur. N. Amer. Bryin. 2: 246, 1897. Type: *Barbula acuta* (Brid.) Brid.

Barbula subsect. *Rigidulae* Chen, Hedwigia 80: 193, 1941.

Sect. *Asteriscium* (C. Müll.) Zand., Cryptogamie, Bryol. Lichénol. 2: 383, 1981 [1982]. Type: *Barbula umbrosa* C. Müll.

Husnotiella Card., Rev. Bryol. 36: 71, 1909. Type: *Husnotiella revoluta* Card.

Trichostomopsis Card., Rev. Bryol. 36: 73, 1909. Type: *Trichostomopsis crispifolia* Card.

Asteriscium (C. Müll.) Hilp., Beih. Bot. Centralbl. 50(2): 618, 1933, *hom. illeg. non* Cham. & Schlecht., 1826.

Barbula sect. *Asteriscium* C. Müll., Linnaea 42: 342, 1879.

Didymodon sect. *Craspedophyllon* Card., Rev. Bryol. 36: 81, 1909.

Sect. *Fallaces* (De Not.) Zand., Phytologia 44: 209, 1979. Type: *Barbula fallax* Hedw.

Geheebia Schimp., Syn. ed. 2: 233, 1876. Type: *Geheebia cataractarum* Schimp.

Dactylhymenium Card., Rev. Bryol. 36: 72, 1909. Type: *Dactylhymenium pringlei* Card.

Limneria Stirt., Trans. Bot. Soc. Edinburgh 26: 428, 1915. Type: *Limneria viridula* Stirt.

Prionidium Hilp., Beih. Bot. Centralbl. 50(2): 640, 1933. Type: *Prionidium setschwanicum* (Broth.) Hilp.

Trichostomum subg. *Zygotrichodon* Schimp., Syn. ed. 2: 169, 1876. Type: *Trichostomum tophaceum* Brid.

Barbula subg. *Geheebia* (Schimp.) Szafr., Fl. Polska Mchy 1: 213, 1957 [1958].

Tortula sect. *Fallaces* De Not., Mem. Roy. Acc. Sci. Torino 40: 287, 1838. Type: *Tortula fallax* (Hedw.) Turn.

Barbula sect. *Graciles* Milde, Bryol. Siles. 117, 1869. Lectotype: *Barbula rigidicaulis* C. Müll. *fide* Saito, J. Hattori Bot. Lab. 39: 601, 1975.

Barbula sect. *Pseudodidymodon* Kindb., Eur. N. Amer. Bryin. 2: 246, 1897, *nom. illeg. incl. sect. prior.*

Barbula sect. *Reflexae* Mönk., Laubm. Eur. 280, 1927, *nom. illeg. incl. sect. prior.*

Barbula sect. *Fallaces* (De Not.) Steere in Grout, Moss Fl. N. Amer. 1: 174, 1938.

Didymodon sect. *Graciles* (Milde) Saito, J. Hattori Bot. Lab. 39: 501, 1975, see Zander, Phytologia 41: 24, 1978.

Barbula subsect. *Fallaciformes* Kindb., Eur. N. Amer. Bryin. 2: 246, 1897. Type: *Barbula fallax* Hedw.

Barbula subsect. *Reflexae* (Mönk.) Chen, Hedwigia 80: 203, 1941, *nom. illeg. incl. sect. prior.*

Sect. *Rufidulus* (Chen) Zand. (a *comb. nov.* made below).

Barbula sect. *Rufidula* Chen, Hedwigia 80: 210, 1941.

Sect. *Vineales* (Steere) Zand., Phytologia 41: 24, 1978. Lectotype nov.: *Didymodon vinealis* (Brid.) Zand.

Barbula sect. *Vineales* Steere in Grout, Moss Fl. N. Amer. 1: 174, 1938.

Barbula sect. *Rubiginosae* Steere in Grout, Moss Fl. N. Amer. 1: 174, 1938. Type: *Barbula rubiginosa* Mitt.

Barbula subsect. *Vinealiformes* Kindb., Eur. N. Amer. Bryin. 2: 246, 1897. Type: *Barbula vinealis* Brid.

From δίδυμος, double, twin + ὀδους (ὀδων), ὀδόντος, tooth; the twin divisions of the peristome teeth.

Plants gregarious or more usually forming turfs or cushions, light to blackish, olive or reddish green above, brown to reddish brown or tan below. *Stems* branching seldom to often, to 2(–9) cm in length, transverse section rounded-pentagonal, occasionally rounded-triangular, *central strand usually distinct*, seldom absent, sclerodermis usually present, hyalodermis only occasionally present; *axillary hairs* of ca. 5 cells, *basal 1–2 cells brownish*; indumentum usually absent. *Leaves* often crowded, appressed-incurved and occasionally twisted or curled when dry, spreading when moist, ovate or more *usually lanceolate or long-lanceolate* to occasionally long-triangular, ca. 0.8–3.0(–6.0) mm in length, upper lamina usually broadly concave, occasionally narrowly channeled or keeled, margins seldom plane or more usually recurved or occasionally revolute, entire or occasionally weakly dentate or crenulate, occasionally bistratose in patches or entirely so; apex narrowly acute to rounded, seldom cucullate; base weakly differentiated to ovate, occasionally oblong and half-sheathing the stem, occasionally decurrent, shoulders rarely present; *costa* ending several cells below apex to excurrent as a blunt awn, superficial cells quadrate to elongate ventrally, *usually short-rectangular dorsally above midleaf*, smooth or occasionally papillose, 2–4(–8) rows of cells across costa ventrally at midleaf, costal transverse section ovate, semicircular or reniform, stereid bands usually weak, occasionally absent ventrally, ventral epidermis present or seldom absent, dorsal present but usually weak, guide cells 2–6(–8) in 1(–2) layers, *hydroid strand seldom present*; upper laminal cells subquadrate to hexagonal or rounded angular, occasionally short-rectangular or rhomboidal, usually 8–13 µm in width, 1:1, occasionally bistratose in patches or entirely, walls thin to thickened, lumens sometimes angular, occasionally trigonous, superficially weakly to strongly convex on both surfaces; *papillae usually low, simple to bifid, usually solid, not obscuring the lumens*, occasionally absent or multiplex; *basal cells usually weakly differentiated* to occasionally strongly differentiated across leaf or extending higher medially, occasionally little different from upper cells, quadrate to rectangular, seldom bulging, usually little wider than the upper, ca. 2–4:1, walls usually rather thin, occasionally porose, smooth to papillose. *Propagula* occasionally present, *green, usually comparatively small, spherical to elliptical, of 1–10 cells, usually borne in leaf axils*, occasionally on ventral surface of costa or on basal rhizoids; occasionally the leaf apex swollen and fragile. Dioicous (occasionally possibly rhizautoicous). Perichaetia terminal, inner leaves ovate to long-lanceolate, occasionally enlarged, not or occasionally sheathing in lower 1/2, seldom convolute-sheathing, lower cells rhomboidal-rectangular in lower 1/2. Perigonia terminal or occasionally as small buds on protonema near archegoniophore. Seta mostly 0.5–2.0 cm in length, 1(–2) per perichaetium, yellowish to reddish brown, twisted clockwise below, occasionally counterclockwise above; theca ca. 1–3 mm in length, yellowish to reddish brown, elliptical to cylindrical, exothecial cells rectangular, thin-walled, stomates phaneropore, at base of theca, annulus of 1–3 rows of hexagonal, often vesiculose cells, often deciduous in pieces or revoluble; *peristome teeth* 16, or 32 and grouped in pairs, occasionally rudimentary or rarely absent, oblong to linear or long-triangular, often perforate or cleft medially, papillose to spiculose or spirally striate, to 700(–1300) µm, often of many articulations, *usually straight or weakly twisted counterclockwise*, but sometimes strongly twisted, *basal membrane absent or low*, occasionally to 70 µm in height, papillose or spiculose. Operculum short- to long-conic or conic-rostrate, ca. 0.5–1.2 mm in length, cells in straight rows or weakly twisted counterclockwise, occasionally to twice twisted. Calyptra cucullate, smooth, ca. 2.0–2.5 mm in length. Spores ca. 7–15 µm in diameter, light brown, smooth to papillose. Laminal KOH color reaction yellow or red. Reported chromosome number n = 12, 12+m, 13, 14.

Found on a variety of substrates, mostly rock or soil; a cosmopolitan genus widely diversified in temperate and montane regions.

Didymodon and *Barbula*, although often treated as one genus, are here distinguished along the lines set out by Saito (1975a) and summarized by Zander (1978e; see also treatment of *Barbula*). In addition, *Didymodon* as presented here remains an apparent potpourri of phyletic lines; Steere (1947) pointed out that *Didymodon* "is a synthetic genus composed of discordant elements originating in several genera of the Pottiaceae, agreeing among themselves only in their (supposedly!) [sic] untwisted peristome teeth." Just as *Barbula* and *Bryoerythrophyllum* are closely related groups now separated as fairly homogeneous genera by a variety of characters, chief among them KOH color reactions, *Didymodon*, which is suspiciously heterogeneous in KOH color reactions, may prove to be profitably split by recognizing various sections as genera. Obvious candidates are described below.

Didymodon sect. *Vineales*

This section is characterized by the leaves spreading to widely spreading and occasionally recurved when moist, concave

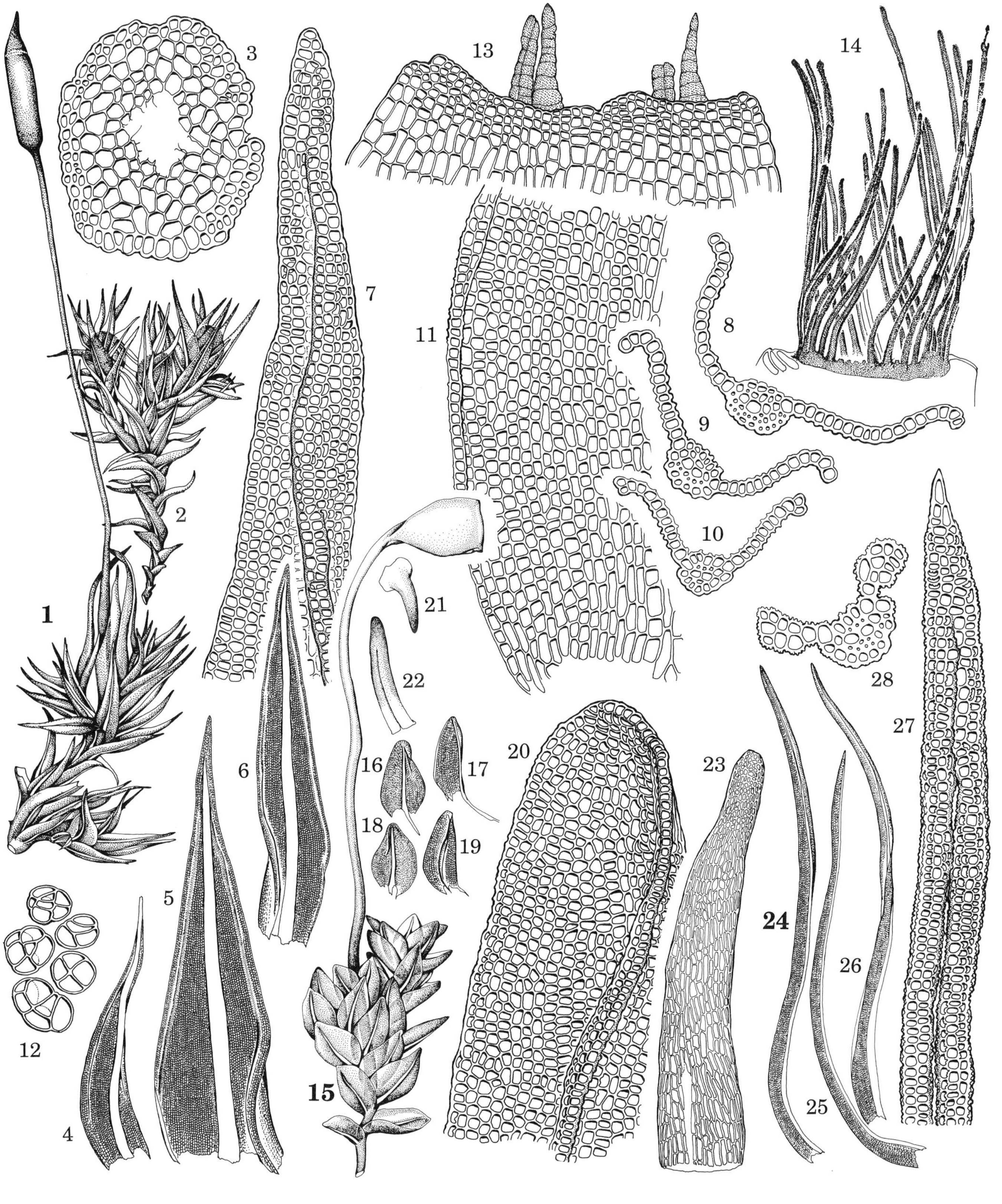

Plate 49. ***Didymodon*. 1–14. *D. rigidulus*.** 1. Habit of sporophyte-bearing plant. 2. Perigoniate plant. 3. Transverse section of stem. 4–6. Three leaves. 7. Leaf apex. 8–10. Three transverse sections at midleaf. 11. Leaf base. 12. Propagula. 13–14. Two peristomes. **15–22. *D. amblyophyllus*.** 15. Habit. 16–19. Four leaves. 20. Leaf apex. 21. Operculum. 22–23. Calyptra. **24–28. *D. bartramii*.** 24–26. Three leaves. 27. Leaf apex. 28. Transverse section at midleaf.

across the leaf to keeled and narrowly channeled along the adaxial surface of the costa, margins weakly decurrent to strongly so in robust plants, weakly recurved below to recurved or revolute to near the apex, often apiculate by a conical cell, the costa usually percurrent to short-excurrent in a broad mucro, the upper laminal cells occasionally bistratose along the leaf margins, epapillose to papillae simple or irregular to more often spiculose-multiplex, 1–4 over each lumen, the adaxial superficial cells of the costa quadrate in the upper half of the leaf, the guide cells occasionally in two layers, and the adaxial stereid band often absent (often replaced by substereid cells); the peristome is absent or rudimentary to well developed and twisted up to 2.5 turns; spores released usually in spring, also summer; KOH color reaction usually red to red-orange (high magnification might be needed to ascertain the exact hue of the internal upper laminal cell walls). A "marker" character, not always present but otherwise distinctive, is the absence of the quadrate ventral costal cells at the extreme leaf apex, resulting in a short, boat-shaped groove bottomed by epapillose elongate cells. Some characteristic species of sect. *Vineales* are *D. brachyphyllus*, *D. cordatus*, *D. herzogii*, *D. luehmanii* (Pl. 50, f. 8–11), *D. luridus*, *D. nicholsonii*, *D. occidentalis*, *D. reedii*, *D. sinuosus*, *D. tectorum* and *D. vinealis*.

Didymodon sect. *Fallaces*
This section is distinguished by the leaves spreading to often strongly recurved when moist, concave to keeled, margins weakly to strongly decurrent, plane to recurved in lower 2/3, not apiculate, the costa ending below the apex to short-excurrent, the upper laminal cells unistratose, epapillose to papillae simple, hemispherical or occasionally conic-apiculate, usually 1–2 over each lumen, the adaxial superficial cells of the costa short-rectangular to elongate in the upper half of the leaf (except *D. asperifolius*) and the adaxial stereid band usually present; peristome rudimentary to well developed and twisted up to 2 turns; spores mature usually in winter or spring; KOH color reaction usually reddish orange. The correct name at the generic level would be *Geheebia*. This section is morphologically similar to *Hymenostylium* by the commonly exposed ventral stereid band and often angular upper medial cell lumens.

Although Sollman (1983) has synonymized *D. constrictus* with *D.* (sect. *Vineales*) *vinealis*, the type of the former at NY ("170") is a mixed collection but largely composed of plants identical with other NY specimens ("Walanchoon" and "Lachen") labeled in Mitten's hand as "*Barbula constricta*." These are not *D. vinealis* but have clearly elongate ventral cells; the Asian *D. constrictus* is near to or the same as the dark-red colored *D.* (sect. *Fallaces*) *laevigatus* of the Andes. Specimens identified in various herbaria as *D. constrictus* but with quadrate ventral costal cells are usually *D.* (sect. *Didymdon*) *rigidulus* var. *icmadophilus*, which has a similar very short leaf base filled with quadrate cells and a very long-acuminate upper leaf.

Although Syed and Crundwell (1973) indicated that *Didymodon maxima* (as *Barbula maxima*) lacked a central strand, it usually has one, albeit weakly developed, even in European plants (e.g. Ireland: Crundwell & Warburg 1962, NY). Trigones, similar to those of *Didymodon giganteus*, are present in the leaves of most specimens. Hill (1981) and H. Crum and D. Hall (in press) have pointed out that *Didymodon ferrugineus* (Schimp. ex Besch.) Hill is the correct name for *D. fallax* var. *reflexus* at the species level. Some characteristic species of sect. *Fallaces* are *D. asperifolius*, *D. calycinus* (Pl. 50, f. 1–3), *D. ceratodonteus* (Pl. 50, f. 4–7), *D. constrictus*, *D. fallax*, *D. ferrugineus*, *D. erosodenticulatus*, *D. giganteus*, *D. hastatus*, *D. inundatus*, *D. johansenii*, *D. laevigatus*, *D. maxima*, *D. michiganensis*, *D. nigrescens*, *D. spadiceus* (Pl. 50, f. 12–16), *D. tomaculosus* (differs from *D. fallax* only in the presence of rhizoidal propagula), *D. tophaceus* and *D. waymouthii* (Pl. 50, f. 17–21).

Didymodon sect. *Asteriscium*
This section is characterized by the stem occasionally with a hyalodermis, leaves spreading to spreading-recurved when moist, occasionally squarrose from a shortly sheathing base, broadly channeled, leaf margins not decurrent, plane to broadly recurved throughout, the costa ending 1–6 cells below apex to short-excurrent, often rather broad at midleaf, adaxial surface convex, upper laminal cells unistratose to bistratose evenly or in patches along the leaf margins, papillae absent to large, low, simple to bifid 1(–4) per cell lumen, adaxial superficial cells of costa quadrate to elongate, adaxial stereid band absent or very small, hydroid groups often present; KOH color reaction usually yellow or yellowish orange. If recognized at the genus level, the correct name would be *Husnotiella* or *Trichostomopsis*, both published at the same time. This section represents at least superficially a morphological intermediate, as noted by Hilpert (1933), between *Didymodon* and *Erythrophyllopsis* in that the upper lamina of *D. challaense* is bistratose in patches medially or completely. Some characteristic species are *D. australasiae*, *D. bartramii* (Pl. 49, f. 24–26), *D. challaense*, *D. revolutus* and *D. umbrosus*. Probably also belonging here is *D. marginatum*.

Hydroid strands are of scattered occurrence in the genus *Didymodon*, having been seen in *D. australasiae*, *D. ceratodonteus*, *D. luehmanii*, *D. revolutus* and *D. xanthocarpus*. This feature is evidently of little taxonomic utility at the present time.

Didymodon tophaceus, a relatively common species of wet habitats, may lack a peristome in some specimens, which occasionally appear as taxonomic types of certain synonyms (e.g. *Gymnostomum knightii* Schimp. in Knight, BM) originally described as members of other, characteristically eperistomate genera.

The type of *Didymodon occidentalis* (NY) lacks a peristome entirely (it is not retained in the operculum on dehiscence). The sect. *Vineales* is similar to *Bryoerythrophyllum* in red coloration in KOH but is distinguishable from the latter by the usually sharply acute leaves, smaller upper laminal papillae, and often more than one layer of guide cells in the costa, which also often lacks a ventral stereid band.

Didymodon lindigii of the Andes has bistratose upper margins and apex, and rudimentary peristome, and is apparently close to *Didymodon rigidulus* (Pl. 49, f. 1–14). The several apparent isotypes at NY consist of considerable admixture with *D. tophaceus* and *D. australasiae*, while the isotype at FH is solely of the last two; lectotypification is sorely needed, with attention to possible confusion with *Astomum lindigii* (*cf.* Mitten's 1859 treatment of this last species).

Weissia waymouthii (Pl. 50, f. 17–21) is here transferred to *Didymodon* (sect. *Fallaces*) reflecting its recurved lower leaf margins, elongate ventral cells of costa, lack of a differentiated ventral costal epidermis, upper laminal cells pellucid, thick-walled, with rounded lumens and medially longitudinally

Plate 50. ***Didymodon.*** **1–3.** ***D. calycinus.*** 1–2. Two leaves. 3. Leaf apex. **4–7.** ***D. ceratodonteus.*** 4–5. Two leaves. 6. Leaf apex. 7. Transverse section at midleaf. **8–11.** ***D. luehmanii.*** 8–9. Two leaves. 10. Leaf apex, dorsal view. 11. Transverse section at midleaf. **12–16.** ***D. spadiceus.*** 12–13. Two leaves. 14. Leaf apex. 15. Transverse section at midleaf. 16. Peristome. **17–21.** ***D. waymouthii.*** 17–19. Three leaves. 20. Leaf apex. 21. Transverse section at midleaf. **22–26.** ***D. xanthocarpus.*** 22–23. Two leaves. 24. Leaf apex. 25. Peristome. 26. Operculum.

elongate, and papillae low and simple. The absence of a central strand in *D. waymouthii* is unusual in the genus, and might indicate a relationship with *Hymenostylium*, but the species is retained at least tentatively in *Didymodon* because of the presence of a peristome and costa ending a few cells below the apex. The KOH color reaction is reddish orange, not the characteristic yellow of *Weissia* species.

Additional, selected literature: Abramova et al. (1987), Allen (1992), Andrews (1941), Bartram (1926a,c), Conard (1945a, 1951a), Corley et al (1987), Crum (1965b, 1969a), Crundwell (1976), Crundwell and Nyholm (1965), Crundwell and Whitehouse (1978), Dismier (1905), Düll (1984), Düll-Hermanns (1984), Düll-Hermanns and Düll (1985), Eckel (1986a), Ellis and Smith (1983), Guerra and Ros (1987), Herzog (1905), Hill (1978), Hillier (1931), Lal and Parihar (1980), Matteri (1988a), Mitten (1867), Philibert (1885), Preston and Whitehouse (1985), Robinson (1968, 1970), Savicz-Ljubitzkaja (1965b), Sollman (1983), Steere (1938b), Vashistha and Chopra (1984), Werner (1982), Williams (1913), Zander (1978b,h, 1981c), Zander and Delgadillo (1984).

Number of accepted species: 122, plus 1 in *Husnotiella*, and 1 in *Trichostomopsis*.

Species examined: *D. alticaulis* (DUKE, TENN), *D. amblyophyllus* (NY), *D. anserinocapitatus* (NY), *D. asperifolius*, *D. australasiae*, *D. bartramii* (F, US), *D. brachyphyllus* (BUF), *D. calycinus* (BM, NY), *D. cardotii*, *D. ceratodonteus* (NY, PC), *D. challaense* (H), *D. constrictus* (BUF, NY), *D. cordatus* (BUF, F), *D. crassicostatus* (FH), *D. deciduus* (NY), *D. fallax*, *D. ferrugineus*, *D. giganteus*, *D. herzogii* (JE, L), *D. hampei* (BM, BUF, FH, TENN), *D. hastatus* (NY), *D. humidus* (NY), *D. imperfectus* (H), *D. incrassatolimbatus* (BM, FH, NY, TENN), *D. inundatus* (NY), *D. japonicus* (H), *D. johansenii* (ALA, MICH), *D. laevigatus* (BUF, NY), *D. lamyanus* (F), *D. leskeoides* (ALTA, BUF, NY, UBC), *D. lindigii* (NY), *D. lingulatus* (NY, BM), *D. luehmanii* (BUF), *D. luridus*, *D. jackvancei* (NY), *D. marginatum* (NY), *D. maxima* (NY), *D. michiganensis* (DUKE, NY, TENN), *D. minusculus* (NY), *D. nicholsonii* (BUF, HSC, UBC, US), *D. nigrescens* (COLO, DUKE, FH, NY, US), *D. occidentalis* (NY), *D. patagonicus* (NY), *D. perobtusus* (H, NY), *D. pruinosus* (NY, BUF), *D. reedii* (US), *D. revolutus*, *D. rigidulus*, *D. rivicola* (BUF), *D. rufidulus* (BUF), *D. sinuosus* (BUF, NY), *D. spadiceus*, *D. stewartii* (NY), *D. subandreaeoides* (ALA, CANM, NY), *D. subtorquatus* (US), *D. taylori* (NY), *D. tectorum* (H), *D. tomaculosus* (BUF), *D. tophaceopsis* (BUF, NY, US), *D. tophaceus*, *D. torquatus* (HSC), *D. umbrosus*, *D. uruguayensis* (US), *D. vinealis*, *D. waymouthii* (NY), *D. xanthocarpus* (NY).

New heterotypic synonymy: *Gyroweisia boliviana* Williams = *Didymodon tophaceus* (Brid.) Lisa. *Husnotiella baueri* Bartr. in Bauer = *Didymodon tophaceus* (Brid.) Lisa. *Gymnostomum knightii* Schimp. in Knight = *Didymodon tophaceus* (Brid.) Lisa.

New combinations and names:

Didymodon sect. *Rufidulus* (Chen) Zand., *comb. nov.* (*Barbula* sect. *Rufidula* Chen, Hedwigia 80: 210, 1941).

Didymodon anserinocapitatus (X.-j. Li) Zand., *comb. nov.* (*Barbula anserinocapitata* X.-j. Li, Acta Bot. Yunnan. 3: 103, 1981).

Didymodon bartramii Zand., *nom. nov.* (*Didymodon angustifolius* Bartr. *non* Warnst. *nec* Herz., Fieldiana Bot. 28: 4, 1951).

Didymodon cardotii (Dus.) Zand., *comb. nov.* (*Barbula cardotii* Dus., Bot. Not. 1905: 299, 1905).

Didymodon challaense (Broth. in Herz.) Zand., *comb. nov.* (*Trichostomum challaense* Broth. in Herz., Biblioth. Bot. 87: 30, 1916; *Erythrophyllopsis challaensis* (Broth. in Herz.) Hilp.).

Didymodon crassicostatus (Bartr.) Zand., *comb. nov.* (*Barbula crassicostata* Bartr., Bryologist 49: 114, 1946).

Didymodon deciduus Zand., *nom. nov.* (*Barbula decidua* C. Müll., Linnaea 43: 425, 1882, *hom. illeg.*; *Barbula uruguensis* Par., *nom. nov.*, Ind. Bryol. 101, 1894).

Didymodon herzogii Zand., *nom. nov.* (*Trichostomum ferrugineum* Herz., Biblioth. Bot. 87: 31, 1916; *Gertrudiella ferruginea* (Herz.) Hilp.; *non Didymodon ferrugineus* (Schimp. ex Besch.) Hill).

Didymodon hampei Zand., *nom. nov.* (*Trichostomum obtusifolium* Hampe, Bot. Zeit. 28: 49, 1870; *Gyroweisia obtusifolia* Broth.; *non Didymodon obtusifolius* Schkuhr).

Didymodon hastatus (Mitt.) Zand., *comb. nov.* (*Barbula hastata* Mitt., J. Linn. Soc. Bot. Suppl. 1: 34, 1859).

Didymodon humidus (Mitt.) Zand., *comb. nov.* (*Tortula humida* Mitt., J. Linn. Soc. Bot. 12: 162, 1869).

Didymodon imperfectus (C. Müll.) Zand., *comb. nov.* (*Trichostomum imperfectum* C. Müll., Linnaea 42: 214, 1879; *Barbula imperfecta* (C. Müll.) Broth.).

Didymodon lindigii (Hampe) Zand., *comb. nov.* (*Hyophila lindigii* Hampe Ann. Sc. Nat. Bot. ser. 5, 3: 343, 1865; *Barbula lindigii* Hampe, Bot. Zeit. 27: 867, 1869).

Didymodon jackvancei Zand., *nom. nov.* (*Husnotiella plicata* Magill, Fl. S. Afr. I. Bryophyta 1: 222, 1981; *non Didymodon plicatus* (C. Müll.) Mont.).

Didymodon marginatum (Robins.) Zand., *comb. nov.* (*Trichostomum marginatum* Robins., Phytologia 12: 389, 1971).

Didymodon minusculus (Williams) Zand., *comb. nov.* (*Tortula minuscula* Williams, Bull. Torrey Bot. Cl. 42: 399, 1915).

Didymodon pruinosus (Mitt.) Zand., *comb. nov.* (*Tortula pruinosa* Mitt., J. Linn. Soc. Bot. 12: 152, 1869; *Barbula pruinosa* (Mitt.) Jaeg.).

Didymodon rigidulus var. *ditrichoides* (Broth.) Zand., *comb. et stat. nov.* (*Barbula ditrichoides* Broth., Sitzungsber. Ak. Wiss. Wien Math. Nat. Kl. 133: 566, 1924).

Didymodon stewartii (Bartr.) Zand., *comb. nov.* (*Barbula stewartii* Bartr., Bull. Torr. Bot. Cl. 82: 23, 1955).

Didymodon taylorii Zand., *nom. nov.* (*Tortula campylocarpa* Tayl., London J. Bot. 7: 187, 1848; *Barbula campylocarpa* (Tayl.) C. Müll.; *non Didymodon campylocarpus* (C. Müll.) Broth.).

Didymodon tophaceopsis Zand., *nom. nov.* (*Gyroweisia latifolia* Dix., S. Afr. J. Sci. 18: 309, 1922; *Husnotiella latifolia* (Dix.) Zand. & Magill; *non Didymodon latifolius* Wahlenb. in Web. & Mohr).

Didymodon uruguayensis Zand., *nom. nov.* (*Barbula uruguayensis* Broth. is an *hom. illeg.* of *Barbula uruguensis* Par. *fide* Margadant in litt., itself a *nom. nov.* for *Barbula decidua* C. Müll. a quite different species here recognised as *Didymodon deciduus* (C. Müll.) Zand., Bih. K. Svensk. Vet. Ak. Foerh. 36 afd. 3(7): 18, 1900).

Didymodon waymouthii (R. Br. ter) Zand., *comb. nov.* (*Weissia waymouthii* R. Br. ter, Trans. New Zealand Inst. Bull. 31: 439, 1899; *Weissia lancifolia* var. *waymouthii* (R. Br. ter) Wijk & Marg.).

Subfamily POTTIOIDEAE

Pottioideae (Limpr.) Broth., Nat. Pfl. 1(3): 381, 1902 "Pottieae."
Hyophilaceae Hampe, Linnaea 20: 68, 1847, *nom. nud.*, *nom. rejic.* Type: *Hyophila* Brid.
Astomaceae Schimp., Coroll. Bryol. Eur. 7, 1855 [1856]. Type: *Astomum* Hampe.
Phascaceae Schimp., Coroll. Bryol. Eur. 4, 1855 [1856]. Type: *Phascum* Hedw.
Weissiaceae Schimp., Coroll. Bryol. Eur. 7, 1855 [1856] "Weisiaceae." Type: *Weissia* Hedw.
Eupottiaceae Hampe, Flora 50: 67, 1867, *nom. illeg.*
Euweisiaceae Hampe, Flora 50: 67, 1867, *nom. illeg.*
Astomataceae Magill, Taxon 26: 597, 1977, *orth. err. pro* Astomaceae Schimp., 1855 [1856].
Tortuloideae Visotska, Citol. Genet. (Kiev) 1(4): 38, 1963, *nom. inval. typ. non. cit.*

This subclade is distinguished from subclades lower in the tree (Timmielloideae, Erythrophyllopsoideae, Gertrudielloideae, Chionolomoideae and Trichostomoideae) by the advanced characters of stem hyalodermis absent and perichaetial leaves sheathing (*cf.* Cladogram 15). The features of the immediate ancestral node are: stem sclerodermis not or little differentiated, leaf base little differentiated in shape, and seta nearly absent to short, less than 1 cm in length.

The Pottioideae is clearly discernible, both without including the *Weissia* group in Cladograms 8–10 and 14 and with the *Weissia* group as a basal part of the lineage in Cladograms 2, 3 and 4. Apparently the more highly evolved of the Pottioideae (those with only one stereid band, laminal margins mostly recurved, and upper laminal cells mostly equally bulging on both surfaces of the lamina) were ultimately derived from ancestors of *Leptobarbula* and *Weissia* (with two stereid bands and lanceolate leaves) through the loss of the ventral stereid band (beginning with the shared ancestor of *Luisierella* and *Weisiopsis*). Thus, from a side branch with spathulate leaves and plane or incurved leaf margins, with ventrally bulging laminal cells and mostly two costal stereid bands (e.g. ancestors of *Hyophila* and its relatives), the somewhat reduced group of *Stegonia, Crossidium, Pterygoneurum, Globulinella, Aloina* and *Aloinella* was derived.

The following salient traits have developed in the more highly evolved genera of this subclade: spathulate leaves, loss of the ventral stereid band, and development of distinct infrageneric sporophyte reduction series; stem central strand and leaf hydroid strand usually present; stem-borne stalked clavate propagula very rare; the leaf margins usually recurved below, plane or recurved above; the upper laminal cells generally rather large and equally convex on both superficial faces; elaborations of specialized photosynthetic tissue not uncommon; and the basal membrane of the peristome occasionally very high.

The gametophytic characters of many species of this subfamily, especially of *Tortula* sect. *Pottia*, are strikingly like those of species of Bryaceae. A separate derivation of these species from the Bryaceae is improbable because the twisted peristome would have had to have been derived in separate families. Note, however, that a twisted peristome is present, albeit rarely, in the Ditrichaceae. An examination of various *Bryum* species shows they generally have a stout, little twisted seta. The seta of Ditrichaceae species, on the other hand is like that of the Pottiaceae in being relatively thin, generally twisted clockwise above and counterclockwise below.

The presence of a small ventral stereid band in various species (e.g. *Tortula rubra* and *Globulinella benoistii*) of genera characteristically having only one stereid band (the dorsal) indicates that the traditional combination of characters for this subfamily was never entirely consistent. This, plus (1) the inclusion of the many genera with two stereid bands in the Pottioideae (see Cladograms 14–16) at the base of the Pottieae subclade, and (2) certain single stereid-banded genera commonly are removed from the Pottieae upon cladistic analysis (e.g., *Calyptopogon* placed near *Tortella*, and *Gymnostomiella* and *Scopelophila* near *Didymodon* in Cladogram 14) demonstrates that the number of stereid bands is as easily influenced by evolutionary pressures as many other characters in the Pottiaceae. The traditional Pottioideae (these genera here treated as the more highly evolved members of the Pottieae and of the Hyophileae) is a paraphyletic group.

Tribe HYOPHILEAE

Hyophileae Chen, Hedwigia 80: 179, 1941.
Pleuroweisieae Limpr., Laubm. Deutschl. 1: 240, 1888 (rank not given, inoper. in priority).
Pleuroweisioideae (Limpr.) Broth., Nat. Pfl. ed. 2, 10: 243, 1924.
Pleuroweisieae (Limpr.) Chen, Hedwigia 80: 41, 1941.

This tribe is distinguished at the immediate ancestral node by the features: leaves tubulose when dry, leaf margins incurved or involute, and upper laminal cell walls ventrally bulging and dorsally nearly flat. *Molendoa*, which includes as a synonym *Pleuroweisia*, the type of the tribe Pleuroweisieae, is included in this group. Although the Pleuroweisieae has been often recognized at the tribal level in the past (see review by Saito 1975a), the name Hyophileae (typified by *Hyophila*), which was conveniently published at the same time, is chosen here (I.C.B.N., Art. 57.2) as tribal name because *Molendoa* is placed with the Barbuleae in several cladograms with various outgroups and must be considered of doubtful position. The distribution of the tribe Hyophileae is essentially worldwide, but with a concentration of significant taxa in tropical areas (*cf.* Cladogram 14). Genera with single stereid bands in their costa apparently have also lost the ventrally bulging and dorsally flat laminal cell feature, though these genera may each have one or more species with some degree of this one-sided bulging.

38. HYPODONTIUM **Plate 51.**

Hypodontium C. Müll., Hedwigia 38: 96, 1899. Type: *Hypodontium dregei* (Hornsch.) C. Müll. (lectotype by Magill 1981).

From ὑπο, under, beneath, less than + ὀδους (ὀδων), ὀδόντος, tooth + *-ium*, characteristic; referring to the poorly developed peristome teeth.

Plants robust, growing in a dense turf, light or dark green, often glaucous above, light tan to blackish brown below. Stems branching often, to 7 cm in length, transverse section rounded-pentagonal, central strand present, sclerodermis weak, hyalodermis present, often weak; axillary hairs long and narrow, ca. 5 cells in length, basal 1–2 cells thick-walled; radiculose or red-tomentose. *Leaves tubulose, incurved, often contorted when dry*, widely spreading from a sheathing base when moist, *lanceolate*, 3.5–6.0 mm in length, *upper lamina broadly channeled, margins plane to broadly incurved or involute*, occasionally tubulose near apex, *entire* above but sometimes denticulate along basal margins, *bordered by 1–5 rows of narrow, thick-walled cells* to midleaf or near apex; apex rounded-acute, occasionally broadly mucronate; *base sheathing in the lower 1/4 of leaf*, with *distinct shoulders*; *costa percurrent or ending in a short, broad mucro*, occasionally ventrally strongly bulging, ventral superficial cells quadrate, dorsally elongate, *papillose to spinose*, 7–12 rows of cells across costa ventrally at midleaf, costal transverse section ovate or semicircular, 2 stereid bands present, the ventral occasionally larger, dorsally lunate, epidermis usually strongly differentiated ventrally, absent or weakly differentiated dorsally, guide cells 4–6 in 1 layer, *hydroid strand absent*; *upper laminal cells* rounded-quadrate, 8–11 µm in width, 1:1, walls evenly thickened, *superficially ventrally bulging, dorsally weakly convex*; *papillae thick, columnar*, to 20 µm in height, one per lumen on each side of the lamina, apex of papilla low-spiculose, papillae larger medially and ventrally on the leaf; basal cells differentiated across leaf, rising higher marginally in a vee, rectangular, 15–25 µm in width, 2–4:1, wider medially, walls hyaline, with distinct pores or these grading into transverse slits. Dioicous. *Perichaetia* terminal, *inner leaves linear-lanceolate, long-awned*, fragile, 6–7 mm in length, tubulose-sheathing in lower 1/2–2/3, lower cells rectangular and hyaline except at base of awn. *Perigonia consisting of flattened axillary buds* in clusters near apex of perigoniate plants similar in stature to the perichaetiate. Seta 5–8 mm in length, 1 per perichaetium, yellowish or reddish brown, twisted counterclockwise; *theca fleshy*, 1.7–2.0 mm in length, yellowish or reddish brown, ovate to cylindrical, *exothecial cells short-rectangular*, 30–40 µm in width, 1–3:1, *very thick-walled*, stomates phaneropore, at base of capsule, *annulus weakly differentiated*, of smaller, transversely elongated cells; *peristome teeth 16, flat and long-triangular or broadly lanceolate, smooth*, 150–220 µm, with of 8–10 articulations, straight, *strongly incurved*, basal membrane absent or very low, smooth; fragmentary remains of an exostome remaining attached to the inside of the operculum or as scattered transverse, narrowly elliptical plates attached externally to the joints of the peristome teeth. Operculum long-rostrate, ca. 1 mm in length, cells straight. *Calyptra cucullate, rough* in upper 1/2 with forward-pointing papillae, 2.7–3.0 mm in length. *Spores large*, ca. 30–40 µm in diameter, brown, finely to very coarsely papillose. Laminal KOH color reaction yellow.

The genus is endemic to southern Africa, where it is found on soil, rock or tree trunks.

Magill's (1981) treatment of the genus is detailed and up to date. Reese et al. (1986), in an instructive review of the genera of Calymperaceae, where *Hypodontium* has heretofore been put, suggested this genus might be pottiaceous. Edwards (1980a) accepted *Hypodontium* as Calymperaceae, and did not discuss it extensively. Edwards (1979, p. 328) pointed out the fragmentary remains of 32 narrow exostome teeth that remain attached to the inside of the operculum of *Hypodontium* and fall with it; scattered, narrowly elliptical, transversely oriented plates attached to the external faces of joints of the peristome were seen in this study. An examination of a series of collections of *H. dregei* (Pl. 51, f. 1–12) and *H. pomiformis* (Pl. 51, f. 13–16) from PRE indicates that *Hypodontium* lacks many of the central characters of genera of Calymperaceae, e.g. the campanulate calyptra of *Calymperes*, a stem central strand, the intramarginal bands of elongated cells, and the enlarged but short-rectangular to quadrate basal cells of *Calymperes*, *Mitthyridium* and *Syrrhopodon*. The elongate, hyaline basal cells of *Hypodontium* are similar to those of many of the genera of Pottiaceae. Characters of *Hypodontium* most characteristic of the Calymperaceae are the peristome teeth (Pl. 51, f. 12) being short and incurved, 16, deep orange, triangular, smooth or weakly verrucose (compare those of *Syrrhopodon*, or even *Octoblepharum cf.* Magill 1981); upper basal cells of at least the outer perichaetial leaves sharply differentiated and inflated-quadrate like the cancellinae of *Calymperes*; cauline leaves lanceolate, bordered by much-elongate cells with thick walls, apex broadly acute to rounded and margins plane to incurved and often spinose, base with high shoulders; the upper laminal cells usually bulging ventrally and weakly convex dorsally; and upper laminal and dorsal costal papillae very thick (Pl. 51, f. 7, 16); none of these characters is, however, unique to the Calymperaceae. The peristome teeth, in being red, flat and triangular, are quite like those of *Oreoweisia* (Dicranaceae), a genus, which, however, has a rather dissimilar gametophyte morphology.

Unusual in both the Pottiaceae and Calymperaceae are the inner perichaetial leaves (Pl. 51, f. 9) sheathing below but narrowly subulate or awned apically (as in *Bryobartramia* of the Encalyptaceae, or species of *Diphyscium*, Diphysciaceae, and approached in the awned perichaetial leaves of some species of *Pseudocrossidium*). *Calymperes* has a persistent calyptra, opening by slits, which is unique to Calymperaceae, while the cucullate calyptra of *Syrrhopodon* often has a distinct collar on the lower margin that opens late. The calyptra of *Hypodontium* (Pl. 51, f. 11) is comparatively small, cucullate, and perched on the rostrum of the operculum, and is more typical of Pottiaceae,

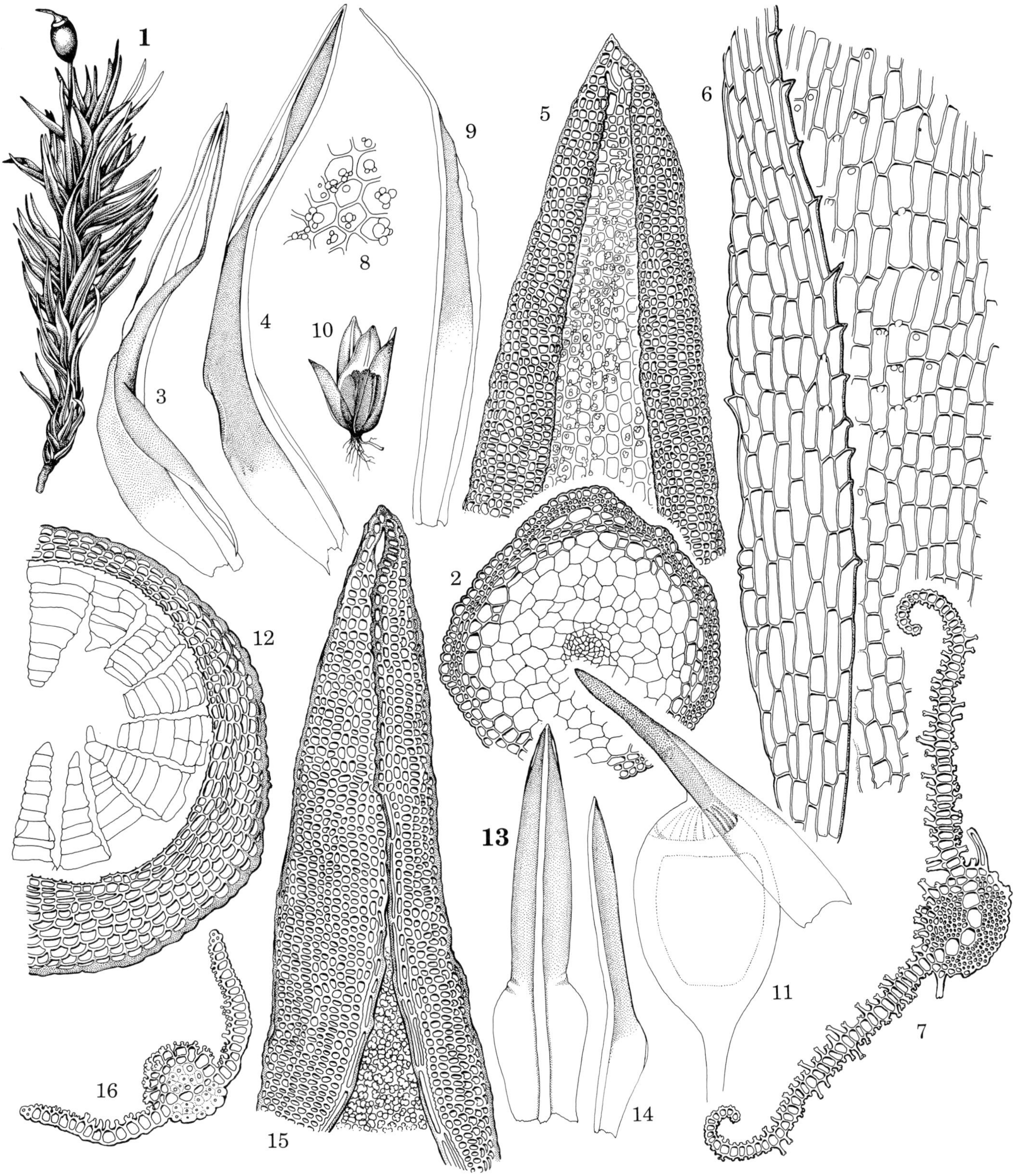

Plate 51. ***Hypodontium.*** **1–12.** ***H. dregei.*** 1. Habit. 2. Transverse section at midleaf. 3–4. Two leaves. 5. Leaf apex. 6. Basal cells. 7. Transverse section at midleaf. 8. Upper laminal papillae. 9. Perichaetial leaf. 10. Perigoniate bud. 11. Sporophyte with calyptra. 12. Vertical view of peristome teeth. **13–16.** ***H. pomiformis.*** 13–14. Two leaves. 15. Leaf apex. 16. Transverse section at midleaf.

although also found in *Mitthyridium*, Calymperaceae (Nowak 1980).

Except for the peristome, *Hypodontium* has the same capsule morphology of fleshy appearance; large, thick-walled exothecial cells; annulus largely undifferentiated; and spores relatively large, as has *Tridontium* (here removed to the Grimmiaceae near *Scouleria*) and *Tetracoscinodon* (Merceyoideae), the last with similarly sheathing, distally narrowly subulate (but not quite awned) perichaetial leaves. The peristome teeth of these three genera are also somewhat alike, being broadly triangular-lanceolate, while *Hypodontium* and *Tetracoscinodon* both have smooth teeth, but the gametophytes are disparate in appearance. If there were some distinctive autapomorphy or compelling combination of unusual characters, *Hypodontium* might be placed in a family of its own.

Number of accepted species: 2.
Species examined: *H. dregei* (PRE), *H. pomiformis* (PRE).

39. HYMENOSTYLIELLA **Plate 52.**

Hymenostyliella Bartr., Philippine J. Sci. 68: 108, 1939. Type: *Hymenostyliella involuta* (Card. & Thér.) Bartr.

From *Hymenostylium*, a genus + i + *-ella*, diminutive; resembling the genus *Hymenostylium*.

Plants in cushions, yellow-green above, yellow-brown below. Stems branching irregularly, to 3 cm in length, transverse section rounded-pentagonal, central strand strong, sclerodermis present, hyalodermis absent; axillary hairs of 7–10 cells, basal 1–2 yellowish brown; red-tomentose. *Leaves involute,*

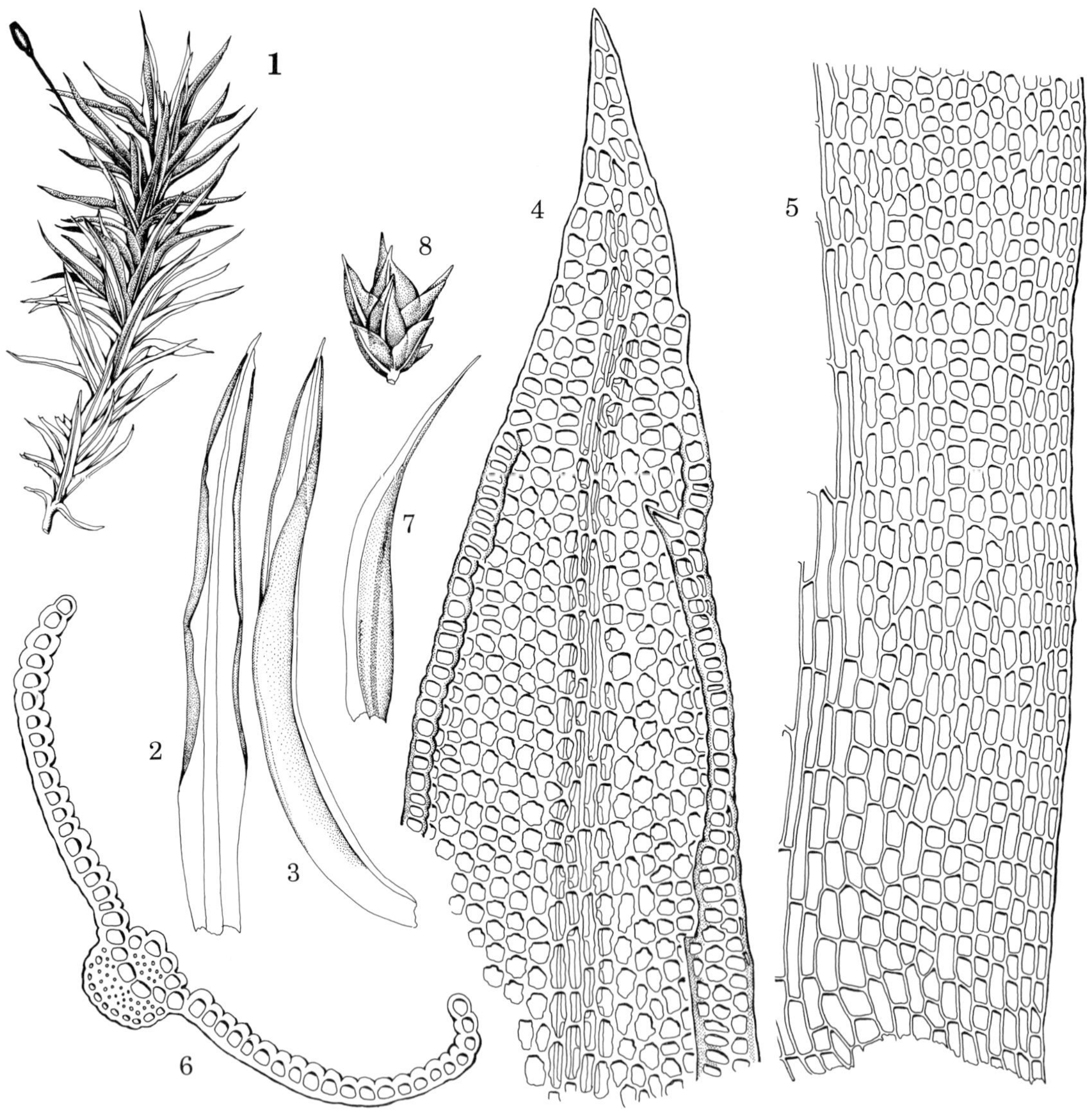

Plate 52. ***Hymenostyliella.* 1–8. *H. llanosii.*** 1. Habit. 2–3. Two leaves. 4. Leaf apex. 5. Basal cells. 6. Transverse section at midleaf. 7. Perichaetial leaf. 8. Perigoniate branchlet.

incurved, curled when dry, spreading-recurved when moist, *long-lanceolate, widest just below midleaf, ca. 4.5 mm in length, upper lamina broadly channeled across leaf, margins involute in upper 4/5, plane below*, entire or more commonly *distantly dentate in upper 1/4*; *apex narrowly acute*; *base not differentiated in shape*, somewhat decurrent; *costa short-excurrent as a mucro*, occasionally flexuose or dentate, *superficial cells* ventrally quadrate to short-rectangular, *bulging*, once *prorulose* at distal ends, dorsally elongate, 4–5 rows of cells across costa ventrally at midleaf, costal transverse section semicircular, 2 stereid bands present, epidermis present ventrally, weakly developed dorsally, guide cells 4 in 1 layer, hydroid strand absent; *upper laminal cells rounded-hexagonal or somewhat longitudinally elongate*, 10–13 μm in width, 1(–2):1, *walls thick, trigonous, superficially strongly bulging ventrally, flat dorsally*; *papillae absent*; basal cells weakly differentiated across leaf or somewhat higher along margins, rectangular, similar to upper cells in width, 2–4:1, walls moderately thick-walled, somewhat porose. Dioicous. *Perichaetia on short lateral branchlets*, inner leaves lanceolate, to 3.0 mm in length, strongly sheathing in lower half, cells long-rhomboidal in lower 4/5. Perigonia lateral, small, gemmate, occurring singly or in clusters. *Seta* ca. 5–6 mm in length, 1 per perichaetium, *reddish brown*, twisted clockwise; *theca* ca. 1.2–1.5 mm in length, *reddish brown*, obovoid to elliptical, exothecial cells rhomboidal, thick-walled, stomates phaneropore, at base of theca, annulus of 2–3 rows of dark, weakly vesiculose cells; *peristome absent. Operculum long-rostrate, longer than the theca*, to 1.5 mm in length, cells in straight rows. Calyptra cucullate, smooth, ca. 1.7 mm in length. Spores ca. 10–13 μm in diameter, light brown, essentially smooth. Laminal KOH color reaction yellow. [This description is based on *H. llanosii* collections at PC and US. Robinson (1971a) recently transferred *Timmiella alata* Herz. to this genus (without examination of authentic material but probably correctly), saying this species is closely related but is distinguished from *H. llanosii* by a dorsal costal surface winged with two ridges up to 12 cells in height and a cucullate leaf apex.]

A rare taxon found on rock in wet areas in the Philippines, India and Brazil.

This genus bears an immediate resemblance to *Timmiella* by the elongate, involute leaf and ventrally bulging-mamillose upper laminal cells (Pl. 52, f. 6), but as Chen (1941) pointed out, the latter genus has a peristome, as well as a bistratose upper lamina. Unusual characters in *Hymenostyliella* are the strongly involute leaf margins, ventrally bulging-mamillose and trigonous (Pl. 52, f. 4) upper laminal cells, and sporophytes borne laterally on short branches (Pl. 52, f. 1). Saito (1975a) noted the monopodial branching of *Hymenostyliella* in his discussion of *H. japonica* (as a synonym of *Didymodon japonicus*). The long-lanceolate leaf shape and distant teeth of this genus are reminiscent of *Tuerckheimia*, which differs by the distinctive massive laminal papillae and terminal perichaetium. Robinson (1971a) indicated a similarity to *Hyophila* (Cladogram 15 bears out a relationship) through the ventrally bulging-mamillose upper laminal cells. *Hyophila* differs, however, in its spathulate leaves, generally has a hydroid strand, and is acrocarpous; *Ganguleea*, with laterally borne sporophytes is seemingly more closely related, but see Cladogram 15. The characteristic trigonous areolation is found to greater or lesser degree in genera of other subfamilies, such as *Hymenostylium*, *Leptodontium*, *Reimersia*, *Trichostomum* and *Tuerckheimia*.

Number of accepted species: 3.

Species examined: *H. llanosii* (PC, US).

40. MOLENDOA **Plate 53.**

Molendoa Lindb., Utkast Nat. Grupp. Eur. Bladmoss. 29, 1878. Type: *Molendoa hornschuchiana* (Hook.) Lindb. ex Limpr.

Pleuroweisia Limpr. ex Schlieph., Flora 68: 359, 1885. Type: *Pleuroweisia schliephackei* Limpr. ex Schlieph.

Ozobryum Smith Merrill, Novon 2: 255, 1992. Type: *Molendoa ogalalensis* (Smith Merrill) Zand.

Anoectangium subg. *Molendoa* (Lindb.) Kindb., Eur. N. Amer. Bryin. 2: 317, 1897.

Named for Ludwig Molendo, 1833–1902, a German muscologist.

Plants in a compact turf, dark to light green, *occasionally glaucous* above, brown, often tan below. Stems branching seldom to often, to 4.0 cm in length, transverse section rounded-triangular to pentagonal, central strand present, usually strong, sclerodermis present, usually weak, hyalodermis absent or occasionally weakly developed; axillary hairs of 3–15 cells, usually all hyaline or occasionally basal 1–2 cells thick-walled; light brown to reddish tomentum sometimes present. *Leaves usually crowded*, appressed incurved to weakly spreading, usually twisted, occasionally tubulose when dry, spreading to spreading-recurved when moist, variously oval, ligulate, long-oblong, linear, ovate- to linear-lanceolate, (0.3–)1.0–4.0 mm in length, upper lamina flat to broadly or occasionally narrowly channeled, *margins plane to weakly recurved in lower 1/2, entire or occasionally sinuolate above, occasionally denticulate at shoulders* of the base (when dilated), *upper margins often bistratose* entirely or in patches; apex broadly to narrowly acute, occasionally broadly rounded and somewhat cucullate; base scarcely differentiated to elliptical and sheathing; costa ending 1–3(–6) cells below apex, percurrent, or excurrent as a stout mucro, superficial cells quadrate to elongate on both sides, 2–7 rows of cells across costa ventrally at midleaf, costal transverse section circular, semicircular to flattened reniform, *stereid bands absent to weak ventrally, present dorsally and flattened in section*, epidermis present ventrally, usually present but weak dorsally, *guide cells 2–4(–7) in 1 layer*, hydroid strand absent; *upper laminal cells often irregular in shape*, rounded-quadrate, oval, rounded-triangular, ca. 8–10 μm in width, 1:1, walls evenly thickened to weakly trigonous, superficially flat to bulging; *papillae usually crowded, low, irregularly scablike*, occasionally simple to bifid or massively multifid; basal cells differentiated across leaf or reaching higher along costa or margins, rectangular, little wider than the upper cells, (1–)2–5:1, walls thin, evenly thickened to porose. *Propagula rarely present*, obovoid to spindle-shaped, ca. 35–50 μm long, multicellular, borne in leaf axils. Dioicous. *Perichaetia terminal on short lateral branches*, inner leaves ovate-lanceolate, often marginally serrulate, to 3 mm in length, sheathing the seta, lower cells rhomboidal in lower 1/2 to throughout. Perigonia lateral. Seta 0.2–0.7 cm in length, 1 per perichaetium, yellow to brown, twisted clockwise below; *theca* 0.6–1.5 mm in length, yellow-brown, ovoid to cylindrical, *often macrostomous*, exothecial cells thin-walled, stomates phaneropore, at base of theca, annulus of 2–3 rows of transversely elongated, hexagonal cells, weakly vesiculose; peristome teeth absent. *Operculum very*

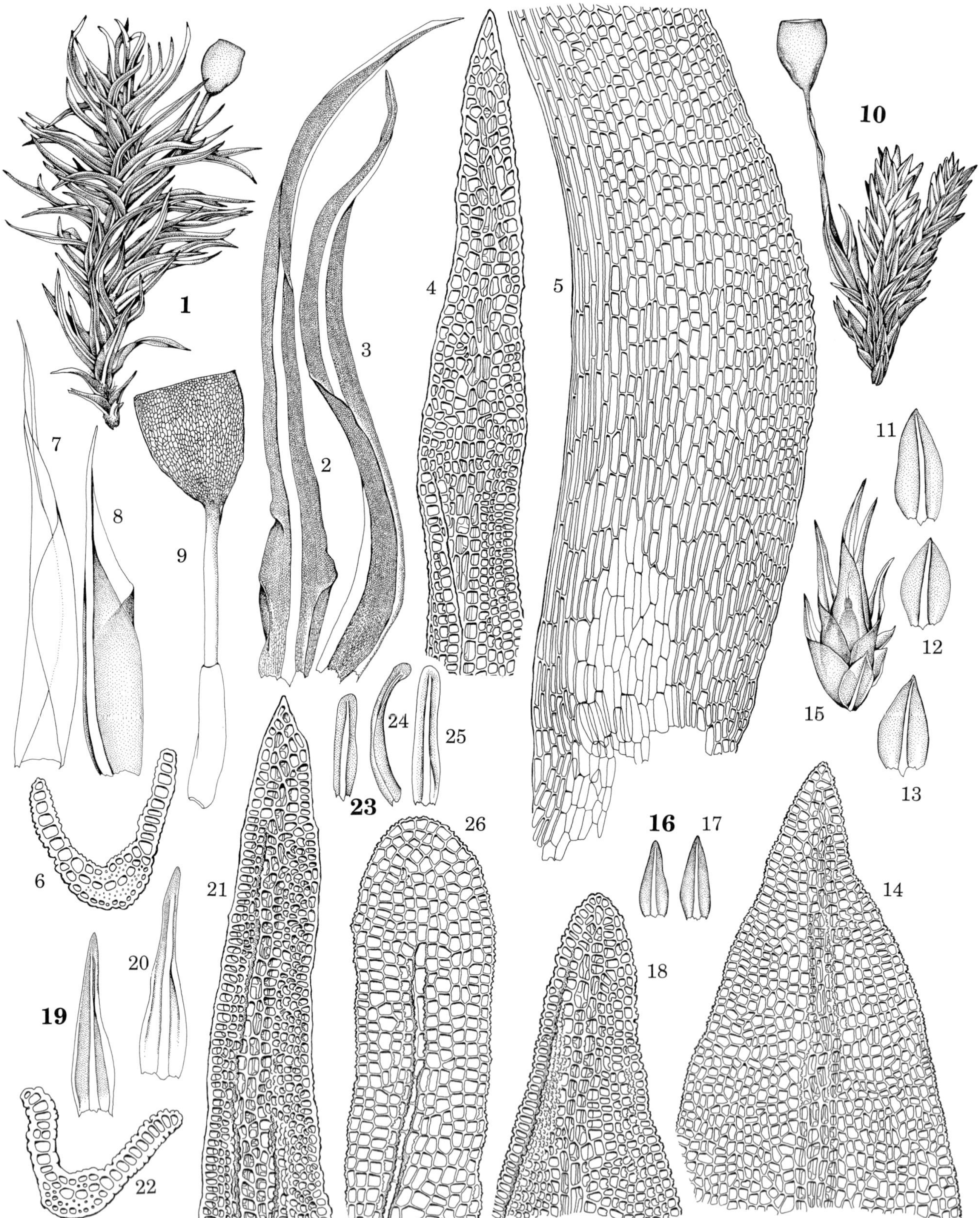

Plate 53. ***Molendoa.*** **1–9.** ***M. duthei.*** 1. Habit. 2–3. Two leaves. 4. Leaf apex. 5. Basal cells. 6. Transverse section at midleaf. 7–8. Perichaetial leaves. 9. Sporophyte on short lateral branch. **10–15.** ***M. platyphyllum.*** 10. Habit. 11–13. Three leaves. 14. Leaf apex. 15. Perichaetium. **16–18.** ***M. warburgii.*** 16–17. Two leaves. 18. Leaf apex. **19–22.** ***M. seravschanica.*** 19–20. Two leaves. 21. Leaf apex. 22. Transverse section at midleaf. **23–26.** ***M. schliephackei.*** 23–25. Three leaves. 26. Leaf apex.

long-rostrate, oblique, ca. 0.8–1.5 mm in length, cells straight. Calyptra cucullate, smooth, 1.0–2.5 mm in length. Spores ca. 9–12 µm in diameter, brown, essentially smooth or weakly to weakly papillose. *Laminal KOH reaction light yellow.* Reported chromosome number n = 13.

This genus is found on all continents except Australia and Antarctica, in montane or arctic areas, growing mainly on soil or rock (often calcareous) in wet or seasonally moist places.

Diagnostic characters are few and somewhat variable. For this reason species of *Molendoa* are often confused with *Anoectangium* species, which also are eperistomate and pleurocarpous, and also often have triangular stem sections. *Molendoa* differs from the latter principally by the presence of a ventral costal stereid band (Pl. 53, f. 6, 22), which, unfortunately, is often absent in *Molendoa* specimens of small size. Other characters that are found in some but not all species of *Molendoa* (or which occur only in well-developed populations of certain species), but which are not found in *Anoectangium,* may be used to refer specimens to the correct genus. These occasional indicator features of *Molendoa* include glaucous color of the upper leaves, becoming light tan in lower parts of the plant; leaves linear-lanceolate or not keeled; apex rounded, not apiculate; upper margins bistratose (*Anoectangium* may be bistratose in transverse spaced rows across upper leaf); leaf base much dilated and shoulders sometimes denticulate; superficial cells of the costa quadrate; costa ending before the apex or very broad (up to 7 rows of cells across costa ventrally at midleaf and up to 7 guide cells in the costa); costal transverse section flattened reniform, or showing two stereid bands; laminal cells heterogeneous in shape (Pl. 53, f. 14), or superficial walls (as seen in section) flat and distinctly thicker than the contiguous, anticlinal walls (Pl. 53 6, 22); papillae low, scab-like, crowded; KOH reaction pale yellow (versus yellow-orange); theca macrostomous (Pl. 53, f. 9).

The leaf shape, leaf section, and propagula are similar to those of *Didymodon* (*sensu* Saito 1975a), but *Molendoa* differs from that genus by the rather distinctive scab-like papillae (in most species) and, of course, the lateral gametoecia. *Molendoa clavuligera* (not seen) is reported to have clavate leaf apices similar to those of *Didymodon johansenii. Molendoa* is, on analysis, more closely related to *Didymodon* than to *Anoectangium* in some cladograms and is placed in the Hyophileae of the Pottioideae in Cladograms 14–16..

Pleuroweisia is here treated as a synonym of *Molendoa* (following the suggestion of Hilpert 1933) in that the key characters of the former are also found in *Molendoa.* The recognition of the genus *Pleuroweisia* by several modern authors (Brotherus 1924–25; Hilpert 1933; Chen 1941; Savicz-Ljubiskaja & Smirnova 1970) is probably due to acceptance of technical characters advanced by Limpricht (1890) that in fact have no substance. The calyptrae, alleged (Limpricht 1890; Chen 1941) to cover only the rostrum of the operculum, are also found perched on the rostrum in *Molendoa sendtneriana* in both microstomous (e.g. Mexico: Müller 1834, MICH, type of *Anoectangium glaucescens*) and macrostomous (e.g. Mexico: Le Sueur E6a, FH) collections. The type specimen (leaves figured in Pl. 53, f. 23–26 from an isotype, Switzerland, Pontresina, Roseg-Gletschers, Graf, 9-7-1883, H) of *Molendoa schliephackei* (= *Pleuroweisia schliephackei*) has calyptrae present only on immature capsules, and, in fact, these partly cover the thecae as well as the opercula. Other characters, such as the recurved leaf margins, rounded apex, rather thick-walled laminal cells, indistinct stem central strand, ovoid perichaetial leaves that sheath the seta, ovoid capsules, and somewhat vesiculose annular cells are also found in various combinations in *Molendoa,* specifically the highly variable (Pilous 1958; Zander 1977c) *M. sendtneriana.* The chromosome number of *M. schliephackei* has been reported (Fritsch 1972) to be the same as that of *Molendoa sendtneriana,* n = 13. Characters that may prove of significance for *M. schliephackei* at the species level are the strongly recurved laminal margins and the rather large leaf cells (12–14 µm in diameter), although these are features seen so far only in the type specimen. Other specimens examined that were identified as this species are within the range of variation accepted (Zander 1977c) for *M. sendtneriana* in the New World, and none match the type of the species in all significant respects. Perhaps Geissler (1985) has a better solution to the problem of variation among these taxa; she viewed *M. sendtneriana, M. schliephackei, M. tenuinerve* and *M. taeniatifolia* as synonyms of *M. hornschuchiana* (as *Anoectangium hornschuchianum*).

Previously placed in *Anoectangium, Molendoa platyphyllum* (Pl. 53, f. 10–15) of the Peruvian Andes and *M. warburgii* (Pl. 53, f. 16–18) of the United Kingdom are similar in the small plant size, much-branched stems, leaf shape (ovate-triangular with a constricted apex) and costa thin (the former has two stereid bands, but none have yet been detected in the latter). The two differ in papillae shape (the former has low, scattered scabs, while the latter has massive simple or less thickened bifid papillae), but these two species are evidently closely related. Long (1982c) found propagula in the latter species, and he rightly suggested these indicate a closer relationship to *Molendoa* than to *Anoectangium.* The mid-North American *M. ogalalensis,* type of *Ozobryum,* is similar to these but differs in the lack of a stem sclerodermis, the broader leaf apex, single massive papilla over each bulging lumen, and superficial walls of upper laminal cells not distinctly thicker than the internal walls; this species is distinctive but, in my opinion, not at the generic level. *Molendoa ogalalensis* is similar in stem section, leaf shape and papillae morphology to *Quaesticula navicularis,* which differs, however, in the incurved upper leaf margins and terminal perichaetium.

The Bryotheca E. Levier specimen of *Molendoa roylei* at NY is *Hymenostylium recurvirostrum* and other distributed duplicates may also be this species. *Hymenostylium* characteristically (but not always) lacks a central strand. This and the rectangular, often irregularly thickened median leaf cells are good distinguishing characters for sterile specimens.

Additional literature: Castelli (1966, 1968), Györffy (1910, 1912, 1914, 1946), Herzog (1943), Iwatsuki & Sharp (1958—this paper actually refers to *Tuerckheimia svihlae* (Bartr.) Zand. as noted by Saito 1972b and Zander 1979b, both as *Tuerckheimia angustifolia*).

Number of accepted species: 15.

Species examined: *M. duthei* (NY), *M. hornschuchiana, M. ogalalensis* (BUF), *M. platyphyllum* (F, NY), *M. schliephackei* (BP, H, NY, PC), *M. seravschanica* (H), *M. sendtneriana, M. warburgii* (BUF).

New heterotypic synonymy: *Hymenostylium secundum* C. Müll. = *Molendoa sendtneriana* (BSG) Limpr. *Hymenostylium validinerve* Dix. & P. Varde = *Molendoa sendtneriana* (BSG) Limpr.

New combinations: *Molendoa ogalalensis* (Smith Merrill) Zand., *comb. nov.* (*Ozobryum ogalalense* Smith Merrill, Novon 2: 255, 1992). *Molendoa platyphyllum* (Williams) Zand., *comb. nov.* (*Anoectangium platyphyllum* Williams, Field Mus. Nat. Hist. 4(5): 130, 1927. *Molendoa schliephackei* (Limpr. ex Schlieph.) Zand., *comb. nov.* (*Pleuroweisia schliephackei* Limpr. ex Schlieph., Flora 68: 359, 1885 [*Pleuroweissia schliephackei* Limpr., Jahresber. Schles. Ges. Vaterl. Cult. 61: 224, 1884, *nom. inval. sin. descr.*]). *Molendoa warburgii* (Crundw. & Hill) Zand., *comb. nov.* (*Anoectangium warburgii* Crundw. & Hill, J. Bryol. 9: 435, 1977 [1978]).

41. HYOPHILA Plates 54–56.

Hyophila Brid., Bryol. Univ. 1(Suppl.): 760, 1827. Lectotype: *Hyophila javanica* (Nees & Blume) Brid. *fide* Hampe in Bot. Zeit. 4: 266, 1846.

Rottleria Brid., Bryol. Univ. 1: 105, 1826, *hom. illeg. non* Wild., 1797.

Hygrophila Syd., Bot. Jahresber. 39(1): 99, 1912, *nom. inval. error pro Hyophila* Brid.

Hyophila subg. *Gymnohyophila* Card. in Grand., Hist. Madag. 39: 208, 1915, *nom. illeg. incl. type. gen.*

Gymnostomum sect. *Hyophila* (Brid.) Reichenb., Consp. Regn. Veg. 1: 33, 1828.

Pottia sect. *Hyophila* (Brid.) C. Müll., Syn. 1: 558, 1849.

Weissia sect. *Hyophila* (Brid.) Mitt., J. Linn. Soc. Bot. 12: 135, 1869.

From ὕω, to wet, rain + φίλος, love, dear, beloved; referring to a preference for wet habitats.

Plants turf-forming or occasionally loosely caespitose, green above, red to reddish brown or dark green below. *Stems* branching occasionally, to 1.0 cm in length, transverse section rounded-pentagonal or triangular, central strand usually strong, occasionally absent or central portion of stem hollow, *central cylinder often of rather thick-walled cells grading into a sclerodermis,* sclerodermis usually present, often strong, in several layers, hyalodermis absent or present (often only weakly differentiated); axillary hairs 6–10 cells in length, these all hyaline; radiculose. *Leaves* often rosulate, *tubulose-twisted, incurved and occasionally contorted when dry,* spreading, occasionally rather fragile when moist, *commonly spathulate or ligulate,* often ovate, oblong or elliptical, *usually narrowed to the base,* to 2.5 mm in length, upper lamina broadly channeled, shallowly grooved along costa, occasionally concave, *margins plane to broadly incurved,* occasionally narrowly recurved in lower 2/3, entire or denticulate to dentate in upper 1/4 or just at apex, *apex broadly acute to rounded,* rarely cucullate or emarginate; base not differentiated in shape or half-sheathing, occasionally weakly auricled; costa subpercurrent or percurrent, ending in an apiculus or occasionally a mucro, superficial cells ventrally quadrate or more commonly short-rectangular, often bulging, dorsally elongate, 2–6 rows of cells across costa ventrally at midleaf, costal transverse section semicircular, 2 stereid bands present, ventral and dorsal epidermis present, guide cells 4(–6) in 1 layer, *hydroid strand occasionally present; upper laminal cells rounded-quadrate or hexagonal, usually small,* 6–13 µm in width, 1:1, *walls thin to evenly thickened, either superficially ventrally strongly bulging and dorsally weakly convex, or bulging equally on both sides; papillae absent or simple, solid, often only on dorsal surface of lamina,* occasionally weakly bifid; *basal cells* differentiated across leaf or medially, *usually restricted to small area near insertion,* short-rectangular, 10–20 µm in width, 1–4:1, walls thin to evenly thickened. *Propagula often present, clavate, stellate or dentate-elliptical,* often to 300 µm in length, *borne on stout, branching stalks in leaf axils. Dioicous or monoicous* (autoicous, paroicous). Perichaetia terminal and inner leaves little different from or smaller than the cauline, half-sheathing, lower cells thin-walled and rhomboidal in lower half. Perigonia as lateral buds on perichaetiate plants or terminal on perigoniate plants, inner leaves little differentiated. Seta 0.4–0.7 cm in length, 1(–2) per perichaetium, reddish to yellowish brown, twisted clockwise; theca 1.0–2.3 mm in length, reddish to yellowish brown, long-ovoid to cylindrical, *exothecial cells* quadrate to rectangular, 20–45 µm in width, 1–5:1, *walls thin or much thickened both on exposed and contiguous sides,* stomates phaneropore, at base of theca, annulus 1–3 rows of vesiculose cells, deciduous in pieces or persistent on theca or operculum; *peristome teeth absent.* Operculum conic to long-conic or rostrate, 0.5–0.8 mm in length, cells not twisted. Calyptra cucullate, often twisted about the seta when mature, smooth, 2–3 mm in length. Spores 7–24 µm in diameter, light brown, papillose. *Laminal KOH color reaction yellow, occasionally with a red blush medially above midleaf,* occasionally cells near leaf insertion red. Reported chromosome number n = 7, 13, 13+m.

Found on rock, soil and trees, generally in moist or wet areas, throughout the tropic and temperate zones.

Abundant synonymy and combinations in other genera will certainly be necessary on revision for presently accepted correct names in *Hyophila*. Recent synonymy by various authors includes several broad-leaved sterile or eperistomate taxa in *Hyophila* transferred to or synonymized with species in such genera as *Barbula, Didymodon, Bryoerythrophyllum, Gymnostomum, Gyroweisia, Plaubelia, Scopelophila, Tortula, Trichostomum* and *Weissia*. Many correct names in *Hyophila* provisionally accepted during this study are based on type or authentic specimens expected, on revision, to prove to be *H. involuta* (Pl. 54, f. 1–11); this species requires comprehensive evaluation, however, and extensive synonymy within the genus was not attempted. The species illustrated here are mostly those fairly different from *H. involuta*.

Many species of the genus are morphologically similar to *Trichostomum*; these seem to intergrade with *Trichostomum* species through such similar, intermediate taxa as *Plaubelia, Hyophila nymaniana* (Pl. 55, f. 18–25) and *T. planifolium. Trichostomum* itself certainly shows tendencies towards the *Hyophila* morphotype (e.g. Magill 1981, p. 262 refers *Hyophila zeyheri* of South Africa to the synonymy of *Trichostomum brachydontium* because of peristome variation from rudimentary to absent, see also Sérgio 1985). The upper laminal cells ventrally bulging and dorsally flattened are developed only medially in *H. bartramiana* and *H. subcucullata* (Pl. 56, f. 6–10), two closely related monoicous species that also have laminal papillae. Ventrally bulging upper laminal cells (usually also ornamented with papillae) are also found in many species of *Weissia*, but *Weissia* is apparently not closely related to *Trichostomum* (see Cladogram 14 and others) on analysis using the full character set, although it is related to *Hyophila*. *Hyophila* appears to be a weakly segregated end point of a parallel

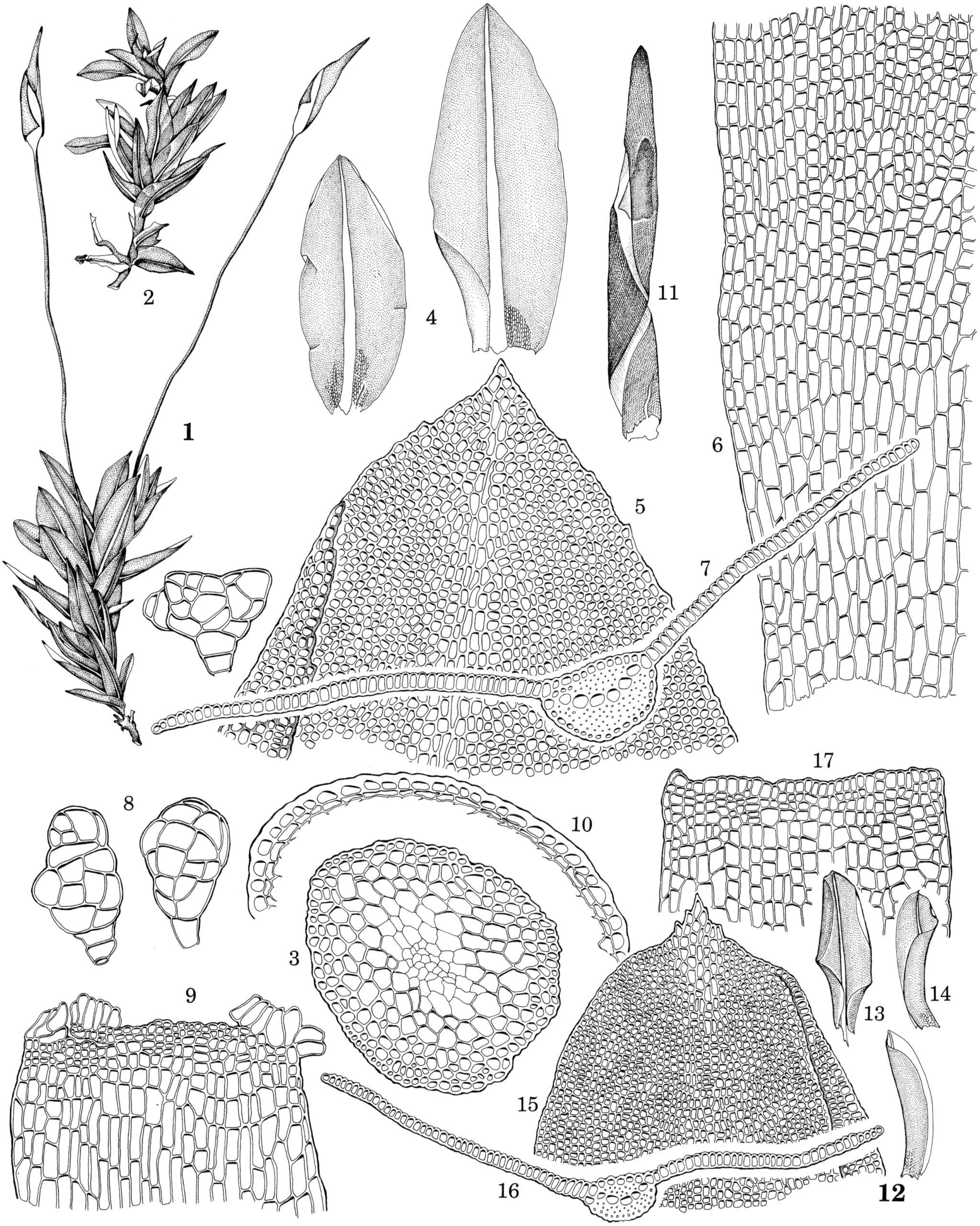

Plate 54. ***Hyophila.*** **1–11.** ***H. involuta.*** 1. Habit of sporangiate plant. 2. Perigoniate plant. 3. Transverse section of stem. 4. Two leaves. 5. Leaf apex. 6. Basal cells. 7. Transverse section at midleaf. 8. Propagula. 9. Thecal mouth. 10. Portion of transverse section of thecal wall. 11. Calyptra, twisted about operculum. **12–17.** ***H. acutifolia.*** 12–14. Three leaves. 15. Leaf apex. 16. Transverse section at midleaf. 17. Thecal mouth.

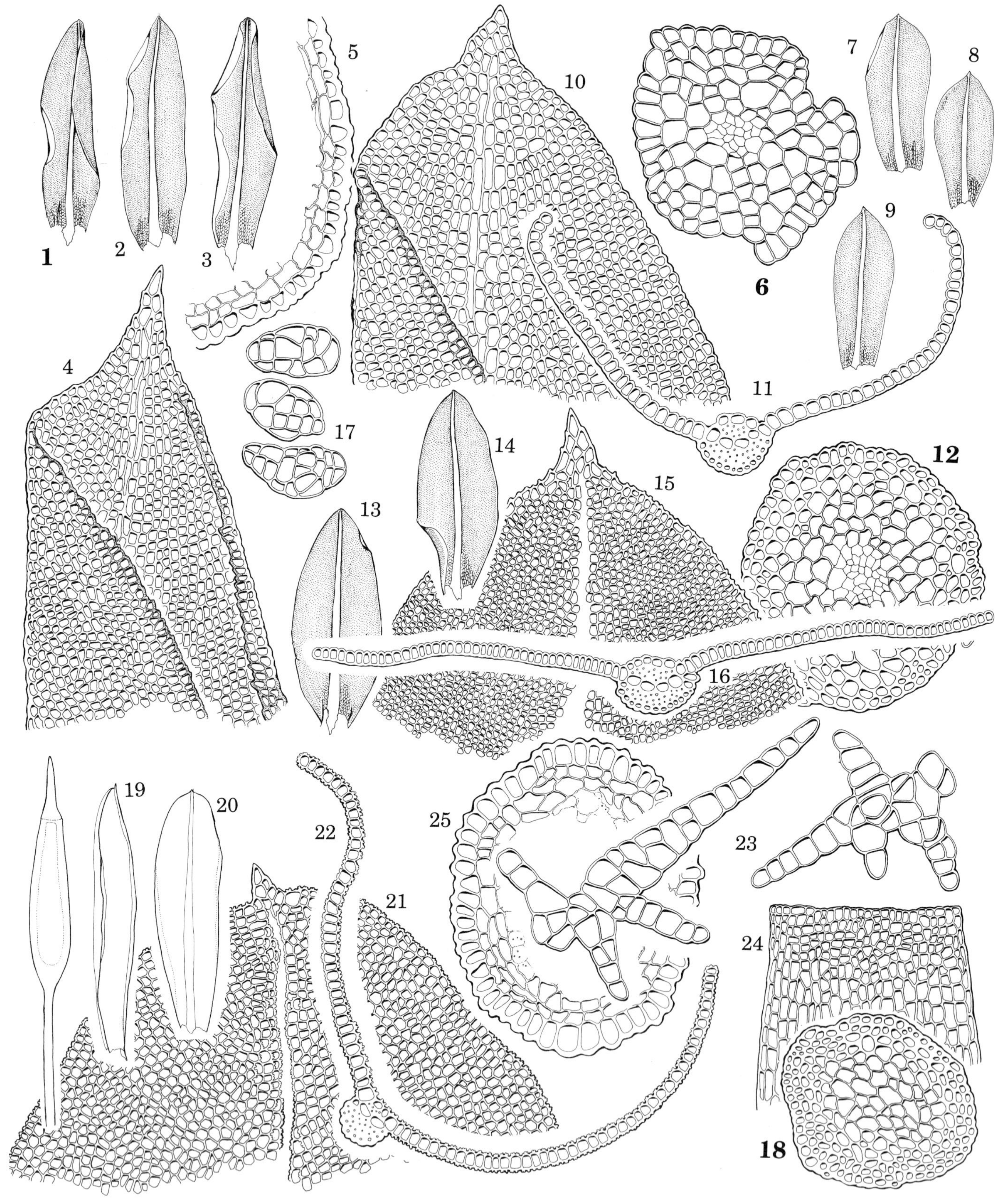

Plate 55. ***Hyophila.*** **1–5.** ***H. apiculata.*** 1–3. Three leaves. 4. Leaf apex. 5. Portion of thecal wall in section. **6–11.** ***H. beruensis.*** 6. Section of stem. 7–9. Three leaves. 10. Leaf apex. 11. Section at midleaf. **12–17.** ***H. propagulifera.*** 12. Section of stem. 13–14. Two leaves. 15. Leaf apex. 16. Section at midleaf. 17. Propagula. **18–25.** ***H. nymaniana.*** 18. Section of stem. 19–20. Two leaves. 21. Leaf apex. 22. Section at midleaf. 23. Propagula. 24. Mouth of theca. 25. Section of theca.

line of evolution towards *Trichostomum* species (sect. *Laminanchium*) with ventrally bulging upper laminal cells, or else the latter is better seen as a separate taxon at the genus level in the Hyophileae or even transferred to *Hyophila*.

Some species of *Hyophila*, including the generitype, *H. involuta*, have certain of the characters of *Barbula* (Norris and Koponen 1989 felt that *Hyophila* is closely related to *Barbula*) including recurved lower laminal margins, medially differentiated basal laminal cells, ventrally colliculate but dorsally smooth upper lamina (e.g. *B. javanica*) and armed, often caltrop-shaped propagula (as in *Barbula indica* and *Barbula* sect. *Hydrogonium*). Also, the capsules of some species (e.g., *H. involuta* and *H. nymaniana*) are large, dark brown and thick-walled, reminiscent of capsules of *Barbula* or *Tortula*, while others (e.g., *H. acutifolia*) are short, yellowish and thin-walled, like those of *Trichostomum* or *Weissia*; intermediates are few. These features indicate that *Hyophila* as presently recognized is polyphyletic. Of some significance is that *H. siamensis* (Pl. 56, f. 1–5) could be placed with *Barbula* by its narrowly recurved marginal laminal cells and other characters, except that the spathulate leaf shape and lack of a peristome at the present time militate against it; the same is true for *H. nymaniana*, which differs widely from other *Hyophila* species in the presence of punctate papillae (curiously like those of certain species of Rhachitheciaceae) on the upper laminal cells, which bulge on both sides of the leaf, the absence of a stem central strand, and the bright red KOH color reaction of the basal laminal cells at the leaf insertion (the last two characters are unusual but duplicated in *Leptodontium viticulosoides* and *Tortula cuneifolia* var. *blissii*). On the other hand, *H. apiculata* (Pl. 55, f. 1–5) appears to be a phenocopy of *B. agraria* in the short-ovate leaf shape, acute apex and sharply excurrent costa. Additional study is needed for satisfactory disposition of these *Hyophila* species.

Distinctive characters often but not always found in *Hyophila* species in various combinations and which must serve to distinguish *Hyophila* from *Trichostomum* and *Barbula* include: leaves ligulate, narrowed to the insertion, broadly concave in transverse section; upper laminal cell surfaces epapillose, ventrally bulging and dorsally weakly convex (Pl. 54, f. 7); basal laminal cells short and poorly differentiated or differentiated cells restricted to a small area near the insertion; propagula present, armed (Pl. 54, f. 8; 55, f. 23); monoicous; capsule eperistomate, and exothecial cells thick-walled (Pl. 55, f. 25). No species of *Hyophila* has all these characters and few have most. *Hyophila* differs from *Plaubelia* mainly in being eperistomate and the ventral surface of the costa usually consisting of elongate cells. The leaf hydroid strand is also often absent in *Hyophila*, but these characters may ultimately prove insufficient to separate the two taxa at the generic level. Two species previously placed in *Hyophila* probably because of their lingulate leaves and eperistomate capsules are here referred to *Tisserantiella* of the Rhachitheciaceae (one as a synonym, the second as a *comb. nov.*, see Excluded Taxa) because their costa mostly end below the leaf apex and the upper laminal papillae are

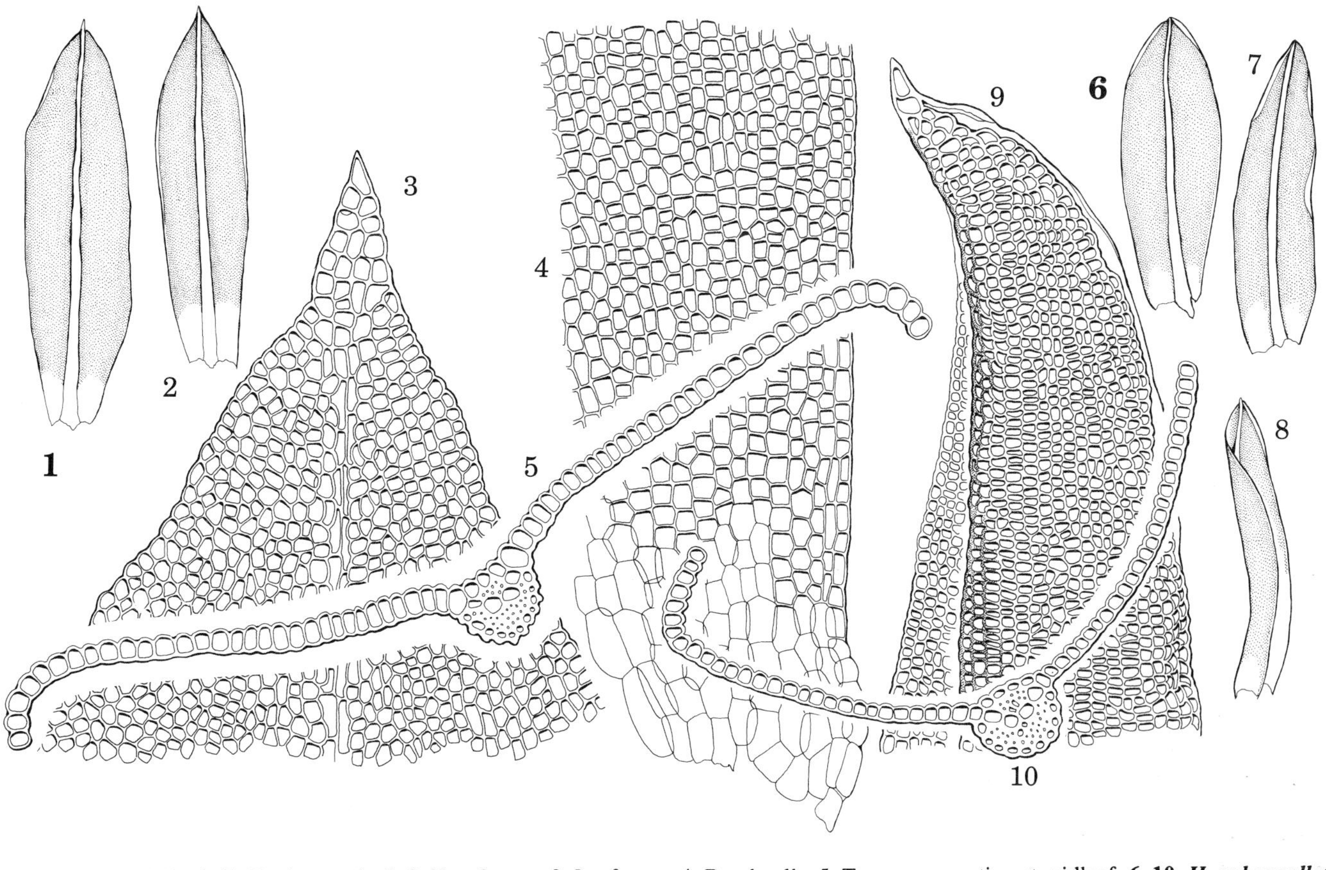

Plate 56. *Hyophila*. 1–5. *H. siamensis*. 1–2. Two leaves. 3. Leaf apex. 4. Basal cells. 5. Transverse section at midleaf. **6–10. *H. subcucullata*.** 6–8. Three leaves. 9. Leaf apex, lateral view. 10. Transverse section at midleaf.

extremely small, solid, punctiform to short-spiculose, with the appearance of tiny specks of light under the compound lens (these characters are unique only in combination).

Additional literature: Andrews and Redfearn (1965), Deguchi et al. (1991), Gao et al. (1991), Long (1978), Mehra (1984, 1988), Nawawi and Mohamed (1989), Olarinmoye (1981), Rahbar and Chopra (1980), Sakurai (1954), Sharma and Chopra (1986), Sharp (1955), Smith and Whitehouse (1974), Yang (1965).

Number of accepted species: 88.

Species examined: *H. acutifolia* (TNS), *H. apiculata* (FH), *H. bartramiana* (BUF, MICH, TENN), *H. beruensis* (H), *H. blanda* (NY), *H. grossidens* (H), *H. involuta*, *H. latifolia* (H), *H. mattogrossensis* (H), *H. nymaniana* (BUF, NY, TENN), *H. potierii* (FH), *H. propagulifera* (H, NY, TNS), *H. siamensis* (BM), *H. spathulata* (NY), *H. subcucullata* (NY), *H. subflaccida* (NY), *H. usambarica* (H), *H. viridula* (BUF).

New heterotypic synonymy: *Gymnostomum denticulatum* Griff. (NY) = *Hyophila involuta* (Hook.) Jaeg.

42. PLAUBELIA Plate 57.

Plaubelia Brid., Bryol. Univ. 1: 522, 1826 (*nom. rejic. vs. Trichostomum* Bruch, 1829). Type: *Plaubelia tortuosa* Brid.

Hyophilopsis Crum, Bryologist 68: 69, 1965, *non* Card. & Dix. 1911, *hom. illeg.*

Neohyophila Crum, Bryologist 68: 470, 1965, *nom. nov.* for *Hyophilopsis* Crum, *hom. illeg.* Type: *Neohyophila sprengelii* (Schwaegr.) Crum.

Tortula sect. *Plaubelia* (Brid.) Mitt., J. Linn. Soc. Bot. 143, 154, 1869.

Named for a mycologist surnamed Plaubel, a specialist of the 1820's in *Puccinia* Pers. and other Uredinales. Major mycological references provide no further information on this person.

Plants forming turf or loosely caespitose, green above, green or sometimes brown below. Stems often branching, to 4 mm in length, in transverse section rounded-pentagonal, *central strand strong*, sclerodermis present, often weak, hyalodermis absent; axillary hairs of up to 5 uniseriate cells, the basal 1–2 yellow; rhizoids sparse. *Leaves* incurved to spreading, *often tubulose when dry*, widely spreading when moist, usually *rosulate*, *spathulate to oblong*, to 3.2 mm in length, ventral surface flat to broadly concave across leaf; *margins incurved, involute, or occasionally plane*, sometimes broadly involute at the apex, entire to distantly denticulate above; *apex rounded acute to broadly obtuse*, *apiculate* or occasionally entire; *base little differentiated in shape*; *costa percurrent or ending up to 4 cells below the apex*, *adaxial surface cells bulging*, rounded-hexagonal, ca. 4 rows of cells across ventral surface of costa at midleaf, dorsal cells elongate, costa in transverse section rounded, showing one or two stereid bands, the dorsal semicircular in section except for a ventral indentation at the hydroid strand, ventral epidermis differentiated, 2–4 guide cells in 1 layer, dorsal epidermis slightly differentiated, *hydroid strand present* (this occasionally difficult to demonstrate in small plants); *upper laminal cells rounded-hexagonal*, 8–10 µm in width, 1:1, walls evenly thickened, *ventrally strongly convex and dorsally nearly flat*; upper laminal papillae often absent, when present solid, small, simple, 1–2 per lumen dorsally or occasionally present on both sides of lamina; basal cells not differentiated or often forming a small group medially or weakly differentiated across the leaf, quadrate to short-rectangular, to 15 µm wide, 2–3:1, hyaline to somewhat yellowish; 1–2 rows of inflated, hyaline cells across insertion sometimes also present, occasionally forming small auricles. Propagula in leaf axils, clavate, often multi-branched at the wider end, mostly 100–300 µm in length, with occasional internal walls. *Dioicous*. Perichaetia terminal, inner leaves ovate-lanceolate to ligulate, slightly larger or somewhat shorter than outer leaves, weakly or strongly sheathing, cells long-rhomboidal or rectangular in lower half. Perigonia terminal, inner leaves ovate, outer leaves large. Seta to 0.8 cm in length, yellow- to red-brown, twisted clockwise below, often counter-clockwise above, 1 per perichaetium; theca ca. 1.0–1.5 mm in length, red- or yellow-brown, ellipsoidal or oblong, exothecial cells somewhat bulging, rectangular, ca. 15–35 µm in width, 2–5:1, walls thin to weakly porose and thickened, stomates phaneropore, present at base of capsule above a comparatively well developed neck; annulus of strongly vesiculose cells, persistent on the capsule mouth, detaching in pieces or revoluble; *peristome of 16 spiculose, lanceolate to long-linear teeth*, cleft to near base or variously cleft or perforate, often only perforate at base and entire, to 180 µm in length, up to 9 articulations, straight, low-spiculose, basal membrane absent or to 25 µm in height, low-spiculose. Operculum rostrate, to 0.9 mm in length, cells in straight rows. Calyptra cucullate, smooth, 0.6–2.0 mm in length. Spores yellow, weakly papillose, small, ca. 8–10 µm in diameter. *Laminal KOH color reaction yellow in upper leaves, often orange-brown in lower.*

Found on rock, soil, especially calcareous substrates, in extreme southwestern and southeastern U.S.A., Mexico, Central America, the West Indies, Venezuela, Brazil, South Africa and Burma.

Bridel (1826–27) established *Plaubelia* to include the single species *Plaubelia tortuosa* Brid. (= *Desmatodon sprengelii* (Schwaegr.) Williams, a taxonomic synonym *fide* Williams 1919). The name *Plaubelia* fell from use after being rejected against the later (1829) generic name *Trichostomum* Bruch; M. Crosby (in litt.) noted that *Plaubelia* is acceptable when it is not a taxonomic synonym of *Trichostomum* Bruch. Crum (1965a) created *Hyophilopsis* (and later the *nomen novum Neohyophila*, a heterotypic synonymn of *Plaubelia*) as a segregate of *Desmatodon*, citing, as a unique combination of features, the spathulate, rosulate leaves (Pl. 57, f. 1, 9, 10) with erect or incurved margins, laminal cells ventrally bulging, and the ventral epidermis of bulging cells (Pl. 57, f. 5). He further referred the genus to the Pottioideae, citing a single stereid band, but I agree with Delgadillo and Cárdenas (1982) that this character is variable, as one or two bands may be present in different leaves of the same plant. Actually *Plaubelia* is apparently most closely related to *Hyophila*, from which it differs most saliently in the presence of a peristome (Pl. 57, f. 8, 18). The strongly involute margins are reminiscent of *Weissia*. The presence of a hydroid strand (Pl. 57, f. 7, 16) and of rounded-quadrate ventral epidermal cells (as seen from above) on the costa are also good characters distinguishing *Plaubelia* from the similar and very commonly distributed *Hyophila involuta*, as well as several other *Hyophila* species but not all (*Hyophila* greatly needs revision). *Globulinella* is very similar in many respects, but is distinguishable by the costa not bulging dorsally and the usually cucullate leaf apex. Crum and Anderson (1981)

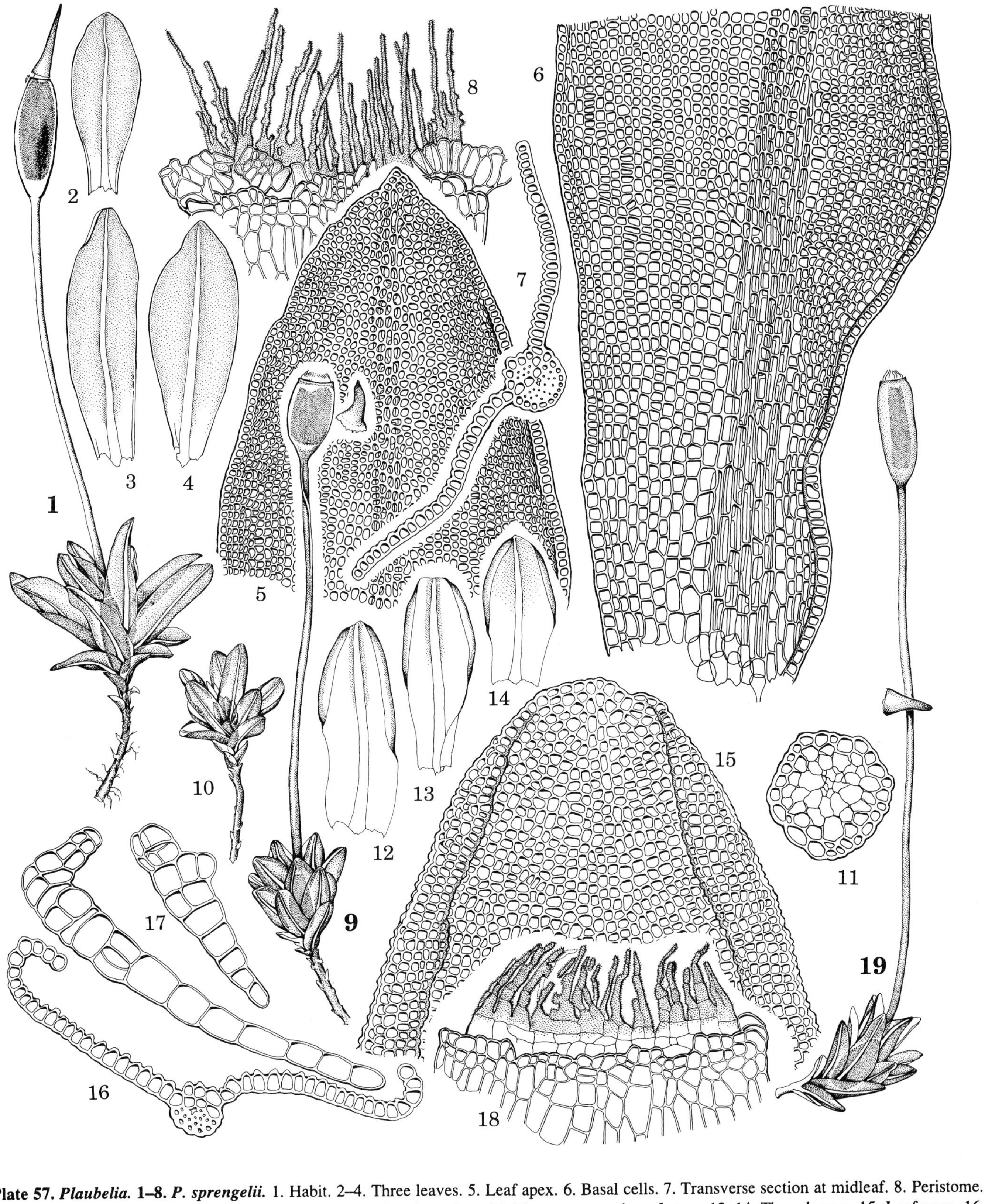

Plate 57. *Plaubelia.* 1–8. *P. sprengelii.* 1. Habit. 2–4. Three leaves. 5. Leaf apex. 6. Basal cells. 7. Transverse section at midleaf. 8. Peristome. **9–18. *P. involuta.*** 9. Habit of sporangiate plant. 10. Perigoniate plant. 11. Transverse section of stem. 12–14. Three leaves. 15. Leaf apex. 16. Transverse section at midleaf. 17. Propagula. 18. Peristome. **19. *P. perinvoluta.*** 19. Habit.

placed the species *Merceyopsis angulosa* with *Plaubelia* (without actually making a combination), but that species is better viewed as the monotypic genus, *Ganguleea* ((*q.v.*).

Two new combinations are here added to *Plaubelia* from the Old World, where they were previously recognized in *Desmatodon* and *Weisiopsis*. These additional species differ from the American *P. sprengelii* in their strongly involute upper laminal margins. They are otherwise little different from each other and may prove synonymous.

Additional literature: Crum (1965c), Delgadillo & Zander (1985), Saito (1973d), Zander (1983c).

Number of accepted species: 3.

Species examined: *P. involuta* (NY), *P. perinvoluta* (FH), *P. sprengelii*.

New combinations: *Plaubelia involuta* (Magill) Zand., *comb. nov.* (*Weisiopsis involuta* Magill, Fl. S. Afr. Mosses 1: 225, 1981 [1982]). *Plaubelia perinvoluta* Zand., *nom. nov.* (*Desmatodon involutus* Bartr., *hom. illeg.*, Farlowia 1: 175, 1943). *Plaubelia sprengelii* (Schwaegr.) Zand., *comb. nov.* (*Barbula sprengelii* Schwaegr., Spec. Musc. Suppl. 2(1): 64, 1823; *Desmatodon sprengelii* (Schwaegr.) Williams; *Trichostomum sprengelii* (Schwaegr.) Lindb.). *Plaubelia sprengelii* var. *stomatodonta* (Card.) Zand., *comb. nov.* (*Hyophila stomatodonta* Card., Rev. Bryol. 36: 76, 1909; *Desmatodon stomatodontus* (Card.) Williams; *Neohyophila stomatodonta* (Card.) Crum).

43. TENIOLOPHORA — Plate 58.

Teniolophora Reese, Bryologist 65: 67, 1962. Type: *Teniolophora fluviatile* (Williams) Reese.

Teniola Reese, Bryoloigst 62: 221, 1959 [1960], *nom. inval.*

From *teniola*, a band of intramarginal differentiated cells + o + φόρος, that which is brought in, from φορέω, to bear or carry, to possess; referring to the differentiated border of elongated cells otherwise characteristic of the Calymperaceae.

Plants growing in a thick turf, *dark green above*, brown below. Stems to 3 cm in length, often branching, in transverse section pentagonal, central strand strong, sclerodermis present, hyalodermis absent, axillary hairs ca. 10 cells in length, basal 1–2 brown. *Leaves appressed, incurved and tubulose when dry, spreading when moist, elliptical*, to 2.5 mm in length, broadly channeled across ventral surface; *margins plane to incurved*, entire to denticulate near apex, *with a bistratose intramarginal border of 5–6 rows of thick-walled, rectangular or rhomboidal cells extending from leaf base to near apex*; apex rounded to broadly acute, often apiculate or sharply reflexed; base scarcely differentiated in shape; costa percurrent, 6–10 cells across ventral surface at midleaf, superficial cells quadrate to short-rectangular ventrally, elongate dorsally, transverse section elliptical, *two strong stereid bands present*, epidermal cells differentiated on both surfaces, guide cells 4–8 in 1 layer, hydroids absent; *upper laminal cells* short-rectangular to hexagonal, *small*, 7–9 µm in width, 1:1, walls evenly thickened, *ventrally bulging, dorsally nearly flat*, homogeneous; papillae apparently absent or low and broad, evident only in leaf section and only present on dorsal surface, simple, one per lumen; basal laminal cells differentiated across leaf, rectangular, little wider than upper laminal cells, walls evenly thickened, 2–3:1. *Propagula borne on stout stalks in upper leaf axils, multicellular, large*, to 350 µm long, obovoid to elliptical. Apparently dioicous, perichaetia terminal, inner leaves not differentiated. Perigonia and sporophyte not known. *Laminal KOH color reaction deep yellow* in upper leaves, deep yellow-orange in lower leaves.

Found on moist rocks near streams or other moist situations; presently only known from the West Indies in Puerto Rico and Haiti.

This genus was established largely on the basis of the differentiated laminal intramarginal border (Pl. 58, f. 5, 8), similar to the teniolae of *Calymperes*. Excepting this band of thick-walled cells and firm-walled basal cells of the axillary hairs, characters of *Teniolophora* are much the same as those of *Hyophila involuta*. Specimens of this latter taxon (e.g. Brazil, Rio de Janeiro, Glaziou 9228, NY) may be found with similar large, obovoid, unarmed propagula (Pl. 58, f. 9, 10) rather than the more usual armed propagula.

Literature: Reese (1959, 1962).

Number of accepted species: 1.

Species examined: *T. fluviatile* (NY).

44. WEISSIA — Plates 59–62.

Weissia Hedw., Sp. Musc. 168, 1801. Lectotype: *Weissia controversa* Hedw. *fide* Mitten, Kew J. Bot. 8: 258, 1856.

Cavanillea Borkh., Disp. Pl. 251, 1809, *hom. illeg. non* Medikus, 1787, *non* Desf. in Lam., 1792.

Subg. *Weissia*

Simophyllum Lindb., Act. Soc. Sci. Fenn. 10: 74, 1871, *nom. illeg. incl. gen. prior.*

Rechingerella Fröhl., Ann. Naturhist. Mus. Wien 66: 36, 1962 [1963], *hom. illeg. non* Petrak. Type: *Rechingerella macedonica* J. Fröhl.

Weissia subg. *Weisia* BSG, Bryol. Eur. 1: 5, 1851 (fasc. 47. Consp. vol. 1: VII), *nom. illeg.*

Weissia subg. *Microweisia* BSG, Bryol. Eur. 1: 5, 1851 (fasc. 46–47 Consp. vol. 1: VII).

Weissia subg. *Euweisia* Schimp., Syn. 54, 1860, *nom. illeg.*

Hymenostomum subg. *Weisia* (Hedw.) Andrews, Bryologist 23: 31, 1920, *nom. illeg. prior. ut gen.*

Tortella subg. *Nanotortella* C. Jens., Danm. Moss. 2: 318, 1923, *nom. illeg. incl. typ. gen. prior., p.p.*

Bryum sect. *Weissia* (Hedw.) Relh., Fl. Cantabr. 425, 1802. Type: *Bryum virens* Relh.

Weissia sect. *Controversae* Nees & Hornsch., Bryol. Germ. 2(2): 25, 31, 1831.

Weissia sect. *Viridulae* BSG, Bryol. Eur. 1: 66, 1846 (fasc. 33–36 Mon. 6).

Weissia sect. *Euweisia* C. Müll., Syn. 1: 651, 1849, *nom. illeg.*

Didymodon sect. *Pusilli* Kindb., Eur. N. Amer. Bryin. 2: 272, 1897. Type: *Didymodon triumphans* (De Not.) Kindb.

Barbula sect. *Edentella* C. Müll., Gen. Musc. Fr. 453, 1900. Type: *Barbula schweinfurthiana* C. Müll.

Weissia subsect. *Sphalerostomae* Nees & Hornsch., Bryol. Germ. 2(2): 24, 25, 1831, *nom. illeg.*

Subg. *Astomum* (Hampe) Kindb., Eur. N. Amer. Bryin. 2: 283, 1897.

Astomum Hampe, Flora 20: 285, 1837. Lectotype: *Astomum crispum* (Hedw.) Hampe *fide* Margadant, Acta Bot. Neerl. 8: 274, 1959.

Sphaerangium Schimp., Syn. 12, 1860, *nom. illeg. incl. gen.*

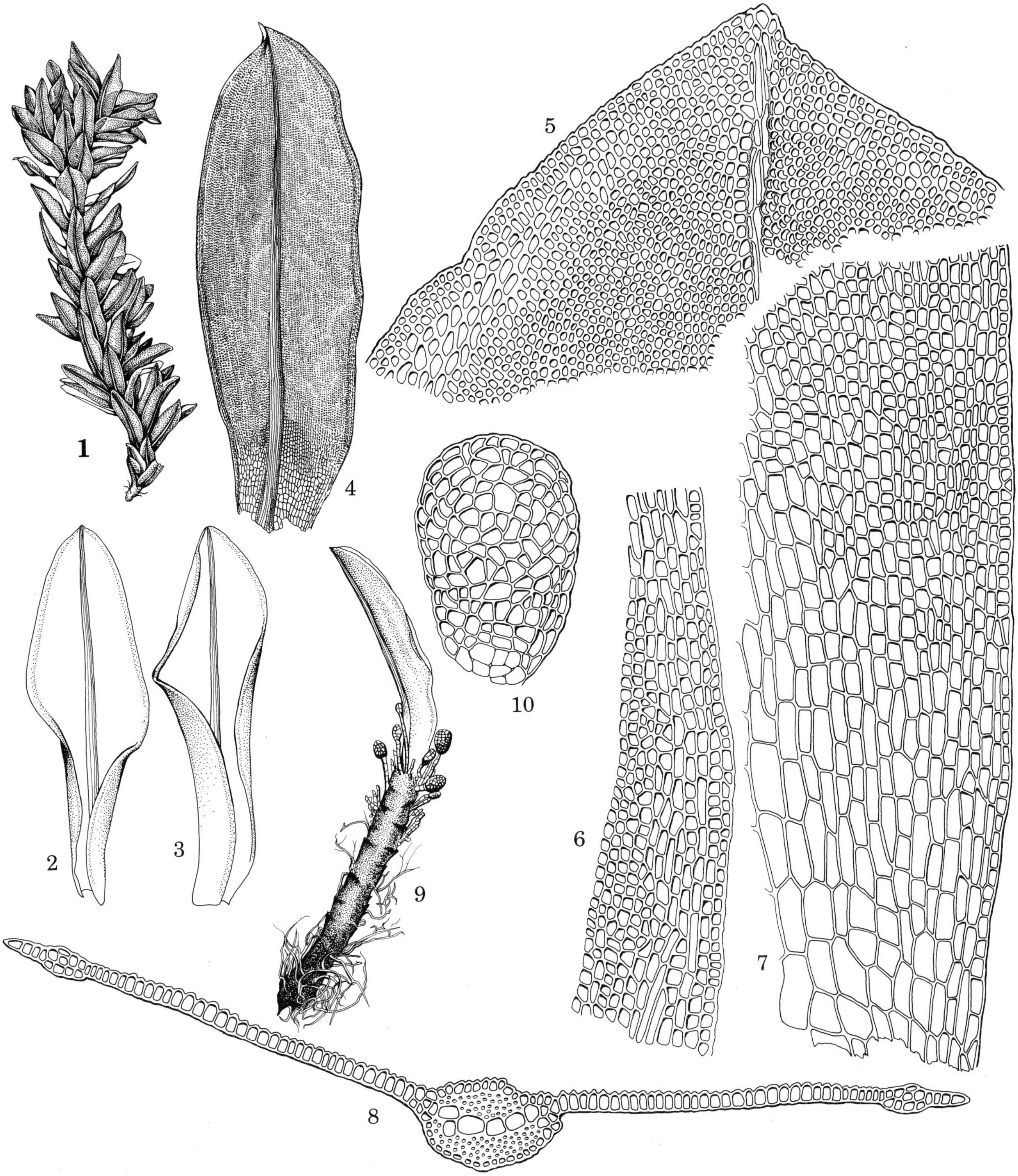

Plate 58. ***Teniolophora.*** **1–10.** ***T. fluviatile.*** 1. Habit. 2–4. Three leaves. 5. Leaf apex. 6. Upper marginal cells, showing intramarginal border. 7. Basal cells. 8. Transverse section at midleaf. 9. Stalked propagula on stem. 10. Propagulum.

prior.

Systegium Schimp., Syn. 30, 1860. Lectotype: *Systegium crispum* (Hedw.) Schimp.

Astomum subg. *Euastomum* (C. Müll.) Broth., Nat. Pfl. 1(3): 384, 1901, *nom. illeg.*

Hymenostomum subg. *Astomum* (Hampe) Andrews, Bryologist 23: 31, 1920.

Tortella subg. *Systegium* C. Jens., Danm. Moss. 2: 317, 1923, *hom. illeg.* Type: *Tortella crispa* (Hedw.) C. Jens.

Phascum sect. *Astomum* (Hampe) Ångstr. in Fries, Summ. Veg. Scand. 1: 97, 1846.

Weissia sect. *Astomum* (C. Müll.) C. Müll., Bot. Zeit. 5: 98, 1847, as autonym.

Weissia sect. *Systegium* Lindb., Öfv. K. Vet. Ak. Förh. 21: 230, 1864. Lectotype: *Weissia crispa* (Hedw.) Mitt.

Acaulon sect. *Macrobryum* C. Müll., Linnaea 43: 353, 1882. Type: *Acaulon lorentzii* C. Müll.

Phascum sect. *Systegium* C. Müll., Hedwigia 37: 76, 1898, *nom. illeg. incl. sect. prior.*

Subg. *Hymenostomum* (R. Br.) Limpr., Laubm. Deutschl. 1: 225, 1886, as autonym.

Hymenostomum R. Br., Trans. Linn. Soc. London 12(2): 572, 1818. Type: *Hymenostomum microstomum* (Hedw.) R. Br.

Gymnostomum subg. *Hymenostomum* (R. Br.) Schimp., Syn. 33, 1860.

Mollia subg. *Hymenostomum* (R. Br.) Lindb., Musci Scand. 21, 1879.

Weissia subg. *Hymenostomum* Limpr., Laubm. Deutschl. 1: 225, 1886, *nom. illeg.*

Hymenostomum subg. *Kleioweisia* Limpr., Laubm. Deutschl. 1: 224, 1886. Type: *Hymenostomum rostellatum* (Brid.) Schimp.

Weissia subg. *Hymenostomum* (R. Br.) Kindb., Eur. N. Amer. Bryin. 2: 283, 1897, *nom. illeg.*

Hymenostomum subg. *Euhymenostomum* Andrews, Bryologist 23: 31, 1920, *nom. illeg.*

Weissia sect. *Hymenostomum* (R. Br.) BSG, 1: 56, 1846 (fasc. 33–36 Mon. 6), as autonym.

Hymenostomum sect. *Microstoma* B.&S. in BSG, Bryol. Eur. 1: 56, 1846 (fasc. 33–36 Mon. 6), *nom. illeg. incl. typ. gen.* Type: *Hymenostomum microstomum* (Hedw.) R. Br.

Hymenostomum sect. *Tortilia* B.&S. in BSG, Bryol. Eur. 1: 56, 1846 (fasc. 33–36 Mon. 6). Type: *Hymenostomum tortile* (Schwaegr.) B.&S.

Weissia sect. *Hymenostomum* (R. Br.) C. Müll., Syn. Musc. 1: 660, 1849.

Mollia sect. *Hymenostomum* (R. Br.) Braithw., Brit. Moss Fl. 1: 230, 1885.

Subg. *Phasconica* (C. Müll.) Zand. see below.

Phasconica C. Müll., Linnaea 43: 438, 1882. Type: *Phasconica lorentzii* C. Müll., *lectotyp. nov.*

Subg. *Pseudopottia* Kindb., Eur. N. Amer. Bryin. 2: 283, 284, 1897.

Sect. *Gymnostomum* Mitt., J. Linn. Soc. Bot. 12: 41, 129, 131, 1869.

Sect. *Tortularia* Mitt., J. Linn. Soc. Bot. 12: 15, 130, 139, 1869.

Subsect. *Hapalostomae* Nees & Hornsch., Bryol. Germ. 2(2): 24, 25, 1831.

Named for Friedrich Wilhelm Weiss (1744–1826) of Göttingen, Germany, a lichenologist.

Plants in low cushions or turfs or loosely caespitose, green above, brown to tan or yellow below. Stems branching irregularly, to 1.0 cm in length, transverse section rounded-pentagonal, *central strand present*, occasionally hollow or very thick, sclerodermis weakly differentiated in 1–2 layers, hyalodermis weakly differentiated to distinct, seldom absent; axillary hairs to 10 cells in length, basal 1–2 cells thicker walled; sparsely radiculose. *Leaves* incurved, tubulose, often contorted or spiralled when dry, spreading and occasionally sharply reflexed above a sheathing base when moist, *long-ligulate, oblong or triangular to lanceolate*, 1.5–2.5(–4.0) mm in length, upper lamina broadly channeled across leaf, seldom narrowly channeled along costa, *margins sharply incurved (or occasionally tightly involute) above the leaf base*, seldom merely erect-incurved near apex, *entire*, occasionally fragile and breaking off in large rectangles; apex sharply acute to subulate, occasionally broadly acute, obtuse, or weakly cucullate; base scarcely differentiated to ovate or rectangular, occasionally half-sheathing; *costa shortly and sharply mucronate*, seldom subpercurrent, *superficial cells quadrate* or occasionally short-rectangular to elongate ventrally, *elongate dorsally*, 4–8(–10) rows of cells across costa ventrally at midleaf, costal transverse section ovate, occasionally circular or semicircular, *two stereid bands present, usually of about equal size*, differentiated epidermis present ventrally, absent or occasionally weakly differentiated dorsally, guide cells 4–6(–8) in 1 layer, *hydroid strand absent* or occasionally present; upper laminal cells subquadrate to hexagonal, 7–13 µm in width, 1:1, *walls* thin to evenly thickened, *superficially strongly bulging on both exposed surfaces or more strongly protuberant on ventral surface*; *papillae bifid*, 2–6 per lumen, occasionally fused into a capitulate papilla covering the lumen, occasionally spiculose-branching, seldom absent; basal cells differentiated across leaf, occasionally rising higher along margins in a vee, rectangular, occasionally rhomboidal, 2–5:1, walls thin to evenly thickened. Rhizoid-borne tuber-like propagula reported for *W. controversa*. *Monoicous, occasionally dioicous*. Perichaetia terminal, *inner leaves little different from cauline leaves or somewhat larger*, occasionally weakly sheathing the seta, lower cells long-rhomboidal to midleaf. Perigonia appearing as stalked lateral buds on perichaetiate plants (but variably present) or terminal on usually smaller perigoniate plants. Seta 0.1–1.3 cm in length, 1(–2) per perichaetium, yellowish brown, twisted clockwise; theca 1.0–2.2 mm in length, occasionally inclined, yellowish to reddish brown, elliptical to ovate or cylindrical, occasionally nearly spherical or urceolate, exothecial cells rectangular, walls thin or becoming rather thick, rarely superficially bulging, stomates phaneropore at base of capsule, *annulus* of 2–6 rows of persistent, vesiculose cells, *often on a differentiated collar*, or annulus not differentiated; *cleistocarpous or stegocarpous, peristome teeth 16, short or rudimentary or absent*, occasionally lifts with operculum, oblong-truncate to long-triangular, *often irregularly cleft into two rami* or variously perforate, papillose to spiculose or spirally ridged, to 125(–250) µm, with up to 10 articulations, straight or twisted weakly counterclockwise, basal membrane absent or low, low-spiculose, *mouth of capsule occasionally closed by a hymenium, this internal to the peristome* (when peristome present), seldom macrostomous. Operculum conic to rostrate, 0.3–0.8(–1.5) mm in length, cells straight or rarely somewhat twisted counterclockwise.

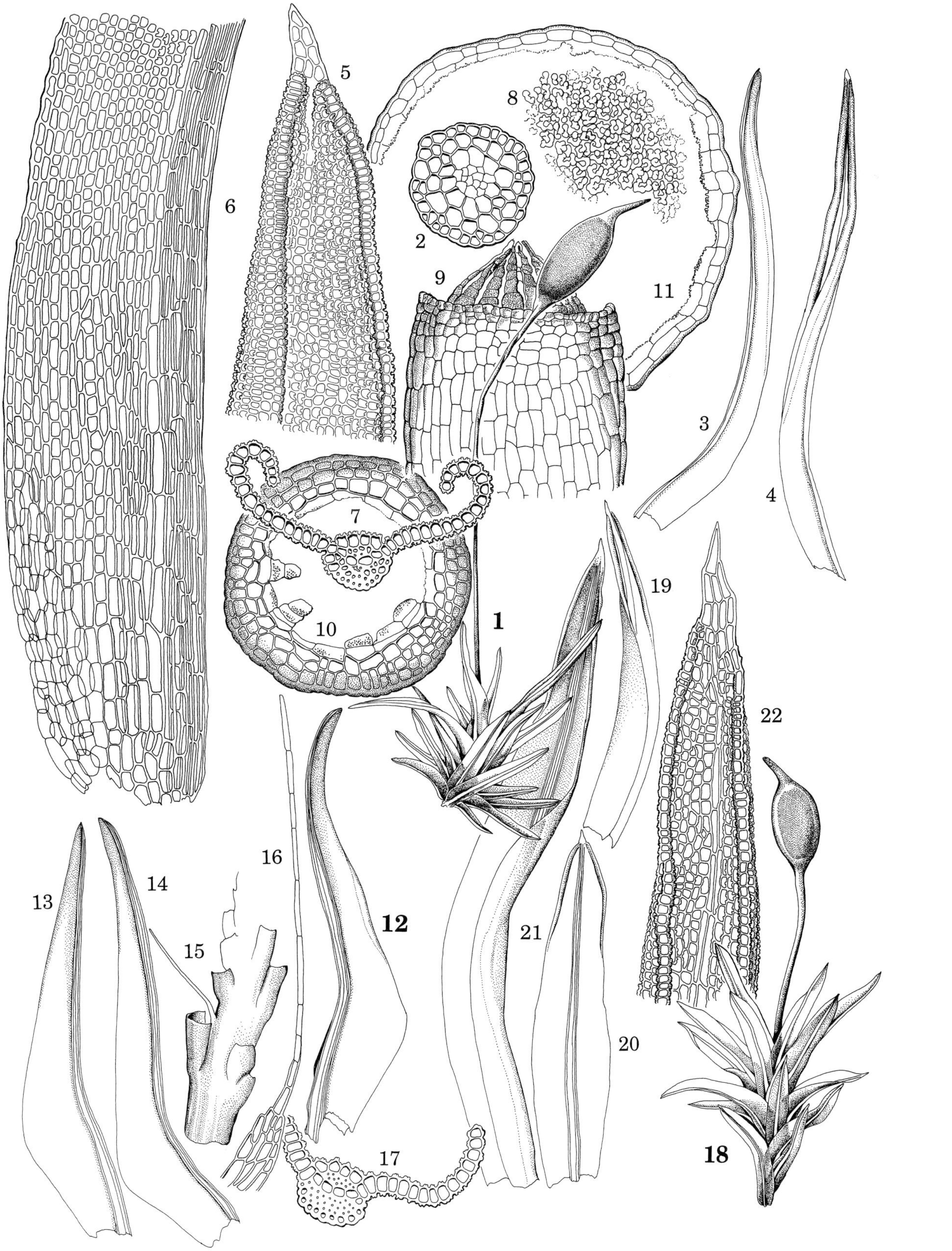

Plate 59. ***Weissia.*** **1–11.** ***W. controversa.*** 1. Habit. 2. Transverse section of stem. 3–4. Two leaves. 5. Leaf apex. 6. Basal cells. 7. Transverse section at midleaf. 8. Upper laminal papillae. 9–10. Two peristomes. 11. Transverse section of capsule wall. **12–17.** ***W. artocosana.*** 12–14. Three leaves. 15–16. Axillary hairs. 17. Transverse section at midleaf. **18–22.** ***W. brachycarpa.*** 18. Habit. 19–21. Three leaves. 22. Leaf apex.

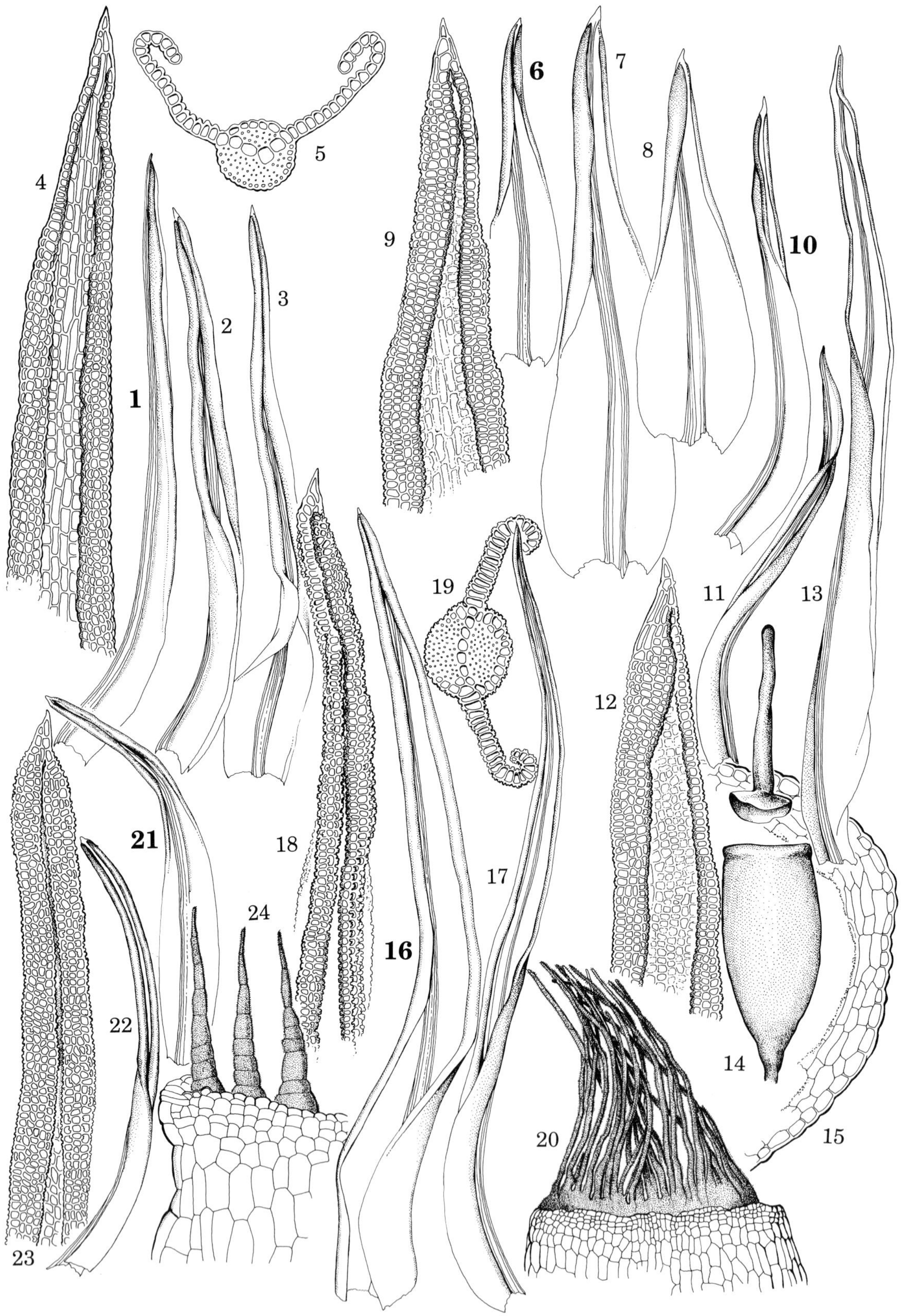

Plate 60. ***Weissia.*** **1–5.** ***W. breutelii.*** 1–3. Three leaves. 4. Leaf apex. 5. Transverse section at midleaf. **6–9.** ***W. crispa.*** 6–8. Three leaves. 9. Leaf apex. **10–15.** ***W. edentula.*** 10–11. Two leaves. 12. Leaf apex. 13. Perichaetial leaf. 14. Capsule with operculum. 15. Transverse section of capsule wall. **16–20.** ***W. jamaicensis.*** 16–17. Two leaves. 18. Leaf apex. 19. Transverse section at midleaf. 20. Peristome teeth. **21–24.** ***W. longidens.*** 21–22. Two leaves. 23. Leaf apex. 24. Peristome teeth.

Calyptra cucullate (to short-triangular on cleistocarpous capsules), smooth, 0.4–2.5 mm in length. Spores 14–28 µm in diameter, brown to yellowish brown, low-papillose or occasionally highly papillose. *Laminal KOH color reaction yellow, usually pale yellow.* Reported chromosome number n = 13, 13+m, 13+0–4m, 14, 26 (subg. *Weissia*); 13, 13+m, 26 (subg. *Astomum*); 8, 13, 14, 26 (subg. *Hymenostomum*).

A large genus found on all continents except Antarctica, mostly growing on soil.

Weissia is superficially not well distinguished from *Trichostomum*. Andrews (1945) felt that the *Weissia* complex should be "extended to cover *Trichostomum*." In *Weissia*, the central strand nearly universally present (Pl. 59, f. 2; 62, f. 7) at least in the specimens examined), upper laminal margins are usually narrowly and sharply incurved (Pl. 59, f. 7; 60, f. 5) as is the case with *Chionoloma* and *Weissiodicranum* (those of *Weissia termitidarum* and certain others are loosely involute), the upper laminal cells are often more strongly bulging ventrally than dorsally (although usually overlain with papillae), the hydroid strand is usually absent (but present in *W. controversa*), the sexual condition is commonly monoicous (position of antheridia often varying between or within collections of a particular species), perichaetial leaves are often larger than the cauline, peristomes are generally short with the teeth more or less flattened (Pl. 59, f. 9–10) or absent, and spores are rather large, 14–28 µm in diameter. When present, the ventrally colliculate condition, i.e. bulging upper laminal cells associated with slightly thicker, flattened dorsal superficial walls, has not been emphasized by previous authors, but is common among the species examined, being present to various degrees in, e.g. *W. artocosana* (Pl. 59, f. 17), *W. breutelii* (Pl. 60, f. 5), *W. canaliculata*, *W. condensa*, *W. controversa* (Pl. 59, f. 7), *W. ghatensis*, *W. glaziouii*, *W. latiuscula*, *W. ligulaefolia*, *W. macrospora*, *W. occidentalis*, *W. striata* (Pl. 61, f. 19), *W. termitidarum*, *W. triumphans*, *W. veviridis* and *W. wimmeriana*.

The best-developed species of *Weissia*, *W. jamaicensis* (Pl. 60, f. 16–20), includes a potpourri of features characteristic of other genera. It has the basal vee of hyaline cells of *Tortella*, the much-enlarged ventral stereid band and dilated leaf base of *Pseudosymblepharis*, and a peristome of 16 straight, linear-lanceolate teeth cleft to a basal membrane into two spiculose rami quite like that of *Trichostomum crispulum* (which, itself, occasionally develops a basal vee and margins incurved near the apex, but lacks the swollen ventral stereid band). In this case, the species are distinctive, but the generic definitions break down. *Trichostomum crispulum* and *T. perligulatum* have leaf margins that are often incurved near the apex but generally have plane or broadly incurved margins throughout most of the leaf. Some species, such as *W. crispa* (Pl. 60, f. 6–9) and the doubtfully distinct *W. muhlenbergiana* (Pl. 61, f. 1–6, see Crum & Anderson's 1981 discussion), have perichaetial leaves with plane to erect-incurved margins but cauline leaves (and those on sterile stems) with weakly but narrowly incurved upper margins, and are perhaps ultimately better assigned to *Trichostomum* in the *T. crispulum* relationship. *Weissia rutilans* (Pl. 61, f. 7–11) has upper laminal margins not easily assigned to either plane or narrowly incurved status. In any case, to provide an interim classification that better reflects perceived relationships, all *Weissia* taxa with clearly plane upper cauline leaf margins examined in the course of this study are here transferred to *Trichostomum*.

The fact that peristome teeth may occasionally be represented in some species only by very much reduced, truncate and colorless fragments, these often hidden below the rim of the capsule mouth, indicates that past generic level distinctions between peristomate and gymnostomous species of *Weissia* with similar leaves may not be worthwhile. There is some correlation between thick capsule walls and lack of a peristome (reflected in distinctions between *Weissia* taxa of my treatments in the Moss Flora of Mexico, in press), but this must be investigated in greater detail. Stoneburner (1985) demonstrated significant variation in peristome expression in species of *Weissia* from the southwestern United States, and even found rudimentary teeth in a species (*W. ligulaefolia*) hitherto considered gymnostomous. She further discussed the "evanescent membrane" of hymenostomous species as a poor taxonomic character, explaining it as the upper surface of a somewhat persistent columella in both eperistomate and peristomate taxa. The present study confirms this in that a "hymenium" can often be seen below the peristome teeth of recently dehisced capsules of, for instance, *W. controversa*. This is an old problem. Referring to Brown's (1818) description of the then new genus *Hymenostomum*, Greville and Arnott (1824b) wrote: "In Dr. Hooker's excellent figure of *Gymnostomum xanthocarpum* in the Musci Exotici, the membrane which closes the sporular bag is admirably delineated.... It ought to be observed, that Palisot has described this membrane in the Mosses without a peristome, but denied it to the others" and (1824a) again asserted: "Now, *every* moss possesses this 'membranula indivisa,' arising from the lining of the theca; but in all the species of *Gymnostomum*, it is peculiarly evident at some particular stage." However, Greville and Arnott (1824b) admitted that: "The membrane to which we have already alluded, as closing the mouth of the sporular bag, there is every reason to suppose, is stronger and more durable in all such as have a naked peristome."

A further taxonomic complication is that the peristome teeth of occasional specimens, e.g. of *W. triumphans*, are somewhat twisted counterclockwise, possibly an indication of a relationship of these species to *Tortella*.

Regarding the synonymy of *Weissia* and *Astomum*, a review of the problem and new evidence (existence of sporophytes of intermediate morphology) was offered by Stoneburner (1985). Many combinations at the genus level in *Astomum* and *Hymenostomum* remain as "correct" names in the list of species and infraspecies given at the end of this treatment. A mass transfer of these names to *Weissia* was considered but rejected because many names, on examination of authentic material, will prove to be more appropriately placed in *Trichostomum* (as emended here). Recognition here of subgenera of *Weissia* that are based largely on sporophyte morphology reflects recent studies by other authors (excepting subg. *Phasconica*). Quite probably, intraspecific lines of evolution involving sporophytic features of species with essentially identical gametophytes will be found that cut across presently recognized subgeneric limits. Unusual characters of sporophytes, such as the thick thecal walls of species of subg. *Hymenostomum* or the strongly bulging exothecial cells of *W. macrospora*, however, should also be evaluated and taken into account.

A monotypic genus now in the Viridivelleraceae, *Viridivellus* Stone, may actually represent reduction of the *Weissia* gametophyte to an extreme, leaving only the stem and

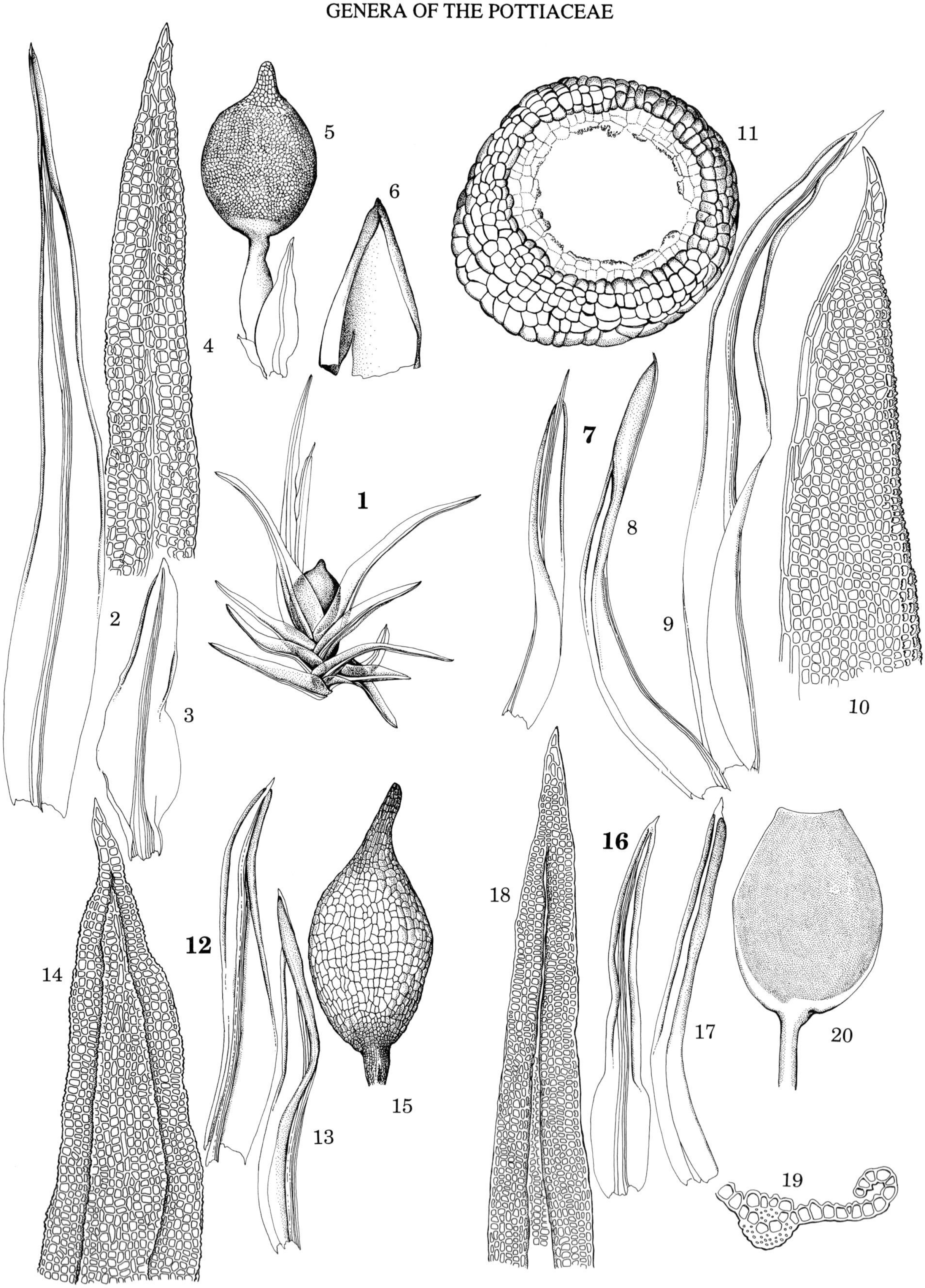

Plate 61. ***Weissia.*** **1–6.** ***W. muhlenbergiana.*** 1. Habit. 2–3. Two leaves. 4. Leaf apex. 5. Sporophyte. 6. Calyptra. **7–11.** ***W. rutilans.*** 7–9. Three leaves. 10. Leaf apex. 11. Peristome, rudimentary, vertical view. **12–15.** ***W. semidiaphana.*** 12–13. Two leaves. 14. Leaf apex. 15. Capsule. **16–20.** ***W. striata.*** 16–17. Two leaves. 18. Leaf apex. 19. Transverse section at midleaf. 20. Capsule.

perichaetial leaves on a persistent protonema; Stone (1980b) indicated that this genus may be better placed in the Pottiaceae. It should be examined during future revisions of the genus *Weissia*.

Brotherus (1924–25) placed *Phasconica* between *Tetrapterum* and *Hymenostomum*, among genera he felt intermediate between *Astomum* and *Weissia*. Hilpert (1933), who saw material of the generitype, *Phasconia lorentzii*, noted the closely inrolled upper laminal margins and suggested a relationship with *Astomum*. Crundwell and Nyholm (1972a, 1974) included *Phasconica* in *Weissia* subg. *Astomum*. Stone (1980b) carefully described and illustrated what is surely *Phasconica balansae* from Australia, showing the strongly incurved upper laminal margins characteristic of *Weissia*; however, she indicated a reluctance to follow Crundwell and Nyholm (1972a, 1974) in assigning the genus to *Weissia* subg. *Astomum* because of its macrostomous capsule with dehiscent flattened operculum, stout columella, and ventral surface of laminal cells distinctly mamillose and strongly papillose, among other characters. Magill (1981) suggested that because of similarities of gametophyte and dissimilarities of sporophyte (ex descr.) this genus "is reminiscent of the relationship between *Weissia* and the segregate genera *Astomum* Hampe and *Hymenostomum* R. Br." Saito (1975, p. 417) reduced *Phasconica* to synonymy with *Weissia*, with no further explanation. Material of *Phasconica* C. Müll. *s. str.*, seen during this study was found mixed in the type of *Trachycarpidium verrucosum* (*P. tisserantii*, PRE!, is referred to *Trichostomum* as a synonym of *T. unguiculatum* see Crundwell & Nyholm 1974), and the genus *Phasconica* is here given a new status as a subgenus of *Weissia* differing from other subgenera mainly by the macrostomous, eperistomate capsule with a short seta and a well differentiated, dehiscent operculum. Because of the unusual sporophyte morphology, subg. *Phasconia* may not be part of a reduction series in *Weissia*, but may instead be a remnant of a more primitive and now largely extinct lineage; further analysis should be done at the species level. *Weissia platystegia* (illustrated and discussed as *Astomum platystegium* by Eddy (1990) and Norris and Koponen (1989) belongs in subg. *Phasconica*.

The singular *Pseudosymblepharis socotrana*, including the synonym *Barbula* (sect. *Edentella*) *schweinfurthiana* (isotype at BM) is transferred to *Weissia* (as *W. artocosana*, a *nom. nov.*) on account of the weakly papillose, ventrally bulging upper laminal cells, ventral stereid band smaller than the dorsal (Pl. 59, f. 17), and the lack of a peristome, all characters found in *Weissia*. The strongly sheathing leaf bases are found in both genera, and are variably demonstrated among species of *Pseudosymblepharis*. This species is gametophytically rather

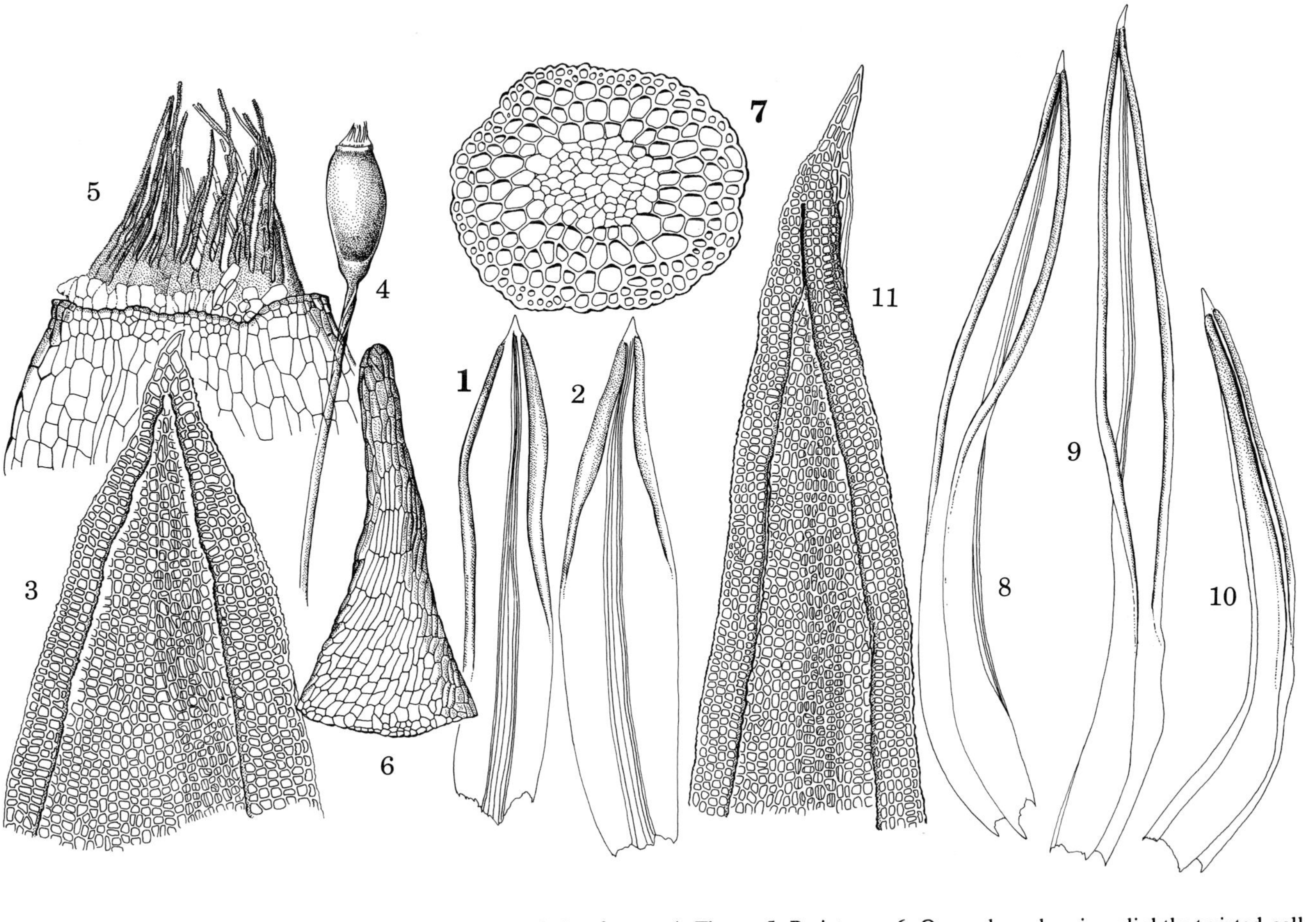

Plate 62. *Weissia*. 1–6. *W. triumphans*. 1–2. Two leaves. 3. Leaf apex. 4. Theca. 5. Peristome. 6. Operculum showing slightly twisted cells. **7–11. *W. veviridis*.** 7. Transverse section of stem. 8–10. Three leaves. 11. Leaf apex.

similar to *Rhamphidium* but differs in the entire leaves and lack of a hydroid strand, as well as absence of a peristome.

Hybrids have not only been reported between species of different subgenera of *Weissia*, but also between *Weissia crispa* and *Tortella flavovirens* (Nicholson 1910). The sporophytes in this last case apparently had short setae; the thecae were ovoid, partially cleistocarpous, rudimentarily peristomate, and covered by very large calyptrae (characteristic of *Tortella flavovirens*). The spores were also reported as remaining adherent in tetrads.

Additional literature: Anderson and Lemmon (1972, 1973, 1974), Andrews (1920, 1922a, 1924, 1933), Bryan (1956), Crundwell (1971b), Dietert (1979), Eckel (1986b), Flowers (1973b), Grout (1900), Khanna (1960, 1964), Lemmon (1968), Lewinski (1983), Mizushima (1957), Nicholson (1905a, 1906), Reese (1988, 1991), Reese and Lemmon (1965), Robinson (1966), Shaw (1987a), Stoneburner (1981, 1986), Stoneburner and Wyatt (1985), Williams (1966b), Zander (1985a), Zuttere et al. (1984).

Number of accepted species: 97, plus 11 in *Astomum* and 16 in *Hymenostomum*.

Species examined: *W. abbreviata* (NY), *W. andersoniana* (BUF, TENN), *W. argentinica* (NY), *W. artocosana* (BM, NY), *W. balansae* (PC), *W. balansaeana* (MO), *W. bizotii* (PC), *W. brachycarpa* (BUF, NY), *W. breutelii* (BUF, NY), *W. canaliculata* (NY), *W. condensa* (BUF, TENN, NY), *W. controversa*, *W. crispa* (BUF, DUKE), *W. diffidentia* (US), *W. edentula* (MICH, NY), *W. fallax* (NY), *W. felipponei* (PC), *W. ghatensis* (BM), *W. glazioui* (NY), *W. jamesonii* (TENN), *W. jamaicensis* (BUF, NY, TENN), *W. latiuscula* (BM), *W. ligulaefolia* (BUF, FH, NY), *W. lineaefolia* (NY), *W. longidens* (DUKE, NY), *W. longifolia* (NY), *W. ludoviciana* (DUKE), *W. macrospora* (NY), *W. muhlenbergianum* (DUKE), *W. newcomeri* (FH), *W. occidentalis* (COLO), *W. opaca* (NY), *W. ovalis* (NY), *W. platystegia* (BM), *W. riograndensis* (H), *W. rutilans* (BUF, NY), *W. semidiaphana* (BUF, FH), *W. sharpii* (BUF, DUKE), *W. subacaulis* (NY), *W. submicacea* (NY), *W. termitidarum* (NY), *W. triumphans* (NY), *W. unguiculata* (NY), *W. veviridis* (H), *W. wimmeriana* (NY).

New heterotypic synonymy: *Barbula* (sect. *Edentella*) *schweinfurthiana* C. Müll. = *Weissia artocosana* (Mitt.) Zand. *Trichostomum bermudanum* (Mitt.) Par. = *Weissia jamaicensis* (Mitt.) Grout. *Trichostomum ekmanii* Thér. = *Weissia jamaicensis* (Mitt.) Grout. *Trichostomum linealifolium* C. Müll. = *Weissia jamaicensis* (Mitt.) Grout.

New names, statuses and combinations:

Weissia subg. *Phasconica* (C. Müll.) Zand., *comb. et stat. nov.* (*Phasconica* C. Müll., Linnaea 43: 438, 1882).

Weissia abbreviata (Thwait. & Mitt.) Zand., *comb. nov.* (*Systegium abbreviatum* Thwait. & Mitt., J. Linn. Soc. Bot. 13: 299, 1873; *Astomum abbreviatum* (Thwait. & Mitt.) Fleisch.).

Weissia artocosana (Mitt.) Zand., *nom. nov.* (*Symblepharis socotrana* Mitt., Trans. R. Soc. Edinburgh 31: 331, 1888; *Pseudosymblepharis socotrana* (Mitt.) Thér.).

Weissia balansae (C. Müll.) Zand., *comb. nov.* (*Phasconica balansae* C. Müll., Linnaea 43: 438, 1882).

Weissia (subg. *Phasconica*) *bizotii* Zand., *nom. nov.* (*Kleioweisiopsis involuta* Biz., Rev. Bryol. Lichénol. 40: 119, 1974 invalid, single element not cited, *fide* Crosby et al., 1992). Lectotype: Herb. Bizot; paratype: Herb. Pocs.

Weissia diffidentia Zand., *nom. nov.* (*Phascum recurvirostum* C. Müll., Flora 71: 5, 1888; *Tetrapterum recurvirostrum* (C. Müll.) Broth.).

Weissia glazioui Zand., *nom. nov.* (*Hymenostomum striatum* Geh. & Hampe, Vid. Medd. Naturh. For. Kjoebenh. ser. 4,1: 84, 1879).

Weissia (subg. *Phasconica*) *lorentzii* (C. Müll.) Zand., *comb. nov.* (*Phasconica lorentzii* C. Müll., Linnaea 43: 438, 1882).

Weissia neocaledonica (Thér.) Zand., *comb. nov.* (*Aschisma neocaledonicum* Thér., Diagn. Esp. Var. Nouv. Mouss. 8: 4, 1910; *Astomum neocaledonicum* (Thér.) Andrews).

Weissia riograndensis (Broth.) Zand., *comb. nov.* (*Hymenostomum riograndense* Broth., Bih. K. Svensk. Vet. Ak. Handl. 26 Afd. 3(7): 19, 1900).

Weissia triumphans var. *monspeliensis* (Schimp.) Zand., *comb. nov.* (*Trichostomum monspeliense* Schimp., Syn. ed. 2, 175, 1876; *Trichostomum triumphans* var. *monspeliense* (Schimp.) Husn.).

Weissia veviridis Zand., *nom. nov.* (*Trichostomum perviride* Broth., Dansk. Bot. Ark. 2(9): 3, 1918).

45. WEISSIODICRANUM Plate 63.

Weissiodicranum Reese in Reese & Buck, Bryologist 94: 308, 1991. Type: *Weissiodicranum insularum* Reese in Reese & Buck.

From *Weissia*, a genus + o + *Dicranum*, a genus (Dicranaceae); this name refers to genera of two different families, characters of which are shared by *Weissiodicranum*.

Plants forming low cushions or turfs, green above, tan below. Stems branching occasionally, ca.0.5(–1.0) cm in length, transverse section rounded-pentagonal, *central strand distinct and rather large*, sclerodermis absent, hyalodermis absent; *axillary hairs* 4–10 cells in length, hyaline or weakly thick-walled basally; rhizoids sparse. *Leaves incurved, tubulose, contorted when dry*, spreading when moist, longly ligulate to lanceolate, ca. 2.5 mm in length, upper lamina broadly channeled across leaf, *margins sharply incurved to involute*, entire, apex narrowly acute; base scarcely differentiated or rectangular; costa shortly and sharply mucronate, superficial cells quadrate, elongate and smooth dorsally, with 4 rows of cells across costa ventrally at midleaf, *costal transverse section nearly round to elliptical or ovate*, 2 stereid bands present, *differentiated epidermis present both ventrally and dorsally*, guide cells (2–)4 in 1 layer, *hydroid strand absent*; *upper laminal cells* subquadrate, ca. 8–10 µm in width, 1:1, walls evenly thickened, *superficially strongly bulging ventrally, nearly flat dorsally*; *low molariform papillae present on upper part of costa ventrally*; basal cells differentiated across leaf, rectangular, 10–13 µm in width, 2–4:1, walls thin to evenly thickened, *several "alar" cells at insertion inflated in 1–3 rows, extending from margin about 2/3 way to costa, decurrent as a pad on stem.* Probably dioicous, only archegonia known. Perichaetia terminal, inner leaves little different. Perigonia and sporophytes unknown. *Laminal color reaction to KOH yellow.*

A monotypic genus found on soil at low to medium elevations in the West Indies (Puerto Rico and Jamaica) and the Galapagos Islands of Ecuador.

As Reese and Buck (1991) indicated, this genus is distinctive in its inflated alar cells (Pl. 63, f. 8), similar to those of the genus *Dicranum* Hedw. (Dicranaceae), differing as well from a very similar species, *Weissia breutelii*, by the absence of a dorsal costal epidermis and by the bulging dorsal surface of the costa (Pl. 63, f. 9). The alar cells are enlarged and banana-shaped, grading below into a unique pad of inflated cells that are 1–2 cells in thickness on the stem. This is apparently an extreme development of a tendency already present in *Weissia* (a synapomorphy not used in the Data Set), in which leaf insertions consist of one or two rows of swollen cells that occasionally partially detach from the stem laterally when the leaf is removed for examination and giving the appearance of differentiated alar cells. Examined species of *Weissia* with small groups of distinctly differentiated but inconspicuous alar cells include: *W. abbreviata*, *W. balsansae*, *W. crispum*, *W. rutilans*, and *W. veviridis*; there are presumably others. It is almost certain, because of the complex and characteristic costal anatomy, and because somewhat differentiated alar cells are found in many species of *Weissia*, that *Weissiodicranum* is derived from ancestors of pottiaceous stock, not of *Dicranaceae*.

Number of accepted species: 1.

Species examined: *Weissiodicranum insulanum* (BUF).

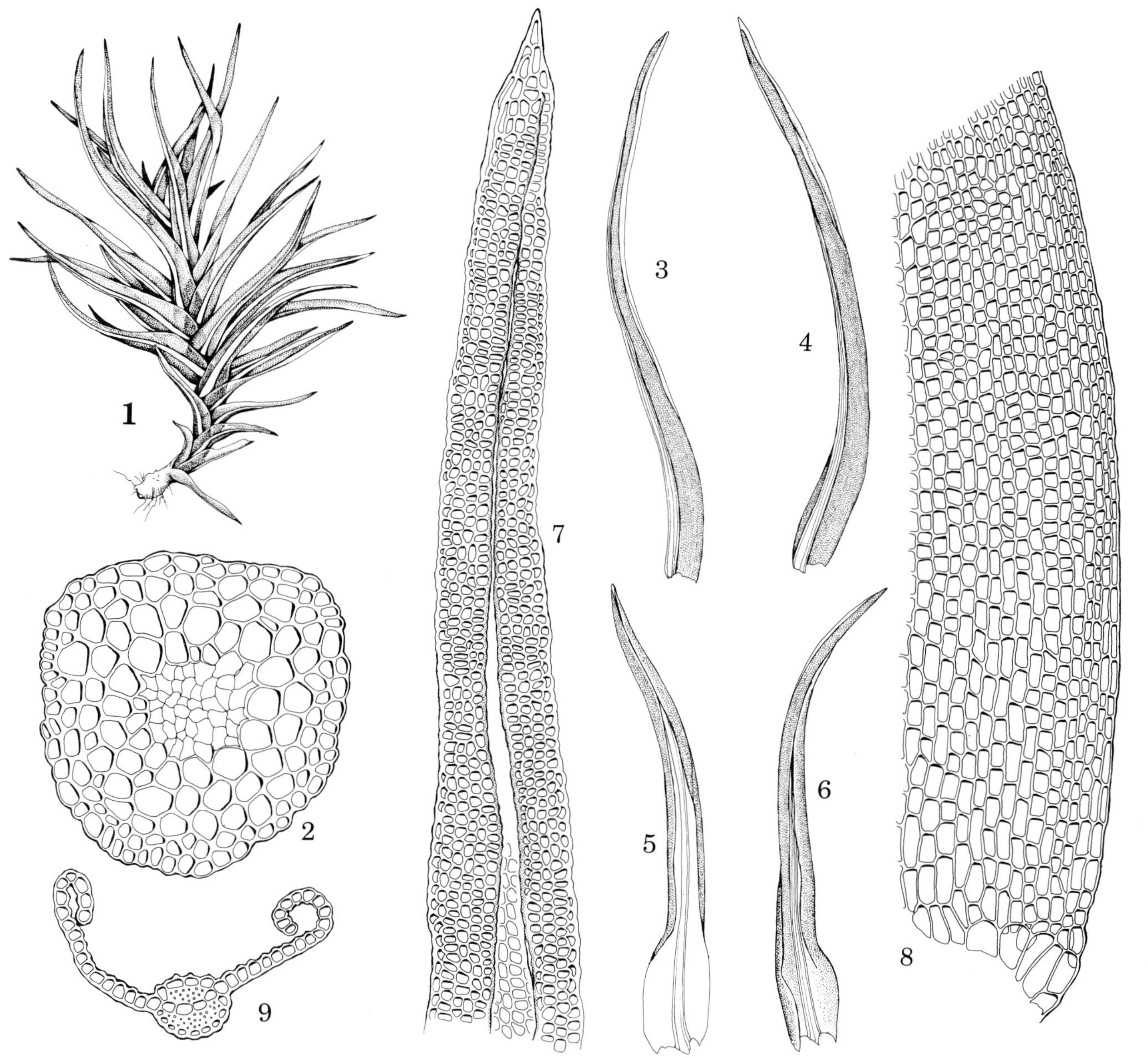

Plate 63. *Weissiodicranum*. 1–9. *W. insulanum*. 1. Habit. 2. Transverse section of stem. 3–6. Four leaves. 7. Leaf apex. 8. Basal cells. 9. Transverse section at midleaf.

46. QUAESTICULA Plate 64.

Quaesticula Zand., *gen. nov.*

Type: Cuba, "moist rocky ledges," Wright, Cuban Mosses 29 (as "*Barbula linearis* Swartz"), NY, lectotype of *Weissia navicularis* Mitt.

Plantae interdum in regione superna glaucae. Caules filum centralem validum evolventes, sclerodermide carentes, hyalodermide praediti. Folia in statu sicco incurva tubulosaque, ligulatolanceolata, marginibus incurvis vel late involutis, integris, apice rotundato, base vix in forma distincto, costa subpercurrenti, in regione mediana cellulis superficialibus ventraliter valde tumescentibus, stratis stereidarum plerumque duobus, cellulis laminalibus supernis rotundatoquadratis vel hexagonis, parietibus crassitie aequis evolventibus, in superficie tumescentibus, papillis magnis, plerumque luminem obtegentibus, simplicibus vel bifidis, plerumque per luminem unicis praedita.

The name refers to the long and careful evaluation required for placing the new genus in proper taxonomic perspective, and is patterned after the hepatic name *Pinguicula*.

Plants forming a low, dense turf, dark to light green, *occasionally glaucous* above, tan below. *Stems* branching occasionally, to 0.6 mm in length, transverse section rounded-pentagonal, *central strand strong, sclerodermis absent, hyalodermis absent;* axillary hairs conspicuous, elongate, of 6–8 hyaline cells; sparsely radiculose. *Leaves* contorted, *incurved and tubulose when dry,* spreading when moist, *ligulate-lanceolate,* ca. 1.5–2.0 mm in length, upper lamina broadly channeled across leaf, *margins incurved to broadly involute,* entire (crenulate by projecting cell walls); *apex rounded,* sometimes weakly cucullate; *base scarcely differentiated; costa* strong, often up to 1/3 leaf width at base, *subpercurrent,* ending 2–4 cells below apex, costa with lamina inserted nearly laterally, *superficial cells* rounded-quadrate, *strongly bulging ventrally,* dorsally elongate and smooth, 3–6 rows of cells across costa ventrally at midleaf, *costal transverse section circular or ovate,* stereid bands ventrally distinct or occasionally absent, dorsally strong and semicircular in shape, ventral and dorsal epidermises well differentiated, guide cells (2–)4 in 1 layer, hydroid strand absent; *upper laminal cells somewhat bulliform, rounded-quadrate or hexagonal,* 10–14 µm in width, 1:1, *walls evenly thickened, superficially strongly bulging ventrally, weakly convex dorsally to almost as strongly convex as ventrally, papillae massive,* usually covering the lumens, *simple to occasionally bifid, usually one over each lumen;* basal cells differentiated across lower 1/5 of leaf, rectangular, 10–14 µm, the marginal cells narrower in width, 3–4:1, walls thin. Dioicous. Perichaetia terminal, inner leaves long-ligulate, to 3 mm in length, sheathing in lower 1/4, lower cells long-rectangular. Perigonia terminal, budlike. Seta 0.9–1.2 cm in length, 1 per perichaetium, brown, twisted clockwise; theca 0.7–1.3 mm in length, brown, ellipsoidal, exothecial cells rectangular, 20–25 µm in width, 3–4:1, thin-walled, stomates phaneropore, at base of capsule, annulus of 3 rows of strongly vesiculose, persistent cells; *peristome teeth 32, longly filamentous,* to 700 µm in length, spiculose, *twisted almost once counterclockwise,* basal membrane 10 to 35 µm, papillose. Operculum conic, 0.7–1.0 mm in length, cells twisted counterclockwise nearly once. Calyptra long, cucullate, smooth, 2.5–3.0 mm in length. Spores 10–13 µm in diameter, light brown, essentially smooth. *Laminal KOH color reaction yellow.*

Found on rocks (and once on a palm tree trunk) near mountain streams and waterfalls in Mexico and the West Indies.

The incurved upper laminal margins and the upper laminal cells (lumens examined in transverse section) bulging more strongly ventrally than dorsally (Pl. 64, f. 8–9) are also characteristic traits of *Weissia*, *Plaubelia* and *Hyophila*, but, in addition to the twisted peristome (Pl. 64, f. 10), the narrowly rounded leaf apices distinguish *Q. navicularis* from the first, the lack of a costal hydroid strand from the second, and the leaves are far too narrow for inclusion of this species in the third. The significant gametophytic morphology of *Quaesticula* is, in fact, nearly that of *Trichostomum crispulum*, especially the ligulate leaves, apex often cucullate, margins broadly incurved, costa subpercurrent and protuberant dorsally, but the upper laminal cells differ in being evenly thick-walled and otherwise quasibulliform, somewhat more strongly convex dorsally than ventrally, and papillae massive, apparently one broad lens-like papilla over each lumen in the type and most known collections, but commonly two over each lumen in one collection (Mexico: Eckel 8511141 BUF); thus approaching the papilla morphology of *Tuerckheimia*. The distinctly twisted peristome is unlike that known for *Tuerckheimia* and *Trichostomum*. One might hypothesize that, of genera with superficially strongly bulging upper laminal cells and massive papillae, *Tuerckheimia* is derived from ancestors of *Trichostomum* subg. *Oxystegus*, while *Quaesticula* is derived from ancestors of *Trichostomum* subg. *Crispuliformes* (see subgeneric descriptions as given in the treatment of *Trichostomum*). *Luisierella* may have been derived from ancestors of *Quaesticula* through reduction of sporophyte (weakly differentiated peristome) and gametophyte (loss of ventral costal stereid band), and elaboration of sexual condition and the strongly differentiated vee-shaped basal cell region. Some of this speculation is supported by the consensus tree of Cladogram 14; perhaps if *Trichostomum* subg. *Crispuliformes* were treated as a separate genus, it might be referred to the Hyophileae. Analysis at the species level is needed.

Only one specimen of *Weissia navicularis* Mitt. is referred to in the protologue; of the variously distributed isotypes of Wright 29, the lower left specimen, from Mitten's herbarium, on the NY sheet is here designated as lectotype of *Weissia navicularis* Mitt. and has been marked as such.

Literature: Brotherus (1910), Iwatsuki & Sharp (1958), Zander (1978f), Zander & Eckel (1987).

Number of accepted species: 1.

Species examined: *Quaesticula navicularis* (BUF, NY).

New heterotypic synonymy: *Tuerckheimia calculosa* Zand. & Eckel = *Quaesticula navicularis* (Mitt.) Zand.

New combination: *Quaesticula navicularis* (Mitt.) Zand., *comb. nov.* (*Weissia navicularis* Mitt., J. Linn. Soc. Bot. 12: 139, 1869; *Tortula navicularis* (Mitt.) Broth.).

47. GANGULEEA Plate 65.

Ganguleea Zand., Phytologia 65: 426, 1989. Type: *Ganguleea angulosa* (Broth. & Dix.) Zand.

Named for H. C. Gangulee, 1957–, author of "Mosses of Eastern India and Adjacent Regions" (1969–1980).

Plants in loose turf, green above, brown to blackish brown

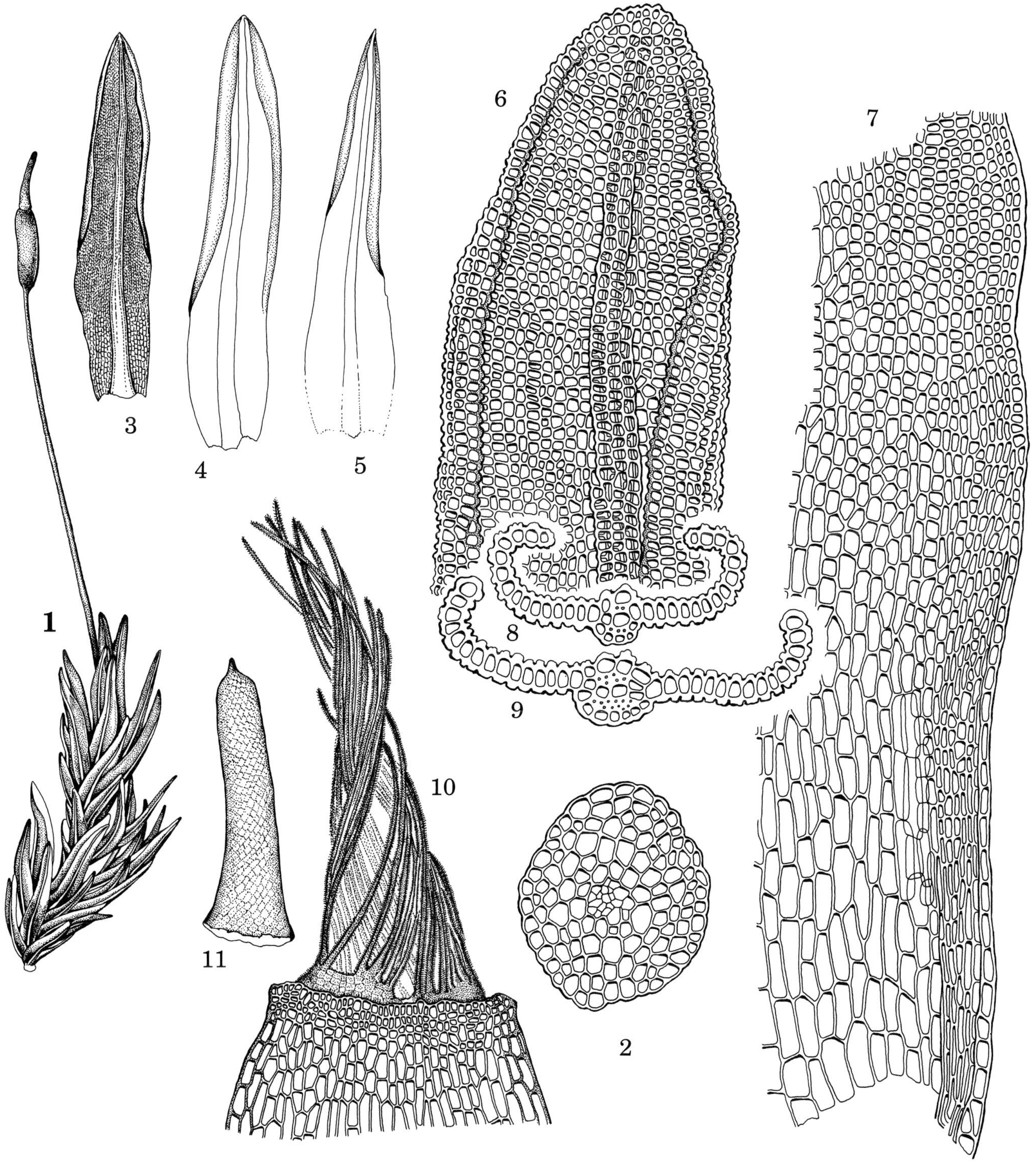

Plate 64. ***Quaesticula.*** **1–11.** ***Q. navicularis.*** 1. Habit. 2. Transverse section of stem. 3–5. Three leaves. 6. Leaf apex. 7. Basal cells. 8–9. Transverse sections of leaves. 10. Peristome teeth. 11. Operculum.

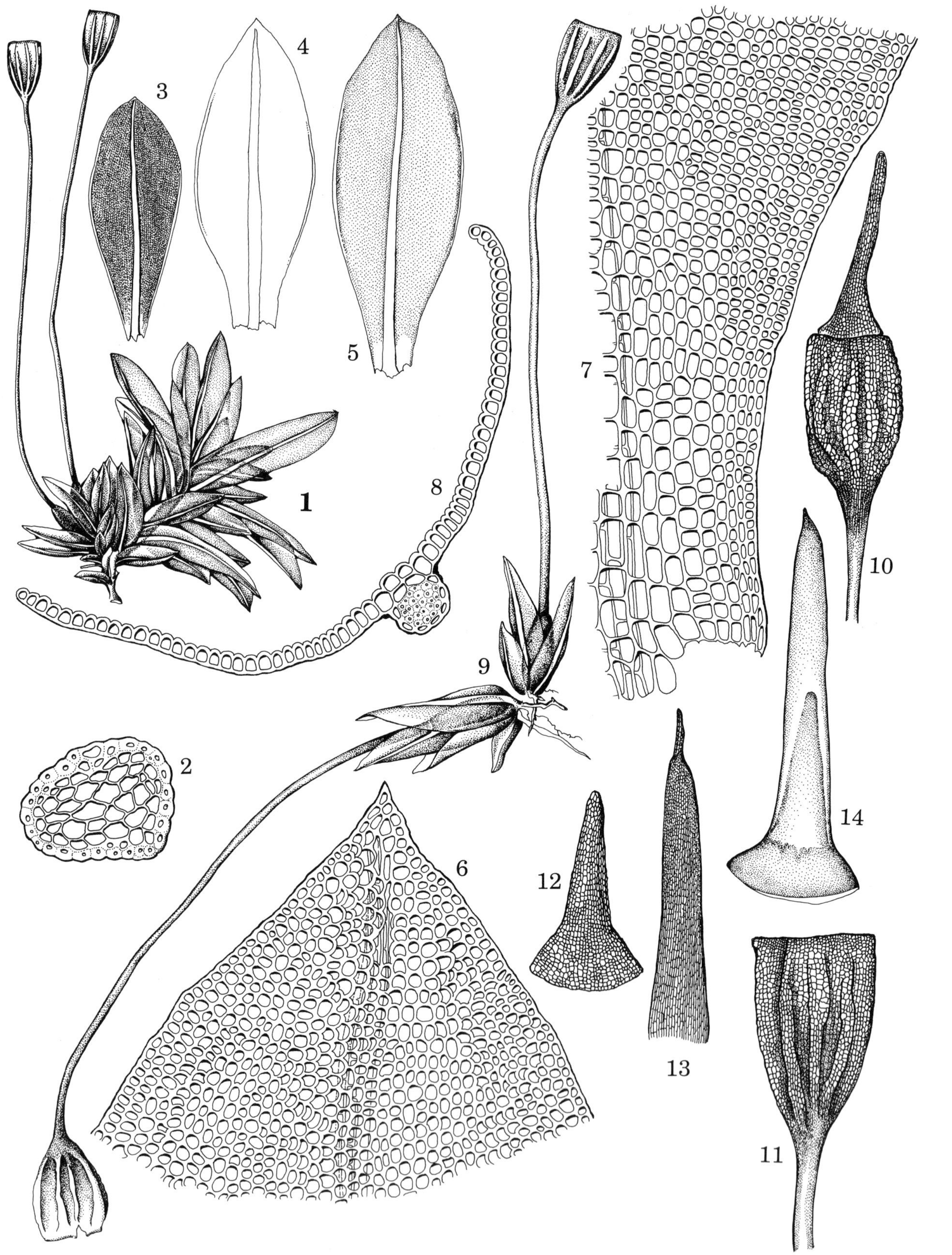

Plate 65. ***Ganguleea.*** **1–14.** ***G. angulosa.*** 1. Habit. 2. Transverse section of stem. 3–5. Three leaves. 6. Leaf apex. 7. Basal cells. 8. Transverse section at midleaf. 9. Cluster of sporophyte-bearing perichaetia. 10. Capsule. 11. Exothecial cells with plicae. 12. Operculum. 13. Calyptra. 14. Calyptra inserted on operculum.

below, *generally rosulate*. *Stems* branching rather often, to 0.3 cm in length, transverse section rounded elliptical to triangular, *central strand absent*, sclerodermis absent or present (variable in same stem), hyalodermis absent; axillary hairs 4–6 cells in length, the basal cell firm-walled; rhizoids sparse. *Leaves incurved and tubulose when dry*, spreading when moist, *spathulate to oblong-elliptical*, 1.5–2.0 mm in length, *upper lamina broadly channeled, margins incurved*, entire; *apex broadly rounded and apiculate; base very narrow, constricted above the insertion;* costa subpercurrent to short-excurrent as a sharp mucro, costa with lamina inserted ventrally but divergent at 180°, superficial cells rounded-quadrate, bulging ventrally, dorsally elongate, 2–4 rows of cells across costa ventrally at midleaf, costal transverse section circular, stereid band single, strong and circular in shape, ventral and dorsal epidermis present, cells flattened, *guide cells 2, flattened-elliptical*, in 1 layer, hydroid strand absent; *upper laminal cells* rounded-quadrate to hexagonal, 8–12 µm in width but rather heterogeneous in size, 1:1, lumens usually rounded, walls usually evenly thickened, *superficially strongly bulging ventrally, nearly flat dorsally; papillae absent;* basal cells differentiated in a small, triangular, juxtacostal area near the insertion, bulging-rectangular, ca. 18 µm in width, 2–3:1, walls thin to thickened and trigonous. Autoicous, paroicous, occasionally probably rhizautoicous. *Pleurocarpous. Perichaetia borne on a short lateral branch, inner perichaetiate leaves* lanceolate, entire to weakly serrulate, *short*, to 0.8 mm in length, not sheathing, lower cells short-rectangular to rhomboidal, walls thin. Perigonia borne in clusters just below perichaetia, occasionally terminal on a separate plant. Seta 2.5–5.5 mm in length, 1 per perichaetium, brown, twisted clockwise; *theca* 0.5–0.8 mm in length, brown, often glistening, ovate, macrostomous, *with 8 plicae*, exothecial cells thin-walled, quadrate to short-rectangular, ca. 18 µm in width, *8 longitudinal rows of thick-walled cells in two ranks each along ridges of plicae*, stomates phaneropore, at base of capsule, annulus of 2 rows of thin-walled, weakly vesiculose cells; *peristome teeth absent*. Operculum long-conic to rostrate, erect, ca. 0.7 mm in length, cells straight. *Calyptra long-conic, sometimes not cleft*, inserted on operculum, smooth to somewhat rough apically with projecting cell walls, ca. 1 mm in length. Spores 10–13 µm in diameter, light yellow, essentially smooth. *Laminal KOH color reaction yellow*.

A rarely collected monotypic genus found on soil over rock, Himalayas of India and Nepal and mountains above Rio de Janeiro, southeastern Brazil (Schäfer-Verwimp 8403, BUF).

Like *Anoectangium*, *Molendoa* and *Pleurochaete*, this genus bears its sporophytes laterally on the main stem axis, at the ends of very short branches bearing highly modified perichaetial leaves (Pl. 65, f. 1, 9). *Ganguleea* differs from these genera, however, in being monoicous. There is considerable resemblance to *Weisiopsis*, in which it was placed by Hilpert (1933), especially in the plicate capsule (Pl. 65, f. 10–11), leaves with narrowed base (Pl. 65, f. 3–5, 7), colliculate ventral surface, margins incurved, and costal section with a usually very strong and rounded stereid band (Pl. 65, f. 8). *Ganguleea* may be derived from ancestors of that genus through the loss of the stem central strand (Pl. 65, f. 2), further narrowing of the leaf base, loss of peristome, and development of pleurocarpy. An isotype at H has the sporophytes fallen with the short perichaetial branchlets (with their much different leaf morphology) attached and giving the appearance of two different species in the packet. The Brazilian specimen mentioned above differs from that from Nepal (Norkett 6118, BM) in having a long-rostrate, slightly inclined operculum with a cleft calyptra, while the Nepal specimen has an erect-conic operculum and long-conic, uncleft calyptra; it may be possible to distinguish the two at the specific level on this basis, but more specimens need to be examined.

Additional literature: Gangulee (1972), Vohra and Wadhwa (1966).

Number of accepted species: 1.

Species examined: *Ganguleea angulosa* (BM, BUF, H).

48. WEISIOPSIS — Plate 66–67.

Weisiopsis Broth., Öfv. Finsk. Vet. Soc. Förh. 62A(9): 7, 1921.
Lectotype: *Weisiopsis anomala* (Broth. & Par.) Broth. *fide* Saito, J. Hattori Bot. Lab. 39: 525, 1975.

Hyophila subg. *Hyophilodonta* Card. in Grand., Hist. Madag. 39: 214, 1915. Type: *Hyophila subplicata* Ren. & Card.

From *Weis[s]ia*, a genus + ὄψις, -εως, appearance; resembling the genus *Weissia*.

Plants *small*, in a low turf, light green above, brown below. Stems branching occasionally, to 3 mm in length, transverse section rounded-pentagonal, *central strand present*, sclerodermis absent or of substereid or occasionally stereid cells, hyalodermis absent; axillary hairs of ca. 3–5 cells, the basal 1 thick-walled; radiculose below. *Leaves usually tubulose and erect-incurved when dry, tubulose and spreading when moist, long-ligulate to spathulate*, to 2.2 mm in length, upper lamina flat or narrowly grooved along costa, *margins plane or broadly and weakly incurved, entire or minutely crenulate by projecting cell walls*, occasionally serrulate at extreme apex; *apex rounded to broadly acute, often apiculate*; base oblong or elliptical; costa tapering, ending 3–6 cells below apex or percurrent or short-excurrent into an apiculus, superficial cells elongate and smooth or rounded-quadrate and bulging ventrally, elongate and smooth dorsally, 2(–3) rows of cells across costa ventrally at midleaf, *costal transverse section nearly round, stereid band one, usually round in section*, ventral epidermis often absent, dorsal epidermis present, guide cells 2(–4) in 1 layer, *often flattened and lenticular in section, hydroid strand absent*; *upper laminal cells rounded-hexagonal to quadrate*, 6–13 µm in width, 1:1, walls evenly thickened to somewhat trigonous, *superficially ventrally strongly bulging, weakly convex dorsally*; *papillae absent*; basal cells differentiated across leaf or larger medially up to 1/3 leaf length, *often abruptly enlarged* and sharply differentiated from the upper laminal cells, rectangular, often bulging, 20–25 µm in width, 3–5:1, walls thin and hyaline or yellowish brown and thick-walled, marginal cells narrower. *Monoicous* (autoicous or occasionally heteroicous or rhizautoicous). *Perichaetia leaves little differentiated* or smaller, terminal. Perigonia lateral and budlike, occasionally terminal on equal-sized, entirely perigoniate plants. Seta 4–6 mm in length, 1 per perichaetium, yellowish brown, twisted clockwise; *theca* 0.6–1.0 mm in length, yellowish brown, ovoid to cylindrical, occasionally plicate with 8 low ridges (best seen in transverse section) with 2 rows of somewhat thick-walled exothecial cells on the crests, other exothecial cells rectangular to rhomboid, 25–35 µm in width, 2–4:1, thin-walled, rarely superficially bulging, stomates phaneropore at base of capsule, annulus of 2–3 rows of strongly

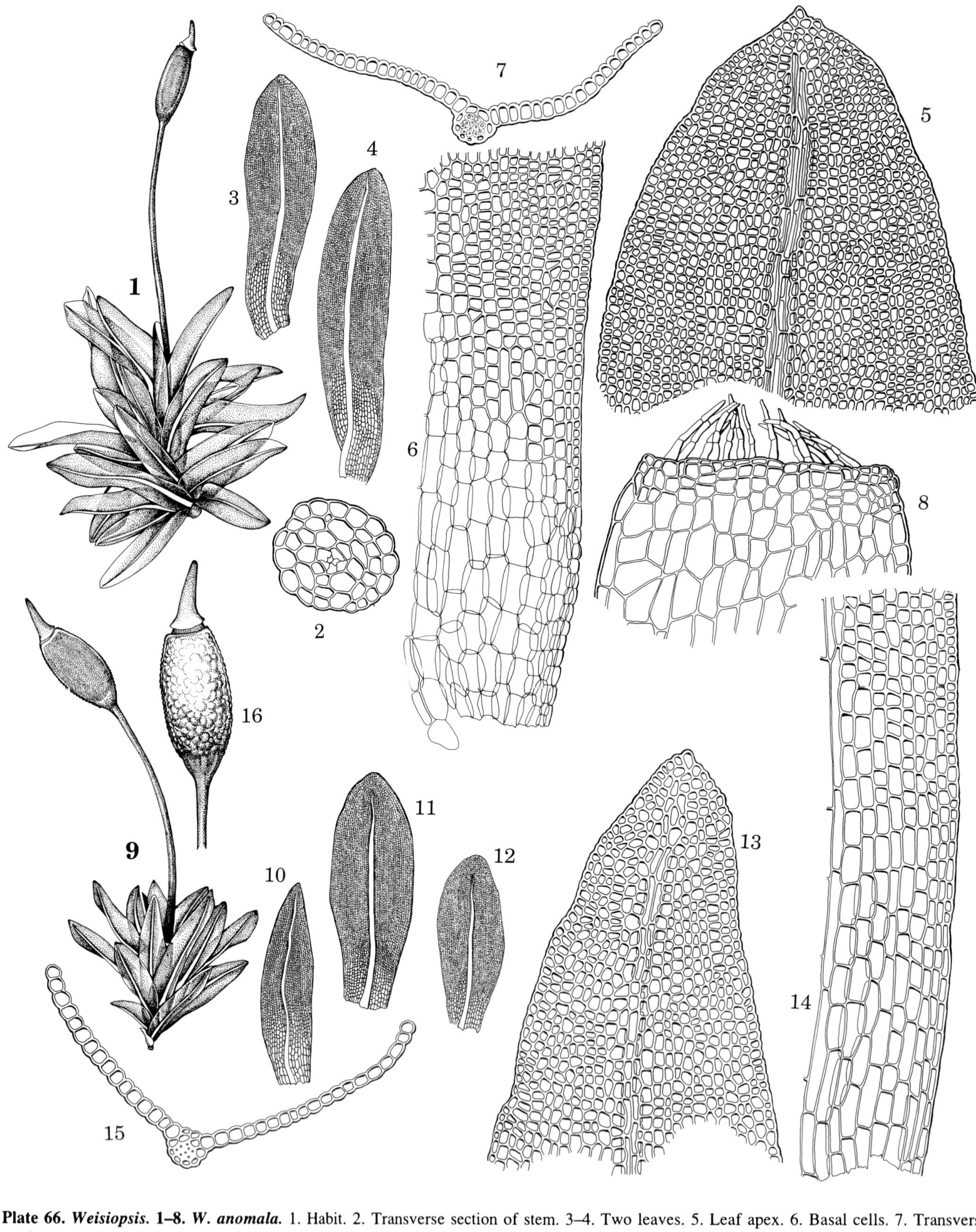

Plate 66. ***Weisiopsis.*** **1–8.** ***W. anomala.*** 1. Habit. 2. Transverse section of stem. 3–4. Two leaves. 5. Leaf apex. 6. Basal cells. 7. Transverse section at midleaf. 8. Peristome. **9–16.** ***W. nigeriana.*** 9. Habit. 10–12. Three leaves. 13. Leaf apex. 14. Basal cells. 15. Transverse section at midleaf. 16. Capsule.

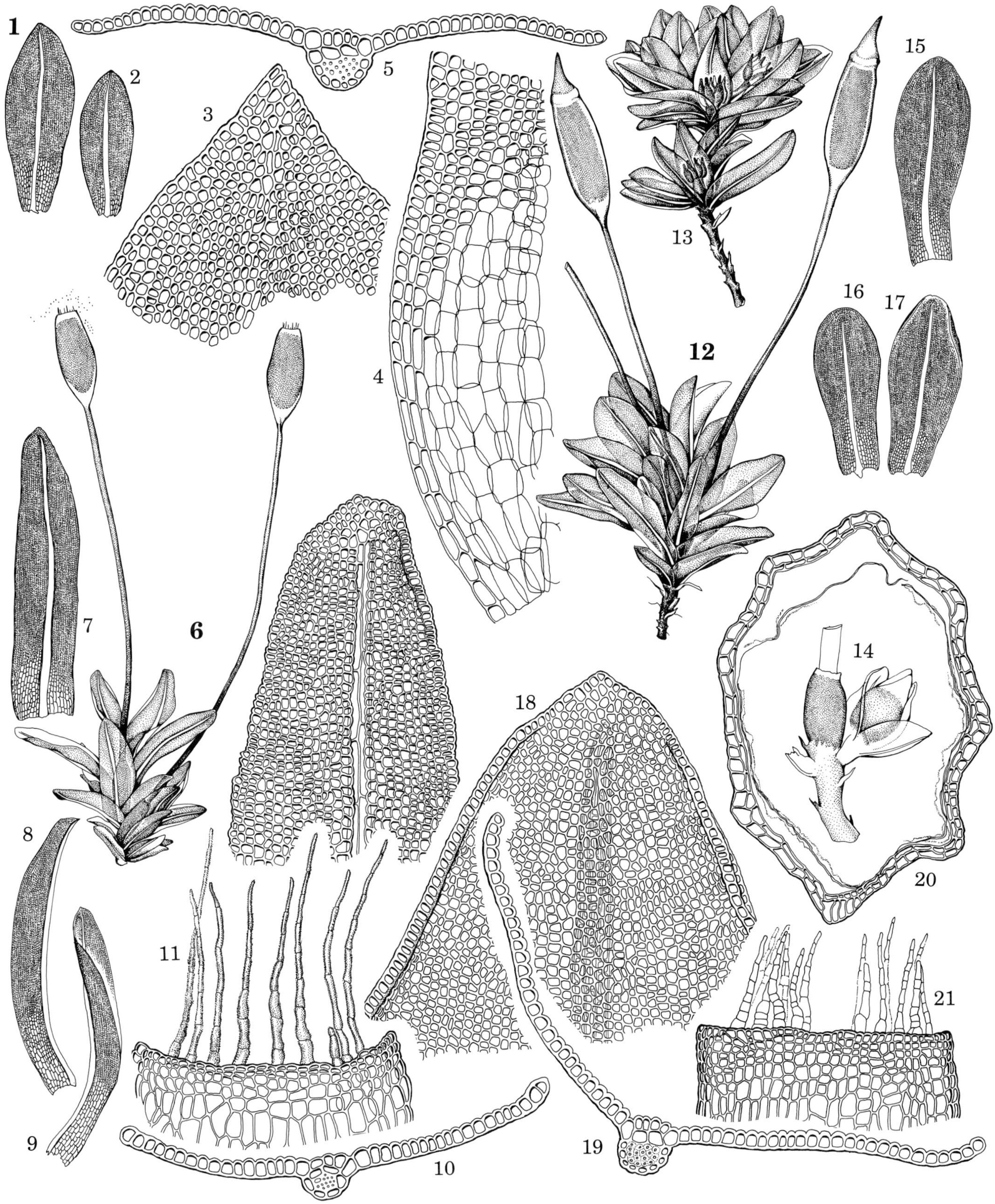

Plate 67. ***Weisiopsis.*** **1–11.** ***W. norrisii.*** 1–2. Two leaves. 3. Leaf apex. 4. Basal cells. 5. Transverse section at midleaf. **6–11.** ***W. oblonga.*** 6. Habit. 7–9. Three leaves. 10. Transverse section at midleaf. 11. Peristome. **12–21.** ***W. plicata.*** 12. Habit of sporophyte-bearing plant. 13. Entirely perigoniate plant. 14. Autoicous bud. 15–17. Three leaves. 18. Leaf apex. 19. Transverse section at midleaf. 20. Transverse section of theca. 21. Peristome.

vesiculose cells, occasionally smaller and not vesiculose; *peristome teeth* absent or more usually *16*, *inserted below the mouth of the theca*, *not contiguous at base*, linear-subulate, entire or occasionally branching above or perforate at base, finely spiculose to smooth, to 200 µm, with 5–7 articulations, straight, basal membrane absent. Operculum long-conic to rostrate from a low-conic base, 0.3–0.7 mm in length, cells straight. Calyptra cucullate, 1.0–1.5 mm in length. Spores 8–13 µm in diameter, yellow, essentially smooth. Laminal KOH color reaction yellow.

A small genus of southern and eastern Asia, Mexico, Central America, Brazil, central and southern Africa, and Madagascar; growing on rocks and soil.

Important characters for *Weisiopsis* are the broad, tubulose leaves with plane or incurved margins; colliculate ventral surface of the lamina (only weakly developed in *W. nigeriana*); guide cells often lenticular in section (Pl. 67, f. 5, 10); single, cylindrical stereid band (Pl. 66, f. 7); occasionally plicate thecae (Pl. 67, f. 20), theca occasionally with bulging exothecial cells (Pl. 66, f. 16—*W. nigeriana*); basal laminal cells abruptly enlarged and rather thick-walled above (Pl. 66, f. 6, 14; 67, f. 4); somewhat distant peristome teeth (Pl. 67, f. 11, 21) when present; and essentially monoicous sexuality. *Ganguleea* is its closest relative. *Weisiopsis* has much the same costal section and laminal basal cells as species of *Gyroweisia* but differs in the ventrally bulging upper laminal cells and the spaced peristome teeth. *Weisiopsis* is possibly composed of species from different lines of reduction, but ultimate relationships cannot be distinguished without rigorous analysis at the species level.

Weisiopsis plicata (type NY!) has thecae with plicae almost as strongly developed as those of *Ganguleea* ((*q.v.*), and it may

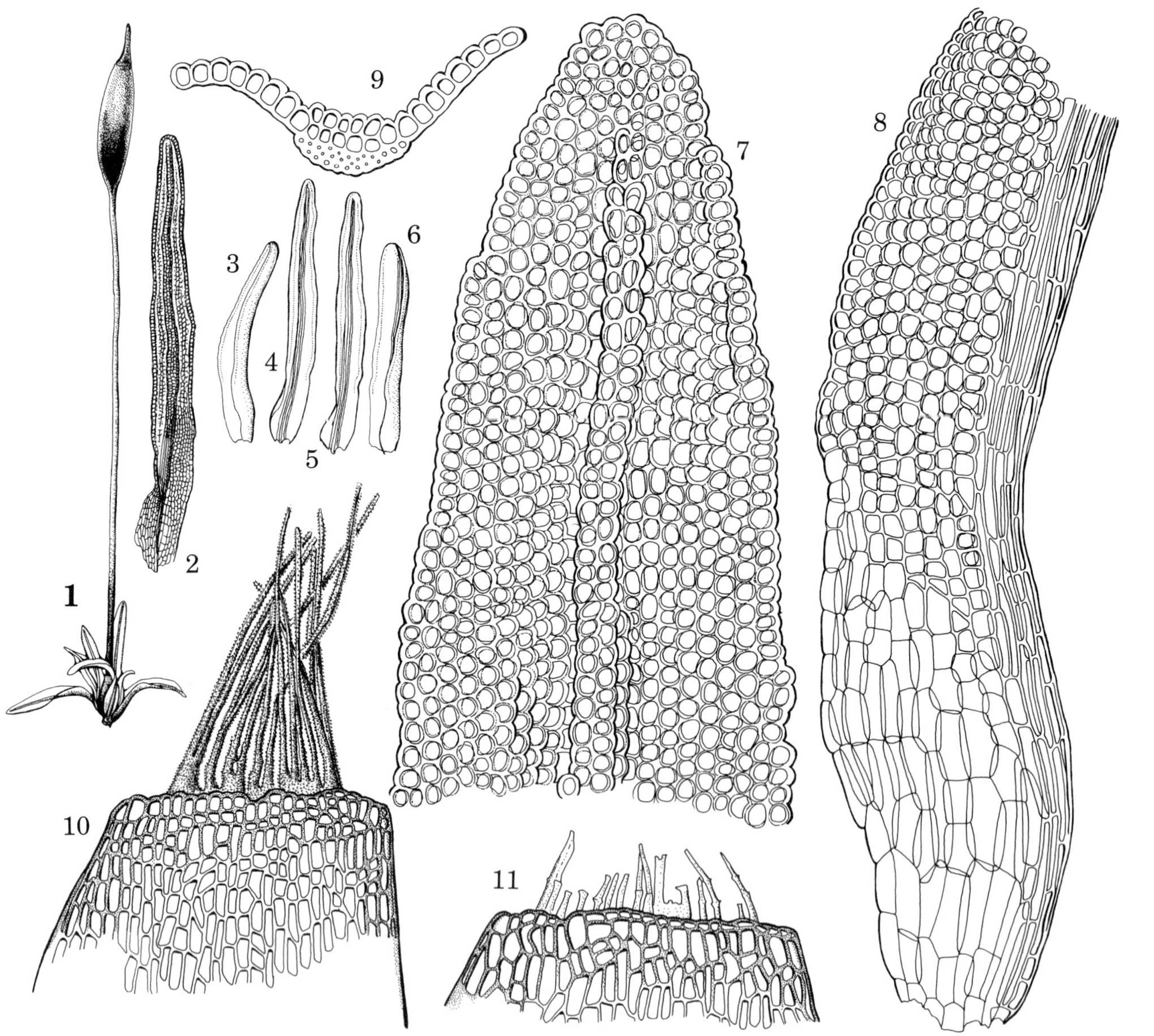

Plate 68. *Luisierella*. 1–11. *L. barbula*. 1. Habit. 2–6. Five leaves, one enlarged. 7. Leaf apex, ventral view. 8. Basal cells. 9. Transverse section at midleaf. 10–11. Two peristomes, one comparatively rudimentary.

be a transitional species between *Ganguleea* and *Weisiopsis*. In addition, *W. plicata* has entirely archegoniate plants, autoicous plants (perigonia as a bud just below the perichaetium), and entirely perigoniate plants (perigonia terminal) of the same size as the sporophyte-bearing gametophytes. Much patience is required to adequately determine the sexual condition of certain species of Pottiaceae.

A specimen at NY labelled "*Weisiopsis bahiensis*...Burchell, Catalogus Geographicus Plantarum Brasiliae Tropicae, No. 7737–2" is not the type of that species; the specimen is taxonomically *Trichostomum brachydontium* Bruch.

Additional literature: Hilpert (1933), Noguchi (1951).

Number of accepted species: 7.

Species examined: *W. anomala* (DUKE, NY), *W. nigeriana* (NY, UIH), *W. norissii* (BUF), *W. oblonga* (BUF, PC, TENN), *W. plicata* (NY).

New heterotypic synonymy: *Gyroweisia pocsii* Biz. = *Weisiopsis nigeriana* Egun. & Olar.

New combinations: *Weisiopsis cucullatifolia* (Gao, Jia & Cao) Zand., *comb. nov.* (*Hyophila cucullatifolia* Gao, Jia & Cao, Bull. Bot. Res. (Harbin) 11(2): 29. 1991), ex descr. et icon. *Weisiopsis nigeriana* (Egun. & Olar.) Zand., *comb. nov.* (*Gyroweisia nigeriana* Egun. & Olar., Bryologist 81: 443, 1978). *Weisiopsis norrisii* (Zand.) Zand., *comb. nov.* (*Scopelophila norrisii* Zand., Bryologist 88: 353, 1985 [1986]).

49. LUISIERELLA — Plate 68.

Luisierella Thér. & P. Varde, Bull. Soc. Bot. France 83: 73, 1936.
Type: *Luisierella pusilla* Thér. & P. Varde.

Named for Rév. Père Alphonse Luisier, 1872–1957, a French muscologist + *-ella*, diminutive.

Plants dull or blackish green, scattered or gregarious, often growing in a thin crust associated with cyanobacteria. Stems occasionally branching, to 2 mm in length, in transverse section rounded-pentagonal, central strand absent, outer cortex not differentiated, hyalodermis apparently absent; axillary hairs ca. 5 cells in length, basal cell mostly yellowish. *Leaves tubulose*, contorted when dry, sometimes rosulate, spreading-recurved from the base when moist, *ligulate-lanceolate to long-elliptical*, ca. 1.0–2.0 mm in length; *margins plane to erect-incurved, crenulate above by projecting cells*; apex rounded to obtusely acute; base elliptical; *costa* thin above, *rather wide and much flattened below, ending ca. 4–5 cells below the apex, ventral superficial cells similar to the laminal cells* and 4–6 cells across at midleaf, the dorsal elongate, costal transverse section showing a single stereid band, this crescent-shaped, 2–4 guide cells in one layer, epidermis present only ventrally, hydroids lacking; *upper laminal cells with a persistent green color, irregularly rounded-hexagonal*, 8–12 µm in width, 1:1(–2), often somewhat transversely elongate at the leaf margins, *superficially ventrally bulging and dorsally nearly flat*, lumens rounded; *papillae absent* or perhaps represented by weakly developed, broad, low, flat thickenings above the lumens; *basal laminal cells sharply differentiated in a vee, inflated, rectangular*, 20–28 µm in width, ca. 3:1, walls hyaline, thin. *Gynodioicous*; plants synoicous, or different gametoecia on a single plant synoicous or only archegoniate, or some plants entirely archegoniate but not producing sporophytes. Perichaetia terminal, enclosing a *highly swollen vaginula*, inner perichaetial leaves smaller than the cauline. Seta 1(–2) per perichaetium, ca. 4–5 mm in length, red-orange, twisted clockwise above; theca 1–2 mm in length, narrowly cylindrical, occasionally slightly curved, exothecial cells ca. 20 µm in width, 3–4:1, thick-walled, stomates phaneropore, at base of capsule; annulus of 2 rows of strongly vesiculose cells, persistent; *peristome extremely variable in development*, absent, rudimentary or to 500 µm in length, when elongate often broken off with the operculum, teeth 16, linear to lanceolate, cleft to near base and often perforate below, straight, red when well developed, finely papillose to densely spiculose, basal membrane absent or weakly developed, weakly spiculose or papillose. Operculum long-conic, to 1 mm in length, cells straight. Calyptra cucullate, smooth, ca. 2 mm in length. Spores light brown, essentially smooth, small, ca. 8 µm in diameter. Laminal KOH color reaction light yellow to orange, occasionally negative.

Growing on calcareous rock in southeastern U.S.A., Mexico, Central America, the West Indies, Brazil, Japan and Java.

This monotypic genus is easily recognized by the blackish coloration of the ligulate, tubulose (when dry) leaves with crenulate upper margins, upper laminal cells bulging ventrally and flat dorsally, and ventral surface of the costa covered with bulging, rounded-hexagonal cells (Pl. 68, f. 7). The basal cells are inflated and arranged in a sharply differentiated vee (Pl. 68, f. 8), similar to those of *Tortella*, a genus differing in its generally acute leaf apices with costa short-excurrent, and two stereid bands in the costa, among other characters. Eperistomate specimens might be confused with *Hyophila*, *Gyroweisia* or *Weisiopsis* species, but none of these has the basal vee of inflated hyaline cells. Javan specimens labeled as the synonym *Gyroweisia brevicaulis* differ weakly from American collections in being somewhat more robust and less commonly associated with the thin turf of cyanobacteria almost universal in New World collections.

Most descriptions of this taxon refer to a dioicous sexual condition. A thorough examination of collections indicates that sporophyte-bearing plants in any particular collection lack antheridia, but other plants are variously entirely archegoniate, synoicous or both archegoniate and synoicous in different gametoecia on the same plant. This difference is probably the reason for past incorrect ascription of dioicous sexual condition. Using a modified form of Wyatt's (1985) nomenclature for sexuality, *Luisierella* is actually gynodioicous (with both monoicous and gynoicous plants), and individual plants are either perigoniate or synoicous or gynomonoicous (with both perigoniate and, in this case, synoicous gametoecia). The lack of sporophytes on synoicous plants may be due to dichogamy, but this needs further investigation. Deguchi (1987) recently described a monoicous (synoicous and cryptoicous) and dioicous or rhizautoicous condition for *L. barbula* in Asia.

Luisierella may be confused with species of Calymperaceae, which may share the bulging, lens-like upper laminal cells, but the vee-shaped area of enlarged basal cells is diagnostic.

Additional literature: Crum & Anderson (1961), Harada and Deguchi (1990), Potier de la Varde (1936), Steere (1945).

Number of species recognized: 1.

Species examined: *L. barbula* (BP, BM—as *Gyroweisia brevicaulis*, BUF, DUKE, FH, NY, TENN).

New heterotypic synonymy: *Trichostomum brevicaule*

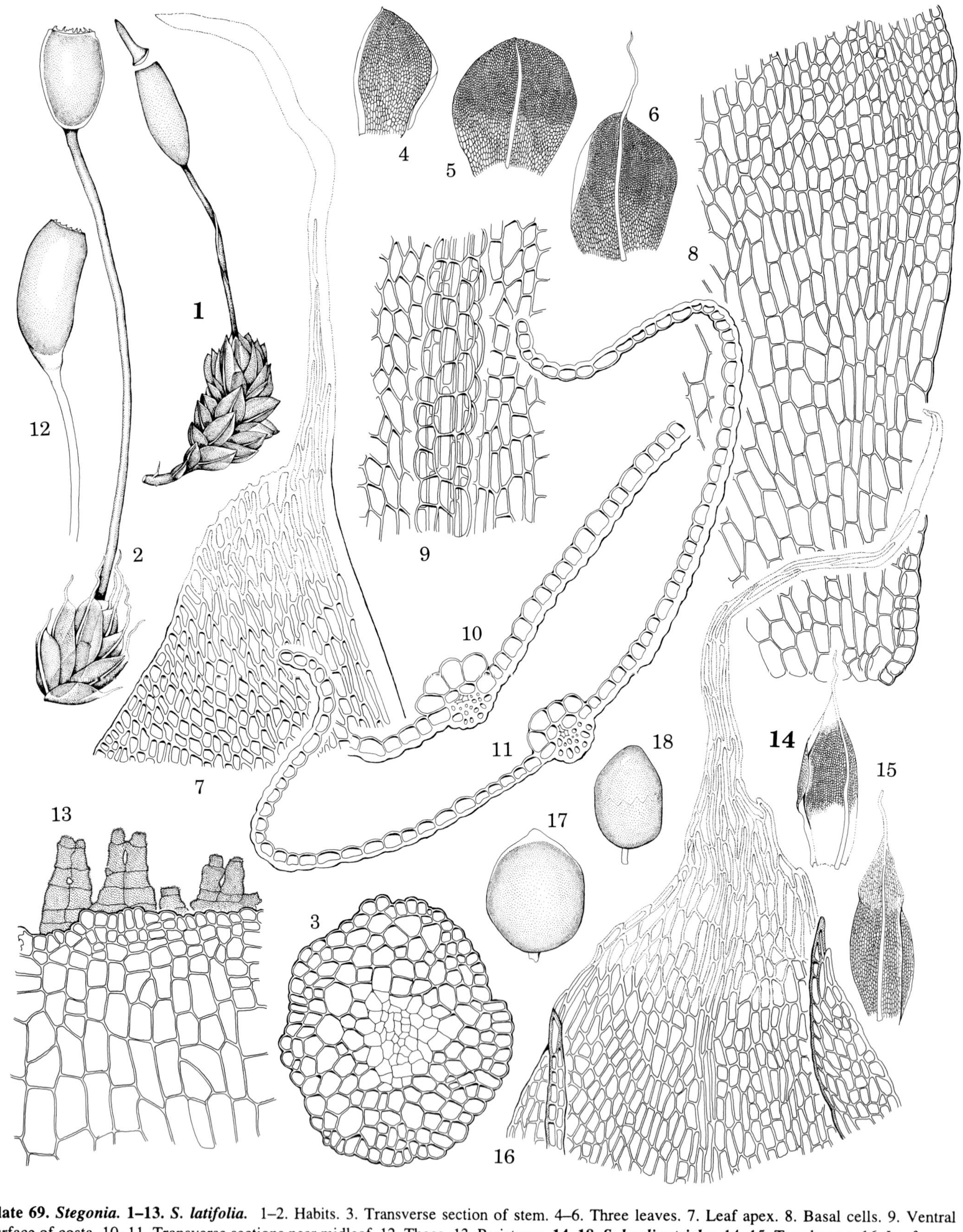

Plate 69. ***Stegonia.*** **1–13.** ***S. latifolia.*** 1–2. Habits. 3. Transverse section of stem. 4–6. Three leaves. 7. Leaf apex. 8. Basal cells. 9. Ventral surface of costa. 10–11. Transverse sections near midleaf. 12. Theca. 13. Peristome. **14–18.** ***S. hyalinotricha.*** 14–15. Two leaves. 16. Leaf apex. 17–18. Capsules.

Hampe ex C. Müll. (*Gyroweisia brevicaulis* (Hampe ex C. Müll.) Broth.) = *Luisierella barbula* (Schwaegr.) Steere. *Weisiopsis hollandii* Bartr. = *Luisierella barbula* (Schwaegr.) Steere.

50. STEGONIA **Plate 69.**

Stegonia Vent., Rev. Bryol. 10: 96, 1883. Type: *Stegonia latifolia* (Schwaegr.) Vent.

Hyalophyllum Warnst., Hedwigia 53: 284, 1913, *nom. illeg. incl. gen. prior.*

Didymodon subg. *Stegonia* (Vent.) Kindb., Eur. N. Amer. Bryin. 2: 272, 1897.

Pottia sect. *Stegonia* (Vent.) C. Müll., Gen. Musc. Fr. 386, 1900.

From στέγω, to cover closely, so as to keep out wet, to cover, protect, to contain; a reference to the concave, protecting leaves.

Plants gregarious, bulbiform, white to whitish green or occasionally just green above, brown below, usually unbranched, *short,* to 3 mm in length, transverse section rounded, central strand present, sclerodermis absent, hyalodermis absent; axillary hairs ca. 3–6 cells in length, the basal 2 cells firm-walled; rhizoids scarce. *Leaves* appressed when dry, *appressed or loosely involute above and weakly spreading below when moist, nearly circular to very broadly ovate, sometimes elliptical,* ca. 1.5–2.0 mm in length, *upper lamina deeply concave, margins broadly recurved to revolute, occasionally plane and seemingly broadly incurved because of broad leaf concavity, commonly serrulate near apex,* occasionally entire, *upper 1/3 to 1/4 of leaf usually hyaline, of rhomboidal thick-walled cells;* apex broadly acute or occasionally rounded; base not differentiated in shape; *costa thin, excurrent as a flexuose hyaline awn or occasionally percurrent or subpercurrent, costa with lamina inserted laterally, superficial cells ventrally short-rectangular, smooth and bulliform,* dorsally rectangular and smooth, ca. 3–4 rows of cells across costa ventrally at midleaf, *costal transverse section round,* stereid band present dorsally (occasionally of substereids), ventral epidermis present, strongly bulging, dorsal epidermis absent above midleaf but present below, guide cells 2 in 1 layer (often as two thick-walled cells larger than stereid cells but smaller than ventral epidermal cells) but often absent above, hydroid strand present and large (occasionally absent above), *ventral epidermal cells forming a bulging ridge on leaf; upper laminal cells rhomboidal or hexagonal and more elongate towards the leaf apex, often hyaline in upper 1/3 to 1/4 of leaf, 13–22 µm in width, 2–3:1, walls usually thin medially but thick-walled near apex and dorsally superficially thick-walled and ventrally thin-walled, about equally convex on both free surfaces; papillae absent;* basal cells differentiated across leaf, rectangular, occasionally bulging, little differentiated or to 30 µm in width and 4:1, walls often somewhat thicker than those of upper cells. Monoicous (paroicous or autoicous). Perichaetia terminal, inner leaves little differentiated. Setae various in length, very short (70–100 µm in length) or 0.2 to 1.2 cm in length, 1(–2) per perichaetium, brown, straight or twisted clockwise below and occasionally counterclockwise above; *capsule cleistocarpous, immersed* and ca. 0.7 mm in length, ovate and very shortly apiculate, *or stegocarpous and theca exerted* and 1–2 mm in length, brown, shiny, elliptical to cylindrical and somewhat curved, exothecial cells 18–30 µm in width, 1–2:1, thin-walled, stomates phaneropore, at base of capsule, annulus when present of 2–3 rows of persistent, vesiculose cells; *peristome teeth absent in stegocarpous capsules or rudimentary or 16 cleft to near base,* occasionally perforate, ligulate, spiculose, to 125 µm in length, with up to 5 articulations, straight, basal membrane absent. Operculum when differentiated short-rostrate, ca. 0.7 mm in length, cells straight or twisted very weakly counterclockwise. Calyptra not seen, reported as cucullate and smooth. *Spores 20–45 µm in diameter,* light to dark brown, essentially smooth to papillose. *Laminal KOH color reaction yellow.* Reported chromosome number n = 26.

Found on soil in Europe, Asia, North Africa, western North America, Mexico.

The two species, *Stegonia latifolia* and *S. hyalinotrichum*, have essentially identical gametophytes. Characters of importance are the bulbiform habit; deeply concave leaves with bases not differentiated in shape; upper laminal cells smooth and rhomboidal to hexagonal, cells of the upper 1/3–1/4 of leaf (in most specimens) hyaline and thick-walled (Pl. 69, f. 7, 16), dorsal laminal cell walls thick-walled (as seen in section or in optical sections of folded portions of the leaf) and the ventral thin-walled in the upper portion of the leaf (Pl. 69, f. 10–11); costa thin, in section dorsally flattened, the single stereid band represented by stereid, substereid or thin-walled cells in various specimens or even in parts of the same leaf (the lower portions of leaf with less thickened cell walls), and the two guide cells not strongly differentiated but, unlike *Tortula* sect. *Schizophascum*, are not absent.

This genus, like *Tortula*, shows a reduction series in sporophyte characters, *S. latifolia* having a peristome (Pl. 69, f. 13—this poorly developed and often apparently absent), *S. mourettii* (not seen) said to be eperistomate, and *S. hyalinotrichum* cleistocarpous (Pl. 69, f. 17–18). The character of dorsally thickened laminal walls is present in *Hilpertia* (which has much the same costal section and hyaline leaf apex) and some species of *Acaulon*, and may be absent in *Stegonia latifolia* in the lower part of the lamina (Flowers 1973a) or almost throughout the leaf (Mosses N. Amer. 484, BUF, has cell walls little thickened on both exposed sides except in the extreme apex), and dorsal wall thickening in leaves of *S. hyalinotrichum* is only present at the medial apical cells. Of pottiaceous genera, *Hilpertia* is most similar in general morphology, but is apparently only distantly related (see Cladograms 9–10, 12–16). *Hilpertia* differs saliently in the revolute and strongly photosynthetic leaf margins, differentiated perichaetial leaves, and red laminal KOH color reaction. One specimen labeled "*Phascum hyalinotrichum*" at MICH (Steere 17728) is a nematode-infested *Bryum* sp.

Number of accepted species: 3.

Species examined: *S. hyalinotrichum* (MICH), *S. latifolia*.

New combinations: *Stegonia hyalinotrichum* (Card. & Thér.) Zand., *comb. nov.* (*Phascum hyalinotrichum* Card. & Thér., Bot. Gaz. 37: 363, 1904).

51. CROSSIDIUM **Plates 70–71.**

Crossidium Jur., Laubmoosfl. Oest.-Ungarn 127, 1882, *nom. cons.* Type: *Crossidium squamiferum* (Viv.) Jur.

Chloronotus Vent., Fauna Fl. Venet. 1(3): 124, 1868, *nom. rejic.*

Pseudaloina Delg., Bryologist 85: 401, 1982 [1983]. Type: *Pseudaloina woodii* Delg.

Barbula sect. *Chloronotae* BSG, Bryol. Eur. 2: 74, 1842 (fasc. 13–15 Mon. 12).

Barbula sect. *Argyrobarbula* C. Müll., Syn. 1: 597, 1849.

Tortula sect. *Chloronotae* (BSG) Spruce, Ann. Mag. Nat. Hist. ser. 2, 3: 374, 1849.

Tortula sect. *Squamigerae* Lindb., Oefv. K. Vet. Ak. Foerh. 21: 214, 1864.

Sect. *Pseudocrossidium* Holz. & Bartr., Bryologist 26: 69, 1923. Type: *Crossidium aberrans* Holz. & Bartr., *lectotyp. nov.*

Crossidium sect. *Paenecrossidium* Holz. & Bartr., Bryologist 27: 54, 1924, *nom. illeg. incl. sect. prior.*

From κροσσός, fringe, tassel + -ιδιον, diminutive; the "tassle-like" fringe on the ventral surface of the costa.

Plants low, gregarious or forming a thin turf, green above, reddish brown to tan below. *Stems mostly buried in soil,* branching occasionally, to 4 mm in length, transverse section rounded-pentagonal to elliptical, *central strand present,* occasionally hollow, *sclerodermis absent, hyalodermis absent;* axillary hairs of 2–6 cells, all clear or basal 1–2 thicker walled; *rhizoids* sparse above, *usually pale and densely clothing the buried portions of the stem. Leaves appressed-incurved, occasionally spiralled about the stem when dry,* weakly spreading when moist, *ovate to short-triangular, occasionally spathulate,* 0.7–1.5 mm in length, *upper lamina broadly concave, margins recurved to revolute below or to near apex* (but occasionally incurved above to infolded, often also narrowly recurved), entire or occasionally denticulate or serrulate near apex or above midleaf, *marginal cells often less papillose than medial* or forming a smooth or hyaline membranous border; *apex broadly acute to rounded, usually concave,* occasionally emarginate or cucullate; base short-ovate, occasionally rectangular; costa usually ending in an apiculus or a short or long (to 300 µm), smooth or denticulate awn, seldom merely percurrent, *superficial cells of upper ventral half of costa forming a pad of separate filaments, these often branching, 1–9 cells in length,* apical cell smooth or papillose, costal superficial cells elongate dorsally, 3–6 rows of cells across costa ventrally at midleaf, costal transverse section round to elliptical, *a single stereid band present dorsally, mostly oval in section, differentiated epidermis absent to weakly discernible dorsally,* guide cells 2–5 in 1(–2) layers, *hydroid strand present, strong,* occasionally encircled by stereid cells; *upper laminal cells subquadrate to hexagonal or rhomboidal,* ca. 13–19 µm in width, 1(–2):1, walls thin or evenly thickened, occasionally somewhat trigonous and lumens rounded, superficially equally weakly convex or strongly convex ventrally and nearly flat dorsally, dorsal superficial walls occasionally thickened; epapillose or with hollow, simple "cee-shaped" papillae, usually occurring medially on the leaf, also often mostly dorsally, 1–4 papillae per lumen; *basal cells differentiated across leaf, rising higher medially,* rectangular, occasionally quadrate, to 25 µm in width, ca. (1–)2–3:1, walls thin, hyaline, transverse walls often thickened. Dioicous (possibly occasionally rhizautoicous) or monoicous. Perichaetial terminal, leaves similar to the cauline or broadly ovate, occasionally rudimentary, inner leaves triangular to broadly ovate or occasionally spathulate, little or distinctly sheathing below, lower cells occasionally hyaline, rectangular to rhomboidal throughout. Perigonia terminal, sometimes cladautoicous or appearing as autoicous buds in upper leaf axils. Seta elongate, ca. 0.4–1.6 cm in length, 1 per perichaetium, orange to reddish brown, occasionally yellow, twisted clockwise below, counterclockwise above; *theca* ca. 0.8–1.5 mm in length, reddish brown, *long-ellipsoidal to cylindrical,* exothecial cells rectangular, ca. 20–30 µm in width, 2–4:1, thin-walled, stomates phaneropore, at base of theca, annulus of 1–2 layers of somewhat vesiculose cells; *peristome teeth 16,* cleft to near base into 32 paired rami, long-subulate to linear, spiculose, to 700 µm, with several articulations, twisted weakly counterclockwise, basal membrane low or to ca. 35 µm in height, spiculose. *Operculum* conic, *occasionally retaining fragments of peristome,* ca. 500–900 µm in length, cells weakly twisted counterclockwise. Calyptra cucullate, smooth, ca. 1.5–3.0 mm in length. Spores ca. 10–15 µm in diameter, yellowish brown, essentially smooth to weakly papillose. *Laminal KOH color reaction yellow,* orange in older leaves, occasionally with red patches medially above. Reported chromosome number n = 24.

Found in arid regions of most continents, on soil or rock.

The genus is easily distinguished by its dry habitat; leaves concave, ovate to spathulate, upper margins recurved or plane (occasionally broadly incurved but then also marginally recurved—Pl. 71, f. 4, 14); costa with a ventral cushion of multicellular, branching filaments of photosynthetic tissue inserted only on the costa, with one stereid band and a prominent hydroid strand (this often encircled by stereid cells, Pl. 70, f. 6, 12, 71, f. 14, 18); laminal cells unistratose; seta elongate; capsule peristomate (Pl. 70, f. 7); and KOH color reaction yellow (occasionally lamina medially orange, especially dorsally at thickened superficial walls). It differs from *Pseudocrossidium* most consistently by the oval (rather than reniform) section of the dorsal stereid band, upper laminal margins never differentiated as photosynthetic tissue (rather than occasionally so in *Pseudocrossidium*), medial upper leaf cells equally weakly convex or strongly convex ventrally and nearly flat dorsally (not strongly bulging on both sides as is usually the case with *Pseudocrossidium*), and the absence of a ventral stereid band (occasionally present in *Pseudocrossidium*).

Crossidium squamiferum and related taxa of sect. *Crossidium* have a characteristic areolation (Pl. 71, f. 13) similar to that of *Aloina* (upper cells rhomboidal) and the upper margins are membranous and incurved over the pad of filaments, but the lamina is entirely unistratose, the filaments are inserted only on the costa (not also on a portion of the lamina), and the reaction to KOH is yellow. *Crossidium woodii* (Pl. 71, f. 15–18), previously recognized in the monotypic genus *Pseudaloina*, belongs here and is very similar to *C. apiculatum*. Species of *Crossidium* sect. *Pseudocrossidium*, such as *C. aberrans* (Pl. 70, f. 8–13), *C. rosei*, *C. geheebii* (Pl. 71, f. 5–7) and *C. seriatum* (Pl. 71, f. 8–10), may be more closely related to *Tortula* sect. *Tortula* (through such species as *T. atherodes* and *T. protobryoides* with their flask-shaped protuberant ventral costal cells) or *Tortula* sect. *Tortula* (through *Tortula atrovirens* and *T. revolvens* with their ventral costal pads of vertically elongate cells) than to *Crossidium s. str.*, lacking the distinctive cylindrical terminal

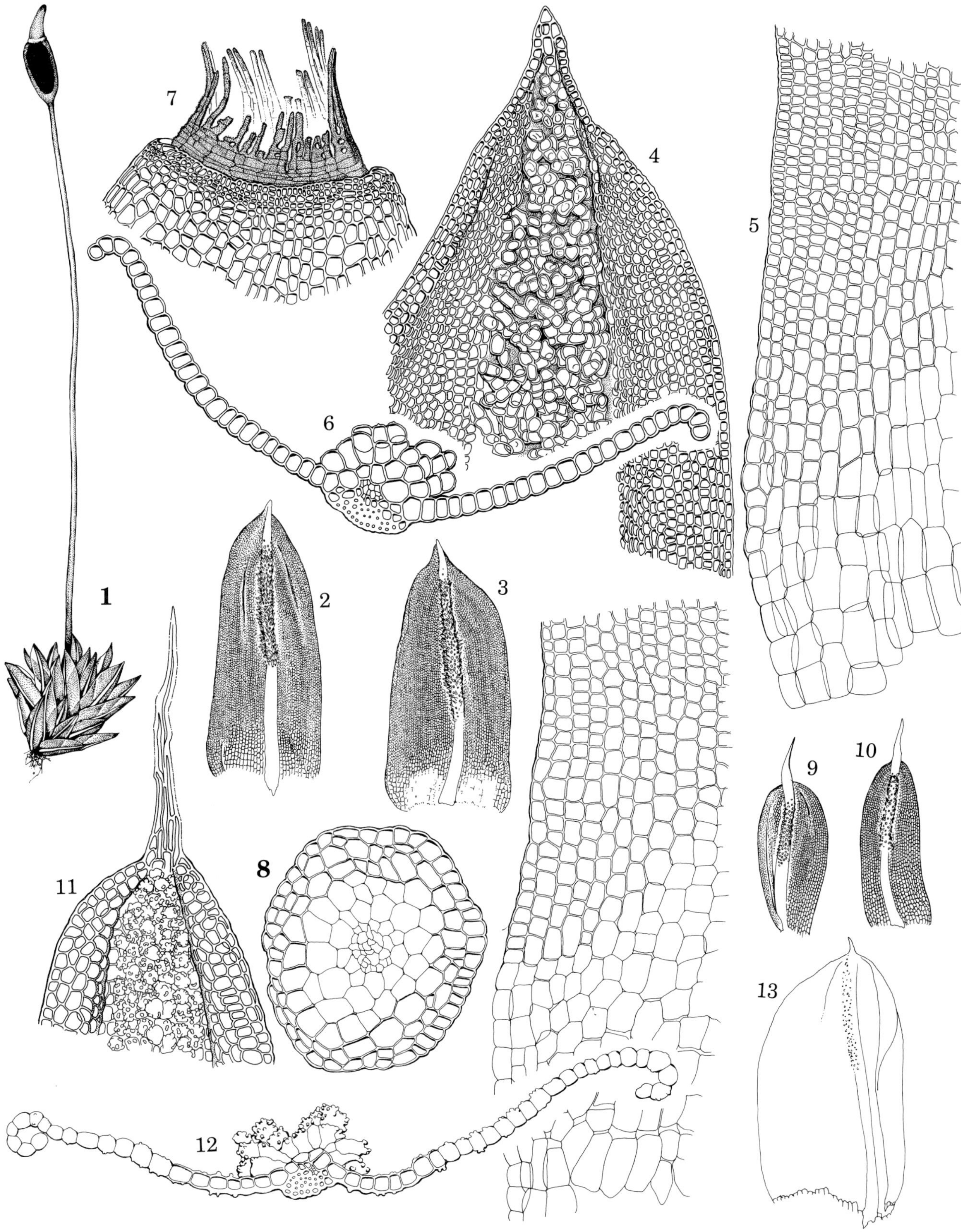

Plate 70. ***Crossidium.*** **1–7.** ***C. spiralifolium.*** 1. Habit. 2–3. Two leaves. 4. Leaf apex. 5. Basal cells. 6. Transverse section at midleaf. 7. Peristome. **8–13.** ***C. aberrans.*** 8. Transverse section of stem. 9–10. Three leaves. 11. Leaf apex. 12. Transverse section at midleaf. 13. Perichaetial leaf.

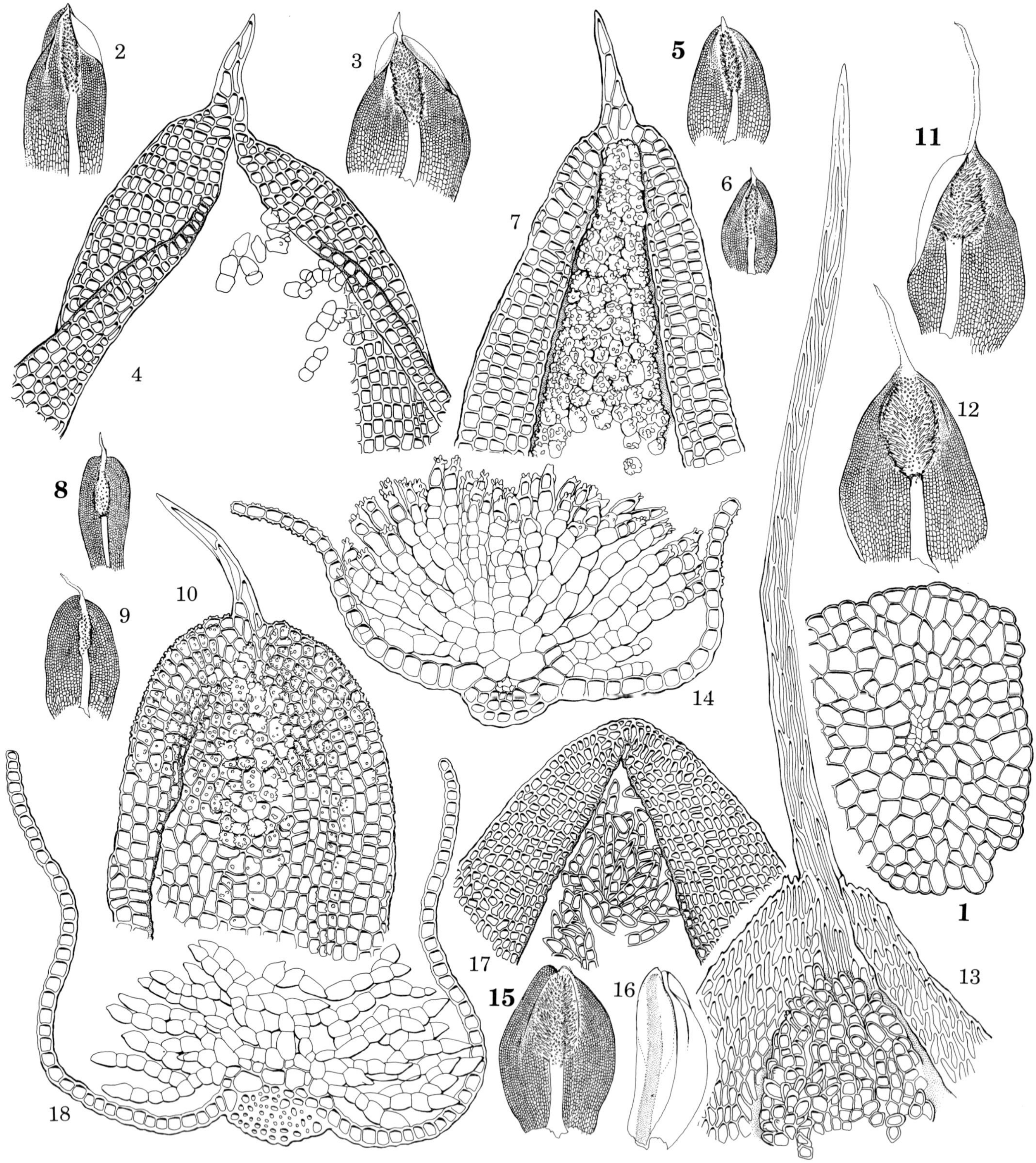

Plate 71. ***Crossidium*. 1–4. *C. crassinerve*.** 1. Transverse section of stem. 2–3. Two leaves. 4. Leaf apex. **5–7. *C. geheebii*.** 5–6. Two leaves. 7. Leaf apex. **8–10. *C. seriatum*.** 8–9. Two leaves. 10. Leaf apex. **11–14. *C. squamiferum*.** 11–12. Two leaves. 13. Leaf apex. 14. Transverse section at midleaf. **15–18. *C. woodii*.** 15–16. Two leaves. 17. Leaf apex. 18. Transverse section at midleaf.

cell of the costal filaments and strongly differentiated, epapillose upper laminal cells of species of sect. *Crossidium*. Although *Crossidium* is placed with the Hyophileae in Cladograms 14–16, perhaps sect. *Pseudocrossidium* would be referred to the Pottieae if recognized as a separate genus.

Crossidium is clearly related to *Pterygoneurum* (see Cladogram 14 and others); the photosynthetic tissue of the former, however, is probably derived from progressive elaboration of a ventral costal pad of cells (such as that of *Tortula muralis*) into filaments, while the latter may have been derived through elaboration of ridges of ventral costal cells like those in *T. atherodes*. The filaments may be considered merely deeply laciniate lamellae, but on dissection, there is no immediate evidence of this in *Crossidium* but see discussion of *Pterygoneurum*. Stark and Castetter (1987) mention both *Crossidium* and *Pterygoneurum* as probably perennial in arid habitats.

Additional literature: Cano et al. (1992), Delgadillo (1973a,b,c, 1975a, 1982), Frey and Kürschner (1984, 1988b, 1991), Fuertes-Lasala (1983), Holzinger and Bartram (1923a,b), McIntosh (1989).

Number of accepted species: 13.

Species examined: *C. aberrans* (BUF, DUKE), *C. apiculatum* (BUF, PRE), *C. crassinerve* (BUF, DUKE, NY, TENN) *C. geheebii* (H), *C. seriatum* (BUF, DUKE), *C. spiralifolium* (PRE), *C. squamiferum* (BUF, DUKE), *C. woodii* (E, MEXU).

New combinations: *Crossidium woodii* (Delg.) Zand., *comb. nov.* (*Pseudaloina woodii* Delg., Bryologist 85: 401, 1982 [1983]).

52. PTERYGONEURUM Plate 72–73.

Pterygoneurum Jur., Laubm. Oest. Ungarn 95, 1882, *nom. et orth. cons.*, as "*Pterigoneurum*". Lectotype: *Pterygoneurum cavifolium* Jur., *nom. illeg.*

Fiedleria Rabenh., Flora 31: 252, 1848, *hom. illeg. non* Reichenb., 1844. Type: *Fiedleria subsessilis* (Brid.) Rabenh.

Pharomitrium Schimp., Syn. 120, 1860, *nom. rejic.*

Pterigoneurum Jur., Hedwigia 21: 154, 1882, *nom. illeg.*

Pterogoneuron Jur. in Kindb., Eur. N. Amer. Bryin. 2: 272, 1897, orthogr. var.

Pottia subg. *Pterygoneurum* (Jur.) Boul., Muscin. France 469, 1884.

Barbula sect. *Lamelligerae* Schimp., Syn. ed. 2: 193, 1876. Type: *Barbula cavifolia* Schimp.

Tortula sect. *Pterygoneuron* (Jur.) Dix., Stud. Handb. Brit. Moss. 176, 1896.

Pottia sect. *Lamellipottia* C. Müll., Gen. Musc. Fr. 386, 1900. Type: *Pottia cavifolia* Fürnr.

Tortula subsect. *Pterygoneuron* (Jur.) Braithw., Brit. Moss. Fl. 1: 206, 1885.

From πτέρυξ, -ύγος, wing + o + νεῦρον, nerve, sinew, tendon; referring to the ridge-like lamellae of the dorsal surface of the costa.

Plants short or bulbiform, gregarious or forming a thin turf, green above, light brown below. *Stems* branching occasionally, 0.5–5.0 mm in length, transverse section rounded-pentagonal, *central strand present*, distinct, *sclerodermis absent*, *hyalodermis absent*; axillary hairs ca. 7 cells in length, all hyaline or basal 1–2 cells brown; *buried stem densely clothed with pale, brown rhizoids*. *Leaves* appressed when dry, weakly spreading when moist, *ovate to ligulate*, lamina 0.5–1.5 mm in length, awn to 2.0 mm in length, *upper lamina broadly concave, weakly cucullate, margins weakly to broadly incurved to near apex*, plane below, occasionally broadly recurved above, entire or denticulate above, occasionally dentate to erose near costa at apex; *apex obtuse to rounded*; base not differentiated in shape or somewhat widened; *costa* long-excurrent as a smooth or denticulate *hyaline awn*, but awn occasionally absent in lower cauline leaves or innermost perichaetial bracts, superficial cells of costa in medial portion of lamina differentiated as longitudinal lamellae ventrally, elongate dorsally, 4–6 dorsal rows of cells across costa dorsally at midleaf, costal transverse section circular, *one stereid band present dorsally* but this often weak, differentiated epidermis usually present dorsally, guide cells 2 in 1 layer, *hydroid strand present, usually large, (2–)3 lamellae longitudinally inserted on ventral surface of costa*, ca. 12 cells in height, occasionally lobed, smooth or papillose, cells bulging superficially; *upper laminal cells quadrate to short-rectangular*, often transversely elongated, 10–13(–20) µm in width, (2–)1:1(–2), walls thin to weakly thickened and often weakly trigonous, superficially flat or somewhat bulging on both sides or only ventrally; *papillae absent or small, simple, solid or hollow*, present dorsally in medial portion of leaf, 2–3 per lumen; basal cells differentiated across leaf base, occasionally rising higher medially, rectangular, little wider than upper cells or weakly inflated, ca. 2–5:1, walls thin. *Autoicous, paroicous or cladautoicous*. Perichaetia terminal, inner leaves elliptical, little differentiated, to 1.5 mm in length, not sheathing, lower cells hyaline, rhomboidal-rectangular in lower 1/2. Perigonia occurring as small buds in upper leaf axils or terminal on short branches from base of perichaetiate plant. Seta 0.4–3.5 mm in length, 1 per perichaetium, dark reddish to yellowish brown, twisted counterclockwise; capsule usually stegocarpous but cleistocarpous in *P. kozlovii* (not seen); *theca 0.5–1.5 mm in length*, dark reddish brown to yellowish brown, *short-cylindrical, often macrostomous, variously sulcate or ridged when dry*, exothecial cells rectangular, outer exposed walls occasionally much thickened (as seen in transverse section), stomates phaneropore, at base of theca, annulus of ca. 1–3 rows of weakly vesiculose cells; *peristome teeth absent or seldom present*, usually removed with the dehiscence of the operculum, *rudimentary*, consisting of a latticework of elliptical fragments borne on a conical hyaline membrane, weakly papillose, to 300 µm in length, with several articulations, weakly twisted counterclockwise, basal membrane low, weakly papillose. Operculum rostrate to conic-rostrate, ca. 0.6–1.4 mm in length, cells straight or twisted counterclockwise. *Calyptra cucullate or mitrate and lobed*, smooth, 0.9–2.0 mm in length. *Spores large*, 25–38 µm in diameter, brown, occasionally hyaline, papillose, occasionally smooth. *Laminal KOH color reaction yellow*, occasionally with red or orange patches medially near apex or on leaf base. Reported chromosome number n = 26, 52.

Found largely in areas of dry climate on soil or occasionally rock, on most continents.

Pterygoneurum is easily identified by the short, often bulbiform habit (Pl. 72, f. 1) with concave, ovate to spathulate leaves with lamellae on the ventral surface of the costa (Pl. 72, f. 7) and capsule stegocarpous. Species with short setae and short capsules have mitrate calyptrae, those with elongate setae

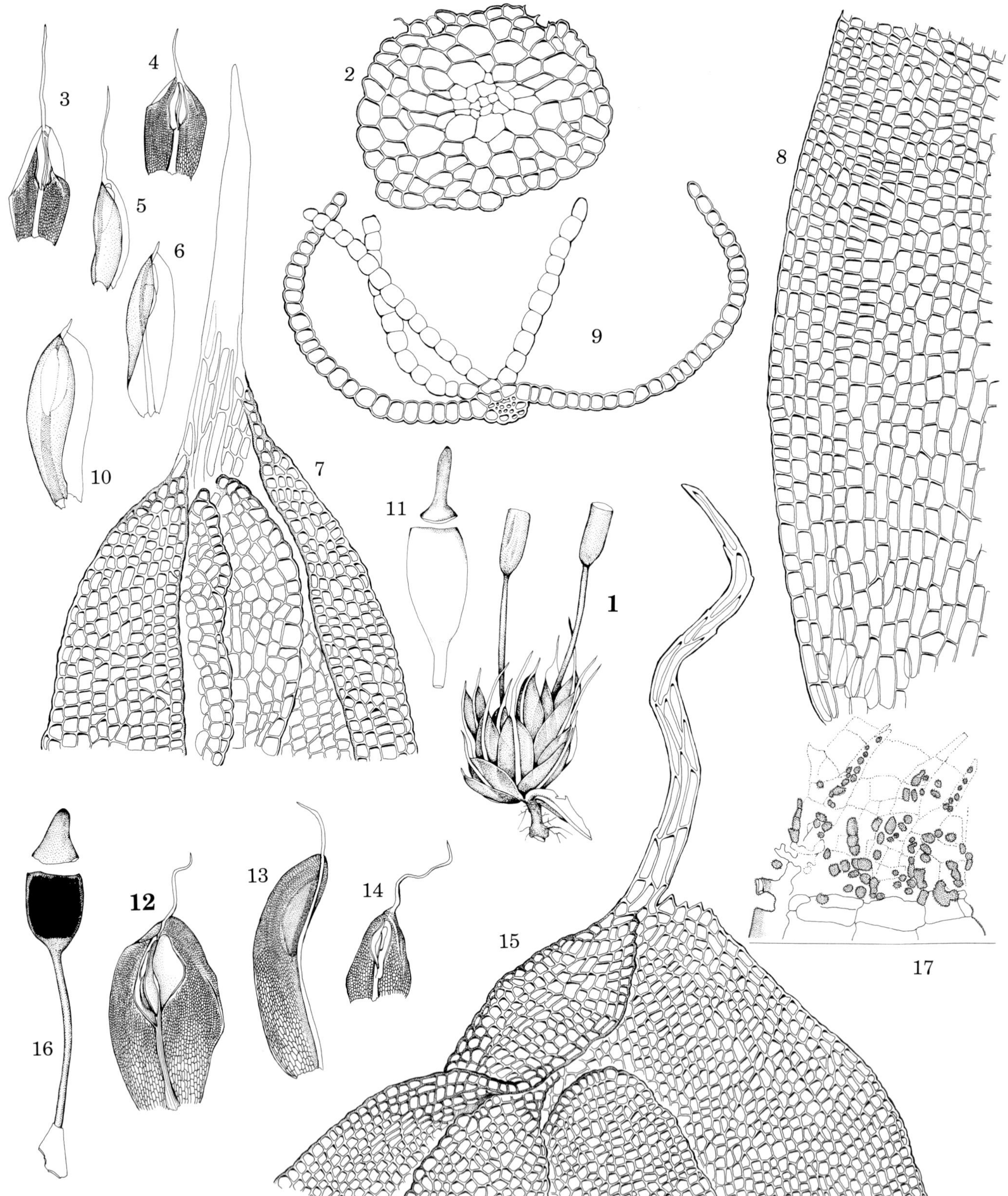

Plate 72. ***Pterygoneurum.*** **1–11.** ***P. ovatum.*** 1. Habit. 2. Transverse section of stem. 3–6. Four leaves. 7. Leaf apex. 8. Basal cells. 9. Transverse section at midleaf. 10. Perichaetial leaf. 11. Capsule. **12–17.** ***P. lamellatum.*** 12–14. Three leaves. 15. Leaf apex. 16. Sporophyte. 17. Peristome (rudimentary).

and capsules have cucullate calyptrae. The species of *Pterygoneurum* examined have extraordinarily similar gametophytes but a range of sporophyte morphology ranging from long-exerted, elliptical, peristomate capsules (Pl. 72, f. 16–17) to short-cylindrical or obovate, eperistomate capsules on short setae (Pl. 73, f. 7). Although one might expect that other species with lamellate leaves and cleistocarpous capsules might also belong here, the candidate taxon *Acaulon* subg. *Alaticosta* truly belongs with *Acaulon* (Pottieae) because of the characteristically nearly circular, more deeply concave leaves; larger, rectangular to rhomboid medial laminal cells (20 μm in width, 2–3:1); calyptra shorter (ca. 0.4 mm); and leaves with an orange to red KOH color reaction throughout the leaf blade. *Pterygoneurum* is quite like *Tortula* sect. *Pottia*, and is approached in most characters by *T. cuneifolia*, *T. grandiretis* and *T. californica*. It is surprising that *Pterygoneurum* appears to be rather distantly related to *Tortula* in Cladograms 14–16, which indicate a derivation from *Hyophila*-like ancestors. Cladograms 9, 10 and 13 show, on the other hand, a close relationship; these hypotheses, of course, must be tested through additional study, preferably at the species level.

The neatly arranged trellis-like peristome illustrated for *P.*

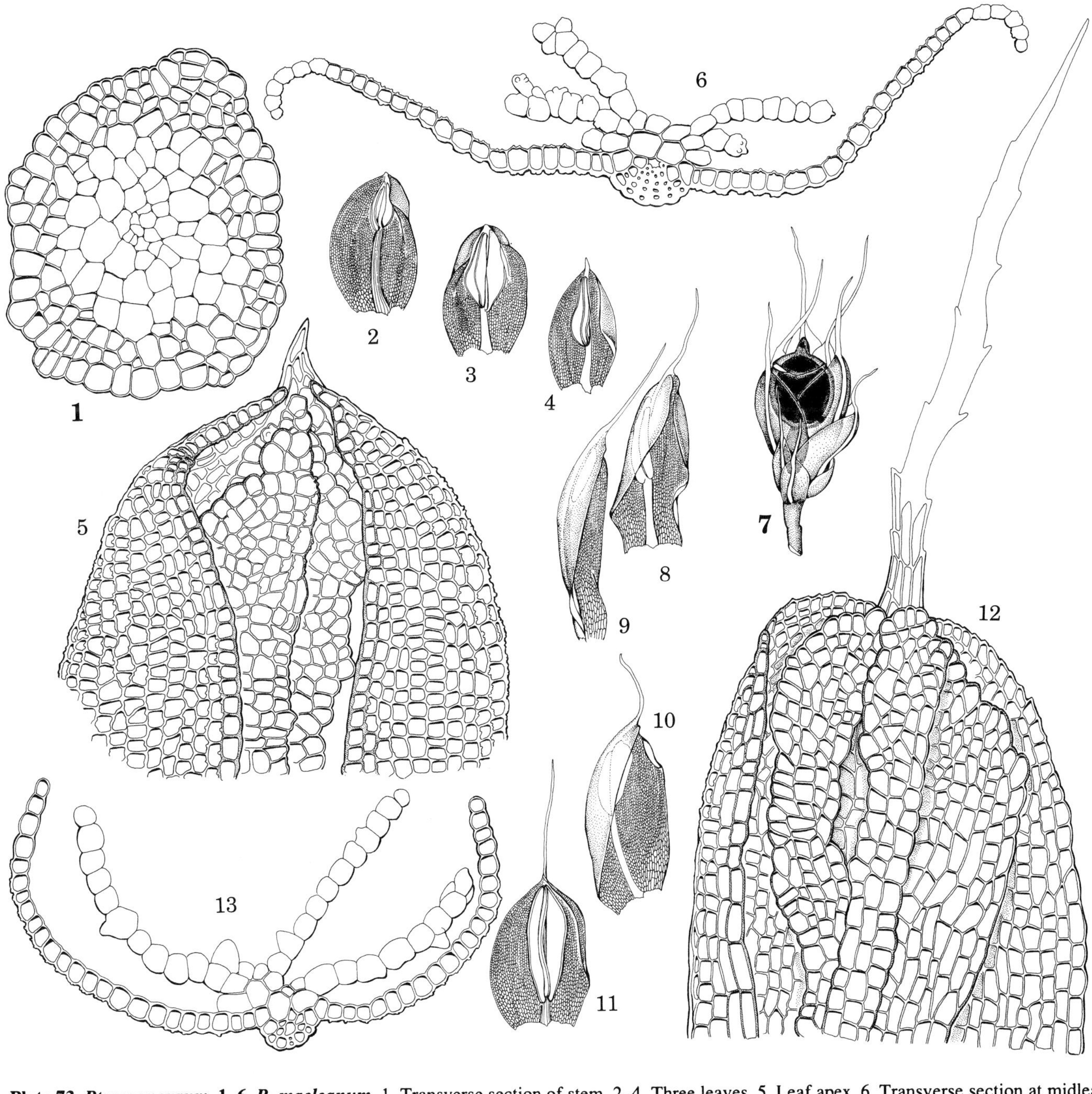

Plate 73. *Pterygoneurum*. 1–6. *P. macleanum*. 1. Transverse section of stem. 2–4. Three leaves. 5. Leaf apex. 6. Transverse section at midleaf. **7–13. *P. subsessile*.** 7. Habit. 8–10. Three upper cauline leaves. 11. Lower cauline leaf. 12. Leaf apex. 13. Transverse section at midleaf.

lamellatum by Flowers (1973a) and others was not seen in the course of this study; a more rudimentary arrangement of small plates over a membrane was seen instead (Pl. 72, f. 17). The calyptra of *P. californicum* is actually not cucullate as was indicated in the original description, but is more accurately described as "long-mitrate," with one long split and 2–4 smaller clefts basally forming small lobes. Both Flowers (1973a) and Wareham (1939c) note that *P. subsessile* has a mitrate calyptra that may at times appear cucullate through uneven splitting. McIntosh (1989) reviewed the possibility (Boros 1953, Corley et al. 1981) that *P. kozlovii* might represent be a hybrid of *Tortula atherodes* (Pl. 84, f. 16–22, discussed as the synonym *Phascum cuspidatum*) and another species of *Pterygoneurum*; the problem, apparently, is the reduced capsule of *P. kozlovii*, which is here accepted as a common and natural outcome of evolution in arid land mosses and requires no postulation of hybridization to explain it away.

Stegonia is similar to *Pterygoneurum* in the strongly concave ovate leaves, piliferous leaf apex at the base of which the leaf margins are commonly erose, laminal papillae absent, and inflated bulging ventral costal cells, but lacks the ventral costal photosynthetic lamellae. Frey et al. (1990) reviewed the literature on the phylogenetic derivation of *Pterygoneurum*, which indicates a derivation from (ancestors of) *Crossidium*. The presence on the ventral costal lamellae of filaments similar to those of *Crossidium* was cited as important evidence of this. The cladistic analysis of the present study supports such a derivation.

Additional literature: Abramova et al. (1973), Pospíšil (1975), Smith (1985).

Number of accepted species: 12.

Species examined: *P. californicum* (MICH), *P. lamellatum* (BUF, DUKE, NY), *P. macleanum* (NY), *P. ovatum* (BUF, DUKE, MICH, NY).

53. GLOBULINELLA — Plate 74.

Globulinella Steere, J. Washington Acad. Sci. 36: 221, 1946.
Lectotype: *Globulinella globifera* (Hampe) Steere *fide* van der Wijk et al. (1959–69).

Seligeria subg. *Globulina* C. Müll., Gen. Musc. Fr. 306, 1800.
Type: *Globulinella globifera* (Hampe) Steere.

From *globuline*, this from *globulus*, diminutive of *globus*, ball + *-ella*, diminutive; the bead-like shape of these tiny plants.

Plants small, scattered, gregarious or forming a turf, green above, tan below. *Stems* branching seldom, then by subperichaetial innovation, ca. 0.4–0.6 cm in length, transverse section rounded-pentagonal, *central strand strong,* sclerodermis absent, hyalodermis absent or weakly differentiated; axillary hairs clavate, short, of 2–3 cells, the basal 1 often yellowish. *Leaves* appressed when dry, weakly spreading when moist, *spathulate or ovate to orbicular,* 0.7–1.3 mm in length, *upper lamina concave to broadly channeled, margins plane or incurved, entire*; *apex broadly rounded, usually cucullate;* base not differentiated in shape; *costa ending 4–6 cells below apex, often laterally spurred, ventrally protuberant, dorsally flat or bulging,* superficial cells ventrally quadrate to short-rectangular, bulging, dorsally elongate, 2–6 rows of cells across costa ventrally at midleaf, costal transverse section round to elliptical, *stereid band usually absent ventrally,* dorsally present and round or flat in shape, ventral and dorsal epidermises present, the dorsal sometimes only present laterally, guide cells 2–4 in 1 layer, hydroid strand indistinct, but apparently present; *upper laminal cells subquadrate to short-rectangular or rhomboidal, lumens rounded,* 9–13 µm in width, 1–2:1, walls irregularly thickened, superficially weakly convex on both sides or occasionally somewhat more bulging ventrally; *papillae absent;* basal cells weakly differentiated across leaf, short-rectangular, ca. 10–13 µm in width, 2–3:1, walls thin to evenly thickened or indistinctly porose. Propagula rare, small, of several cells, short-clavate to spherical, ca. 40 µm in length, borne in leaf axils. Dioicous. Perichaetia terminal, inner leaves little differentiated, ovate, less cucullate, somewhat enlarged, usually sheathing the seta, cells elliptical to rhomboidal in lower 1/4–3/4. Perigonia terminal. Seta 0.4–0.8 cm in length, 1 per perichaetium, yellow to brown, twisted clockwise above or straight; theca 0.6–1.6 mm in length, brown, ovoid to elliptical, exothecial cells rectangular, walls thin to evenly thickened, stomates phaneropore, at base of theca, annulus of 2 rows of vesiculose cells; *peristome teeth 16, irregularly cleft or anastomosing,* ligulate to lanceolate, densely spiculose, 65–115 µm, with 3–6 articulations, straight, basal membrane absent or low, spiculose. Operculum rostrate, inclined, 0.6–0.7 mm in length, cells straight. Calyptra cucullate, smooth, ca. 1.0 mm in length. Spores 8–13 µm in diameter, brown, essentially smooth to weakly papillose. *Laminal KOH color reaction yellow.*

A small genus known from southwestern U.S.A. south through Mexico and Central America to Ecuador, found on soil in dry, montane areas.

Globulinella (see also discussion of *Globulina* in treatment of *Bryoerythrophyllum*) can be recognized by the following combination of distinctive character states: small, julaceous plants growing on soil, leaves usually broadly rounded and deeply concave, often cucullate (Pl. 74, f. 2–6), upper laminal cell walls incrassate, costa bulging ventrally (Pl. 74, f. 7, 10), seta elongate, and peristome teeth 16, irregularly cleft and densely spiculose (Pl. 74, f. 8). *Saitoella peruviana*, known only from sterile material, has previously been recognized in *Globulinella* (sharing the features of rounded, cucullate leaves with upper cells mainly rhomboidal and thick-walled, and costa ending before the apex) and, in the light of Cladogram 14, may represent convergent evolution. *Saitoella* differs in its upper laminal cells being papillose only medially in the leaf, the complete lack of costal guide cells (the single layer of ventrally exposed costal parenchyma is papillose, and therefore probably best interpreted as epidermal tissue), and its red reaction to KOH solution. Naked, lateral archegonia borne singly on the stems of both *G. globifera* and *S. peruviana* were seen, but this is also the case in certain species of other genera of Pottiaceae.

Propagula were found in only one specimen (of *G. globifera* from Mexico: Zander 4900, BUF). The two stereid bands seen in sections of the costae of a recent collection of *G. benoistii* (Pl. 74, f. 10—Ecuador, Steere E35, NY) indicate a possible relationship of this species (and the similar, but singly stereid-banded *G. globifera*) with *Plaubelia*, with which it shares a short leaf with incurved upper margins, rounded leaf apex, smooth (in majority of *Plaubelia* specimens) upper laminal cells, general lack of a ventral stereid band, and 16 irregularly cleft peristome teeth. *Plaubelia* differs in its oblong leaves, ventrally colliculate areolation, and non-spurred costa.

Literature: Magill (1977a), Steere & Chapman (1946), Zander (1976).

Number of accepted species: 2.

Species examined: *Globulinella benoistii* (NY, PC), *Globulinella globifera* (BUF, FH, TENN).

54. ALOINA Plates 75–76.

Aloina Kindb., Bih. K. Svensk. Vet. Ak. Handl. 6(19): 22, 1882, *nom. cons.* Lectotype: *Aloina aloides* (Schultz) Kindb.

Aloidella Vent., Comm. Fauna Fl. Venet. 1(3): 124, 1868, *nom. rejic.*

Barbula subg. *Tortula* BSG, Bryol. Eur. 2: 1, 1851 (fasc. 45–47 Consp. 2: III).

Barbula subg. *Aloidella* (Vent.) Schimp., Syn. ed. 2: 188, 1876.

Tortula subg. *Aloidella* (Vent.) C. Jens., Medd. Groenland 3: 379, 1887.

Barbula sect. *Aloideae* Fürnr., Flora 12: 598, 1829. Type: *Tortula aloides* (Schultz) De Not.

Tortula sect. *Aloideae* (Fürnr.) De Not., Mem. R. Acc. Sc. Torino 40: 287, 1838.

Barbula sect. *Aloina* C. Müll., Syn. 1: 596, 1849, *nom. illeg. incl. sect. prior.*

Tortula sect. *Aloidella* De Not., Musc. Ital. 1: 3, 14, 1862, *nom. illeg. incl. sect. prior.*

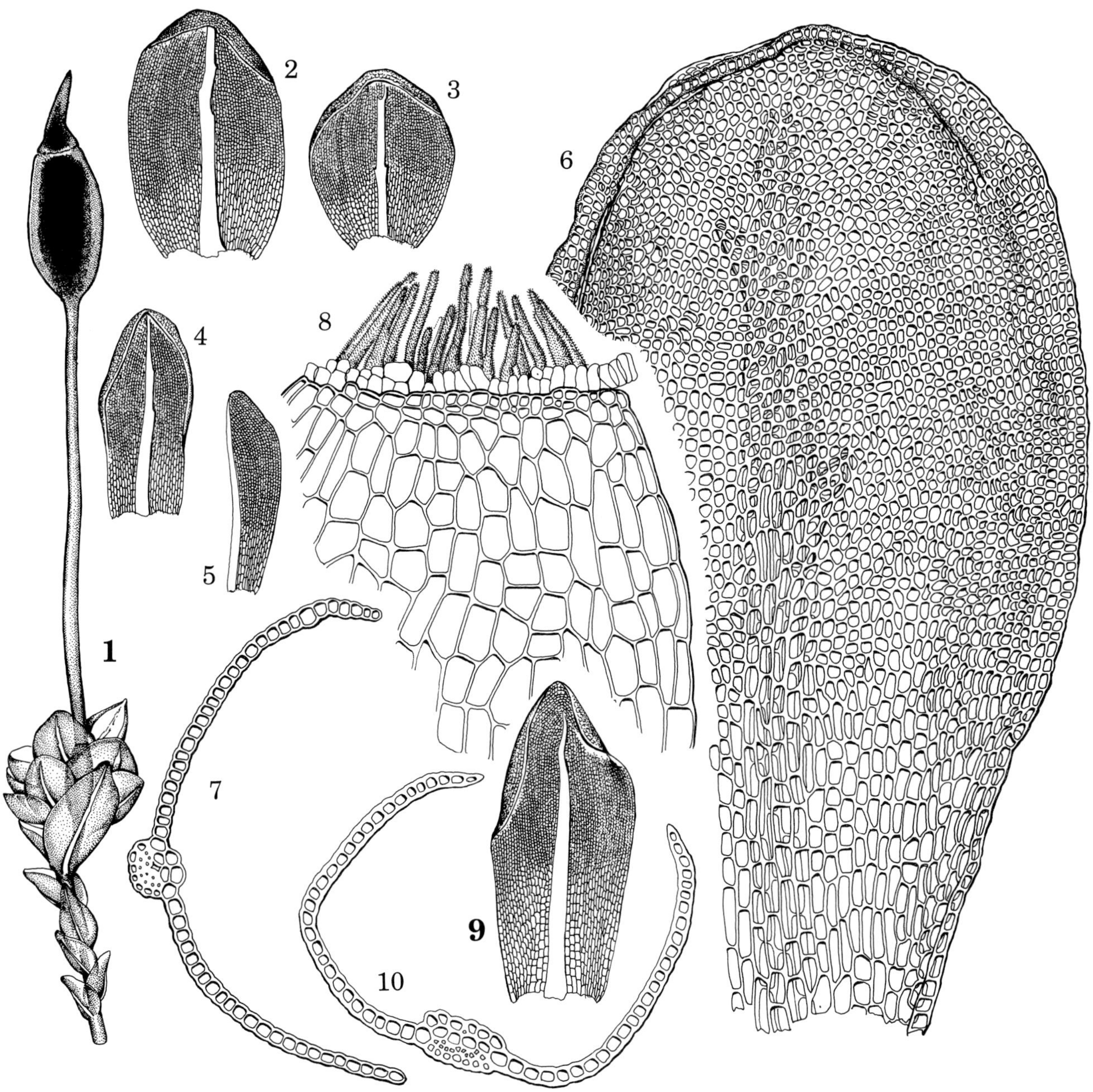

Plate 74. ***Globulinella.*** **1–8.** ***G. globifera.*** 1. Habit. 2–5. Four leaves. 6. Leaf areolation. 7. Transverse section at midleaf. 8. Peristome. **9–10.** ***G. benoistii.*** 9. Leaf. 10. Transverse section at midleaf.

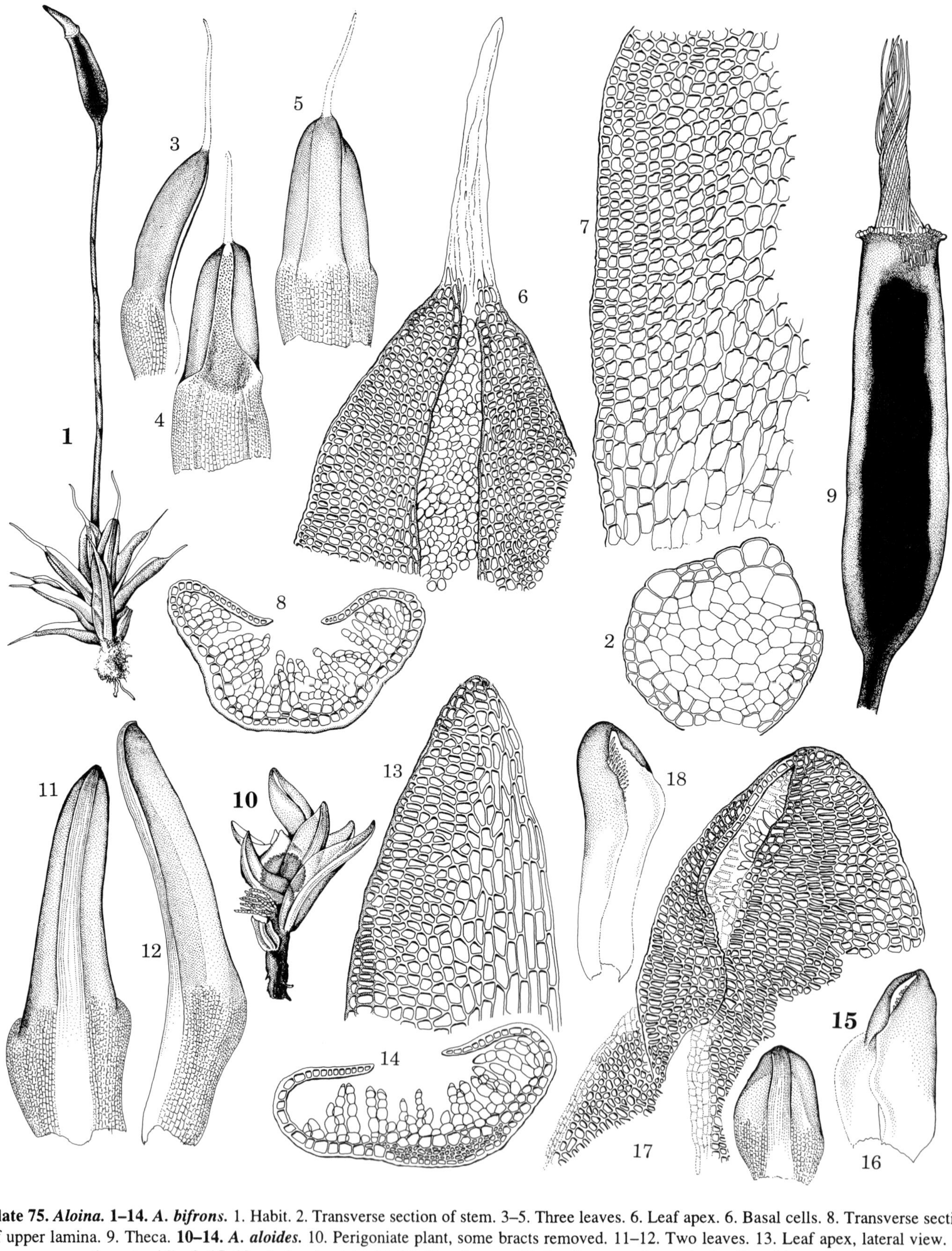

Plate 75. ***Aloina.*** **1–14.** ***A. bifrons.*** 1. Habit. 2. Transverse section of stem. 3–5. Three leaves. 6. Leaf apex. 6. Basal cells. 8. Transverse section of upper lamina. 9. Theca. **10–14.** ***A. aloides.*** 10. Perigoniate plant, some bracts removed. 11–12. Two leaves. 13. Leaf apex, lateral view. 14. Transverse section at midleaf. **15–18.** ***A. brevirostris.*** 15–16. Two leaves. 17. Leaf apex and marginal cells with membranous border. 18. Perichaetial leaf.

Tortula sect. *Aloina* Mitt., J. Linn. Soc. Bot. 12: 143, 157, 1869, *nom. illeg. incl. sect. prior.*
Barbula sect. *Aloidella* (Vent.) Lesq. & James, Man. N. Amer. Moss. 115, 1884, *nom. illeg. incl. sect. prior.*
Tortula subsect. *Aloideae* (Fürnr.) Braithw., Brit. Moss Fl. 1: 208, 1855.

From *Aloë* + *-ina*, characteristic of; fleshy leaved like *Aloë* (Liliaceae).

Plants gregarious or in a thin turf, deep green above, brown below. *Stems* branching irregularly, *fleshy*, *buried in soil*, ca. 0.5 mm exposed above soil surface, transverse section rounded-pentagonal, cells thin-walled, central strand absent or present and distinct, sclerodermis and hyalodermis absent; axillary hairs ca. 6 cells in length, all clear; *buried stem densely covered with pale rhizoids. Leaves* incurved and somewhat tubulose when dry, spreading-incurved when moist, *obovate or ovate to ligulate*, ca. 3.0–3.5 mm in length, *upper lamina broadly and deeply channeled, margins strongly infolded*, entire to weakly sinuose or denticulate above, *marginal cells often longitudinally elongate or forming a membranous border*; *apex rounded, cucullate*; *base broadly rectangular, sheathing*, hyaline, often with distinct shoulders; *costa* brownish, subpercurrent to percurrent or mucronate or long-excurrent as a smooth or serrate awn, *ventrally covered from midleaf to near laminal apex with photosynthetic filaments, which are also inserted on the medial portion of lamina*, dorsally with elongate superficial cells or these occasionally quadrate to short-rectangular near

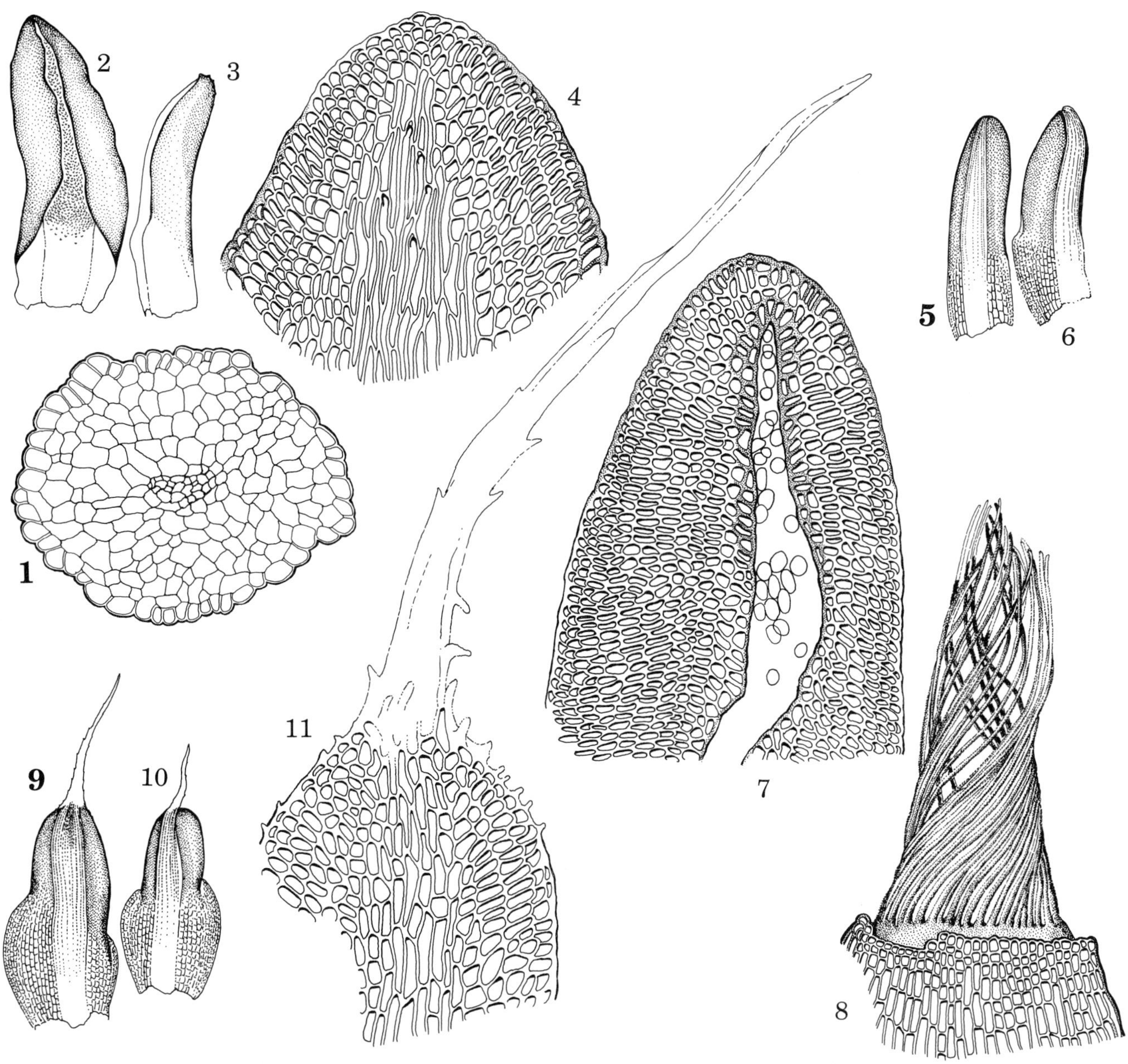

Plate 76. ***Aloina.*** **1–4.** ***A. hamulus.*** 1. Transverse section of stem. 2–3. Two leaves. 4. Leaf apex, dorsal view. **5–8.** ***A. rigida.*** 5–6. Two leaves. 7. Leaf apex. 8. Peristome. **9–11.** ***A. roseae.*** 9–10. Two leaves. 11. Leaf apex, dorsal view.

apex, 10–20 rows of cells across costa dorsally at midleaf, *costal transverse section broadly reniform*, stereid band present only dorsally, crescent-shaped, occasionally weak or absent, dorsal epidermis weakly differentiated, guide cells ca. 6–12 in 1(–2) layers, hydroid strand present, ventral epidermis differentiated as a series of uniseriate filaments ca. 3–4 cells in length, extending over bistratose portion of lamina, occasionally branching, apical cell of filament occasionally thickened and papillose; *upper laminal cells transversely elongate*, rectangular to rhomboidal, *medially bistratose*, ca. 8–13 µm in width, 1–2:1, *walls thickened, lumens rounded*, superficially flat to weakly convex, thickened dorsally; *papillae absent* (cells near apex occasionally rough when awn is rough); *basal cells* differentiated across leaf, *medially bistratose*, rectangular, ca. 25 µm in width, 3–5:1, walls thin, occasionally transversely slightly thickened. Dioicous or monoicous. *Perichaetia* terminal, *inner leaves usually little different from cauline leaves*, occasionally with margins plane and not infolded, innermost leaves serrulate on margins and cells lax throughout. Perigonia terminal, clustered, inner leaves triangular, paraphyses filamentous, uniseriate. Seta ca. 0.8–1.5 cm in length, 1(–2) per perichaetium, brown to reddish brown, twisted clockwise below, counterclockwise above; theca ca. 2–3 mm in length, reddish brown, cylindrical, exothecial cells rectangular, ca. 25–30 µm in width, 2–3:1, thin-walled, stomates phaneropore at base of capsule, annulus of ca. 2 rows of vesiculose cells, persistent; peristome teeth 16, linear, variously cleft above or more usually to basal membrane, spiculose, to ca. 1000 µm in length, with several articulations, twisted 0.5–1.0 times counterclockwise, basal membrane absent or present, to 60 µm in height, spiculose. Operculum conic to rostrate, ca. 0.5–1.0 mm in length, cells spiralling counterclockwise. Calyptra cucullate, smooth, ca. 2.5 mm in length. Spores ca. 10–15 µm in diameter, light brown, smooth or papillose. *Laminal KOH color reaction red.* Reported chromosome number n = 23+m, 25+2m, 24, 26, 28, 48.

A widespread genus found in North, Central and South America, Europe, Asia, northern and southern Africa, Australia and New Zealand, growing on soil, occasionally walls and rock.

Delgadillo (1973b, 1975a) recently produced a useful revision of this genus, supplying workable distinctions between *Aloina* and genera with similar morphology, and discussing his conclusions about their phylogenetic relationships. Delgadillo (1975a) and Brotherus (1924–25) both used the absence of a central strand in stem sections to assist in separating *Aloina* from related genera; however, this may be misleading in that some species of *Aloina* (e.g. *A. brevirostris*, Pl. 75, f. 15–18, and *A. hamulus*, Pl. 76, f. 1–4) may in fact have a distinct central strand. These species, however, lack the stem satellite hydroid strands of *Aloinella*. The taxonomically most important features of *Aloina* are the fleshy, buried stem (like that of *Crossidium*), the upper laminal margins strongly folded over the pad of photosynthetic filaments, which itself covers the upper costa and the bistratose medial portion of the lamina (Pl. 75, f. 8), and the KOH reaction red (*Aloinella* is golden yellow in KOH). The pad of filaments may not be homologous in derivation with those of *Aloinella* and *Crossidium*, because they arise from a portion of the lamina. The genus *Pseudaloina* is treated here as a synonym of *Crossidium* ((*q.v.*).

Additional literature: Baczurina (1972), Barker (1904), Craig (1939), Delgadillo (1973c, 1982), Koponen and Oittinen (1967), Mitten (1874), Persson (1944), Steere (1950), Weber (1979).

Number of accepted species: 12.

Species examined: *A. aloides* (BUF, NY), *A. bifrons* (NY), *A. brevirostris* (BUF), *A. hamulus* (BUF, NY), *A. rigida* (BUF), *A. rosei* (NY).

55. ALOINELLA — Plate 77.

Aloinella Card., Rev. Bryol. 36: 76, 1909. Type: *Aloinella catenula* Card.

Barbula sect. *Climacocaulon* C. Müll., Linnaea 42: 329, 1879. Type: *Aloinella galeata* (C. Müll.) Broth. see van der Wijk et al. (1959–69, 5: 301).

From *Aloina*, a genus + *-ella*, diminutive; resembling the genus *Aloina*.

Plants growing in a turf or somewhat mixed with other species, yellowish green above, tan to dark or reddish brown below. *Stems* branching often and irregularly, to 2 cm in length, transverse section rounded-pentagonal, cells of central cylinder often large, to 25 µm in diameter, thin-walled, *central strand strong, occasionally additional small hydroid strands embedded in outer cortex*; sclerodermis absent, outer cortex of substereid or occasionally stereid cells, hyalodermis absent; axillary hairs to 9 cells in length, basal 1–3 cells thick-walled; sparsely radiculose. *Leaves* appressed to incurved-catenulate when dry, *appressed to weakly spreading when moist, short- to long-ovate or obovate*, occasionally circular or oblong, 0.4–1.0 mm in length, *upper lamina deeply concave, margins weakly and broadly incurved*, entire or weakly denticulate at apex or weakly serrulate above or throughout, occasionally bordered by 2–3 rows of thin-walled, hyaline cells; *apex rounded, deeply cucullate*; base not differentiated in shape; *costa broad, flat*, ending a few to several cells below the apex, superficial cells elongate dorsally, ca. 12 rows of cells across costa dorsally at midleaf, *costa often strongly and broadly decurrent at leaf base*, costal transverse section broadly reniform, stereid band single, crescent-shaped, *epidermis modified ventrally as a pad of crowded filaments restricted to the costa*, each filament uniseriate and ca. 2–6 cells in length, erect or slanted distally, occasionally hollow-papillose, epidermis not differentiated or weakly differentiated dorsally, guide cells 4–8 in 1 layer, often differentiated only medially on the costa, hydroid strand present; *upper laminal cells rectangular to rhomboidal*, 9–15 µm in width, 1(–2):1, walls evenly thickened, lumens angular, superficially weakly convex on both exposed surfaces, occasionally bulging slightly more ventrally than dorsally, *dorsal superficial walls often thickened*; papillae absent or lamina papillose medially with 3–6 bifid, scattered papillae per lumen, often papillose only dorsally; *basal cells weakly differentiated across base*, rectangular, to 18 µm in width, 1–4:1, walls thin to evenly thickened. Dioicous. Perichaetia terminal, inner leaves ovate to short-elliptical, not cucullate, to 1.9 mm in length, sheathing, lower cells lax, thin-walled, rectangular-rhomboidal throughout. Perigonia terminal, swollen-gemmate, paraphyses occasionally of cells with internal longitudinal walls. Seta 0.5–1.1 cm in length, 1 per perichaetium, yellowish to reddish brown, twisted clockwise; theca 0.9–2.0 mm in length, reddish brown, long-ellipsoidal to cylindrical, exothecial cells rectangular, thin-walled, ca. 18–28 µm in width, 2–3:1, stomates phaneropore, at base of theca, annulus of 1–3 rows of vesiculose cells; *peristome teeth 16, short*, entire or cleft into 2

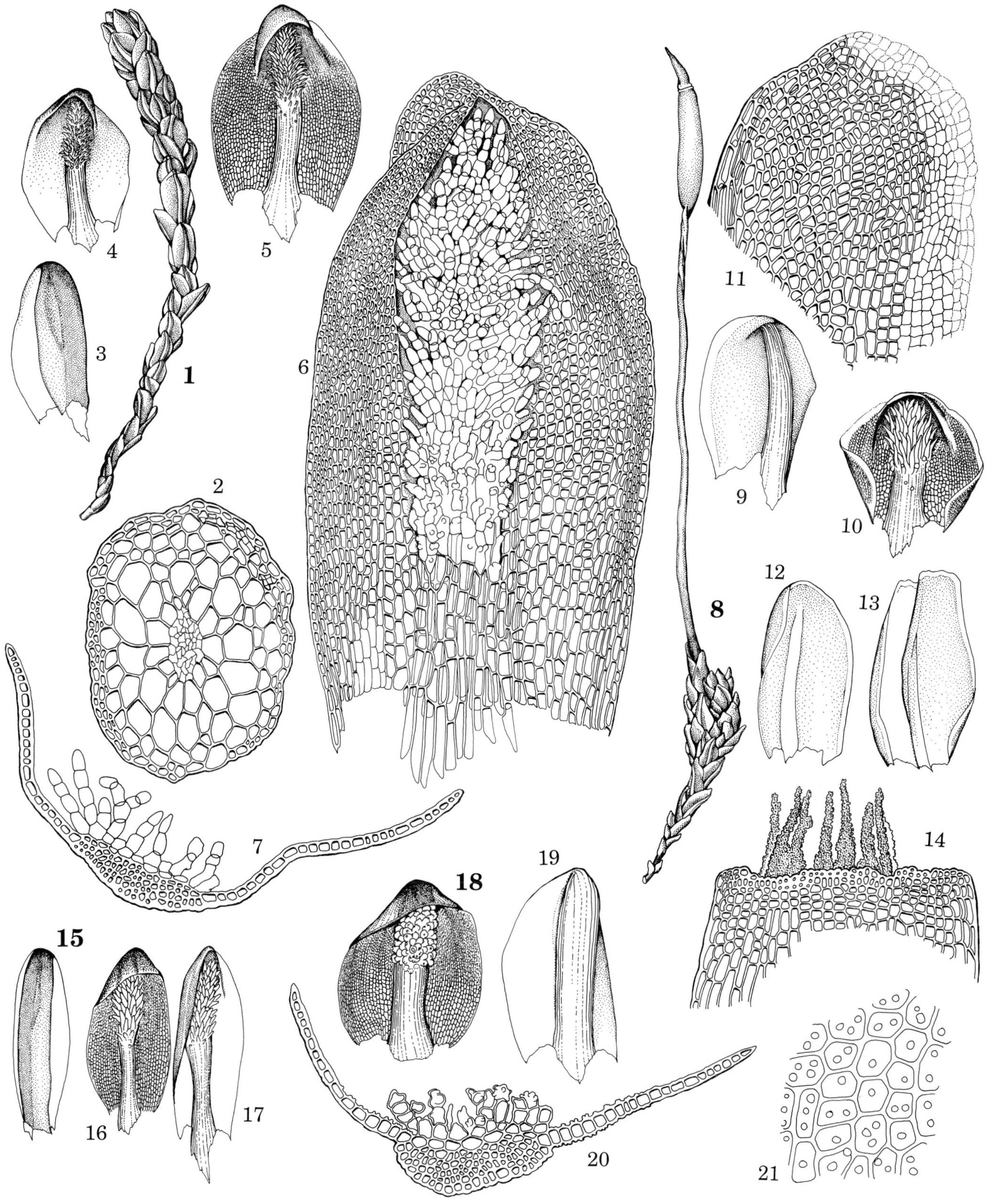

Plate 77. ***Aloinella.*** **1–7.** ***A. catenula.*** 1. Habit. 2. Transverse section of stem showing subepidermal satellite hydroid strands. 3–5. Three leaves. 6. Leaf areolation. 7. Transverse section at midleaf. **8–14.** ***A. boliviana.*** 8. Habit. 9–10. Two leaves. 11. Upper marginal laminal cells. 12–13. Perichaetial leaves. 14. Peristome. **15–17.** ***A. cucullifera.*** 15–17. Three leaves. **18–21.** ***A. venezuelana.*** 13–19. Two leaves. 20. Transverse section at midleaf. 21. Upper laminal papillae.

rami, occasionally anastomosing, 110–165 µm in length, with several articulations, straight or nearly so, densely spiculose, basal membrane lacking. Operculum conic-rostrate, ca. 0.4–1.0 mm in length, cells in straight rows. Calyptra cucullate, smooth, ca. 1 mm in length. Spores 11–22 µm in diameter, brown, lightly papillose. *Laminal KOH color reaction golden yellow.*

Restricted to South America and Mexico, growing on soil at high elevations.

This genus is relatively homogeneous (the species differ by few but probably not insignificant characters) and has been revised recently (by Delgadillo 1975a). The striking features of *Aloinella* are the appressed, deeply cucullate leaves (Pl. 77, f. 3–5, 15–17, 9–10) with ventral photosynthetic filaments arising from and restricted to the broad, flat costa (Pl. 77, f. 7, 20). The genus is unusual for the multiseriate-celled paraphyses of the perigonia and the small satellite hydroid strands in the stem (Pl. 77, f. 2), these latter probably associated with the long decurrency of the costa. Satellite strands have also been found in the stem of *Timmia* (Mastracci 1993).

Additional literature: Brotherus (1911), Delgadillo (1973b,c), Griffin (1975).

Number of accepted species: 7.

Species examined: *A. boliviana* (BUF, NY), *A. catenula* (BUF, PC, TENN), *A. cucullifera* (NY), *A. venezuelana* (FLAS).

Tribe POTTIEAE

Pottieae (Limpr.) Dix., Stud. Handb. Brit. Moss. 174, 1924.
Pottieae Limpr., Laubm. Deutschl. 1: 518, 1888, rank not given.

This subclade is distinguished at the immediate ancestral node by the character states: stem short, less than 1 cm in length and peristome of 32 similar rami, distinctly twisted. There is a clear evolutionary transformation from taxa with lanceolate leaves and two stereid bands in the costa to spathulate leaves and one stereid band in the costa, paralleling that in the Hyophileae. The Pottieae is the only suprageneric group with many genera having the seta twisted counterclockwise. The distribution of species is generally worldwide, with a concentration in South America.

56. LEPTOBARBULA — Plate 78.

Leptobarbula Schimp., Rev. Bryol. 2: 17, 1875. Lectotype: *Leptobarbula meridionalis* Schimp. *fide* Philibert, Rev. Bryol. 9: 18, 1882.
Barbula sect. *Leptobarbula* (Schimp.) Kindb., Eur. N. Amer. Bryin. 2: 246, 1897.

From λεπτός, peeled, fine, small, thin, delicate + o + *Barbula*, a genus; a small *Barbula*-like moss.

Plants loosely caespitose, *in a thin turf*, yellow-green above, brown below. Stems seldom branching, to 3 mm in length, rounded-pentagonal in transverse section, central strand strong, sclerodermis absent or weakly developed, hyalodermis absent; *axillary hairs* of ca. 6–7 cells, the *basal 1–2 brown*. Cauline leaves much different from the more obvious perichaetial leaves, *cauline leaves* small, spreading-recurved to incurved and twisted when dry, spreading-recurved when wet, *ligulate*, ca. 0.7 mm in length; *margins plane*, entire; apex acute to obtuse; *base scarcely differentiated in shape*; *costa percurrent* or ending 1–2 cells below the apex, 2–3 rows of cells in width at midleaf, *superficial cells* papillose, *ventrally quadrate*, dorsally short-rectangular, costal transverse section semicircular to ovate, *stereid bands two*, the ventral often weak or absent, *epidermis present on both sides*, guide cells 2–4 in 1 layer, *hydroid strand absent*; *upper laminal cells quadrate, ca. 10 µm in diameter*, 1:1, walls thin, superficially flat to weakly convex; *papillae bifid to multiplex, 2–4 per lumen, hollow*; *basal cells differentiated across leaf, reaching higher medially*, rectangular, ca. 10 µm in width, 2–3:1, walls moderately thickened to porose. Dioicous. *Perichaetia terminal, inner leaves long-ligulate to long-lanceolate, convolute-sheathing*, to 1.5 mm in length, *with distinct, often denticulate shoulders*, basal cells long rhomboidal and thick-walled, filling most of leaf. Perigonia terminal, gemmate, inner leaves ovate-triangular. Seta short, 0.6 to 0.9 cm in length, yellow-brown, occasionally red below, straight or twisted clockwise above, 1 per perichaetium; theca 0.7–2.0 mm in length, red-brown, elliptical to cylindrical; exothecial cells rhomboidal, 40–45 µm in width, 2–4:1, walls thin; stomates phaneropore, at base of theca; *annulus strongly vesiculose, persistent or revoluble*; *peristome of 32 filamentous, yellow, spiculose teeth*, to ca. 300 µm in length, of several articulations, *twisted counterclockwise* about once, basal membrane low, 25–50 µm in height, sometimes hidden by a persistent annulus, low-spiculose. Operculum long-conic to rostrate, 0.5–0.7 mm in length, cells in counterclockwise twisted rows. Calyptra cucullate, smooth, ca. 1.6 mm in length. Spores ca. 8 µm in diameter, light brown, essentially smooth. *Laminal KOH color reaction yellow.*

The genus is of local occurrence in the Mediterranean region of Europe, the Middle East and north Africa.

Schimper at first took this taxon to be a *Seligeria* by reason of its tiny size and generally slender habit (Pl. 78, f. 1–4), but its large, filamentous, somewhat twisted peristome (Pl. 78, f. 13) shows a clear relationship to the Pottiaceae. Characters diagnostic of *Leptobarbula* are the small size of the plants, twisted peristome teeth, convolute-sheathing perichaetial leaves (Pl. 78, f. 12), ligulate cauline leaves with plane margins, little differentiated base, two costal stereid bands (Pl. 78, f. 9–10), small quadrate upper laminal cells with bifid to multifid papillae (Pl. 78, f. 11), basal cells differentiated mostly medially, and laminal color in KOH yellow. Plants of

Leptobarbula grow in close mixtures of perichaetiate, perigoniate and sterile stems, which gave Casares-Gil (1932) cause to postulate possible rhizautoicy.

Hilpert (1933) recognized a relationship with *Barbula* when he made the combination *Streblotrichum bericum* (De Not.) Hilp. ("in nächste Nähe von *Str. bicolor*"). It is indeed similar to *Barbula* sect. *Convolutae* by the yellow setae (at least above); twisted, spiculose peristome; the highly differentiated perichaetial leaves; reflexed, plane-margined cauline leaves with upper laminal cells having thin walls and bifid to multifid papillae; basal laminal cells differentiated more highly in the leaf medially than marginally; and yellow color in KOH. Appleyard et al. (1985), however, felt that *Leptobarbula* was "allied to *Gymnostomum* and *Gyroweisia*, differing apparently only in the revoluble annulus and well-developed peristome" (*Gyroweisia* usually has a revoluble annulus). The cladistic analysis indicates that *Leptobarbula* is probably a relict of a now much depleted lineage inserted near the base of the Pottioideae.

Additional literature: Corbière (1890), Dglevskaja (1963), Loeske (1916), Philibert (1882a,c), Pilous (1952), Whitehouse and During (1987), Zander and Hermann (1986).

Number of accepted species: 1.

Species examined: *L. berica* (BUF, CU, DUKE, FI, NY, WTU).

57. TETRAPTERUM Plate 79.

Tetrapterum Hampe ex Jaeg., Ber. S. Gall. Naturw. Ges. 1868–1869: 85, 1869. Type: *Tetrapterum australe* Hampe.

From τετρά-, four + πτερόν, feather, wing, fin, leaf; referring to the strongly ridged capsules.

Plants in low cushions or turfs, green above, brown below. *Stems* branching occasionally, to 4.5 mm in length, transverse section rounded-pentagonal, *central strand very strong, sclerodermis absent* or rarely weakly developed, hyalodermis absent or seldom weakly differentiated; axillary hairs rather long, of 8–10 cells, the basal 2 thicker walled. *Leaves* incurved when dry, spreading when moist, *oblong-lanceolate to long-lanceolate*, occasionally ligulate, ca.

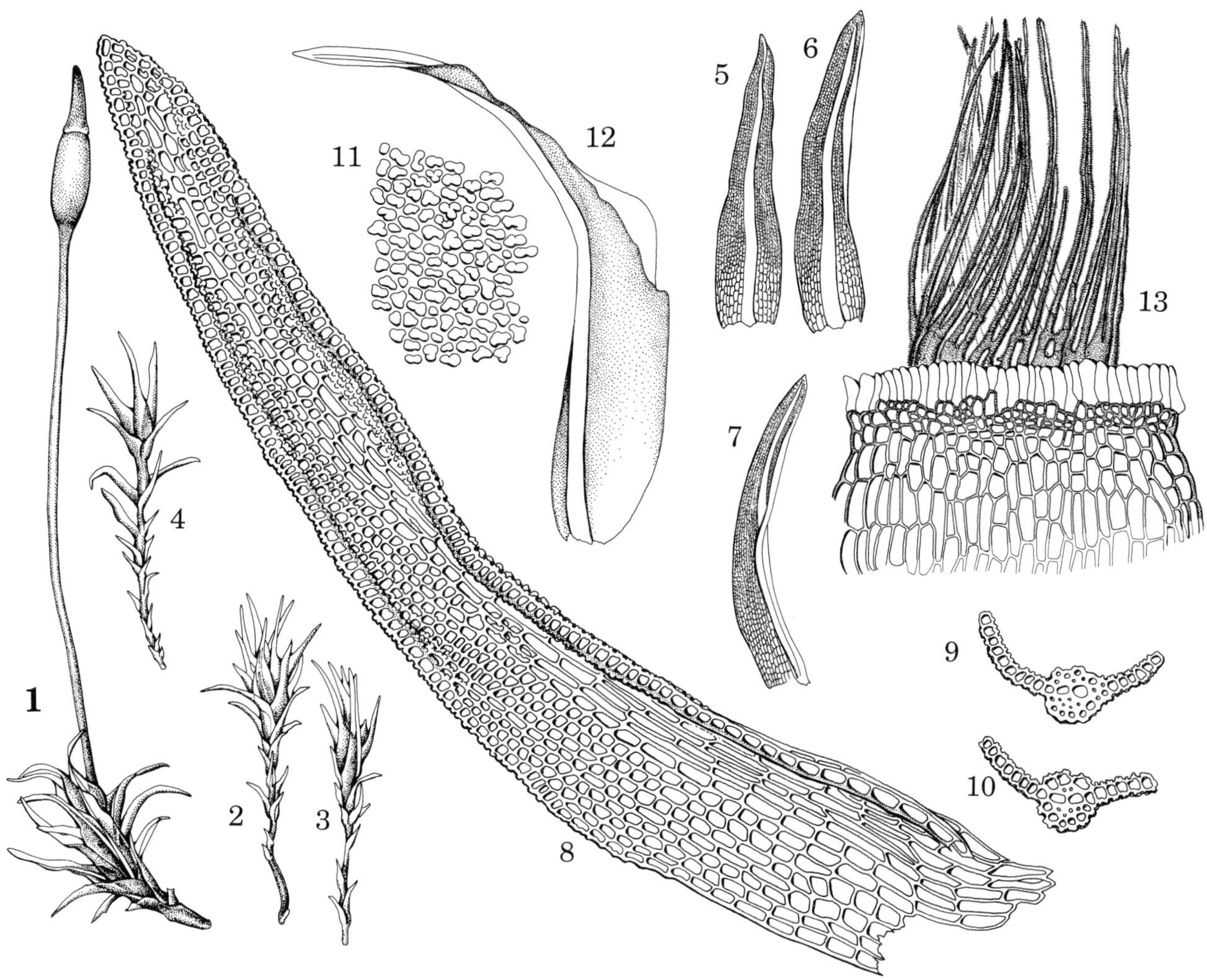

Plate 78. *Leptobarbula*. 1–13. *L. berica*. 1. Habit of sporophyte-bearing plant. 2–. Perigoniate plants. 4. Sterile plant. 5–7. Three leaves. 8. Leaf areolation. 9–10. Transverse sections near midleaf. 11. Papillae. 12. Perichaetial leaf. 13. Peristome.

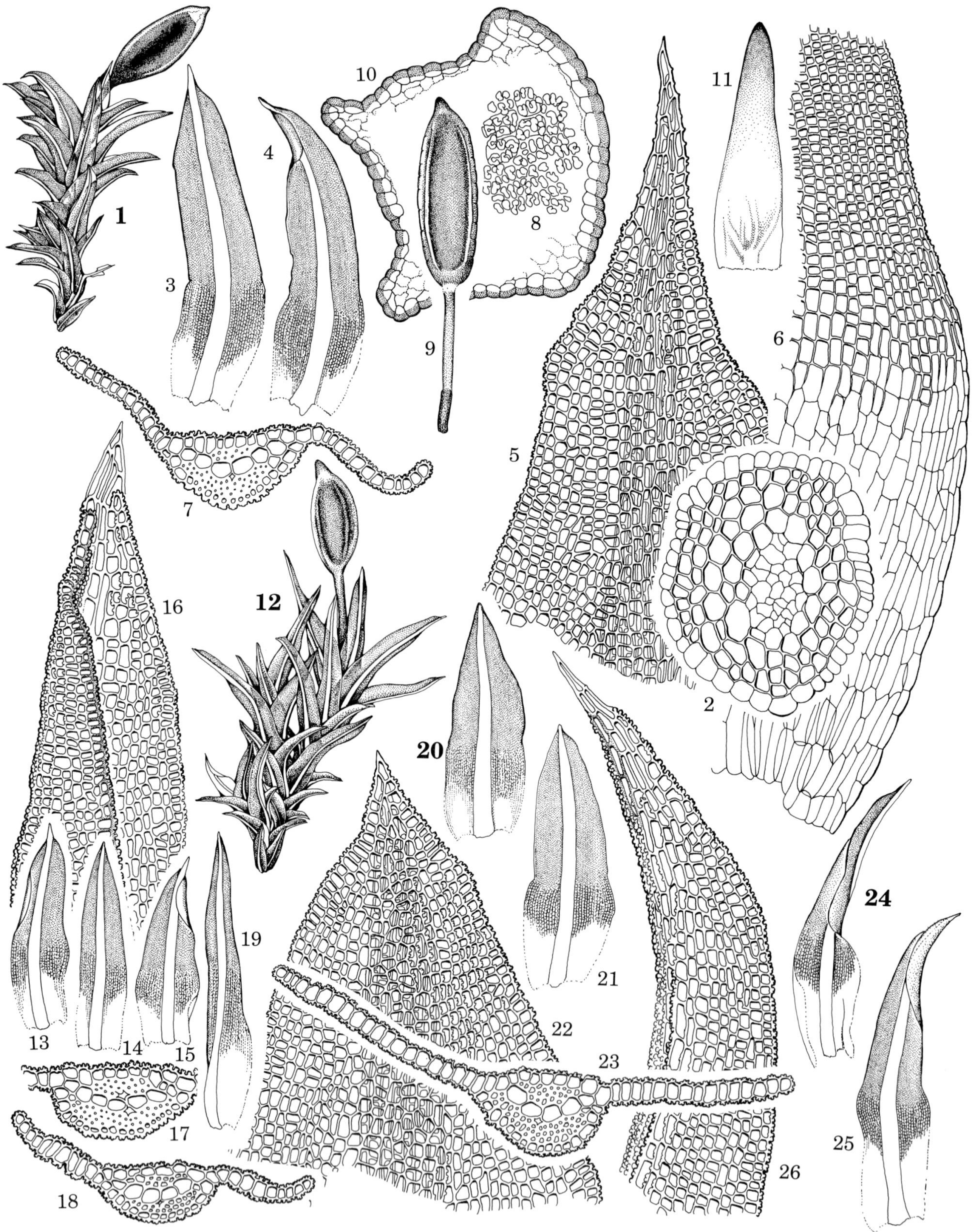

Plate 79. ***Tetrapterum*****. 1–11.** ***T. tetragonum*****.** 1. Habit. 2. Transverse section of stem. 3–4. Two leaves. 5. Leaf apex. 6. Basal cells. 7. Transverse section at midleaf. 8. Upper laminal papillae. 9. Sporophyte. 10. Transverse section of capsule. 11. Calyptra. **12–19.** ***T. cylindricum*****.** 12. Habit. 13–15. Three leaves. 16. Leaf apex. 17–18. Transverse sections near midleaf. 19. Perichaetial leaf. **20–23.** ***T. sullivanii*****.** 20–21. Two leaves. 22. Leaf apex. 23. Transverse section at midleaf. **24–26.** ***T. weymouthii*****.** 24–25. Two leaves. 26. Leaf apex.

1.5–2.0 mm in length, upper lamina broadly channeled, *margins plane, entire*; apex broadly to narrowly acute; base oblong to broadly elliptical; *costa excurrent as a sharp, smooth mucro, usually rather broad below midleaf, superficial cells ventrally quadrate, papillose, dorsally elongate below grading to short-rectangular above, papillose, 6–12 rows of cells across costa ventrally at midleaf,* costal transverse section semicircular to reniform, two stereid bands present, ventral and dorsal epidermises present, the dorsal often only weakly developed, guide cells 4–6 in 1 layer, *hydroid strand present, often multiple* and occasionally on both sides of guide cells; *upper laminal cells quadrate,* occasionally hexagonal or rectangular, 8–13 µm in width, 1(–2):1, *walls evenly thickened,* lumens quadrate, superficially flat to convex; *papillae hollow or solid, crowded, low, 4–6 per lumen, bi-(tri-)fid*; basal cells differentiated across leaf, rectangular, slightly wider than upper cells, 3–5:1, walls thin. *Monoicous* (autoicous). Perichaetia terminal, *inner leaves long-lanceolate from a high, often shouldered base,* to 2.5 mm in length, *strongly sheathing in lower half,* lower cells long-rectangular to long-rhomboidal. Perigonia gemmate, as lateral buds or terminal on a branch. *Seta short,* 0.5–1.5 mm in length, 1 per perichaetium, pale yellowish brown, straight, *distal end of seta with a narrow, pale band of cells forming an abscission layer*; *capsule cleistocarpous,* 1.5–1.9 mm in length, yellowish translucent brown, *ovate to elliptical, nearly cylindrical in transverse section or with 4(–8) flat sides and then often with sharp longitudinal folds or wings when dry, exothecial* cells rectangular, ca. 20–24 µm in width, 2–3:1, thin-walled anticlinally but superficially thick-walled, *spore sac easily visible and surrounded by a large air space,* stomates phaneropore, at base of theca, annulus absent, operculum and peristome undifferentiated. *Calyptra cucullate,* smooth, 1.4–1.6 mm in length. *Spores rather large,* ca. 25–28 µm in diameter, yellowish, *strongly spiculose to verrucose. Laminal KOH color reaction deep yellow to orange.*

Found on bare soil in dry areas of South America, Australia, Tasmania and South Africa.

Tetrapterum is similar in many gametophytic characters to *Barbula* sect. *Convolutae,* particularly those species with broad costae having quadrate ventral cells, but *Tetrapterum* may be distinguished by the short seta with a distal abscission layer in a pale band just below the capsule (Pl. 79, f. 9), and the capsule cleistocarpous, apically blunt, often with several flat sides, and the spore sac surrounded with a large air space. Like *Aschisma* and *Tortula* sect. *Schizophascum,* the capsule tends to rupture along distinct encircling lines of apparent weakness at the transverse exothecial cell walls. It is possible that *Barbula calycina, B. microcalycina* and *B. subcalycina* belong with *Tetrapterum* in that the gametophytes are nearly identical and no propagula typical of *Barbula* sect. *Convolutae* are known for these. Final disposition requires a revision of *Barbula.*

Andrews (1945) discussed *Tetrapterum,* recognizing only *T. tetragonum.* He was of the opinion that of the 11 species recognized by Brotherus (1924–25), those with cylindrical capsules are not *Tetrapterum* but instead belong to *Astomum.* I disagree with this (which in any case would have excluded the generitype) in part and emphasize other characters (above), thus retaining *T. cylindricum* among other species. Catcheside (1980) likewise recognized *T. cylindricum.* Andrews (1945) felt that *Tetrapterum* was ultimately related to *Trichostomum,* but cladistic study indicates a somewhat more distant relationship. The yellow KOH reaction does not support Hilpert's (1933) suggestion of a relationship with *Bryoerythrophyllum* (as *Erythrophyllum*), although it is true that the costal sections are quite alike. The large superficial cells that are lateral to the costa and just dorsal of the laminal insertion (Pl. 79, f. 7, 17, 18, 23) are reminiscent of those of *Pseudocrossidium.*

The distal end of the capsule (Pl. 79, f. 1, 9) of all species of *Tetrapterum* recognized here is broadly rounded-acute or very bluntly apiculate, lacking the distinct rostrum or apiculus of capsules of the reduced, cleistocarpous *Trichostomum* species placed in *Tetrapterum* by Brotherus (1924–25). Moistened capsules, even when quite mature, contain a substantial air space around the spore sac. After abscission of the capsule at the distal end of the seta, the capsule apparently disperses as an integral unit possibly by floating in seasonal floods; free capsules with a seed-like appearance are commonly found in the loose debris within collection packets. This condition is, however, not unique to *Tetrapterum,* but is most highly developed in this genus. Brotherus (1924–25) made several combinations in *Tetrapterum* reflecting his emendation of the genus as including all those species of Trichostomoideae with undifferentiated opercula and long-elliptical capsules. It is apparent that most (*Tetrapterum lilliputanum* is here transferred to *Tortella*) of Brotherus' combinations were of species better viewed as species of *Trichostomum* at the ends of one or more evolutionary series involving reduction of the sporophyte, though this needs detailed evaluation at the species level. *Tetrapterum* species may be distinguished from cleistocarpous *Trichostomum* species by a hydroid strand being visible in costal sections. The capsules of *T. tetragonum* are actually 4–8 sided when dry, and the corners may form longitudinal, flattened ridges or low wings (Pl. 79, f. 10).

In light of the discussion above, the genus may well be represented by only a few taxa, since the Australian species seen during this study are rather similar, but synonymy, if any, should await revision.

Number of accepted species: 8.

Species examined: *T. cylindricum* (NY), *T. sullivanii* (H), *T. tetragonum* (BM, NY), *T. weymouthii* (H).

58. TRACHYCARPIDIUM Plate 80.

Trachycarpidium Broth., Nat. Pfl. 1(3): 383, 1901. Type: *Trachycarpidium verrucosum* (Besch.) Broth.

Astomum subg. *Pycnocaulon* (C. Müll.) Broth., Nat. Pfl. 1(3): 384, 1901.

Acaulon sect. *Pycnocaulon* C. Müll., Linnaea 37: 144, 1872. Type: *Acaulon brisbanicum* C. Müll.

From τράχύς, rough + καρπός, fruit + -ιδιον, diminutive + *-ium,* characteristic of.

Plants gregarious or forming a thin turf, light green above, green or brownish below. Stems sometimes branching,

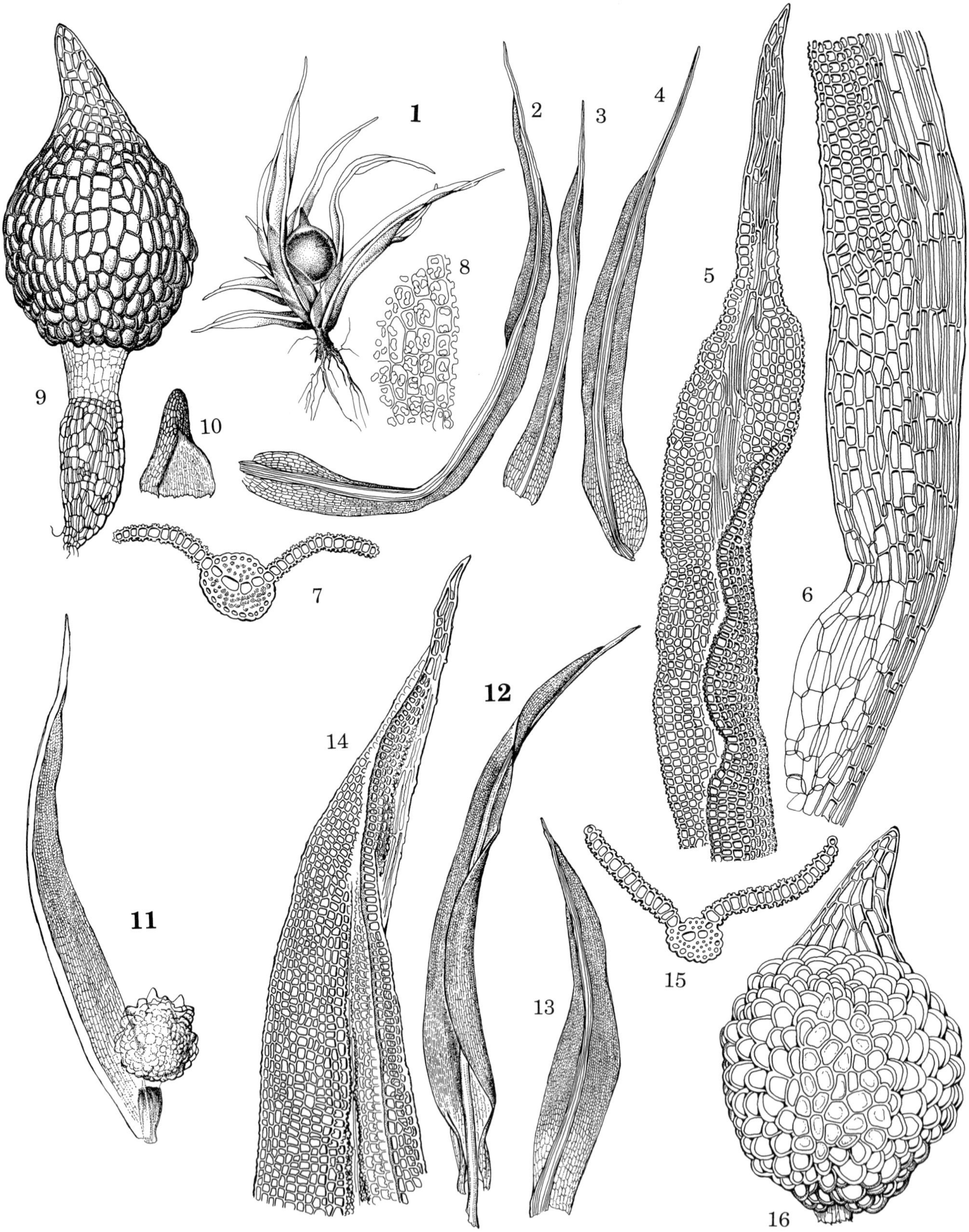

Plate 80. ***Trachycarpidium.*** **1–10.** ***T. brisbanicum.*** 1. Habit. 2–4. Three leaves. 5. Leaf apex. 6. Basal cells. 7. Transverse section at midleaf. 8. Laminal papillae. 9. Sporophyte. 10. Calyptra. **11.** ***T. echinatum.*** 11. Sporophyte and perichaetial leaf. **12–16.** ***T. tisserantii.*** 12–13. Two leaves. 14. Leaf apex. 15. Transverse section at midleaf. 16. Mature capsule fallen from seta.

ca. 0.3 mm in length, transverse section rounded-pentagonal, central strand present but weak, sclerodermis absent, hyalodermis absent; axillary hairs ca. 3 cells in length, basal cell yellowish. *Leaves* erect-spreading, weakly incurved and occasionally somewhat twisted when dry, erect-spreading and weakly recurved when moist, *long-lanceolate*, 1.5–2.0 mm in length, upper lamina shallowly but narrowly grooved along costa, this sometimes masked when costa is stout, *margins plane*, *entire* (but minutely crenulate by projecting papillae), *sometimes marginal cells rectangular and epapillose near apex or throughout; apex narrowly acute*; base short-ovate to oblong; *costa stout, excurrent as a thick, sharp, smooth mucro or short awn, superficial cells elongate ventrally* and elongate dorsally, ca. 4–6 rows of cells across costa ventrally at midleaf, costal transverse section semicircular to ovate, with two stereid bands, weak or strong ventrally, strong dorsally, epidermis present and weakly developed ventrally, absent or present and weakly developed dorsally, guide cells 2–4 in 1 layer, hydroid strand variably absent or present, small; upper laminal cells hexagonal to subquadrate, occasionally short-rectangular, ca. 7–10 µm in width, 1(–2):1, walls thin to weakly evenly thickened, superficially weakly convex or strongly bulging on both sides; *papillae bifid, 2–3 per lumen, solid or hollow*, crowded or somewhat centered over lumens; *basal cells differentiated across leaf base and rising distinctly higher along margins*, inflated-rectangular, ca. 15–20 µm in width, 2–6:1, walls thin, hyaline. *Autoicous or rhizautoicous*. Perichaetial leaves little different than the cauline, lower cells rhomboidal to long-rectangular in lower 1/4. Perigonia terminal on small buds or plantlets at the base of perichaetiate plants. *Seta very short*, ca. 75 µm in length, 1 per perichaetium, *hyaline*, straight; *theca cleistocarpous, falling off distal end of seta*, ovate to short-elliptical, 300–400 µm in length plus a 150–200 µm sharp apiculus (theca occasionally very broad at base), *with protuberances basally or throughout, walls very light yellowish brown, translucent*, columella apparently absent in mature capsules, *exothecial cells with superficial walls considerably and evenly thickened, cells* hexagonal to short-rectangular, *strongly projecting-mamillose at base or throughout* (except the apiculus), stomates phaneropore, at base of theca, annulus absent. Calyptra campanulate to short-conic-cucullate, smooth to slightly rough, 350–400 µm in length. Spores ca. 25 µm in diameter, brown, low-spiculose. Laminal KOH color reaction yellow.

Found on bare soil; Brazil, central Africa, Australia, New Caledonia, New Guinea.

Trachycarpidium is a genus of pygmy mosses (Pl. 80, f. 1) characterized by long-lanceolate, plane-margined, entire leaves with a stout costa ending in a short awn, the basal cells differentiated in a vee up the margins (Pl. 80, f. 6) as in *Tortella*, and bulging, strongly protuberant cells of the body (not the apiculus) of the immersed, cleistocarpous capsule (Pl. 80, f. 9, 11, 16). It is similar to *Tortella eckendorffii* and *Tortella fruchartii*, and may be viewed as an apparent reduction product with further elaboration of the capsule; on detailed analysis, however, the cladograms do not reflect this hypothesis. Like *Bryoceuthospora* and *Uleobryum*, the exothecial walls of the capsule are mamillose. They are only very light yellow in color, easily transmitting the darker brown color of the spores, but are not definitely hyaline as in *Uleobryum*. *Trachycarpidium* also differs from *Bryoceuthospora* and *Uleobryum* in the even thickening of the superficial walls, that is, these are without a central lens-like spot. Superficially bulging exothecial cells were also observed in the operculate taxa *Weisiopsis nigeriana* and *Weissia macrospora*, apparently as a convergent development. The capsule is, in *Trachycarpidium*, easily broken off the end of the short, weak seta (Pl. 80, f. 16).

Bryobartramia Sainsb. (Encalyptaceae, see Excluded Taxa) is similar in its nearly spherical capsule, hyaline exothecial cells, large spores, short seta, papillose calyptra, and yellow KOH reaction, but differs most significantly in the calyptra not detaching from the vaginula and inflated as an epigonium (Sainsbury 1948; Stone 1977a), the brown seta, the cauline leaves with only one stereid band, and the perichaetial leaves subulate, often consisting only of a costa.

Type material of *Trachycarpidium verrucosum* loaned by PC was *Weissia* (subg. *Phasconica*) *balsanae*; this material, however, was not in the original packet and was evidently segregated from the true type material, presumably still at PC, which is a *Trachycarpidium* from the illustration by Brotherus (1924–25).

Additional literature: Dixon (1942b), Potier de la Varde (1928), Stone (1975).

Number of accepted species: 5.

Species examined: *T. brisbanicum* (BUF, HSC, NY, MICH), *T. echinatum* (BM), *T. lonchophyllum* (NY), *T. tisserantii* (BM, US).

New combination: *Trachycarpidium lonchophyllum* (Roth) Zand., *comb. nov.* (*Astomum lonchophyllum* Roth, Aussereur. Laubm. 182, 1910).

59. ASCHISMA — Plate 81.

Aschisma Lindb., Utkast Nat. Grupp. Eur. Bladm. 28, 1878.
Type: *Aschisma carniolicum* (Web. & Mohr) Lindb.
Phascum subg. *Aschisma* (Lindb.) Kindb., Eur. N. Amer. Bryin. 2: 403, 1897.

From the alpha privative + σχίσμα, -ατος, a split; referring to the capsule lacking an operculum.

Plants very small, gregarious or forming a thin green turf. Stems seldom branching, ca. 0.2 mm in length, transverse section rounded-pentagonal, central strand present, weak, sclerodermis not well differentiated, hyalodermis absent; axillary hairs ca. 3 cells in length, basal cell thicker walled. *Leaves* incurved and tubulose when dry, weakly spreading when moist, *oblong or triangular to short-lanceolate, often falcate and plicate on one side*, 0.6–1.0 mm in length, upper lamina broadly channeled, *margins plane to occasionally weakly incurved*, entire to sharply serrulate above midleaf, *sometimes bordered above, below or throughout by 1–3 rows of weakly papillose, elongate cells*; apex broadly or rounded-acute; *base little differentiated in shape*; *costa excurrent as a sharp mucro, this occasionally rough or denticulate, superficial cells elongate both ventrally and dorsally*, 2–4 rows of cells

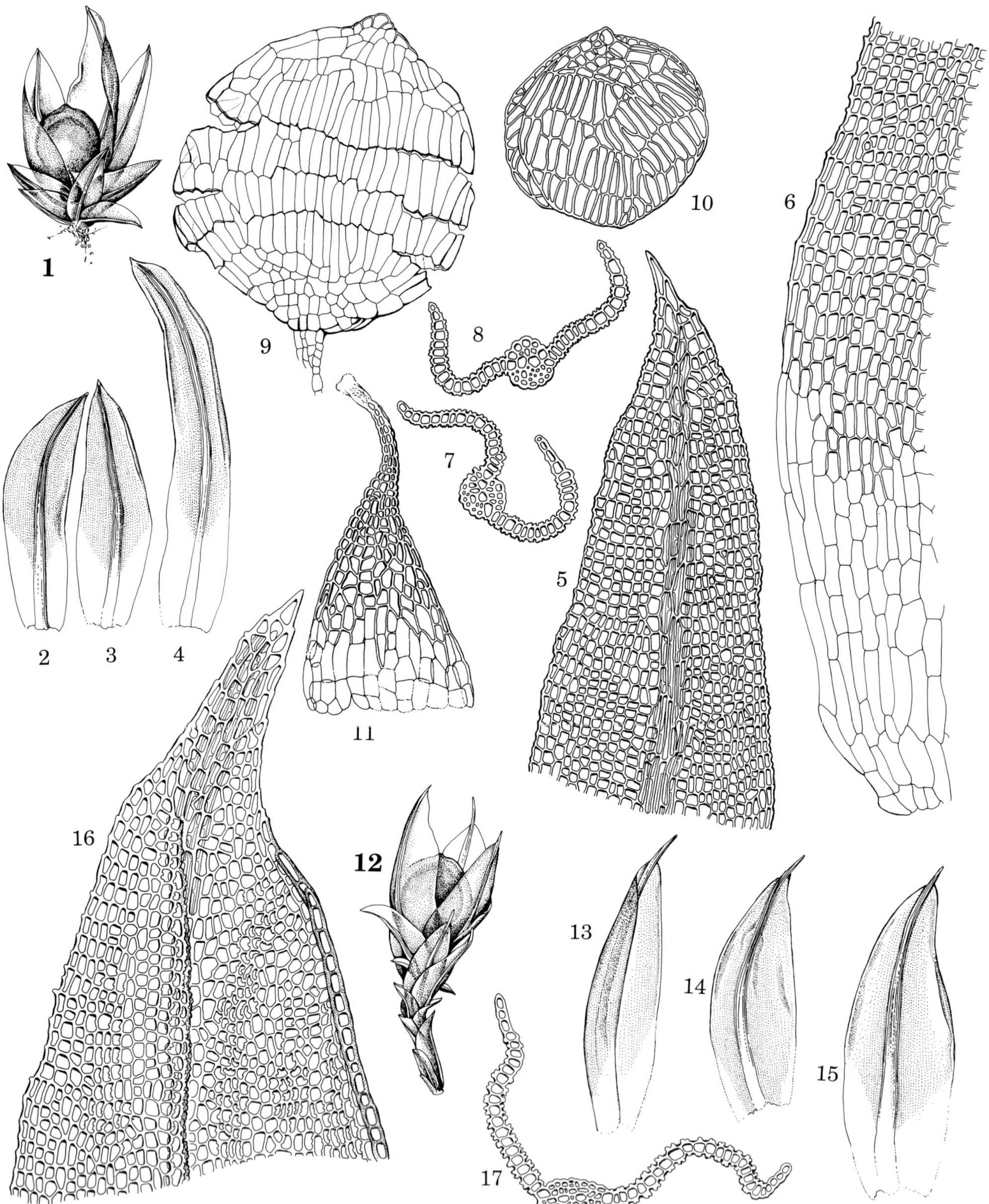

Plate 81. ***Aschisma.*** **1–11.** ***A. carniolicum.*** 1. Habit. 2–4. Three leaves. 5. Leaf apex, dorsal view. 6. Basal cells. 7. Transverse section through leaf base. 8. Transverse section through upper leaf. 9. Sporophyte. 10. Exothecial cells. 11. Calyptra. **12–17.** ***A. kansanum.*** 12. Habit. 13–15. Three leaves. 16. Leaf apex. 17. Transverse section at midleaf.

across costa ventrally at midleaf, costal transverse section circular to elliptical, *two stereid bands present*, mostly substereid ventrally, often also substereid dorsally, *epidermis not differentiated*, guide cells 2–4 in 1 layer, hydroid strand apparently absent or occasionally present; upper laminal cells quadrate to occasionally short-rectangular, 9–12 µm in width, 1:1(–2), walls evenly thickened, superficially flat or bulging only ventrally or on both exposed surfaces; *papillae stout, bifid*, crowded, scattered or centered over lumens, ca. 4 per lumen; *basal cells strongly differentiated, rising higher along margins in a weak vee*, rectangular, little wider than upper cells, 3–5:1, walls very thin to evenly thickened. Monoicous. Perichaetia terminal, inner leaves long-elliptical, to 1.5 mm in length, concave or not sheathing, lower cells oblong. Antheridia paroicous in axils of bracts or in stalked autoicous buds at base of the perichaetiate plant. *Seta very short*, ca. 50 µm in length, 1 per perichaetium, *hyaline*, straight; *capsule cleistocarpous*, 0.3 mm in diameter, *yellowish brown*, *spherical*, *with a small blunt apiculus* (ca. 30 µm in length), *exothecial cells rectangular, mostly* ca. 13–18 µm in width, *4–5:1*, *circling capsule longitudinally in several even bands with the appearance of a palisade, stomates absent, annulus absent*; *peristome teeth absent*. *Calyptra conic*, split once or twice, rough, 0.2–0.3 mm in length. *Spores large*, 18–24 µm in diameter, yellowish brown, essentially smooth to spiculose-papillose. *Laminal KOH color reaction yellow*.

A genus widely distributed in the Mediterranean area and disjunctive to central North America; growing in dry climates on soil, occasionally under translucent rocks in exposed situations.

The most prominent traits of this distinctive genus are the dry habitat; small size of the plants; spherical, cleistocarpous capsules with yellow, rectangular cells arranged in neat encircling bands, lacking stomates (Pl. 81, f. 9–10); the weak, hyaline seta; the falcate leaves usually with a distinct border of epapillose cells (Pl. 81, f. 5, 6, 16) and the leaf base not differentiated in shape; and the costa with usually merely substereid cells in two bands (Pl. 81, f. 7, 8, 17). As is the

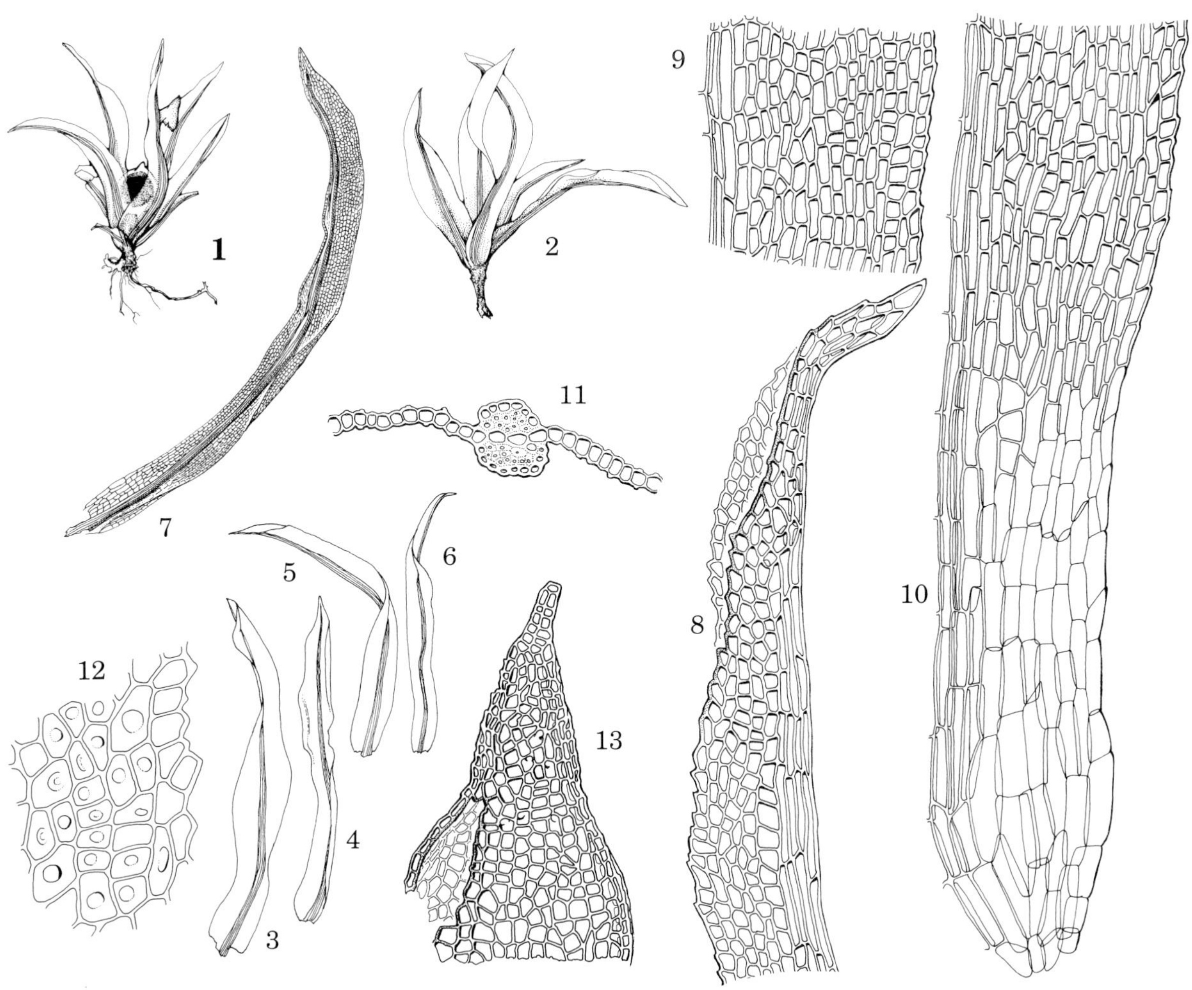

Plate 82. *Bryoceuthospora*. 1–13. *B. mexicana*. 1. Habit of sporophyte-bearing plant. 2. Perigoniate plant. 3–7. Five leaves, one enlarged. 8. Leaf apex. 9. Upper marginal cells. 10. Basal cells. 11. Transverse section at midleaf. 12. Papillae. 13. Calyptra.

case with *Tetrapterum* and *Tortula* sect. *Schizophascum*, both Pottioideae but rather distantly related, the capsule tends to rupture along the transverse walls at the butt ends of the exothecial cells, which are superficially rather thick-walled.

Aschisma carniolicum and *A. kansanum* are fairly distinct in the material seen. The latter (Pl. 81, f. 12–17) has more strongly bordered and serrulate margins, more strongly bulging upper laminal cells with centered (not scattered) papillae, a larger, more strongly denticulate mucro, and spores less ornamented and toward the small end of the size range for the genus. The protonema of *A. kansanum* survives arid conditions under translucent pebbles, ultimately producing leafy axes and sporophytes peripherally (Cridland 1959); this scenario is also the case with *Bruchia brevifolia* (Bruchiaceae) (*cf.* Rushing 1989) and, at least occasionally, *Syntrichia inermis* (Weger & During 1989).

Additional literature: Andrews (1915), Cárdenas (1988), Sérgio (1972a).

Number of accepted species: 2.

Species examined: *A. carniolicum* (BUF, NY), *A. kansanum* (NY).

60. BRYOCEUTHOSPORA — Plate 82.

Bryoceuthospora Crum & Anders., Bryologist 62: 66, 1959.
Type: *Bryoceuthospora mexicana* (Bartr.) Crum & Anders.
Ceuthospora Crum & Anders., J. Elisha Mitchell Sci. Soc. 74: 31, 1958, *hom. illeg. non* Fries, 1825.

From βρύον, a moss + κεῦθος, hidden + o + σπορά, seed, spore; referring to the immersed capsules.

Plants small, gregarious, often with a persistent protonema, yellow-green. Stems not branching, 0.5–4.0 mm in length, transverse section rounded-pentagonal, central strand absent or weak, outer cortex cells similar to those of central cylinder, hyalodermis absent; axillary hairs ca. 4 cells in length, the basal 1 yellowish. *Leaves* few (to 12), crowded and larger above, widely spreading and weakly contorted when dry, patent and recurved above midleaf when moist, *oblong-lanceolate to long-ligulate*, grading to the long-lanceolate perichaetial leaves, to 1.5 mm in length, upper lamina broadly channeled above, *margins broadly incurved to plane, irregularly dentate to closely serrate above midleaf*, sometimes entire except near the apex, somewhat undulate; apex acute; base scarcely differentiated in shape; *costa strong*, to 40–45 µm in width at midleaf, *ending 2–4 cells below apex, superficial cells long-rectangular and smooth on both sides*, with several rows of cells across costa at midleaf, costal transverse section circular to ovoid, *stereid bands two, strong, epidermis absent or weakly differentiated on both sides*, guide cells 2–4 in 1 layer, hydroid strand present but often weak or apparently absent; *upper laminal cells quadrate to hexagonal or short-rectangular, heterogeneous in shape*, (7–)9–12 µm in width, 1(–2):1, walls thin, superficially weakly bulging; *papillae large, simple, 1(–2) over each lumen*, hollow to solid, often weak or present on only a few of the medial upper laminal cells, basal cells differentiated across leaf base or rising higher marginally in a vee, rectangular, mostly 9–12 µm in width, 2–4:1, walls thin, somewhat bulging. Dioicous (or possibly rhizautoicous) or monoicous. *Perichaetia* terminal, *inner leaves about twice the length of the cauline leaves*. Paroicous or apparently rhizautoicous and perigoniate plants smaller than the perichaetiate, associated on the protonema. *Seta very short*, 35–40 µm in length, 1(–2) per perichaetium, yellow-green; *theca* 230–360 µm (apiculus 45–90 µm) in length, translucent yellow to yellowish brown, the brown spore mass evident, *cleistocarpous, spherical and conic-apiculate, surface smooth to colliculate*, exothecial cells hexagonal to rhomboidal, *thin-walled but with superficial, medial lens-like thickenings*, stomates at base of capsule, annulus not differentiated. *Calyptra* conic-campanulate and distinctly 4-lobed, *cells smooth to bulging conic-mamillose*, especially on the lobes, ca. 255 µm in length. *Spores large*, oval to elliptical, 25–30(–35) µm in longest diameter, dark brown, densely spiculose. Laminal KOH color reaction yellow.

This genus is known only from Mexico and Angola, growing on soil.

Bryoceuthospora is quite similar in general appearance to *Trachycarpidium* of Africa and the Australasian region (see Stone 1975) and to *Aschisma* of central Europe and North American prairie states. It differs gametophytically from both in the longer leaves (Pl. 82, f. 8) and low, unipapillose upper laminal cells (Pl. 82, f. 12). *Trachycarpidium* has similarly mamillose cells of the calyptra (Pl. 82, f. 13), but the cells of the exothecium of that genus are more strongly protuberant. Important features of *Bryoceuthospora* (holotype: Mexico, Sinaloa, Bartram 505, FH) are the dentate, broadly channeled leaves, margins nearly plane (not strongly and narrowly incurved as in most *Weissia* species), the strong costa, the cleistocarpous capsule with convex exothecial cells, and the (often) rough calyptra. The rather translucent exothecial cell walls relate it to *Uleobryum*, which has entirely colorless cell walls but a rather different gametophyte. *Ephemerum* species (both species of *Bryoceuthospora* were originally described in *Ephemerum*) look much like *Bryoceuthospora*, and may even have similarly unipapillose leaves (e.g. *E. cohaerens* (Hedw.) Hampe) with two (faint) stereid bands, but the leaves of that genus are more strongly serrate, the costa is rather thin, the exothecial cells are not bulging, and the calyptrae are smooth, distinctly campanulate rather than conic-campanulate. *Bryoceuthospora aethiopica*, recently reported from Mexico (Cárdenas 1988), is distinguishable from *B. mexicana* by its paroicous sexual condition, weak central strand, cauline leaves about half the length of the perichaetial and more strongly papillose (pluripapillose in the lower leaves), and smooth calyptrae; in spite of the species' name, the type (BM) was apparently collected near Golungo Alto, now in Angola.

Additional literature: Crum & Anderson (1958b).

Number of accepted species: 2.

Species examined: *B. aethiopica* (BM), *B. mexicana* (FH).

New combination: *Bryoceuthospora aethiopica* (Welw. & Dub.) Zand., *comb. nov.* (*Ephemerum aethiopicum* Welw. & Dub., Mém. Soc. Phys. Hist. Nat. Genève 21: 443, 1871; *Aschisma aethiopica* (Welw. & Dub.) Lindb.).

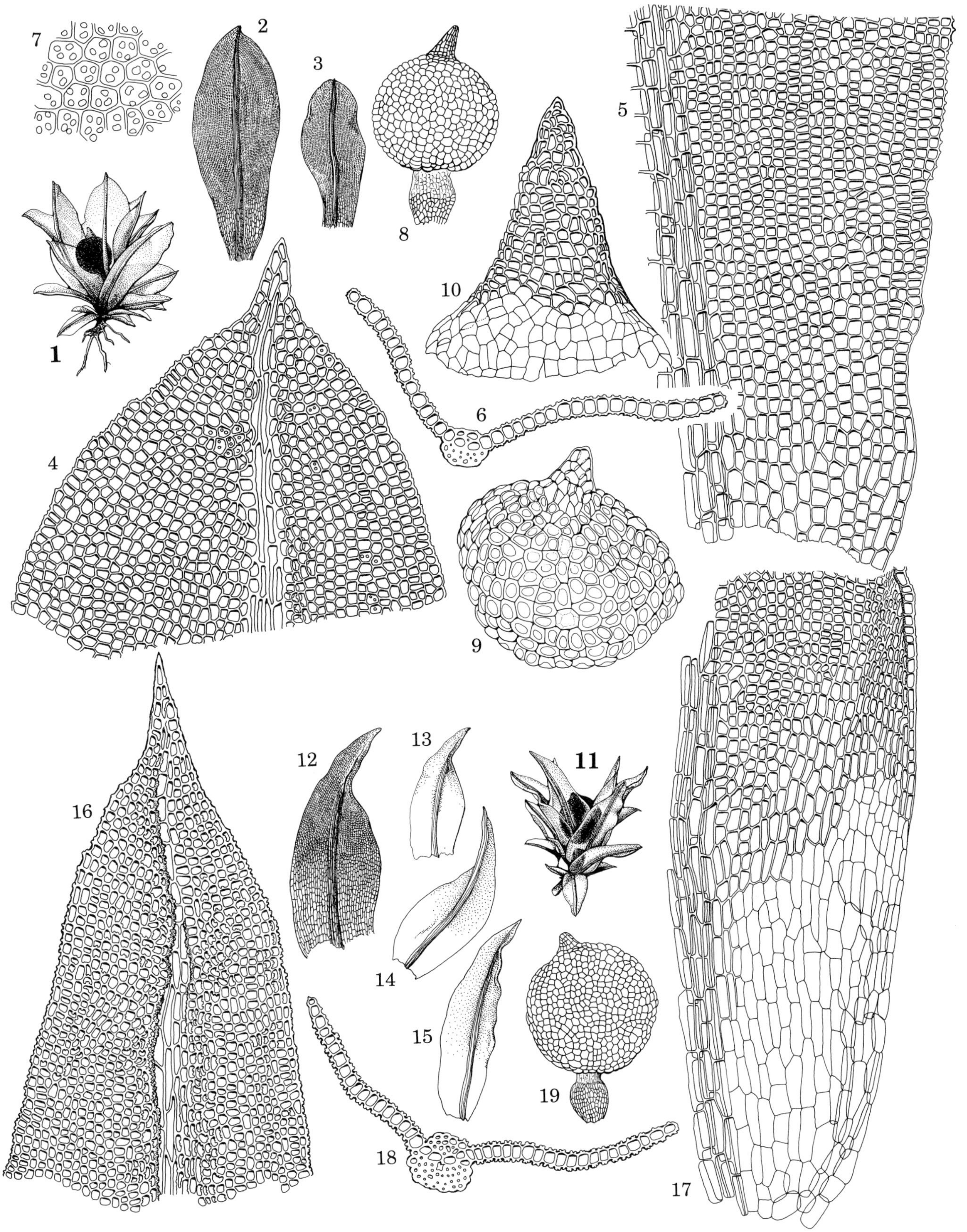

Plate 83. ***Uleobryum*. 1–10. *U. peruvianum*.** 1. Habit. 2–3. Two leaves. 4. Leaf apex. 5. Basal cells. 6. Transverse section at midleaf. 7. Laminal papillae. 8. Sporophyte. 9. Exothecial cells. 10. Calyptra. **11– 19. *U. curtisii*.** 11. Habit. 12–15. Four leaves. 16. Leaf apex. 17. Basal cells. 18. Transverse section at midleaf. 19. Sporophyte.

61. ULEOBRYUM Plate 83.

Uleobryum Broth., Hedwigia 45: 271, 1906. Type: *Uleobryum peruvianum* Broth.

Named for Ernst H. G. Ule, 1854–1915, a German collector of South American plants + o + βρύον, a moss.

Plants very short, gregarious or forming a thin turf, green above, brown below. Stems branching occasionally, 1–2 mm in length, transverse section rounded-pentagonal, central strand present, sclerodermis absent, hyalodermis absent; axillary hairs ca. 4 cells, basal 1 brownish. *Leaves* appressed-incurved when dry, spreading when moist, *spathulate to obovate or elliptical, ca. 1 mm in length, upper lamina narrowly grooved along costa, margins plane, minutely crenulate-serrulate by projecting papillae and cell walls*, marginal row of cells often smooth and rectangular; *apex* acute to rounded, *mucronate*; *base scarcely differentiated in shape*; *costa* percurrent to short-excurrent, occasionally ending 1–2 cells below apex, *superficial cells elongate ventrally* and dorsally, 2–6 rows of cells across costa ventrally at midleaf, costal transverse section round to semicircular, with two stereid bands (the ventral occasionally absent), epidermis variably present or absent ventrally, absent dorsally, guide cells 2–4 in 1 layer, hydroid strand usually present, often small; upper laminal cells hexagonal or subquadrate to short-rectangular, 8–13 µm in width, 1(–2):1, walls thin to evenly thickened, superficially convex on both sides, sometimes more convex ventrally; papillae simple or bifid, 2–4 per cell, occasionally indistinct, mostly hollow; *basal cells differentiated across leaf base, somewhat higher along margins*, rectangular, slightly inflated, to 20 µm in width, mostly 3–4:1, walls thin, hyaline. *Autoicous*. Perichaetia terminal, *inner leaves* with base often oblong, otherwise *little different from the cauline leaves*, not or little sheathing, lower cells little different from those of the cauline leaves. Perigonia present as small axillary buds on perichaetiate plants. *Seta very short*, ca. 30–40 µm in length, 1(–2) per perichaetiate plant, *hyaline, straight; theca spherical with a strong, conical apiculus*, ca. 350–400 µm (plus 110–150 µm for apiculus) in diameter, *cleistocarpous, glistening*, transmitting the brown color of the contained spores, *exothecial cells hexagonal, hyaline* (entirely transparent, colorless), *mamillose, superficial walls centrally thickened and lens-like*, columella apparently absent at maturity, stomates ca. 4, phaneropore, at base of theca, *annulus absent, operculum not differentiated. Calyptra conic-campanulate*, split once or twice, *cells slightly rough to strongly* mamillose, ca. 250–300 µm in length. *Spores large*, ca. 25–33 µm in diameter, brown, closely low-spiculose. *Laminal KOH color reaction yellow*.

Found on bare soil; West Indies, Mexico, Peru, Brazil and Australia.

Uleobryum is quite like *Bryoceuthospora* in its small size, spherical capsules with hyaline setae (Pl. 83, f. 8, 19), the transparent capsule walls that are thickened on the surface to form bulging lenses (Pl. 83, f. 9) and which transmit the color of the large spores, the conic-campanulate calyptrae that are usually rough or pustulate apically (Pl. 83, f. 10), the leaves serrulate (very weakly so in *Uleobryum*), and the costa with two stereid bands (Pl. 83, f. 6, 18) and elongate superficial cells ventrally. It differs significantly, albeit not altogether satisfactorily, from *Bryoceuthospora* in the shorter, lingulate to elliptical leaves with pluripapillose leaf cells (Pl. 83, f. 7). The two genera are quite close; see also Cladogram 14. *Trachycarpidium* has very lightly yellow-brown colored exothecial cells and the gametophyte is similar in areolation to that of *U. curtisii* (Pl. 83, f. 11–19), but it differs in a teardrop-shaped capsule that is strongly pustulate, and the exothecial cells are evenly thickened superficially, not just medially as in *Uleobryum*. *Uleobryum occultum* of Brazil differs little from the Australian *U. curtisii*, being slightly more strongly serrulate on the leaf margins by projecting cell walls.

Additional literature: Cárdenas (1988), Stone (1984, 1985).

Number of accepted species: 3.

Species examined: *U. curtisii* (MELU), *U. occultum* (SPA), *U. peruvianum* (H, NY).

New heterotypic synonymy: *Phascum brittoniae* Crum & Steere (*nom. nov.* for *Phascum sessile* E. Britt.) = *Uleobryum peruvianum* Broth.

New combination: *Uleobryum occultum* (Roth) Zand., *comb. nov.* (*Aschisma occultum* Roth, Aussereur. Laubm. 173, 1911).

62. TORTULA Plates 84–89.

Tortula Hedw., Sp. Musc. 122, 1801, *nom. cons. non* Roxburgh, 1800. Lectotype: *Tortula subulata* Hedw.

Beccaria C. Müll., Nuov. Giorn. Bot. Ital. 4: 11, 1872.

Bauriella Warnst., Hedwigia 57: 88, 1915, *nom. inval. prov.* Type: *Tortula polyseta* (C. Müll.) Warnst.

Tortula sect.? *Piliferae* De Not., Mem. R. Acc. Sc. Torino 40: 287, 1838, rank not indicated; inoperative in piority I.C.B.N. Art. 35.2.

Barbula sect. *Amphidiopsis* C. Müll., Linnaea 42: 332, 1879. Type: *Barbula amphidiifolia* C. Müll.

Barbula sect. *Pilifera* Lázaro é Ibiza, Bot. Descr. Comp. Fl. Esp. 1: 586, 1896.

Barbula sect. *Orthopodiae* Kindb., Eur. N. Amer. Bryin. 2: 245, 1897.

Barbula sect. *Catillaria* C. Müll., Gen. Musc. Fr. 425, 1900. Type: *Barbula pellata* Schimp.

Pottia sect. *Beccaria* (C. Müll.) C. Müll., Gen. Musc. Fr. 389, 1900.

Pottia subsect. *Acutae* C. Jens., Skand. Bladmfl. 203, 1939, *nom. inval. descr. suec.*

See sectional synonymy for additional nomenclature.

From *tortus*, twisted + *-ula*, diminutive; referring to the twisted peristome teeth.

Plants forming cushions or turfs, green or occasionally blackish green above, yellow-brown to dark brown below. *Stems* branching occasionally, to 2 cm in length, transverse section rounded-pentagonal, *central strand present* or very rarely absent, *sclerodermis absent, hyalodermis absent;* axillary hairs ca. 5–8 cells in length, basal 1–3 cells thicker walled; rhizoids often dense. *Leaves* appressed-incurved to lax when dry, weakly to widely spreading when moist, *usually obovate to spathulate,* occasionally ovate to elliptical or ligulate, 1–4(–6) mm in length, upper lamina nearly flat to concave, broadly channeled, occasionally grooved along costa, *margins recurved below or rarely plane, entire or occasionally weakly serrulate near apex,* marginal 1–4 rows of cells often less papillose and smaller than the medial or walls thicker, occasionally marginal cells elongate, rarely bistratose; apex broadly acute to rounded; base scarcely differentiated in shape to elliptical, rarely weakly auricled; costa short- to long-excurrent as an awn, occasionally percurrent or subpercurrent, costa with lamina inserted laterally or to 45°, superficial cells quadrate or occasionally short-rectangular and papillose or smooth ventrally, dorsally short-rectangular to elongate and papillose or smooth, *3–4(–5) rows of cells across costa ventrally at midleaf, costal transverse section circular to semicircular,* ventral stereid band absent or occasionally small and represented by a few cells, *dorsally present and round, elliptical or semicircular in shape,* epidermis present ventrally and dorsally or occasionally only laterally on the dorsal side, rarely absent dorsally, guide cells 2(–3) in 1(–2) layers or rarely absent, *hydroid strand usually present,* often large, very rarely absent; rarely an elliptical pad of cells bulging from ventral surface of the costa; *upper laminal cells* rounded-quadrate to hexagonal, occasionally rhomboidal, *ca. 15–19 µm in width,* 1–2:1, *walls* thin or seldom evenly thickened, *superficially convex; papillae usually hollow, simple or bifid, 4–6 per lumen, occasionally on a conical salient, rarely absent;* basal cells differentiated across leaf or higher medially, rectangular, often rather lax, 18–25 µm in width, 2–5:1, walls thin, hyaline, rarely little differentiated. *Propagula absent.* Dioicous or monoicous (commonly autoicous or paroicous). Perichaetia terminal, inner leaves little differentiated or somewhat larger than the cauline. Perigonia terminal or as autoicous buds in subperichaetial or lower leaf axils. *Seta very short or to 2.5 cm in length,* 1 (very rarely 2) per perichaetium, yellowish brown to brown, *twisted counterclockwise, clockwise or not twisted; theca stegocarpous or else cleistocarpous,* 0.5–3.0(–7.0) mm in length, yellowish brown to dark brown, *spherical, ovate, elliptical or cylindrical, occasionally inclined,* occasionally macrostomous, exothecial cells rectangular, 25–30 µm, ca. 2–3:1, rarely 4–5:1, walls thin or evenly thickened, stomates present at base of theca, phaneropore, annulus of 1–2 rows of vesiculose cells, persistent or very rarely revoluble, occasionally absent or rarely with up to 8 circumferential weak lines of dehiscence; *peristome teeth of 32 filaments or 16 triangular teeth or rudimentary or absent,* long or shortly triangular, cleft to near base, spiculose, up to 2000 µm in length, with many articulations, straight to twisted counterclockwise, basal membrane absent or low or up to 1000 µm in height, tessellated and spiculose. *Operculum when differentiated long-conic, occasionally shortly rostrate,* 0.5–2.5 mm in length, cells twisted counterclockwise. Calyptra cucullate, smooth, 2.5–6.0 mm in length. Spores 13–30(–50) µm in diameter, light brown, papillose, rarely densely spiculose. *Laminal KOH color reaction usually yellow,* occasionally red medially, occasionally negative, rarely reddish orange. Reported chromosome numbers: Sect. *Tortula*: n = 13+m, 14, 24, 26, 27, 28, 30, 39, 40, 48, 48+m, 50, 52, 60, 66. Sect. *Pottia*: n = 12, 13, 15, 20, 21, 24, 25, 26, 26+m, 27, 28+1–2acc, 30, 32, 42, 52. The most often reported number for both sections is n = 26.

Found on most continents in various habitats, mainly soil.

With the segregation of various genera (Zander 1989: *Chenia, Dolotortula, Hennediella, Hilpertia, Sagenotortula, Stonea, Syntrichia*), *Tortula* becomes a fairly homogeneous group with very similar gametophytes and a characteristic tendency to reduction in the sporophyte. Thus, the correlations of season of sporophyte maturation (Zander 1979d) with genus (*Pottia* having spring and winter sporophyte maturation dates and *Tortula* in the traditional sense having dates mainly in the spring and summer) is probably a reflection of "life strategy" (*cf.* During 1979) rather than phylogeny, though this needs to be tested. Major characters of *Tortula* as presented here include presence of stem central strand (rarely absent or present in different stems of same collection, e.g. *T. brevissima*) and absence of sclerodermis and hyalodermis (Pl. 84, f. 2, 17); leaves usually obovate to spathulate, margins usually narrowly recurved below and entire; costal stereid band usually semicircular to rounded in section, hydroid strand present, dorsal epidermis usually present (Pl. 84, f. 7, 8, 21); upper laminal cells usually rather large and clear (i.e. walls usually relatively unobscured by the papillae, Pl. 84, f. 9); propagula absent; upper laminal KOH reaction usually yellow.

Tortula is distinguishable from *Syntrichia* by the semicircular to rounded stereid band (not crescent-shaped) and yellow KOH reaction of the upper laminal cells (not red). Unlike *Syntrichia*, a dorsal costal epidermis is usually differentiated, either completely or occasionally only laterally (Pl. 89, f. 15—as is also the case in *Hennediella*). The yellow KOH reaction is usually present, but some few taxa have no color reaction (e.g. *T. entosthodontacea*), or the basal cells may be brick red (e.g. *T. raucopapillosa*), or the leaves may blush red medially in the upper part of the leaf (e.g. *T. atrovirens*), or the older leaves may be red and the younger yellow (e.g. *T. lingulata*), or all leaves may have a reddish orange cast (e.g. *T. nevadensis*). But the characters of these taxa otherwise are those of *Tortula* as here emended. *Hennediella* is distinguished by its red KOH reaction, commonly dentate or serrate and plane upper laminal margins, and superficially flattened upper laminal cells. Papillae may be absent in some *Tortula* species or variously expressed in different specimens of the same species.

Chamberlain (1978) recognized *Pottia caespitosa*, but the leaves have plane margins, the upper laminal cells are rather small and thick-walled, and costal sections show two stereid bands. This small-statured species is actually a *Trichostomum* (as witness the combination *Trichostomum caespitosum* (Bruch ex Brid.) Jur.), differing somewhat from other species of that genus by the broadly sheathing perichaetial leaves and quite neckless capsule.

Upon examination of the sporophytes of a considerable range of species of Pottieae, there was found to be no sharp

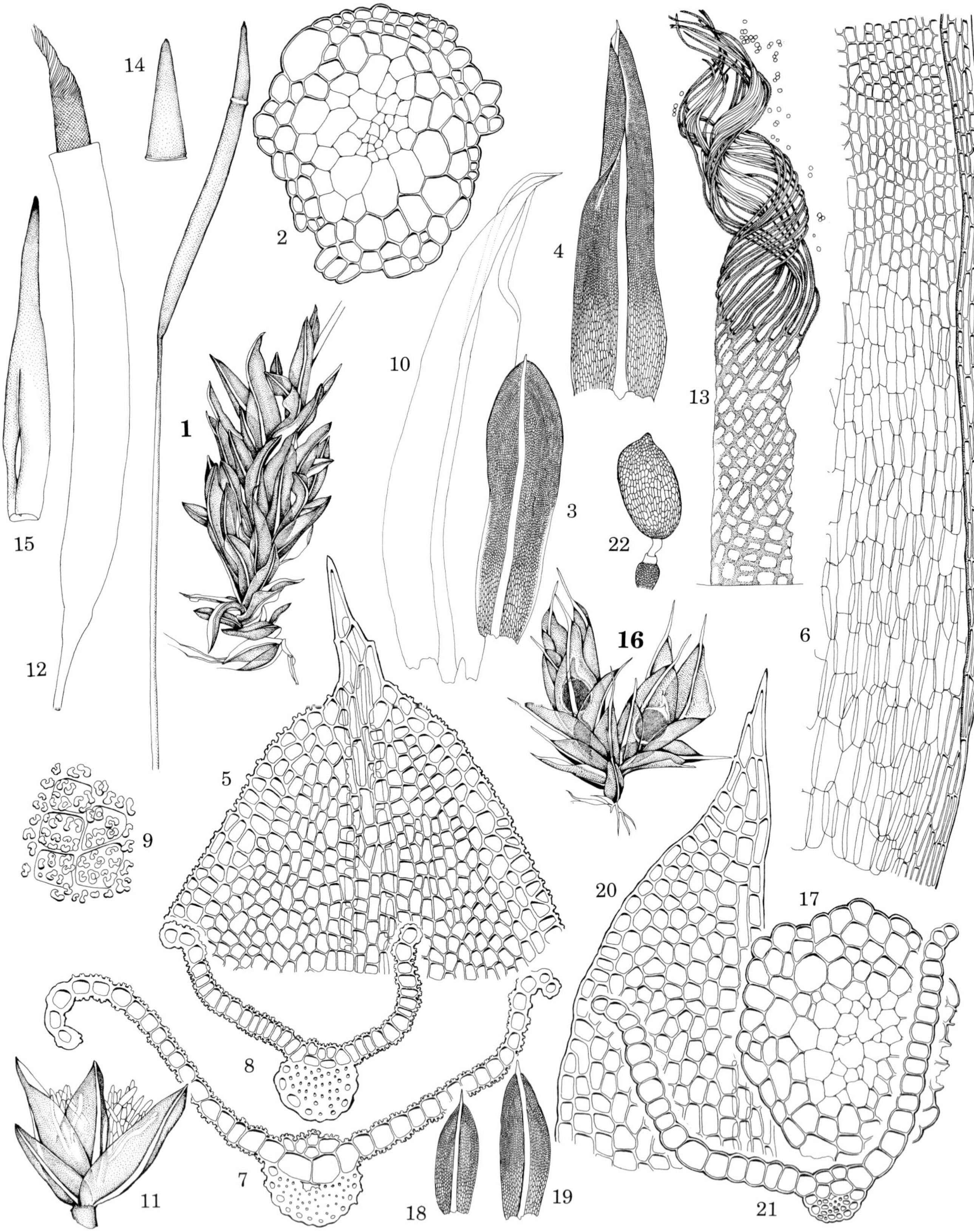

Plate 84. ***Tortula.*** **1–15.** ***T. subulata.*** 1. Habit, with sporophyte. 2. Transverse section of stem. 3–4. Two leaves. 5. Leaf apex. 6. Basal cells. 7. Transverse section at midleaf. 8. Transverse section near leaf apex. 9. Upper laminal papillae. 10. Perichaetial leaf. 11. Autoicous bud. 12. Theca. 13. Portion of basal membrane and peristome. 14. Operculum. 15. Calyptra. **16–22.** ***T. atherodes.*** 16. Habit. 17. Transverse section of stem. 18–19. Two leaves. 20. Leaf apex, lateral view. 21. Transverse section at midleaf. 22. Sporophyte.

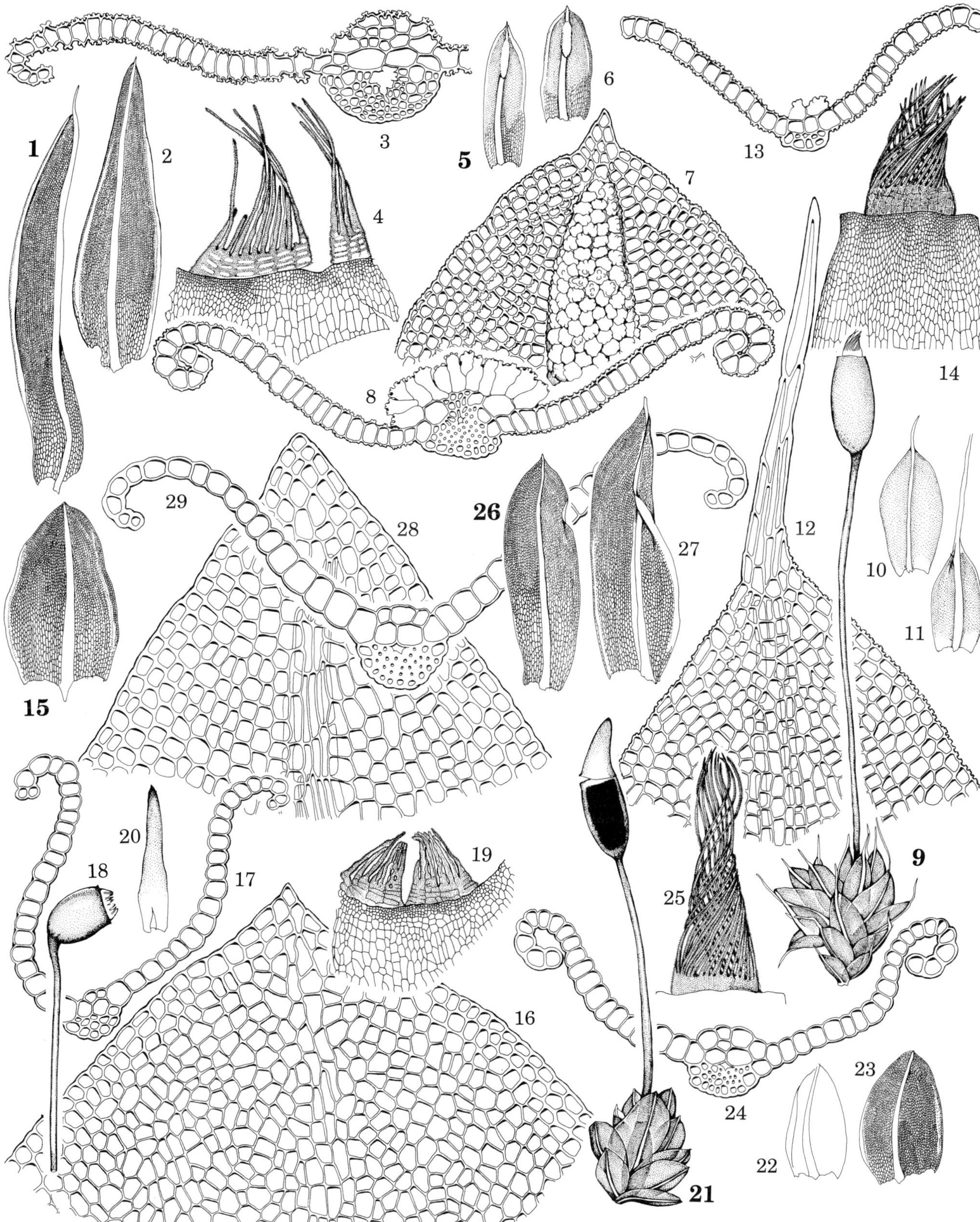

Plate 85. ***Tortula.*** **1–4.** ***T. altipes.*** 1–2. Two leaves. 3. Transverse section at midleaf. 4. Portion of peristome. **5–8.** ***T. atrovirens.*** 5–6. Two leaves. 7. Leaf apex. 8. Transverse section at midleaf. **9–14.** ***T. canescens.*** 9. Habit. 10–11. Two leaves. 12. Leaf apex. 13. Transverse section at midleaf. 14. Peristome. **15–20.** ***T. cernua.*** 15. Leaf. 16. Leaf apex. 17. Transverse section at midleaf. 88. Theca with peristome. 19. Portion of peristome. 20. Calyptra. **21–25.** ***T. cuneifolia* var. *blissii.*** 21. Habit. 22–23. Two leaves. 24. Transverse section at midleaf. 25. Peristome. **26–29.** ***T. deciduidentata.*** 26–27. Two leaves. 28. Leaf apex. 29. Transverse section at midleaf.

difference between traditional "*Tortula*" peristomes with 32 similar rami (Pl. 84, f. 12–13) and "*Desmatodon*" peristomes with 16 teeth cleft to near the base or to a basal membrane (i.e., 32 paired rami, Pl. 86, f. 16; 87, f. 9–10). Even if some difference was statistically demonstrable, it would cut across observed (and as taxonomically recognized here) clearly defined generic groupings based on several gametophytic characters. It is simpler to entertain convergence of one character (short, probably reduced peristomes having teeth paired) than convergent evolution of several distinctive gametophyte morphotypes two or more times. The flattened basal portion of "*Desmatodon*" peristome teeth may be explained as a cleft portion of the basal membrane. The proximally flat teeth of *Desmatodon* species are also different from the almost terete filaments of *Tortula* with long, twisted peristomes simply because the long filaments of *Tortula* have distal regions little wider than their thickness. A cladistic evaluation at the species level may clarify this.

Visotska's (1967) proposal of a subfamily Tortuloideae (no type cited), based on a chromosome number of x = 12 and intended to contain *Tortula*, *Aloina* and *Crossidium*, was criticized on cytological grounds by Nyholm and Wigh (1973) because several species of *Tortula* have a basic chromosome number of x = 13. Evaluation of chromosome counts given by Fritsch (1982, 1991) gave both 12 and 13 as basic numbers for both *Tortula s. str.* and *Syntrichia* as conceived in the present study; however, Newton (1972) found n = 7 for *S. robusta* and Ramsay (1974) found n = 6+m for *S. papillosa*.

An electron microscopical study by Lewinsky (1974) of spore ornamentation in 10 European species of *Tortula s. lat.* showed differences between the spores of the specimens studied (only one or two collections were examined for each species although 20–30 spores from two to five capsules were studied in each species), which probably represent differences between the species, but she found no evidence of differences between traditional sections of the genus.

Mishler's (1986a) cladogram of postulated phylogenetic relationships of several species of *Tortula s. lat.* recognized *Tortula s. str.* (as recognized here) as a primitive group (he listed *T. subulata*, *T. mucronifolia* and *T. muralis*) distinct from several other species (all recognized here as *Syntrichia*) by the upper laminal cells not strongly mammilose.

Corley et al. (1981) gave an up-to-date presentation of the sections of the genus as represented in Europe. Their apprehension of *Desmatodon* as a rather small assemblage of the type species and closely related taxa presages the present study. They stated that *Desmatodon* "is not defined by sound technical characters, and there has been much confusion about which species should be assigned to it. As with *Didymodon* there has been too much emphasis placed on the peristome, which is not a conservative character in Pottiaceae."

Tortula muralis, *T. leucostoma* and *T. altipes* occasionally have a small ventral (sub)stereid band (Pl. 85, f. 3). The costa is, however, rounded in section and generally unlike that of *Barbula*.

Selected literature on *Tortula s. str.* (here including *Phascum*, *Desmatodon* and *Pottia*): Arts (1987a,b, 1988), Bachelot (1813), Bennett (1965), Brown (1894b,c), Bryan (1956), Carrión et al. (1990), Chamerlain (1978), Crundwell (1953, 1955, 1956), Dixon (1927b), Favali and Gianni (1973), Guerra et al. (1988, 1991, 1992), Häusler (1984), Hernnstadt and Heyn (1989), Holzinger (1925), Hughes (1969, 1979, 1982), Hughes and Wiggin (1969), Jiménez et al. (1990), Kanda (1981), Lazarenko (1969, 1974), Lazarenko and Lesnyak (1972), Lazarenko et al. (1961), Lewinsky (1974), Lightowlers (1984, 1985a,c, 1986a,b,c), Lobachevskaya et al. (1986), Matteri (1977a,b), Mishler (1985b, 1986a, 1990), Mishler and Newton (1988), Ripetskij (1978, 1979), Ripetskij et al. (1983), Risse (1985), Rumsey (1992), Rungby (1957), Sainsbury (1936), Saito (1973a), Savicz-Ljubitzkaya and Smirnova (1963b, 1965), Sérgio (1972a, 1978), Springer (1935), Steere (1939a, 1940), Stone (1989), Toth (1987), Ulycna (1977), Wareham (1939a), Wareham and Whitney (1939), Warnstorf (1912, 1916), Zander (1989).

Number of accepted species: 163, none remaining in *Desmatodon*, plus 10 as yet undistributed in *Phascum*, plus 30 undistributed in *Pottia*.

Species examined: see below.

New heterotypic synonymy: *Didymodon schimperi* (Mont.) Broth. = *Tortula atrovirens* (Sm.) Lindb.

New taxa, combinations, statuses and names:

Tortula altipes (Broth.) Zand., *comb. nov.* (*Desmatodon altipes* Broth., Act. Hort. Bot. Ac. Sc. U.R.S.S. 42: 154, 1931).

Tortula argentinica (Broth.) Zand., *comb. nov.* (*Desmatodon argentinicus* Broth., Ark. Bot. 15(6): 5, 1918), not seen.

Tortula atherodes Zand., *nom. nov.* (*Phascum cuspidatum* Schreb. ex Hedw., Spec. Musc. 22, 1801).

Tortula atherodes var. *arcuata* (Herrnstadt & Heyn) Zand., *comb. nov.* (*Phascum cuspidatum* var. *arcuatum* Herrnstadt & Heyn, Bryologist 94: 175, 1991), not seen.

Tortula atherodes var. *affinis* (Nees & Hornsch.) Zand., *comb. nov.* (*Phascum affine* Nees & Hornsch., Bryol. Germ. 1: 74, 1823; *Phascum cuspidatum* var. *affine* (Nees & Hornsch.) Hampe), not seen.

Tortula atherodes var. *curviseta* (Dicks.) Zand., *comb. nov.* (*Phascum curvisetum* Dicks., Pl. Crypt. Brit. 4: 2, 1801; *Phascum cuspidatum* var. *curvisetum* (Dicks.) Nees & Hornsch.), not seen.

Tortula atherodes var. *diaphora* (Hag.) Zand., *comb. nov.* (*Phascum acaulon* var. *diaphorum* Hag., K. Norsk. Vid. Selsk. Skrift. 1928(3): 19, 1929; *Phascum cuspidatum* var. *diaphorum* (Hag.) C. Jens.), not seen.

Tortula atherodes var. *elata* (Brid.) Zand., *comb. nov.* (*Phascum elatum* Brid., J. Bot. (Schrader) 1800(1): 269, 1801; *Phascum cuspidatum* var. *elatum* (Brid.) Drumm.), not seen.

Tortula atherodes var. *intertexta* (Brid.) Zand., *comb. nov.* (*Phascum intertextum* Brid., Mant. Musc. 8, 1819; *Phascum cuspidatum* var. *intertextum* (Brid.) Brid., not seen.)

Tortula atherodes var. *marginata* (Hernnstadt & Heyn) Zand., *comb. nov.* (*Phascum cuspidatum* var. *marginatum* Hernnstadt & Heyn, Bryologist 94: 175, 1991), not seen.

Tortula atherodes var. *mitraeformis* (Limpr.) Zand., *comb. nov.* (*Phascum cuspidatum* var. *mitraeforme* Limpr., Laubm. Deutschl. 1: 187, 1885), not seen.

Tortula atherodes var. *papillosa* (Lindb.) Zand., *comb. nov.* (*Phascum papillosum* Lindb., Oefv. K. Vet. Ak. Foerh. 21: 217, 1864; *Phascum cuspidatum* var. *papillosum* (Lindb.) Roth); *Phascum cuspidatum* ssp. *papillosum* Guerra & Ros in Guerra, Jiménez, Ros & Carrión), not seen.

Tortula atherodes var. *pilifera* (Hedw.) Zand., *comb. nov.* (*Phascum piliferum* Scherb. ex Hedw., Spec. Musc. 20, 1801; *Phascum cuspidatum* var. *piliferum* Scherb. ex Hedw.) Hook. & Tayl.).

Tortula atherodes var. *retortifolia* (Guerra & Ros in Guerra, Jiménez, Ros & Carrión) Zand., *comb. nov.* (*Phascum cuspidatum* var. *retortifolium* Guerra & Ros in Guerra, Jiménez, Ros & Carrión, Cryptogamie, Bryol. Lichénol. 12: 390, 1991),˙not seen.

Tortula atherodes var. *schreberiana* (Dicks.) Zand., *comb. nov.* (*Phascum schreberianum* Dicks., Pl. Crypt. Brit. 4: 2, 1801; *Phascum cuspidatum* var. *schreberianum* (Dicks.) Brid.), not seen.

Tortula atrovirens var. *leucodonta* (Corb.) Zand., *comb. nov.* (*Barbula atrovirens* var. *leucodonta* Corb., Mém. Soc. Sc. Nat. Cherbourg 26: 244, 1889; *Desmatodon convolutus* var. *leucodontus* (Corb.) Wijk & Marg.), not seen.

Tortula bogosica (C. Müll.) Zand., *comb. nov.* (*Desmatodon bogosicus* C. Müll., Nuov. Giorn. Bot. Ital. 4: 12, 1872).

Tortula capillaris (Chen) Zand., *comb. nov.* (*Desmatodon capillaris* Chen, Hedwigia 80: 287, 1941), not seen.

Tortula cernua var. *xanthopus* (Kindb.) Zand., *comb. nov.* (*Desmatodon cernuus* var. *xanthopus* Kindb., Ottawa Natural. 4: 61, 1890), not seen.

Tortula cuneifolia var. *blissii* Zand., *var. nov.* (Pl. 85, f. 21–25.) *A varietate typica gametophytis subatris, in solutione KOH colorem profundiorem evolventibus, foliis rigide appressis, seta curta, crassa, longitudine 4–5 mm, latitudine 0.2–0.3 mm differt.*

Differs from the typical variety by the blackish gametophytes, these more strongly colored in KOH; leaves stiffy appressed; seta short, ca. 4–5 in length, stout, 0.2–0.3 mm in width. Type: Canada, Northwest Territories, Cornwallis Island, Resolute Bay area, L. C. Bliss, 1977, holotype, BUF; isotype, ALTA. The leaves of var. *blissii* are identical in morphology to those of muticous-leaved forms of the European species *T. cuneifolia*, but show a very strong color reaction to KOH: bright yellow upper lamina and deep brick-red basal cells, colors which are pale in European specimens at BUF. The short, thick seta is apparently unique to this arctic variety; European specimens have setae 0.7–1.5 mm in length and ca. 0.15 mm in width. Like European specimens, the capsule is variably macro- and microstomous, and the operculum is broadly to narrowly conic. This specimen was incorrectly reported (Vitt & Zander 1978) as a second known station for *Crumia deciduidentata* (here treated as a species of *Tortula* near *T. cuneifolia*), which has a similar short, thick seta and smooth, weakly bordered leaves that are bright yellow in KOH except for brick-red basal cells, but which differs in its capsule about twice as long (2.7–3.3 mm vs. 1.3–1.6 mm) and proportionately thicker; operculum pushed off by the elongating columella; spores larger (ca. 18 µm vs. ca. 13 µm); leaves ovate to clearly spathulate (vs. short- to long-ovate); constricted leaf apex; and much enlarged basal cells. A section across the leaf base of var. *blissii* shows what appear to be two stereid cell groups of equal size separated by a group of hydroid cells (also seen in sections of the basal portion of the costa of *Tortula muralis*, and see discussion of *Phascopsis*).

Tortula entosthodontacea (Card. & Dix.) Zand., *comb. nov.* (*Hyophilopsis entosthodontacea* Card. & Dix., J. Bot. 49: 137, 1911.).

Tortula euryphylla Zand., *nom. nov.* (*Dicranum latifolium* Hedw., Spec. Musc. 140, 1801; *Desmatodon latifolius* (Hedw.) Brid.).

Tortula euryphylla ssp. *brevifolia* (Kindb.) Zand., *comb. nov.* (*Tortula latifolia* (Hedw.) Lindb. ssp. *brevifolia* Kindb., Bih. K. Svensk. Vet. Ak. Handl. 7(9): 135, 1883; *Desmatodon latifolius* ssp. *brevifolius* (Kindb.) Kindb.).

Tortula euryphylla var. *eucalyptrata* (Lindb.) Zand., *comb. nov.* (*Tortula eucalyptrata* Lindb., Bot. Not. 1886: 100, 1886; *Desmatodon latifolius* varr *eucalyptratus* (Lindb.) Kaur.).

Tortula euryphylla var. *flavescens* (Brid.) Zand., *comb. nov.* (*Dicranum latifolium* var. *flavescens* Brid., Spec. Musc. 140, 1801; *Desmatodon latifolius* var. *flavescens*).

Tortula euryphylla var. *spelaea* (Amann) Zand., *comb. nov.* (*Desamtodon spelaeus* Amann, Bull. Murithienne 40: 46, 1919; *Desmatodon latifolius* var. *spelaeus* (Amann) Podp.).

Tortula euryphylla var. *subobliqua* (Lindb.) Zand., *comb. nov.* (*Desmatodon latifolius* var. *suboliquus* Lindb., Oefv. K. Vet. Ak. Foerh. 23: d553, 1867).

Tortula chungtienia Zand., *nom. nov.* (*Desmatodon yuennanensis* Broth., Symb. Sin. 4: 44, 1929).

Tortula deciduidentata (Sharp & Iwats.) Zand., *comb. nov.* (*Crumia deciduidentata* Sharp & Iwats., J. Hattori Bot. Lab. 32: 95, 1969).

Tortula kabir-khanii (Broth.) Zand., *comb. nov.* (*Desmatodon kabir-khanii* Broth., Mitteil. Inst. Allg. Bot. Hamburg 8: 400, 1931), not seen.

Tortula lanceola Zand., *nom. nov.* (*Encalypta lanceolata* Hedw., Spec. Musc. 63, 1801; *Pottia lanceolata* (Hedw.) C. Müll.; *non Tortula lanceolata* (Hedw.) P. Beauv.).

Tortula lanceola var. *albidens* (Corb.) Zand., *comb. nov.* (*Pottia lanceolata* var. *albidens* Corb., Rev. Bryol. 22: 35, 1895), not seen.

Tortula lanceola var. *angustata* (B.&S. in BSG) Zand., *comb. nov.* (*Anacalypta lanceolata* var. *angustata* B.&S. in BSG, Bryol. Eur. 2: 48, 1843; *Pottia lanceolata* var. *angustata* (B.&S. in BSG) C. Müll.), not seen.

Tortula lanceola var. *lejolisii* (Corb.) Zand., *comb. nov.* (*Pottia lanceolata* var. *lejolisii* Corb., Mem. Soc. Sci. Nat. Cherbourg 26: 238, 1889), not seen.

Tortula lanceola var. *leucodonta* (Schimp.) Zand., *comb. nov.*

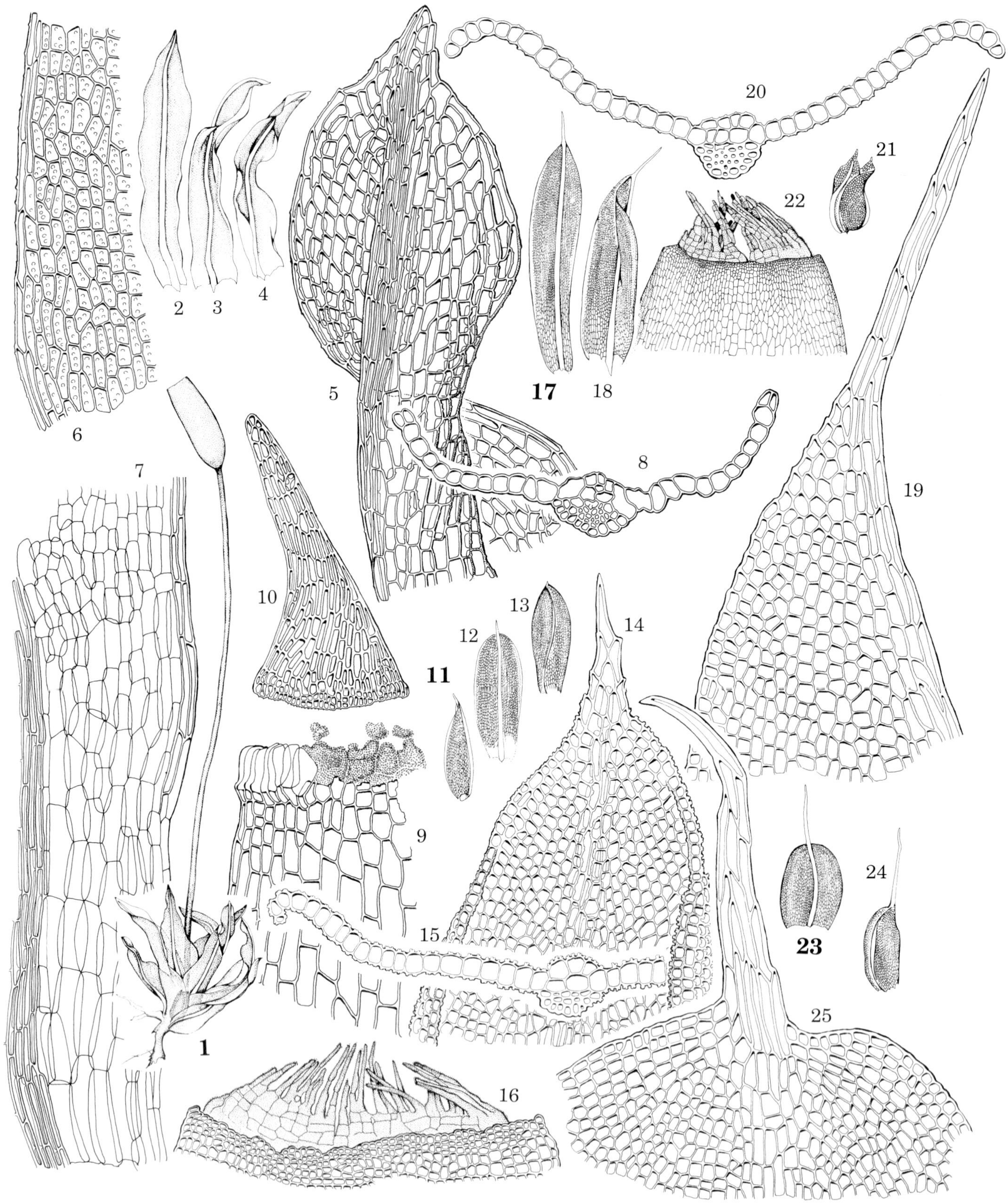

Plate 86. ***Tortula.*** **1–10** ***T. entosthodontacea.*** 1. Habit. 2–4. Two leaves. 5. Leaf apex. 6. Upper marginal cells showing papillae. 7. Basal cells. 8. Transverse section at midleaf. 9. Portion of rudimentary peristome. 10. Operculum. **11–16.** ***T. euryphylla.*** 11–13. Three leaves. 14. Leaf apex. 15. Transverse section at midleaf. 16. Peristome. **17–22.** ***T. lanceola.*** 17–18. Two leaves. 19. Leaf apex, lateral view. 20. Transverse section at midleaf. 21. Autoicous bud. 22. Peristome. **23–25.** ***T. transcaspica.*** 23–24. Two leaves. 25. Leaf apex.

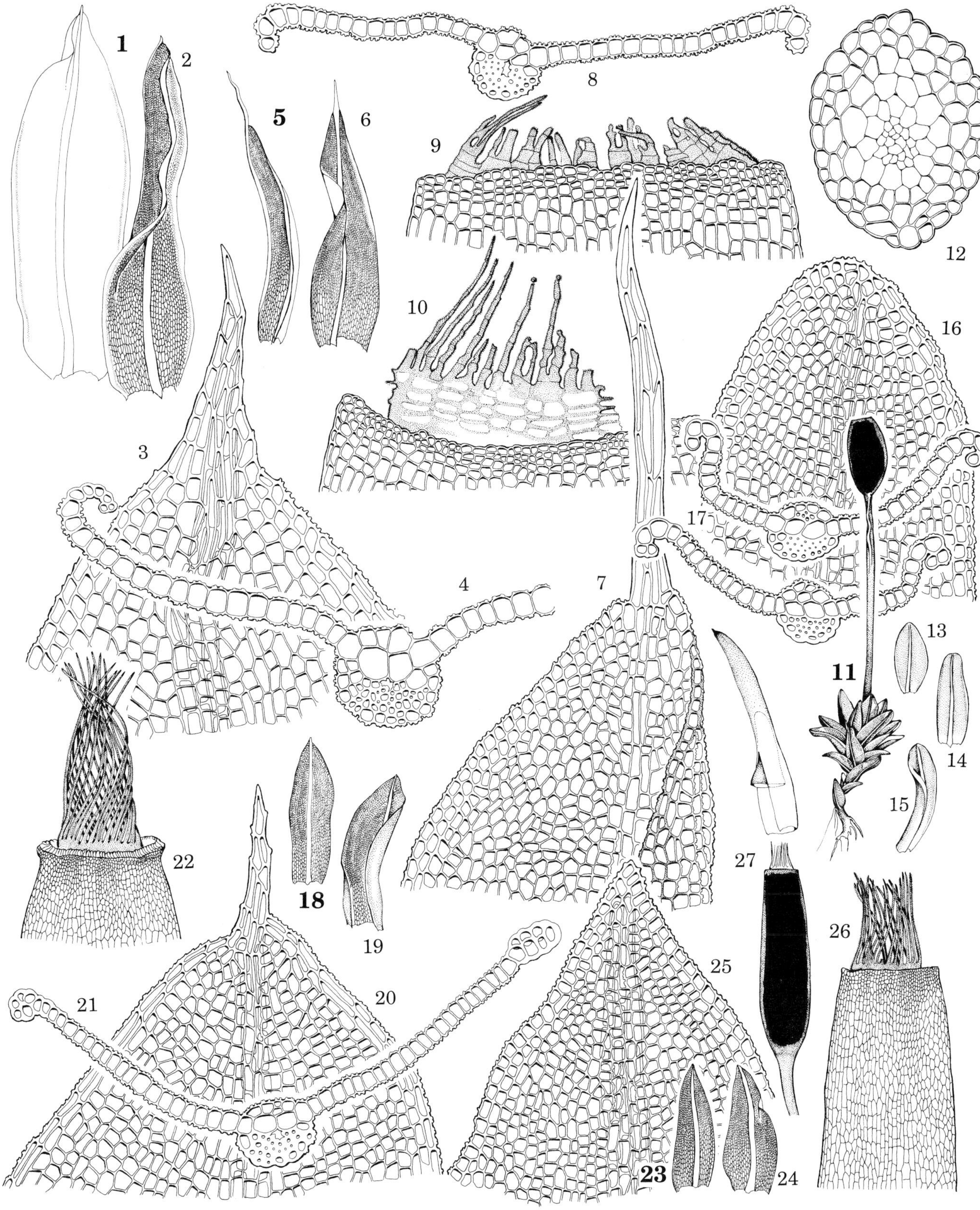

Plate 87. *Tortula.* 1–4. *T. laureri.* 1–2. Two leaves. 3. Leaf apex. 4. Transverse section at midleaf. **5–10. *T. leucostoma.*** 5–6. Two leaves. 7. Leaf apex. 8. Transverse section at midleaf. 9–10. Two examples of peristome teeth. **11–17. *T. lingulata.*** 11. Habit. 12. Transverse section of stem. 13–15. Two leaves. 16. Leaf apex. 17. Two transverse sections at midleaf. **18–22. *T. marginata.*** 18–19. Two leaves. 20. Leaf apex. 21. Transverse section at midleaf. 22. Peristome. **23–27. *T. porteri.*** 23–24. Two leaves. 25. Leaf apex. 26. Peristome. 27. Theca, operculum and calyptra.

(*Pottia lanceolata* var. *leucodonta* Schimp., Syn. ed. 2: 158, 1876), not seen.

Tortula lanceola var. *macrophylla* (Warnst.) Zand., *comb. nov.* (*Pottia lanceolata* var. *macrophylla* Warnst., Hedwigia 58: 133, 1916), not seen.

Tortula lanceola var. *microphylla* (Warnst.) Zand., *comb. nov.* (*Pottia lanceolata* var. *microphylla* Warnst., Hedwigia 58: 134, 1916), not seen.

Tortula lanceola var. *mucronata* (Amann) Zand., *comb. nov.* (*Pottia lanceolata* var. *mucronata* Amann, Bull. Soc. Vaudoise Sci. Nat. 53: 85, 1920), not seen.

Tortula lanceola var. *ovalifolia* (Warnst.) Zand., *comb. nov.* (*Pottia lanceolata* var. *ovalifolia* Warnst., Hedwigia 58: 131, 1916), not seen.

Tortula lanceola var. *papillosa* (Corb.) Zand., *comb. nov.* (*Pottia lanceolata* var. *papillosa* Corb., Mém. Soc. Sci. Nat. Cherbourg 26: 237, 1889), not seen.

Tortula lanceola var. *rigidior* (Schwaegr.) Zand., *comb. nov.* (*Encalypta lanceolata* var. *rigidior* Schwaegr., Spec. Musc. Suppl. 1(1): 61, 1811.)

Tortula laureri var. *setschwanica* (Broth.) Zand., *comb. nov.* (*Desmatodon setschwanicus* Broth., Symb. Sin. 4: 43, 1929; *Desmatodon laureri* var. *setschwanicus* (Broth.) Chen), not seen.

Tortula maritima (R. Br. ter) Zand., *comb. nov.* (*Dendia maritima* R. Br. ter, Trans. New Zealand Inst. 30: 411, 1898; *Pottia maritima* (R. Br. ter) Broth.).

Tortula minor (C. Müll.) Zand., *comb. nov.* (*Beccaria minor* C. Müll., Nuov. Giorn. Bot. Ital. 4: 11, 1872; *Pottia minor* (C. Müll.) Wijk & Marg.), not seen.

Tortula minor var. *elatior* (C. Müll.) Zand., *comb. nov.* (*Beccaria elatior* C. Müll., Nuov. Giorn. Bot. Ital. 4: 11, 1872; *Pottia minor* var. *elatior* (C. Müll.) Wijk & Marg.).

Tortula modica Zand., *nom. nov.* (*Gymnostomum intermedium* Turn., Musc. Hib. 7, 1804; *Pottia intermedia* (Turn.) Fuernr.).

Tortula modica var. *corsa* (Fleisch. & Warnst.) Zand., *comb. nov.* (*Pottia intermedia* var. *corsa* Fleisch. & Warnst., Bot. Centralbl. 65: 299, 1896), not seen.

Tortula modica var. *gymnandra* (Schiffn.) Zand., *comb. nov.* (*Pottia intermedia* var. *gymnandra* Schiffn., Oesterr. Bot. Zeitschr. 47: 55, 1897), not seen.

Tortula modica var. *gymnogyna* (Schiffn.) Zand., *comb. nov.* (*Pottia intermedia* var. *gymnogyna* Schiffn., Oesterr. Bot. Zeitschr. 48: 389, 1898), not seen.

Tortula modica var. *revoluta* (Schiffn.) Zand., *comb. nov.* (*Pottia intermedia* var. *revoluta* Schiffn., Oesterr. Bot. Zeitschr. 47: 55, 1897), not seen.

Tortula modica var. *stenocarpa* (Velen.) Zand., *comb. nov.* (*Pottia intermedia* var. *stenocarpa* Velen., Rozpravy Cesk. Ak. Ved. Tr. 2, 6(6): 148, 1897), not seen.

Tortula modica var. *tenuis* (Vent.) Zand., *comb. nov.* (*Pottia intermedia* var. *tenuis* Vent., Muscin. Trent. 31, 1899), not seen.

Tortula nevadensis (Card. & Thér.) Zand., *comb. nov.* (*Pottia nevadensis* Card. & Thér., Bot. Gaz. 37: 365, 1904).

Tortula pallida (Lindb.) Zand., *comb. nov.* (*Pottia pallida* Lindb., Oefv. K. Vet. Ak. Foerh. 21: 220, 1864).

Tortula pallida var. *longicuspis* (Warnst.) Zand., *comb. nov.* (*Pottia pallida* var. *longicuspis* Warnst., Hedwigia 58: 113, 1916).

Tortula porteri (Jam. in Aust.) Zand., *comb. nov.* (*Desmatodon porteri* Jam. in Aust., Musci Appal. 123, 1870).

Tortula protobryoides Zand., *nom. nov.* (*Phascum bryoides* Dicks., Pl. Crypt. Brit. 4: 3, 1801; *Pottia bryoides* (Dicks.) Mitt.).

Tortula protobryoides var. *brevifolia* (De Not.) Zand., *comb. nov.* (*Phascum bryoides* var. *brevifolium* De Not., Atti Univ. Genova 1: 734, 1869; *Pottia bryoides* var. *brevifolia* (De Not.) Wijk & Marg.), not seen.

Tortula protobryoides var. *thornhillii* (Wils.) Zand., *comb. nov.* (*Phascum bryoides* var. *thornhillii* Wils., Bryol. Brit. 33, 1855; *Pottia bryoides* var. *thornhillii* (Wils.) Braithw.), not seen.

Tortula randii (Kenn.) Zand., *comb. nov.* (*Pottia randii* Kenn., Rhodora 1: 78, 1899; *Desmatodon randii* (Kenn.) Laz.).

Tortula raucopapillosa (X.-j. Li) Zand., *comb. nov.* (*Desmatodon raucopapillosus* X.-j. Li, Acta Bot. Yunnan. 3: 105, 1981 "*raucopapillosum*").

Tortula rhodonia Zand., *nom. nov.* (*Desmatodon wilczekii* Meyl., Bull. Soc. Vaudoise Sc. Nat. 52: 383, 1919), not seen.

Tortula sainsburyana Zand., *nom. nov.* (*Pottia stevensii* R. Br. ter, Trans. N. Z. Inst. 26: 291, 1894).

Tortula solomensis (Broth.) Zand., *comb. nov.* (*Desmatodon solomensis* Broth., Rev. Bryol. n. ser. 2: 2, 1929), not seen.

Tortula splachnoides (Hornsch.) Zand., *comb. nov.* (*Phascum splachnoides* Hornsch., Horae Phys. Berol. 57, 1820; *Pottia splachnoides* (Hornsch.) Broth.).

Tortula thompsonii (C. Müll.) Zand., *comb. nov.* (*Trichostomum thompsonii* C. Müll., Bot. Zeit. 22: 359, 1854; *Desmatodon thompsonii* (C. Müll.) Jaeg.).

Tortula tonkinensis (Besch.) Zand., *comb. nov.* (*Desmatodon tonkinensis* Besch., Bull. Soc. Bot. France 41: 80, 1894), not seen, *cf.* Brotherus (1924: 297).

Tortula truncata var. *brevirostris* (Lisa) Zand., *comb. nov.* (*Gymnostomum truncatum* var. *brevirostre* Lisa, Elenco Muschi Torino 16, 1837; *Pottia truncata* var. *brevirostris* (Lisa) De Not.), not seen.

Tortula truncata var. *illyrica* (Latz.) Zand., *comb. nov.* (*Pottia illyrica* Latz., Beih. Bot. Centralbl. 48(2): 481, 1931; *Pottia truncata* var. *illyrica* (Latz.) Podp.), not seen.

Tortula truncata var. *littoralis* (Mitt.) Zand., *comb. nov.* (*Pottia littoralis* Mitt., J. Bot. 9: 4, 1871; *Pottia truncata* var. *littoralis* (Mitt.) Warnst.), not seen.

Tortula truncata var. *minutissima* (Warnst.) Zand., *comb. nov.* (*Pottia truncata* var. *minutissima* Warnst., Hedwigia 58: 117, 1916), not seen.

Tortula ucrainica (Laz.) Zand., *comb. nov.* (*Desmatodon ucrainicus* Laz., Bull. Jard. Bot. Kieff 4: 34, 1926).

Tortula willisiana Zand., *nom. nov.* (*Phascum drummondii*

Wils., London J. Bot. 7: 26, 1848; *Pottia drummondii* (Wils.) Willis).

Tortula willisiana var. *obscura* (Willis) Zand., *comb. nov.* (*Pottia drummondii* var. *obscura* Willis, Vict. Nat. 70: 171, 1954), not seen.

Tortula wilsonii (Hook.) Zand., *comb. nov.* (*Gymnostomum wilsonii* Hook., Bot. Misc. 1: 143, 41, 1829; *Pottia wilsonii* (Hook.) BSG).

Tortula wilsonii var. *asperula* (Mitt.) Zand., *comb. nov.* (*Pottia asperula* Mitt., J. Bot. 9: 4, 1871; *Pottia wilsonii* ssp. *asperula* (Mitt.) Kindb.), not seen.

Tortula wilsonii var. *crinita* (Wils. ex B.&S.) Zand., *comb. nov.* (*Pottia crinita* Wils. ex B.&S. in BSG, Bryol. Eur. 2: 43, 1843; *Pottia wilsonii* var. *crinita* (Wils. ex B.&S.) Warnst.), not seen.

Tortula wilsonii var. *mucronifolia* (Bruch in F. A. Müll.) Zand., *comb. nov.* (*Entosthymenium mucronifolium* Bruch in F. A. Müll., Flora 12: 387, 1829; *Pottia wilsonii* var. *mucronifolia* (Bruch in F. A. Müll.) Warnst.), not seen.

Tortula zoddae Zand., *nom. nov.* (*Pottia cuneifolia* Solms ex Schimp., Syn. ed 2, 154, 1876).

TORTULA Sect. TORTULA

Tortula sect. *Tortula* Hedw., Sp. Musc. 122, 1801, *nom. cons. non* Roxburgh, 1800. Lectotype: *Tortula subulata* Hedw.

Desmatodon Brid., Mant. Musc. 86, 1819. Lectotype: *Desmatodon latifolius* (Hedw.) Brid. *fide* Venturi, Comm. Fauna Fl. Gea Venezia 1: 123, 1868.

Zygotrichia Brid., Bryol. Univ. 1: 520, 1826. Type: *Zygotrichia leucostoma* (R. Br.) Brid.

Dermatodon Hüb., Musc. Germ. 14: 109, 1833, *p.p.*

Pachyneurum Amann, F. Mouss. Suisse 2: 112, 1912.

Tortula subg. *Desmatodon* (Brid.) Lindb., Musci Scand. 20, 1879.

Tortula subg. *Zygotrichia* (Brid.) Lindb., Musci Scand. 20, 1879.

Tortula subg. *Tortula* C. Jens., Medd. Groenland 3: 379, 1887, *nom. illeg.*

Didymodon subg. *Desmatodon* (Brid.) Kindb., Eur. N. Amer. Bryin. 2: 273, 1897.

Barbula subg. *Zygotrichia* Kindb., Eur. N. Amer. Bryin. 2: 245, 1897.

Barbula subg. *Tortula* Kindb., Eur. N. Amer. Bryin. 2: 246, 1897, *hom. illeg.*

Tortula subg. *Pachyneurum* (Amann) C. Jens., Skand. Bladmfl. 200, 1939.

Pottia subg. *Pseudodesmatodon* Medel. in C. Jens., Skand. Bladmfl. 207, 1939, *nom. inval. descr. suec.*

Tortula subg. *Eutortula* C. Jens., Skand. Bladmfl. 199, 1939, *nom. illeg.*

Barbula sect. *Tortulae* Rebent., Prodr. Fl. Neomarch. 257, 1804.

Tortula sect. *Subulatae* De Not., Mem. R. Acc. Sc. Torino 40: 287, 1838. Type: *Tortula subulata* Hedw.

Barbula sect. *Subulatae* B.&S. in BSG, Bryol. Eur. 2: 98, 1842 (fasc. 13–15 Mon. 36).

Trichostomum sect. *Desmatodon* (Brid.) C. Müll., Syn. 1: 588, 1849.

Barbula sect. *Crassinerves* (De Not.) Milde, Bryol. Soles. 112, 1869. Type: *Barbula nervosa* (BSG) Milde.

Tortula sect. *Desmatodon* (Brid.) Mitt., J. Linn. Soc. Bot. 12: 145, 164, 1869.

Tortula sect. *Zygotrichia* (Brid.) Mitt., J. Linn. Soc. Bot. 12: 145, 168, 1869.

Barbula sect. *Crassicostatae* Schimp., Syn. ed. 2: 194, 1876.

Desmatodon sect. *Eudesmatodon* Jur., Laubmfl. Öst. Ungarn 129, 1882, *nom. illeg.*

Desmatodon sect. *Crassicostati* Jur., Laubmfl. Öst.-Ungarn 135, 1882. Type: *Desmatodon atrovirens* (Sm.) Jur.

Desmatodon sect. *Subulati* Jur., Laubmfl. Öst. Ungarn 129, 1882. Lectoype nov.: *Desmatodon subulatus* (Hedw.) Jur.

Pottia sect. *Didyctium* C. Müll. in Broth., Hedwigia 34: 124, 1895. Type: *Pottia asperula* C. Müll. in Broth.

Barbula sect. *Crassinervia* Lázaro é Ibiza, Bot. Descr. Comp. Fl. Esp. 1: 506, 1896, *p.p. Crossidium* & *p.p. Aloina.*

Barbula sect. *Muraliformes* Kindb., Eur. N. Amer. Bryin. 2: 246, 1897. Type: *Barbula muralis* (Hedw.) Crome.

Barbula sect. *Canescentes* Kindb., Eur. N. Amer. Bryin. 2: 245, 1897. Type: *Barbula canescens* (Mont.) B.&S.

Barbula sect. *Limbatae* Kindb., Eur. N. Amer. Bryin. 2: 246, 1897. Type: *Barbula marginata* B.&S.

Barbula sect. *Subulataeformes* Kindb., Eur. N. Amer. Bryin. 2: 245, 1897. Type: *Barbula subulata* (Hedw.) P. Beauv.

Barbula sect. *Tortula* Hérib., Mém. Ac. Sc. Clermont-Ferrand ser. 2, 14: 364, 1899, *nom. illeg.*

Tortula sect. *Tortula* Broth., Nat. Pfl. 1(3): 429, 1902, *nom. illeg.*

Tortula sect. *Eutortula* Broth., Act. Hort. Gothoburg 1: 191, 1924, *nom. illeg.*

Syntrichia sect. *Zygotrichia* (Brid.) Mönk., Laubm. Eur. 306, 1927.

Tortula sect. *Crassicostatae* (Schimp.) Podp., Consp. Musc. Eur. 242, 1954.

Tortula sect. *Crassinerves* (Milde) Wijk & Marg., Taxon 8: 75, 1959. Type: *Barbula nervosa* (De Not.) Milde, *hom. illeg.*

Tortula subsect. *Desmatodon* (Brid.) Braithw., Brit. Moss. Fl. 3: 212, 1885.

Upper laminal cells usually densely papillose and relatively small (ca. 10–13 µm in width), ventral costal cells 3–6 in transverse section, usually forming a convex pad and usually densely hollow papillose with simple or bifid papillae, leaves usually with rounded apices and commonly with long hyaline awns; costal sections with guide cells; peristomes usually present.

Lindberg (1864) treated *T. subulata* and *T. mucronifolia* as a single taxon. Steere (1940) pointed out that, although the variable (Warnstorf 1912) *T. subulata* and *T. mucronifolia* intergrade extensively in Europe, the two species appear to have have distinct ranges and morphology in North America.

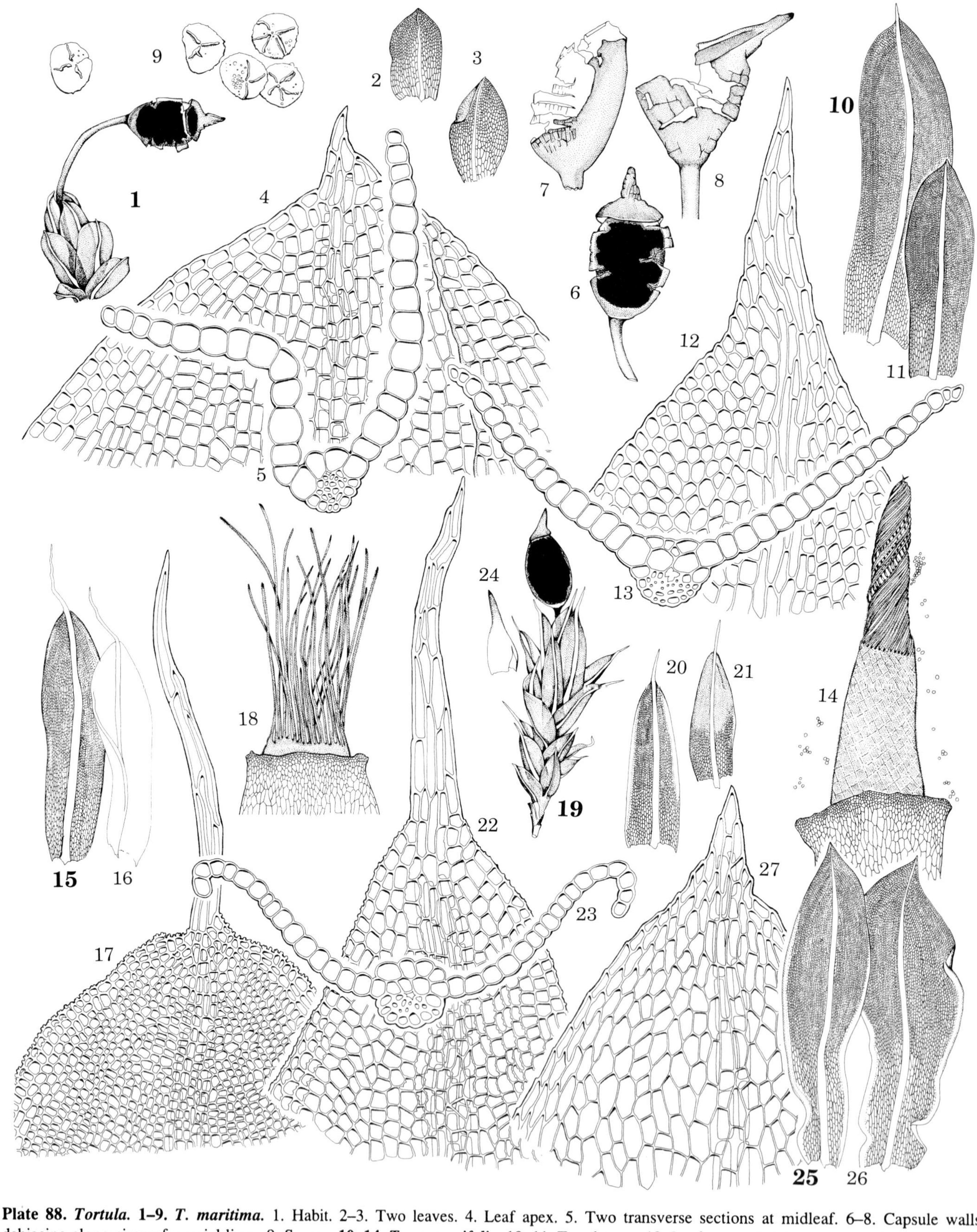

Plate 88. *Tortula.* 1–9. *T. maritima.* 1. Habit. 2–3. Two leaves. 4. Leaf apex. 5. Two transverse sections at midleaf. 6–8. Capsule walls dehiscing along circumferencial lines. 9. Spores. **10–14. *T. mucronifolia.*** 10–11. Two leaves. 12. Leaf apex. 13. Transverse section at midleaf. 14. Peristome. **15–18. *T. platyphylla.*** 15–16. Two leaves. 17. Leaf apex. 18. Peristome. **19–25. *T. protobryoides.*** 19. Habit. 20–21. Two leaves. 22. Leaf apex. 23. Transverse section at midleaf. 24. Calyptra. **25–27. *T. randii.*** 25–26. Two leaves. 27. Leaf apex.

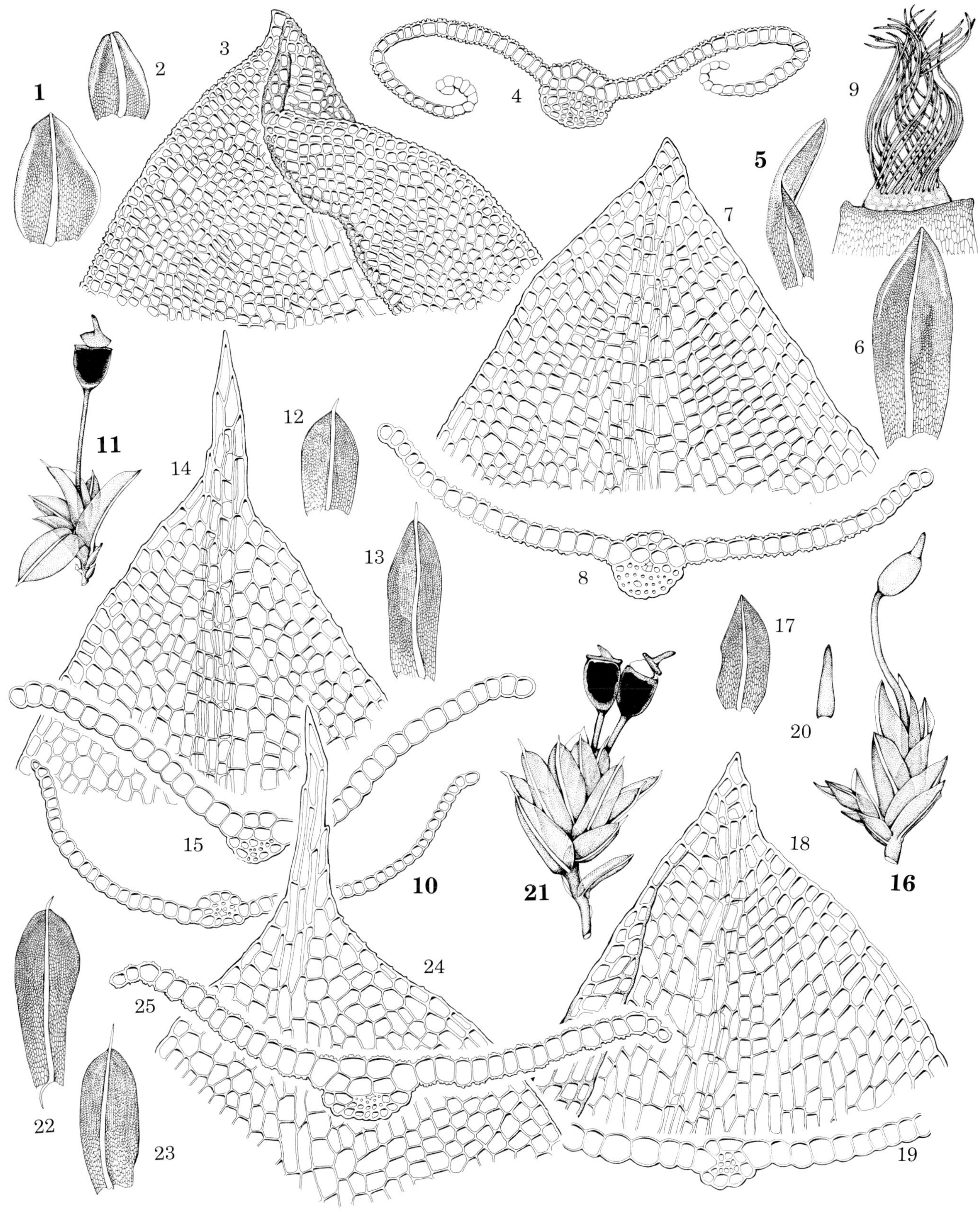

Plate 89. ***Tortula.*** **1–4.** ***T. revolvens.*** 1–2. Two leaves. 3. Leaf apex. 4. Transverse section at midleaf. **5–9.** ***T. solmsii.*** 5–6. Two leaves. 7. Leaf apex. 8. Transverse section at midleaf. 9. Peristome. **10.** ***T. splachnoides.*** 10. Transverse section at midleaf. **11–15.** ***T. truncata.*** 11. Habit. 12–13. Two leaves. 14. Leaf apex. 15. Transverse section at midleaf. **16–20.** ***T. willisiana.*** 16. Habit. 17. Leaf. 18. Leaf apex. 19. Transverse section at midleaf. 20. Calyptra. **21–25.** ***T. wilsonii.*** 21. Habit. 22–23. Two leaves. 24. Leaf apex. 25. Transverse section at midleaf.

Chen (1941) commented, significantly as it turns out, that *T. mucronifolia* (as *Syntrichia mucronifolia*) has a leaf morphology quite like that of many *Pottia* species. This is similarly unfortunate for nomenclatural purposes, since the closely related *T. subulata* is the type of the genus and therefore of the sect. *Tortula*. In the present study, the variation between *T. subulata* and *T. mucronifolia* was found to parallel major differences in laminal structure between sect. *Tortula* and sect. *Pottia*. Thus, in spite of an apparent close relationship, these two species are placed in different sections of the genus. This is not unacceptable if the two are considered taxa whose ancestors are inserted near the base of the lineage of *Tortula* species; this needs to be analysed at the species level.

Tortula sect. *Tortula* includes species with papillose leaves and small cells (e.g. *T. muralis*) that are nearly identical with species with larger, smooth laminal cells (e.g. *T. californica* and *T. transcaspica*, Pl. 86, f. 23–25), but these last are distinctly awned and placed in sect. *Tortula* on that account. Perhaps *T. subulata*, with its short awn, ought to be placed with the species of section *Pottia* (a name change at the sectional level would be necessary), but *T. subulata* otherwise has the bulk of characters given above for the section. However, many species of sect. *Tortula* with the laminal morphology of *T. muralis*, the species probably most representative of sect. *Tortula*, also have short awns or lack them altogether. The evidence that sect. *Pottia* demonstrates far more sporophyte reduction than sect. *Tortula* may be interpreted either as an additional character supporting sect. *Pottia* as a distinctive line of evolution (just as *Syntrichia* characteristically lacks sporophyte reduction) or that smooth cells and reduced sporophytes are correlated at the end of several separate reduction series from sect. *Tortula*. In any case, this should be further analysed at the species level.

Steere (1939a) felt that *Tortula leucostoma* (Pl. 87, f. 5–10) and *T. guepinii* would better be regarded as subspecies of *T. euryphylla* (all as *Desmatodon* spp., the last as *D. latifolius*).

Tortula revolvens (Pl. 89, f. 1–4) has the innermost cells of its revolute upper margins hollow-papillose, the group somewhat differentiated as a photosynthetic organ like that of *Hilpertia* and species of *Pseudocrossidium*; *T. revolvens* is, however, closely related to *T. muralis* by the costa round in section, with a single strong stereid band, and upper lamina highly papillose, cells superficially flat, and KOH reaction yellow.

Tortula atrovirens (Pl. 85, f. 5–8) is near *T. revolvens* in short-ovate leaf shape, thickened costa, and upper marginal cells epapillose. *Tortula muralis* has a similarly thickened costa, likewise often with two apparent stereid bands. *Tortula atrovirens* is similar to *Crossidium* by its habit of low-growing, gemmiform plants with spiralling ovate to obovate leaves, the development of specialized ventral costal photosynthetic tissue (the cells of which have lumens vertically elongate in section, as was nicely illustrated by Flowers 1973a), the rounded dorsal stereid band, the medial placement of the simple, hollow papillae on the leaf, upper lamina yellow but with a reddish blush medially on the leaf in KOH, and a general similarity to *Crossidium aberrans* and *C. seriatum* of the areolation (upper cells ca. 15–18 µm in width), the thickened dorsal superficial upper median cell walls, and other features. Delgadillo (1975a, p. 282) proposed that *Crossidium* was derived, along with *Aloina*, from a *Tortula* or *Desmatodon*-like ancestor through elaboration of ventral costal outgrowths and reduction of leaves and sporophytes. The cladograms do not support this otherwise cogent inference, either placing *Crossidium* in a different tribe (the Hyophileae, Cladograms 14–16) or having the immediate ancestors of *Crossidium* inserted deeper in the cladogram than those of *Tortula* (see Cladograms 9, 10 and 13). The costal sections of *C. aberrans* and *T. atrovirens* are extraordinarily similar. *Tortula atrovirens* has sporophyte maturation dates (Zander 1979d) more characteristic of *Crossidium* than species previously placed in *Desmatodon*. As in the extension of the *Pseudocrossidium* evolutionary series (Zander 1979f) to include *P. revolutum*, which has no specialized ventral photosynthetic tissue, it may be possible to include *T. atrovirens* in *Crossidium* to fill it out as a natural genus. On the other hand, *Crossidium* sect. *Pseudocrossidium* (no nomenclatural relation to the genus *Pseudocrossidium*), which includes *C. aberrans* and other species with similar costal filaments, may be more appropriately included in *Tortula*, perhaps as a section or recognized as a separate genus in its own right. This needs further evaluation.

Tortula bogosica, because of its rather deep ventral costal groove and lack of dorsal costal epidermal cells, actually may well be *Barbula indica* or a closely related species lacking the ventral stereid band; this also requires further study. The little-papillose *Tortula californica* is placed in sect. *Tortula* largely because of its evident relationship with *T. transcaspica*, especially through the rounded leaf apex and long, hyaline awn, but has the rather large upper laminal cells of sect. *Pottia*; the same is true for *T. grandiretis*, which has somewhat longer leaves than the first two.

Species of sect. *Tortula* examined: *T. altipes* (H), *T. atrovirens*, *T. bogosica* (NY), *T. brevipes* (BUF), *T. brevissima* (US), *T. californica* (BUF), *T. canescens* (BUF), *T. euryphylla*, *T. grandiretis* (NY), *T. guepinii* (CANM), *T. leucostoma* (NY), *T. lingulata* (NY), *T. peruviana* (NY), *T. marginata* (BUF), *T. muralis*, *T. obtusifolius*, *T. platyphylla* (NY), *T. plinthobia*, *T. raucopapillosa* (NY), *T. revolvens* (NY), *T. sublimbata* (NY), *T. subulata* (BUF), *T. thianschanica* (H), *T. trachyphylla* (H), *T. transcaspica* (H), *T. vahliana* (NY), *T. wilsonii* (BUF, NY).

TORTULA Sect. POTTIA

Tortula sect. *Pottia* (Ehrh. ex Reichenb.) Kindb., Bih. K. Svensk. Vet. Ak. Handl. 7(9): 131, 1883. Lectotype: *Pottia truncata* (Hedw.) BSG, see Wareham in Grout, Moss Fl. N. Amer. 1: 197, 1939.

Phascum L. ex Hedw., Spec. Musc. 19, 1801. Lectotype: *Phascum cuspidatum* Hedw., see Wareham in Grout, Moss Fl. N. Amer. 1: 195, 1939.

Anacalypta Röhl. ex Leman, Dict. Sci. Nat. ed. 2, 2 Suppl. 38, 1816. Type: *Anacalypta lanceolata* (Hedw.) Nees & Hornsch.

Physedium Brid., Bryol. Univ. 1: 51, 1826. Type: *Physedium splachnoides* (Hornsch.) Brid.

Pottia (Ehrh. ex Reichenb.) Ehrh. ex Fürnr., Flora 12(2 Erg.): 10, 1829.

Mildeella Limpr., Laubm. Deutschl. 1: 191, 1885, *hom. illeg. non* Trevisan, 1876. Type: *Mildeella bryoides* (Dicks.) Limpr.

Mildea Warnst., Krypt. Fl. Brandenburg 2: 82, 1904, *hom. illeg. non* Grisebach, 1866.

Phascum subg. *Phascum* Sull. in A. Gray, Man. Bot. N. U.S. ed. 2: 615, 1856.

Pottia subg. *Anacalypta* (Röhl. ex Leman) Boul., Fl. Crypt. Est Muscin. 509, 1872.

Tortula subg. *Pottia* (Ehrh. ex Reichenb.) Lindb., Musci Scand. 21, 1879.

Pottia subg. *Eupottia* Boul., Muscin. France 471, 1884, *nom. illeg.*

Phascum subg. *Euphascum* Limpr., Laubm. Deutschl. 7: 185, 1885, *nom. illeg.*

Phascum subg. *Mildeella* Kindb., Eur. N. Amer. Bryin. 2: 403, 1897. Type: *Phascum bryoides* Dicks.

Pottia subg. *Pottia* (Ehrh. ex Reichenb.) Broth., Nat. Pfl. 1(3): 423, 1902, *hom. illeg.*

Pottia subg. *Mildeella* (Kindb.) Broth., Nat. Pfl. 1(3): 423, 1902.

Gymnostomum sect. *Pottia* Ehrh. ex Reichenb., Consp. Regn. Veg. 1: 33, 1828.

Barbula sect. *Cuneifoliae* BSG, Bryol. Eur. 2: 93, 1842 (fasc. 13–14 Mon. 31).

Phascum sect. *Annua* Ångstr. in Fries, Summ. Veg. Scand. 1: 97, 1846.

Pottia sect. *Eupottia* C. Müll., Syn. 1: 550, 1849, *nom. illeg.*

Pottia sect. *Anacalypta* (Röhl. ex Leman) C. Müll., Syn. 1: 547, 1849, *nom. illeg.* prior. ut gen.

Tortula sect. *Cuneifoliae* (B.&S. in BSG) Spruce, Ann. Mag. Nat. Hist. ser. 2, 3: 375, 1849. Type: *Tortula cuneifolia* (With.) Turn.

Trichostomum sect. *Anacalypta* (Röhl. ex Leman) C. Müll., Linnaea 42: 312, 316, 1879.

Desmatodon sect. *Cuneifolii* (BSG) Jur., Laubmfl. Oest. Ungarn 129, 1882. *Lectotyp. nov.*: *Desmatodon cuneifolius* (With.) Jur.

Barbula sect. *Camptopodiae* Kindb., Eur. N. Amer. Bryin. 2: 245, 1897. Type: *Barbula laureri* (Schultz) Kindb.

Pottia sect. *Mildeella* (Kindb.) Mönk., Laubm. Eur. 324, 1927.

Phascum sect. *Euphascum* Podp., Consp. Musc. Eur. 222, 1954, *nom. illeg.* Basionym: *Phascum* subg. *Euphascum* Limpr., *nom. illeg.*

Pottia sect. *Pseudodesmatodon* Podp., Consp. Musc. Eur. 232, 1954, *nom. inval.* Type: *Pottia randii* Kenn.

Pottia sect. *Pottia* (Ehrh. ex Reichenb.) Nyholm, Ill. Fl. Nordic Mosses 2: 81, 1989, *nom. superfl.*

Upper laminal cells usually smooth or weakly papillose and large (ca. 15–20 µm in width), ventral costal cells 2–3 in transverse section, commonly bulging individually or arranged in longitudinal rows as low lamellae and usually smooth or simply once- or twice-papillose, leaves usually with acute apices and short yellow-brown awns (rarely hyaline awned); costal section with guide cells; peristomes often rudimentary or absent.

Magill (1981) pointed out that *Acaulon* and *Phascum* are not easily distinguishable in South Africa by the traditional characters of size of capsule beak, curvature of leaf margins and the papillosity of the laminal cells. He defined *Phascum* by its narrower, plane-margined, erect to spreading leaves, and emergent capsule and cucullate calyptra, distinguishing *Acaulon* by the bulbiform habit, broad, concave leaves and small, mitrate calyptra. *Acaulon* is recognized here as a distinctive genus. Chamberlain (1978) suggested that *Phascum* and *Pottia* are linked through *Pottia recta* and *Pottia bryoides* and that it would be taxonomically more satisfactory if the two genera were united. The two genera are here synonymized with *Tortula s. str.*, but *Pottia recta* belongs with *Microbryum*.

Holzinger (1925) thought *Pottia randii* to be a depauperate form of *Desmatodon cernuus*; this is also reflected in the combination *D. randii* (Kenn.) Laz. of Lazarenko (1963b). These two (Pl. 85, f. 15–20; 88, f. 25–27) are kept together here in sect. *Pottia*. These and other species, like *T. laureri* (Pl. 87, f. 1–4) and *T. thompsonii*, with bordered, mostly plane and weakly serrulate margined leaves and large, superficially flat upper laminal cells are quite like species of *Hennediella*, a genus otherwise only distinguished from *Tortula* by commonly plane and serrate laminal borders and red laminal KOH reaction. These species of sect. *Pottia* (often placed in *Desmatodon* but not including the type of that genus) commonly have summer sporophyte maturation dates (Zander 1979d) rather than the more usual winter and spring dates of traditional *Tortula* species, and, on revision, may prove worthy of a section of their own.

Tortula lanceola (Pl. 86, f. 17–22) is similar to various *Crossidium* species in the weakly twisted peristome of 16 split, triangular teeth, the short, bulging cells of the ventral surface of the upper costa, the smooth areolation, and the laminal KOH reaction yellow with a red blush in the upper medial portion of the leaf. It differs mainly in the narrow recurvature of the laminal margins and very weak development of the ventral costal bulge, not quite sufficient for one to consider it a specialized photosynthetic organ. Another species similar to *Crossidium*, *T. atrovirens*, is here placed with sect. *Tortula*.

Certain specimens of small-statured species with basal laminal cells commonly reddish in KOH (e.g. *Tortula truncata*, Pl. 89, f. 11–15) may react red throughout the lamina giving the impression of the genus *Microbryum*; these are apparently from unusually harsh environments, and usually may be assigned to *Tortula* by a lack of laminal papillae.

Species of sect. *Pottia* examined: *T. atherodes*, *T. cernua* (NY), *T.* chungtienia (H, NY), *T. cuneifolia* (ALTA, BUF, NY), *T. deciduidentata* (NY), *T. modica* (BUF), *T. lanceola* (BUF), *T. laureri* (BUF, NY), *T. minor* (NY), *T. mucronifolia* (BUF), *T. nevadensis* (US), *T. pallida* (NY), *T. paulsenii* (H), *T. planifolia* (NY), *T. protobryoides* (BUF), *T. randii* (H, NY), *T. sainsburyana* (NY), *T. solmsii* (BM), *T. systylia* (NY), *T. thompsonii* (NY), *T. truncata*, *T. ucrainica* (NY), *T.*

websteri (US), *T. zoddae* (NY).

TORTULA Sect. SCHIZOPHASCUM

Tortula sect. *Schizophascum* (C. Müll.) Zand., *comb. nov.*
Phascum sect. *Schizophascum* C. Müll., Flora 71: 6, 1888. Type: *Phascum disrumpens* C. Müll.
Dendia R. Br. ter, Trans. New Zealand Inst. 30: 411, 1898. Type: *Dendia maritima* R. Br. ter
Pottia subg. *Schizophascum* (C. Müll.) Broth., Nat. Pfl. 1(3): 423, 1901.
Pottia sect. *Schizophascum* (C. Müll.) Wareham in Grout, Moss Fl. N. Amer. 1(4): 197, 1939.

Costal section lacking guide cells, and stereid band distinctly central in the section, otherwise gametophyte similar to that of sect. *Pottia*; sporophyte with elongate seta and elliptical capsule, this being cleistocarpous or dehiscing along up to 8 circumferential lines, usually along weakened cell walls at butt ends of longitudinally elongate rectangular cells (ca. 3–5:1) arranged in even rows (or "pallisades").

Found on soil, often near the sea; South Africa, Australia and New Zealand.

Brotherus (1902–09) sunk the monotypic *Dendia* (type species *Dendia maritima* R. Br. ter = *Tortula maritima*, Pl. 88, f. 1–9) into *Pottia* subg. *Schizophascum*, a move approved by Dixon (1923), who regarded the curious tattered capsule dehiscence of the two known collections as a merely "unhealthy, not to say thoroughly rotten condition" associated with "more or less complete immersion of the capsules from time to time in sea-water." *Tortula willisiana*, Pl. 89, f. 16–20, from Australia, however, has similar wall weakenings in its exothecial cells. Characters allowing recognition at least at the section level are the elongate, palisade-like exothecial cells that have weak walls along the butt ends allowing dehiscence along up to 8 circumferencial lines in at least two of the three species, and the lack of guide cells in the costa in all species. The hydroid strand is sometimes absent in some leaves of species in this section. Species of other sections of *Tortula*, if cleistocarpous, dehisce along irregular lines, these not distinctly arranged circumferencially, and have guide cells in their costae.

The type of *Tortula* sect. *Schizophascum*, *Pottia disrumpens* (C. Müll.) Broth., was referred to *T. willisiana* (as *Pottia drummondii*) by Willis (1954), a species very similar or the same as *T. maritima*. *Tortula* sect. *Schizophascum* may be apprehended as an end member of a reduction series through sect. *Pottia*. This may possibly involve ancestors of *T. cuneifolia*, which has nearly identical gametophytes but does have guide cells, but not ancestors of the cleistocarpous *T. atherodes* (Pl. 84, f. 16–22) (which, although it has a minute seta and spherical capsule, has more strongly awned, narrower, little concave leaves with a less lax areolation). Involved would be loss of the costal guide cells and elaboration of a unique dehiscence feature in some species. The spores are very large (ca. 35–40 µm diam.), but this is a common condition of other cleistocarpous (e.g. *T. atherodes*) and eperistomate species (e.g. *T. nevadensis*). Note, however, that the various cladograms demonstrate that reduced austral taxa are commonly relicts of lineages inserted deeply on the tribal or subfamilial subclade. Thus, sect. *Schizophascum* may be better recognized as a separate genus. Again, further analysis at the species level is called for.

Tortula maritima (syntype NY! New Zealand, "Godley Heads"), of which only two collections are known, is not operculate, but neither is it exactly cleistocarpous since capsules are cleanly ruptured along the central one or two of five to eight weak annular lines encircling the capsule from near the base to near the apex. In nearly mature but undehisced capsules, these annular lines are easily seen as minute cleavages between the cells in lateral view around the outline of the somewhat flattened capsule on a microscope slide. No exothecial cells are differentiated, however, as smaller, specialized annular cells as in the case with stegocarpous mosses. The exothecial cells are rectangular and have much the same "pallisade" arrangement as those of *Aschisma* and *Tetrapterum*, (see Cladograms 14–16, but compare with other cladograms in the study). In *Aschisma* rectangular exothecial cells (ca. 3–4:1) are considered unusual because the capsule is spherical, and, in taxa with spherical capsules, the exothecial cells are generally short-rhomboidal or short-rectangular (e.g. as in *T. atherodes*).

A significant character of all species here referred to sect. *Schizophascum* is the costal section showing a central region of stereid cells, this often including a ventrally or centrally situated hydroid strand, the stereid band surrounded by a ring of substereid cells, then surrounded again by an epidermis of thin-walled and often bulliform cells; thus, the guide cells are either absent or somehow hidden amoung similar cells of different developmental origin. The costal section is somewhat like that of *Stegonia* in the bulliform ventral cells and flattened dorsal surface, and similar to that of *Saitoella*, which lacks guide cells and has bulliform ventral epidermal cells, but also differs in lacking a hydroid strand.

Magill (1981, p. 209) suggested that *Tortula maritima* of New Zealand, *T. splachnoides* (Pl. 89, f. 10) of South Africa (as *Pottia*) and *T. willisiana* of Australia (as *Pottia drummondii*) may constitute a single, albeit variable circumantarctic taxon (see also Catcheside 1980). He did not, however, synonymize them, but referred South African specimens previously identified as *T. maritima* to *T. splachnoides* (as *Pottia*). An authentic specimen of *T. willisiana* (NY!) has the characters of *T. maritima*: concave leaves, upper laminal cells either epapillose or papillose medially, the costal section with centrally located stereid band, and the sporophytes, though immature, show a pallisade arrangement of cells with weak breaks along circumferencial lines. On the other hand, an isotype (NY! one plant) of *Tortula splachnoides* has exothecial cells ca. 2:1 in dimension, not in a pallisade arrangement, and breaks in the capsule wall are across cells, not along cell walls; the costal section, however, is like that of the other two species in the stereid (substereid) band positioned centrally and lacking guide cells. *Tortula splachnoides* is included here with other species (pending further study) because of the like costal section, the fact that immature capsules of *T. maritima* have rather short exothecial cells, and in view of the short, concave, nearly smooth leaves. The non-type specimens identified as *T. splachnoides* at BM have plane leaves with rather papillose laminae (except for a smooth border), and a distinct guide cell layer with the stereid band not central, but rather situated dorsally, thus there may be more than one taxon referred to *T. splachnoides* in herbaria; these non-type and as yet unidentified specimens, although cleistocarpous, lack a pallisade arrangement of exothecial cells but have similarly large spores. The number of accepted species in this section is three.

Additional literature: Brown (1898b).

Species of sect. *Schizophascum* examined: *Tortula maritima* (NY), *T. splachnoides* (NY), *T. willisiana* (NY).

TORTULA Sect. HYOPHILOPSIS

Tortula sect. *Hyophilopsis* (Card. & Dix.) Zand., *comb. et stat. nov.*

Hyophilopsis Card. & Dix., J. Bot. 49: 137, 1911. Type: *Hyophilopsis entosthodontacea* Card. & Dix.

Near *Tortula* sect. *Pottia* but differing by the annulus being revoluble, or persistent but strongly developed and bulging-reflexed when well wetted. Additional characters of value in identification are upper laminal cells 2–3:1, usually weakly papillose, costa percurrent, margins narrowly bordered by elongate cells, autoicous or rhizautoicous sexual condition; perigonia terminal on small, loosely foliate stems at base of perichaetiate plants; seta elongate and capsule cylindrical but peristome teeth rudimentary, consisting of several nubbins barely rising above a low basal membrane which itself scarcely exceeds the annulus; operculum conic, cells very weakly twisted counterclockwise.

A single rare taxon found on walls, mortar, and lateritic rocks in India's Western Ghats.

Hyophilopsis was found to be weakly distinguishable from *Tortula* sect. *Pottia*, mainly by the peristome reduction unaccompanied by seta and capsule reduction (Pl. 86, f. 1–10). The gametophyte of this monotypic genus (type: India, Western Ghats, Sedgwick 119, isotypes, NY, PC) is much like that of *Tortula laueri*.

Literature: Dixon (1911a).

Species examined: *T. entosthodontacea* (NY, PC, US).

63. WILLIA **Plates 90–91.**

Willia C. Müll. in Neum., Deutsch. Exp. Int. Polarforsch. 2: 311, 1890. Type: *Willia grimmioides* C. Müll.

Willia subg. *Euwillia* (C. Müll.) Broth., Nat. Pfl. 1(3): 417, 1902, *nom. illeg.*

Willia sect. *Euwillia* C. Müll., Gen. Musc. Fr. 424, 1900, *nom. illeg.*

Named for Elie Jean François Guillou, 1806–? of the voyage of the French ships Astrolabe and Zélée, 1837–1840), whose collections are at NSW and P.

Plants forming cushions or turfs, green above, occasionally blackened, reddish brown below. *Stems* branching occasionally, ca. 1–2 cm in length, transverse section rounded-pentagonal, *central strand usually present but weak, sclerodermis absent, hyalodermis absent;* axillary hairs 5–10 cells in length, basal 2–3 cells with thickened walls; rhizoids sparse. *Leaves* appressed when dry, spreading when moist, *ligulate to spathulate, sometimes constricted medially,* ca. 2–5 mm in length, *upper lamina concave or broadly channeled across leaf,* narrowly grooved along costa, *margins plane,* entire or occasionally weakly denticulate near apex, often bordered with 2–4 rows of cells with thicker walls, cells often elongate-rhomboidal and hyaline near apex; apex broadly rounded to rounded-acute; base weakly or strongly differentiated, elliptical to rhomboidal, occasionally clearly sheathing; *costa excurrent in a short or long awn,* in immature branch leaves represented only by an apiculus, costa with lamina inserted laterally or ventrally, superficial cells quadrate to short-rectangular and papillose ventrally, dorsally elongate and smooth to somewhat roughened, 3–5 rows of cells across costa ventrally at midleaf, costal transverse section elliptical to round, *stereid band* present dorsally, *large and reniform to semicircular in shape, ventral epidermis present, dorsal absent,* ca. 6 guide cells in 2 layers (ventrally 2 large cells, dorsally ca. 4 smaller), *hydroid strand present, distinct; upper laminal cells subquadrate, rather small,* 10–15 µm in width, 1(–2):1, walls thin to evenly thickened, superficially convex; papillae solid or hollow, bifid, 4–8 per lumen; basal cells strongly differentiated medially or occasionally across leaf in lower 1/4–1/2 of leaf, rectangular, 15–23 µm in width, mostly 3–4:1, walls thin, often bordered by 3–4 rows of narrower, more sturdy cells. Dioicous. *Perichaetia* terminal, *inner leaves highly differentiated,* long-ligulate to broadly lanceolate, 4–8 mm (inclusive of awn) in length, often secund, *loosely sheathing the seta,* lower cells rectangular to inflated-rhomoidal, differentiated in lower 1/2 to throughout leaf. Perigonia terminal, gemmate. *Seta mostly rather short,* 1–7 mm in length, 1 per perichaetium, yellow-brown, straight or twisted clockwise; *theca often short,* 0.8–3.5 mm in length, yellow-brown, very short-elliptical to cylindrical, exothecial cells rectangular, ca. 15–25 µm in width, 2–6:1, walls thin or somewhat thickened, stomates phaneropore, at base of theca, *annulus of 2–4 rows of highly vesiculose cells, often everted, persistent; peristome absent or teeth of 32 short anastomosing rami,* filamentous, low-spiculose, ca. 800 µm, with several articulations, twisted very weakly counterclockwise, *basal membrane comparatively high,* 400 µm in height, low-spiculose. Operculum short- to long-conic, 0.6–1.6 mm in length, cells twisted weakly counterclockwise. *Calyptra cucullate to long-mitrate, often lobed below, plicate below or smooth,* 1.5–4.0 mm in length. Spores 8–10 µm in diameter, light brown, essentially smooth to papillose. *Laminal KOH color reaction red.*

Found mainly on rock and tree trunks in austral regions, often near the coast.

The austral genus *Willia* is near *Syntrichia* in gross characteristics but differs largely in a tendency towards reduced sporophytes (Pl. 90, f. 1, 12; 91, f. 1) and highly differentiated perichaetial leaves (Pl. 90, f. 10, 19, 20; 91, f. 8). Associated

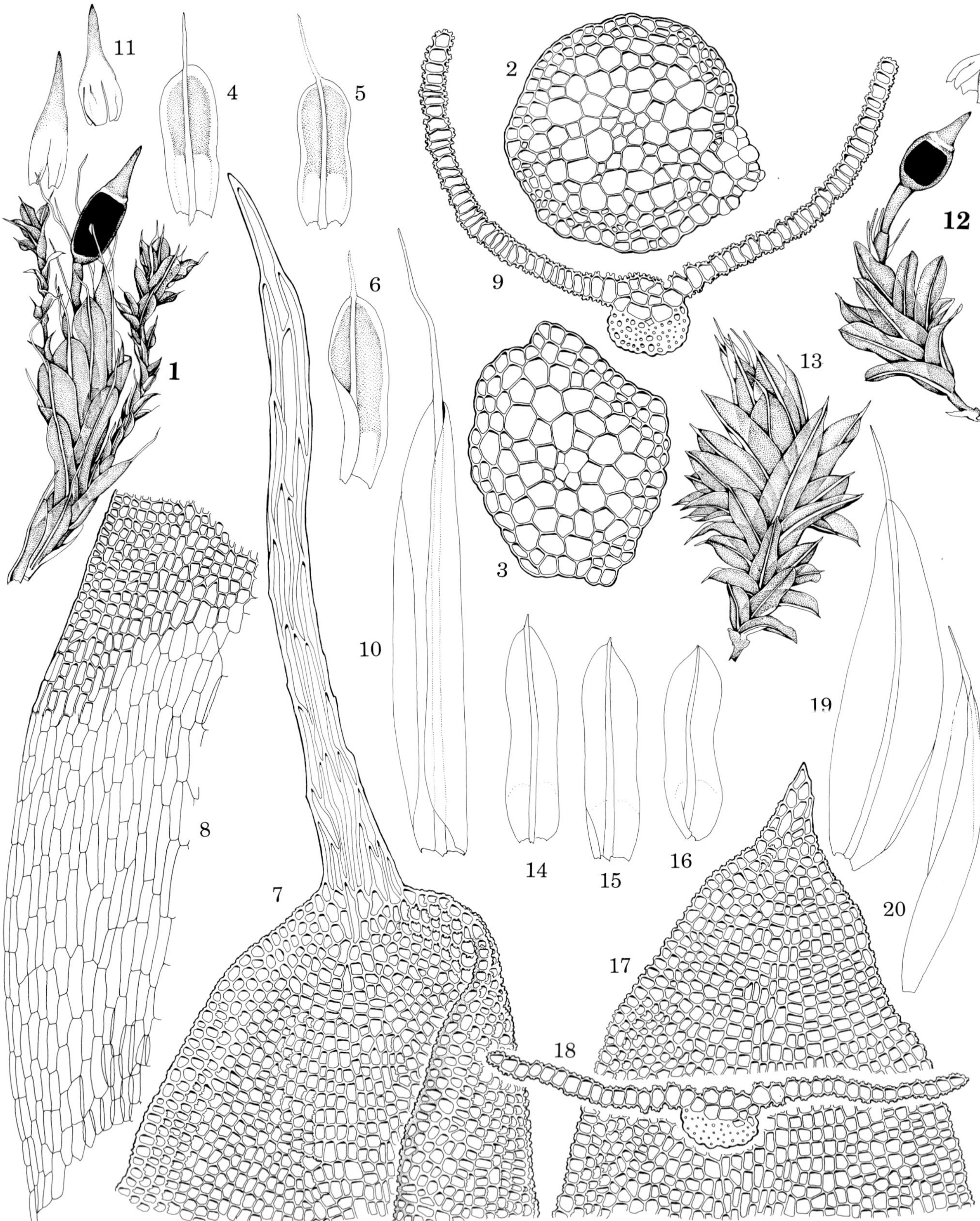

Plate 90. *Willia.* 1–11. *W. austroleucophaea.* 1. Habit. 2–3. Two transverse sections of stem. 4–6. Three leaves. 7. Leaf apex. 8. Basal cells. 9. Transverse section at midleaf. 10. Perichaetial leaf. 11. Calyptra. **12–20. *W. calobolax.*** 12–13. Two habits. 14–16. Three leaves. 17. Leaf apex. 18. Transverse section at midleaf. 19–20. Two perichaetial leaves.

characters are commonly panduriform cauline leaves often with elongate (occasionally hyaline) laminal cells near the apex; the leaf margins plane, weakly bordered by somewhat thickened cells; the peristome absent or short but with a high basal membrane (Pl. 91, f. 9); the annulus in some species everted on capsule dehiscence; and the calyptra is usually large and lobed below (Pl. 90, f. 11–12, but *cf.* 91, f. 11).

Lightowlers (1985c) placed *Tortula calobolax* (Pl. 90, f. 12–20) in *Willia* because of its immersed, eperistomate capsules, the rigid, oblong, somewhat pandurate leaves with plane margins, and the perichaetial leaves with long-acuminate apices and differentiated marginal cells. A specimen (Prince Edward I., Zinderen Bakker 671, NY) indicated by him (1985c) to be this species additionally has the characteristic short, fleshy seta and short macrostomous capsules also characteristic of *W. austroleucophaea*, and clearly

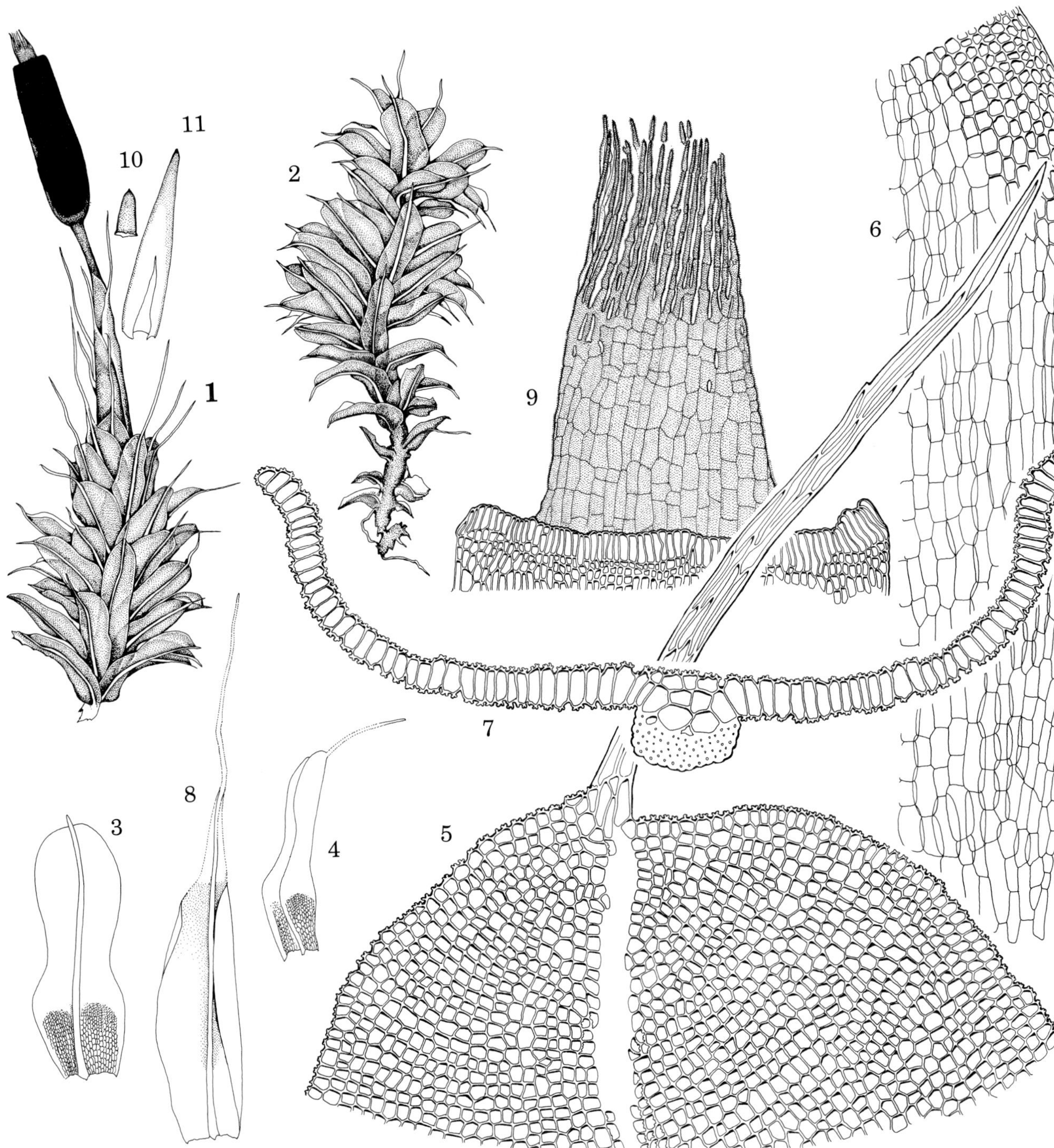

Plate 91. ***Willia.*** **1–11.** ***W. brachychaete.*** 1. Habit of sporophyte-bearing plant. 2. Perigoniate plant. 3–4. Two leaves. 5. Leaf apex. 6. Basal cells. 7. Transverse section at midleaf. 8. Perichaetial leaf. 9. Peristome. 10. Operculum. 11. Calyptra.

belongs with *Willia*. The genus is expanded here with the addition of *W. brachychaete* (Pl. 91, f. 1–11, discussed by Kramer 1988). Clearly, *W. brachychaete* approaches *Syntrichia* by its cucullate calyptra, presence of a peristome (albeit consisting mostly of a high basal membrane), less strongly reduced seta and capsule, and is itself approached by *S. cavallii* (Pl. 108, f. 7–12), which has a strongly differentiated perichaetium and weakly pandurate leaves. *Syntrichia cavallii* differs from *Willia* in the recurved upper leaf margins, somewhat cucullate leaf apices, and well developed sporophyte and peristome. *Willia* may ultimately be acceptable as only a section of *Syntrichia*, but more intensive study is needed to demonstrate this. Of especial value would be an evaluation of the relationship of *Willia* to plane-margined species of *Syntrichia* (sect. *Aesiotortula*). See Cladograms 13 and 15 for character state changes of phylogenetic hypotheses of relationships based on a global evaluation of all characters.

Additional literature: Bell (1974).

Number of accepted species: 3.

Species examined: *W. austroleucophaea* (BM, H, NY), *W. brachychaete* (BUF), *W. calobolax* (NY).

New combination: *Willia brachychaete* (Dus.) Zand., *comb. nov.* (*Tortula brachychaete* Dus., Bot. Not. 1905: 300, 1905).

64. SAITOELLA — Plate 92.

Saitoella Menzel, J. Hattori Bot. Lab. 71: 239, 1992.

Saitoa Zand., Phytologia 65: 430, 1989, *hom. illeg. non Saitoa* Rajendran & Muthappa, Proc. Indian Acad. Sci. sect. B, 89: 185, 1980. Type: *Saitoella peruviana* (Williams) Menzel.

Named for Japanese bryologist Kamezo Saito, author of "A Monograph of Japanese Pottiaceae (Musci)" (1975a) and many other papers on the Pottiaceae.

Small terete plants growing in dense clumps, deep reddish brown above, reddish tan below. *Stems branching occasionally, pseudodichotomous,* to 4 mm in length, transverse section rounded-pentagonal, central strand distinct, sclerodermis absent, hyalodermis absent; axillary hairs 2–4 cells in length, all cells hyaline; sparsely radiculose. *Leaves* closely appressed when dry, weakly spreading when moist, *broadly oblong-obovate, elliptic or suborbicular, 0.5–0.6 mm in length, upper lamina concave* and apex therefore somewhat cucullate, margins usually recurved at midleaf and below, crenulate in upper half by projecting cell walls, *bordered by ca. 15–18 rows of smooth, rhomboidal, thick-walled cells with somewhat rounded lumens; apex broadly rounded to weakly*

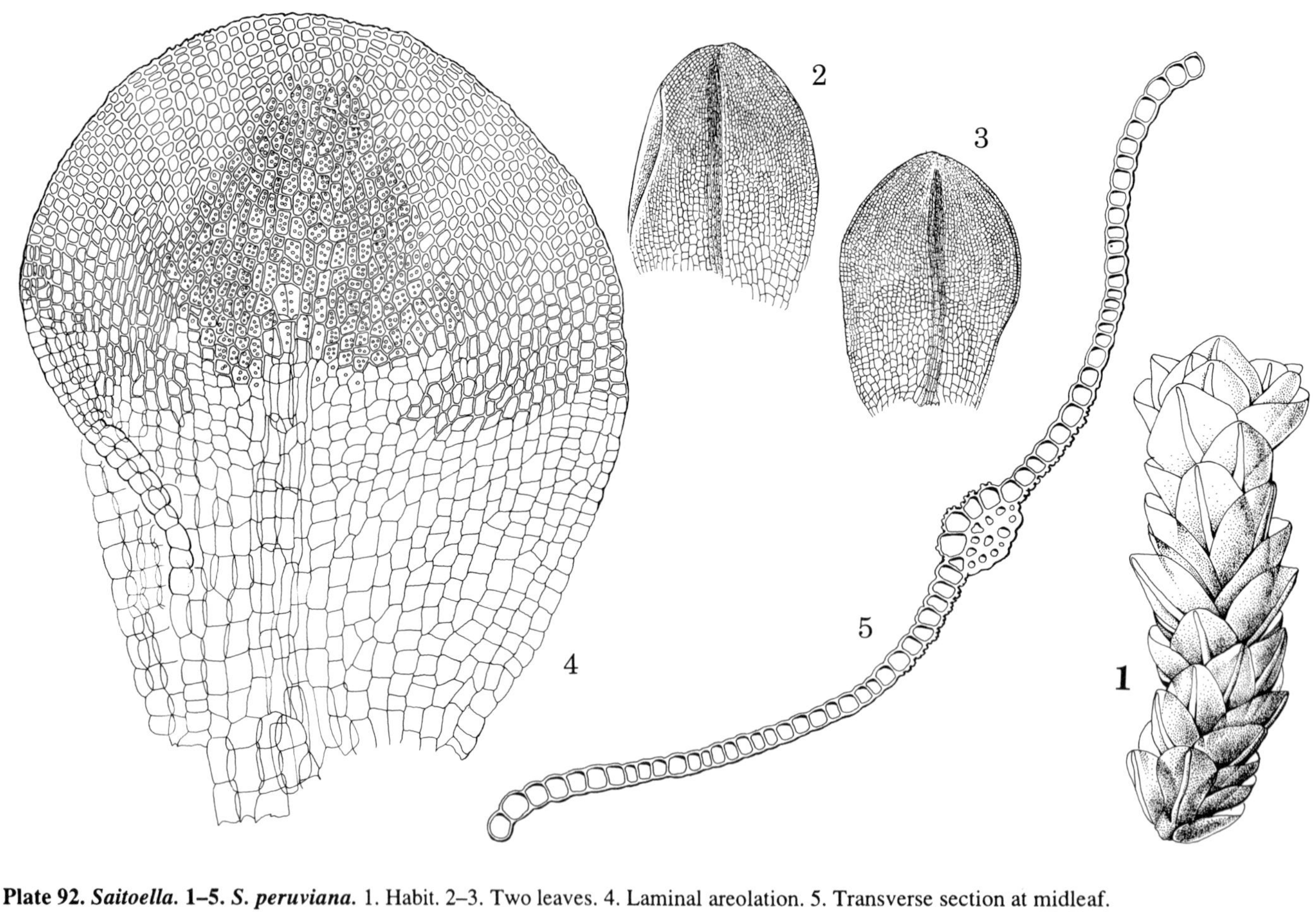

Plate 92. *Saitoella*. 1–5. *S. peruviana*. 1. Habit. 2–3. Two leaves. 4. Laminal areolation. 5. Transverse section at midleaf.

emarginate; base not differentiated in shape; leaf widest above middle; costa ending 2–3 cells below apex, thickest distally, superficial cells ventrally quadrate to short-rectangular, papillose, dorsally elongate, smooth or papillose, 4–5 rows of cells across costa ventrally at midleaf, *ventral surface of upper costa forming a bulging pad of 1 layer of papillose cells, costal transverse section elliptical to reversed semicircular* (i.e. flattened dorsally), stereid band present dorsally, strong, elliptical, ventral epidermis strongly differentiated, of thin-walled, papillose cells, dorsal epidermis absent or of thick-walled bulging cells, *guide cells absent, hydroid strand absent; upper laminal cells* subquadrate to short-rectangular, 6–10 µm in width, 1–2:1, walls thin to thickened, superficially equally weakly bulging on both sides of lamina; *papillae restricted to a small area of the lamina near the costa,* small, mostly bifid, hollow or solid, about 4–6 per lumen; basal cells differentiated across the leaf base, quadrate to short-rectangular, ca. 16–18 µm in width, 1–2:1, walls thin. Sexual structures and sporophyte unknown. *Laminal KOH color reaction deep red.*

Found on soil over volcanic rock at high elevations; Peru, Ecuador, Mexico.

This genus contains only *S. peruviana*, which has been recognized in *Globulinella* for some time, being similar in the concave, ovate to rounded leaf shape and rather thick costa ending before the apex. *Saitoella* differs considerably, however, in the upper lamina bordered by many rows of rhomboidal, thick-walled cells, upper laminal cells papillose in a medial patch (Pl. 92, f. 4), costa lacking guide cells (Pl. 92, f. 5), and red color KOH reaction. The branching is pseudodichotomous (Zander 1976), a pattern apparently rare in mosses and as yet inadequately surveyed. The naked archegonia found in leaf axils in *S. peruviana* are uncommon in Pottiaceae; they are also present in *Globulinella globifera* and have been found in several other genera of Pottiaceae and in other families (Zander 1976).

Saitoella is considered here to lack differentiated guide cells, the superficial ventral parenchyma being of papillose quadrate cells, and therefore probably not interpretable as guide cells. *Phascopsis* is similar in lacking guide cells in at least some leaf sections—in that genus cells in the guide cell position become narrower and more thick-walled from the base of the leaf to the apex. *Scopelophila ligulata* either lacks guide cells or, more probably, a ventral epidermis, since the ventral cell layer is of inflated, smooth cells, but *Scopelophila* may be distinguished because it also lacks laminal papillae and a stem central strand. In *Saitoella*, putative guide cells cannot be distinguished anywhere in the costa. *Bryoerthrophyllum columbianum* has a similar ventrally bulging costa with the lamina inserted laterally, but guide cells are present and two stereid bands may be distinguished in at least some leaves. *Tortula maritima* has a similar medial concentration of upper papillose cells in the leaf and nearly identical costal section (guide cells absent and epidermal cells bulliform), and *Stegonia latifolia* has similar rhomboidal upper laminal cells and a costal section (epidermal cells bulliform but guide cells present) approaching that of *Saitoella*; these two taxa are easily distinguished by their yellow KOH reaction. *Aloinella* species may likewise have papillae located only medially on the leaf, with a broad border of non-papillose rectangular cells, but ventral costal filaments are differentiated. *Tortula revolvens* is somewhat similar to *Saitoella* in having upper laminal cells papillose only medially, but may be distinguished by the presence of guide cells in the costa, strongly revolute upper laminal margins, and yellow KOH reaction.

A dorsal costal epidermis is clearly differentiated in some leaves of *S. peruviana* but not in others. The epidermal presence, even partial, is of some value in separating it from *Syntrichia*, which lacks a dorsal costal epidermis.

Additional literature: Steere & Chapman (1946).

Number of accepted species: 1.

Species examined: *S. peruviana* (BUF, FH, MICH, TENN, NY).

65. MICROBRYUM — Plate 93.

Microbryum Schimp., Syn. 10, 1860. Type: *Microbryum floerkeanum* (Web. & Mohr) Schimp.

Bryella Berk., Handb. Brit. Mosses 16: 300, 1863, Type: *Bryella recta* (With.) Berk.

Cycnea Berk., Handb. Brit. Moss. 60, 301, 1863, *hom. illeg. non Cycnia* Griff., 1854. Type: *Cycnea curvicolla* (Hedw.) Berk.

Pottiella (Limpr.) Gams, Krypt. Fl. Mitteleur. ed. 2, 1: 101, 1948.

Lydiaea Laz., Not. Syst. Sect. Cryptog. Inst. Bot. Nom. Komar. Acad. Sci. URSS 12: 280, 1959. Type: *Lydiaea vlassovii* (Laz.) Laz.

Phascum subg. *Microbryum* (Schimp.) Limpr., Laubm. Deutschl. 1(3): 182, 1885.

Phascum subg. *Pottiella* Limpr., Laubm. Deutschl. 1: 188, 1885. Type: *Phascum curvicollum* Hedw., *lectotyp. nov.*

Pottia subg. *Pottiella* (Limpr.) Broth., Nat. Pfl. 1(3): 423, 1902.

Phascum sect. *Pottiella* (Limpr.) Par., Actes Soc. Linn. Bordeaux 51[1]: 66, 1897; Ind. Bryol. [4]: 1030, 1898.

Acaulon sect. *Microbryum* (Schimp.) C. Müll., Gen. Musc. Fr. 20, 1900.

Phascum sect. *Microbryum* (Schimp.) Podp., Consp. Musc. Eur. 221, 1954.

Pottia sect. *Pottiella* (Limpr.) Nyholm, Ill. Fl. Nordic Mo. 2: 81, 1989.

Pottia subsect. *Muticae* C. Jens., Skand. Bladmfl. 210, 1939, *nom. inval. descr. suec.*

From μῖκρός, small + o + *bryum* from βρύον, a moss.

Plants forming a low turf, scattered or gregarious, *occasionally bulbiform,* reddish brown above, brown below. *Stems* seldom branching, *extremely short,* 0.2–0.4 mm in length, transverse section round to rounded-pentagonal, central strand present or absent, sclerodermis absent or weakly differentiated, hyalodermis absent; axillary hairs 3–6 cells in length, the basal 1–2 usually with thicker walls; rhizoids sparse. *Leaves* appressed when dry, weakly spreading and tips occasionally reflexed when moist, lanceolate, elliptical or ovate, occasionally spathulate, *short,* 0.6–1.8 mm in length,

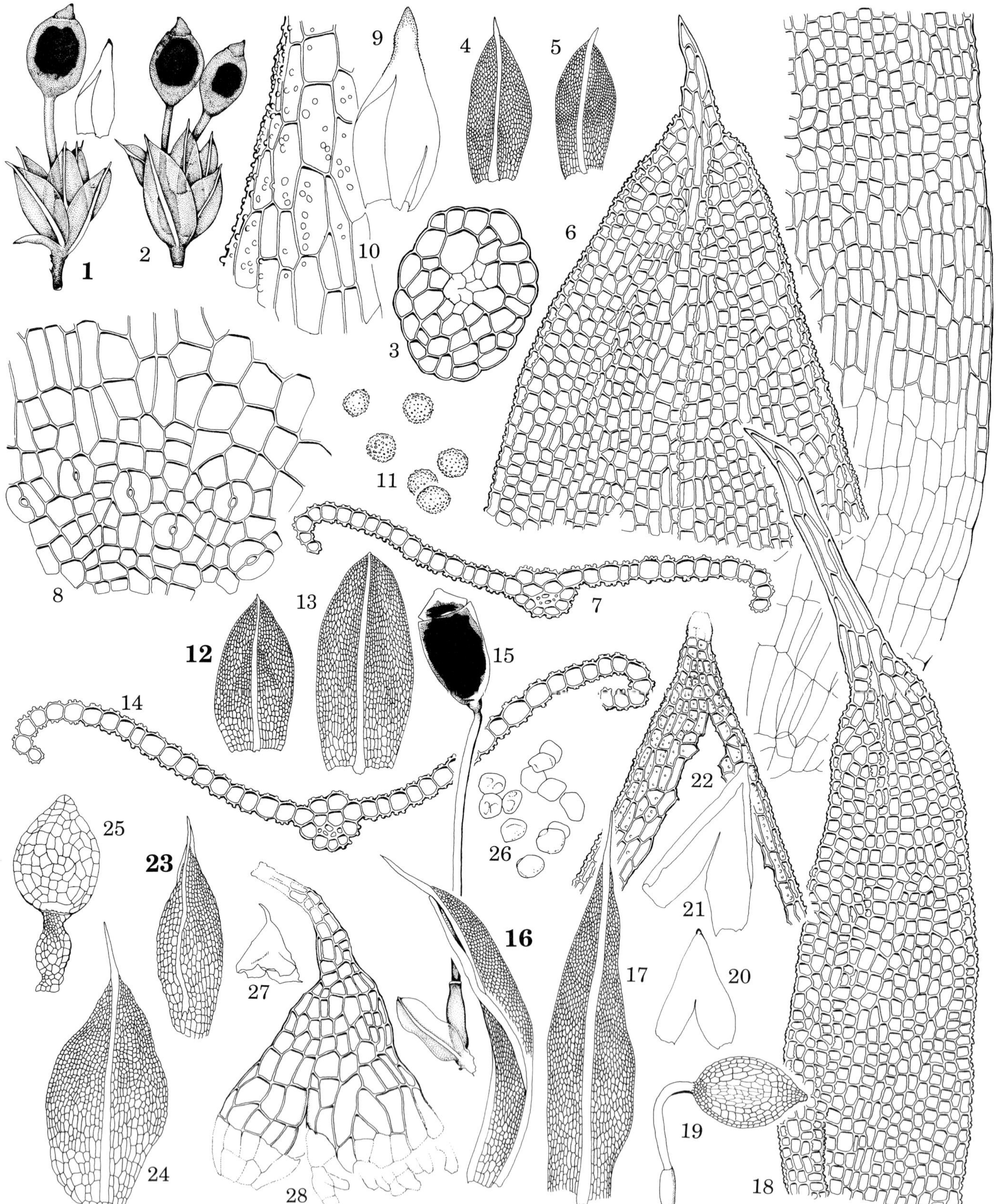

Plate 93. ***Microbryum.*** **1–11.** ***M. rectum.*** 1–2. Habits. 3. Transverse section of stem. 4–5. Two leaves. 6. Leaf apex. 7. Transverse section at midleaf. 8. Stomates. 9. Calyptra. 10. Detail of calyptra. 11. Spores. **12–15.** ***M. brevicaule.*** 12–13. Two leaves. 14. Transverse section at midleaf. 15. Sporophyte. **16–19.** ***M. curvicolum.*** 16–17. Two leaves. 18. Leaf apex. 19. Sporophyte. 20–22. Calyptrae and detail. **23–28.** ***M. floerckeanum.*** 23–24. Two leaves. 25. Sporophyte. 26. Spores. 27. Calyptra. 28. Areolation of calyptra.

upper lamina weakly concave to broadly channeled, *margins recurved at midleaf, commonly narrowly recurved to near apex,* entire or rarely serrulate near apex, marginal cells often less papillose and somewhat thicker walled than the medial; apex broadly acute; base not differentiated; *costa* excurrent as an apiculus or forming a mucro or short awn, occasionally only percurrent, costa with lamina inserted laterally, superficial cells smooth or papillose, ventrally quadrate or short-rectangular or elongate, dorsally short-rectangular to elongate, 2(4–6) rows of cells across costa ventrally at midleaf, costal transverse section usually round, stereid band present dorsally, round to semicircular in shape, *ventral and dorsal epidermises present,* guide cells 2(–4) in 1 layer, hydroid strand present, occasionally centrally located in the stereid band, occasionally the ventral epidermis differentiated as a pad of enlarged parenchymatic cells; *upper laminal cells* quadrate to hexagonal or short-rectangular, occasionally rhomboidal, rather large, 11–20 µm in width, 1–2:1, walls thin to moderately and evenly thickened, superficially convex on both sides; *papillae usually simple, seldom bifid, hollow,* 1–6 per lumen, occasionally branching and tall; basal cells differentiated across leaf or higher medially, rectangular, ca. 18–30 µm in width, 2–4:1, walls usually thin. Monoicous, usually paroicous, occasionally synoicous. Perichaetia terminal, inner leaves often somewhat enlarged, otherwise little different from the cauline. Seta nearly absent to 4 mm in length, 1(–2) per perichaetium, yellow-brown, twisted straight or counterclockwise below and clockwise above; *theca cleistocarpous or stegocarpous,* ca. 0.5–1.1 mm in length, brown or reddish or yellowish brown, ovate to short-elliptical, *apiculate when cleistocarpous,* exothecial cells short-rectangular, ca. 18–25 µm in width, mostly 2–5:1, thin-walled, stomates phaneropore, at base of theca, annulus absent or of 1–2 rows of weakly vesiculose cells, persistent; eperistomate or peristome teeth 16, irregular, often rudimentary, often apically truncate, ligulate to triangular, spiculose, short, with few articulations, straight, basal membrane absent. *Operculum when differentiated low-conic,* ca. 0.1–0.2 mm in length, cells straight. *Calyptra mitrate to conic-cucullate, smooth or finely papillose,* ca. 0.2–0.8 mm in length. Spores ca. 20–30 µm in diameter, light brown, essentially smooth to warty or spiculose or hollow-tuberculate. *Laminal KOH color reaction red.* Reported chromosome numbers: n = 26, 27+m, 28, 30.

Found in the temperate zones worldwide, especially in somewhat arid situations, mainly on soil.

Microbryum is distinctive in the combination of the small habit, red KOH color reaction of the upper lamina, single round to semicircular costal stereid band (Pl. 93, f. 7, 14), capsules apiculate when cleistocarpous (Pl. 93, f. 19), peristomes when present commonly apically truncate and seemingly large in comparison with the size of the capsule; and calyptrae often papillose (Pl. 93, f. 10, 22). If one considers the possibility that taxa with single stereid bands evolved independently from taxa with two stereid bands, then *Microbryum* might have been derived through reduction from ancestors of *Bryoerythrophyllum,* which has similar areolation. Evidence for this is that there are no other highly reduced taxa related to *Bryoerythrophyllum* (see also discussion of *Saitoella* and *Acaulon*). Otherwise, *Microbryum* could have come from progenitors similar to *Tortula* sect. *Tortula.* Cladograms 11 and 14, however, indicate a different derivation.

Chamberlain's (1969, 1978) treatment of *Pottia* species with erostrate opercula recognized *M. starckeanum* as a single species including both tuberculate and papillose-spiculose spored plants. The present study, however, supports the traditional arrangement of Corley et al. (1982) that distinguishes material differing by the two spore ornamentations at the species level. Chamberlain's infraspecies are all recognized at the varietal level, with the addition of *M. davallianum* var. *commutatum* and *M. starckeanum* var. *fosbergii.* All varieties of these two species that were recognized by Chamberlain (1978) for Gt. Britain are also present in U.S.A. in California. One collection, California, Pasadena, *s.n.,* "g. 9", US, includes *M. starckeanum* var. *starckeanum,* var. *fosbergii,* and an intermediate with a capsule having a differentiated operculum and short peristome that is indehiscent even when boiled in KOH solution. The intermediate is also found in other collections (e.g. California, Ikenberry 369, CANM, *comm.* T. McIntosh). The type of *Pottia arizonica* (= *M. starkeanum* var. *starckeanum*) has spores that appear to be both papillose and tuberculate, but the "papillae" are loose in the spore sac as well as partially coating the spore. The type of *Pottia fosbergii* is more probematic, with spores that are slightly wrinkled (very weakly tuberculate) and also weakly papillose. In this case, the sporophytes and spores (if unreduced) may be of hybrid origin; other specimens clearly of var. *fosbergii* (operculum not differentiated at all) have the spores typical of *M. starckeana. Microbryum davallianum* var. *conicum* may have spores that are epapillose, but these are never wrinkled. The two species may be seen as two series of infraspecific peristome reduction.

The calyptrae are roughened with low, simple papillae in many of the species with comparatively large calyptrae (e.g. *M. commutatum, M. rectum*); calyptral papillae are apparently absent in those species with much reduced sporophytes and tiny calyptrae. The presence of calyptral papillae helps distinguish this genus from *Acaulon* (which, like highly reduced members of *Microbryum,* has strongly bulging vaginulae), *Syntrichia* and *Tortula* sect. *Tortula.* The new combination *Microbryum rufochaete* reflects the strongly apiculate capsules and recurved upper laminal margins of this species. *Microbryum tasmanicum* is similar but has simple laminal papillae. In *M. rufochaete* the perigoniate plants are about a quarter to a third the size of the perichaetiate plants, and are situated near the base of the perichaetiate plants (possibly rhizautoicous).

McIntosh (1989) discussed *M. vlassovii* for North America (as *Phascum*). Both *M. vlassovii* and *M. raddei* have an enlarged pad of parenchymatic cells ventrally on the costa. Carrión et al. (1990) described the spore morphology of several species of *Microbryum* (as *Phascum*), and indicated that the spore surface of *M. vlassovii* was rather different from that of *M. curvicolle* (Pl. 93, f. 16–19) and *M. floerkeanum* (Pl. 93, f. 23–28), being more like that of *Tortula atherodes* (discussed as *P. cuspidatum*).

Additional literature: Guerra et al. (1991, 1992).

Number of accepted species: 13.

New heterotypic synonymy: *Pottia arizonica* Wareham in Grout = *Microbryum starckeanum* (Hedw.) Zand. var. *starckeanum*. *Pottia arizonica* var. *mucronulata* Wareham in Grout = *Microbryum starckeanum* var. *brachyodus* (BSG) Zand.

Species examined: *M. brevicaule* (NY), *M. curvicolle* (NY), *M. davallianum* (BUF, CANM, FH, US), *M. floerkeanum* (NY), *M. longipes* (BUF), *M. raddei* (H), *M. rectum* (BUF), *M. rufochaete* (NY), *M. starckeanum* (BUF, CANM), *M. subplanomarginatum* (BUF), *M. tasmanicum* (BM), *M. vlassovii* (BUF), *M. zeelandiae* (NY).

New combinations and statuses:

Microbryum brevicaule (Tayl.) Zand., *comb. nov.* (*Gymnostomum brevicaule* Tayl., London J. Bot. 5: 42, 1846; *Pottia brevicaulis* (Tayl.) C. Muell.).

Microbryum curvicolle (Hedw.) Zand., *comb. nov.* (*Phascum curvicolle* Hedw., Spec. Musc. 21, 1801; *Pottia curvicollis* (Hedw.) Mitt.).

Microbryum davallianum (Sm.) Zand., comb. nov. (*Gymnostomum davallianum* Sm., Ann. Bot. 1: 577, 1805; *Pottia davalliana* (Sm.) C. Jens.).

Microbryum davallianum var. *commutatum* (Limpr.) Zand., *comb. nov.* (*Pottia commutata* Limpr., Laubm. Deutschl. 1: 537, 1888; *Pottia davalliana* subsp. *commutata* (Limpr.) Podp.).

Microbryum davallianum var. *conicum* (Schleich. ex Schwaegr.) Zand., *comb. nov.* (*Gymnostomum conicum* Schleich. ex Schwaegr., Sp. Musc. Suppl. 1(1) 25, 1811; *Pottia starckeana* ssp. *conica* (Schleich. ex Schwaegr.) Chamberl.).

Microbryum floerkeanum var. *arbense* (Loitl.) Zand., *comb. nov.* (*Phascum arbense* Loitl., Verh. Zool. Bot. Ges. Wien 59: 55, 1909; *Phascum floerkeanum* var. *arbense* (Loitl.) Podp.), not seen.

Microbryum longipes (Guerra, Martínez & Ros) Zand., *comb. nov.* (*Phascum longipes* Guerra, Martínez & Ros, J. Bryol. 16: 55, 1990 as "*longipedis*" typographical error *cf.* J. Bryol. 16: 335, 1991).

Microbryum raddei (Broth.) Zand., *comb. nov.* (*Tortula raddei* Broth., Bot. Centralbl. 34: 26, 1888).

Microbryum rectum (With.) Zand., *comb. nov.* (*Phascum rectum* With., Syst. Arr. Britt. Pl. ed. 4: 771, 1801; *Pottia recta* (With.) Mitt.).

Microbryum rufochaete (Magill) Zand., *comb. nov.* (*Acaulon rufochaete* Magill, Fl. S. Afr. I. Bryophyta 1: 201, 1981 [1982]).

Microbryum starckeanum (Hedw.) Zand., *comb. nov.* (*Weisia starckeana* Hedw., Spec. Musc. 65, 1801; *Pottia starckeana* (Hedw.) C. Müll.).

Microbryum starckeanum var. *brachyodus* (BSG) Zand., *comb. nov.* (*Anacalypta starckeana* var. *brachyodus* BSG, Bryol. Eur. 2: 47, 1843; *Pottia starckeana* var. *brachyodus* (BSG) C. Müll.), commonly "*brachyoda*."

Microbryum starckeanum var. *brevidens* (Latz.) Zand., *comb. nov.* (*Pottia starckeana* var. *brevidens* Latz., Beih. Bot. Centralbl. 48(2): 483, 1931), not seen.

Microbryum starckeanum var. *fosbergii* (Bartr.) Zand., *comb. nov.* (*Pottia fosbergii* Bartr., Bryologist 33: 18, 1930; *Pottia starckeanum* var. *fosbergii* (Bartr.) Zand.).

Microbryum starckeanum var. *leiostoma* (Corb. in Corb. & Pitard) Zand., *comb. nov.* (*Pottia starckeana* var. *leiostoma* Corb. in Corb. & Pitard, Bull. Soc. Bot. France 56: ccxxiv, 1909), not seen.

Microbryum starckeanum var. *subgymnostoma* (De Not.) Zand., *comb. nov.* (*Anacalypta starckeana* var. *subgymnostoma* De Not., Atti Univ. Genova 1: 583, 1869; *Pottia starckeana* var. *subgymnostoma* (De Not.) Grav.), not seen.

Microbryum starckeanum var. *submutica* (Latz.) Zand., *comb. nov.* (*Pottia starckeana* var. *submutica* Latz., Beih. Bot. Centralbl. 48(2): 483, 1931), not seen.

Microbryum starckeanum fo. *brevifolium* (Limpr.) Zand., *comb. nov.* (*Pottia starckeana* fo. *brevifolia* Limpr., Laubm. Deutsch. 1: 536, 1888), not seen.

Microbryum starckeanum fo. *dextrorsum* (Limpr.) Zand., *comb. nov.* (*Pottia starckeana* fo. *dextrorsa* Limpr., Laubm. Deutsch. 1: 536, 1888), not seen.

Microbryum starckeanum fo. *microphyllum* (Warnst.) Zand., *comb. nov.* (*Pottia starckeana* fo. *microphylla* Warnst., Hedwigia 58: 144, 1917), not seen.

Microbryum subplanomarginatum (Dix.) Zand., *comb. nov.* (*Pottia subplanomarginata* Dix., Trans. R. Soc. S. Afr. 18: 253, 1929).

Microbryum tasmanicum (Dix. & Rodw.) Zand., *comb. nov.* (*Phascum tasmanicum* Dix. & Rodw., Pap. Proc. R. Soc. Tasmania 1923: 25, 1923).

Microbryum vlassovii (Laz.) Zand., *comb. nov.* (*Phascum vlassovii* Laz., J. Inst. Bot. Ac. Sc. R.S.S. Ukraine 26–27: 196, 1938).

Microbryum zeelandiae (R. Br. ter) Zand., *comb. nov.* (*Anacalypta zealandiae* R. Br. ter, Trans. Proc. New Zealand Inst. 30: 413, 1898; *Pottia zealandiae* (R. Br. ter) Par.).

66. CRUMIA **Plate 94.**

Crumia Schof., Canad. J. Bot. 44: 609, 1966. Type: *Crumia latifolia* (Kindb.) Schof.

Named for Howard Alvin Crum, 1922–, author of numerous valuable bryological publications, including "Mosses of Eastern North America" (1981) with Lewis E. Anderson.

Plants forming cushions, green to reddish above, *blackish green to iridescent tan below*. *Stems* branching often, to 3.0 cm in length, transverse section rounded-pentagonal, *central strand absent or very weak, sclerodermis absent, hyalodermis absent*; axillary hairs ca. 6 cells in length, basal 1–2 cells yellowish; sparsely radiculose. *Leaves* appressed, weakly contorted when dry, spreading when moist, *spathulate*, to 4.0 mm in length, upper lamina flat, shallowly channeled along costa, margins recurved along 1 or both margins below midleaf,

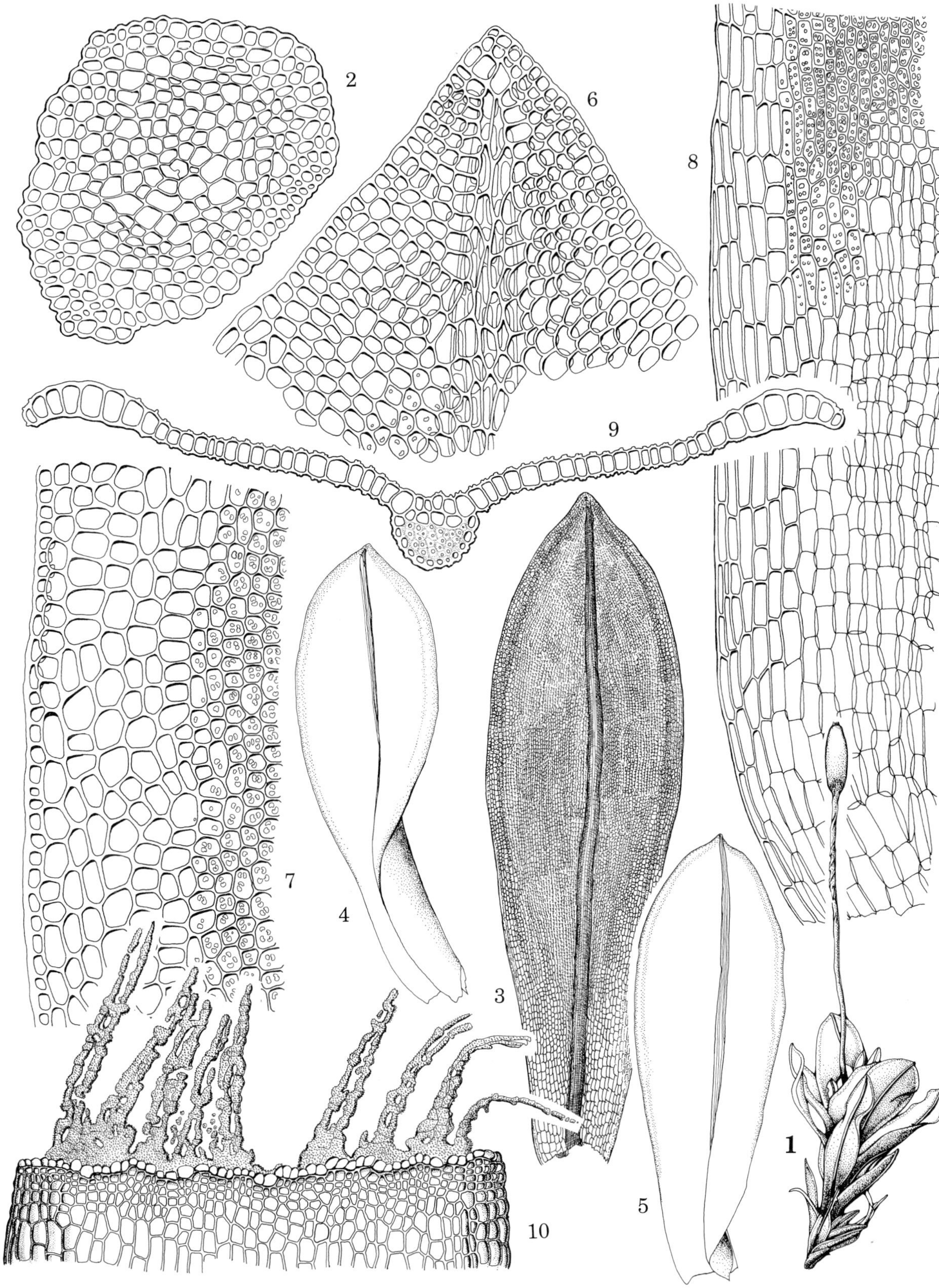

Plate 94. ***Crumia.* 1–10. *C. latifolia.*** 1. Habit. 2. Transverse section of stem. 3–5. Three leaves, one enlarged. 6. Leaf apex. 7. Upper marginal cells. 8. Basal cells. 9. Transverse section at midleaf. 10. Portion of peristome.

entire, *bordered by ca. 6 rows of enlarged, thicker walled, rounded rhomboidal to rectangular cells, except the extreme 1–2 marginal rows*; *apex* broadly acute to rounded-acute, *usually broadly apiculate*; base scarcely differentiated in shape, bordered by a few rows of long-rectangular cells; *costa percurrent or ending 1–2 cells below the apex*, superficial cells elongate on both sides, 5–6 rows of cells across costa ventrally at midleaf, costal transverse section semicircular, one stereid band present, strong, epidermis present ventrally, weakly developed dorsally, guide cells 3–4 in 1 layer, *hydroid strand absent*; upper laminal cells in medial portion of leaf hexagonal to short-rectangular, 13–20 µm in width, 1–2:1, walls thin, superficially weakly convex on both sides; *papillae small but distinct, simple to bifid, 8–12 per lumen, scattered*; basal cells differentiated medially, inflated, rectangular, wider than the upper cells, 3–5:1, walls thin, brown. Dioicous. *Perichaetia* terminal, *inner leaves little different from cauline leaves, not sheathing*. Perigonia reported as terminal, bracts not sheathing. Seta ca. 4–15 mm in length, 1(–2) per perichaetium, reddish brown, twisted clockwise below, counterclockwise above; theca 2.5–3.0 mm in length, reddish brown, cylindrical, exothecial cells rectangular, thin-walled, stomates phaneropore, at base of theca, annulus of ca. 2 rows of vesiculose cells; *peristome teeth 16, cleft to near base into two branches, linear, rami much perforated and anastomosing* (but articulations held together by a hyaline membrane), *densely spiculose*, ca. 200 µm in length, with ca. 7 articulations, *twisted very weakly counterclockwise*, basal membrane low, crazed. *Operculum* conic, 0.8–1.1 mm in length, *cells weakly twisted counterclockwise*. Calyptra cucullate, smooth, ca. 3.0 mm in length. Spores ca. 15–18 µm in diameter, light brown, weakly papillose. *Laminal KOH color reaction reddish orange*.

Found on soil or rock, usually in wet areas, especially along streams, often on calcareous substrates; found in southwestern Canada and western United States.

Crumia is much like *Scopelophila* in the often blackish coloration of the plants, the lack of both a sclerodermis and a hyalodermis in the stem (Pl. 94, f. 2), leaves spathulate, often with a broad apiculus, costa percurrent, stereid band single, hydroid strand absent (Pl. 94, f. 9), and perichaetial leaves little differentiated from the cauline. It differs significantly from *Scopelophila* in the presence of a stem central strand (though generally only weakly developed or occasionally absent), a more strongly differentiated laminal border of thick-walled cells with one or two rows of much smaller cells on the extreme margin (Pl. 94, f. 7), upper laminal cells papillose (rarely mostly smooth), and peristome (Pl. 94, f. 10) present, though rarely fruiting. Schofield (1966) added a few other distinctions for *Crumia*, such as always entire leaf margins, fruiting in spring (rather than late autumn for *Scopelophila*), stomata well developed (versus rudimentary or obscure), exothecial cells elongate, and annular cells weakly developed, but these are either minor characters or are based too strongly on only one species of *Scopelophila*. His observation, however, that *Crumia* is found on alkaline substrates poor in metallic ions, while *Scopelophila* is largely restricted to acid habitats often rich in sulfur and iron, copper or other metallic ions, has considerable weight, given the other features. The leaf shape of *Crumia* is similar to that of *Tortula* but the upper marginal cells of the former are short-elliptical and swollen in section, with lumens larger than those of the medial cells, while the marginal cells of *Tortula* species are either undifferentiated or nearly so or rectangular and lumens smaller (as seen in section) than those of the medial cells.

The peristome of *Crumia* is quite interesting, being weakly twisted counterclockwise and composed of fragmented articulations of the two rami, these held together by a very thin, hyaline membrane reminiscent of that of *Cinclidotus* (but here the 16 teeth are quite separate). The peristome is quite unlike that of *Tortula deciduidentata*, previously placed in *Crumia*, which is strongly twisted, composed of 32 integral rami above a distinct basal membrane, and comes off with the operculum (the columella elongates and pushes the operculum, with the peristome inside it, up and away from the capsule mouth). Although the cauline leaves of *T. deciduidentata* are similar to those of *Crumia* or *Scopelophila*, the perichaetial leaves bear an awn, which is curiously blunt or flattened apically. Other characters that would be anomalous in *Crumia* for *Tortula deciduidentata* are the latter's strong central strand, presence of a hydroid strand in the costa, and the paroicous or autoicous sexual condition.

Additional literature: Sharp & Iwatsuki (1969), Vitt and Zander (1978).

Number of accepted species: 1.

Species examined: *C. latifolia* (BUF, NY).

67. HENNEDIELLA — Plates 95–98.

Hennediella Par., Ind. Bryol. 557, 1896, *nom. nov.* for *Hennedia* R. Br. ter. Type: *Hennediella macrophylla* (R. Br. ter) Par.

Hennedia R. Br. ter, Trans. New Zealand Inst. 25: 285, 1893, *hom. illeg. non* Harv., 1860. Type: *Hennedia macrophylla* R. Br. ter, *lectotyp. nov.*

Beckettia C. Müll., Hedwigia 37: 77, 1898. Type: *Beckettia bruchioides* C. Müll.

Neobarbula Dus., Bot. Not. 1905: 299, 1905. Type: *Neobarbula magellanica* Dus.

Bauriella Warnst., Hedwigia 57: 88, 1915, *nom. inval. prov.* Type: *Tortula polyseta* (C. Müll.) Warnst.

Henediella Paris ex Fleisch., Musci Fl. Buitenzorg 4: 1697, 1923, *orthogr. var.*

Willia subg. *Schistidiella* (C. Müll.) Broth., Nat. Pfl. 1(3): 417, 1902. Type: *Willia marginata* (Hook. f. & Wils.) C. Müll.

Hennediella subg. *Beckettia* (C. Müll.) Roth, Aussereur. Laubm. 219, 1911.

Hennediella subg. *Hennedia* Roth, Aussereur. Laubm. 220, 1911, *nom. illeg. incl. typ. gen.*

Willia sect. *Schistidiella* C. Müll., Gen. Musc. Fr. 424, 1900. Type: *Willia marginata* (Hook. f. & Wils.) C. Müll.

Tortula sect. *Neobarbula* (Dus.) Card., Wiss. Ergebn. Schwed. Südpol. Exp. Nordenskjöld 4(8): 95, 1908. Type: *Tortula magellanica* Dus.

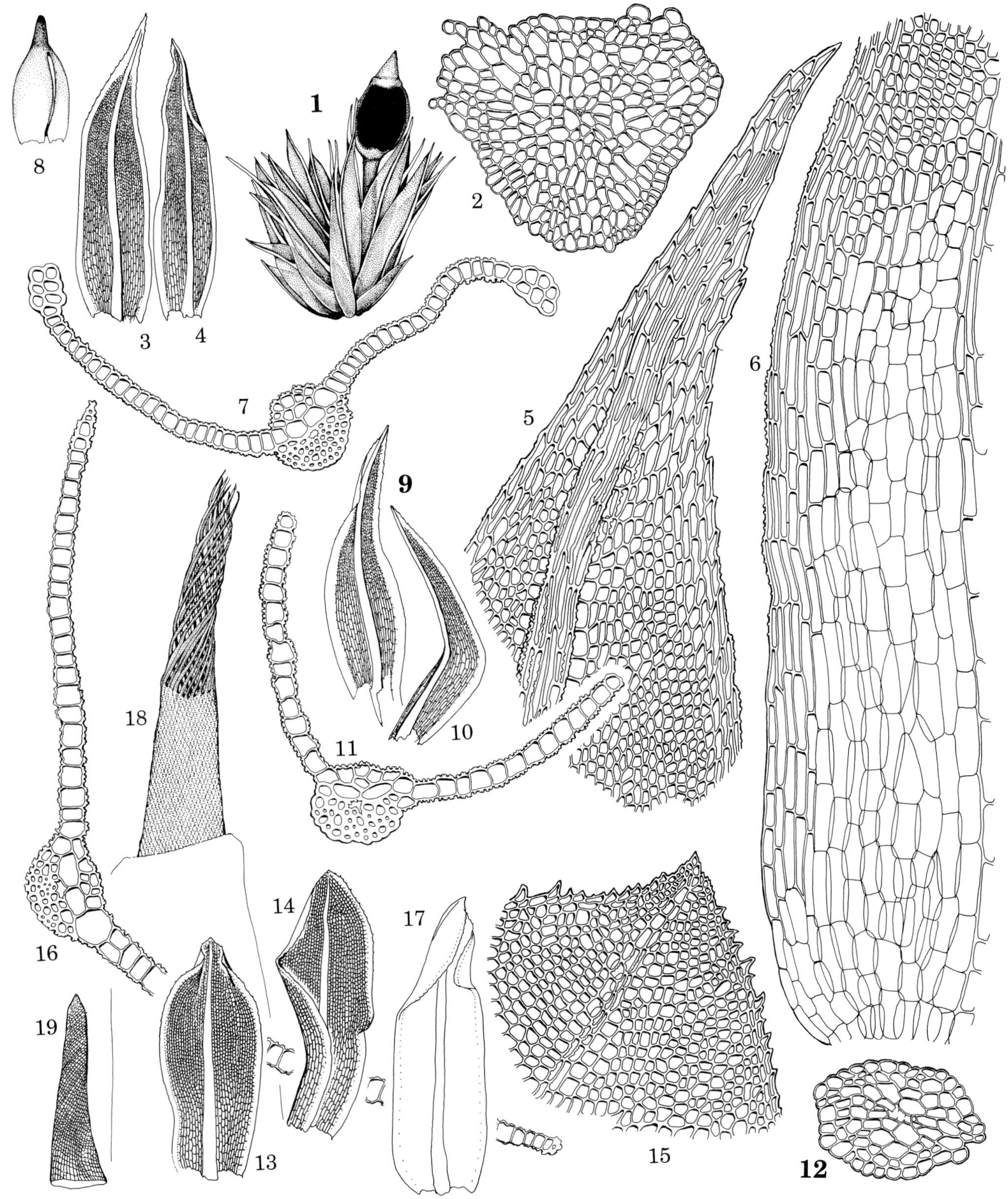

Plate 95. ***Hennediella*. 1–8. *H. macrophylla*.** 1. Habit. 2. Transverse section of stem. 3–4. Two leaves. 5. Leaf apex, dorsal view. 6. Basal cells. 7. Transverse section of costa at midleaf. 8. Calyptra. **9–11. *H. acutidentata*.** 9–10. Two leaves. 11. Transverse section of costa at midleaf. **12–19. *H. acletoi*.** 12. Transverse section of stem. 13–14. Two leaves. 15. Leaf apex. 16. Transverse section of costa at midleaf. 17. Perichaetial leaf. 18. Peristome. 19. Operculum.

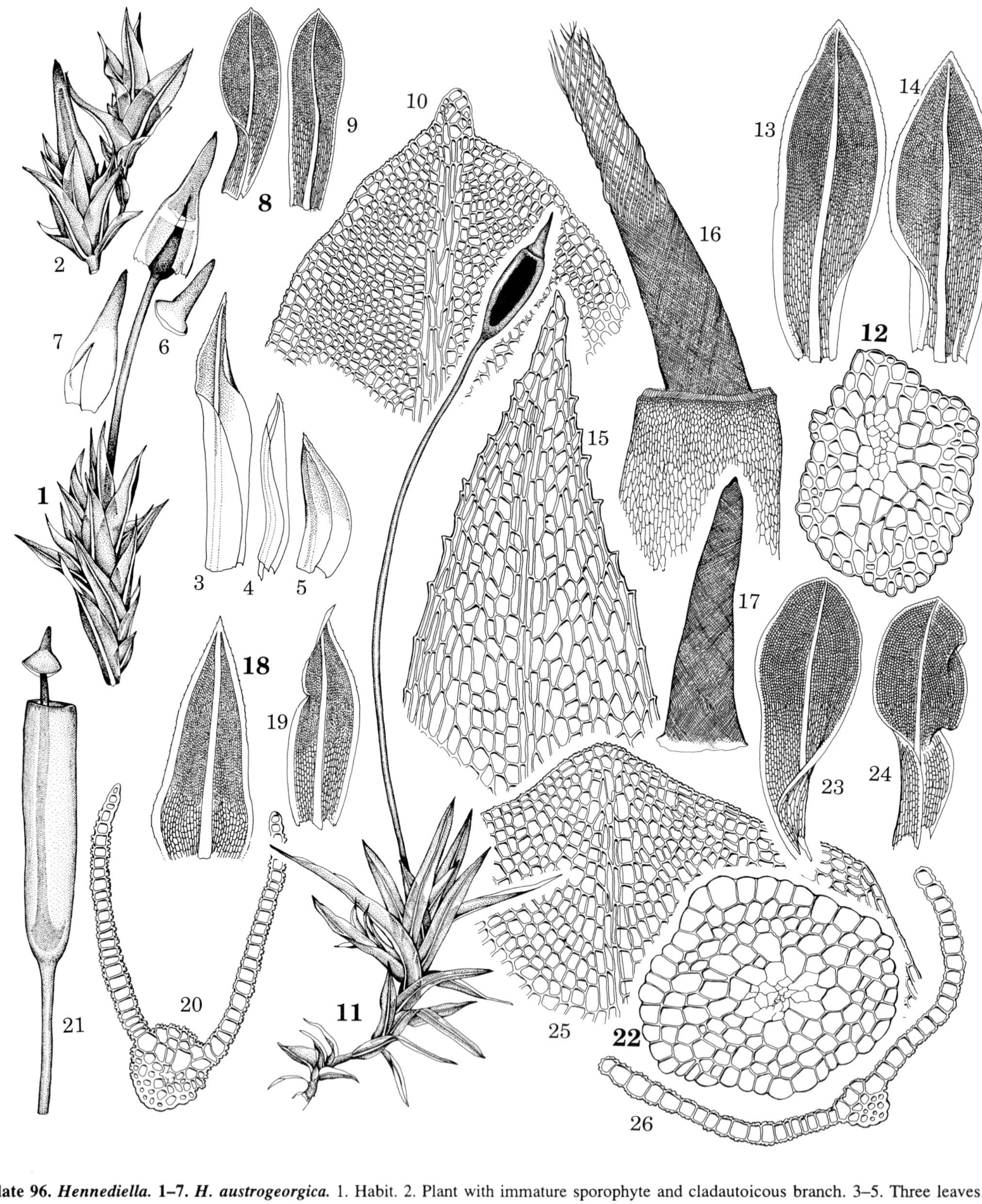

Plate 96. *Hennediella.* 1–7. *H. austrogeorgica.* 1. Habit. 2. Plant with immature sporophyte and cladautoicous branch. 3–5. Three leaves. 6. Operculum. 7. Calyptra. **8–10. *H. bellei.*** 8–9. Two leaves. 10. Leaf apex. **11. *H. densifolia.*** 11. Habit. **12–17. *H. denticulata.*** 12. Transverse section of stem. 13–14. Two leaves. 15. Leaf apex. 16. Peristome. 17. Operculum. **18–21. *H. heimei.*** 18–19. Two leaves. 20. Transverse section of costa at midleaf. 21. Systylious capsule. **22–26. *H. heteroloma.*** 22. Transverse section of stem. 23–24. Two leaves. 25. Leaf apex. 26. Transverse section of costa at midleaf.

Named for Roger Hennedy, 1809–1877, Glascow phycologist, who was Brown's professsor + *-ella*, diminutive.

Plants forming loose turf or cushions, green above, reddish brown below. *Stems* branching irregularly, ca. 0.5–1.0 cm in length, transverse section rounded-pentagonal or elliptical, central strand present, *sclerodermis absent or represented as an outer cortex of substereid cells, hyalodermis absent or weakly developed, not collapsed*; axillary hairs of ca. 5 cells, the basal 1–2 cells occasionally somewhat thick-walled; rhizoids usually common. *Leaves* appressed or incurved when dry, spreading and sometimes reflexed at midleaf when moist, *ovate to long-lanceolate, occasionally ligulate or spathulate, comparatively large*, (1.5–)3–4(–7.0) mm in length, *upper lamina plane or broadly channeled across leaf, margins plane or only rarely narrowly recurved, often dentate above*, less commonly entire, *with a border (occasionally intramarginal) of short-rectangular to elongate cells*, these usually less papillose or smooth and occasionally thick-walled but rarely bi- or tristratose; *apex acute or rarely obtuse*, rarely fragile; base not differentiated in shape or ovate and somewhat sheathing, commonly bordered by cells similar to those of the upper border; *costa percurrent or short-excurrent as a flattened, denticulate mucro*, rarely ending 3–4 cells below apex, costa with lamina inserted laterally, *superficial cells quadrate to short-rectangular and papillose ventrally, dorsally elongate and smooth or papillose*, 4–5 rows of cells across costa ventrally at midleaf, costal transverse section round to elliptical, *stereid band* one, of substereid cells and *elliptical to round in shape*, ventral epidermis present, *dorsal epidermis usually present* (occasionally present only laterally on the costa or sometimes absent, guide cells 2–4 per layer in (1–)2 layers, *hydroid strand present, usually large*; *upper laminal cells relatively large*, quadrate to hexagonal or short-rectangular, ca. 18–24 µm in width, 1–2:1, *walls thin to evenly thickened, superficially flat to weakly convex; papillae hollow, simple to bifid, usually 6–8 or more per lumen*, occasionally absent; basal cells differentiated across leaf, rectangular, 18–30 µm in width, (2–)4–6:1, walls thin. Autoicous, paroicous, cladautoicous, or dioicous. Perichaetial leaves somewhat larger than the cauline leaves. Perigonia in dioicous species borne terminally on plants somewhat smaller than the perichaetiate. *Seta extremely various in size among species, 0.05–2.5 cm in length*, 1 per perichaetium, yellow-brown to dull brown, twisted counterclockwise; capsule occasionally systylious, *theca 0.8–3.5 mm in length*, yellow-brown to dark brown, *ovate to cylindrical*, occasionally microstomous or macrostomous, exothecial cells rectangular, 25–35 µm in width, 3–5:1, walls thin to irregularly thickened, stomates phaneropore, at base of capsule, annulus of 2–4 rows of vesiculose cells; *peristome teeth absent, rudimentary, or of 32 long, filamentous teeth*, densely spiculose, when well developed to 1.5 mm in length, with many articulations, twisted counterclockwise, *basal membrane occasionally to 500 µm in height*, spiculose. *Operculum conic to rostrate, often narrowly so*, (0.6–)1.2–1.8 mm in length, cells twisted counterclockwise. *Calyptra cucullate, occasionally flaring below and not split*, smooth, 2.0–4.5 mm in length. *Spores sometimes quite large*, 8–30 µm in diameter, yellowish, essentially smooth, papillose or tuberculate. *Laminal KOH color reaction red.* Reported chromosome numbers: n = 25+m, 26, 26+m, 26+2m, 50.

Found on soil and rock, commonly in wet areas, nearly worldwide but most diverse in temperate areas of the austral region.

This previously monotypic genus has been treated as a synonym of *Pottia* (Dixon 1923; Sainsbury 1955; van der Wijk et al. 1959–69); however, a clear sporophyte reduction series involving other species (here transferred to *Hennediella*) and the distinctive morphology of gametophyte warrant recognition at the generic level. Support for recognition of distinctive characters of at least the gametophyte includes Sainsbury's (1955) comment that *Hennediella* be treated as "a section of *Pottia* distinguished by the toothed and bordered leaves and by the very large, usually mitriform, calyptra..."; A. Fife's comment (pers. comm.) that *Pottia macrophylla* (Pl. 95, f. 1–8) deserves recognition (as *Hennediella*); Corley et al. (1981) suggested *Tortula stanfordensis* (Pl. 98, f. 7–11) should be in a separate section or genus; and Matteri (1977a, 1977b, 1988a) pointed out that *Pottia austrogeorgica* (Pl. 96, f. 1–7), which she regarded as a good species, was vegetatively similar both to the *P. heimii* (Pl. 96, f. 18–21) complex and to *Hennediella macrophylla*. Recently, Mishler (1990), in a phylogenetic analysis of 23 *Tortula s. lat.* species from Mexico, found that *T. stanfordensis*, *T. polyseta* (Pl. 98, f. 1–6) and *T. leiostoma*, all here recognized in *Hennediella*, are "likely to be monophyletic." Blockeel (1991) appreciated the relationships of *H. macrophylla*, *T. stanfordensis* and *P. austrogeorgica*, and provided a detailed evaluation recognizing *Hennediella* as a good genus for at least these three species, making the appropriate combinations (and mentioned my sketchy generic key that included *Hennediella* as conceived here, Zander 1989, as support).

Characters here considered of most importance are the often large and commonly broadly lanceolate leaves (Pl 95, f. 3–4) bordered by a band of usually elongate thick-walled cells, with usually serrulate to dentate and almost always plane (narrowly recurved in *H. limbata*, Pl. 97, f. 1–6) upper laminal margins, upper medial laminal cells large, superficially rather flat (Pl. 95, f. 7, 11, 16), thin-walled and strongly hollow-papillose (rarely epapillose, as in *H. denticulata*, Pl. 96, f. 12–14), costa usually with a dorsal epidermis, stereid band elliptical, or semicircular to rounded (Pl. 95, f. 7, 11, 16—like *Tortula* and unlike *Syntrichia*, which has lunate dorsal stereid band sections), sporophytes in a reduction series (also like *Tortula* and unlike *Syntrichia*), operculum often long and narrowly rostrate (Pl. 98, f. 6), calyptra enlarged, sometimes inflated (Pl. 95, f. 8; 97, f. 7; 98, f. 17), and KOH color reaction red. The perigonia are borne terminally on lateral branchlets on the archegoniate stems of *H. macrophylla* and *H. longirostris*. *Hennediella densifolia* is also monoicous but has gemmate perigonia sessile on the main axis. Dixon (1923) noted that the calyptra of *H. macrophylla* is campanulate but tears basally as it matures, while Sainsbury (1955) observed that in this species the inflated calyptra may appear campanulate in some collections and cucullate in others with much variability and intergradation in shape.

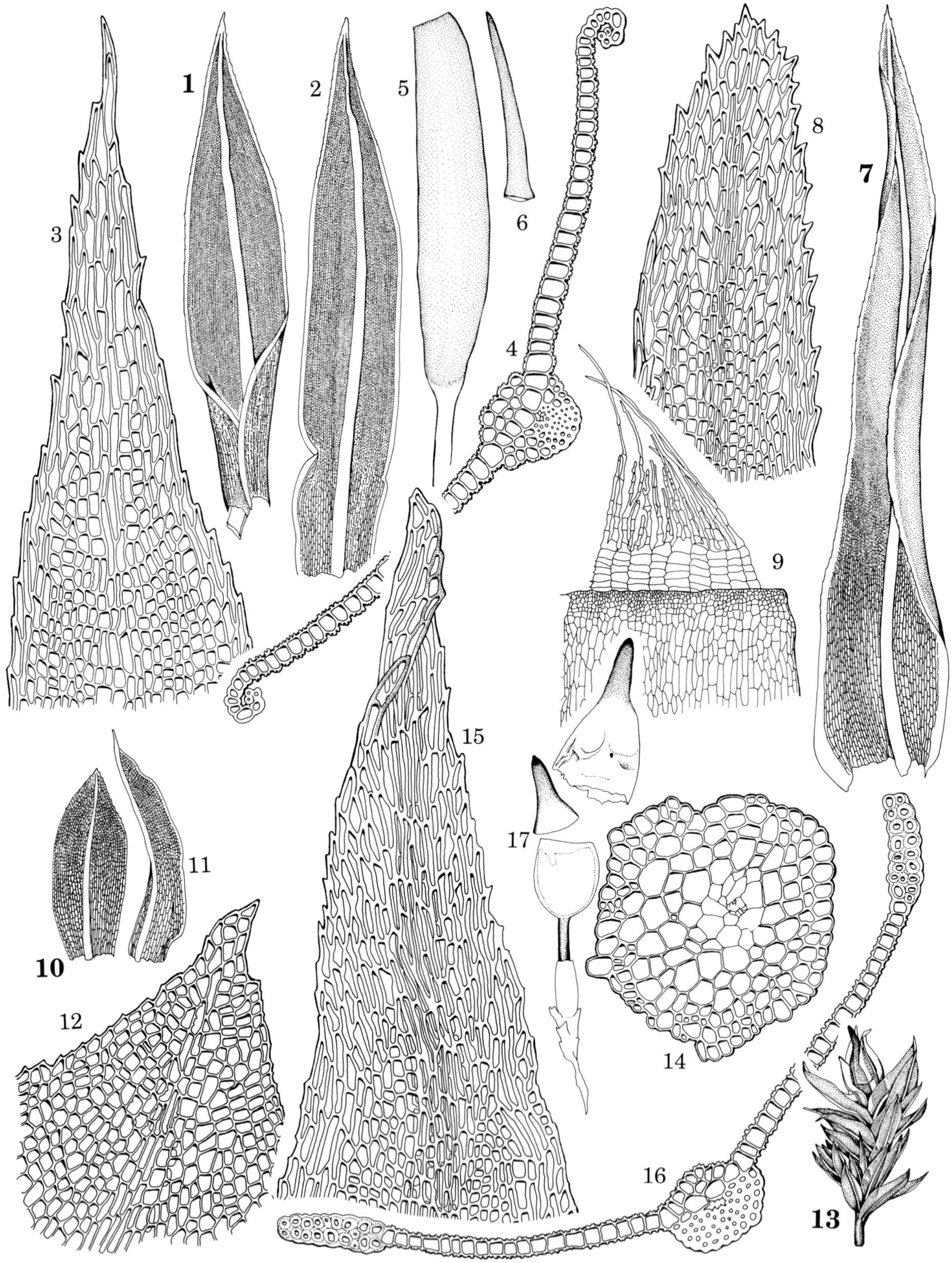

Plate 97. ***Hennediella.*** **1–6.** ***H. limbata.*** 1–2. Two leaves. 3. Leaf apex. 4. Transverse section of costa at midleaf. 5. Theca. 6. Operculum. **7–9.** ***H. magellanica.*** 7. Leaf. 8. Leaf apex. 9. Portion of peristome. **10–12.** ***H. oedipodioides.*** 10–11. Two leaves. 12. Leaf apex. **13–17.** ***H. marginata.*** 13. Habit. 14. Transverse section of stem. 15. Leaf apex. 16. Transverse section at midleaf. 17. Sporophyte, with operculum and calyptra.

The costal guide cells are in what may be interpreted as two layers, the ventral layer being of smaller and somewhat thicker walled cells than those of the dorsal guide cell layer. Because the stereid band is made up of substereid cells, the ventral layer of guide cells, although intermediate in appearance between the larger layer of guide cells and the dorsal substereids, may be interpreted as indication of a second stereid band. This, plus the common lanceolate leaf shape may imply a relationship to the Barbuleae. Its closest relationship, however, is probably with *Tortula* species having leaves with similar lax areolation and differentiated border (e.g. *Tortula solmsii*), but which differ by the generally broader leaves, more convex superficial surface of the laminal cells, lack of marginal teeth, and yellow KOH reaction (see, however, Cladograms 11 and 14 for alternative hypotheses). The upper laminal cells of *Hennediella* are rather large, but the genus may be distinguished from *Chenia*, *Dolotortula* and *Sagenotortula*, similar genera with red KOH reactions and very large upper laminal cells, by the papillose upper laminal cells, except in the case of the epapillose *H. denticulata*, which may be distinguished by its unistratose border of elongate cells that are denticulate in the upper portion of the leaf.

Like *Pseudocrossidium*, *Hennediella* is an essentially austral genus with extensions of some species into the north temperate zone probably through migration routes along north-south cordilleras.

Number of accepted species: 20.

Species examined: *H. acletoi* (US), *H. acutidentata* (NY), *H. angustifolia* (NY), *H. arenae* (NY, PC), *H. austrogeorgica* (NY), *H. bellei* Bartr. (FH), *H. densifolia* (NY), *H. denticulata* (NY, BM), *H. heimii*, *H. heteroloma* (BUF, TENN), *H. kunzeana* (NY), *H. limbata* (NY), *H. longipedunculata* (NY), *H. longirostris* (BM), *H. macrophylla* (BM), *H. marginata* (BM), *H. oedipodioides* (NY), *H. polyseta* (BUF, FH, TENN), *H. serrulata* (NY), *H. stanfordensis* (BUF, NY), *H. steereana* (BUF).

New combinations:
Hennediella acletoi (Robins.) Zand., *comb. nov.* (*Tortula*

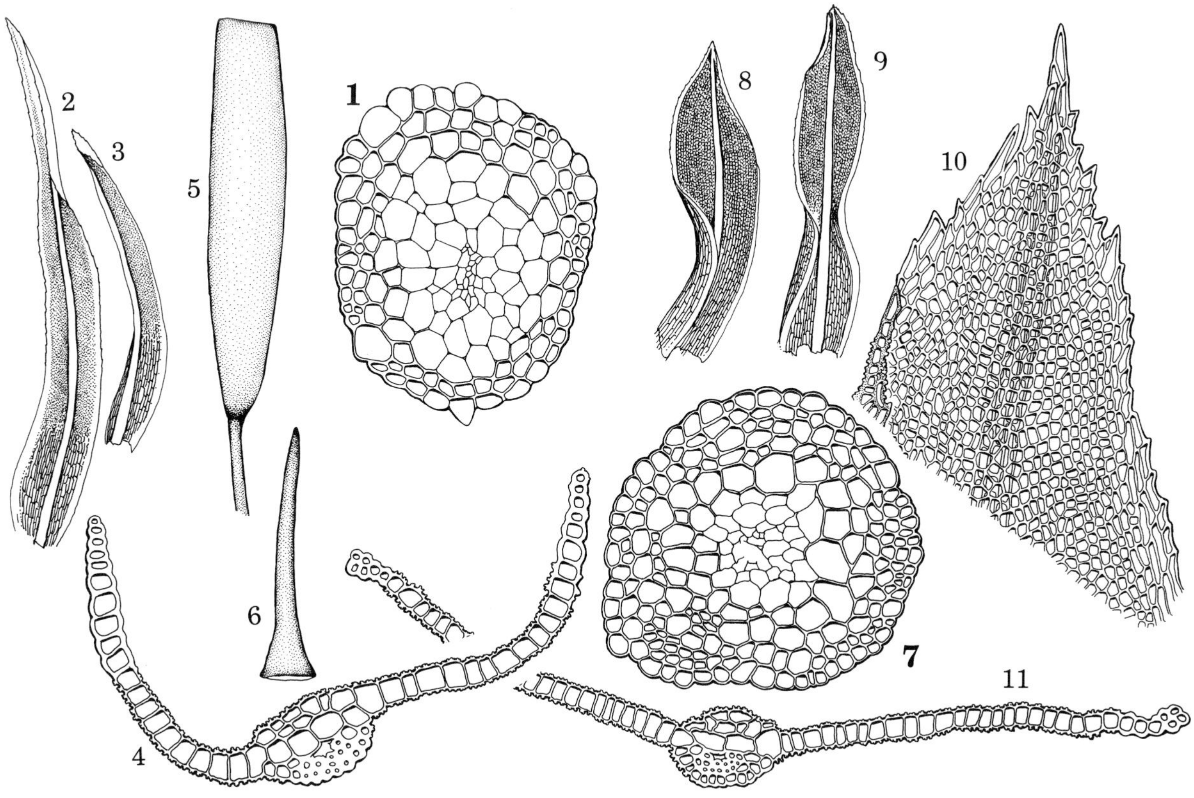

Plate 98. ***Hennediella.*** **1–6.** ***H. polyseta.*** 1. Transverse section of stem. 2–3. Two leaves. 4. Transverse section of costa at midleaf. 5. Theca. 6. Operculum. **7–11.** ***H. stanfordensis.*** 7. Transverse section of stem. 8–9. Two leaves. 10. Leaf apex. 11. Transverse section of costa at midleaf.

acletoi Robins., Phytologia 12: 390, 1971).

Hennediella acutidentata (Card. & Thér.) Zand., *comb. nov.* (*Pottia acutidentata* Card. & Thér., Bull. Mus. Hist. Nat. Paris 22: 338, 1916).

Hennediella angustifolia (Herz.) Zand., *comb. nov.* (*Calyptopogon angustifolius* Herz., Beih. Bot. Centralbl. 26(2): 63, 1910; *Tortula angustifolia* (Herz.) Herz., *nom. illeg.*; *Tortula herzogii* Zand.).

Hennediella arenae (Besch.) Zand., *comb. nov.* (*Barbula arenae* Besch., Bull. Soc. Bot. France 32: 59, 1885; *Tortula arenae* (Besch.) Broth.).

Hennediella arenae var. *petriei* (Broth. ex Beck.) Zand., *comb. et stat. nov.* (*Tortula petriei* Broth. ex Beck., Trans. New Zealand Inst. 29: 441, 1897; *Tortula arenae* subsp. *petriei* (Broth.) Lightowlers, J. Bryol. 13: 371, 1985), not seen.

Hennediella bellei (Bartr.) Zand., *comb. nov.* (*Desmatodon bellii* Bartr., Bull. Brit. Mus. Nat. Hist. Bot. 2: 56, 1955).

Hennediella densifolia (Hook. f. & Wils.) Zand., *comb. nov.* (*Barbula densifolia* London J. Bot. 3: 543, 1844; *Tortula densifolia* (Hook. f. & Wils.) Hook. f. & Wils.).

Hennediella denticulata (Wils. in Mitt.) Zand., *comb. nov.* (*Barbula denticulata* Wils. in Mitt., Kew J. Bot. 3: 50, 1851; *Tortula denticulata* (Wils. in Mitt.) Mitt.). This peristomate species is probably the same as *Tortula densifolia* (not seen).

Hennediella heimii (Hedw.) Zand., *comb. nov.* (*Gymnostomum heimii* Hedw., Spec. Musc. 32, 1801; *Pottia heimii* (Hedw.) Hampe).

Hennediella heimii var. *alpina* (Amann) Zand., *comb. nov.* (*Pottia heimii* var. *alpina* Amann Bull. Soc. Vaudoise Sc. Nat. 54: 42, 1921), not seen.

Hennediella heimii var. *arctica* (Lindb.) Zand., *comb. nov.* (*Pottia heimii* var. *arctica* Oefv. K. Vet. Ak. Foerh. 23: 551, 1867), not seen.

Hennediella heimii var. *brachyphylla* (Warnst.) Zand., *comb. nov.* (*Pottia heimii* var. *brachyphylla* Warnst., Hedwigia 58: 100, 1916), not seen.

Hennediella heimii var. *brevicuspis* (Warnst.) Zand., *comb. nov.* (*Pottia heimii* var. *brevicuspis* Warnst., Hedwigia 58: 98, 1916), not seen.

Hennediella heimii var. *breviseta* (Warnst.) Zand., *comb. nov.* (*Pottia heimii* var. *breviseta* Warnst., Hedwigia 58: 99, 1916), not seen.

Hennediella heimii var. *drummondii* (Warnst.) Zand., *comb. nov.* (*Pottia heimii* var. *drummondii* Warnst., Hedwigia 58: 96, 1916), not seen.

Hennediella heimii var. *eurystoma* (Card. & Broth.) Zand., *comb. nov.* (*Pottia heimii* var. *eurystoma* Card. & Broth., K. Svensk. Vet. Ak. Handl. 63(10): 20, 1923), not seen.

Hennediella heimii var. *guessfeldtii* (Schlieph.) Zand., *comb. nov.* (*Pottia guessfeldtii* Schlieph., Ber. Deutsch. Bot. Ges. 2: 461, 1884; *Pottia heimii* var. *guessfeldtii* (Schlieph.) Warnst.), not seen.

Hennediella heimii var. *lanceolata* (Warnst.) Zand., *comb. nov.* (*Pottia heimii* var. *lanceolata* Warnst., Hedwigia 58: 98, 1916), not seen.

Hennediella heimii var. *magellanica* (Warnst.) Zand., *comb. nov.* (*Pottia heimii* var. *magellanica* Warnst., Hedwigia 58: 95, 1916).

Hennediella heimii var. *maxima* (Card.) Zand., *comb. nov.* (*Pottia heimii* var. *maxima* Car., Bull. Herb. Boiss. ser. 2, 5: 1002, 1905), not seen.

Hennediella heimii var. *spegazzinii* (C. Müll.) Zand., *comb. nov.* (*Pottia spegazzinii* C. Müll., Flora 68: 414, 1885; *Pottia heimii* var. *spegazzinii* (C. Müll.) Warnst.), not seen.

Hennediella heimii var. *thaxteri* (Card. & Thér.) Zand., *comb. nov.* (*Pottia heimii* var. *thaxteri* Card. & Thér., Rev. Bryol. n. ser. 2: 165, 1930), not seen.

Hennediella heteroloma (Card.) Zand., *comb. nov.* (*Tortula heteroloma* Card., Rev. Bryol. 37: 127, 1910).

Hennediella heteroloma var. *eckeliae* (Zand.) Zand., *comb. nov.* (*Tortula eckeliae* Zand., Bryologist 88: 354, 1985 [1986]).

Hennediella kunzeana (C. Müll.) Zand., *comb. nov.* (*Barbula kunzeana* C. Müll., Linnaea 17: 586, 1843; *Tortula kunzeana* (C. Müll.) Mont. in Gay).

Hennediella limbata (Mitt.) Zand., *comb. nov.* (*Barbula limbata* Mitt., Kew J. Bot. 3: 354, 1851; *Tortula limbata* (Mitt.) Mitt., *hom. illeg.*).

Hennediella longipedunculata (C. Müll.) Zand., *comb. nov.* (*Barbula longipedunculata* C. Müll., Syn. Musc. 1: 630, 1849).

Hennediella longirostris (Hampe ex C. Müll.) Zand., *comb. nov.* (*Pottia longirostris* Hampe ex C. Müll., Syn. 1: 552, 1849; *Tortula longirostris* (Hampe ex C. Müll.) Broth. in Par., *hom. illeg.*).

Hennediella marginata (Hook. f. & Wils.) Zand., *comb. nov.* (*Schistidium marginatum* Hook. f. & Wils., London J. Bot. 3: 539, 1844; *Streptopogon marginatus* (Hook. f. & Wils.) Mitt.; *Willia marginata* (Hook. f. & Wils.) C. Müll.)."

Hennediella oedipodioides (C. Müll.) Zand., *comb. nov.* (*Pottia oedipodioides* C. Müll., Bot. Jahrb. 5: 79, 1883).

Hennediella polyseta (C. Müll.) Zand., *comb. nov.* (*Barbula polyseta* C. Müll., Nuov. Giorn. Bot. Ital. n. ser. 4: 414, 1897; *Tortula polyseta* (C.Müll.) Warnst.).

Hennediella serrulata (Hook. & Grev.) Zand., *comb. nov.* (*Tortula serrulata* Hook. & Grev., Edinburgh J. Sc. 1: 299, 1824).

Hennediella steereana (Zand. & Crum) Zand., *comb. nov.* (*Desmatodon steereanus* Zand. & Crum, Bryologist 80: 638, 1977).

68. DOLOTORTULA — Plate 99.

Dolotortula Zand., Phytologia 65: 425, 1989. Type: *Dolotortula mniifolia* (Sull.) Zand.

From *dolus*, from δόλος, artifice, deceit, guile + o + *Tortula*, a genus; mimicking the genus *Tortula*.

Plants in a loose turf, green above, brownish green below. Stems seldom branching, to 1.5 cm in length, transverse section round, central strand present, distinct, sclerodermis and hyalodermis absent; axillary hairs of clear cells, several cells in length; weakly rhizoidiferous. *Leaves* contorted, appressed when dry, widely spreading and somewhat rosulate when moist, *spathulate*, 2.5–4.0 mm in length, upper lamina flat, very weakly channeled along the costa, *margins* weakly recurved below midleaf, *entire, with a narrow cartilaginous border of stereid cells to 4 cells in thickness and ending at or just before apex;* apex rounded or emarginate, often bluntly apiculate; base little differentiated in shape; *costa thin, percurrent or ending up to 4 cells below apex,* costa with lamina inserted laterally, superficial cells ventrally long-rectangular, dorsally narrowly elongate, 2 rows of cells across costa ventrally at midleaf, costal transverse section round, *stereid band weak and rounded in shape, ventral and dorsal epidermis present*, the latter only laterally, guide cells 2 in 1 layer, *hydroid strand present*, small; *upper laminal cells hexagonal to short-rectangular, large, ca. 22–28 µm in width,* 1–2:1, *walls thin, often weakly trigonous,* superficially weakly convex on both sides; *papillae absent;* basal cells scarcely differentiated, short-rectangular, ca. 30 µm in width, 2–3:1, walls thin. Dioicous. Perichaetia terminal, inner leaves slightly larger than the cauline. Perigonia not seen. Seta 1.0–1.3 cm in length, 1 per perichaetium, red-brown, twisted clockwise; theca ca. 1.7–2.0 mm in length, red- or yellow-brown, cylindrical, exothecial cells thin-walled, hexagonal to short-rectangular, 25–35 µm in width, stomates at base of theca, phaneropore, annulus of 3–4 rows of vesiculose cells, persistent; peristome teeth ca. 32, filamentous, anastomosing at base, densely spiculose, ca. 1000 µm in length, with many articulations, weakly twisted counterclockwise, basal membrane ca. 70 µm in height, densely spiculose-papillose. Operculum not seen (blunt-conic *ex descr.*). Calyptra not seen. Spores 10–13 µm in diameter, yellowish, weakly papillose. *Laminal KOH color reaction red.*

A rare taxon of scattered distribution, being found on soil, often calcareous, at moderate elevations; Mexico, West Indies, Central America and the Andes of South America.

Among the three genera with epapillose, very large upper laminal cells (*Chenia*, *Dolotortula* and *Sagenotortula*), *Dolotortula* is unusual in its strong, multistratose border of stereid cells (Pl. 99, f. 6, 8). Species of *Tortula* sect. *Pottia* and *Tortula* sect. *Hyophilopsis* with bordered leaves and large upper laminal cells have somewhat the same appearance, but these have (broadly) acute leaf apices, uni- to bistratose borders of substereid cells, at least weakly papillose upper laminal cells, and have a yellow laminal KOH reaction. Quite possibly, through elaboration of the marginal border, loss of upper laminal papillae, and modification of cell wall chemistry to enable the red color reaction (all apparently advanced character states), *Dolotortula* was derived from shared ancestors of such taxa (see Cladogram 14). It appears unrelated to *Sagenotortula* because of the *Syntrichia*-like costal section of the latter genus, while *Chenia* may be derived from borderless, serrate-margined species of *Tortula* sect. *Pottia* (but if so then distantly, see Cladograms 11 and 14). *Tortula domingensis*, which Crum and Steere (1957) indicated was "doubtfully distinct" from *D. mniifolia* (discussed as *Tortula*), is actually a *Brachymenium* (new combination made in section on Excluded Taxa).

Additional literature: Bartram (1949), Zander (1989).
Number of accepted species: 1.
Species examined: *D. mniifolia* (BM, BUF, FH, TENN).

69. PHASCOPSIS — Plate 100.

Phascopsis Stone, J. Bryol. 11: 17, 1980. Type: *Phascopsis rubicunda* Stone.

From *Phascum*, a genus + ὄψις, -εως, appearance; resembling the genus *Phascum*.

Plants in dense clumps, green above, brown below. *Stems* branching often, 1–5 mm in length, cells of stem apex filled with oil globules, transverse section rounded-pentagonal, *central strand present* but generally very weak, sclerodermis absent, hyalodermis absent or possibly present but weak; axillary hairs of 3–4 cells, basal 1–2 cells somewhat thicker-walled; rhizoids sparse. *Leaves* appressed, somewhat incurved when dry, weakly spreading when moist, ca. 1.3 mm in length, *oblong to spathulate*, upper lamina flat to broadly channeled across leaf, *margins plane, entire, weakly bordered by 1(–4) rows of slightly thicker-walled cells*; *apex rounded to broadly acute*; base scarcely differentiated in shape; *costa short-excurrent as a mucro*, superficial cells quadrate to short-rectangular ventrally, elongate dorsally, ca. 4 rows of cells across costa ventrally at midleaf, costal transverse section semicircular to round, *stereid band strong and mostly circular in section*, epidermis present ventrally, absent dorsally, *guide cells 2–4 (occasionally absent) in 1(–2) layers, hydroid strand very strong or absent, often encircled by stereid cells*, costa often greatly bulging ventrally; *upper laminal cells* quadrate to irregularly rhomboidal or occasionally short-rectangular, *ca. 13–15 µm in width, 1(–2):1*, walls thin, superficially weakly convex on both sides; *papillae hollow, oh-shaped, simple or occasionally bifid, rather large, 4–6 per lumen*; basal cells differentiated across leaf, rising higher medially, short-rectangular, somewhat wider than upper cells, 2–3:1, walls thin, hyaline. Cauline leaves with unusually thickened costae apparently act as propagula. *Dioicous*. Perichaetia weakly differentiated, inner leaves oblong, apex emarginate, somewhat shorter in length, not sheathing, cells long-rhomboidal in lower 4/5. Perigonia terminal (but overtopped by subperigonial innovations on smaller plants). *Seta very short*, ca. 0.2 mm in length, 1 per perichaetium, yellowish brown, not twisted; *capsule cleistocarpous*, ca. 1 mm in length, yellowish brown, short-elliptical, *apiculate*, exothecial cells thin-walled, hexagonal to short-rhomboidal to rectangular, stomates phaneropore, at base of theca, annulus absent. *Calyptra mitrate, lobed*, smooth, ca. 0.7 mm in length. Spores round to elliptical, 20–23 µm in longest diameter, hyaline, weakly papillose. *Laminal KOH color reaction red.*

Known to date from only a few collections in arid southern and western Australia and in New Zealand, on clay, calcareous or sandy soil.

Phascopsis is characterized by the cleistocarpous, nearly globose, apiculate capsule (Pl. 100, f. 1); lobed, mitrate calyptra; broad, entire leaves with stout costa (Pl. 100, f. 3–5) and large, hollow-papillose upper laminal cells (Pl. 100, f. 11); transverse section of costa showing only one stereid band, a very strong hydroid strand, and occasional absence of distinct guide cells (Pl. 100, f. 8–10); and dioicous sexual condition. Stone (1980a) described, illustrated and discussed this taxon exhaustively. There is a certain resemblance to *Aschisma*, but that genus clearly has two stereid bands. *Phascopsis* may be closely related, as the end of a reduction series, to *Syntrichia*

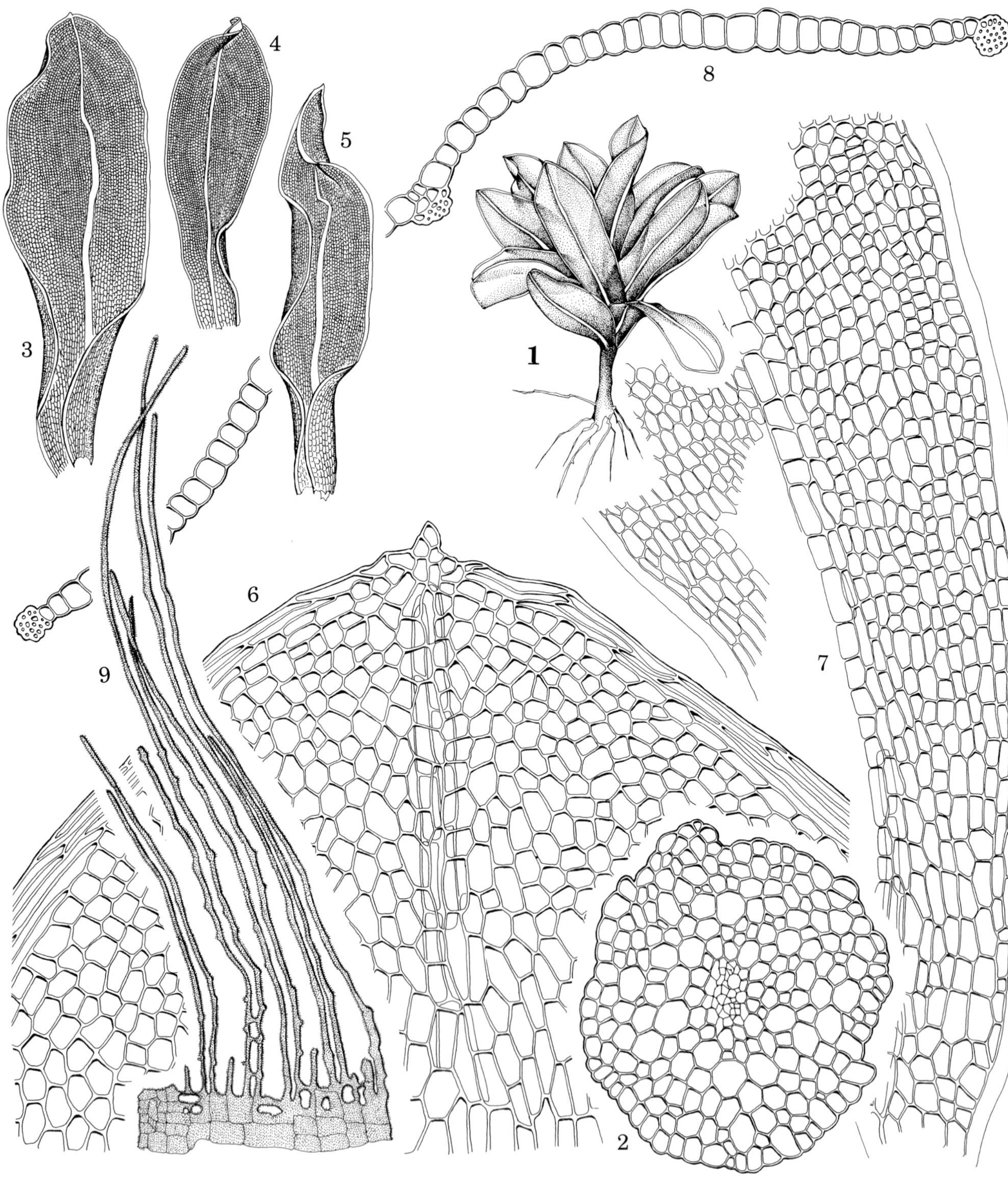

Plate 99. ***Dolotortula.* 1–9. *D. mniifolia*.** 1. Habit. 2. Transverse section of stem. 3–5. Three leaves. 6. Leaf apex. 7. Basal cells. 8. Transverse section at midleaf. 9. Portion of peristome.

sect. *Aesiotortula*, but Cladograms 13 and 14 indicate its ancestors are more primitive (inserted deeper in the Pottieae subclade) than those of *Syntrichia*.

The morphology exhibited in the transverse section of the costa of *Phascopsis* is curious in its variability. A build up of stereid and parenchymatous cells ventral to the hydroid strand, possibly associated ultimately with occasional production of propaguloid cauline leaves with much swollen costae (Pl. 100, f. 12), results in encirclement of the hydroid strand (Pl. 100, f. 9) so that it appears in the center of the stereid band. This also may result in apparent absence of guide cells. Sections made below midleaf in larger leaves (Pl. 100, f. 8) and at midleaf in smaller leaves, however, demonstrate morphology consistent with other Pottieae. *Crossidium aberrans*, *C. convolutum* and *Tortula muralis* may have a similar encirclement of the hydroid strand by stereid cells, though there is no immediate phylogenetic relationship between *Phascopsis* and these taxa (Cladograms 13 and 14). Stone (1980) observed a weak line of dehiscence of an operculum in some capsules of *Phascopsis*.

New Zealand collections of *P. rubicunda* include one labeled as *Pottia stevensii* [R. Br. ter], New Zealand, North I., Hawke Bay, near Wairoa, Whakamahia, sandy soil adjoining sea beach, Sainsbury, 3 Oct. 1931, US ex F and another determined as *Pottia longifolia* R. Br. ter by H. Dixon, North I., Wellington, Berggren, 1874, NY. These names predate that of *P. rubicunda* and their types should be examined. The upper costa is only weakly swollen in collections other than the type and the combination of plane, unbordered laminal margins and red KOH reaction is here considered diagnostic.

Number of accepted species: 1.

Species examined: *P. rubicunda* (MELU, NY, US).

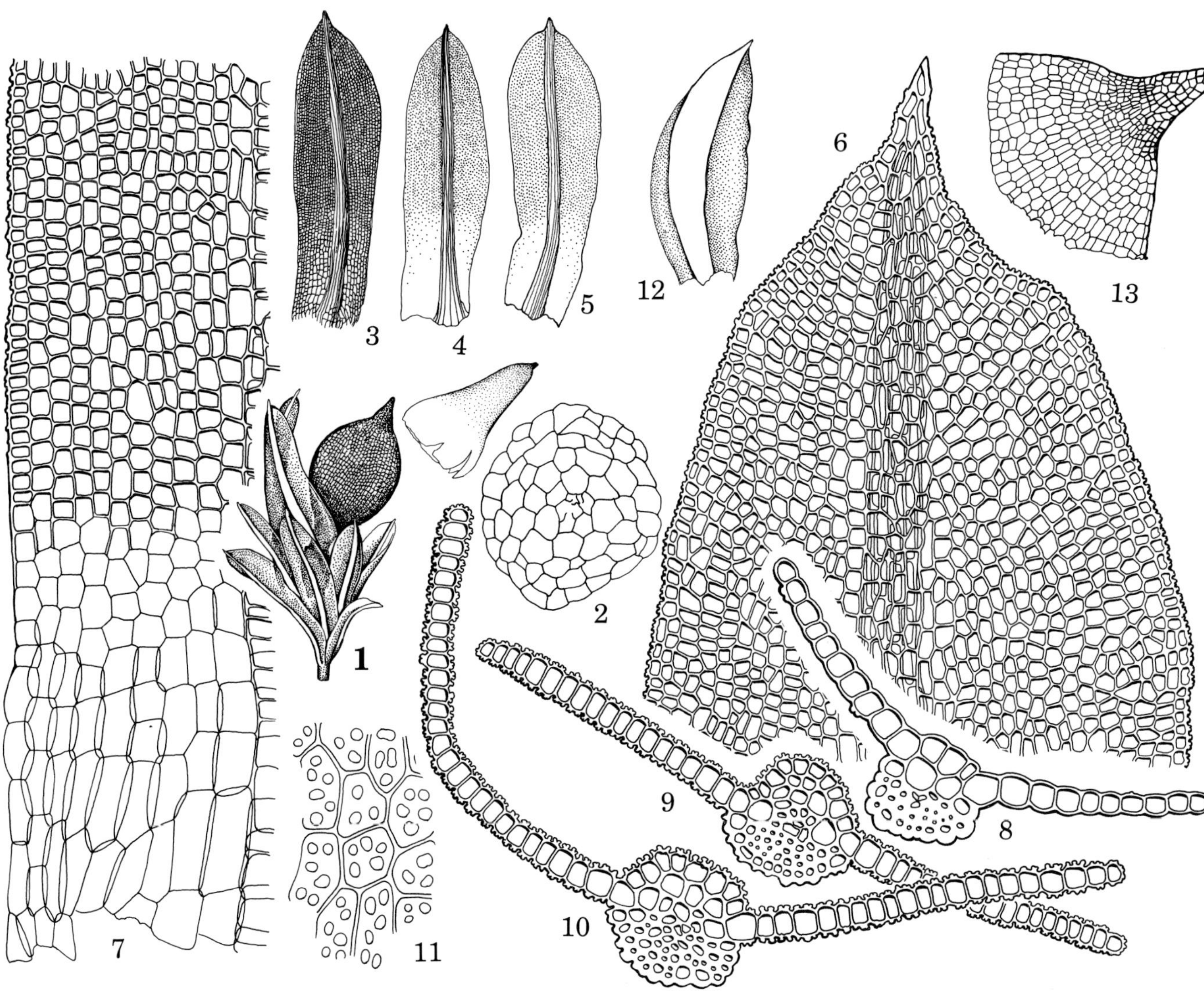

Plate 100. ***Phascopsis.*** **1–13.** ***P. rubicunda.*** 1. Habit (and calyptra). 2. Transverse section of lower stem. 3–5. Three leaves. 6. Leaf apex. 7. Basal cells. 8. Transverse section of leaf base. 9–10. Transverse sections near midleaf. 11. Upper laminal papillae. 12. Propaguloid cauline leaf. 13. Apiculus of capsule.

70. STONEA Plate 101.

Stonea Zand., Phytologia 65: 431, 1989. Type: *Stonea oleaginosa* (Stone) Zand.

Named for Ilma G. Stone, an Australian bryologist well known for her detailed morphological and anatomical studies of Australasian arid habitat mosses.

Plants gregarious, mostly buried in soil, green above, reddish brown below. *Stems* seldom branching, *very short,* to 0.3 mm in length, transverse section rounded, central strand absent, sclerodermis absent, hyalodermis absent; axillary hairs 2–4 cells in length, basal cells firm-walled; lower stem clothed with fine rhizoids. *Leaves* incurved when dry, weakly spreading when moist, *obovate or short-lingulate, occasionally wider than long, short,* ca. 0.4–0.5 mm in length, *upper lamina broadly and deeply concave, margins plane,* entire or dentate at apex; *apex usually broadly and often also sharply apiculate, usually cucullate;* base not differentiated in shape; *costa percurrent* or occasionally ending 1–3 cells below apex, costa with lamina inserted nearly laterally, *superficial cells papillose, quadrate and usually bulging ventrally,* dorsally elongate, *sharply papillose near apex dorsally,* ca. 3 rows of cells across costa ventrally at midleaf, costal transverse section semicircular to circular, *stereid band weak and rounded in shape, ventral epidermis present, often strongly bulging, dorsal epidermis present or absent,* guide cells 2 in one layer or absent, hydroid strand absent, *costa often expanded as a large, ventrally bulging, rounded, oil-rich excrescence* nearly as wide as the leaf but this affecting only the deciduous, uppermost leaves; upper laminal cells quadrate, ca. 13 µm in width, 1:1, walls thin, superficially weakly convex on both sides of lamina; *laminal papillae present only dorsally near costa at leaf apex,* 1–3 per lumen, simple, hollow to solid; basal cells only weakly differentiated, quadrate to short-rectangular, ca. 15 µm in width, 1–2:1, walls thin. Apparently dioicous (naked axillary archegonia reported in original description). Sporophytes and androgametophytes unknown. *Laminal KOH color reaction red.* (Because of paucity of available material, details of the costal section are from the original description and illustrations.)

Found on soil or thin soil over limestone in dry areas of southern Australia.

Stonea oleaginosa has a swollen, oil-rich lenticular knob on the ventral surface of the upper costa of the very uppermost leaves on the stem. Leaves farther down on the stem lack this excrescence, but instead have protruding, bottle-shaped, papillose cells on the ventral costal surface much like those on the costa of *Crossidium aberrans. Stonea* is entirely red in KOH and has plane margins, features absent in *Crossidium.* As in the case of *Gymnostomiella* ((*q.v.*), the leaves of *Stonea* may have attained their general lack of distinctive characters through paedomorphosis. This genus is probably closely related to *Syntrichia*; *S. caninervis* has a similar spinose-papillose dorsal surface of the costa, a very small immature habit, and similar oil globules in its upper laminal cells, but *Stonea* differs by its obovate leaves with plane

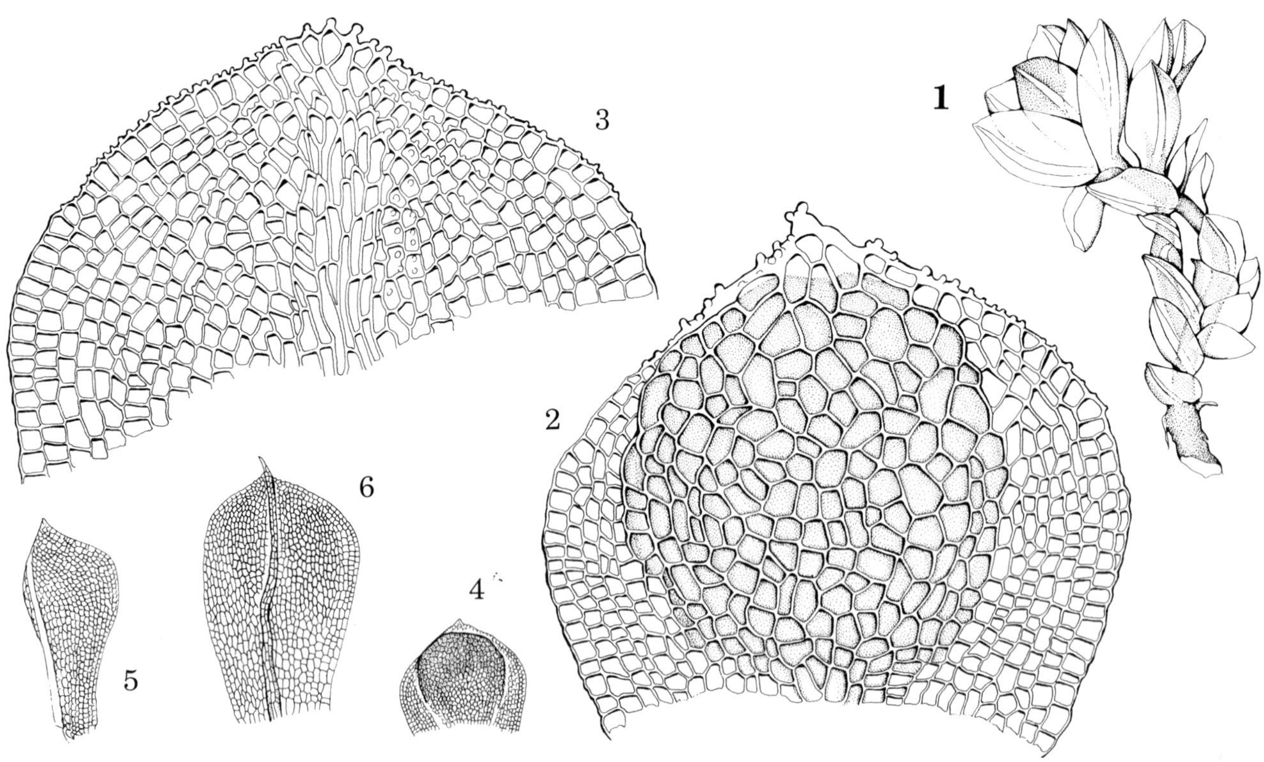

Plate 101. ***Stonea.* 1–6. *S. oleaginosa.*** 1. Habit. 2. Ventral surface of a propaguloid leaf from near plant apex. 3. Apex of lower leaf, dorsal view. 4. Propaguloid leaf. 5–6. Lower leaves, dorsal and ventral views.

margins and upper laminal cells smooth except for the extreme upper marginal cells and the tip of the costa. There is some similarity with *Chenia leptophylla*, which also has a strongly papillose and sharply apiculate leaf apex, but which has larger laminal cells and a very narrow costa.

Literature: Stone (1978).

Number of accepted species: 1.

Species examined: *Stonea oleaginosa* (MELU).

71. ACAULON **Plate 102.**

Acaulon C. Müll., Bot. Zeit. 5: 99, 1847. Lectotype: *Acaulon muticum* (Hedw.) C. Müll. see Limpr., Laubm. Deutschl. 1: 178. 1885.

Sphaerangium Schimp., Syn. 12, 1860, *nom. illeg. incl. gen. prior.*

Phascum subg. *Acaulon* (C. Müll.) Wils. in Hook. f., Fl. Nov. Zel. 2: 58, 1854.

Acaulon subg. *Alaticosta* Stone, J. Bryol. 9: 213, 1976, *nom. inval. holotyp. non cit.*

Acaulon sect. *Sphaerangium* C. Müll., Gen. Musc. Fr. 20, 1900.

Subg. *Alaticosta* Stone, J. Bryol. 9: 573, 1977. Type: *Acaulon chrysacanthum* Stone.

Subg. *Acaulonopsis* Stone, J. Bryol. 15: 746, 1989. Type: *Acaulon robustum* Broth. ex Roth.

From the alpha privitive + καυλός, stalk, stem; stemless.

Plants very small, gemmate, gregarious or scattered, reddish or occasionally yellowish brown above, brown below. *Stems* not branching, *very short,* to 0.5 mm in length, transverse section rounded, central strand absent, sclerodermis absent, hyalodermis absent; axillary hairs ca. 5 cells in length, proximal cell walls sometimes thickened; sparsely radiculose. *Leaves strongly appressed and apices usually reflexed when dry,* appressed to weakly spreading when moist, *ovate,* small, 0.5–1.75 mm in length, *lamina broadly channeled and usually deeply concave, margins plane,* entire to serrulate or dentate; apex broadly acute; base not differentiated in shape; *costa excurrent in a sharp apiculus or stout mucro,* occasionally only percurrent or as a short, sometimes dentate awn, costa with lamina inserted laterally, superficial cells elongate and smooth ventrally, dorsally elongate and usually smooth, 3–4 rows of cells across costa ventrally at midleaf, costal transverse section round, stereid bands ventrally generally absent, dorsally present (but usually weak) and rounded in shape, ventral and dorsal epidermises present, guide cells 2–4 in 1 layer or rarely absent, hydroid strand usually present, occasionally apparently centered in the stereid band, *costal outgrowths sometimes present, of ca. 2 longitudinal lamellae formed on ventral surface of the costa; upper laminal cells* rounded-quadrate to rhomboidal, ca. 13–15 µm in width, 1–4:1, walls evenly thickened, *occasionally highly thickened on dorsal walls,* superficially convex on both sides of lamina; *papillae absent or occasionally large and simple, 1(–2) per lumen;* leaf base not differentiated in shape, basal cells rectangular, little wider than upper cells, 3–4:1, walls thin. Dioicous and perigoniate plants smaller than the perichaetiate (often much reduced), or else monoicous and usually paroicous. Perichaetia terminal, inner leaves somewhat enlarged. *Seta very short,* to 0.2 mm in length, 1 per perichaetium, light brown; *theca cleistocarpous, spherical, apiculus lacking,* ca. 0.4–0.7 mm in diameter, light brown, exothecial cells quadrate to rhomboidal, 25–50 µm in width, 1:1, thin-walled, stomates phaneropore, occurring at base of capsule. Operculum absent. Calyptra mitrate, often lobed, smooth, ca. 0.1–0.4 mm in length. *Spores rather large,* ca. 25–35 µm in diameter, spherical to weakly elliptical, light brown, *lightly papillose to irregularly warty or spiculose. Laminal KOH color reaction red.* Reported chromosome number n = 26.

Found on soil, a widely distributed genus mainly found in temperate regions of low rainfall.

Acaulon is clearly similar to *Microbryum*, differing in the even smaller habit size, capsules spherical and lacking an apiculus (Pl. 102, f. 28), upper laminal margins plane, and papillae lacking in most species. *Acaulon schimperianum* (Pl. 102, f. 29) has upper laminal papillae somewhat like those of *Microbryum vlassovii* and *M. floerckeanum* (one to two over each lumen, simple or occasionally branching apically, hollow, often rather tall). Those species of *Microbryum* with short-ovate leaves and no papillae also, like *Acaulon*, have rectangular ventral quadrate cells. *Acaulon* may be an end member of a series including *Tortula* sect. *Tortula* and *Microbryum* through reduction of gametophyte and sporophyte size and complexity, and differentiation of red coloration. This is supported somewhat by Cladograms 12 and 14 to the extent that *Tortula* is lower on the tree than the other genera.

In *A. eremicola* (Pl. 102, f. 14–15), *A. muticum* and *A. schimperianum*, the perigoniate plants are about a quarter to a third the size of the perichaetiate plants, and are situated near the base of the perichaetiate plants (possibly rhizautoicous).

The costa may appear, in section, to have two stereid bands in species (e.g. *A. chrysacanthum*, Pl. 102, f. 11–13) that develop strong awns. Stone (1976b, 1977b, 1979, 1988, 1989) has discussed the morphology of Australasian *Acaulon* species in a particularly incisive manner. Casas et al. (1990) gave a key to the five species of the genus on the Iberian Peninsula, while Sérgio (1992) noted that the European *A. piligerum* belongs in the lamellate subgenus *Alaticosta*, previously known only from Australia.

Additional literature: Bryan (1956), Cárdenas (1988), Crum and Anderson (1965), Grout (1945), Hill (1982), Sérgio (1972a, 1992).

Number of accepted species: 15.

Species examined: *A. chrysacanthum* (NY), *A. eremicola* (MELU), *A. integrifolium* (BUF, NY), *A. leucochaete* (BUF, NY), *A. muticum* (BUF), *A. robustum* (NY), *A. schimperianum*, *A. triquetrum* (BUF), *A. uleanum* (BM).

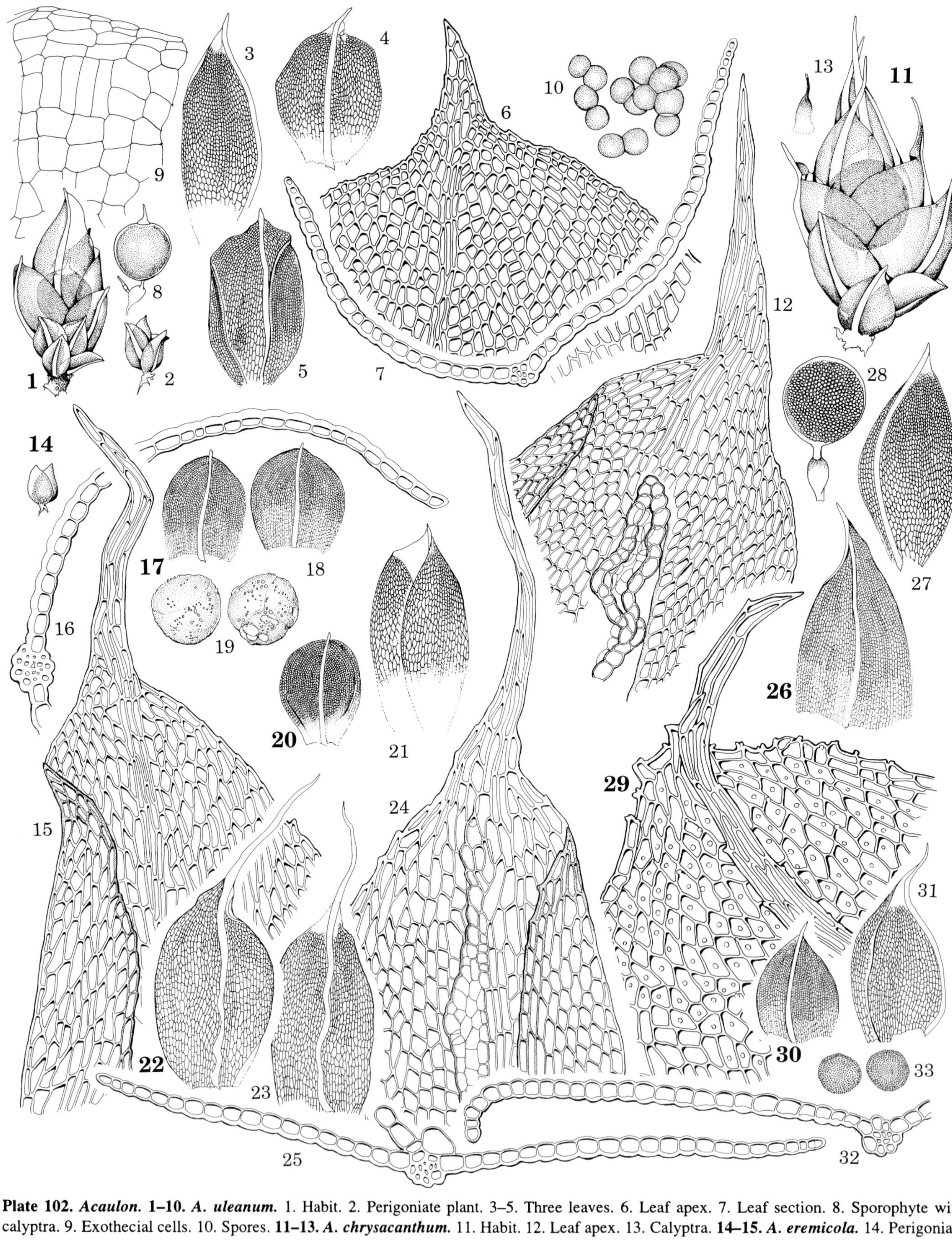

Plate 102. ***Acaulon.*** **1–10.** ***A. uleanum.*** 1. Habit. 2. Perigoniate plant. 3–5. Three leaves. 6. Leaf apex. 7. Leaf section. 8. Sporophyte with calyptra. 9. Exothecial cells. 10. Spores. **11–13.** ***A. chrysacanthum.*** 11. Habit. 12. Leaf apex. 13. Calyptra. **14–15.** ***A. eremicola.*** 14. Perigoniate plant. 15. Leaf apex. 16. Leaf section. **17–19.** ***A. inaequalifolium.*** 17–18. Two leaves. 19. Spores. **20–21.** ***A. integrifolium.*** 20–21. Two leaves. **22–25.** ***A. leucochaete.*** 22–23. Two leaves. 24. Leaf apex. 25. Leaf section. **26–28.** ***A. robustum.*** 26–27. Two leaves. 28. Sporophyte with vaginula. **29.** ***A. schimperianum.*** 29. Leaf apex. **30–33.** ***A. triquetrum.*** 30–31. Two leaves. 32. Leaf section. 33. Spores.

72. SARCONEURUM Plate 103.

Sarconeurum Bryhn, Nyt Mag. Naturvid. 40: 204, 1902. Type: *Sarconeurum antarctium* Bryhn.

From σάρξ, σαρκόσ, flesh + o + νεῦρον, nerve, sinew, tendon; the costa is thickened, appearing meaty to some.

Plants forming dense cushions, blackish green above, reddish brown below. Stems often branching, to 3 cm in length, rounded-pentagonal in transverse section, central strand absent to strong, sclerodermis absent, hyalodermis indistinct; axillary hairs of ca. 10 cells, the basal 2–3 firm-walled, yellow. *Leaves* appressed-incurved when dry, spreading-recurved when moist, *ligulate-lanceolate,* to 1.5 mm in length, broadly channeled across the ventral surface; margins plane, entire; *lamina constricted below apex, which ends in a deciduous, cylindrical, sharply apiculate propagulum*; base scarcely differentiated in shape to ovate; *costa* running into the apical propagulum, epidermal cells quadrate to short-rectangular on both leaf surfaces, ca. 4 rows of cells across costa ventrally at midleaf, *in transverse section with one stereid or substereid band dorsally (this generally lacking in small leaves),* 2–3 guide cells in 1(–2) layers, hydroid strand usually present, *a ventral stereid band occasionally present near the leaf base in large leaves,* epidermis present ventrally, absent dorsally; upper laminal cells quadrate to rectangular, often transversely elongated, especially along the margins, 15–20 µm wide, 1–2:1, walls evenly thickened or collenchymatous, *superficially flattened, papillae low, small, punctiform or bifid, apparently solid, several per lumen*; *basal cells differentiated medially, thin-walled, bulging-rectangular,* to 23 µm in width, 3–5:1. Sexual organs and sporophyte unknown. *Laminal KOH color reaction red.*

Found on sandy and volcanic soil or on lava at low elevations in Antarctica, and, rarely, southernmost South America.

A report (Matteri 1982) of sporophytes of *Sarconeurum glaciale* is based on specimens (Argentina: trunks of *Nothofagus*, TBPA B109, TBPA 3486, BA!) that can be referred to *Tortula pygmaea* Dix., which is a good species (BM! and see Lightowlers 1985b) found in southern South America and New Zealand. Greene (1975) reported *Sarconeurum* from South America apparently from specimens of *T. pygmaea*, a species curiously similar to *S. glaciale* in general habit and the constricted leaf apex bearing a deciduous, sharply apiculate propagulum (Pl. 103, f. 3–6), but differs in the smaller (ca. 10 µm in width) leaf cells with large, solid bifid papillae and the deep medial groove along the ventral surface of the costa. On the other hand, Lightowlers (1985b) demonstrated that the type ("Fuegia septentrionalis") of the South American *T. lithophila* at S is actually *S. glaciale.*

Sarconeurum tortelloides has been transferred to *Tortella*, leaving *Sarconeurum* monotypic.

Hilpert (1933) referred *Sarconeurum* to the Pottieae near *Tortula* (without further discussion) but probably because of the generally single stereid band in the costa (Pl. 103, f. 8–10) and the large upper laminal cells. There is considerable similarity to *Bryoerythrophyllum*, however, in the essentially oblong-lanceolate leaf shape, the presence of apical laminal propagulum in the related genus *Mironia* (but *cf. Tortella tortelloides*, which has caducous leaf apices), the bifid papillae, the occasional second stereid band, the medially differentiated basal cells (Pl. 103, f. 7) and the red color in KOH. The cladistic analysis places *Sarconeurum* in the Pottieae (see Cladograms 13 and 14), surprisingly, near *Acaulon.*

Additional literature: Greene et al. (1970), Savicz-Ljubitzkaja & Smirnova (1961), Zander (1978h), Zander & Hoe (1979).

Number of accepted species: 1.

Species examined: *S. glaciale* (BM, NY, US).

73. CHENIA Plate 104.

Chenia Zand., Phytologia 65: 424, 1989. Type: *Chenia subobliqua* (Williams) Zand.

Phascum sect. *Leptophascum* C. Müll., Flora 71: 7, 1888. Type: *Phascum leptophyllum* C. Müll.

Phascum subg. *Leptophascum* (C. Müll.) Roth, Aussereur. Laubm. 2: 214, 1911.

Named for Chen Pan-chieh, 1907–1970, author of "Studien über die ostasiatischen Arten der Pottiaceae" (1941) and also author or editor of "Genera Muscorum Sinicorum" (1963, 1978).

Plants forming turfs, occasionally rosulate, green above, brownish below. Stems seldom branching, 0.3–1.0 cm in length, transverse section rounded, central strand weak to strong, sclerodermis present, hyalodermis absent; axillary hairs small, of 3–4 cells, basal cell firm-walled; rhizoids usually few. *Leaves* appressed and somewhat contorted when dry, spreading when moist, *ligulate to spathulate, 1.5–2.5 mm in length,* upper lamina occasionally grooved along costa, plane or broadly channeled across leaf, margins plane above, weakly recurved below, *sharply crenulate to irregularly dentate above with sharp mid-marginal wall projections usually ending in a weak simple papilla,* marginal cells often smaller than the medial; *apex* rounded to broadly acute, *often sharply apiculate by a distinctive thick-walled cell or cells*; base rectangular or not differentiated in shape; *costa weak,* ending several (6–9) cells before the apex or percurrent, costa with lamina inserted ventrally to laterally, ventral and dorsal superficial cells short-rectangular, narrower than laminal cells, 2 rows of cells across costa ventrally at midleaf, costal transverse section rounded to elliptical, *stereid band very weak or occasionally absent,* rounded in shape, ventral and dorsal epidermis present, guide cells 2 in 1 layer, hydroid strand present, small to large; *upper laminal cells large, bulging-hexagonal, 15–18 µm in width,* 1:1, *walls thin, weakly trigonous,* superficially convex on both sides; *papillae absent* (upper marginal teeth may be interpreted as ending in sharp, simple papillae); basal cells differentiated across the leaf base (except for one row of marginal cells similar to the upper cells), rectangular, ca. 20–25 µm in width, 2–4:1, walls thin, weakly trigonous. *Propagula when present borne on rhizoids in soil, irregularly rounded to clavate,* ca. 100–130 µm in longest dimension. Dioicous. Perichaetia terminal, inner leaves little different from the cauline, slightly larger.Seta ca. 0.1–1.2 cm in length, 1 per perichaetium, reddish brown, twisted clockwise; when present theca 0.7–2.0 mm in length, brown, nearly spherical (then with a narrow beak to 0.25 mm in length) or short-ovate or cylindrical, exothecial cells rectangular, 20–45 µm in width, 1–4:1, thin-walled, stomates at base of theca, phaneropore, cleistocarpous or stegocarpous, annulus when present of 2–3 layers of strongly vesiculose cells, persistent; when present peristome teeth 32, filamentous, somewhat anastomosing, densely branching-spiculose, ca. 400 µm in length, with ca. 5 articulations, nearly straight or weakly twisted counterclockwise, basal membrane 25–35 µm in height, low spiculose. Operculum

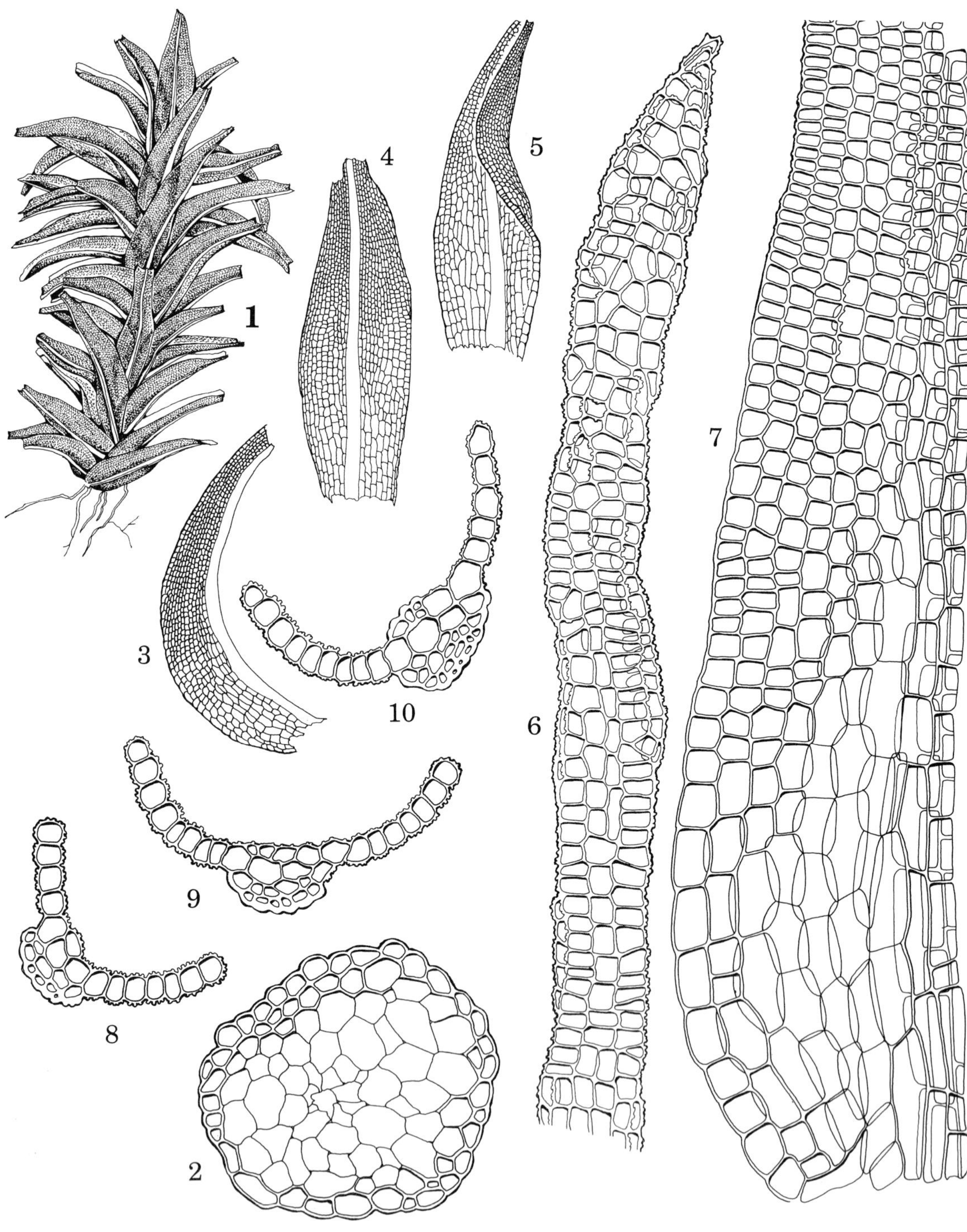

Plate 103. ***Sarconeurum*****. 1–10.** ***S. glaciale*****.** 1. Habit. 2. Transverse section of stem. 3–5. Three leaves. 6. Propaguloid leaf apex. 7. Basal cells. 8–10. Transverse sections near midleaf.

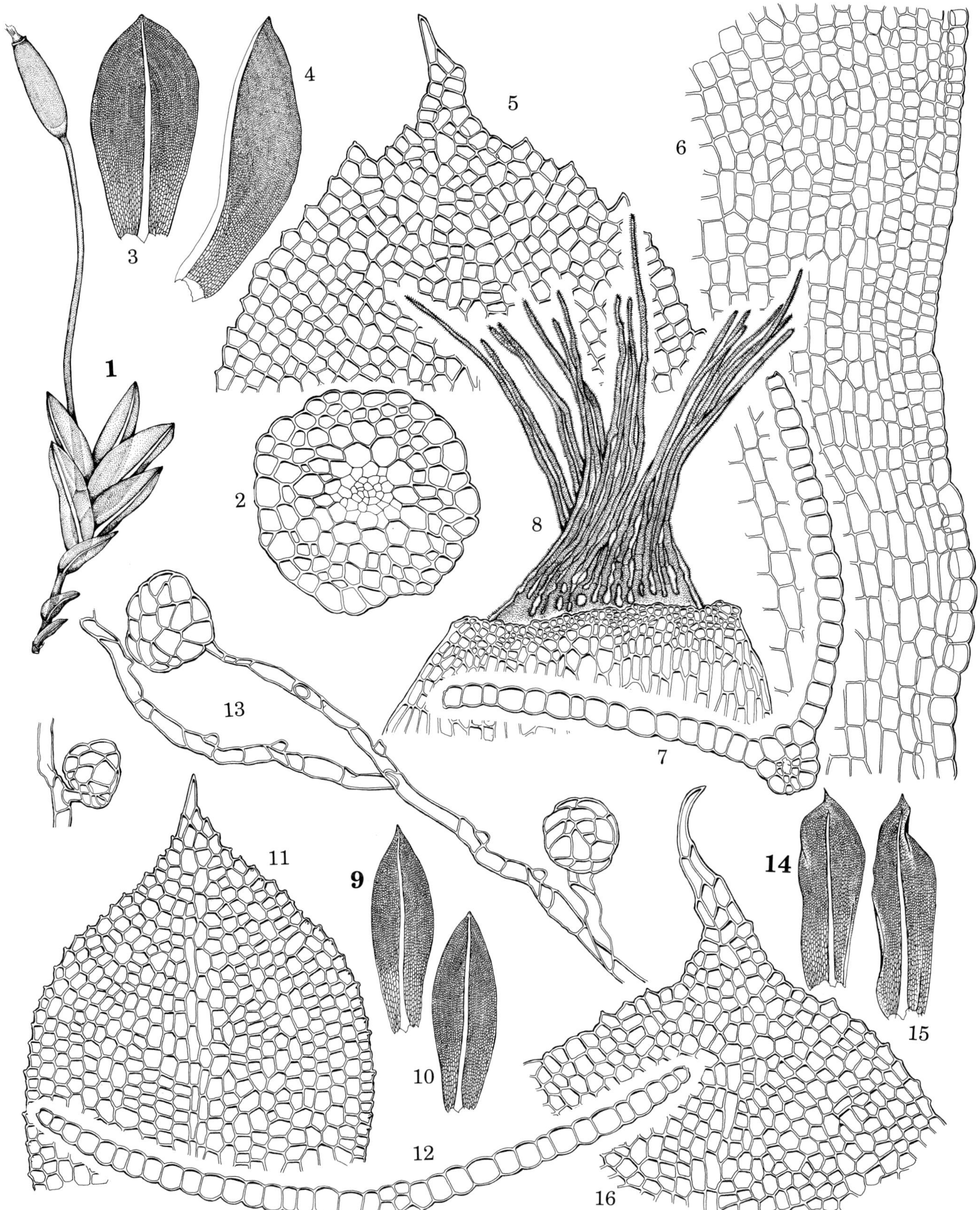

Plate 104. ***Chenia.*** **1–8.** ***C. subobliqua.*** 1. Habit. 2. Transverse section of stem. 3–4. Two leaves. 5. Leaf apex. 6. Basal cells. 7. Transverse section at midleaf. 8. Peristome. **9–13.** ***C. leptophylla.*** 9–10. Two leaves. 11. Leaf apex. 12. Transverse section at midleaf. 13. Propagula. **14–16.** ***C. lorentzii.*** 14–15. Two leaves. 16. Leaf apex.

when differentiated short- to long-conic, 500–650 µm in length, cells twisted weakly counterclockwise. Calyptra cucullate or mitriform and then three-lobed, smooth, 1.5–2.0 mm in length. Spores 10–20 µm in diameter, light brown, nearly smooth to finely papillose. *Laminal KOH color reaction red.* [Information on the cleistocarpous sporophyte of *C. leptophylla* from Arts and Sollman 1991]

Found on soil from sea level to nearly 3000 meters elevation; North and South America, Europe, eastern Asia and Australia.

Corley et al. (1981) suggested that *Chenia leptophylla* (as *Tortula rhizophylla*) "is completely anomalous in *Tortula*." The Andean *C. obliqua* (Pl. 104, f. 1–8) is closely related and differs from *C. rhizophylla* in the peristomate sporophyte, broad leaf apex, margins strongly dentate in the upper 1/3 of the leaf and lacking a leaf apiculus of thick-walled cells. Although these two species are rather different, a third species, *C. lorentzii* (Pl. 104, f. 14–18), is sterile but otherwise intermediate in character, with narrower leaves, not or moderately dentate, and the apiculus is occasionally differentiated and thick-walled. Small, sterile plants mixed in with the type of *C. lorentzii* are quite like *C. leptophylla*. The autoicous condition ascribed to *C. lorentzii* in the original description cannot be confirmed as no antheridia were found in the NY isotype.

Like *Stegonia*, *Chenia* is a small genus that includes both peristomate and cleistocarpous species. *Chenia* is easily distinguished from other *Tortula*-like species by the combination of the dentate upper leaf margins, large, epapillose upper laminal cells, a thin costa (Pl. 104, f. 12), and red coloration in KOH. The last may have to be determined by examination of the upper laminal cell walls under high magnification because the dense yellow-green of the chlorophyll overwhelms the color of the very thin cell walls. The leaf teeth are each usually tipped with a single, simple papilla. *Phascum leptophyllum* of central and southern Africa, an earlier name than *Physcomitrium rhizophyllum* Sak., was reported by Arts and Sollman (1991) to be the same as *C. rhizophylla*, along with *Pottia denticulata* Dix. & Varde of India and *P. splachnobryoides* C. Müll. of Asia; they describe and illustrate the sporophyte with its cleistocarpous capsule, and review the literature on this widely distributed species. *Chenia leptophylla* is, as is the case with *Didymodon australasiae* var. *umbrosus* (*cf.* Crundwell & Whitehouse 1978, Eckel 1986a, Preston & Whitehouse 1985, Synnott & Robinson 1990, and others), apparently spread by human agency through rhizoid-borne propagula (Pl. 104, f. 13) in the soil. *Chenia* is similar to *Hennediella*, especially *H. serrulata* (which has a quite similar gametophytic appearance); also, *Tortula paulsenii* has somewhat similar morphology. Phylogenetic analysis indicates a close relationship, however, to *Syntrichia* (see Cladograms 13 and 14).

Additional literature: Cortini and Aleffi (1989), Iwatsuki and Saito (1972), Martinez et al. (1989), Neumann (1972), Pedrotti and Aleffi (1989), Reese (1967, 1968), Smith and Whitehouse (1974), Sollman (1979), Stone (1980a).

Number of accepted species: 3.

Species examined: *C. lorentzii* (NY), *C. leptophylla* (BUF, PAC), *C. subobliqua* (NY).

New combinations: *Chenia leptophylla* (C. Müll.) Zand., *comb. nov.* (*Phascum leptophyllum* C. Müll., Flora 71: 6, 1888). *Chenia lorentzii* (C. Müll.) Zand., *comb. nov.* (*Barbula lorentzii* C. Müll., Linnaea 42: 346, 1879 ["-*i*"]; *Tortula lorentzii* (C. Müll.) Broth.

74. SYNTRICHIA — Plates 105–111.

Syntrichia Brid., J. Bot. (Schrader) 1(2): 299, 19 April 1801 (*vide* Sayre 1959). Type: *Syntrichia ruralis* (Hedw.) Web. & Mohr, lectotype *fide* Zander, Phytologia 65: 432, 1989.

Tortula subg. *Syntrichia* (Brid.) Chev., Fl. Gén. Env. Paris 2: 52, 1827.

Barbula subg. *Syntrichia* (Brid.) BSG, Bryol. Eur. 2:10, 1851 (fasc. 46–47 Consp. 2: III).

Tortula sect. *Syntrichia* (Brid.) Lam. & Cand., Syn. 100, 1806.

Tortula sect. *Rurales* De Not., Mem. R. Acc. Sc. Torino 40: 286, 1838. Type: *Tortula ruralis* (Hedw.) Gaetrn., Meyer & Scherb.

Barbula sect. *Rurales* BSG, Bryol. Eur. 2: 101, 1842 (fasc. 13–15 Mon. 39).

Barbula sect. *Syntrichia* (Brid.) C. Müll., Syn. 1: 632, 1849.

Barbula sect. *Vallidens* C. Müll., Linnaea 42: 347, 1879, *nom. inval.* Type: *Barbula percarnosa* C. Müll.

Barbula sect. *Syntrichiae* Lesq. & James, Man. N. Amer. Moss. 130, 1884, *nom. illeg.*

Barbula sect. *Ruraliformes* Kindb., Eur. N. Amer. Bryin. 2: 245, 1897, *nom. illeg. incl. sect. prior.* Type: *Barbula ruralis* Hedw.

Syntrichia sect. *Eusyntrichia* Moenk., Laubm. Eur. 306, 1927, *nom. illeg.*

From σύν, together, with + θρίχ, τρίχός, hair; the peristome teeth are fused at the base into a cone.

Plants usually coarse, forming often deep turf, green, occasionally reddish or blackish above, reddish brown below. Stems branching occasionally, 1–4(–12) cm in length, transverse section rounded-pentagonal, central strand present or absent, *sclerodermis absent or occasionally of 1–2 layers of substereid cells,* hyalodermis absent or present, sometimes collapsed; axillary hairs ca. 3–10 cells in length, basal 1–3 cells thicker walled or all hyaline; rhizoids common, often dense. *Leaves* appressed to weakly spreading when dry, *spreading to squarrose when moist, narrowly ligulate to broadly spathulate, occasionally broadly lanceolate,* (1.5–)4–7 mm (inclusive of awn) in length, *upper lamina broadly channeled to keeled, often narrowly grooved along costa, margins* plane to recurved, rarely broadly incurved, *usually entire,* occasionally serrulate to dentate above, *occasionally bordered with thick-walled, less papillose cells, these occasionally intramarginal or rarely elongate; apex rounded to broadly acute,* occasionally narrowly acute or cucullate; base not differentiated in shape to elliptical; *costa often stout, commonly prominent dorsally,* ending several cells before apex to excurrent as a long, often hyaline, often serrate, occasionally flattened awn, costa with lamina inserted at 45°,

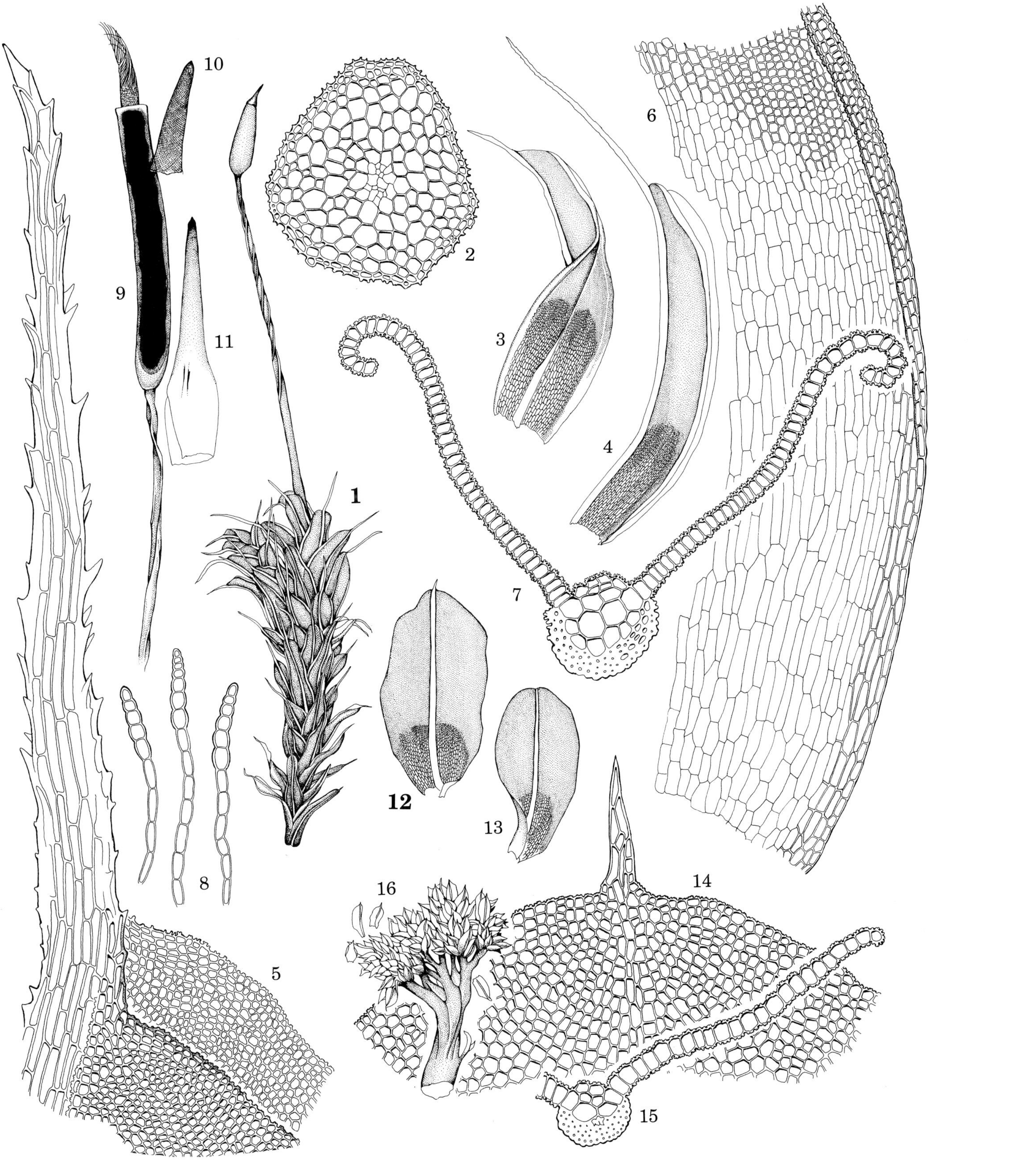

Plate 105. ***Syntrichia.*** **1–11.** ***S. ruralis.*** 1. Habit. 2. Transverse section of stem. 3–4. Two leaves. 5. Leaf apex. 6. Basal cells. 7. Transverse section at midleaf. 8. Paraphyses. 9. Capsule with peristome. 10. Operculum. 11. Calyptra. **12–16.** ***S. ammondsiana.*** 12–13. Two leaves. 14. Leaf apex. 15. Transverse section at midleaf. 16. Cluster of propagula.

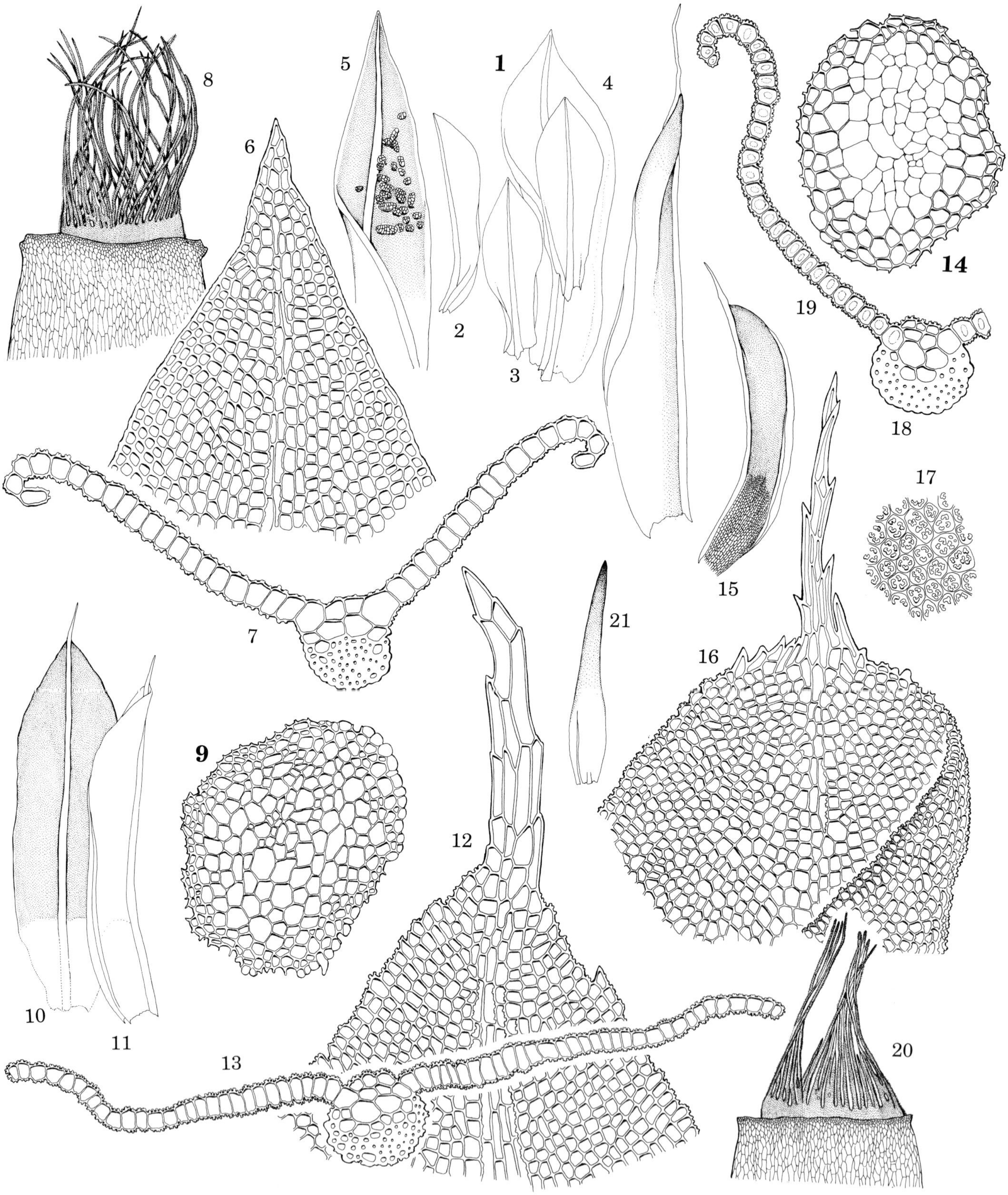

Plate 106. ***Syntrichia.*** **1–8.** ***S. amphidiacea.*** 1–4. Four leaves. 5. Leaf with propagula. 6. Leaf apex. 7. Transverse section at midleaf. 8. Peristome. **9–13.** ***S. anderssonii.*** 9. Transverse section of stem. 10–11. Two leaves. 12. Leaf apex. 13. Transverse section at midleaf. **14–21.** ***S. andicola.*** 14. Transverse section of stem. 15. Leaf. 16. Leaf apex. 17. Upper laminal papillae. 18. Transverse section at midleaf (internally pored laminal cells). 18. Perichaetial leaf. 20. Peristome. 21. Calyptra.

costal superficial cells rounded-quadrate, papillose, occasionally elongate and smooth near apex ventrally, dorsally elongate, papillose to denticulate or spinose, occasionally nearly smooth, 2–4 rows of cells across costa ventrally at midleaf, *costal transverse section round, occasionally elliptical or semicircular, stereid bands* ventrally absent or very rarely of a few substereid or stereid cells, *dorsally present and crescent-shaped, occasionally semicircular, ventral epidermis present, the dorsal usually absent, guide cells 2–4 per layer in 1–4 layers,* hydroid strand absent or present, occasionally multiple; *upper laminal cells* rounded-quadrate, ca. 11–15*(–20) µm in width, 1(–2):1,* occasionally bistratose entirely or in patches, *walls thin or occasionally collenchymatous and interiorly porose, superficially strongly convex on both sides; papillae ca. (1–)4–8 per lumen, usually bifid,* occasionally simple, solid or hollow, rarely simple; basal cells differentiated across base, rising higher medially, often sharply differentiated from the upper cells, occasionally inflated medially, 18–30 µm in width, 2–6:1, walls thin, sometimes irregularly porose, 3–6 rows of basal marginal cells usually narrowly rectangular and thicker walled. Propagula occasionally present, variously clusters of small, deciduous leaves; rhizoid-borne tubers; very fragile cauline leaves or leaf tips (rarely deciduous); obovate or clavate or raspberry-shaped propagula found on ventral surface of costa or medially or marginally on upper lamina. Dioicous or monoicous (then usually autoicous or synoicous). Perichaetia terminal, inner leaves usually little differentiated in size and shape, rarely enlarged and sheathing, lower cells rhomboidal, rarely throughout leaf. Perigonia gemmate, terminal or lateral as stalked buds, paraphyses often clavate. *Seta elongate,* (0.8–)*1.0–3.0 cm in length,* 1(–2) per perichaetium, reddish brown, twisted counterclockwise; *theca 2–6 mm in length, reddish brown, cylindrical,* commonly slightly curved, exothecial cells rectangular, 20–30 µm in width, ca. 4–6:1, walls thin to evenly thickened, capsule neck occasionally distinct, stomates phaneropore, at base of theca, annulus of 2–4 rows of vesiculose cells, persistent; *peristome teeth 32, filamentous, densely spiculose, ca. 1000–2000 µm in length, with many articulations,* twisted counterclockwise ca. once, *basal membrane usually present,* ca. 100–800 µm in height, papillose to spiculose. *Operculum conic, rather large,* 1.2–2.2 mm in length, cells twisted counterclockwise. *Calyptra cucullate, smooth, rather large,* 3.0–4.5 mm in length. Spores 8–15 µm in diameter, light brown, lightly papillose. *Laminal KOH color reaction brick red.* Reported chromosome number n = 6+m, 12, 12+m, 13, 13+m, 24, 24+2m, 26, 28, 32+2m, 48.

Found on rock, soil and bark on all continents, most commonly in temperate areas but also characteristic of dry climates.

Syntrichia is segregated from *Tortula s. lat.* by the combination of red KOH reaction of the upper laminal cells, lack of narrowly elongate upper laminal marginal cells, the crescent-shaped transverse section of the stereid band, and the general lack of differentiated dorsal costal epidermal cells. This concept is essentially that of Kramer (1980, 1988), who emphasized the exposed dorsal stereid band (Pl. 105, f. 7, 15), not covered dorsally by parenchymatous or otherwise differentiated epidermal cells. Recently Ochyra (1992) supported this concept (Zander 1989) as "a natural group that deserves recognition as a genus of its own" with a series of new combinations appropriate for the Polish flora. *Syntrichia* is distinguished from *Hennediella* (likewise KOH red) by the leaves ligulate to spathulate or very seldom lanceolate, margins usually recurved and seldom dentate or bordered, upper laminal cells smaller and superficially strongly convex, costa commonly excurrent as an awn, almost always lacking a dorsal costal epidermis, stereid band semicircular to crescent-shaped, and sporophytes not in a reduction series within the genus.

Note that in some few species (e.g. *S. brandisii*) a dorsal epidermis is very weakly differentiated (seen as somewhat wider lumens in costal transverse section), but these species are placed here rather than with *Hennediella* because of their recurved, unbordered margins and small upper laminal cells. *Syntrichia costesii* (Pl. 107, f. 17–19), like certain species of *Hennediella*, has an intramarginal border of differentiated cells, but those cells are isodiametric and the costa lacks a differentiated dorsal epidermis. The existence of a weakly differentiated dorsal epidermis in some species may indicate a derivation of *Syntrichia* from *Bryoerythrophyllum*, which differs, in traditional characters, in the presence of a second stereid band and usually narrower (broadly lanceolate to ligulate) leaves. Most cladograms in the phylogenetic analysis show *Syntrichia* and *Bryoerythrophyllum* to be quite distant. A few species, like *S. laevipila* (Pl. 109, f. 4–5), may occasionally and perhaps abnormally have one or two stereid or substereid cells present ventrally between the guide cells and the ventral epidermis as a tiny second stereid band, while in the type (isotype, NY) of *S. rubra* (Pl. 111, f. 5–8) and a second specimen seen (as *Tortula rubra* var. *subantarctica*, Campbell I., Sorensen 1946, isotype, NY), the ventral stereid band is of several stereid cells, the ventral epidermis is lacking, and superficially, the ventral surface of the costa is similar to the dorsal surface, being of superficially elongate, simply papillose cells. *Syntrichia percarnosa* (Pl. 109, f. 20–22) has much the same leaf morphology as *Trichostomum crispulum*, and the former species may well belong with that genus; the substereid cells of the costa are, however, arranged more like those of *Syntrichia.*

Certain species of *Didymodon* sect. *Vineales* (KOH red) may lack a ventral stereid band, but have lanceolate leaves and narrowly rectangular basal cells. *Bryoerythrophyllum* and *Mironia* species are also KOH red and may have long-ligulate leaves, but the transverse section of the costa is reniform in section and the stem has a sclerodermis. *Syntrichia geheebiaeopsis* (Pl. 108, f. 13–17, *cf.* discussion by Lightowlers 1985a as *Tortula*) has many of the characters of *Bryoerythrophyllum* and *Mironia*, including broadly lanceolate, dentate leaves with a sheathing leaf base, small upper laminal cells with low, crowded papillae. It further differs from *Syntrichia* by the very narrowly rectangular basal cells. *Bryoerythrophyllum* and *Mironia* have semicircular to reniform costal sections, however, with usually two distinct stereid bands and a differentiated dorsal epidermis, but *S. geheebiaeopsis* has the typical costal morphology of *Syntrichia*: a nearly circular costal section, with a single stereid band and no differentiated dorsal epidermal cells. It may

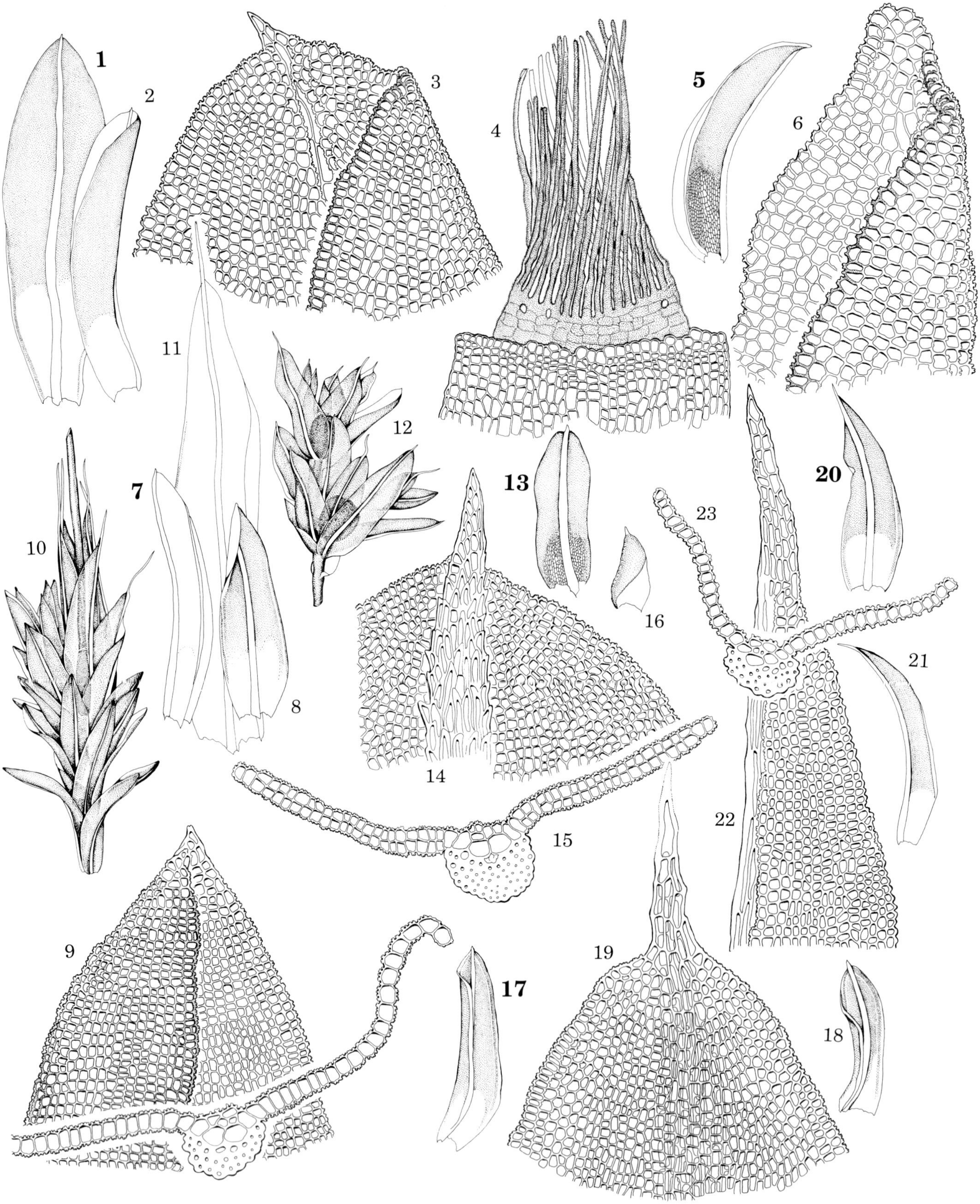

Plate 107. ***Syntrichia.*** **1–4.** ***S. bipedicellata.*** 1–2. Two leaves. 3. Leaf apex. 4. Peristome. **5–6.** ***S. cainii.*** 5. Leaf. 6. Leaf apex. **7–12.** ***S. cavalii.*** 7–8. Two leaves. 9. Leaf apex. 10. Perichaetium. 11. Perichaetial leaf. 12. Perigonia. **13–16.** ***S. chisosa.*** 13. Leaf. 14. Leaf apex, dorsal view. 15. Transverse section at midleaf. 16. Propagulum. **17–19.** ***S. costesii.*** 17–18. Two leaves. 19. Leaf apex. **20–23.** ***S. didymodontoides.*** 20–21. Two leaves. 22. Leaf apex. 23. Transverse section at midleaf.

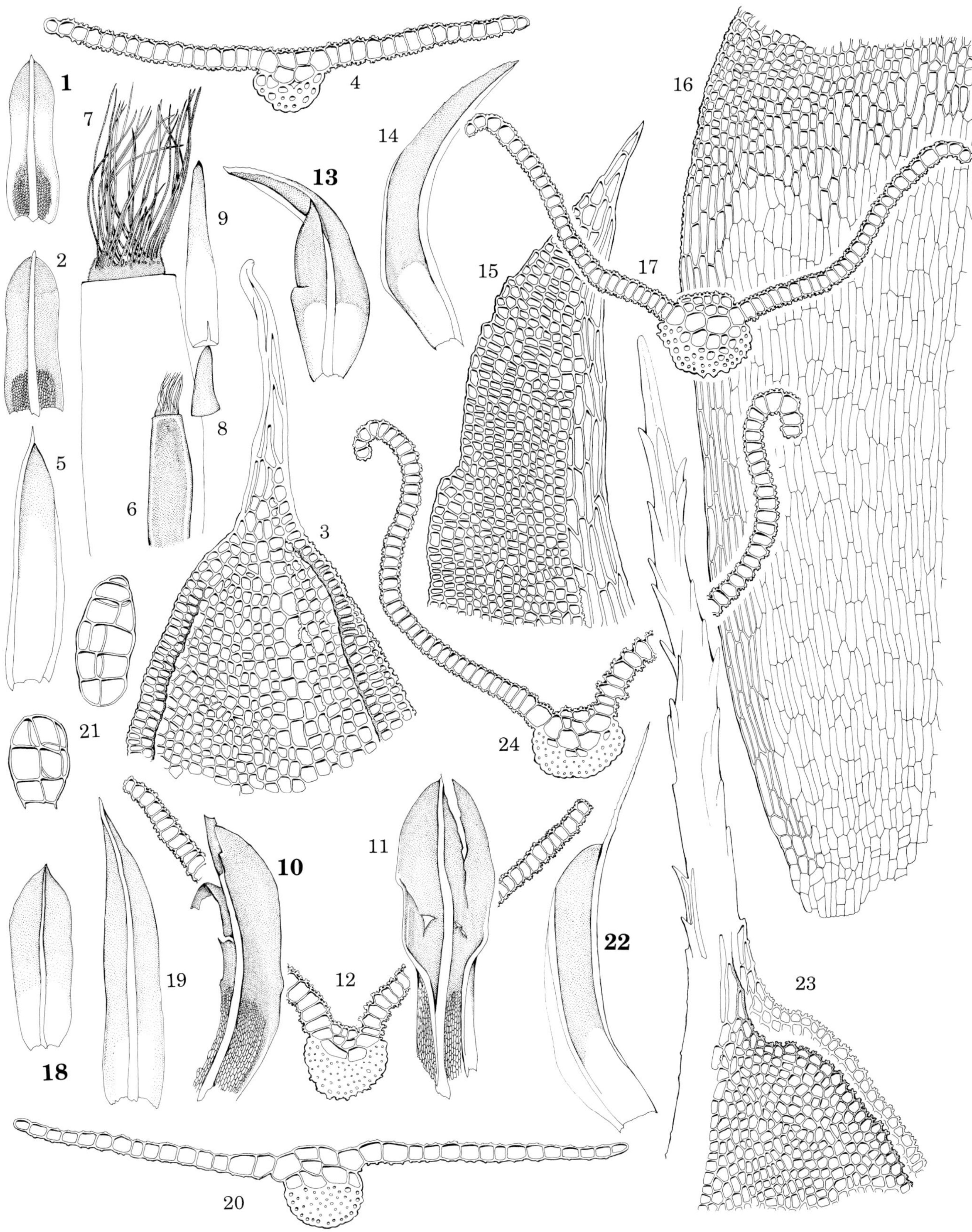

Plate 108. ***Syntrichia.*** **1–9.** ***S. epilosa.*** 1–2. Two leaves. 3. Leaf apex. 4. Transverse section at midleaf. 5. Perichaetial leaf. 6–7. Peristomes. 8. Operculum. 9. Calyptra. **10–12.** ***S. fragilis.*** 10–11. Two leaves. 12. Transverse section at midleaf. **13–17.** ***S. geheebiaeopsis.*** 13–14. Two leaves. 15. Leaf apex. 16. Basal cells. 17. Transverse section at midleaf. **18–21.** ***S. gemmascens.*** 18–19. Two leaves. 20. Transverse section at midleaf. 21. Propagula. **22–24.** ***S. obtusissima.*** 22. Leaf. 23. Leaf apex. 24. Transverse section at midleaf.

prove to be a good genus in the Merceyoideae apparently having lost, like *Streptopogon*, the ventral stereid band.

Syntrichia cavallii (Pl. 107, f. 7–12) is unusual in its strongly differentiated perichaetial leaves that are similar to those of *S. papillosa* (sect. *Collototortula*); it is, however, apparently related to *S. percarnosa*, instead. *Syntrichia cavallii* differs from *Calyptopogon*, in which perichaetial leaves are differentiated, in that propagula are lacking and the upper laminal papillae are multiplex. *Willia* also has strongly differentiated perichaetial leaves, but these are usually somewhat secund, and the sporophyte is distinctively reduced in length and complexity. *Syntrichia cavallii* is here placed in sect. *Syntrichia*, not sect. *Collototortula*, because of the lack of collenchymatous thickenings in the upper laminal cell walls. Eventually, perhaps, further study may require that a separate section or genus be devised for *S. cavallii* (see also discussion under *Willia*).

Syntrichia sect. *Collototortula* Zand., *sect. nov.* Type: *Syntrichia andicola* (Mont.) Zand.

A sectione typica cellulis laminalibus in regione mediana superna plerumque distincte in angulis incrassatis vel tumescentitrigonis, marginibus folii plerumque recurvis, papillis interdum simplicibus, propagulis si effectis clavatis vel ellipticis differt.

Differs from the typical section by the following combination of characters: medial upper laminal cells usually somewhat thickened at the corners or even trigonous (i.e. with bulging knots), leaf margins usually recurved, papillae sometimes simple, and propagula when present clavate or elliptical. Examination of transverse leaf sections indicates that the interior upper laminal cell walls may have large, round central pores (staining helps define this) in at least some species (*Syntrichia aculeata*, *S. amphidiacea*, *S. andicola* (Pl. 106, f. 19), *S. bogotensis*, *S. gemmascens*, *S. gromschii* and *S. papillosa*), while in other species, including those of other sections of *Syntrichia*, the pores are absent or what appear to be interior pores are occasional, irregular and possibly artifacts. Features correlated with laminal cell collenchymatous thickenings are the upper leaf margin usually serrate and the stem central strand often lacking, but these are not unique characters. The section may not be sharply distinct from sect. *Syntrichia* because certain species of that section may have weakly thickened cell corners (appearing as bright triangular points of light under high magnification). The counterclockwise-twisted seta of *Syntrichia papillosa* commented upon by Dixon (1923) as unusual is not taxonomically significant, since long setae in the Pottiaceae are generally twisted clockwise below and counterclockwise above, and short setae are untwisted or merely twisted clockwise. The seta of *S. papillosa* (Pl. 109, f. 19—from a specimen misidentified as *Tortula panduraefolia*, Tasmania, Weymouth 2821, NY) is only about 5 mm in length. Species (some surely synonyms of *S. andicola*) belonging to sect. *Collototortula* include: *S. aculeata*, *S. alpestris*, *S. amphidiacea* (Pl. 106, f. 1–8), *S. andicola* (Pl. 106, f. 14–20), *S. antarctica* (possibly belongs here, but laminal cell walls are only weakly collenchymatous), *S. bogotensis*, *S. cainii* (Pl. 107, f. 5–6), *S. ciliata*, *S. conferta*, *S. fontana* (laminal cells flat superficially, as in *Hennediella*), *S. gemmascens* (Pl. 106, f. 18–21), *S. goudotii*, *S. gromschii*, *S. mollis*, *S. papillosa* (Pl. 109, f. 14–19), *S. rivularis* (Pl. 109, f. 20–23), *S. robusta* (Pl. 111, f. 1–4), *S. rubra* (Pl. 111, f. 5–8), and *S. subaristata*. Curiously, *S. rigescens* has propagula (Pl. 110, f. 18–19) borne ventrally on the costa, as in sect. *Collototortula*, but is clearly in sect. *Syntrichia* near *S. caninervis*. Note also *Syntrichia leucostega*, not seen, for which ventral costal propagula are illustrated by Kramer (1988).

Syntrichia sect. *Aesiotortula* Zand., *sect. nov.* Type: *S. pagorum* (Milde) Amann.

A sectione typica plantis parvioribus quam congeneribus, foliis ligulatis vel spathulatis, marginibus omnino planis, papillis bifidis, propagulis si effectis foliaceis, in termino caulis vel in apicibus foliorum deciduis portatis, costa supra regione folii mediana perincrassata et in sectione transversalis semicircularibus differt.

Differs from the typical section by the following combination of characters: plants rather small for the genus, leaves ligulate to spathulate, margins plane throughout, papillae bifid, and propagula when present leaf-like, borne terminally on the stem or as deciduous leaf apices. The costa is often very strongly thickened above midleaf and semicircular in transverse section. This group of species includes *S. ammonsiana* (Pl. 105, f. 12–16), *S. baileyi*, *S. bartramii*, *S. chisosa* (Pl. 107, f. 13–16), *S. epilosa* (Pl. 108, f. 1–9), *S. pagorum* (Pl. 109, f. 10–11), *S. phaea* (Pl. 109, f. 23–27) and *S. pygmaea* (Pl. 110, f. 10–16). *Syntrichia papillosa*, although here placed with sect. *Collototortula* because of the clearly trigonous upper laminal cells and clavate to spherical propagula, also has plane margins and a much thickened costa.

Section *Aesiotortula* (as the "*Tortula laevipila-Tortula pagorum*-Gruppe") was not dealt with by Kramer (1980), who viewed it as a complex requiring special study. Certain of the infraspecific taxa of *S. laevipila* may also belong here, but not var. *laevipila* (Pl. 109, f. 1–9) itself. *Syntrichia pygmaea* is only tentatively placed in this section; it is propaguliferous, however, by a deciduous leaf apex and has plane margins. On the other hand, species of similarly small stature in many genera have a tendency to lose or exhibit less distinctly some important characters, including recurvature of the lower margins. *Syntrichia pygmaea* is also unusual for its ventral leaf surface very narrowly grooved along the costa and ventral costal cells elongate, which may be an indication of a relationship with the Bryoerythrophylleae for this species. The upper laminal cells of the most highly derived of the species of sect. *Aesiotortula* are relatively large for the genus, generally 13–15 µm in diameter; species with upper laminal cells small (ca. 8–10 µm in diameter), such as *S. bartramii* and *S. chisosa*, are less distinctive. The cleistocarpous genus *Phascopsis* could be seen as having been derived from this section of *Syntrichia*, but is apparently a relict of a more primitive lineage (see Cladograms 13 and 14).

The New World austral species *Syntrichia epilosa* is much the same as *S. bartramii* of western North America and may

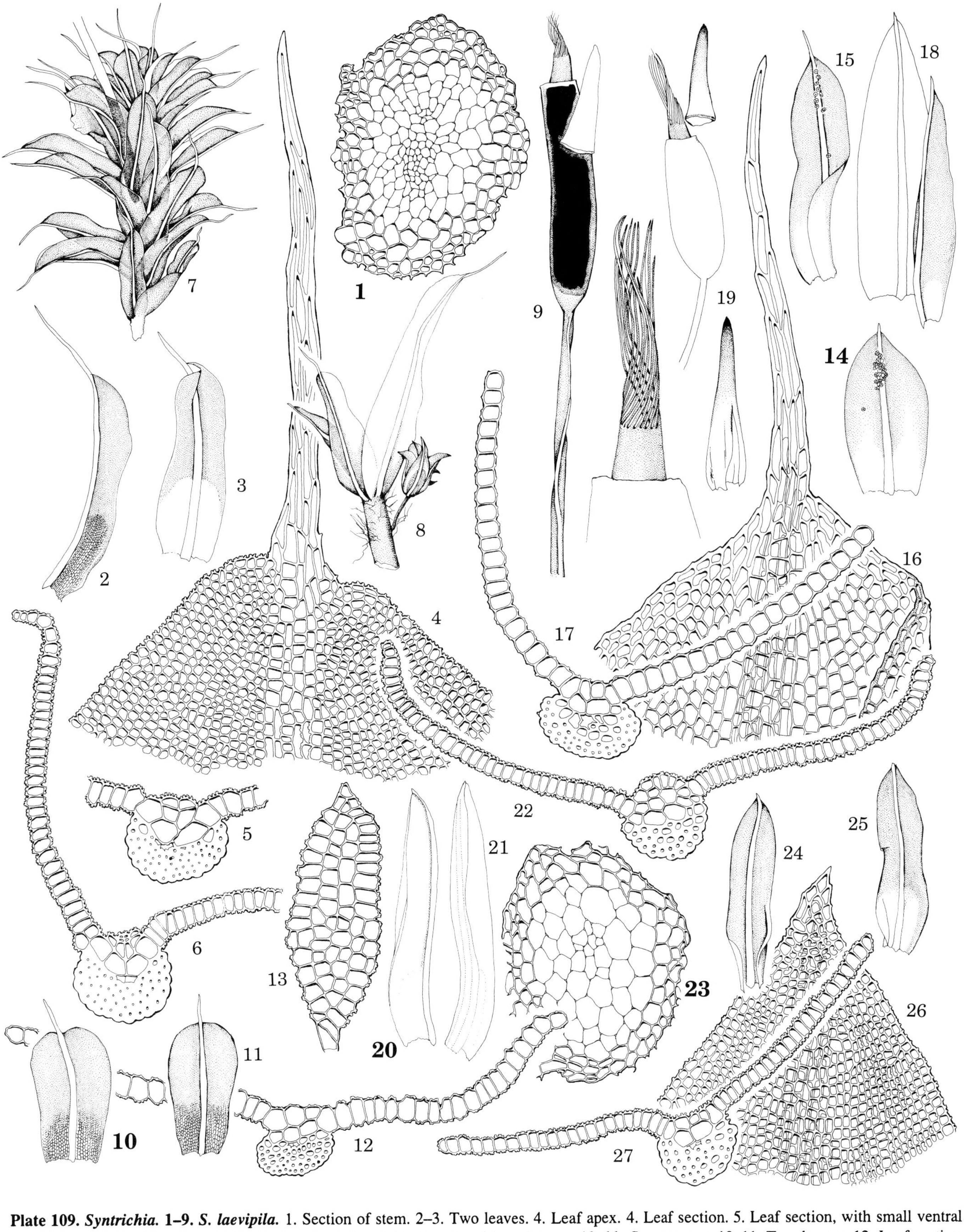

Plate 109. ***Syntrichia.*** **1–9.** ***S. laevipila.*** 1. Section of stem. 2–3. Two leaves. 4. Leaf apex. 4. Leaf section. 5. Leaf section, with small ventral stereid band. 7. Perichaetium. 8. Stalked perigonial bud. 9. Capsule with peristome. **10–11.** ***S. pagorum.*** 10–11. Two leaves. 12. Leaf section. 13. Propagula. **14–19.** ***S. papillosa.*** 14–15. Two leaves. 16. Leaf apex. 17. Leaf section. 18. Perichaetial leaves. 19. Capsule, calyptra. **20–22.** ***S. percarnosa.*** 20–21. Two leaves. 22. Leaf section. **23–27.** ***S. phaea.*** 23. Section of stem. 24–25. Two leaves. 26. Leaf apex. 27. Leaf section.

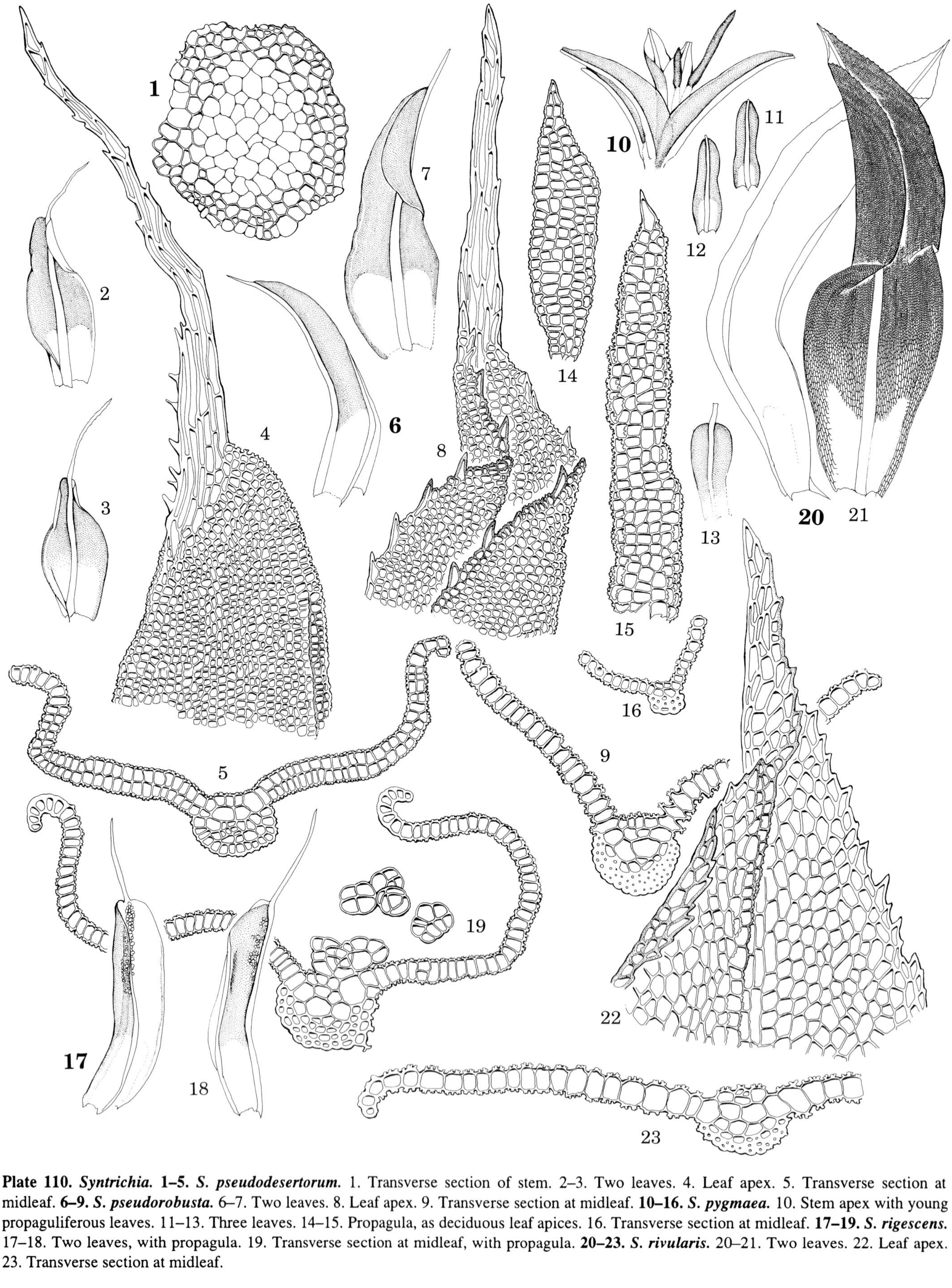

Plate 110. ***Syntrichia.*** **1–5.** ***S. pseudodesertorum.*** 1. Transverse section of stem. 2–3. Two leaves. 4. Leaf apex. 5. Transverse section at midleaf. **6–9.** ***S. pseudorobusta.*** 6–7. Two leaves. 8. Leaf apex. 9. Transverse section at midleaf. **10–16.** ***S. pygmaea.*** 10. Stem apex with young propaguliferous leaves. 11–13. Three leaves. 14–15. Propagula, as deciduous leaf apices. 16. Transverse section at midleaf. **17–19.** ***S. rigescens.*** 17–18. Two leaves, with propagula. 19. Transverse section at midleaf, with propagula. **20–23.** ***S. rivularis.*** 20–21. Two leaves. 22. Leaf apex. 23. Transverse section at midleaf.

be synonymous; a parallel southern South America and southwestern North America distribution is that of the austral *Pseudocrossidium crinitum* and its Arizona population previously known as *Tortula aurea* (see treatment of *Pseudocrossidium*).

Additional, selected literature: Allorge (1938), Anderson (1943), Barkman (1963), Bartram (1924b, 1926b), Bewley (1972, 1973a,b), Bewley et al. (1974), Bizot (1954, 1956), Blomquist (1930), Boudier (1992), Casas de Puig (1975a), Casas de Puig and Molinas (1975), Catcheside (1980, 1992), Dedkov et al. (1989), Doei et al. (1985), El-Oqlah et al. (1988), Hedenäs (1989b), Kalenov (1977), Kramer (1978), Lazarenko (1959), Lightowlers (1985a, 1986a,b,c), Magill et al. (1983), Maya (1986), Mishler (1984a, 1985a,b, 1986a, 1987a, 1990), Mishler and Newton (1987, 1988), Mishler and Oliver (1991), Mishler and Scheirer (1983), Oliver and Mishler (1988, 1990), Ovezova (1989), Saito (1973a), Side and Whitehouse (1974), Steere (1939a, 1940), Stone (1971), Studlar et al. (1984), Toth (1987), Tuba (1984, 1985), Willis (1964), Zander (1989).

Number of accepted species: 82.

New heterotypic synonymy: *Tortula abruptinervis* Dix. = *Syntrichia pygmaea* (Dus.) Zand. *Tortula kingii* Robins. = *Syntrichia percarnosa* (C. Müll.) Zand. *Tortula lemniscata* Zand. = *Syntrichia aculeata* (Wils.) Zand. *Tortula nigra* Zand. = *Syntrichia percarnosa* (C. Müll.) Zand. *Tortula tanganyikae* Dix. = *Syntrichia amphidiaceus* (C. Müll.) Zand.

Species examined: *S. aculeata* (BUF, NY), *S. alpestris* (BM), *S. ammonsiana* (BUF), *S. amphidiaceus* (BM, BUF, TENN), *S. amplexa* (CU), *S. anderssonii* (NY), *S. andicola*, *S. antarctica* (NY), *S. baileyi* (NY), *S. bartramii* (BUF), *S. bipedicellata* (NY), *S. bogotensis* (NY), *S. bolanderi* (BUF), *S. brandisii* (NY), *S. cainii* (BUF, TRTC), *S. caninervis* (BUF, H), *S. cavallii* (NY, US), *S. chisosa* (BUF), *S. ciliata* (H), *S. conferta* (NY), *S. costesii* (NY), *S. didymodontoides* (H), *S. epilosa* (COLO), *S. filaris* (H, US), *S. flagellaris* (BUF), *S. fontana* (S) as *Tortula rivularis*, *S. fragilis*, *S. fuscoviridis* (NY), *S. geheebiaeopsis* (NY), *S. gemmascens* (NY), *S. gromschii* (US), *S. inermis*, *S. intermedia* (BUF), *S. jaffuelii* (NY), *S. lacerifolia* (NY), *S. laevipila* (BUF), *S. latifolia*, *S. limensis* (NY), *S. linguifolia* (L), *S. longimucronata* (NY), *S. mongolica* (NY), *S. mollis* (US), *S. norvegica*, *S. obtusissima* (BM, BUF, TENN), *S. papillosa* (BUF, NY), *S. percarnosa* (BUF, H, US), *S. phaea* (BUF), *S. pichinchensis* (US), *S. princeps*, *S. prostrata* (US), *S. pseudodesertorum* (S), *S. pseudorobusta* (NY), *S. pygmaea* (BM, BUF, MICH, NY), *S. ramosissima* (NY), *S. reflexa* (NY), *S. rigescens* (FU), *S. robusta* (NY), *S. rubella* (NY), *S. rubra* (NY), *S. ruralis*, *S. saxicola* (NY), *S. scabrella* (NY), *S. scabrinervis* (US), *S. schnyderi* (NY), *S. serrata* (NY), *S. serripungens* (NY), *S. sinensis* (NY), *S. socialis* (S), *S. subaristata* (NY), *S. virescens* (BUF), *S. viridula* (NY).

New combinations and statuses:

Syntrichia aculeata (Wils.) Zand., *comb. nov.* (*Barbula aculeata* Wils., Kew J. Bot. 3: 51, 1851; *Tortula aculeata* (Wils.) Mitt.).

Syntrichia alpestris (Dix. in Herz.) Zand., *comb. nov.* (*Tortula alpestris* Dix. in Herz., Rep. Spec. Nov. Reg. Veg. 38: 103, 1935).

Syntrichia amphidiacea (C. Müll.) Zand., *comb. nov.* (*Barbula amphidiacea* C. Müll., Linnaea 42: 332, 1879; *Tortula amphidiacea* (D. Müll.) Broth.).

Syntrichia amplexa (Lesq.) Zand., *comb. nov.* (*Barbula amplexa* Lesq., Trans. Am. Phil. Soc. n. ser. 13: 5, 1865; *Tortula amplexa* (Lesq.) Steere in Grout).

Syntrichia anderssonii (Ångstr.) Zand., *comb. nov.* (*Tortula anderssonii* Ångstr., Oefv. K. Vet. Ak. Foerh. 29(4): 6, 1872).

Syntrichia anderssonii var. *fagicola* (C. Müll.) Zand., *comb. nov.* (*Barbula conotricha* var. *fagicola* C. Müll., Flora 68: 416, 1885), not seen.

Syntrichia antarctica (Hampe in C. Müll.) Zand., *comb. nov.* (*Barbula antarctica* Hampe in C. Müll., Syn. 1: 642, 1849; *Tortula antarctica* (Hampe in C. Müll.) Wils. in Hook. f.).

Syntrichia austroafricana (Kramer) Zand., *comb. nov.* (*Tortula austroafricana* Kramer, J. Hattori Bot. Lab. 65: 92, 1988), not seen but *cf.* Kramer, J. Hattori Bot. Lab. 65: 84, 1988.

Syntrichia baileyi (Broth.) Zand., *comb. nov.* (*Tortula baileyi* Broth., Oefv. Finsk. Vet. Soc. Foerh. 33: 97, 1891).

Syntrichia bartramii (Steere in Grout) Zand., *comb. nov.* (*Tortula bartramii* Steere in Grout, Moss Fl. N. Amer. 1: 241, 1939).

Syntrichia bipedicellata (Britt.) Zand., *comb. nov.* (*Tortula bipedicellata* Britt., Bull. Torr. Bot. Cl. 23: 431, 1896).

Syntrichia bogotensis (Hampe) Mitt. (*Barbula bogotensis* Hampe, Ann. Sc. Nat. Bot. ser. 5, 3: 349, 1865; *Tortula bogotensis* (Hampe) Mitt.).

Syntrichia bolanderi (Lesq. & James) Zand., *comb. nov.* (*Barbula bolanderi* Lesq. & James, Trans. Am. Phil. Soc. n. ser. 13: 5, 1865; *Tortula bolanderi* (Lesq. & James) Howe).

Syntrichia brachyclada (Card.) Zand., *comb. nov.* (*Tortula brachyclada* Card., Bull. Herb. Boiss. ser. 2, 5: 1002, 1905), not seen but *cf.* Kramer, J. Hattori Bot. Lab. 65: 84, 1988.

Syntrichia brandisii (C. Müll.) Zand., *comb. nov.* (*Barbula brandisii* C. Müll., Flora 61: 82, 1878; *Tortula brandisii* (C. Müll.) Broth.).

Syntrichia brevisetacea (F. Müll.) Zand., *comb. nov.* (*Barbula brevisetacea* F. Müll., Analyt. Draw. Austral. Moss. Tab. 4, 1864; *Tortula brevisetacea* (F. Müll.) Thér.), not seen but *cf.* Kramer, J. Hattori Bot. Lab. 65: 84, 1988.

Syntrichia cainii (Crum & Anders.) Zand., *comb. nov.* (*Tortula cainii* Crum & Anders., J. Elisha Mitchell Sci. Soc. 74: 35, 1958).

Syntrichia campestris (Dus.) Zand., *comb. nov.* (*Tortula campestris* Dus., Ark. Bot. 6(8): 25, 1906), not seen but *cf.* Kramer, J. Hattori Bot. Lab. 65: 84, 1988.

Syntrichia caninervis var. *spuria* (Amann) Zand., *comb. nov.* (*Tortula spuria* Amann, Bull. Murithienne 39: 351, 1916; *Tortula caninervis* ssp. *spuria* (Amann) Kramer), not seen.

Syntrichia caninervis var. *abranchesii* (Luis.) Zand., *comb.*

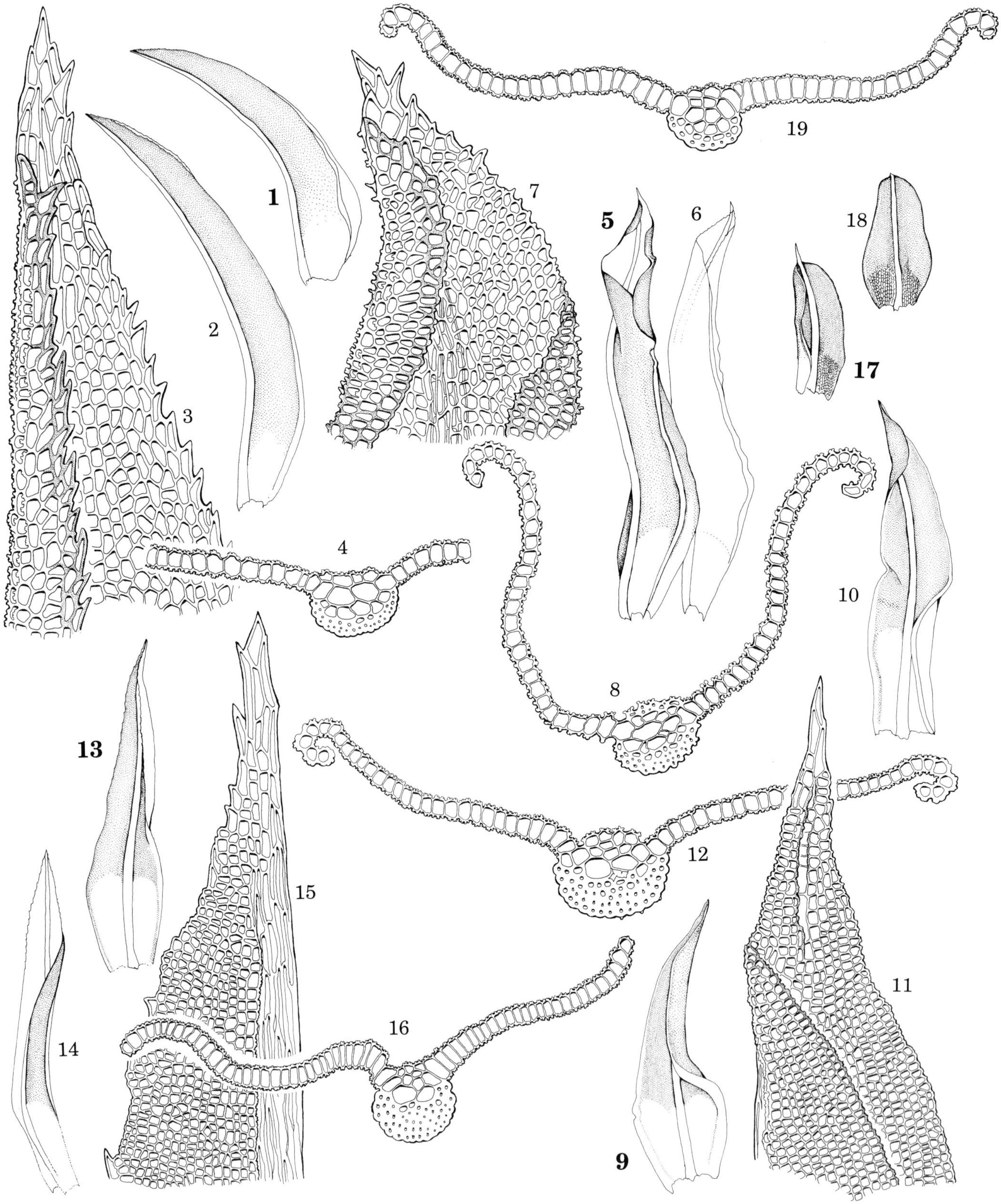

Plate 111. ***Syntrichia.*** **1–4.** ***S. robusta.*** 1–2. Two leaves. 3. Leaf apex. 4. Transverse section at midleaf. **5–8.** ***S. rubra.*** 5–6. Two leaves. 7. Leaf apex. 8. Transverse section at midleaf. **9–12.** ***S. saxicola.*** 9–10. Two leaves. 11. Leaf apex. 12. Transverse section at midleaf. **13–16.** ***S. serrata.*** 13–14. Two leaves. 15. Leaf apex, lateral view. 16. Transverse section at midleaf. **17–19.** ***S. virescens.*** 17–18. Two leaves. 19. Transverse section at midleaf.

nov. (*Tortula abranchesii* Luis., Broteria ser. Bot. 14: 115, 1916; *Tortula caninervis* var. *abranchesii* (Luis.) Kramer), not seen.

Syntrichia caninervis var. *spuria* (Amann) Zand., *comb. nov.* (*Syntrichia spuria* Amann, Fl. Mouss. Suisse 2: 119, 384, 1918).

Syntrichia chisosa (Magill, Delg. & Stark) Zand., *comb. nov.* (*Tortula chisosa* Magill, Delg. & Stark, Ann. Missouri Bot. Gard. 70: 200, 1983).

Syntrichia ciliata (Broth.) Zand., *comb. nov.* (*Tortula ciliata* Broth., Bot. Jahrb. 49: 175, 1912).

Syntrichia conferta (Bartr.) Zand., *comb. nov.* (*Tortula conferta* Bartr., Bryologist 60: 140, 1957; *Tortula princeps* var. *conferta* (Bartr.) Lightowlers).

Syntrichia costesii (Thér.) Zand., *comb. nov.* (*Tortula costesii* Thér., Rev. Chil. Hist. Nat. 25: 298, 1921).

Syntrichia didymodontoides (Broth. in Dryg.) Zand., *comb. nov.* (*Tortula didymodontoides* Broth. in Dryg., Deutsch. Südpolar Exp. 8: 86, 1906).

Syntrichia epilosa (Broth. ex Dus.) Zand., *comb. nov.* (*Tortula epilosa* Dus., Ark. Bot. 6(8): 25, 1906.

Syntrichia epilosa var. *pilifera* (Thér.) Zand., *comb. nov.* (*Tortula epilosa* var. *pilifera* Thér., Rev. Chil. Hist. Nat. 30: 343, 1926), not seen.

Syntrichia filaris (C. Müll. in Neum.) Zand., *comb. nov.* (*Barbula filaris* C. Müll. in Neum., Deutsch. Exp. Int. Polarforsch. 2: 309, 1890; *Tortula filaris* (C. Müll. in Neum.) Broth.).

Syntrichia flagellaris (Schimp.) Zand., *comb. nov.* (*Barbula flagellaris* Schimp., Ann. Sci. Nat. Bot. ser. 2, 6: 146, 1836; *Tortula flagellaris* Mont. in Gay).

Syntrichia flagellaris var. *densiretis* (Thér.) Zand., *comb. nov.* (*Tortula flagellaris* var. *densiretis* Thér., Recueil Publ. Soc. Havraise Etud. Div. 1917: 7, 1917), not seen.

Syntrichia fontana (C. Müll. in Neum.) Zand., *comb. nov.* (*Barbula fontana* C. Müll. in Neum., Deutsch. Exp. Int. Polarforsch. 2: 308, 1890; *Tortula fontana* (C. Müll.) Broth.).

Syntrichia fuscoviridis (Card.) Zand., *comb. nov.* (*Tortula fuscoviridis* Card., Bull. Herb. Boiss. ser. 2, 6: 6, 1906).

Syntrichia geheebiaeopsis (C. Müll.) Zand., *comb. nov.* (*Barbula geheebiaeopsis* C. Müll., Bot. Jahrb. 5; 80, 1883; *Tortula geheebiaeopsis* C. Müll.) Broth.).

Syntrichia gemmascens (Chen) Zand., *comb. nov.* (*Desmatodon gemmascens* Chen, Hedwigia 80: 297, 1941; *Didymodon gemmascens* Broth., Symb. Sin. 4: 38, 1929, *hom. illeg.*).

Syntrichia gemmascens var. *hopeiensis* (Chen) Zand., *comb. nov.* (*Desmatodon gemmascens* var. *hopeiensis* Chen, Hedwigia 80: 229, 1941).

Syntrichia glacialis (Kunze ex C. Müll.) Zand., *comb. nov.* (*Barbula glacialis* Kunze ex C. Müll., Syn. 1: 634, 1849; *Tortula glacialis* (Kunze ex C. Müll.) Mont. in Gay).

Syntrichia gromschii (Thér.) Zand., *comb. nov.* (*Tortula gromschii* Thér., Rev. Bryol. Lichénol. 7: 173, 1935).

Syntrichia jaffuelii (Thér.) Zand., *comb. nov.* (*Tortula jaffuelii* Thér., Rev. Chil. Hist. Nat. 27: 9, 1923).

Syntrichia lacerifolia (Williams) Zand., *comb. nov.* (*Tortula lacerifolia* Williams, Bull. Torr. Bot. Cl. 43: 326, 1916).

Syntrichia leucostega (C. Müll.) Zand., *comb. nov.* (*Barbula leucostega* C. Müll., Syn. 1: 641, 1894; *Tortula leucostega* (C. Müll.) Broth.), not seen but *cf.* Kramer, J. Hattori Bot. Lab. 65: 106, 1988.

Syntrichia leucostega var. *trachyneura* (Dix.) Zand., *comb. nov.* (*Tortula trachyneura* Dix., Trans. R. Soc. S. Africa 8: 195, 1920; *Tortula leucostega* var. *trachyneura* Kramer), not seen but *cf.* Kramer, J. Hattori Bot. Lab. 65: 107, 1988.

Syntrichia limensis (Williams) Zand., *comb. nov.* (*Tortula limensis* Williams, Bull. Torr. Bot. Cl. 42: 398, 1915).

Syntrichia linguifolia (Herz.) Zand., *comb. nov.* (*Tortula linguifolia* Herz., Biblioth. Bot. 87: 49, 1916).

Syntrichia longimucronata (X.-j. Li) Zand., *comb. nov.* (*Tortula longimucronata* X.-j. Li, Act. Bot. Yunnan. 3: 107–109, 1981).

Syntrichia magellanica (Mont. in Gay) Zand., *comb. nov.* (*Tortula magellanica* Mont. in Gay, Hist. Fis. Polit. Chile Bot. 7: 145, 1850; *Tortula princeps* var. *magellanica* (Mont. in Gay) Lightowlers), not seen but *cf.* Kramer, J. Hattori Bot. Lab. 65: 110, 1988.

Syntrichia mollis (B.&S. ex C. Müll.) Zand., *comb. nov.* (*Barbula mollis* B.&S. ex C. Müll., Syn. 1: 637, 1849; *Tortula mollis* (B.&S. ex C. Müll.) Broth.).

Syntrichia obtusissima (C. Müll.) Zand., *comb. nov.* (*Barbula obtusissima* C. Müll., Syn. 1: 640, 1849; *Tortula obtusissima* (C. Müll.) Mitt.).

Syntrichia papillosa var. *chilensis* (Thér.) Zand., *comb. nov.* (*Tortula papillosa* var. *chilensis* Thér., Rev. Chil. Hist. Nat. 25: 297, 1921), not seen.

Syntrichia percarnosa (C. Müll.) Zand., *comb. nov.* (*Barbula percarnosa* C. Müll., Linnaea 42: 347, 1879; *Tortula percarnosa* (C. Müll.) Broth.).

Syntrichia phaea (Hook. f. & Wils.) Zand., *comb. nov.* (*Trichostomum phaeum* Hook. f. & Wils., Fl. Nov. Zel. 2: 72, 1854; *Tortula phaea* (Hook. f. & Wils.) Dix.).

Syntrichia pichinchensis (Tayl.) Zand., *comb. nov.* (*Tortula pichinchensis* Tayl., London J. Bot. 6: 333, 1847).

Syntrichia princeps var. *brachycarpa* (De Not.) Zand., *comb. nov.* (*Tortula princeps* var. *brachycarpa* De Not., Atti Univ. Genova 1: 538, 1869).

Syntrichia princeps var. *echinata* (Shiffn.) Zand., *comb. nov.* (*Tortula echinata* Schiffn., Oesterr. Bot. Zeitschr. 65: 4–5, 1915; *Tortula princeps* ssp. *echinata* (Schiffn.) Kramer).

Syntrichia prostrata (Mont.) Zand., *comb. nov.* (*Tortula prostrata* Mont., Ann. Sc. Nat. Bot. ser. 3, 4: 104, 1845).

Syntrichia pseudorobusta (Dus.) Zand., *comb. nov.* (*Tortula pseudorobusta* Dus., Ark. Bot. 6(8): 19, 1906).

Syntrichia pygmaea (Dus.) Zand., *comb. nov.* (*Tortula pygmaea* Dus., Ark. Bot. 6(10): 8, 1907).

Syntrichia ramosissima (Thér.) Zand., *comb. nov.* (*Tortula ramosissima* Thér., Rev. Chil. Nat. 33: 136, 1926).

Syntrichia reflexa Zand., *nom. nov.* (*Tortula reflexa* X.-j. Li, Acta Bot. Yunnan. 3: 109, 1981, *hom. illeg. non* Brid.).

Syntrichia robusta (Hook. & Grev.) Zand., *comb. nov.* (*Tortula robusta* Hook. & Grev., Edinburgh J. Sc. 1: 299,

1824).

Syntrichia robusta var. *recurva* (Lightowlers) Zand., *comb. nov.* (*Tortula robusta* var. *recurva* Lightowlers, Brit. Ant. Surv. Bull. 64: 64, 1984).

Syntrichia rubella (Hook. f. & Wils.) Zand., *comb. nov.* (*Tortula rubella* Hook. f. & Wils., Fl. Tasman. 2: 176, 1859).

Syntrichia rubra (Mitt. in Hook. f.) Zand., *comb. nov.* (*Tortula rubra* Mitt. in Hook. f., Handb. New Zealand Fl. 419, 1867).

Syntrichia rubra var. *subantarctica* (Sainsb.) Zand., *comb. nov.* (*Tortula subantarctica* Sainsb., Svensk. Bot. Tidskr. 44: 72, 1950; *Syntrichia rubra* var. *subantarctica* (Sainsb.) Lightowlers).

Syntrichia ruralis var. *gigantea* (Lesq.) Zand., *comb. nov.* (*Barbula ruralis* var. *gigantea* Lesq., Mem. California Acad. Sci. 1: 13, 1868; *Tortula ruralis* var. *gigantea* (Lesq.) L. Koch.), not seen.

Syntrichia ruralis var. *gracilis* (C. Jens.) Zand., *comb. nov.* (*Tortula ruralis* var. *gracilis* C. Jens., Medd. Groenland 15: 409, 1898), not seen.

Syntrichia ruralis var. *spiralis* (Herz.) Zand., *comb. nov.* (*Tortula ruralis* var. *spiralis* Herz., Biblioth. Bot. 87: 49, 1916), not seen.

Syntrichis ruralis var. *submamillosa* (Kramer) Zand., *comb. nov.* (*Tortula ruralis* var. *submamillosa* Kramer, Bryophyt. Biblioth. 21: 127, 1980), not seen.

Syntrichia ruralis var. *subpapillosissima* (Biz. & Pierr.) Zand., *comb. nov.* (*Tortula ruralis* var. *subpapillosissima* Biz. & Pierr., Acta Bot. Acad. Sci. Hung. 18: 10, 1973; *Tortula ruraliformis* var. *subpapillosissima* (Biz. & Pierr.) Kramer), not seen.

Syntrichia saxicola (Card.) Zand., *comb. nov.* (*Tortula saxicola* Card., Bull. Herb. Boiss. ser. 2, 5: 1002, 1905).

Syntrichia scabrella (Dus.) Zand., *comb. nov.* (*Tortula scabrella* Dus., Ark. Bot. 6(10): 4, 1907).

Syntrichia scabrinervis (C. Müll.) Zand., *comb. nov.* (*Barbula scabrinervis* C. Müll., Syn. 1: 634, 1849; *Tortula scabrinervis* (C. Müll.) Mitt.).

Syntrichia schnyderi (C. Müll.) Zand., *comb. nov.* (*Barbula schnyderi* C. Müll., Linnaea 43: 434, 1882; *Tortula schnyderi* (C. Müll.) Broth.).

Syntrichia serrata Zand., *comb. nov.* (*Tortula serrata* Dix., New Zealand Inst. Bull. 3(3): 146, 1923, *nom. legit.* contrary to Index Muscorum, also *cf.* Lightowlers, J. Bryol. 13: 373, 1985).

Syntrichia serripungens (Lor. & C. Müll.) Zand., *comb. nov.* (*Barbula serripungens* Lor. & C. Müll., Linnaea 42: 360, 1879; *Tortula serripungens* (Lor. & C. Müll.) Broth.).

Syntrichia serripungens var. *excesa* (C. Müll.) Zand., *comb. nov.* (*Barbula serripungens* var. *excesa* C. Müll., Linnaea 42: 351, 1879.

Syntrichia socialis (Dus.) Zand., *comb. nov.* (*Tortula socialis* Dus., Ark. Bot. 6(10): 6, 1907).

Syntrichia subaristata (B.&S. ex C. Müll.) Zand., *comb. nov.* (*Barbula subaristata* B.&S. ex C. Müll., Syn. 1: 644, 1849; *Tortula subaristata* (B.&S. ex C. Müll.) Broth.).

Syntrichia virescens var. *bizotiana* (Kramer) Zand., *comb. nov.* (*Tortula virescens* ssp. *bizotiana* Kramer, J. Hattori Bot. Lab. 65: 123, 1988, *nom. nov.* for *Tortula virescens* ssp. *bizotii* (Laz.) Kramer, Bryophyt. Biblioth. 21: 102, 1980, *comb. inval. basion. inval.*; *Tortula bizotii* Laz., Vopr. Evol. Biogeogr. Genet. Sel. 145, 1960, *nom. inval.*), not seen but *cf.* Kramer, J. Hattori Bot. Lab. 65: 84, 1988.

Syntrichia virescens var. *iranica* (Kramer) Zand., *comb. nov.* (*Tortula virescens* var. *iranica* Kramer, Bryophyt. Biblioth. 21: 101, 1980), not seen.

Syntrichia viridula (C. Müll.) Zand., *comb. nov.* (*Barbula viridula* C. Müll., Nuov. Giorn. Bot. Ital. n. ser. 4: 114, 1897; (*Tortula viridula* (C. Müll.) Broth.; the isotype at NY is probably *S. fragilis*).

75. HILPERTIA Plate 112.

Hilpertia Zand., Phytologia 65: 427, 1989. Type: *Hilpertia velenovskyi* (Schiffn.) Zand.

Named for Friedrich Hilpert, 1907–, originally of Hildburghausen, Germany, student of T. Herzog and author of "Studien zur Systematik der Trichostomaceen" (1933) as a doctoral dissertation.

Plants growing in loose cushions, greenish brown above, light brown below. Stems branching irregularly, to 1.0 cm in length, transverse section rounded-pentagonal, central strand distinct, sclerodermis not or weakly differentiated, hyalodermis absent; axillary hairs to 8 cells in length, all hyaline or the basal one yellow-brown; rhizoids rare. *Leaves* crowded, larger above, appressed and tightly spiralled when dry, weakly spreading when moist, *ovate to circular,* 1.3–2.0 mm in length (including awn), *upper lamina* flat to more *usually quite concave, margins strongly revolute* (to 2 times), entire or broadly toothed at or near the base of the awn; *apex* broadly acute, *hyaline in an apical patch or triangle;* base not differentiated in shape; *costa narrow but broader above, excurrent as a hyaline awn,* costa with lamina inserted laterally, superficial cells long-rectangular and smooth on both sides, 2–4 rows of cells across costa ventrally at midleaf, costal transverse section rounded, *stereid band rounded in shape, ventral epidermis present but dorsally absent,* guide cells 2 in 1 layer, hydroid strand present; *upper laminal cells hexagonal to short-rectangular or rhomboidal,* 14–25 µm in width, 2–4:1, internal walls thin to thickened and porose, *dorsal superficial walls much thickened,* weakly convex superficially on both sides, *cells of leaf apex rhomboidal to fusiform, smooth, cells of revolute margin enlarged, strongly chlorophyllose; papillae absent medially, usually hollow-papillose on revolute margins;* basal cells weakly differentiated, rectangular, ca. 16–18 µm in width, 2–3:1, walls thin. Propagula when present (1–)3–4 celled, brown, spherical to elliptical, mostly 30–50 µm in length, borne on basal rhizoids. Synoicous, paroicous, autoicous or apparently dioicous but probably rhizautoicous. *Perichaetia* terminal, *inner leaves usually differentiated,* long-oval, margins usually little differentiated, to 1.7 mm in length, basal portion of leaf sheathing, lower cells rectangular, very thin-walled. Perigonia lateral or occasionally terminal on a separate plant. Seta 3.5–4.0 mm in length, 1 per perichaetium, yellow-brown, twisted counterclockwise above, clockwise below; theca 1.2–1.5 mm in length, yellow-brown, elliptical, occasionally weakly ventricose, exothecial cells ca. 16–23 µm in width, 2–3:1, thin-walled, stomates phaneropore, on capsule neck, annulus of 3 rows of smaller, quadrate, highly vesiculose cells; peristome teeth 32, linear, densely branching-spiculose, 300–700 µm in length, with many articulations, twisted counterclockwise about 1/2 turn, basal membrane short, to 45 µm in height, papillose-spiculose. Operculum broadly short-conic to long-conic, 0.4–1.0 mm in length, cells twisted 1/2 turn counterclockwise. Calyptra cucullate, smooth, ca. 2.8 mm in length. Spores 13–16 µm in diameter, light brown, indistinctly papillose. *Laminal KOH color reaction red.*

Found on soil in Canada (Northwest Territories) and countries of eastern central Europe.

Corley et al. (1981) suggested that *Tortula velenovskyi* is probably best placed in a separate section of *Tortula*. It is here treated, together with *T. scotteri*, as a separate genus, *Hilpertia*.

Schiffner (1893) reviewed the extensive variation in sexuality in the type species. The modification of the upper laminal margins of *Hilpertia* into tubes of photosynthetic tissue (Pl. 112) is paralleled in species of *Pseudocrossidium* and to a lesser extent in *Tortula revolvens* (Pl. 89, f. 4). *Hilpertia*

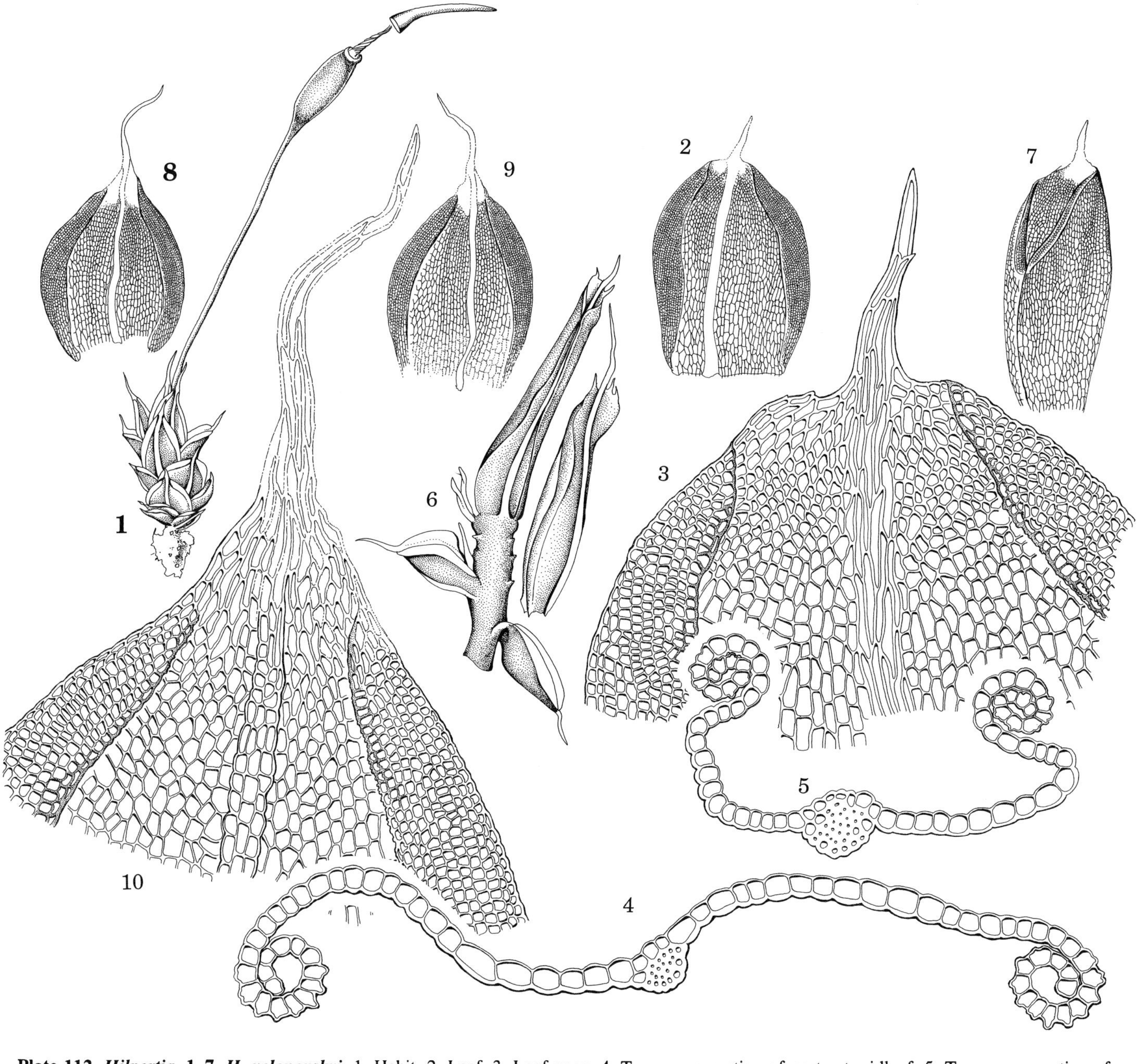

Plate 112. ***Hilpertia.*** **1–7.** ***H. velenovskyi.*** 1. Habit. 2. Leaf. 3. Leaf apex. 4. Transverse section of costa at midleaf. 5. Transverse section of costa near apex. 6. Dissected plant apex. 7. Perichaetial leaf. **8–10.** ***H. scotteri.*** 8–9. Two leaves. 10. Leaf apex.

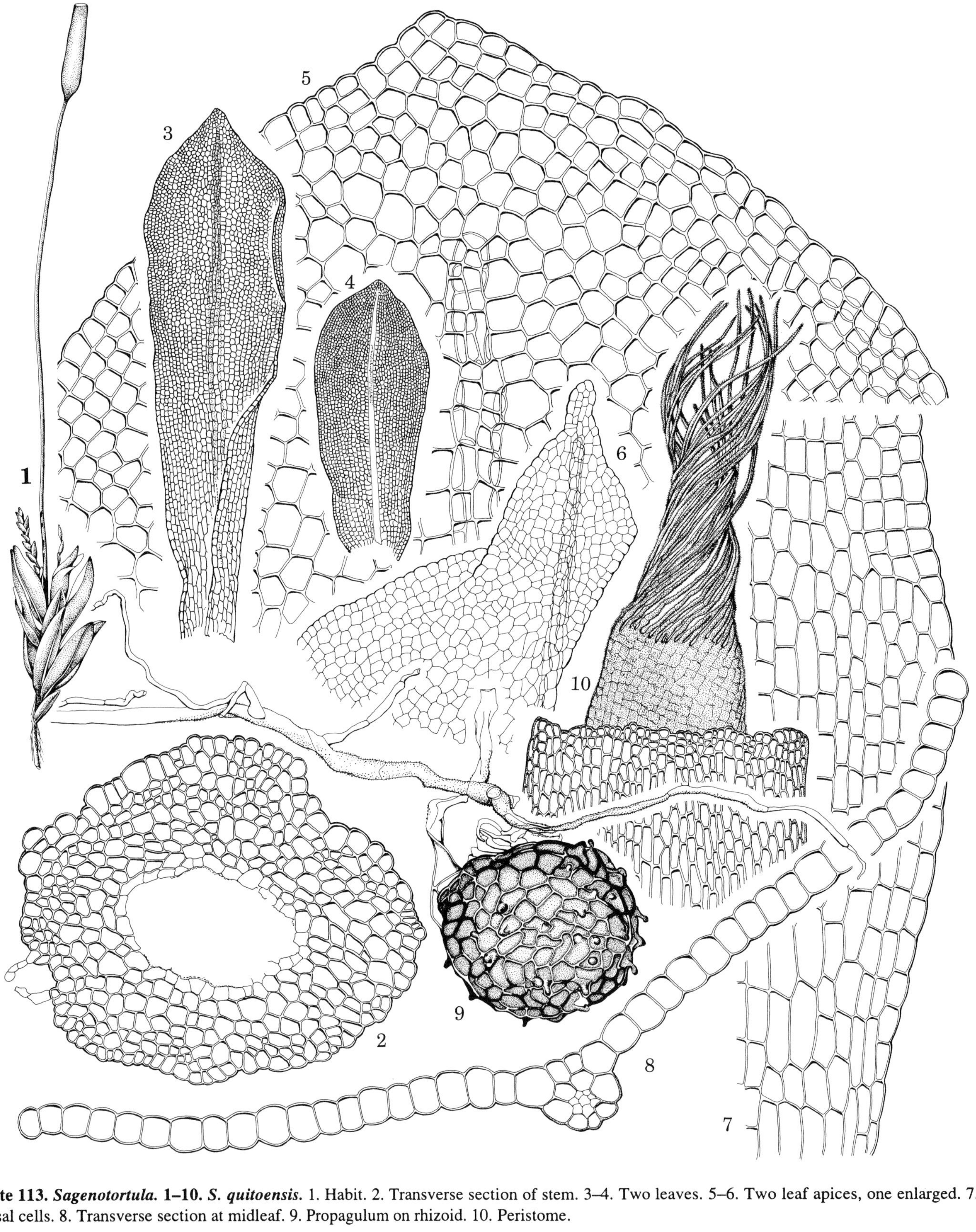

Plate 113. ***Sagenotortula.*** **1–10.** ***S. quitoensis.*** 1. Habit. 2. Transverse section of stem. 3–4. Two leaves. 5–6. Two leaf apices, one enlarged. 7. Basal cells. 8. Transverse section at midleaf. 9. Propagulum on rhizoid. 10. Peristome.

has many of the features of species of *Acaulon* including the ovate (rather concave) leaf shape, thin costa, upper laminal cells often hyaline apically and dorsally superficially thick-walled (as easily seen in section, Pl. 112, f. 4–5), a hyaline awn, and red KOH color reaction. The concave leaves may indicate an ancestry of taxa with a somewhat bulbiform habit, thus *Hilpertia* may well be derived from *Acaulon*-like ancestors through desuppression of the sporophyte and elaboration of the leaf margins (but *cf.* Cladograms 13 and 14). *Hilpertia* is easily distinguished from *Acaulon* by its revolute leaf margins, elongate stems, elongate setae, and peristomate, cylindrical capsules. On the other hand, the leaves of *Stegonia latifolia* and *S. hyalinotricha* (Pl. 69) are quite similar morphologically to those of *Hilpertia*, thus *Hilpertia* may have been derived from ancestors shared with *Stegonia* by development of photosynthetically specialized laminal margins, differentiated perichaetial leaves, and red laminal KOH reaction. Cladistic analysis (see Cladogram 14), contrarily, indicates that *Stegonia* is best hypothesized as rather distantly related, and that *Acaulon* does not share immediate ancestors with *Hilpertia*.

Additional literature: Boros (1941), Karczmarz (1961), Kuc (1960), Peciar (1960b), Pospíšil (1977), Waclawska (1958), Zander (1989), Zander and Steere (1978).

Number of accepted species: 2.

Species examined: *Hilpertia scotteri* (BUF), *H. velenovskyi* (FH, NY).

76. SAGENOTORTULA — Plate 113.

Sagenotortula Zand., Phytologia 65: 429, 1989. Type: *Sagenotortula quitoensis* (Tayl. in Hook.) Zand.

From *sagena,-ae*, fish-net + o + *Tortula*, a genus; a *Tortula*-like moss with a net-like areolation.

Plants loosely caespitose or turf-forming, green above, greenish brown below. Stems branching occasionally, especially from just below the perichaetia, to 2.0 cm in length, transverse section round to elliptical, *central strand present, often strong or stem centrally hollow*, sclerodermis absent, hyalodermis absent; axillary hairs several cells in length, all hyaline; stem thickly matted with rhizoids. *Leaves* appressed, little changed to incurved but *little contorted when dry*, erect to weakly spreading when moist, *broadly ligulate to spathulate, 3–5 mm in length*, upper lamina flat or very broadly and shallowly channeled, *margins plane*, occasionally broadly incurved above, *entire* (occasionally distantly bluntly toothed, especially evident in young leaves); apex broadly acute, occasionally with a broad, blunt apiculus; base not differentiated in shape; *costa ending ca. 4 cells below apex to percurrent*, costa with lamina inserted ventrally at about 45°, superficial cells quadrate to hexagonal or short-rectangular ventrally, dorsally rectangular, 2(–3) rows of cells across costa ventrally at midleaf, costal transverse section somewhat rounded, *stereid band little developed, weakly substereid and reniform in shape*, ventral epidermis present, dorsal absent, guide cells 2 in 1 layer, hydroid strand present, often large; *upper laminal cells hexagonal*, occasionally quadrate or short-rectangular, *very large*, (25–)40–60(–75) µm in width, 1(–2):1, *walls thin*, occasionally evenly thickened, *often weakly trigonous*, superficially convex on both sides; *papillae absent;* basal cells differentiated in lower 1/4 of leaf across leaf or just medially, rectangular, ca. 35–60 µm in width, 2–4:1, walls thin to somewhat thickened. Propagula consisting of *red brood bodies borne on tomentum of perigoniate plants*, elliptical to spherical, 50–65 µm in longest dimension, *bearing superficially scattered short peg-like projections*. Dioicous. Perichaetia terminal, inner leaves slightly larger than the cauline, otherwise little different, to 6 mm in length. Perigonia terminal, comparatively large, inner leaves ovate. Seta ca. 2.5 cm in length, 1–3 per perichaetium, red-brown, twisted counterclockwise above, occasionally clockwise below; theca 3–4 mm in length, red-brown, cylindrical, exothecial cells long-rectangular, thin-walled, 20–25 µm in width, stomates at base of theca, phaneropore, annulus of 2–3 rows of vesiculose cells, apparently deciduous in pieces; *peristome teeth 32*, filamentous, low-spiculose, ca. 1000 µm, with many articulations, twisted counterclockwise once, basal membrane to 200 µm in height, low-spiculose to crazed. Operculum conic, 1.5–1.7 mm in length, cells twisted counterclockwise. Calyptra cucullate, smooth, ca. 4.5 mm in length. Spores 10–13 µm in diameter, tan, essentially smooth. *Laminal KOH color reaction red.*

Found on rocks and soil at rather high elevations; Mexico and the Andes of South America.

Sagenotortula is distinguished from other genera of Pottiaceae by the unique red brood bodies (Pl. 113, f. 9) that have numerous short, blunt, papilla-like projections (unlike the armed or caltrop-like propagula of *Hyophila* species that have cell walls protruding). The brood bodies were found only on perigoniate plants, being uncolored and transparent when young, and are distributed in the tomentum along the entire length of the stem. The peg-like projections develop one per superficial cell and resemble simple papillae when small; rhizoids are occasionally produced apically from the longer protrusions. The genus differs from other genera that have large, epapillose upper laminal cells by the mostly entire, unbordered margins and the extremely large upper laminal cells (Pl. 113, f. 5).

The apparent lack of a dorsal costal epidermis (Pl. 113, f. 8) indicates that this genus could have immediate ancestors shared with *Syntrichia*, with derivation through loss of papillae, inflation of the laminal cells, and elaboration of the unique brood body. Cladograms 13 and 14 support this.

Additional literature: Brotherus (1924–25).

Number of accepted species: 1.

Species examined: *Sagenotortula quitoensis* (NY, TENN).

EXCLUDED AND UNTREATED TAXA

GENERA

ARVILDIA

Arvildia Ignatov, Palaeontograph. B. 217: 150, 1990. Type: *Arvildia elenae* Ignatov.

The two species of this Permian fossil genus were referred to either the Pottiaceae or Trichostomaceae by Ignatov (1990), although he indicated that other, closely related families were also possibile dispositions. The published figures, however, are reminiscent of the Dicranaceae because of the very broad costae and the bulging, thin-walled laminal cells, which lack papillae. Additional study is needed.

BRYOBARTRAMIA

Bryobartramia Sainsb., Bryologist 51: 10, 1948. Type: *Bryobartramia robbinsii* Sainsb.

Bryobartramia is a genus of arid areas of Australia and South Africa. The single species, *B. novae-valesiae* (Broth.) Stone, has been placed in its own family, the Bryobartramiaceae Sainsb., but is here referred to the Encalyptaceae because of the large calyptra, yellow KOH reaction of the upper lamina, and large, coarsely papillose upper laminal cells. Stone (1977a) indicated that *Bryobartramia* had more characters in common with the Encalyptaceae than with any other family and described these in detail. *Bryobartramia* differs from other genera of Encalyptaceae by, among other features, the non-dehiscing calyptra (epigonium); stomates scattered over the capsule, which is spherical and cleistocarpous; exothecial cells transparent and nearly hyaline (very weakly tinted yellow); seta very short; and perichaetial leaves subulate, not papillose, elaminate above (rather like the long-awned perichaetial leaves of *Diphyscium*). The inflated, non-dehiscing calyptra may act as a flotation device as does, perhaps, the thecal air sac of *Tetrapterum*.

CLADOPHASCUM

Cladophascum Dix. ex Sim, Bryoph. S. Afr. 143, 1926. Type: *Cladophascum gymnomitrioides* (Dix.) Dix. ex Sim.

This genus was referred to the Pottiaceae in Crosby and Magill's (1977, 1978, 1981) Dictionary of the Mosses, but has been placed in the Dicranaceae subf. Trematodontoideae by Magill (1981) in his treatment of the bryoflora of South Africa.

ENTOSTHYMENIUM

Entosthymenium Brid., Bryol. Univ. 1: 761. 1827. Type: *Entosthymenium tristichum* Brid.

Paris (1906) and van der Wijk et al. (1959–69) recognized this genus. Of the four combinations, the Index Muscorum (van der Wijk et al. 1959–1969) reported two names assigned to the synonymy of *Desmatodon convolutus* (Brid.) Grout, and one to *Pottia wilsonii* var. *mucronifolia* (Bruch) Warnst. Nothing seems known of the generitype, *E. tristichum*. The type of *E. tristichum*, collected in southeastern France, is absent from the De Candolle Herbarium at G (P. Geissler, pers. comm.). C. Müller (1849–51) and W. Schultze-Motel (pers. comm.) have indicated that the type is not to be found in the Bridel Herbarium (B); the former referred the genus to "Pottiacea suspectissima." L. Bertrand (pers. comm.) wrote me that it is also not in the Grateloup herbarium at Montpellier (MPU). Brotherus (1902–09, 1924–1925, Corley et al. (1981), Husnot (1884–1894), Jaeger and Sauerbeck (1870–1880), Lindberg (1864), Müller (1901), Podpěra (1954) and Reed and Robinson (1972), perhaps understandably, neither acknowledged nor disposed of the genus in their treatments, which include the Pottiaceae of France. The original description is extremely vague. The genus is here set aside pending further study that might lead to adequate lecto- or neotypification.

ERPODIOPSIS

Erpodiopsis C. Müll., Flora 73: 470, 1890. Type: *Erpodiopsis kilimandscharica* C. Müll. (= *Phascum kilimandscharicum* (C. Müll.) Roth).

Brotherus (1902–09: 1239) placed the type of this genus very near *Bryum argenteum* Hedw. (Bryaceae), while Roth (1911) indicated that it is close to *Phascum hyalinotrichum* (here transferred to *Stegonia* and in fact with much the appearance of *B. argenteum*). Inasmuch as the type (not seen) is, according to both aforementioned authors, sterile and Roth's (1911) illustration is strikingly bryaceous, a new combination in *Bryum* would be an effective disposition pending evaluation by a specialist in that group.

New combination: *Bryum kilimandscharicum* (C. Müll.) Zand., comb. nov. (*Erpodiopsis kilimandscharica* C. Müll., Flora 73: 470, 1890; *Phascum kilimandscharicum* (C. Müll.) Roth).

KLEIOWEISIOPSIS

Kleioweisiopsis Dix., Smithsonian Misc. Coll. 72(3): 18, 1920. Type: *Kleioweisiopsis denticulata* Dix., Kenya, Aberdare Mts., Allan 395a, holotype, BM, isotype, US.

Because of the lack of characteristic papillae, the rectangular, evenly thickened upper laminal cell walls, and the subulate perichaetial leaves, this genus is excluded from the Pottiaceae. For the time being, it should be recognized in the Orthotrichaceae near *Amphidium. Kleioweisiopsis denticulata* has dentate upper laminal margins reminiscent of *A. cyathicarpum* but laminal cell ornamentation consists of longitudinal lines of indistinct, minute verrucae rather than small elliptical verrucae. The capsule of *Kleioweisiopsis* is spherical and the operculum is differentiated but persistent. There is also some resemblance to Dicranaceae subfam. Rhabdoweisioideae as noted by Dixon in the original description, but the upper leaf cells of that group are generally only as long as wide, and verrucae are apparently absent. The holotype at BM is annotated by Saito as a possible combination in *Pleuridium*; a second species *K. involuta* (PC! an invalid name fide Crosby et al. 1992) proved to be *Weissia* subg. *Phasconica*, see treatment of *Weissia*.

MELOPHYLLUM

Melophyllum Herz., Rep. Soc. Nov. Regn. Veg. 74: 98, 1939. Type: *Melophyllum radiculosum* Herz.

This genus is a synonym of *Astomiopsis* C. Müll. (Ditrichaceae) according to Buck and Zander (1980).

PSEUDOHYOPHILA

Pseudohyophila Hilp., Hedwigia 73: 68, 1933. Type: *Pseudohyophila peruviana* (Williams) Hilp., Peru, Juliaca, Williams 2874, holotype, NY.

Originally published as *Hyophila peruviana* Williams (1903), this monotypic genus has been recognized in the Grimmiaceae. Churchill (1981), in a treatment of the Grimmiaceae, indicated that *Pseudohyophila* does not fit well in that family. The somewhat tomentose stems; entire, lanceolate leaves with epapillose, quadrate upper laminal cells, often bistratose in patches on the upper margins; weak, flattened, deeply grooved costa; sheathing perichaetial leaves; autoicous sexual condition; and capsule with distinct, furrowed neck are all characters shared by *Dicranoweisia* Lindb. (cf. *D. cirrata* (Hedw.) Lindb. ex Milde), and, although *Pseudohyophila* is eperistomate, it clearly should be placed near that genus in the Dicranaceae.

PSEUDOTIMMIELLA

Pseudotimmiella Biz., Cryptogamie Bryol. Lichénol. 1: 425, 1980. Type: *Pseudotimmiella pocsii* Biz., Tanzania, Uluguru Mts., above Morogoro, Pocs, Mwanjabe & Sharma 6549–H, isotype, NY. = *Diphyscium* Mohr.

Characters indicating correct placement of this taxon (the type collection lacks sporophytes) in *Diphyscium* (Diphysciaceae) are the rigid, subulate perichaetical leaves, the long-ligulate cauline leaves that develop an excurrent costa just below the perichaetium, and the lamina entirely bistratose, not unistratose marginally as in *Timmiella*.

New combination: *Diphyscium pocsii* (Biz.) Zand., comb. nov. (*Pseudotimmiella pocsii* Biz., Cryptogamie Bryol. Lichénol. 1: 425, 1980).

RHAMPHIDIUM

Rhamphidium Mitt., J. Linn. Soc. Bot. 12: 45, 1869. Type: *Rhamphidium macrostegium* (Sull.) Mitt.

Trichostomum subg. *Rhamphidium* (Mitt.) Besch., Ann. Sc. Nat. Bot., ser. 6(3): 198, 1876.

Norris and Koponen (1989) noted that the gametophyte of *Rhamphidium* was very similar to that of *Dicranella* (Dicranaceae), and encouraged a restudy of familial placement. Eddy (1990) found that *Rhamphidium* "appears to combine Pottioid sporophytic features with gametophytes of more Dicranoid character." According to Hilpert (1933), *Rhamphidium* is better placed in the Ditrichaceae and he suggested a close relationship to *Cheilothela* (Lindb.) Broth. of that family. *Cheilothela chilensis* (Mont.) Broth., in face, has the characteristic sheathing leaf base of *Rhamphidium*. In addition, the closely spirally ridged and often spiculose peristome teeth rising from a low, weakly ornamented and interiorly thickened plate of *Rhamphidium* are closely approached if not matched by those of *Saelania* Lindb. (Ditrichaceae). Although the Ditrichaceae and the Pottiaceae are similar in many gametophyte and peristome traits, the relationship of *Rhamphidium* is clearly with the former family.

New heterotypic synonymy: *Rhamphidium mussuriense* Dix. = *Barbula amplexifolia* (Mitt.) Jaeg.

SEBILLEA

Sebillea Biz., Rev. Bryol. Lichénol. 40: 120, 1974, nom. inval. fide Crosby et al., 1992. Type: *Sebillea brasiliensis* Biz., nom. inval.

Neither the genus and species names were validly published (I.C.B.N., Art. 37 and 42 fide Crosby et al. 1992); the original description merely cited "Type: Herbiers Sébille et Bizot" without indication of holotype. Although the herbarium of Sébille has been incorporated into that of Bizot now at PC, according to H. Bischler (in litt.), the type is not present in PC or at Dijon, and may have been lost with certain other material during transfer of Bizot's herbarium to Paris. The illustration of this taxon is reminiscent of *Rhabdoweisia* (Dicranaceae).

SERPOTORTELLA

Serpotortella Dix., J. Bot. 80: 44, 1942. Type: *Serpotortella madagassa* Dix., Malagassy Republic, sin coll., "Herb. Mus. Brit.," "Ref. No. 15," BM, lectotype nov.

Serpotortella consists of two species of the island of Madagascar and is extraordinary for a number of features otherwise characteristic of many species of *Macromitrium* (Orthotrichaceae). These features (some pointed out by Dixon 1942 in the original description) include: the creeping main axis with distant leaves; erect branches with crowded leaves; leaves crisped when dry, plicate, yellow in KOH, tendency towards a border of elongate cells at the leaf base; upper laminal cells rounded in section (with the appearance of a chain of beads), with columnar, simple to bifid papillae (although rather massive in *Serpotortella*). *Serpotortella* is similar to *Tortella* in the following combination of characters: stem central strand present (absent in the Orthotrichaceae fide Brotherus 1924–25 and cf. Zander & Vitt 1979); costa with one layer of guide cells and a ventral stereid band (*Macromitrium* has two layers of guide cells and ventral stereid band absent or substereid at most); basal cells inflated and sharply differentiated in a vee reaching up the leaf margins. The strongly corrugate upper lamina of one species is reminiscent of that of a variant of *Anoectangium aestivum* (Zander 1976) and of *Macromitrium incrustatifolium* Robins. (Robinson 1968; Gangulee 1969–80) and other *Macromitrium* species, but the cells are not bistratose as they are in these latter species, and the extra cell growth involved may be a malformation. Although *Tortella* species may sometimes send up new shoots from an occasional prostrate shoot, this is always secondary, occurring after damage such as crushing of a turf. *Serpotortella* has a definite prostrate axis with distant leaves somewhat differently shaped than those of the crowded branches. Although *Mitthyridium* Robins. of the Calymperaceae has such a prostrate main axis, the differentiated, sheathing perichaetial leaves and the stem central strand do not occur in the Calymperaceae.

Several characteristics that in combination led W. D. Reese and me to place *Serpotortella* in its own monogeneric family, Serpotortellaceae Reese & Zand. (1987, 1988b), include creeping main stems, with central strand; leaves with only a few porose basal cells; sheathing perichaetial leaves; peristome teeth smooth, with transverse bars, reflexed when dry; gemmae absent. The lectotype should be based on material seen by the original author, and so is changed above from *S. chenagonii* (Ren. & Card.) Reese & Zand.

New heterotypic synonymy: *Serpotortella madagassa* Dix. = *Serpotortella chenagonii* (Ren. & Card.) Reese & Zand., Bryologist 90: 234, 1987 [1988]. *Serpotortella marginata* Dix. = *Serpotortella chenagonii* (Ren. & Card.) Reese & Zand., Bryologist 90: 234, 1987 [1988].

SPLACHNOBRYUM

Splachnobryum C. Müll., Verh. Zool. Bot. Ges. Wien 19: 503, 1869. Lectotype: *Splachnobryum obtusum* (Brid.) C. Müll.

This genus has been removed (A. Koponen 1981) from the Pottiaceae to the Splachnobryaceae A. Kop., Ann. Bot. Fenn. 18: 128, 1981, type: *Splachnobryum* C. Müll. There has been considerable support for this. Andrews (1949) first used the term "Splachnobryaceae" but did not actually accept it as a family. Corley et al. (1981) also felt that *Splachnobryum* "would be best placed in a new family, Splachnobryaceae" but did not take the necessary steps to do so.

TRIDONTIUM

Tridontium Hook. f. in Hook., Icon. Pl. Rar. 3: 148, 1840. Type: *Tridontium tasmanicum* Hook. f., Australia, Tasmania, ex herb. Hooker, isotype, NY.

Plants forming turfs or cushions, greenish brown above, *blackish brown to dark brown below*. Stems branching irregularly, ca. 3.0 cm in length, transverse section rounded-pentagonal, central strand present, strong, sclerodermis present, hyalodermis absent; axillary hairs to 5 cells in length, basal cell brown; occasionally sparsely radiculose. *Leaves* incurved, contorted when dry, spreading when moist, *ligulate-lanceolate*, ca. 3.0 mm in length, *upper lamina broadly channelled, margins plane to erect above, recurved in lower 1/4, entire, ca. 4 rows of enlarged cells forming intramarginal border* in lower 1/3–1/2, border sometimes absent; *apex broadly rounded to occasionally rounded-acute*, somewhat cucullate; base scarcely differentiated to long-elliptical; *costa strong*, flattened, ending ca. 6 cells below apex, *superficial cells elongate on both sides*, ca. 6 rows of cells across costa ventrally at midleaf, costal transverse section semicircular to elliptical, stereid band absent to weak ventrally, present dorsally, epidermis present dorsally and absent ventrally, guide cells 4(–6) in 1 layer, hydroid strand absent; *upper laminal cells irregularly hexagonal, heterogeneous in shape*, ca. 9–13 µm in width, approximately 1:1, *walls often irregularly thickened, superficially flat*, lumens usually round to elliptical; papillae absent; basal cells differentiated across leaf from intramarginal border to costa, rhomboidal to rectangular, ca. 15 µm in width, 3–4:1, walls evenly thickened. Dioicous. Perichaetia terminal, archegonia long, to 1.5 mm, inner perichaetial leaves long-oblong, to 4 mm in length, sheathing seta, cells long-rhomboidal and inflated in lower 1/2. Perigonia terminal, gemmate, paraphyses filiform, of uniseriate cells. Seta ca. 1 cm in length, 1 per perichaetium, dark brown, twisted clockwise above; theca fleshy, ca. 1.2–1.5 mm in length, dark brown, urceolate to short-ovoid, often with a circumstomal ring, exothecial cells short-rectangular, ca. 20 µm in width, 2:1, walls evenly thickened, stomates phaneropore, at base of theca, usually sunken in pits, annulus 4–6 rows of vesiculose, persistent cells; *peristome teeth 16, lanceolate, cleft 2–3 times* halfway or to base, often variously perforate, yellow, low-papillose to densely spiculose, to 450 µm in length, with several articulations, straight, basal membrane absent. Operculum rostrate, oblique, ca. 2 mm in length, cells straight. Calyptra cucullate, smooth, ca. 3.8 mm in length. Spores ca. 25–30 µm in diameter, yellow, obscurely papillose. Laminal KOH color reaction orange.

Found in wet, generally limy areas in New Zealand, Tasmania, and Auckland and Macquarie Islands.

This critical monotypic genus is treated here in detail to give some perspective on its familial relationship.

Clavate propagula in leaf axils were reported for *Tridontium* by Norris and Koponen (1989). Hilpert (1933) emphasized a relationship with the Pleuroweisieae (a tribe whose members are here assigned elsewhere in the Pottiaceae in other subfamilial groups), specifically with *Eucladium* (now in the Trichostomoideae), but a careful comparison of the characters italicized in the discriptions given here demonstrates that the similarity of habitat has probably been given undue weight. Although variation in appearance of the leaf apex is considerable in many taxa of Pottiaceae as correlated with variation in ratio of length and width of the leaf, only a high level of character similarity in other respects or a clear phyletic series of related species can replace utility of this feature in taxonomic evaluations in the family. *Tridontium* has no apparent relationship with *Eucladium*. I agree with Dixon (1923) and Sainsbury (1955) that the marginal border may be practically absent in some specimens (e.g. type of *Weissia lancifolia* (BM, see below), and I also find, as did Sainsbury, that the peristome is not so regularly trifid as figured by Brotherus (1924–25).

Although Dixon appreciated a resemblance to *Cinclidotus* P. Beauv. (Cinclidotaceae fide Saito 1975a), *Tridontium* is usually placed with the Pottiaceae, near *Leptodontium* (note the absence of a costal ventral epidermis, this a characteristic of the *Leptodontium* group), *Erythrophyllopsis* or *Eucladium*. Eddy (1990) noted that *Tridontium* "stands well apart from the other Pottiaceae...." Although (1) the peristome of *Tridontium* is quite unlike the net-like peristomial cone of *Cinclidotus*, (2) the opercular cells are rectangular, not isodiametric as in *Cinclidotus*, and (3) stomates are present, it is because of the fleshy theca, large spores, and above attributes of the gametophyte that there is considerable resemblance to that similarly aquatic genus.

Tridontium has a striking resemblance to *Scouleria* (Grimmiaceae, or Scouleriaceae Churchill in Funk & Brooks, Advances Cladistics 143, 1981, type: *Scouleria* Hook. in Drumm. see Churchill 1985), including the hydric habitat, thick-walled exothecial cells, lanceolate leaves with intramarginal border, rounded hexagonal upper laminal cells, and cucullate calyptra. Character states listed by Murray (1984) for certain other genera of Grimmiaceae (subfam. Coscinodontoideae), *Indusiella*, *Aligrimmia* and *Coscinodontella*, such as macrostomous capsules with cleft or perforate peristome teeth, strong stem central strand, ovate-lanceolate leaves with obtuse apices, leaves with involute upper margins and leaf cells with angular lumina, also occur in *Tridontium*. These taxa of the Grimmiaceae are distinct, however, in the plicate calyptrae (cucullate calyptrae are rare in Grimmiaceae) and monoicous sexuality. It is quite possible that both *Tridontium* and *Cinclidotus* are taxa bridging the Pottiaceae and the Grimmiaceae.

Tridontium is here placed in the Grimmiaceae in the Scoulerioideae pending some future evaluation of the relationship of the two families.

Number of accepted species: 1.

Species examined: *T. tasmanicum* (ALTA, BM, BUF, DUKE, MO, NY).

New heterotypic synonymy: *Weissia lancifolia* (R. Br. ter) Wijk & Marg. (*Dicranum lancifolium* R. Br. ter) = *Tridontium tasmanicum* Hook. f. in Hook.

TISSERANTIELLA

Tisserantiella P. Varde, Bull. Soc. Bot. France 88: 469, 1941. Type: *Tisserantiella spathulata* P. Varde, Cameroon, LeMauf, 1937, holotype, PC.

Macroglossum Hilp., Beih. Bot. Centralbl. 50(2): 670, 1933, nom. illeg. non Copeland, 1909. Type: *Macroglossum pulchellum* Thér. & Hilp., Congo, Overlaet, 1923, isotype, CU.

Potier de la Varde (1941) in the original description of this genus noted its similarity to papillose species of *Uleastrum* Buck (as *Ulea* C. Müll.). *Tisserantiella* lacks a peristome but nonetheless clearly belongs to the Rhachitheciaceae by the arborescent substrate, presence of a stem central strand, sclerodermis absent; leaves spathulate with costa ending 15–30 cells below apex; laminal papillae very small, simple, solid; basal cells much differentiated, elongate, smooth; seta rather short, 0.6–1.0 mm in length; annulus strongly vesiculose; and laminal KOH color reaction yellow. The type of *Weissia minutissima* (NY, see below) from Brazil is also a member of this genus, differing in the longer costa.

The family Rhachitheciaceae Robins. is distinguished from the Pottiaceae (emending Robinson's 1964 description, also cf. Iwatsuki & Sharp 1976 and Allen & Pursell 1991) by the following combination of characters: small corticolous plants usually of low elevations in tropical areas; without stem sclerodermis, small or absent central strand; leaves ligulate, upper laminal cells often with solid or hollow papillae, basal cells inflated or much elongate, smooth; costa usually ending several cells below the apex, often reaching only to 2/3 or 3/4 the leaf length; perichaetial leaves usually sheathing; seta stout, often kinked just below the capsule; theca ovoid, smooth or 8–ribbed; peristome teeth seldom absent, spreading when wet, strongly incurved when dry, 16 and paired, smooth, articulations generally close; calyptra cucullate, smooth; and KOH laminal color reaction yellow. Multi-ribbed thecae are also found in the Pottiaceae in *Ganguleea angulosa* and in *Weisiopsis plicata*. The genera of the Rhachitheciaceae are an apparent link between the Pottiaceae and the Orthotrichaceae. Three genera are here recognized in the Rhachitheciaceae: *Hypnodontopsis* Iwats. & Nog., *Rhachithecium* Broth. ex Le Jolis, and *Tisserantiella*. *Jonesiobryum* Biz. & Pocs ex Allen & Pursell apparently belongs in this family, too (cf. Allen & Pursell 1991).

New heterotypic synonymy: *Tisserantiella spathulata* P. Varde = *Tisserantiella pulchellum* (Thér. & Hilp.) Zand.

New combinations in *Tisserantiella*: *Tisserantiella pulchella* (Thér. & Hilp.) Zand., comb. nov. (*Macroglossum pulchellum* Thér. & Hilp., Beih. Bot. Centralbl. 50(2): 671, 1933; *Hyophila pulchella* (Thér. & Hilp.) Wijk & Marg.). *Tisserantiella minutissima* (Mitt.) Zand., comb. nov. (*Weissia minutissima* Mitt., J. Linn. Soc. Bot. 12: 138, 1869; *Hyophila minutissima* (Mitt.) Jaeg.; *Macroglossum minutissimum* (Mitt.) Hilp.).

New combination in *Rhachithecium*: *Rhachithecium welwitchii* (Duby) Zand., comb. nov. (*Zygodon welwitchii* Duby, Mem. Soc. Phys. Hist. Nat. Genève 21: 44, 1871; *Ulea welwitschii* (Duby) Broth. in Par.).

ULEASTRUM

Uleastrum Buck, Candollea 40: 203, 1985. Type: *Ulea palmicola* C. Müll.

Ulea C. Müll., Hedwigia 36: 102, 1897, hom. illeg. non Schroeter, 1892. Type: *Ulea palmicola* C. Müll.

Spruceella C. Müll., Gen. Musc. Fr. 396, 1900, hom. illeg. non Pierre, 1890. Type: *Spruceella octoblepharis* (Jaeg.) C. Müll.

Except for the sharply differentiated basal cells, the combination of the the ligulate leaves with acute apices, smooth bistratose upper laminal cells and other gametophyte characters are much like *Ptychomitrium incurvum* (Schwaegr.) Spruce of the Ptychomitriaceae. *Uleastrum* differs from other genera of the Ptychomitriaceae, however, in the cucullate calyptra and smooth peristome teeth. Chen (1941: 39) suggested that this genus is closely related to *Rhachithecium* and there is indeed some evidence of relationship. The costa is relatively weak and generally ends below the leaf apex. *Uleastrum palmicola*, the generitype, has propagula borne on the dorsal side of the basal cells, as has *Rhachithecium perpusillum*. *Uleastrum octoblepharis* has a kinked seta similar to that of species of *Rhachithecium* and *Hypnodontopsis*, but other species of *Uleastrum* lack this, and, in fact, probably should belong in another genus. Pending further study, *Uleastrum* can be recognized in the Orthotrichaceae, a disposition making the least necessary emendation of recognized family limits.

Species examined: *U. octoblephare* (BUF, H, NY), *U. nitidum* (PC), *U. palmicola* (H, NY), *U. paraguense* (H).

New heterotypic synonymy: *Zygodon palmarum* C. Müll. (isotype H!) = *Uleastrum palmicola* (C. Müll.) Zand.

New combinations: *Uleastrum nitidum* (Thér. in Felipp.) Zand., comb. nov. (*Ulea nitida* Thér. in Felipp., Rev. Bryol. n. ser. 2: 214, 1930). *Uleastrum octoblephare* (Spruce ex Jaeg.) Zand., comb. nov. (*Pottia octoblepharis* Spruce ex Jaeg., Ber. S. Gall. Naturw. Ges. 1871–72: 343, 1873 (Ad. 1: 191); *Weissia octoblepharis* Spruce ex Mitt., J. Linn. Soc. Bot. 12: 40, 1869, hom. illeg.; *Ulea octoblepharis* (Spruce ex Jaeg.) C. Müll., Hedwigia 37: 234, 1898). *Uleastrum palmicola* (C. Müll.) Zand., comb. nov. (*Ulea palmicola* C. Müll., Hedwigia 36: 102, 1897).

ULEOPSIS

Uleopsis Thér., Rev. Bryol. Lichénol. 9: 20, 1936. Type: *Uleopsis mamillosa* Thér..

The type of this species (Ecuador, rochers du Condorguachana, Benoist 3154, PC) is *Oreoweisia brasiliensis* Hampe (sensu Griffin 1986a). *Uleopsis* Thér. thus becomes a synonym of *Oreoweisia* De Not.

SPECIES

Tortula montana Mitt., J. Linnean Soc. Bot. 12: 156, 1869 (*Didymodon montanus* (Mitt.) Broth.) = *Rhamphidium montanus* (Mitt.) Zand., comb. nov.

Gymnostomum lessonii Besch. = *Racopilum* sp. (Racopilaceae).

Phascum carinatum Hampe, Vid. Medd. Naturh. For. Kjoebenh. ser. 4, 1: 76, 1879 = *Bruchia carinata* (Hampe) Zand., comb. nov. The type (BM) of this Brazilian species consists of (a very few) plants with the characteristic enlarged capsule neck and rectangular laminal cells of *Bruchia* (Bruchiaceae).

Pottia macrocarpa Schimp., Ann. Sc. Nat. Bot. ser. 2, 6: 145, 1836 = *Funaria macrocarpa* (Schimp.) Zand., comb. nov. A fragmentary type at NY is apparently a *Funaria* by the large, eperistomate capsule, subpercurrent costa, bluntly dentate upper leaf margins, lax upper laminal cells, and yellow KOH reaction.

Pottia mirabilis Broth. & Par., Rev. Bryol. 31: 118, 1904 = *Physcomitrium mirabile* (Broth. & Par.) Zand., comb. nov. The gametophytes of the type at H are soft, glossy and otherwise funariaceous or bryaceous in general appearance. The specimen keys (Brotherus 1924–25) to *Physcomitrium* sect. *Cryptopyxis* C. Müll. (Funariaceae), and a new combination in *Physcomitrium* is appropriate here.

Tortula domingensis Thér., Rev. Bryol. Lichénol. 14: 12, 1944 = *Brachymenium domingense* (Thér.) Zand., comb. nov. Isotypes at NY and US have the characters of *Brachymenium* Schwaegr. (Bryaceae). One may especially note the dense red tomentum, lack of differentiated basal laminal cells, and upright capsule with an apparently bryaceous peristome (hyaline and poorly developed in the rather young capsules). gametophytes of this species are similar to those of *Tortula* sect. *Pottia* but are more densely yellow in KOH solution. The operculum is large for the genus, being stoutly long-conic.

AN UPDATED LIST OF GENERA, SPECIES AND INFRASPECIFIC TAXA OF THE POTTIACEAE

This is a list of presently recognised genera and specific and infraspecific taxa and their distributions, as an extension to Index Muscorum and its first supplement (Van der Wijk et al. 1959–1969), and later supplements (Crosby 1977, 1979; Crosby & Bauer 1981, 1983, 1986; Crosby, Magill & Bauer 1992). Other supraspecific taxa and thier synonyms are given in the family and generic treatments. Only italicized (correct) names were extracted from the Index Muscorum for inclusion here—this list does not provide all synonymy; included is new synonymy made since publication of that work, plus citations of formae, largely from Podpěra (1954). Synonyms with which the author disagrees or which are superseded are placed in square brackets.

Series of new combinations at the species level that would put all epithets into some other genus were usually not made. Combinations at the species level were made only after seeing authentic material or illustrations of authentic material that show distinctive characters. For example, all epithets of correct names in *Astomum* and *Hymenostomum* (now in synonymy with *Weissia*) were not transferred to *Weissia* because, although the generitypes belong in *Weissia*, many species probably belong with *Trichostomum*. Even the Index Muscorum has many "orphaned" specific and infraspecific epithets languishing in combinations with otherwise synonymized generic and specific names. Revisionists are encouraged to examine types of these names and make the appropriate combinations or synonymy. Complete series of new combinations for many infraspecific epithets were made, however, with the assumption that their relationship with the typical variety is secure.

In the past, I have used varietal names for widely distributed taxa (e.g. segregates of *Didymodon australasiae s. lat.*, *D. rigidulus s. lat.*, *D. vinealis s. lat.*, *Leptodontium viticulosoides s. lat.*) that intergrade in morphology in major parts of their geographic range (for further discussion see Zander 1981c). Since there is apparently a resistence of the part of other bryologists to using these admittedly ungainly three-part names for widely distributed taxa, and because in some parts of the range of these species they apparently do not intergrade, the names of such taxa are given here as correct at the species level. Widespread infraspecific taxa of *D. rigidulus s. lat.* are excepted since the abundant variation in these populations precludes morphological sorting of such names so as to assign types to even vaguely circumscribed species limits. This argument is also true for weakly distinguishable local populations or "facies" of New World *Anoectangium aestivum* (see Zander 1977c), which are not recognized as separate taxa.

In accordance with Art. 73.9 of the 1987 I.C.B.N. (Greuter 1988), hyphens are eliminated from all epithets unless the connected words may stand independently. Also included here are the names of hybrid formulas, with authors given for each name and the reporting author after "*fide*"; the usage of the Index Muscorum is apparently not in consonance with the Code. The abbreviation used here for rank not cited (see Art. 35.1; also W. Margadant pers. comm.) is "*nom. inval. dispon. non cit.*"

The floristic distribution codes of the Index Muscorum were updated at the correct name whenever new synonymy was added, but, inasmuch as the taxonomy of the Pottiaceae was the priority for the time allotted to complete this study, no general survey of the floristics literature since the publication of Index Muscorum was made to update all distributions. Afr1 includes North Africa, Madeira, Azores, Canary Islands; Afr2 includes Central Africa, St. Helena; Afr3 includes Madagascar, Mauritius, Réunion; Afr4 includes South Africa, Kerguelen; Am1 includes North America, Greenland, Aleutians, Bermuda; Am2 is Central America; Am3 is the West Indies; Am4 includes Venezuela, Colombia, Peru, Bolivia, Ecuador, Galapagos; Am5 includes Brazil, Paraguay, Guiana, Trinidad, Tobago; Am6 includes Chili, Argentina, Uruguay, Falklands; Ant is Antarctica; As1 includes Northern Asia, Sakhalin; As2 includes China, Mongolia, Japan, Korea, Formosa; As3 includes India, Pakistan, Ceylon, Burma, Thailand, Indochina; As4 includes Indonesia, Malaya, Philippines, New Guinea; As5 includes Asiatic portion of Middle East, Cyprus; Austr1 includes Australia, Tasmania; Austr2 is New Zealand and nearby islands; Oc includes the Pacific islands.

This list recognizes as correct names in the Pottiaceae a total of 1457 species, 31 subspecies, 536 varieties, 339 formae and 7 subformae.

ACAULON C. Müll.

Acaulon capense C. Müll. [see Magill, Fl. S. Afr. I. Mosses 1: 201, 1981 (1982)] Afr4

Acaulon casasianum Brugués & Crum, Lindbergia 10: 1, 1984 Eur

Acaulon chrysacanthum Stone, J. Bryol. 9: 213, 1976 [1977] Austr1

Acaulon crassinervium C. Müll. Austr1 Austr2

Acaulon dertosense Casas, Sérgio, Cros & Brugués, Anales Jard. Bot. Madrid 42: 299, 1986 Eur

Acaulon eremicola Stone, J. Bryol. 10: 467–474. 1979 [1980] Austr1

Acaulon fontiquerianum Casas & Sérgio, Cryptogamie Bryol. Lichénol. 11: 61, 1990 Eur

Acaulon granulosum Stone, J. Bryol. 15: 257. Austr1

Acaulon integrifolium C. Müll. Austr1

var. *aristatum* (Willis) Willis Austr1

Acaulon leucochaete Stone, J. Bryol. 9: 217, 1976 [1977] Afr4 Austr1

[*Acaulon mediterraneum* Limpr. Eur ≡ *Acaulon muticum* var. *mediterraneum* (Limpr.) Sérgio *fide* Sérgio, Bol. Soc. Brot. 46: 460, 1972]

[*Acaulon minus* (Hook. & Tayl.) Jaeg. Eur Afr1 ≡ *Acaulon muticum* (Hedw.) C. Müll. *fide* Hill, J. Bryol. 12: 11, 1982]

Acaulon muticum (Hedw.) C. Müll. Eur Am1 As1 As5 Afr1 Afr2 Afr4

var. *mediterraneum* (Limpr.) Sérgio *fide* Sérgio, Bol. Soc. Brot. 46: 460, 1972

[var. *minus* (Hook. & Tayl.) B.&S. in BSG Eur Afr1 = *Acaulon muticum* (Hedw.) C. Müll. *fide* Hill, J. Bryol. 12: 11, 1982]

var. *rufescens* (Jaeg.) Crum, Bryologist 72: 240, 1969 Am1

fo. *subintegrifolium* C. Jens., Skand. Bladmfl. 216, 1939 *nom. inval. descr. suec.*

Acaulon nanum C. Müll. Am5

[*Acaulon piligerum* (De Not.) Limpr. Eur = *Acaulon triquetrum* (Spruce) C. Müll. *fide* Corley et al., J. Bryol. 11: 621, 1981

(1982)]
Acaulon recurvatum Magill, Fl. S. Afr. I. Mosses 1: 199, 1981 [1982] Afr4
Acaulon robustum Broth. ex Roth Austr1
[*Acaulon rufescens* Jaeg. Am1 ≡ *Acaulon muticum* var. *rufescens* (Jaeg.) Crum *fide* Crum, Steere & Anderson 1973]
[*Acaulon rufochaete* Magill, Fl. S. Afr. I. Mosses 1: 201, 1981 [1982] Afr4 ≡ *Microbryum rufochaete* (Magill) Zand., see treatment of *Microbryum*]
Acaulon schimperianum (Sull.) Sull. in Sull. & Lesq. Am1 Am2
Acaulon sphaericum Shaw (Lsee Magill, Fl. S. Afr. I. Mosses 1: 201, 1981 [1982]) Afr4
Acaulon triquetrum (Spruce) C. Müll. Eur Afr1 Am1 As1 Austr1
var. *desertorum* (Besch.) Jelenc Afr1
Acaulon uleanum C. Müll. Am5
Acaulon vesiculosum C. Müll. Am6

ALOINA Kindb.
Aloina aloides (Schultz) Kindb. Eur As1 As2 As5 Afr1 Am1
var. *ambigua* (BSG) Craig in Grout Eur As1 As2 As5 Afr1 Am1 Am2 Austr1
[*Aloina ambigua* (BSG) Limpr. ≡ *Aloina rigida* var. *ambigua* (BSG) Craig]
fo. *microphylla* (Latz.) Podp., Consp. Musc. Eur. 238, 1954 (*Aloina ericaefolia* fo.) Eur
Aloina apiculata (Bartr) Delgad. Am3
Aloina bifrons (De Not.) Delgad. Eur As1 As2 As5 Afr1 Afr4 Am1 Am2 Am3
Aloina brevirostris (Hook. & Grev.) Kindb. Eur As1 As2 As5 Am1
[var. *breidleri* (Limpr.) Limpr. Eur = *Aloina brevirostris* (Hook. & Grev.) Kindb. *fide* Delgadillo, 1973]
[var. *elongata* (Ångstr.) Par. Eur = *Aloina brevirostris* (Hook. & Grev.) Kindb. *fide* Delgadillo, 1973]
[var. *rotundifolia* Warnst. Eur = *Aloina brevirostris* (Hook. & Grev.) Kindb. *fide* Delgadillo, 1973]
[*Aloina calceolifolia* (Mitt.) Broth. Am2 Am4 = *Aloina rigida* (Hedw.) Limpr. var. *rigida fide* Delgadillo, 1975]
Aloina catillum (C. Müll.) Broth. Am2 Am4
Aloina cornifolia Delgad., Bryologist 78: 265, 1975 As2
[*Aloina ericaefolia* (Lindb.) Kindb. Eur = *Aloina rigida* var. *ambigua* (BSG) Craig]
[fo. *microphylla* Latz., Bot. Centralbl. Beih. 48(2): 484, 1931 ≡ *Aloina ambigua* fo. *microphylla* (Latz.) Podp., Consp. Musc. Eur. 238, 1954]
Aloina hamulus (C. Müll.) Broth. Am1 Am2
[*Aloina longirostris* Card. *hom. illeg.* Am2 = *Aloina hamulus* (C. Müll.) Broth. *fide* Delgadillo 1973]
[*Aloina obliquifolia* (C. Müll.) Broth. As2 ≡ *Aloina rigida* var. *obliquifolia* (C. Müll.) Delgad.]
[*Aloina pilifera* (De Not.) Crum & Steere *nom. superfl.* Eur As5 Am1 Am2 = *Aloina bifrons* (De Not.) Delgad. see Corley et al., J. Bryol. 11: 620, 1981 (1982)]
Aloina recurvipatula (C. Müll.) Broth. Am6
Aloina rigida (Hedw.) Limpr. Eur Afr1 Afr4 Am1 Am2 As1 As2 As3 As5 Austr1
[var. *ambigua* (BSG) Craig Eur As1 As2 As5 Afr1 Am1 ≡ *Aloina aloides* var. *ambigua* (BSG) Craig in Grout *fide* Crum, Steere & Anderson 1973]
[var. *longirostris* (Torka) Podp. Eur = *Aloina rigida* (Hedw.) Limpr. *fide* Delgadillo, Bryologist 78: 160, 1975]
var. *mucronulata* (BSG) Limpr. Eur (see Delgadillo, Bryologist 78: 262, 1975
var. *obliquifolia* (C. Müll.) Delgad., Bryologist 78: 264, 1975 (*Barbula*) As2
fo. *mucronata* Mönk., Laubm. Eur. 317, 1927 Eur
[fo. *mucronulata* (BSG) Podp., Consp. Musc. Eur. 237, 1954 (*Tortula rigida* var.) Eur ≡ *Aloina rigida* fo. *mucronata* Mönk.]
fo. *obtusa* (Jur.) Mönk., Laubm. Eur. 317, 1927 (*Tortula rigida* var.) Eur
[fo. *pilifera* (De Not.) Mönk., 317, 1927 (*Tortula rigida* var.) Eur As5 ≡ *Aloina pilifera* (De Not.) Crum & Steere *nom. superfl.* = *Aloina bifrons* (De Not.) Delgad. see Corley et al., J. Bryol. 11: 620, 1981 (1982)]
Aloina rosei (Williams) Delgad., Bryologist 76: 273, 1973 Am1 Am4
Aloina sedifolia (C. Müll.) Broth. Am6
Aloina sullivaniana (C. Müll.) Broth. Austr1 Austr2 [= *Aloina bifrons fide* Delgadillo, 1973, but see Scott & Stone, Mo. Southern Australia 454, 1976 and *fide* Catcheside, Mosses S. Austr. 155, 1980]

ALOINELLA Card.
Aloinella andina Delgad. Am4
[*Aloinella apiculata* Bartr. Am3 ≡ *Aloina apiculata* (Bartr.) Delgad.]
Aloinella boliviana Broth. in Herz. Am4
Aloinella catenula Card. Am2 Am4
Aloinella cucullatifolia (C. Müll.) Broth. Am6
Aloinella cucullifera (Mitt.) Steere Am4
Aloinella galeata (C. Müll.) Broth. Am6
[*Aloinella hamulus* (C. Müll.) Bartr. Am2 ≡ *Aloina hamulus* (C. Müll.) Broth.]
Aloinella venezuelana Griffin, Bull. Torrey Bot. Club 102: 26, 1975 Am4

[**ANICTANGIUM** Hedw. *nom. rejic.* ≡ *Anoectangium* Schwaegr. *nom. cons.*]
[*Anictangium ciliatum* Hedw. ≡ *Hedwigia ciliata* (Hedw.) P. Beauv.]
var. *incanum* Sw. Eur
Anictangium orthotrichoides Gill. ex Grev. Am6

ANOECTANGIUM Schwaegr.
Anoectangium abyssinicum Hampe ex Geh. Afr2
Anoectangium aestivum (Hedw.) Mitt. Eur As2 As3 As4 Afr1 Afr2 Am1 Am2 Am4 Am5 Austr2 Oc
var. *glaciale* (Lor. & Mol.) Wijk & Marg. Eur
var. *pellucidum* (Wils.) Braithw. Eur.
Anoectangium afrocompactum C. Müll. ex Dus. Afr1
[*Anoectangium angustifolium* Mitt. Afr1 = *Anoectangium aestivum* (Hedw.) Mitt. *fide* Dirkse et al., Cryptogamie Bryol. Lichénol. 14: 15, 1993]
[*Anoectangium anomalum* Bartr. As4 = *Hymenostylium recurvirostrum* (Hedw.) Dix. *fide* Norris & Koponen, Acta Bot. Fenn. 137: 102, 1989]
[var. *trifarium* Bartr., Lloydia 5: 255, 1942 As4 = *Hymenostylium recurvirostrum* (Hedw.) Dix. *fide* Norris & Koponen, Acta Bot. Fenn. 137: 102, 1989]
[*Anoectangium apiculatum* Schimp. in Besch. Am2 = *Anoectangium aestivum* (Hedw.) Mitt. *fide* Zander, Bryologist 80: 243, 1977]
[*Anoectangium arizonicum* Bartr. ex Grout Am1 Am2 = *Gymnostomum aeruginosum* Sm. *fide* Zander, Bryologist 80: 259, 1977]
[*Anoectangium balfourei* Mitt. Afr2 = *Semibarbula orientalis* (Web.) Wijk & Marg. *fide* Frey & Kürschner, Nova Hedw. 46: 94, 1988 = *Barbula indica* (Hook.) Spreng]

Anoectangium bellii Broth. ex Dix. Austr1 Austr2
Anoectangium bicolor Ren. & Card. As3
Anoectangium borbonense Besch. Afr3
Anoectangium brachyphyllum Broth. in Herz. As4
[*Anoectangium breutelianum* BSG ex Besch. Am2 = ?*Hymenostylium recurvirostrum* (Hedw.) Dix. *fide* Zander, Bryologist 80: 265, 1977]
Anoectangium brotherusii Kis, Mosses South-east Trop. Afr., Inst. Ecol. Bot. Hungarian Acad. Sci. 46, 1985 (*nom. nov.* for *Anoectangium torquatum* Broth. *hom. illeg.*) Afr2
[*Anoectangium calidum* Mitt. Am4 = *Anoectangium aestivum* (Hedw.) Mitt. *fide* Zander, Bryologist 80: 243, 1977]
Anoectangium clarum Mitt. As2 As3
[*Anoectangium compactum* Schwaegr. = *Anoectangium aestivum* (Hedw.) Mitt.]
[var. *alaskanum* Card. & Thér. Am1 = *Anoectangium aestivum* (Hedw.) Mitt., see treatment of *Anoectangium*]
var. *madeirense* Geh. in Geh. & Herz. Afr1
Anoectangium contortum Broth. As1
Anoectangium crassinervium Mitt. As2
[*Anoectangium crustatum* J. Fröhl., Ann. Naturh. Mus. Wien 66: 35–36, 1962 (1963) Eur = *Gymnostomum viridulum* Brid. *fide* Corley & Crundwell, J. Bryol. 16: 346, 1991]
[*Anoectangium handelii* Schiffn. As5 = *Molendoa sendtneriana* (BSG) Limpr. *fide* Zander, Bryologist 80: 248, 1977]
Anoectangium hanningtonii Mitt. Afr2
Anoectangium harttiae Bartr. Oc
Anoectangium herzogii Broth. in Herz. Am4
Anoectangium hobsonii Mitt. As5
Anoectangium humblotii Ren. & Card. in Ren. Afr3
Anoectangium hymenodontoides (C. Müll.) Jaeg. As3
Anoectangium impressum Hampe Afr3
[*Anoectangium incrassatum* Broth. in Boerg. Am3 = *Anoectangium aestivum* (Hedw.) Mitt. *fide* Zander, Bryologist 80: 243, 1977]
[*Anoectangium incurvans* (Schimp. ex Besch.) Bartr. Am2 = *Molendoa sendtneriana* (BSG) Limpr. *fide* Zander, Bryologist 80: 248, 1977]
[*Anoectangium jamaicense* (C. Müll.) Par. Am3 = *Anoectangium aestivum* (Hedw.) Mitt. *fide* Zander, Bryologist 80: 243, 1977]
Anoectangium kashmiriense Aziz & Vohra, Bull. Bot. Surv. India 35: 239, 1983 As3
Anoectangium keniae (P. Varde) Zand. (*Gymnostomum*), see treatment of *Anoectangium* Afr2
[*Anoectangium kilimandscharicum* Broth. *hom. illeg.*]
var. *minutum* Broth. Afr2
[*Anoectangium laetevirens* Besch. & Card. As2 = *Anoectangium thomsonii* Mitt. *fide* Saito, J. Hattori Bot. Lab. 39: 458, 1975]
[*Anoectangium laetum* Ren. & Card. = *Anoectangium stracheyanum* Mitt.]
fo. *henryi* Tix., Rev. Bryol. Lichénol. 34: 132, 1966 As3
[*Anoectangium lechlerianum* Mitt. Am4 = *Molendoa sendtneriana* (BSG) Limpr. *fide* Zander, Bryologist 80: 247, 1977]
var. *laetius* Hampe Am4
var. *limbatulum* Bartr. in Bauer Am6
[*Anoectangium liebmannii* Schimpr. ex Besch. Am2 = *Anoectangium aestivum* (Hedw.) Mitt. *fide* Zander, Bryologist 80: 243, 1977]
[var. *viride* Card. Am2 = *Molendoa sendtneriana* (BSG) Limpr. *fide* Zander, Bryologist 80: 247, 1977]
Anoectangium lineare (C. Müll.) Kindb. Am4 Am6
[*Anoectangium lombokense* Broth. As4 = *Anoectangium aestivum* (Hedw.) Mitt. *fide* Touw, J. Hattori Bot. Lab. 71: 341, 1992]
Anoectangium madagassum Ren. & Par. Afr3
Anoectangium mafatense Ren. & Card. Afr2 Afr3
Anoectangium magnirete Ren. & Card. Afr2 Afr3
Anoectangium microphyllum Card. As2
Anoectangium nigerianum Broth. & Par. Afr2
Anoectangium papuanum Fleisch. As4
Anoectangium patagonicum Card. & Broth. Am6
[*Anoectangium peckii* (Sull.) Sull. ex Aust. Am1 = *Anoectangium aestivum* (Hedw.) Mitt. *fide* Crum, Steere & Anders., Bryologist 76: 109, 1973, see also Zander, Bryologist 80: 243, 1977]
[*Anoectangium peruvianum* Sull. Am4 = *Molendoa sendtneriana* (BSG) Limpr. *fide* Zander, Bryologist 80: 247, 1977]
Anoectangium pflanzii Broth. Am4
[*Anoectangium platyphyllum* Williams Am4 ≡ *Molendoa platyphyllum* (Williams) Zand., see treatment of *Molendoa*]
Anoectangium pleuroweisioides J. Frölich, Ann. Naturhist. Mus. Wien 67: 151, 1964 As3
Anoectangium rhaphidostegium C. Müll. ex Broth. Afr3
Anoectangium rivale Card. As2
[*Anoectangium rubrigemmium* Hoe & Crum Oc ≡ *Zygodon rubrigemmius* (Hoe & Crum) Zand. & Vitt., Canad. J. Bot. 57: 296, 1979]
Anoectangium rufoviride Besch. As2
var. *eucollum* Besch. Afr3
Anoectangium schimperi Mitt. Afr2
Anoectangium sellae Negri Afr2
Anoectangium shepherdae (Card. & Dix.) Zand. (*Hymenostylium*), see treatment of *Anoectangium* As3
Anoectangium sikkimense Aziz & Vohra, Bull. Bot. Surv. India 30: 185, 1988 [1990] As 3
Anoectangium spathulatum Mitt. Afr2
Anoectangium stracheyanum Mitt. As2 As3
[var. *gymnostomoides* (Broth. & Yas.) Wijk & Marg. As2 = *Anoectangium aestivum* (Hedw.) Mitt. *fide* Saito, J. Hattori Bot. Lab. 39: 457, 1975]
[*Anoectangium sublaetevirens* Card. As2 = *Anoectangium thomsonii* Mitt. *fide* Saito, J. Hattori Bot. Lab. 39: 458, 1975]
[*Anoectangium taeniatifolium* (Herz.) Hill, J. Bryol. 11: 600, 1981 (1982) ≡ *Molendoa taeniatifolia* Herz.]
Anoectangium tapes Besch. Oc
Anoectangium tasmanicum Broth. Austr1
[*Anoectangium termale* Card. As2 = *Gymnostomum aeruginosum* Sm. *fide* Saito, J. Hattori Bot. Lab. 39: 450, 1975, as "*thermale*"]
Anoectangium thomsonii Mitt. As 2 As3
[*Anoectangium torquatum* Broth. *hom. illeg.* Afr2 ≡ *Anoectangium brotherusii* Kis, Mosses South-east Trop. Afr., Inst. Ecol. Bot. Hungarian Acad. Sci. 46, 1985]
Anoectangium walkeri Broth. As3
[*Anoectangium warburgii* Crundw. & Hill, J. Bryol. 9: 435, 1977 [1978] Eur ≡ *Molendoa warburgii* (Crundw. & Hill) Zand., see treatment of *Molendoa*]
Anoectangium weisioides C. Müll. Am4
Anoectangium wilmsianum (C. Müll.) Par. Afr4

ASCHISMA Lindb.
[*Aschisma aethiopicum* (Welw. & *dub.*) Lindb. Afr2 ≡ *Byroceuthospora aethiopica* (Welw. & *dub.*) Zand., see treatment of *Byroceuthospora*]
Aschisma carniolicum (Web. & Mohr) Lindb. Eur Afr1
var. *speciosum* Limpr. Eur
Aschisma kansanum Andrews Am1
[*Aschisma occultum* Roth Am3 ≡ *Uleobryum occultum* (Roth) Zand., see treatment of *Uleobryum*]

[**ASTOMUM** Hampe = *Weissia* Hedw. *fide* Saito, J. Hattori Bot. Lab. 39: 417, 1975]

[*Astomum abbreviatum* (Thwait. & Mitt.) Fleisch. As3 As4 ≡ *Weissia abbreviata* (Thwait. & Mitt.) Zand., see treatment of *Weissia*]

[*Astomum acuminatum* Dix. & Thér. As2 = *Weissia crispa* (Hedw.) Mitt. *fide* Saito, J. Hattori Bot. Lab. 39: 418, 1975 ≡ *Weissia longifolia* Mitt. *fide* Crundwell & Nyholm, J. Bryol. 7: 13, 1972]

Astomum alternifolium Spruce *hom. illeg.* Eur

[*Astomum austrocrispum* (Beckett) Broth. Austr1 Austr2 ≡ *Weissia austrocrispa* (Beckett) Stone, J. Bryol. 11: 236, 1980 ≡ *Trichostomum austrocrispum* (Beck.) Zand., see treatment of *Trichostomum*]

[var. *longifolium* (R. Br. ter) Dix. Austr2 ≡ *Trichostomum austrocrispum* var. *longifolium* (R. Br. ter) Zand., see treatment of *Trichostomum*]

[*Astomum borbonicum* Biz. & Onraedt in Biz., Rev. Bryol. Lichénol. 40: 116, 1974 *nom. inval. holoty. non cit.* Afr3 ≡ *Astomum borbonicum* Biz. & Onraedt ex Onraedt, Bull. Jard. Bot. Natl. Belgique 46: 356, 1976]

Astomum borbonicum Biz. & Onraedt ex Onraedt, Bull. Jard. Bot. Natl. Belgique 46: 356, 1976 Afr3

[*Astomum brisbanicum* (C. Müll.) Broth. ≡ *Trachycarpidium brisbanicum* (C. Müll.) Stone]

[*Astomum chilense* Williams Am4 ≡ *Trichostomum williamsii* Zand. *nom. nov.*, see treatment of *Trichostomum*]

[*Astomum crispum* (Hedw.) Hampe Eur As1 As2 As3 As5 Afr1 ≡ *Weissia crispa* (Hedw.) Mitt. ≡ *Weissia longifolia* Mitt. *fide* Crundwell & Nyholm, J. Bryol. 7: 13, 1972]

[var. *aciculatum* (Mitt.) Podp. Eur ≡ *Weissia crispa* var. *aciculata* (Mitt.) Dix. = *Weissia longifolia* Mitt. *fide* Crundwell & Nyholm, J. Bryol. 7: 13, 1972]

[var. *angustifolium* Baumg. in Ginzb. Eur ≡ *Weissia longifolia* var. *angustifolia* (Baumg. in Ginzb.) Crundw. & Nyh., J. Bryol. 7: 14, 1972]

[var. *brevifolium* Card. & Copp. Eur = *Weissia longifolia* var. *angustifolia* (Baumg. in Ginzb.) Crundw. & Nyholm, J. Bryol. 7: 14, 1972]

var. *exsertum* Nog. As3

[var. *laubacense* (Roth) Podp. Eur = *Weissia longifolia* Mitt. *fide* Crundwell & Nyholm, J. Bryol. 7: 13, 1972]

[var. *philibertii* (Husn.) Wijk & Marg. Eur Afr1 = *Weissia levieri* (Limpr.) Kindb. *fide* Crundwell & Nyholm, J. Bryol. 7: 16, 1972]

[var. *sterile* (Nich.) Moenk. Eur ≡ *Weissia crispa* subsp. *sterilis* (Nich.) Dix. ≡ *Weissia sterilis* Nich. *fide* Crundwell & Nyholm, J. Bryol. 7: 11, 1972]

[fo. *major-planifolia* Breidl. in Limpr., Laubm. Deutsch. 3: 638, 1901 Eur = *Astomum crispum* var. *philibertii* (Husn.) Wijk & Marg. *fide* Podpěra, Consp. Musc. Eur. 184, 1954 = *Weissia levieri* (Limpr.) Kindb. *fide* Crundwell & Nyholm, J. Bryol. 7: 16, 1972]

Astomum crispum (Hedw.) Hampe × *Astomum crispata* (Nees. & Hornsch.) C. Müll. *fide* Nich., Rev. Bryol. 32: 20, 1905 Eur

Astomum crispum (Hedw.) Hampe × *Astomum microstoma* Hornsch. ex Nees & Hornsch. *fide* Nich., Rev. Bryol. 33: 1, 1906 Eur

[*Astomum cryptocarpum* Broth. Am5 ≡ *Tortella cryptocarpa* (Broth.) Zand., see treatment of *Tortella*]

[*Astomum fruchartii* (C. Müll.) Broth. Am5 Am6 ≡ *Tortella fruchartii* (C. Müll.) Zand., see treatment of *Tortella*]

Astomum japonicum Roth. As2

[*Astomum latifolium* Broth. in Roth Am5 = *Tortella fruchartii* (C. Müll.) Zand., see treatment of *Tortella*]

[*Astomum lindigii* (Hampe) Jaeg. Am4 ≡ *Trichostomum lindigii* (Hampe) Zand., see treatment of *Trichostomum*]

[*Astomum lonchophyllum* Roth Am5 ≡ *Trachycarpidium lonchophyllum* (Roth) Zand., see treatment of *Trachycarpidium*]

Astomum lorentzii (C. Müll.) Broth. Am6

[*Astomum ludovicianum* (Sull.) Sull. Am1 ≡ *Weissia ludoviciana* (Sull.) Reese & Lemmon, Bryologist 68: 282, 1965]

Astomum minutum Dix. & P. Varde As3

[*Astomum mittenii* BSG Eur ≡ *Weissia mittenii* (BSG) Mitt.]

Astomum mollifolium (C. Müll.) Broth. Am5

[*Astomum muehlenbergianum* (Sw.) Grout As2 Am1 ≡ *Weissia muehlenbergiana* (Sw.) Reese & Lemmon, Bryologist 68: 282, 1965]

[*Astomum multicapsulare* (Sm.) BSG Eur As1 ≡ *Weissia multicapsularis* (Sm.) Mitt.]

[*Astomum neocaledonicum* (Thér.) Andrews in Broth. Oc ≡ *Weissia neocaledonica* (Thér.) Zand., see treatment of *Weissia*]

Astomum nicholsonii Roth [= *Weissia crispata* (Nees & Hornsch.) C. Müll. × *Astomum crispum* (Hedw.) Hampe] Eur

[*Astomum nitidulum* (C. Müll.) Sull. & Lesq. = *Weissia muehlenbergianum* (Sw.) Grout]

var. *pygmaeum* Lesq. & Jam. Am1

Astomum novae-valesiae Broth. ex Roth Eur Afr1

[*Astomum occidentale* Flow. ex Crum, Bryologist 76: 286, 1973 Am1 ≡ *Weissia occidentalis* (Flow. ex Crum) Stoneburner, Bryologist 88: 310, 1985]

[*Astomum phascoides* (Hook.) Grout Am1 Eur ≡ *Weissia rostellata* var. *phascoides* (Hook.) Reese & Lemmon, Bryologist 68: 283, 1965]

[*Astomum platystegium* Dix. As4 ≡ *Weissia platystegia* (Dix.) Zand., see treatment of *Weissia*]

[*Astomum unguiculatum* (Mitt.) Broth. Afr2 ≡ *Weissia unguiculata* (Mitt.) Crundw. & Nyholm, J. Bryol. 8: 69, 1974 ≡ *Trichostomum unguiculatum* (Mitt.) Zand., see treatment of *Trichostomum*]

[*Astomum viride* C. Müll. Austr1 ≡ *Pleuridium viride* (C. Müll.) Kindb.]

Astomum wattsii Broth. ex Roth Austr1

BARBULA Hedw.

[*Barbula aaronis* (Lor.) Hilp. Eur As5 Afr1 ≡ Trichostomopsis aaronis (Lor.) Agnew & Towns., Israel J. Bot. 19: 258, 1970 ≡ *Didymodon aaronis* (Lor.) Guerra in Guerra & Ros, Cryptogamie Bryol. Lichénol. 8: 55, 1987]

[*Barbula abbonii* Thér. Am2 = *Didymodon tophaceus* (Brid.) Lisa *fide* Zander, Cryptogamie Bryol. Lichénol. 2: 406, 1981 (1982)]

[*Barbula aciphylla* BSG = *Tortula ruralis* var. *alpina* Wahlenb.]

var. *mucronata* Sendtn. in Gerb. Eur

Barbula acrophylla C. Müll. Austr1

[*Barbula acuta* (Brid.) Brid. Eur As1 As3 As5 Afr1 Am1 Am2 = *Didymodon rigidulus* var. *gracilis* (Hook. & Grev.) Zand. *fide* Zander, Cryptog. Bryol. Lichénol. 2: 393, 1981 (1982)]

[subsp. *abbreviatifolia* (H. Muell.) Kindb. Eur ≡ *Barbula acuta* var. *abbreviatifolia* (H. Müll.) Podp.]

[subsp. *icmadophila* (Schimp. ex C. Müll.) Amann Eur As1 As3 As5 Am1 Am2 ≡ *Didymodon rigidulus* var. *icmadophilus* (Schimp. ex C. Müll.) Zand., Cryptogamie Bryol. Lichénol. 2: 394, 1981]

[var. *bescherellei* (Sauerb. ex Jaeg. & Sauerb.) Crum, Bryologist 72: 241, 1969 Am2 = *Didymodon rigidulus* Hedw. *s. lat. fide* Zander, Cryptogamie Bryol. Lichénol. 2: 389, 1981 (1982)]

var. *abbreviatifolia* (H. Müll.) Podp. Eur

[var. *icmadophila* (Schimp. ex C. Müll.) Crum, Bryologist 72:

241, 1969 Am1 ≡ *Didymodon rigidulus* var. *icmadophilus* (Schimp. ex C. Müll.) Zand. *fide* Zander, Cryptogamie Bryol. Lichénol. 2: 394, 1981 (1982)]

fo. *brevifolia* (Roth) Podp., Consp. Musc. Eur. 209, 1954 (*Barbula gracilis* var.) Eur

fo. *calabrica* (Roth) Podp., Consp. Musc. Eur. 209, 1954 (*Barbula gracilis* var.) Eur

fo. *irrigata* (H. Müll.) Podp., Consp. Musc. Eur. 209, 1954 (*Barbula gracilis* var.) Eur

fo. *multiseta* (Limpr.) Podp., Consp. Musc. Eur. 209, 1954 (*Barbula gracilis* var.) Eur

fo. *patens* (Głow.) Podp., Consp. Musc. Eur. 209, 1954 (*Barbula gracilis* var.) Eur

fo. *pulveriplena* (Loeske) Podp., Consp. Musc. Eur. 209, 1954 (*Barbula icmadophila* fo.)

fo. *rufescens* (Limpr.) Podp., Consp. Musc. Eur. 209, 1954 (*Barbula gracilis* var.) Eur

fo. *viridis* (BSG) Podp., Consp. Musc. Eur. 209, 1954 (*Barbula gracilis* var.) Eur

[*Barbula acutata* C. Müll. Afr4 = *Pseudocrossidium replicatum* (Tayl.) Zand. *fide* Frey & Kürschner, Cryptogamie Bryol. Lichénol. 9: 99, 1988 also Sollman, Lindbergia 16: 22, 1990]

[*Barbula afroduriuscula* C. Müll. Afr3 = *Serpotortella chenagonii* (Ren. & Card.) Reese & Zand., Bryologist 90: 234, 1987 (1988)]

Barbula afrofontana (C. Müll.) Broth. Afr4

var. *acutiuscula* P. Varde Afr2

Barbula agraria Hedw. Am1 Am2 Am3 Am5

[fo. *involuta* Biz. & Thér. Am3 = *Barbula agraria* Hedw. *fide* Zander, Phytologia 44: 202, 1979]

Barbula alpicola C. Müll. Am6

Barbula altipapillosa Bartr. As3

[*Barbula altiseta* Card. Am2 = *Didymodon rigidulus* Hedw. *s. lat. fide* Zander, Cryptogamie Bryol. Lichénol. 2: 390, 1981 (1982)]

Barbula amoena C. Müll. *hom. illeg.* Austr1

Barbula amplexifolia (Mitt.) Jaeg. As3 As4 Am1

Barbula anastomosans C. Müll. Am4 Am6

Barbula anceps Card. As2

Barbula angustifolia C. Müll. & Kindb. in Macoun, Cat. Canad. Pl. 6: 264, 1892 *hom. illeg.*

Barbula arcuata Griff. As3 As4 Oc

[*Barbula anderssonii* (Ångstr.) Jaeg. Am6 ≡ *Syntrichia anderssonii* (Aongstr.) Zand., see treatment of *Syntrichia*]

[*Barbula andreaeoides* Kindb. Am1 = *Didymodon subandreaeoides fide* Zander, Phytologia 41: 23, 1978]

Barbula aneitensis Broth. & Watts. Oc

[*Barbula anserinocapitata* X.-j. Li, Acta Bot. Yunnan. 3: 103, 1981 As2 ≡ *Didymodon anserinocapitatus* (X.-j. Li) Zand., see treatment of *Didymodon*]

[*Barbula apiculata* Hedw. ≡ *Barbula unguiculata* fo. *apiculata* (Hedw.) Mönk.]

var. *minor* Mart. Eur

Barbula appressifolia (Mitt.) Jaeg. Am4

Barbula arctoamericana C. Müll. Am1

Barbula arcuata Griff. As3 As4 Oc Am2 Am3 Am5

[*Barbula arenicola* Dus. Am6 = *Pseudocrossidium crinitum* (Schultz) Zand., see treatment of *Pseudocrossidium*]

[*Barbula asperifolia* Mitt.e Eur As1 As2 As3 Am1 ≡ *Didymodon asperifolius* (Mitt.) Crum, Steere & Ander.]

[var. *gorodkovii* (A. Abr. & I. Abr.) L. Savicz, Novosti Sist. Niz. Rast. 6: 148, 1969 [1970] As1 ≡ *Didymodon rufus* var. *gorodkovii* A. Abr. & I. Abr. in I. Abr., Trudy Arktic. Antarktic. Naucno-Issl. Inst. 224: 220, 1963]

var. *gracilis* (Amann & Meyl.) Wijk & Marg. Eur

var. *grauhauptiana* (De Not.) Wijk & Marg. Eur

var. *kneuckeri* (Loeske & Osterw.) Wijk & Marg. Eur

var. *spitsbergensis* (Jones) Wijk & Marg. Eur

[*Barbula aurea* (Bartr.) Zand. in Zand. & Steere, Bryologist 81: 466, 1978 Am1 Am2 ≡ *Pseudocrossidium aureum* (Bartr.) Zand. *fide* Zander, Phytologia 44: 207, 1979 = *Pseudocrossidium crinitum* (Schultz) Zand.]

Barbula aureola C. Müll. Austr2

[*Barbula australasiae* (Hook. & Grev.) Brid. Am4 Am6 Austr1 Austr2 ≡ *Trichostomopsis australasiae* (Hook. & Grev.) Robins., Phytologia 20: 187, 1970 ≡ *Didymodon australasiae* (Hook. & Grev.) Zand., Phytologia 41: 21, 1978]

Barbula austrogracilis Dus. Am6

[*Barbula austrorevoluta* Besch. in Britt. Am4 ≡ *Pseudocrossidium austrorevolutum* (Besch.) Zand., see treatment of *Pseudocrossidium*]

[*Barbula barbuloides* (Herz.) U. Miz., J. Jap. Bot. 46: 320. 1971 = *Didymodon erosodenticulatus* (C. Müll.) Saito *fide* Saito J. Hattori Bot. Lab. 39: 504, 1975]

Barbula bagelensis Fleisch. As4

[*Barbula barbuloides* (Herz.) Mizushima, Jap. J. Bot. 46: 320, 1971 (*Erythrophyllum*) As2 = *Didymodon erosodenticulatus* (C. Müll.) Saito *fide* Saito, J. Hattori Bot. Lab. 39: 504, 1975]

[*Barbula bescherellei* Sauerb. in Jaeg. Am1 Am2 Am3 = *Didymodon rigidulus* Hedw. *s. lat. fide* Zander, Cryptogamie Bryol. Lichénol. 2: 389, 1981 (1982)]

[var. *crassinervia* Thér. Am2 = *Didymodon rigidulus* Hedw. *s. lat. fide* Zander, Cryptogamie Bryol. Lichénol. 2: 390, 1981 (1982)]

[var. *stenocarpa* Card. Am2 = *Didymodon rigidulus* Hedw. *s. lat. fide* Zander, Cryptogamie Bryol. Lichénol. 2: 390, 1981 (1982)]

Barbula bicolor (BSG) Loeske Eur

Barbula bistrata Rungby, Bot. Not. 112: 81, 1959 Am4

Barbula boliviana (Broth.) Hilp. Am4

Barbula bolleana (C. Müll.) Broth. Afr2

[*Barbula bourgaeana* Besch. Am2 = *Didymodon vinealis* (Brid.) Zand. *fide* Zander, Cryptogamie Bryol. Lichénol. 2: 407, 1981 (1982)]

Barbula brachymenia (Mitt.) Jaeg. Am4

[*Barbula brachyphylla* Sull. in Whippl. Am1 ≡ *Didymodon brachyphyllus* (Sull. in Whippl.) Zand., Phytologia 41: 24, 1978]

[*Barbula brunneola* C. Müll. Am2 = *Didymodon nigrescens* (Mitt.) Saito *fide* Zander, Phytologia 41: 22, 1978]

Barbula bulbiformis Brid. As5

[*Barbula byrdii* Bartr. Ant ≡ *Bryoerythrophyllum byrdii* (Bartr.) Zand., see treatment of *Bryoerythrophyllum*]

[*Barbula calcarea* Thér. Am2 ≡ *Bryoerythrophyllum calcareum* (Thér.) Zand., Bryologist 83: 232, 1980]

Barbula calcicola (Hampe) Broth. Austr1

Barbula calodictyon Broth. in Schum. & Lauterb. As4

Barbula calycina Schwaegr. [see Catcheside, Mosses S. Austr. 185, 1980 and Magill, Fl. S. Afr. I. Mosses 1: 241, 1981 (1982)] As3 Afr4 Am6 Austr1 Austr2

Barbula calyculosa (Mitt.) Jaeg. Am4 Am5

[*Barbula campylocarpa* (Tayl.) C. Müll. Am4 ≡ *Didymodon taylorii* Zand. *nom. nov.*, see treatment of *Didymodon*]

[*Barbula canaliculata* (Dix.) R. S. Chopra, Taxon. Indian Mosses 138, 1975 (*Didymodon*) As3 ≡ *Didymodon canaliculatus* Dix.]

[*Barbula cancellata* C. Müll. Am1 = *Barbula indica* (Hook.) Spreng. *fide* Zander, Phytologia 44: 1979 and, see treatment of *Barbula*]

Barbula capillipes Broth. Am5
[*Barbula cardotii* Dus. Am6 ≡ *Didymodon cardotii* (Dus.) Zand., see treatment of *Didymodon*]
Barbula catenulata Dix., Anniv. Vol. Bot. Gard. Calcutta 181, 1942 As3
[*Barbula chloronotos* (Brid.) Brid. ≡ *Crossidium chloronotos* (Brid.) Limpr. Eur As2 As3 As5 Afr1 Am1 Austr2 ≡ *Crossidium squamiferum fide* Delgadillo, Bryologist 78: 276, 1975]
[fo. *nivea* Besch., Cat. Moun. Alg. 12, 1882 Afr1 ≡ *Crossidium chloronotos* var. *niveum* (Besch.) Jelenc ≡ *Crossidium chloronotos* fo. *nivea* (Besch.) Par.]
Barbula chlorotricha (Broth. & Geh.) Par. Austr1
Barbula chocayensis Broth. & Herz. Am4
Barbula chrysochaete C. Müll. Austr1
Barbula chrysopus C. Müll. Austr1
Barbula clavicostata (Ren. & Card.) Zand. (*Hyophila*), see treatment of *Barbula* Afr3
[*Barbula columbiana* (Herm. & Lawt.) Herm. & Lawt., Bull. Torrey Bot. Club 99: 309, 1972 ≡ *Bryoerythrophyllum columbianum* (Herm. & Lawt.) Zand., Bryologist 81: 548, 1978]
[*Barbula comosa* Dozy & Molk. = *Barbula arcuata* Griff. *fide* Saito, J. Hattori Bot. Lab. 39: 496, 1977]
[var. *japonica* Broth. As2 = *Barbula arcuata* Griff. *fide* Saito, J. Hattori Bot. Lab. 39: 496, 1975]
[*Barbula commutata* Jur. ≡ *Streblotrichum convolutum* subsp. *commutatum* (Jur.) Giac. ≡ *Barbula convolutum* subsp. *commutata* (Jur.) Boul.]
var. *erosa* Corb. Afr1
[*Barbula concava* Dietr., Filic. Jenens. 49, 1827 *nom. illeg. incl. Barbula cuneifolia* (Dicks.) Web. & Mohr Eur ≡ *Tortula cuneifolia* (Dicks.) Turn.]
Barbula confertifolia Mitt. As3
Barbula congoana Thér. Afr2
[*Barbula consanguinea* (Thwait. & Mitt.) Jaeg. As3 As4 = *Barbula javanica* Dozy & Molk. *fide* Saito, J. Hattori Bot. Lab. 39: 495, 1975]
[*Barbula constricta* Mitt. As2 As3 As4 ≡ *Didymodon constrictus* (Mitt.) Saito *fide* Saito, J. Hattori Bot. Lab. 39: 514, 1975 (= *Didymodon vinealis* (Brid.) Zand. *fide* Sollman, Bryologist 86: 271, 1983)]
[var. *flexicuspis* Chen As2 ≡ *Didymodon constrictus* var. *flexicuspis* (Chen) Saito *fide* Saito, J. Hattori Bot. Lab. 39: 516, 1975 (= *Didymodon vinealis* (Brid.) Zand. *fide* Sollman, Bryologist 86: 272, 1983)]
[*Barbula convolutifolia* Dix. = *Barbula javanica* Dozy & Molk. *fide* K. Saito, J. Hattori Bot. Lab. 39: 495, 1975]
Barbula convoluta Hedw. Eur As1 As2 Afr1 Am1 Am2 Austr2
subsp. *commutata* (Jur.) Boul. Eur As5 Afr1
[subsp. *austriaca* (Schiffn. & Baumg.) Podp., Consp. Musc. Eur. 206, 1954 Eur ≡ subsp. *austriacus* (Schiffn. & Baumg.) Wijk & Marg. Eur]
var. *gallinula* Zand., Phytologia 44: 195, 1979 Am1
var. *obtusata* C. Müll. & Kindb. In Macoun Am1
[var. *propagulifera* Głow. = *Gymnostomum aeruginosum* Sm. *fide* R. Zander, Phytologia 44: 211, 1979]
[var. *robusta* Schimp. in Husn., Musci Gall. 11: 513A, 1875 = *Barbula revolvens* Schimp. *fide* Boul., Rev. Bryol. 2(2): 20, 1875 ≡ *Tortula revolvens* (Schimp.) Roth]
var. *sardoa* BSG Eur
fo. *brevifolia* Podp., Cas. Morav. Mus. Zemsk. 13: 52, 1913 Eur
[fo. *brunnescens* Podp., Cas. Morav. Mus. Zemsk. 13: 52, 1913 Eur ≡ *Barbula convoluta* fo. *rufescens* Loeske & Quelle]
fo. *insolata* Latz., Bot. Centralbl. Beih. 48(2): 480, 1931 Eur
fo. *rufescens* Loeske & Quelle, Moosfl. Harz. 175, 1903 Eur
fo. *rufipes* Bauer, Musc. Eur. Exs. 1586a, 1923 Eur
fo. *uliginosa* Limpr. in Cohn, Krypt. Fl. Schles. 1: 172, 1876 Eur
[*Barbula cordata* (Jur.) Loeske ≡ *Didymodon cordatus* Jur.]
fo. *brevicaulis* (Röll) Podp., Consp. Musc. Eur. 205, 1954 *nom. illeg.* (*Didymodon cordatus* fo.) Eur ≡ *Didymodon cordatus* var. *brevicaulis* Röll]
fo. *gracilis* (Röll) Podp., Consp. Musc. Eur. 205, 1954 (*Didymodon cordatus* var.) Eur ≡ *Didymodon cordatus* var. *gracilis*]
[fo. *longicaulis* (Röll) Podp., Consp. Musc. Eur. 206, 1954 (*Didymodon cordatus* fo.) Eur ≡ *Didymodon cordatus* fo. *longicaulis* Röll]
[fo. *ramosa* (Röll) Podp., Consp. Musc. Eur. 206 (*Didymodon cordatus* fo.) ≡ *Didymodon cordatus* fo. *ramosus* Röll]
[fo. *robusta* (Röll) Podp., Consp. Musc. Eur. 206, 1954 (*Didymodon cordatus* fo.) ≡ *Didymodon cordatus* fo. *robustus* Röll]
[fo. *stricta* (Röll) Podp., Consp. Musc. Eur. 206, 1954 (*Didymodon cordatus* fo.) ≡ *Didymodon cordatus* fo. *strictus* Röll]
[fo. *typica* (Röll) Podp., Consp. Musc. Eur. 205, 1954 *nom. illeg.* (*Didymodon cordatus* fo.) Eur ≡ *Didymodon cordatus* Jur. fo. *cordatus*
Barbula coreensis (Card.) Saito, J. Hattori Bot. Lab. 39: 484, 1975 (*Barbula paludosa* var.) As2
Barbula costa-ricensis Ren. & Card. Am2
Barbula costata (Mitt.) Jaeg. Am4
Barbula costesii Thér. Am6
[*Barbula crassicostata* Bartr. Am2 ≡ *Didymodon crassicostatus* (Bartr.) Zand., see treatment of *Didymodon*]
[*Barbula crassicuspis* Robins. Am2 ≡ *Morinia crassicuspis* (Robins.) Zand. *fide* Zander, Bryologist 81: 556, 1978 ≡ *Mironia crassicuspis* (Robins.) Zand., see treatment of *Mironia*]
[*Barbula crinita* Schultz (good species *fide* Catcheside, Mosses S. Austr. 180, 1980 and Magill, Fl. S. Afr. I. Mosses 1: 237, 1981 [1982]) Afr2 Afr4 Am6 Austr1 Austr2 ≡ *Pseudocrossidium crinitum* (Schultz) Zand., see treatment of *Pseudocrossidium*]
Barbula crocea (Brid.) Web. & Mohr Eur As2 Afr1
var. *funckiana* (Schultz) Margadant, Lindbergia 1: 124, 1972 (*Barbula*) Eur
Barbula crozalsii (Philib.) Broth. Eur
[*Barbula cruegeri* Sond. ex C. Müll. Am1 Am2 Am3 Am4 Am5 = *Barbula indica* (Hook.) Spreng. *fide* Zander, Phytologia 44: 185, 1979]
var. *laevinervis* Broth. Am3
Barbula cucullata J. Frölich, Ann. Naturhist. Mus. Wien 67: 153, 1964 As3
[*Barbula cucullifera* (Mitt.) Jaeg. Am4 ≡ *Aloinella cucullifera* (Mitt.) Steere]
[*Barbula curvipes* C. Müll. Am6 ≡ *Trichostomopsis curvipes* (C. Müll.) Robins., Phytologia 20: 186, 1970]
[*Barbula curvirostris* Lindb. ≡ *Hymenostylium recurvirostre* (Hedw.) Dix.]
[fo. *commutata* (Mitt.) Lindb., Musci Scand. 22, 1879 Eur ≡ *Hymenostylium recurvirostre* var. *commutatum* (Mitt.) Podp.]
[fo. *laeviuscula* Lindb., Musci Scand. 22, 1879 Eur ≡ *Hymenostylium recurvirostre* fo. *breviusculum* (Lindb.) Podp.]
[fo. *scabra* Lindb., Musci Scand. 22, 1879 Eur As1 Am1 ≡ *Hymenostylium recurvirostrum* var. *scabrum* (Lindb.) Podp = *Hymenostylium recuvirostrum* var. *latifolium* (Zett.) Wijk & Marg.]
Barbula cylindrangia C. Müll. Austr1
[*Barbula cylindrica* (Tayl.) Schimp. in Boul. Eur As1 As5 Afr1 Am1 ≡ *Didymodon vinealis* var. *flaccidus* (BSG) Zand., Phytologia 41: 26, 1978 = *Didymodon vinealis* (Brid.) Zand.

var. *vinealis fide* Sollman, Bryologist 86: 272, 1983]
[fo. *rivularis* Warnst., Krypt. Fl. Brandenb. 2(2): 253, 1904 ≡ *Barbula vinealis* fo. *rivularis* (Loeske) Podp., Consp. Musc. Eur. 210, 1954 (*Barbula*) Eur]
[fo. *robusta* Loeske, Moosfl. Harz. 174, 1903 Eur ≡ *Barbula vinealis* fo. *robusta* (Loeske) Podp., Consp. Musc. Eur. 210, 1954]
[fo. *rubella* (Schiffn.) Herz., Krypt. Forsch. 4: 279, 1919 (*Barbula cylindrica* var.) Eur ≡ *Barbula vinealis* fo. *rubella* (Schiffn.) Podp., Consp. Musc. Eur. 210, 1954]
[fo. *rufescens* C. Jens., Bot. Faeröes 155, 1901 Eur = *Barbula vinealis* fo. *rubella* (Schiffn.) Podp. *fide* Podpěra, Consp. Musc. Eur. 210, 1954]
fo. *viridis* Warnst., Krypt. Fl. Brandenb. 2(2): 253, 1904 Eur
Barbula declivium C. Müll. Am6
[*Barbula decolorans* Hampe Am4 = *Trichostomopsis australasiae* (Hook. & Grev.) Robins., Phytologia 20: 187, 1970 ≡ *Didymodon australasiae* (Hook. & Grev.) Zand., Phytologia 41: 21, 1978]
Barbula decurrens Laz. As1
[*Barbula denticulata* Dix. & P. Varde As3 *hom. illeg.* = *Barbula vardei* R.S. Chopra *nom. nov.*, Bryologist 80: 544, 1977]
Barbula dharvarensis Dix. As3
Barbula dioritica C. Müll. Afr2
Barbula dissita C. Müll. Oc
[*Barbula ditrichoides* Broth. As3 Am1 ≡ *Didymodon acutus* var. *ditrichoides* (Broth.) Zand., Phytologia 41: 20, 1978 (*Barbula ditrichoides*) ≡ *Didymodon rigidulus* var. *ditrichoides* (Broth.) Zand., see treatment of *Didymodon*]
[*Barbula divergens* Broth. in Hall. As4 = *Barbula inaequalifolia* Tayl. *fide* Sollman, Lindbergia 10: 54, 1984 ≡ *Bryoerythrophyllum inaequalifolium* (Tayl.) Zand.]
[*Barbula dixonii* R. S. Chopra, Taxon. Indian Mosses 139, 1975 (*Didymodon obtusifolius* Card. ex Dix. & P. Varde *hom. illeg. non* Schkurh) As3 = *Barbula inaequalifolia* Tayl. *fide* Sollman, Lindbergia 10: 54, 1984 ≡ *Bryoerythrophyllum inaequalifolium* (Tayl.) Zand.]
Barbula dorrii Ren. & Card. Afr3
[*Barbula dregeana* C. Müll. Afr4 = *Aloina bifrons* (C. Müll.) Kindb. *fide* Magill, Fl. S. Afr. I. Mosses 1: 193, 1981 (1982)]
Barbula dusenii C. Müll. ex Broth. Afr2
[*Barbula ecuadoriensis* Broth. Am4 = *Trichostomopsis australasiae* (Hook. & Grev.) Robins., Phytologia 20: 187, 1970 ≡ *Didymodon australasiae* (Hook. & Grev.) Zand., Phytologia 41: 21, 1978]
Barbula ehrenbergii (Lor.) Fleisch. Eur As2 As5 Afr1 Am1 Am2 Am3 Austr1
var. *algeriae* (C. Müll.) Vent. & Bott. Eur Afr1
[var. *mexicana* Thér. Am2 = *Barbula ehrenbergii* (Lor.) Fleisch. *fide* Zander, Phytologia 44: 198, 1979]
Barbula elata Dur. & Mont. ex C. Müll. Afr1
[*Barbula elbertii* Broth. in Hall. As4 = *Didymodon vinealis* (Brid.) Zand. *fide* Sollman, Lindbergia 10: 54, 1984]
Barbula elliottii Broth. Afr2
Barbula enderesii Garov. Eur As1
[*Barbula erosa* Hampe in C. Müll. Am2 Am4 = *Barbula indica* (Hook.) Spreng. *fide* Zander, Phytologia 44: 185, 1979]
[*Barbula erythropoda* Schimp. ex Besch. Am2 = *Didymodon rigidulus* Hedw. *s. lat. fide* Zander, Cryptogamie Bryol. Lichénol. 2: 389, 1981 (1982)]
Barbula eubryum C. Müll. (good species *fide* Magill, Fl. S. Afr. I. Mosses 1: 245, 1981 [1982]) Afr2 Afr4
Barbula eustegia Card. & Thér. Am1
[*Barbula falcifolia* C. Müll. = *Didymodon rigidicaulis* (C. Müll.) Saito *fide* K. Saito, J. Hattori Bot. Lab. 39: 502, 1975 = *Didymodon ferrugineus* (Schimp. ex Besch.) Hill, J. Bryol. 11: 599, 1981 (1982)]
[*Barbula fallax* Hedw. Eur As1 As2 As3 As5 Afr1 Am1 ≡ *Didymodon fallax* (Hedw.) Zand., Phytologia 41: 28, 1978]
subsp. *brevifolia* (With.) Kindb. Eur Afr1 Am1
var. *atroviridis* Hüb. Eur
[var. *brevifolia* (Dicks. ex With.) Schultz Eur Af1 Am1 ≡ *Didymodon fallax* var. *brevifolius* (Dicks. ex With.) Ochyra, Fragm. Fl. Geobot. 28: 449, 1982 (1984)]
var. *filescens* Röll. Eur
var. *laevifolia* Hess. Eur
var. *longifolia* Warnst. & Fleisch. Eur
var. *obtusifolia* Amann Eur
[var. *recurvifolia* (Wils.) Husn. = *Didymodon rigidicaulis* (C. Müll.) Saito *fide* K. Saito, J. Hattori Bot. Lab. 39: 502, 1975 = *Didymodon ferrugineus* (Schimp. ex Besch.) Hill, J. Bryol. 11: 599, 1981 (1982)]
var. *stricta* Schultz Eur As5
var. *tristicha* Brid. Am1
var. *vinealoides* March. Eur
fo. *alpina* Loeske & Paul, Krypt. Forschung. Bayer. Bot. Ges. 5: 354, 1920 Eur
fo. *atrata* Röll, Jahrb. Ak. Wiss. Erfurt n.ser. 41: 123, 1915 Eur
fo. *biseta* Györf., Magyar Bot. Lapok 5: 340, 1906 Eur
fo. *brevicaulis* (Schwaegr.) Podp., Consp. Musc. Eur. 206, 1954 (*Barbula*) Eur Am1
fo. *crispula* (Warnst.) Podp., Consp. Musc. Eur. 206, 1954 (*Barbula fallax* var.) Eur
fo. *elata* Loeske in Bauer, Musc. Eur. Exs. 1590, 1923 Eur
fo. *fastigiata* (Warnst.) Podp., Consp. Musc. Eur. 206, 1954 (*Barbula fallax* var.) Eur
fo. *robusta* Podp., Consp. Musc. Eur. 207, 1954 (*Barbula fallax* var. *robusta* Warnst. *hom. illeg.*) Eur
Barbula farriae Crum & Bartr. Am3
Barbula fendleri C. Müll. Am4
[*Barbula ferrinervis* C. Müll. = *Barbula arcuata* Griff. *fide* Zander, Phytologia 44: 199, 1975]
[var. *eggersiana* C. Müll. Am3 = *Barbula arcuata* Griff. *fide* Zander, Phytologia 44: 199, 1975]
Barbula fidelis Crum & Steere Am3
[*Barbula flaccidiseta* Lor. Am2 = *Didymodon rigidulus* Hedw. *s. lat. fide* Zander, Cryptogamie Bryol. Lichénol. 2: 389, 1981 (1982)]
[*Barbula flaviseta* (Herz.) Wijk & Marg. Am6 = *Trichostomopsis umbrosa* (C. Müll.) Robins., Phytologia 20: 185, 1970 (≡ *Didymodon australasiae* (C. Müll.) Zand., Cryptogamie Bryol. Lichénol. 2: 400, 1981 [1982]) ≡ *Didymodon umbrosus* (C. Müll.) Zand.]
[*Barbula fontana* (C. Müll.) Broth. *hom. illeg.* Afr2 ≡ *Barbula meidensis* Cufodontis, *nom. nov.*]
Barbula francii Thér. Oc
Barbula frigida C. Müll. Am4
Barbula funalis Dix. & Badhw. As3
Barbula furvofusca C. Müll. Oc
[*Barbula fusca* C. Müll. Am4 Am6 = *Didymodon vinealis* (Brid.) Zand. *fide* Sollman, Bryologist 86: 271, 1983; Dusèn collections of this sp. at NY are also *Didymodon vinealis*]
Barbula fuscescens Wallich As2 As3
[*Barbula fuscinervia* (Mitt.) Jaeg. Am6 ≡ *Bryoerythrophyllum fuscinervium* (Mitt.) Zand., see treatment of *Byroerythrophyllum*]
Barbula fuscovirens Bartr. As3
Barbula fuscoviridis Broth. ex Thér. Am6

[*Barbula gangetica* C. Müll. As3 = *Barbula arcuata* Griff. *fide* Gangulee, Mosses E. India 3: 725, 1972]
Barbula gattefossei P. Varde Afr1
Barbula geminata C. Müll. Austr1
Barbula geniculata (Mont.) C. Müll. Am6
Barbula glaucescens Hampe Am4
var. *latifolia* Herz. Am4
Barbula glaucula C. Müll. Oc
Barbula goniospora C. Müll. Oc
Barbula gracilenta Mitt. As3
[*Barbula gracilescens* Schimp. ex Besch. Am2 = *Didymodon rigidulus* Hedw. *s. lat. fide* Zander, Cryptogamie Bryol. Lichénol. 2: 389, 1981 (1982)]
[*Barbula graciliformis* Schimp. ex Besch. Am2 = *Didymodon rigidulus* Hedw. *s. lat. fide* Zander, Cryptogamie Bryol. Lichénol. 2: 389, 1981 (1982)]
[*Barbula graminicolor* C. Müll. Am2 Am6 = *Didymodon australasiae* (Hook. & Grev.) Zand. *s. lat. fide* Zander, Cryptogamie Bryol. Lichénol. 2: 397, 1981 (1982)]
subsp. *subgraminicolor* Thér. Am6
Barbula granulosa Thér. Am4
[*Barbula gregaria* (Mitt.) Jaeg. As2 As3 ≡ *Barbula indica* var. *gregaria* (Mitt.) Zand., Cryptogamie Bryol. Lichénol. 2: 6, 1981]
Barbula grimmiacea C. Müll. Am6
Barbula gymnostoma C. Müll. As3
Barbula hampeana Par. Austr1
[*Barbula haringae* Crum Am1 = *Barbula amplexifolia* (Mitt.) Jaeg. *fide* Zander, Phytologia 44: 193, 1979]
[*Barbula hastata* Mitt. As3 ≡ *Didymodon hastatus* (Mitt.) Zand., see treatment of *Didymodon*]
Barbula hiroshii Saito, J. Hattori Bot. Lab. 39: 499, 1975 As2
Barbula hispaniolensis Buck & Steere, Moscosoa 2(1): 30, 1983 (*nom. nov.* for *Barbula leptodontioides* Crum & Steere) Am3
[*Barbula hookeri* Steud. Afr4 = *Barbula calycina* Schwaegr. *fide* Magill, Fl. S. Afr. I. Mosses 1: 241, 1981 (1982)]
[*Barbula hornschuchiana* Schultz Eur As5 Afr1 Afr4 Am1 ≡ *Pseudocrossidium hornschuchianum* (Schultz) Zand. *fide* Zander, Phytologia 44: 205, 1979]
var. *incrassata* Podp. Eur
[var. *obtusula* (Lindb.) Podp. Eur Am1 ≡ *Pseudocrossidium revolutum* var. *obtusulum* (Lindb.) Tan, Zand. & T. Tayl., Lindbergia 7: 41, 1981]
fo. *brevifolia* Reim., Hedwigia 79: 268, 1940 Eur
[*Barbula horrinervis* Saito, J. Hattori Bot. Lab. 39: 486, 1975 As5 = *Barbula indica* (Hook.) Spreng. *fide* Zander, Phytologia 44: 186, 1979]
Barbula hosseusii Thèr. Am4
Barbula horricomis C. Müll. ex Gangulee, Nov. Hedw. 12: 423, 1966 [1967] As2
Barbula humboldtii Herz. Am4 ≡ *Didymodon humboldtii* (Herz.) Hegew. & Hegew.]
[*Barbula husnotii* Schimp. ex Besch. Am3 = *Barbula agraria* Hedw. *fide* Zander, Phytologia 44: 202, 1979]
Barbula hyalinobasis Broth. Am4
Barbula hymenostylioides Broth. in Urban Am3
[*Barbula icmadophila* Schimp. ex C. Müll. Eur As2 As3 As5 Am1 Am2 ≡ *Didymodon rigidulus* var. *icmadophilus* (Schimp. ex C. Müll.) Zand. *fide* Zander, Cryptogamie Bryol. Lichénol. 2: 394, 1981 (1982)]
[fo. *pulveriplena* Loeske, Hedwigia 49: 29, 1910 Eur ≡ *Barbula acuta* fo. *pulveriplena* (Loeske) Podp., Consp. Musc. Eur. 209, 1954]
[*Barbula imbricata* Herrnstadt & Heyn, Bryologist 94: 174, 1991 (*nom. nov.* for *Didymodon luridus* Hornsch. ex Spreng. Eur As5 Afr1 Afr2 ≡ *Didymodon luridus* Hornsch. ex Spreng.]
[*Barbula imperfecta* (C. Müll.) Broth. Am6 ≡ *Didymodon imperfectum* (C. Müll.) Zand., see treatment of *Didymodon*]
[*Barbula imshaugii* Vitt, Bryologist 74: 464, 1971 (1972) Austr2 ≡ *Trichostomum imshaugii* (Vitt) Zand., see treatment of *Trichostomum*]
[*Barbula inaequalifolia* Tayl. Am1 Am2 Am4 As2 As4 ≡ *Bryoerythrophyllum inaequalifolium* (Tayl.) Zand., Bryologist 83: 232, 1980]
Barbula incerta Dix. *hom. illeg.* Austr1
Barbula inclinans Schimp. ex Besch. Afr3
[*Barbula inclinata* (Hedw. f.) Schwaegr. ≡ *Tortella inclinata* (Hedw. f.) Limpr.]
var. *magellanica* Card. Am6
[fo. *acuminata* Farn., Muschi Prov. Pavia, Atti Istituto Bot. Univ. Pavia 3: 17, 1891 ≡ *Tortella inclinata* fo. *acuminata* (Farn.) Par.]
Barbula indica (Hook.) Spreng. As2 As3 As4 Afr2 Afr3 Afr4 Am1 Am2 Am3 Am4 Austr1
var. *gregaria* (Mitt.) Zand., Cryptogamie Bryol. Lichénol. 2: 6, 1981 (*Barbula*) As3 Am1 Am2
var. *scaberrima* Dix. Afr2
Barbula inflexa (Duby) C. Müll. As2 As3 As4 Oc
Barbula integrifolia (Williams) Zand., Bryologist 75: 277, 1972 (*Leptodontium*) Am4
Barbula isoindica Zand. (*nom. nov.* for *Pottia papillinervis* Lor.), see treatment of *Barbula* Afr2
[*Barbula jacksharpii* Crum, Bryologist 87: 204, 1984 Am3 ≡ *Gymnostomum jacksharpii* (Crum) Allen, Bryologist 93: 207, 1991]
Barbula javanica Dozy & Molk. As2 As3 As4
var. *epapillosa* Fleisch. As4
var. *robusta* Dix. As3
[*Barbula johansenii* Williams Am1 ≡ *Didymodon johansenii* (Williams) Crum]
Barbula juniperoidea C. Müll. Am6
Barbula kiaerii Broth. Afr3
Barbula kivuensis Leroy & Potier de la Varde. in Demar. & Leroy, Expl. Parc Nation. Albert 4 Mission Lebrun 6: 15, 1944 Afr2
[*Barbula laevifolia* Broth. & Yas. = *Barbula subcomosa* Broth. *fide* K. Saito, J. Hattori Bot. Lab. 39: 493. 1975]
[*Barbula laevipila* (Brid.) Garov. ≡ *Tortula laevipila* (Brid.) Schwaegr.]
var. *populina* Brockm. Eur.
fo. *pagorum* (Milde) Mol., Ber. Naturhis. Ver. Passau 10: 100, 1875 Eur
[*Barbula lagunicola* C. Müll. Am2 = *Didymodon rigidulus* Hedw. *s. lat. fide* Zander, Cryptogamie Bryol. Lichénol. 2: 389, 1981 (1982)]
Barbula lambarenensis P. Varde Afr2 [≡ *Semibarbula lambarenensis* (P. Varde) Biz., Sv. Bot. Tidsk. 63: 453, 1969]
Barbula lamprocalyx C. Müll. Am6
Barbula lanceolata Schum *hom. illeg.* Eur
Barbula laureriana Lor. Afr4
Barbula laxiretis Broth. in Hall. As4
Barbula leiophylla Tix., Rev. Bryol. Lichénol. 34: 136, 1966 As3
[*Barbula leptocarpa* Besch. Am2 = *Didymodon rigidulus* Hedw. *s. lat. fide* Zander, Cryptogamie Bryol. Lichénol. 2: 389, 1981 (1982)]
[*Barbula leptodontoides* Crum & Steere *hom. illeg.* Am3 ≡ *Barbula hispaniolensis* Buck & Steere *nom. nov.*]
Barbula leucobasis Dix. in Dix. & Greenwood, Proc. Linn. Soc. N. South Wales 55: 277, 1930 Oc

Barbula leucodontoides C. Müll. in Par. ex Gangulee, Nov. Hedw. 12: 425, 1966 As3

[*Barbula ligularis* (Mitt.) Jaeg. Am4 ≡ *Bryoerythrophyllum ligulare* (Mitt.) Zand. see treatment of *Bryoerythrophyllum*]

[*Barbula lindigii* Hampe ≡ *Didymodon lindigii* (Hampe) Zand., see treatment of *Didymodon* Am4

[*Barbula linguaefolia* Bartr. Am2 = *Barbula calcarea* Thér. *fide* Zander, Phytologia 44: 183, 1979 ≡ *Bryoerythrophyllum calcareum* (Thér.) Zand., Bryologist 83: 232, 1980]

[*Barbula linearis* Web. & Mohr Am3 Am5 ≡ *Tortella linearis* (Web. & Mohr.) Zand., see treatment of *Tortella*]

Barbula linguaecuspis Broth. Am4

Barbula lobayetensis Williams As4

[*Barbula lombokensis* Broth. in Hall. As4 = *Bryoerythrophyllum ferruginascens* (Stirt.) Giac. *fide* Sollman in Touw, J. Hattori Bot. Lab. 71: 342, 1992]

Barbula lonchodonta C. Müll. Am6

[*Barbula lonchostega* C. Müll. Am2 = *Barbula indica* (Hook.) Spreng. *fide* Zander, Cryptogamie Bryol. Lichénol. 2: 418, 1981 (1982)]

Barbula longicostata X.-j. Li, Acta Bot. Yunnan. 3: 105, 1981 As2

[*Barbula longirostris* Hampe Am2 Am4 = *Tortula leiostoma* Herz. = *Hennediella limbata* (Mitt.) Zand., see treatment of *Hennediella* and cf. Mishler in Sharp et al., Moss Fl. Mex.]

[*Barbula louisiadum* Broth. Oc = *Barbula consanguinea* (Thw. & Mitt.) Jaeg. *fide* Eddy, Handb. Males. Mosses 2: 178, 1991 = *Barbula javanica* Dozy & Molk. *fide* Saito, J. Hattori Bot. Lab. 39: 495, 1975]

[*Barbula lozanoi* Card. Am2 = *Didymodon vinealis* (Brid.) Zand. *fide* Zander, Cryptogamie Bryol. Lichénol. 2: 408, 1981 (1982)]

[*Barbula luehmannii* Broth. & Geh. Austr1 ≡ *Didymodon luehmannii* (Broth. & Geh.) Catcheside, Mosses S. Austr. 175, 1980]

Barbula lurida Hornsch. Am2 Am4 Am5

[*Barbula lurida* (Hornsch.) Lindb. *hom. illeg.* Eur As5 Afr1 Afr2 Am1 ≡ *Didymodon luridus* Hornsch. ex Spreng.]

[fo. *cuspidata* (Schimp.) Mönk., Laubm. Eur. 294, 1927 (*Didymodon luridus* var.) ≡ *Didymodon trifarius* var. *cuspidata* (Schimp.) Wijk & Marg.]

Barbula luteola (Mitt.) Par. (good species *fide* Catcheside, Mosses S. Austr. 182, 1980) Austr1

[*Barbula mamillosa* Crundw., J. Bryol. 9: 163, 1976 [1977] ≡ *Didymodon mamillosus* (Crundw.) Hill]

[*Barbula macrogonia* Besch. Am3 = *Barbula arcuata* Griff. *fide* Zander, Phytologia 44: 199, 1979]

Barbula macassarensis Fleisch. As3 As4

Barbula majuscula C. Müll. As2

Barbula malagana Crum, Bryologist 70: 235, 1967 Am4

Barbula marginans C. Müll. Eur

Barbula marginatula C. Müll. ex Gangulee, Nov. Hedw. 12: 424, 1966 [1967] As3

[*Barbula maschalogena* Ren. & Card. As3 ≡ *Didymodon maschalogena* (Ren. & Card.) Broth.]

[*Barbula maxima* Syed & Crundw., J. Bryol. 7: 527. 1973 [1974] nov. nov. for *Barbula reflexa* var. *robusta* Braithw. ≡ *Didymodon maximus* (Syed & Crundw.) M. Hill]

Barbula meidensis Cufodontis, Oesterr. Bot. Z. 98: 225, 1951 (*nom. nov.* for *Barbula fontana* (C. Müll.) Broth.) Afr2

[*Barbula mendozensis* (Mitt.) Jaeg. Am6 ≡ *Pseudocrossidium mendozense* (Mitt.) Zand., see treatment of *Pseudocrossidium*]

[*Barbula michiganensis* Steere in Grout Am1, Am2, As2 ≡ *Didymodon michiganensis* (Steere in Grout) Saito, J. Hattori Bot. Lab. 39: 517. 1975]

Barbula microcalycina Magill, Fl. S. Afr. I. Mosses 1: 241, 1981 [1982] Afr4

[*Barbula microglottis* C. Müll. Am3 = *Barbula indica* (Hook.) Spreng. *fide* Zander, Phytologia 44: 186, 1979]

Barbula microstoma (Dix. & Badhw.) R. S. Chopra, Bryologist 80: 544, 1977 (*Didymodon*) As3

[*Barbula mobilis* C. Müll. Am4 = *Didymodon rigidulus* Hedw. *s. lat. fide* Zander, Cryptogamie Bryol. Lichénol. 2: 389, 1981 (1982)]

[*Barbula mosis* (Lor.) Hilp. As5 ≡ *Gymnostomum mosis* (Lor.) Jur.]

Barbula mucronulata Ren. & Card. *hom. illeg.* Afr3

Barbula mucronulata Lilj. *hom. illeg.* Eur

[*Barbula muenchii* Card. Am2 = *Barbula indica* (Hook.) Spreng. *fide* Zander, Phytologia 44: 186, 1979]

Barbula munyensis Williams Am4

[*Barbula muralis* (Hedw.) Crom. ≡ *Tortula muralis* Hedw.]

var. *breviseta* Opiz Eur

var. *microcarpa* Hüb. Eur

var. *tenuis* Opiz Eur

[fo. *stenocarpa* Baker, Pacif. Slope Bryoph. 360, 1902 = *Tortula brevipes* (Lesq.) Broth. *fide* Steere in Grout, Moss Fl. N. Am. 1(4): 232, 1939]

[*Barbula mutica* (Schultz) Kindb., Eur. N. Am. Bryin. 2: 247, 1897 *hom. illeg.* (*Syntrichia laevipila* var.) Am1 = *Tortula latifolia* Hartm.]

Barbula nana C. Müll. Afr2

[*Barbula nigrescens* Mitt. As2 As3 Am1 Am2 ≡ *Didymodon nigrescens* (Mitt.) Saito *fide* Saito, J. Hattori Bot. Lab. 39: 510, 1975]

[*Barbula nipponica* Nog. As2 = *Didymodon constrictus* (Mitt.) Saito *fide* Saito, J. Hattori Bot. Lab. 39: 514, 1975 (= *Didymodon vinealis* (Brid.) Zand. *fide* Sollman, Bryologist 86: 271, 1983)]

[var. *gracilis* Nog. As2 = *Didymodon constrictus* (Mitt.) Saito *fide* Saito, J. Hattori Bot. Lab. 39: 514, 1975 (= *Didymodon vinealis* (Brid.) Zand. *fide* Sollman, Bryologist 86: 271, 1983)]

Barbula novae-caledoniae C. Müll. Oc

Barbula novoguinensis Broth. As4

Barbula novogranatensis Hampe Am4

var. *gracilior* Hampe Am4

[*Barbula obscuriretis* Dix. As3 As4 Oc = *Barbula consanguinea* (Thw. & Mitt.) Jaeg. *fide* Eddy, Handb. Males. Mosses 2: 178, 1991 = *Barbula javanica* Dozy & Molk. *fide* Saito, J. Hattori Bot. Lab. 39: 495, 1975]

Barbula obtusissima Broth. & Par. *hom. illeg.* Oc

Barbula occidentalis (Mitt.) Broth. Am4

[*Barbula ochrocarpa* Toyama As2 = *Barbula indica* (Hook.) Spreng. *fide* Saito, J. Hattori Bot. Lab. 39: 488, 1975]

[*Barbula olivacea* (mitt.) Besch. Am2 = *Didymodon vinealis* (Brid.) Zand. *fide* Zander, Cryptogamie Bryol. Lichénol. 2: 407, 1981 (1982)]

Barbula omissa Thér. Afr3

[*Barbula orientalis* (Web.) Broth. = *Barbula indica* (Hook.) Spreng. *fide* Saito, J. Hattori Bot. Lab. 39: 488, 1975]

var. *scaberrima* Dix.

Barbula orizabensis C. Müll. Am1 Am2 Am3

[*Barbula pachydictyon* Broth. in Hall. As4 = *Didymodon vinealis* (Brid.) Zand. *fide* Sollman, Lindbergia 10: 54, 1984]

Barbula pachyloma Broth. [≡ *Cinclidotus involutus* Hilp.] As4

Barbula pallidobasis Dix. As1

[*Barbula papillinervis* (Lor.) Broth. *hom. illeg.* Afr2 ≡ *Barbula isoindica* Zand. *nom. nov.*, see treatment of *Barbula*]

[*Barbula perlinearis* C. Müll. = *Pseudocrossidium replicatum* (Tayl.) Zand. *fide* Zander, Phytologia 44: 206, 1975]

Barbula pernana C. Müll. Am6
[*Barbula perobtusa* (Broth.) Chen As2 ≡ *Didymodon perobtusus* Broth. *fide* Zander, Phytologia 41: 23, 1978]
[*Barbula perrevoluta* C. Müll. Am6 ≡ *Pseudocrossidium perrevolutum* (C. Müll.) Zand., see treatment of *Pseudocrossidium*]
[var. *acutifolia* (C. Müll.) Par. Am6 ≡ *Pseudocrossidium perrevolutum* var. *acutifolium* (C. Müll.) Zand., see treatment of *Pseudocrossidium*]
[var. *linearifolia* (C. Müll.) Par. Am6 ≡ *Pseudocrossidium perrevolutum* var. *linearifolium* (C. Müll.) Zand., see treatment of *Pseudocrossidium*]
Barbula pertorquescens Broth. Am6
Barbula peruviana (Mitt.) Jaeg. Am4
Barbula pflanzii (Broth.) Herz. Am4
var. *falcatula* Broth. in Herz. Am4
[*Barbula pilifera* (Hook.) Sim. As4 Afr2 Afr4 Am6 = *Barbula crinita* Schultz *fide* Catcheside, Mosses S. Austr. 180, 1980 and Magill, Fl. S. Afr. I. Mosses 1: 237, 1981 [1982] ≡ *Pseudocrossidium crinitum* (Schultz) Zand., see treatment of *Pseudocrossidium*]
var. *gracilis* Hornsch. Afr4 As4
fo. *elata* Thér., Rev. Bryol. Lichénol. 7: 173, 1934 [1935] *nom. inval. descr. gall.* Am4
[*Barbula planifolia* Broth. & Yas. As2 = *Trichostomum brachydontium* Bruch *fide* Saito, J. Hattori Bot. Lab. 39: 431, 1975]
[*Barbula platydictyon* Broth. in Hall., Meded. Rijks Herb. Leiden 14, Elbert's Sunda Exp. 19, 1912 As4 ≡ *Didymodon vinealis* (Brid.) Zand. *fide* Sollman, Lindbergia 10: 54, 1990]
[*Barbula platyneura* C. Müll. & Kindb. ex Macoun & Kindb. Am1 (= *Pseudocrossidium revolutum* (Brid. in Schrad.) Zand. sensu Zander, Phytologia 44: 204, 1975) = *Pseudocrossidium revolutum* var. *obtusulum fide* Tan, Zand. & T. Tayl., Lindbergia 7: 41, 1981]
Barbula plebeja C. Müll. Am6
[*Barbula porphyreoneura* C. Müll. Afr2 Afr4 ≡ *Tortula porphyreoneura* (C. Müll.) Townsend, J. Bryol. 10: 576, 1979]
Barbula potaninii Broth. ex C. Müll. As2
[*Barbula prionophylla* Saito, J. Jap. Bot. 46: 142, 1971 = *Barbula barbuloides fide* Iwatski & Nog. 1973 = *Didymodon erosodenticulatus* (C. Müll.) Saito *fide* Saito, J. Hattori Bot. Lab. 39: 504, 1975]
Barbula prschewalzkii Broth. in C. Müll. As2
[*Barbula pruinosa* (Mitt.) Jaeg. Am4 ≡ *Didymodon pruinosus* (Mitt.) Zand., see treatment of *Didymodon*]
Barbula pseudoehrenbergii Fleisch. As3 As4 Afr2
var. *aspera* Baumg. & Fröhl. As4
[*Barbula pseudogracilis* C. Müll. Afr2 = *Barbula acuta* (Brid.) Brid. *fide* Townsend, Lindbergia 10: 177, 1985 = *Didymodon rigidulus* var. *gracilis* (Schleich. ex Hook. & Grev.) Zand., Cryptogamie Bryol. Lichénol. 2: 393, 1981]
Barbula pseudonigrescens Tix., Ann. Hist.-Nat. Mus. Natl. Hungarici 66: 88, 1974 As3
[*Barbula pseudopilifera* C. Müll. & Hampe Austr1 Austr2 = *Barbula pilifera* (Hook.) Brid. (*hom. illeg.*) *fide* Weber, Lindbergia 1: 214–216, 1972 = *Barbula crinita* Schultz *fide* Catcheside, Mosses S. Austr. 180, 1980 ≡ *Pseudocrossidium crinitum* (Schultz) Zand., see treatment of *Pseudocrossidium*]
[var. *obscura* (Dix.) Sainsb. Austr2 = *Barbula pilifera* (Hook.) Brid. (*hom. illeg.*) *fide* Weber, Lindbergia 1: 214–216, 1972 = *Barbula crinita* Schultz *fide* Catcheside, Mosses S. Austr. 180, 1980 ≡ *Pseudocrossidium crinitum* (Schultz) Zand., see treatment of *Pseudocrossidium*]
Barbula punae Herz. Am4
Barbula punctulata (Ren. & Par.) Broth. *hom. illeg.* Afr3
[*Barbula pungens* (Hook. f. & Wils.) Jaeg. Austr1 = *Barbula luteola* (Mitt.) Par. *fide* Catcheside, Mosses S. Austr. 182, 1980]
Barbula purpurascens Dus. Am6
[*Barbula purpuripes* C. Müll. Am3 = *Barbula indica* (Hook.) Spreng. *fide* Zander, Phytologia 44: 186, 1979]
Barbula pycnophylla Card. Am6
Barbula pygmaea C. Müll. Afr2
Barbula rechingeri Broth. Oc
[*Barbula rectifolia* Tayl. Am2 Am4 = *Didymodon vinealis* (Brid.) Zand. *fide* Zander, Cryptogamie Bryol. Lichénol. 2: 407, 1981 (1982)]
[*Barbula recurva* (Griff.) R. S. Chopra, Bryologist 80: 544, 1977 (*Gymnostomum*) *inval. basion. non cit.* ≡ *Bryoerythrophyllum recurvum* (Griff.) Saito, Bull. Univ. Mus. Univ. Tokyo 8: 254, 1975 ≡ *Bellibarbula recurva* (Griff.) Zand., see treatment of *Bellibarbula*]
Barbula recurvopungens C. Müll. Am4
[*Barbula recurvirostris* (Hedw.) Dix. ≡ *Bryoerythrophyllum recurvirostrum* (Hedw.) Chen]
[var. *latinervia* Holmen Am1 ≡ *Bryoerythrophyllum recurvirostrum* var. *latinervium* (Holmen) Murray, Bryobrothera 1: 14, 1992.]
[*Barbula recurvis* (Griff.) R. S. Chopra, Taxon. Indian Mosses 139, 1975 (*Gymnostomum*) *comb. inval. basion. non cit.* As3 ≡ *Bryoerythrophyllum recurvum* (Griff.) Saito ≡ *Bellibarbula recurva* (Griff.) Zand., see treatment of *Bellibarbula*]
[*Barbula recurvis* (Griff.) R. S. Chopra ex R. S. Chopra, Bryologist 80: 544, 1977 (*Gymnostomum*) As3 ≡ *Bryoerythrophyllum recurvum* (Griff.) Saito ≡ *Bellibarbula recurva* (Griff.) Zand., see treatment of *Bellibarbula*]
[*Barbula reflexa* (Brid.) Brid. Eur As1 As2 Am1 Am2 = *Didymodon rigidicaulis* (C. Müll.) Saito *fide* Saito, J. Hattori Bot. Lab. 39: 502, 1975 = *Didymodon ferrugineus* (Schimp. ex Besch.) Hill, J. Bryol. 11: 599, 1981 (1982)]
var. *obtusata* Mönk. Eur
[var. *robusta* Braithw. Eur ≡ *Barbula maxima* Syed & Crundw., J. Bryol. 7: 527. 1973 nov. nov. ≡ Didymodon maximus (Syed & Crundw.) M. Hill]
fo. *robusta* Röll, Hedwigia 56: 152, 1915 Eur
[*Barbula reflexifolia* Fleisch. As4 = *Barbula inaequalifolia* Tayl. *fide* Zander, Bryologist 71: 421, 1968 ≡ *Bryoerythrophyllum inaequalifolium* (Tayl.) Zand., Bryologist 83: 232, 1980]
Barbula rehmannii C. Müll. [good species *fide* Magill, Fl. S. Afr. I. Mosses 1: 245, 1981 (1982)] Afr2 Afr4
[*Barbula replicata* Tayl. Am4 Am6 ≡ *Pseudocrossidium replicatum* (Tayl.) Zand. *fide* Zander, Phytologia 44: 206, 1979.]
[*Barbula revoluta* Brid. in Schrad. Eur As5 Afr1 Afr4 ≡ *Pseudocrossidium revolutum* (Brid.) Zand. *fide* Zander, Phytologia 44: 204, 1979]
var. *propagulifera* Amann Eur
fo. *elata* Herz., Ber. Zürich. Bot. Ges. 15: 46, 1905 Eur
fo. *macrophylla* Warnst., Krypt. Fl. Brandenb. 2(2): 244, 1904 Eur
fo. *mucronata* Loeske, Moosfl. Harz. 175, 1903 Eur
fo. *umbrosa* Loeske in Bauer, Musc. Eur. Exs. 978, 1913 Eur
[fo. *mucronata* Loeske, Moosfl. Harz. 175, 1903 Eur = *Barbula hornschuchiana* var. *obtusula* (Lindb.) Podp. *fide* Podp., Consp. Musc. Eur. 212, 1954 ≡ *Pseudocrossidium revolutum* var. *obtusulum* (Lindb.) Tan, Zand. & T. Tayl.]
Barbula riebeckii C. Müll. Afr2
[*Barbula rigida* Hedw. ≡ *Aloina rigida* (Hedw.) Limpr.]

var. *desertorum* Fröhl. As5
var. *rostellifolia* Brid. Eur
[*Barbula rigidula* (Hedw.) Mild. ≡ *Didymodon rigidulus* Hedw.]
[var. *desertorum* Fröhl., Ann. Naturhist. Mus. Wien 63: 31, 1959 As5 ≡ *Barbula trifaria* var. *desertorum* (Fröhl.) S. Agnew, Feddes Rep. 86: 366, 1975]
var. *perobtusa* Broth. As2
[fo. *biseta* (Pét.) Podpěra, Consp. Musc. Eur. 202, 1954 (*Didymodon rigidulus* fo.) Eur ≡ *Didymodon rigidulus* fo. *biseta* Pét., Magyar Bot. Lapok 2: 292, 1903]
[fo. *brevicaulis* Röll, Deutsche Bot. Monatsschr. 3: 163, 1885 ≡ *Didymodon rigidulus* fo. *brevicaulis* (Röll) Röll]
[fo. *brevifolia* (Röll) Podp., Consp. Musc. Eur. 202, 1954 ≡ *Didymodon rigidulus* fo. *brevifolius* Röll, Hedwigia 42: 299, 1903]
fo. *densa* (BSG) Podp., Consp. Musc. Eur. 203, 1954 (*Trichostomum rigidulum* var.) Eur
[fo. *gemmipara* Herz., Wiener Bot. Z. 93: 39, 1944 Eur = *Barbula rigidula* (Hedw.) Milde *fide* Crundwell & Nyholm, Svensk Bot. Tidskr. 59(2): 211, 1965 ≡ *Didymodon rigidulus* Hedw.]
[fo. *laxa* (Mol.) Mönk., Laubm. Eur. 1927 (*Didymodon rigidulus* var.) Eur ≡ *Didymodon rigidulus* fo. *laxus* Mol.]
[fo. *longicaulis* Röll, Deutsche Bot. Monatsschr. 3: 163, 1885 ≡ *Didymodon rigidulus* fo. *longicaulis* (Röll) Röll]
[fo. *longifolia* Podp., Consp. Musc. Eur. (*Didymodon rigidulus* var.) Eur. 203, 1954 Eur]
[fo. *major* (Podp.) Podp., Consp. Musc. Eur. 203, 1954 (*Didymodon rigidulus* var.) Eur ≡ *Didymodon rigidulus* var. *major* Podp.]
[fo. *propagulifera* (Schiffn.) Podp., Consp. Musc. Eur. 203, 1954 (*Didymodon rigidulus* fo.) Eur ≡ *Didymodon rigidulus* fo. *propaguliferus* (Schiffn.) Limpr.]
[fo. *tenuis* (Hammerschm.) Podp., Consp. Musc. Eur. 203, 1954 (*Didymodon rigidulus* var.) Eur ≡ *Didymodon rigidulus* var. *tenuis* Hammerschm.]
[fo. *viridis* Röll, Deutsche Bot. Monatsschr. 3: 164, 1885 ≡ *Didymodon rigidulus* fo. *brevicaulis* (Röll) Röll *fide* Podpěra, Consp. Musc. Eur. 203, 1954]
Barbula riograndensis Bartr. Am5
Barbula riparia C. Müll. Am6
[*Barbula rivicola* Broth. As2 ≡ *Didymodon rivicola* (Broth.) Zand. in Kop., Gao, Lou & Järvinen]
Barbula robbinsii Bartr. As4
Barbula rothii Herz. Am4
Barbula rottensis Weyl. Eur, fossil
[*Barbula rubella* (Hüb.) Mitt. in Lindb. *hom. illeg.* Eur ≡ *Bryoerythrophyllum recurvirostrum* (Hedw.) Chen]
fo. *minor* Arnell in Lindb. & Arnell, K. Svensk. Vet.-Ak. Handl. 23(10): 72, 1890 As1
[*Barbula rubiginosa* Mitt. Am1 ≡ *Didymodon occidentalis* Zand., Phytologia 41: 26, 1978 *nom. nov.* (≡ *Didymodon vinealis* var. *rubiginosus* (Mitt.) Zand., Cryptogamie Bryol. Lichénol. 2: 417, 1981)]
[*Barbula rubricaulis* Thér. Am2 = *Barbula arcuata* Griff. *fide* Zander, Phytologia 44: 200, 1979]
Barbula rubriseta Bartr. in Bauer Am6
[*Barbula rufa* (Lor.) Jur. ≡ *Didymodon asperifolius* (Mitt.) Crum, Steere & Anders.]
fo. *laevis* C. Jens., Bot. Faeröes 155, 1901 [Skand. Bladmfl. 238, 1939] Eur
fo. *sublaevigata* (Herz.) Podp., Consp. Musc. Eur. 211, 1954 (*Didymodon rufus* fo.) Eur
[*Barbula rufofusca* Lawt. & Herm., Bull. Torrey Bot. Club 99: 307, 1972 Am1 = *Didymodon nigrescens* (Mitt.) Saito *fide* Zander, Phytologia 41: 22, 1978]
Barbula scaberrima Broth. & Par. As3
Barbula semirosulata Zand. (*nom. nov.* for *Gymnostomiella rosulata* P. Varde), see treatment of *Barbula* Afr1
[*Barbula sobolifera* Fleisch. As4 = *Barbula arcuata* Griff. *fide* Saito, J. Hattori Bot. Lab. 39: 496, 1975]
Barbula solfatariensis Fleisch. As4
Barbula somaliae C. Müll. Afr2
Barbula sordida Besch. As3
[*Barbula spadicea* (Mitt.) Braithw. Eur As1 As5 ≡ *Didymodon spadiceus* (Mitt.) Limpr.]
var. *squarrosa* Latz. Eur
[var. *vaginans* (Lindb.) G. Roth in Bauer, Lotos 54(5): 15, 1906 (*Barbula*) Eur ≡ *Barbula spadicea* fo. *vaginans* (Lindb.) Podp.]
fo. *bernensis* (Culm.) Podp., Consp. Musc. Eur. 208, 1954 (*Barbula spadicea* var.) Eur
fo. *brevifolia* Latz., Magyar Bot. Lapok 29: 117, 1930 Eur
fo. *debilis* Latz., Magyar Bot. Lapok 29: 117, 1930 Eur
fo. *leptoderma* Podp., Consp. Musc. Eur. 208, 1954 (*Barbula spadicea* var. *leptoderma* Jens. *nom. inval.*) Eur
fo. *mollis* (Burchard) Podp., Consp. Musc. Eur. 208, 1954 (*Didymodon spadiceus* var.) Eur
fo. *obtusifolia* (Röll) Podp., Consp. Musc. Eur. 208, 1954 (*Didymodon spadiceus* var.) Eur
fo. *vaginans* (Lindb.) Podp., Consp. Musc. Eur. 208, 1954 (*Didymodon*) Eur
[*Barbula sparsifolia* Ren. & Card. Afr3 ≡ *Trichostomum sparsifolium* (Ren. & Card.) Card. in Grand.]
Barbula spathulifolia (Dix. & P. Varde) Zand., see treatment of *Barbula* (*Merceyopsis*) As3
Barbula speirostega C. Müll. Austr1
[*Barbula spiralis* Schimp. ex C. Müll. Am1 Am2 = *Pseudocrossidium replicatum* (Tayl.) Zand. *fide* Zander, Phytologia 44: 206, 1979]
[var. *emarginata* Card. Am2 ≡ *Pseudocrossidium replicatum* (Tayl.) Zand. *fide* Zander, Phytologia 44: 206, 1979]
Barbula squarrosa Schultz *hom. illeg.* Eur
Barbula stenocarpa Hampe Am4
[*Barbula stenotheca* Thér. Am2 (= *Barbula orizabensis* C. Müll. *fide* Zander, Phytologia 44: 183, 1979) ≡ *Morinia stenotheca* (Thér.) Zand., Bryologist 86: 156, 1983 ≡ *Mironia stenotheca* (Thér.) Zand., see treatment of *Mironia*]
[*Barbula stewartii* Bartr. As3 ≡ *Didymodon stewartii* (Bartr.) Zand., see treatment of *Didymodon*]
[*Barbula stillicidiorum* Card. Am2 = *Barbula arcuata* Griff. *fide* Zander, Phytologia 44: 200, 1979]
[*Barbula strictidens* C. Müll. Am2 = *Didymodon rigidulus* Hedw. *s. lat. fide* Zander, Cryptogamie Bryol. Lichénol. 2: 390, 1981 (1982)]
[*Barbula strictifolia* (Dix. & P. Varde) R. S. Chopra, Taxon. Indian Mosses 140, 1975 As3 ≡ *Didymodon strictifolius* Dix. & P. Varde As3 = *Didymodon recurvus* (Griff.) Broth. *fide* Robinson, Bryologist 71: 85, 1968 ≡ *Bryoerythrophyllum recurvum* (Griff.) Saito ≡ *Bellibarbula recurva* (Griff.) Zand., see treatment of *Bellibarbula*]
Barbula stulhmannii (Broth.) Broth. Afr2 Afr4
[*Barbula subanomala* C. Müll. As3 = *Timmiella anomala* (BSG) Limpr.]
Barbula subcespitosa (Hampe) Broth. Am4
Barbula subcalycina C. Müll. Austr1
Barbula subcernua Schimp. in Besch. Afr4
Barbula subcomosa Broth. As2
[*Barbula subcontorta* As2 = *Didymodon vinealis* (Brid.) Zand. *fide* Sollman, Bryologist 86: 272, 1983]

Barbula subdenticulata Dix. As3
[*Barbula suberythropoda* C. Müll. Am2 = *Barbula arcuata* Griff. *fide* Zander, Phytologia 44: 199, 1979]
Barbula subglaucescens C. Müll. Am4
[*Barbula subglaucula* Dix. ex Sainsb. *nom. inval. descr. angl.* Austr2 ≡ *Barbula unguiculata* Hedw. *fide* Sollman, Lindbergia 10: 54, 1984]
Barbula subgracilis C. Müll. & Kindb. in Macoun Am1
Barbula subgrimmiacea Thér. in Felipp. Am6
[*Barbula sublaevifolia* Toyama As2 = *Barbula javanica* Dozy & Molk. *fide* Saito, J. Hattori Bot. Lab. 39: 495, 1975]
Barbula subobtusa Thér. Afr3
Barbula subpellucida Mitt. As2 As3
Barbula subreflexifolia C. Müll. Am4
Barbula subreplicata Broth. Am4
[*Barbula subrivicola* Chen As2 = *Didymodon nigrescens* (Mitt.) Saito *fide* Saito, J. Hattori Bot. Lab. 39: 510, 1975]
[var. *densifolia* Chen As2 = *Didymodon nigrescens* (Mitt.) Saito *fide* Saito, J. Hattori Bot. Lab. 39: 510, 1975]
Barbula subrufa Broth. & C. Müll. As1
Barbula subrunciata C. Müll. Am6
[*Barbula subscabrinervis* Dix. & Nav. Afr2 = *Bryoerythrophyllum recurvum* (Griff.) Saito *fide* Sollman, Lindbergia 16: 22, 1990 ≡ *Bellibarbula recurva* (Griff.) Zand., see treatment of *Bellibarbula*]
[*Barbula subteretiuscula* Card. Am2 Am3 = *Didymodon rigidulus* Hedw. *s. lat. fide* Zander, Cryptogamie Bryol. Lichénol. 2: 390, 1981 (1982)]
[*Barbula subtorquata* C. Müll. & Hampe Austr1 ≡ *Didymodon subtorquatus* (C. Müll. & Hampe) Catcheside, Mosses S. Austr. 174, 1980]
[*Barbula subulifolia* Sull. Am2 Am3 Am4 Am5 = *Barbula arcuata* Griff. *fide* Zander, Phytologia 44: 199, 1979]
Barbula sulcata Geh. in Geh. & Herz. Afr2
Barbula sumatrana Baumg. & Dix. As4
[*Barbula svihlae* Bartr. As3 = *Hymenostylium recurvirostrum* (Hedw.) Dix. var. *recurvirostrum*, see treatment of *Hymenostylium*]
Barbula swartziana C. Müll. Am3
[*Barbula tamakii* Broth. As2 = *Dichodontium pellucidum* (Hedw.) Schimp. *fide* Saito, J. Hattori Bot. Lab. 39: 528, 1975]
Barbula taylorii Bartr. & Steere *hom. illeg.* Am4 [≡ *Hydrogonium taylorii* Weber *nom. nov.*, Lindbergia 3: 81, 1975 (1976)]
[*Barbula tectorum* C. Müll. As2 ≡ *Didymodon tectorum* (C. Müll.) Saito, J. Hattori Bot. Lab. 39: 517, 1975]
[*Barbula tenii* Herz. As2 = *Barbula inaequalifolia* Tayl. *fide* Zander, Bryologist 71: 41, 1968 ≡ *Bryoerythrophyllum inaequalifolium* (Tayl.) Zand., Bryologist 83: 232, 1980]
Barbula tenuicoma C. Müll. ex Broth. Am5
Barbula tenuirostris Brid. As3 As4 Oc
[*Barbula teretiuscula* Schimp. ex C. Müll. Am2 = *Didymodon rigidulus* Hedw. *s. lat. fide* Zander, Cryptogamie Bryol. Lichénol. 2: 389, 1981 (1982)]
Barbula thelimitria C. Müll. As4
Barbula tisserantii (P. Varde) P. Varde Afr2
[*Barbula tokyensis* Besch. As2 = *Barbula unguiculata* Hedw. *fide* Saito, J. Hattori Bot. Lab. 39: 491, 1975 as "*tokyoensis*"]
[*Barbula tonkinensis* (Besch.) Broth. As3 = *Barbula indica* (Hook.) Spreng., see treatment of *Barbula*]
[*Barbula tomaculosa* Blockeel, J. Bryology 11: 583, 1981 (1982) Eur ≡ *Didymodon tomaculosus* (Blockeel) Corley in Corley et al., J. Bryol. 11: 649, 1981 (1982)]
[*Barbula tophacea* (Brid.) Mitt. ≡ *Didymodon tophaceus* (Brid.) Lisa]
[fo. *acutifolia* (Schimp.) Mönk., Laubm. Eur. 295, 1927 (*Trichostomum tophaceum* var.) Eur ≡ *Didymodon tophaceus* fo. *acutifolius* Mönk., Laubm. Eur. 295, 1927 Eur]
[fo. *anatina* (Hammerschm.) Podp., Consp. Musc. Eur. 201, 1954 (*Didymodon tophaceus* var.) Eur ≡ *Didymodon tophaceus* var. *anatinus* Hammerschm.]
[fo. *bosniaca* (Głow.) Mönk. Laubm. Eur. 296, 1927 (*Didymodon*) Eur = *Didymodon tophaceus* var. *decurrens* Card. & Thér. *fide* Podp., Consp. Musc. Eur. 201, 1954]
fo. *cylindrica* (Boul.) Podp., Consp. Musc. Eur. 201, 1954 (*Trichostomum tophaceum* fo.) Eur
[fo. *humilis* (Schimp.) Mönk., Laubm. Eur. 295, 1927 (*Didymodon tophaceus* var.) Eur ≡ *Didymodon tophaceus* var. *humilis* Schimp.]
[fo. *laxa* (Kindb.) Podp., Consp. Musc. Eur. 201, 1954 (*Didymodon tophaceus* fo.) Eur ≡ *Didymodon tophaceus* fo. *laxus* Kindb., Boll. Soc. Bot. Ital. 7: 15, 1896]
[fo. *linearis* (De Not.) Podp., Consp. Musc. Eur. 201, 1954 (*Trichostomum tophaceum* var.) Eur ≡ *Didymodon tophaceus* var. *linearis* (De Not.) Limpr.]
fo. *lingulata* (Boul.) Mönk., Laubm. Eur. 295, 1927 (*Trichostomum tophaceum* fo.) Eur
[fo. *propagulifera* (Amann) Mönk., Laubm. Eur. 296, 1927 (*Didymodon tophaceus* fo.) Eur ≡ *Didymodon tophaceus* fo. *propaguliferus* Amann]
[fo. *recurvifolia* (Boul.) Demar., Bull. Jard. Bot. Natl. Belgique 17: 335, 1945 (*Trichostomum tophaceum* fo.) Eur ≡ *Didymodon tophaceus* fo. *recurvifolius* De Willd., Prodr. 434, 1899]
[fo. *scabrinervis* (Podp.) Podp., Consp. Musc. Eur. 202, 1954 (*Didymodon tophaceus* fo.) Eur ≡ *Didymodon tophaceus* fo. *scabrinervis* Podp.]
fo. *thermarum* Boros in Bauer, Musc. Eur. Exs. 2011, 1928 Eur
fo. *torrentium* (Loeske) Podp., Consp. Musc. Eur. 202, 1954 (*Barbula tophacea* var.) Eur
fo. *truncata* (Boul.) Podp., Consp. Musc. Eur. 202, 1954 (*Trichostomum tophaceum* fo.) Eur
[*Barbula torquata* Tayl. Austr1 Austr2 (= *Trichostomopsis australasiae* (Hook. & Grev.) Robins. *fide* Robinson, Phytologia 20: 184, 1970) ≡ *Didymodon torquata* (Tayl.) Catcheside, Mosses S. Austr. 174, 1980]
[*Barbula torquatifolia* Geh. Afr2 Afr4 = *Tortula porphyreoneura* (C. Müll.) Townsend *fide* Magill, Fl. S. Afr. I. Mosses 1: 213, 1981 [1982] ≡ *Pseudocrossidium porphyreoneurum* (C. Müll.) Zand., see treatment of *Pseudocrossidium*]
Barbula tortelloides C. Müll. Am6
[*Barbula tortuosa* (Hedw.) Web. & Mohr ≡ *Tortella tortuosa* (Hedw.) Limpr.]
[fo. *alpina* Boul., Muscin. France 421, 1886 Eur ≡ *Tortella tortuosa* fo. *alpina* (Boul.) Podp.]
[fo. *dentata* Farn., Muschi Prov. Pavia, Atti Istituto Bot. Univ. Pavia 3: 18, 1891 Eur ≡ *Tortella tortuosa* fo. *dentata* (Farn.) Podp.]
[fo. *gracilescens* Zett., K. Svensk Vet. Ak. Handl. 5(10): 25, 1865 Eur ≡ *Tortella tortuosa* fo. *gracillescens* (Zett.) Par.]
[*Barbula tosaënsis* Broth., Oefv. Finsk. Vet. Sco. Foerh. 62A(9): 10, 1921 As2 = *Barbula subcomosa* Broth. *fide* Saito, J. Jap. Bot. 47: 11, 1972]
Barbula translucens Salzm. ex Bruch Afr2
Barbula trichomanoides Broth. ex Ihs. As2
[*Barbula trichostomacea* C. Müll. Afr4 = *Trichostomopsis australasiae* (Hook. & Grev.) Robins., Phytologia 20: 187, 1970 ≡ *Didymodon australasiae* (Hook. & Grev.) Zand., Phytologia 41: 21, 1978]

[*Barbula trifaria* (Hedw.) Mitt. Eur As1 As2 As3 As5 Afr1 Am1 Am2 Oc = *Saelania glaucescens* (Hedw.) Broth. in Bomanss & Broth. *fide* Zander, Cryptog. Bryol. Lichénol. 2: 412, 1981 (1982)]
  var. *desertorum* (Fröhl.) S. Agnew, Feddes Rep. 86: 366, 1975 (*Barbula rigidula* var.) As2
  [fo. *brevifolia* (Latz.) Podp., Consp. Musc. Eur. 204, 1954 Eur (*Didymodon luridus* fo.) Eur ≡ *Didymodon luridus* fo. *brevifolius* Latz.]
  [fo. *rubella* (Loeske in Bauer) Podp., Consp. Musc. Eur. 204, 1954 (*Didymodon luridus* fo.) Eur ≡ *Didymodon luridus* fo. *rubella* Loeske in Bauer]
  [fo. *subscabra* (Linder) Podp., Consp. Musc. Eur. 204, 1954 (*Didymodon luridus* fo.) Eur ≡ *Didymodon luridus* fo. *subscabrus* Linder]
  [fo. *tophacea* (Amann) Podp., Consp. Musc. Eur. 204, 1954 (*Didymodon luridus*) ≡ *Didymodon luridus* fo. *tophaceus* Amann, Fl. Mousses Suisse Ad. 3: 26, 1935]
  [fo. *cuspidata* (Schimp.) Mönk., Laubm. Eur. 294, 1927 (*Didymodon luridus* var.) Eur ≡ *Didymodon trifarius* var. *cuspidatus* (Schimp.) Wijk & Marg.]
[*Barbula trivialis* C. Müll. Afr4 ≡ *Trichostomopsis trivialis* (C. Müll.) Robins., Phytologia 20: 187, 1970 ≡ *Didymodon trivialis* (C. Müll.) Guerra in Guerra & Ros, Cryptogamie Bryol. Lichénol. 8: 64, 1987]
Barbula tuberculosa (Ren. & Par.) Card. in Grand. Afr3
[*Barbula umbrosa* C. Müll. Afr6 ≡ *Trichostomopsis umbrosa* (C. Müll.) Robins., Phytologia 20: 185, 1970 ≡ *Didymodon australasiae* var. *umbrosus* (C. Müll.) Zand. *fide* Zander, Cryptogamie Bryol. Lichénol. 2: 400, 1981 (1982) ≡ *Didymodon umbrosus* (C. Müll.) Zand.]
Barbula umtaliensis Magill in Magill & Schelpe, Mem. Bot. Surv. S. Afr. 43: 5, 1979 (*nom. nov.* for *Tortella obtusifolia* Dix.) Afr2
Barbula unguiculata Hedw. Eur As1 As2 As3 As5 Afr1 Am1 Am2 Am6 Austr1
  var. *elongata* (Schultz) Brid. Eur
  var. *patagonica* C. Müll. Am6
  [var. *proligera* Broth. As2 = *Barbula indica* (Hook.) Spreng. *fide* Saito, J. Hattori Bot. Lab. 39: 488, 1975]
  var. *rigidula* Röll Eur
  var. *rupestris* Hüb. Eur
  var. *simplex* Brid. Eur
  [var. *trichostomifolia* (C. Müll.) Chen As2 = *Barbula unguiculata* Hedw. *fide* Saito, J. Hattori Bot. Lab. 39: 490, 1975]
  fo. *apiculata* (Hedw.) Mönk., Laubm. Eur. 286, 1927 (*Barbula*) Eur Am1
  fo.? *brachypus* (Brid.) Podp., Consp. Cusc. Eur. 213, 1954 *nom. inval. dispon. non cit.* (*Barbula*) Eur
  fo.? *breviseta* (Farn.) Podp., Consp. Musc. Eur. 213, 1954 *nom. inval. dispon. non cit.* (*Barbula unguiculata* var.) Eur
  fo.? *bulbifera* (Schiff.) Podp., Consp. Musc. Eur. 213, 1954 *nom. inval. dispon. non cit.* (*Barbula unguiculata* var.) Eur
  fo. *cuspidata* (Schultz) Mönk., Laubm. Eur. 286, 1927 (*Barbula cuspidata*) Eur
  fo. *fastigiata* (Schultz) Mönk., Laubm. Eur. 286, 1927 (*Barbula*) Eur
  fo. *flaccida* Röll, Jahrb. Ak. Wiss. Erfurt 41: 123, 1915 Eur
  fo.? *lanceolata* (Hedw.) Podp., Consp. Musc. Eur. 213, 1954 *nom. inval. dispon. non cit.* (*Barbula*) Eur Am1
  fo.? *microcarpa* (Schultz) Podp., Consp. Musc. Eur. 213, 1954 *nom. inval. dispon. non cit.* (*Barbula*) Eur
  fo. *minus* Tosco, Webbia 28: 284, 1973 "*minor*" Eur
  fo.? *nitidocostata* (Farn.) Podp., Consp. Musc. Eur. 213, 1954 *nom. inval. dispon. non cit.* (*Barbula unguiculata* var.) Eur
  fo. *obtusifolia* Mönk., Laubm. Eur. 286, 1927 (*Barbula obtusifolia* Schultz *hom. illeg.*) Eur!
  fo.? *paludosa* (Röll) Podp., Consp. Musc. Eur. 214, 1954 *nom. inval. dispon. non cit.* (*Barbula unguiculata* var.) Eur
  fo. *polyseta* Peterfi, Mag. Bot. Lapok 1: 49, 1902 Eur
  [fo. *propagulosa* Crum, Bryologist 72: 241, 1969 Am1 = *Barbula indica* (Hook.) Spreng. *fide* Zander, Photologia 44: 186, 1979]
  fo. *robusta* (Lindb. ex Hag.) Podp. in Sap., Bot. Jahrb. Syst. 46 (Beibl. 105): 12, 1911 (*Barbula unguiculata* var.) Eur
  fo. *subsquarrosa* Latz., Bot. Centralbl. Beih. 48: 2, 480, 1931 Eur
Barbula unguiculatula C. Müll. Am4 Am6
[*Barbula uruguayensis* Broth. (*hom. illeg.* of *Barbula uruguensis* Par. *fide* Margadant in litt.) Am6 ≡ *Didymodon uruguayensis* (Broth.) Zand., see treatment of *Didymodon*]
[*Barbula uruguensis* Par., Act. Soc. Linn. Bordeaux 46: 111, 1893 [1894] Am6 ≡ *Didymodon deciduus* Zand. *nom. nov.*, see treatment of *Didymodon*]
Barbula vaginata Warnst. Am6
[*Barbula valida* (Limpr.) Möll. Eur = *Barbula acuta* (Brid.) Brid. *fide* Crundwell & Nyholm, Svensk Bot. Tidskr. 59(2): 214, 1965 = *Didymodon rigidulus* var. *gracilis* (Hook. & Grev.) Zand. *fide* Zander, Cryptog. Bryol. Lichénol. 2: 393, 1981 (1982)]
Barbula validinervia C. Müll. Afr4
Barbula vardei R.S. Chopra, Bryologist 80: 544, 1977 (*nom. nov.* for *Barbula denticulata* Dix. & P. Varde) As3
Barbula ventanica C. Müll. Am6
[*Barbula vinealis* Brid. Eur As1 As2 As3 As5 Afr1 Am1 Am2 Oc ≡ *Didymodon vinealis* (Brid.) Zand., Phytologia 41: 25, 1978]
  [subsp. *cylindrica* (Tayl.) Podp. Eur As1 As5 Afr1 Am1 Am6 Oc = *Didymodon vinealis* var. *flaccidus* (BSG) Zand., Phytologia 41: 25, 1978 ≡ *Didymodon vinealis* (Brid.) Zand. var. *vinealis* *fide* Sollman, Bryologist 86: 272, 1983]
  var. *decipiens* Meyl. in Amann Eur
  [var. *flaccida* BSG Eur As1 As5 Afr1 Am1 Am6 Oc ≡ *Didymodon vinealis* var. *flaccidus* (BSG) Zand., Phytologia 41: 25, 1978 ≡ *Didymodon vinealis* (Brid.) Zand. var. *vinealis* *fide* Sollman, Bryologist 86: 272, 1983]
  var. *propagulifera* Amann Eur
  fo. *compacta* Herz., Ber. Zürich. Bot. Ges. 15: 45, 1905 Eur
  fo. *propagulifera* (Amann) Podp., Consp. Musc. Eur. 210, 1954 (*Didymodon vinealis* var.) Eur
  fo. *rivularis* (Loeske) Podp., Consp. Musc. Eur. 210, 1954 (*Barbula*) Eur
  fo. *robusta* (Loeske) Podp., Consp. Musc. Eur. 210, 1954 (*Barbula cylindrica* fo.) Eur
  fo. *rubella* (Schiffn.) Podp., Consp. Musc. Eur. 210, 1954 (*Barbula cylindrica* var.) Eur
  fo. *viridis* Podp., Consp. Musc. Eur. 210, 1954 Eur
Barbula vulcanica Lor. Am4
[*Barbula whitehouseae* Crum Am1 = *Barbula eustegia* Card. & Thér. *fide* Zander, Phytologia 44: 197, 1979]
Barbula williamsii (Chen) Iwats. & Tan, Kilikasan, Philippine J. Biol. 8: 186, 1979 (*Hydrogonium*) As4
[*Barbula wisselii* Dix. As4 ≡ *Didymodon wisselii* (Dix.) Norris & T. Kop., Acta Bot. Fenn. 137: 127, 1989]
[*Barbula wrightii* Saurb. in Jäg. Am3 = *Barbula indica* (Hook.) Spreng. *fide* Zander, Phytologia 44: 186, 1979]
[*Barbula xanthocarpa* C. Müll. Afr2 Afr4 ≡ *Didymodon xanthocarpus* (C. Müll.) Magill, Fl. So. Afr. Bryoph. 1: 235, 1981]
Barbula yunnanensis Copp. As2

Barbula zambesiaca Magill in Magill & Schelpe, Mem. Bot. Surv. S. Afr. 43: 5, 1979 (*nom. nov.* for *Semibarbula elongata* Hilp.) Afr2

[*Barbula zollingeri* (Fleisch.) Broth. As4 ≡ *Barbula indica* (Hook.) Spreng. var. *indica fide* Norris & Koponen, Acta Bot. Fenn. 137: 114, 1989]

[**BARNESIA** Card. = *Streptocalyptra* C. Müll. *fide* Zander, Lindbergia 8: 162, 1982 (1983)]

[*Barnesia tortelloides* Card. Am2 ≡ *Streptocalypta tortelloides* (Card.) Zand., Lindbergia 8: 163, 1982 (1983)]

BELLIBARBULA Chen

Bellibarbula kurziana Chen As3

var. *purpurascens* Gangulee, Nov. Hedw. 8: 148, 1964 As3

[*Bellibarbula obtusicuspis* (Besch.) Chen As2 = *Bellibarbula recurva* (Griff.) Zand., see treatment of *Bellibarbula*]

Bellibarbula recurva (Griff.) Zand., see treatment of *Bellibarbula* (*Gymnostomum*) As2 As3 Am1 Am2

BYROCEUTHOSPORA Crum & Anders.

Byroceuthospora aethiopica (Welw. & *dub.*) Zand. (*Ephemerum*), see treatment of *Byroceuthospora* Afr2

Byroceuthospora mexicana (Bartr.) Crum & Anders. Am2

BRYOERYTHROPHYLLUM Chen

Bryoerythrophyllum alpigenum (Vent.) Chen Eur As1 As2 As3 Am1 Austr2

Bryoerythrophyllum andersonianum Zand. & Sharp, Bryologist 84: 545, 1981 Am2

[*Bryoerythrophyllum angustulum* (Herz.) Robins., Bryologist 70: 22, 1970 (*Didymodon*) Am4 = *Trichostomopsis australasiae* (Hook. & Grev.) Robins. *fide* Robinson, Smiths. Contr. Bot. 27: 30, 1975 ≡ *Didymodon australasiae* (Hook. & Grev.) Zand. *fide* Zander, Phytologia 41: 21, 1978]

[*Bryoerythrophyllum arcuatum* (Mitt.) Crum Am4 (= *Bryoerythrophyllum jamesonii* (Tayl.) Crum *fide* Zander, Bryologist 81: 549, 1978 [1979]) = *Bryoerythrophyllum campylocarpum* (C. Müll.) Crum *fide* Zander, Bryologist 89: 15, 1986]

[*Bryoerythrophyllum afrorubellum* (Broth. & Wag.) De Sloover, Bull. Jard. Bot. Nat. Belg. 49: 398, 1979 Afr4 (= *Bryoerythrophyllum jamesonii* (Tayl.) Crum *fide* Magill, Fl. S. Afr. I. Mosses 1: 248, 1981 [1982]) = *Bryoerythrophyllum campylocarpum* (C. Müll.) Crum from illustration by Magill, Fl. S. Afr. I. Mosses 1: 248, 1981 (1982); see Zander, Bryologist 89: 13, 1986]

[*Bryoerythrophyllum atrorubens* (Besch.) Chen As2 As3 = *Bryoerythrophyllum wallichii* (Mitt.) Chen *fide* Saito, J. Hattori Bot. Lab. 39: 479, 1975]

Bryoerythrophyllum binnsii (R. Brown ter) Wijk & Marg. Austr1 Austr2 (good species *fide* Zander, Bryologist 89: 15, 1986)

Bryoerythrophyllum bolivianum (C. Müll.) Zand., Bryologist 81: 545, 1978 [1979] (*Globulina*) Am2 Am4

Bryoerythrophyllum brachystegium (Besch.) Saito, J. Jap. Bot. 47: 14, 1972 As2

Bryoerythrophyllum byrdii (Bartr.) Zand. (*Barbula*), see treatment of *Bryoerythrophyllum* Ant

Bryoerythrophyllum calcareum (Thér.) Zand., Bryologist 83: 232, 1980 (*Barbula*) Am2

Bryoerythrophyllum campylocarpum (C. Müll.) Crum Eur As2 As3 Afr2 Afr4 Am2 Am3 Am4 Am6 [= *Bryoerythrophyllum jamesonii* (Tayl.) Crum *fide* Zander, Bryologist 81: 549, 1978 (1979)] = *Bryoerythrophyllum campylocarpum* (C. Müll.) Crum *fide* Zander, Bryologist 89: 15, 1986]

[*Bryoerythrophyllum cavernarum* (Mol.) Podp. = *Bryoerythrophyllum rubrum* (Jur.) Chen]

fo. *brevifolium* (Herz.) Podp., Consp. Musc. Eur. 219, 1954 (*Erythrophyllum rubrum* f.) Eur

Bryoerythrophyllum chimborazense (Mitt.) Zand. (*Tortula*), see treatment of *Bryoerythrophyllum* Am4

Bryoerythrophyllum columbianum (Herm. & Lawt.) Zand., Bryologist 81: 548, 1978 (*Didymodon*) Am1

[var. *atacamense* Zand. & Lewis in Lewis, Bryologist 84: 536, 1981 [1982] "*atacamensis*" Am4 = *Bryoerythrophyllum fuscinervium* (Mitt.) Zand., see treatment of *Byroerythrophyllum*]

[*Bryoerythrophyllum dentatum* (Mitt.) Chen As3 As4 = *Leptodontium flexifolium* (Dicks.) Hampe in Lindb. *fide* Zander, Bryologist 75: 231, 1972 cf. Zander, Bryologist 84: 546, 1972]

Bryoerythrophyllum ferruginascens (Stirt.) Giac. Eur As1 As4 Am1 Am2

Bryoerythrophyllum ferrugineum Gangulee, Nov. Hedwigia 8: 147, 1964 As3

Bryoerythrophyllum fuscinervium (Mitt.) Zand. (*Tortula*), see treatment of *Byroerythrophyllum* Am4 Am6

Bryoerythrophyllum gymnostomum (Broth.) Chen As2

Bryoerythrophyllum hostile (Herz.) Chen As2

Bryoerythrophyllum inaequalifolium (Tayl.) Zand., Bryologist 83: 232, 1980 (*Barbula*) As1 As2 As3 As4 Afr1 Am1 Am2 Am4

[*Bryoerythrophyllum linearifolium* Saito, J. Hattori Bot. Lab. 39: 481, 1975 As2 = *Bryoerythrophyllum recurvum* (Griff.) Saito *fide* Sollman, Lindbergia 16: 22, 1990 ≡ *Bellibarbula recurva* (Griff.) Zand.,, see treatment of *Bellibarbula*]

[*Bryoerythrophyllum lusitanicum* (Card. & Dix.) Hill, J. Bryol. 11: 600, 1981 [1982] (*Hyophila*) Eur = *Bryoerythrophyllum campylocarpum* (C. Müll.) Crum *fide* Zander, Bryologist 89: 15, 1986]

Bryoerythrophyllum jamesonii (Tayl.) Crum Am2 Am4

Bryoerythrophyllum ligulare (Mitt.) Zand. (*Tortula*), see treatment of *Bryoerythrophyllum* Am4

Bryoerythrophyllum machadoanum (Sérgio) Hill, J. Bryol. 11: 601, 1981 [1982] (*Hyophila*) Eur

Bryoerythrophyllum noguchianum (Gangulee) Saito, Bull. Univ. Mus. Univ. Tokyo 8: 254, 1975 (*Bryoerythrophyllum yunnanense* var.) As2

[*Bryoerythrophyllum obtussissimum* (Broth.) Chen As2 = *Didymodon brachystegius* (Besch.) Broth. *fide* Iwats. & Nog., J. Hattori Bot. Lab. 37: 316, 1973 ≡ *Bryoerythrophyllum brachystegium* (Besch.) Saito *fide* Saito, J. Jap. Bot. 47: 14, 1972 and J. Hattori Bot. Lab. 39: 477, 1975]

[*Bryoerythrophyllum pergemmascens* (Broth.) Chen As2 = *Leptodontium flexifolium* (Dicks. ex. With.) Hampe in Lindb. *fide* Zander, Bryologist 75: 231, 1972]

[*Bryoerythrophyllum recurvifolium* (Tayl.) Zand., Bryologist 75: 277, 1972 Eur Am1 ≡ *Oxystegus recurvifolius* (Tayl.) Zand., Lindbergia 8: 187, 1982 ≡ *Trichostomum recurvifolium* (Tayl.) Zand., see treatment of *Trichostomum*]

Bryoerythrophyllum recurvirostrum (Hedw.) Chen Eur As1 As2 As3 As4 As5 Afr1 Afr2 Afr4 Am1 Am2 Austr1 Oc

var. *aeneum* (Ren. & Card.) Zand., Bryologist 79: 229, 1977 (*Trichostomum*) Am2

var. *antarcticum* Sav.-Ljub. & Z. Smirn. Ant

[var. *angustifolium* (P. Varde) Wijk & Marg. Afr2 = *Bryoerythrophyllum afrorubellum* (Broth. & Wag.) De Sloover *fide* De Sloover, Bull. Jard. Bot. Natl. Belg. 49: 398, 1979]

var. *brevifolium* (Lindb. & Arn.) Podp. As1 Am1

[var. *dentatum* (Schimp.) Crum, Steere & Anders., Bryologist 67: 163, 1964 Am1 = *Bryoerythrophyllum recurvirostrum* (Hedw.) Chen *fide* Crum, Steere & Anderson, Bryologist 68: 418, 1973]

var. *crassinerve* (Herz.) Podp. Eur
var. *latinervium* (Holmen) Murray, Bryobrothera 1: 14, 1992 (*Bryoerythrophyllum recurvirostrum* var.) Am1
var. *robustum* Saito, J. Hattori Bot. Lab. 39: 474, 1975 As2
var. *serratum* (Röll) Podp. Eur
fo. *alpinum* (Herz.) Podp., Consp. Musc. Eur. 218, 1954 (*Erythrophyllum rubellum* f.) Eur
fo. *apiculatum* (Röll) Podp., Consp. Musc. Eur. 218, 1954 (*Didymodon rubellus* var.) Eur
fo. *brevirostre* (Warnst.) Podp., Consp. Musc. Eur. 218, 1954 (*Didymodon rubellus* f.) Eur
fo. *confertum* (Velen.) Podp., Consp. Musc. Eur. 218, 1954 (*Didymodon rubellus* var.) Eur
fo. *flaccidum* (Röll) Podp., Consp. Musc. Eur. 218, 1954 (*Didymodon rubellus* var.) Eur
fo. *gracile* (Limpr.) Podp., Consp. Musc. Eur. 218, 1954 (*Didymodon rubellus* f.) Eur
fo. *latifolium* (Herz.) Podp., Consp. Musc. Eur. 218, 1954 (*Erythrophyllum rubellum* f.) Eur
fo. *longirostre* (Warnst.) Podp., Consp. Musc. Eur. 218, 1954 (*Didymodon rubellus* f.) Eur
fo. *maius* (Breidl. ex Matous.) Podp., Consp. Musc. Eur. 218, 1954 (*Didymodon rubellus* f.] Eur
fo. *obtusifolium* (Röll) Podp., Consp. Musc. Eur. 218, 1954 (*Didymodon rubellus* var.) Eur
fo. *pallens* (Ryan & Hag.) Podp., Consp. Musc. Eur. 218, 1954 (*Didymodon rubellus* var.) Eur
fo. *pygmaeum* (Meyl.) Podp., Consp. Musc. Eur. 218, 1954 (*Didymodon rubellus* f.) Eur
fo. *pulvinatum* (Herz.) Podp., Consp. Musc. Eur. 218, 1954 (*Erythrophyllum rubellum* f.) Eur
fo. *viride* (Schlieph. ex Limpr.) Podp., Consp. Musc. Eur. 218, 1954 (*Didymodon rubellus* f.) Eur
[*Bryoerythrophyllum recurvum* (Griff.) Saito, Bull. Univ. Mus. Univ. Tokyo 8: 254, 1975 (*Didymodon*) As3 Am1 Am2 ≡ *Bellibarbula recurva* (Griff.) Zand.,, see treatment of *Bellibarbula*]
Bryoerythrophyllum rotundatum (Lindb. & Arn.) Chen As1
Bryoerythrophyllum rubrum (Geh.) Chen Eur As1 As2
[var. *minus* Saito, J. Hattori Bot. Lab. 39: 476, 1975 As2 = *Bryoerythrophyllum ferruginascens* (Stirt.) Giac. *fide* Sollman, Bryologist 86: 271, 1983]
Bryoerythrophyllum sharpii Zand., Bryologist 89: 13, 1986 Am2
[*Bryoerythrophyllum tenerrimum* (Broth.) Chen As2 = *Bryoerythrophyllum recurvum* (Griff.) Saito *fide* Sollman, Bryologist 86: 271, 1983 ≡ *Bellibarbula recurva* (Griff.) Zand.,, see treatment of *Bellibarbula*]
Bryoerythrophyllum wallichii (Mitt.) Chen As2 As3
Bryoerythrophyllum yichunense Ch. Gao, Fl. Musc. Chinae Bor.-Or. 379, 1977 As2
Bryoerythrophyllum yunnanense (Herz.) Chen As2
[var. *noguchianum* Gangulee, Mosses of India and Adj. Reg., Fasc. 3: 750, 1972 As2 = *Bryoerythrophyllum noguchianum* Saito, Bull. Univ. Mus. Univ. Tokyo 8: 254, 1975]
var. *pulvinans* (Herz.) Chen As2

CALYMPERASTRUM Stone, J. Bryol. 14: 315, 1986 [1987]
Calymperastrum latifolium Stone, J. Bryol. 14: 315, 1986 [1987] Austr1

CALYPTOPOGON (Mitt.) Broth.
Calyptopogon mnioides (Schwaegr.) Broth. Am4 Am6 Austr1 Austr2
var. *anguste-limbatus* Salm. Austr2

CHENIA Zand.
Chenia lorentzii (C. Müll.) Zand. (*Barbula*), see treatment of *Chenia* Am6
Chenia leptophylla (C. Müll.) Zand. (*Phascum*) Eur As2 As4 Oc Afr1 Afr2 Afr4 Am1 Am2 Am4 Am5 Am6 Austr1
[*Chenia rhizophylla* (Sak.) Zand., Phytologia 65: 425, 1989 (*Physcomitrium*) Eur As2 As4 Oc Afr2 Afr4 Am1 Am2 Am4 Am5 Am6 Austr1 = *Chenia leptophylla* (C. Müll.) Zand., see treatment of *Chenia*]
Chenia subobliqua (Williams) Zand., Phytologia 65: 425, 1989 (*Tortula*) Am4

CHIONOLOMA Dix. [= *Pseudosymblepharis* Broth. *fide* Eddy, Handb. Males. Mosses 2: 153, 1991]
[*Chionoloma angustata* (Mitt.) Menzel, Willdenowia 22: 197, 1992 ≡ *Pseudosymblepharis angustata* (Mitt.) Hilp.]
[*Chionoloma bartramii* (Thér. ex Bartr.) Menzel, Willdenowia 22: 198, 1992 ≡ *Pseudosymblepharis bartramii* Thér. ex Bartr.] = *Pseudosymblepharis schimperiana* (Par.) Crum, see treatment of *Pseudosymblepharis*]
[*Chionoloma duriuscula* (Mitt.) Menzel, Willdenowia 22: 198, 1992 ≡ *Pseudosymblepharis duriuscula* (Mitt.) Chen]
Chionoloma induratum Dix. As3 [= *Pseudosymblepharis subduriuscula* (C. Müll.) Chen *fide* Eddy, Handb. Males. Mosses 2: 153, 1991]
Chionoloma latifolium Dix. As3 [= *Pseudosymblepharis subduriuscula* (C. Müll.) Chen *fide* Eddy, Handb. Males. Mosses 2: 153, 1991]
Chionoloma longifolium Dix. As4 [= *Pseudosymblepharis subduriuscula* (C. Müll.) Chen *fide* Eddy, Handb. Males. Mosses 2: 153, 1991]
[*Chionoloma schimperiana* (Par.) Menzel, Menzel, Willdenowia 22: 198, 1992 ≡ *Pseudosymblepharis schimperiana* (Par.) Crum]
[*Chionoloma socotrana* (Mitt.) Menzel, Willdenowia 22: 198, 1992 ≡ *Pseudosymblepharis socotrana* (Mitt.) Thér. Afr2 ≡ *Weissia artocosana* (Mitt.) Zand. *nom. nov.*, see treatment of *Weissia*]
[*Chionoloma subduriuscula* (C. Müll.) Menzel, Willdenowia 22: 198, 1992 ≡ *Pseudosymblepharis subduriuscula* (C. Müll.) Chen As2 As3 As4 = *Pseudosymblepharis angustata* (Mitt.) Hilp. *fide* Norris & Koponen, Acta Bot. Fenn. 137: 94, 1987]

[**CINCLIDOTUS** P. Beauv. referred to Cinclidotaceae]

CROSSIDIUM Jur.
Crossidium aberrans Holz. & Bartr. Afr1 Am1 Am2
fo. *epilosum* Flow., Bryologist 76: 289, 1973 Am1
Crossidium apiculatum Magill, Fl. S. Afr. I. Mosses 1: 195, 1981 [1982] Afr4
Crossidium asirense Frey & Kürschner, J. Bryol. 13: 25, 1984 As5
[*Crossidium chloronotos* (Brid.) Limpr. Eur As2 As3 As5 Afr1 Am1 Austr2 = *Crossidium squamiferum* (Viv.) Jur. *fide* Delgadillo, Bryologist 78: 276, 1975]
Crossidium crassinerve (De Not.) Jur. Eur As3 Afr1 Am1 Am2
[var. *laevipilum* (Thér. & Trab.) Delg., Bryologist 78: 275, 1975 (*Crossidium*) As5 Afr1 ≡ *Crossidium laevipilum* Thér. & Trab. *fide* Frey & Kürchner, Cryptogamie Bryol. Lichénol. 12: 441–450, 1991]
Crossidium davidai Catcheside, Mosses S. Australia 152, 1980 Austr1
Crossidium deserti Frey & Kürschner, Nova Hedw. 46: 87, 1988 (nom. nov. for *Crossidium desertorum* Frey & Kürschner) As5

[*Crossidium desertorum* Holz. & Bartr. Am1 Am2 = *Crossidium crassinerve* (De Not.) Jur. *fide* Delgadillo, Sharp et al., Moss Fl. Mex.]

[*Crossidium desertorum* Frey & Kürschner, Nova Hedw. 45: 132, 1987 *hom. illeg.* As5 ≡ *Crossidium deserti* Frey & Kürschner, Nova Hedw. 46: 87, 1988]

[*Crossidium elatum* Williams Am4 ≡ *Pseudocrossidium elatum* (Williams) Delg., Bryologist 78: 278, 1975]

[*Crossidium erosum* Holz. & Bartr. Am1 = *Crossidium crassinerve* (De Not.) Jur. *fide* Delgadillo, Sharp et al., Moss Fl. Mex.]

Crossidium geheebii (Broth.) Broth. Austr1 Austr2

Crossidium laevipilum Thér. & Trab. Afr1 As5 [≡ *Crossidium crassinerve* var. *laevipilum* (Thér. & Trab.) Delg. *fide* Delgadillo, Bryologist 78: 275, 1975 but good species *fide* Frey & Kürchner, Cryptogamie Bryol. Lichénol. 12: 441–450, 1991]

Crossidium laxefilamentosum Frey & Kürschner, Nova Hedw. 45: 130, 1987 As5

[*Crossidium peruvianum* Broth. Am4 = *Aloina rosei* (Williams) Delg.]

[*Crossidium rosei* Williams Am4 Am6 ≡ *Aloina rosei* (Williams) Delg.

Crossidium seriatum Crum & Steere Eur Am1 Am2

[*Crossidium spatulaefolium* Holz. & Bartr. Am4 Am6 = *Crossidium aberrans* Holz. & Bartr. *fide* Delgadillo, Bryologist 78: 270, 1975]

Crossidium spiralifolium Magill, Fl. S. Afr. I. Mosses 1: 197, 1981 Afr4

Crossidium squamiferum (Viv.) Jur. Eur As1 As2 As3 As5 Afr1 Am1 Austr2

var. *brevisetum* (Besch.) Par. Afr1

var. *longipilum* Amann Eur

var. *pottioideum* (De Not.) Mönk. Eur As3 Afr1 Am1 Am2

var. *pumilum* Amann Eur

Crossidium woodii (Delgad.) Zand. (*Pseudaloina*), see treatment of *Crossidium* As5

CRUMIA Schof.

[*Crumia deciduidentata* Sharp & Iwats., J. Hattori Bot. Lab. 32: 95, 1969 Am1 ≡ Tortula deciduidentata (Sharp & Iwats.) Zand., see treatment of *Tortula*]

Crumia latifolia (Kindb. ex Macoun) Schof., Canad. J. Bot. 44: 610, 1966 (*Merceya*) Am1

[**DESMATODON** Brid. = *Tortula* Hedw., see treatment of *Tortula*]

[*Desmatodon africanus* P. Varde Afr2 = *Scopelophila cataractae* (Mitt.) Broth., see treatment of *Scopelophila*]

[*Desmatodon altipes* Broth. As1 ≡ *Tortula altipes* (Broth.) Zand., see treatment of *Tortula*]

[*Desmatodon argentinicus* Broth. Am6 ≡ *Tortula argentinica* (Broth.) Zand., see treatment of *Tortula*]

[*Desmatodon austrogeorgicus* (Card.) Ochyra in Ochyra, Vitt & Hort., Cryptog. Bryol. Lichénol. 7: 57, 1986 ≡ *Hennediella austrogeorgica* (Card.) Blockeel, J. Bryol. 16: 191, 1991]

[*Desmatodon bellii* Bartr. Am4 ≡ *Hennediella bellei* (Bartr.) Zand., see treatment of *Hennediella*]

[*Desmatodon bogosicus* C. Müll. Afr1 Afr2 Afr4 ≡ *Tortula bogosica* (C. Müll.) Zand., see treatment of *Tortula*]

[*Desmatodon capillaris* Chen As2 ≡ *Tortula capillaris* (Chen) Zand., see treatment of *Tortula*]

[*Desmatodon cernuus* (Hüb.) BSG Eur As1 As2 Am1 ≡ *Tortula cernua* (Hueb.) Lindb., see treatment of *Tortula*]

[var. *xanthopus* Kindb. Am1 ≡ *Tortula cernua* var. *xanthopus* (Kindb.) Zand., see treatment of *Tortula*]

[*Desmatodon coloradensis* Grout Am1 = *Desmatodon obtusifolius* (Schwaegr.) Schimp. *fide* Crum & Anderson, Mo. E. N. Amer. 371, 1981 ≡ *Tortula obtusifolia* (Schwaegr.) Math. *fide* Corley et al., J. Bryol. 11: 620, 1981 (1982)]

[*Desmatodon convolutus* (Brid.) Grout Eur Am1 Am2 Am3 Am4 Am6 Austr1 Austr2 Oc Afr1 Afr2 Afr3 Afr4 = *Tortula atrovirens* (Sm.) Lindb. *fide* Corley et al., J. Bryol. 11: 620, 1981 (1982)]

[var. *edentulus* (BSG) Grout Eur Afr4 ≡ *Tortula atrovirens* var. *edentula* (BSG) Schimp.]

[var. *gasilienii* (Vent.) Wijk & Marg. Eur ≡ *Tortula atrovirens* var. *gasilienii* (Vent.) Limpr.]

[var. *leucodontus* (Corb.) Wijk & Marg. Eur ≡ *Tortula atrovirens* var. *leucodonta* (Corb.) Zand., see treatment of *Tortula*]

[*Desmatodon ellesmerensis* Brass., Bryologist 74: 208, 1971 Am1 (= *Pseudocrossidium revolutum* (Brid. in Schrad.) Zand. *fide* Zander, Phytologia 44: 204, 1979) = *Pseudocrossidium revolutum* var. *obtusulum* (Lindb.) Tan, Zand. & T. Tayl., Lindbergia 7: 41, 1981]

[*Desmatodon fisherae* Crum, Bryologist 75: 360, 1972 Am1 = *Desmatodon porteri* Jam. in Aust. *fide* Crum & Anderson, Mo. E. N. Amer. 371, 1981 ≡ *Hennediella porteri* (Jam. in Aust.) Zand., see treatment of *Hennediella*]

[*Desmatodon gemmascens* Chen As2 ≡ *Syntrichia gemmascens* (Chen) Zand., see treatment of *Syntrichia*]

[var. *hopeiensis* Chen As2 ≡ *Syntrichia gemmascens* var. *hopeiensis* (Chen) Zand., see treatment of *Syntrichia* As2]

[*Desmatodon guepinii* BSG Eur Am1 Am2 ≡ *Tortula guepinii* (B.&S. in BSG) Broth., see treatment of *Tortula*]

[*Desmatodon haussknechtii* (Jur. & Milde) Fröhl., Ann. Nat. Mus. Wien 67: 155, 1964 As5 ≡ *Didymodon haussknechtii* (Jur. & Mild.) Broth.]

[*Desmatodon heimii* (Hedw.) Mitt. ≡ *Pottia heimii* (Hedw.) Fürnr. in Hampe ≡ *Hennediella heimii* (Hedw.) Zand., see treatment of *Hennediella*]

[var. *arcticus* (Lindb.) Crum, Bryologist 72: 242, 1969 "*arctica*" Am1 ≡ Hennediella heimii var. *arctica* (Lindb.) Zand., see treatment of *Hennediella*]

[*Desmatodon hendersonii* (Ren. & Card.) Williams in Millsp. & Nutt. Am1 = *Didymodon tophaceus* (Brid.) Lisa *fide* Anderson et al., Bryologist 93: 478, 1990]

[*Desmatodon involutus* Bartr. *hom. illeg.* As3 ≡ *Plaubelia perinvoluta* Zand., see treatment of *Plaubelia*]

[*Desmatodon kabir-khanii* Broth. As3 ≡ *Tortula kabir-khanii* (Broth.) Zand., see treatment of *Tortula*]

[*Desmatodon latifolius* (Hedw.) Brid. Eur As1 As2 As3 As5 Afr1 Am1 ≡ *Tortula euryphylla* Zand. *nom. nov.*, see treatment of *Tortula*]

[subsp. *brevifolius* (Kindb.) Kindb. Eur ≡ *Tortula euryphylla* subsp. *brevifolia* (Kindb.) Zand., see treatment of *Tortula*]

[var. *eucalyptratus* (Lindb.) Kaur. Eur ≡ *Tortula euryphylla* var. *eucalyptrata* (Lindb.) Zand., see treatment of *Tortula*]

[var. *flavescens* (Brid.) Steud. Eur ≡ *Tortula euryphylla* var. *flavescens* (Brid.) Zand., see treatment of *Tortula*]

[var. *muticus* (Brid.) Brid. Eur As1 Am1 Am3 = *Desmatodon latifolius* fo. *fide* Anderson et al. 93: 478, 1990 ≡ *Tortula euryphylla* Zand., see treatment of *Tortula*]

[var. *pilifer* (Dicks.) Rabenh. Eur = *Desmatodon latifolius* (Hedw.) Brid. var. *latifolius fide* Lawton, Moss Fl. Pacif. Northw. 95, 1971 ≡ *Tortula euryphylla* Zand., see treatment of *Tortula*]

[var. *spelaeus* (Amann) Podp. Eur ≡ *Tortula euryphylla* var. *spelaeus* (Amann) Zand., see treatment of *Tortula*]

[var. *subobliquus* Lindb. Eur ≡ *Tortula euryphylla* var. *subobliqua* (Lindb.) Zand., see treatment of *Tortula*]
fo. *eucalyptratus* (Lindb.) Hag., Skand. Bladmfl. 204, 1939 (*Tortula*) Eur
[*Desmatodon laureri* (Schultz) BSG Eur As1 As2 Afr4 Am1 ≡ *Tortula laureri* (Schultz) Lindb., see treatment of *Tortula*]
[var. *setschwanicus* (Broth.) Chen As2 ≡ *Tortula laureri* var. *setschwanica* (Broth.) Zand., see treatment of *Tortula*]
[*Desmatodon leucostoma* (R. Br.) Berggr. Eur Am1 As1 As2 ≡ *Tortula leucostoma* (R. Br.) Hook. & Grev., see treatment of *Tortula*]
var. *muticus* (Lindb.) Sav.-Ljubits., Novosti Sist. Niz. Rast. 6: 248, 1969 [1970] (*Desmatodon obliquus* var.) Eur
[*Desmatodon lingulatus* (Hook. f. & Wils.) Sainsb. Austr2 ≡ *Didymodon lingulatus* (Hook. f. & Wils.) Broth., see treatment of *Didymodon*]
[*Desmatodon longipedunculatus* (C. Müll.) Magill, Fl. S. Afr. I. Mosses 1: 210, 1981 (1982) (*Barbula*) Afr4 ≡ *Hennediella longipedunculata* (C. Müll.) Zand., see treatment of *Hennediella*]
[*Desmatodon meridionalis* Luis. Eur ≡ *Tortula meridionalis* (Luis.) Mach.]
[*Desmatodon obtusifolius* (Schwaegr.) Schimp. Eur As1 As2 As5 Afr1 Am1 ≡ *Tortula obtusifolia* (Schwaegr.) Math. *fide* Corley et al., J. Bryol. 11: 620, 1981 (1982)]
var. *brevifolius* Schimp. Eur
var. *limbatus* (Herz.) Wijk & Marg. Eur
[*Desmatodon plinthobius* Sull. & Lesq. Am1 ≡ *Tortula plinthobia* (Sull. & Lesq.) Broth. *fide* Zander, Bryologist 82: 551, 1979]
[*Desmatodon porteri* Jam. in Aust. Am1 ≡ *Tortula porteri* (Jam. in Aust.) Zand., see treatment of *Tortula*]
[*Desmatodon randii* (Kenn.) Laz. Eur Am1 ≡ *Tortula randii* (Kenn.) Zand., see treatment of *Tortula*]
[*Desmatodon raucopapillosus* X.-j. Li, Acta Bot. Yunnan. 3: 105, 1981 As2 ≡ *Tortula raucopapillosa* (X.-j. Li) Zand., see treatment of *Tortula*]
[*Desmatodon recurvatus* (Hook.) Mitt. (good species *fide* Catcheside, Mosses S. Austr. 150, 1980) Afr4 Austr1 ≡ *Tortula recurvata* Hook.]
[*Desmatodon reflexidens* (Hampe) Jaeg. Austr1 = *Desmatodon recurvatus* (Hook.) Mitt. *fide* Catcheside, Mosses S. Austr. 148,1980]
[*Desmatodon solomensis* Broth. As2 ≡ *Tortula solomensis* (Broth.) Zand., see treatment of *Tortula*]
[*Desmatodon spathulifolius* Bartr. Am2 = *Neohyophila sprengelii* (Schwaegr.) Crum *s. lat. fide* Zander, Bryologist 86: 135, 1983 ≡ *Plaubelia sprengelii* (Schwaegr.) Zand., see treatment of *Plaubelia*]
[*Desmatodon sprengelii* (Schwaegr.) Williams Am1 Am2 Am3 ≡ *Neohyophila sprengelii* (Schwaegr.) Crum, Bryologist 68: 470, 1965 ≡ *Plaubelia sprengelii* (Schwaegr.) Zand., see treatment of *Plaubelia*]
[*Desmatodon steereanus* Zand. & Crum, Bryologist 80: 638, 1977 Am4 ≡ *Hennediella steereana* (Zand. & Crum) Zand., see treatment of *Hennediella*]
[*Desmatodon stomatodontus* (Card.) Williams Am2 Am3 Am5 ≡ *Neohyophila sprengelii* var. *stomatodonta* (Card.) Zand., Bryologist 86: 138, 1983 ≡ *Plaubelia sprengelii* var. *stomatodonta* (Card.) Zand., see treatment of *Plaubelia*]
[*Desmatodon suberectus* (Hook.) Limpr. Eur As1 As2 Am1 = *Tortula leucostoma* (R. Brown) Berggr.]
[var. *limbatus* Amann Eur = *Desmatodon wilczekii* Meyl.]
fo. *apiculatus* (Lindb.) Podp., Consp. Musc. Eur. 241, 1954 (*Desmatodon obliquus* var.) Eur
fo. *muticus* (Lindb.) Podp., Consp. Musc. Eur. 241, 1954 (*Desmatodon obliquus* var.) Eur
fo. *pilifer* (Lindb.) Podp., Consp. Musc. Eur. 241, 1954 (*Desmatodon obliquus* var.) Eur
[*Desmatodon systylius* Schimp. Eur As1 As3 Am1 ≡ *Tortula systylia* (Schimp.) Lindb., see treatment of *Tortula*]
[*Desmatodon thomsonii* (C. Müll.) Jaeg. As2 As3 ≡ *Tortula thompsonii* (C. Müll.) Zand., see treatment of *Tortula*]
[*Desmatodon tonkinensis* Besch. As3 ≡ *Tortula tonkinensis* (Besch.) Zand., see treatment of *Tortula*]
[*Desmatodon ucrainicus* Laz. Eur ≡ *Tortula ucrainica* (Laz.) Zand., see treatment of *Tortula*]
[*Desmatodon viridipilus* (Dix. & Sainsb.) Sainsb. Austr2 ≡ *Tortula viridipila* Dix. & Sainsb.]
[*Desmatodon wilczekii* Meyl. Eur ≡ *Tortula rhodonia* Zand. *nom. nov.*, see treatment of *Tortula*]
[*Desmatodon yuennanensis* Broth. As2 ≡ *Tortula chungtienia* Zand. *nom. nov.*, see treatment of *Tortula*]

DIALYTRICHIA (Schimp.) Limpr.
[*Dialytrichia brebissonii* (Brid.) Limpr. = *Dialytrichia mucronata* (Brid.) Broth.]
var. *pumila* Amann Eur
[*Dialytrichia fragillima* Dix. & Sak. As2 = *Barbula subcomosa* Broth. *fide* Saito, J. Hattori Bot. Lab. 39: 493, 1975]
Dialytrichia mucronata (Brid.) Broth. Eur As2 As5 Afr1
var. *conferta* (Corb.) Corb. in Pitard Eur Afr1
var. *fragilifolia* Biz. & Roux, Rev. Bryol. Lichénol. 36: 110, 1968 [1969] *inval. holotyp. non cit.* Eur

DIDYMODON Hedw.
Didymodon aaronis (Lor.) Guerra in Guerra & Ros, Cryptogamie Bryol. Lichénol. 8: 55, 1987 (*Trichostomum*) Eur As5 Afr1
[*Didymodon acutus* (Brid.) Saito, J. Hattori Bot. Lab. 39: 519, 1975. Eur As1 As2 Afr1 Am1 = *Didymodon rigidulus* var. *gracilis* (Schleich. ex Hook. & Grev.) Zand. *fide* Zander, Cryptogamie Bryol. Lichénol. 2: 393, 1981 (1982)]
[var. *icmadophilus* (Schimp. ex C. Müll.) Zand., Phytologia 41: 20, 1978 (*Barbula*) ≡ *Didymodon rigidulus* var. *icmadophilus* (Schimp. ex C. Müll.) Zand. *fide* Zander, Cryptogamie Bryol. Lichénol. 2: 394, 1981 (1982)]
[var. *ditrichoides* (Broth.) Zand., Phytologia 41: 20, 1978 (*Barbula ditrichoides*) As3 Am1 ≡ *Didymodon rigidulus* var. *ditrichoides* (Broth.) Zand., see treatment of *Didymodon*]
[*Didymodon afrorubellus* Broth. & Wag. ex Dix. Afr4 ≡ *Bryoerythrophyllum afrorubellum* (Broth. & Wag.) DeSloover, Bull. Jard. Bot. Natl. Belg. 49: 398, 1979]
Didymodon alticaulis Bartr. Am2
Didymodon amblyophyllus (Hook.) Broth., Nat. Pfl. 1(3): 406, 1902 Am5 Am6
Didymodon ampliretis Card. & Broth. Am6
Didymodon andreaeoides Card. & Broth. Am6
[*Didymodon angustifolius* Bartr. *hom. illeg.* Am4 ≡ *Didymodon bartramii* Zand. *nom. nov.*, see treatment of *Didymodon*]
[*Didymodon angustulus* Herz. Am4 ≡ *Bryoerythrophyllum angustulum* (Herz.) Robins. *fide* Robinson, Bryologist 70: 22, 1967 = *Bryoerythrophyllum campylocarpum* (C. Müll.) Crum *fide* Robinson, Smithsonian Contr. Bot. 27: 31, 1975]
Didymodon anserinocapitatus (X.-j. Li) Zand. (*Barbula*), see treatment of *Didymodon* As2
[*Didymodon arcticus* (Kaal.) Broth. Eur As1 Am1 ≡ *Trichostomum arcticum* Kaal., cf. Frisvoll, J. Bryol. 13: 435, 1985]
Didymodon argentinicus (Par.) Par. Am4 Am6
Didymodon argentiniensis Warnst. Am6

Didymodon asperifolius (Mitt.) Crum, Steere & Anders., Bryologist 67: 163, 1964 (*Barbula*) Eur As1 As2 As3 Am1

Didymodon australasiae (Hook. & Grev.) Zand., Phytologia 41: 21, 1978 (*Trichostomopsis*) Eur Afr1 Afr4 Am1 Am2 Am3 Am4 Am6 Austr1 Austr2

[var. *umbrosus* (C. Müll.) Zand., Cryptogamie Bryol. Lichénol. 2: 400, 1981 [1982] (*Barbula*) Eur Am1 Am2 Am4 Am6 ≡ *Didymodon umbrosus* (C. Müll.) Zand.]

fo. *propaguliferus* Brugués ex Guerra in Guerra & Ros, Cryptogamie Bryol. Lichénol. 8: 55, 1987 Eur

Didymodon austroalpigena (C. Müll.) Broth. Afr4

[*Didymodon atrorubens* (Besch.) Broth. = *Bryoerythrophyllum wallichii* (Mitt.) Chen *fide* Saito, J. Hattori Bot. Lab. 39: 479, 1975]

Didymodon barbuloides Libert ex March. Eur

Didymodon bartramii Zand. *nom. nov.* (*Didymodon angustifolius* Bartr. *hom. illeg.*), see treatment of *Didymodon* Am4

Didymodon berthoanus Thér. Am6

Didymodon brachyphyllus (Sull. in Whippl.) Zand., Phytologia 41: 24, 1978 (*Barbula*) Am1 [≡ *Didymodon vinealis* var. *brachyphyllus* (Sull. in Whippl.) Zand., Cryptogamie Bryol. Lichénol. 2: 411, 1981 (1982)]

[*Didymodon brachystegius* (Besch.) Broth. As2 ≡ *Bryoerythrophyllum brachystegium* (Besch.) Saito, J. Jap. Bot. 47: 14, 1972]

Didymodon brotheri (Ren. & Par.) Zand. (*Trichostomum*), see treatment of *Didymodon* Afr3

Didymodon brunneus (C. Müll.) Warnst. in Broth. Am6

Didymodon calycinus Dix. Austr2

[*Didymodon calymperidictyon* Broth. in Skottsb. Am6 = *Bryoerythrophyllum campylocarpum* (Tayl.) Crum *fide* Robinson, Smithsonian Contr. Bot. 27: 30, 1975]

Didymodon canaliculatus Dix. As3 [≡ *Barbula canaliculata* (Dix.) R. S. Chopra, Taxon. Indian Mosses 138, 1975]

[*Didymodon capillaceus* (Hedw.) Web. & Mohr ≡ *Distichium capillaceum* (Hedw.) BSG]

var. *elongatus* Brid. *ignot.*

Didymodon capitatus Dix. ex Sappa & Piov. Eur?

Didymodon cardotii (Dus.) Zand., see treatment of *Didymodon* Am6

Didymodon catenulatus Dix. As3

Didymodon ceratodonteus (C. Müll.) Dix. Afr2 Afr4

Didymodon challaense (Broth. in Herz.) Zand., see treatment of *Didymodon* Am4

[*Didymodon chimborazensis* (Mitt.) Broth. in Par. Am4 ≡ *Bryoerythrophyllum chimborazense* (Mitt.) Zand., see treatment of *Bryoerythrophyllum*]

Didymodon coffeanus Norris & T. Kop., Acta Bot. Fenn. 137: 125, 1989 As4

[*Didymodon columbianus* Herm. & Lawt. Am1 ≡ *Bryoerythrophyllum columbianum* (Herm. & Lawt.) Zand. *fide* Zander, Bryologist 81: 548, 1987 (1979)]

Didymodon constrictus (Mitt.) Saito, J. Hattori Bot. Lab. 39: 514, 1975 As2 [= *Didymodon vinealis* (Brid.) Zand. *fide* Sollman, Bryologist 86: 271, 1983]

var. *flexicuspis* (Chen) Saito, J. Hattori Bot. Lab. 39: 516, 1975 As2 [= *Didymodon vinealis* (Brid.) Zand. *fide* Sollman, Bryologist 86: 272, 1983]

Didymodon contortus Herz. Am4

Didymodon cordatus Jur. Eur

subsp. *austriacus* (Schiffn. & Baumg.) Wijk & Marg. Eur

var. *austriacus* (Schiffn. & Baumg.) Latz. Eur

var. *latifolius* (Röll) Röll Eur

var. *longifolius* Röll Eur

fo. *brevicaulis* Röll, Deutsche Bot. Monatsschr. 3: 47, 1885 Eur

fo. *gracilis* Röll, Deutsche Bot. Monatsschr. 3: 47, 1885 Eur

[fo. *latifolius* Röll, Deutsch. Bot. Monatsh. 3: 46, 1885 Eur ≡ *Didymodon cordatus* var. *latifolius* (Röll) Röll]

[fo. *latifolius* Röll, Deutsche Bot. Monatsschr. 3: 46, 1885 Eur ≡ *Barbula cordata* var. *latifolia* (Röll) Podp. *fide* Podpěra, Consp. Musc. Eur. 205, 1954 ≡ *Didymodon cordatus* var. *latifolius* (Röll) Röll]

fo. *longicaulis* Röll, Hedwigia 56: 145, 1915 Eur

fo. *ramosus* Röll, Deutsche Bot. Monatsschr. 3: 57, 1885 Eur

fo. *robustus* Röll, Deutsche Bot. Monatsschr. 3: 58, 1885 Eur

fo. *strictus* Röll, Deutsche Bot. Monatsschr. 3: 57, 1885 Eur

fo. *tenellus* Podp. in Latz., Bot. Centralbl. Beih. 48(2): 477, 1931 Eur

[fo. *typicus* Röll, Deutsche Bot. Monatsschr. 3: 47, 1885 *nom. illeg.* Eur ≡ *Didymodon cordatus* Jur. var. *cordatus*]

[*Didymodon craspedophyllus* Card. Am2 = *Didymodon australasiae* (Hook. & Grev.) Zand. *s. lat. fide* Zander, Cryptogamie Bryol. Lichénol. 2: 397, 1981 (1982)]

Didymodon crassicostatus (Bartr.) Zand. (*Barbula*), see treatment of *Didymodon* Am2

Didymodon deciduus Zand. (*nom. nov.* for *Barbula decidua* C. Müll.), see treatment of *Didymodon* Am6

[*Didymodon decolorans* (Hampe) Williams Am4 = *Trichostomopsis australasiae* (Hook. & Grev.) Robins., Phytologia 20: 187, 1970 ≡ *Didymodon australasiae* (Hook. & Grev.) Zand., Phytologia 41: 21, 1978]

var. *obtusus* Herz. Am4

[*Didymodon diaphanobasis* Card. Am1 = *Trichostomopsis australasiae* (Hook. & Grev.) Robins., Phytologia 20: 187, 1970 ≡ *Didymodon australasiae* (Hook. & Grev.) Zand., Phytologia 41: 21, 1978]

[var. *angustifolius* Thér. in Bartr. Am1 = *Trichostomopsis australasiae* (Hook. & Grev.) Robins., Phytologia 20: 187, 1970 ≡ *Didymodon australasiae* (Hook. & Grev.) Zand., Phytologia 41: 21, 1978]

[*Didymodon dimorphus* (C. Müll.) Broth. (not *Gymnostomum dimorphum* (C. Müll.) Sim *fide* Magill & Schelpe, 1979) Afr4 = *Didymodon ceratodonteus* (C. Müll.) Dix. *fide* Magill, Fl. S. Afr. I. Mosses 1: 233, 1981 (1982)]

[*Didymodon distans* (Hampe in C. Müll.) Jaeg. Afr2 ≡ *Trichostomum distans* Hampe in C. Müll., see treatment of *Trichostomum*]

[*Didymodon dixonii* Wadhwa & Vohra, Curr. Sci. 32: 483, 1963 As3 (*nom. nov.* for *Didymodon obtusifolius* Card. ex Dix. & P. Vard.) As3 ≡ *Barbula dixonii* R. S. Chopra, Taxon. Indian Mosses 139, 1975 *nom. nov.* = *Barbula inaequalifolia* Tayl. *fide* Sollman, Lindbergia 10: 54, 1984 ≡ *Bryoerythrophyllum inaequalifolium* (Tayl.) Zand.]

Didymodon dubius (Schwaegr.) Par. Austr1

Didymodon eckendorfii P. Varde Afr2

[*Didymodon ehrenbergii* (Lor.) Kindb. ≡ *Barbula ehrenbergii* (C. Müll.) Wijk & Marg.]

[fo. *densiretis* Amann, Fl. Mousses Suisse 3: 27, 1933 Eur ≡ *Hydrogonium ehrenbergii* fo. *densirete* (Amann) Podp., Consp. Musc. Eur. 199, 1954]

[fo. *laxiretis* Amann, Rev. Bryol. 51: 10, 1924 Eur ≡ *Hydrogonium ehrenbergii* fo. *laxirete* (Amann) Podp., Consp. Musc. Eur. 199, 1954]

Didymodon erosodenticulatus (C. Müll.) Saito, J. Hattori Bot. Lab. 39: 504, 1975 (*Barbula*) As2

Didymodon fallax (Hedw.) Zand., Phytologia 41: 28, 1978 (*Barbula*) Eur As1 As2 As3 As5 Afr1 Am1

var. *brevifolius* (Dicks. ex With.) Ochyra, Fragm. Fl. Geobot. 28:

449, 1982 [1984] (*Bryum*) Eur

[var. *reflexus* (Brid.) Zand., Bryologist 83: 230, 1980 (*Barbula*) Eur As1 As2 Am1 Am2 = *Didymodon rigidicaulis* (C. Müll.) Saito *fide* Saito, J. Hattori Bot. Lab. 39: 502, 1975 ≡ *Didymodon ferrugineus* (Schimp. ex Besch.) Hill, J. Bryol. 11: 599, 1981 (1982)]

Didymodon ferrugineus (Schimp. ex Besch.) Hill, J. Bryol. 11: 599, 1981 (1982) (*Barbula*) As1 As2 Am1 Am2 Am3 Eur Afr [= *Didymodon fallax* var. *reflexus* (Brid.) Zand. *fide* Zander, Bryologist 83: 230, 1980]

[*Didymodon filicaulis* Card. Am2 = *Bryoerythrophyllum recurvirostrum* (Hedw.) Chen var. *recurvirostrum fide* Zander, Cryptogamie Bryol. Lichénol. 2: 418, 1981 (1982)]

Didymodon fragilicuspis Broth. As2

[*Didymodon fuscoviridis* Card. Am1 Am2 Am3 = *Didymodon rigidulus* Hedw. *s. lat. fide* Zander, Cryptogamie Bryol. Lichénol. 2: 390, 1981 (1982)]

Didymodon gelidus Card. Ant

[*Didymodon gemmascens* Mitt. in Hunt, Mem. Litt. Phil. Soc. Manchester ser. 3, 3: 235, 1868 [as "*gemmescens*"] Eur ≡ *Leptodontium gemmascens* (Mitt. in Hunt) Braithw. good species *fide* Zander, Bryologist 75: 236, 1972]

Didymodon giganteus (Funck) Jur. Eur As2 As3 Am1

[var. *japonicus* Broth. ex Ihs. As2 = *Barbula gigantea* Funck *fide* Iwats. & Nog., J. Hattori Bot. Lab. 37: 308, 1973 ≡ *Didymodon giganteus* (Funck) Jur.]

Didymodon glaucoviridis (C. Müll.) Broth. Am5

Didymodon glaucus Lindb. ex Roth Eur *hom. illeg.*

[*Didymodon godmanianus* (C. Müll.) Bartr. Am2 = *Didymodon rigidulus* Hedw. *s. lat. fide* Zander, Cryptogamie Bryol. Lichénol. 2: 390, 1981 (1982)]

Didymodon gracilescens Bartr. in Bauer *hom. illeg.* Am6

Didymodon guineensis Broth. & Par. Afr2

Didymodon gymnus (C. Müll.) Broth. Am4 Am6

Didymodon hampei Zand. (*nom. nov.* for *Gyroweisia obtusifolia* Hampe), see treatment of *Didymodon* Am2

Didymodon hastatus (Mitt.) Zand. (*Barbula*), see treatment of *Didymodon* As3

Didymodon haussknechtii (Jur. & Mild.) Broth. As5

[*Didymodon heribaudii* Card. Am2 = *Didymodon rigidulus* Hedw. *s. lat. fide* Zander, Cryptogamie Bryol. Lichénol. 2: 390, 1981 (1982)]

[*Didymodon homomallus* Hedw. ≡ *Ditrichum heteromallum* (Hedw.) Britt.]

var. *elongatus* Hüb. Eur

Didymodon humboldtii (Herz.) Hegew. & Hegew., Nov. Hedw. 28: 747, 1976 (*Barbula*) Am4

Didymodon humidus (Mitt.) Zand. (*Tortula*), see treatment of *Didymodon* Am4

[*Didymodon icmadophilus* Schimp. ex C. Müll.) Saito, J. Hattori Bot. Lab. 39: 519, 1975 (as "*icmadophyllus*") (*Barbula*) As1 As2 As3 Eur Am1 Am2 ≡ *Didymodon rigidulus* var. *icmadophilus* (Schimp. ex C. Müll.) Zand. *fide* Zander, Cryptogamie Bryol. Lichénol. 2: 394, 1981 (1982)]

Didymodon imperfectus (C. Müll.) Zand. (*Trichostomum*), see treatment of *Didymodon* Am4

Didymodon incrassatolimbatus Card. Am2

Didymodon incrassatus (Broth.) Broth. Eur

[*Didymodon insulanus* (De Not.) Hill, J. Bryol. 11: 599, 1981 (1982) (*Tortula*) Eur = *Didymodon vinealis* var. *flaccidus* (BSG) Zand. cf. Corley et al., J. Bryol. 11: 622, 1981 (1982) (= *Didymodon vinealis* (Brid.) Zand. *fide* Sollman, Bryologist 86: 272, 1983)]

Didymodon integrifolius Broth. in Mildbr. Afr2

[var. *paucidentatus* Thér. Afr2 = *Bryoerythrophyllum campylocarpum* (C. Müll.) Crum, see treatment of *Bryoerythrophyllum*]

Didymodon inundatus (Mitt.) Broth. Am4 Am6

Didymodon japonicus (Broth.) Saito, J. Hattori Bot. Lab. 39: 508, 1975 (*Molendoa*) As2

Didymodon johansenii (Williams) Crum Eur Am1

Didymodon juniperinus (C. Müll.) Broth. Am4

[*Didymodon killipii* Williams Am4 = *Morinia ehrenbergiana* (C. Müll.) Thér., see treatment of *Morinia* ≡ *Mironia ehrenbergiana* (C. Müll.) Zand., see treatment of *Mironia*]

[*Didymodon knysnae* Rehm. ex Sim Afr4 = *Didymodon xanthocarpus* (C. Müll.) Magill in Magill & Schelpe, Mem. Bot. Surv. S. Afr. 43: 5, 1979]

Didymodon laevigatus (Mitt.) Zand., Phytologia 41: 29, 1978 (*Tortula*) Am4 Am6

Didymodon lamyanus (Schimp.) Thér. Eur

Didymodon leskeoides Saito, J. Hattori Bot. Lab. 39: 508, 1975 As2 Am1

[*Didymodon linearis* Broth. in Skottsb. *hom. illeg.* Am5 = *Bryoerythrophyllum campylocarpum* (C. Müll.) Crum *fide* Robinson, Smiths. Contr. Bot. 27: 30, 1975]

Didymodon lindigii (Hampe) Zand. (*Hyophila*), see treatment of *Didymodon* Am4

Didymodon lingulatus (Hook. f. & Wils.) Broth. Austr2

[*Didymodon lingulatus* (Rehm. ex Sim) Magill, Mem. Bot. Surv. S. Afr. 43: 5, 1979 (*Gymnostomum*) *hom. illeg.* Afr4 ≡ *Gymnostomum lingulatum* Rehm. ex Sim *fide* Magill, Fl. S. Afr. I. Mosses 1: 184, 1981 (1982)]

Didymodon loeskei Fleisch. As4

[*Didymodon longifolius* (Brid.) Hook. ≡ *Leucoloma longifolia* (Brid.) Wijk. & Marg.]

var. *curvifolius* Hook. f. & Wils. Am6

Didymodon lorentzianus (C. Müll.) Broth. Am6

Didymodon luehmannii (Broth. & Geh.) Catcheside, Mosses S. Australia 175, 1980 (*Barbula*) Austr1

Didymodon luridus Hornsch. in Spreng. Eur As1 As2 As5 Afr1 Afr2 Am2 Am3 [≡ *Didymodon vinealis* var. *luridus* (Hornsch. in Spreng.) Zand. *fide* Zander, Cryptogamie Bryol. Lichénol. 2: 412, 1981 (1982)]

var. *latifolius* Hill Eur

fo. *brevifolius* Latz., Bot. Centralb. Beih. 48(2): 477, 1931 Eur

fo. *cuspidatus* (Schimp.) Röll, Hedwigia 56: 144, 1915 Eur

fo. *rubellus* Loeske in Bauer, Musc. Eur. Exs. 1581, 1923 Eur

fo. *subscabrus* Linder, Mitt. Bad. Landesvereins Naturk. 234–235: 265, 1909 Eur

fo. *tophaceus* Amann, Fl. Mousses Suisse Ad. 3: 26, 1935 Eur

[*Didymodon luzonensis* Bartr. As4 = *Bryoerythrophyllum ferruginascens* (Stirt.) Giac., see treatment of *Bryoerythrophyllum*]

Didymodon macrophyllus Broth. in Herz. Am4

Didymodon jackvancei Zand. *nom. nov.* (*Husnotiella plicata* Magill), see treatment of *Didymodon*] Afr4

Didymodon mamillosus (Crundw.) Hill, J. Bryol. 11: 599, 1981 [1982] (*Barbula*) Eur

Didymodon marginatum (Robins.) Zand. (*Trichostomum*), see treatment of *Didymodon* Am4

Didymodon maschalogena (Ren. & Card.) Broth. As3

Didymodon maximus (Syed & Crundw.) Hill, J. Bryol. 11: 599, 1981 [1982] (*Barbula*) Am1 Eur

[*Didymodon merceyoides* Broth. in Herz. Am4 = *Bryoerythrophyllum campylocarpum* (C. Müll.) Crum, see treatment of *Bryoerythrophyllum*]

[*Didymodon mexicanus* Besch. Am1 Am2 = *Didymodon rigidulus*

Hedw. *s. lat. fide* Zander, Cryptogamie Bryol. Lichénol. 2: 389, 1981 (1982)]
[var. *subulatus* Thér. & Bartr. Am1 Am2 ≡ *Didymodon rigidulus* var. *subulatus* (Thér. & Bartr.) Zand. *fide* Zander, Cryptogamie Bryol. Lichénol. 2: 395, 1981 (1982)]
Didymodon michiganensis (Steere in Grout) Saito, J. Hattori Bot. Lab. 39: 517, 1975 (*Barbula*) As2 As3 Am1 Am2
[*Didymodon microstomus* Dix. & Badhwar As3 = *Bryoerythrophyllum recurvirostrum* (Hedw.) Chen *fide* Gangulee, Mosses E. India 3: 742, 1972 also *fide* Sollman, Lindbergia 16: 22, 1990]
Didymodon microthecius (C. Müll.) Broth. Am6
Didymodon minusculus (Williams) Zand. (*Tortula*), see treatment of *Didymodon* Am4
Didymodon mittenii Gangulee, Nov. Hedw. 8: 149, 1964. As3
[*Didymodon montanus* (Mitt.) Broth. Am4 ≡ *Rhamphidium montanus* (Mitt.) Zand. see Excluded Taxa]
Didymodon montevidensis Broth. in Felipp. Am5
Didymodon moritzianus (C. Müll.) Broth. Am4
Didymodon nicholsonii Culm. Eur Am1
Didymodon nigrescens (Mitt.) Saito, J. Hattori Bot. Lab. 39: 510, 1975 (*Barbula*) As2 Am1 Am2
[*Didymodon obscurus* (Mitt.) Dix. As3 *hom. illeg.* = *Didymodon mittenii* Gangulee *fide* Gangulee, Mosses E. India 758, 1972]
[*Didymodon obtusifolius* Card. ex Dix. & P. Varde As3 *hom. illeg. non* Schkuhr ≡ *Barbula dixonii* R. S. Chopra, Taxon. Indian Mosses 139, 1975 *nom. nov.* = *Barbula inaequalifolia* Tayl. *fide* Sollman, Lindbergia 10: 54, 1984 ≡ *Bryoerythrophyllum inaequalifolium* (Tayl.) Zand.]
Didymodon obtussissimus Broth. = *Bryoerythrophyllum brachystegium* (Besch.) Saito *fide* Saito, J. Jap. Bot. 47: 14, 1972 and J. Hattori Bot. Lab. 39: 477, 1975]
var. *japonicus* Broth. As2
Didymodon occidentalis Zand., Phytologia 41: 26, 1978 (*nom. nov.* for *Barbula rubiginosa*) Am1 [= *Didymodon vinealis* var. *rubiginosus* (Mitt.) Zand., Cryptogamie Bryol. Lichénol. 2: 417, 1981]
Didymodon orbignyanus (C. Müll.) Broth. Am6
[*Didymodon ovatus* (Mitt.) Jaeg. As3 = *Bryoerythrophyllum recurvum* (Griff.) Saito *fide* Sollman, Lindbergia 16: 22, 1990 ≡ *Bellibarbula recurva* (Griff.) Zand., see treatment of *Bellibarbula*]
[*Didymodon papillinervis* (Dix. & Herz.) Demar. Afr2 = *Bryoerythrophyllum recurvirostrum* (Hedw.) Chen *fide* Sollman, Lindbergia 16: 22, 1990]
[*Didymodon parvulus* (Kindb.) Britt. in Williams Am1 = *Distichium inclinatum* (Hedw.) BSG *fide* Steere & Crum, Mem. New York Bot. Gard. 28(2): 76, 1977]
Didymodon patagonicus (Mitt.) Broth. Am6
[*Didymodon patentifolius* Thér. Am2 = *Didymodon australasiae* (Hook. & Grev.) Zand. *s. lat. fide* Zander, Cryptogamie Bryol. Lichénol. 2: 397, 1981 (1982)]
[*Didymodon paucidentatus* (C. Müll.) Broth. in Par. Afr2 = *Bryoerythrophyllum campylocarpum* (C. Müll.) Crum *fide* Sollman, Lindbergia 16: 23, 1990]
[*Didymodon pelichucensis* Williams Am4 = *Bryoerythrophyllum campylocarpum* (C. Müll.) Crum, see treatment of *Bryoerythrophyllum*]
[*Didymodon percarinatus* Dix. & Sak. As2 = *Grimmia* sp. *fide* Saito, J. Hattori Bot. Lab. 39: 528, 1975]
Didymodon perexilis (C. Müll.) Broth. Am4
Didymodon perobtusus Broth. As2 Am1
[*Didymodon perrevolutus* P. Varde Afr2 = *Didymodon ceratodonteus* (C. Müll.) Dix. *fide* Magill, Fl. S. Afr. I. Mosses 1: 235, 1981 (1982)]
[*Didymodon planifolius* P. Varde & Thér. Am3 = *Trichostomum brachydontium* Bruch in F. Müll. *fide* Zander, Cryptogamie Bryol. Lichénol. 2: 419, 1981 (1982)]
Didymodon planotophaceus Fröhl. As5
Didymodon polycephalus Mont. Am6
Didymodon pruinosus (Mitt.) Zand., see treatment of *Didymodon* (*Tortula*) Am4
[*Didymodon pusillus* Card. *hom. illeg.* Am2 = *Didymodon rigidulus* Hedw. *s. lat. fide* Zander, Cryptogamie Bryol. Lichénol. 2: 390, 1981 (1982)]
[*Didymodon ramulosus* (Schimp. ex Besch.) Card. in Broth. Am2 = *Didymodon rigidulus* Hedw. *s. lat. fide* Zander, Cryptogamie Bryol. Lichénol. 2: 389, 1981 (1982)]
[*Didymodon recurvus* (Griff.) Broth. As3 ≡ *Bryoerythrophyllum recurvum* (Griff.) Saito ≡ *Bellibarbula recurva* (Griff.) Zand., see treatment of *Bellibarbula*]
Didymodon reedii Robins., Bryologist 70: 323, 1967 Am1 Eur
Didymodon reflexus Thér. Am6
Didymodon reticulatus Gillies ex Grev. Am6
Didymodon revolutus (Card.) Williams *fide* Zander, Cryptogamie Bryol. Lichénol. 2: 401, 1981 [1982] (*Husnotiella*) Am1 Am2 Am4
[*Didymodon rigidicaulis* (C. Müll.) Saito, J. Hattori Bot. Lab. 39: 502, 1975 (*Barbula*) As1 As2 Am1 Am2 Am3 Eur Afr = *Didymodon fallax* var. *reflexus* (Brid.) Zand. *fide* Hill, J. Bryol. 11: 599, 1981 (1982) = *Didymodon ferrugineus* (Schimp. ex Besch.) Hill, J. Bryol. 11: 599, 1981 (1982)]
[*Didymodon rigidifolius* Dix. Afr2 = *Bryoerythrophyllum recurvum* (Griff.) Saito *fide* Sollman, Lindbergia 16: 22, 1990 ≡ *Bellibarbula recurva* (Griff.) Zand., see treatment of *Bellibarbula*]
Didymodon rigidulus Hedw. Eur As1 As2 As3 As5 Afr1 Am1 Am6
subsp. *andreaeoides* (Limpr.) Wijk & Marg. Eur As2
[subsp. *validus* (Limpr.) Loesk. Eur ≡ *Didymodon rigidulus* var. *validus* (Limpr.) R. Düll, J. Hattori Bot. Lab. 55: 263, 1984 = *Didymodon rigidulus* var. *gracilis* (Schleich. ex Hook. & Grev.) Zand. *fide* Zander, Cryptogamie Bryol. Lichénol. 2: 393, 1981 (1982)]
[subsp. *verbanus* (Nichol. & Dix.) Loesk. Eur = *Didymodon glaucus* Ryan *fide* Corley et al., J. Bryol. 11: 622, 1981 ≡ *Didymodon rigidulus* var. *glaucus* (Ryan) Wijk & Marg.]
var. *acutus* Biz., Acta Bot. Acad. Sci. Hung. 18: 21, 1973 Afr2
var. *brevifolius* Zodd. *illeg. hom.* Afr1
var. *campicola* (Broth.) Wijk & Marg. As1
var. *compactus* Röll Eur
[var. *excurrens* Hag. Eur ≡ *Didymodon rigidulus* fo. *excurrens* (Hag.) Düll, J. Hattori Bot. Lab. 55: 259, 1984]
var. *flaccidus* (Röll) Röll Eur
var. *glaucus* (Ryan) Wijk & Marg. Eur As1
var. *gracilis* (Schleich. ex Hook. & Grev.) Zand., Cryptogamie Bryol. Lichénol. 2: 393, 1981 (*Tortula*) Eur As1 As3 As5 Afr1 Am1 Am2
var. *icmadophilus* (Schimp. ex C. Müll.) Zand., Cryptogamie Bryol. Lichénol. 2: 394, 1981 (*Barbula*) Eur As1 As3 As5 Am1 Am2
var. *insidiosus* (Jur. & Mild.) Röll, Hedwigia 56(1–3): 149, 1915 (*Barbula rigidulus* var.) Eur = *Didymodon spadiceus* (Mitt.) Limpr. *fide* Limpricht, Laubm. 556, 1890 (1888)]
var. *major* Podp., Sitzungsber. K. Boehm. Ges. Wiss. 1899(44): 14, 1899 Eur
var. *paludosus* (Lang.) Par. Eur Am1
var. *rigidus* (Röll) Röll Eur
var. *subulatus* (Thér. & Bartr.) Zand., Cryptogamie Bryol.

Lichénol. 2: 395, 1981 (*Didymodon mexicanus* var.) Am1 Am2
var. *tenuis* Hamm. Eur
var. *trifarius* Hartm. Eur
[var. *validus* (Limpr.) R. Düll, J. Hattori Bot. Lab. 55: 263, 1984 Eur = *Didymodon rigidulus* var. *gracilis* (Schleich. ex Hook. & Grev.) Zand. *fide* Zander, Cryptogamie Bryol. Lichénol. 2: 393, 1981 (1982)]
fo. *bisetus* Pét. Magyar Bot. Lapok 2: 292, 1903 Eur
fo. *brevicaulis* (Röll) Röll, Hedwigia 56: 147, 1915 (*Barbula rigidula* fo.) Eur
fo. *brevifolius* Röll, Hedwigia 42: 299, 1903 Eur
fo. *densus* (B.&S. in BSG) Limpr., Laubm. Deutsch. 1: 556, 1888 (*Trichostomum rigidulum* var.) Eur
fo. *excurrens* (Hag.) Düll, J. Hattori Bot. Lab. 55: 259, 1984 (*Didymodon rigidulus* var.) Eur
fo. *laxus* Mol., Auläu-Studien 65, 1864 Eur
fo. *longicaulis* (Röll) Röll, Hedwigia 56: 147, 1915 (*Barbula rigidula* fo.) Eur
fo. *propaguliferus* (Schiffn.) Limpr., Laubm. Deutsch. 3: 691, 1901 Eur
subfo. *viridis* (Röll) Röll, Hedwigia 56: 147, 1915 (*Barbula rigidula* fo.) Eur
Didymodon rivicola (Broth.) Zand. in Kop., Gao, Lou & Järvinen, Ann. Bot. Fenn. 20: 222, 1983 (*Barbula*) As2
[*Didymodon rubellus* BSG = *Bryoerythrophyllum recurvirostrum* (Hedw.) Chen]
var. *angustifolius* P. Varde Afr2 = *Bryoerythrophyllum afrorubellum* (Broth. & Wag.) De Sloover, Bull. Jard. Bot. Natl. Belg. 49: 398, 1979]
[var. *subdentatus* Thér., Rev. Bot. 9: 486, 1892 Eur ≡ *Didymodon rubellus* fo. *subdentatus* (Thér.) Hérib., Mém. Ac. Sc. Clermont-Ferrand ser. 2, 14: 382, 1899]
[fo. *brevirostris* Warnst., Krypt. Fl. Brandenburg 224, 1904 Eur ≡ *Bryoerythrophyllum recurvirostrum* fo. *brevirostre* (Warnst.) Podp.]
[fo. *flaccidus* Röll, Hedwigia 38: 261, 1899 Eur ≡ *Bryoerythrophyllum recurvirostrum* fo. *flaccidum* (Röll) Podp.]
[fo. *gracilis* Limpr., Laubm. Deutsch. 1: 547, 1888 Eur ≡ *Bryoerythrophyllum recurvirostrum* fo. *gracile* (Limpr.) Podp.]
[fo. *longirostris* Warnst., Krypt. Fl. Brandenburg 224, 1904 Eur ≡ *Bryoerythrophyllum recurvirostrum* fo.]
[fo. *maior* Breidl. ex Matous., Hedwigia 44: 29, 1904 Eur ≡ *Bryoerythrophyllum recurvirostrum* fo. *maius*]
[fo. *obtusifolius* Röll, Hedwigia 38: 261, 1899 Eur ≡ *Bryoerythrophyllum recurvirostrum* fo. *obtusifolium* (Röll) Podp.]
[fo. *pygmaeus* Meyl., Bull. Soc. Vaud. 57: 123, 1929 ≡ *Bryoerythrophyllum recurvirostrum* fo. *pygmaeum* (Meyl.) Podp.]
fo. *subdentatus* Thér. in Hérib., Mém. Ac. Sc. Clermont-Ferrand ser. 2, 14: 382, 1899 Eur
[fo. *viridis* Schlieph. ex Limpr., Laubm. Deutsch. 1: 547, 1888 Eur ≡ *Bryoerythrophyllum recurvirostrum* fo. *viride* (Schlieph. ex Limpr.) Podp]
Didymodon rubiginosus (C. Müll.) Broth. Austr1
Didymodon rufidulus (C. Müll.) Broth. As2
[*Didymodon rufus* Lor. in Rabenh. ≡ *Didymodon asperifolius* (Mitt.) Crum, Steere & Anders.)]
var. *gorodkovii* A. Abr. & I. Abr. in I. Abr., Trudy Arktic. Antarktic. Naucno-Issl. Inst. 224: 220, 1963 As1
fo. *sublaevigatus* Herz., Weiner Bot. Z. 93: 39, 1914 Eur
Didymodon schilleri Herz. & Thér. Am6
[*Didymodon schimperi* (Mont.) Broth. Am6 = *Tortula atrovirens* (Sm.) Lindb., see treatment of *Tortula*]
[*Didymodon saxicola* Broth. ex Gangulee, Mosses E. India 774, 1972 *nom. nud. in syn.* = *Desmatodon gemmascens* Chen *fide* Gangulee, Mo. E. India 3: 774, 1972]
[*Didymodon semivaginatus* (Britt.) Broth. Am4 = *Erythrophyllopsis boliviana* Broth. in Herz., see treatment of *Erythrophyllopsis*]
Didymodon sinuosus (Mitt.) Delogn. Eur
Didymodon soaressii Luis. Eur
Didymodon spadiceus (Mitt.) Limpr. Eur As1 As5
subsp. *poeninus* (Amann) Wijk & Marg. Eur
var. *siluricus* Velen. Eur
Didymodon spathulatolinearis (C. Müll.) Broth. Am6
[*Didymodon stenopyxis* Card. Am2 = *Bryoerythrophyllum recurvirostrum* var. *aeneum* (C. Müll.) Zand. *fide* Zander in Sharp, Moss Fl. Mexico]
Didymodon stewartii (Bartr.) Zand. (*Barbula*), see treatment of *Didymodon* As3
[*Didymodon strictifolius* Dix. & P. Varde As3 ≡ *Barbula strictifolia* (Dix. & P. Varde) R. S. Chopra, Taxon. Indian Mosses 140, 1975 = *Didymodon recurvus* (Griff.) Mitt. *fide* Robinson, Bryologist 71: 86, 1968 ≡ *Bellibarbula recurva* (Griff.) Zand., see treatment of *Bellibarbula*]
Didymodon subandreaeoides (Kindb.) Zand., Phytologia 41: 23, 1978 (*Barbula*) Am1
Didymodon subfontanus Dix. in Sim Afr4
[*Didymodon sublingulatus* Dix. Afr2 = *Bryoerythrophyllum campylocarpum* (C. Müll.) Crum *fide* Sollman, Lindbergia 16: 22, 1990]
Didymodon subrevolutus (Hampe) Broth. Afr3
Didymodon subtriquetrus Robins., Bryologist 70: 319, 1967 Am4
[*Didymodon subtophaceus* Williams Am4 Am6 = *Trichostomopsis australasiae* (Hook. & Grev.) Robins., Phytologia 20: 187, 1970 ≡ *Didymodon australasiae* (Hook. & Grev.) Zand., Phytologia 41: 21, 1978]
Didymodon subtorquatus (C. Müll. & Hampe) Catcheside, Mosses S. Australia 174, 1980 (*Barbula*) Austr1
Didymodon subulatus Cambessèdes *hom. illeg.* Eur
Didymodon taylorii Zand. (*nom. nov.* for *Tortula campylocarpa* Tayl.) Am4
Didymodon tenellus Hedw. f. ex Brid. *ignot.*
Didymodon tectorum (C. Müll.) Saito, J. Hattori Bot. Lab. 39: 516, 1975 (*Barbula*) As2
Didymodon tomaculosus (Blockeel) Corley in Corley et al., J. Bryol. 11: 649, 1981 [1982] (*Barbula*) Eur
Didymodon tophaceopsis Zand. (*nom. nov.* for *Gyroweisia amplexicaulis* Sim), see treatment of *Didymodon* Afr4
Didymodon tophaceus (Brid.) Lisa Eur As1 As3 As3 As5 Afr1 Am1 Am2 Am4
var. *anatinus* Hammerschm. Eur
var. *bosniacus* (Głow.) Latz. Eur = *Didymodon tophaceus* var. *decurrens* Card. & Thér. *fide* Podp., Consp. Musc. Eur. 201, 1954]
var. *breidleri* Bauer Eur
var. *brevifolius* (BSG) Warnst. Eur
var. *decurrens* Card. & Thér. Eur
var. *excurrens* (Broth.) Wijk & Marg. As1
var. *humilis* (Schimp.) Warnst. Eur
var. *linearis* (De Not.) Limpr. Eur
var. *riparius* Amann Eur
fo. *acutifolius* (Boul.) Zodda, Malphighia 22: 509, 1908 (*Trichostomum tophaceum* fo.) Eur
[fo. *anatinus* (Hammerschm.) Marg., Lindbergia 1: 127, 1972 ≡ *Didymodon tophaceus* var. *anatinus* Hammerschm.]
fo. *cylindrica* (Boul.) Limpr., Laubm. Deutsch. 1: 554, 1888 (*Trichostomum tophaceum* fo.) Eur

fo. *elatus* (Boul.) Artaria in E. Bauer, Musci. Eur. Am. Exsicc. 37: n. 1823, 1925 (*Trichostomum tophaceum* fo.) Eur
fo. *laxus* Kindb., Bull. Soc. Bot. Ital. 7: 15, 1896 Eur
[fo. *humilis* (Schimp.) Marg., Lindbergia 1: 128, 1972 ≡ *Didymodon tophaceus* var. *humilis* (Schimp.) Warnst.]
fo. *lingulatus* (Boul.) Mönk., Süsswasserfl. 14: 67, 1914 (*Trichostomum tophaceum* fo.) Eur
fo. *lingulifolius* (Roth in Zodda) Marg., Lindbergia 1: 128, 1972 Eur
fo. *propaguliferus* Amann, Bull. Soc. Vaud. Sc. Nat. 53: 86, 1920 Eur
fo. *recurvifolius* (Boul.) De Willd., Prodr. 434, 1899 (*Trichostomum tophaceum* fo.) Eur
fo. *scabrinervis* Podp., Acta Bot. Bohem. 1: 11, 1922 Eur
fo. *truncatus* (Boul.) Limpr., Laubm. Deutsch. 1: 554, 1888 (*Trichostomum tophaceum* fo.) Eur
fo. *vulgaris* Mönk., Süsswasserfl. 14: 67, 1914 Eur
Didymodon torquatus (Tayl.) Catcheside, Mosses of South Australia 174, 1980 (*Barbula*) Austr1 Austr2
[*Didymodon tosaënsis* Card. As2 = *Barbula subcomosa* Broth. *fide* Saito, J. Jap. Bot. 47: 11, 1972]
[*Didymodon trifarius* (Hedw.) Röhl., the type, not the concept of modern authors which is correctly *Didymodon luridus* (Hornsch. in Spreng.) = *Saelania glaucescens* (Hedw.) Broth. in Bomanss & Broth. *fide* Zander, Bryologist 81: 421, 1978]
[subsp. *nicholsonii* (Culm.) Wijk & Marg. Eur ≡ *Didymodon vinealis* var. *nicholsonii* (Culm.) Zand., Cryptogamie Bryol. Lichénol. 2: 416, 1981 (1982) ≡ *Didymodon nicholsonii* Culm.]
var. *angustifolius* (Warnst.) Wijk & Marg. Eur
var. *cuspidatus* (Schimp.) Wijk & Marg. Eur Am1
var. *intermedius* (Ruth.) Wijk & Marg. Eur
var. *krimensis* (Warnst.) Wijk & Marg. Eur
Didymodon trivialis (C. Müll.) Guerra in Guerra & Ros, Cryptogamie Bryol. Lichénol. 8: 64, 1987 (*Barbula*) Eur Afr4
Didymodon umbrosus (C. Müll.) Zand., Phytologia 41: 22, 1978 (*Barbula*) Eur Am1 Am2 Am4 Am6 [≡ *Didymodon australasiae* var. *umbrosus* (C. Müll.) Zand. *fide* Zander, Cryptogamie Bryol. Lichénol. 2: 400, 1981 (1982)]
Didymodon uruguayensis Zand. (*nom. nov.* for *Barbula uruguayensis* Broth. *hom. illeg.* of *Barbula uruguensis* Par. *fide* Margadant in litt.), see treatment of *Didymodon* Am6
[*Didymodon viridissimus* Card. Am2 = *Didymodon rigidulus* Hedw. *s. lat. fide* Zander, Cryptogamie Bryol. Lichénol. 2: 390, 1981 (1982)]
Didymodon vinealis (Brid.) Zand., Phytologia 41: 25, 1978 (*Barbula*) Eur As1 As2 As3 As5 Afr1 Am1 Am2 Am4 Am6 Oc
[var. *brachyphyllus* (Sull. in Whippl.) Zand., Cryptogamie Bryol. Lichénol. 2: 411, 1981 (*Barbula*) Am1 ≡ *Didymodon brachyphyllus* (Sull. in Whippl.) Zand., Phytologia 41: 24, 1978]
[var. *flaccidus* (BSG) Zand., Phytologia 41: 25, 1978 (*Barbula*) Eur As1 As5 Afr1 Am1 Am6 Oc [= *Didymodon vinealis* (Brid.) Zand. var. *vinealis fide* Sollman, Bryologist 86: 272, 1983]
[var. *luridus* (Hornsch. in Spreng.) Zand., Cryptogamie Bryol. Lichénol. 2: 412, 1981 (*Didymodon*) Eur As1 As2 As5 Afr1 Afr2 Am2 Am3 ≡ *Didymodon luridus* Hornsch. ex Spreng.]
[var. *nicholsonii* (Culm.) Zand., Cryptogamie Bryol. Lichénol. 2: 416, 1981 (*Didymodon*) Eur Am1 ≡ *Didymodon nicholsonii* Culm.]
[var. *rubiginosus* (Mitt.) Zand., Cryptogamie Bryol. Lichénol. 2: 417, 1981 (*Barbula*) Am1 ≡ *Didymodon occidentalis* Zand., Phytologia 41: 26, 1978]
Didymodon waymouthii (R. Br. ter) Zand. (*Weissia*), see treatment of *Didymodon* Austr2
Didymodon wildii (Broth.) Broth. Austr1
Didymodon wisselii (Dix.) Norris & T. Kop., Acta Bot. Fenn. 137: 127, 1989 (*Barbula*) As4
[*Didymodon wollei* (Aust.) Aust. Am1 = *Trichostomum tenuirostre* var. *holtii* (Braithw.) Dix., see treatment of *Trichostomum*]
Didymodon xanthocarpus (C. Müll.) Magill in Magill & Schelpe, Mem. Bot. Surv. S. Afr. 43: 5, 1979 (*Barbula*) Afr4

DOLOTORTULA Zand.
Dolotortula mniifolia (Sull.) Zand., Phytologia 65: 426, 1989 (*Barbula*) Am2 Am3 Am4

ENTOSTHYMENIUM Brid.
Entosthymenium tristichum Brid., Bryol. Univ. 1: 761, 1827 *ignot. nom. dub.* Eur

ERYTHROPHYLLASTRUM Zand., see treatment of *Erythrophyllastrum*
Erythrophyllastrum andinum (Sull.) Zand. (*Trichostomum*), see treatment of *Erythrophyllastrum* Am4

ERYTHROPHYLLOPSIS Broth. in Herz.
[*Erythrophyllopsis andina* (Sull.) Zand., Bryologist 80: 159, 1977 Am4 (*Trichostomum*) ≡ *Erythrophyllastrum andinum* (Sull.) Zand., see treatment of *Erythrophyllastrum*]
[*Erythrophyllopsis boliviana* Broth. in Herz. Am4 Am6 (= *Erythrophyllopsis andina* (Sull.) Zand. *fide* Zander, Bryologist 80: 159, 1977, but, see treatment of *Erythrophyllastrum*) = *Erythrophyllopsis fuscula* (C. Müll.) Hilp. *fide* Hilpert, Beih. Bot. Centralbl. 50(2): 639, 1933]
[*Erythrophyllopsis challaensis* (Broth. in Herz.) Hilp. Am4 ≡ *Didymodon challaense* (Broth. in Herz.) Zand., see treatment of *Didymodon*]
Erythrophyllopsis fuscula (C. Müll.) Hilp. Am4 Am6

[**ERYTHROPHYLLUM** (Lindb.) Loeske ≡ *Bryoerythrophyllum* Chen]
[*Erythrophyllum rubellum* Hilp. = *Bryoerythrophyllum recurvirostrum* (Hedw.) Chen]
[fo. *alpinum* Herz., Weiner Bot. Z. 93: 38, 1944 Eur ≡ *Bryoerythrophyllum recurvirostrum* fo.]
[fo. *latifolium* Herz., Wiener Bot. Z. 93: 38, 1944 Eur ≡ *Bryoerythrophyllum recurvirostrum* fo. *latifolium* (Herz.) Podp.]
[fo. *pulvinatum* Herz., Wiener Bot. Z. 93: 38, 1944 Eur ≡ *Bryoerythrophyllum recurvirostrum* fo.]
[*Erythrophyllum recurvirostrum* (Hedw.) Loeske ≡ *Bryoerthrophyllum recurvirostrum* (Hedw.) Chen]
[fo. *serratum* (Schimp. ex Roell) Lazar., Opr. Listv. Mchov Ukr. 168. 1955 (*Didymodon rubellus* var.) Eur ≡ *Bryoerythrophyllum recurvirostrum* var. *serratum* (Röll) Podp.]
[fo. *viride* (Schlieph. ex Limpr.) Lazar., Opr. Listv. Mchov Ukr. 168, 1955 Eur ≡ *Bryoerythrophyllum recurvirostrum* fo. *viride* (Schlieph. ex Limpr.) Podp.]

EUCLADIUM BSG
[*Eucladium irroratum* (Mitt. in Hook.) Jaeg. Austr2 ≡ *Tetracoscinodon irrorata* (Mitt. in Hook. f.) Zand. see treatments of *Tetracoscinodon* and *Eucladium*]
Eucladium verticillatum (Brid.) BSG Eur As1 As2 As3 As5 Afr1 Afr4 Am1 Am2
subsp. *crassinervium* Podp. Eur
subsp. *styriacum* (Glow.) Amman Eur
var. *acuminatum* Glow. Eur

var. *angustifolium* Lindb. Eur Afr1 As2
var. *brevifolium* Warnst. Eur
var. *clinotheca* (Besch.) Par. Afr1
var. *commutatum* (Glow.) Podp. Eur
var. *crassinervium* (Podp.) Podp. Eur
var. *dalmaticum* Par. Eur As2
var. *latebricola* Lamarl. & Maheu Eur
var. *michelettii* Fleisch. in Warnst. Eur
var. *obtusifolium* Warnst. Eur
var. *penicilliforme* Farn. Eur
var. *recurvatum* Dunk & Dunk, Herzogia 2: 419, 1973 Eur
var. *recurvifolium* Lindb. As5
var. *styriacum* (Glow.) Glow. Eur
fo. *crispum* (Röll) Röll, Hedwigia 56: 139, 1915 (*Eucladium verticillatum* var.) Eur
fo. *gracile* Röll, Hedwigia 56: 139, 1915 Eur
fo. *inundatum* (Farn.) Podp., Consp. Musc. Eur. 174, 1954 (*Eucladium verticillatum* var.) Eur
[fo. *lacustre* Amann, Fl. Mousses Suisse II: 37, 1918 Eur = fo. *inundatum* (Farn.) Podp. *fide* Podp., Consp. Musc. Eur. 174, 1954]
fo. *laetevirens* (Zett.) Podp., Consp. Musc. Eur. 174, 1954 (*Eucladium verticillatum* var.) Eur
fo. *thermale* Boros in Bauer, Musc. Eur. Exs. 2057, Sched. 42: 2, 1930 Eur
fo. *viridissimum* Bauer, Musc. Eur. Exs. 2059, Sched. 42: 2, 1930 Eur

GANGULEEA Zand.
Ganguleea angulosa (Broth. & Dix.) Zand., Phytologia 65: 427, 1989 (*Merceyopsis*) As3 Am5

[**GEHEEBIA** Schimp. = *Didymodon* Hedw. *fide* Saito, J. Hattori Bot. Lab. 39: 501, 1975]
[*Geheebia gigantea* (Funck) Boul. Eur As2 As3 Am1 ≡ *Didymodon giganteus* (Funck) Jur. *fide* Zander, Phytologia 41: 29, 1978 ≡ *Didymodon fallax* var. *giganteus* (Funck) Boul., see treatment of *Didymodon*]

[**GERTRUDIA** Herz. ≡ *Gertrudiella* (Herz.) Broth.]
[*Gertrudia validinervis* var. *serrato-pungens* Herz. Am4 ≡ *Gertrudiella validinervis* var. *serrato-pungens* (Herz.) Zand., see treatment of *Gertrudiella*]

GERTRUDIELLA Broth.
[*Gertrudiella ferruginea* (Herz.) Hilp. Am4 ≡ *Didymodon herzogii* Zand. *nom. nov.*, see treatment of *Didymodon*]
Gertrudiella validinervis (Herz.) Broth. Am4
var. *serratopungens* (Herz.) Zand., see treatment of *Gertrudiella* Am4

[**GLOBULINA** (C. Müll.) Broth. = *Bryoerythrophyllum* Chen *fide* Zander, Bryologist 81: 541, 1978]
[*Globulina boliviana* C. Müll., Nuovo Giorn. Bot. Ital. 4(1): 39, 1897 Am4 ≡ *Bryoerythrophyllum bolivianum* (C. Müll.) Zand., Bryologist 81: 545, 1978]

GLOBULINELLA Steere in Steere & Chapman
Globulinella benoistii (Thér.) Magill, Bryologist 80: 79, 1977 Am4
Globulinella globifera (Hampe) Steere Am1 Am2 Am3
[*Globulinella peruviana* (Williams) Steere Am2 Am4 ≡ *Saitoa peruviana* (Williams) Zand., Phytologia 65: 431, 1989 ≡ *Saitoella peruviana* (Williams) Menzel]

GYMNOSTOMIELLA Fleisch.
Gymnostomiella burmensis Bartr. As3
Gymnostomiella longinervis Broth. As2 As4
Gymnostomiella monodii P. Varde Afr2
Gymnostomiella orcuttii Bartr. in Orcutt Am1 Am2 Am3 Am5
[*Gymnostomiella rosulata* P. Varde Afr1 ≡ *Barbula semirosulata* Zand. *nom. nov.*, see treatment of *Barbula*]
Gymnostomiella tanganyikae Sloover, Bull. Jard. Bot. Natl. Belgique 47: 146, 1977 Afr2
Gymnostomiella vernicosa (Hook.) Fleisch. As2 As3 As4 Austr1

GYMNOSTOMUM Nees & Hornsch.
Gymnostomum aeruginosum Sm. Eur As1 As2 As3 As5 Afr1 Afr4 Am1 Am2 Am3 Am6 Austr1 Austr2 Oc
var. *cochlearifolium* Karczmarz, Lindbergia 7: 127, 1981 [1982] As3
fo. *compactum* (BSG) Wijk & Marg., Lindbergia 1: 128, 1972 Eur
fo. *ramosissimum* (BSG) Wijk & Marg., Lindbergia 1: 128, 1972 (*Gymnostomum rupestre* var.) Eur
fo. *rigidum* (Schimp.) Wijk & Marg., Lindbergia 1: 128, 1972 Eur
fo. *stelligerum* (Schimp.) Lazar.(Saviz.-Lub. & Smirnova 1970) Eur
[*Gymnostomum angustifolium* Saito, J. Hattori Bot. Lab. 36: 163, 1972 *nom. inval.* Am1 Am2 As2 ≡ *Tuerckheimia angustifolia* (Saito) Zand., Misc. Bryol. Lichenol. 8: 27, 1978 = *Tuerckheimia svihlae* (Bartr.) Zand., see treatment of *Tuerckheimia*]
[*Gymnostomum aurantiacum* (Mitt.) Jaeg. ≡ *Hymenostylium recurvirostrum* var. *aurantiacum* (Mitt.) Gangulee, Mosses E. India 648, 1972 (*Hymenostylium*) *hom. illeg. incl.* var. prior. = *Hymenostylium recurvirostrum* var. *luzonense* (Broth.) Bartr. *fide* Gangulee, Mosses E. India 3: 648, 1972 = *Hymenostylium recurvirostrum* var. *cylindricum* (Bartr.) Zand., see treatment of *Hymenostylium*]
Gymnostomum bewsii Dix. Afr4 [good species, see Magill, Fl. S. Afr. 1: 185, 1981 (1982)]
Gymnostomum bescherellei Broth. & Geh. ex Herz. Afr2
Gymnostomum boreale Nyholm & Hedenäs, Lindbergia 12: 41, 1986 Eur
[*Gymnostomum brachystegium* Besch., J. Bot. (Morot) 12: 281, 1898 As2 ≡ *Bryoerythrophyllum brachystegium* (Besch.) Saito, J. Jap. Bot. 47: 14, 1972]
Gymnostomum calcareum Nees & Hornsch. Eur As1 As2 As3 As5 Afr1 Afr4 Am1 Am2 Am6 Austr1 Austr2 Oc [= *Gymnostomum aeruginosum* Sm. *fide* Zander, Bryologist 80: 259, 1977; Magill, 1981, Fl. S. Afr. 1: 183, 1981 (1982)]
var. *australe* Broth. & Geh. Austr1 Austr2
[var. *brevifolium* Schimp. Eur Afr1 = *Gymnostomum luisieri* (Sérgio) Sérgio ex Crundw. *fide* Sérgio, Anales Biol. (Univ. Murcia) 2, Sec. Esp. 2) 1984: 361, 1984 = *Gymnostomum viridulum* Brid. *fide* Whitehouse & Crundwell, J. Bryol. 16: 563, 1991]
var. *longifolium* Dix. Austr1 Austr2
var. *muticum* Boul. Eur Afr1
var. *perpusillum* Sull. Am1
[var. *viridulum* (Brid.) B.&S. in BSG Eur ≡ *Gymnostomum viridulum* Brid. *fide* Whitehouse & Crundwell, J. Bryol. 16: 563, 1991]
fo. *cavernarum* Grom, Acta Bot. Croat. 26–27: 249, 1967–68 [1968] Eur
fo. *gracile* (Breidl.) Podp., Consp. Musc. Eur. 170, 1954 (*Gymnostomum calcareum* var.) Eur

fo. *intermedium* (Schimp.) Podp., Consp. Musc. Eur. 170, 1954 (*Gymnostomum calcareum* var.) Eur
fo. *longifolium* Meylan, Bull. Soc. Vaud. Sci. Nat. 60(249): 270, 1939 Eur
fo. *subrotundifolium* Latz., Beih. Bot. Centralbl. 48(2): 473, 1931 Eur
fo. *tenellum* (BSG) Podp., Consp. Musc. Eur. 170, 1954 (*Gymnostomum calcareum* var.) Eur
Gymnostomum carthusianum (Brid.) P. Beauv. Eur
Gymnostomum chenii Saito, J. Jap. Bot. 48: 164, 1973 As3
Gymnostomum cirrhatum De Not. Eur
[*Gymnostomum curvirostre* Hedw. ex Brid. ≡ *Hymenostylium recurvirostrum* (Hedw.) Dix.]
[var. *anoectangioides* Thér. Eur ≡ *Hymenostylium recurvirostrum* var. *anoectangioides* (Thér.) Wijk & Marg.]
[*Gymnostomum denticulatum* Griff. As3 = *Hyophila involuta* (Hook.) Jaeg., see treatment of *Hyophila*]
[*Gymnostomum dimorphum* (C. Müll.) Sim Afr4 ≡ *Didymodon dimorphus* (C. Müll.) Broth. *fide* Magill & Schelpe, Mem. Bot. Surv. S. Afr. 43: 23, 1979]
Gymnostomum foliosum Röhl. Eur
[*Gymnostomum gracile* Dix. *hom. illeg.* Afr2 Afr4 = *Didymodon ceratodonteus* (C. Müll.) Dix. *fide* Magill, Fl. S. Afr. I. Mosses 1: 233, 1981 (1982)]
Gymnostomum hymenostylioides (Broth. & Dix.) Zand. (*Merceyopsis*), see treatment of *Gymnostomum* As3
[*Gymnostomum insigne* (Dix.) A. Sm., J. Bryol. 9: 279, 1976 (1977) Eur As2 As3 Am1 ≡ *Hymenostylium recurvirostrum* var. *insigne* (Dix.) Bartr. *fide* Zander & Eckel, Canad. J. Bot. 60: 1596, 1982]
Gymnostomum jacksharpii (Crum) Allen, Bryologist 93: 207, 1991 (*Barbula*) Am3
[*Gymnostomum keniae* P. Varde Afr2 ≡ *Anoectangium keniae* (P. Varde) Zand., see treatment of *Anoectangium*]
[*Gymnostomum knightii* Schimp. in Knight, Trans. New Zealand Inst. 7: 354, 1875 Austr2 = *Didymodon tophaceus* (Brid.) Lisa, see treatment of *Didymodon*]
Gymnostomum laxirete (Broth.) Chen As2
Gymnostomum leptostomum Brid. *hom. illeg.* Afr3
[*Gymnostomum lessonii* Besch. Afr2 = *Racopilum* sp. see Excluded Taxa]
Gymnostomum lingulatum Rehm. ex Sim Afr4 [≡ *Didymodon lingulatus* (Sim) Magill, Mem. Bot. Surv. S. Afr. 43: 5, 1979 = good species *fide* Magill, Fl. S. Afr. I. Mosses 1: 184, 1981 (1982)]
Gymnostomum ludovicae Broth. & Par. Oc
[*Gymnostomum luisieri* (Sérgio) Sérgio ex Crundw., J. Bryol. 11: 603, 1981 [1982] (*Gyroweisia*) Eur = *Gymnostomum viridulum* Brid. *fide* Whitehouse & Crundwell, J. Bryol. 16: 563, 1991]
[*Gymnostomum microcarpon* Nees & Hornsch. = *Hymenostylium recurvirostrum* (Hedw.) Dix.]
var. *elongatum* Nees & Hornsch. Eur
Gymnostomum minutulum Saporta *hom. illeg.*, fossil Eur
Gymnostomum mosis (Lor.) Jur. As5
Gymnostomum mucronulatum Hedw. f. in Mohr *nom. inval.* Eur
[*Gymnostomum ovatum* Hedw. ≡ *Pterygoneurum ovatum* (Hedw.) Dix.]
var. *longicapsulum* Chev. Eur
[*Gymnostomum pottsii* (Dix.) Sim = *Didymodon dimorphus* (C. Müll.) Broth. *fide* Magill & Schelpe, Mem. Bot. Surv. So. Afr. 43: 23, 1979 = *Didymodon ceratodonteus* (C. Müll.) Dix. *fide* Magill, Fl. S. Afr. I. Mosses 1: 233, 1981 (1982)]
[*Gymnostomum recurvirostrum* Hedw. = *Hymenostylium recurvirostrum* (Hedw.) Dix.]
[var. *latifolium* (Zett.) Flow. ex Crum, Bryologist 72: 243, 1969. Eur As1 Am1 = *Hymenostylium recurvirostrum* (Hedw.) Dix. *fide* Zander, Bryologist 80: 253, 1977]
[*Gymnostomum rupestre* Schleich. ex Schwaegr. = *Gymnostomum aeruginosum* Sm.]
var. *minus* De Not. Eur
var. *obtusifolium* Kindb, Nuov. Giorn. Bot. Ital. 25: 116, 1893 Eur
fo. *arboreum* Geh., Bot. Centralb. Beih. 22(2): 98, 1907 Eur
fo. *cavernicola* Boros., Bot. Közlem. 32: 108, 1935 Eur
[fo. *compactum* (BSG) Mönk., Laubm. Eur. 251, 1927 (*Gymnostomum rupestre* var.) Eur ≡ *Gymnostomum aeruginosum* fo. *compactum* (BSG) Wijk & Marg.]
fo. *crispatulum* Röll, Hedwigia 38: 260, 1899 Eur
fo. *curvisetum* (Amann) Podp., Consp. Musc. Eur. 170, 1954 (*Gymnostomum rupestre* var.) Eur
fo. *elatum* Šmarda, Cas. Morav. Mus. Zemsk. 32: 14, 1949 Eur
fo. *intermedium* Limpr., Laubm. 233, 1886 Eur
fo. *ramosissimum* (BSG) Mönk., Laubm. Eur. 250, 1927 (*Gymnostomum rupestre* var.) Eur
fo. *rigidum* (Schimp.) Podp., Consp. Musc. Eur. 171, 1954 (*Gymnostomum rupestre* var.) Eur
[fo. *stelligerum* (BSG) Mönk., Laubm. Eur. 251, 1927 (*Gymnostomum rupestre* var.) Eur Am1 ≡ *Gymnostomum aeruginosum* fo. *stelligerum* (BSG) Lazar.]
[*Gymnostomum setifolium* Hook. & Arn. Am4 ≡ *Bartramidula setifolia* (Hook. & Arn.) Fransen, Lindbergia 14: 31, 1988 ≡ *Flowersia setifolia* (Hook. & Arn.) Griffin & Buck, Bryologist 92: 372, 1989]
Gymnostomum simplicissimum (P. Beauv.) Brid. As3
Gymnostomum splachnobryoides Biz., Cryptogamie Bryol. Lichénol. 1: 425, 1980 Afr2
[*Gymnostomum subrigidulum* (Broth.) Chen As2 = *Hymenostylium recurvirostrum* var. *insigne* (Dix.) Bartr. *fide* Zander & Eckel, Canad. J. Bot. 60: 1596, 1982]
Gymnostomum tenerrimum (C. Müll.) Wijk & Marg., Taxon 17: 467, 1968 (*Anoectangium*) Am6
[*Gymnostomum truncatum* (With.) Hedw. ex Schum. ≡ *Pottia truncata* (Hedw.) BSG]
var. *intermedium* Lilj. Eur
Gymnostomum unguiculatum Philib. in Husn., Musci Gall. Herb. 12: n. 551, 1879 Eur
[*Gymnostomum valerianum* (Bartr.) Zand., Bryologist 80: 262, 1977 (*Leptodontium*) Am2 ≡ *Tuerckheimia valeriana* (Bartr.) Zand. *fide* Zander, Misc. Bryol. Lichenol. 8: 27, 1978]
[*Gymnostomum venezuelense* (C. Müll.) Kindb. Am4 = *Hymenostylium recurvirostrum* (Hedw.) Dix., see treatment of *Hymenostylium*]
Gymnostomum viridulum Brid. good species *fide* Whitehouse & Crundwell, J. Bryol. 16: 563, 1991 Eur Afr1
[*Gymnostomum wageri* Schelpe in Magill & Schelpe, Mem. Bot. Surv. S. Afr. 43: 5, 1979 (*Gymnostomum gracile* Dix. *non* (R. Br.) Hook.) Afr4 = *Didymodon ceratodonteus* (C. Müll.) Dix. *fide* Magill, Fl. S. Afr. I. Bryoph. 1: 235, 1981 (1982)]

GYROWEISIA Schimp.
[*Gyroweisia amplexicaulis* Sim Afr4 = *Husnotiella latifolia* (Dix.) Zand. & Magill in Magill & Schelpe, Mem. Bot. Surv. S. Afr. 43: 23, 1979 ≡ *Didymodon tophaceopsis* Zand. *nom. nov.*, see treatment of *Didymodon*]
Gyroweisia barbulacea (C. Müll.) Broth. Am2 (ignot., type cannot be not located *fide* Zander, Bryologist 80: 265, 1977)
[*Gyroweisia benoistii* Thér. Am4 ≡ *Globuilinella benoistii* (Thér.)

Magill, Bryologist 80: 79, 1977]

[*Gyroweisia boliviana* Williams (= *Barbula* sp. *fide* Hilpert 1933) = *Didymodon tophaceus* (Brid.) Lisa, see treatment of *Gyroweisia*]

[*Gyroweisia brevicaulis* (C. Müll.) Broth. As4 Oc = *Luisierella barbula* (Schwaegr.) Steere *fide* Zander, Bryologist 80: 266, 1977]

[*Gyroweisia hildebrandtii* (C. Müll.) Kindb. Afr2 ≡ Hymenostylium hildebrandtii (C. Müll.) Zand., see treatment of *Hymenostylium*]

[*Gyroweisia latifolia* Dix. Afr2 ≡ *Husnotiella latifolia* (Dix.) Zand. & Magill in Magill & Schelpe, Mem. Bot. Surv. S. Afr. 43: 7, 1979 ≡ *Didymodon tophaceopsis* Zand. *nom. nov.*, see treatment of *Didymodon*]]

var. *tanneri* Biz., Rev. Bryol. Lichénol. 40: 120, 1974 Afr2

[*Gyroweisia lindigii* (Hampe) Broth. Am4 ≡ *Didymodon lindigii* (Hampe) Zand., see treatment of *Didymodon*]

[*Gyroweisia luisieri* Sérgio, Bol. Soc. Port. Ciênc. Nat. 14: 82, 1972 Eur (= *Gymnostomum aeruginosum fide* Zander, Bryologist 80: 259, 1977) ≡ *Gymnostomum luisieri* (Sérgio) Sérgio ex Crundw., J. Bryol. 11: 603, 1981 (1982) = *Gymnostomum viridulum* Brid. *fide* Whitehouse & Crundwell, J. Bryol. 16: 563, 1991]

Gyroweisia monterreia Zand. & Herm., Bryologist 89: 227, 1986 [1987] Am2

[*Gyroweisia nigeriana* Egun. & Olar., Bryologist 81: 443, 1978 Afr2 ≡ *Weisiopsis nigeriana* (Egun. & Olar.) Zand., see treatment of *Weisiopsis*]

[*Gyroweisia obtusifolia* Broth. Am2 ≡ *Husnotiella obtusifolia* (Hampe) Zand., Bryologist 80: 265, 1977 ≡ *Gyroweisia obtusifolia* Broth. *fide* Zander, Cryptogamie Bryol. Lichénol. 2: 419, 1981 (1982) ≡ *Didymodon hampei* Zand. *nom. nov.*, see treatment of *Didymodon*]

[*Gyroweisia papillosa* Thér. Am2 = *Husnotiella obtusifolia* (Broth.) *fide* Zander, Bryologist 80: 265, 1977 ≡ *Gyroweisia obtusifolia* Broth. *fide* Zander, Cryptogamie Bryol. Lichénol. 2: 419, 1981 (1982)] ≡ *Didymodon hampei* Zand. *nom. nov.*, see treatment of *Didymodon*]

[*Gyroweisia pocsii* Biz., Cryptogamie Bryol. Lichénol. 1: 424, 1980 Afr2 = *Weisiopsis nigeriana* Egun. & Olar. Zand., see treatment of Weisiopsis]

Gyroweisia pusilla Broth. Am1

Gyroweisia reflexa (Brid.) Schimp. Eur Afr1 Am1

Gyroweisia rohlfsiana (C. Müll.) Par. Afr1

Gyroweisia shansiensis Sak. As2

Gyroweisia tenuis (Hedw.) Schimp. Eur As5 Afr1 Am1

var. *badia* Limpr. Eur Afr1

var. *cuspidata* Röll Eur

var. *lacustris* Amann Eur

var. *schisticola* Roth in Zodda Eur

fo. *compacta* (Hag.) Podp., Consp. Musc. Eur. 171, 1954 (*Gyroweisia tenuis* var.) Eur

[fo. *propagulifera* Limpr. Eur = *Gyroweisia tenuis* var. *compacta* (Hag.) Podp., Consp. Musc. Eur. 171, 1954]

[*Gyroweisia tophicola* (C. Müll.) Kindb. Afr2 = *Hymenostylium recurvirostrum* (Hedw.) Dix., see treatment of *Hymenostylium*]

Gyroweisia yuennanensis Broth. As2

HENNEDIELLA Par.

Hennediella acletoi (Robins.) Zand. (*Tortula*), see treatment of *Hennediella* Am4

Hennediella acutidentata (Card. & Thér.) Zand. (*Pottia*), see treatment of *Hennediella* Afr4

Hennediella angustifolia (Herz.) Zand. (Calyptopogon), see treatment of Hennediella Am4

Hennediella arenae (Besch.) Zand. (*Barbula*), see treatment of *Hennediella* Austr2 Ant

var. *petriei* (Broth.) Zand. (*Tortula*), see treatment of *Hennediella* Austr2

Hennediella austrogeorgica (Card.) Blockeel, J. Bryol. 16: 191, 1991 (*Pottia*) Am6

Hennediella bellei (Bartr.) Zand. (*Desmatodon*), see treatment of *Hennediella* Am4

Hennediella densifolia (Hook. f. & Wils.) Zand. (*Barbula*), see treatment of *Hennediella* Am6 Ant

Hennediella denticulata (Wils. in Mitt.) Zand. (*Barbula*), see treatment of *Hennediella* Am3 Am4

Hennediella heimii (Hedw.) Zand. (*Gymnostomum*), see treatment of *Hennediella* Eur As1 As2 Am1 Am6 Austr1 Austr2 Ant

var. *alpina* (Amann) Zand. (*Pottia heimii* var.), see treatment of *Hennediella* Eur

var. *arctica* (Lindb.) Zand. (*Pottia heimii* var.), see treatment of *Hennediella* Eur As1 Am1

var. *brachyphylla* (Warnst.) Zand. (*Pottia heimii* var.), see treatment of *Hennediella* Eur

var. *brevicuspis* (Warnst.) Zand. (*Pottia heimii* var.), see treatment of *Hennediella* Eur

var. *breviseta* (Warnst.) Zand. (*Pottia heimii* var.), see treatment of *Hennediella* Am1

var. *drummondii* (Warnst.) Zand. (*Pottia heimii* var.), see treatment of *Hennediella* Am1

var. *eurystoma* (Card. & Broth.) Zand. (*Pottia heimii* var.), see treatment of *Hennediella* Am6

var. *guessfeldtii* (Schlieph.) Zand. (*Pottia*), see treatment of *Hennediella* Eur

var. *lanceolata* (Warnst.) Zand. (*Pottia heimii* var.), see treatment of *Hennediella* Eur Am1

var. *magellanica* (Warnst.) Zand. (*Pottia heimii* var.), see treatment of *Hennediella* Am6

var. *maxima* (Card.) Zand. (*Pottia heimii* var.), see treatment of *Hennediella* Am6

var. *spegazzinii* (C. Müll.) Zand. (*Pottia*), see treatment of *Hennediella* Am6

var. *thaxteri* (Card. & Thér.) Zand. (*Pottia heimii* var.), see treatment of *Hennediella* Am6

Hennediella heteroloma (Card.) Zand. (*Tortula*), see treatment of *Hennediella* Am2

var. *eckeliae* (Zand.) Zand., see treatment of *Hennediella* Am2

Hennediella kunzeana (C. Müll.) Zand. (*Barbula*), see treatment of *Hennediella* Afr4 Am6

Hennediella limbata (Mitt.) Zand. (*Barbula*), see treatment of *Hennediella* Am2 Am4

Hennediella longipedunculata (C. Müll.) Zand. (*Barbula*), see treatment of *Hennediella* Afr4

Hennediella longirostris (Hampe ex C. Müll.) Zand. (*Pottia*), see treatment of *Hennediella* Afr2

Hennediella macrophylla (R. Br. ter) Par. Am6 Austr2

Hennediella marginata (Hook. f. & Wils.) Zand. (*Schistidium*), see treatment of *Hennediella* Afr4

Hennediella oedipodioides (C. Müll.) Zand. (*Pottia*), see treatment of *Hennediella* Afr4

Hennediella polyseta (C. Müll.) Zand. (*Barbula*), see treatment of *Hennediella* Am2 Am4

Hennediella serrulata (Hook. & Grev.) Zand. (*Tortula*), see treatment of *Hennediella* Am6

Hennediella stanfordensis (Steere) Blockeel J. Bryol. 16: 191, 1991 (*Tortula*) Eur Am1 Austr1

Hennediella steereana (Zand. & Crum) Zand. (*Desmatodon*), see treatment of *Hennediella* Am4

HILPERTIA Zand.

Hilpertia scotteri (Zand. & Steere) Zand., Phytologia 65: 428, 1989 (*Tortula*) Am1

Hilpertia velenovskyi (Schiffn.) Zand., Phytologia 65: 428, 1989 (*Tortula*) Eur Am6

[**HUSNOTIELLA** Card. = *Didymodon fide* Zander, Cryptogamie Bryol. Lichénol. 2: 384, 1981 (1982)]

[*Husnotiella baueri* Bartr. in Bauer Am6 = *Didymodon tophaceus* (Brid.) Lisa see treatments of *Didymodon* and *Bryoerythrophyllum*]

Husnotiella glossophylla Herz. Am4 (type apparently lost, see treatment of *Bryoerythrophyllum*)

[*Husnotiella latifolia* (Dix.) Zand. & Magill in Magill & Schelpe, Mem. Bot. Surv. S. Afr. 43: 7, 1979 (*Gyroweisia*) Afr4 ≡ *Didymodon tophaceopsis* Zand. *nom. nov.*, see treatment of *Didymodon*]

[*Husnotiella obtusifolia* (Hampe) Zand., Bryologist 80: 265, 1977 (*Trichostomum*) Am2 ≡ *Gyroweisia obtusifolia* Broth. *fide* Zander, Cryptogamie Bryol. Lichénol. 2: 419, 1981 (1982) ≡ *Didymodon hampei* Zand. *nom. nov.*, see treatment of *Didymodon*]

[*Husnotiella plicata* Magill, Fl. S. Afr. I. Mosses 1: 222, 1981 [1982] Afr4 ≡ *Didymodon jackvancei* Zand. *nom. nov.*, see treatment of *Didymodon*]

[*Husnotiella revoluta* Card. Am1 Am2 Am4 ≡ *Didymodon revolutus* (Card.) Williams *fide* Zander, Cryptogamie Bryol. Lichénol. 2: 401, 1981 (1982)]

[var. *palmeri* (Card.) Thér. Am1 = *Didymodon revolutus* (Card.) Williams *fide* Zander, Cryptogamie Bryol. Lichénol. 2: 401, 1981 (1982)]

fo. *elata* Thér., Smiths. Misc. Coll. 85: 7, 1931 Am2

[*Husnotiella torquescens* (Card.) Bartr. Am1 Am2 Am4 = *Didymodon australasiae* (Hook. & Grev.) Zand. *s. lat. fide* Zander, Cryptogamie Bryol. Lichénol. 2: 397, 1981 (1982)]

[**HYDROGONIUM** (C. Müll.) Jaeg. = *Barbula fide* Saito, J. Hattori Bot. Lab. 39: 492, 1975]

[*Hydrogonium afrofontanum* (C. Müll.) Hilp. Afr4 ≡ *Barbula afrofontana* (C. Müll.) Broth.]

[*Hydrogonium amplexifolium* (Mitt.) Chen As3 ≡ *Barbula amplexifolia* (Mitt.) Jaeg.]

[*Hydrogonium anceps* (Card.) Herz. & Nog. As2 ≡ *Barbula anceps* Card.]

[*Hydrogonium aneitense* (Broth. & Watts) Schultze-Motel, Willdenowia 7: 55, 1973 ≡ *Barbula aneitensis* Broth. & Watts.]

[*Hydrogonium arcuatum* (Griff.) Wijk & Marg. As3 As4 Oc ≡ *Barbula arcuata* Griff.]

[*Hydrogonium bolleanum* (C. Müll.) Jaeg. Afr2 ≡ *Barbula bolleana* (C. Müll.) Broth.]

[*Hydrogonium brotheri* (Ren. & Par.) Hilp. Afr3 ≡ *Didymodon brotheri* (Ren. & Par.) Zand. (*Trichostomum*), see treatment of *Didymodon*]

[*Hydrogonium clavicostatum* (Ren. & Card.) Hilp. Afr3 ≡ *Barbula clavicostata* (Ren. & Hilp.) Zand., see treatment of *Barbula*]

[*Hydrogonium comosum* (Dozy & Molk.) Hilp. As4 = *Barbula arcuata* Griff. *fide* Saito, J. Hattori Bot. Lab. 39: 496, 1977]

[var. *japonica* Broth. As2 = *Barbula arcuata* Griff. *fide* Saito, J. Hattori Bot. Lab. 39: 496, 1977]

[*Hydrogonium consanguineum* (Thwait. & Mitt.) Hilp. As3 As4 ≡ *Barbula consanguinea* (Thwait. & Mitt.) Jaeg. = *Barbula javanica* Dozy & Molk. *fide* Saito, J. Hattori Bot. Lab. 39: 495, 1975]

[*Hydrogonium crozalsii* (Philib.) Podp. Eur ≡ *Barbula crozalsii* (Philib.) Broth.]

[*Hydrogonium decolyi* Broth. ex Gangulee, Nov. Hedw. 12: 426, 1966 As3 ≡ *Bryoerythrophyllum recurvum* (Griff.) Saito *fide* Sollman, Lindbergia 16: 23, 1990]

Hydrogonium dicranelloides Gangulee, Nov. Hedw. 12: 427, 1966 As3

Hydrogonium dixonianum Chen As2

[*Hydrogonium dorrii* (Ren. & Card.) Hilp. Afr3 ≡ *Barbula dorrii* Ren. & Card.]

[*Hydrogonium ehrenbergii* (Lor.) Jaeg. Eur As2 As5 Afr1 Am1 Am2 ≡ *Barbula ehrenbergii* (Lor.) Fleisch.]

[var. *algeriae* (C. Müll.) Wijk & Marg. Eur Afr1 ≡ *Hydrogonium ehrenbergii* var. *algeriae* (C. Müll.) Vent. & Bott.]

fo. *densirete* (Amann) Podp., Consp. Musc. Eur. 199, 1954 (*Didymodon ehrenbergii* fo.) Eur

fo. *laxirete* (Amann) Podp., Consp. Musc. Eur. 199, 1954 (*Didymodon ehrenbergii* fo.) Eur

[*Hydrogonium fontanum* (C. Müll.) Jaeg. Afr2 ≡ *Barbula fontana* (C. Müll.) Broth. ≡ *Barbula meidensis* Cufodontis *nom. nov.*]

[*Hydrogonium gangeticum* (C. Müll.) Chen As3 ≡ *Barbula gangetica* C. Müll. = *Barbula arcuata* Griff. *fide* Gangulee, Mosses E. India 3: 725, 1972]

[*Hydrogonium gracilentum* (Mitt.) Chen As3 ≡ *Barbula gracilenta* Mitt.]

Hydrogonium heterophyllum Fröhl., Ann. Naturhist. Mus. Wien 67: 154, 1964 As3

Hydrogonium hygrophilum Hilp. As4

[*Hydrogonium javanicum* (Dozy & Molk.) Hilp. As2 As3 As4 ≡ *Barbula javanica* Dozy & Molk.]

[var. *convolutifolium* (Dix.) Chen As2 = *Barbula javanica* Dozy & Molk. *fide* Saito, J. Hattori Bot. Lab. 39: 495, 1975]

[var. *epapillosum* (Fleisch.) Hilp. As4 ≡ *Barbula javanica* var. *epapillosa* Fleisch.]

var. *kurzii* (C. Müll.) Gangulee, Mosses E. India 3: 738, 1972 As3

[*Hydrogonium laevifolium* (Broth. & Yas.) Chen As2 ≡ *Barbula laevifolia* Broth. & Yas.]

[*Hydrogonium leucobasis* (Dix. in Dix. & Greenwood) Schultze-Motel, Willdenowia 7: 371, 1974 Oc (*Barbula*) ≡ *Barbula leucobasis* Dix. in Dix. & Greenwood]

Hydrogonium leucodontoides Gangulee, Nov. Hedw. 12: 425, 1966 [1967] As3

[*Hydrogonium leucodontoides* (C. Müll. in Par. ex Gangulee) Chopra, Taxon. Indian Mosses 145. 1975 (*Barbula*) As3 ≡ *Barbula leucodontioides* C. Müll. in Par. ex Gangulee, Nov. Hedw. 12: 425, 1966]

[*Hydrogonium louisiadum* (Broth.) Schultze-Motel, Willdenowia 7: 55, 1973 (*Barbula*) Oc = *Barbula consanguinea* (Thw. & Mitt.) Jaeg. *fide* Eddy, Handb. Males. Mosses 2: 178, 1991 = *Barbula javanica* Dozy & Molk. *fide* Saito, J. Hattori Bot. Lab. 39: 495, 1975]

[*Hydrogonium majusculum* (C. Müll.) Chen As2 ≡ *Barbula majuscula* C. Müll.]

Hydrogonium mamatkulovii Laz., Dopov. Akad. Nauk Ukr. RSR, Ser. B 8: 753, 1967

Hydrogonium mussoorianum Vohra, J. Bombay Nat. Hist. Soc. 63: 464, 1967. As3

[*Hydrogonium novoguinense* (Broth.) Chen As4 ≡ *Barbula novoguinense* Broth.]

Hydrogonium patulifolium Fröhl., Ann. Naturhist. Mus. Wien 67: 154, 1964 As3

[*Hydrogonium pseudoehrenbergii* (Fleisch.) Chen As2 As4 Afr2 ≡ *Barbula pseudoehrenbergii* Fleisch.]

[*Hydrogonium rechingeri* (Broth.) Schultze-Motel, Willdenowia 7: 55, 1973 Oc ≡ *Barbula rechingeri* Broth.]

[*Hydrogonium setschwanicum* (Broth.) Chen As2 ≡ *Barbula setschwanica* Broth. = *Barbula indica* (Hook.) Spreng. *fide* Saito, J. Hattori Bot. Lab. 39: 488, 1975]

[*Hydrogonium soboliferum* (Fleisch.) Hilp. As4 ≡ *Barbula sobolifera* Fleisch. = *Barbula arcuata* Griff. *fide* Saito, J. Hattori Bot. Lab. 39: 496, 1975]

[*Hydrogonium sordidum* (Besch.) Chen As3 ≡ *Barbula sordida* Besch.]

[*Hydrogonium subcomosum* (Broth.) Chen As2 ≡ *Barbula subcomosa* Broth.]

[*Hydrogonium subpellucidum* (Mitt.) Hilp. ≡ *Barbula subpellucida* (Mitt.) Broth.]

var. *hyaloloma* Herz. As2

[*Hydrogonium taylorii* W. Weber, Lindbergia 3: 81, 1975 [1976] ≡ *Barbula taylorii* Bartr. & Steere *hom. illeg. non* Lindb.]

[*Hydrogonium tisserantii* (P. Varde) Schultze-Motel, Willdenowia 7: 491, 1975 (*Didymodon*) Afr2 ≡ *Barbula tisserantii* (P. Varde) P. Varde]

[*Hydrogonium williamsii* Chen As4 ≡ *Barbula williamsii* (Chen) Iwats. & Tan, Kilikasan 8: 186, 1979]

[**HYMENOSTOMUM** R. Br. = *Weissia* Hedw. *fide* Saito, J. Hattori Bot. Lab. 39: 417, 1975]]

[*Hymenostomum anomalum* Broth. in Herz. Am4 ≡ *Trichostomum ovatifolium* Zand. *nom. nov.*, see treatment of *Trichostomum*]

Hymenostomum aristatulum Par. & Broth. Oc

[*Hymenostomum ayresii* (Schimp.) Broth. Afr2 Afr3 ≡ *Weissia ayresii* Schimp. in Besch.]

[*Hymenostomum balansaeanum* Besch. Am5 Am6 ≡ *Weissia balansaeana* (Besch.) C. Müll.]

var. *densirete* Thér. Am6

[*Hymenostomum brachypelma* (C. Müll.) Kindb. Afr2 ≡ *Weissia brachypelma* C. Müll. Afr2

[*Hymenostomum breutelii* (C. Müll.) Kindb. Am2 Am3 Am5 ≡ *Weissia breutelii* C. Müll. Am2 Am3 Am4 Am5

[*Hymenostomum castaneum* Crum & Steere, Amer. Midland Nat. 60: 12, 1959 Am3 ≡ *Trichostomum castaneum* (Crum & Steere) Zand., see treatment of *Trichostomum*]

[*Hymenostomum chloropus* (Besch.) Broth. Afr2 Afr3 = *Hymenostomum ayresii* (Schimp. ex Besch.) Broth. *fide* Een, Lindbergia 3: 125, 1976 ≡ *Weissia ayresii* Schimp. in Besch.]

[*Hymenostomum cucullatum* (C. Müll.) Kindb. Afr4 ≡ *Weissia cucullata* C. Müll. Afr4

Hymenostomum densirete Thér. Am4

[*Hymenostomum eckendorffii* P. Varde Afr2 ≡ *Tortella eckendorffii* (P. Varde) Zand., see treatment of *Tortella*]

[*Hymenostomum edentulum* (Mitt.) Besch. As3 As4 Oc ≡ *Weissia edentula* Mitt. As3 As4 Oc

[*Hymenostomum eurybasis* Dix. Afr2 = *Weissia latiuscula* C. Müll. *fide* Magill, Fl. S. Afr. I. Mosses 1: 267, 1981 (1982)]

[*Hymenostomum exsertum* (Broth.) Broth. As2 ≡ *Weissia exserta* (Broth.) Chen]

Hymenostomum fasciculatum Hampe Am5

[*Hymenostomum flavescens* Britt. in N. Britt. & Millsp. Am1 Am3 ≡ *Weissia flavescens* (Britt. in N. Britt. & Millsp.) Reese, Bryologist 94: 54, 1991 ≡ *Trichostomum brittonianum nom. nov.*, see treatment of *Trichostomum*]

Hymenostomum francii Thér. Oc

Hymenostomum goyazense (Broth.) Broth. Am5

Hymenostomum guineense Broth. & Par. Afr2

[*Hymenostomum guyazensis* (Broth.) Broth., Nat. Pfl. 1(3): 386, 1902 err. pro *Hymenostomum goyazense* (Broth.) Broth.]

[*Hymenostomum gymnostomum* (Besch.) Nog. As2 = *Weissia edentula* Mitt. *fide* Saito, J. Hattori Bot. Lab. 39: 421, 1975]

[*Hymenostomum humicola* (C. Müll.) Par. Afr2 Afr4 ≡ *Weissia humicola* C. Müll.]

[*Hymenostomum inoperculatum* Crum, Madroño 14: 74, 1957 Am1 ≡ *Weissia inoperculata* (Crum) Crum, Steere & Anders., Bryologist 67: 164, 1973]

[*Hymenostomum jamesonii* (Arnott) Hampe Am2 Am5 ≡ *Weissia jamesonii* Tayl.]

[*Hymenostomum krassavinii* Laz. As1 ≡ *Weissia krassavinii* (Laz.) Laz. ex Ochyra]

[*Hymenostomum kunzeanum* (C. Müll.) Broth. Am6 ≡ *Weissia kunzeana* C. Müll.]

Hymenostomum latifolium Nog. As2

Hymenostomum laxirete (Broth.) Broth. Am6

Hymenostomum leratii Par. & Broth. Oc

var. *acuminatum* Thér. Oc

[*Hymenostomum lineaefolium* (C. Müll.) Par. Afr2 ≡ *Weissia lineaefolia* C. Müll.]

[*Hymenostomum malayense* Fleisch. As2 As4 ≡ *Weissia malayensis* (Fleisch.) Manuel, Fed. Mus. Jour. 26: 161, 1981 = *Barbula indica* (Hook.) Spreng. *fide* Iwatsuki & Noguchi, J. Hattori Bot. Lab. 37: 355, 1973 as *Barbula cruegeri* Sond. ex C. Müll.]

[*Hymenostomum mexicanum* Card. Am2 = *Weissia controversa* Hedw. *fide* Zander in Sharp et al., Moss Fl. Mex.]

[*Hymenostomum meylanii* Amann = *Hymenostomum squarrosum* Nees & Hornsch. *fide* Crundwell & Nyholm, J. Bryology 7: 9, 1972 ≡ *Weissia squarrosa* (Nees & Hornsch.) C. Müll.]

[*Hymenostomum micaceum* (Schlecht.) Hampe As2 As3 As4 Am2 Am3 Am5 Oc ≡ *Weissia micacea* (Schlecht.) C. Müll.]

[*Hymenostomum microstomum* (Hedw.) R. Br. in Nees & Hornsch. Eur As1 As2 As3 As5 Afr1 Am1 ≡ *Weissia microstoma* (Hedw.) C. Müll. Eur As1 As2 As3 As5 Afr1 Am1 *hom. illeg.* = *Weissia brachycarpa* (Nees & Hornsch.) Jur. var. *brachycarpa fide* Corley et al., J. Bryol. 11: 650, 1981 (1982) and Koponen, Isoviita & Lammes, Flora Fennica 6: 61, 1977]

[var. *brachycarpum* (Nees & Hornsch.) Hüb. Eur ≡ *Weissia brachycarpa* (Nees & Hornsch.) Jur. *fide* Corley et al., J. Bryol. 11: 650, 1981 (1982) and Koponen, Isoviita & Lammes, Flora Fennica 6: 61, 1977]

[var. *obliquum* (Nees in Nees & Hornsch.) Hüb. Eur ≡ *Weissia brachycarpa* var. *obliqua* (Nees in Nees & Hornsch.) Hill *nom. illeg.*]

fo. *elatum* (B.&S. in BSG) Podp., Consp. Musc. Eur. 186, 1954 Eur

fo. *planifolium* Fleisch., Beitr. Laubmoosfl. Ligur. Atti Congr. Botan. Intern. 1892: 268, 1893 Eur

[*Hymenostomum newcomeri* Bartr. As2 ≡ *Weissia newcomeri* (Bartr.) Saito]

Hymenostomum noumeanum Thér. Oc

Hymenostomum obscurissimum Dix. As3

[*Hymenostomum obtusatum* (C. Müll.) Broth. Am5 ≡ *Weissia obtusata* C. Müll.]

Hymenostomum olivaceum C. Müll. ex Geh. Austr1

[*Hymenostomum opacum* Wag. & Dix. Afr4 = *Barbula indica* (Hook.) Spreng. *fide* Magill, Fl. S. Afr. I. Mosses 1: 243, 1981 (1982)]

[*Hymenostomum ovale* Williams Oc ≡ *Weissia ovalis* (Williams) Bartr.]

[*Hymenostomum papillosissimum* Lazar., Dopov. Akad. Nauk Ukr. R.S.R. ser. B 8: 752, 1967 *nom. inval.* in syon. Eur ≡ *Weissia papillosissima* Lazar.]

[*Hymenostomum papillosissimum* (Lazar.) Savicz-Ljubitsk., Novosti Sist. Niz. Rast. 6: 248, 1969 [1970] Eur ≡ *Weissia papillosissima* Lazar.]

Hymenostomum patulum (Knight) Dix. Austr2

[*Hymenostomum perpusillum* (C. Müll.) Par., Acta Soc. Linn. Bordeaux 49: 272, 1895 [1896], Ind. Bryol. [2]: 596, 1896 Austr1 ≡ *Weissia perpusilla* (C. Müll.) Stone, J. Bryol. 11: 231, 1980]

[*Hymenostomum pulicare* Besch. Afr3 ≡ *Trichostomum pulicare* (Besch.) Zand., see treatment of *Trichostomum*]

[*Hymenostomum riograndense* Broth. Am5 ≡ *Weissia riograndensis* (Broth.) Zand., see treatment of *Weissia*]

[*Hymenostomum rostellatum* (Brid.) Schimp. Eur ≡ *Weissia rostellata* (Brid.) Lindb.]

[subsp. *meylanii* (Amann) Podp. Eur = *Hymenostomum squarrosum* Nees & Hornsch. *fide* Crundwell & Nyholm, J. Bryology 7: 9, 1972 ≡ *Weissia squarrosa* (Nees & Hornsch.) C. Müll.]

[var. *meylanii* (Amann) Podp. Eur = *Hymenostomum squarrosum* Nees & Hornsch. *fide* Crundwell & Nyholm, J. Bryology 7: 9, 1972 ≡ *Weissia squarrosa* (Nees & Hornsch.) C. Müll.]

[var. *phascoides* (Hook.) Card. Eur Am1 ≡ *Weissia rostellata* var. *phascoides* (Hook.) Reese & Lemmon, Bryologist 68: 283, 1965]

[*Hymenostomum semidiaphanum* Thér. Am2 ≡ *Weissia semidiaphana* (Thér.) Zand., Monogr. Syst. Bot. Missouri Bot. Gard. 11: 197, 1985]

[*Hymenostomum semiinvolutum* (C. Müll.) Kindb. Am6 ≡ *Weissia semiinvoluta* C. Müll.]

Hymenostomum siamense Dix. As3

[*Hymenostomum socotranum* (Mitt.) Broth. Afr2 ≡ *Weissia socotrana* Mitt.]

[*Hymenostomum squarrosum* Nees & Hornsch. Eur ≡ *Weissia squarrosa* (Nees & Hornsch.) C. Müll.]

[*Hymenostomum striatum* Geh. & Hampe Am6 ≡ *Weissia glaziouii* Zand. *nom. nov*., see treatment of *Weissia*]

[*Hymenostomum strictifolium* Dix. & Sak. As2 = *Weissia edentula* Mitt. *fide* Saito, J. Hattori Bot. Lab. 39: 421, 1975]

[*Hymenostomum subacaule* (Mitt.) Par., Acta Soc. Linn. Bordeaux 49: 272, 1895 (1896), Ind. Bryol. (2): 596, 1896 Am4 ≡ *Weissia subacaulis* (Mitt.) Par.]

[*Hymenostomum submicaceum* (C. Müll.) Par., Acta Soc. Linn. Bordeaux 49: 273, 1895 [1896], Ind. Bryol. [2]: 597, 1896 Am5 ≡ Weissia submicacea C. Müll.]

Hymenostomum subrostellatum Schimp. ex Basch. Afr1

Hymenostomum sullivanii C. Müll. ex Geh. Austr1

[*Hymenostomum termitarum* (C. Müll.) Broth. Am5 ≡ *Trichostomum termitarum* (C. Müll.), see treatment of *Trichostomum*]

[*Hymenostomum termitidarum* (C. Müll.) Par., Acta Soc. Linn. Bordeaux 49: 273, 1895 (1896), Ind. Bryol. (2): 597, 1896 Afr2 ≡ *Weissia termitidarum* C. Müll.]

[*Hymenostomum tortile* (Schwaegr.) B.&S. in BSG Eur As1 As3 As5 Afr1 Austr2 Am1 Am2 = *Weissia condensa* (Voit) Lindb. *fide* Corley et al., J. Bryol. 11: 623, 1981 (1982)]

var. *brevifolium* Schiffn. Eur

[var. *intermedium* (Mönk.) Podp. Eur ≡ *Weissia tortilis* var. *intermedia* Mönk.]

var. *pseudocrispatum* Podp. Eur

var. *subalpinum* Kern Eur

[var. *subcylindricum* BSG Eur ≡ *Weissia tortilis* var. *subcylindrica* (BSG) Dix.]

var. *tunetana* Besch. in Besch. & Pat. Afr1

[fo. *brevifolium* Amann, Fl. Mouss. Suisse 3: 7, 1923 *hom. illeg. non Hymenostomum tortile* var. *brevifolium* Schiffn. Eur]

fo. *rufidulum* Podp., Bul. Grăd. Bot. Univ. Cluj 11: 55, 1931 Eur

[*Hymenostomum urceolare* (Hampe) Hampe Am5 ≡ *Trichostomum urceolare* (Hampe) Zand., see treatment of *Trichostomum*]

HYMENOSTYLIELLA Bartr.

Hymenostyliella alata (Herz.) Robins., Phytologia 21: 3, 1971 (*Timmiella*) Am4

Hymenostyliella calcarea (Dix.) Iwats., J. Hattori Bot. Lab. 40: 146, 1976 (*Diphyscium*) As3

[*Hymenostyliella involuta* (Card. & Thér.) Bartr. As4 = *Hymenostyliella llanosii* (Broth.) Robins. *fide* Robinson, Phytologia 21: 2, 1971]

[*Hymenostyliella japonica* (Broth.) Saito, J. Jap. Bot. 46: 145, 1971 As2 ≡ *Didymodon japonicus* (Broth.) Saito, J. Hattori Bot. Lab. 39: 508, 1975]

Hymenostyliella llanosii (Broth.) Robins., Phytologia 21: 2, 1971 (*Barbula*) As2 As3

HYMENOSTYLIUM Brid.

[*Hymenostylium annotinum* Mitt. ex Dix. As3 = *Hymenostylium recurvirostrum* var. *insigne* (Dix.) Bartr. *fide* Zander & Eckel, Canad. J. Bot. 60: 1596, 1982]

[*Hymenostylium anoectangioides* (C. Müll.) Broth. As2 = *Hymenostylium recurvirostrum* (Hedw.) Dix. *fide* Chen, Hedwigia 80: 62, 1941]

[*Hymenostylium aurantiacum* Mitt. As2 As3 As4 ≡ *Hymenostylium recurvirostrum* var. *aurantiacum* (Mitt.) Gangulee, Mosses E. India 648, 1972 (*Hymenostylium*) *hom. illeg. incl.* var. prior. = *Hymenostylium recurvirostrum* var. *luzonense* (Broth.) Bartr. *fide* Gangulee, Mosses E. India 3: 648, 1972 = *Hymenostylium recurvirostrum* var. *cylindricum* (Bartr.) Zand., see treatment of *Hymenostylium*]

[*Hymenostylium collenchymaticum* Baumg. & Fröhl. As4 = *Hymenostylium recurvirostrum* (Hedw.) Dix. *fide* Sollman in Touw, J. Hattori Bot. Lab. 71: 343, 1992]

Hymenostylium congoanum Dix. & Nav. Afr2

Hymenostylium contextum Herz. Am4

[*Hymenostylium courtoisii* Broth. & Par. As2 = *Hymenostylium recurvirostrum* var. *aurantiacum* (Mitt.) Gangulee, Mosses E. India 648, 1972 *hom. illeg. incl.* var. prior = *Hymenostylium recurvirostrum* var. *luzonense* (Broth.) Bartr. *fide* Gangulee, Mosses E. India 3: 648, 1972 = *Hymenostylium recurvirostrum* var. *cylindricum* (Bartr.) Zand., see treatment of *Hymenostylium*]

Hymenostylium crassinervium Broth. & Dix. Afr2 Afr4

Hymenostylium crispulum Broth. & Par. Afr2

[*Hymenostylium curvirostre* Mitt. = *Hymenostylium recurvirostrum* (Hedw.) Dix.]

[var. *bicolor* Dix. As3 ≡ *Hymenostylium recurvirostrum* var. *bicolor* (Dix.) Crum]

[fo. *commutatum* (Mitt.) Hag., Musci Norv. Bor. 2, 1899 (*Hymenostomum*) Eur ≡ *Hymenostylium recurvirostrum* var. *commutatum* (Mitt.) Podp.]

[fo. *megalosporum* Latz., Hedwigia 66: 138, 1926 Eur ≡ *Hymenostylium recurvirostrum* fo. *megalosporum* (Latz.) Podp.]

[fo. *scabrum* (Lindb.) Hag., Musci Norv. Bor. 2, 1899 Eur = *Hymenostylium recurvirostrum* var. *latifolium* (Zett.) Wijk & Marg. = *Hymenostylium recurvirostrum* (Hedw.) Dix. *fide* Zander in Sharp et al., Moss Fl. Mex.]

[fo. *serrulatum* Röll, Hedwigia 38: 260, 1899 Eur ≡ *Hymenostylium recurvirostrum* fo. *serrulatum* (Röll) Podp.]

Hymenostylium dicranelloides Broth. ex Dix. As3

Hymenostylium diversifolium Fröhl., Ann. Naturhist. Mus. Wien 67: 152, 1964 As3

Hymenostylium filiforme Dix. As3

[*Hymenostylium firmum* (C. Müll.) Broth. in Bartr. Oc = *Hymenostylium recurvirostrum* (Hedw.) Dix. var. *recurvirostrum*, see treatment of *Hymenostylium*]
[*Hymenostylium glaucum* (C. Müll.) Broth. = *Hymenostylium recurvirostrum* (Hedw.) Dix.]
[var. *cylindricum* Bartr. Am3 (= *Hymenostylium recurvirostrum* (Hedw.) Dix. *fide* Zander, Bryologist 80: 253, 1977) ≡ *Hymenostylium recurvirostrum* var. *cylindricum* (Bartr.) Zand. in Zand. & Eckel, Canad. J. Bot. 60: 1982]
Hymenostylium grandirete Dix., Anniv. Vol. Bot. Gard. Calcutta 178, 1942 As3
Hymenostylium hildebrandtii (C. Müll.) Zand. (*Weissia*) Afr2
[*Hymenostylium insigne* (Dix.) Podp. Eur ≡ *Hymenostylium recurvirostrum* var. *insigne* (Dix.) Bartr. *fide* Zander & Eckel, Canad. J. Bot. 60: 1596, 1982]
Hymenostylium kunzeanum (C. Müll.) C. Müll. *nom. illeg.* Am6
[*Hymenostylium luzonense* Broth. = *Hymenostylium recurvirostrum* var. *cylindricum* (Bartr.) Zand., see treatment of *Hymenostylium*]
[var. *minus* Broth. in Hall. As4 = *Hymenostylium recurvirostrum* (Hedw.) Dix. *fide* Sollman in Touw, J. Hattori Bot. Lab. 71: 343, 1992]
Hymenostylium papillinerve Dix. Afr2
[*Hymenostylium pellucidum* Broth. & Yas. As2 = *Hymenostylium recurvirostrum* (Hedw.) Dix. *fide* Chuang, J. Hattori Bot. Lab. 37: 469, 1973]
Hymenostylium recurvirostrum (Hedw.) Dix. Eur As1 As2 As3 As4 As5 Afr1 Afr2 Afr4 Am1 Am2 Am3 Am4 Am6 Austr1 Austr2
[var. *aurantiacum* (Mitt.) Gangulee, Mosses E. India 648, 1972 *hom. illeg. incl.* var. prior = *Hymenostylium recurvirostrum* var. *luzonense* (Broth.) Bartr. *fide* Gangulee, Mosses E. India 3: 648, 1972 = *Hymenostylium recurvirostrum* var. *cylindricum* (Bartr.) Zand. (based on *Hymenostylium glaucum* var. *cylindricum* Bartr., 1936), see treatment of *Hymenostylium*]
var. *anoectangioides* (Thér.) Wijk & Marg. Eur
var. *bicolor* (Dix.) Crum As3
var. *cataractarum* (Schimp.) Podp. Eur As1 As2 As3
var. *commutatum* (Mitt.) Podp. Eur As1 As2 As3 Am1
var. *cylindricum* (Bartr.) Zand. in Zander & Eckel, Canad. J. Bot. 60: 1599, 1982 (*Hymenostylium glaucum* var. *cylindricum* Bartr., 1936) As2 As3 Am2 Am3 Oc
var. *insigne* (Dix.) Bartr. Eur As2 As3 Am1
[var. *latifolium* (Zett.) Wijk & Marg. Eur As1 Am1 = *Hymenostylium recurvirostrum* (Hedw.) Dix. *fide* Zander in Sharp et al., Moss Fl. Mex.]
[var. *luzonense* (Broth.) Bartr. As4 = *Hymenostylium recurvirostrum* var. *cylindricum* (Bartr.) Zand., see treatment of *Hymenostylium*]
var. *sendtnerianaeforme* (Györf.) Podp. Eur
fo. *crassicoligum* Vaněk in Šmarda, Cas. Morav. Mus. Zemsk. 32: 15, 1948 Eur
fo. *laeviuscula* (Lindb.) Podp., Consp. Musc. Eur. 173, 1954 (*Barbula curvirostris* var.) Eur As1
fo. *megalosporum* (Latz.) Podp., Consp. Musc. Eur. 172, 1954 (*Hymenostylium curvirostre* fo.) Eur
fo. *microcarpum* (Nees & Hornsch.) Podp., Consp. Musc. Eur. 172, 1954 (*Gymnostomum*) Eur As2
fo. *minimum* (Amann) Podp., Consp. Musc. Eur. 172, 1954 (*Hymenostylium curvirostre* var.) Eur
fo. *pallidisetum* (Nees & Hornsch.) Podp., Consp. Musc. Eur. 172, 1954 (*Gymnostomum*) Eur
fo. *serrulatum* (Röll) Podp., Consp. Musc. Eur. 172, 1954 (*Hymenostylium curvirostre* fo.) Eur
Hymenostylium rigescens (C. Müll.) Broth. Afr2 Afr4
Hymenostylium scaturiginosum (C. Müll.) Broth. Afr2 Afr4
[*Hymenostylium secundum* C. Müll. Afr2 = *Molendoa sendtneriana* (BSG) Limpr., see treatment of *Molendoa*]
[*Hymenostylium shepheardae* Card. & Dix. As3 ≡ *Anoectangium shepherdae* (Card. & Dix.) Zand., see treatment of *Anoectangium*]
Hymenostylium sinense Sak. As2
[*Hymenostylium sordidum* Card. As2 = *Gymnostomum recurivrostrum* Hedw. *fide* Saito, J. Hattori Bot. Lab. 39: 39: 452, 1975 ≡ *Hymenostylium recurvirostrum* (Hedw.) Dix.]
Hymenostylium subcrispulum Thér. Afr3
Hymenostylium xanthocarpum (Hook.) Brid. [good species *fide* Aziz & Vohra, Bull. Bot. Surv. India 30: 185, 1988] As3
[*Hymenostylium validinerve* Dix. & P. Varde As3 = *Molendoa sendtneriana* (BSG) Limpr., see treatment of *Molendoa*]

HYOPHILA Brid.
Hyophila acuminata Broth. & P. Varde Afr2 Afr3
Hyophila acuminata Bartr. *hom. illeg.* As4
Hyophila acutifolia Saito, J. Hattori Bot. Lab. 39: 470, 1975 As2
[*Hyophila acutiuscula* Broth. Afr2 ≡ *Trichostomum acutiusculum* (Broth.) Zand., see treatment of *Trichostomum*]
[*Hyophila afrophaea* (C. Müll.) Warnst. Afr4 = *Trichostomum brachydontium* Bruch ex F. Müll. *fide* Magill, Fl. S. Afr. I. Mosses 1: 260, 1981]
[*Hyophila amblyphylla* Card. As2 = *Merceya ligulata* (Spruce) Schimp. *fide* Noguchi, J. Jap. Bot. 35(10): 316, 1960 ≡ *Scopelophila ligulata* (Spruce) Spruce]
[*Hyophila angustifolia* Card. *hom. illeg.* As2 = *Trichostomum platyphyllum* (Broth. ex Ihs.) Chen *fide* Saito, J. Hattori Bot. Lab. 39: 434, 1975]
Hyophila angustifolia Par. & Ren. Afr3
Hyophila angustiuscula Baumg. & Dix. As4
Hyophila anoectangioides C. Müll. ex Dus. Afr2
Hyophila apiculata Fleisch. As4
[*Hyophila arborea* (Mitt.) Jaeg. Am5 ≡ *Trichostomum arboreum* (Mitt.) Zand., see treatment of *Trichostomum*]
Hyophila argentinica Thér. Am6
[*Hyophila asanoi* Sak. As2 = *Barbula indica* (Hook.) Spreng. *fide* Saito, J. Hattori Bot. Lab. 39: 488, 1975]
Hyophila ascensionis Card. Afr2
Hyophila assimilis Broth. Am5
[*Hyophila atrovirens* (C. Müll.) Broth. Afr2 Afr4 = *Hyophila involuta* (Hook.) Jaeg. *fide* Magill, Fl. S. Afr. I. Mosses 1: 228, 1981 (1982)]
var. *oubanguiensis* Thér. & P. Varde Afr2
Hyophila baginsensis C. Müll. Afr2 Afr4
Hyophila bartramiana Steere Am2
[*Hyophila basutensis* Sim Afr4 = *Didymodon certatodonteus* (C. Müll.) Dix. *fide* Magill, Fl. S. Afr. I. Mosses 1: 235, 1981 (1982)]
Hyophila beruensis Dix. Oc
[*Hyophila biloinsularis* Sak. As2 = *Barbula indica* (Hook.) Spreng. *fide* Saito, J. Hattori Bot. Lab. 39: 488, 1975 as "*bilo-insulare*"]
Hyophila bingeri Broth. & Par. Afr2
Hyophila blanda (Hook. & Wils.) Jaeg. Am5
Hyophila brevifolia Hampe Am5
[*Hyophila calymperoides* Thér. & Nav. Afr2 = *Bryoerythrophyllum campylocarpum* (C. Müll.) Crum., see treatment of *Bryoerythrophyllum*]
Hyophila combae Broth. Oc
[*Hyophila comosa* Dix. & P. Varde As3 = *Hyophila rosea* Williams *fide* Zander in Sharp et al., Moss Fl. Mex. = *Hyophila*

nymaniana (Fleisch.) Menzel, Willdenowia 22: 198, 1992]
Hyophila compacta (Welw. & *dub.*) Jaeg. Afr2
Hyophila congolensis Thér. & Nav. Afr2
[*Hyophila contorta* (Kunz.) Jaeg. Afr1 ≡ *Trichostomum contortum* (Kunze) Sérgio, Portug. Acta Biol. (B) 14: 169, 1985]
[*Hyophila crenulata* C. Müll. ex Dus. Afr2 = *Hyophila involuta* (Hook.) Jaeg. *fide* Sollman, Lindbergia 10: 55, 1984]
var. *brevifolia* Biz., Svensk. Bot. Tidsk. 63: 444, 1969 *nom. inval.* Afr2
[*Hyophila crenulata* Mach. *hom. illeg.* Eur = *Hyophila machadoana* Sérgio, Rev. Bryol. Lichénol. 36: 628, 1969 *nom. nov.* ≡ *Bryoerythrophyllum machadoanum* (Sérgio) Hill *fide* Hill, J. Bryol. 11: 601, 1981 (1982) = *Bryoerythrophyllum campylocarpum* (C. Müll.) Crum *fide* Sollman, Lindbergia 16: 22, 1990]
Hyophila crenulatula C. Müll. ex Par. Afr2
Hyophila crispula (Sak.) Sak.
[*Hyophila cucullatifolia* Gao, Jia & Cao, Bull. Bot. Res. (Harbin) 11(2): 29. 1991 As2 ≡ *Weisiopsis cucullatifolia* (Gao, Jia & Cao) Zand., see treatment of *Weisiopsis*]
Hyophila cuspidatissima Par. & Broth. Afr2
[*Hyophila cyathiformis* (Dix.) Sim Afr2 Afr4 = *Hypodontium dregei* (Hornsch.) C. Müll. *fide* Magill & Schelpe, Mem. Bot. Surv. S. Afr. 43: 23, 1979]
Hyophila dendroides Dix. As3
Hyophila elliptica Baumg. & Fröhl. As4
[*Hyophila erosa* Sim As4 = *Oreoweisia erosa* (C. Müll.) Kindb. *fide* Magill & Schelpe, Mem. Bot. Surv. S. Afr. 43: 23, 1979]
Hyophila excurrentinervis Par. & Broth. Afr2
[*Hyophila flavipes* Broth. As2 = *Hyophila involuta* (Hook.) Jaeg. *fide* Saito, J. Hattori Bot. Lab. 39: 468, 1975]
Hyophila fouta-djalloni Par. & Broth. Afr2
Hyophila girodii Ren. & Card. Afr3
Hyophila glaucoviridis Par. & Broth. Afr2
Hyophila grandiretis Sak. *hom. illeg.* As2
Hyophila grossidens Broth. Am4
Hyophila guaraya Herz. Am4
Hyophila gymnostomoides (Welw. & *dub.*) Jaeg. Afr2
Hyophila holstii Broth. Afr2
Hyophila incurva Jaeg. Am2 Am5
Hyophila integrifolia Dix. & Thér. Afr2
Hyophila involuta (Hook.) Jaeg. Eur As2 As3 As4 Afr4 Am1 Am2 Am3 Am4 Am5 Oc
fo. *circinata* (C. Müll.) Chen, Hedwigia 80: 190, 1941 (*Pottia*) As3
fo. *flaccida* (Warnst.) Podp., Consp. Musc. Eur. 199, 1954 (*Trichostomum warnstorfii* var.) Eur
fo. *serrata* (Amann) Podp., Consp. Musc. Eur. 199, 1954 (*Hyophila riparia* var.) Eur
Hyophila involutifolia (C. Müll.) Jaeg. Am4
Hyophila javanica (Nees & Blum.) Brid. As1 As2 As3 As4
[*Hyophila khartoumensis* (Pettet) A. Sm. & H. Whiteh., J. Bryol. 8: 14, 1974 (*Tortula*) Afr2 = *Desmatodon bogosicus* C. Müll. *fide* Corley et al., J. Bryol. 11: 620, 1981 (1982)]
Hyophila kurziana Gangulee, Nov. Hedw. 12: 422, 1966 As3
Hyophila laete-virens Broth. Am5
Hyophila lanceolata Ren. & Card. Afr3
Hyophila latifolia Broth. Afr2
[*Hyophila lauterbachii* Broth. in Schum. & Lauterb. As4 ≡ *Hyophila involuta* (Hook.) Jaeg. *fide* Norris & Koponen, Acta Bot. Fenn. 137: 109, 1987]
Hyophila leikipaiae (C. Müll.) Broth. Afr2
Hyophila leioneura Ren. & Par. Afr3
[*Hyophila leprieurii* (Mont.) Jaeg. Am5 = *Hyophila tortula* (Schwaegr.) Hampe *fide* Florsch., Mosses Suriname 1: 173, 1964 = *Hyophila involuta* (Hook.) Jaeg.
[*Hyophila ligulaefolia* Broth. & Par. Afr2 ≡ *Trichostomum ligulaefolium* (Broth. & Par.) Zand., see treatment of *Trichostomum*]
Hyophila linguaeformis Broth. ex Ihs. As2
[*Hyophila lingulata* Card. Am2 = *Neohyophila sprengelii* var. *stomatodonta* (Card.) Zand. *fide* Zander, Bryologist 86: 137, 1983 ≡ *Plaubelia sprengelii* var. *stomatodonta* (Card.) Zand., see treatment of *Plaubelia*]
[*Hyophila liukiensis* Sak. As2 = *Hyophila involuta* (Hook.) Jaeg. *fide* Saito, J. Hattori Bot. Lab. 39: 468, 1975]
[*Hyophila lombokensis* Broth. in Hall. As4 = *Trichostomum brachydontium* Bruch fidy Sollman in Touw, J. Hattori Bot. Lab. 71: 345, 1992]
Hyophila loxorhyncha C. Müll. ex Ånstr. Am5
[*Hyophila lusitanica* Card. & Dix. Eur ≡ *Bryoerythrophyllum lusitanicum* (Card. & Dix.) Hill, J. Bryol. 11: 600, 1981 (1982)]
[*Hyophila machadoana* Sérgio, Rev. Bryol. Lichénol. 36: 628, 1969 *nom. nov.* Eur ≡ *Bryoerythrophyllum machadoanum* (Sérgio) Hill *fide* Hill, J. Bryol. 11: 601, 1981 [1982] = *Bryoerythrophyllum campylocarpum* (C. Müll.) Crum *fide* Sollman, Lindbergia 16: 22, 1990]
Hyophila mattogrosensis Broth. Am5
Hyophila melanostoma (Mitt.) Jaeg. Am5
[*Hyophila mexicana* Thér. Am2 = *Trichostomum brachydontium* Bruch ex F. Müll. *fide* Zander in Sharp et al., Moss Fl. Mex.]
[*Hyophila machadoana* Sérgio, Rev. Bryol. Lichénol. 36: 628, 1969 *nom. nov.* for *Hyophila crenulata* Mach. = *Bryoerythrophyllum machadoanum* (Sérgio) Hill *fide* Hill, J. Bryol. 11: 601, 1981 (1982)]
[*Hyophila micholitzii* Broth. = *Hyophila involuta* (Hook.) Jaeg.]
var. *sterilis* Fleisch. As3 As4 Oc
[*Hyophila microcarpa* (Schimp. ex Besch.) Broth. Am2 Am3 Am4 Am5 = *Weissia sinaloensis* Bartr. *fide* Zander in Sharp et al., Moss Fl. Mex. ≡ *Trichostomum sinaloense* (Bartr.) Zand., see treatment of *Trichostomum*]
[*Hyophila minutissima* (Mitt.) Jaeg. Am4 Am5 ≡ *Tisserantiella minutissima* (Mitt.) Zand., see treatment of *Tisserantiella* (Rhachitheciaceae), Excluded Taxa]
Hyophila millifolia Dix. & P. Varde As3
Hyophila mosenii Both. Am5
Hyophila muelleri (Dub.) Jaeg. As3
[*Hyophila naganoi* Sak. As2 = *Hyophila propagulifera* Broth. *fide* Saito, J. Hattori Bot. Lab. 39: 471, 1975]
[*Hyophila nakayamae* Sak. As2 = *Barbula indica* (Hook.) Spreng. *fide* Saito, J. Hattori Bot. Lab. 39: 488, 1975]
[*Hyophila nakayamae* Sak. As2 = *Barbula indica* (Hook.) Spreng. *fide* Saito, J. Hattori Bot. Lab. 39: 488, 1975]
Hyophila neocaledonica Broth. & Par. Oc
Hyophila niam-niamiae C. Müll. Afr2
Hyophila novae-guineae Broth. ex C. Müll. As4
Hyophila novae-seelandiae Dix. & Sainsb. Austr2
Hyophila nymaniana (Fleisch.) Menzel, Willdenowia 22: 198, 1992 (*Glyphomitrium*) As3 As4 Am2
Hyophila obtusifolia (C. Müll.) Jaeg. Am5 [= *Gyroweisia* sp. cf. Brotherus 1902]
Hyophila ochracea Broth. Am5
[*Hyophila okamurae* Broth. As2 = *Hyophila propagulifera* Broth. *fide* Saito, J. Hattori Bot. Lab. 39: 471, 1975]
Hyophila ovalifolia (Hampe) Hampe Am5
Hyophila pampanini Zodda Afr1
Hyophila paraguensis Broth. Am5
Hyophila parietalis Card. in Grand. Afr3

[*Hyophila perannulata* Ren. & Card. As3 ≡ *Trichostomum criotum* Zand. *nom. nov*., see treatment of *Trichostomum*]
Hyophila perpendiculata Dix. Afr2
Hyophila perpusilla Thér. & Trab. Afr1
Hyophila potieri Besch., Rev. Bryol. 7(2): 21, 1880 Afr2 Afr3
var. *atroviridis* Ren. & Card. Afr3
var. *denticulta* Broth. Afr2
Hyophila procera Par. & Broth. Afr2
Hyophila propagulifera Broth. As2
[var. *elata* Sak. As2 ≡ *Hyophila propagulifera* Broth. *fide* Saito, J. Hattori Bot. Lab. 39: 471, 1975]
[*Hyophila pulchella* (Thér. & Hilp.) Wijk & Marg. Afr2 ≡ *Tisserantiella pulchella* (Thér. & Hilp.) Zand., see treatment of *Tisserantiella* (Rhachitheciaceae)]
Hyophila punctulata (Mitt.) Kindb. Afr3
Hyophila regnellii C. Müll. ex Ångstr. Am5
[*Hyophila rosea* Williams As2 As3 As4 Am2 = *Hyophila nymaniana* (Fleisch.) Menzel, Willdenowia 22: 198, 1992]
Hyophila rubiginosa Hampe Am5
Hyophila sakalavensis Par. & Ren. Afr3
Hyophila samoana Mitt. Oc
Hyophila setschwanica (Broth.) Hilp. ex Chen As2
Hyophila siamensis Dix. As3
[*Hyophila sieboldii* Besch. As2 = *Barbula unguiculata* Hedw. *fide* Sollman, Lindbergia 10: 54, 1984]
[*Hyophila sinaloensis* (Bartr.) Bartr. Am2 Am3 Am4 ≡ *Weissia sinaloensis* Bartr. *fide* Zander in Sharp et al., Moss Fl. Mex.]
[*Hyophila somaliae* C. Müll. Afr2 ≡ *Tortella somaliae* (C. Müll.) Zand., see treatment of *Tortella*]
Hyophila spathulata (Harv.) Jaeg. As3
[*Hyophila stanfordensis* (Steere) A. Sm. & Whiteh., J. Bryol. 8: 13, 1974. Eur Am1 ≡ *Tortula stanfordensis* Steere ≡ *Hennediella stanfordensis* (Steere) Blockeel, J. Bryol. 16: 191, 1991]
[*Hyophila stenophylla* Card. As2 = *Trichostomum platyphyllum* (Broth. ex Ihs.) Chen *fide* Saito, J. Hattori Bot. Lab. 39: 434, 1975]
[*Hyophila stomatodonta* Card. Am2 ≡ *Neohyophila stomatodonta* (Card.) Crum ≡ *Neohyophila sprengelii* var. *stomatodonta* (Card.) Zand., Bryologist 86: 138, 1983 ≡ *Plaubelia sprengelii* var. *stomatodonta* (Card.) Zand., see treatment of *Plaubelia*]
Hyophila streimannii Norris & T. Kop., Acta Bot. Fenn. 137: 111, 1989 As4
[*Hyophila styriaca* Głow. Eur = *Gymnostomum aeruginosum* Sm. *fide* Corley et al., J. Bryol. 11: 649, 1981 (1982)]
Hyophila subacutiuscula P. Varde & Thér. Afr2
[*Hyophila subangustifolia* Thér. Am2 ≡ *Weissia subangustifolia* (Thér.) Zand., Bryologist 86: 156, 1983 ≡ *Trichostomum subangustifolium* (Thér.) Zand., see treatment of *Trichostomum*]
Hyophila subcucullata Williams Am3
Hyophila subflaccida Broth. & Dix. As3
[*Hyophila subspathulata* Sak. As2 = *Trichostomum platyphyllum* (Broth. ex Ihs.) Chen *fide* Saito, J. Hattori Bot. Lab. 39: 434, 1975]
Hyophila tisserantii P. Varde Afr2
[*Hyophila tortula* (Schwaegr.) Hampe Am1 Am2 Am3 Am4 Am5 Am6 = *Hyophila involuta* (Hook.) Jaeg.]
Hyophila treleasii Card. Afr1
Hyophila uleana C. Müll. Am5
[*Hyophila usambarica* Broth. Afr2 = *Trichostomum brachydontium* Bruch, see treatment of *Trichostomum*]
[*Hyophila validinervis* Card. & P. Varde As3 = *Hyophila involuta* (Hook.) Jaeg. *fide* Sollman, Lindbergia 10: 55, 1984]
Hyophila variegata Ångstr. Am5
Hyophila victoriae C. Müll. ex Dus. Afr2
Hyophila viridula Card. & P. Varde As3
Hyophila vitiana (C. Müll.) Jaeg. Oc
[*Hyophila walkeri* Broth. As3 ≡ *Tortella walkeri* (Broth.) Zand., see treatment of *Tortella*]
Hyophila warmingii Hampe Am5
var. *angustifolia* Broth. Am5
[*Hyophila zeyheri* (Hampe) Jaeg. Afr2 Afr4 = *Trichostomum brachydontium* Bruch ex F. Müll. *fide* Magill, Fl. S. Afr. I. Mosses 1: 260, 1981 (1982)]
var. *brevimucronata* P. Varde Afr2
var. *lanceolata* Sim Afr4

[**HYOPHILOPSIS** Card. & Dix. ≡ *Tortula* sect. *Hyophilopsis* (Card. & Dix.) Zand., see treatment of *Tortula*]
[*Hyophilopsis entosthodontacea* Card. & Dix. As3 ≡ *Tortula entosthodontacea* (Card. & Dix.) Zander, see treatment of *Tortula*]

[**HYOPHILOPSIS** Crum *non* Card. & Dix., 1911 *hom. illeg.* ≡ *Neohyophila* Crum = *Plaubelia* Brid.]
[*Hyophilopsis lingulata* (Card.) Crum Am2 = *Neohyophila sprengelii* var. *stomatodonta* (Card.) Zand. *fide* Zander, Bryologist 86: 137, 1983 ≡ *Plaubelia sprengelii* var. *stomatodonta* (Card.) Zand., see treatment of *Plaubelia*]
[*Hyophilopsis spathulifolia* (Bartr.) Crum Am2 = *Neohyophila sprengelii* (Schwaegr.) Crum *s. lat. fide* Zander, Bryologist 86: 135, 1983 ≡ *Plaubelia sprengelii* (Schwaegr.) Zand., see treatment of *Plaubelia*]
[*Hyophilopsis sprengelii* (Schwaegr.) Crum Am1 Am2 Am3 Am4 Am5 ≡ *Neohyophila sprengelii* (Schwaegr.) Crum ≡ *Plaubelia sprengelii* (Schwaegr.) Zand., see treatment of *Plaubelia*]
[*Hyophilopsis stomatodonta* (Card.) Crum Am2 Am3 ≡ *Neohyophila sprengelii* var. *stomatodonta* (Card.) Zand., Bryologist 86: 135, 1983 ≡ *Plaubelia sprengelii* var. *stomatodonta* (Card.) Zand., see treatment of *Plaubelia*]

HYPODONTIUM C. Müll., Hedwigia 38: 96, 1899
Hypodontium dregei (Hornsch.) C. Müll., Hedwigia 38: 97, 1899 Afr4
Hypodontium pomiforme (Hook.) C. Müll., Hedwigia 38: 91, 1899 Afr4

[**KLEIOWEISIOPSIS** Dix. referred to Orthotrichaceae, see treatment of *Kleioweisiopsis*, Excluded Taxa]
[*Kleioweisiopsis involuta* Biz., Rev. Bryol. Lichénol. 40: 119, 1974 Afr2 ≡ *Weissia bizotii* Zand. *nom. nov.*—see treatment of *Weissia*]

LEPTOBARBULA Schimp.
Leptobarbula berica (De Not.) Schimp. Eur As3 As5 Afr1
var. *meridionalis* (Schimp.) Limpr. Eur
var. *winteri* (Schimp.) Limpr. Eur

LEPTODONTIELLA Zand. & Hegew.
Leptodontiella apiculata (Zand.) Zand. & Hegew., Bryologist 79: 16, 1976 Am4

LEPTODONTIUM (C. Müll.) Hampe
[*Leptodontium abyssinicum* Broth. Afr2 = *Leptodontium capituligerum* C. Müll. *fide* Sloover, Bull. Jard. Bot. Natl. Belgique 57(3/4): 448, 1987]
[*Leptodontium acutifolium* Mitt. Am2 Am4 = *Leptodontium*

pungens (Mitt.) Kindb. *fide* Zander, Bryologist 75: 256, 1972]
[var. *grimmioides* (Britt.) Herz. = *Leptodontium pungens* (Mitt.) Kindb. *fide* Zander, Bryologist 75: 256, 1972]
[*Leptodontium acutissimum* Bartr. Am4 = *Leptodontium pungens* (Mitt.) Kindb. *fide* Zander, Bryologist 75: 256, 1972]
Leptodontium aggregatum (C. Müll.) Kindb. As4
[subsp. *hyalinum* Fleisch., Musci Fl. Buitenz. 1: 369, 1904 As4 ≡ *Leptodontium aggregatum* (C. Müll.) Kindb. *fide* Norris & Koponen, Acta Bot. Fenn. 137: 106, 1987]
[var. *hyalinum* (Fleisch.) Broth., Nat. Pfl. 1(3): 1190, 1909 As4 ≡ *Leptodontium aggregatum* (C. Müll.) Kindb. *fide* Norris & Koponen, Acta Bot. Fenn. 137: 106, 1987]
[*Leptodontium albovaginatum* Herz. = *Bryoerythrophyllum jamesonii* (Tayl.) Crum *fide* Zander, Bryologist 75: 278, 1972]
[*Leptodontium allorgei* Biz. Afr2 = *Cynodontium* sp. *fide* Sloover, Bull. Jard. Bot. Natl. Belgique 57(3/4): 449, 1987]
[*Leptodontium angustinerve* Thér. (= *Anoectangium compactum* Schwaegr. *fide* Zander, Bryologist 75: 278, 1972) = *Hymenostylium recurvirostrum* (Hedw.) Dix. *fide* Zander, 80: 243, 1977]
[*Leptodontium anoectangiaceum* (C. Müll.) Broth. Am5 = *Leptodontium stellatifolium* (Hampe) Broth. *fide* Zander, Bryologist 75: 240, 1972]
[*Leptodontium anomalum* Dix. & Thér. Am4 = *Leptodontium wallisii* (C. Müll.) Kindb. *fide* Zander, Bryologist 75: 259, 1972]
[*Leptodontium apiculatum* Zand., Bryologist 75: 238, 1972 Am4 ≡ *Leptodontiella apiculata* (Zand.) Zand. & Hegew., Bryologist 79: 16, 1976]
[*Leptodontium arachnoideum* C. Müll. Am6 = *Leptodontium capituligerum* C. Müll. *fide* Zander, Bryologist 75: 273, 1972]
Leptodontium araucarieti (C. Müll.) Par. Am4 Am5
[*Leptodontium arsenii* Thér. = *Bryoerythrophyllum ferruginascens* (Stirt.) Giac. *fide* Zander, Bryologist 75: 278, 1972]
Leptodontium brachyphyllum Broth. & Thér. Am2 Am4 Afr4
[*Leptodontium brasiliense* Mitt. Am5 = *Leptodontium viticulosoides* (P. Beauv.) Wijk & Marg. var. *viticulosoides fide* Zander, Bryologist 75: 244, 1972]
[*Leptodontium braunioides* C. Müll. Am6 = *Leptodontium pungens* (Mitt.) Kindb. *fide* Zander, Bryologist 75: 256, 1972]
[*Leptodontium brevicaule* Bartr. Oc = *Leptodontium flexifolium* (Dicks.) Hampe in Lindb. *fide* Zander, Bryologist 75: 231, 1972]
[*Leptodontium buesii* Williams Am4 = *Leptodontium luteum* (Tayl.) Mitt. *fide* Zander, Bryologist 75: 264, 1972]
[*Leptodontium calymperoides* Thér. Am4 (= *Leptodontium capituligerum* C. Müll. *fide* Zander, Bryologist 75: 273, 1972) = *Leptodontium tricolor* (Williams) Zand. *fide* Zander, Bryologist 79: 20, 1976]
Leptodontium capituligerum C. Müll. Am2 Am4 Am5 Am6 Afr2 Afr3
[*Leptodontium chrysobaseum* (C. Müll.) Broth. Am5 = *Leptodontium stellatifolium* (Hampe) Broth. *fide* Zander, Bryologist 75: 240, 1972]
[*Leptodontium citrinum* (Hampe) Hampe Am5 = *Leptodontium viticulosoides* var. *panamense* (Lor.) Zand., Bryologist 75: 250, 1972 = *Leptodontium viticulosoides* var. *sulphureum* (Lor.) Zand. *fide* Zander, Bryologist 86: 156, 1983]
Leptodontium debatii (Husn.) Hag. Eur
[*Leptodontium densifolium* (Mitt.) Mitt. Am4 = *Leptodontium viticulosoides* (P. Beauv.) Wijk & Marg. var. *viticulosoides fide* Zander, Bryologist 75: 244, 1972]
[*Leptodontium dentatum* (Mitt.) Kindb. As3 As4 = *Leptodontium flexifolium* (Dicks.) Hampe in Lindb. *fide* Zander, Bryologist 75: 231, 1972 cf. Zander, Bryologist 84: 546, 1972]
[*Leptodontium erectifolium* Dix. As4 ≡ *Leptodontium flexifolium* (Dicks.) Hampe in Lindb. *fide* Norris & Koponen, Acta Bot. Fenn. 137: 105, 1987]
Leptodontium erythroneuron Herz. Am4
[*Leptodontium exasperatum* Card. Am2 ≡ *Leptodontium viticulosoides* var. *exasperatum* (Card.) Zand. *fide* Zander, Bryologist 75: 254, 1972]
[*Leptodontium excelsum* (Sull.) Britt. Am1 Am2 = *Leptodontium viticulosoides* var. *panamense* (Lor.) Zand., Bryologist 75: 251, 1972 = *Leptodontium viticulosoides* var. *sulphureum* (Lor.) Zand. *fide* Zander, Bryoloist 86: 156, 1983]
[*Leptodontium felipponei* Broth. Am6 = *Leptodontium capituligerum* C. Müll. *fide* Zander, Bryologist 75: 273, 1972]
[*Leptodontium fernandezianum* Broth. in Skottsb. Am6 = *Leptodontium longicaule* var. *microruncinatum* (Dus.) Zand. *fide* Zander, Bryologist 75: 269, 1972]
[*Leptodontium ferrugineum* Broth. Am4 = *Leptodontium wallisii* (C. Müll.) Kindb. *fide* Zander, Bryologist 75: 259, 1972]
[*Leptodontium filescens* (Hampe) Mitt. Am2 Am4 = *Leptodontium flexifolium* (Dicks.) Hampe in Lindb. *fide* Zander, Bryologist 75: 231, 1972.
[var. *denticulatum* Bartr. Am2 = *Leptodontium filicola* Herz. *fide* Zander, Bryologist 75: 241, 1972]
[*Leptodontium filicaule* Dix. Afr2 = *Leptodontium flexifolium* (Dicks.) Hampe in Lindb. *fide* Sloover, Bull. Jard. Bot. Natl. Belgique 57(3/4): 428, 1987]
Leptodontium filicola Herz. Am2 Am4 Am5 Am6
[*Leptodontium filiformis* (Lor.) Steere Am4 = *Leptodontium flexifolium* (Dicks.) Hampe in Lindb. *fide* Zander, Bryologist 75: 231, 1972]
Leptodontium flexifolium (Dicks.) Hampe in Lindb. (see Karttunen, Taxon 37:156–157, 1988 for discussion of Dickson as authority) Am1 Am2 Am4 Afr2 As2 As3 As4
[var. *americanum* (Grout) Grout Am1 Am2 = *Leptodontium flexifolium* (Dicks.) Hampe in Lindb. *fide* Zander, Bryologist 75: 231, 1972]
fo. *compacta* Hessel. ex Rosevinge, Bot. Iceland 2: 452, 1918 Eur
[fo. *gemmifera* Mönk., Laubm. Eur. 274, 1927 Eur ≡ *Leptodontium gemmascens* (Mitt. in Hunt) Braithw.]
fo. *gemmipara* Frahm, Nov. Hedw. 24: 418, 1973 [1975] Eur
Leptodontium fuhrmannii Broth. & Irmsch. (type not available for study *fide* Zander, Bryologist 75: 278, 1972) Am4
Leptodontium fuscescens Bartr. Am5 = *Leptodontium capituligerum* C. Müll. *fide* Zander, Bryologist 75: 273, 1972]
Leptodontium gambaragarae Negri Afr2
Leptodontium gemmascens (Mitt. in Hunt) Braithw. [good species *fide* Zander, Bryologist 75: 236, 1972] Eur Afr3
[*Leptodontium gemmigerum* Broth. in Mildbr. Afr2 = *Leptodontium longicaule* Mitt. var. *longicaule fide* Zander, Bryologist 75: 268, 1972]
[*Leptodontium gracile* C. Müll. ex Britt. *hom. illeg.* Am4 = *Leptodontium capituligerum* C. Müll. *fide* Zander, Bryologist 75: 273, 1972]
[var. *gemmascens* Bartr. in Bauer Am5 = *Leptodontium capituligerum* C. Müll. *fide* Zander, Bryologist 75: 273, 1972]
[*Leptodontium gracilescens* C. Müll. Am4 = *Leptodontium capituligerum* C. Müll. *fide* Zander, Bryologist 75: 273, 1972]
[*Leptodontium gracillimum* Nog. As2 = *Leptodontium flexifolium* (Dicks. ex With.) Hampe in Lindb. *fide* Zander, Bryologist 75: 232, 1972]
Leptodontium handelii Thér. As2
[*Leptodontium humillimum* Broth. in Hall. As4 = *Leptodontium flexifolium* (Dick. ex With.) Hampe in Lindb. *fide* Zander, Bryologist 75: 231, 1972]

[*Leptodontium hyalinum* (Fleisch.) Fleisch. As4 ≡ *Leptodontium aggregatum* (C. Müll.) Kindb. *fide* Norris & Koponen, Acta Bot. Fenn. 137: 106, 1987]

[*Leptodontium insolitum* Thér. & P. Varde Afr2 = *Barbula eubryum* C. Müll. *fide* Sloover, Bull. Jard. Bot. Natl. Belgique 57(3/4): 449, 1987]

var. *perundulatum* Thér. & P. Varde Afr3

[*Leptodontium integrifolium* Williams ≡ *Barbula integrifolia* (Williams) Zand. *fide* Zander, Bryologist 75: 277, 1972]

Leptodontium interruptum (Mitt.) Broth. Afr4 Austr2

[*Leptodontium japonicum* Sak. = *Didymodon rigidicaulis* (C. Müll.) Saito *fide* Saito, J. Hattori Bot. Lab. 39: 502, 1975 = *Didymodon ferrugineus* (Schimp. ex Besch.) Hill, J. Bryol. 11: 599, 1981 (1982)]

[*Leptodontium joannis-meyeri* C. Müll. Afr2 = *Leptodontium pungens* (Mitt.) Kindb. *fide* Sloover, Bull. Jard. Bot. Natl. Belgique 57(3/4): 434, 1987]

[var. *cameruniae* Broth. in Mildbr. Afr2 = *Leptodontium pungens* (Mitt.) Kindb. *fide* Sloover, Bull. Jard. Bot. Natl. Belgique 57(3/4): 434, 1987]

[*Leptodontium kinabaluense* Dix. As4 = *Leptodontium flexifolium* (Dicks.) Hampe in Lindb. *fide* Zander, Bryologist 75: 231, 1972]

[*Leptodontium laevigatum* Herz. Am5 = *Leptodontium viticulosoides* (P. Beauv.) Wijk & Marg. var. *viticulosoides fide* Zander, Bryologist 75: 245, 1972]

[*Leptodontium laticuspis* Broth. Am4 = *Leptodontium syntrichioides* (C. Müll.) Kindb. *fide* Zander, Bryologist 75: 271, 1972]

Leptodontium latifolium Broth. Afr2

[*Leptodontium laxifolium* Broth. Am4 = *Leptodontium araucarieti* (C. Müll.) Par. *fide* Zander, Bryologist 75: 265, 1972]

Leptodontium leptoprion C. Müll. *nom. inval.* Afr2

[*Leptodontium limbatulum* Fleisch. As4 = *Leptodontium flexifolium* (Dicks.) Hampe in Lindb. *fide* Zander, Bryologist 75: 231, 1972]

Leptodontium longicaule Mitt. Am2 Am4 Afr2 Afr3 Afr4

subsp. *stellatum* (Brid.) Sloover, Bull. Jard. Bot. Natl. Belgique 57(3/4): 444, 1987 (*Leptodontium*) Afr3

var. *microruncinatum* (Dus.) Zand., Bryologist 75: 269, 1972 Am4 Am6 Afr2 Afr3 Afr4

Leptodontium luteum (Tayl.) Mitt. Am4 Afr2

[*Leptodontium mandonii* C. Müll. Am4 = *Leptodontium longicaule* Mitt. var. *longicaule fide* Zander, Bryologist 75: 268, 1972]

[*Leptodontium matucamense* Besch. Am4 = *Leptodontium pungens* (Mitt.) Kindb. *fide* Zander, Bryologist 75: 256, 1972]

[*Leptodontium microruncinatum* Dus. Am6 ≡ *Leptodontium longicaule* var. *microruncinatum* (Dus.) Zand. *fide* Zander, Bryologist 75: 269, 1972]

[*Leptodontium nakaii* Okam. As2 = *Leptodontium flexifolium* (Dicks.) Hampe in Lindb. *fide* Saito, J. Hattori Bot. Lab. 39: 463, 1975]

[*Leptodontium norvegicum* Kaal. Eur = *Bartramia pomiformis* Hedw. *fide* Frisvoll, J. Bryol. 12: 186, 1982]

Leptodontium novae-seelandiae C. Müll. Austr2

[*Leptodontium orthotrichoides* (C. Müll.) Par. Am6 = *Zygodon palmarum* C. Müll. (Orthotrichaceae) *fide* Malta, Gatt. Zygodon 170, 1926 see Zander, Bryologist 75: 278, 1972]

[*Leptodontium papillosum* [Hampe] Kindb. Am4 = *Leptodontium longicaule* Mitt. var. *longicaule fide* Zander, Bryologist 75: 268, 1972]

Leptodontium paradoxicum Stone & Scott, J. Bryol. 11: 701, 1981 Austr1

[*Leptodontium pergemmascens* Broth. As2 = *Leptodontium flexifolium* (Dicks.) Hampe in Lindb. *fide* Zander, Bryologist 75: 231, 1972]

[*Leptodontium persquarrosum* Broth. in Mildbr. Afr2 = *Leptodontium wallisii* (C. Müll.) Kindb. *fide* Zander, Bryologist 75: 259, 1972]

Leptodontium planifolium Herz. Am4

[*Leptodontium procumbens* C. Müll. Am4 = *Leptodontium viticulosoides* var. *panamense* (Lor.) Zand., Bryologist 75: 251, 1972 = *Leptodontium viticulosoides* var. *sulphureum* (Lor.) Zand. *fide* Zander, Bryologist 86: 156, 1983]

Leptodontium proliferum Herz. Am2 Am4

[*Leptodontium pumilum* (C. Müll.) Kindb. Afr2 = *Leptodontium pungens* (Mitt.) Kindb. *fide* Sloover, Bull. Jard. Bot. Natl. Belgique 57(3/4): 434, 1987]

Leptodontium pungens (Mitt.) Kindb. Afr2 Am2 Am4 Am5 Am6

[*Leptodontium quennoae* C. Müll. Am6 = *Leptodontium viticulosoides* (P. Beauv.) Wijk & Marg. var. *viticulosoides fide* Zander, Bryologist 75: 244, 1972]

[*Leptodontium ramosum* Crum & Richards, J. Bryol. 13: 194, 1984 Am2 = *Leptodontium stoloniferum* Zand., see treatment of *Leptodontium*]

[*Leptodontium recurvifolium* (Tayl.) Lindb. Eur Am1 ≡ *Bryoerythrophyllum recurvifolium* (Tayl.) Zand., Bryologist 75: 277, 1972 ≡ *Paraleptodontium recurvifolium* (Tayl.) Long, J. Bryology 12: 181, 1982 ≡ *Oxystegus recurvifolius* (Tayl.) Zand., Lindbergia 8: 187, 1982 ≡ *Trichostomum recurvifolium* (Tayl.) Zand., see treatment of *Trichostomum*]

Leptodontium repens (C. Müll.) Kindb. Afr2

[*Leptodontium rhacomitrioides* Lor. & C. Müll. Am4 Am6 = *Leptodontium viticulosoides* (P. Beauv.) Wijk & Marg. var. *viticulosoides fide* Zander, Bryologist 75: 244, 1972]

[*Leptodontium rhynchophorum* Dix. Afr2 = *Zygodon* sp. *fide* Sloover, Bull. Jard. Bot. Natl. Belgique 57(3/4): 449, 1987]

[*Leptodontium rigidum* Broth. Am5 = *Leptodontium viticulosoides* (P. Beauv.) Wijk & Marg. var. *viticulosoides fide* Zander, Bryologist 75: 245, 1972]

[*Leptodontium rufescens* Broth. in Herz. Am4 = *Leptodontium longicaule* var. *microruncinatum* (Dus.) Zand. *fide* Zander, Bryologist 75: 269, 1972]

Leptodontium saxicola [C. Müll.] C. Müll. ex Par. (type not available for study *fide* Zander, Bryologist 75: 278, 1972) Am5

Leptodontium scaberrimum Broth. As2

[*Leptodontium schiffneri* Broth. Am5 = *Leptodontium wallisii* (C. Müll.) Kindb. *fide* Zander, Bryologist 75: 259, 1972]

[*Leptodontium serrae* (C. Müll.) Par. Am5 = *Leptodontium viticulosoides* var. *panamense* (Lor.) Zand., Bryologist 75: 251, 1972 = *Leptodontium viticulosoides* var. *sulphureum* (Lor.) Zand. *fide* Zander, Bryologist 86: 156, 1983]

[*Leptodontium serrifolium* (C. Müll.) Broth. Afr2 = *Leptodontium pungens* (Mitt.) Kindb. *fide* Sloover, Bull. Jard. Bot. Natl. Belgique 57(3/4): 434, 1987]

[*Leptodontium setschwanicum* Broth. As2 = *Didymodon erosodenticulatus* (C. Müll.) Saito *fide* Saito, J. Hattori Bot. Lab. 39: 504, 1975]

[*Leptodontium sikokianum* Sak. in Oti As2 = *Dichodontium pellucidum* (Hedw.) Schimp. (Dicranaceae) *fide* Saito, J. Hattori Bot. Lab. 39: 528, 1975]

[*Leptodontium skottsbergii* Bartr. Am6 = *Leptodontium pungens* (Mitt.) Kindb. *fide* Zander, Bryologist 75: 256, 1972]

[*Leptodontium spinosum* Williams Am4 = *Zygodon pichinchensis* (Tayl.) Mitt. (Orthotrichaceae) *fide* Zander, Bryologist 75: 278, 1972]

[*Leptodontium spongiosum* Herz. Am4 = *Leptodontium flexifolium* (Dicks.) Hampe in Lindb. *fide* Zander, Bryologist 75: 231, 1972]

[*Leptodontium squamifolium* (C. Müll.) Broth. Am5 = *Leptodont-*

ium stellatifolium (Hampe) Broth. *fide* Zander, Bryologist 75: 240, 1972]
Leptodontium stellaticuspis Bartr. Am4
Leptodontium stellatifolium (Hampe) Broth. Am5
[*Leptodontium stellatum* (Brid.) Ren. Afr3 ≡ *Leptodontium longicaule* subsp. *stellatum* (Brid.) Sloover, Bull. Jard. Bot. Natl. Belgique 57(3/4): 444, 1987]
Leptodontium stoloniferum Zand., Bryologist 75: 239, 1972 Am2 Am4
Leptodontium styriacum (Jur.) Limpr. Eur As1
[*Leptodontium subalpinum* (De Not.) Lindb. Eur = *Dichodontium pellucidum* (Hedw.) Schimp. (Dicranaceae) *fide* Frahm, Lindbergia 12: 81, 1987]
[*Leptodontium subcirrhifolium* (C. Müll.) Kindb. Am4 = *Leptodontium viticulosoides* var. *panamense* (Lor.) Zand., Bryologist 75: 251, 1972 = *Leptodontium viticulosoides* var. *sulphureum* (Lor.) Zand. *fide* Zander, Bryologist 86: 156, 1983]
[*Leptodontium subfilescens* Thér. & Nav. Afr2 = *Leptodontium flexifolium* (Dicks.) Hampe in Lindb. *fide* Sloover, Bull. Jard. Bot. Natl. Belgique 57(3/4): 428, 1987]
[*Leptodontium subgracile* Ren. & Card. Am2 Am4 = *Leptodontium longicaule* Mitt. var. *longicaule fide* Zander, Bryologist 75: 268, 1972]
[*Leptodontium subgrimmioides* Broth. & Thér. Am4 = *Leptodontium pungens* (Mitt.) Kindb. *fide* Zander, Bryologist 75: 256, 1972]
Leptodontium subintegrifolium Thér. ex Herz., Fedde Rep. Spec. Nov. Reg. Veg. 45: 45, 1938. Am4
Leptodontium sublaevifolium Broth. in Mildbr. Afr2
[*Leptodontium subplanifolium* Thér. = *Bryoerythrophyllum jamesonii* (Tayl.) Crum *fide* Zander, Bryologist 75: 278, 1972]
[*Leptodontium sulphureum* (C. Müll.) Mitt. Am2 Am4 (= *Leptodontium viticulosoides* var. *panamense* (Lor.) Zand., Bryologist 75: 250, 1972) ≡ *Leptodontium viticulosoides* var. *sulphureum* (Lor.) Zand. *fide* Zander, Bryologist 86: 156, 1983]
[var. *flagellaceum* Bartr. Am2 ≡ *Leptodontium viticulosoides* var. *flagellaceum* (Bartr.) Zand. fide Zander, Bryologist 75: 255, 1972]
[var. *motelayi* (Ren. & Card.) Bartr. Am2 = *Leptodontium viticulosoides* var. *panamense* (Lor.) Zand., Bryologist 75: 251, 1972 = *Leptodontium viticulosoides* var. *sulphureum* (Lor.) Zand. *fide* Zander, Bryologist 86: 156, 1983]
[var. *panamense* Lor. Am2 ≡ *Leptodontium viticulosoides* var. *panamense* (Lor.) Zand., Bryologist 75: 250, 1972 = *Leptodontium viticulosoides* var. *sulphureum* (Lor.) Zand., Bryologist 86: 156, 1983]
Leptodontium syntrichioides (C. Müll.) Kindb. Am2 Am4
Leptodontium taiwanense Nog. As2
[*Leptodontium tenerascens* Broth. in Mildbr. Afr2 = *Leptodontium flexifolium* (Dicks.) Hampe in Lindb. *fide* Zander, Bryologist 75: 231, 1972]
[var. *majus* Broth. in Mildbr. Afr2 = *Leptodontium flexifolium* (Dicks.) Hampe in Lindb. *fide* Zander, Bryologist 75: 231, 1972]
var. *planifolium* P. Varde & Thér. Afr2
[var. *subfilescens* (Thér. & Nav.) Thér. Afr2 = *Leptodontium flexifolium* (Dicks.) Hampe in Lindb. *fide* Sloover, Bull. Jard. Bot. Natl. Belgique 57(3/4): 428, 1987]
Leptodontium tricolor (Williams) Zand. in Zand. & Hegew., Bryologist 79: 20, 1976 (*Williamsiella*) Am4
Leptodontium trifarium Broth. [type not available for study *fide* Zander, Bryologist 75: 278, 1972] Am5
[*Leptodontium turgidum* Herz. Am4 = *Leptodontium viticulosoides* (P. Beauv.) Wijk & Marg. var. *viticulosoides fide* Zander, Bryologist 75: 244, 1972]
[*Leptodontium ulocalyx* (C. Müll.) Mitt. Am2 Am4 = *Leptodontium viticulosoides* var. *panamense* (Lor.) Zand., Bryologist 75: 250, 1972 = *Leptodontium viticulosoides* var. *sulphureum* (Lor.) Zand. *fide* Zander, Bryologist 86: 156, 1983]
[var. *cirrhifolium* (Mitt.) Bartr. Am2 Am4 = *Leptodontium viticulosoides* var. *panamense* (Lor.) Zand., Bryologist 75: 251, 1972 = *Leptodontium viticulosoides* var. *sulphureum* (Lor.) Zand. *fide* Zander, Bryologist 86: 156, 1983]
[*Leptodontium undulatum* Herz. Am4 = *Leptodontium longicaule* var. *microruncinatum* (Dus.) Zand. *fide* Zander, Bryologist 75: 269, 1972]
[*Leptodontium vaginatum* Herz. Am4 = *Leptodontium wallisii* (C. Müll.) Kindb. *fide* Zander, Bryologist 75: 259, 1972]
[*Leptodontium valerianum* Bartr. ≡ *Gymnostomum valerianum* (Bartr.) Zand., Bryologist 75: 277, 1972 ≡ *Tuerckheimia valeriana* (Bartr.) Zand., Misc. Bryol. Lichenol. 8: 27, 1978]
[*Leptodontium variegatum* Herz. Am5 = *Leptodontium wallisii* (C. Müll.) Kindb. *fide* Zander, Bryologist 75: 259, 1972]
Leptodontium viticulosoides (P. Beauv.) Wijk & Marg. Am2 Am4 Am5 Am6 Afr2 Afr3 Afr4 As2 As3 As4
var. *abbreviatum* (Dix.) Wijk & Marg. As3
var. *exasperatum* (Card.) Zand., Bryologist 75: 254, 1972 Am2 Am4
var. *flagellaceum* (Bartr.) Zand., Bryologist 75: 255, 1972 Am2
[var. *panamense* (Lor.) Zand., Bryologist 75: 250, 1972 Am1 Am2 Am3 Am4 Am5 = *Leptodontium viticulosoides* var. *sulphureum* (Lor.) Zand. *fide* Zander, Bryologist 86: 156, 1983, I.C.B.N. Art. 57.3]
var. *subdenticulatum* (C. Müll.) Wijk & Marg. As4
var. *sulphureum* (Lor.) Zand., Bryologist 86: 86, 1983 Am1 Am2 Am3 Am4 Am5
[*Leptodontium volkensii* Broth. Afr2 = *Leptodontium viticulosoides* (P. Beauv.) Wijk & Marg. var. *viticulosoides fide* Zander, Bryologist 75: 244, 1972]
Leptodontium wallisii (C. Müll.) Kindb. Am4
[*Leptodontium warnstorfii* Fleisch. As4 = *Leptodontium flexifolium* (Dicks.) Hampe in Lindb. *fide* Zander, Bryologist 75: 231, 1972]
Leptodontium zygodontoides C. Müll. (type not available for study *fide* Zander, Bryologist 75: 278, 1972) Am6

LUISIERELLA Thér. & P. Varde
Luisierella barbula (Schwaegr.) Steere As2 As4 Am1 Am2 Am3 Am5 Oc

[**LYDIAEA** Laz. = *Microbryum* Schimp.]
[*Lydiaea vlassovii* (Laz.) Laz. As1 ≡ *Phascum vlassovii* Laz. *fide* Savicz-Ljubitzkaja, L. I. & Z. N. Smirnova, Handb. Mosses U.S.S.R. Acrocarp. 1970 ≡ *Microbryum vlassovii* (Laz.) Zand., see treatment of *Microbryum*]

[**MELOPHYLLUM** Herz. = *Bryomanginia* Thér. *fide* Zander, Bryologist 75: 333, 1978 = *Astomiopsis* C. Müll. (Ditrichaceae) *fide* Buck & Zander, Bryologist 83: 255, 1980]
[*Melophyllum radiculosum* Herz. Am4 (= *Bryomanginia saintpierrei* Thér. *fide* Zander, Bryologist 75: 333, 1978) ≡ *Astomiopsis radiculosa* (Herz.) Buck & Zander (Ditrichaceae) *fide* Buck and Zander, Bryologist 83: 255, 1980]

[**MERCEYA** Schimp. = *Scopelophila* (Mitt.) Lindb. *fide* Zander, Bryologist 70: 406, 1967]
[*Merceya difficilis* Herz. & Thér., Fedde Rep. Spec. Nov. Reg. Veg. 45: 46, 1938 Am4 = *Scopelophila ligulata* (Spruce) Spruce *fide*

Zander, Bryologist 70: 412, 1967]

[*Merceya gedeana* (Lac.) Nog. As2 As3 As4 = *Scopelophila cataractae* (Mitt.) Broth. *fide* Zander, Bryologist 70: 408, 1967]

[*Merceya latifolia* Kindb. Am1 ≡ *Crumia latifolia* (Kindb. ex Macoun) Schof.]

[*Merceya ligulata* (Spruce) Schimp. Eur As2 As3 As4 As5 Afr1 Am1 Am2 Am4 ≡ *Scopelophila ligulata* (Spruce) Spruce *fide* Zander, Bryologist 70: 411, 1967]

var. *acutiuscula* (Broth.) Chen Eur

[*Merceya longirostris* (Griff.) Wijk & Marg. As3 ≡ *Trichostomum contractum* Zand. *nom. nov.*, see treatment of *Trichostomum*]

Merceya repandula Baumg. & Fröhl. As4

[*Merceya serratinervis* Tak. As2 = *Encalypta streptocarpa* Hedw. *fide* Iwats. & Nog., J. Hattori Bot. Lab. 37: 368, 1973]

[**MERCEYOPSIS** Broth. & Dix. ex Dix. = *Scopelophila* (Mitt.) Lindb., see treatment of *Scopelophila*]

[*Merceyopsis angulosa* Broth. & Dix. ex Dix. As3 Am5 ≡ *Weisiopsis angulosa* (Broth. & Dix. in Dix.) Hilp. ≡ *Ganguleea angulosa* (Broth. & Dix.) Zand., Phytologia 65: 427, 1989]

[*Merceyopsis crassinervis* (Broth.) Bartr., Bishop Mus. Occ. Pap. 15: 98, 1939 Oc ≡ *Molendoa crassinervis* Broth. *fide* Nog. in Hoe, Lyonia 1: 28, 1974 = *Hymenostylium recurvirostrum* var. *cylindricum* (Bartr.) Zand. *fide* Zander & Eckel, Canad. J. Bot. 60: 1599, 1982]

[*Merceyopsis excavata* Sak. As2 = *Gymnostomum aeruginosum* Sm. *fide* Saito, J. Hattori Bot. Lab. 39: 450, 1975]

[*Merceyopsis hymenostylioides* Broth. & Dix. ex Dix. As3 ≡ *Gymnostomum hymenostylioides* (Broth. & Dix.) Zand., see treatment of *Gymnostomum*]

[*Merceyopsis longirostris* (Griff.) Broth. & Dix. ex Dix. As3 ≡ *Merceya longirostris* (Griff.) Wijk & Marg. ≡ *Trichostomum contractum* Zand. *nom. nov.*, see treatment of *Trichostomum*]

[*Merceyopsis robusta* Dix. As3 ≡ *Tuerckheimia robusta* (Dix.) Zand. *comb. nov.*, see treatment of *Tuerckheimia*]

[*Merceyopsis spathulifolia* Dix. & P. Varde As3 ≡ *Barbula spathulifolia* (Dix. & P. Varde) Zand., see treatment of *Barbula*]

MICROBRYUM Schimp.

Microbryum brevicaule (Tayl.) Zand. (*Gymnostomum*), see treatment of *Microbryum* Austr1

Microbryum curvicolle (Hedw.) Zand. (*Phascum*), see treatment of *Microbryum* Eur Afr1 As5

Microbryum davallianum (Sm.) Zand. (*Gymnostomum*), see treatment of *Microbryum* Eur As5 Afr1 Am1 Austr1

var. *commutatum* (Limpr.) Zand. (*Pottia*), see treatment of *Microbryum* Eur As5 Afr1

var. *conicum* (Schleich. ex Schwaegr.) Zand. (*Gymnostomum*), see treatment of *Microbryum* Eur Afr1 As5 Am2

Microbryum floerkeanum (Web. & Mohr) Schimp. (*Phascum*), see treatment of *Microbryum* Eur Afr1 Am1 As5

var. *arbense* (Loitl.) Zand. (*Phascum*), see treatment of *Microbryum* Eur

var. *badium* BSG Eur Am1 As5

Microbryum longipes (Guerra, Martínez & Ros) Zand. (*Phascum*), see treatment of *Microbryum* Eur

Microbryum raddei (Broth.) Zand. (*Tortula*) As1

Microbryum rectum (With.) Zand. (*Phascum*), see treatment of *Microbryum* Eur Afr1

Microbryum rufochaete (Magill) Zand. (*Acaulon*), see treatment of *Microbryum* Afr4

Microbryum starckeanum (Hedw.) Zand. (*Weisia*), see treatment of *Microbryum* Eur As5 Afr1 Am1 Am2 Austr1 Austr2

var. *brachyodus* (BSG) Zand. (*Anacalypta starckeana* var.), see treatment of *Microbryum* Eur As5 Afr1

var. *brevidens* (Latz.) Zand. (*Pottia starckeana* var.), see treatment of *Microbryum* Eur

var. *fosbergii* (Bartr.) Zand. (*Pottia*), see treatment of *Microbryum* Am1 Am2

var. *leiostoma* (Corb in Corb. & Pitard) Zand. (*Pottia starckeana* var.) Afr1

var. *subgymnostoma* (De Not.) Zand. (*Anacalypta starckeana* var.), see treatment of *Microbryum* Eur

var. *submutica* (Latz.) Zand. (*Pottia starckeana* var.), see treatment of *Microbryum* Eur

fo. *brevifolium* (Limpr.) Zand. (*Pottia starckeana* fo.), see treatment of *Microbryum* Eur

fo. *dextrorsum* (Limpr.) Zand. (*Pottia starckeana* fo.), see treatment of *Microbryum* Eur

fo. *microphyllum* (Warnst.) Zand. (*Pottia starckeana* fo.), see treatment of *Microbryum* Eur

Microbryum subplanomarginatum (Dix.) Zand. (*Pottia*), see treatment of *Microbryum* Afr4

Microbryum tasmanicum (Dix. & Rodw.) Zand. (*Phascum*), see treatment of *Microbryum* Austr1

Microbryum vlassovii (Laz.) Zand. (*Phascum*), see treatment of *Microbryum* As1 Am1

Microbryum zeelandiae (R. Br. ter) Zand. (*Anacalypta*), see treatment of *Microbryum* Austr2

[**MILDEELA** Limpr. ≡ *Pottia* subg. *Mildeella* (Kindb.) Broth. = *Tortula* sect. *Tortula*, see treatment of *Tortula*]

[*Mildeella bryoides* (Dicks.) Limpr. ≡ *Pottia bryoides* (Dicks.) Mitt. ≡ *Tortula protobryoides* Zand., see treatment of *Tortula*]

[fo. *longiseta* Röll, Hedwigia 56: 157, 1915 ≡ *Pottia bryoides* fo. *longiseta* (Röll) Podp.]

MIRONIA Zand. (*nom. nov.* for *Morinia* Card.,, see treatment of *Mironia*)

Mironia crassicuspis (Robins.) Zand., see treatment of *Mironia* (*Barbula*) Am2

Mironia ehrenbergiana (C. Müll.) Zand. (*Barbula*), see treatment of *Mironia nbergiana* (C. Müll.) Thér. Am2 Am4

var. *elongata* (Wils. in Mitt.) Zand., see treatment of *Mironia* (*Barbula*) Am4

Mironia stenotheca (Thér.) Zand., see treatment of *Mironia* (*Barbula*) Am2

MOLENDOA Lindb.

Molendoa andina (Mitt.) Broth. Am4

[*Molendoa boliviana* Both. in Herz. Am4 = *Molendoa sendtneriana* (BSG) Limpr. *fide* Zander, Bryologist 80: 248, 1977]

[var. *brevifolia* Herz. Am4 = *Molendoa sendtneriana* (BSG) Limpr. *fide* Zander, Bryologist 80: 248, 1977]

[*Molendoa burmensis* Bartr. As3 = *Hymenostylium recurvirostrum* var. *cylindricum* (Bartr.) Zand. in Zand. & Eckel;, see treatment of *Molendoa*]

[*Molendoa clavuligera* Castelli, Rev. Bryol. Lichénol. 34: 716, 1966 (1967)] Eur = *Molendoa taeniatifolia* Herz. (see Corley et al., J. Bryol. 11: 623, 1981 (1982)]

[*Molendoa crassinervis* Broth. Oc = *Hymenostylium recurvirostrum* var. *cylindricum* (Dix.) Bartr. *fide* Zander & Eckel, Canad. J. Bot. 60: 1599, 1982]

Molendoa cucullata (Herz.) Hilp. Am4

Molendoa duthiei (Broth.) Broth. As3

[*Molendoa excelsa* (C. Müll.) Broth. Am6 = *Molendoa*

sendtneriana (BSG) Limpr. *fide* Zander, Bryologist 80: 247, 1977]

Molendoa fuegiana Bartr. Am6

[*Molendoa guentheri* Broth. Am4 = *Molendoa sendtneriana* (BSG) Limpr. *fide* Menzel, J. Hattori Bot. Lab. 71: 219, 1992]

[*Molendoa herzogii* Broth. in Herz. Am4 = *Molendoa sendtneriana* (BSG) Limpr. *fide* Zander, Bryologist 80: 248, 1977]

Molendoa hornschuchiana (Hook.) Lindb. ex Limpr. Eur As1 As2 Afr1 Am1

var. *minor* (Mol.) Limpr. Eur

[*Molendoa japonica* Broth. As2 ≡ *Molendoa sendtneriana* var. *japonica* (Broth.) Iwats. ≡ *Hymenostyliella japonica* (Broth.) Saito, J. Jap. Bot. 46: 145, 1971 ≡ *Didymodon japonicus* (Broth.) Saito *fide* Saito, J. Hattori Bot. Lab. 39: 508, 1975]

Molendoa kitaibelana Györf. in Bauer Eur

[*Molendoa obtusifolia* Broth. & Par. Am1 Am2 = *Molendoa sendtneriana* (BSG) Limpr. *fide* Zander, Bryologist 80: 248, 1977]

[var. *densissima* Thér. Am2 = *Molendoa sendtneriana* (BSG) Limpr. *fide* Zander, Bryologist 80: 248, 1977]

[var. *incrassata* Thér. Am2 = *Molendoa sendtneriana* (BSG) Limpr. *fide* Zander, Bryologist 80: 248, 1977]

Molendoa ogalalensis (Smith Merrill) Zand. (*Ozobryum*), see treatment of *Molendoa* Am1

Molendoa platyphyllum (Williams) Zand. (*Anoectangium*), see treatment of *Molendoa* Am4]

Molendoa roylei (Mitt.) Broth. As2

Molendoa schliephackei (Limpr.) Zand. (*Pleuroweisia*), see treatment of *Molendoa*) Eur As2

Molendoa sendtneriana (BSG) Limpr. Eur As1 As2 As3 Am1 Am2 Am4 Am5 Am6

[var. *japonica* (Broth.) Iwats. As2 ≡ *Hymenostyliella japonica* (Broth.) Saito, J. Jap. Bot. 46: 145, 1971 ≡ *Didymodon japonicus* (Broth.) Saito *fide* Saito, J. Hattori Bot. Lab. 39: 508, 1975]

var. *limprichtii* Györf. Eur

var. *sudetica* Podp. Eur

[var. *tenuinervis* (Limpr.) Pilous Eur Am1 = *Molendoa sendtneriana* (BSG) Limpr. *fide* Zander, Bryologist 80: 247, 1977]

var. *transcaspica* Györf. in Fedch. As1

var. *yunnanica* Györf in Thér.

fo. *andreaeoides* (Limpr.) Pilous, Preslia 30: 168, 1958 (*Grimmia*) Eur

fo. *propagulifera* Podp., Cas. Morav. Mus. Zemsk. 13: 42, 1913 Eur

Molendoa seravschanica Broth. & Györf. in Fedch. As2

[*Molendoa sordida* (Mitt.) Steere Am4 = *Hymenostylium recurvirostrum* (Hedw.) Dix., see treatment of *Molendoa*]

Molendoa sublaevis Demar. & P. Varde Afr2

Molendoa subobtusifolia Broth., Rev. Bryol. 47: 7, 1921 (not in Index Muscorum) Am4

Molendoa taeniatifolia Herz. Eur

[*Molendoa tenuinervis* Limpr. Eur ≡ *Molendoa sendtneriana* var. *tenuinervis* (Limpr.) Pilous = *Molendoa sendtneriana* (BSG) Limpr. *fide* Zander, Bryologist 80: 247, 1977]

[fo. *arctica* Györffy, Bryologist 15: 1912 Am1 = *Molendoa sendtneriana* var. *tenuinervis* Limpr. *fide* Grout, Moss Fl. N. Am. 1(3): 150, 1933 = *Molendoa sendtneriana* (BSG) Limpr. *fide* Zander, Bryologist 80: 247, 1977]

[*Molendoa warburgii* (Crundw. & Hill) Zand. (*Anoectangium*), see treatment of *Molendoa*]

[**MORINIA** Card. ≡ *Mironia*]

[*Morinia crassicuspis* (Robins.) Zand., Bryologist 81: 556, 1978 (*Barbula*) Am2 ≡ *Mironia crassicuspis* (Robins.) Zand., see treatment of *Mironia*]

[*Morinia ecuadorensis* Bartr. Am4 = *Morinia ehrenbergiana* var. *elongata* (Wils.) in Mitt.) Zand. *fide* Zander, Bryologist 81: 557, 1978 ≡ *Mironia ehrenbergiana* var. *elongata* (Wils. in Mitt.) Zand., see treatment of *Mironia*]

[*Morinia ehrenbergiana* (C. Müll.) Thér. Am2 Am4 ≡ *Mironia ehrenbergiana* (C. Müll.) Zand., see treatment of *Mironia*]

[var. *elongata* (Wils. in Mitt.) Zand., Bryologist 81: 557, 1973 (*Barbula*) Am4 ≡ *Mironia ehrenbergiana* var. *elongata* (Wils. in Mitt.) Zand., see treatment of *Mironia*]

[*Morinia saitoana* Zand, Delg. & Eckel, Bryologist 83: 510, 1980 (1981) Am2 ≡ *Morinia stenotheca* (Thér.) Zand., Bryologist 86: 156, 1983 ≡ *Mironia stenotheca* (Thér.) Zand., see treatment of *Mironia*]

[*Morinia setschwanica* (Broth.) Hilp. As2 = *Didymodon eroso-denticulatus* (C. Müll.) Saito *fide* Saito, J. Hattori Bot. Lab. 39: 504, 1975]

[*Morinia stenotheca* (Thér.) Zand., Bryologist 86: 156, 1983 (*Barbula*) Am2 ≡ *Mironia stenotheca* (Thér.) Zand., see treatment of ≡ *Mironia stenotheca* (Thér.) Zand., see treatment of *Mironia*]

[**NEOCARDOTIA** Thér. & Bartr. = *Rhexophyllum* Herz. *fide* Hilpert, Bot. Centralb. Beih. 50(2): 684, 1933]

[*Neocardotia subnigra* (Mitt.) Thér. & Bartr. Am1 Am2 Am4 ≡ *Rhexophyllum subnigrum* (Mitt.) Thér. ex Hilp. *fide* Hilpert, Bot. Centralb. Beih. 50(2): 684, 1933]

NEOHYOPHILA Crum

[*Neohyophila lingulata* (Card.) Crum Am2 = *Neohyophila sprengelii* var. *stomatodonta* (Card.) Zand., Bryologist 86: 138, 1983 ≡ *Plaubelia sprengelii* var. *stomatodonta* (Card.) Zand., see treatment of *Plaubelia*]

[*Neohyophila spathulifolia* (Bartr.) Crum Am2 = *Neohyophila sprengelii* (Schwaegr.) Crum *s. lat. fide* Zander, Bryologist 86: 135, 1983 ≡ *Plaubelia sprengelii* (Schwaegr.) Zand., see treatment of *Plaubelia*]

[*Neohyophila sprengelii* (Schwaegr.) Crum Am1 Am2 Am3 Am4 Am5 ≡ *Plaubelia sprengelii* (Schwaegr.) Zand., see treatment of *Plaubelia*]

[var. *stomatodonta* (Card.) Zand., Bryologist 86: 138, 1983 (*Hyophila*) Am1 Am2 Am3 ≡ *Plaubelia sprengelii* var. *stomatodonta* (Card.) Zand., see treatment of *Plaubelia*]

[*Neohyophila stomatodonta* (Card.) Crum Am1 Am2 Am3 ≡ *Neohyophila sprengelii* var. *stomatodonta* (Card.) Zand., Bryologist 86: 138, 1983]

[**OXYSTEGUS** (Limpr.) Hilp. = *Trichostomum* Bruch]

[*Oxystegus burmensis* (Bartr.) Gangulee, M. E. India 653, 1972 "*burmense*" (*Trichostomum*) As3 = *Trichostomum tenuirostre* (Hook. & Tayl.) Lindb. *fide* Sollman, Lindbergia 10: 55, 1984]

Oxystegus crassicostatus Norris & T. Kop., Acta Bot. Fenn. 137: 90, 1989 As4

[*Oxystegus cuspidatus* (Doz. & Molk.) Chen As2 As4 ≡ *Trichostomum cuspidatum* (Doz. & Molk.) Doz. & Molk. = *Trichostomum tenuirostre* (Hook. & Tayl.) Lindb. *fide* Sollman, Lindbergia 10: 55, 1984]

[*Oxystegus cylindricus* (Brid.) Hilp. Eur As1 As3 Afr1 Afr2 Afr3 Afr4 Am1 Am2 Am4 = *Oxystegus tenuirostris* (Hook. & Tayl.)

A. Sm. *fide* Smith, J. Bryol. 9: 393, 1977 ≡ *Trichostomum tenuirostre* (Hook. & Tayl.) Lindb.]

[subsp. *hibernicus* (Mitt.) Podp. Eur ≡ *Oxystegus hibernicus* (Mitt.) Hilp. *fide* Corley et al., J. Bryol. 11: 623, 1981 (1982) ≡ *Trichostomum hibernicum* (Mitt.) Dix.]

var. *daldinianus* (De Not.) Podp Eur

[var. *holtii* (Braithw.) Podp. Eur ≡ *Oxystegus tenuirostris* var. *holtii* (Braithw.) A. Sm. *fide* Smith, J. Bryol. 9: 393, 1977 ≡ *Trichostomum tenuirostre* var. *holtii* (Braithw.) Dix.]

var. *irriguus* (Limpr.) Podp. Eur

var. *norvegicus* (Hag.) Podp. Eur

var. *robustus* (Schimp.) Podp. Eur

[fo. *asper* (Podp.) Podp., Consp. Musc. Eur. 176, 1954 (*Trichostomum cylindricum* var.) Eur ≡ *Trichostomum cylindricum* var. *asperum* Podp.]

fo. *filifer* (Herz.) Podp., Consp. Musc. Eur. 176, 1954 (*Oxystegus cylindricus* var.) Eur

fo. *longifolius* Podp., Consp. Musc. Eur. 176, 1954 (*Trichostomum cylindricum* fo. *longifolius* Latz. *hom. illeg.*) Eur

[*Oxystegus cylindrothecus* (Mitt.) Gangulee, Nov. Hedw. 12: 430, 1966 (*Tortula*) As3 ≡ *Trichostomum cylindrothecum* (Mitt.) Broth. = *Trichostomum bombayense* C. Müll. *fide* Townsend, J. Bryol. 12: 561, 1983]

[*Oxystegus fallax* (Herz.) Hilp. Am4 ≡ *Trichostomum fallax* Herz.]

[*Oxystegus hibernicus* (Mitt.) Hilp. Eur ≡ *Trichostomum hibernicum* (Mitt.) Dix.]

[*Oxystegus indicus* (Dix. & P. Varde) Hilp. As3 ≡ *Pseudosymblepharis indica* Dix. & P. Varde]

[*Oxystegus khasianus* (Mitt.) Gangulee, Nov. Hedw. 8: 149, 1964 (*Tortula*) As3 ≡ *Psuedosymblepharis khasiana* (Mitt.) Zand., see treatment of *Pseudosymblepharis*]

[*Oxystegus lignicola* (Herz.) Hilp. Am5 ≡ *Trichostomum lignicola* Herz.]

[*Oxystegus linearis* (Web. & Mohr) Hilp. Am3 Am5 ≡ *Tortella linearis* (Web. & Mohr.) Zand., see treatment of *Tortella*]

Oxystegus melanostoma (Mitt.) Hilp. Am5

[*Oxystegus recurvifolius* (Tayl.) Zand., Lindbergia 8: 187, 1982 (*Bryum*) Eur Am1 ≡ *Trichostomum recurvifolium* (Tayl.) Zand., see treatment of *Trichostomum*]

[*Oxystegus rhodesiae* (Broth.) Hilp. Afr2 ≡ *Trichostomum rhodesiae* Broth.]

[*Oxystegus soulae* (C. Müll. in Ren. & Card.) Wijk & Marg. Afr3 ≡ *Trichostomum soulae* (C. Müll. in Ren. & Card.) Zand., see treatment of *Trichostomum*]

[var. *corticicola* (Ren. & Card.) Wijk & Marg. Afr3 ≡ *Trichostomum soulae* var. *corticicola* (Ren. & Card.) Zand., see treatment of *Trichostomum*]

[*Oxystegus spiralis* (Grout) Crum & Anderson, Bryologist 92: 533, 1989 ≡ *Trichostomum spirale* Grout, see treatment of *Trichostomum*]

[*Oxystegus stenophyllus* (Mitt.) Gangulee, Nov. Hedw. 12: 429, 1966 As3 = *Oxystegus cylindricus* (Brid.) Hilp. *fide* Saito, J. Hattori Bot. Lab. 39: 437, 1975 = *Oxystegus tenuirostris* (Hook. & Tayl.) A. Sm. *fide* Smith, J. Bryol. 9: 393, 1977 ≡ *Trichostomum tenuirostre* (Hook. & Tayl.) Lindb.]

[*Oxystegus svihlae* (Bartr.) Gangulee, M. E. India 654, 1972 (*Trichostomum*) As2 ≡ *Tuerckheimia svihlae* (Saito) Zand., see treatment of *Tuerckheimia*]

[*Oxystegus syrrhopodontoides* (Herz.) Hilp. Am4 ≡ *Trichostomum syrrhopodontoides* Herz. = *Trichostomum tenuirostre* (Hook. & Tayl.) Lindb. *fide* W. Steere, in litt., having seen type at JE]

[*Oxystegus verrucosus* (Broth. & Par.) Hilp. Oc ≡ *Pseudosymblepharis verrucosa* (Broth. & Par.) Zand., see treatment of *Pseudosymblepharis*]

[*Oxystegus tenuirostris* (Hook. & Tayl.) A. Sm., J. Bryol. 9: 393, 1977 Eur As1 As3 Afr1 Afr2 Afr3 Afr4 Am1 Am2 Am4 ≡ *Trichostomum tenuirostre* (Hook. & Tayl.) Lindb.]

[var. *gemmiparus* (Schimp.) Zand., Lindbergia 4: 285, 1978 (*Didymodon cylindricus* var.) Eur As3 Am2 Am3 Am5 ≡ *Trichostomum tenuirostre* var. *gemmiparum* (Schimp.) Zand., see treatment of *Trichostomum*]

[var. *holtii* (Braithw.) A. Sm., J. Bryol. 9: 393, 1977 Eur Am1 ≡ *Trichostomum tenuirostre* var. *holtii* (Braithw.) Dix.]

[var. *stenocarpus* (Thér.) Zand., Misc. Bryol. Lichenol. 9: 73, 1982 (*Weisiopsis*) As2 Am1 Am2 = *Trichostomum spirale* Grout, see treatment of *Trichostomum*]

[**OZOBRYUM** Smith Merrill, Novon 2: 255, 1992 = *Molendoa* Lindb., see treatment of *Molendoa*]

[*Ozobryum ogalalense* Smith Merrill, Novon 2: 255, 1992 Am1 ≡ *Molendoa ogalalensis* (Smith Merrill) Zand., see treatment of *Molendoa*]

PACHYNEUROPSIS H. Mill.

Pachyneuropsis bartlettii (Bartr.) H. Mill., Taxon 19: 822, 1970 (*Pachyneurum* Bartr. *hom. illeg.*) As4

[**PARALEPTODONTIUM** Hill., J. Bryology 12: 181, 1982 = *Oxystegus* (Limpr.) Hilp. *fide* Zander, Lindbergia 8: 187, 1982 = *Trichostomum* Hedw., see treatment of *Trichostomum*]

[*Paraleptodontium recurvifolium* (Tayl.) Long, J. Bryology 12: 181, 1982 ≡ *Oxystegus recurvifolius* (Tayl.) Zand., Lindbergia 8: 187, 1982 ≡ *Trichostomum recurvifolium* (Tayl.) Zand., see treatment of *Trichostomum*]

[**PHASCONICA** C. Müll. ≡ *Weissia* subg. *Phasconica* (C. Müll.) Zand., see treatment of *Weissia*]

[*Phasconica balansae* C. Müll. Austr1 Oc ≡ *Weissia balansae* (C. Müll.) Zand., see treatment of *Weissia*]

[*Phasconica lorentzii* C. Müll. Am6 ≡ *Weissia lorentzii* (C. Müll.) Zand., see treatment of *Weissia*]

[*Phasconica tisserantii* P. Varde Afr2 Afr4 (good species of *Phasconica fide* Magill, Fl. S. Afr. I. Mosses 1: 251, 1981 [1982]) = *Weissia unguiculata* (Mitt.) Crundw. & Nyh., J. Bryol. 8: 69, 1974 ≡ *Trichostomum unguiculatum* (Mitt.) Zand., see treatment of *Trichostomum*]

PHASCOPSIS Stone, J. Bryol. 11: 17, 1980

Phascopsis rubicunda Stone, J. Bryol. 11: 17, 1980 Austr1

[**PHASCUM** Hedw. = *Tortula* Hedw., see treatment of *Tortula*]

Phascum beaugleholei Stone, J. Bryol. 15: 766, 1989 Austr1

[*Phascum brittoniae* Crum & Steere Am3 = *Uleobryum peruvianum* Broth., see treatment of *Uleobryum*]

Phascum calodictyon C. Müll. Am6

[*Phascum carinatum* Hampe Am5 ≡ *Bruchia carinata* (Hampe) Zand. (Ditrichaceae) see Excluded Taxa]

[*Phascum crispum* Hedw. ≡ *Astomum crispum* (Hedw.) Hampe = *Weissia longifolia* Mitt. *fide* Crundwell & Nyholm, J. Bryol. 7: 13, 1972]

var. *minus* De Not. Eur

[*Phascum curvicolle* Hedw. Eur Afr1 As5 ≡ *Microbryum curvicolle* (Hedw.) Zand., see treatment of *Microbryum*]

[*Phascum cuspidatum* Hedw. Eur As1 As2 As5 Afr1 Am1 Am2 Am4 ≡ *Tortula atherodes* Zand. *nom. nov.*, see treatment of *Tortula*]

[subsp. *elatum* (Brid.) C. Jens. Eur ≡ *Tortula atherodes* var. *elata* (Brid.) Zand., see treatment of *Tortula*]
[subsp. *mitraeforme* (Limpr.) C. Jens. Eur ≡ *Tortula atherodes* var. *mitraeformis* (Limpr.) Zand., see treatment of *Tortula*]
[subsp. *papillosum* (Lindb.) Guerra & Ross in Guerra, Jiménez, Ros & Carrión, Cryptogamie Bryol. Lichénol. 12: 391, 1991 (*Phascum p.*) Eur ≡ *Tortula atherodes* var. *papillosa* (Lindb.) Zand., see treatment of *Tortula*]
[subsp. *piliferum* (Hedw.) Amann Eur As1 As5 Afr2 Am1 ≡ *Tortula atherodes* var. *pilifera* (Hedw.) Zand., see treatment of *Tortula*]
[var. *affine* (Nees & Hornsch.) Hampe Eur ≡ *Tortula atherodes* var. *affinis* (Nees & Hornsch.) Zand., see treatment of *Tortula*]
[var. *arcuatum* Herrnstadt & Heyn, Bryologist 94: 175, 1991 As5 ≡ *Tortula atherodes* var. *arcuata* (Herrnstadt & Heyn) Zand., see treatment of *Tortula*)
[var. *curvisetum* (Dicks.) Nees & Hornsch. Eur ≡ *Tortula atherodes* var. *curviseta* (Dicks.) Zand., see treatment of *Tortula*]
[var. *diaphorum* (Hag.) C. Jens. Eur ≡ *Tortula atherodes* var. *diaphora* (Hag.) C. Jens., see treatment of *Tortula*]
[var. *elatum* (Brid.) Drumm. Eur ≡ *Tortula atherodes* var. *elata* (Brid.) Zand., see treatment of *Tortula*]
[var. *flagellaceum* Ruthe in Warnst. Eur ≡ *Phascum cuspidatum* fo. *flagellaceum* (Ruthe in Warnst.) Podp., Consp. Musc. Eur. 222, 1954]
[var. *grandiusculum* (Brid.) Röll, Hedwigia 56(1–3): 156, 1915 (*Phascum grandiusculum*) Eur = *Phascum cuspidatum* var. *elatum* (Brid.) Drumm. *fide* Web. & Mohr 1807 ≡ *Tortula atherodes* var. *elata* (Brid.) Zand., see treatment of *Tortula*]
[var. *henrici* (Ren. & Card.) Wijk & Marg. Am1 = *Phascum cuspidatum fide* Crum & Anders., Mosses E. N. Amer. 1: 350, 1981 ≡ *Tortula atherodes* Zand., see treatment of *Tortula*]
[var. *intertextum* (Brid.) Brid. Eur ≡ *Tortula atherodes* var. *intertexta* (Brid.) Zand., see treatment of *Tortula*]
[var. *marginatum* Hernnstadt & Heyn, Bryologist 94: 175, 1991 As5 ≡ *Tortula atherodes* var. *marginata* (Hernnstadt & Heyn) Zand., see treatment of *Tortula*]
[var. *mitraeforme* Limpr. Eur ≡ *Tortula atherodes* var. *mitraeformis* (Limpr.) Zand., see treatment of *Tortula*]
[var. *papillosum* (Lindb.) C. Hartm. in Hartm., Handb. Skand. Fl. ed. 10, 2: 119, 1871 (*Phascum papillosum*) Eur ≡ *Tortula atherodes* var. *papillosa* (Lindb.) Zand., see treatment of *Tortula*]
[var. *piliferum* (Hedw.) Hook. & Tayl. Eur As1 As5 Afr2 Am1 ≡ *Tortula atherodes* var. *pilifera* (Hedw.) Zand., see treatment of *Tortula*]
[var. *retortifolium* Guerra & Ros in Guerra, Jiménez, Ros & Carrión, Cryptogamie Bryol. Lichénol. 12: 390, 1991 Eur ≡ *Tortula atherodes* var. *retortifolia* (Guerra & Ros in Guerra, Jiménez, Ros & Carrión) Zand., see treatment of *Tortula*]
[var. *schreberianum* (Dicks.) Brid. Eur Am1 ≡ *Tortula atherodes* var. *schreberiana* (Dicks.) Zand., see treatment of *Tortula*]
[var. *varium* Steud. Eur = *Phascum cuspidatum* var. *schreberianum* (Dicks.) Brid. *fide* Brid., Mant. Musc. 8, 1819 ≡ *Tortula atherodes* var. *schreberiana* (Dicks.) Zand., see treatment of *Tortula*]
fo. *affine* (Nees & Hornsch.) Casares Gil, Fl. Iber. Musg. 256, 1932 (*Phascum affine*) Eur
fo. *flagellaceum* (Ruthe in Warnst.) Podp., Consp. Musc. Eur. 222, 1954 (*Phasum cuspidatum* var.) Eur
Phascum cuynettii Biz. & Pier. ex Guerra, Jiménez, Ros & Carrión, Cryptogamie Bryol. Lichénol. 12: 401, 1991 (*Phascum cuynettii* Biz. & Pier., Rev. Bryol. Lichénol. 36: 506, 1969 *nom. inval.*, single element not cited as holotype *fide* Crosby et al. 1992) Eur
[*Phascum floerkeanum* Web. & Mohr Eur Afr1 Am1 As5 ≡ *Microbryum floerkeanum* (Web. & Mohr) Schimp., see treatment of *Microbryum*]
[var. *arbense* (Loitl.) Podp. Eur ≡ *Microbryum floerkeanum* var. *arbense* (Loitl.) Zand., see treatment of *Microbryum*]
[var. *badium* (BSG) Schimp. Eur Am1 As5 ≡ *Microbryum floerkeanum* var. *badium* BSG]
[*Phascum fosbergii* (Bartr.) Guerra in Guerra, Martínez & Ros, J. Bryol. 16: 59, 1990 Am1 Am2 ≡ *Microbryum starckeanum* var. *fosbergii* (Bartr.) Zand.]
Phascum galilaeum Hernnstadt & Heyn, Bryologist 94: 175, 1991 As5
[*Phascum halophilum* Smarda Eur = *Phascum cuspidatum* Hedw. *fide* Corley et al., J. Bryol. 11: 621, 1981 [1982] ≡ *Tortula atherodes* Zand., see treatment of *Tortula*]
[*Phascum hyalinotrichum* Card. & Thér. Am1 Am2 ≡ *Stegonia hyalinotrichum* (Card. & Thér.) Zand., see treatment of *Stegonia*]
[*Phascum kilimandscharicum* (C. Müll.) Roth Afr2 ≡ *Bryum kilimandscharicum* (C. Müll.) Zand. (Bryaceae) see Excluded Taxa]
Phascum laticostum Stone, J. Bryol. 15: 753, 1989 Austr1
[*Phascum leptophyllum* C. Müll. Afr2 Afr4 ≡ *Chenia leptophylla* (C. Müll.) Zand., see treatment of *Chenia*]
[*Phascum longipes* Guerra, Martínez & Ros, J. Bryol. 16: 55, 1990 Eur ["*longipedis*" typographical error cf. A. Smith, J. Bryol. 16: 335, 1991] ≡ *Microbryum longipes* (Guerra, Martínez & Ros) Zand., see treatment of *Microbryum*]
Phascum longipilum Stone, J. Bryol. 15: 755, 1989 Austr1
Phascum nepalense Brid. As3
[*Phascum nitidum* Hedw. ≡ *Pseudephemerum nitidum* (Hedw.) Reim.]
var. *bulbiferum* Husn. Eur
[var. *strictum* (Dicks.) Wils., Bryol. Brit. 35, 1855 *nom. illeg. incl. Phascum nitidum* var. *minus* Bayrh. = *Pseudephemerum nitidum* var. *minimum* (Rabenh.) Wijk & Marg. cf. Wilson 1855]
Phascum peraristatum C. Müll. Afr4
[*Phascum piliferum* Schreb. ex Hedw. ≡ *Phascum cuspidatum* var. *piliferum* (Hedw.) Hook. & Tayl. ≡ *Tortula atherodes* var. *pilifera* (Hedw.) Zand., see treatment of *Tortula*]
var. *subacaule* Schwaegr. ex Steud. Eur
[var. *imberbe* Hartm., Handb. Skand. Fl. ed. 3: 261, 1838 ≡ *Phascum floerkeanum* Web. & Mohr. 1807]
Phascum piptocarpum Dur. & Mont. in Mont. Afr1
Phascum readeranum Stone, J. Bryol. 15: 757, 1989 Austr1
[*Phascum rectum* With. ≡ *Pottia recta* (With.) Mitt. ≡ *Microbryum rectum* (With.) Zand., see treatment of *Microbryum*]
var. *luxurians* De Not. Eur
[*Phascum robustum* (Broth.) ex Roth) Stone, J. Bryol. 15: 747, 1989 Austr1 ≡ *Acaulon robustum* Broth. ex Roth]
var. *crassinervium* (C. Müll.) Stone, J. Bryol. 15: 751, 1989 Austr1
[*Phascum schimperianum* Sull. in A. Gray, Man. Bot. E. N. Am. ed. 2: 615, 1856 ≡ *Acaulon schimperianum* (Sull. in A. Gray) Sull.]
[*Phascum tasmanicum* Dix. & Rodw. Austr1 ≡ *Microbryum tasmanicum* (Dix. & Rodw.) Zand., see treatment of *Microbryum*]
var. *unguiculatum* Stone, J. Bryol. 15: 761, 1989 Austr1
[*Phascum vlassovii* Laz. As1 Am1 ≡ *Microbryum vlassovii* (Laz.) Zand., see treatment of *Microbryum*]
[*Phascum vogesiacum* Moug. & Nestl., Stirp. Crypt. Vog.-Rhen.

fasc. 7: n. 706, 1823 *inval.* Eur ≡ *Bruchia vogsesiaca* Schwaegr.]

PLAUBELIA Brid.

Plaubelia involuta (Magill) Zand. (*Weisiopsis*), see treatment of *Plaubelia* Afr4

Plaubelia perinvoluta Zand. (*nom. nov.* for *Desmatodon involutus* Bartr. *hom. illeg.*), see treatment of *Plaubelia* As3

Plaubelia sprengelii (Schwaegr.) Zand. (*Barbula*), see treatment of *Plaubelia* Am1 Am2 Am3 Am4 Am5

var. *stomatodonta* (Card.) Zand. (*Hyophila*), see treatment of *Plaubelia* Am1 Am2 Am3

PLEUROCHAETE Lindb.

Pleurochaete beccarii Vent. Afr2

Pleurochaete ecuadoriensis Broth. Am4 Am6

[*Pleurochaete luteola* (Besch.) Thér. Am1 Am2 Am3 Am4 ≡ *Pleurochaete squarrosa* var. *luteola* (Besch.) Zand., see treatment of *Pleurochaete*]

Pleurochaete malacophylla (C. Müll.) Broth. As5

Pleurochaete squarrosa (Brid.) Lindb. Eur As1 As2 As3 As5 Afr1 Afr2 Am1 Am2 Am3 Am4

[var. *crispifolia* Nog. As2 = *Pseudosymblepharis angustata* (Mitt.) Chen *fide* Saito, J. Hattori Bot. Lab. 39: 439, 1975]

var. *densifolia* Amann Eur

var. *luteola* (Besch.) Zand., see treatment of *Pleurochaete* (*Pleurochaete luteola*) Am1 Am2 Am3 Am4

var. *nitida* (Farn.) Podp. Eur

fo. *brevifolia* (Amann) Podp., Consp. Musc. Eur. 183, 1954 (*Pleurochaete squarrosa* var.) Eur

fo. *subintegra* Latz., Bot. Centralb. Beih. 48: 477, 1931 Eur

[**PLEUROWEISIA** Limpr. = *Molendoa* Lindb. (see treatment of *Molendoa*)]

[*Pleuroweisia schliephackei* Limpr. Eur ≡ *Molendoa schliephackei* (Limpr.) Zand., see treatment of *Molendoa*]

[**POTTIA** (Reichenb.) Ehrh. ex Fürnr. = *Tortula* Hedw., see treatment of *Tortula*]

[*Pottia acutidentata* Card. & Thér. Afr4 ≡ *Hennediella acutidentata* (Card. & Thér.) Zand., see treatment of *Hennediella*]

[*Pottia affinis* (Hook. & Tayl.) Herrnstadt & Heyn, Bryologist 94: 175, 1991 (*Weissia*) As5 = *Pottia mutica* Vent. in De Not. Eur As5 Afr1 = *Pottia starckeana* var. *brachyoda* (BSG) C. Müll. *fide* Moenkemeyer, Laub. Eur. 327, 1927 ≡ *Microbryum starckeanum* var. *brachyodus* (BSG) Zand., see treatment of *Microbryum*]

Pottia alpicola Dix. As3

Pottia altipes Broth. Am5

Pottia antarctica (Ångstr.) C. Müll. Am6

Pottia appertii Warnst. Afr1

Pottia areolata (Knight) R. Br. ter Austr2

[*Pottia arizonica* Wareh. in Grout Am1 = *Microbryum starckeanum* (Hedw.) Zand. var. *starckeanum*, see treatment of *Microbryum*]

[var. *mucronulata* Wareh. in Grout Am1 = *Microbryum starckeanum* var. *brachyodus* (BSG) Zand., see treatment of *Microbryum*]

[*Pottia aurantiaca* (Mitt.) C. Müll. in Chen, Hedwigia 80: 62, 1941 *inval.* ≡ *Hymenostylium aurantiacum* Mitt. = *Hymenostylium recurvirostrum* var. *luzonense* (Broth.) Bartr. *fide* Gangulee, Mosses E. India 3: 648, 1972 = *Hymenostylium recurvirostrum* var. *cylindricum* (Bartr.) Zand., see treatment of *Hymenostylium*]

[*Pottia austrogeorgica* Card. (good species *fide* Matteri 1977) Am6 ≡ *Hennediella austrogeorgica* (Card.) Blockeel, J. Bryol. 16: 191, 1991]

[*Pottia brevicaulis* (Tayl.) C. Müll. (see Catcheside, Mosses S. Austr. 128, 1980) Austr1 ≡ *Microbryum brevicaule* (Tayl.) Zand., see treatment of *Microbryum*]

[*Pottia bryoides* (Dicks.) Mitt. Eur As5 Am1 Am2 ≡ *Tortula protobryoides* Zand. (*nom. nov.* for *Phascum bryoides* Dicks.), see treatment of *Tortula*]

[var. *brevifolia* (De Not.) Wijk & Marg. Eur ≡ *Tortula protobryoides* var. *brevifolia* (De Not.) Zand., see treatment of *Tortula*]

[var. *thornhillii* (Wils.) Braithw. Eur ≡ *Tortula protobryoides* var. *thornhillii* (Wils.) Zand., see treatment of *Tortula*]

fo. *atroviridis* (Schimp.) Podp., Consp. Musc. Eur. 224, 1954 (*Phascum bryoides* var.) Eur

fo. *brachycarpa* Mönk., Laubm. Eur. 326, 1927 (*Phascum bryoides* var.) Eur

fo. *curviseta* (Mönk.) BSG, Laub. Eur. 326, 1927 (*Phascum bryoides* var.) Eur

fo. *longiseta* (Röll) Podp., Consp. Musc. Eur. 224, 1954 (*Mildeella bryoides* fo.) Eur

fo. *minor* (Nees & Hornsch.) Podp., Consp. Musc. Eur. 224, 1954 (*Phascum bryoides* var.) Eur

fo. *pilifera* (Schultz) Mönk., Bryol. Eur. 326, 1927 (*Phascum bryoides* var.) Eur

[*Pottia caespitosa* (Bruch ex Brid.) C. Müll. cf. Corley et al., J. Bryol. 11: 620, 1981 [1982] Eur ≡ *Trichostomum caespitosum* (Bruch ex Brid.) Jur., see treatment of *Trichostomum*]

Pottia caucasica (Lindb.) Par. Eur

[*Pottia charcotii* Card. Ant ≡ *Pottia heimii* var. *charcotii* (Card.) Sav. & Smirn., Novosti Sist. Niz. Rast. 1965: 258, 1965 = *Pottia heimei* (Hedw.) Fürnr. in Hampe var. *heimei fide* Matteri, Brit. Antarct. Surv. Bull. 46: 141, 1977]

Pottia chucutensis Card. & Broth. Am6

[*Pottia ciliatiseta* Sak. = *Rhabdoweisia crispata* (With.) Lindb. *fide* Saito, J. Hattori Bot. Lab. 39: 528, 1975]

[*Pottia conica* (Schleich. ex Schwaegr.) Nyh., Ill. Fl. Nord. Mosses 2: 82, 1989 Eur As5 Afr1 ≡ *Microbryum davallianum* var. *conicum* (Schleich. ex Schwaegr.) Zand., see treatment of *Microbryum*]

Pottia cucullata (Hampe) Jaeg. Am4

[*Pottia cuneifolia* Solms ex Schimp. Eur ≡ *Tortula zoddae* Zand. *comb. nov.*, see treatment of *Tortula*]

[*Pottia davalliana* (Sm.) C. Jens. Eur As5 Afr1 Am1 Austr1 (= *Pottia starckeana* subsp. *minutulum* (Schleich. ex Schwaegr.) Chamberl. *fide* Chamberlain, Roy. Bot. Gard. Edinburg Not. 29: 403, 1969) ≡ *Microbryum davallianum* (Sm.) Zand., see treatment of *Microbryum*]

[subsp. *commutata* (Limpr.) Podp. Eur As5 Afr1 ≡ *Microbryum davallianum* var. *commutatum* (Limpr.) Zand., see treatment of *Microbryum*]

var. *angustifolia* (Par.) Podp. Eur

[var. *charcotii* (Card.) Sav. & Smirn., Novosti Sist. Niz. Rast. 1965: 258, 1965 = *Pottia heimei* (Hedw.) Fürnr. in Hampe var. *heimei fide* Matteri, Brit. Antarct. Surv. Bull. 46: 141, 1977]

[var. *conica* (Schleich. ex Schwaegr.) Podp. Eur As5 Afr1 ≡ *Pottia starckeana* subsp. *conica* (Schleich. ex Schwaegr.) Chamberl. *fide* Chamberlain, Roy. Bot. Gard. Edinburg Not. 29: 403, 1969 ≡ *Microbryum davallianum* var. *conicum* (Schleich. ex Schwaegr.) Zand., see treatment of *Microbryum*]

var. *microphylla* (Warnst.) Podp. Eur

var. *oblonga* (B.&S. in BSG) Podp. Eur

[fo. *conica* (Schleich. ex Schwaegr.) Pilous, Preslia 10: 130, 1969 Eur As5 Afr1 *nom. inval. basion. non cit.* ≡ *Microbryum davallianum* var. *conicum* (Schleich. ex Schwaegr.) Zand., see treatment of *Microbryum*]

[*Pottia davallii* Wareham in Grout, Moss Fl. N. Am. 1(4): 201, 1929 *nom. illeg. incl. typon. Gymnostomum davallianum* Sm. ≡ *Pottia davalliana* (Sm.) C. Jens. (= *Pottia starckeana* subsp. *minutulum* (Schleich. ex Schwaegr.) Chamberl. *fide* Chamberlain, Roy. Bot. Gard. Edinburg Not. 29: 403, 1969) ≡ *Microbryum davallianum* (Sm.) Zand., see treatment of *Microbryum*]

[*Pottia denticulata* Dix. & P. Varde *hom. illeg.* As3 = *Chenia leptophylla* (C. Müll.) Zand., see treatment of *Chenia* and vide Arts & Sollman, Lindbergia 17: 22, 1991]

[*Pottia drummondii* (Wils.) Willis Austr1 ≡ *Tortula willisiana* Zand. *nom. nov.*, see treatment of *Tortula*]

[var. *obscura* Willis Austr1 ≡ *Tortula willisiana* var. *obscura* (Willis) Zand., see treatment of *Tortula*]

Pottia flavipes Mont. Am6

[*Pottia fosbergii* Bartr. Am1 Am2 ≡ *Microbryum starckeanum* var. *fosbergii* (Bartr.) Zand., see treatment of *Microbryum*]

Pottia globosa Catcheside, Mosses S. Austr. 124, 1980 Austr1

Pottia groenlandica (Kindb.) Par. Am1

[*Pottia heimii* (Hedw.) Fürnr. in Hampe Eur As1 As2 Am1 Am6 Austr1 Austr2 Ant ≡ *Hennediella heimii* (Hedw.) Zand., see treatment of *Hennediella*]

[var. *alpina* Amann Eur ≡ *Hennediella heimii* var. *alpina* (Amann) Zand., see treatment of *Hennediella*]

[var. *arctica* Lindb. Eur As1 Am1 ≡ *Hennediella heimii* var. *arctica* (Lindb.) Zand., see treatment of *Hennediella*]

[var. *brachyphylla* Warnst. Eur ≡ *Hennediella heimii* var. *brachyphylla* (Warnst.) Zand., see treatment of *Hennediella*]

[var. *brevicuspis* Warnst. Eur ≡ *Hennediella heimii* var. *brevicuspis* (Warnst.) Zand., see treatment of *Hennediella*]

var. *brevinervis* Sav.-Ljubits., Bot. Mater. Otd. Sporov. Rast. Bot. Inst. Korarova Akad. Nauk SSSR 16: 193, 1963 Ant

[var. *breviseta* Warnst. Am1 ≡ *Hennediella heimii* var. *breviseta* (Warnst.) Zand., see treatment of *Hennediella*]

[var. *drummondii* Warnst. Am1 ≡ *Hennediella heimii* var. *drummondii* (Warnst.) Zand., see treatment of *Hennediella*]

[var. *eurystoma* Card. & Broth. Am6 ≡ *Hennediella heimii* var. *eurystoma* (Card. & Broth.) Zand., see treatment of *Hennediella*]

[var. *guessfeldtii* (Schlieph.) Warnst. Eur ≡ *Hennediella heimii* var. *guessfeldtii* (Schlieph.) Zand., see treatment of *Hennediella*]

[var. *krausei* (Warnst.) Warnst. Eur = *Pottia heimii* (Hedw.) Hampe *fide* Podpěra, Consp. Musc. Eur. 1954 ≡ *Hennediella heimii* (Hedw.) Zand.]

[var. *lanceolata* Warnst. Eur Am1 ≡ *Hennediella heimii* var. *lanceolata* (Warnst.) Zand., see treatment of *Hennediella*]

[var. *magellanica* Warnst. Am6 ≡ *Hennediella heimii* var. *magellanica* (Warnst.) Zand., see treatment of *Hennediella*]

[var. *maxima* Card. Am6 ≡ *Hennediella heimii* var. *maxima* (Card.) Zand., see treatment of *Hennediella*]

[var. *spegazzinii* (C. Müll.) Warnst. Am6 ≡ *Hennediella heimii* var. *spegazzinii* (C. Müll.) Zand., see treatment of *Hennediella*]

[var. *thaxteri* Card. & Thér. Am6 ≡ *Hennediella heimii* var. *thaxteri* (Card. & Thér.) Zand., see treatment of *Hennediella*]

fo. *cylindrica* (BSG) Podp., Consp. Musc. Eur. 232, 1954 (*Pottia heimii* var.) Eur

fo. *longiseta* (Arnell) Podp., Consp. Musc. Eur. 232, 1954 (*Pottia heimii* var.) Eur

fo. *ryanii* (Philib.) Warnst., Hedwigia 58: 101, 1916 Eur = *Pottia heimii* var. *arctica* Lindb. cf. Par., Ind. Bryol. ed. 2, 4: 90, 1905]

fo. *tschuctschica* Warnst., Hedwigia 58: 102, 1916 (*Pottia tschuctschica* C. Müll. syn. *inval.*) As1

Pottia humillima (Ångstr.) Par. Am5

[*Pottia inconspicua* (Griff.) C. Müll. ex Chen, Hedwigia 80: 63, 1941 syn. *inval.* As2 As3 As4 ≡ *Reimersia inconspicua* (Griff.) Chen.]

[*Pottia intermedia* (Turn.) Fürnr. Eur As2 Afr1 Am1 ≡ *Tortula modica* Zand. *nom. nov.*, see treatment of *Tortula*]

[var. *corsa* Fleisch. & Warnst. Eur ≡ *Tortula modica* var. *corsa* (Fleisch. & Warnst.) Zand., see treatment of *Tortula*]

[var. *gymnandra* Schiffn. Eur ≡ *Tortula modica* var. *gymnandra* (Schiffn.) Zand., see treatment of *Tortula*]

[var. *gymnogyna* Schiffn. Eur ≡ *Tortula modica* var. *gymnogyna* (Schiffn.) Zand., see treatment of *Tortula*]

[var. *revoluta* Schiffn. Eur ≡ *Tortula modica* var. *revoluta* (Schiffn.) Zand., see treatment of *Tortula*]

[var. *stenocarpa* Velen. Eur ≡ *Tortula modica* var. *stenocarpa* (Velen.) Zand., see treatment of *Tortula*]

[var. *tenuis* Vent. Eur ≡ *Tortula modica* var. *tenuis* (Vent.) Zand., see treatment of *Tortula*]

fo. *cylindrica* Röll, Hedwigia 56: 158, 1915 Eur

fo. *robusta* Podp., Vslyedky II. V. Kl. Přírod. Prostějov 8: 32, 1906 Eur

[*Pottia lanceolata* (Hedw.) C. Müll. Eur As2 As5 Afr1 Am1 ≡ *Tortula lanceola* Zand. *nom. nov.*, see treatment of *Tortula*]

[subsp. *leucodonta* (Schimp.) Boul. Eur = *Pottia lanceolata* var. *angustata* (B.&S. in BSG) C. Müll. *fide* Husnot, Musc. Gall. 1(3): 76, 1885 ≡ *Tortula lanceola* var. *angustata* (B.&S. in BSG) Zand., see treatment of *Tortula*]

[var. *albidens* Corb. Eur ≡ *Tortula lanceola* var. *albidens* (Corb.) Zand., see treatment of *Tortula*]

[var. *angustata* (B.&S. in BSG) C. Müll. Eur As5 Afr1 ≡ *Tortula lanceola* var. *angustata* (B.&S. in BSG) Zand., see treatment of *Tortula*]

[var. *lejolisii* Corb. Eur Afr1 ≡ *Tortula lanceola* var. *lejolisii* (Corb.) Zand., see treatment of *Tortula*]

var. *leucocephala* Schimp. in Paques *nom. nud.? fide* Index Muscorum 4: 195, 1967 Eur

[var. *leucodonta* Schimp. Eur ≡ *Tortula lanceola* var. *leucodonta* (Schimp.) Zand., see treatment of *Tortula*]

[var. *macrophylla* Warnst. Eur ≡ *Tortula lanceola* var. *macrophylla* (Warnst.) Zand., see treatment of *Tortula*]

[var. *microphylla* Warnst. Eur ≡ *Tortula lanceola* var. *microphylla* (Warnst.) Zand., see treatment of *Tortula*]

[var. *mucronata* Amann Eur ≡ *Tortula lanceola* var. *mucronata* (Amann) Zand., see treatment of *Tortula*]

[var. *ovalifolia* Warnst. Eur ≡ *Tortula lanceola* var. *ovalifolia* (Warnst.) Zand., see treatment of *Tortula*]

[var. *papillosa* Corb. Eur ≡ *Tortula lanceola* var. *papillosa* (Corb.) Zand., see treatment of *Tortula*]

[var. *rigidior* (Schwaegr.) Wijk & Marg. Eur ≡ *Tortula lanceola* var. *rigidior* (Schwaegr.) Zand., see treatment of *Tortula*]

var. *trabeculata* Podp. Eur

fo. *breviseta* (Röll) Podp., Consp. Musc. Eur. 229, 1954 (*Pottia lanceolata* var.) Eur

fo. *conica* Röll, Hedwigia 56: 159, 1915 Eur

fo. *major* Röll, Hedwigia 56: 158, 1915 Eur

[fo. *minor* Geh. in Warnst., Hedwigia 58: 134, 1916 syn. *inval.* Eur = *Pottia lanceolata* var. *microphylla* Warnst. *fide* Warnstorf 1916]

[fo. *ovata* Röll, Hedwigia 56(1–3): 159, 1915 Eur ≡ *Pottia lanceolata* subfo. *ovata* (Röll) Podp.]

fo. *pilifera* Röll, Hedwigia 56: 159, 1915 Eur

fo. *stenocarpa* Podp., Vy'sledky VII. Sborn. Klubu Přír. v Brno 5: 13, 1923 Eur

fo. *tortuosa* Röll, Hedwigia 56: 158, 1915 Eur

subfo. *ovata* (Röll) Podp., Consp. Musc. Eur. 229, 1956

subfo. *conica* (Röll) Podp., Consp. Musc. Eur. 229, 1954 (*Pottia lanceolata* fo.) *nom. inval.* Eur

[*Pottia latifolia* (Schwaegr.) C. Müll. ≡ *Stegonia latifolia* (Schwaegr.) Vent.]

[fo. *pilifera* (Brid.) Mönk., Laubm. Eur. 333, 1927 Eur As1 Am1 ≡ *Stegonia latifolia* var. *pilifera* (Brid.) Broth.]

Pottia latzii Catcheside, Mosses S. Austr. 126, 1980 Austr1

Pottia ligularifolia C. Müll., Hedwigia 34: 123, 1895 Am5

Pottia longifolia R. Br. ter Austr2

[*Pottia longirostris* Hampe ex C. Müll. Afr2 ≡ *Hennediella longirostris* (Hampe ex C. Müll.) Zand., see treatment of *Hennediella*]

[*Pottia macowaniana* C. Müll. Afr4 [see Magill, Fl. S. Afr. I. Mosses 1: 207, 1981 (1982)] = *Pottia starckeana* subsp. *conica* (Schleich. ex Schwaegr.) Chamberl., Roy. Bot. Gard. Edinburg Not. 29: 403, 1969 ≡ *Microbryum davallianum* var. *conicum* (Schleich. ex Schwaegr.) Zand., see treatment of *Microbryum*]

[*Pottia macrocarpa* Schimp. Am6 ≡ *Funaria macrocarpa* (Schimp.) Zand. (Funariaceae) see Excluded Taxa]

[*Pottia macrophylla* (R. Br. ter) Sainsb. Am6 Austr2 ≡ *Hennediella macrophylla* (R. Br. ter) Par., see treatment of *Hennediella*]

[*Pottia maritima* (R. Br. ter) Broth. Austr2 ≡ *Tortula maritima* (R. Br. ter) Zand, see treatment of *Dendia*]

[*Pottia minor* (C. Müll.) Wijk & Marg. As3 Afr2 ≡ *Tortula minor* (C. Müll.) Zand., see treatment of *Tortula*]

[var. *elatior* (C. Müll.) Wijk & Marg. As3 Afr2 ≡ *Tortula minor* var. *elatior* (C. Müll.) Zand., see treatment of *Tortula*]

[*Pottia mirabilis* Broth. & Par. Afr2 ≡ *Physcomitrium mirabile* (Broth. & Par.) Zand. (Funariaceae) see Excluded Taxa]

[*Pottia mutica* Vent. in De Not. Eur As5 Afr1 = *Pottia starckeana* var. *brachyoda* (BSG) C. Müll. *fide* Mönkemeyer, Laub. Eur. 327, 1927 ≡ *Microbryum starckeanum* var. *brachyodus* (BSG) Zand., see treatment of *Microbryum*]

var. *brachyphylla* Warnst. Austr1

var. *gymnostoma* Corb. Afr1

var. *leucodonta* Corb. Afr1

var. *parvifolia* Warnst. Eur

[var. *salina* (Warnst.) Podp. Eur = *Pottia starckeana* subsp. *conica* (Schleich. ex Schwaegr.) Chamberl., Roy. Bot. Gard. Edinburg Not. 29: 403, 1969 ≡ *Microbryum davallianum* var. *conicum* (Schleich. ex Schwaegr.) Zand., see treatment of *Microbryum*]

Pottia namaquensis Magill, Fl. S. Afr. I. Mosses 1: 206, 1981 [1982] Afr4

Pottia naumanii C. Müll. Afr4

Pottia neocaledonia Thér. *nom. inval.* Oc

[*Pottia nevadensis* Card. & Thér. Am1 ≡ *Tortula nevadensis* (Card. & Thér.) Zand., see treatment of *Tortula*]

Pottia notarisii Schimp. Eur

var. *cyclopica* Zodda, Malphighia 24: 274, 1911 Eur

Pottia obliqua R. Br. ter Austr2

[*Pottia obtusifolia* (C. Müll.) C. Müll. Am5 ≡ *Pottia gardneriana* C. Müll., Syn. 2(9–10): 754, 772, 1851 ≡ *Hyophila obtusifolia* (C. Müll.) Jaeg.]

[*Pottia oedipodioides* C. Müll. Afr4 ≡ *Hennediella oedipodioides* (C. Müll.) Zand., see treatment of *Hennediella*]

[*Pottia pallida* Lindb. Eur Afr1 As5 ≡ *Tortula pallida* (Lindb.) Zand., see treatment of *Tortula*]

[var. *longicuspis* Warnst. Eur As5 Afr1 ≡ *Tortula pallida* var. *longicuspis* (Warnst.) Zand., see treatment of *Tortula*]

Pottia pellata (Schimp.) Broth. Am2

Pottia propagulifera Herz. Eur As2

[*Pottia pusilla* Warnst. Afr2 *hom. illeg.* = *Pottia minor* (C. Müll.) Wijk & Marg. ≡ *Tortula minor* (C. Müll.) Zand., see treatment of *Tortula*]

[*Pottia randii* Kenn. Eur Am1 ≡ *Tortula randii* (Kenn.) Zand., see treatment of *Tortula*]

[*Pottia recta* (Sm.) Mitt. Eur Afr1 ≡ *Microbryum rectum* (With.) Zand., see treatment of *Microbryum*]

Pottia recurvifolia Warnst. Afr1

Pottia sampaiana Mach. Eur

Pottia scabrifolia Bartr. Austr1

Pottia serrata R. Br. ter Austr2

[*Pottia splachnobryoides* C. Müll. As2 = *Chenia leptophylla* (C. Müll.) Zand., see treatment of *Chenia* and vide Arts & Sollman, Lindbergia 17: 22, 1991]

[*Pottia splachnoides* (Hornsch.) Broth. Afr4 ≡ *Tortula splachnoides* (Hornsch.) Zand., see treatment of *Tortula*]

[*Pottia starckeana* (Hedw.) C. Müll. Eur As5 Afr1 Am1 Am2 Austr1 Austr2 ≡ *Microbryum starckeanum* (Hedw.) Zand., see treatment of *Microbryum*]

[subsp. *conica* (Schleich. ex Schwaegr.) Chamberl., Roy. Bot. Gard. Edinburg Not. 29: 403, 1969 Eur Afr1 As5 Am2 ≡ *Microbryum davallianum* var. *conicum* (Schleich. ex Schwaegr.) Zand., see treatment of *Microbryum*]

[subsp. *minutula* (Schleich. ex Schwaegr.) Bouvet, Bull. Soc. Études Sci. Angers 26: 87, 1896 (*Gymnostomum*) Eur Afr1 Austr1 = *Microbryum davallianum* (Hedw.) Zand. var. *davallianum*, see treatment of *Microbryum*]

[subsp. *minutula* (Schleich. ex Schwaegr.) Chamberl., Roy. Bot. Gard. Edinburg Not. 29: 403, 1969 (*Gymnostomum*) *hom. illeg.* Eur Afr1 Austr1 ≡ *Pottia starckeana* subsp. *minutula* (Schleich. ex Schwaegr.) Bouvet = *Microbryum davallianum* (Hedw.) Zand. var. *davallianum*, see treatment of *Microbryum*

[var. *brachyodus* (BSG) C. Müll. Eur As5 Afr1 ≡ *Microbryum starckeanum* var. *brachyodus* (BSG) Zand., see treatment of *Microbryum*]

[var. *brevidens* Latz. Eur ≡ *Microbryum starckeanum* var. *brevidens* (Latz.) Zand. (*Pottia starckeana* var.), see treatment of *Microbryum*]

[var. *fosbergii* (Bartr.) Zand., Novon 3: 92, 1993 (*Pottia*) ≡ *Microbryum starckeanum* var. *fosbergii* (Bartr.) Zand., see treatment of *Microbryum*]

[var. *leiostoma* Corb. in Corb. & Pitard Afr1 ≡ *Microbryum starckeanum* var. *leiostoma* (Corb in Corb. & Pitard) Zand., see treatment of *Microbryum*]

[var. *subgymnostoma* (De Not.) Grav. Eur ≡ *Microbryum starckeanum* var. *subgymnostoma* (De Not.) Zand., see treatment of *Microbryum*]

[var. *submutica* Latz. Eur ≡ *Microbryum starckeanum* var. *submutica* (Latz.) Zand., see treatment of *Microbryum*]

[fo. *brevifolia* Limpr., Laubm. Deutsch. 1: 536, 1888 Eur ≡ *Microbryum starckeanum* fo. *brevifolium* (Limpr.) Zand., see treatment of *Microbryum*]

[fo. *dextrorsa* Limpr., Laubm. Deutsch. 1: 536, 1888 Eur ≡ *Microbryum starckeanum* fo. *dextrorsum* (Limpr.) Zand., see treatment of *Microbryum*]

[fo. *microphylla* Warnst., Hedwigia 58: 144, 1917 Eur ≡ *Microbryum starckeanum* fo. *microphyllum* (Warnst.) Zand., see treatment of *Microbryum*]

[fo. *minima* Latz., Bot. Centralbl. Beih. 2: 483, 1931 Eur = *Pottia starckeana* (Hedw.) C. Müll. ≡ *Microbryum starckeanum* (Hedw.) Zand., see treatment of *Microbryum*]

Pottia stenocarpa P. Varde Afr2

[*Pottia stevensii* R. Br. ter Austr2 ≡ *Tortula sainsburyana* Zand. *nom. nov.*, see treatment of *Tortula*]

[*Pottia subplanomarginata* Dix. Afr4 ≡ *Microbryum*

subplanomarginatum (Dix.) Zand., see treatment of *Microbryum*]
Pottia tasmanica Broth. Austr1
[*Pottia texana* Wareh. in Grout Am1 Am2 = *Pottia starckeana* subsp. *conica* (Schleich. ex Schwaegr.) Chamberl. *fide* Chamberl., Roy. Bot. Gard. Edinburg Not. 29: 403, 1969 ≡ *Microbryum davallianum* var. *conicum* (Schleich. ex Schwaegr.) Zand., see treatment of *Microbryum*]
Pottia thraustophylla (Ångstr.) Par. Am5
[*Pottia truncata* (Hedw.) B.&S. Eur As1 As2 As3 Afr1 Am1 Am6 Austr1 Austr2 ≡ *Tortula truncata* (Hedw.) Mitt. in Godm., see treatment of *Tortula*]
[subsp. *intermedia* (Turn.) Bouvet, Bull. Soc. Études Sci. Angers 26: 86, 1896 (*Gymnostomum*) Eur As2 Afr1 ≡ *Pottia intermedia* (Turn.) Fürnr. ≡ *Tortula modica* Zand. *nom. nov.*, see treatment of *Tortula*]
[subsp. *littoralis* (Mitt.) Giac. Eur Afr1 ≡ *Tortula truncata* var. *littoralis* (Mitt.) Zand., see treatment of *Tortula*]
[var. *brevirostris* (Lisa) De Not. Eur ≡ *Tortula truncata* var. *brevirostris* (Lisa) Zand., see treatment of *Tortula*]
[var. *illyrica* (Latz.) Podp. Eur ≡ *Tortula truncata* var. *illyrica* (Latz.) Zand., see treatment of *Tortula*]
[var. *littoralis* (Mitt.) Warnst. Eur Afr1 ≡ *Tortula truncata* var. *littoralis* (Mitt.) Zand., see treatment of *Tortula*]
[var. *major* (Web. & Mohr) B.&S. ≡ *Pottia intermedia* (Turn.) Fürnr. ≡ *Tortula modica* Zand. *nom. nov.*, see treatment of *Tortula*]
[var. *minutissima* Warnst. Eur ≡ *Tortula truncata* var. *minutissima* (Warnst.) Zand., see treatment of *Tortula*]
fo. *angustata* Warnst., Hedwigia 58: 116, 1916 Eur
fo. *angustifolia* Wanst., Hedwigia 58: 122, 1916 Eur
fo. *brevicuspis* Warnst., Hedwigia 58: 119, 1916 Eur
fo. *compacta* Röll, Hedwigia 56(1–3): 158, 1915 Eur
fo. *elongata* Hammerschm., Mitt. Bayer. Bot. Ges. 2: 184, 1908 Eur
fo. *latifolia* Warnst., Hedwigia 58: 120, 1916 Eur
fo. *longicuspis* Warnst., Hedwigia 58: 118, 1916 Eur
fo. *longifolia* Warnst., Hedwigia 58: 121, 1916 Am2
fo. *major* Röll, Hedwigia 56(1–3): 158, 1915 *nom. illeg. non Pottia truncata* var. *major* (Web. & Mohr) B.&S. Eur
fo. *pusilla* Röll, Hedwigia 56(1–3): 158, 1915 Eur
fo. *serrulata* Röll, Hedwigia 56(1–3): 158, 1915 Eur
fo. *spathulata* (Warnst.) Warnst., Hedwigia 58: 115, 1916 (*Pottia truncatula* var.) Eur
[*Pottia truncatula* (With.) Bus. = *Pottia truncata* (Hedw.) BSG ≡ *Tortula truncata* (Hedw.) Mitt. in Godm.]
var. *macrocarpa* Bus. Eur
Pottia uleana Par. Am5
[*Pottia verrucosa* Warnst. ex Rehm. Afr4 = *Pottia starckeana* var. *conica* (Schleich. ex Schwaegr.) Chamberl., Bot. Gard. Edinburg Not. 29: 403, 1969 ≡ *Microbryum davallianum* var. *conicum* (Schleich. ex Schwaegr.) Zand., see treatment of *Microbryum*]
[*Pottia viridifolia* Mitt. Eur = *Pottia crinita* Wils. ex BSG *fide* Corley et al., J. Bryol. 11: 620, 1981 [1982] ≡ *Pottia wilsonii* var. *crinita* (Wils. ex BSG) Warnst. ≡ *Tortula wilsonii* var. *crinita* (Wils. ex B.&S.) Zand., see treatment of *Tortula*]
[var. *flavescens* (Corb.) Podp. Eur ≡ *Tortula wilsonii* var. *crinita* (Wils. ex B.&S.) Zand.]
[*Pottia watsonii* R. S. Chopra, Taxon, Indian Mosses 155, 1975 (*nom. nov.* for *Pottia denticulata* Dix. & P. Varde *hom. illeg.*) = *Chenia leptophylla* (C. Müll.) Zand., see treatment of *Chenia*]
[*Pottia willisiana* Sainsb., Rev. Bryol. Lichénol. 25: 237, 1956 Austr1 ≡ *Weissia willisiana* (Sainsb.) Catcheside, Mosses S. Austr. 194, 1980]
[*Pottia wilsonii* (Hook.) B.&S. Eur As1 As5 Afr1 Am1 ≡ *Tortula wilsonii* (Hook.) Zand., see treatment of *Tortula*]
[subsp. *asperula* (Mitt.) Kindb. Eur ≡ *Tortula wilsonii* var. *asperula* (Mitt.) Zand., see treatment of *Pottia*]
[var. *crinita* (Wils. ex B.&S.) Warnst. Eur As5 Afr1 ≡ *Tortula wilsonii* var. *crinita* (Wils. ex B.&S.) Zand., see treatment of *Pottia*]
[var. *mucronifolia* (Bruch) Warnst. As1 As5 ≡ *Tortula wilsonii* var. *mucronifolia* (Bruch in F. Müll.) Zand., see treatment of *Pottia*]
fo. *pilifera* (Limpr.) Podp., Consp. Musc. Eur. 228, 1954 (*Pottia wilsonii* var.) Eur
[*Pottia zealandiae* (R. Br. ter) Par. Austr2 ≡ *Microbryum zeelandiae* (R. Br. ter) Zand., see treatment of *Microbryum*]
[*Pottia zealandioides* Dix. & Sainsb. Austr2 = *Pottia starckeana* var. *conica* (Schleich. ex Schwaegr.) Chamberl., Roy. Bot. Gard. Edinburg Not. 29: 403, 1969 ≡ *Microbryum davallianum* var. *conicum* (Schleich. ex Schwaegr.) Zand., see treatment of *Microbryum*]

[**PRIONIDIUM** Hilp. = *Didymodon* Hedw. *fide* Saito, J. Hattori Bot. Lab. 39: 500, 1975]
[*Prionidium erosodenticulatum* (C. Müll.) Chen As2 ≡ *Didymodon erosodenticulatus* (C. Müll.) Saito *fide* Saito, J. Hattori Bot. Lab. 39: 504, 1975]
[*Prionidium setschwanicum* (Broth.) Hilp. As2 = *Didymodon erosodenticulatus* (C. Müll.) Saito *fide* Saito, J. Hattori Bot. Lab. 39: 504, 1975]

[**PSEUDALOINA** Delgad., Bryologist 85: 401, 1982 (1983) = *Crossidium* Jur., see treatment of *Crossidium*]
[*Pseudaloina woodii* Delgad., Bryologist 85: 401, 1982 [1983] As5 ≡ *Crossidium woodii* (Delgad.) Zand., see treatment of *Crossidium*]

PSEUDOCROSSIDIUM Williams
Pseudocrossidium apiculatum Williams Am4 Am6
[*Pseudocrossidium aureum* (Bartr.) Zand., Phytologia 44: 207, 1979 (*Barbula*) Am1 Am2 = *Pseudocrossidium crinitum* (Schultz) Zand., see treatment of *Pseudocrossidium*]
Pseudocrossidium austrorevolutum (Besch. in Britt.) Zand. (*Barbula*), see treatment of *Pseudocrossidium* Am4
Pseudocrossidium carinatum (Gill. ex Grev.) Zand. (*Tortula*), see treatment of *Pseudocrossidium* Am6
Pseudocrossidium chilense Williams Am6
Pseudocrossidium crinitum (Schultz) Zand., see treatment of *Pseudocrossidium* (*Barbula*) Afr2 Afr4 Am1 Am2 Am6 Austr1 Austr2
Pseudocrossidium elatum (Williams) Delg., Bryologist 78: 278, 1975 (*Crossidium*) Am4
Pseudocrossidium excavatum (Mitt.) Williams Am4
Pseudocrossidium hornschuchianum (Schultz) Zand., Phytologia 44: 205, 1979 (*Barbula*) Eur As5 Afr1 Afr4 Am1 Austr1
Pseudocrossidium leucocalyx (Mont.) Thér. Am6
[*Pseudocrossidium obtusulum* (Lindb.) Crum & Anderson, Bryologist 92: 533, 1989 Am1 ≡ *Pseudocrossidium revolutum* var. *obtusulum* (Lindb.) Tan, Zand. & T. Tayl., see treatment of *Pseudocrossidium*]
Pseudocrossidium mendozense (Mitt.) Zand. (*Tortula*), see treatment of *Pseudocrossidium* Am6
Pseudocrossidium pachygastrellum (Herz.) Broth. Am4
Pseudocrossidium perrevolutum (C. Müll.) Zand. (*Barbula*), see treatment of *Pseudocrossidium* Am6
var. *acutifolium* (C. Müll.) Zand. (*Barbula perrevoluta* var.), see

treatment of *Pseudocrossidium* Am6
var. *linearifolium* (C. Müll.) Zand. (*Barbula perrevoluta* var.), see treatment of *Pseudocrossidium* Am6
Pseudocrossidium porphyreoneurum (C. Müll. ex Vent.) Zand. (*Barbula*), see treatment of *Pseudocrossidium* Afr2 Afr4
Pseudocrossidium replicatum (Tayl.) Zand., Phytologia 44: 206, 1979 (*Barbula*) Afr2 Afr4 Am1 Am2 Am4 Am6
var. *longicuspidatum* Zand., Willdenowia 16: 258, 1986 Am4
Pseudocrossidium revolutum (Brid. in Schrad.) Zand., Phytologia 44: 204, 1979 (*Barbula*) Eur As5 Afr1 Afr4
var. *obtusulum* (Lindb.) Tan, Zand. & T. Tayl., Lindbergia 7: 41, 1981 (*Barbula hornschuchiana* var.) Eur Am1
Psuedocrossidium steerei Churchill, Bryologist 93: 353, 1990 Am4

PSEUDOSYMBLEPHARIS Broth.
Pseudosymblepharis angustata (Mitt.) Hilp. As2 As3 As4
Pseudosymblepharis angustifolia (Crum & Steere) Zand. (*Trichostomum*), see treatment of *Pseudosymblepharis* Am3
[*Pseudosymblepharis bartramii* Thér. ex Bartr. Am2 = *Pseudosymblepharis schimperiana* (Par.) Crum, see treatment of *Pseudosymblepharis*]
Pseudosymblepharis cavernarum (Broth.) Zand. (*Trichostomum*), see treatment of *Pseudosymblepharis* Am5
Pseudosymblepharis circinnatula (Broth. in Völtzk) Zand. (*Trichostomum*), see treatment of *Pseudosymblepharis* Afr3
Pseudosymblepharis duriuscula (Mitt.) Chen As2 As3
Pseudosymblepharis indica Dix. & P. Varde As3
Psuedosymblepharis khasiana (Mitt.) Zand. (*Tortula*), see treatment of *Pseudosymblepharis* As3
[*Pseudosymblepharis mauiensis* (C. Müll.) Broth. As4 Oc = *Pseudosymblepharis angustata* (Mitt.) Hilp. *fide* Norris & Koponen, Acta Bot. Fenn. 137: 94, 1987]
[*Pseudosymblepharis pallidens* Dix. As3 ≡ *Trichostomum pallidens* (Dix.) Zand., see treatment of *Trichostomum*]
[*Pseudosymblepharis papillosula* (Card. & Thér.) Broth. As2 As3 = *Pseudosymblepharis angustata* (Mitt.) Hilp. *fide* Saito, J. Hattori Bot. Lab. 39: 439, 1975]
var. *robusta* Tix., Rev. Bryol. Lichénol. 34: 134, 1966 As3
Pseudosymblepharis perlongifolia (Fröhl.) Zand. (*Trichostomum*), see treatment of *Trichostomum* As4
Pseudosymblepharis schimperiana (Par.) Crum Am2 Am3 Am4 Am5
[*Pseudosymblepharis socotrana* (Mitt.) Thér. Afr2 ≡ *Weissia artocosana* Zand. *nom. nov.*, see treatment of *Weissia*]
[*Pseudosymblepharis subduriuscula* (C. Müll.) Chen As2 As3 As4 ≡ *Pseudosymblepharis angustata* (Mitt.) Hilp. *fide* Norris & Koponen, Acta Bot. Fenn. 137: 94, 1987]
Pseudosymblepharis syrrhopodontoides (Dix.) Zand., see treatment of *Pseudosymblepharis* Afr2
Pseudosymblepharis verrucosa (Broth. & Par.) Zand. (*Trichostomum*), see treatment of *Pseudosymblepharis* Oc

[**PSEUDOTIMMIELLA** Biz., Cryptogamie Bryol. Lichénol. 1: 425, 1980 = *Diphyscium* Mohr (Diphysciaceae) see Excluded Taxa]
[*Pseudotimmiella pocsii* Biz., Cryptogamie Bryol. Lichénol. 1: 425, 1980 Afr2 ≡ *Diphyscium poscii* (Biz.) Zand. (Diphysciaceae) see Excluded Taxa]

PTERYGONEURUM Jur.
[*Pterygoneurum arcticum* Steere, Bryologist 62: 217, 1960 Am1 = *Pterygoneurum lamellatum* (Lindb.) Jur. *fide* Steere, 1976]
Pterygoneurum californicum Crum, Madroño 19: 92, 1967 Am1
[*Pterygoneurum cavifolium* Jur. ≡ *Pterygoneurum ovatum* (Hedw.) Dix.]
var. *muticum* Schiffn. As5
[fo. *crossidioides* Sapegin, Bot. Jahrb. Syst. 46(Beibl. 105): 13, 1911 Eur ≡ *Pterygoneurum ovatum* fo. *crossidioides* (Sapegin) Podp., Consp. Musc. Eur. 235, 1954]
[fo. *longiseta* Amann, Bull. Soc. Vaudoise Sci. Nat. 57: 123, 1929 Eur ≡ *Pterygoneurum ovatum* fo. *longiseta* (Amann) Podp.]
Pterygoneurum chotticum (Trab.) Broth. Afr1
Pterygoneurum crossidioides Frey, Herrnstadt & Kürschner, Nov. Hedw. 50: 239–244, 1990 As5
Pterygoneurum kemsleyi Willis Austr1
Pterygoneurum kozlovii Laz., Vizn. List. Moch. Ukr. S.R.R. 104, 1936 *nom. inval. sin. descr. lat.* Eur ≡ *Pterygoneurum kozlovii* Laz. ex Laz.
Pterygoneurum kozlovii Laz. ex Laz., Bot. Zhurn. Ak. Nauk Uk. RSS 3: 61, 1946 Eur Am1
Pterygoneurum lamellatum (Lindb.) Jur. Eur As1 As2 Am1
Pterygoneurum macleanum Warnst. Afr4
Pterygoneurum medium (Salm.) Broth. As1 As5
Pterygoneurum ovatum (Hedw.) Dix. Eur As1 As5 Afr1 Am1 Am6 Austr1
var. *humile* (Amann) Podp. Eur
var. *perraldieri* (Besch.) Wijk & Marg. Afr1
fo. *crossidioides* (Sapegin) Podp., Consp. Musc. Eur. 235, 1954 Eur
fo. *epilosum* (Brid.) Podp., Consp. Musc. Eur. 235, 1954 (*Gymnostomum ovatum* var.) Eur
fo. *incanum* (Nees & Hornsch.) Podp., Consp. Musc. Eur. 235, 1954 (*Gymnostomum ovatum* var.) Eur As5 Afr1 Am1
fo. *longisetum* (Amann) Podp., Consp. Musc. Eur. 235, 1954 (*Pterygoneurum cavifolium* fo.) Eur
fo. *majus* Röll, Hedwigia 56: 160, 1915 Eur
fo. *minimum* Röll, Hedwigia 56: 160, 1915 Eur
fo. *polycarpum* (Györf.) Podp., Consp. Musc. Eur. 235, 1954 (*Pterygoneurum cavifolium* var.) Eur
[*Pterygoneurum rosei* Williams Am4 ≡ *Aloina rosei* (Williams) Delgadillo, Bryologist 76: 273, 1973]
[*Pterygoneurum smardaeanum* Vanek Eur (= *Phascum cuspidatum* Hedw. but see discussion by Corley et al., J. Bryol. 11: 647, 1981 [1982]) = *Pterygoneurum koslovii* Laz. *fide* Abramova et al., Notyl. Syst. Plant *non* Vascular. Acad. Sci. URSS Inst. Bot. V. L. Komarovii 10: 314, 1973]
Pterygoneurum subsessile (Brid.) Jur. Eur As1 Afr1 Am1 Am6
var. *henrici* (Rau) Wareh. in Grout Am1
var. *kieneri* Hab. Am1
fo. *robustum* Röll, Hedwigia 56: 160, 1915 Eur
fo. *tenellum* Röll, Hedwigia 56: 160, 1915 Eur
Pterygoneurum subsessile (Brid.) Jur. ×*P. ovatum* (Hedw.) Dix. *fide* Podp., Consp. Musc. Eur. 235, 1954 As1
fo. *apterum* Lazar., Opr. Listv. Mchov Ukr. 181, 1955 *nom. inval. sin. descr. lat.* Eru

QUAESTICULA Zand. (see treatment of *Quaesticula*)
Quaesticula navicularis (Mitt.) Zand. (*Weissia*), see treatment of *Quaesticula* Am2 Am3

[**RECHINGERELLA** Fröhl., Ann. Naturh. Mus. Wien 66: 36, 1962 (1963) *hom. illeg. non* Petrak. = *Weissia* Hedw. *fide* H. Robinson, Bryologist 69: 112, 1966]
[*Rechingerella macedonica* Fröhl., Ann. Naturh. Mus. Wien 66: 36, 1962 (1963) = *Weissia controversa* Hedw. *fide* H. Robinson, Bryologist 69: 112, 1966]

REIMERSIA Chen
Reimersia inconspicua (Griff.) Chen As2 As3 As4

[**RHAMPHIDIUM** Mitt. referred to Ditrichaceae in Excluded Taxa]

RHEXOPHYLLUM Herz.
[*Rhexophyllum laciniatum* Herz. = *Rhexophyllum subnigrum* (Mitt.) Thér. ex Hilp. *fide* Zander in Sharp et al., Mo. Fl. Mexico]
Rhexophyllum subnigrum (Mitt.) Thér. ex Hilp. Am1 Am2 Am4

SAGENOTORTULA Zand.
Sagenotortula quitoensis (Tayl. in Hook.) Zand., Phytologia 65: 430, 1989 (*Tortula*) Am2 Am4 Am6

[**SAITOA** Zand. ≡ *Saitoella* Menzel, J. Hattori Bot. Lab. 71: 239, 1992]
[*Saitoa peruviana* (Williams) Zand., Phytologia 65: 431, 1989 (*Globulina*) Am2 Am4 ≡ *Saitoella peruviana* (Williams) Menzel, J. Hattori Bot. Lab. 71: 240, 1992]

SAITOELLA Menzel, J. Hattori Bot. Lab. 71: 239, 1992
Saitoella peruviana (Williams) Menzel, J. Hattori Bot. Lab. 71: 240, 1992 (*Globulina*) Am2 Am4

SARCONEURUM Bryhn
[*Sarconeurum antarcticum* Bryhn, Nyt Mag. Naturvidensk. 40: 205, 1902 = *Sarconeurum glaciale* (C. Müll.) Card. & Bryhn *fide* Robinson in Llano, Antarctic Terrestrial Biology, Antarctic Res. Ser. 20: 171, 1972)
Sarconeurum glaciale (C. Müll.) Card. & Bryhn Am6 Ant
[*Sarconeurum tortelloides* Greene, Sci. Rep. Brit. Antarct. Surv. 64: 38, 1970 Ant ≡ *Tortella tortelloides* (Greene) Robinson *fide* Robinson in Llano, Antarctic Terrestrial Biology, Antarctic Res. Ser. 20: 171, 1972 = *Tortella alpicola* Dix.]

SCOPELOPHILA (Mitt.) Lindb.
Scopelophila cataractae (Mitt.) Broth. Eur As2 As3 As4 As5 Afr2 Am1 Am2 Am4
Scopelophila infericola Hoe, Bryologist 76: 192, 1973 Oc
Scopelophila ligulata (Spruce) Spruce Am1 Am2 Am4 Eur Afr1 As2 As3 As4 As5
[*Scopelophila norrisii* Zand., Bryologist 88: 353, 1986 Am2 ≡ *Weisiopsis norrisii* (Zand.) Zand., see treatment of *Weisiopsis*]

[**SEBILLEA** Biz., Rev. Bryol. Lichénol. 40: 120, 1974 *nom. inval. fide* Crosby et al. 1992, probably referable to Dicranaceae, see Excluded Taxa]
[*Sebillea brasiliensis* Biz., Rev. Bryol. Lichénol. 40: 120, 1974 *nom. inval. fide* Crosby et al. 1992 Am5 = Dicranaceae? see Excluded Taxa]

[**SEMIBARBULA** Herz. ex Hilp. = *Barbula* Hedw.]
[*Semibarbula congoana* (Thér.) Biz., Rev. Bryol. Lichénol. 38: 549, 1971–72 (1973) Afr2 ≡ *Barbula congoana* Thér.]
[*Semibarbula elongata* Hilp. Afr2 ≡ *Barbula zambesiaca* Magill in Magill & Schelpe, Mem. Bot. Surv. S. Afr. 43: 21, 1979 *nom. nov.*]
[*Semibarbula lambarenensis* (P. Varde) Biz., Rev. Bryol. Lichénol. 38: 594, 1971–72 (1973) Afr2 ≡ *Barbula lambarenensis* P. Varde]
[*Semibarbula orientalis* (Web.) Wijk & Marg. As2 As3 As4 Afr2 Afr3 Afr4 Oc = *Barbula indica* (Hook.) Spreng. *fide* Saito, J. Hattori Bot. Lab. 39: 488, 1975]
Semibarbula ranuii Gangulee, Nova Hedw. 8: 149, 1964 As3
[*Semibarbula rufipes* (Schimp. ex Besch.) Hilp. Am2 = *Barbula indica* (Hook.) Spreng. *fide* Zander, Phytologia 44: 185, 1979]
[*Semibarbula scaberrima* (Broth. & Par.) Hilp. As3 ≡ *Barbula scaberrima* Broth. & Par.]
[*Semibarbula stulhmannii* (Broth.) Hilp. Afr2 Afr4 ≡ *Barbula stulhmannii* (Broth.) Broth.]
[*Semibarbula trachyphylla* (Broth.) Laz. in Lazar et al., Biul. Mosk. Obsh. Ispyt. Prir. Otd. Biol. 75(3): 147, 1970 (*Tortula*) *comb. inval. basion. non cit.* As1 ≡ *Tortula trachyphylla* Broth.]

[**SERPOTORTELLA** Dix. referred to Serpotortellaceae, see Excluded Taxa]

[**SPLACHNOBRYUM** C. Müll. referred to Splachnobryaceae, see Excluded Taxa]

STEGONIA Vent.
Stegonia hyalinotrichum (Card. & Thér.) Zand. (*Phascum*), see treatment of *Stegonia* Am1 Am2
Stegonia latifolia (Schwaegr.) Vent. ex Broth. Eur As1 As2 As5 Am1
var. *pilifera* (Brid.) Broth. Eur As1 Am1 [≡ *Stegonia pilifera* (Brid.) Crum & Anderson, Bryologist 92: 533, 1989]
[fo. *pilifera* (Brid.) Nyh., Ill. Moss Fl. Fenno. 98, 1956 *nom. inval. ref. incompl.* (*Coscinodon*) Eur As1 Am1 ≡ *Stegonia pilifera* (Brid.) Crum & Anderson, Bryologist 92: 533, 1989 ≡ *Stegonia latifolia* var. *pilifera* (Brid.) Broth.]
Stegonia mouretii (Corb.) Broth. Afr1
var. *crinita* Corb. ex Jelenc. Afr1
[*Stegonia pilifera* (Brid.) Crum & Anderson, Bryologist 92: 533, 1989 Eur As1 Am1 ≡ *Stegonia latifolia* var. *pilifera* (Brid.) Broth.]

[**STEPHANODICTYON** Dix. = *Trichostomum* Bruch, see treatment of *Trichostomum*]
[*Stephanodictyon angustinerve* Fröhl. As4 ≡ *Trichostomum finukamactum* Zand. *nom. nov.*, see treatment of *Trichostomum*]
[*Stephanodictyon borneense* Dix. As4 (= *Pseudosymblepharis subduriuscula* (C. Müll.) Chen *fide* Eddy, Handb. Males. Mosses 2: 153, 1991) ≡ *Trichostomum borneense* (Dix.) Zand., see treatment of *Trichostomum*]
[*Stephanodictyon obscurirete* Dix. As4 (= *Pseudosymblepharis angustata* (Mitt.) Chen *fide* Eddy, Handb. Males. Mosses 2: 153, 1991) ≡ *Trichostomum brachydontium* Bruch, see treatment of *Trichostomum*]

STONEA Zand.
Stonea oleaginosa (Stone) Zand., Phytologia 65: 431, 1989 (*Tortula*) Austr1

[**STREBLOTRICHUM** P. Beauv. = *Barbula* Hedw. *fide* Saito, J. Hattori Bot. Lab. 39: 499, 1975]
[*Streblotrichum bicolor* (BSG) Loeske Eur ≡ *Barbula bicolor* (BSG) Lindb.]
[*Streblotrichum canaliculata* (Dix.) R. S. Chopra, Taxon. Indian Mosses 138, 1975 As3 ≡ *Didymodon canaliculatus* Dix.]
[*Streblotrichum convolutum* (Hedw.) P. Beauv. Eur As1 As2 As5 Afr1 Am1 Am2 Austr2 ≡ *Barbula convoluta* Hedw.]
[subsp. *commutatum* (Jur.) Giac. Eur As5 Afr1 ≡ *Barbula convolutum* subsp. *commutata* (Jur.) Boul.]
var. *commutatum* (Jur. Amann Eur As5 Afr1
var. *girodii* (Thér.) Podp. Eur

[var. *sardoa* (BSG) Podp. Eur ≡ *Barbula convoluta* var. *sardoa* BSG]
fo.? *angustifolium* (Warnst.) Podp., Consp. Musc. Eur. 216, 1954 *nom. inval. dispon. non cit.* (*Barbula convoluta* var.) Eur
[fo.? *brevifolium* (Podp.) Podp., Consp. Musc. Eur. 216, 1954 *nom. inval. dispon. non cit.* (*Barbula convoluta* fo.) Eur ≡ *Barbula convoluta* fo. *brevifolia* Podp.]
fo.? *filiformis* (Hag.) Podp., Consp. Musc. Eur. 216, 1954 *nom. inval. dispon. non cit.* (*Barbula convoluta* var.) Eur
[fo.? *insolatum* (Latz.) Podp., Consp. Musc. Eur. 216, 1954 *nom. inval. dispon. non cit.* (*Barbula convoluta* fo.) ≡ *Barbula convoluta* fo. *insolata* Latz.]
fo.? *latifolium* (Amann) Podp., Consp. Musc. Eur. 216, 1954 *nom. inval. dispon. non cit.* (*Streblotrichum convolutum* var.) Eur
fo.? *obtusatum* (Artr. & Loeske in Bauer) Podp., Consp. Musc. Eur. 216, 1954 *nom. inval. dispon. non cit.* (*Barbula convoluta* var.) Eur
fo.? *propaguliferum* (Głow.) Podp., Consp. Musc. Eur. 216, 1954 *nom. inval. dispon. non cit.* (*Barbula convoluta* var.) Eur
[fo.? *rufescens* (Loeske & Quelle) Podp., Consp. Musc. Eur. 216, 1954 *nom. inval. dispon. non cit.* (*Barbula convoluta* fo.) ≡ *Barbula convoluta* fo. *rufescens* Loeske & Quelle]
[fo.? *rufipes* (Bauer) Podp., Consp. Musc. Eur. 216, 1954 *nom. inval. dispon. non cit.* (*Barbula convoluta* fo.) Eur ≡ (*Barbula convoluta* fo. *rufipes* Bauer]
fo.? *stockumii* (Warnst.) Podp., Consp. Musc. Eur. 216, 1954 *nom. inval. dispon. non cit.* (*Barbula convoluta* var.) Eur
fo.? *turfaceum* (Warnst.) Podp., Consp. Musc. Eur. 216, 1954 *nom. inval. dispon. non cit.* (*Barbula convoluta* var.) Eur
fo.? *uliginosum* (Limpr.) Podp., Consp. Musc. Eur. 216, 1954 *nom. inval. dispon. non cit.* (*Barbula convoluta* var.) ≡ *Barbula convoluta* fo. *uliginosa* (Limpr.) Mönk.]
[*Streblotrichum croceum* (Brid.) Loeske Eur As2 Afr1 ≡ *Barbula crocea* (Brid.) Web. & Mohr]
var. *carpaticum* Pilous, Preslia 41: 120, 1969 Eur
[var. *coreense* (Card.) Wijk & Marg. ≡ *Barbula coreense* (Card.) Saito, J. Hattori Bot. Lab. 39: 484, 1975]
var. *dichodontioides* Pilous, Preslia 41: 121, 1969 Eur
var. *funckianum* (Schultz) Podp. Eur
var. *grimmioides* Pilous, Preslia 41: 121, 1969 Eur
var. *lingulatum* Pilous, Preslia 41: 121, 1969 Eur
fo. *cataractarum* (Loeske) Podp., Consp. Musc. Eur. 217, 1954 (*Streblotrichum croceum* var.) Eur
[*Streblotrichum enderesii* (Garov.) Loeske Eur As1 ≡ *Barbula enderesii* Garov.]
Streblotrichum gracillimum Herz. As2
[*Streblotrichum hypselostegium* (Card.) Hilp. Am2 = *Barbula indica* (Hook.) Spreng. *fide* Zander, Phytologia 44: 196, 1979]
Streblotrichum obtusifolium (Hilp.) Chen As2
[*Streblotrichum pringlei* (Card.) Hilp. Am2 = *Barbula indica* (Hook.) Spreng. *fide* Zander, Phytologia 44: 186, 1979]
Streblotrichum propaguliferum Li & Zhang, Acta Bot. Yunnan. 5(4): 385, 1983 As2

STREPTOCALYPTA C. Müll.
Streptocalypta lorentziana C. Müll. Am6
Streptocalypta santosii (Bartr.) Zand., Lindbergia 8: 165, 1982 [1983] (*Tortella*) Am2
Streptocalypta tortelloides (Card.) Zand., Lindbergia 8: 163, 1982 [1983] (*Barnesia*) Am2

STREPTOPOGON Wils. in Mitt.
[*Streptopogon australis* Mitt. Afr4 [(= *Didymodon* sp. *fide* Salmon 1903: 109) = *Daltonia* cf. *angustifolia* Dozy & Molk. (Daltoniaceae), see treatment of *Streptopogon*]
Streptopogon calymperes C. Müll. ex Geh. Afr3 Am2 Am3 Am4
Streptopogon calymperoides Demar. & P. Varde Afr2
Streptopogon cavifolius Mitt. Am2 Am4
Streptopogon clavipes (Spruce) Spruce ex Mitt. Am4
Streptopogon elimbatus Card. in P. Varde Afr3
Streptopogon erythrodontus (Tayl.) Wils. Afr3 Am2 Am3 Am4 Oc
var. *intermedius* Salm. Am4
var. *rutenbergii* (Geh.) Salm. Afr2 Afr3
Streptopogon heterophyllus Herz. Am4
Streptopogon juarezii Sharp, Phytologia 61: 372, 1986 Am2
Streptopogon lindigii Hampe Am4
Streptopogon matudianus Crum Am2
Streptopogon peruvianus Broth. ex Herz. *nom. inval. sin. descr.* Am4
[*Streptopogon rzedowskii* Cárdenas, Phytologia 61: 297, 1986 Am2 = *Leptodontium proliferum* Herz. *fide* Cárdenas, Anales Inst. Biol. Univ. Nac. Auton. México 58, Ser. Bot. 1: 94, 1987 (1989)]
Streptopogon spathulatus Herz. Am4
Streptopogon stenophyllus P. Varde & Thér. Afr2
Streptopogon subelimbatus Card. in Grand. Afr3

STREPTOTRICHUM Herz.
Streptotrichum ramicola Herz. Am4

SYNTRICHIA Brid.
[*Syntrichia abranchesii* (Luis.) Ochyra, Fragm. Florist. Geobot. 37: 212, 1992 (*Tortula*) Eur ≡ *Syntrichia caninervis* var. *abranchesii* (Luis.) Zand., see treatment of *Syntrichia*]
Syntrichia abruptinervis (Dix.) Zand. (*Tortula*) see treatment of *Syntrichia* Austr2
[*Syntrichia aciphylla* (B.&S.) Jur. = *Tortula ruralis* var. *alpina* Wahlenb.]
[var. *calva* Amann Eur ≡ *Syntrichia norvegica* var. *calva* (Amann) Ochyra, Fragm. Florist. Geobot. 37: 212, 1992]
Syntrichia aculeata (Wils.) Zand. (*Barbula*), see treatment of *Syntrichia* Am4
Syntrichia alpestris (Dix. in Herz.) Zand. (*Tortula*) see treatment of *Syntrichia* Afr2
Syntrichia ammonsiana (Crum & Anders.) Ochyra, Fragm. Florist. Geobot. 37: 212, 1992 (*Tortula*) Am1 Afr4
Syntrichia amplexa (Lesq.) Zand. (*Barbula*), see treatment of *Syntrichia* Eur Am1
Syntrichia amphidiacea (C. Müll.) Zand. (*Barbula*), see treatment of *Syntrichia* Afr2 Am1 Am2 Am3
Syntrichia anderssonii (Ångstr.) Zand. (*Tortula*), see treatment of *Syntrichia* Am6 Austr2 Ant
var. *fagicola* (C. Müll.) Zand. (*Barbula conotricha* var.), see treatment of *Syntrichia* Am6
Syntrichia andicola (Mont.) Ochyra, Fragm. Florist. Geobot. 37: 212, 1992 (*Tortula*) Am2 Am4
Syntrichia antarctica (C. Müll. & Hampe) Zand. (*Barbula*), see treatment of *Syntrichia* Afr4 Am6 Austr1 Austr2
Syntrichia baileyi (Broth.) Zand. (*Tortula*), see treatment of *Syntrichia* Austr1
Syntrichia bartramii (Steere in Grout) Zand. (*Tortula*), see treatment of *Syntrichia* Am1 Am2
Syntrichia bipedicellata (Britt.) Zand. (*Tortula*), see treatment of *Syntrichia* Am4
Syntrichia bogotensis (Hampe) Mitt. (*Barbula*), see treatment of *Syntrichia* [I.M. citation of Am3 is incorrect as Mitten's (1869)

"Guadelupe" refers to an area in Colombia, not the West Indies] Am4

Syntrichia bolanderi (Lesq. & James) Zand. (*Barbula*), see treatment of *Syntrichia* Eur Afr1 Am1

Syntrichia brachyclada (Card.) Zand. (*Tortula*), see treatment of *Syntrichia* Am6

Syntrichia brandisii (C. Müll.) Zand. (*Barbula*), see treatment of *Syntrichia* As3

Syntrichia brevisetacea (F. Müll.) Zand. (*Barbula*), see treatment of *Syntrichia* Austr1

Syntrichia cainii (Crum & Anders.) Zand. (*Tortula*), see treatment of *Syntrichia* Am1

Syntrichia calcicola Amann Eur

Syntrichia campestris (Dus.) Zand. (*Tortula*), see treatment of *Syntrichia* Am6

Syntrichia caninervis Mitt. Eur As2 Am1

var. *abranchesii* (Luis.) Zand. (*Tortula*), see treatment of *Syntrichia* Eur

var. *gypsophila* (Amann ex Roth) Ochyra, Fragm. Florist. Geobot. 37: 212, 1992 (*Tortula ruralis* var.) Eur As2

var. *spuria* (Amann) Zand. (*Tortula*), see treatment of *Syntrichia* Eur

Syntrichia cavallii (Negri) Ochyra, Fragm. Florist. Geobot. 37: 212, 1992 (*Tortula*) Afr2

Syntrichia chisosa (Magill, Delg. & Stark) Zand. (*Tortula*), see treatment of *Syntrichia* Am1 Am2 Afr4

Syntrichia ciliata (Broth.) Zand. (*Tortula*), see treatment of *Syntrichia* Am4

Syntrichia conferta (Bartr.) Zand. (*Tortula*), see treatment of *Syntrichia* Ant

Syntrichia costesii (Thér.) Zand. (*Tortula*), see treatment of *Syntrichia* Am6

Syntrichia didymodontoides (Broth. in Dryg.) Zand. (*Tortula*), see treatment of *Syntrichia* Afr4

[*Syntrichia echinata* (Schiffn.) Herrnstadt & Ben-Sasson, Bryologist 85: 216, 1982 (*Tortula*) Eur As5 Tortula princeps subsp. *echinata* (Schiffn.) Kramer, Bryophyt. Biblioth. 21: 86, 1980 ≡ *Syntrichia princeps* var. *echinata* (Shiffn.) Zand. (*Tortula*), see treatment of *Syntrichia*]

Syntrichia epilosa (Broth. ex Dus.) Zand. (*Tortula*), see treatment of *Syntrichia* Am6

var. *pilifera* (Thér.) Zand. (*Tortula epilosa* var.), see treatment of *Syntrichia* Am6

[*Syntrichia erythroneura* (C. Müll.) Sim Afr4 = *Tortula antarctica* (Hampe in C. Müll.) Wils. in Hook. f. *fide* Kramer, J. Hattori Bot. Lab. 65: 89, 1988 ≡ *Syntrichia antarctica* (Hampe in C. Müll.) Zand., see treatment of *Syntrichia*]

Syntrichia filaris (C. Müll. in Neum.) Zand. (*Barbula*), see treatment of *Syntrichia* Am6 Ant

Syntrichia flagellaris (Schimp.) Zand. (*Barbula*), see treatment of *Syntrichia* Am6

var. *densiretis* (Thér.) Zand. (*Tortula flagellaris* var.), see treatment of *Syntrichia*] Am6

Syntrichia fontana (C. Müll. in Neum.) Zand. (*Barbula*), see treatment of *Syntrichia* Am6 Ant

Syntrichia fragilis (Tayl.) Ochyra, Fragm. Florist. Geobot. 37: 212, 1992 (*Tortula*) Am1 Am2 Am3 Am4 As3 Afr3 Afr4

Syntrichia fuscoviridis (Car.) Zand. (*Tortula*), see treatment of *Syntrichia* Am6

Syntrichia geheebiaeopsis (C. Müll.) Zand. (*Barbula*) Afr4 Ant

Syntrichia gemmascens (Chen) Zand. (*Desmatodon*)see treatment of *Syntrichia* As2

Syntrichia gemmascens var. *hopeiensis* (Chen) Zand. (*Desmatodon gemmascens* var.), see treatment of *Syntrichia* As2

Syntrichia glacialis (Kunze ex C. Müll.) Zand. (*Barbula*), see treatment of *Syntrichia* Am4 Am6

Syntrichia gromschii (Thér.) Zand. (*Tortula*), see treatment of *Syntrichia* Am6

[*Syntrichia hadacii* Vondr., Bull. Soc. Amis Sci. Lett. Poznon, ser. D, 6: 121, 1965 As2 = *Tortula caninervis* var. *gypsophila* (Roth) Kramer, Bryoph. Biblioth. 21: 106, 1980 ≡ *Syntrichia caninervis* var. *gypsophila* (Roth.) Ochyra, Fragm. Florist. Geobot. 37: 212, 1992]

Syntrichia handelii (Schiffn.) Bach. Eur As5

var. *ferganensis* (Laz.) Ochyra, Fragm. Florist. Geobot. 37: 212, 1992 (*Tortula*) As1

[var. *pseudodesertorum* Vondr., Bull. Soc. Amis Sci. Lett. Poznon, ser. D, 6: 121, 1965 As5 ≡ *Syntrichia pseudohandelii* Fröhl. *fide* Agnew & Vondráček, Feddes Rep. 86: 401, 1975]

Syntrichia inermis (Brid.) Bruch in Hueb. Eur As1 As2 As3 As5 Afr1 Am1 Am2

var. *submarginata* (Schiffn.) Podp. Eur

Syntrichia intermedia Brid. Eur As1 As3 As5 Afr1 Afr4 Am1 Am2

var. *calva* (Dur. & Sag.) Delogn. Eur

Syntrichia jaffuelii (Thér.) Zand. (*Tortula*), see treatment of *Syntrichia* Am6

Syntrichia lacerifolia (Williams) Zand. (*Tortula*), see treatment of *Syntrichia* Am4

Syntrichia laevipila Brid. Eur As1 As3 As5 Afr1 Am1 Am6

var. *meridionalis* (Schimp.) Jur. Eur

Syntrichia latifolia (Hartm.) Hüb. Eur Am1

var. *propagulifera* (Milde) Amann Eur

fo. *perfragilis* Amann, Fl. Mousses Suisse Add. 3: 36, 1933 Eur

[*Syntrichia leptopyxis* (C. Müll.) Lazar., J. Bot. Acad. Sci. Ukraine 1(3–4): 95, 1940 (*Barbula*) As2 ≡ *Tortula leptopyxis* (C. Müll.) Lindb. & Arn.]

Syntrichia leucostega (C. Müll.) Zand. (*Barbula*), see treatment of *Syntrichia* Afr4 Austr1

var. *trachyneura* (Dix.) Zand. (*Tortula*) Afr4

Syntrichia limensis (Williams) Zand. (*Tortula*) see treatment of *Syntrichia* Am4

Syntrichia linguifolia (Herz.) Zand. (*Tortula*), see treatment of *Syntrichia* Am4

Syntrichia longimucronata (X.-j. Li) Zand. (*Tortula*), see treatment of *Syntrichia* As2

Syntrichia magellanica (Mont. in Gay) Zand. (*Tortula*), see treatment of *Syntrichia* Am6 Ant

[*Syntrichia mniadelphus* (C. Müll.) Herz. ≡ *Tortula mniadelphus* (C. Müll.) Broth. = *Tortula quitoensis* Tayl. in Hook. *fide* Mishler in Sharp et al., Moss Fl. Mex. ≡ *Sagenotortula quitoensis* (Tayl. in Hook.) Zand., Phytologia 65: 430, 1989]

var. *cochlearifolia* Herz. Am6?

Syntrichia mollis (B.&S. ex C. Müll.) Zand. (*Barbula*), see treatment of *Syntrichia* Afr2

[*Syntrichia mongolica* Boros, Trans. Brit. Bryol. Soc. 6: 70, 1970 As1 = *Syntrichia submontana* (Broth.) Ochyra, Fragm. Florist. Geobot. 37: 212, 1992]

[*Syntrichia montana* Nees in Raab = *Tortula intermedia* (Brid.) Berk.]

var. *brevifolia* C. Jens. Eur Am1

[var. *gelida* Podp. Eur ≡ *Tortula intermedia* var. *gelida* (Amann) Wijk & Marg.]

var. *nivalis* Amann Eur

var. *rufipila* Amann Eur

[fo. *brevifolia* (Arn.) Podp., Consp. Musc. Eur. 258, 1954 (*Tortula ruralis* fo.) Eur ≡ *Tortula ruralis* fo. *brevifolia* Arn. in

Rabenh.]
[*Syntrichia mucronifolia* (Schwaegr.) Brid. ≡ *Tortula mucronifolia* Schwaegr.]
var. *systilia* Amann Eur
fo. *perpusilla* (Warnst.) Podp., Consp. Musc. Eur. 251, 1954 (*Tortula mucronifolia* var.) Eur
fo. *pygmaea* Mart. ex Warnst., Hedwigia 52: 74, 1912 Eur
Syntrichia norvegica Web. Am1 Eur Afr1 Afr4 As1 As2 As3
var. *calva* (Amann) Ochyra, Fragm. Florist. Geobot. 37: 212, 1992 Eur
Syntrichia obtusissima (C. Müll.) Zand. (*Barbula*), see treatment of *Syntrichia* Am1 Am2
Syntrichia pagorum (Milde) Amann, see treatment of *Syntrichia* Eur Afr4 Am1 Am2 Austr1
Syntrichia papillosa (Wils.) Jur. Eur Afr4 Am1 Am2 Am4 Am5 Am6 Austr1 Austr2
var. *chilensis* (Thér.) Zand. (*Tortula papillosa* var.), see treatment of *Syntrichia* Am4
var. *meridionalis* (Warnst.) Zand. Eur
fo. *saxatilis* (Warnst.) Podp., Consp. Musc. Eur. 253, 1954 (*Tortula papillosa* var.) Eur
Syntrichia percarnosa (C. Müll.) Zand. (*Barbula*), see treatment of *Syntrichia* Am4 Am6
Syntrichia perichaetialis Herz. *nom. inval.* Am6
Syntrichia phaea (Hook. f. & Wils.) Zand. (*Trichostomum*) see treatment of *Syntrichia* Austr2
Syntrichia pichinchensis (Tayl.) Zand. (*Tortula*), see treatment of *Syntrichia* Am3 Am4
Syntrichia princeps (De Not.) Mitt. Eur As1 As3 As5 Afr1 Afr4 Am1 Am2 Am4 Am6 Austr1 Austr2 Oc Ant
var. *brachycarpa* (De Not.) Zand. (*Tortula princeps* var.), see treatment of *Syntrichia* Eur
var. *parnassica* (Schiffn.) Podp. Eur
Syntrichia prostrata (Mont.) Zand. (*Tortula*), see treatment of *Syntrichia* Am4 Am6
[*Syntrichia pseudodesertorum* (Fröhl.) Agnew & Vondr., Feddes Rep. 86(6–8): 402, 1975 (*Tortula*) As3
Syntrichia pseudohandelii (Fröhl.), Agnew & Vondr., Feddes Rep. 86(6–8): 401, 1975 (*Tortula*) As3
Syntrichia pseudorobusta (Dus.) Zand. (*Tortula*), see treatment of *Syntrichia* Am6
[*Syntrichia pulvinata* (Jur.) Jur. Eur Afr1 Am1 ≡ *Tortula pulvinata* (Jur.) Limpr. = *Tortula virescens* (De Not.) De Not. *fide* Kramer, Bryoph. Biblioth. 21: 99, 1980 ≡ *Syntrichia virescens* (De Not.) Ochyra, Fragm. Florist. Geobot. 37: 212, 1992]
fo. *macrophylla* (Warnst.) Podp., Consp. Musc. Eur. 255, 1954 (*Tortula pulvinata* var.) Eur
fo. *microphylla* (Warnst.) Podp., Consp. Musc. Eur. 255, 1954 (*Tortula pulvinata* var.) Eur
fo. *versispora* (Warnst.) Podp., Consp. Musc. Eur. 255, 1954 (*Tortula pulvinata* var.) Eur
Syntrichia pygmaea (Dus.) Zand. (*Tortula*) see treatment of *Syntrichia* Am6 Austr2
Syntrichia ramosissima (Thér.) Zand. (*Tortula*), see treatment of *Syntrichia* Am4
Syntrichia reflexa Zand. (*nom. nov.* for *Tortula reflexa* Li *hom. illeg.*), see treatment of *Syntrichia* As2
Syntrichia rigescens (Broth. & Geh. in Broth.) Ochyra, Fragm. Florist. Geobot. 37: 212, 1992 As5
Syntrichia robusta (Hook. & Grev.) Zand. (*Tortula*) see treatment of *Syntrichia* Am5 Am6 Austr1 Ant
var. *recurva* (Lightowlers) Zand. (*Tortula robusta* var.) see treatment of *Syntrichia* Ant
Syntrichia rubella (Hook. f. & Wils.) Zand. (*Tortula*), see treatment of *Syntrichia* Afr4 Austr1 Austr2
Syntrichia rubra (Mitt. in Hook. f.) Zand. (*Tortula*) see treatment of *Syntrichia* Am6 Austr1 Austr2 Ant
var. *subantarctica* (Sainsb.) Zand. (*Tortula*) see treatment of *Syntrichia* Austr2
[*Syntrichia ruraliformis* var. *subpapillosissima* (Biz. & Pierrot in Biz.) Ochyra, Fragm. Florist. Geobot. 37: 212, 1992 Afr1 ≡ *Syntrichia ruralis* var. *subpapillosissima* (Biz. & Pierr.) Zand. (*Tortula ruralis* var.), see treatment of *Syntrichia*]
Syntrichia ruralis (Hedw.) Web. & Mohr Eur As1 As2 As3 As5 Afr1 Afr2 Afr4 Am1 Am2 Am4 Am5 Austr1 Oc
var. *aetnensis* Reim. Eur
var. *arenicola* (Braithw.) Amann Eur As5 Afr1 Am1
[var. *calcicola* (Amann) Moenk. Eur ≡ *Syntrichia calcicola* Amann]
var. *gigantea* (Lesq.) Zand. (*Barbula ruralis* var.), see treatment of *Syntrichia* Am1
var. *gracilis* (C. Jens.) Zand. (*Tortula ruralis* var.), see treatment of *Syntrichia* Am1
var. *hirsuta* (Vent.) Podp. Eur As2 Am1
var. *pontresinae* (Geh. & Warnst.) Podp. Eur
var. *pseudodesertorum* Podp. Eur
var. *spiralis* (Herz.) Zand. (*Tortula ruralis* var.), see treatment of *Syntrichia* Am4
var. *submamillosa* (Kramer) Zand. (*Tortula ruralis* var.), see treatment of *Syntrichia* As5
var. *subpapillosissima* (Biz. & Pierr.) Zand. (*Tortula ruralis* var.), see treatment of *Syntrichia* Afr1
var. *substereidosa* (Kramer) Zand. (*Tortula ruralis* var.), see treatment of *Syntrichia* As5
fo. *brevipila* (Warnst.) Podp., Consp. Musc. Eur. 255, 1954 (*Tortula ruralis* var.) Eur
[fo. *calva* (Amann) Podp., Consp. Musc. Eur. 257, 1954 (*Syntrichia aciphylla* var.) Eur ≡ *Syntrichia norvegica* var. *calva* (Amann) Zand., see treatment of *Syntrichia*]
fo. *compacta* (Głow.) Podp., Consp. Musc. Eur. 257, 1954 (*Tortula aciphylla* var.) Eur
[fo. *contorta* (Podp.) Podp., Consp. Musc. Eur. 255, 1954 (*Tortula ruralis* fo.) Eur ≡ *Tortula ruralis* fo. *contorta* Podp.]
fo. *fallax* (Herz.) Podp., Consp. Musc. Eur. 255, 1954 (*Tortula ruralis* fo.) Eur
fo. *gigantea* (Farn.) Podp., Consp. Musc. Eur. 256, 1954 (*Barbula ruraliformis* fo.) Eur
fo. *gracilis* (Meyl.) Podp., Consp. Musc. Eur. 255, 1954 (*Tortula ruralis* fo.) Eur
fo. *longipila* Papp, Bul. Grăd. Bot. Univ. Cluj 26: 13, 1946 Eur
fo. *planifolia* (Warnst.) Podp., Consp. Musc. Eur. 256, 1954 (*Tortula ruralis* var.) Eur
fo. *rubicundula* (Kindb.) Podp., Consp. Musc. Eur. 256, 1954 (*Tortula ruralis* subsp.) Eur
fo. *rufipila* Herz. ex Rübel, Bot. Jahrb. Syst. 47: 478, 1912 Eur
fo. *rufoneura* (Podp.) Podp., Consp. Musc. Eur. 256, 1954 (*Tortula ruralis* fo.) Eur
fo. *subrufa* (Podp.) Podp., Consp. Musc. Eur. 256, 1954 (*Tortula ruralis* fo.) Eur
fo. *viridis* (Matous.) Podp., Consp. Musc. Eur. 256, 1954 (*Tortula ruralis* fo.) Eur
Syntrichia saxicola (Card.) Zand. (*Tortula*), see treatment of *Syntrichia* Am6
Syntrichia scabrella (Dus.) Zand. (*Tortula*), see treatment of *Syntrichia* Am6
Syntrichia scabrinervis (C. Müll.) Zand. (*Barbula*), see treatment of

Syntrichia Am4 Am6
Syntrichia serrata (Dix.) Zand. (*Tortula serrata* Dix. *nom. legit.* contrary to Index Muscorum; cf. Lightowlers, J. Bryol. 13: 373, 1985) Austr2
Syntrichia serripungens (Lor. & C. Müll.) Zand. (*Barbula*) Am4 Am6
var. *excesa* (C. Müll.) Zand. (*Barbula serripungens* var.), see treatment of *Syntrichia* Am6
Syntrichia sinensis (C. Müll.) Ochyra, Fragm. Florist. Geobot. 37: 212, 1992 (*Barbula*) Eur As1 As2 As3 As5 Afr1
Syntrichia socialis (Dus.) Zand. (*Tortula*), see treatment of *Syntrichia* Am6
Syntrichia subaristata (B.&S. ex C. Müll.) Zand. (*Barbula*), see treatment of *Syntrichia* Afr2
Syntrichia submontana (Broth.) Ochyra, Fragm. Florist. Geobot. 37: 212, 1992 (*Tortula*) As1
[*Syntrichia subulata* (Hedw.) Web. & Mohr. ≡ *Tortula subulata* Hedw.]
var. *major* Hüb. Eur
fo. *brevifolia* Herz., Kryptog. Forsch. 4: 280, 1919 Eur
fo. *calcarea* (Velen.) Podp., Consp. Musc. Eur. 250, 1954 (*Tortula subulata* var.) Eur
fo. *compacta* (Schiffn.) Podp., Consp. Musc. Eur. 250, 1954 (*Tortula subulata* var.) Eur
fo. *dentata* (Boul.) Podp., Consp. Musc. Eur. 250, 1954 (*Barbula subulata* var.) Eur Afr1
[fo. *denticulata* (Latz.) Podp., Consp. Musc. Eur. 250, 1954 (*Tortula subulata* fo.) Eur ≡ *Tortula subulata* fo. *denticulata* Latz.]
[fo. *graeffii* (Warnst.) Podp., Consp. Musc. Eur. 250, 1954 (*Barbula*) Eur ≡ *Tortula subulata* var. *graeffii* Warnst.]
fo. *inframarginata* (Głow.) Podp., Consp. Musc. Eur. 250, 1954 (*Tortula subulata* var.) Eur
fo. *lapidicola* (C. Jens.) Podp., Consp. Musc. Eur. 250, 1954 "lapicidicola" (*Tortula subulata* var.) Eur
fo. *mucronata* (Farn.) Podp., Consp. Musc. Eur. 250, 1954 (*Barbula subulata* var.) Eur
fo. *recurvomarginata* (Breidl. ex Limpr.) Podp., Consp. Musc. Eur. 250, 1954 (*Tortula subulata* var.) Eur
fo. *robusta* (Warnst.) Podp., Consp. Musc. Eur. 250, 1954 (*Tortula subulata* var.) Eur
fo. *saxicola* (Lamy) Podp., Consp. Musc. Eur. 250, 1954 (*Barbula*) Eur
fo. *transiens* (Velen.) Podp., Consp. Musc. Eur. 250, 1954 (*Tortula subulata* var.) Eur
Syntrichia virescens (De Not.) Ochyra, Fragm. Florist. Geobot. 37: 212, 1992 (*Tortula ruralis* var.) (cf. *Syntrichia virescens* (De Not.) Boros in Soó, Magyar Fl. 1: 425, 1964 *inval. basion. non cit.*) Eur As5 Afr1 Am1
var. *bizotiana* (Kramer) Zand. (*Tortula virescens* subsp.), see treatment of *Syntrichia* As5
var. *iranica* (Kramer) Zand., see treatment of *Syntrichia* As5
var. *minor* (Biz.) Ochyra, Fragm. Florist. Geobot. 37: 212, 1992 (*Tortula laevipila* var.) As5

Syntrichia viridula (C. Müll.) Zand. (*Barbula*), see treatment of *Syntrichia* Am4

TENIOLOPHORA Reese
Teniolophora fluviatile (Williams) Reese Am3

TETRACOSCINODON R. Br. ter
Tetracoscinodon irroratus (Mitt.) Zand., see treatment of *Tetracoscinodon* Austr2

TETRAPTERUM Hampe
Tetrapterum brachypelma (C. Müll.) Broth. Austr1
Tetrapterum cylindricum (Tayl.) Jaeg. Austr1
Tetrapterum lamprocarpum (C. Müll.) Broth. Am6
Tetrapterum lamprothecium (C. Müll.) Broth. Am6
[*Tetrapterum lilliputanum* (C. Müll. ex Roth) Broth. Am6 ≡ *Tortella lilliputana* (C. Müll. ex Roth) Zand., see treatment of *Tortella*]
[*Tetrapterum recurvirostre* (C. Müll.) Broth. Am5 Am6 ≡ *Weissia diffidentia* Zand. (*nom. nov.* for *Phascum recurvirostre* C. Müll.), see treatment of *Weissia*]
Tetrapterum sullivanii (C. Müll.) Broth. Austr1
Tetrapterum tetragonum (Hook.) Andrews Afr4
Tetrrapterum vernicosum (Roth) Broth. Am5 ≡ *Trichostomum exulatum* Zand. *nom. nov.*, see treatment of *Trichostomum*]
Tetrapterum weymouthii (C. Müll.) Broth. Austr1

TIMMIELLA (De Not.) Limpr.
Timmiella acaulon (C. Müll.) Zand., see treatment of *Timmiella* Am6
[*Timmiella alata* Herz. Am5 ≡ *Hymenostyliella alata* (Herz.) Robins.]
Timmiella anomala (BSG) Limpr. Eur As1 As2 As3 As4 As5 Afr1 Am1 Am2
[*Timmiella argentinica* Broth. Am4 = *Timmiella acaulon* (C. Müll.) Zand., see treatment of *Timmiella*]
[*Timmiella barbula* Limpr. = *Timmiella barbuloides* (Brid.) Moenk.]
var. *minor* Schimp. ex Luis. Afr1
Timmiella barbuloides (Brid.) Mönk. Eur As1 As3 As5 Afr1 Afr2 Am4
var. *longiseta* (Brid.) Wijk & Marg. Eur
Timmiella brevidens Dix. Afr2
Timmiella cameruniae Broth. Afr2
Timmiella corniculata (Wahlenb.) Broth. As1
Timmiella crassinervis (Hampe) Koch Am1
Timmiella diminuta (C. Müll.) Chen As2
Timmiella flexiseta (Bruch) Limpr. Eur Afr1
[var. *vancouveriensis* (Broth.) Grout Am1 = *Timmiella crassinervis* (Hampe) Koch *fide* Crum, Steere & Anderson, Bryologist 68: 431, 1973]
Timmiella grosseserrata Schiffn. As5
[*Timmiella japonica* Iwas. As2 = *Timmiella anomala* (BSG) Limpr. *fide* Iwatsuki & Noguchi, J. Hattori Bot. Lab. 37: 412, 1973]
Timmiella pelindaba Magill in Magill & Schelpe, Mem. Bot. Surv. S. Afr. 43: 3, 1979 Afr4
[*Timmiella subanomala* (Besch.) Broth. Am2 = *Timmiella anomala* (BSG) Limpr. *fide* Zander in Sharp et al., Moss Fl. Mex.]
Timmiella subintegra Dix. As3
Timmiella umbrosa (C. Müll.) Broth. Am6

TORTELLA (Lindb.) Limpr.
[*Tortella acaulon* (C. Müll.) Broth. Am6 ≡ *Timmiella acaulon* (C. Müll.) Zand., see treatment of *Timmiella*]
Tortella alpicola Dix. Am1 As3 Oc Ant
Tortella aprica (C. Müll.) Broth. Afr2
[*Tortella arctica* (Arn.) Crundw. & Nyh., Trans. Brit. Bryol. Soc. 4: 187, 1963 (1964) Am1 As1 As3 ≡ *Tortella tortuosa* var. *arctica* (Arn.) Broth. in Fedch.]
Tortella brotheri (Broth.) Broth. Eur
Tortella tetrapteroides (C. Müll.) Broth. Am5
[*Tortella bryotropica* Zand. in Schultze-Motel & Menzel, Beih.

Nov. Hedw. 88: 21, 1987 Am4
[*Tortella caespitosa* (Schwaegr.) Limpr. ≡ *Tortella humilis* (Hedw.) Jenn.]
var. *longirostris* Papp Eur
[*Tortella calycina* (Schwaegr.) Dix. As3 Afr4 Am6 Austr1 Austr2 ≡ *Barbula calycina* Schwaegr. *fide* Magill, Fl. S. Afr. I. Mosses 1: 241, 1981 (1982)]
[*Tortella ceylonensis* Fleisch. ex Dix. As3 = *Tortella humilis* (Hedw.) Jenn. *fide* Sollman, Lindbergia 16: 24, 1990]
[*Tortella cirrhata* Broth. Austr1 ≡ *Trichostomum eckelianum* Zand. *nom. nov.* see treatment of *Trichostomum*]
Tortella cirrifolia (Mitt.) Broth. Afr1
Tortella contortifolia (Mitt.) Broth. in Par. Am4
Tortella cryptocarpa (Broth.) Zand. (*Astomum*), see treatment of *Tortella* Am5
[*Tortella cylindrica* (Brid.) Loeske = *Trichostomum tenuirostre* (Hook. & Tayl.) Lindb.]
var. *perpapillosa* (Ihs.) Nog. As2
[fo. *irrigua* (Limpr.) Mönk. Eur ≡ *Oxystegus cylindricus* var. *irriguus* (Limpr.) Podp., Consp. Musc. Eur. 176, 1954]
Tortella cyrtobasis Dix. As3
Tortella dakinii Willis Austr1
Tortella densa (Lor. & Mol.) Crundw. & Nyh. Eur
Tortella eckendorffii (P. Varde) Zand. (*Hymenostomum*), see treatment of *Tortella* Afr2
Tortella elkantarensis Thér. & Trab. Afr1
Tortella erosodentata Sak. As2
[*Tortella eutrichostomum* (C. Müll.) Broth. Afr4 = *Tortella humilis* (Hedw.) Jenn. *fide* Magill, Fl. S. Afr. I. Mosses 1: 256, 1981 (1982)]
Tortella flavovirens (Bruch) Broth. Eur As5 Afr1 Am1
subsp. *esterelensis* (Roth) Giac. Eur
subsp. *limosella* (Stirt.) Podp. Eur
subsp. *viridiflava* (De Not.) Giac. Eur
var. *glareicola* (T. Christ.) Crundw. & Nyh. Eur
var. *laevis* Luis. Afr1
var. *papillosissima* Sérgio & Cas. de Puig, Portugaliae Acta Biol., ser. B., Sist. 13: 116, 1981 (*"papilosissima"*) Eur
var. *viridiflava* (De Not.) Cas. Gil. Eur
fo. *brevifolia* (Schiffn.) Podp., Consp. Musc. Eur. 178, 1954 (*Trichostomum flavovirens* var.) Eur
fo. *nitidicostatum* (Zodda) Podp., Consp. Musc. Eur. 178, 1954 (*Trichostomum flavovirens* var.) Eur
Tortella flavovirens (Bruch) Broth. × *Weissia crispa* (Hedw.) Mitt. *fide* Nicholson, Rev. Bryol. 37: 23, 1910 Eur
Tortella fragilis (Drumm.) Limpr. Eur As1 As2 As3 Afr4 Am1
var. *moravica* Podp. Eur
var. *setacea* (Farn.) Par. Eur
[var. *tortelloides* (Greene) Zand. & Hoe, Bryologist 82: 84, 1979 Am1 As3 Oc Ant Am1 As3 Oc Ant (*Sarconeurum*) ≡ *Tortella tortelloides* (Greene) Robins. in Llano = *Tortella alpicola* Dix.]
fo. *paludosa* Amann, Bull. Soc. Vaudoise Sci. Nat. 57: 123, 1929 Eur
fo. *riparia* (Amann) Podp., Consp. Musc. Eur. 182, 1954 (*Tortella tortuosa* var.) Eur
Tortella fragillima P. Varde Afr2
Tortella fristedtii (C. Müll.) Broth. Austr2
Tortella fruchartii (C. Müll.) Zand. (*Phascum*), see treatment of *Tortella* Am5 Am6
Tortella germainii (C. Müll.) Broth. Am4
Tortella goniospora (C. Müll.) Zand. (*Barbula*), see treatment of *Tortella* Oc
Tortella goughii Dix. As3
[*Tortella grossiretis* Bartr. Am5 = *Pseudosymblepharis schimperiana* (Par.) Crum, see treatment of *Pseudosymblepharis*]
[*Tortella guatemalensis* Bartr. Am2 = *Tortella tortuosa* (Hedw.) Limpr. *fide* Zander in Sharp et al., Moss Fl. Mex.]
Tortella hosseusii Herz. Am6
Tortella humilis (Hedw.) Jenn. Eur As2 As5 Afr1 Afr2 Afr3 Afr4 Am1 Am2 Am3 Am4 Am5 Am6 Oc
Tortella inclinata (Hedw. f.) Limpr. Eur As5 Afr1 Afr4 Am1 Am4 Austr1
var. *brachypoda* (Besch.) Par. Afr1
var. *leptotheca* (Brid.) Par. Afr4
fo. *acuminata* (Farn.) Par., Index Bryol. ed. 2, 30, 1906 (*Barbula inclinata* fo.) Eur
[fo. *actutifolia* Gand. in Dalla Torre, Moose v. Tirol 212, 1904 Eur ≡ *Tortella inclinata* fo. *acuminata* (Farn.) Par. *fide* Podpěra, Consp. Musc. Eur. 179, 1954]
fo. *alpina* Tosco, Webbia 28: 284, 1973 Eur
fo. *fragilifolia* Herz., Wiener Bot. Z. 93: 37, 1944 Eur
fo. *saltans* Loeske, Herbarium 62: 130, 1922 Eur
fo. *mutica* (Latz.) Podp., Consp. Musc. Eur. 179, 1954 (*Tortella inclinata* var.) Eur
Tortella inflexa (Bruch) Broth. Eur As2 As5 Afr1
fo. *elata* (Glow.) Podp., Consp. Musc. Eur. 177, 1954 Eur
Tortella involutifolia Dix. Afr2
Tortella japonica (Besch.) Broth. As2 Am2
Tortella jugicola (Dub.) Par. Am5
Tortella kmetiana Pilous Eur
Tortella knightii (Mitt.) Broth. Austr1 Austr2
Tortella lilliputana (C. Müll. ex Roth) Zand. (*Phascum*), see treatment of *Tortella* Am6
Tortella limbata (Schiffn.) Geh. & Herz. Afr1
Tortella lindmaniana Broth. Am5
Tortella linearis (Web. & Mohr) Zand. (*Barbula*), see treatment of *Tortella* Am3 Am5
[*Tortella mollissima* Broth. ex Bartr. Am1 Am2 Am3 Am5 ≡ *Trichostomum mollissimum* (Broth. ex Bartr.) Crum, Bryologist 72: 245, 1969 = *Psuedosymblepharis schimperiana* (Par.) Crum *fide* Zander in Sharp et al., Moss Fl. Mex.]
Tortella mooreae Sainsb. Austr2
Tortella nitida (Lindb.) Broth. Eur As5 Afr1
var. *irrigata* Winter, Hedwigia 55: 82, 1914 Afr1
var. *media* (Boul.) Corb. Eur Afr1
var. *obtusa* (Boul.) Jelenc Eur Afr1
var. *subtortuosa* (Boul.) Jelenc Eur Afr1
fo. *alpigena* Grom, Acta Bot. Croat. 26–27: 249, 1967–68 [1969] Eur
fo. *brachyphylla* Latz., Bot. Centralbl. Beih. 48: 476, 1931 Eur
fo. *media* (Boul.) Podp., Consp. Musc. Eur. 177, 1954 (*Trichostomum nitidum* var.) Eur
fo. *obtusa* (Boul.) Podp., Consp. Musc. Eur. 177, 1954 (*Trichostomum nitidum* var.) Eur
Tortella novae-valesiae Broth. Austr1
[*Tortella obtusifolia* Dix. Afr2 ≡ *Barbula umtaliensis* Magill *nom. nov. fide* Magill in Magill & Schelpe, Mem. Bot. Surv. S. Afr. 43: 5, 1979]
[*Tortella opaca* Dix. Afr2 ≡ *Weissia opaca* (Dix.) Magill in Magill & Schelpe, Mem. Bot. Surv. S. Afr. 43: 7, 1979]
Tortella perrufula (C. Müll.) Broth. Am6
[*Tortella petrieana* Sim Afr4 = *Tortella humilis* (Hedw.) Jenn. *fide* Magill, Fl. S. Afr. I. Mosses 1: 256, 1981 (1982)]
Tortella pileomayica Herz. Am4
Tortella pseudocaespitosa (C. Müll.) Broth. Am6
var. *brachybasis* (C. Müll.) Par. Am6

var. *pungens* (C. Müll.) Par. Am6

[*Tortella richardsii* Bartr. Am2 Am3 = *Pseudosymblepharis schimperiana* (Par.) Crum *fide* Zander in Sharp et al., Moss Fl. Mex.]

Tortella rigens Alb. Eur Am1

Tortella rubripes (Mitt.) Broth. Austr2

[*Tortella rufiseta* (C. Müll.) Broth. Afr4 = *Tortella xanthocarpa* (C. Müll.) Broth. *fide* Magill, Fl. So. Afr. I. Mosses 1: 255, 1982]

[*Tortella santosii* Bartr. Am2 ≡ *Streptocalypta santosii* (Bartr.) Zand., Lindbergia 8: 165, 1982 (1983)]

Tortella satoi Sak. As2

Tortella simplex Robins. in Robinson, Holm-Nielsen & Løjtnant, Lindbergia 4: 109, 1977 [1978] Am4

Tortella smithii Towns., Trans. Brit. Bryol. Soc. 6: 791, 1969 Afr2

Tortella somaliae (C. Müll.) Zand. (*Hyophila*), see treatment of *Tortella* Afr2

[*Tortella spathulata* Sak. As2 = *Trichostomum brachydontium* Bruch *fide* Saito, J. Hattori Bot. Lab. 39: 432, 1975]

Tortella subflavovirens Broth. & Watts Austr1

[*Tortella subfragilis* Crum & Steere Am3 = *Pseudosymblepharis schimperiana* (Par.) Crum *fide* Zander in Sharp et al., Moss Fl. Mex.]

[*Tortella syrrhopodontoides* Dix. Afr2 ≡ *Pseudosymblepharis syrrhopodontoides* (Dix.) Zand., see treatment of *Pseudosymblepharis*]

[*Tortella theriotii* Broth. & P. Varde Afr2 = *Tortella humilis* (Hedw.) Jenn. *fide* Sollman, Lindbergia 16: 24, 1990]

[var. *angustata* Dix. & P. Varde Afr2 = *Tortella humilis* (Hedw.) Jenn. *fide* Sollman, Lindbergia 16: 24, 1990]

[*Tortella tortelloides* (Greene) Robins. in Llano, Antarctic Terrestrial Biology, Antarctic Res. Ser. 20: 170, 1972 Am1 As3 Oc Ant [≡ *Tortella fragilis* var. *tortelloides* (Greene) Zand. & Hoe, Bryologist 82: 84, 1979 = *Tortella alpicola* Dix.]

Tortella tortuosa (Hedw.) Limpr. Eur As1 As2 As3 As5 Afr1 Am1 As2 Am6

subsp. *fasciculata* Culm Eur

var. *angustifolia* (Jur.) Limpr. Eur

var. *arctica* (Arn.) Broth. in Fedch. As1 As3 Am1

var. *brevifolia* Breidl. in Limpr. Eur

var. *brevirostris* Papp Eur

var. *cucullata* Amann Eur

var. *curvula* (Hartm.) Hag. Eur

var. *dicranoidea* (Ferg.) Limpr. Eur

var. *fleischeri* (Bauer) Latz. Eur

[var. *gracilescens* (Zett.) Podp. Eur = *Tortella tortuosa* var. *tenella* (Walth. & Mol.) Limpr. *fide* Podpěra, Consp. Musc. Eur. 181, 1954]

var. *hamifolia* Herz. Eur

var. *longifolia* (Ren.) Par. Eur

var. *nitida* Pilous Eur

var. *rauligera* Latz. Eur

var. *rigida* (Boul.) Limpr. Eur

var. *riparia* Amann Eur

var. *robusta* (Pfeff.) Limpr. Eur

var. *rotaeana* (De Not.) Limpr. Eur

var. *setifera* Velen. Eur

var. *tenella* (Walth. & Mol.) Limpr. Eur

var. *viridis* (Delogn.) Wild. Eur

fo. *alpina* Giac., Ann. Bryol. 11: 72, 1938 Eur

fo. *alpina* (Boul.) Podp., Consp. Musc. Eur. 180, 1954 *hom. illeg.* (*Barbula tortuosa* fo.) Eur

fo. *atrovirens* Latz., Hedwigia 66: 138, 1926 Eur

[fo. *brevifolia* (Breidl. in Limpr.) Herz. in Rübel, Bot. Jahrb. Syst. 47: 476, 1912 Eur ≡ *Tortella tortuosa* var. *brevifolia* Breid. in Limpr. *fide* Podpěra, Consp. Musc. Eur. 180, 1954]

fo. *dentata* (Farn.) Podp., Consp. Musc. Eur. 180, 1954 (*Barbula tortuosa* fo.) Eur

[fo. *gracilescens* (Zett.) Par., Index Bryol. ed 2, 32, 1906 (*Barbula tortuosa* var.) Eur = *Tortella tortuosa* var. *tenella* (Walth. & Mol.) Limpr. *fide* Podpěra, Consp. Eur. 181, 1954]

fo. *latifolia* Herz., Weiner Bot. Z. 93: 38, 1944 Eur

fo. *spinidens* (Roth ex Zodd.) Bauer, Fedde Repert. 50: 341, 1941 (*Tortella*) Eur

fo. *subrecurva* Latz., Magyar Bot. Lapok 33: 174, 1934 Eur

fo. *superalpina* Podp., Consp. Musc. Eur. 181, 1954 *nom. illeg.* Eur

Tortella undulatifolia Dix., Anniv. Vol. Bot. Gard. Calcutta 180, 1942 As3

Tortella vernicosa (Ren. & Card.) Broth. Afr3

Tortella walkeri (Broth.) Zand. (*Hyophila*), see treatment of *Tortella* As3

Tortella xanthocarpa (C. Müll.) Broth. Afr4

TORTULA Hedw.

[*Tortula abranchesii* Luis. Eur ≡ *Tortula caninervis* var. *abranchesii* (Luis.) Kramer, Bryophyt. Biblioth. 21: 108, 1980 ≡ *Syntrichia caninervis* var. *abranchesii* (Luis.) Zand., see treatment of *Syntrichia*]

[*Tortula abruptinervis* Dix. Austr2 ≡ *Syntrichia pygmaea* (Dus.) Zand. see treatment of *Syntrichia*]

Tortula abyssinica De Not. Afr2

[*Tortula aciphylla* (B.&S. in BSG) Hartm. ≡ *Tortula norvegica* Web.) Lindb. *fide* Kramer, Bryoph. Biblioth. 21: 128, 1980 ≡ *Syntrichia norvegica* Web.]

var. *mucronata* (Sendtn.) Limpr. Eur

[*Tortula acletoi* Robins., Phytologia 12: 390, 1971 Am4 ≡ *Hennediella acletoi* (Robins.) Zand., see treatment of *Hennediella*]

[*Tortula aculeata* (Wils.) Mitt. Am4 ≡ *Syntrichia aculeata* (Wils.) Zand., see treatment of *Syntrichia*]

Tortula aculeonervis (C. Müll.) Broth. Am6

[*Tortula adusta* (Mitt.) Mitt. Austr1 = *Barbula hornschuchiana* Schultz *fide* Catcheside, Mosses S. Austr. 178, 1980] ≡ *Pseudocrossidium hornschuchianum* (Schultz) Zand., Phytologia 44: 205, 1979]

[*Tortula aestiva* (Hedw.) P. Beauv. ≡ *Tortula muralis* var. *aestiva* Hedw.]

[fo. *terrestris* Podp., Sborn. Klubu Přír. v. Brno 5: 13, 1923 Eur ≡ *Tortula muralis* fo. *terrestris* (Podp.) Podp., Consp. Musc. Eur. 248, 1954]

Tortula afanassievii Laz. As2

[*Tortula afroruralis* (C. Müll.) Broth. Afr4 = *Tortula leucostega* (C. Müll.) Broth. *fide* Kramer, J. Hattori Bot. Lab. 65: 84, 1988 ≡ *Syntrichia leucostega* (C. Müll.) Zand., see treatment of *Syntrichia*]

[*Tortula alpestris* Dix. in Herz. Afr2 ≡ *Syntrichia alpestris* (Dix. in Herz.) Zand. see treatment of *Syntrichia*]

[*Tortula alpina* (BSG) Bruch in Breut. = *Tortula sinensis* (C. Müll.) Broth. ≡ *Syntrichia sinensis* (C. Müll.) Ochyra, Frag. Florist. Geobot. 37: 212, 1992]

[var. *inermis* (Mild.) De Not. Eur As2 Afr1 Am1 Oc = *Tortula fragilis* Tayl. *fide* Corley et al., J. Bryol. 11: 619, 1981 (1982)] ≡ *Syntrichia fragilis* (Tayl.) Ochyra, Fragm. Florist. Geobot. 37: 212, 1992]

[fo. *pagorum* (Milde) Squiv., Rev. Bryol. Lichénol. 30: 215, 1962 (1963) Eur Am1 Am2 Austr1 ≡ *Tortula pagorum* (Milde) De Not. ≡ *Syntrichia pagorum* (Milde) Amann, see treatment of

Syntrichia]

Tortula altipes (Broth.) Zand. (*Desmatodon*), see treatment of *Tortula* As1

[*Tortula amphidiacea* (C. Müll.) Broth. Afr2 Am1 Am2 Am3 ≡ *Syntrichia amphidiacea* (C. Müll.) Zand., see treatment of *Syntrichia*]

[*Tortula ammonsiana* Crum & Anders., Bryologist 82: 469, 1979 Am1 Afr4 ≡ *Syntrichia ammonsiana* (Crum & Anders.) Ochyra]

Tortula amphidiifolia (C. Müll.) Broth. Am6

[*Tortula amplexa* (Lesq.) Steere in Grout Eur Am1 ≡ *Syntrichia amplexa* (Lesq.) Zand., see treatment of *Syntrichia*]

Tortula ampliretis Crundwell & Long, J. Bryol. 10: 104, 1978 Afr1

Tortula anacamptophylla (C. Müll.) Broth. Am5 Am6 = *Tortula robusta* Hook. & Grev. *fide* Lightowlers, Brit. Antarct. Surv. Bull. 67: 67, 1985 ≡ *Syntrichia robusta* (Hook. & Grev.) Zand. see treatment of *Syntrichia*]

[*Tortula anderssonii* Ångstr. Am6 Austr2 Ant ≡ *Syntrichia anderssonii* (Aongstr.) Zand., see treatment of *Syntrichia*]

[var. *fagicola* (C. Müll.) Card. Am4 ≡ *Syntrichia anderssonii* var. *fagicola* (C. Müll.) Zand., see treatment of *Syntrichia*]

[*Tortula angustifolia* (Herz.) Herz. *nom. illeg.* Am4 ≡ *Tortula herzogii* Zand. *nom. nov.* Bryologist 82: 631, 1979 ≡ *Hennediella angustifolia* (Herz.) Zand., see treatment of *Hennediella*]

[*Tortula andicola* Mont. Am2 Am4 ≡ *Syntrichia andicola* (Mont.) Ochyra, Fragm. Florist. Geobot. 37: 212, 1992]

[*Tortula androgyna* (C. Müll.) Broth. Austr1 = *Tortula antarctica* (Hampe in C. Müll.) Wils. in Hook. f. *fide* Kramer, J. Hattori Bot. Lab. 65: 84, 1988 ≡ *Syntrichia antarctica* (C. Müll. & Hampe) Zand., see treatment of *Syntrichia*]

[*Tortula antarctica* (Hampe in C. Müll.) Wils. in Hook. f. Am6 (good species *fide* Kramer, J. Hattori Bot. Lab. 65: 84, 1988) ≡ *Syntrichia antarctica* (Hampe in C. Müll.) Zand., see treatment of *Syntrichia*]

Tortula appressa Mitt. Am6

[*Tortula arenae* (Besch.) Broth. Am6 Ant ≡ *Hennediella arenae* (Besch.) Zand., see treatment of *Hennediella*]

[subsp. *petriei* (Broth.) Lightowlers, J. Bryol. 13: 371, 1985 (*Tortula petriei*) Austr2 ≡ *Hennediella arenae* var. *petriei* (Broth. ex Beck.) Zand., see treatment of *Hennediella*]

Tortula argentinica (Broth.) Zand. (*Desmatodon*), see treatment of *Tortula* Am6

Tortula astoma Schiffn. As5

Tortula atherodes Zand. (*nom. nov.* for *Phascum cuspidatum* Schreb. ex Hedw.), see treatment of *Tortula* Eur As1 As2 As5 Afr1 Am1 Am2 Am4

var. *affinis* (Nees & Hornsch.) Zand. (*Phascum affine*), see treatment of *Tortula* Eur

var. *atherodes* var. *arcuata* (Herrnstadt & Heyn) Zand. (*Phascum cuspidatum* var.) Eur

var. *curviseta* (Dicks.) Zand. (*Phascum curvisetum*), see treatment of *Tortula* Eur

var. *diaphora* (Hag.) Zand. (*Phascum acaulon* var.), see treatment of *Tortula* Eur

var. *elatum* (Brid.) Zand. (*Phascum elatum*), see treatment of *Tortula* Eur

var. *intertexta* (Brid.) Zand. (*Phascum intertextum*), see treatment of *Tortula* Eur

var. *mitraeformis* (Limpr.) Zand. (*Phascum cuspidatum* var.), see treatment of *Tortula* Eur

var. *papillosa* (Lindb.) Zand. (*Phascum papillosum*), see treatment of *Tortula* Eur

var. *pilifera* (Hedw.) Zand. (*Phascum piliferum*), see treatment of *Tortula* Eur As1 As5 Afr2 Am1

var. *retortifolia* (Guerra & Ros in Guerra, Jiménez, Ros & Carrión) Zand. (*Phascum cuspidatum* var.), see treatment of *Tortula* Eur

var. *schreberiana* (Dicks.) Zand. (*Phascum schreberianum*), see treatment of *Tortula* Eur Am1

Tortula atrata Thér. Am4

Tortula atrovirens (Sm.) Lindb. Eur As4 As5 Am1 Am2 Am3 Am4 Am6 Austr1 Austr2 Oc Afr1 Afr2 Afr3 Afr4

var. *brevifolia* Thér. Am6

var. *edentula* (BSG) Schimp. Eur Afr4

var. *gasilienii* (Vent.) Limpr. Eur

var. *leucodonta* (Corb.) Zand. (*Barbula atrovirens* var.), see treatment of *Tortula* Eur

var. *subrevolvens* Amann Eur

[*Tortula aurea* Bartr. Am1 Am2 ≡ *Barbula aurea* (Bartr.) Zand. in Zand. & Steere, Bryologist 81: 466, 1978 ≡ *Pseudocrossidium aureum* (Bartr.) Zand., Phytologia 44: 207, 1979 = *Barbula crinita* Schultz *fide* Sollman, Lindbergia 16: 22, 1990 ≡ *Pseudocrossidium crinitum* (Schultz) Zand., see treatment of *Pseudocrossidium* and cf. Sollman, Lindbergia 16: 22, 1990]

[*Tortula austroafricana* Kramer, J. Hattori Bot. Lab. 65: 92, 1988 Afr4 ≡ *Syntrichia austroafricana* (Kramer) Zand., see treatment of *Syntrichia*]

Tortula austroruralis (C. Müll.) Broth. Am6 Austr1

Tortula baetica (Casas & Oliva) Guerra & Ros in Guerra, Ros & Carrión, J. Bryol. 17: 281, 1992 (*Tortula muralis* var.) Eur

[*Tortula baileyi* Broth. Austr1 ≡ *Syntrichia baileyi* (Broth.) Zand., see treatment of *Syntrichia*]

[*Tortula bartramii* Steere in Grout Am1 Am2 ≡ *Syntrichia bartramii* (Steere in Grout) Zand., see treatment of *Tortula*]

Tortula bauriana Warnst. ex Baur Eur

[*Tortula bealeyensis* R. Br. ter Austr2 = *Tortula anderssonii*Ångstr. *fide* Lightowlers, Brit. Antarct. Surv. Bull. 67: 45, 1985 & J. Bryol. 13: 369, 1985 ≡ *Syntrichia anderssonii* (Aongstr.) Zand., see treatment of *Syntrichia*]

Tortula berthoana Thér. Am6

[*Tortula bipedicellata* Britt. Am4 ≡ *Syntrichia bipedicellata* (Britt.) Zand., see treatment of *Syntrichia*]

[*Tortula bistratosa* Flow. Am1 = *Tortula caninervis* (Mitt.) Broth. *fide* Corley et al., J. Bryol. 11: 619, 1981 (1982)] ≡ *Syntrichia caninervis* Mitt.]

[*Tortula bizotii* Laz., Vopr. Evol. Biogeogr. Genet. Sel. 145, 1960 *nom. inval. typon. non cit.* As5 ≡ *Tortula virescens* subsp. *bizotii* (Laz.) Kramer, Bryophyt. Biblioth. 21: 102, 1980 *comb. inval. basion. inval.*]

Tortula bogosica (C. Müll.) Zand. (*Desmatodon*), see treatment of *Tortula* Afr1 Afr2 Afr4

[*Tortula bogotensis* (Hampe) Mitt. Am4 (I.M. citation of Am3 is incorrect as Mitten's [1869] "Guadelupe" refers to an area in Colombia, not the West Indies) ≡ *Syntrichia bogotensis* (Hampe) Mitt., see treatment of *Syntrichia*]

[*Tortula bolanderi* (Lesq. & James) Howe Eur Afr1 Am1 ≡ *Syntrichia bolanderi* (Lesq. & James) Zand., see treatment of *Syntrichia*]

[*Tortula brachychaete* Dus. Am6 ≡ *Willia brachychaete* (Dus.) Zand., see treatment of *Willia*]

[*Tortula brachyclada* Card. Am6 ≡ *Syntrichia brachyclada* (Card.) Zand., see treatment of *Syntrichia*]

[*Tortula brachydontia* (Bruch in F. Müll.) Mitt., J. Linn. Soc. Bot. 12: 148, 1869 Am4 ≡ *Trichostomum brachydontium* Bruch in F. Müll.]

[*Tortula brachypelma* Des. Am6 = *Tortula magellanica* Mont. in Gay fide Kramer, J. Hattori Bot. Lab. 65: 111, 1988 ≡ *Syntric-*

hia magellanica (Mont.) Zand.]

[*Tortula brachytricha* (C. Müll.) Broth. Austr1 = *Tortula antarctica* (Hampe in C. Müll.) Wils. in Hook. f. *fide* Kramer, J. Hattori Bot. Lab. 65: 89, 1988 ≡ *Syntrichia antarctica* (Hampe in C. Müll.) Zand.]

[*Tortula brandisii* (C. Müll.) Broth. As3 ≡ *Syntrichia brandisii* (C. Müll.) Zand., see treatment of *Syntrichia*]

Tortula brevimucronata (C. Müll.) Broth. Afr4

Tortula brevipes (Lesq.) Broth. Am1 Am2

[*Tortula brevis* Whitehouse & Newton, J. Bryol. 15: 83, 1988 Eur ≡ *Hennediella brevis* (Whitehouse & Newton) Blockeel, J. Bryol. 16: 191, 1991]

Tortula breviseta Mont. Am6

[*Tortula brevisetacea* (F. Müll.) Thér. Austr1 ≡ *Syntrichia brevisetacea* (F. Müll.) Zand., see treatment of *Syntrichia*]

Tortula brevissima Schiffn. Eur As5

Tortula brunnea (C. Müll.) Broth. Am4

Tortula brunnea Broth. & Watts *hom. illeg.* Austr1

Tortula buchtienii Herz. Am4

Tortula bullata Herz. Am6 *hom. illeg.*

Tortula bushii (Card. & Thér.) Steere *nom. inval. in syn.* Am1

Tortula buyssonii (Philib.) Broth. Eur

[*Tortula cabulica* J. Fröhl., Mitt. Thür. Bot. Ges. 1: 61, 1955 As5 = *Trichostomopsis aaronis* (Lor.) Agnew & Towns., Israel J. Bot. 19: 258, 1970 ≡ *Didymodon aaronis* (Lor.) Guerra in Guerra & Ros, Cryptogamie Bryol. Lichénol. 8: 55, 1987]

[*Tortula cainii* Crum & Anders. Am1 ≡ *Syntrichia cainii* (Crum & Anders.) Zand., see treatment of *Syntrichia*]

[*Tortula calcicolens* Kramer, Bryophyt. Biblioth. 21: 90, 1980 (*Tortula calcicola* Greebe *hom. illeg.*, *Tortula ruralis* subsp. *calcicola* (Amann) Giac.) Eur (= *Tortula ruralis* (Hedw.) Gaertn., Meyer & Scherb. *fide* Corley et al., J. Bryol. 11: 619, 1981 [1982]) ≡ *Syntrichia calcicola* Amann]

Tortula californica Bartr. Am1 Am2

[*Tortula calobolax* (C. Müll.) Broth. Afr4 ≡ *Willia calobolax* (C. Müll.) Lightowlers, J. Bryol. 13: 370, 1985]

[var. *angustinervia* (C. Müll.) Par. Afr4 ≡ *Willia calobolax* var. *angustinervia* (C. Müll.) Lightowlers, J. Bryol. 18: 370, 1985]

[*Tortula campestris* Dus. Am6 ≡ *Syntrichia campestris* (Dus.) Zand., see treatment of *Syntrichia*]

Tortula canescens Mont. Eur As1 As5 Afr1

var. *longipila* (Carb.) Jelenc Afr1

[*Tortula caninervis* (Mitt.) Broth. Eur As2 Am1 ≡ *Syntrichia caninervis* Mitt.]

[subsp. *spuria* (Amann) Kramer, Bryophyt. Biblioth. 21: 106, 1980 (*Tortula*) Eur ≡ *Syntrichia caninervis* var. *spuria* (Amann) Zand., see treatment of *Syntrichia*]

[var. *abranchesii* (Luis.) Kramer, Bryophyt. Biblioth. 21: 108, 1980 (*Tortula*) Eur ≡ *Syntrichia caninervis* var. *abranchesii* (Luis.) Zand.]

[var. *gypsophila* (G. Roth) Kramer, Bryophyt. Biblioth. 21: 108, 1980 (*Tortula ruralis* var.) Eur ≡ *Syntrichia caninervis* var. *gypsophila* (Roth.) Ochyra, Fragm. Florist. Geobot. 37: 212, 1992]

Tortula capillaris (Chen) Zand. (*Desmatodon*), see treatment of *Tortula* As2

Tortula cardotii Thér. & Nav. Afr2

[*Tortula carinata* Gill. ex Grev. Am6 ≡ *Pseudocrossidium carinatum* (Gill. ex Grev.) Zand., see treatment of *Pseudocrossidium*]

[*Tortula caroliniana* Andrews Am1 Am2 Am3 = *Tortula amphidiacea* (C. Müll.) Broth. *fide* Mishler in Sharp et al., Moss Fl. Mex. ≡ *Syntrichia amphideacea* (C. Müll.) Zand., see treatment of *Syntrichia*]

[*Tortula cavallii* Negri Afr2 ≡ *Syntrichia cavallii* (Negri) Ochyra, Fragm. Florist. Geobot. 37: 212, 1992]

Tortula cernua (Hueb.) Lindb. Eur As1 As2 Am1

var. *xanthopus* (Kindb.) Zand. (*Desmatodon cernuus* var.), see treatment of *Tortula* Am1

Tortula characodonta (C. Müll.) Broth. Am6

[*Tortula chisosa* Magill, Delg. & Stark, Ann. Missouri Bot. Gard. 70: 200, 1983 Am1 Am2 Afr4 ≡ *Syntrichia chisosa* (Magill, Delg. & Stark) Zand., see treatment of *Syntrichia*]

Tortula chrysopila (C. Müll.) Par. Am6

Tortula chubutensis Dus. Am6

Tortula chubutensis Broth. Am6 *hom. illeg.*

Tortula chungtienia Zand. (*nom. nov.* for *Desmatodon yuennanensis* Broth.), see treatment of *Tortula* As2

[*Tortula ciliata* Broth. Am4 ≡ *Syntrichia ciliata* (Broth.) Zand.]

Tortula cochlearifolia P. Varde Afr2

[*Tortula conferta* Bartr. Ant (= *Tortula grossiretis* Card. *fide* Robins., Antarct. Res. Ser. 20: 172, 1972) ≡ *Tortula princeps* var. *conferta* (Bartr.) Lightowlers, Brit. Antarct. Surv. Bull. 67: 61, 1985 (= *Tortula princeps* De Not. *fide* Lightowlers, J. Bryol. 14: 290, 1986 [1987] ≡ *Syntrichia princeps* (De Not.) Mitt.) ≡ *Syntrichia conferta* (Bartr.) Zand., see treatment of *Syntrichia* Ant]

[*Tortula conica* Turton, Gen. Syst. Nat. Veg. Kingd. 2: 1720, 1806 (*Barbula conica* Brid.) Am1 = *Barbula agraria* Hedw. *fide* Brid., Bryol. Univ. 1: 532, 1827]

Tortula contorta (C. Müll.) Mont. in Gay Am6

[*Tortula convoluta* (Hedw.) Gaertn., Meyer & Scherb. ≡ *Barbula convoluta* Hedw.]

var. *major* De Not. Eur

[*Tortula costesii* Thér. Am6 ≡ *Syntrichia costesii* (Thér.) Zand., see treatment of *Syntrichia*]

Tortula crawfordii (Par.) Watts in Watts & Whitel. Austr1

Tortula crenata Mitt. Am4

[*Tortula crenulata* Warnst = *Tortula subulata* var. *serrulata* Warnst.]

var. *latifolia* (Warnst.) Warnst. Eur

Tortula cucullifolia Fröhl., Mitt. Thürin. Bot. Ges. 1(2–3): 62, 1955 Eur

Tortula cuneifolia (Dicks.) Turn. Eur As5 Afr1 [for discussion of Dickson as authority see Karttunen, Taxon 37: 156–157, 1988]

subsp. *freibergii* (Dix. & Loeske) Giac. Eur

var. *blissii* Zand. var. nov., see treatment of *Tortula* Am1

var. *luteomarginata* Höhn. Eur

var. *marginata* Fleisch. Eur

var. *piligera* Latz. Eur

fo. *brevifolia* Fleisch. in E. Bauer, Musc. Eur. Exs. n. 178, 1906 Eur

[*Tortula cuspidata* (Schultz) Chev. Eur Afr1 ≡ *Barbula unguiculata* fo. *cuspidata* (Schultz) Mönk., Laub. Eur. 286, 1927]

var. *elongata* Chev. Eur

[*Tortula cuspidata* Hook. f. & Wils. Austr1 = *Tortula anderssonii*Ångstr. *fide* Kramer, J. Hattori Bot. Lab. 65: 84, 1988 ≡ *Syntrichia anderssonii* (Aongstr.) Zand., see treatment of *Syntrichia*]

Tortula cuspidatissima (C. Müll.) Broth. Afr2

Tortula deciduidentata (Sharp & Iwats.) Zand. (*Crumia*), see treatment of *Tortula* Am1

Tortula demawendica Schiffn. As5

[*Tortula densifolia* (Hook. f. & Wils.) Hook. f. & Wils. Am6 Ant ≡ *Hennediella densifolia* (Hook. f. & Wils.) Zand., see treatment of *Hennediella*]

[*Tortula denticulata* (Wils. in Mitt.) Mitt. Am3 Am4 ≡ *Hennediella denticulata* (Wils. in Mitt.) Zand., see treatment of *Hennediella*]

[*Tortula deserta* (C. Müll.) Broth. Afr4 = *Desmatodon longipedunculatus* (C. Müll.) Magill, Fl. S. Afr. I. Mosses 1: 210, 1981 (1982) ≡ *Hennediella longipedunculata* (C. Müll.) Zand., see treatment of *Hennediella*]

[*Tortula desertorum* Broth. Eur As1 As3 As5 Am1 = *Tortula caninervis* (Mitt.) Broth. *fide* Corley et al., J. Bryol. 11: 619, 1981 (1982)] ≡ *Syntrichia caninervis* Mitt.]

[*Tortula dicksoniana* (Schultz) Podp. = *Tortula cuneifolia* (Dicks.) Turn.]

[fo. *piligera* (Latz.) Podp., Consp. Musc. Eur. 245, 1954 (*Tortula cuneifolia* var.) Eur ≡ *Tortula cuneifolia* var. *piligera* Latz.]

[fo. *brevifolia* (Fleisch. in Bauer) Podp., Consp. Musc. Eur. 245, 1954 Eur ≡ *Tortula cuneifolia* fo. *brevifolia* Fleisch. in Bauer]

[*Tortula didymodontoides* Broth. in Dryg. Afr4 ≡ *Syntrichia didymodontoides* (Broth. in Dryg.) Zand., see treatment of *Syntrichia*]

[*Tortula domingensis* Thér. Am3 ≡ *Brachymenium domingensis* (Thér.) Zand. see Excluded Taxa]

[*Tortula eckeliae* Zand., Bryologist 88: 354, 1985 (1986) Am2 ≡ *Hennediella heteroloma* var. *eckeliae* (Zand.) Zand., see treatment of *Hennediella*]

[*Tortula elongata* (Wils. in Mitt.) Mitt. *hom. illeg.* Am2 Am4 ≡ *Morinia ehrenbergiana* var. *elongata* (Wils. in Mitt.) Zand., Bryologist 81: 557, 1978 (1979) ≡ *Mironia ehrenbergiana* var. *elongata* (Wils. in Mitt.) Zand., see treatment of *Mironia*]

Tortula entosthodontacea (Card. & Dix.) Zander (*Hyophilopsis*), see treatment of *Tortula* As3

Tortula epilosa Broth. ex Dus. Am6

var. *pilifera* Thér. Am6

[*Tortula erythrodonta* (Tayl.) Wils. ≡ *Streptopogon erythrodontus* (Tayl.) Wils.]

[var. *clavipes* Spruce Am4 ≡ *Streptopogon clavipes* (Spruce) Spruce ex Mitt.]

[*Tortula erythroneura* (C. Müll.) Broth. Afr4 = *Tortula antarctica* (Hampe in C. Müll.) Wils. in Hook. f. *fide* Kramer, J. Hattori Bot. Lab. 65: 89, 1988 ≡ *Syntrichia antarctica* (Hampe in C. Müll.) Zand., see treatment of *Syntrichia*]

[*Tortula eubryum* (C. Müll.) Broth. in Engl. Afr2 Afr4 ≡ *Barbula eubryum* C. Müll. *fide* Magill, Fl. S. Afr. I. Mosses 1: 245, 1981 (1982)]

Tortula euryphylla Zand. (*nom. nov.* for *Dicranum latifolium* Hedw.), see treatment of *Tortula* Eur As1 As2 As3 As5 Afr1 Am1

subsp. *brevifolia* (Kindb.) Zand. (*Tortula latifolia* (Hedw.) Lindb. subsp. *brevifolia* Kindb.), see treatment of *Tortula* Eur

var. *eucalyptrata* (Lindb.) Zand. (*Tortula*) Eur

var. *flavescens* (Brid.) Zand. (*Dicranum latifolium* var.), see treatment of *Tortula* Eur

var. *spelaea* (Amann) Zand. (*Desmatodon*), see treatment of *Tortula* Eur

var. *subobliqua* (Lindb.) Zand. (*Desmatodon latifolius* var.), see treatment of *Tortula* Eur

Tortula evanescens Broth. Austr1

[*Tortula excelsa* Card. Ant = *Tortula filaris* (C. Müll. in Newm.) Broth. *fide* Lightowlers, Brit. Antarct. Surv. Bull. 67: 49, 1985 ≡ *Syntrichia filaris* (C. Müll. in Neum.) Zand., see treatment of *Tortula*]

[*Tortula fallax* (Hedw.) Schrad. ex Turn. ≡ *Barbula fallax* Hedw.]

var. *brevifolia* Chev. Eur

Tortula felipponei Thér. in Felipp. Am6

[*Tortula ferganensis* Laz. As1 ≡ *Tortula handelii* var. *ferganensis* (Laz.) Kramer, Bryologist 81: 385, 1978 ≡ *Syntrichia handelii* var. *ferganensis* (Laz.) Ochyra, Fragm. Florist. Geobot. 37: 212, 1992 (*Tortula*) As1]

Tortula ferruginea Bartr., Rev. Bryol. Lichénol. 33: 325, 1964–65 [1965] Am4

[*Tortula filaris* (C. Müll. in Neum.) Broth. Am6 Ant ≡ *Syntrichia filaris* (C. Müll. in Neum.) Zand., see treatment of *Tortula*]

[*Tortula flagellaris* (Schimp.) Mont. in Gay Am6 ≡ *Syntrichia flagellaris* (Schimp.) Zand., see treatment of *Syntrichia*]

[var. *densiretis* Thér. Am6 ≡ *Syntrichia flagellaris* var. *densiretis* (Thér.) Zand., see treatment of *Syntrichia*]

[*Tortula flavinervis* Dix. Austr1 Austr2 = *Barbula crinita* Schultz *fide* Weber, Lindbergia 1: 216, 1972 ≡ *Pseudocrossidium crinitum* (Schultz) Zand., see treatment of *Pseudocrossidium*]

[var. *gigantea* Dix. & Sainsb. Austr2 = *Barbula crinita* Schultz *fide* Weber, Lindbergia 1: 216, 1972 ≡ *Pseudocrossidium crinitum* (Schultz) Zand., see treatment of *Pseudocrossidium*]

[var. *obscura* Dix. Austr2 = *Barbula crinita* Schultz *fide* Weber, Lindbergia 1: 216, 1972 ≡ *Pseudocrossidium crinitum* (Schultz) Zand., see treatment of *Pseudocrossidium*]

[var. *parviretis* Sainsb. Austr1 = *Barbula crinita* Schultz *fide* Weber, Lindbergia 1: 216, 1972 ≡ *Pseudocrossidium crinitum* (Schultz) Zand., see treatment of *Pseudocrossidium*]

Tortula flavipes Broth. *hom. illeg.* Am6

[*Tortula flexomarginata* (C. Müll. & Hampe) Mitt. Austr1 = *Tortula antarctica* (Hampe in C. Müll.) Wils. in Hook. f. *fide* Kramer, J. Hattori Bot. Lab. 65: 84, 1988 ≡ *Syntrichia antarctica* (C. Müll. & Hampe) Zand., see treatment of *Syntrichia*]

[*Tortula flexuosa* Brid. Am4 = *Macromitrium* sp. *fide* C. Müll., Syn. 1(5): 645, 1849]

[*Tortula flexuosa* Hook. *hom. illeg.* Afr4 = *Barbula calycina* Schwaegr. *fide* Magill, Fl. S. Afr. I. Mosses 1: 240, 1981 (1982)]

[*Tortula fontana* (C. Müll.) Broth. Am6 Ant ≡ *Syntrichia fontana* (C. Müll. in Neum.) Zand., see treatment of *Syntrichia*]

[*Tortula fragilis* Tayl. Am1 Am2 Am3 Am4 ≡ *Syntrichia fragilis* (Tayl.) Ochyra, Fragm. Florist. Geobot. 37: 212, 1992 (*Tortula*) Am1 Am2 Am3 Am4 As3 Afr3 Afr4]

Tortula fragillima Herz. Am4

Tortula fuegiana (Mitt.) Mitt. (= *Tortula princeps* var. *magellanica* (Mont. in Gay) Lightowlers, Brit. Antarct. Surv. Bull. 67: 61, 1985 ≡ *Syntrichia magellanica* (Mont. in Gay) Zand., see treatment of *Syntrichia*) a good species *fide* cf. Kramer, J. Hattori Bot. Lab. 65: 84, 1988 Afr4 Am6 Ant

[*Tortula fuscomucronata* (C. Müll.) Broth. Afr2 = *Pseudocrossidium replicatum* C. Müll. *fide* Townsend, Lindbergia 10: 178, 1985]

[*Tortula fuscoviridis* Card. Am6 ≡ *Syntrichia fuscoviridis* (Car.) Zand., see treatment of *Syntrichia*]

[*Tortula geheebiaeopsis* (C. Müll.) Broth. Afr4 Ant ≡ *Syntrichia geheebiaeopsis* (C. Müll.) Zand., see treatment of *Syntrichia*]

[*Tortula glacialis* (Kunze ex C. Müll.) Bomt. in Gay Am4 Am6 ≡ *Syntrichia glacialis* (Kunze ex C. Müll.) Zand., see treatment of *Syntrichia*]

Tortula goudotii (Hampe) Mitt. Am4

var. *boliviana* Broth. in Herz. Am4

[*Tortula graeffii* (Warnst.) Warnst. ≡ *Syntrichia subulata* fo. *graeffii* (Warnst.) Podp., Consp. Musc. Eur. 250, 1954 ≡ *Tortula subulata* var. *graeffii* Warnst.]

[var. *angustifolia* Warnst. Eur ≡ *Tortula subulata* fo. *angustifolia* (Wanst.) Sav.-Ljub., Novosti Sist. Niz. Rast. 6: 248, 1969 (1970)]

var. *latifolia* Warnst.

fo. *denticulata* Warnst., Hedwigia 52: 77, 1912 Eur

[*Tortula gregaria* Mitt. As 3 Am1 Am2 ≡ *Barbula indica* var. *gregaria* (Mitt.) Zand. *fide* Zander, Phytologia 44: 185, 1979]

Tortula grandiretis Broth. Eur As1

[*Tortula gromschii* Thér. Am6 ≡ *Syntrichia gromschii* (Thér.) Zand., see treatment of *Syntrichia*]

[*Tortula grossiretis* Card. Am6 Ant = *Tortula princeps* De Not. *fide* Lightowlers, Brit. Antarct. Surv. Bull. 67: 58, 1985) ≡ *Syntrichia princeps* (De Not.) Mitt.]

[var. *atrata* Card. Am6 = *Tortula princeps* De Not. *fide* Lightowlers, Brit. Antarct. Surv. Bull. 67: 58, 1985 ≡ *Syntrichia princeps* (De Not.) Mitt.]

[*Tortula guatemalensis* Bartr. Am2 = *Tortula bogotensis* (Hampe) Mitt. *fide* Mishler in Sharp et al., Moss Fl. Mex.]

Tortula guepinii (B.&S. in BSG) Broth. Eur Am1 Am2

[*Tortula handelii* Schiffn. Eur As5 ≡ *Syntrichia handelii* (Schiffn.) Bach.]

[var. *ferganensis* (Laz.) Kramer, Bryologist 81: 385, 1978 As1 ≡ *Syntrichia handelii* var. *ferganensis* (Laz.) Ochyra, Fragm. Florist. Geobot. 37: 212, 1992 (*Tortula*) As1]

[*Tortula hellenica* Schiffn. & Baumg. Eur = *Didymodon australasiae* (Hook. & Grev.) Zand. *fide* Sollman, Lindbergia 10: 54, 1984]

[var. *brevifolia* Schiffn. & Baumg. Eur = *Didymodon australasiae* (Hook. & Grev.) Zand. *fide* Sollman, Lindbergia 10: 54, 1984]

[*Tortula herzogii* Zand. Bryologist 82: 631, 1979 (*Calyptopogon angustifolius* Herz.; *nom. nov.* for *Tortula angustifolia* Herz. *hom. illeg.*) Am4 ≡ *Syntrichia angustifolia* (Herz.) Zand., see treatment of *Syntrichia*]

[*Tortula heteroloma* Card. Am2 ≡ *Hennediella heteroloma* (Card.) Zand., see treatment of *Hennediella*]

[*Tortula heteroneura* Card. Ant = *Tortula princeps* De Not. *fide* Lightowlers, Brit. Antarct. Surv. Bull. 67: 58, 1985 ≡ *Syntrichia princeps* (De Not.) Mitt.]

[*Tortula hildebrandtii* (C. Müll.) Broth. Afr2 Afr3 Afr4 = *Tortula schmidii* (C. Müll.) Broth. *fide* Magill, Fl. S. Afr. I. Mosses 1: 215, 1981 (1982) = *Tortula fragilis* Tayl. *fide* Frey & Kürschner, Nova Hedw. 46: 96, 1988 ≡ *Syntrichia fragilis* (Tayl.) Ochyra, Fragm. Florist. Geobot. 37: 212, 1992]

var. *papillosa* (Broth.) Wijk & Marg. Afr2

[*Tortula hirsuta* (Vent.) Laz. in Laz., Visotsk., & Mamatk., Biul. Mosk Obsh. Ispyt Prir. Otd. Biol. 73(3): 145, 1968 Eur ≡ *Syntrichia ruralis* var. *hirsuta* (Vent.) Podp.]

[*Tortula humida* Mitt. Am4 ≡ *Didymodon humidus* (Mitt.) Zand., see treatment of *Didymodon*]

Tortula humillima Card. & Copp. Afr1

[*Tortula husnotii* (Besch.) Broth. Am3 = *Barbula agraria* Hedw. *fide* Zander, Phytologia 44: 202, 1979]

[*Tortula hyalinotricha* (C. Müll.) Broth. Afr4 (= *Tortula princeps* De Not. *fide* Lightowlers, J. Bryol. 14: 290, 1986 (1987) ≡ *Syntrichia princeps* (De Not.) Mitt.) = *Tortula magellanica* Mont. in Gay fide Kramer, J. Hattori Bot. Lab. 65: 111, 1988 ≡ *Syntrichia magellanica* (Mont.) Zand.]

[*Tortula incerta* (Mitt.) Broth. Afr3 ≡ *Trichostomum incertum* (Mitt.) Zand., see treatment of *Trichostomum*]

[*Tortula inermis* (Brid.) Mont. Eur As1 As2 As3 As5 Afr1 Am1 Am2 ≡ *Syntrichia inermis* (Brid.) Bruch in Hueb.]

[var. *submarginata* Schiffn. Eur ≡ *Syntrichia inermis* var. *submarginata* (Schiffn.) Podp.]

[*Tortula intermedia* (Brid.) Berk., Handb. Brit. Mosses 134, 1855 Eur As1 As3 As5 Afr1 Afr4 Am1 Am2 ≡ *Syntrichia intermedia* Brid.]

[subsp. *handelii* (Schiffn.) Wijk & Marg. Eur As5 ≡ *Tortula handelii* Schiffn. *fide* Kramer, Bryoph. Biblioth. 21: 109, 1980]

[var. *calva* (Dur. & Sag.) Wijk & Marg. Eur ≡ *Syntrichia intermedia* var. *calva* (Dur. & Sag.) Delogn.]

var. *gelida* (Amann) Wijk & Marg. Eur

var. *planifolia* (Fleisch. & Warnst.) Wijk & Marg. Eur

var. *subcalcicola* (Giac.) Wijk & Marg. Eur

[*Tortula irregularis* Sim Afr4 = *Tortula hildebrandtii* (C. Müll.) Broth. *fide* Magill & Schelpe, Mem. Bot. Surv. S. Afr. 43: 25, 1979 = *Tortula schmidii* (C. Müll.) Broth. *fide* Magill, Fl. S. Afr. I. Mosses 1: 215, 1981 (1982) = *Tortula fragilis* Tayl. *fide* Frey & Kürschner, Nova Hedw. 46: 96, 1988 ≡ *Syntrichia fragilis* (Tayl.) Ochyra, Fragm. Florist. Geobot. 37: 212, 1992]

[*Tortula jaffuelii* Thér. Am6 ≡ *Syntrichia jaffuelii* (Thér.) Zand., see treatment of *Syntrichia*]

Tortula kabir-khanii (Broth.) Zand. (*Desmatodon*), see treatment of *Tortula* As3

[*Tortula khartoumensis* Pettet, Trans. Brit. Bryol. Soc. 5: 322, 1966 (1967) Afr2 ≡ *Hyophila khartoumensis* (Pettet) A. Sm. & H. Whiteh., J. Bryol. 8: 14, 1974 = *Desmatodon bogosicus* C. Müll. *fide* Townsend & Whitehouse, J. Bryol. 10: 475, 1979 and Corley et al., J. Bryol. 11: 620, 1981 (1982)]

[*Tortula kingii* Robins., Bryologist 70: 24, 1967 Am4 = *Tortula percarnosa* (C. Müll.) Broth. ≡ *Syntrichia percarnosa* (C. Müll.) Zand., see treatment of *Tortula*]

[*Tortula kunzeana* (C. Müll.) Mont. in Gay Afr4 Am6 ≡ *Hennediella kunzeana* (C. Müll.) Zand., see treatment of *Hennediella*]

[*Tortula lacerifolia* Williams Am4 ≡ *Syntrichia lacerifolia* (Williams) Zand., see treatment of *Syntrichia*]

Tortula laevinervis Broth. ex Dus. Am6

[*Tortula laevipila* (Brid.) Schwaegr. Eur As1 As3 As5 Afr1 Am1 Am6 Ant Austr2 ≡ *Syntrichia laevipila* Brid., see treatment of *Syntrichia*]

var. *gemmifera* Squiv. Eur

[var. *meridionalis* (Schimp.) Wijk & Marg. As1 As3 As5 Afr1 Am1 Am6 ≡ *Syntrichia laevipila* Brid. var. *meridionalis* (Schimp.) Jur.]

[var. *minor* (Biz.) Biz. As5 ≡ *Syntrichia virescens* var. *minor* (Biz.) Ochyra, Fragm. Florist. Geobot. 37: 212, 1992]

var. *notarisii* Barkm. Eur

var. *saccardoana* (De Not.) Barkm. Eur

var. *wachteri* Barkm. Eur

Tortula lanceola Zand. (*nom. nov.* for *Encalypta lanceolata* Hedw. *non Tortula lanceolata* (Hedw.) P. Beauv.), see treatment of *Tortula* Eur As2 As5 Afr1 Am1

var. *albidens* (Corb.) Zand. (*Pottia lanceolata* var.), see treatment of *Tortula* Eur

var. *angustata* (B.&S. in BSG) Zand. (*Anacalypta lanceolata* var.), see treatment of *Tortula* Eur As5 Afr1

var. *lejolisii* (Corb.) Zand. (*Pottia lanceolata* var.), see treatment of *Tortula* Eur Afr1

var. *leucodonta* (Schimp.) Zand. (*Pottia lanceolata* var.), see treatment of *Tortula* Eur

var. *macrophylla* (Warnst.) Zand. (*Pottia lanceolata* var.), see treatment of *Tortula* Eur

var. *microphylla* (Warnst.) Zand. (*Pottia lanceolata* var.), see treatment of *Tortula* Eur

var. *mucronata* (Amann) Zand. (*Pottia lanceolata* var.), see treatment of *Tortula* Eur

var. *ovalifolia* (Warnst.) Zand. (*Pottia lanceolata* var.), see treatment of *Tortula* Eur

var. *papillosa* (Corb.) Zand. (*Pottia lanceolata* var.), see treatment of *Tortula*]

var. *rigidior* (Schwaegr.) Zand. (*Encalypta lanceolata* var.), see treatment of *Tortula* Eur

[*Tortula latifolia* Bruch ex Hartm. Eur Am1 ≡ *Syntrichia latifolia* (Hartm.) Hüb.]

[var. *propagulifera* (Milde) Limpr., Laubm. Deutschl. 1(11): 678, 1888 Eur ≡ *Syntrichia latifolia* var. *propagulifera* (Milde) Amann]

[fo. *propagulifera* (Milde) Baumgartn. in Degen, Fl. Veleb. 3: 415, 1938 Eur ≡ *Tortula latifolia* var. *propagulifera* (Milde) Limpr. ≡ *Syntrichia latifolia* var. *propagulifera* (Milde) Amann]

[*Tortula latrobeana* (C. Müll.) Mitt. Austr1 = *Tortula antarctica* (Hampe in C. Müll.) Wils. in Hook. f. *fide* Kramer, J. Hattori Bot. Lab. 65: 84, 1988 ≡ *Syntrichia antarctica* (C. Müll. & Hampe) Zand., see treatment of *Syntrichia*]

Tortula laureri (Schultz) Lindb. Eur As1 As2 Afr4 Am1

var. *setschwanica* (Broth.) Zand. (*Desmatodon*), see treatment of *Tortula* As2

Tortula lazarenkoi Sav. in Sav.-Ljub. & Smirn., Handb. Mosses U.S.S.R. 330, 1970 (*nom. nov.* for *Tortula scabrinervis* Laz.) As1

[*Tortula lechleri* (C. Müll.) Broth. Am6 = *Tortula magellanica* Mont. in Gay *fide* Kramer, J. Hattori Bot. Lab. 65: 84, 1988 ≡ *Syntrichia magellanica* (Mont. in Gay) Zand., see treatment of *Syntrichia*]

[*Tortula leiostoma* Herz. Am2 Am4 = *Hennediella limbata* (Mitt.) Zand., see treatment of *Hennediella* and Long, J. Bryol. 10: 381, 1979]

Tortula leiostomoides P. Varde Afr2

[*Tortula lemniscata* Zand., Willdenowia 16: 253, 1986 Am4 = *Syntrichia aculeata* (Wils.) Zand., see treatment of *Tortula*]

Tortula leptopyxis (C. Müll.) Lindb. & Arn. As1

[*Tortula leptosyntrichia* (C. Müll.) Broth. Am6 = *Tortula robusta* Hook. & Grev. *fide* Lightowlers, Brit. Antarct. Surv. Bull. 67: 67, 1985 ≡ *Syntrichia robusta* (Hook. & Grev.) Zand. see treatment of *Syntrichia*]

Tortula leptotheca (Broth.) Chen As2

Tortula le-testui P. Varde Afr2

Tortula leucochlora (C. Müll.) Broth. Afr2

[*Tortula leucostega* (C. Müll.) Broth. Afr4 ≡ *Syntrichia leucostega* (C. Müll.) Zand., see treatment of *Syntrichia*]

[var. *trachyneura* (Dix.) Kramer, J. Hattori Bot. Lab. 65: 107, 1988 ≡ *Syntrichia leucostega* var. *trachyneura* (Dix.) Zand. (*Tortula*), see treatment of *Syntrichia*]

Tortula leucostoma (R. Br.) Hook. & Grev. Eur Am1 As1 As2

Tortula ligulata Herz. Am4

[*Tortula limensis* Williams Am4 ≡ *Syntrichia limensis* (Williams) Zand. see treatment of *Syntrichia*]

[*Tortula linearifolia* P. Beauv., Prodr. 92, 1805 (1806) (*Bryum acuminatum* Sw. *nom. inval.*; *Barbula acuminata* Brid. *nom. inval.*) = *Barbula agraria* Hedw. *fide* C. Müll., Syn. 1: 604, 1849]

[*Tortula linguifolia* Herz. Am4 ≡ *Syntrichia linguifolia* (Herz.) Zand., see treatment of *Syntrichia*]

[*Tortula lingulaefolia* Card. & Broth. Am6 = *Tortula arenae* (Besch.) Broth. *fide* Lightowlers, Brit. Antarct. Surv. Bull. 67: 46, 1985]

Tortula lingulata Lindb. Eur

subsp. *montenegrina* (Breidl. & Szysz.) Podp. Eur

[*Tortula lithophila* Dus. Am6 = *Sarconeurum glaciale* (C. Müll.) Card. & Bryhn *fide* Green, Brit. Antarct. Surv. Bull. 41–42: 187, 1975 and Lightowlers, Bryologist 88: 365, 1985]

Tortula litorea Card. in Card. & Broth. Am6

[*Tortula longimucronata* X.-j. Li, Acta Bot. Yunnan. 3: 107, 1981 As2 ≡ *Syntrichia longimucronata* (X.-j. Li) Zand., see treatment of *Syntrichia*]

[*Tortula longipedunculata* (C. Müll.) Broth. Afr4 ≡ *Desmatodon longipedunculatus* (C. Müll.) Magill, Fl. S. Afr. I. Mosses 1: 210, 1981 [1982] ≡ *Hennediella longipedunculata* (C. Müll.) Zand., see treatment of *Hennediella*]

[*Tortula lorentzii* (C. Müll.) Broth. Am6 ≡ *Chenia lorentzii* (C. Müll.) Zand., see treatment of *Chenia*]

[*Tortula luteola* Mitt. Austr1 = *Barbula crinita* Schultz *fide* Weber, Lindbergia 1: 216, 1972 ≡ *Pseudocrossidium crinitum* (Schultz) Zand., see treatment of *Pseudocrossidium*]

Tortula madagassa Dix. Afr3

[*Tortula magellanica* Mont. in Gay Am6 ≡ *Tortula princeps* var. *magellanica* (Mont.) Lightowlers, Brit. Antarct. Surv. Bull. 67: 61, 1985 ≡ *Syntrichia magellanica* (Mont.) Zand., see treatment of *Syntrichia*]

Tortula mairei Thér. & Trab. Afr1

Tortula marginata (B.&S. in BSG) Spruce Eur As3 As5 Afr1 Am1 Am3

subsp. *limbata* (Lindb.) Podp. Eur Afr1

fo. *ragusina* Latz., Bot. Centralbl. Beih. 48(2): 484, 1931 Eur

Tortula maritima (R. Br. ter) Zand. (*Dendia*), see treatment of *Tortula* Austr2

[*Tortula maudii* R. Brown ter Austr2 = *Tortula rubella* Hook. f. & Wils. *fide* Kramer, J. Hattori Bot. Lab. 65: 84, 1988 ≡ *Syntrichia rubella* (Hook. f. & Wils.) Zand., see treatment of *Syntrichia*]

Tortula meridionalis (Luis.) Mach. Eur

Tortula minima Herz. Am4

Tortula minor (C. Müll.) Zand. (*Pottia*), see treatment of *Tortula* As3 Afr2

var. *elatior* (C. Müll.) Zand. (*Pottia minor* var.), see treatment of *Tortula* As3 Afr2

[*Tortula minuscula* Williams Am4 ≡ *Didymodon minuscula* (Williams) Zand., see treatment of *Didymodon*]

Tortula minutirosula (C. Müll.) Broth. Am6

[*Tortula mniadelphus* (C. Müll.) Broth. Am2 Am4 Am6 = *Tortula quitoensis* Tayl. in Hook. *fide* Mishler in Sharp et al., Moss Fl. Mex. ≡ *Sagenotortula quitoensis* (Tayl. in Hook.) Zand., Phytologia 65: 430, 1989]

[*Tortula mniifolia* (Sull.) Mitt. Am2 Am3 Am4 ≡ *Dolotortula mniifolia* (Sull.) Zand., Phytologia 65: 426, 1989]

Tortula modica Zand. (*nom. nov.* for *Gymnostomum intermedium* Turn.), see treatment of *Tortula* Eur As2 Af1 Am1

var. *corsa* (Fleisch. & Warnst.) Zand. (*Pottia intermedia* var.), see treatment of *Tortula* Eur

var. *gymnandra* (Schiffn.) Zand. (*Pottia intermedia* var.), see treatment of *Tortula* Eur

var. *gymnogyna* (Schiffn.) Zand. (*Pottia intermedia* var.), see treatment of *Tortula* Eur

var. *revoluta* (Schiffn.) Zand. (*Pottia intermedia* var.), see treatment of *Tortula* Eur

var. *stenocarpa* (Velen.) Zand. (*Pottia intermedia* var.), see treatment of *Tortula* Eur

var. *tenuis* (Vent.) Zand. (*Pottia intermedia* var.), see treatment of *Tortula* Eur

[*Tortula mollis* (B.&S. ex C. Müll.) Broth. Afr2 ≡ *Syntrichia mollis* (B.&S. ex C. Müll.) Zand., see treatment of *Syntrichia*]

[*Tortula mongolica* (Boros) Ochyra & Pacyna, Fragm. Florist. & Geobot. 26: 73, 1980 (*Syntrichia*) As1 = *Tortula submontana* Broth. *fide* Kramer, Bryoph. Biblioth. 21: 89, 1980) ≡ *Syntrichia submontana* (Broth.) Ochyra, Fragm. Florist. Geobot. 37: 212, 1992]

[*Tortula monoica* Card. Am6 Ant = *Tortula princeps* var. *magellanica* (Mont.) Lightowlers, Brit. Antarct. Surv. Bull. 67: 61, 1985 also see Kramer, J. Hattori Bot. Lab. 65: 84, 1988 ≡ *Syntrichia magellanica* (Mont.) Zand., see treatment of *Syntrichia*]

[*Tortula montana* (Nees) Lindb. *hom. illeg.* = *Tortula intermedia* (Brid.) Berk.]

var. *paludosa* Röll, Hedwigia 56(1–3): 164, 1915 Eur
fo. *brevifolia* Arnold in Rabenh. Bryoth. Eur. 22: n. 1069, 1869 Eur
Tortula mucronifolia Schwaegr. Eur As1 As2 Afr1 Am1
var. *arctica* Hook. & Grev. Am1 Am6
var. *aristata* (Warnst.) C. Müll. ex Warnst. Eur
var. *brevifolia* (Warnst.) Warnst. As1
var. *emucronata* Arn. As1
var. *hyperborea* (Hag.) C. Jens. Eur
var. *mucronata* (Warnst.) Warnst. Eur
fo. *angustifolia* Warnst., Hedwigia 52: 79, 1912 Eur
fo. *breviseta* Warnst., Hedwigia 52: 79, 1912 Eur
fo. *latifolia* Warnst., Hedwigia 52: 79, 1912 Eur
fo. *longifolia* Warnst., Hedwigia 52: 79, 1912 Eur
fo. *perpusilla* Warnst., Hedwigia 52: 80, 1912 Eur
fo. *recurvata* Warnst., Hedwigia 52: 79, 1912 Am1
subfo. *crassiseta* Warnst., Hedwigia 52: 80, 1912 Eur
Tortula muralis Hedw. Cosm
subsp. *aestiva* (Brid. ex Hedw.) Meyl. Eur As2 Afr1 Am1 Am6
var. *aestiva* Brid. ex Hedw. Eur As2 Afr1 Am1 Am6
[var. *albida* Podp. Eur ≡ *Tortula muralis* fo. *albida* (Podp.) Podp., Consp. Musc. Eur. 247, 1954]
[var. *baetica* Cas. & Oliva, Acta Bot. Malacitana 7: 104, 1932 Eur ≡ *Tortula baetica* (Casas & Oliva) Guerra & Ros in Guerra, Ros & Carrión, J. Bryol. 17: 281, 1992]
var. *brevipedunculata* Rehm. ex Sim Afr4
var. *emarginata* Broth. Eur
var. *heribaudii* (Corb.) Corb. in Culm. Eur
var. *humilis* Papp. Eur
var. *israelis* (Biz. & Bil.) Biz. As5
var. *longipila* Dus. Am6
var. *madagascariensis* Thér. Afr3
var. *obcordata* (Schimp.) Limpr. (good taxon *fide* Iwatsuki & Noguchi, J. Hattori Bot. Lab. 37: 413, 1973) Eur As2 As5 Afr1
var. *sublaevis* Latz. Eur
var. *vulcanicola* (Schiffn.) Podp. Eur
fo. *aestiva* (Brid. ex Hedw.) Koppe, Mitth. Thüring. Bot. Vereins N.F. 50: 137, 1943 (*Tortula muralis* var.) Eur As2 Afr1 Am1 Am6
fo. *albida* (Podp.) Podp., Consp. Musc. Eur. 247, 1954 (*Tortula muralis* var.) Eur
fo. *brachyrhyncha* (Warnst.) Podp., Consp. Musc. Eur. 248, 1954 (*Tortula aestiva* var.) Eur
fo. *brevifolia* (Schiffn.) Podp., Consp. Musc. Eur. 248, 1954 (*Tortula aestiva* var.) As5
fo.? *calcarea* Podp., Consp. Musc. Eur. 247, 1954 *nom. inval. dispon. non cit.* (*Tortula muralis* var.) Eur
fo. *dentata* Loeske, Moosfl. Harz 178, 1903 Eur
fo.? *heterophylla* Podp., Consp. Musc. Eur. 247, 1954 *nom. inval. dispon. non cit.* (*Barbula muralis* var.) Eur
fo. *incana* Sapegen, Bot. Jahrb. System. 46(Beibl. 105): 14, 1911 (*Tortula muralis* var.) Eur Afr1 Afr4
fo. *melanocarpa* Podp., Sitzungsber. Köngl. Böhm. Ges. Wiss. Prag 1899(46): 16, 1899 Eur
fo. *obcordata* (Schimp.) Mönk., Laubm. Eur. 304, 1927 (*Tortula muralis* var.) Eur As2 As5 Afr1
fo.? *ovata* Podp., Consp. Musc. Eur. 248, 1954 *nom. inval. dispon. non cit.* (*Tortula muralis* var.) Eur
fo. *rupestris* (Schultz) Sapěhin, Bot. Jahrb. 46(Beibl. 104): 14, 1911 (*Tortula muralis* var.) Eur
fo. *terrestris* (Podp.) Podp., Consp. Musc. Eur. 248, 1954 (*Tortula aestiva* fo.) Eur
Tortula murina (C. Müll.) Broth. Austr1
[*Tortula nankomontana* Nog. = *Desmatodon latifolius* (Hedw.) Brid. *fide* Saito, J. Hattori Bot. Lab. 39: 71, 1975 ≡ *Tortula euryphylla* Zand.]
Tortula napoana De Not. Am4
[*Tortula navicularis* (Mitt.) Broth. Am3 ≡ *Quaesticula navicularis* (Mitt.) Zand., see treatment of *Quaesticula*]
Tortula nevadensis (Card. & Thér.) Zand. (*Pottia*), see treatment of *Tortula* Am1
[*Tortula nigra* Zand., Willdenowia 16: 255, 1986 = *Syntrichia percarnosa* (C. Müll.) Zand., see treatment of *Tortula*]
[*Tortula nigrescens* Griff. in Gangulee, Mosses E. India 3: 700, 1972 *nom. inval.* sin *descr.* As3 ≡ *Barbula nigrescens* Mitt. ≡ *Didymodon nigrescens* (Mitt.) Saito]
[*Tortula norvegica* (Web. & Mohr) Wahlenb. Am1 Eur Afr1 As1 As2 ≡ *Syntrichia norvegica* Web.]
[var. *calva* (Amonn) Kramer, Bryoph. Biblioth. 21: 132, 1980 (*Syntrichia aciphylla* var.) ≡ *Syntrichia norvegica* var. *calva* (Amann) Zand., see treatment of *Syntrichia*]
[*Tortula novoguinensis* Bartr. As4 = *Tortula caroliniana* Andrews *fide* Eddy, Handb. Males. Mosses 2: 246, 1991 = *Tortula amphidiacea* (C. Müll.) Broth. *fide* Mishler in Sharp et al., Moss Fl. Mex. ≡ *Syntrichia amphideacea* (C. Müll.) Zand., see treatment of *Syntrichia*]
Tortula oamaruana R. Br. ter Austr2
Tortula obscuriretis Thér. Am6
Tortula obtusifolia (Schwaegr.) Math. Eur As1 As2 As5 Afr1 Am1
[*Tortula obtusissima* (C. Müll.) Mitt. Am1 Am2 ≡ *Syntrichia obtusissima* (C. Müll.) Zand. (*Barbula*), see treatment of *Syntrichia*]
[var. *connectens* (Card.) Thér. Am2 = *Tortula obtusissima* (C. Müll.) Mitt. *fide* Crum, Appalachian-Ozarkian Element Moss Fl. Mexico (dissertation, Univ. Microfilms) 165, 1951 ≡ *Syntrichia obtusissima* (C. Müll.) Zand.]
[*Tortula oleaginosa* Stone, J. Bryol. 10: 117, 1978 Austr1 ≡ *Stonea oleaginosa* (Stone) Zand., Phytologia 65: 432, 1989]
[*Tortula ovata* (Hedw.) Dix., Rep. Brit. Bryol. Soc. 4: 118, 1939 (*Gymnostomum*) Eur As1 As5 Afr1 Am1 Am6 Austr1 ≡ *Pterygoneurum ovatum* (Hedw.) Dix.]
[*Tortula pachyneura* Dix. & Sainsb. Austr2 = *Willia calobolax* (C. Müll.) Lightowlers, J. Bryol. 13: 370, 1985
[*Tortula pagorum* (Milde) De Not. Eur Afr4 Am1 Am2 Austr1 (≡ *Tortula alpina* fo. *pagorum* (Milde) Squiv. *fide* Squivet de Carondelet, Rev. Bryol. Lichénol. 30: 215, 1962 [1963]) ≡ *Syntrichia pagorum* (Milde) Amann, see treatment of *Syntrichia*]
Tortula pallida (Lindb.) Zand. (*Pottia*), see treatment of *Tortula* Eur Afr1 As5
var. *longicuspis* (Warnst.) Zand. (*Pottia pallida* var. see treatment of *Tortula* Eur As5 Afr1
[*Tortula panduraefolia* (C. Müll. & Hampe) Broth. Austr1 = *Tortula antarctica* (Hampe in C. Müll.) Wils. in Hook. f. *fide* Kramer, J. Hattori Bot. Lab. 65: 84, 1988 ≡ *Syntrichia antarctica* (C. Müll. & Hampe) Zand., see treatment of *Syntrichia*]
[*Tortula panduriformis* R. Brown ter Austr2 = *Tortula rubella* Hook. f. & Wils. *fide* Kramer, J. Hattori Bot. Lab. 65: 84, 1988 ≡ *Syntrichia rubella* (Hook. f. & Wils.) Zand., see treatment of *Syntrichia*]
[*Tortula papillosa* Wils. in Spruce Eur Afr4 Am1 Am2 Am4 Am5 Am6 Austr1 Austr2 ≡ *Syntrichia papillosa* (Wils. in Spruce) Jur., see treatment of *Syntrichia*]
[var. *chilensis* Thér. Am4 ≡ *Syntrichia papillosa* var. *chilensis* (Thér.) Zand., see treatment of *Syntrichia*]
[var. *meridionalis* Warnst. Eur ≡ *Syntrichia papillosa* var. *meridionalis* (Warnst.) Zand.]

[*Tortula papillosissima* (Copp.) Broth. Am1 Eur (= *Tortula ruralis* (Hedw.) Gaertn., Meyer & Scherb. *fide* Corley et al., J. Bryol. 11: 619, 1981 [1982]) = *Tortula ruralis* subsp. *hirsuta* (Vent.) W. Kramer ≡ *Syntrichia ruralis* var. *hirsuta* (Vent.) Podp.]

Tortula parramattana Mitt. Austr1

Tortula parva Card. Am2

[var. *latifolia* Thér. Am2 = *Tortula fragilis* Tayl. *fide* Crum, Appalachian-Ozarkian Element Moss Fl. Mexico (dissertation, Univ. Microfilms) 166, 1951 ≡ *Syntrichia fragilis* (Tayl.) Ochyra, Fragm. Florist. Geobot. 37: 212, 1992]

Tortula paulsenii Broth. As1

Tortula perarmata Broth. Am6

[*Tortula percarnosa* (C. Müll.) Broth. Am4 Am6 ≡ *Syntrichia percarnosa* (C. Müll.) Zand., see treatment of *Tortula*]

[*Tortula perlimbata* Geh. ex Card. in Luis. *hom. illeg.* Afr1 ≡ *Tortula solmsii* fo. *perlimbata* Düll, Cryptogamie Bryol. Lichénol. 1: 179, 1980 (*Tortula perlimbata* Geh. ex Herz. *non* Geh. ex Card. in Luis. see Crosby & Bauer 1983) ≡ *Hennediella solmsii* fo. *perlimbata* (Düll) Zand., see treatment of *Hennediella*]

Tortula perpusilla (C. Müll.) Broth. Am6

[*Tortula peruviana* Mitt. Am4 ≡ *Barbula peruviana* (Mitt.) Jaeg., see treatment of *Barbula*]

[*Tortula petriei* Broth. ex Beck. Austr2 ≡ *Tortula arenae* subsp. *petriei* (Broth.) Lightowlers, J. Bryol. 13: 371, 1985 ≡ *Syntrichia arenae* subsp. *petriei* (Broth. ex Beck.) Zand., see treatment of *Syntrichia*]

[*Tortula phaea* (Hook. f. & Wils.) Dix. Austr2 ≡ *Syntrichia phaea* (Hook. f. & Wils.) Zand. see treatment of *Syntrichia*]

[*Tortula pichinchensis* Tayl. Am3 Am4 ≡ *Syntrichia pichinchensis* (Tayl.) Zand., see treatment of *Syntrichia*]

Tortula pierrotii Biz., Acta Bot. Acad. Sci. Hung. 18: 21, 1973 "*pierrotti*" Afr2

[*Tortula pilifera* Hook. As4? Afr2 Afr4 Am6 = *Barbula crinita* Schultz *fide* Magill, Fl. S. Afr. I. Mosses 1: 237, 1981 (1982) ≡ *Pseudocrossidium crinitum* (Schultz) Zand., see treatment of *Pseudocrossidium*]

[var. *denticulata* (Dus.) Wijk & Marg. Am6 = *Barbula crinita* Schultz *fide* Weber, Lindbergia 1: 216, 1972 ≡ *Pseudocrossidium crinitum* (Schultz) Zand., see treatment of *Pseudocrossidium*]

[var. *gracilis* (Dus.) Wijk & Marg. Am6 = *Barbula crinita* Schultz *fide* Weber, Lindbergia 1: 216, 1972 ≡ *Pseudocrossidium crinitum* (Schultz) Zand., see treatment of *Pseudocrossidium*]

[var. *longifolia* Sim Afr4 = *Barbula crinita* Schultz *fide* Weber, Lindbergia 1: 216, 1972 and *fide* Magill, Fl. S. Afr. I. Mosses 1: 237, 1981 (1982)]

[var. *oliviensis* (Card.) Wijk & Marg. Am6 = *Barbula crinita* Schultz *fide* Weber, Lindbergia 1: 216, 1972 ≡ *Pseudocrossidium crinitum* (Schultz) Zand., see treatment of *Pseudocrossidium*]

Tortula planicosta Herz. Am6

Tortula planifolia X.-j. Li, Acta Bot. Yunnan. 3: 109, 1981 As2

Tortula platyphylla Mitt. Am6

Tortula plinthobia (Sull. & Lesq.) Aust., Musci Appalach. Suppl. 1:10, 1878 cf. Zander, Bryologist 82: 551, 1979 Am1

Tortula podocarpi (C. Müll.) Broth. Am6

Tortula polycarpa Dus. Am6

Tortula polylepidis Herz. Am4

[*Tortula polyseta* (C. Müll.) Warnst. Am2 Am4 ≡ *Hennediella polyseta* (C. Müll.) Zand., see treatment of *Hennediella*]

[*Tortula porphyreoneura* (C. Müll.) Townsend, J. Bryol. 10: 576, 1979 (*Barbula*) Afr2 Afr4 ≡ *Psuedocrossidium porphyreoneurum* (C. Müll.) Zand., see treatment of *Pseudocrossidium*]

Tortula porteri (Jam. in Aust.) Zand., see treatment of *Tortula* Am1

[*Tortula preissiana* (C. Müll.) Broth. Austr1 = *Tortula antarctica* (Hampe in C. Müll.) Wils. in Hook. f. *fide* Kramer, J. Hattori Bot. Lab. 65: 84, 1988 ≡ *Syntrichia antarctica* (Hampe in C. Müll.) Zand., see treatment of *Syntrichia*]

[*Tortula princeps* De Not. Eur As1 As3 As5 Afr1 Afr4 Am1 Am2 Am4 Am6 Austr1 Austr2 Oc Ant ≡ *Syntrichia princeps* (De Not.) Mitt., see treatment of *Syntrichia*]

[subsp. *echinata* (Schiffn.) Kramer, Bryophyt. Biblioth. 21: 86, 1980 (*Tortula*) Eur As5 (= *Tortula princeps* De Not. *fide* Corley et al., J. Bryol. 11: 618, 1981 [1982]) ≡ *Tortula princeps* subsp. *echinata* (Schiffn.) Kramer, Bryophyt. Biblioth. 21: 86, 1980 ≡ *Syntrichia princeps* var. *echinata* (Shiffn.) Zand., see treatment of *Syntrichia*]

[subsp. *parnassica* (Schiffn.) Kramer, Bryophyt. Biblioth. 21: 87, 1980 (*Tortula muelleri* var.) Eur ≡ *Syntrichia princeps* var. *parnassica* (Schiffn.) Podp.]

[var. *brachycarpa* De Not. Eur ≡ *Syntrichia princeps* var. *brachycarpa* (Zand.) De Not., see treatment of *Syntrichia*]

[var. *conferta* (Bartr.) Lightowlers, Brit. Antarct. Surv. Bull. 67: 61, 1985 (*Tortula*) Ant (= *Tortula princeps* De Not. *fide* Lightowlers, J. Bryol. 14: 290, 1986 [1987] ≡ *Syntrichia princeps* (De Not.) Mitt., see treatment of *Syntrichia*) ≡ *Syntrichia conferta* (Bartr.) Zand., see treatment of *Syntrichia*]

[var. *echinata* (Schiffn.) Biz. Eur As5 (= *Tortula princeps* De Not. *fide* Corley et al., J. Bryol. 11: 619, 1981 [1982]) ≡ *Tortula princeps* subsp. *echinata* (Schiffn.) Kramer, Bryophyt. Biblioth. 21: 86, 1980 ≡ *Syntrichia princeps* var. *echinata* (Shiffn.) Zand., see treatment of *Syntrichia*]

[var. *magellanica* (Mont.) Lightowlers, Brit. Antarct. Surv. Bull. 67: 61, 1985 (*Tortula*) Am6 Ant ≡ *Syntrichia* magellanica (Mont.) Zand., see treatment of *Syntrichia*]

[var. *parnassica* (Schiffn.) Wijk & Marg. Eur ≡ *Syntrichia princeps* var. *parnassica* (Schiffn.) Podp.]

[*Tortula propagulata* Laz., J. Bot. Acad. Sci. Ukr. 3(3–4): 61, 1946 As1 = *Tortula laevipila* (Brid.) Schwaegr. var. *propagulifera* Lindb. *fide* Sav.-Ljub. & Smirn., Handb. Mosses USSR 352, 1970]

[*Tortula propagulosa* Sharp Am1 = *Rhachithecium perpusillum* (Thwait. & Mitt.) Broth. *fide* Zander, Bryologist 81: 458, 1978]

Tortula propinqua (C. Müll.) Broth. Austr1

Tortula protobryoides Zand. (*nom. nov.* for *Phascum bryoides* Dicks.), see treatment of *Tortula* Eur As5 Am1 Am2

var. *brevifolia* (De Not.) Zand. (*Phascum bryoides* var.), see treatment of *Tortula* Eur

var. *thornhillii* (Wils.) Zand. (*Phascum bryoides* var.), see treatment of *Tortula*] Eur

[*Tortula prostrata* Mont. Am4 Am6 ≡ *Syntrichia prostrata* (Mont.) Zand. see treatment *Syntrichia*]

fo. *angustofolia* Herz., Arch. Es. Farm. Fac. Ci. Méd. Córdoba (Secc. Ci.) 7: 41, 1938 *nom. inval. descr.* hisp. Am6

fo. *latifolia* Herz., Arch. Es. Farm. Fac. Ci. Méd. Córdoba (Secc. Ci.) 7: 41, 1938 *nom. inval. descr.* hisp. Am6

[*Tortula pseudaciphylla* (Kindb.) Broth. Am1 = *Tortula intermedia* (Brid.) Berk. *fide* Steere in Grout, Moss Fl. N. Am. 1: 246, 1939]

[*Tortula pseudodesertorum* Fröhl., Ann. Naturhist. Mus. Wien 67: 155, 1964 As3 ≡ *Syntrichia pseudohandelii* (Fröhl.), Agnew & Vondr., Feddes Rep. 86(6–8): 401, 1975]

[*Tortula pseudohandelii* Fröhl., Ann. Naturhist. Mus. Wien 67: 155, 1964 As3 ≡ *Syntrichia pseudohandelii* (Fröhl.), Agnew & Vondr., Feddes Rep. 86(6–8): 401, 1975]

Tortula pseudolatifolia Card. Am6

Tortula pseudoprinceps Dix. As3

[*Tortula pseudorobusta* Dus. Am6 ≡ *Syntrichia pseudorobusta*

(Dus.) Zand., see treatment of *Syntrichia*]

Tortula pugionata (C. Müll.) Broth. As2

[*Tortula pulvinata* (Jur.) Limpr. Eur Afr1 Am1 = *Tortula virescens* (De Not.) De Not. *fide* Kramer, Bryoph. Biblioth. 21: 99, 1980 ≡ *Syntrichia virescens* (De Not.) Ochyra, Fragm. Florist. Geobot. 37: 212, 1992]

subsp. *papillinervis* (C. Müll. & Kindb.) Par. *nom. dub.* Am1

Tortula pulvinatula Dus. Am6

[*Tortula purpureovelutina* Herz. Am6 = *Barbula crinita* Schultz *fide* Sollman, Lindbergia 16: 22, 1990 ≡ *Pseudocrossidium aureum* *fide* Zander, Phytologia 44: 207, 1979 = *Pseudocrossidium crinitum* (Schultz) Zand., see treatment of *Pseudocrossidium*]

[*Tortula pusilla* Ångst. *hom. illeg.* Am6 = *Tortula princeps* var. *magellanica* (Mont.) Lightowlers, Brit. Antarct. Surv. Bull. 67: 61, 1985 ≡ *Syntrichia magellanica* (Mont.) Zand., see treatment of *Syntrichia*]

[*Tortula pygmaea* Dus. Am6 ≡ *Syntrichia pygmaea* (Dus.) Zand. see treatment of *Syntrichia*]

[*Tortula quitoensis* Tayl. Am2 Am4 Am6 ≡ *Sagenotortula quitoensis* (Tayl. in Hook.) Zand., Phytologia 65: 430, 1989]

[*Tortula raddei* Broth. As1 ≡ *Microbryum raddei* (Broth.) Zand., see treatment of *Microbryum*]

Tortula rallieri Card. Afr4

[*Tortula ramosissima* Thér. Am4 ≡ *Syntrichia ramosissima* (Thér.) Zand., see treatment of *Syntrichia*]

Tortula randii (Kenn.) Zand. (*Pottia*), see treatment of *Tortula* Eur Am1

Tortula raucopapillosa (X.-j. Li) Zand. (*Desmatodon*), see treatment of *Tortula* As2

Tortula readeri (C. Müll.) Broth. Austr1

Tortula recurvata Hook. (good species *fide* Catcheside, Mosses S. Austr. 150, 1980 as *Desmatodon*) Afr4 Austr1

[*Tortula recurvifolia* (Schimp.) Aust., Musci Appalach. Suppl. 1: 10, 496, 1878 *hom. illeg.* (*Barbula*) Eur As2 As3 Am1 = *Barbula reflexa* (Brid.) Brid. *fide* Limpricht 1888 ≡ *Didymodon fallax* var. *reflexus* (Brid.) Zand., Bryologist 83: 230, 1980 = *Didymodon rigidicaulis* (C. Müll.) Saito *fide* Saito, J. Hattori Bot. Lab. 39: 502, 1975 = *Didymodon ferrugineus* (Schimp. ex Besch.) Hill, J. Bryol. 11: 599, 1981 (1982)]

[*Tortula reflexa* X.-j. Li, Acta Bot. Yunnan. 3: 109, 1981 *hom. illeg. non* Brid. As2 ≡ *Syntrichia reflexa* Zand. *nom. nov.*, see treatment of *Syntrichia*]

[*Tortula rehmannii* (C. Müll.) Sim Afr2 Afr4 ≡ *Barbula rehmannii* C. Müll. *fide* Magill, Fl. S. Afr. I. Mosses 1: 245, 1981 (1982)]

[*Tortula remotifolia* Tak. As2 = *Tortula norvegica* (Web. & Mohr) Wahlenb. *fide* Kramer, Bryoph. Biblioth. 21: 128, 1980 ≡ *Syntrichia norvegica* Web.]

[*Tortula reticularia* (C. Müll.) Broth. Afr4 = *Tortula papillosa* Wils. in Spruce *fide* Magill & Schelpe, Mem. Bot. Surv. S. Afr. 43: 25, 1979 and see Magill, Fl. S. Afr. I. Mosses 1: 218, 1981 (1982)]

Tortula revolutifolia Laz. As1

Tortula revolvens (Schimp.) Roth Eur Afr1

var. *obtusata* Reim. Eur As5

[*Tortula rhizophylla* (Sak.) Iwats. & Saito Eur As2 Am1 Am2 Austr1 ≡ *Chenia rhizophylla* (Sak.) Zand., Phytologia 65: 425, 1989 = *Chenia leptophylla* (C. Müll.) Zand., see treatment of *Chenia*]

Tortula rhodonia Zand. (*nom. nov.* for *Desmatodon wilczekii* Meyl.), see treatment of *Tortula* Eur

[*Tortula rigescens* Broth. & Geh. in Broth. As5 ≡ *Syntrichia rigescens* (Broth. & Geh. in Broth.) Ochyra, Fragm. Florist. Geobot. 37: 212, 1992]

Tortula ripicola Thér. Am2

[*Tortula rivularis* Dus. Am6 = *Tortula fontana* (C. Müll.) Broth. *fide* Lightowlers, Brit. Antarct. Surv. Bull. 67: 53, 1985 ≡ *Syntrichia fontana* (C. Müll. in Neum.) Zand., see treatment of *Syntrichia*]

[*Tortula robusta* Hook. & Grev. Am5 Am6 Austr1 Ant ≡ *Syntrichia robusta* (Hook. & Grev.) Zand. see treatment of *Syntrichia*]

[var. *laxa* Bartr. Am6 = *Tortula robusta* Hook. & Grev. *fide* Lightowlers, Brit. Antarct. Surv. Bull. 67: 67, 1985 ≡ *Syntrichia robusta* (Hook. & Grev.) Zand. see treatment of *Syntrichia*]

[var. *recurva* Lightowlers, Brit. Antarct. Surv. Bull. 64: 64, 1984 Ant ≡ *Syntrichia robusta* var. *recurva* (Lightowlers) Zand. (*Tortula robusta* var.) see treatment of *Syntrichia*]

[var. *runcinata* (C. Müll.) Broth. Am4 Am6 = *Tortula robusta* Hook. & Grev. *fide* Lightowlers, Brit. Antarct. Surv. Bull. 67: 67, 1985 ≡ *Syntrichia robusta* (Hook. & Grev.) Zand. see treatment of *Syntrichia*]

[*Tortula robustula* Card. Am6 = *Tortula filaris* (C. Müll.) Broth. *fide* Lightowlers, Brit. Antarct. Surv. Bull. 67: 49, 1985 ≡ *Syntrichia filaris* (C. Müll. in Neum.) Zand., see treatment of *Tortula*]

[*Tortula rubella* Hook. f. & Wils. good species *fide* Kramer, J. Hattori Bot. Lab. 65: 84, 1988 Afr4 Austr1 Austr2 = *Syntrichia rubella* (Hook. f. & Wils.) Zand., see treatment of *Syntrichia*]

[*Tortula rubra* Mitt. in Hook. f. Am6 Austr1 Austr2 Ant ≡ *Syntrichia rubra* (Mitt. in Hook. f.) Zand. see treatment of *Syntrichia*]

[var. *subantarctica* (Sainsb.) Lightowlers, J. Bryol. 13: 373, 1985 (*Tortula subantarctica*) *inval. basion. non cit.* Austr2 ≡ *Syntrichia rubra* var. *subantarctica* (Sainsb.) Zand.]

[*Tortula rubripila* Dix. As3 = *Tortula norvegica* (Web. Lindb. *fide* Kramer, Bryoph. Biblioth. 21: 128, 1980 ≡ *Syntrichia norvegica* Web.]

Tortula rufa (Besch.) Broth. *hom. illeg.* Afr3

[*Tortula ruraliformis* (Besch.) Grout, Mosses Handl. Micr. 167, 1903 Eur As5 Afr1 Am1 ≡ *Tortula ruralis* var. *ruraliformis* (Besch.) Wild. = *Syntrichia ruralis* var. *arenicola* (Braithw.) Amann, cf. Kramer, Bryoph. Biblioth. 21: 116, 1980]

[*Tortula ruraliformis* (Besch.) Ingh. (inval. *fide* Crundwell, Trans. Brit. Bryol. Soc. 6: 325, 1971) Eur As5 Afr1 Am1 ≡ *Tortula ruralis* var. *ruraliformis* (Besch.) Wild. = *Syntrichia ruralis* var. *arenicola* (Braithw.) Amann, cf. Kramer, Bryoph. Biblioth. 21: 116, 1980]

[var. *subpapillosissima* (Biz. & Pierr.) Kramer, Bryophyt. Biblioth. 21: 120, 1980 Afr1 ≡ *Syntrichia ruralis* var. *subpapillosissima* (Biz. & Pierr.) Zand., see treatment of *Syntrichia*]

[*Tortula ruralis* (Hedw.) Gaertn., Meyer & Scherb. Eur As1 As2 As3 As5 Afr1 Afr2 Afr4 Am1 Am2 Am4 Am5 Austr1 Oc ≡ *Syntrichia ruralis* (Hedw.) Web. & Mohr]

[subsp. *calcicola* (Amann) Giac. Eur ≡ *Tortula calcicolens* Kramer, Bryophyt. Biblioth. 21: 90, 1980 (≡ *Tortula ruralis* (Hedw.) Gaertn., Meyer & Scherb. *fide* Corley et al., J. Bryol. 11: 619, 1981 [1982]) ≡ *Syntrichia calcicola* Amann]

[subsp. *hirsuta* (Vent.) Kramer, Bryophyt. Biblioth. 21: 126, 1980 Eur As2 Am1 ≡ *Syntrichia ruralis* var. *hirsuta* (Vent.) Podp.]

[subsp. *ruraliformis* (Besch.) Dix. Eur As5 Afr1 Am1 = *Syntrichia ruralis* var. *arenicola* (Braithw.) Amann cf. Kramer, Bryoph. Biblioth. 21: 116, 1980]

[var. *aciphylloides* Podp. Eur = *Tortula ruraliformis* (Besch.) Ingh. *fide* Kramer, Bryoph. Biblioth. 21: 116, 1980 = *Syntrichia ruralis* var. *arenicola* (Braithw.) Amann cf. Kramer, Bryoph. Biblioth. 21: 116, 1980]

[var. *alpina* Wahlenb. Eur As1 Afr1 Am1 = *Tortula norvegica* (Web.) Wahlenb. ex Lindb. *fide* Corley et al., J. Bryol. 11: 619, 1981 (1982)]

var. *alpina* De Not., Mem. R. Acc. Sc. Torino Cl. Sc. Fis. Mat. 40: 291, 1838 *hom. illeg.* Eur
var. *brevifolia* Warnst., Krypt. Fl. Brandenburg 2: 277, 1904 *hom. illeg.* (*Tortula ruralis* fo. *brevifolia* Arn. in Rabenh., 1869) Eur
[var. *calcicola* (Amann) Barkm. Eur ≡ *Tortula calcicolens* Kramer, Bryophyt. Biblioth. 21: 90, 1980 (= *Tortula ruralis* (Hedw.) Gaertn., Meyer & Scherb. *fide* Corley et al., J. Bryol. 11: 619, 1981 [1982]) ≡ *Syntrichia ruralis* var. *calcicola* (Ammann) Mönk. ≡ *Syntrichia calcicola* Amann]
[var. *densa* Velen. Eur = *Tortula ruralis* (Hedw.) Gaertn., Meyer & Scherb. var. *ruralis fide* Kramer, Bryoph. Biblioth. 21: 122, 1980 ≡ *Syntrichia ruralis* (Hedw.) Web. & Mohr]
[var. *gigantea* (Lesq.) Koch Am1 ≡ *Syntrichia ruralis* var. *gigantea* (Lesq.) Zand., see treatment of *Syntrichia*]
[var. *gracilis* C. Jens. Am1 ≡ *Syntrichia ruralis* var. *gracilis* (C. Jens.) Zand., see treatment of *Syntrichia*]
[var. *gypsophila* Amann ex Roth Eur ≡ *Tortula caninervis* var. *gypsophila* (G. Roth) Kramer, Bryophyt. Biblioth. 21: 108, 1980 ≡ *Syntrichia caninervis* var. *gypsophila* (Roth.) Ochyra, Fragm. Florist. Geobot. 37: 212, 1992]
[var. *hirsuta* (Vent.) Par. Eur As2 Am1 ≡ *Syntrichia ruralis* var. *hirsuta* (Vent.) Podp.]
[var. *latifolia* Arnott, Disp. Méth. Esp. Mouss. 38, 1825 [1826] Eur ≡ *Tortula latifolia* (Arnott) Hartm.]
[var. *pontresinae* (Geh. & Warnst.) Wijk & Marg. Eur ≡ *Syntrichia ruralis* var. *pontresinae* (Geh. & Warnst.) Podp.]
[var. *pseudoaciphylla* Wint. Eur = *Tortula norvegica* (Web.) Lindb. var. *norvegica fide* Kramer, Bryoph. Biblioth. 21: 128, 1980 ≡ *Syntrichia norvegica* Web.]
[var. *rufipila* Herz. Eur = *Tortula norvegica* (Web.) Lindb. var. *norvegica fide* Kramer, J. Hattori Bot. Lab. 65: 84, 1988]
[var. *ruraliformis* (Besch.) Wild. Eur As5 Afr1 Am1 = *Syntrichia ruralis* var. *arenicola* (Braithw.) Amann cf. Kramer, Bryoph. Biblioth. 21: 116, 1980]
[var. *spiralis* Herz. Am4 ≡ *Syntrichia ruralis* var. *spiralis* (Herz.) Zand., see treatment of *Syntrichia*]
[var. *submamillosa* Kramer, Bryophyt. Biblioth. 21: 127, 1980 As5 ≡ *Syntrichia ruralis* var. *submamillosa* (Kramer) Zand., see treatment of *Syntrichia*]
[var. *subpapillosissima* Biz. & Pierr., Acta Bot. Acad. Sci. Hungaricae 18: 11, 1973 Afr1 ≡ *Tortula ruraliformis* var. *subpapillosissima* (Biz. & Pierr.) Kramer, Bryoph. Biblioth. 21: 120, 1980 ≡ *Syntrichia ruralis* var. *subpapillosissima* (Biz. & Pierr.) Zand., see treatment of *Syntrichia*]
[var. *substereidosa* Kramer, Bryophyt. Biblioth. 21: 125, 1980 As5 ≡ *Syntrichia ruralis* var. *substereidosa* (Kramer) Zand., see treatment of *Syntrichia*]
fo. *brevifolia* Arn. in Rabenh., Bryoth. Eur. 1069, 1869 Eur
fo. *contorta* Podp., Sitzungsber. Köngl. Böhm. Ges. Wiss. Prag 1899(46): 16, 1899 Eur
[fo. *fallax* Herz., Bot. Z. 93: 41, 1944 Eur ≡ *Syntrichia ruralis* fo. *fallax* (Herz.) Podp.]
[fo. *fuscipila* Herz., Kryptog. Forsch. 4: 280, 1920 Eur ≡ *Syntrichia ruralis* fo. *rufipila* Herz. ex Rübel *fide* Podp., Consp. Musc. Eur. 256, 1954]
[fo. *gracilis* Meyl., Rev. Bryol. 52: 54, 1925 Eur ≡ *Syntrichia ruralis* fo. *gracilis* (Meyl.) Podp.]
fo. *pseudoruraliformis* Tosco, Webbia 28: 286, 1973 Eur
[fo. *rufoneura* Podp., Sitzungsber. Köngl. Böhm. Ges. Wiss. Prag 1899(46): 16, 1899 Eur ≡ *Syntrichia ruralis* fo. *rufoneura* (Podp.) Podp.]
[fo. *subrufa* Podp., Veröff. Geobot. Inst. Rübel Zürich 1: 244, 1924 Eur ≡ *Syntrichia ruralis* fo. *subrufa* (Podp.) Podp.]
[fo. *viridis* Matous., Hedwigia 44: 31, 1904 Eur ≡ *Syntrichia ruralis* fo. *viridis* (Matous.) Podp.]
Tortula sabinae Townsend, Lindbergia 10: 176, 1985 Afr2
[*Tortula saharae* Trab. Afr1 = *Tortula caninervis* (Mitt.) Broth. *fide* Kramer, Bryoph. Biblioth. 21: 103, 1980 ≡ *Syntrichia caninervis* Mitt.]
Tortula sainsburyana Zand. (*nom. nov.* for *Pottia stevensii*) Austr2
Tortula santiagensis Broth. Am6
Tortula santorinensis Schiffn. Eur
fo. *typica* Schiffn., Verh. Zool.-Bot. Ges. Wien 69: 329, 1920 *nom. illeg.* ≡ *Tortula santorinensis* Schiffn. fo. *santorinensis*]
var. *apiculata* Schiffn. Eur
Tortula satoi Sak. As2
Tortula savatieri (Besch.) Broth. Am4
[*Tortula saxicola* Card. Am6 ≡ *Syntrichia saxicola* (Card.) Zand., see treatment of *Syntrichia*]
[*Tortula scabrella* Dus. Am6 ≡ *Syntrichia scabrella* (Dus.) Zand., see treatment of *Syntrichia*]
[*Tortula scabrinervis* (C. Müll.) Mont. in Gay Am4 Am6 ≡ *Syntrichia scabrinervis* (C. Müll.) Zand., see treatment of *Syntrichia*]
[*Tortula scabrinervis* Laz. As1 *hom. illeg.* ≡ *Tortula lazarenkoi* L. Sav. in Sav.-Ljub. & Smirn., Handb. Mosses U.S.S.R. 330, 1970 *nom. nov.*]
[*Tortula schmidii* (C. Müll.) Broth. As3 Afr3 Afr4 = *Tortula fragilis* Tayl. *fide* Frey & Kürschner, Nova Hedw. 46: 96, 1988) ≡ *Syntrichia fragilis* (Tayl.) Ochyra, Fragm. Florist. Geobot. 37: 212, 1992]
[*Tortula schnyderi* (C. Müll.) Broth. Am6 ≡ *Syntrichia schnyderi* (C. Müll.) Zand., see treatment of *Syntrichia*]
[*Tortula scotteri* Zand. & Steere, Bryologist 81: 463, 1978 Am1 ≡ *Hilpertia scotteri* (Zand. & Steere) Zand., Phytologia 65: 428, 1989]
[*Tortula searlii* R. Brown ter Austr2 = *Tortula rubella* Hook. f. & Wils. *fide* Kramer, J. Hattori Bot. Lab. 65: 84, 1988 ≡ *Syntrichia rubella* (Hook. f. & Wils.) Zand., see treatment of *Syntrichia*]
[*Tortula semirubra* (C. Müll.) Broth. Afr4 = *Tortula princeps* De Not. *fide* Lightowlers, J. Bryol. 14: 290, 1986 (1987) ≡ *Syntrichia princeps* (De Not.) Mitt.]
[*Tortula serrata* Dix. (*nom. legit.* contrary to Index Muscorum; cf. Lightowlers, J. Bryol. 13: 373, 1985) Austr2 ≡ *Syntrichia serrata* (Dix.) Zand., see treatment of *Syntrichia*]
[*Tortula serripungens* (Lor. & C. Müll.) Broth. Am4 Am6 ≡ *Syntrichia serripungens* (Lor. & C. Müll.) Zand., see treatment of *Syntrichia*]
[var. *exesa* (C. Müll.) Herz. Am6 ≡ *Syntrichia serripungens* var. *excesa* (C. Müll.) Zand., see treatment of *Syntrichia*]
[*Tortula serrulata* Hook. & Grev. (see Lightowlers, J. Bryol. 13, 373, 1985) Am6 ≡ *Hennediella serrulata* (Hook. & Grev.) Zand., see treatment of *Hennediella*]
[*Tortula serrulata* Warnst. *hom. illeg.* ≡ *Tortula subulata* var. *serrulata* Warnst.]
fo. *tenuiseta* Warnst., Hedwigia 52: 76, 1912 Eur
fo. *minor* (A. Braun) Warnst., Hedwigia 52: 76, 1912 (*Syntrichia subulata* var.) Eur
[*Tortula sinensis* (C. Müll.) Broth. in Levier Eur As1 As2 As3 As5 Afr1 ≡ *Syntrichia sinensis* (C. Müll.) Ochyra, Fragm. Florist. Geobot. 37: 212, 1992]
Tortula sinuata Bartr., Rev. Bryol. Lichénol. 33: 325, 1964–65 [1965] Am6
[*Tortula socialis* Dus. Am6 ≡ *Syntrichia socialis* (Dus.) Zand., see treatment of *Tortula*]
Tortula solmsii (Schimp.) Limpr. Eur Afr1

var. *minor* Roth Afr1
fo. *perlimbata* Düll, Cryptogamie Bryol. Lichénol. 1: 179, 1980 (*Tortula perlimbata* Geh. ex Herz. *non* Geh. ex Card. in Luis. see Crosby & Bauer 1983) Afr1
Tortula solomensis (Broth.) Zand. (*Desmatodon*), see treatment of *Tortula* As2
Tortula sordida Herz. Am4
Tortula splachnoides (Hornsch.) Zand. (*Phascum*), see treatment of *Tortula* Afr4
[*Tortula spuria* Amann Eur ≡ *Tortula caininervis* subsp. *spuria* (Amann) Kramer, Bryophyt. Biblioth. 21: 106, 1980 ≡ *Syntrichia caninervis* var. *spuria* (Amann) Zand., see treatment of *Syntrichia*]
Tortula squarripila Thér. Am6
[*Tortula stanfordensis* Steere Eur Am1 Austr1 ≡ *Hyophila stanfordensis* (Steere) A. Sm. & Whiteh. ≡ *Hennediella stanfordensis* (Steere) Blockeel, J. Bryol. 16: 191, 1991]
[*Tortula stenophylla* Mitt. As2 = *Oxystegus tenuirostris* (Hook. & Grev.) A. Sm. (as *Oxystegus cylindricus* (Brid.) Hilp.) *fide* Saito, J. Hattori Bot. Lab. 39: 437, 1975 ≡ *Trichostomum tenuirostre* (Hook. & Tayl.) Lindb.]
Tortula stenophylla Card. & Broth. *hom. illeg.* Am6
Tortula streptopogoniacea (C. Müll.) Broth. Austr1
[*Tortula subantarctica* Sainsb. Austr2 ≡ *Tortula rubra* var. *subantarctica* (Sainsb.) Lightowlers, J. Bryol. 13: 373, 1985]
[*Tortula subaristata* (B.&S. ex C. Müll.) Broth. Afr2 ≡ *Syntrichia subaristata* (B.&S. ex C. Müll.) Zand., see treatment of *Syntrichia*]
Tortula subbrunnea Broth. & Watts Austr1
Tortula subcaroliniana Biz., Svensk Bot. Tidskr. 63: 446, 1969 Afr2
Tortula sublimbata (Mitt.) Broth. As2
[*Tortula submontana* Broth. As1 ≡ *Syntrichia submontana* (Broth.) Ochyra, Fragm. Florist. Geobot. 37: 212, 1992]
[*Tortula subobliqua* Williams Am4 = *Chenia subobliqua* (Williams) Zand., Phytologia 65: 425, 1989]
Tortula subrufa Card. in Grand. Afr3
[*Tortula subspathulata* (C. Müll.) Broth. Afr2 = *Tortula fragilis* Tayl. *fide* Townsend, Lindbergia 10: 177, 1985 ≡ *Syntrichia fragilis* (Tayl.) Ochyra, Fragm. Florist. Geobot. 37: 212, 1992]
[*Tortula subspiralis* (Hampe) Broth. Austr1 = *Barbula hornschuchiana* Schultz *fide* Catcheside, Mosses S. Austr. 178, 1980] ≡ *Psuedocrossidium hornschuchianum* (Schultz) Zand. Phytologia 44: 205, 1979]
[*Tortula subtorquatifolia* Dix. Afr4 = *Desmatodon convolutus* (Brid.) Grout *fide* Magill, Fl. S. Afr. I. Mosses 1: 210, 1981 [1982] = *Tortula atrovirens* (Sm.) Lindb.]
Tortula subtranscaspica Fröhl., Mitt. Thüring. Bot. Ges. 1(2–3): 63, 1955 Eur?
Tortula subulata Hedw. Eur As1 As2 As3 As5 Afr1 Am1 Am3
subsp. *angustata* (Schimp.) Kindb. Eur Afr1 Am1
subsp. *serrulata* (Warnst.) Giac. Eur
var. *angustata* (Schimp.) Limpr. Eur Afr1 Am1
var. *angustifolia* (Schimp. ex Milde) Röll, Hedwigia 56(1–3): 162, 1915 Eur
var. *bifrons* Hag. Eur
var. *buergeneri* Warnst. Eur
var. *elongata* (Lang.) Par. Eur Am1
var. *flaviseta* Warnst. Eur
var. *graeffii* Warnst. Eur
var. *laxifolia* Warnst. Eur
var. *microphylla* Warnst. Eur
var. *minor* Wahlenb. Eur
var. *mutica* (Schimp.) Limpr. Eur Afr1 Am1
var. *serrulata* Warnst. Eur
var. *subinermis* (B.&S. in BSG) Wils. Eur As5 Afr1 Am1
fo. *angustifolia* (Wanst.) Sav.-Ljub., Novosti Sist. Niz. Rast. 6: 248, 1969 [1970] (*Tortula graeffii* var.) Eur
fo. *denticulata* Latz., Magyar Bot. Lapok 29: 119, 1930 Eur
fo. *pygmaea* Mart. ex Warnst., Hedwigia 52: 74, 1912 Eur
[*Tortula synecia* R. Brown ter Austr2 = *Tortula rubella* Hook. f. & Wils. *fide* Kramer, J. Hattori Bot. Lab. 65: 84, 1988 ≡ *Syntrichia rubella* (Hook. f. & Wils.) Zand., see treatment of *Syntrichia*]
Tortula systylia (Schimp.) Lindb. Eur As1 As3 Am1
[*Tortula tanganyikae* Dix. Afr2 = *Syntrichia amphidiaceus* (C. Müll.) Zand. see treatment of *Syntrichia*]
[*Tortula tenella* Broth. Afr4 Austr1 Austr2 = *Tortula rubella* Hook. f. & Wils. *fide* Kramer, J. Hattori Bot. Lab. 65: 84, 1988 ≡ *Syntrichia rubella* (Hook. f. & Wils.) Zand., see treatment of *Syntrichia*]
Tortula thianschanica Broth. As1
Tortula thompsonii (C. Müll.) Zand. (*Trichostomum*), see treatment of *Tortula* As2 As3
Tortula tonkinensis (Besch.) Zand. (*Desmatodon*), see treatment of *Tortula* As3
[*Tortula tortuosa* Hedw. ≡ *Tortella tortuosa* (Hedw.) Limpr.]
var. *microcarpa* Hartm. Eur
Tortula toutonii Biz., Acta Bot. Acad. Sci. Hung. 18: 26, 1973 Afr2
[*Tortula trachyneura* Dix. Afr4 ≡ *Tortula ruralis* fo. *trachyneura* (Dix.) Sim = *Tortula ruralis* (Hedw.) Gaetrn., Meyer & Scherb. *fide* Magill, Fl. S. Afr. I. Mosses 1: 219, 1981 (1982)]
Tortula trachyphylla Broth. As1
Tortula transcaspica Broth. As1
Tortula truncata (Hedw.) Mitt. in Godm. Eur As1 As2 As3 Afr1 Am1 Am6 Austr1 Austr2
var. *littoralis* (Mitt.) Zand. (*Pottia littoralis*), see treatment of *Tortula* Eur Afr1
var. *brevirostris* (Lisa) Zand. (*Gymnostomum truncatum* var.), see treatment of *Tortula* Eur
var. *illyrica* (Latz.) Zand. (*Pottia*), see treatment of *Tortula* Eur
var. *minutissima* (Warnst.) Zand. (*Pottia truncata* var.), see treatment of *Tortula* Eur
[*Tortula tutigae* Sak. As2 = *Pottia intermedia* (Turn.) Fürnr. *fide* Saito, J. Hattori Bot. Lab. 39: 522, 1975 ≡ *Tortula modica* Zand. *nom. nov.*, see treatment of *Tortula*]
Tortula ucrainica (Laz.) Zand. (*Desmatodon*), see treatment of *Tortula* Eur
Tortula umbrosa Dus. Am6
[*Tortula unguiculata* (Hedw.) A. Roth ex P. Beauv. *hom. illeg.* ≡ *Barbula unguiculata* Hedw.]
var. *compacta* De Not. Eur
var. *robusta* Lindb. in Hartm., Handb. Skand. Fl. ed. 10, 2: 91, 1871 Eur
Tortula vahliana (Schultz) Mont. in Gay Eur As5 Afr1 Am1 Am6
var. *minor* (Husn.) Par. Afr1
[*Tortula vectensis* Warb. & Crundw., Trans. Brit. Bryol. Soc. 4: 763, 1965 Eur Am1 = *Tortula rhizophylla* (Sak.) Iwats. & Saito, Misc. Bryol. Lichenol. 6: 59, 1972 = *Chenia leptophylla* (C. Müll.) Zand., see treatment of *Chenia*]
[*Tortula velenovskyi* Schiffn. Eur Am6 ≡ *Hilpertia velenovskyi* (Schiffn.) Zand., see treatment of *Hilpertia*]
Tortula vesiculosa (C. Müll.) Broth. Austr1
var. *involucrata* (C. Müll.) Par. Austr1
[*Tortula virescens* (De Not.) De Not. Eur Afr1 Am1 ≡ *Syntrichia virescens* (De Not.) Ochyra, Fragm. Florist. Geobot. 37: 212, 1992]

[subsp. *bizotii* (Laz.) Kramer, Bryophyt. Biblioth. 21: 102, 1980 *comb. inval. basion. inval.* (*Tortula*) As5 ≡ *Tortula virescens* subsp. *bizotiana* Kramer, J. Hattori Bot. Lab. 65: 123, 1988 ≡ *Syntrichia virescens* var. *bizotiana* (Kramer) Zand., see treatment of *Syntrichia*]

[var. *iranica* Kramer, Bryophyt. Biblioth. 21: 101, 1980 As5 ≡ *Syntrichia virescens* var. *iranica* (Kramer) Zand., see treatment of *Syntrichia*]

Tortula viridipila Dix. & Sainsb. Austr2

[*Tortula viridula* (C. Müll.) Broth. Am4 ≡ *Syntrichia viridula* (C. Müll.) Zand., see treatment of *Syntrichia*]

Tortula websteri Robins., Bryologist 68: 316, 1965 As3

Tortula wilczekii Meyl. Afr1

[*Tortula williamsii* Bartr. Am1 = *Scopelophila cataractae* (Mitt.) Broth. *fide* Zander, Bryologist 70: 408, 1967 (1968)]

Tortula willisiana Zand. (*nom. nov.* for *Phascum drummondii* Wils.), see treatment of *Tortula* Austr1

var. *obscura* (Willis) Zand. (*Phascum drummondii* var.), see treatment of *Tortula* Austr1

Tortula wilsonii (Hook.) Zand. (*Gymnostomum*) Eur As1 As5 Afr1

var. *asperula* (Mitt.) Zand. (*Pottia*), see treatment of *Pottia* Eur

var. *crinita* (Wils. ex B.&S.) Zand. (*Pottia*), see treatment of *Pottia* Eur As5 Afr1

var. *mucronifolia* (Bruch in F. Müll.) Zand. (*Entosthymenium*), see treatment of *Pottia* As1 As5

Tortula xerophila Herz. Am4

Tortula yuennanensis Chen As2

Tortula zoddae Zand. (*nom. nov.* for *Pottia cuneifolia* Solms ex Schimp.), see treatment of *Tortula* Eur

TRACHYCARPIDIUM Broth.

Trachycarpidium brisbanicum (C. Müll.) Stone, Muelleria 3: 122, 1975 (*Acaulon*) As4 Austr1

Trachycarpidium echinatum Dix. As4

Trachycarpidium lonchophyllum (Roth) Zand. (*Astomum*), see treatment of *Trachycarpidium* Am5

[*Trachycarpidium novae-valesiae* Broth. ex Roth. Austr1 ≡ *Bryobartramia novae-valesiae* (Broth.) Stone & Scott, J. Bryol. 7: 604, 1973, Encalyptaceae]

Trachycarpidium tisserantii Dix. & P. Varde Afr2

Trachycarpidium verrucosum (Besch.) Broth. Oc

TRACHYODONTIUM Steere, Bryologist 89: 17, 1986

Trachyodontium zanderi Steere, Bryologist 89: 17, 1986 Am4

[**TRICHOSTOMOPSIS** Card. = *Didymodon* Hedw. *fide* Zander, Phytologia 41: 14, 1978]

[*Trichostomopsis aaronis* (Lor.) Agnew & Towns., Israel J. Bot. 19: 258, 1970 Eur As5 Afr1 ≡ *Didymodon aaronis* (Lor.) Guerra in Guerra & Ros, Cryptogamie Bryol. Lichénol. 8: 55, 1987]

[*Trichostomopsis australasiae* (Hook & Grev.) Robins., Phytologia 20: 187, 1970 (*Tortula*) Eur Afr4 Am1 Am2 Am4 Am6 Austr1 Austr2 ≡ *Didymodon australasiae* (Hook. & Grev.) Zand. *fide* Zander, Phytologia 41: 21, 1978]

[*Trichostomopsis brevifolia* Bartr. Afr1 Am1 Am2 = *Trichostomopsis australasiae* (Hook. & Grev.) Robins., Phytologia 20: 187, 1970 ≡ *Didymodon australasiae* (Hook. & Grev.) Zand. *fide* Zander, Phytologia 41: 21, 1978]

[*Trichostomopsis crispifolia* Card. Am2 = *Trichostomopsis umbrosus* (Card.) Robins., Phytologia 20: 185, 1970 ≡ *Didymodon umbrosus* (C. Müll.) Zand., Phytologia 41: 22, 1978]

fo. *crassiretis* Thér., Smiths. Misc. Coll. 85(4): 9, 1931 Am1

Trichostomopsis curvipes (C. Müll.) Robins., Phytologia 20: 186, 1970 (*Barbula*) Am6

[*Trichostomopsis diaphanobasis* (Card.) Grout Am1 Am2 = *Trichostomopsis australasiae* (Hook. & Grev.) Robins., Phytologia 20: 187, 1970 ≡ *Didymodon australasiae* (Hook. & Grev.) Zand. *fide* Zander, Phytologia 41: 21, 1978]

[*Trichostomopsis fayae* Grout Am1 Am2 = *Didymodon australasiae* (Hook. & Grev.) Zand. *fide* Zander, Phytologia 41: 21, 1978]

[*Trichostomopsis haussknechtii* (Jur. & Milde) Agnew & Towns., Israel J. Bot. 19: 258, 1970 (*Barbula*) As5 ≡ *Didymodon haussknechtii* (Jur. & Mild.) Broth.]

[*Trichostomopsis trivialis* (C. Müll.) Robins., Phytologia 20: 187, 1970 (*Barbula*) Eur Afr4 ≡ *Didymodon trivialis* (C. Müll.) Guerra in Guerra & Ros, Cryptogamie Bryol. Lichénol. 8: 64, 1987]

[*Trichostomopsis umbrosa* (C. Müll.) Robins., Phytologia 20: 185, 1970 (*Barbula*) Am2 ≡ *Didymodon australasiae* var. *umbrosus* (C. Müll.) Zand. *fide* Zander, Cryptogamie Bryol. Lichénol. 2: 400, 1981 (1982) ≡ *Didymodon umbrosus* (C. Müll.) Zand.]

fo. *propagulifera* Brugués, Acta Phytotax. Barcinon. 21: 15, 1976 [1977] *nom. inval. sin. descr. lat. et holotyp. non cit.* Eur

TRICHOSTOMUM Bruch

Trichostomum abyssinicum (Thér.) Zand. (*Weissia*), see treatment of *Trichostomum* Afr2

Trichostomum acutiusculum (Broth.) Zand. (*Hyophila*), see treatment of *Trichostomum* Afr2

Trichostomum acutum Herz. Am4

Trichostomum aduncum Par. Oc

Trichostomum aequatoriale Spruce ex Dix. Afr4 Am4

Trichostomum affine Warnst. *hom. illeg.* Eur

Trichostomum alpinum Kindb. Am1

[*Trichostomum andinum* Sull. Am4 ≡ *Erythrophyllopsis andina* (Sull.) Zand., Bryologist 80: 159, 1977 ≡ *Erythrophyllastrum andinum* (Sull.) Zand., see treatment of *Erythrophyllastrum*]

[*Trichostomum angustifolium* Crum & Steere Am3 ≡ *Pseudosymblepharis angustifolia* (Crum & Steere) Zand., see treatment of *Pseudosymblepharis*]

Trichostomum apophysatulum Herz. Am4

Trichostomum arboreum (Mitt.) Zand. (*Weissia*), see treatment of *Trichostomum* Am5

Trichostomum arcticum Kaal. Eur As1 Am1

[*Trichostomum ardjunense* Fleisch. As4 = *Trichostomum crispulum* Bruch in F. Müll. *fide* Norris & Koponen, Acta Bot. Fenn. 137: 96, 1989]

[*Trichostomum aristatulum* Broth. Austr1 = *Trichostomum brachydontium* Bruch in F. Müll. *fide* Norris & Koponen, Acta Bot. Fenn. 137: 96, 1989]

Trichostomum aristatulum (Broth.) Hilp. ex Chen *hom. illeg.* As2

Trichostomum atrocaule (Saito) Zand. (*Weissia*), see treatment of *Trichostomum* As2

Trichostomum austrocrispum (Beck.) Zand. (*Phascum*), see treatment of *Trichostomum* Austr1 Austr2

var. *longifolium* (R. Br. ter) Zand. (*Phascum longifolium*), see treatment of *Trichostomum* Austr2

[*Trichostomum atrorubens* Besch. As2 As3 = *Bryoerythrophyllum wallichii* (Mitt.) Chen *fide* Saito, J. Hattori Bot. Lab. 39: 479, 1975]

Trichostomum ayresianum Schimp. ex Besch. Afr3

Trichostomum barbuloides (Broth.) Chen *hom. illeg.* As2

[*Trichostomum bartramii* H. Miller, J. Hattori Bot. Lab. 30: 272, 1967 (*Trichostomum mauiense* Broth. *hom. illeg.*) Oc = *Trichostomum crispulum* Bruch in F. Müll. *fide* Norris & Koponen,

Acta Bot. Fenn. 137: 96, 1987]

[*Trichostomum bartramii* Iwats. & Tan, Misc. Bryol. Lichenol. 7: 152, 1977 *hom. illeg.* As4 ≡ *Trichostomum brevifolium* Bartr. *hom. illeg.*) ≡ *Trichostomum philippenense* Iwats. & Tan, Kalikasan 8: 195, 1979]

[*Trichostomum baurianum* Warnst. ex Amann Eur = *Barbula spadicea* (Mitt.) Braithw. *fide* Geissler, Candollea 40: 198, 1985 ≡ *Didymodon spadiceus* (Mitt.) Limpr.]

Trichostomum bellii Bartr. Am4

[*Trichostomum bermudanum* (Mitt.) Par. Am1 = *Weissia jamaicensis* (Mitt.) Grout, see treatment of *Weissia*]

Trichostomum bombayense C. Müll. As3

Trichostomum borbonicum Een, Lindbergia 3: 217, 1976 (*nom. nov.* for *Trichostomum glaucoviride* Ren. & Card. *hom. illeg. non* C. Müll. 1882) *nom. illeg. incl.* spec. prior., *Trichostomum cardotii* Biz. [≡ *Trichostomum cardotii* Biz. *nom. inval. basion. non cit.*]

Trichostomum borneense (Dix.) Zand. (*Stephanodictyon*), see treatment of *Trichostomum* Dix. As4

Trichostomum brachydontium Bruch in F. Müll. Eur As1 As2 As3 As4 As5 Afr1 Afr2 Afr3 Afr4 Am1 Am2 Am3 Am4 Am6

subsp. *cuspidatum* (Braithw.) Giac. Eur

subsp. *densum* (B.&S. in BSG) Giac. Eur Afr1

subsp. *mutabile* (Bruch) Giac. Eur As1 As2 As5 Afr1 Am2 Am3 Am4

var. *angustifolium* (Lindb.) Wijk & Marg. Eur

var. *antillarum* Bartr. in Orcutt Am3

var. *cophocarpum* (Schimp.) P. Cout. Eur

var. *cuspidatum* (Braithw.) Sav. Eur

var. *cylindricum* (Schimp.) P. Cout. Eur

var. *devonicum* (Podp.) Wijk & Marg. Eur

var. *esquirolii* (Thér.) Chen As2

var. *littorale* (Mitt.) C. Jens. in Weim. Eur Afr1

var. *longifolium* (Warnst.) Wijk & Marg. Eur

var. *lutescens* (Lindb.) Wijk & Marg. Eur

var. *nigroviride* (Ren. & Card.) Luis. Afr1

var. *robustum* (Ren. & Card.) During, Lindbergia 7: 121, 1981 Afr2

var. *sibiricum* (Podp.) Laz. As1

var. *unguiculatum* (Philib.) Corb. & Jah. Eur

[*Trichostomum brevifolium* Bartr. As4 *hom. illeg.* ≡ *Trichostomum bartramii* Iwats. & Tan, Misc. Bryol. Lichenol. 7: 152, 1977 *hom. illeg.* ≡ *Trichostomum philippenense* Iwats. & Tan, Kalikasan 8: 195, 1979]

Trichostomum brittonianum Zand. (*nom. nov.* for *Hymenostomum flavescens* E. Britt. in N. Britt. & Millsp.), see treatment of *Trichostomum* Am3

[*Trichostomum bulbilliferum* Herz. Am5 = *Oxystegus tenuirostris* var. *gemmiparus* (Schimp.) Zand., Lindbergia 4: 285, 1978 ≡ *Trichostomum tenuirostre* var. *gemmiparum* (Schimp.) Zand., see treatment of *Trichostomum*]

[*Trichostomum burmense* Bartr. As4 ≡ *Oxystegus burmensis* (Bartr.) Gangulee, Moss. E. India 653, 1972 = *Trichostomum tenuirostre* (Hook. & Tayl.) Lindb. *fide* Sollman, Lindbergia 10: 55, 1984]

Trichostomum caespitosum (Bruch ex Brid.) Jur. [≡ *Pottia caespitosa* (Bruch ex Brid.) C. Müll. *fide* Corley et al., J. Bryol. 11: 620, 1981 (1982)], see treatment of *Trichostomum* Eur

var. *mucronatum* Pilous Eur

Trichostomum calymperaceum Broth. & Par. Afr2

[*Trichostomum cardotii* Biz., Rev. Bryol. Lichénol. 40: 120, 1974 *nom. inval. basion. non cit.* (*nom. nov.* for *Trichostomum glaucoviride* Ren. & Card.) Afr3 ≡ *Trichostomum borbonicum* Een, Lindbergia 3: 217, 1976]

Trichostomum carinatum Bartr. As4

Trichostomum castaneum (Crum & Steere) Zand. (*Hymenostomum*), see treatment of *Trichostomum* Am3

[*Trichostomum cavernarum* Broth. Am5 ≡ *Pseudosymblepharis cavernarum* (Broth.) Zand., see treatment of *Pseudosymblepharis*]

[*Trichostomum chlorphyllum* C. Müll. Am2 = *Weissia jamaicensis* (Mitt.) Grout *fide* Zander in Sharp et al., Moss Fl. Mex.]

[var. *brevifolium* Thér. Am2 = *Trichostomum crispulum* Bruch in F. Müll. *fide* Zander in Sharp et al., Moss Fl. Mex.]

[*Trichostomum circinnatulum* Broth. in Völtzk. Afr3 ≡ *Pseudosymblepharis circinnatula* (Broth. in Völtzk) Zand., see treatment of *Pseudosymblepharis*]

Trichostomum clavinerve (Card. & P. Varde) H. O. Whitt. in H. O. Whitt. & B. Whitt., Bryologist 77: 433, 1974 (*Weissia*) Oc

[*Trichostomum clintonii* C. Müll. Am2 = *Oxystegus tenuirostris* (Hook. & Tayl.) A. Sm. *fide* Zander in Sharp et al., Moss Fl. Mex. ≡ *Trichostomum tenuirostre* (Hook. & Tayl.) Lindb.]

Trichostomum compactulum C. Müll. Am6

Trichostomum connivens (Broth.) Par. Eur

Trichostomum contortum (Kunze) Sérgio, Portug. Acta Biol. (B) 14: 169, 1985 (*Hymenostomum*) Afr1

Trichostomum contractum Zand., see treatment of *Trichostomum* (*nom. nov.* for *Gymnostomum longirostre* Griff.) As2

Trichostomum criotum Zand. (*nom. nov.* for *Hyophila perannulata* Ren. & Card.), see treatment of *Trichostomum* As3

Trichostomum crispulum Bruch in F. Müll. Eur As1 As2 As5 Afr1 Afr2 Am1 Am2

subsp. *brevifolium* (B.&S. in BSG) Giac. Eur Afr1

subsp. *hammerschmidii* (Loeske & Paul) Amann Eur

var. *acuminatum* Meyl. Eur

var. *algarvicum* Schimp. Eur Afr1

var. *armatum* (Thér. & Trab.) Biz. Afr1

var. *brevifolium* B.&S. Eur Afr1

var. *cucullatum* (Roth) Podp. Eur Am1

var. *elatum* Schimp. Eur As5 Afr1

var. *hammerschmidii* (Loeske & Paul) Blumr. Eur

var. *mucronatulum* (Card.) Podp. Eur Afr1

var. *muticum* Podp. Eur

var. *nigroviride* (Braithw.) Dix. Eur

var. *pseudoweisia* Schimp. Eur Afr1

var. *sudeticum* Velen. Eur

fo. *brevissimum* Reim., Hedwigia 79: 261, 1940 Eur

fo. *longifolium* Bouvet, Bull. Soc. Et. Sc. Angers 26: 84, 1896 (*Trichostomum crispulum* var. *longifolium* Schimp. *nom. illeg.*) Eur

fo. *tophaceum* Latz., Bot. Centralbl. Beih. 48(2): 242, 1931 Eur

[*Trichostomum crispulum* Besch. Am2 *hom. illeg.* = *Oxystegus tenuirostris* (Hook. & Tayl.) A. Sm. (as *Oxystegus cylindricus* (Brid.) Hilp.) *fide* Crum, Appalachian-Ozarkian Element Moss Fl. Mexico (dissertation, Univ. Microfilms) 137, 1951 ≡ *Trichostomum tenuirostre* (Hook. & Tayl.) Lindb.]

[*Trichostomum cuspidatissimum* Card. & Thér. Eur As1 Am1 = *Trichostomum arcticum* Kaal. *fide* Frisvoll, Bryologist 81: 156, 1978]

[*Trichostomum cuspidatum* (Doz. & Molk.) Doz. & Molk. As2 As4 = *Trichostomum tenuirostre* (Hook. & Tayl.) Lindb. *fide* Sollman, Lindbergia 10: 55, 1984]

[*Trichostomum cylindricum* (Brid.) C. Müll. *nom. inval.* = *Oxystegus tenuirostris* (Hook. & Tayl.) A. Sm., J. Bryol. 9: 393, 1977 ≡ *Trichostomum tenuirostre* (Hook. & Tayl.) Lindb.]

var. *asperum* Podp., Sborn. Klub. Přirod. Brno 5: 11, 1923 Eur

var. *cataractarum* (Loeske) Culm. (*Streblotrichum croceum* var.) Eur

[fo. *longifolia* Latz., Hedwigia 66: 138, 1926 *hom. illeg.* Eur ≡

Oxystegus cylindricus fo. *longifolia* Podp.]
[*Trichostomum cylindrotheca* (Mitt.) Broth. As3 ≡ *Oxystegus cylindrothecus* (Mitt.) Gangulee, Nov. Hedw. 12: 430, 1966 = *Trichostomum bombayense* C. Müll. *fide* Townsend, J. Bryol. 12: 561, 1983]
Trichostomum deciduaefolium (Saito) Zand. (*Weissia*), see treatment of *Trichostomum* As2
Trichostomum decurvifolium Dix. Afr2
Trichostomum distans Hampe in C. Müll. Afr2
[*Trichostomum dubium* Thér. Oc ≡ *Pseudosymblepharis angustata* (Mitt.) Hilp. *fide* Norris & Koponen, Acta Bot. Fenn. 137: 94, 1989]
Trichostomum duidense Bartr. Am5 [= *Trichostomum brachydontium* Bruch in F. Muell. *fide* Sollman, Lindbergia 10: 55, 1984 but recognized here as a good species]
Trichostomum edentulum Broth. in Herz. Am4
[*Trichostomum ehrenbergii* Lor. ≡ *Barbula ehrenbergii* (Lor.) Fleisch.]
var. *denticuspis* Broth., Bot. Jahrb. 30: 261, 1901 loc?
[*Trichostomum ekmanii* Thér. Am3 = *Weissia jamaicensis* (Mitt.) Grout, see treatment of *Trichostomum*]
Trichostomum elliottii Broth. ex Dus. Am6
[*Trichostomum etessei* Broth. & Par. Oc. ≡ *Trichostomum crispulum* Bruch in F. Müll. *fide* Norris & Koponen, Acta Bot. Fenn. 137: 96, 1989]
Trichostomum exulatum (C. Müll. ex Roth) Zand. (*Tetrapterum*), see treatment of *Trichostomum* Am5
Trichostomum fallaciosum Welch & Crum Am3
Trichostomum fallax Herz. Am4
fo. *minutum* Herz., Biblioth. Bot. 87: 92, 1916 Am4
Trichostomum flavescens Dix. Afr2
Trichostomum flavisetum DC. in Lam. & DC. Eur [= *Ditrichum* sp. *fide* W. Margadant in litt. DC]
Trichostomum fragilifolium Dix. Afr2
[*Trichostomum fuscomucronatum* C. Müll. Afr2 = *Pseudocrossidium replicatum* (Tayl.) Zand. *fide* Townsend, Lindbergia 10: 178, 1985]
[*Trichostomum glaucoviride* Ren. & Card. Afr3 *hom. illeg.* ≡ *Trichostomum cardotii* Biz., Rev. Bryol. Lichénol. 40: 120, 1974 *nom. inval. basion. non cit.* (*nom. nov.* for *Trichostomum glaucoviride* Ren. & Card.) ≡ *Trichostomum borbonicum* Een, Lindbergia 3: 217, 1976]
Trichostomum gracillimum C. Müll. Am6
Trichostomum hibernicum (Mitt.) Dix. Eur
Trichostomum hyalinoblastum (Broth.) Broth. As3
Trichostomum imshaugii (Vitt) Zand., see treatment of *Trichostomum* (*Barbula*) Austr2
Trichostomum incertum (Mitt.) Zand. (*Weissia*), see treatment of *Trichostomum* Afr3
[*Trichostomum inclinans* Schimp. ex Besch., Mém. Soc. Nat. Sci. Natl. 16: 176, 1872 = *Rhamphidium dicranoides* (C. Müll.) Par. *fide* Zander in Sharp et al., Moss Fl. Mex.]
Trichostomum insulare (Besch.) Broth. Oc
Trichostomum involutum Sull. Am3 Am4
var. *minus* Sull. Am3
Trichostomum involutum Broth. As2 *hom. illeg.*
[*Trichostomum involvens* Card. Am2 = *Trichostomum crispulum* Bruch in F. Müll. *fide* Zander in Sharp et al., Moss Fl. Mex.]
[*Trichostomum jamaicense* (Mitt.) Jaeg. Am1 Am2 Am3 = *Weissia jamaicensis* (Mitt.) Grout *fide* Zander in Sharp et al., Moss Fl. Mex.]
Trichostomum kanieriense R. Br. ter Austr2
[*Trichostomum khasianum* (Mitt.) Broth. As3 ≡ *Oxystegus khasianus* (Mitt.) Gangulee, Nov. Hedw. 8: 149, 1964 ≡ *Psuedosymblepharis khasiana* (Mitt.) Zand., see treatment of *Pseudosymblepharis*]
Trichostomum knightii Hampe ex C. Müll. Am2
Trichostomum lambii Bartr., Rev. Bryol. Lichénol. 33: 324, 1964–65 [1965] Am5
[*Trichostomum lamprothecium* C. Müll. Am2 = *Trichostomum crispulum* Bruch in F. Müll. *fide* Zander in Sharp et al., Moss Fl. Mex.]
[*Trichostomum lanuginosum* Hedw. ≡ *Rhacomitrium lanuginosum* (Hedw.) Brid.]
var. *incanum* Hornsch. Am1
Trichostomum laticostatum Thér. Oc
Trichostomum leptocylindricum C. Müll. Am5
[*Trichostomum leiodontium* C. Müll. = (*Tortella xanthocarpa* (C. Müll.) Broth. *fide* Broth, Nat. Pfl. 1(3): 397, 1902) = *Tortella humilis* (Hedw.) Jenn. *fide* Sollman, Lindbergia 16: 22, 1990]
Trichostomum leptotheca C. Müll. Austr1
Trichostomum lignicola Herz. Am5
Trichostomum ligulaefolium (Broth. & Par.) Zand. (*Hyophila*), see treatment of *Trichostomum* Afr2
Trichostomum lillei Dix. As2
Trichostomum lindigii (Hampe) Zand. (*Systegium*), see treatment of *Trichostomum* Am4
[*Trichostomum linealifolium* C. Müll. Am4 = *Weissia jamaicensis* (Mitt.) Grout, see treatment of *Weissia*]
[*Trichostomum lingulatum* Hook. f. & Wils., Fl. N. Zeal. 2: 71, 1854 = *Didymodon tophaceus* (Brid.) Lisa, see treatment of *Didymodon*]
Trichostomum lorifolium Broth. & Par. Afr2
[*Trichostomum makanuiense* R. Br. ter, Trans. N. Zeal. Inst. 35: 332, 1903 *nom. inval.* in Robinson, Phytologia 20: 187, 1970 = *Trichostomopsis australasiae* (Hook. & Grev.) Robins. *loc. cit.* ≡ *Didymodon australasiae* (Hook. & Grev.) Zand. var. *australasiae*]
[*Trichostomum marginatum* Robins., Phytologia 12: 389, 1971 Am4 ≡ *Didymodon marginatum* (Robins.) Zand., see treatment of *Didymodon*]
[*Trichostomum mauiense* Broth. *hom. illeg.* Oc ≡ *Trichostomum bartramii* H. Miller, J. Hattori Bot. Lab. 30: 272, 1967 = *Trichostomum crispulum* Bruch in F. Müll. *fide* Norris & Koponen, Acta Bot. Fenn. 137: 96, 1989]
Trichostomum melanostomum (Mitt.) Zand. (*Weissia*) Am5
Trichostomum mildeanum Jur. As5
Trichostomum minusculum Dix. & P. Varde As3
Trichostomum minutissimum Sak. Oc
Trichostomum mitteneanum Zand. *nom. nov.* (*Weissia umbrosa* Mitt.), see treatment of *Trichostomum* Am4
[*Trichostomum molariforme* Zand., Bryologist 85: 126, 1982 Am1 Am2 Am3 = *Trichostomum portoricense* Crum & Steere, see treatment of *Trichostomum*]
[*Trichostomum mollissimum* (Broth. ex Bartr.) Crum Am1 Am2 Am3 Am5 = *Pseudosymblepharis schimperiana* (Par.) Crum *fide* Zander in Sharp et al., Moss Fl. Mex.]
Trichostomum mouense Broth. & Par. Oc
[*Trichostomum mutabile* Bruch in De Not. ≡ *Trichostomum brachydontium* subsp. *mutabile* (Bruch) Giac.]
var. *alpinum* Amann Eur
var. *brevifolium* Schiffn. Eur
var. *majus* Podp. Eur
[var. *robustum* Ren. & Card. Afr1 ≡ *Trichostomum brachydontium* var. *robustum* (Ren. & Card.) During, Lindbergia 7: 121, 1981]

var. *subalpinum* BSG *nom. inval. prov.* Eur

fo. *litorale-mutabile* Herz., Abh. Leop.-Car. Ak. Naturf. 73(3): 471, 1907 Eur

fo. *mutabile-cuspidatum* Herz., Abh. Leop.-Car. Ak. Naturf. 73(3): 476, 1907 Eur

subfo.? *crispulomutabile* Herz., Abh. Leop.-Car. Ak. Naturf. 73(3): 471, 1907 Eur

subfo.? *flaccidolitorale* Herz., Abh. Leop.-Car. Ak. Naturf. 73(3): 471, 1907 Eur

subfo.? *strictolitorale* Herz., Abh. Leop.-Car. Ak. Naturf. 73(3): 470, 1907 Eur

Trichostomum muticum Par. Afr3

[*Trichostomum nigrescens* P. Beauv., Prodr. 47, 1805 (*Bryum nervosum* Vill. 1789 *nom. inval.*) Eur ≡ p.p. *Racomitrium aquaticum* (Schrad.) Brid. (Grimmiaceae) cf. Brid., Musc. Rec. 2(3): 51, 1803)]

[*Trichostomum nitidum* (Lindb.) Schimp. ≡ *Tortella nitida* (lindb.) Broth.]

var. *irrigatum* Wint. Afr1

Trichostomum nordenskioeldii Schimp. in Heer, Fl. Foss. Arct. 2(3): 88, 1870 Eur

Trichostomum novogranatense Broth. & Irmsch. Am4

[*Trichostomum oblongifolium* Bartr. Oc = *Trichostomum brachydontium* Bruch in F. Muell. *fide* Sollman, Lindbergia 10: 55, 1984]

[*Trichostomum obtusifolium* (Schwaegr.) Boul., Fl. Crypt. Est. Muscin. 491, 1872 *hom. illeg.* Eur As1 As2 As5 Afr1 Am1 ≡ *Desmatodon obtusifolius* (Schwaegr.) Schimpr. ≡ *Tortula obtusifolia* (Schwaegr.) Math. *fide* Corley et al., J. Bryol. 11: 620, 1981 (1982)]

[*Trichostomum orientale* Web. = *Barbula indica* (Hook.) Spreng.]

fo. *sterile* Fleisch., Musci Fl. Buitenz. 1: 346, 1902 [1904] As4

Trichostomum orthodontum (C. Müll.) Broth. As3

Trichostomum ovatifolium Zand. (*nom. nov.* for *Hymenostomum anomalum* Broth. in Herz.), see treatment of *Trichostomum* Am4

Trichostomum pallidens (Dix.) Zand. (*Pseudosymblepharis*), see treatment of *Trichostomum* As3

Trichostomum paludicola (Broth.) Hilp. Am4

Trichostomum eckelianum Zand., see treatment of *Trichostomum* (*nom. nov.* for *Trichostomum cirrhatum* Hampe *hom. illeg.*) Austr1

Trichostomum pennequinii Ren. & Par. Afr3

Trichostomum perangustum Besch. Afr4

Trichostomum perannulatum Dix. & P. Varde As3

Trichostomum perinvolutum Tix., Rev. Bryol. Lichénol. 34: 132, 1966 As3

Trichostomum perligulatum (Flow. ex Crum) Zand. (*Weissia*), see treatment of *Trichostomum* Am1

[*Trichostomum perlongifolium* J. Fröhl. *fide* As4 ≡ *Pseudosymblepharis perlongifolia* (Fröhl.) Zand., see treatment of *Pseudosymblepharis*]

Trichostomum perplexum P. Varde Afr2

Trichostomum perpusillum C. Müll. ex Warnst. Am2

Trichostomum perrieri Thér. Afr2

Trichostomum persicum Jur. & Milde As5

[*Trichostomum perviride* Broth. Am3 Am4 ≡ *Weissia veviridis* Zand. *nom. nov.*, see treatment of *Weissia*]

Trichostomum philippenense Iwats. & Tan, Kalikasan 8: 195, 1979 (*nom. nov.* for *Trichostomum brevifolium* Bartr.) As4

Trichostomum planifolium (Dix.) Zand. (*Weissia*), see treatment of *Trichostomum* As1 As2 Am2

Trichostomum platyphyllum (Ihs.) Chen As2

Trichostomum plicatum C. Müll. Am6

Trichostomum pomangium Broth. & Herz. in Herz. Am4

Trichostomum portoricense Crum & Steere Am1 Am2 Am3

Trichostomum prionodon C. Müll. Am5

Trichostomum pulicare (Besch.) Zand. (*Hymenostomum*), see treatment of *Trichostomum* Afr3

Trichostomum pygmaeum Bartr. Am2

[*Trichostomum quitense* Hampe in C. Müll. *nom. nud.* = *Tortula pichinchensis* Tayl.]

var. *longifolium* Herz. *comb. inval.* Am4

Trichostomum recurvifolium (Tayl.) Zand., see treatment of *Trichostomum* Eur Am1

[*Trichostomum rehmannii* Sim Afr4 = *Trichostomum tortuloides* Sull. & Lesq. *fide* Magill & Schelpe, Mem. Bot. Surv. S. Afr. 43: 25, 1979 = *Tortella xanthocarpa* (C. Müll.) Broth. *fide* Magill, Fl. S. Afr. I. Mosses 1: 255, 1981]

Trichostomum rhodesiae Broth. Afr2

Trichostomum rigidulum (Hedw.) Turn. ≡ *Didymodon rigidulus* Hedw.]

var. *flaccidum* Röll *nom. inval. prov.* Eur

var. *rigidum* Röll *nom. inval. prov.* Eur

Trichostomum robustum Broth. ex Ihs. As2

Trichostomum ruvenzorense (Broth.) Broth. Afr2

Trichostomum sarawakense Dix. As4

[*Trichostomum schlimii* C. Müll. Am4 (= ?*Oxystegus tenuirostris* (Hook. & Tayl.) A. Sm. (as *Oxystegus cylindricus* (Brid.) Hilp.) *fide* Crum, Appalachian-Ozarkian Element Moss Fl. Mexico (dissertation, Univ. Microfilms) 137, 1951) ≡ *Trichostomum tenuirostre* (Hook. & Tayl.) Lindb., see treatment of *Trichostomum*]

Trichostomum siamense Broth. in Hosseus, Beih. Bot. Centralbl. 28: 362, 1911 As3

Trichostomum sinaloense (Bartr.) Zand. (*Weissia*), see treatment of *Trichostomum* Am2 Am3

[*Trichostomum sitkanum* Card. & Thér. Am1 = *Tortella tortuosa*

var. *arctica* (Arnell) Broth. in Fedch., see treatment of *Tortella*]

Trichostomum soulae (C. Müll. in Ren. & Card.) Zand. (*Ptychomitrium*), see treatment of *Trichostomum* Afr3

var. *corticicola* (Ren. & Card.) Zand. (*Barbula*), see treatment of *Trichostomum* Afr3

Trichostomum sparsifolium (Ren. & Card.) Card. in Grand. Afr3

Trichostomum spirale Grout Am1 Am2 As2 (= *Oxystegus tenuirostris* (Hook. & Tayl.) A. Sm. var. *stenocarpus* (Thér.) Zand., Misc. Bryol. Lichenol. 9: 73, 1982) ≡ *Oxystegus spiralis* (Grout) Crum & Anderson, Bryologist 92: 533, 1989)

Trichostomum sporaphyllum (Ren. & Card.) Card. In Grand. Afr3

Trichostomum stanilandsii R. Br. ter Austr2

[*Trichostomum stenocarpum* Wils. in Seem., Bot. Voyage Herald 344, 1857 Am2 ≡ *Atractylocarpus stenocarpus* (Wils. in Seem.) Zand., Bryologist 85: 128, 1982]

[*Trichostomum stenocarpum* (Thér.) Crum, Bryologist 84: 390, 1981 *hom. illeg.* (*Weisiopsis*) Am2 ≡ *Oxystegus tenuirostris* var. *stenocarpus* (Thér.) Zand., Misc. Bryol. Lichenol. 9: 73, 1982 = *Trichostomum spirale* Grout, see treatment of *Trichostomum*]

Trichostomum stenophyllum (Mitt.) Broth. [see discussion of Townsend, J. Bryol. 12: 563, 1983] As2

[*Trichostomum striatum* (P. Beauv.) Arn. Am1 = p.p. *Racomitrium patens* (Hedw.) Hüb. cf. Arnott *loc. cit.*]

Trichostomum subangustifolium (Thér.) Zand. (*Hyophila*), see treatment of *Trichostomum* Am2

Trichostomum subcirrhatum Hampe Am5

Trichostomum subconnivens Thér. Am3

Trichostomum subdenticulatum Aust. Am1 *hom. illeg.*

Trichostomum subintegrum (Broth.) Broth. Afr2
Trichostomum sublamprothecium Par. Am3
Trichostomum subminusculum Dix. & P. Varde As3
[*Trichostomum subulifolium* Bartr. As4 ≡ *Pseudosymblepharis angustata* (Mitt.) Hilp. *fide* Norris & Koponen, Acta Bot. Fenn. 137: 94, 1989]
Trichostomum sumatranum Baumg. in J. Fröhl. As4
[*Trichostomum svihlae* Bartr. As4 ≡ *Oxystegus svihlae* (Bartr.) Gangulee, M. E. India 654, 1972 ≡ *Tuerckheimia svihlae* (Bartr.) Zand., see treatment of *Tuerckheimia*]
[*Trichostomum syrrhopodontoides* Herz. Am4 = *Trichostomum tenuirostre* (Hook. & Tayl.) Lindb. *fide* W. Steere, in litt. having seen type at JE]
Trichostomum tenuirostre (Hook. & Tayl.) Lindb. Eur As1 As3 Afr1 Afr2 Afr3 Afr4 Am1 Am2 Am4 Oc
var. *gemmiparum* (Schimp.) Zand., see treatment of *Trichostomum* Eur As3 Am2 Am3 Am5
var. *holtii* (Braithw.) Dix. Eur Am1
Trichostomum termitarum (C. Müll.) Zand. (*Weissia*), see treatment of *Trichostomum* Am5
[*Trichostomum tophaceum* Brid. ≡ *Didymodon tophaceus* (Brid.) Lisa]
var. *brevifolium* Lindb. *hom. illeg.* Eur
var. *leucotrichion* Brid. Eur
[fo. *acutifolium* Boul., Mouss. France 449, 1884 Eur ≡ *Didymodon tophaceus* fo. *acutifolius* (Boul.) Zodda, Malphighia 22: 509, 1908]
fo. *brevicaule* (Schimp.) Boul., Muscin. France 449, 1884 (*Trichostomum tophaceum* var.) Eur
[fo. *cylindricum* Boul., Muscin. France 449, 1884 Eur ≡ *Didymodon tophaceus* fo. *cylindrica* (Boul.) Limpr., Laubm. Eur. 1: 554, 1888]
[fo. *elatum* Boul., Muscin. France 449, 1884 Eur ≡ *Didymodon tophaceus* fo. *elatus* (Boul.) Artaria in E. Bauer, Musci Eur. Am. Exsicc. 37: n. 1823, 1925]
[fo. *lingulatum* Boul., Muscin. France 449, 1884 Eur ≡ *Didymodon tophaceus* fo. *lingulatus* (Boul.) Mönk., Süsswasserfl. 14: 67, 1914]
[fo. *recurvifolium* Boul., Muscin. France 449, 1884 Eur ≡ *Didymodon tophaceus* fo. *recurvifolius* (Boul.) De Willd., Prodr. 434, 1899]
[fo. *truncatum* Boul., Muscin. France 449, 1884 Eur ≡ *Didymodon tophaceus* fo. *truncatus* (Boul.) Limpr., Laubm. Deutsch. 1: 554, 1888]
Trichostomum tortelloides (Broth. & Dix.) Zand. (*Calymperes*), see treatment of *Trichostomum* As3
[*Trichostomum tortuloides* Sull. & Lesq. Afr4 = *Tortella xanthocarpa* (C. Müll.) Broth. *fide* Magill, Fl. S. Afr. I. Mosses 1: 255, 1981 (1982)]
[*Trichostomum trifarium* (Hedw.) Sm. ≡ *Didymodon trifarius* (Hedw.) Röhl. *nom. dub. fide* Zander, Cryptogamie Bryol. Lichénol. 2: 414, 1981 (1982)]
var. *sterile* Mazz. Eur
Trichostomum trirete Dix. Afr2
[*Trichostomum triumphans* De Not. ex Schimp. Eur As5 Afr1 ≡ *Weissia triumphans* (De Not.) Hill, J. Bryol. 11: 600, 1981 (1982)]
[subsp. *monspeliense* (Schimp.) Boul. Eur Afr1 ≡ *Weissia triumphans* var. *monspeliensis* (Schimp.) Zand., see treatment of *Weissia*]
subsp. *pallidisetum* (H. Müll.) Giac. Eur As5 Afr1
subsp. *philibertii* (Schimp.) Boul. Eur Afr1
var. *azoricum* (Card.) Podp. Eur Afr1
var. *brachyodon* (Spindl.) Podp. Eur
[var. *monspeliense* (Schimp.) Boul. Eur Afr1 ≡ *Weissia triumphans* var. *monspeliensis* (Schimp.) Zand., see treatment of *Weissia*]
var. *pallidisetum* (H. Müll.) Husn. Eur As5 Afr1
var. *philibertii* (Schimp.) Husn. Eur Afr1
Trichostomum tucumanense Bartr., Rev. Bryol. Lichénol. 33: 324, 1964–65 [1965] As5
Trichostomum unguiculatum (Mitt.) Zand., see treatment of *Trichostomum* Afr2 Afr4
Trichostomum uncifolium Dix., Anniv. Vol. Bot. Gard. Calcutta 180, 1942 As3
Trichostomum urceolare (Hampe) Zand. (*Hyophila*), see treatment of *Trichostomum* Am5
Trichostomum usambaricum (Broth.) Broth. Afr2
Trichostomum villaumei Thér. Afr3
[*Trichostomum viridulum* Bruch in F. Müll. Eur As5 Afr1 [= *Trichostomum crispulum* fo. *longifolium* Bouvet] = *Trichostomum crispulum* Bruch in F. Müll. *fide* Corley et al., J. Bryol. 11: 623, 1981 (1982)]
Trichostomum wagneri (C. Müll.) Broth. Am4
Trichostomum weisioides C. Müll. Am5
Trichostomum whittonii R. Br. ter Austr2
Trichostomum williamsii Zand. (*nom. nov.* for *Astomum chilense* Williams), see treatment of *Trichostomum* Am4

[**TRIDONTIUM** Hook. f. in Hook. referred to Grimmiaceae, see Excluded Taxa]
[*Tridontium tasmanicum* Hook. f. in Hook. As4 Austr1 Austr2 = Grimmiaceae, see Excluded Taxa]

TRIQUETRELLA C. Müll.
Triquetrella arapilensis Luis. Eur
Triquetrella californica (Lesq.) Grout Am1
[*Triquetrella ferruginea* (Besch.) Thér. Am2 = *Didymodon fallax* var. *reflexus* (Brid.) Zand. *fide* Zander, Bryologist 83: 230, 1980 = *Didymodon rigidicaulis* (C. Müll.) Saito *fide* Saito, J. Hattori Bot. Lab. 39: 502, 1975 = *Didymodon ferrugineus* (Schimp. ex Besch.) Hill, J. Bryol. 11: 599, 1981 (1982)]
[*Triquetrella filicaulis* Dus. Am6 Afr4 ≡ *Triquetrella patagonica* fo. *filicaulis* (Dus.) Herz.]
Triquetrella fragilis C. Müll. Austr1
[*Triquetrella japonica* Dix. in Sak. As2 *nom. inval. descr. angl.* = *Didymodon rigidicaulis* (C. Müll.) Saito *fide* Saito, J. Hattori Bot. Lab. 39: 502, 1975 = *Didymodon ferrugineus* (Schimp. ex Besch.) Hill, J. Bryol. 11: 599, 1981 (1982)]
[*Triquetrella nipponensis* Sak. As2 = *Dichodontium pellucidum* (Hedw.) Schimp. (Dicranaceae) *fide* Saito, J. Hattori Bot. Lab. 39: 502, 1975]
Triquetrella papillata (Hook. f. & Wils.) Broth. Austr1 Austr2
Triquetrella patagonica C. Müll. Am6
fo. *filicaulis* (Dus.) Herz., Arch. Esc. Farm. Fac. Ci. Méd. Córdoba (Secc. Ci.) 7: 40, 1938 Am6 Afr4
[*Triquetrella preissiana* (Hampe) C. Müll. Austr1 = *Triquetrella papillata* (Hook. f. & Wils.) Broth. *fide* Watts & Whitelegge, Proc. Linn. Soc. N.S.W. 27: 1–90, 1902 cf. Stone & Scott, Mo. So. Australia 220, 1976, and, see treatment of *Triquetrella*]
[*Triquetrella recurvifolia* Dix. & Sak. = *Didymodon rigidicaulis* (C. Müll.) Saito *fide* Saito, J. Hattori Bot. Lab. 39: 502, 1975 = *Didymodon ferrugineus* (Schimp. ex Besch.) Hill, J. Bryol. 11: 599, 1981 (1982)]
Triquetrella richardsiae C. Müll. Austr1
Triquetrella spiculosa Thér. Am4
Triquetrella tasmanica (Broth.) Granzow-de la Cerda, Bryologist

92: 383, 1989 Austr1

[*Triquetrella tenuicaulis* Sak. As2 = *Didymodon rigidicaulis* (C. Müll.) Saito *fide* Saito, J. Hattori Bot. Lab. 39: 502, 1975 = *Didymodon ferrugineus* (Schimp. ex Besch.) Hill, J. Bryol. 11: 599, 1981 (1982)]

Triquetrella tristicha (C. Müll.) C. Müll. Afr4

TUERCKHEIMIA Broth.

[*Tuerckheimia angustifolia* (Saito) Zand., Misc. Bryol. Lichenol. 8: 27, 1978 *nom. inval.* (*Gymnostomum*) Am1 Am2 As2 = *Tuerckheimia svihlae* (Bartr.) Zand., see treatment of *Tuerckheimia*]

[*Tuerckheimia angustinervis* (Card.) Broth. Am1 Am2 Am3 Am5 = *Weissia jamaicensis* (Mitt.) Grout *fide* Zander in Sharp et al., Moss Fl. Mex.]

[*Tuerckheimia calculosa* Zand. & Eckel, Mem. New York Bot. Gard. 45: 293, 1987 Am2 = *Quaesticula navicularis* (Mitt.) Zand., see treatment of *Quaesticula*]

Tuerckheimia guatemalensis Broth. Am2

[*Tuerckheimia linearis* (Web. & Mohr) Britt. Am3 Am5 ≡ *Oxystegus linearis* (Web. & Mohr) Hilp. *fide* Zander, Bryologist 85: 128, 1982 ≡ *Tortella linearis* (Web. & Mohr) Zand., see treatment of *Tortella*]

[*Tuerckheimia longifolia* Thér. Am3 = *Oxystegus tenuirostris* var. *gemmiparus* (Schimp.) Zand., Lindbergia 4: 285, 1978 ≡ *Trichostomum tenuirostre* var. *gemmiparum* (Schimp.) Zand., see treatment of *Trichostomum*]

Tuerckheimia robusta (Dix.) Zand. *comb. nov.*, see treatment of *Tuerckheimia* (*Merceyopsis*) As3

Tuerckheimia svihlae (Bartr.) Zand. (see treatment of *Tuerckheimia*) Thér. Afr2 ≡ *Trichostomum abyssinicum* (Thér.) Zand., see treatment of *Trichostomum*]

Tuerckheimia valeriana (Bartr.) Zand., Misc. Bryol. Lichenol. 8: 27, 1978 (*Leptodontium*) Am2

ULEOBRYUM Broth.

Uleobryum curtisii Stone, J. Bryol. 13: 19, 1984 Austr1

Uleobryum peruvianum Broth. Am2 Am3 Am4 Austr1

Uleobryum occultum (Roth) Zand. (*Aschisma*), see treatment of *Uleobryum* Am3

WEISIOPSIS Broth.

[*Weisiopsis angulosa* (Broth. & Dix.) Hilp. As3 ≡ *Ganguleea angulosa* (Broth. & Dix.) Zand., Phytologia 65: 427, 1989]

Weisiopsis anomala Broth. & Par. As2

Weisiopsis bahiensis (C. Müll.) Broth. Am5

[*Weisiopsis hollandii* Bartr. Oc = *Luisierella barbula* (Schwaegr.) Steere, see treatment of *Luisierella*]

Weisiopsis cucullatifolia (Gao, Jia & Cao) Zand. (*Hyophila*), see treatment of *Weisiopsis* As2

[*Weisiopsis hyophiloides* Dix. & Thér. As2 = *Trichostomum platyphyllum* (Ihs.) Chen p.p. & *Hyophila involuta* (Hook.) Jaeg. p.p. *fide* Saito, J. Jap. Bot. 47: 17, 1972]

[*Weisiopsis involuta* Magill, Fl. S. Afr. I. Mosses 1: 225, 1981 (1982) Afr4 ≡ *Plaubelia involuta* (Magill) Zand., see treatment of *Plaubelia*]

Weisiopsis nigeriana (Egun. & Olar.) Zand. (*Gyroweisia*), see treatment of *Weisiopsis* Afr2

Weisiopsis norrisii (Zand.) Zand. (*Scopelophila*), see treatment of *Weisiopsis* Am2

Weisiopsis oblonga Thér. Am2

Weisiopsis plicata (Mitt.) Broth. Afr2 Afr3 Afr4

[*Weisiopsis pulchriretis* Dix. Afr4 ≡ *Streptocalypta pulchriretis* (Dix.) Zand., see treatment of *Streptocalypta*]

[*Weisiopsis spathulifolius* Crum & Bartr. Am3 = *Neohyophila sprengelii* (Schwaegr.) Crum *s. lat. fide* Zander, Bryologist 86: 137, 1983 ≡ *Plaubelia sprengelii* (Schwaegr.) Zand., see treatment of *Plaubelia*]

[*Weisiopsis stenocarpa* Thér. Am2 ≡ *Trichostomum stenocarpum* (Thér.) Crum, Bryologist 84: 390, 1981 ≡ *Oxystegus tenuirostris* var. *stenocarpus* (Thér.) Zand., Misc. Bryol. Lichenol. 9: 73, 1982 = *Trichostomum spirale* Grout, see treatment of *Trichostomum*]

WEISSIA Hedw.

Weissia abbreviata (Thwait. & Mitt.) Zand. (*Systegium*), see treatment of *Weissia* As3 As4

[*Weissia abyssinica* Thér. Afr2 ≡ *Trichostomum abyssinicum* (Thér.) Zand., see treatment of *Trichostomum*]

[*Weissia acuta* Hedw. ≡ *Blindia acuta* (Hedw.) BSG]

var. *curvula* Hartm. Eur

Weissia alianuda Tan, Mem. New York Bot. Gard. 68: 5, 1992 (*nom. nov.* for *Weissia nuda* Mitt.) Austr1

[*Weissia amblyphylla* (Dix.) Dix. As3 *hom. illeg. non* Zett 1869 ≡ *Weissia norketii* R. S. Chopra, Taxon. Indian Mosses 125, 1975]

Weissia andersoniana Zand., Monogr. Syst. Bot. Missouri Bot. Gard. 11: 195, 1985 (*nom. nov.* for *Weissia glauca* Bartr. *hom. illeg.*) Am1 Am2

[*Weissia andrewsii* Bartr. Am1 Am2 = *Weissia controversa* Hedw. *fide* Stoneburner & Wyatt, Monogr. Syst. Bot. Missouri Bot. Gard. 11: 184, 1985 and, Bryologist 88: 302, 1985]

Weissia argentinica C. Müll. Am6

Weissia artocosana (Mitt.) Zand. (*nom. nov.* for *Pseudosymblepharis socotrana* (Mitt.) Thér.), see treatment of *Weissia* Afr2

[*Weissia atrocaulis* Saito, J. Hattori Bot. Lab. 39: 425, 1975 As2 ≡ *Trichostomum atrocaule* (Saito) Zand., see treatment of *Trichostomum*]

Weissia austrocrispa (Beckett) Stone, J. Bryol. 11: 236, 1980 Austr2

Weissia ayresii Schimp. in Besch. Afr3

Weissia balansae (C. Müll.) Zand. (*Phasconica*), see treatment of *Weissia* Austr1 Oc

Weissia balansaeana (Besch.) C. Müll. Am5 Am6

Weissia bizotii Zand. (*nom. nov.* for *Kleioweisiopsis involuta* Biz.), see treatment of *Weissia* Afr2

Weissia brachycarpa (Nees & Hornsch.) Jur. (replaces *Weissia microstoma* (Hedw.) C. Müll. *hom. illeg. fide* Koponen et al., Flora Fenn. 6: 61, 1977)

var. *obliqua* (Nees & Hornsch.) Hill, J. Bryol. 11: 601, 1981 [1982]

Weissia brachypelma C. Müll. Afr2

Weissia brachypoma Townsend, J. Bryol. 11: 695, 1981 [1982] Afr2

Weissia brevitheca Chen *nom. inval. descr.* germ. As2

Weissia breutelii C. Müll. Am2 Am3 Am4 Am5

Weissia canaliculata Hampe Am5

[*Weissia cirrata* Hedw. ≡ *Dicranoweisia cirrata* (Hedw.) Lindb.]

var. *atra* Brid. Eur

var. *ovatum* Hüb. Eur

var. *porrigens* Brid. Eur

[*Weissia clavinervis* Card. & P. Varde Oc ≡ *Trichostomum clavinerve* (Card. & P. Varde) Whitt. in Whitt. & Whitt., Bryologist 77: 433, 1974]

Weissia condensa (Voit) Lindb. Eur As1 As3 As5 Afr1 Am1 Austr2 (replaces *Weissia tortilis* (Schwaegr.) C. Müll. *hom. illeg. fide* Corley et al., J. Bryol. 11: 609, 1981 [1982])

Weissia controversa Hedw. Eur As1 As2 As3 As4 As5 Afr1 Afr2 Afr4 Am1 Am2 Am3 Am4 Am5 Am6 Austr1 Austr2
subsp. *perssonii* (Kindb.) Podp. Eur
var. *amblyodon* (Brid.) Sendtn. Eur As2 Afr1 Am1 Am3
var. *amoene-viridis* Nees. & Hornsch. Eur
var. *arenicola* (Limpr.) Podp. Eur
[var. *australis* (Aust.) Schornh. Am1 Am3 = *Weissia controversa* var. *longiseta* (Lesq. & James) Crum, Steere & Anders. *fide* Crum et al., Bryolgist 67: 164, 1973 = *Weissia controversa* Hedw. *fide* Stoneburner, Bryologist 88: 302, 1985]
[var. *brachycarpa* Nees. & Hornsch. Eur ≡ *Weissia brachycarpa* (Nees & Hornsch.) Jur. *fide* Koponen et al., Flora Fenn. 6: 61, 1977
[var. *crispata* (Nees & Hornsch.) Nyholm, Ill. Moss Fl. Fenn. II. Musci 775, 1969 (*Hymenostomum*) *nom. inval.*]
var. *densifolia* (BSG) Demaret in Demaret & Castagne, Fl. Gén. Belgique, Bryoph. 2: 242, 1964 Eur Afr1 Am1
var. *exigua* Schultz ex Nees & Hornsch. Eur
var. *gibbosula* (Amann) Podp. Eur
var. *gymnostoma* (Dix.) Sainsb. Austr1 Austr2
var. *hioramii* Thér. ex Biz, Bull. Soc. Linn. Lyon 34: 311, 1965 *nom. inval.* Am2
var. *latifolia* (Velen.) Podp. Eur
[var. *longiseta* (Lesq. & James) Crum, Steere & Anders., Bryologist 67: 164, 1967, *illeg. hom.* (*Weissia*) Am1 Am3 = *Weissia controversa* Hedw. *fide* Stoneburner, Bryologist 88: 302, 1985]
var. *macrophylla* (Thér. & P. Varde) Wijk & Marg. Afr2
var. *microdonta* (Hedw.) Röhl. Eur Afr1 Am1
var. *microstoma* Hüb. Eur
var. *minutissima* (Par.) Wijk & Marg. As2
var. *nitidifolia* (Farn.) Podp. Eur
var. *pillansii* Schelpe, Trans. Roy. Soc. South Africa 44: 113, 1979 *nom. nov.* (*Weissia viridula* var. *brachycarpa* Dix. *non* Nees & Hornsch.) Afr4
var. *rutilans* Hüb. Eur
var. *stenocarpa* Nees & Hornsch. Eur Afr1 Am1
var. *subglobosa* (Limpr.) Podp. Eur
var. *tenuiseta* (Besch.) Nog. Eur
var. *turfosa* (Podp.) Podp. Eur
[var. *wolffii* (Lesq. & James) Crum, Steere & Anders., Bryologist 67: 164, 1967 "*wolfii*" (*Weissia*) Am1 = *Weissia controversa* Hedw. *fide* Stoneburner, Bryologist 88: 302, 1985]
fo. *fasciculata* (Latz.) Podp., Consp. Musc. Eur. 189, 1954 (*Weissia viridula* fo.) Eur
fo. *protonematica* (Latz.) Podp., Consp. Musc. Eur. 189, 1954 (*Weissia viridula* fo.) Eur
fo. *longiseta* (Latz.) Podp., Consp. Musc. Eur. 189, 1954 (*Weissia viridula* fo.) Eur
fo. *ticinensis* (Farn.) Podp., Consp. Musc. Eur. 190, 1954 (*Weissia viridula* fo.) Eur
[*Weissia crispa* (Hedw.) Mitt. *hom. illeg.* Eur As1 As2 As3 As5 Afr1 = *Weissia longifolia* Mitt. *fide* Corley et al., J. Bryol. 11: 623, 1981 (1982)]
[subsp. *sterilis* (Nich.) Dix. ≡ *Weissia sterilis* Nich. *fide* Crundwell & Nyholm, J. Bryol. 7: 11, 1972]
[var. *aciculata* (Mitt.) Dix. Eur = *Weissia longifolia* Mitt. *fide* Crundwell & Nyholm, J. Bryol. 7: 13, 1972]
Weissia crispa (Hedw.) Mitt. [= *Weissia longifolia* Mitt.] × *W. crispata* (Nees & Hornsch.) C. Müll. *fide* Nicholson, Rev. Bryol. 32: 20, 1905 Eur
Weissia crispa (Hedw.) Mitt. [= *Weissia longifolia* Mitt.] × *W. microstoma* (Hedw.) C. Müll. *fide* Nicholson, Rev. Bryol. 33: 1, 1906 Eur
Weissia crispata (Nees & Hornsch.) C. Müll. = *Weissia fallax* Sehlm.
Weissia crispata (Nees & Hornsch.) C. Müll. [= *Weissia fallax* Sehlm.] × *W. crispa* (Hedw.) Mitt. *fide* Nicholson, Rev. Bryol. 32: 22, 1905 Eur
[*Weissia crispula* Hedw. ≡ *Dicranoweisia crispula* (Hedw.) Mild.]
var. *ambigua* Hook. f. Austr2
var. *longirostris* Nees & Hornsch. Eur
var. *subulata* Nees & Hornsch. Eur
Weissia cucullata C. Müll. Afr4
Weissia cucullata P. Varde Afr2 *hom. illeg.*
[*Weissia cucullifolia* Dix. & Sak. in Sak. As2 = *Weissia planifolia* Dix. *fide* Saito, J. Hattori Bot. Lab. 39: 427, 1975 ≡ *Trichostomum planifolium* (Dix.) Zand., see treatment of *Trichostomum*]
[*Weissia curvirostris* C. Müll. *hom. illeg.* ≡ *Hymenostylium recurvirostrum* (Hedw.) Dix.]
var. *curvipes* Kindb. Am1
[*Weissia deciduaefolia* Saito, J. Hattori Bot. Lab. 39: 429, 1975 As2 ≡ *Trichostomum deciduaefolium* (Saito) Zand., see treatment of *Trichostomum*]
Weissia dieterlenii Thér. Afr4
Weissia diffidentia Zand. (*nom. nov.* for *Phascum recurvirostre* C. Müll.), see treatment of *Weissia* Am5 Am6
Weissia edentula Mitt. As3 As4 Oc
Weissia erythrogona Brid. Am1
[*Weissia euteiches* Zand., Monogr. Syst. Bot. Missouri Bot. Gard. 11: 195, 1985 (*nom. nov.* for *Weissia tortilis* (Schwaegr.) C. Müll.) ≡ *Weissia condensa* (Voit) Lindb. (replaces *Weissia tortilis* (Schwaegr.) C. Müll. *hom. illeg. fide* Corley et al., J. Bryol. 11: 609, 1981 (1982)]
Weissia exserta (Broth.) Chen As2
Weissia fallax Sehlm. Eur As2 As5 Afr1 Am1
var. *alpina* (Schimp.) Podp. Eur
var. *subgymnostoma* (Podp.) Podp. Eur
[*Weissia flavescens* (Britt. in N. Britt. & Millsp.) Reese, Bryologist 94: 54, 1991 Am1 Am3 ≡ *Trichostomum brittonianum nom. nov.*, see treatment of *Trichostomum*]
Weissia felipponei Thér. in Felipp. Am6
Weissia fornicata Brid. Eur
Weissia ghatensis Dix. & P. Varde As3
[*Weissia glauca* Bartr. Am1 Am2 *hom. illeg.* ≡ *Weissia andersoniana* Zand., Monogr. Syst. Bot. Missouri Bot. Gard. 11: 195, 1985]
Weissia glazioudii Zand. (*nom. nov.* for *Hymenostomum striatum* Geh. & Hampe), see treatment of *Weissia* Am5
[*Weissia graeca* Schiffn. Eur = *Weissia controversa* Hedw. *fide* Corley et al., J. Bryol. 11: 623, 1981 (1982)]
[*Weissia hedwigii* Crum, Bryologist 74: 169, 1971 Eur As1 As2 As3 As5 Afr1 Am1 (*nom. nov.* for *Weissia microstoma* (Hedw.) C. Müll. *hom. illeg.*) = *Weissia brachycarpa* (Nees & Hornsch.) Jur. var. *brachycarpa fide* Corley et al., J. Bryol. 11: 650, 1981 (1982) and Koponen et al., Flora Fenn. 6: 61, 1977]
Weissia humicola C. Müll. Afr2 Afr4
Weissia inoperculata (Crum) Crum, Steere & Anders., Bryologist 67: 164, 1973 Am1 (*Hymenostomum*)
Weissia jamaicensis (Mitt.) Grout Am1 Am2 Am3
Weissia jamesonii Tayl. Am4
Weissia kaikouraensis R. Br. ter Austr2
[*Weissia krassavinii* (Lazar.) Lazar. in Lazar. et al., Biul. Mosk. Obsh. Ispyt. Prir, Otd. Biol. 73(2): 142, 1968 *comb. inval. basion. non cit.* (*Hymenostomum*) As1 ≡ *Weissia krassavinii* (Lazar.) Lazar. ex Ochyra]
Weissia krassavinii (Lazar.) Lazar. ex Ochyra, J. Hattori Bot. Lab.

64: 343, 1988 As1
Weissia kunzeana C. Müll. Am6
[*Weissia lancifolia* (R. Br. ter) Wijk & Marg. Austr2 = *Tridontium tasmanicum* Hook. f. in Hook., see treatment of *Tridontium*]
[var. *waymouthii* (R. Br. ter) Wijk & Marg. Austr2 ≡ *Didymodon waymouthii* (R. Br. ter) Zand., see treatment of *Didymodon*]
Weissia laticuscula C. Müll. Afr2 Afr4 (good species *fide* Magill, Fl. S. Afr. I. Mosses 1: 267, 1981 [1982])
Weissia leptocarpa Schimp. ex Besch. in Ler. & Levier Eur Afr1 *hom. illeg.*
Weissia levieri (Limpr.) Kindb. cf. Crundwell & Nyholm, J. Bryol. 7: 16, 1972 Eur
Weissia ligulaefolia (Bartr.) Grout Am1
Weissia lineaefolia C. Müll. Afr2
Weissia longidens Card. As2 (good species *fide* Saito, J. Hattori Bot. Lab. 39: 429, 1975)
Weissia longifolia Mitt. Eur As1 As2 As3 As5 Afr1
var. *angustifolia* (Baumg. in Giznb.) Crundw. & Nyh., J. Bryol. 7: 14, 1972 (*Astomum crispum* var.) Eur
Weissia lorentzii (C. Müll.) Zand. (*Phasconica*), see treatment of *Weissia* Am6
Weissia ludoviciana (Sull.) Reese & Lemmon, Bryologist 68: 282, 1965 (*Astomum*) Am1
Weissia ludoviciana (Sull.) Reese & Lemmon × *W. controversa* Hedw. *fide* Reese & Lemmon, Bryologist 68: 280, 1965 Am1
Weissia macrospora Card. & P. Varde As3
[*Weissia malayensis* (Fleisch.) Manuel, Fed. Mus. Jour. 26: 161, 1981 (*Hymenostomum*) As2 As4 = *Barbula indica* (Hook.) Spreng. *fide* Iwatsuki & Noguchi, J. Hattori Bot. Lab. 37: 355, 1973 as *Barbula cruegeri*]
Weissia melanostoma Mitt. Am5
Weissia micacea (Schlecht.) C. Müll. Am3 Am5
[*Weissia microstoma* (Hedw.) C. Müll. Eur As1 As2 As3 As5 Afr1 Am1 *hom. illeg.* = *Weissia brachycarpa* (Nees & Hornsch.) Jur. var. *brachycarpa fide* Hill, J. Bryol. 11: 601, 1981 (1982) and Koponen et al., Flora Fenn. 6: 61, 1977]
[*Weissia minycarpa* Zand., Monogr. Syst. Bot. Missouri Bot. Gard. 11: 196, 1985 (*nom. nov.* for *Trichostomum microcarpum* Schimp. ex Besch. *non Weissia microcarpa* Hook. f. & Wils.) Am2 Am3 Am4 ≡ *Weissia sinaloensis* Bartr. *fide* Zander in Sharp et al., Moss Fl. Mex. ≡ *Trichostomum sinaloense* (Bartr.) Zand., see treatment of *Trichostomum*]
Weissia mittenii (BSG) Mitt. Eur
Weissia muehlenbergiana (Sw.) Reese & Lemmon, Bryologist 68: 282, 1965 (*Astomum*) As2 Am1
Weissia multicapsularis (Sm.) Mitt. Eur As1
Weissia neocaledonica (Thér.) Zand. (*Aschisma*), see treatment of *Weissia* Oc
Weissia newcomeri (Bartr.) Saito, J. Hattori Bot. Lab. 39: 423, 1975 (*Hymenostomum*) As2
Weissia nitida Reinw. & Hornsch. As4
Weissia norketii R. S. Chopra, Taxon. Indian Mosses 125, 1975 (*nom. nov.* for *Weissia amblyphylla*) (Dix.) Dix. As3
[*Weissia nuda* Mitt. Austr1 *hom. illeg.* ≡ *Weissia alianuda* Tan, Mem. New York Bot. Gard. 68: 5, 1992, *nom. nov.*]
[*Weissia nudiflora* C. Müll. & Hampe Austr1 = *Weissia controversa* Hedw. *fide* Catcheside, Mosses S. Austr. 196, 1980]
Weissia obtusata C. Müll. Am5
Weissia obtusifolia C. Müll. Am5
Weissia occidentalis (Flow. ex Crum) Stoneburner, Bryologist 88: 310, 1985 (*Astomum*) Am1
Weissia occulta Wallr. Eur
Weissia opaca (Dix.) Magill in Magill & Schelpe, Mem. Bot. Surv. S. Afr. 43: 7, 1979 (*Tortella*) Afr2
Weissia ovalis (Williams) Bartr. Oc
Weissia papillosa Dix. & Nav. Afr2
Weissia papillosissima Laz., Dopov. Akad. Nauk Ukr. R.S.R. Ser. B 8: 752, 1967 As1
Weissia patagonica Card. & Broth. Am6
[*Weissia perligulata* Flow. ex Crum, Bryologist 76: 291, 1973 Am1 (= *Trichostomum crispulum* Bruch ex F. Müll. *fide* Stoneburner & Wyatt, Monogr. Syst. Bot. Missouri Bot. Gard. 11: 184, 1985) ≡ *Trichostomum perligulatum* (Flow. ex Crum) Zand., see treatment of *Trichostomum*]
Weissia perpusilla (C. Müll.) Stone, J. Bryol. 11: 231, 1980 (*Hymenostomum*) Austr1
Weissia phascopsis Zand. (*nom. nov.* for *Gymnostomum phascoides* Hook. ex Drumm. *non Weissia phascoides* (BSG) C. Müll.) Eur Am1
[*Weissia planifolia* Dix. As1 As2 Am2 ≡ *Trichostomum planifolium* (Dix.) Zand., see treatment of *Trichostomum*]
[*Weissia platyphylloides* Card. As2 As3 = *Weissia edentula* Mitt. *fide* Saito, J. Hattori Bot. Lab. 39: 421, 1975]
Weissia platystegia (Dix.) Eddy, Handb. Males. Mosses 2: 165, 1991 (*Astomum*) As4
Weissia riograndensis (Broth.) Zand. (*Hymenostomum*), see treatment of *Weissia* Am5
[*Weissia rigescens* Broth. As2 *hom. illeg.* = *Weissia controversa* Hedw. *fide* Saito, J. Hattori Bot. Lab. 39: 426, 1975]
[*Weissia riparia* Hampe Austr1 = *Weissia controversa* Hedw. *fide* Catcheside, Mosses S. Austr. 196, 1980]
Weissia rostellata (Brid.) Lindb. Eur Am1
[var. *phascoides* (Hook.) Reese & Lemmon, Bryologist 68: 283, 1965 (*Astomum*) Eur Am1 ≡ *Weissia phascopsis* Zand. *nom. nov.*, Monogr. Syst. Bot. Missouri Bot. Gard. 11: 196, 1985 *non Weissia phascoides* (BSG) C. Müll.]
Weissia rutilans (Hedw.) Lindb. Eur As1 Afr1 Afr2 Am1 Austr1
subsp. *ganderi* (Jur.) Kindb. Eur
var. *ganderi* (Jur.) Macoun Eur
var. *hillieri* Meyl. Eur
var. *himalayana* Broth. As3
var. *subgymnostomum* Limpr.
Weissia semidiaphana (Thér.) Zand, Monogr. Syst. Bot. Missouri Bot. Gard. 11: 197, 1985 (*Hymenostomum*) Am2
Weissia semiinvoluta C. Müll. Am6
Weissia semipallida C. Müll. As2
Weissia sharpii Anders. & Lemmon, Bryologist 76: 133, 1973 Am1
Weissia simplex Brid. *ignot.*
[*Weissia sinaloensis* Bartr. Am2 Am3 Am4 (see Zander in Sharp et al., Moss Fl. Mex.) ≡ *Trichostomum sinaloensis* (Bartr.) Zand., see treatment of *Trichostomum*]
Weissia socotrana Mitt. Afr2
Weissia splachnum Garov. Eur
Weissia sterilis Nich. see Crundwell & Nyholm, J. Bryol. 7: 11, 1972 Eur
Weissia squarrosa (Nees & Hornsch.) C. Müll. Eur
Weissia subacaulis (Mitt.) Par. Am4
[*Weissia subangustifolia* (Thér.) Zand., Bryologist 86: 156, 1983 (*Hyophila*) Am2 ≡ *Trichostomum subangustifolium* (Thér.) Zand., see treatment of *Trichostomum*]
Weissia submicacea C. Müll. Am5
Weissia sweetii Bartr. Am1 [= *Trichostomum crispulum* Bruch in F. Müll. *fide* Stoneburner & Wyatt, Monogr. Syst. Bot. Missouri Bot. Gard. 11: 184, 1985 and Stoneburner, Bryologist 88: 1985, but retained here because leaf margins narrowly inflexed and monoicous, possibly synonymous with *Weissia perligulata*]

[*Weissia termitarum* C. Müll. Am5 ≡ *Trichostomum termitarum* (C. Müll.) Zand., see treatment of *Trichostomum*]
Weissia termitidarum C. Müll. Afr2 Am5
[*Weissia tortilis* (Schwaegr.) C. Müll. *hom. illeg.* Eur As1 As3 As5 Afr1 Am1 Austr2 = *Weissia condensa* (Voit) Lindb. *fide* Corley et al., J. Bryol. 11: 623, 1981 (1982)]
[subsp. *papillosissima* Lazar., Dopov. Akad. Nauk Ukr. R.S.R. Ser. B 8: 752, 1967 *nom. inval. in syn.* Eur ≡ *Weissia papillosissima* Lazar.]
var. *intermedia* Mönk. Eur
var. *subcylindrica* (BSG) Dix. Eur
[*Weissia tortivelata* Williams Am4 ≡ *Trichostomum tenuirostre* var. *gemmiparum* (Schimp.) Zand., see treatment of *Trichostomum*]
Weissia triumphans (De Not.) Hill, J. Bryol. 11: 600, 1981 [1982] (*Trichostomum*) Eur As5 Afr1
var. *monspeliensis* (Schimp.) Zand. (*Trichostomum*), see treatment of *Weissia* Eur Afr1
Weissia tyrrhena Fleisch. Eur As1
[*Weissia umbrosa* Mitt., J. Linn. Soc. Bot. 12: 133, 1869 Am4 ≡ *Trichostomum mitteneanum* Zand. *nom. nov.*, see treatment of *Trichostomum*]
Weissia unguiculata (Mitt.) Crundw. & Nyholm, J. Bryol. 8: 69, 1974 (*Astomum*) Afr2
[*Weissia vallis-gratiae* C. Müll., Hedwigia 38: 111, 1899 = *Weissia controversa* Hedw. *fide* Magill, Fl. S. Afr. I. Mosses 1: 265, 1981 (1982)]
[*Weissia verticillata* Brid. ≡ *Eucladium verticillatum* (Brid.) BSG]
var. *sterilis* Mazz. Eur
Weissia veviridis Zand. (*nom. nov.* for *Trichostomum perviride* Broth.), see treatment of *Weissia* Am3 Am4
[*Weissia viridula* Hedw. ex Brid. *illeg.* = *Weissia controversa* Hedw.]
subsp. *longirostris* Kindb. in Röll Am1
var. *canaliculata* Hampe Am4
var. *cylindrica* Schimp. in Par. Afr1
[var. *gymnostoma* Dix., New Zealand Inst. Bull. 3: 112, 1923 = *Trichostomum brachydontium* Bruch *fide* Stoneburner, Bryologist 88: 312, 1985]
[var. *hioramii* Thér. *nom. inval. descr. gall.* Am3 ≡ *Weissia controversa* var. *hioramii* Thér. ex Biz, Bull. Soc. Linn. Lyon 34: 311, 1965]
var. *longidens* De Not. Eur
var. *longifolia* Broth. & Wag. Afr4
var. *polycarpa* Chen As2
var. *rugeliana* Ren. & Card. Am1
var. *seligerioides* Amann, Fl. Mouss. Suisse 2(Add. Rectific.):1, 1991 Eur
fo. *fasciculata* Latz., Beih. Bot. Centralb. 48: II. 472, 1931 Eur
fo. *longiseta* Latz., Beih. Bot. Centralb. 48: II. 472, 1931 *hom. illeg.* Eur
fo. *protonematica* Latz., Magyar Bot. Lapok 29: 16, 1930 Eur
fo. *ticinensis* Farn., Muschi Prov. Pavia, Atti Istituto Bot. Univ. Pavia 3: 9, 1891 Eur
[*Weissia waymouthii* R. Br. ter Austr2 ≡ *Didymodon waymouthii* (R. Br. ter) Zand., see treatment of *Didymodon*]
Weissia welwitchii Schimp. Eur
Weissia willisiana (Sainsb.) Catcheside, Mosses S. Austr. 194, 1980 (12 Dec. 1980) (*Pottia*) (comb. also by Stone, J. Bryol. 11: 235, 1980 (27 Mar. 1981) Austr1
Weissia wimmeriana (Sendtn.) BSG Eur As3 Am1
subsp. *pallescens* (Besch.) Giac. Eur Afr1
var. *dalmatica* (Latz.) Podp. Eur
var. *linderi* Broth. & Geh. Eur
var. *muralis* (Spruce) Breidl. Eur
var. *pallescens* (Besch.) Bott. Eur Afr1
[fo. *hymenostomoides* Spruce = *Weissia wimmeriana* var. *muralis* (Spruce) Breidl. *fide* Podp., Consp. Musc. Eur. 190, 1954]
[fo. *subgymnostoma* Limpr., Laubm. Deutsch. 259, 1886 = *Weissia wimmeriana* var. *muralis* (Spruce) Breidl. *fide* Podp., Consp. Musc. Eur. 191, 1954]

WILLIA C. Müll. in Neum.
Willia austroleucophaea (Besch.) Broth. Am6 Ant
Willia brachychaete (Dus.) Zand. (*Tortula*), see treatment of *Willia* Am6
Willia calobolax (C. Müll.) Lightowlers, J. Bryol. 13: 370, 1985 (*Barbula*) Afr4
var. *angustinervia* (C. Müll.) Lightowlers, J. Bryol. 13: 370, 1985 (*Barbula calobolax* var.) Afr4
[*Willia grimmioides* C. Müll. in Neum. Am6 = *Willia austroleucophaea* (Besch.) Broth. *fide* Bell, Brit. Antarct. Surv. Bull. 38: 73, 1974]
[*Willia marginata* (Hook. f. & Wils.) C. Müll. Afr4 ≡ *Hennediella marginata* (Hook. f. & Wils.) Zand., see treatment of *Hennediella*]

[**WILLIAMSIELLA** Britt. = *Leptodontium*]
[*Williamsiella tricolor* (Williams) Britt. Am4 ≡ *Leptodontium tricolor* (Williams) Zand. in Zand. & Hegew., Bryologist 79: 20, 1976]

BIBLIOGRAPHY

This literature list includes selected taxonomic treatments and morphological or physiological discussions of the Pottiaceae in addition to the citations in the text. Published notices of simple range extensions and other minor works are largely eliminated. Readers are directed to a far larger bibliography published electronically (Zander 1993b), which is more suitable for *ad hoc* searches.

A

Abramova, A. L., L. S. Blagodatskich & L. A. Czerepanova. 1973. Conspectus generis *Pterygoneurum* Jur. (Musci) in URSS. Nov. Syst. Pl. non Vasc., Ac. Sc. URSS, Inst. Bot. V. L. Komarovii 10: 305–316.

Abramova, A. L. & U. K. Mamatkulov. 1967. [De proprietatibus biologicis *Barbula johansenii* Williams.] Nov. Syst. Plant. non Vasc., Acad. Sci. USSR, Komarov Inst. 1967: 364–371.

Abramova, A. L. & Ts. Tsegmed. 1987. De positione taxonomica *Trichostomopsis arronis* (Lor.) Agnew et Townsend notula. Novosti Sistematiki Nizshikh Rastenii 24: 169–178. [In Russian.]

Agnew, S. & M. Vondráček. 1975. A moss flora of Iraq. Feddes Repertorium 86(6–8): 341–489.

Akiyama, H. & N. Nishimura. 1993. The *Climacium*-type branch development on the Bartramiaceae and its taxonomic significance. Bryologist 96: 185–191.

Albert, V. A. & B. D. Mishler. 1992. On the rationale and utility of weighting nucleotide sequence data. Cladistics 8: 73–83.

Allen, B. H. 1992. Teratolgical axillary "hairs" in *Didymodon tophaceus*. Bryologist 95: 97–99.

Allen, B. H. & R. E. Magill. 1987. In support of a distinct terminology for bryophyte sexuality. Taxon 36: 57–58.

Allen, B. H. & R. A. Pursell. 1991. A reconsideration of the systematic position of *Jonesiobryum*. Bryologist 94: 438–442.

Allorge, P. 1938. Observations sur *Tortula desertorum* Broth., mousse aralo-caspienne des plateaux castillans. Rev. Bryol. Lichénol. 11: 110–112.

Anderson, L. E. 1943. The distribution of *Tortula pagorum* (Mild.) De Not. In North America. Bryologist 46: 47–66.

Anderson, L. E. 1951. The mosses of North Carolina. VI. Encalyptaceae to Pottiaceae. Bryologist 54: 156–161.

Anderson, L. E. 1954. Hoyer's Solution as a rapid mounting medium for bryophytes. Bryologist 57: 242–244.

Anderson, L. E. & B. E. Lemmon. 1972. Cytological studies of natural intergeneric hybrids and their parental species in the moss genera *Astomum* and *Weissia*. Ann. Missouri Bot. Gard. 59: 382–416.

Anderson, L. E. & B. E. Lemmon. 1973. *Weissia sharpii*, a new species from North America, with cytological and and ecological correlations. Bryologist 76: 131–143.

Anderson, L. E. & B. E. Lemmon. 1974. Gene flow distances in the moss *Weissia controversa* Hedw. J. Hattori Bot. Lab. 38: 67–90.

Anderson, L. E. & R. H. Zander. 1973. The mosses of the Southern Blue Ridge Province and their phytogeographic relationship. J. Elisha Mitchell Scientific Soc. 89: 15–60.

Anderson, L. E. & R. H. Zander. 1986. False anisospory in the moss *Leptodontium viticulosoides* (P.-Beauv.) Wijk & Marg. Am. J. Bot. 73(5): 603 (Abstr.).

Andrews, A. L. 1915. Bryological notes. I. *Aschisma kansanum*, a new species, with remarks upon the genus. Torreya 15: 63–67.

Andrews, A. L. 1920. *Hymenostomum* in North America. I. Delimitation of the genus. Bryologist 23: 28–31.

Andrews, A. L. 1922a. *Hymenostomum* in North America. II. The case of *Astomum sullivantii*. Bryologist 25: 66–71.

Andrews, A. L. 1922b [1923]. The status of *Gyroweisia* in North America. Bryologist 25: 97–100.

Andrews, A. L. 1922c [1923]. The status of *Gyroweisia* in North America. Bryologist 25: 97–100.

Andrews. A. L. 1924. *Hymenostomum* in North America. III. *Astomum ludovicianum*. Bryologist 27: 1–3.

Andrews, A. L. 1926. *Hymenostylium longopulvinatum*. Bryologist 29: 69–70.

Andrews, A. L. 1929. [Review of:] Fritz von Wettstein, Morphologie und Physiologie des Formwechsels der Moose auf genetischer Grundlage, II. Bryologist 32: 14–15.

Andrews, A. L. 1933. *Hymenostomum* in North America. V. *Weisia viridula*. Bryologist 36: 28–31.

Andrews, A. L. 1941. Taxonomic notes. I. The genus *Dactylhymenium* Card. Bryologist 44: 105–107.

Andrews, A. L. 1943. Taxonomic notes, III. Further synonyms of *Hymenostylium curvirostrum*. Bryologist 46: 131–135.

Andrews, A. L. 1945. Taxonomic notes. V. The genus *Tetrapterum*. Bryologist 48: 190–193.

Andrews, A. L. 1949. Taxonomic notes IX. *Splachnobryum kieneri*. Bryologist 52: 78–83.

Andrews, A. L. 1953. The possible type locality of *Anoectangium peckii*. Bryologist 56: 49–50.

Andrews, S. & P. L. Redfearn, Jr. 1965. Observations on the germination of the gemmae of *Hyophila tortula* (Schwaegr.) Hampe. Bryologist 68: 345–347.

Anonymous. 1990a. Leaf lenses: a finer focus. Sci. News 138(6): 85.

Anonymous. 1990b. Why hot plants resort to fetal position. Sci. News 138(6): 93.

Appleyard, J., M. O. Hill & H. L. K. Whitehouse. 1985. *Leptobarbula berica* (De Not.) Schimp. in Britain. J. Bryol. 18: 461–470.

Archie, J. W. 1989a. A randomization test for phylogenetic information in systematic data. Syst. Zool. 38: 239–252.

Archie, J. W. 1989b. Homoplasy excess ratios: new indices for measuring levels of homoplasy in phylogenetic systematics and a critique of the consistency index. Syst. Zool. 38: 253–269.

Arrocha, C. 1992. Musgos de Costa Rica: nuebos registros. Tropical Bryology 5: 1–3.

Arts, T. 1987a. *Pottia bryoides* (Dicks.) Mitt., *P. lanceolata* (Hedw.) C. Müll. and *P. truncata* (Hedw.) B.&S. with rhizoidal tubers recorded from Belgium. Bull. Soc. Bot. Belgique 120: 3–6.

Arts, T. 1987b. *Pottia bryoides* (Dicks.) Mitt., *P. lanceolata* (Hedw.) C. Müll. and *P. truncata* (Hedw.) B.&S. with rhizoidal tubers. Lindbergia 13: 130–132.

Arts, T. 1988. Rhizoidal tubers and protonematal gemmae in *Pseudocrossidium revolutum* (Brid.) Zander var. *revolutum* and *Scopelophila cataractae* (Mitt.) Broth. Lindbergia 14: 59–62.

Arts, T. 1989 [1990]. *Rhamphidium purpuratum* Mitt.: its vegetative propagation and distribution. Lindbergia 15: 106–108.

Arts, T. & P. Sollman. 1991. Remarks on *Phascum leptophyllum* C. Müll., an earlier name for *Tortula rhizophylla* (Sak.) Iwats. & K. Saito. Lindbergia 17: 20–27.

Atchley, W. R. & W. M. Fitch. 1991. Gene trees and origins of inbred strains of mice. Science 254: 554–558.

Azziz, M. N. & J. N. Vohra. 1988 [1990]. Note on the identity of *Hymenostylium xanthocarpum* (Hook.) Brid. Bull. Bot. Surv. India 30: 185–187.

B

Bachelot de La Pylaie, A. 1813. Tableau du genre *Phascum*. J. Bot. Appliquée Agric. Set 2, 2(6): 269–288.

Baczurina, A. F. 1972. On distribution of representatives of *Aloina* Lindb. in bryoflora of the Ukraine. Ukr. Bot. Žurn. 29: 239–241.

Bapna, K. R. & B. L. Choudhary. 1989. Mosses of Rajasthan II. Dicranales and Pottiales. J. Indian Bot. Soc. 68: 379–388.

Barker, T. 1904. Note on *Tortula rigida* and *T. brevirostris*. Rev. Bryol. 31: 23.

Barkman, J. 1963. A contribution to the taxonomy of the *Tortula laevipila-T. pagorum*-complex. Rev. Bryol. Lichénol. 32: 183–192.

Bartram, E. B. 1924a. *Scopelophila ligulata* Spruce in North America. Rev. Bryol. 51: 47–48.

Bartram, E. B. 1924b. Studies in *Tortula* as represented in southern Arizona. Bull. Torrey Bot. Club 51: 335–340.

Bartram, E. B. 1924c. *Weisia sinaloensis* not a *Weisia*. Bryologist 29: 8.

Bartram, E. B. 1926a. Notes on the genus *Husnotiella* Cardot. Bryologist 29: 44–46.

Bartram, E. B. 1926b. *Tortula fragilis* Tayl. and *Tortula bogotensis* (Hpe.) Mitt. Bryologist 29: 53–54.

Bartram, E. B. 1926c. A variety of *Didymodon mexicanus* Besch., in Arizona. Bryologist 29: 1–2.

Bartram, E. B. 1927. *Weisia* and *Hymenostomum* in southwestern United States. Bryologist 80: 77–83 + plates 6–7.

Bartram, E. B. 1939. Mosses of the Philippines. Philippine J. Sci. 68: 1–437 + 29 plates.

Bartram, E. B. 1949. Mosses of Guatemala. Fieldiana: Bot. 25: i-v, 1–442.

Basile, D. V. & M. R. Basile. 1984. Probing the evolutionary history of bryophytes experimentally. J. Hattori Bot. Lab. 55: 173–185.

Basile, D. V., M. R. Basile & Q.-Y. Li. 1984. "Desuppression" of cell division in leaf primordia in *Plagiochila arctica* (Hepaticae) by 3,4–dehydroproline, a selective inhibitor of prolyl hydroxylase. Am. J. Bot. 71 (5 part 2): 18 (Abstr.).

Bauer, C. R. & M. R. Crosby. 1986. Index Muscorum supplementum: 1982–1983. Taxon 35: 416–439.

Bell, G. 1974. A synoptic flora of South Georgia mosses: V. *Willia* and *Racomitrium*. Bull. Brit. Antarct. Surv. 38: 73–101.

Bell, G. 1981. Mosses and their relationship to an arid environment. Australian Syst. Bot. Soc. Newsl. 29. Pp. 11–13.

Bell, G. 1982. Leaf morphology of arid-zone moss species from South Australia. J. Hattori Bot. Lab. 53: 147–151.

Bennett, R. A. 1965. Observations on the growth and maturation of the sporophytes of *Tortula muralis* Hedw. Trans. Bot. Soc. Edinburgh 48: 120–126.

Bewley, J. D. 1972. The conservation of polyribosomes in the moss *Tortula ruralis* during total desiccation. J. Exp. Bot. 23: 692–698.

Bewley, J. D. 1973a. Desiccation and protein synthesis in the moss *Tortula ruralis*. Canad. J. Bot. 51: 203–206.

Bewley, J. D. 1973b. Polyribosomes conserved during desiccation of the moss *Tortula ruralis* are active. Plant Physiol. 51: 285–288.

Bewley, J. D. 1979. Physiological aspects of dessication tolerance. Ann. Rev. Pl. Physiol. 30: 199–238.

Bewley, J. D. & A. Thorpe. 1974. On the metabolism of *Tortula ruralis* following desiccation and freezing: Respiration and carbohydrate oxidation. Physiol. Plant. 32: 147–153.

Bewley, J. D., E. B. Tucker & E. A. Gwóźdź. 1974. The effects of stress on the metabolism of *Tortula ruralis*. Bull. Roy. Soc. New Zealand 12: 395–402.

Bizot, M. 1954. Remarques sur *Tortula papillosissima* (Copp.) Broth. Rev. Bryol. Lichénol. 23: 268–270.

Bizot, M. 1956. Nouvelles remarques sur *Tortula papillosissima* (Copp.) Broth. Rev. Bryol. Lichénol. 25: 268–271.

Bizot, M. & C. Roux. 1968 [1969]. *Dialytrichia mucronata* (Brid.) Broth. var. nov. *fragilifolia*. Rev. Bryol. Lichénol. 36: 109–110. Bryol. 73: 157. 1970.

Blockeel, T. L. 1991. The genus *Hennediella* Par.: a note on the affinities of *Tortula brevis* Whitehouse & Newton and *T. stanfordensis* Steere. J. Bryol. 16: 187–192.

Blomquist, H. L. 1930. Archegonial plants of *Tortula pagorum* (Milde) De Not. in North Carolina. Bryologist 33: 41–43.

Blumrich, J. 1916. Verwandtschaft zwischen *Trichostomum crispulum* und *viridulum*. Bryol. Zeitschr. 1: 62.

Boros, Á. 1953. *Pterygoneurum smardaeanum* Vanek *sp. nov.* Rev. Bryol. Lichénol. 22: 332.

Boros, Á. & S. Polgar. 1941. Die *Tortula velenovskyi* in Ungarn. Bot. Dozlem. 38: 126–130.

Boros, Á. & L. Vajda. 1957. Der systematiser Wert von *Tortula obtusifolia* Schleich. Rev. Bryol. Lichénol. 26: 266.

Boudier, P. 1992. *Tortula rigescens* Broth. & Geh. (Pottiaceae—Musci), nouveau pour la bryoflore européene. Cryptpogamie Bryol. Lichénol. 13: 1–6.

Bowers, M. C. 1964. A water-soluble, rapid, permanent mounting medium for bryophytes. Bryologist 67: 358–359.

Braithwaite, R. 1887. The British Moss Flora. Vol. 1. London.

Brassard, G. R. 1979. The moss genus *Timmia*. 1. Introduction, and revision of *T. norvegica* and allied taxa. Lindbergia 5: 39–53.

Brassard, G. R. 1980. The moss genus *Timmia*. 2. Sect. *Timmiaurea*. Lindbergia 6: 129–136.

Brassard, G. R. 1984. The moss genus *Timmia*. 3. Sect. *Timmia*. Lindbergia 10: 33–40.

Bremer, B. 1981. A taxonomic revision of *Schistidium* (Grimmiaceae, Bryophyta). 3. Lindbergia 7: 73–90.

Bremer, K., C. J. Humphries, B. D. Mishler & S. P. Churchill. 1987. On cladistic relationships in green plants. Taxon 36: 339–349.

Bridel, S. E. 1826–27. Bryologia universa seu systematica ad novam methodum disposito, historia et descriptio omnium muscorum frondosorum hucusque cognitorum cum synonymia ex auctoribus probatissimis. Lipsiae. Vols. 1–2.

Brotherus, V. F. 1902–1909. Musci. In A. Engler & K. Prantl, Die Natürlichen Pflanzenfamilien 1(3). Leipzig.

Brotherus, V. F. 1910. Neue Laubmoosgattungen. Öfvers. Finska Vetensk.-Soc. Förh. 52A(7): 1–12, pl. 1–4.

Brotherus, V. F. 1911. *Aloinella*, eine neue Laubmoosgattung aus Ecuador. Helsingf. Övers. F. Vet. Soc. 53(13): 1–4.

Brotherus, V. F. 1924–1925. Musci. In A. Engler & K. Prantl, Die Natürlichen Pflanzenfamilien, Ed. 2. 10: 1–478; 11: 1–542. Leipzig.

Brown, R. 1818. Characters and description of *Lyellia*, a new genus of mosses, with observations on the section of the order to which it belongs; and some remarks on *Leptostomum* and *Buxbaumia*. Trans. Linn. Soc. London 12: 560–583.

Brown, R. (ter) 1894a. Notes on the genus *Gymnostomum*, with descriptions of new species. Trans. & Proc. New Zealand Inst. 26: 296–301.

Brown, R. (ter) 1894b. Notes on some species of New Zealand Musci: genus *Phascum*. Trans. & Proc. New Zealand Inst. 26: 302–304.

Brown, R. (ter) 1894c. Notes on New-Zealand mosses: genus *Pottia*. Trans. & Proc. New Zealand Inst. 26: 288–296.

Brown, R. (ter) 1896. New Zealand Musci: notes on a new genus: *Tetracoscinodon*. Trans. & Proc. New Zealand Inst. 28: 531.

Brown, R. (ter) 1897. Notes on the New Zealand Musci, and

descriptions of new species. Further notes on the New Zealand Musci: genus *Trichostomum*, with descriptions of some new species. Trans. & Proc. New Zealand Inst. 29: 465–477, 478–491.

Brown, R. (ter) 1898a. New Zealand Musci: Notes on the genus *Streptopogon* Wills. with descrption of a new species. Trans. & Proc. New Zealand Inst. 30: 409–410.

Brown, R. (ter) 1898b. New Zealand Musci: Notes on the new genus *Dendia*. Trans. & Proc. New Zealand Inst. 30: 411–412.

Brown, R. (ter) 1898c. New Zealand Musci: Notes on the genus *Tortula*, with descriptions of new species. Trans. & Proc. New Zealand Inst. 30: 399–409.

Bryan, V. S. 1956. Cytological and taxonomic studies of some species of *Astomum, Acaulon* and *Phascum*. Bryologist 59: 118–129.

Bryhn, N. 1902. *Sarconeurum*, genus muscorum novum. Nyt. Mag. Naturv., Kristiania 40: 204–208, 2 pl.

Buck, W. R. & R. H. Zander. 1980. On *Astomiopsis, Bryomanginia* and *Melophyllum* (Ditrichaceae). Bryologist 83: 254–255.

C

Camus, F. 1904. Sur une anomalie de *Phascum cuspidatum*. Bull. Soc. Bot. France 48: 421–423.

Cano, M. J., J. Guerra & R. M. Ros. 1992. *Crossidium seriatum* (Pottiaceae, Musci) new to Europe. Bryologist 95: 280–283.

Cao, Tong & D. H. Vitt. 1986. A taxonomic revision and phylogenetic analysis of *Grimmia* and *Schistidium* (Bryopsida; Grimmiaceae) in China. J. Hattori Bot. Lab. 61: 123–247.

Cárdenas S., A. 1988. Los musgos cleistocárpicos del Valle de México, México. Bryologist 9: 214–216.

Carlquist, S. 1966. The biota of long-distance dispersal. III. Loss of dispersibility in the Hawaiian flora. Brittonia 18: 310–335.

Carlquist, S. 1974. Island Biology. New York.

Carrión, J. S., J. Guerra & R. M. Ros. 1990. Spore morphology of the European species of *Phascum* Hedw. (Pottiaceae, Musci). Nov. Hedw. 51: 411–433.

Casares-Gil, A. 1932. Flora Ibérica. Briófias. Musgos. Museo Nacional de Ciencias Naturales. Madrid.

Casas de Puig, C. 1975a. Notulae bryologicae VIII. Consideraciones sobre el área de distribución y ecología de *Tortula desertorum* Broth. en España. Acta Phytotaxon. Barcinon. 14: 3–13.

Casas de Puig, C. 1975b. Aportación al estudio de la flora briologica española. Musgos y Hepáticas de las provincias de Soria, Logroño, Burgos y Segovia. Anales Inst. Bot. Cavanilles 32(2): 731–762.

Casas de Puig, C. & M. J. Molinas. 1974. Étude au microscope électronique à balayage de la surface des feuilles de *Tortula ruralis* (Hedw.) Gaertn. var. *hirsuta* (Vent.) Par. Rev. Bryol. Lichénol. 40: 267–270.

Casas de Puig, C. & M. L. Molinas. 1975. Notulae bryologicae IX. Estudio al microscopio electronico de barrido del enves de las hojas de *Tortula desertorum* Broth., procedentes de deferentes localidates de España. Acta Phytotaxon. Barcinon. 15: 14–18.

Casas [de Puig], C., R. M. Cros & J. Muñoz. 1993. *Triquetrella arapilensis* y especies afines: su morfología y distribución geográfica. Bryologist 96: 122–131.

Casas de Puig, C., C. Sérgio, R. M. Cros & M. Brugués. 1990. Datos sobre el género *Acaulon* en la Península Ibérica. Cryptogamie Bryol. Lichénol. 11: 63–70.

Castaldo-Cobianchi, R., S. Giordano & G. Cafiero. 1982. Studies on *Timmiella barbuloides* (Brid.) Moenk., IV. SEM and TEM characterization of spore wall and first germination stages. J. Bryol. 12: 273–278.

Castelli, L. 1955. *Desmatodon milezekii* Meyl. au Mont-Cenis. Rev. Bryol. 24: 371.

Castelli, L. 1966 [1967]. *Molendoa clavuligera* n. sp. Rev. Bryol. Lichénol. 34: 715–719.

Castelli, L. 1968 [1969]. À propos du *Molendoa clavuligera*. Rev. Bryol. Lichénol. 36: 130–131.

Catcheside, D. G. 1967. *Tortula pagorum* in Australia. Muelleria 1: 227–230.

Catcheside, D. G. 1980. Mosses of South Australia. Handbook of the Flora and Fauna of South Australia, South Australian Gov't.

Catcheside, D. G. 1992. *Tortula* section *Rurales* in Australasia. Australas. Bryol. Newsl. 27(Dec. 1992): 6.

Chamberlain, D. F. 1969. New combinations in *Pottia starkeana*. Notes Roy. Bot. Gard. Edinburgh 29: 403–404.

Chamberlain, D. F. 1978. *Pottia*. In A. J. E. Smith, The Moss Flora of Britain and Ireland, pp. 234–242. Cambridge Univ. Press, London.

Changoiwala, M. & P. Chatterjee. 1988. SEM studies of spores of four species of Pottiaceae from West Bengal, India. Cryptogamie Bryol. Lichénol. 9: 63–67.

Chen, P.-c. 1941. Studien über die ostasiatischen Arten der Pottiaceae, I-II. Hedwigia 80: 1–76; 141–322.

Chen, P.-c. 1963, 1978. Genera Muscorum Sinicorum. Parts I & II.

Chopra, R. N. & A. Kapur. 1989. The effect of some chemicals on protonemal growth and bud formation in the moss *Timmiella anomala*. Cryptogamie Bryol. Lichénol. 10: 325–335.

Chopra, R. S. 1975. Taxonomy of Indian Mosses (An Introduction). Chronica Botanica, New Delhi.

Christensen, T. 1959. *Tortella glareicola* n. sp. Bot. Tidsskr. 55: 47–48.

Christy, J. A. 1987 [1988]. New source for wetting agent. Bryologist 90: 281.

Churchill, S. P. 1981. A phylogenetic analysis, classification and synopsis of the genera of the Grimmiaceae (Musci). In V. A. Funk & D. R. Brooks (eds.), Advances in Cladistics. Proceedings of the First Meeting of the Willi Hennig Society, pp. 127–144. New York Botanical Garden, New York.

Churchill, S. P. 1985. The systematics and biogeography of *Scouleria* Hook. (Musci: Scouleriaceae). Lindbergia 11: 59–71.

Churchill, S. P. 1990. *Pseudocrossidium steerei* (Pottiaceae), a new species from Ecuador. Bryologist 93: 353–356.

Churchill, S. P. & P. M. Eckel. 1987. Musci Planitiebus Incolae Exsiccati. Mosses of the Great Plains of North America, First Quarter, Century I. New York Botanical Garden, New York.

Churchill, S. P. & P. M. Eckel. 1987. Musci Planitiebus Incolae Exsiccati. Mosses of the Great Plains of North America, Second Quarter, Century I. New York Botanical Garden, New York.

Clerbois, P. & A. Mansion. 1893. Découverte du *Phascum floerkeanum* (Web. et Mohr) en Belgique. Bull. Soc. Roy. Bot. Belgique 2: 44.

Coats, C. E. & W. F. Mahler. 1985. Distribution of *Barbula whitehouseae* (Pottiaceae), new to Oklahoma, and additional county records. Southwestern Naturalist 30: 151–152.

Collins, J. F. 1904. Some erroneous references (in connection with *Hymenostomum* and *Weisia*). Bryologist 7: 32–33.

Comis, D. 1992. Miracle moss: add water and watch it grow. J. Agric. Res. 40(6): 10–11.

Conard, H. S. 1945a. The decurrent leaves of *Didymodon tophaceus*. Bryologist 48: 27–28.

Conard, H. S. 1945b. *Gyroweisia reflexa* in North America. Bryologist 48: 28–29.

Conard, H. S. 1951a. Decurrent leaves of *Barbula rubinginosa* and *B. cylindrica*. Bryologist 54: 77.

Conard, H. S. 1951b. *Desmatodon obtusifolius* in Iowa. Bryologist

54: 128–130.

Coppey, A. 1908. Sur *Barbula papillosissima* Copp. Rev. Bryol. 35(3): 74–79.

Corbière, L. 1889. *Weisia alberti* Corb. Rev. Bryol. 16: 33–35.

Corbière, L. 1890. *Leptobarbula berica* Schp. trouvé en Normandie. Rev. Bryol. 17: 22.

Corbière, L. 1894. *Didymodon Therioti*. Rev. Bryol. 21: 88–89.

Corbière, L. 1895a. Le *Desmatodon gasilieni* et les *Pottia* du littoral. Rev. Bryol. 22: 6.

Corbière, L. 1895b. Le *Dematodon gasilieni* Vent., est-il une espèce nouvelle?—Quelques mots sur les *Pottia* du littoral. Rev. Bryol. 22: 34–36.

Corley, M. F. V. and A. C. Crundwell. 1991. Additions and amendments to the mosses of Europe and the Azores. J. Bryol. 16: 337–356.

Corley, M. F. V., A. C. Crundwell, R. Düll, M. O. Hill & A. J. E. Smith. 1981 [1982]. Mosses of Europe and the Azores; an annotated list of species, with synonyms from the recent literature. J. Bryol. 11: 609–689.

Corley, M. F. V., M. E. Newton & S. R. Edwards. 1987. *Didymodon nicholsonii* Culm. with sporophytes in England. J. Bryol. 14: 653–657.

Corley, M. F. V. & A. R. Perry. 1985. *Scopelophila cataractae* (Mitt.) Broth. in South Wales, new to Europe. J. Bryol. 13: 323–328.

Cornet, A. 1905. Trois mousses nouvelles pour la flore Belge. *Pterygoneurum lamellatum* Jur., *Dicranum blyttii* B.S. et *Plagiothecium denticulatum* Schpr. Bull. Soc. Roy. Bot. Belgique 41: 143–144.

Corstorphine, M. 1912. *Barbula gracilis* Schwaegr. new to Scotland. Scottish Bot. Rev. p. 46.

Cortes Latorre, C. 1956. Aportaciones a la briologia española. La *Tortula guepini* (Br. eur.) Limpr. musgo nuevo para la flora española. Anales Inst. Bot. A. J. Cavanilles Madrid 14: 171–178.

Cortini Pedrotti, C. & M. Aleffi. 1989. A second record of *Tortula rhizophylla* (Sak.) Iwats. & Saito in Italy. J. Bryol. 15: 803–804.

Cour, P. 1955. Reviviscence chez *Triquetrella arapilensis* Luis. Rev. Bryol. Lichénol. 24: 146.

Craig, E. J. 1939. *Aloina*. In A. J. Grout, Moss Flora of North America North of Mexico. I(4): 211–215.

Cridland, A. A. 1959. The habitat of *Aschisma kansanum*. Bryologist 62: 132–135.

Crosby, M. R. 1974. Toward a revised classification of the Hookeriaceae (Musci). J. Hattori Bot. Lab. 38: 129–141.

Crosby, M. R. 1977. Index Muscorum supplementum: 1974–1975. Taxon 26: 285–307.

Crosby, M. R. 1979. Index Muscorum supplementum: 1976–1977. Taxon 28: 237–264.

Crosby, M. R. 1980. The diversity and relationships of mosses. In R. J. Taylor & A. E. Leviton (eds.), The Mosses of North America. Pac. Div. Am. Assoc. Advan. Sci., San Francisco, pp. 115–129.

Crosby, M. R. & C. R. Bauer. 1983. Index Muscorum supplementum: 1980–1981. Taxon 32: 670–691.

Crosby, M. R. & C. R. Bauer. 1986. Index Muscorum supplementum: 1982–1983. Taxon 35: 416–439.

Crosby, M. R. & R. E. Magill. 1977. A Dictionary of Mosses. Monogr. System. Bot. 3. Missouri Botanical Garden, St. Louis.

Crosby, M. R. & R. E. Magill. 1978. A Dictionary of Mosses, Second Printing. Monogr. Syst. Bot. 3. Missouri Botanical Garden, St. Louis.

Crosby, M. R. & R. E. Magill. 1981. A Dictionary of Mosses, Third Printing. Monogr. Syst. Bot. 3. Missouri Botanical Garden, St. Louis.

Crosby, M. R., R. E. Magill & C. R. Bauer. 1992. Index of Mosses 1963–1989. Monogr. Syst. Bot. 42. Missouri Botanical Garden, St. Louis.

Crum, H. 1949. The discovery of fruiting *Gymnostomiella orcutti* in Mexico. Bryologist 52: 208–211.

Crum, H. 1951. The Appalachian-Ozarkian element in the moss flora of Mexico with a check-list of all known Mexican mosses. Dissert., Univ. Michigan, Ann Arbor, Michigan.

Crum, H. 1952a. *Pseudosymblepharis* in Middle America. Bryologist 55: 137–142.

Crum, H. 1952b. Two new species of Mexican mosses collected by Aaron J. Sharp. Bryologist 55: 51–58.

Crum, H. 1956. Two new species of *Barbula* (Musci) from the Southwest. Southw. Nat. 1: 35–38.

Crum, H. 1957a. A contribution to the moss flora of Ecuador. Svensk Bot. Tidskr. 51(1): 197–206.

Crum, H. 1957b. Some additions to the California moss flora. Madroño 14: 74–79.

Crum, H. 1965a. *Hyophilopsis*, a new genus of Pottiaceae. Bryologist 68: 68–71.

Crum, H. 1965b. *Barbula johansenii*, an arctic disjunct in the Canadian Rocky Mountains. Bryologist 68: 344–345.

Crum, H. 1965c. *Hyophilopsis*—a nomenclatural correction. Bryologist 68: 470.

Crum, H. 1965d. *Barbula eustegia*, a moss new to Canada. Canad. Field-Nat. 79: 157.

Crum, H. 1967a. *Barbula malagana*, a new species from Peru. Bryologist 70: 325–237.

Crum, H. 1967b. A new species of *Pterigoneurum* [sic] from California. Madroño 19: 92–94.

Crum, H. 1969. A reconsideration of the relationship of *Barbula johansenii* (Musci). Canad. Field-Nat. 83: 156–157.

Crum, H. 1972a. Mosses of unusual interest from Baja California. Madroño 21: 403–404.

Crum, H. 1972b. A new *Desmatodon* from Virginia. Bryologist 75: 360–362.

Crum, H. 1973. Mosses of the Great Lakes Forest. Contr. Univ. Michigan Herbarium 10: i-iii, 1–404.

Crum, H. 1976. Mosses of the Great Lakes Forest, Revised Ed. Contr. Univ. Michigan Herbarium 10. 404 pp.

Crum, H. 1986 [1987]. Traditional make-do taxonomy. Bryologist 88: 221–222.

Crum, H. & L. E. Anderson. 1956. Taxonomic studies on North American mosses. I-V. J. Elisha Mitchell Sci. Soc. 72: 276–291.

Crum, H. & L. E. Anderson. 1958a. Taxonomic studies on North American mosses. VI-XIII. J. Elisha Mitchell Sci. Soc. 74: 31–40.

Crum, H. & L. E. Anderson. 1958b. *Byroceuthospora* nom. nov. Bryologist 62: 66.

Crum, H. & L. E. Anderson. 1961 [1962]. *Luisierella barbula* in Georgia. Bryologist 64: 315–320.

Crum, H. & L. E. Anderson. 1965. The taxonomy and distribution of *Acaulon schimperianum*. Bryologist 68: 208–211.

Crum, H. & L. E. Anderson. 1979. *Tortula ammonsiana*, a new species from West Virginia. Bryologist 82: 469–472.

Crum, H. & L. E. Anderson. 1981. Mosses of Eastern North America. Vol. 1 & 2. Columbia University Press, New York.

Crum, H. & E. B. Bartram. 1958. A survey of the moss flora of Jamaica. Inst. Jamaica Bull. Sci. Ser. 8: 1–90.

Crum, H. & W. C. Steere. 1957. The mosses of Porto Rico and the Virgin Islands. New York Acad. Sci., Sci. Surv. Porto Rico Virgin Islands 7(4): 395–599.

Crum, H., W. C. Steere & L. E. Anderson. 1973. A new list of mosses of North America north of Mexico. Bryologist 76: 85–130.

Crundwell, A. C. 1953. *Tortula subulata* Hedw. var. *graeffii* Warnst. in Scotland. Trans. Brit. Bryol. Soc. 2: 292.

Crundwell, A. C. 1955. Further notes on *Tortula subulata* var. *graeffii*. Trans. Brit. Bryol. Soc. 3: 115.

Crundwell, A. C. 1956. Further notes on *Tortula subulata* var. *graeffii*. Trans. Brit. Bryol. Soc. 3: 125–126.

Crundwell, A. C. 1971a. An early English specimen of *Tortella inflexa* (Bruch) Broth. Trans. Brit. Bryol. Soc. 6: 327.

Crundwell, A. C. 1971b. *Weissia perssonii* Kindb., a neglected West European moss. Trans. Brit. Bryol. Soc. 6: 221–224.

Crundwell, A. C. 1976 [1977]. *Barbula mamillosa*, a new species from Scotland. J. Bryol. 9: 163–166.

Crundwell, A. C. 1979. Rhizoids and moss taxonomy. In G. C. S. Clarke & J. G. Duckett (eds.), Bryophyte Systematics, pp. 347–363. Academic Press, New York.

Crundwell, A. C. 1981 [1982]. New combinations in European mosses. II. *Gymnostomum* and *Amblystegium*. J. Bryol. 11: 603–604.

Crundwell, A. C. 1985. The introduced bryophytes of the British Isles. Bull. Brit. Bryol. Soc. 45: 8–9 (Abstr.).

Crundwell, A. C. 1986. *Scopelophila cataractae* in Devonshire. J. Bryol. 14: 387.

Crundwell, A. C., H. J. During & B. P. Koutstaal. 1978. *Tortula grandiretis* Broth. in the Netherlands and in Turkey. Lindbergia 4: 255–257.

Crundwell, A. C. & M. O. Hill. 1977. *Anoectangium warburgii*, a new species of moss from the British Isles. J. Bryol. 9: 435–440.

Crundwell, A. C. & E. Nyholm. 1962. Notes on the genus *Tortella*. I. *T. inclinata*, *T. densa*, *T. flavovirens* and *T. glareicola*. Trans. Brit. Bryol. Soc. 4: 187–193.

Crundwell, A. C. & E. Nyholm. 1963. Notes on the genus *Tortella* II. *Tortella arctica*. Bryologist 66: 184–191.

Crundwell, A. C. & E. Nyholm. 1965. The status of *Barbula valida* and its relationship to *B. acuta*. Svensk Bot. Tidskr. 59: 211–215.

Crundwell, A. C. & E. Nyholm. 1972a. A revision of *Weissia* subgenus *Astomum*. I. The European species. J. Bryol. 7: 7–19.

Crundwell, A. C. & E. Nyholm. 1972b. *Tortula freibergii* Dex. & Loeske in Sussex, new to the British Isles. J. Bryol. 7: 161–164.

Crundwell, A. C. & E. Nyholm. 1974. A revision of *Weissia* subgenus *Astomum*. II. The African species. J. Bryol. 8: 69–70.

Crundwell, A. C. & H. L. K. Whitehouse. 1976. *Tortula bolanderi* (Lesq. & James) Howe in France, new to Europe. J. Bryol. 9: 13–15.

Crundwell, A. C. & H. L. K. Whitehouse. 1978. *Trichostomopsis umbrosa* (C. Müll.) Robins. in England. J. Bryol. 10: 5–8.

Culmann, P. 1907. *Barbula* aut *Didymodon nicholsoni*, *sp. nova*. Rev. Bryol. 34: 100–102.

Culmann, P. 1908. *Barbula Nicholsoni sp. nov.* et *Barbula spadicea* Mitt. var. *bernensis*. Rev. Bryol. 34: 100–102.

Culmann, P. 1924. *Tortella fasciculata*. Rev. Bryol. 51: 22–23.

D

Dalby, D. H. 1966 [1967]. The growth of *Eucladium verticillatum* in a poorly illuminated cave. Rev. Bryol. Lichénol. 34: 288–301; 929–930.

Dallwitz, M. J. 1974. A flexible computer program for generating identification keys. Syst. Zool. 23: 50–57.

Dallwitz, M. J. 1980. A general system for coding taxonomic descriptions. Taxon 29: 41–46.

Dallwitz, M. J. & T. A. Paine. 1986. User's guide to the DELTA system—a general system for processing taxonomic descriptions. Third Edition. CSIRO Aust. Div. Entomol. Rep. 13.

Davies, J. H. 1902. On *Weisia rostellata* in Ireland. Irish Natur., Dublin 11: 289.

Davies, J. H. 1909. *Barbula Hornschuchiana* in counties Down and Armagh. Irish Natur. 18: 23.

Davis, W. B. 1909. Farrant's [sic] Medium for mounting mosses. Bryologist 12: 8.

Debat, M. 1895. *Didymodon debati sp. n.* Rev. Bryol. 22: 79–80.

Dedkov, V. P., P. D. Gunin & V. I. P'Yankov. 1989. Influence of *Tortula desertorum* Broth. on dominants in the grass cover of the Turkmen SSR., U.S.S.R. Probl. Rast. Osvoen. Pust. 1989(1): 38–45.

Deguchi, H. 1987. *Luisierella* (Pottiaceae, Musci), a moss genus with a disjunctive distribution in Neotropics and Japan. J. Jap. Bot. 62: 7–15.

Deguchi, H., K. Oginuma & T. Tada. 1991. Unusual sporogenesis in *Hyophila involuta* (Hook.) Joeg. (Musci, Bryophyta). Mem. Fac. Sci. Kochi Univ., Ser. D, Biology 12: 17–21.

Delgadillo, M., C. 1973a. A Taxonomic Revision of *Aloina*, *Aloinella*, and *Crossidium* (Musci). Ph. D. Dissert., Duke University.

Delgadillo, M., C. 1973b. A new species, nomenclatural changes, and generic limits in *Aloina*, *Aloinella* and *Crossidium* (Musci). Bryologist 76: 271–277.

Delgadillo, M., C. 1973c. A quantitative study of *Aloina*, *Aloinella* and *Crossidium* (Musci). Bryologist 76: 301–305.

Delgadillo M., C. 1975a. Taxonomic revision of *Aloina*, *Aloinella* and *Crossidium* (Musci). Bryologist 78: 245–303.

Delgadillo M., C. 1975b. Literature adicional para las briofitas de México. Bol. Soc. Bot. México 35: 7–12.

Delgadillo, M., C. 1979. Notes on alpine mosses of Mexico. Bryologist 82: 629–631.

Delgadillo M., C. 1982 [1983]. *Pseudaloina* (Pottiaceae, Musci), a new genus from Yemen. Bryologist 85: 401–404.

Delgadillo M., C. 1984. Mosses of the Yucatán Peninsula, Mexico. III. Phytogeography. Bryologist 87: 12–16.

Delgadillo, M., C. 1991. [Abstract.] Cleistocarpic mosses in the neovolcanic belt of Mexico. Am. J. Bot. 78(6, suppl.): 5.

Delgadillo, M., C. 1992. Floristic corridors for moss distribution across the Neovolcanic Belt of Mexico. II. The cleistocarpic species. J. Bryol. 17: 313–316.

Delgadillo M., C. & M. A. Cárdenas S. 1982. Mosses of the Yucatán Peninsula, Mexico. I. Bryologist 85: 253–257.

Delgadillo M., C. & R. H. Zander. 1984 [1985]. The mosses of the Tehuacán valley, Mexico, and notes on their distribution. Bryologist 87: 319–322.

Dillenius, J. J. 1741. Historia Muscorum in qua circiter sexcentae species veteres et novae ad sua genera relatae describuntur et iconibus geminis illustrantur: cum appendice et indice synonymorum. Oxon.

Demaret, F. 1964. *Pottia davalliana* (Sm.) C. Jens. var. *conica* (Schwaegr.) Podp. en Belgique. Bull. Jard. Bot. Etat. Bruxelles 34: 313–314.

Demaret, F. & É. Castagne. 1964. Flore Générale de Belgique. Bryophytes 2(3). Brussels.

Demaret, F. & R. Wilczek. 1980. La présence de *Pohlia obtusifolia* (Brid.) L. Koch douteuse en Belgique. Dumortiera 16: 22–23.

DeMeulder, H. & J. Slembrouk. 1991. *Tortula latifolia* Bruch ex Hartm. (Musci), nieuw voor het Belgisch deel van het Kempens distrikt, zijn huidige verspeiding en oekologie in het gebiece van de "Nieuwe Flora." Dumortiera 47: 19–21.

De Notaris, J. 1838a. Specimen de Tortulis italicus. Mem. Roy. Acad. Sci. Torino 40: 281–326.

De Notaris, J. 1838b. Syllabus muscorum in Italia et in insulis

circumstantibus hucusque cognitorum. Turin.
De Notaris, J. 1862. Musci Italici. Particula 1. Trichostomacei-gen. *Tortula*. Genova.
Dglevskaja, I. V. 1963. *Leptobarbula* Schimp. genus nova pro URSS. Not. Syst. Sect. Crypt. Inst. Bot. Acad. Sci. USSR 16: 199–202.
Dhingra-Babbar, S. 1986. Studies on bud induction in the moss *Hymenostomum edentulum* (Mitt.) Besch. J. Hattori Bot. Lab. 61: 93–99.
Dhingra-Babbar, S. 1988. *In vitro* behavior of gemmae in *Trematodon brevicalyx* and *Semibarbula orientalis*. I. Effects of light and some growth regulators. J. Hattori Bot. Lab. 64: 411–427.
Dhingra-Babbar, S. & R. N. Chopra. 1985. Production of protonemal gemmae in the moss *Semibarbula orientalis* (Web.) Wijk & Marg. J. Bryol. 13: 533–536.
Diehl, W. W. 1929. An improved method for sealing microscopic mounts. Science 49: 276–277.
Dietert, M. F. 1979. Studies on the gametophyte nutrition of the cosmopolitan species *Funaria hygrometrica* and *Weissia controversa*. Bryologist 82: 417–431.
Dilks, T. J. K. & M. C. F. Proctor. 1974. The pattern of recovery of bryophytes after desiccation. J. Bryol. 8: 97–115.
Dilks, T. J. K. & M. C. F. Proctor. 1976a. Effects of intermittent desiccation on bryophytes. J. Bryol. 9: 249–264.
Dilks, T. J. K. & M. C. F. Proctor. 1976b [1977]. Seasonal variation in desiccation tolerance in some British bryophytes. J. Bryol. 9: 239–247.
Dismier, G. 1905. *Didymodon rigidulus* Hedw. et *Didymodon spadiceus* Limpr. Bull. Soc. Bot. (Paris) 52: 184–189.
Dismier, G. 1921. Observations sur le *Didymodon cordatus*. Rev. Bryol. 48: 52–53.
Dismier, G. 1924. Notes sur les *Pottia starkeana*, etc. Rev. Bryol. 51: 17–19.
Dismier, G. 1928. Note sur la repartition en France du *Phascum mitraeforme* (Limpr.) Warnst. Rev. Bryol. 1: 130–131.
Dismier, G. 1932. Note sur la presence du *Tortula obtusifolia* Schleich. dans l'Oisans. Rev. Bryol. 5: 47–48.
Dixon, H. N. 1900. A remarkable form of *Trichostomum tortuosum*. Rev. Bryol. 27: 36–37.
Dixon, H. N. 1910. *Merceyopsis*, a new genus of mosses with further contributions to the byrology of India. J. Bot. 48: 297–310, pl. 507–508.
Dixon, H. N. 1911a. *Hyophilopsis*, a new genus of Pottiaceae, with further contributions to the bryology of India. J. Bot. 49: 137–150.
Dixon, H. N. 1911b. *Leptodontium recurvifolium* (Tayl.) Lindb. in England. J. Bot. 49: 349.
Dixon, H. N. 1911c. *Tortula canescens*. J. Bot. 49: 127–128.
Dixon, H. N. 1912. *Eucladium verbanum* Nicholson and Dixon. n. sp. Rev. Bryol. 39: 89–92.
Dixon, H. N. 1913–29. Studies in the Bryology of New Zealand. Bull. New Zealand Inst. 1913, 3: 1–29; 1914, 3: 31–74; 1923, 3: 75–152, pls. 7–8; 1926, 3: 153–238; 1927, 3: 239–298; 1929, 3: 299–372.
Dixon, H. N. 1915. *Tortula atrovirens* var. *edentula* (B. et S.) J. Bot. 53: 217–218.
Dixon, H. N. 1922a. Some new genera of mosses. J. Bot., Lond. 60: 101–110.
Dixon, H. N. 1922b. Note on a form of *Pottia heimii* from Greenland. Bryologist 25: 92.
Dixon, H. N. 1924. The Student's Handbook of British Mosses, Ed. 3. London.
Dixon, H. N. 1927a. *Hymenostylium xanthocarpum* (Hook.) Brid. Bryologist 30: 106–109.
Dixon, H. N. 1927b. Regeneration of *Tortula muralis*. Bryologist 30: 28–29.
Dixon, H. N. 1932. Classification of mosses. In F. Verdoorn (ed.), Manual of Bryology, pp. 397–412. The Hague.
Dixon, H. N. 1942b. Papuan mosses. J. Bot. 80: 1–11; 25–35.
Doei, H., S. Nakanishi & S. Shinpoh. 1985. Some new localities of *Tortula pagorum*. Proc. Bryol. Soc. Japan 4: 4–6. [In Japanese with English summary.]
Doyle, J. J. 1993. DNA, phylogeny, and the flowering of plant systematics. BioScience 43: 380–389.
Driver, P. J. 1982. *Leptodontium gemmascens* in terrestrial habitats in south-east England. J. Bryol. 12: 113.
Düll, R. 1984. Taxonomy and distribution of some critical taxa of the genus *Didymodon* in Europe. J. Hattori Bot. Lab. 55: 259–266.
Düll-Hermanns, I. 1984. Evaluation of different characters of *Didymodon* taxa by modern statistical methods. J. Hattori Bot. Lab. 55: 253–257.
Düll-Hermanns, I. & R. Düll. 1985. A taxonomical re-investigation of eight critical European taxa of the genus *Didymodon*. Abstr. Bot. 9(Suppl. 2): 33–44.
Duncan, J. B. 1948. *Barbula rubella* var. *dentata* Schp. and *Leptodontium flexifolium* Hampe (forma). Trans. Brit. Bryol. Soc. 1: 113–114.
Dunk, von der, K. & K. von der Dunk. 1973. *Eucladium verticillatum* var. nov. *recurvatum*. Herzogia 2: 419–422.
During, H. J. 1979. Life strategies of bryophytes: a preliminary review. Lindbergia 5: 2–18.
During, H. J., B. Odé & B. F. van Tooren. 1986. New records of *Tortella inflexa* (Bruch) Broth. and *Cephaloziella baumgartneri* Schiffn. in W. Europe. Lindbergia 12: 47–48.

E

Eckel, P. M. 1985a. *Orthotrichum obtusifolium* and *Tortula papillosa* in western New York State. Evansia 2(1): 11–12.
Eckel, P. M. 1985b. *Gymnostomiella orcuttii* (Pottiaceae) new to the western coast of Mexico. Evansia 2: 28–29.
Eckel, P. M. 1986a. *Didymodon australasiae* var. *umbrosus* new to eastern North America. Bryologist 89: 70–72.
Eckel, P. M. 1986b. *Weissia phascopsis* (= *Astomum phascoides*) new to the moss flora of New Mexico with additional county records. Evansia 3: 44–45.
Eckel, P. M. 1987. Mosses new and rare for New York State. Rhodora 89: 375–379.
Eckel, P. M. 1988. *Barbula ehrenbergii* new to the moss flora of New Mexico with additional Eddy County records. Evansia 5: 14–15.
Eckel, P. M. 1990a. Latin glossary of bryophyte terms. In R. E. Magill, Glossarium Polyglottum Bryologiae, pp. 183–214. Monogr. Syst. Bot. 33. Missouri Botanical Garden.
Eckel, P. M. 1990b. *Eucladium verticillatum* (Musci) second Ontario station. Evansia 7(1): 15.
Eckel, P. M. 1990c. *Bryoerythrophyllum ferruginascens* (Musci: Pottiaceae) in eastern North America. Bryologist 93: 208–210.
Eckel, P. M. 1990d. Taxa minora delineata. II. *Bryoerythrophyllum recurvirostrum* (Hedw.) Chen var. *robustum* Saito. Bryologist 93: 386.
Eckel, P. M. 1991. *Tortella tortelloides* (Musci: Pottiaceae) new to North America. Bryologist 94: 84–87.
Eckel, P. M. & M. P. Eckel. 1988. *Weissia hedwigii* on Goat Island, second locality in New York State. Clintonia 3(3): 8–9.
Eddy, A. 1988, 1990. A Handbook of Malesian Mosses. Vol. 1,

Sphagnales to Dicranales; Vol. 2, Leucobryaceae to Buxbaumiaceae. British Museum (Natural History), London.

Edwards, S. R. 1978. Taxonomic implications of cell patterns in haplolepidous moss peristomes. Bull. Brit. Bryol. Soc. 31: 13–14.

Edwards, S. R. 1979. Taxonomic implications of cell patterns in haplolepidous moss peristomes. In G. C. S. Clarke & J. G. Duckett, Bryophyte Systematics, pp. 317–346. Syst. Assoc. Spec. Vol. 14. Academic Press, New York.

Edwards, S. R. 1980a. A revision of west tropical African Calymperaceae. I. Introduction and *Calymperes*. J. Bryol. 11: 49–93.

Edwards, S. R. 1980b. Spore discharge in *Calymperes*. J. Bryol. 11: 95–97.

Edwards, S. R. 1984. Homologies and inter-relationships of moss peristomes. In R. M. Schuster (editor), New Manual of Bryology. Volume 2, Chapter 12, pp. 658–695. Hattori Botanical Laboratory, Nichinan.

Egunyomi, A. & S. O. Olarinmoye. 1978. A new species of *Gyroweisia* (Musci: Pottiaceae) from Nigeria. Bryologist 81: 443–445.

Ehrhart, F. 1787. Beiträge zur Naturkunde. I. Hannover.

Ellis, L. T. & A. C. Smith. 1983. *Barbula cylindrica* (Tayl.) Schimp. with rhizoidal tubers. J. Bryol. 12: 509–510.

El-Oqlah, A. A., W. Frey & H. Kürschner. 1988. *Tortula rigescens* Broth. & Geh. (Pottiaceae), a remarkable species new to the moss flora of Jordan. Lindbergia 14: 27–29.

El-Saadawi, W. & Afaf Badawi. 1977. The moss flora of Egypt. I. Order Pottiales. Publ. Cairo Univ. Herb. 7/8: 181–206.

Engelmark, T.-D. 1974. *Trichostomum cuspidatissimum* Card. & Thér. and *Rhizomnium andrewsianum* (Steere) Kop. discovered in northern Sweden. Lindbergia 2: 218–221.

Erwin, L. T. 1991. An evolutionary basis for conservation strategies. Science 253: 750–752.

F

Faith, D. P. 1991. Cladistic permutation tests for monophyly and nonmonophyly. Syst. Zool. 40: 366–375.

Farr, E. R., J. A. Leussink & F. A. Stafleu. 1979. Index Nominum Genericorum (Plantarum). Vol. I-III. Reg. Veg. 100–102.

Farris, J. S. 1974. Formal definitions of paraphyly and polyphyly. Syst. Zool. 23: 548–554.

Farris, J. S. 1988. "Hennig86" Version 1.5. Software with reference manual. Published by the author.

Farris, J. S. 1989. The retention index and the rescaled consistency index. Cladistics 5: 417–419.

Favali, M. A. & F. Gianni. 1973. Sporophyte ultrastructure in *Tortula muralis* Hedw. Österr. Bot. Zeit. 122: 323–332.

Field, J. H. 1977. Bryophyte notes. Proc. Birmingham Nat. Hist. Soc. 23: 242–244.

Field, J. H. 1990 [1991]. A gemmiferous specimen of *Barbula unguiculata* found in Warwickshire. Proc. Birmingham Nat. Hist. Soc. 26: 110.

Fife, A. J. 1985. A generic revision of the Funariaceae (Bryophyta: Musci). Part 1. J. Hattori bot. Lab. 58: 149–196.

Fleischer, M. 1902–1922. Die Musci der Flora von Buitenzorg. 1–4: 1–1729. Leiden.

Fleischer, M. 1920. Natürliches System der Laubmoose. Hedwigia 61: 390–400.

Flowers, S. 1930. *Pottia Randii* Kennedy. Bryologist 33: 30.

Flowers, S. 1951. A new species of *Tortula* from Utah and Arizona. Bryologist 54: 278–282.

Flowers, S. 1953. *Tortula papillosissima*, new to North America. Bryologist 56: 160–164.

Flowers, S. 1973a. Mosses: Utah & the West. Brigham Young Univ. Press, Provo, Utah.

Flowers, S. 1973b. Mosses of Utah new to science. Bryologist 76: 286–292.

Foster, W. D. 1977. Freehand sectioning of bryophytes. Bull. Brit. Bryol. Soc. 29: 21.

Frahm, J.-P. 1970. *Syntrichia ruraliformis* (Besch.) Dix. in Schleswig-Holstein. Kieler Not. Pflanzenk. Schleswig-Holstein 2: 1–3.

Frahm, J-P. 1973 [1975]. Verbreitung, Systematik und Ökologie von *Leptodontium flexifolium* (Dicks) Hampe. *L. flexifolium fo. nov. gemmipara* Frahm. Nov. Hedw. 24: 413–429.

Frahm, J.-P. 1981. Ein praktisches Einschlussmittle für Micropräparate von Moosen. Herzogia 5: 531–533.

Frahm, J.-P. 1986. Type revision of European mosses. 1. *Leptodontium*. Lindbergia 12: 76–82.

Frahm, J.-P. 1990a. Water-soluble permanent mounting medium for microscopic slides. Bryol. Times 53: 9–10.

Frahm, J.-P. 1990b. *Scopelophila cataractae*, ein neues Moos in unserer Flora. Bryol. Rundbr. 3: 6–7.

Frahm, J.-P. 1991a. *Oxystegus hibernicus* (Mitt.) Hilp. neu für Frankreich. Cryptogamie Bryol. Lichénol. 12: 165–167.

Frahm, J.-P. 1991b. A phenetic and cladistic study of the Campylopodioideae. J. Hattori Bot. Lab. 69: 65–78.

Frey, W. 1971. Blattentwicklung bei *Timmia austriaca* Hedw. Nov. Hedw. 21: 119–136.

Frey, W. & I. Herrnstadt & H. Kürschner. 1990. *Pterygoneurum crossidioides* (Pottiaceae, Musci), a new species to the desert flora of the Dead Sea area. Nov. Hedw. 50: 239–244.

Frey, W. & H. Kürschner. 1983. New records of bryophytes from Transjordan with remarks on phytogeography and endemism in SW Asiatic mosses. Lindbergia 9: 121–132.

Frey, W. & H. Kürschner. 1984. Studies in Arabian bryophytes 3: *Crossidium asirense* (Pottiaceae), a new species from Asir Mountains (Saudi Arabia). J. Bryol. 13: 25–31.

Frey, W. & H. Kürschner. 1987. A desert bryophyte synusia from the Jabal Tuwayq mountain systems (Central Saudi Arabia) with the description of two new *Crossidium* species (Pottiaceae). Nov. Hedw. 45: 119–136.

Frey, W. & H. Kürschner. 1988a. Bryophytes of the Arabian Peninsula and Socotra. Floristics, phytogeography and definition of the Xerothermic Pangaean element. Studies in Arabian bryophytes 12. Nov. Hedw. 46: 37–120.

Frey, W. & H. Kürschner. 1988b. Re-evaluation of *Crossidium geheebii* (Broth.) Broth. (Pottiaceae) from Sinai, a Xerothermic Pangaean element. J. Bryol. 15: 123–126.

Frey, W. & H. Kürschner. 1988c. *Pseudocrossidium replicatum* (Tayl.) Zander replaces *Barbula acutata* C. Müll. A note on its synonymy, distribution and the xeropottioid life syndrome. Cryptogamie Bryol. Lichénol. 9: 95–102.

Frey, W. & H. Kürschner. 1991. *Crossidium laevipilum* Thér. et Trab. (Pottiaceae, Musci), ein eigenständiges, morphologish und standortökologish deutlich unterscheidbares Taxon der Saharo-Arabishen Florenregion. Cryptogamie Bryol. Lichénol. 12: 441–150.

Frisvoll, A. 1978. Nomenclatural note on *Trichostomum cuspidatissimum*. Bryologist 81: 156–159.

Frisvoll, A. 1982. The status of *Leptodontium norvegicum* Kaal. J. Bryol. 12: 185–186.

Frisvoll, A. 1984. The identity of *Trichostomum saxatile* Tayl. and *T. microcarpon* var. *oblongum* Tayl., two names in the *Racomitrium heterostichum* group. J. Bryol. 13: 291–292.

Frisvoll, A. 1985. The identity of *Didymodon arcticus* Broth. of Index Muscorum. J. Bryol. 13: 435.

Fritsch, R. 1972. Chromosomenzahlen der Bryophyten, eine

Übersicht und Diskussion ihres Aussagewertes für das System. Wiss. Zeitschr. Friedrich-Schiller Univ. Jena, Math.-Naturwiss. Reihe 21: 839–944.

Fritsch, R. 1982. Index to Plant Chromosome Numbers—Bryophyta. Reg. Veg. 108.

Fritsch, R. 1991. Index to bryophyte chromosome counts. Bryoph. Biblioth. 40: 1–352.

Frohlich, M. W. 1984. Freehand sectioning with parafilm. Stain Technol. 59: 61–62.

Fuertes-Lasala, E. 1983. El género *Crossidium* Jur. en la Península Ibérica; Islas Baleares, Canarias y Madeira. Anales Jard. Bot. Madrid 40: 29–35.

Fuertes-Lasala, E. 1987. *Desmatodon heimii* (Hedw.) Mitt. in the Iberian Peninsula. Cryptogamie Bryol. Lichénol. 8: 147–148.

Funk, V. A. 1985. Cladistics and generic concepts in the Compositae. Taxon 34: 72–80.

Fürnrohr, A. E. 1829. Noch ein Wort über *Barbula obtusifolia* und *Bryum erythropcarpon*; nebst einigen andern bryologischen Bemerkungen. Bot. Zeit. 38: 593–601.

G

Galambos, I. & S. Orban. 1984. *Crossidium crassinerve* (De Not.) Jur., new member of the Hungarian bryoflora. Bryol. Beitr. 3: 23–27. [German summary.]

Gambardella, R., R. Ligrone & R. Castaldo. 1980. Studies on *Timmiella barbuloides* (Brid.) Moenk. III. Ultrastructural evidence of catalase activity in the phylloid microbodies (peroxisomes). Cryptogamie Bryol. Lichénol. 1: 415–422.

Gangulee, H. C. 1969, 1971, 1972, 1974, 1976 [1978], 1977 [1978], 1978, 1980. Mosses of Eastern India and Adjacent Regions: A Monograph. Fascicles 1–8. Calcutta.

Gao, C., X. Jia & T. Cao. 1991. A new species of *Hyophila* in China. Bull. Bot. Res. (Harbin) 11(2): 29–31.

Garcia G., R. & E. Fuertes L. 1976. Sobre la presencia de *Tortula desertorum* Broth. en Navarra. Anales Inst. Bot. Cavanilles 33: 135–138.

Geheeb, A. 1877. Le *Weisia ganderi*. Rev. Bryol. 4: 4–5.

Geheeb, A. 1880. Note sur le *Weisia welwitschii*. Rev. Bryol. 7: 45.

Geheeb, A. 1906a. Une station étrange du *Gymnostomum rupestre*. Rev. Bryol. 33: 42.

Geheeb, A. 1906b. Le *Gyroweisia reflexa*, espèce nouvelle en Espagne. Rev. Bryol. 33: 58.

Geissler, P. 1985. Notulae Bryofloristicae Helveticae. II. Candollea 40: 193–200.

Gerson, U. 1987. Mites which feed on mosses. Symp. Biol. Hung. 35: 721–724.

Gimingham, C. H. 1948. The role of *Barbula fallax* Hedw. and *Bryum pendulum* Schp. in sand-dune fixation. Trans. Brit. Bryol. Soc. 1: 70–72.

Głowacki, J. 1909. Eine neue Art von *Eucladium* Br. Eur. Österr. Bot. Zeitschr. Wien 59: 222–224.

Głowacki, J. 1913. *Hyophila styriaca* Glow., eine neue Laubmoosart aus Steiermark. Österr. Bot. Zeitschr. Wien 63: 405–406.

Gorton, B. S. & R. E. Eakin. 1958. Development of the gametophyte in the moss *Tortella caespitosa*. Bot. Gaz. 119: 31–38.

Gorton, B. S., C. G. Skinner & R. E. Eakin. 1957. Activity of some 6–substituted purines on the development of the moss *Tortella caespitosa*. Arch. Biochem. Biophys. 66: 493–496.

Gravet, F. 1874a. *Barbula sinuosa*. Rev. Bryol. 1: 5–7.

Gravet, F. 1874b. *Barbula nitida* Lindb. Rev. Bryol. 1: 19–20.

Gray, F. W. 1931. *Tortula fragilis* Tayl. in West Virginia. Bryologist 34: 71–72.

Gray, P. 1954. The Microtomist's Formulary and Guide. Blakiston Co., New York.

Grebe, C. 1909. *Ditrichum julifiliforme* und *Tortula calcicola*, zwei neue Laubmoose. Hedwigia 49: 66–77.

Greene, S. W. 1975. The Antarctic moss *Sarconeurum glaciale* (C. Muell.) Card. et Bryhn in southern South America. Brit. Antarctic Surv. Bull. 41–42: 187–191.

Greene, S. W., D. M. Greene, P. D. Brown & J. M. Pacey. 1970. Antarctic moss flora. 1. The genera *Andreaea*, *Pohlia*, *Polytrichum*, *Psilopilum* and *Sarconeurum*. Brit. Antarct. Surv. Sci. Rep. 64: 1–118.

Greuter, W. (ed.) 1988. International Code of Botanical Nomenclature. Adopted by the Fourteenth International Botanical Congress, Berlin, July-August 1987. Veg. Reg. 118.

Greven, H. C. 1990. *Didymodon nicholsonii* Culm. (Pottiaceae) een nieuwe soort voor de Nederlandse mosflora. Lindbergia 16: 69–72.

Greville, R. K. & G. A. W. Arnott. 1824a. Tentamen methodi Muscorum; or a new arrangement of the genera of mosses, with characters, and observations on their distribution, history, and structure. Memoir I. Gymnostomoideae. Mem. Wernerian Nat. Hist. Soc. 5(2): 42–89.

Greville, R. K. & G. A. W. Arnott. 1824b. Tentamen methodi Muscorum; or a new arrangement of the genera of mosses, with characters, and observations on their distribution, history, and structure. Memoir II. Mem. Wernerian Nat. Hist. Soc. 5(2): 109–150.

Greville, R. K. & G. A. W. Arnott. 1826. Tentamen methodi Muscorum; or a new arrangement of the genera of mosses, with characters, and observations on their distribution, history, and structure. Memoir III. Splachnoideae. Mem. Wernerian Nat. Hist. Soc. 5(2): 442–474.

Griffin, D., III. 1975. *Aloinella venezuelana*, a new species of moss from the Andes of northern South America. Bull. Torrey Bot. Club 102: 26–28.

Griffin, D., III. 1979a. Los Musgos del Estado Mérida: Clave para los Generos. Florida St. Mus., Univ. of Florida, Gainesville. Mimeogr. 54 pp.

Griffin, D., III. 1979b. On the sporophyte of *Streptopogon calymperes* C. Müll. ex Geh. (Musci; Pottiaceae). Misc. Bryol. Lichenol. 8: 70–71.

Griffin, D., III. 1986a. *Oreoweisia* (Dicranaceae, Musci) in tropical America: an annotated key to species. Cryptogamie Bryol. Lichénol. 7: 433–438.

Griffin, D., III. 1986b. An unusual dimorphism in *Streptopogon calymperes* C. Muell. ex Geh. (Musci; Pottiaceae). Evansia 3: 1–3.

Griffin, D., III. 1990a. *Oreoweisia erosa* (C. Muell.) Kindb., an african-neotropical disjunct. Cryptogamie Bryol. Lichénol. 10: 297–300.

Griffin, D., III. 1990b. The use of axillary hairs in the taxonomy of two neotropical Bartramiaceae. J. Bryol. 16: 61–66.

Grims, F. 1991. Ein dritter Nachweis von *Tortula latifolia* (Bruch) Hartm. in Süd-Bayern. Bryol. Rundbr. 5: 5.

Grout, A. J. 1900. *Weissia*. Bryologist 3: 43.

Grout, A. J. 1902. *Tortula pagorum* (Milde) De Not. in Georgia. Bryologist 7: 64–65.

Grout, A. J. 1927. Studies in *Ditrichum*. Bryologist 30: 4–5.

Grout, A. J. 1928–1940. Moss Flora of North America North of Mexico. 3 vols. Newfane, Vt.

Grout, A. J. 1945. Two new species of *Acaulon* from Texas. Bryologist 48: 25–26.

Guerra, J., M. N. Jiménez, R. M. Ros & J. S. Carrión. 1991. El género *Phascum* (Pottiaceae) en la península Ibérica.

Cryptogamie Bryol. Lichénol. 12: 179–423.

Guerra, J., J. J. Martínez, R. M. Ros & J. S. Carrión. 1990. *Phascum longipedis sp. nov.* on gypsum soils from Almería, Spain. J. Bryol. 16: 55–60.

Guerra, J. & R. M. Ros. 1987. Revisión de la Sección *Asteriscium* del género *Didymodon* (Pottiaceae, Musci) (= *Trichostomopsis*) en la Peninsula Iberica. Cryptogamie Bryol. Lichénol. 14: 729–736.

Guerra, J. & R. M. Ros. 1988. Characterización taxonómica, corología y ecología de *Pottia pallida* Lindb. y relaciones con otras especies. Cryptogamie Bryol. Lichénol. 9: 343–352.

Guerra, J., R. M. Ros & J. S. Carrión. 1992. The taxonomic status of *Tortula muralis* var. *baetica* (Musci, Pottiaceae): a comparative study. J. Bryol. 17: 275–283.

Guinet, A. 1902. Note sur *Geheebia cataractarum* Schpr. Bull. Herb. Boissier 2: 1023–1024.

Gwóźdź, E. A. & E. B. Tucker. 1974. Studies on protein synthesis in *Tortula ruralis*. Polyribosome reformation following desiccation. J. Exp. Bot. 25: 599–608.

Györffy, I. 1905a. Fundort von *Hymenostylium curvirostre* var. β *scabrum* in Ungarn; anatomie dieses Mooses. Ungar. Növényt. Közlem 4: 59–61, 95–101.

Györffy, I. 1905b. *Pterygoneurum cavifolium* Jur. Magyar Bot. Lapok 4: 340.

Györffy, I. 1910. A *Molendoa sendtneriana* (Bryol. Eur.) Limpr. újabb hazai elöfordulásáról. (Über die neueren Standorte von *Molendoa sendtneriana* (Bryol. Eur.) Limpr. in Ungarn.) Magyar Bot. Lapok Budapest 9: 194–198.

Györffy, I. 1911. *Dialytrichia Brebissonia* (Brid.) Limpr. Ungarn. Bot. Blatter 10: 85.

Györffy, I. 1912. *Molendoa tenuinervis* Limpr. in America arctica. Bryologist 15: 75–81.

Györffy, I. 1914. Über das *Pleurozygodon sibiricum* Arnell. Ark. Bot. 14(2): 1–3, Taf. 1.

Györffy, I. 1916. Ueber die Entdeckung des *Molendoa tenuinervis* und deren samt *M. sendtneriana* Verbreitung in den Karpathien. Bul. Grad. Bot. Cluj 26: 149–153.

H

Haesler, M. 1973. Beitrag zur *Tortula canescens* (Mont., 1933). Rev. Bryol. Lichénol. 39: 489–491.

Hagen, I. 1929. Forarbeider til en Norsk Lovmosflora. XXI. Pottiaceae. Det. Kongl. Norske Vidensk. Selsk. Skrift. 3: 96 pp.

Hammerschmid, A. 1915. Die Verdickungen auf der Kapselwand von *Trichostomum Hammerschmidii* Lske. et Paul. Mitt. Bayer. Bot. Ges. 3: 215–216.

Harada, H. & H. Deguchi. 1990. The habitat of *Luisierella barbula* (Schwaegr.) Steere in Japan. Proc. Bryol. Soc. Japan 5: 53–54.

Häusler, M. 1984. Die selteneren *Tortula*-Arten der Sektion *Cuneifolia* in Deutschland. Bryol. Beitr. 3: 1–22.

Hébant, C. 1979. Conducting tissues in bryophyte systematics. In G. C. S. Clarke & J. G. Duckett (eds.), Bryophyte Systematics, pp. 365–383. Syst. Assoc. Spec. Vol. 14. Academic Press, New York.

Hedenäs, L. 1989a. Axillary hairs in pleurocarpous mosses—a comparative study. Lindbergia 15: 166–180.

Hedenäs, L. 1989b [1991]. Notes on *Tortula novegica* in southern Sweden. Lindbergia 15: 213–214.

Hedwig, J. 1789–1792. Discriptio et adumbratio microscopico-analytica muscorum frondosorum nec non aliorum vegetantium e classe cryptogamica Linnaei novorum dubiisque vexatorum. Vol. II, 1789, Vol III, 1792. Leipzig.

Hedwig, J. 1801. Species muscorum frondosorum descriptae et tabulis aeneis coloratis illustratae. Leizig.

Hennen, J. A. 1925. À propos de *Desmatodon cernuus* Br. Eur. Bull. Soc. Roy. Belgique 56: 164.

Hennig, W. 1965. Phylogenetic systematics. Ann. Rev. Entom. 10: 97–116.

Hennig, W. 1966. Phylogenetic systematics. Univ. of Illinois Press, Urbana.

Hermann, F. J. & E. Lawton. 1968. A new species of *Didymodon* (Bryophyta) from Oregon and Washington. Bull. Torrey Bot. Club 95: 387–389.

Hernnstadt, I. & C. C. Heyn. 1989 [1991]. Variation in spore texture in the genus *Pottia* (Pottiaceae) as opposed to uniformity in some terrestrial cleistocarpous mosses (Pottiaceae and Dicranaceae), pp. 181–183. In T. Herben & C. B. McQueen (eds.), Proceedings of the Sixth Meeting of the Central and East European Bryological Working Group (CEBWG), Botanical Institute of Czechoslovak Academy of Sciences, Puŭhonice.

Herzog, T. 1905. *Barbula sinuosa*. Bot. Centralbl. Beih. 18(Abt. 2): 115–118.

Herzog, T. 1907. Studien über den Formenkreis des *Trichostomum mutabile* Br. Abh. Kaiserl. Leop.-Carol. Deutsch. Akad. Naturf., Nova Acta 73: 453–481, tab.6–12.

Herzog, T. 1916. Die Bryophyten meiner zweiter Reise durch Bolivia. Biblioth. Bot. 87: 1–347, pl. 1–8, map.

Herzog, T. 1926. Die Geographie der Moose. 1–439, f. 1–8. Jena.

Herzog, T. 1932. *Leptodontium*. Die Pflanzenareale, 3 Reihe, 5: 58–61.

Herzog, T. 1939. *Melophyllum* n. gen., eine neue Laubmoosgattung der Hochanden von Peru. Feddes Repert. 47: 98–100.

Herzog, T. 1943. *Molendoa taeniatifolia* Herz. nov. spec. Eine neue Laubmoosart der Alpen. Flora, N.F. 37: 57–60.

Herzog, T. & I. Thériot. 1933. Bryophyta nova. 8. *Didymodon Schilleri* Herz. et Thér. *sp. nov*. Ann. Bryol. 6: 125.

Heywood, V. H. 1983. The mythology of taxonomy. Trans. Bot. Soc. Edinburgh 44: 79–94.

Higuchi, M. 1989. A new locality of *Tortula pagorum* (Milde) De Not. in Japan. Proc. Bryol. Soc. Japan 5: 9–10. [In Japanese.]

Hill, E. J. 1913. The annulus of *Tortella caespitosa* (Schwaegr.) Limpr. Bryologist 16: 17–18.

Hill, M. O. 1978. The taxonomic position of *Oxystegus sinuosus* (Mitt.) Hilp. in relation to the genus *Barbula* Hedw. J. Bryol. 10: 273–276.

Hill, M. O. 1980. Vice county distributions of segregates of *Pottia starkeana*. Bull. Brit. Bryol. Soc. 35: 9.

Hill, M. O. 1981. New combinations in European mosses, I. Pottiaceae. J. Bryol. 11: 599–602.

Hill, M. O. 1982. A reassessment of *Acaulon minus* (Hook. & Tayl.) Jaeg. in Britain, with remarks on the status of *A. mediterraneum* Limpr. J. Bryol. 12: 11–14.

Hillier, L. 1931. À propos de *Didymodon luridus* Horn. Sa repartition dans les monts Jura. Rev. Bryol. 4: 197–199.

Hillier, L. 1936. Les Trichostomées dans la zone inférieure de la chaîne jurassienne. Bull. Soc. Nat. Doubs 47: 75–103.

Hillis, D. M., J. J. Bull, M. E. White, M. R. Badgett & I. J. Molineux. 1992. Experimental phylogenetics: generation of a known phylogeny. Science 255: 589–592.

Hilpert, F. 1933. Studien zur Systematik der Trichostomaceen. Bot. Centralb. Beih. 50(2): 585–706.

Hobkirk, C. P. 1874. *Tortula sinuosa* in Oxfordshire. J. Bot. 3: 112.

Hobkirk, C. P. 1901. *Tortula cernua* (Lindb.), a moss new to the British flora. Naturalist (London) 1901: 1–3.

Hodgson, E. A. 1930. Notes on the apical gemmae of *Tortula abruptinervis*. Trans. & Proc. New Zealand Inst. 60: 544–546.

Hoe, W. J. 1973. Notes on some Hawaiian species of the Pottiaceae. Bryologist 76: 192–195.

Hoffmann, A. 1957. Untersuchungen über die Geschlechtsdifferenzierung bei haploiden und diploiden Gametophyten von *Barbula unguiculata* (Hudson) Hedwig. Zeitschr. Indukt. Abstammungs- Vererb. Lehre 88: 374–387.

Holden, C. (Ed.) 1992. Miracle moss. Science 257: 322.

Holmen, K. 1957. Three West Arctic moss species in Greenland. On the occurrence of *Cinclidium latifolium*, *Aulacomnium acuminatum* and *Trichostomum cuspidatissimum*. Meddel. Groenland 156: 15 pp.

Holmes, E. M. 1904. *Pottia heimii* Fürnr. J. Bot. 42: 89.

Holmgren, P. K., N. H. Holmgren & L. C. Barnett. 1990. Index Herbariorum. Part I: The herbaria of the World. Eighth Ed. Reg. Veg. 120.

Holzinger, J. M. 1904a. The genus *Hymenostomum* in North America. Bryologist 7: 8–10.

Holzinger, J. M. 1904b, Notes on new or rare mosses: A moss new to North America (*Pterygoneurum cavifolium incanum)* collected near Blackfoot, Idaho). Bryologist 7: 77.

Holzinger, J. M. 1923. *Desmatodon arenaceus* Sull., redivivus. Bryologist 38: 60.

Holzinger, J. M. 1925. *Pottia randii* not a *Pottia*. Bryologist 28: 6.

Holzinger, J. M. & E. B. Bartram. 1923a. The genus *Crossidium* in North America. Bryologist 26: 67–73.

Holzinger, J. M. & E. B. Bartram. 1923b. The genus *Crossidium* in North America (concluded). Bryologist 27: 3–9.

Holzinger, J. M. & E. B. Bartram. 1924. The case of *Barbula manniae* C. M. Bryologist 27: 79.

Hooker, W. J. & R. K. Greville. 1824. On the genus *Tortula*, of the order Musci. Edinb. J. Sci. 1: 287–302, pl. 12.

Horikawa, Y. 1954. Range of *Tortula emarginata* (Dozy & Molk.) Mitt. J. Jap. Bot. 29: 79–32.

Horton, D. G. 1982. A revision of the Encalyptaceae (Musci), with particular reference to the North American taxa. Part I. J. Hattori Bot. Lab. 53: 365–418.

Horton, D. G. 1983. A revision of the Encalyptaceae (Musci), with particular reference to the North American taxa. Part II. J. Hattori Bot. Lab. 54: 353–532.

Hughes, J. G. 1969. Factors conditioning development of sexual and apogamous races of *Phascum*. New Phytologist 68: 883–900.

Hughes, J. G. 1979. The occurrence of polysety in relation to the number of archegonia in the female inflorescence of *Phascum cuspidatum* Hedw. J. Bryol. 10: 553–560.

Hughes, J. G. 1982. Penetration by rhizoids of the moss *Tortula muralis* Hedw. into well-cemented oolitic limestone. Int. Biodeter. Bull. 18(2): 43–46.

Hughes, J. G. & A. J. A. Wiggin. 1969. Light intensity and sexual reproduction in *Phascum cuspidatum* Hedw. Trans. Brit. Bryol. Soc. 5: 823–826.

Husnot, T. 1884–1894. Muscologia Gallica. Paris.

Hyvönen, J. 1991. *Tortella fragilis* (Pottiaceae) reported for southern South America. Bryologist 94: 416–418.

I

Ignatov, M. 1990. Upper Permian mosses from the Russian platform. Palaeobiologica 217: 147–189.

Ignatov, M. 1992. On the occurrence of *Barbula inaequalifolia* Tayl. (Pottiaceae, Musci) in Altai Mountains. Ukr. Botan. Žurn 49: 95–97.

Ingham, W. 1900. *Weissia rupestris* var. *humilis*. J. Bot. 38: 486–489.

Ingham, W. 1904. *Tortula laevipiliformis* De Not. (in vegetative reproduction). *Riccia sorocarpa* Bisch. *Jungermannia minuta* Crantz. Naturalist (London) 1904: 378–389.

Ireland, R. R. 1987. Scanning electron microscope study of the spores of the North American species of *Plagiothecium*. Mem. New York Bot. Gard. 45: 95–110.

Ireland, R. & M. J. Shchepanek. 1993. The spread of the moss *Hyophila involuta* in Ontario. Bryologist 96: 132–137.

Iwatsuki, Z. & K. Saito. 1972. Notes on *Tortula vectensis*. Misc. Bryol. Lichenol. 6: 57–60. (partly in Japanese)

Iwatsuki, Z. & A. J. Sharp. 1958. *Molendoa sendtneriana* in the United States. Bryologist 61: 356–359.

Iwatsuki, Z. & A. J. Sharp. 1976. Notes on *Rhachithecium* in Assam, India. Misc. Bryol. Lichenol. 7: 105–107.

J

Jackson, A. B. 1898. *Tortula intermedia* Berk. in Leicestershire. J. Bot. 36: 149.

Jackson, A. B. 1908. *Leptodontium gemmascens* in Herts. J. Bot. 46: 58.

Jackson, B. D. 1960. A Glossary of Botanic Terms with Their Derivation and Accent. 4th Ed. Hafner Publ., New York.

Jaeger, A. 1869. Musci Cleistocarpi. Übersicht über die Cleistokarpischen Moose. Ber. Tätigk. St. Gallischen Naturwiss. Ges. 1868–69: 1–75.

Jaeger, A. & F. Sauerbeck. 1870–1880. Genera et species muscorum systematice disposita seu adumbratio florae muscorum totius orbis terrarum. Ber. Thätigk. St. Gallischen Naturwiss. Ges. 1869–70: 245–299, 1870; 1870–71: 357–451, 1872; 1871–72: 209–490, 1873; 1872–73: 61–236, 1874; 1873–74: 53–278, 1875; 1874–75: 85–188, 1876; 1875–76: 201–371, 1877; 1876–77: 211–454, 1878; 1877–78: 257–514, 1879; 1878–79: 218–252, 1880.

Janssens, J. A. & R. H. Zander. 1980 [1981]. *Leptodontium flexifolium* and *Pseudocrossidium revolutum* as 60,000–year-old subfossils from the Yukon Territory, Canada. Bryologist 83: 486–496.

Jelenc, F. 1954a. Les bryophytes nord-africains. II. Precisions sur le *Tortula saharae* Trab. Rev. Bryol. Lichénol. 23: 105–106.

Jelenc, F. 1954b. Les bryophytes nordsafricains. III. Le *Tortula desertorum* Broth. existe-t-il en Afrique du Nord? Rev. Bryol. 23: 107–108.

Jennings, O. E. 1935. *Desmatodon arenaceus* versus *Tortula obtusifolia*. Bryologist 38: 60.

Jensen, C. 1928. *Pottia Randii* in Schweden. Ann. Bryol. 1: 112–113.

Jiménez, M. N., R. M. Ros & J. Guerra. 1990. *Phascum vlassovii* Laz. (Pottiaceae, Musci) en Europa. Anales Jard. Bot. Madrid 47: 234–235.

Jones, E. W. 1949. Notes on some British *Barbula*. Trans. Brit. Bryol. Soc. 1: 191–193.

Jones, E. W. 1951. *Didymodon spitsbergensis* and *Bryum pauperidens*, two new mosses from Spitsbergen. Rev. Bryol. 20: 122–125.

Jones, G. N. 1961. The discovery of *Merceya* in Illinois. Bryologist 64:

Juárez, G., G. & A. J. Sharp. 1976. *Weissia rostellata* var. *phascoides* new to Mexico. Bryologist 79: 384.

Juratzka, J. 1870. Notiz über *Pottia mutica* Vent. und *Bryum gemmiparum* De Not. Hedwigia 9: 49–50.

Juratzka, J. 1875. Zwei neue Laubmoose. *Hypnum breidleri* Jur. n. sp. und *Weisia ganderi* Jur. n. sp. Hedwigia 14: 182–184.

K

Kaalaas, B. 1900. *Trichostomum arcticum* n. sp. Bot. Not. 1900: 257–258.

Kaiser, B. 1991. *Bryoerythrophyllum feruginascens* (Stirt.) Giac.

erstfund in Nordbayern. Bryol. Rundbr. 7: 5.

Kalenov, G. S. 1977. Environmental and geographical investigation of *Tortula desertorum* Broth. distribution in middle Asia deserts. Bot. Žurn. (Moscow & Leningrad) 62: 1015–1022.

Kanda, H. 1981. Two moss species of the genus *Pottia* collected from the vicinity of Syowa Station, East Antarctica. Antarct. Rec. 71: 96–108.

Kapur, A. 1989. Studies on protonemal differentiation and bud formation in *Timmiella anomala*, I. Effects of polyamines and amino acids. Phytomorphology 39: 129–133.

Kapur, A. 1991. Studies on protonemal differentiation and bud formation in *Timmiella anomala*, II. The effect of cytokinins and auxins. Cryptogamie Bryol. Lichénol. 12: 63–72.

Karczmarz, K. 1961. *Pottia heimii* (Hedw.) Br. Eur. (*Desmatodon heimii* Lazar.) w Polsce. Ekol. Polska, ser. B, 7: 297–302.

Karczmarz, K. 1980. New localities of *Tortula velenovskyi* Schiffn. in the northern verge of the Podolia region (USSR). Cryptogamie Bryol. Lichénol. 1: 311–313.

Kawai, I. 1968. Taxonomic studies on the midrib in Musci. (1) Significance of the midrib in systematic botany. Sci. Rep. Kanazawa Univ., Biol. 13: 127–157, 15 fig. 4 tab.

Kawai, I. 1976. Systematic studies on the conducting tissues of the gametophyte in Musci. (6) On the essential coördination among the anatomical characteristics of the stem in some species of Hypnaceae. Sci. Rep. Kanazawa Univ. 21: 47–124.

Kawai, I. 1989. Systematic studies on the conducting tissues of the gametophyte in Musci: XVI. Relationships between the anatomical charactersitcs of the stem and the classification system. Asian J. Pl. Sci. 1(2): 19–52.

Kawai, I. & H. Ochi. 1987. Systematic studies on the conducting tissues of the gametophyte in Musci. (15) Relationships between the taxonomic system and anatomical characteristics of stems in some species of Bryaceae. Sci. Rep. Kanazawa Univ. 32: 1–67.

Kawai, I. & K. Walther. 1969. Observation on the midrib of mosses in the Harburg District (1). Sci. Rep. Kanazawa Univ., Biol. 14: 175–215, 1 fig. 7 pl., 6 tab.

Khanna, K. R. 1960. Studies on natural hybridization in the genus *Weisia*. Bryologist 63: 1–16.

Khanna, K. R. 1964. Differential evolutionary activity in bryophytes. Evolution 18: 652–670.

Khanna, K. R. 1976. Generalized distance (D^2) among Himalayan populations of *Gymnostomum*. Rev. Bryol. Lichénol. 42: 683–691.

Koch, L. F. 1955. Mosses of California V. *Pterigoneurum* [sic] *ovatum* and *P. subsessile* Madroño 13: 100–101.

Koch, R. G. 1971. Pottiaceae in Nebraska. Bryologist 74: 206–207.

Koopmans, A. N. 1967. *Tortula*'s op Bizondere Windplaatsen. Buxbaumia 21: 87–88.

Koponen, A. 1981. Splachnobryaceae, a new moss family. Ann. Bot. Fenn. 18: 123–132.

Koponen, T. 1968. Generic revision of Mniaceae Mitt. (Bryophyta). Ann. Bot. Fenn. 5: 117–151.

Koponen, T. 1973. *Rhizomnium* (Mniaceae) in North America. Ann. Bot. Fenn. 10: 1–26.

Koponen, T. 1978. Modern taxonomical methods and the classification of mosses. Bryophytorum Biblioth. 13: 443–481.

Koponen, T. 1982. Generic and family concepts in the Mniaceae. Beih. Nov. Hedw. 71: 249–259.

Koponen, T. & V. Oittinen. 1967. *Aloina brevirostris* (Hook. & Grev.) Kindb. found in Finland. Memoranda Soc. Fauna Flora Fennica 43: 5–7.

Kramer, W. 1978. Contribution to the taxonomy and distribution of several taxa of *Tortula* Hedw. sect. *Rurales* De Not. (Musci) with bistratose lamina [sic]. Bryologist 81: 378–385.

Kramer, W. 1980. *Tortula* Hedw. sect. *Rurales* De Not. (Pottiaceae, Musci) in der östlichen Holarktis. Bryophyt. Biblioth. 21. Cramer, Vaduz.

Kramer, W. 1988. Beiträge zur Systematik und Bryogeographie einiger Sippen von *Tortula* Hedw. sect. *Rurales* De Not. (Pottiaceae, Musci) unter besonderer Berücksightigung der Südhemisphäre. J. Hattori Bot. Lab. 65: 81–144.

Kuc, M. 1959. *Rhynchostegiella tenella* Limpr., *Desmatodon cernuus* Br. Eur. and *Cirriphyllum germanicum* Loeske et Fleischer—in Poland. Ekologia Polska, Ser. B, 4: 47–51.

Kuc, M. 1960. La distribution geographique de *Tortula velenovskyi* Schiff. Rev. Bryol. Lichénol. 29: 92–96.

Kucyniak, J. 1947. Two Pottiaceae new to Quebec.: *Phascum cuspidatum* var. *americanum* and *Astomum phascoides*. Bryologist 50: 178–180.

Kucyniak, J. 1949a. On the occurrence of *Tortula latifolia* in Quebec. Bryologist 52: 23–33.

Kucyniak, J. 1949b. Muscinées nouvelles pour le Québec: *Calypogeia fissa*, *Odontoschisma elongatum*, *Didymodon rufus* et *Hypnum bambergeri*. Ann. ACFAS 15: 97.

Kucyniak, J. 1950. Muscinées nouvelles pour le Québec: *Calypogeia fissa*, *Odontoschisma elongatum*, *Didymodon rufus* et *Hypnum bambergeri*. Naturaliste Canad. 77: 307–312.

Kucyniak, J. 1958. Une mousse inattendue pour le Québec: *Pterigoneurum* [sic] *ovatum*. Naturaliste Canad. 85: 217–224.

Kucyniak, J. 1962. Un autre *Desmatodon* pour le Québec: *D. systylius*. Mém. Jard. Bot. Montréal 56: 12–16.

Kuehnemann, O. 1938. Catalogo de los musgos argentions. Lilloa 2: 37–183.

Kuehnemann, O. 1949. Catalogo de los hepaticas argentions. Lilloa 19: 319–375.

L

Lal, J. & N. S. Parihar. 1980. Production of protonemal and secondary gemmae in *Didymodon michiganensis* (Steere) Saito (Pottiaceae). J. Indian Bot. Soc. 59(Suppl.): 36.

Lambion, J. 1974. La mousse *Dialytrichia mucronata* (Brid.) Broth. dans le Grand Honnelle à Roisin. Nat. Mosana 27: 43.

Lambion, J. & A. Empain. 1973. Les espèces de *Cinclidotus* (Musci) de la Meuse et de la Sambre, en Belgique et das les Ardennes Françaises. Bull. Soc. Roy. Bot. Belgique 106: 175–186.

Lampton, R. K. 1966. *Merceya ligulata* in Georgia. Bryologist 69: 511.

Lane, D. M. 1978a. Chemical test for red-pigmented sections of *Sphagnum:* survey of 17 North American species. Bryologist 81: 602–605.

Lane, D. M. 1978b. A geographical guide to bryophyte floras of the world I. Misc. Bryol. Lichenol. 8: 15–23; 29–34.

Lasala, E. F. 1987. *Desmatodon heimii* (Hedw.) Mitt. in the Iberian Peninsula. Cryptogamie Bryol. Lichénol. 8: 147–148.

Lauridsen, V. B. 1972. A new preparational method for moss capsules. Lindbergia 1: 225.

Lawton, E. 1972. *Barbula rufofusca*, a new species of moss from northwestern North America. Bull. Torrey Bot. Club 99: 307, 1972.

Lazarenko, A. S. 1927. Noch eine neue Art der Gattung *Desmatodon* in der Ukraine. Bull. Jard. Bot. Kiev 5–6: 104–106.

Lazarenko, A. S. 1928. Über eine neue Art *Tortula ferganensis* aus Zentralasien. Ukrain. Bot. Žurn. 4: 64–65.

Lazarenko, A. S. 1959. Observations on the ecology and morphology of the regenerative protonema of *Tortula ruralis* and *T.*

desertorum. Ukrain. Bot. Žurn. 16: 55–64.

Lazarenko, A. S. 1960a. Apogamous sporogonia on two species of *Desmatodon* with polyploid protonema. Dokl. Akad. Nauk SSSR 134: 1239–1244.

Lazarenko, A. S. 1960b. On *Desmatodon* (*Pottia*) *heimii* (Hedw.) Lazar. from the Middle Miocene of the Ukraine. Akad. Sci. Ukraine 17: 97–100.

Lazarenko, A. S. 1961. The apogamic sporogonia on polyploid protonema of two species of bryophyte of genus *Desmatodon*. Dokl. Bot. Sci. 134: 177–179.

Lazarenko, A. S. 1963a. [Apogamous sporogonia in the haplophase of *Pottia intermedia*]. Dopov. Akad. Nauk. Ukr. RSR 11: 1524–1525.

Lazarenko, A. S. 1963b. Dynamics of quantitative variation in the sporophyte of *Desmatodon randii* (Kenn.) Lazar. in a natural population and monosporic cultures. Bull. Moscow Soc. Naturalists, Biol. Series V. 68: 133–148.

Lazarenko, A. S. 1965. The phenomenon of amphomorphism in mosses. Dokl. Bot. Sci. 162(4): 962–964.

Lazarenko, A. S. 1969. Experimental apogamy in *Pottia lanceolata* (Hedw.) C. Müll. Compt. Rend. (Dopov.) Acad. Sci. Ukrainian SSR, ser. B. Nr. 5: 459–462.

Lazarenko, A. S. 1974. Some considerations on the nature and behavior of the relict moss, *Desmatodon randii*. Bryologist 77: 474–477.

Lazarenko, A. S. & E. N. Lesnyak. 1972. A comparative study of two moss sibling species—*Desmatodon cernuus* (Hüb.) BSG—*D. ucrainicus* Laz. (Contribution to the problem of infrastructure of the moss species. Žurn. Obsch. Biol. [J. Gen. Biol.] 33(6): 657–667. [In Ukrainian, English summary.]

Lazarenko, A. S. & Ch. T. Pashuk. 1961. A statistical study of the intraspecific variability of the spores in *Desmatodon heimii* (Hedw.) Lazar. J. Bot. Ukraine 18: 68–80.

Lazarenko, A. S., Ch. T. Pashuk & E. N. Lesnyak. 1961. Apogamy in the haplophase of *Desmatodon randii* (Kenn.) Lazar. Dopov. Akad. Nauk Ukraine 10: 1381–1384.

Lecointe, A. 1976. Un groupement bryo-épiphytique subordonné aux zones inondables: *Tortuletum latifoliae*. Colloques Phytosocial. (Lille) 5: 141–151.

Lecointe, A. & P. Boudier. 1989. *Leptodontium gemmascens* (Mitt. ex Hunt) Braithw. (Pottiaceae, Musci) en Normandie et dans l'Ouest. Bull. Soc. Bot. Centre-Ouest, N.S. 20: 151–156.

Lecointe, A. & R. Schumacker. 1988. *Scopelophila cataractae* (Mitt.) Broth. (Musci, Pottiaceae), nouveau pour la Normandie, dans l'Orne. Bull. Soc. Bot. Centre-Ouest, n.s. 19: 245–248.

Lee, R. D. 1952. Re-discovering the moss *Bryobartramia robbinsii*. Victorian Naturalist 69: 9–11.

Lemmon, B. A. E. 1968. Cytological Investigation in the Genus *Weissia* in Southeastern United States. Dissert., Louisiana State University.

Lepage, E. 1949. Une mousse rare en Amérique du Nord: *Didymodon rufus* Lor. Ann. ACFAS 15: 97.

Lett, H. W. 1901. *Leptodontium recurvifolium* in Ireland. J. Bot. 39: 187.

Lewinsky, J. 1974. An electron microscopical study of the genus *Tortula* Hedw., with special reference to the exine ornamentation. J. Bryol. 8: 269–273.

Lewinsky, J. 1983 [1984]. The genus *Weissia* in the Faroes. Lindbergia 9: 200–203.

Lewis, K. R. 1961. The genetics of bryophytes. Trans. Brit. Bryol. Soc. 4: 111–130.

Lewis, M. 1981. Prodromus Bryologiae Andinae II. *Bryoerythrophyllum columbianum* disjunctively distributed between the steppes of northwestern U.S.A. and Bolivia. Bryologist 84: 536–538.

Li Xingjian (ed.). 1981. Some new taxa of Pottiaceae from Xizang (Tibet). Acta Bot. Yunnan. 3: 101–112. 6 fig.

Ligrone, R. 1986. Structure, development and cytochemistry of mucilage-secreting hairs in the moss *Timmiella barbuloides* (Brid.) Moenk. Ann. Bot. 58: 859–868.

Lightowlers, P. J. 1981. Techniques for mounting fragile tissue in gum chloral solution. J. Bryol. 11: 843–844.

Lightowlers, P. J. 1984. A new variety of the moss *Tortula robusta* from South Georgia. British Antarctic Survey Bull. 64: 63–66.

Lightowlers, P. J. 1985a. A synoptic flora of South Georgian mosses: *Tortula*. Bull. Brit. Antarct. Surv. 67: 41–77.

Lightowlers, P. J. 1985b [1986]. *Tortula lithophila* Dus. is *Sarconeurum glaciale* (C. Müll.) Card. et Bryhn. Bryologist 88: 365–366.

Lightowlers, P. J. 1985c. Taxonomic notes on New Zealand species of *Tortula*. J. Bryol. 13: 369–375.

Lightowlers, P. J. 1986a. [Summary of:] The systematics of austral *Tortula:* unravelling southern hemisphere taxa of a temperate genus. Bull. British Bryol. Soc. 47: 7–8.

Lightowlers, P. J. 1986b. The moss genus *Tortula* from the Antarctic botanical zone. Brit. Antarct. Surv. Bull. 72: 71–76.

Lightowlers, P. J. 1986c [1987]. Taxonomy and distribution of the subantarctic species of *Tortula*. J. Bryol. 14: 281–295.

Ligrone, R. 1983 [1984]. Ultrastructure of mucilage-secreting hairs in *Timmiella barbuloides* (Pottiales, Musci). Giorn. Bot. Ital. 117: 168–169.

Ligrone, R. 1985. Studies on the leaf structure in some species of Leucobryaceae. III. *Leucobryum sanctum* (Brid.) Hampe and *Leucobryum* sp. (Leucobryeae). J. Bryol. 13: 411–416.

Ligrone, R., R. Castaldo & R. Gambardella. 1980a. Studies on *Timmiella barbuloides* (Brid.) Moenk. I. Historical and ultrastructural differentiation of the cauloid. Cryptogamie Bryol. Lichénol. 1: 115–142.

Ligrone, R., R. Gambardella & R. Castaldo. 1980b. Studies on *Timmiella barbuloides* (Brid.) Moenk. II. Ultrastructure and development of unusual spherical bodies in the differentiating cells of the cauloid. Cryptogamie Bryol. Lichénol. 1: 239–246.

Limpricht, K. G. 1890. Die Laubmoose Deutschlands, Oesterreichs und der Schweiz. 1: 1–836. Leipzig.

Lindberg, S. O. 1864. Om de Europeiska Trichostomeae. Öfvers. Förh. Kongl. Svenska Vetensk.-Akad. 21: 213–254.

Lindberg, S. O. 1880. *Tortula lingulata*. Rev. Bryol. 7: 40–41.

Lindsay, E. & A. F. Finocchio. 1970. Stomate morphology of mosses. Sci. Stud. St. Bonaventure Univ. 26: 5–14, 5 fig.

Little, E. L., Jr. 1943. Later generic homonyms among North American mosses. Bryologist 46: 105–125.

Llamas, B. & J. Guerra. 1992. *Leptodontium flexifolium* (Dicks.) Hampe (Pottiaceae) in the Iberian Peninsula. J. Bryol. 17: 160–162.

Lloret, F. 1986. *Tortula sinensis* y *Tortula fragilis*, novedades para la brioflora de la Península Ibérica. Anales Jard. Bot. Madrid 42: 303–308.

Lloret, F. 1987. *Bryoerythrophyllum inaequalifolium* (Tayl.) Zander, new to the European continent. Lindbergia 13: 127–129.

Lobachevskaya, O. V., O. T. Demkiv & R. T. Ripetsky. 1986. The cytophotometric measurement of nuclear DNA amount in polyploid complex of *Tortula muralis* Hedw. Ukraiin. Bot. Žurn. 43(6): 23–26.

Loeske, L. 1914. Neue Prinzipien der systematischen Bryologie. Hedwigia 54: 210–216.

Loeske, L. 1916. Über die systematische Stellung von *Leptobarb-*

ula. Bryol. Zeitschr. 1: 31.

Loeske, L. 1926. Der Einfluss des Wassers auf Papillen und Mamillen. Folia Cryptogamica, Zeitschr. Erforsch. ungarischen Kryptogamenflora 1: 215–219.

Loeske, L. 1934. Über *Tortula freibergii*, *T. obtusifolia* und verwandte Formen. Ann. Bryol. 7: 98–107.

Long, D. G. 1978. The moss *Hyophila standfordensis* in Scotland. Trans. Bot. Soc. Edinb. 43: 25–26.

Long, D. G. 1979a. A reassessment of the systematic position of *Tortula stanfordensis* Steere and *T. khartoumensis*. J. Bryol. 10: 377–381.

Long, D. G. 1979b. Some additions to the bryophyte flora of Iraq. Rev. Bryol. Lichénol. 45: 103–105.

Long, D. G. 1982a [1983]. *Bryoerythrophyllum caledonicum*, a new moss from Scotland. J. Bryol. 12: 141–157.

Long, D. G. 1982b [1983]. *Paraleptodontium*, a new genus of Pottiaceae. J. Bryol. 12: 179–184.

Long, D. G. 1982c. Axillary gemmae in *Anoectangium warburgii* Crundw. & Hill. J. Bryol. 12: 49–51.

Long, D. G. & M. O. Hill. 1982. *Tortula solmsii* (Schimp.) Limpr. in Devon and Cornwall, newly recorded in the British Isles. J. Bryol. 12: 159–169.

Longton, R. E. 1988. Life-history strategies among bryophytes of desert regions. J. Hattori Bot. Lab. 64:15–28.

Lübenau, R. 1991. Zur Aufsatz von Herrn S. Risse: *Bryoerythrophyllum ferruginascens* (Stirt.) Giac. in Deutschland. Bryol. Rundbr. 7: 5.

Luisier, A. 1913. Fragments de Bryologie Ibérique. I. Le genre *Triquestrella* en Europe. Brotéria, série Botánica 11: 135–138.

M

Machado, A. 1917. Uma *Pottia* nova do Algarve. Ann. Sci. Acad. Polytechn. Porto 12: 51–52.

Maddison, D. R. 1991. The discovery and importance of multiple islands of most-parsimonious trees. Syst. Zool. 40: 315–328.

Maddison, W. P., M. J. Donoghue & D. R. Maddison. 1984. Outgroup analysis and parsimony. Sys. Zool. 33: 83–103.

Magill, R. E. 1977a. A reexamination of *Globulinella* (Musci; Pottiaceae). Bryologist 80: 76–82.

Magill, R. E. 1977b [1978]. Towards the conservation of four family names in Musci. Taxon 26: 597–598.

Magill, R. E. 1981. Flora of Southern Africa. Bryophyta. Part 1. Mosses. Fascicle 1. Sphagnaceae-Grimmiaceae. Botanical Res. Inst. Dept. Agri. and Fisheries. Pretoria.

Magill, R. E. (ed.) 1990. Glossarium Polyglottum Bryologiae. Monogr. Syst. Bot. 33. Missouri Botanical Garden. English: R. E. Magill & R. E. Stotler; French: H. Bischler & D. Lamy; German: P. Geissler & J.-P. Frahm; Japanese: Z. Iwatsuki; Latin: P. Eckel; Russian: O. M. Afonia & R. N. Schljakov; Spanish: G. Hässel de Menéndez.

Magill, R. E., C. Delgadillo M. & L. R. Stark. 1983. *Tortula chisosa sp. nov.*, a bistratose-leaved species from the United States, Mexico, and southern Africa. Ann. Missouri Bot. Gard. 70: 200–202.

Maheu, J. 1908. Production expérimetale des propagules dans le genre *Barbula*. Bull. Soc. Bot. France 55: 445–453, pl. 26–27.

Malek, L. & J. D. Bewley. 1978. Effects of various rates of freezing on the metabolism of a drought-tolerant plant, the moss *Tortula ruralis*. Pl. Physiol. 61: 334–338.

Malta, N. 1921. Versuche über die Widerstandsfähigkeit der Moose gegen Austrocknung. Acta Univ. Latv. 1: 125–129.

Malta, N. 1928. *Pottia randii* Kenn. auch im Östbaltichen Gebiet gefunden. Acta Horti Bot. Univ. Latv. 3: 61–62.

Mandoli, D. F. & W. R. Briggs. 1984. Fiber optics in plants. Sci. Am. 251: 90–98.

Mansion, A. 1904–5. Note sur le *Pterygoneurum lamellatum* Jur. et *Breutelia arcuata* Schimp. (espèces nouvelles pour la flore belge). Bull. Soc. Roy. Bot. Belgique 42: 61–63, 66–68.

Mardorf, W. 1916. Über die Lebensweise von *Tortula papillosa*, *T. pulvinata* und *T. laevipila*. Hedwigia 57: 255–256.

Mårtensson, O. & A. Berggren. 1954. Some notes on the ecology of the "copper mosses." Oikos 5: 98–99.

Mårtensson, O. & E. Nilsson. 1974. On the morphological colour of bryophytes. Lindbergia 2: 145–159.

Marshall, J. J. 1903. *Tortula ruralis* var. *arenicola* on the Yorkshire coast. Naturalist (London) 1903: 461.

Martínez Lacal, F., F. D. Mateo & J. Varo. 1989. *Tortula rhizophylla* (Sak.) Iwats. & Saito, musgo nuevo para la Península Ibérica. Fol. Bot. Misc. 6: 81–84.

Mastracci, M. 1993. Taxonomic significance of stem and leaf-sheath anatomy in *Timmia* Hedw. J. Bryol. 17: 481–487.

Matteri, C. M. 1977a. Notes on Antarctic bryophytes: X. The genus *Pottia* from the Antarctic Botanical Zone. Brit. Antarct. Surv. Bull. 46: 23–28.

Matteri, C. M. 1977b. A synoptic flora of South Georgian mosses: VII. *Pottia*. British Antarctic Survey Bull. 46: 23–28.

Matteri, C. M. 1982. Patagonian bryophytes 6. Fruiting *Sarconeurum glaciale* (C. Müll.) Card. et Bryhn newly found in southern Patagonia. Lindberia 8: 105–109.

Matteri, C. M. 1988a. New records for the Fuegian moss flora and the status of *Oedipodium fuegianum* Kühn. et Gonç.-Carr. Lindbergia 14: 63–66.

Matteri, C. M. 1988b. Restos de *Didymodon* sp. (Pottiaceae, Musci) en sedimentos cuaternarios de isla Gable, Argentina. Bol. Soc. LatinoAm. Paleobot. & Palinol. 11: 23–27.

May, R. 1987. Notes on some Macaronesian *Tortella* species. Bryologische Beiträge 6: 58–66.

Maya, F. L. 1986. *Tortula sinensis* (C. Müll.) Broth. y *Tortula fragilis* Tayl., novedades para la brioflora de la Península Ibérica. Anales Jard. Bot. Madrid 42: 303–308.

McIntosh, T. T. 1988. Phytogeographic relationships of steppe bryophytes of British Columbia. Am. J. Bot. 75(6, Part 2): 5 (Abstr.).

McIntosh, T. T. 1989. Bryophyte records from the semiarid steppe of northwestern North America, including four species new to North America. Bryologist 92: 356–362.

Meenks, J. L. D., Z. Tuba & Z. Csintalan. 1991. Ecophysiological responses of *Tortula ruralis* upon transplantation around a power plant in west Hungary. J. Hattori Bot. Lab. 69: 21–35.

Mehra, K. 1984. Effect of ammonium and nitrate nitrogen on callus induction in *Hyophila involuta* (Musci). Beitr. Biol. Pfl. 59: 333–339.

Mehra, P. 1988. Effect of some growth regulators on the formation and behaviour of protonemal gemmae in the moss *Hyophila involuta*. Phytomorphology 38: 283–291.

Melick, H. van. 1975. *Phascum floerkeanum* Web. et Mohr, nieuw voor Nederland. Lindbergia 3: 121–122.

Melick, H. van. 1987. *Scopelophila cataractae* (Mitt.) Broth. ook in Nederland. Lindbergia 12: 163–165.

Menzel, M. 1992. Nomenklatorische Notizen zu Eddys "Handbook of Malesian Mosses" (Band 2). Willdenowia 22: 197–199.

Meylan, C. 1929. *Barbula rufa* (Lor.) et *Barbula Kneuckeri* (Loeske et Osterw.). Bull. Soc. Bot. Genève, ser. 2, 21: 264–267.

Meyland, C. 1936. Note sur un nouveau *Tortula* ap. R. Maire et Wilczek. Florule des îles Haas. Bull. Soc. Hist. Nat. Afr. N. 26: 75–77.

Milde, J. 1869. *Barbula insidiosa* Jur. et Milde spec. Nova. Hedwigia 8: 97–98.

Miller, H. A. 1979. The phylogeny and distribution of the Musci. In G. C. S. Clarke & J. G. Duckett (eds.), Bryophyte Systematics, pp. 11–39. Syst. Assoc. Spec. Vol. 14. Academic Press, New York.

Mishler, B. D. 1984a. Systematic studies in the genus *Tortula* Hedw. (Musci: Pottiaceae). 236 pp. Illustrated. Dissert., Harvard University.

Mishler, B. D. 1984b. A cladistic approach to the phylogeny of the "bryophytes." Brittonia 36: 406–424.

Mishler, B. D. 1985a. Biosystematic studies of the *Tortula ruralis* complex. 1. Variation of taxonomic characters in culture. J. Hattori Bot. Lab. 58: 225–253.

Mishler, B. D. 1985b [1986]. The phylogenetic relationships of *Tortula:* An SEM survey and a preliminary cladistic analysis. Bryologist 88: 388–403.

Mishler, B. D. 1985c. The morphological, developmental, and phylogenetic basis of species concepts in bryophytes. Bryologist 88: 207–214.

Mishler, B. D. 1986a. Ontogeny and phylogeny in *Tortula* (Musci: Pottiaceae). Syst. Bot. 11: 189–208.

Mishler, B. D. 1986b. A Hennigian approach to bryophyte phylogeny. J. Bryol. 14: 71–81.

Mishler, B. D. 1987a. Leaf development in *Tortula papillosissima* (Pottiaceae). Mem. New York Bot. Gard. 45: 48–54.

Mishler, B. D. 1990. Reproductive biology and species distinctions in the moss genus *Tortula*, as represented in Mexico. Syst. Bot. 15: 86–97.

Mishler, B. D. & S. P. Churchill. 1984. A cladistic approach to the phylogeny of the "bryophytes." Brittonia 36: 406–424.

Mishler, B. D. & S. P. Churchill. 1985a. Cladistics and the land plants: A response to Robinson. Brittonia 37: 282–285.

Mishler, B. D. & S. P. Churchill. 1985b. Transition to a land flora: phylogenetic relationships of the green algae and bryophytes. Cladistics 1: 305–328

Mishler, B. D. & S. P. Churchill. 1987. Transition to a land flora: a reply. Cladistics 3: 65–71.

Mishler, B. D. & N. G. Miller. 1983. Distributional studies of Massachusetts bryophytes. Rhodora 85: 421–432.

Mishler, B. D. & A. E. Newton. 1987. [Abstract:] Germination of spores and gametophyte fragments of *Tortula* under a variety of experimental conditions. Am. J. Bot. 74: 598.

Mishler, B. D. & A. E. Newton. 1988. Influences of mature plants and desiccation on germination of spores and gametophytic fragments of *Tortula.* J. Bryology 15: 327–342.

Mishler, B. D. & M. J. Oliver. 1988. [Abstract:] Evolution of desiccation-tolerance in the *Tortula ruralis* complex. I. Distribution, habitats, and water relationships. Am. J. Bot. 75(6), Part 2: 5.

Mishler, B. D. & M. J. Oliver. 1991. Gametophytic phenology of *Tortula ruralis*, a dessication-tolerant moss, in the Organ Mountains of southern New Mexico. Bryologist 94: 143–153.

Mishler, B. D. & D. C. Scheirer. 1983. Development of the leaf of *Tortula obtusissima*: systematic and eological implications. Am. J. Bot. 70 (5 part 2): 6. (Abstr.).

Mitten, W. 1859. Musci indiae orientalis: an enumeration of the mosses of the East Indies. J. Proc. Linn. Soc. Bot. 1(Suppl.): 1–171.

Mitten, W. 1867. A few notes on some British mosses allied to *Tortula fallax* Hedw. J. Bot. 5: 324–329.

Mitten, W. 1868. New or rare British mosses: *Trichostomum flavovirens*, *T. diffractum*, *T. littorale*. J. Bot. 6: 97–99.

Mitten, W. 1869. Musci Austro-americani. J. Linn. Soc. Bot. 12: 1–659.

Mitten, W. 1874. On the *Aloina* section of the genus *Tortula*. J. Bot. 3: 139–142.

Mizushima, U. 1957. On the correct Latin names of *Weissia* Hedwig and *Heterophyllum* (Schimper) C. Müller. J. Jap. Bot. 32(12): 367–369.

Mizushima, U. 1971. On the nomenclature of *Barbula prionophylla* Saito. J. Jap. Bot. 46: 319–320. [In Japanese.]

Mogensen, G. S. 1978. Spore development and germination in *Cinclidium* (Mniaceae, Bryophyta), with special reference to spore mortality and false anisospory. Canad. J. Bot. 56: 1032–1060.

Mogensen, G. S. 1981. The biological significance of morphological characters in bryophytes: the spore. Bryologist 84: 187–207.

Mohan, G. 1984. *Desmatodon cernuus*, ein neuer Laubmoos-Nachweis für die Karpathen und die Bryoflora von Rumänien. Feddes Repertorium 94: 247–250.

Moore, C. J., S. E. Luff & N. D. Hallam. 1982. Fine structure and physiology of the desiccation-tolerant mosses, *Barbula torquata* Tayl. and *Triquetrella papillata* (Hook. f. and Wils.) Borth., during desiccation and rehydration. Bot. Gaz. 143: 358–367.

Mues, R. 1990. The significance of flavonoides for the classification of bryophyte taxa at different taxonomic ranks. In H. D. Zinmeister & R. Mues, eds., Bryophytes: Their Chemistry and Chemical Taxonomy. Proc. Rhytochem. Soc. Europe 29: 421–436. Cladendon Press, Oxford.

Mues, R. & H. D. Zinsmeister. 1988. The chemotaxonomy of phenolic compounds in bryophytes. J. Hattori Bot. Lab. 64: 109–141.

Müller, C. 1849–1851. Synopsis Muscorum Frondosorum. Vols. I & II. Berlin.

Müller, C. 1853. Deutschlands Moose. Halle.

Müller, C. 1873. Sechs neue Laubmoose Nordamerikas: *Mnium nevii*, *Catharinea xanthopelma*, *Bartramia mohriana*, *Barbula cancellata*, *Grimmia nevii*, *Hypnum homolostegium*. Flora 56: 481–484.

Müller, C. 1897a. Prodromus bryologiae Bolivianae. Nuov. Giorn. Bot. Ital., N.S. 4: 4–172.

Müller, C. 1897b. *Triquetrella*, genus Muscorum novum conditum et descriptum. Österr. Bot. Zeit. 47: 420–442.

Müller, C. 1901. Genera Muscorum Frondosorum. Leipzig.

Murray, B. M. 1984. A revision of the monotypic genera *Indusiella*, *Aligrimmia* and *Coscinodontella* (Musci: Grimmiaceae), with comments on convergent xeromorphological features. Bryologist 87: 24–36.

Murray, B. M. 1992. Bryophyte flora of Alaskan steppes. Bryobrothera 1: 9–33.

Murray, J. 1926. Mounting mosses on microscope slides. Bryologist 29: 55–57.

Muyldermans, L. 1959. *Leptodontium flexifolium* in de Beneluxlanden. Buxbaumia 13: 65–73.

N

Nagano, I. 1959. On the distribution and substrata of *Eucladium verticillatum* (Smith) B.S.G. in Miura Peninsula, eastern coast of central Japan. Sci. Rep. Yokosuka City Mus. 4: 34.

Nagano, I. & Y. Shimizu. 1973. Ecological notes of the genus *Scopelophila* (II). Relations between spore germination and hydrogen-ion concentration of *Scopelophila ligulata* and *Mielichhoferia mielichhoferi*. J. Coll. Lib. Arts, Saitama Univ., Nat. Sci. 8: 29–51.

Nagano, I., Yukiko Kato & H. Ezaki. 1969. Ecological notes of genus *Scopelophila (Merceya)* (1). J. Coll. Lib. Arts, Saitama Univ., Nat. Sci. 5: 5–13.

Nawawi, A. & M. A. H. Mohamed. 1989. Multiradiate bryophyte propagules from foam and scum in Malaysian rivers. Malayan Nat. J. 43: 96–99.

Nebel, M. 1990. Zur Verbreitung und Ökologie von *Tortula ruraliformis* (Besch.) Ingh. und *Pleurochaete squarrosa* (Brid.) Lindb. in Sudwestdeutschland. Jahresh. Ges. Naturk. Wurttemberg 145: 163–176.

Nelson, G. J. 1972. Phylogenetic relationship and classification. Syst. Zool. 21: 227–231.

Nelson, G. J. 1974. Classification as an expression of phylogenetic relationships. Syst. Zool. 22: 344–359.

Newton, M. E. 1968. Cyto-taxonomy of *Tortula muralis* Hedw. in Britain. Trans. Brit. Bryol. Soc. 5: 523–535.

Newton, M. E. 1972. Chromosome studies in some South Georgian bryophytes. Brit. Antarct. Surv. Bull. 30: 41–49.

Newton, M. E. 1983. Natural variation and diagnostic characters in *Gymnostomum*, *Anoectangium* and *Molendoa*. J. Bryol. 12: 343–349.

Newton, M. E. & D. Boyce. 1987. Gemmae in British *Leptodontium flexifolium* (With.) Hampe. J. Bryol. 14: 737–740.

Neumann, A. J. 1972 [1973]. Observations on the morphology and biology of the moss *Tortula vectensis* E. Warb. & Crundw. in Louisiana. Bryologist 75: 580–583.

Nicholson, W. E. 1903. *Weisia sterilis, sp. n.* J. Bot. 41: 247–248.

Nicholson, W. E. 1905a. Notes on two forms of hybrid *Weisia*. Rev. Bryol. 32: 19–25.

Nicholson, W. E. 1905b. *Tortula pagorum* (Milde) De Not. Bryologist 8: 70.

Nicholson, W. E. 1906. *Weisia crispa*, Mitt. (female) × *W. microstoma*, C.M. (male). Rev. Bryol. 33: 1–2.

Nicholson, W. E. 1908. *Barbula* (aut *Didymodon) Nicholsoni* spec. nov. J. Bot. 46: 173.

Nicholson, W. E. 1909. *Tortula aciphylla* in Britain. J. Bot. 47: 374–375, pl. 499.

Nicholson, W. E. 1910. A new hybrid moss. Rev. Bryol. 37: 23–24.

Nilsson, E. & G. Bendz. 1975. Flavonoids in bryophytes. Pp. 117–120. In G. Bendz & J. Santesson (eds.), Chemistry in Botanical Classification. Academic Press, New York.

Noguchi, A. 1951. Notes on a genus of mosses, *Weisiopsis* Broth. J. Jap. Bot. 26: 131–134.

Noguchi, A. 1956. On some mosses of *Merceya* with special reference to the variation and ecology. Kumamoto J. Sci. Ser. B, Sect. 2, 2: 239–257.

Noguchi, A. 1988. Illustrated Moss Flora of Japan. Part 2. Supplemented by Zennoske Iwatsuki. Pages i-vi + 243–491. Calymperaceae to Bryaceae. Hattori Botanical Laboratory, Obi, Nichinan, Japan.

Noguchi, A. & H. Furuta. 1956. Germination of spores and regeneration of leaves of *Merceya ligulata* and *M. gedeana*. J. Hattori Bot. Lab. 17: 32–44.

Noguchi, A. & K. Ochi. 1956. Studies on ecology on *Merceya ligulata* and *Merceyopsis sikkimensis*. J. Hattori Bot. Lab. 16: 122–123.

Norris, D. H. & T. Koponen. 1989. Bryophyte flora of the Huon Peninsula, Papua New Guinea. XXVII. Pottiaceae (Musci). Acta Bot. Fennica 137: 81–138.

Norris, D. H. & R. H. Zander. 1981. *Splachnobryum limbatum sp. nov.* from San Cristobal, Solomon Islands. Bryologist 84: 404–407.

Nowak, H. 1980. Revision der Laubmoosgattung *Mitthyridium* (Mitten) Robinson für Ozeanien (Calymperaceae). Bryophyt. Biblioth. 20.

Nyholm, E. 1989. Illustrated Flora of Nordic Mosses. Fasc. 2. Pottiaceae—Splachnaceae—Schistostegaceae. Pp. 75–141. Nordic Bryological Society, Lund.

Nyholm, E. & L. Hedenäs. 1986. A new species of *Gymnostomum*. Lindbergia 12: 41–42.

Nyholm, E. & K. Wigh. 1973. Cytotaxonomical studies in some Turkish mosses. Lindbergia 2: 105–113.

O

Ochyra, R. 1988. New taxa and new combinations of mosses proposed by Andrey S. Lazarenko. J. Hattori Bot. Lab. 64: 335–346.

Ochyra, R. 1992. New combinations in *Syntrichia* and *Warnstorfia* (Musci). Fragm. Flor. Geobot. 37: 211–214.

Odgaard, B. 1978. *Trichostomum crispulum* new to Denmark. Lindbergia 4: 332.

Odgaard, B. 1985 [1986]. A new find of *Leptodontium flexifolium* (With.) Hampe in Denmark. Lindbergia 11: 175–176.

Olarinmoye, S. O. 1981. Regeneration and gemma development in *Hyophila crenulata* C. Müll. ex Dus. Cryptogamie Bryol. Lichénol. 2: 457–460.

Oliver, M. J. & J. D. Bewley. 1984a. Plant desiccation and protein synthesis. IV. RNA synthesis, stability, and recruitment of RNA into protein synthesis during desiccation and rehydration of the desiccation-tolerant moss, *Tortula ruralis*. Pl. Physiol. 74: 21–25. 4 Fig.

Oliver, M. J. & J. D. Bewley. 1984b. Plant desiccation and protein synthesis. V. Stability of poly(a)- and poly(a)+ RNA during desiccation and their synthesis upon rehydration in the desiccation-tolerant moss *Tortula ruralis* and the intolerant moss *Cratoneuron filicinum*. Pl. Physiol. 74: 917–922.

Oliver, M. J. & J. D. Bewley. 1984c. Plant desiccation and protein synthesis. VI. Changes in protein synthesis elicited by desiccation of the moss *Tortula ruralis* are effected at the translational level. Pl. Physiol. 74: 923–927. 3 Fig.

Oliver, M. J. & B. D. Mishler. 1988. Evolution of desiccation-tolerance in the *Tortula ruralis* complex. II. Physiological and biochemical mechanisms. Am. J. Bot. 75(6, Part 2): 6 (Abstr.).

Oliver, M. J. & B. D. Mishler. 1990. Interspecific differences in the control and repair of desiccation damage within the *Tortula ruralis* complex. [Abstract.] Am. J. Bot. 77(6, Suppl.): 2.

Ovezova, A. Kh. 1989. Peculiarities of summer population of *Tortula desertorum* in eastern Karakum. Izvest. Akad. Nauk Turkmen., Ser. Biol. Nauk 1989(5): 19–23.

Osada, T. 1958. On the habitats of *Eucladium verticillatum*. Misc. Bryol. Lichenol. 15: 3.

P

Paris, E. G. 1906. Index Bryologicus. 2nd Ed. Vol. 4. Paris.

Partridge, T. R., M. J. Dallwitz & L. Watson. 1986. A Primer for the DELTA System on VMS, MS-DOS, and PRIMOS. Div. of Entomol. Rep. 38, Commonw. Sci. Industr. Res. Org., Australia, CSIRO, Canberra.

Partyka, L. Ya. 1986. *Anoectangium handelii* Schiffn., a species of moss new for bryoflora of Europe. Ukr. Bot. Žurn. 43(2): 66–67.

Paton, J. A. 1957. The occurrence, structure and functions of the stomata in British bryophytes. Part. I. Occurrence and structure. Trans. Brit. Bryol. Soc. 3: 228–242; 249–252 + pl. 1.

Paton, J. A. 1966. *Tortula stanfordensis* Steere in Yorkshire. Trans. Brit. Bryol. Soc. 5: 147–148.

Patterson, P. M. 1964. Problems presented by bryophytic xerophytism. Bryologist 67: 390–396.

Peciar, V. 1960a. *Funaria mediterranea* Lindb. und *Phascum curvicollum* Hedw. seltene Arten den Slovakischen Bryoflora. Biologia 15: 362–365.

Peciar, V. 1960b. *Tortula velenovskyi* Schiffn. auf neuen slovakischen Lokalitäten. Biologia 10: 790–794.

Pedrotti, C. C. & M. Aleffi. 1989. A second record of *Tortula rhizophylla* (Sak.) Iwats. & Saito in Italy. J. Bryol. 15: 803–805.

Persson, H. 1944. On some species of *Aloina*, with special reference to their dispersal by the wind. Svensk Bot. Tidskr. 38: 260–268.

Persson, H. 1945. *Barbula icmadophila* funnen i Sverige. Svensk Bot. Tidskr. 40: 124–125.

Persson, H. 1948. On the discovery of *Merceya ligulata* in the Azores with a discussion of the so-called "copper mosses." Rev. Bryol. Lichénol. 17: 75–78.

Peterfi, M. 1911. Bryologische Mitteilungen V. Über das Vorkomen von *Tortella squarrosa* Brid. im Ungarischen Alfold. Bot. Kozlem. 10: 14–17.

Pettet, A. 1966. Bryophytes of the Sudan. I. Khartoum Province. Trans. Brit. Bryol. Soc. 5: 316–331.

Philibert, H. 1882a. Sur le *Leptobarbula berica*. Rev. Bryol. 9: 17–20.

Philibert, H. 1882b. *Gyroweisia acutifolia* et l'*Orthotrichum Schawii*. Rev. Bryol. 9: 3–4, 9–11.

Philibert, H. 1882c. *Leptobarbula berica*. Rev. Bryol. 9: 17–20.

Philibert, H. 1883. Le vertitable *Trichostomum nitidum*. Rev. Bryol. 10: 77–80.

Philibert, H. 1885. La fructification du *Didymodon ruber*. Rev. Bryol. 12: 98–94.

Philibert, H. 1886. *Barbula Buyssoni sp. nov.* Rev. Bryol. 13: 36–37.

Philibert, H. 1896. *Pottia Ryani* nov. sp., *Trichostomum Crozalsi sp. n.*, et *Webera rubella sp. n.* Rev. Bryol. 23: 10–14, 28–30, 85–90.

Pierrot, R. B. 1958. *Tortella inflexa* (Bruch) Broth. et *Leptobarbula berica* (De Not.) Schp. in Charente-Maritime. Bull. Féd. Fr. Sci. Nat., Sér. 2, 14: 81–83.

Pierrot, R. B. 1973. *Astomum multicapsulare* (Sm.) B.S.G. dans l'île d'Oléron, espèce nouvelle pour la France. Rev. Bryol. Lichenol. 39: 163–165.

Pierrot, R. B. 1976. *Gymnostomum calcareum* B. G. et *Gyroweissia tenuis* (Schr.) Schp. dans le Centre-Ouest. Bull. Soc. Bot. Centre-Ouest, n.s. 7: 135–137.

Pierrot, R. B. 1986. *Crossidium aberrans* Holz. & Bartr., mousse nouvelle pour la France. Bull. Soc. Bot. Centre-Ouest, n.s. 17: 149–150.

Pierrot, R. B. 1989. Contribution à l'étude de *Gymnostomum luisieri* (Sérgio) Sérgio ex Crundw. Bull. Soc. Bot. Centre-Ouest, n.s. 20: 141–143.

Pilous, Z. 1951a. *Pottia latifolia* C. M. a new moss in the mountains Krkonose (North-East Bohemia). Ceskoslov. Bot. Listy 3: 127–128.

Pilous, Z. 1951b. *Aloina brevirostris* Kindb., novy druh mechu v CSR. Ceskoslov. Bot. Listy 3: 138–139.

Pilous, Z. 1951c. Mech *Tortella nitida* Limpr. v. Ceskoslovensku. Ceskoslov. Bot. Listy 3: 109–110.

Pilous, Z. 1952. Mediterrani mech *Leptobarbula berica* Schpr. v Krkonosich. Preslia 24: 105–106.

Pilous, Z. 1958. Fragmenta bryologica 31–40. Preslia 30(2): 165–178.

Platnick, N. I., C. E. Griswold & J. A. Coddington. 1991. On missing entries in cladistic analysis. Cladistics 7: 337–343.

Podpěra, J. 1950. Roste *Syntrichia desertorum* Amann na Morave? Ceskoslov. Bot. Listy 2: 123–124.

Podpěra, J. 1954. Conspectus Muscorum Europaeorum. Ceskosl. Akad. Věd, Prague.

Pospíšil, V. 1975. Die Bedeutung der Moose *Pterygoneurum subsessile* (Brid.) Jur. und *P. ovatum* (Hedw.) Dix. als Indikatoren der Klimagebiete in der Tschechoslowakei. Čas. Morav. Mus. v. Burně, Vědy Přir. 60: 125–146.

Pospíšil, V. 1977. *Tortula velenovskyi* Schiffn., eine gefährdete Art der Tschechoslowakischen Bryoflora. Čas. Morav. Mus. Zemsk. 62: 47–58.

Pospíšil, V. 1978. Moose *Acaulon muticum* (Hedw.) C. Müll., *A. triquetrum* (Spruce) C. Müll. und *Phascum curvicolle* Ehrh. in der Tschechoslowakei. Čas. Morav. Mus. Zemsk. 63: 57–70.

Pospíšil, V. 1987. Verbreitung, Oekologie und Grade der Gefährdung der Laubmoose *Tortula virescens* (De. Not.) De Not., *T. papillosa* Wils. und *T. latifolia* Bruch ex Hartm. in der Tschechoslowakei. Časopis Moravskeho Musea v Brně, Vědy Prirodni 72: 107–134.

Potier de la Varde, R. 1921. Une correction au nom de *Weisia viridula* var. *longifolia*. Rev. Bryol. 48: 11.

Potier de la Varde, R. 1927. À propos de *Phasconica Balansae* C. M. Arch. Bot. Bull. Mens. 2: 39–40.

Potier de la Varde, R. 1928. Mousses de l'Oubangui. Arch. Bot. 1(Mém.) 3): 1–152.

Potier de la Varde, R. 1934. *Barbula Gattefossei* P. de la V. spec. nov. sect. Helicopogon. Ann. Crypt. Exot. 7: 195.

Potier de la Varde, R. 1936. *Luisierella* Thér. et P. de la V. genus novum familiae Pottiacearum, s. f. Pottioidearum. Bull. Soc. Bot. France 83: 73–76.

Potier de la Varde, R. 1941. *Tisserantiella* P. de la V. genus novum familiae Pottiacearum. Bull. Soc. Bot. France 88: 469–472.

Potier de la Varde, R. 1942. *Weisia rutilans* (Hedw.) Lindb. dans la Manche. Bull. Soc. Linn. Normandie 3: 74.

Potier de la Varde, R. 1953. Découverte du genre *Gymnostomiella* en Mauritanie. Bull. Inst. Fr. Afrique Noire 15: 1387–1389.

Preston, C. D. & H. L. K. Whitehouse. 1985. *Trichostomopsis umbrosa* in semi-natural chalk habitats. J. Bryol. 13: 471–474.

Proctor, M. C. F. 1979a. Structure and eco-physiological adaptation in bryophytes. In G. C. S. Clarke & J. G. Duckett (eds.), Bryophyte Systematics, pp. 479–509. Academic Press, New York.

Proctor, M. C. F. 1979b. Surface wax on the leaves of some mosses. J. Bryol. 10: 531–538.

Q

Quarterman, E. 1956. Contributions to our knowledge of *Pleurochaete squarrosa* in North America. Bryologist 59: 180–182.

Quelle, F. 1906. *Barbula Fiorii*, ein Characktermoos mitteldeutscher Gipsberge. Hedwigia 45: 289–297.

R

Raeymaekers, G. & B. Kruijsen. 1980. *Tortula marginata* (B.S.G.) Spruce in Belgie. Dumortiera 17: 26–28.

Rahbar, K. & R. N. Chopra. 1980. Preliminary studies on callus induction in the moss: *Hyophila involuta*. Zeitschr. Pflanzenphysiol. 99: 199–205.

Ramsay, H. P. 1974. Cytological studies of Australian mosses. Austral. J. Bot. 22: 293–348.

Rashid, A. 1970 [1971]. Spore germination, protonema development and bud formation in *Anoectangium thomsonii*. Phytomorphology 20: 49–57.

Ratkowsky, D. A. 1980. Some further synonyms of *Bryoerythrophyllum jamesonii* (Tayl.) Crum. Bryologist 83: 367.

Redfearn, P. L., Jr. 1962. Notes on the bryophytes of Missouri. V. Distribution of *Tortula pagorum*. Castanea 27: 159–160.

Redfearn, P. L., Jr. 1969. Bryophytes of the Interior Highlands XIV. *Tortella mollissima* new to North America. Bryologist 72: 60.

Redfearn, P. L., Jr. 1976. A *Trichostomum* from Texas new to the United States. Bryologist 79: 80–82.

Redfearn, P. L., Jr. 1991. *Gymnostomiella* (Musci: Pottiaceae) in the Neotropics and eastern Asia. Bryologist 94: 392–395.

Reed, C. F. & H. E. Robinson. 1972. Index to Die Natürlichen Pflanzenfamilien (Musci-Hepaticae), Editions 1 and 2. Contributions of Reed Herbarium 21. Baltimore, Maryland.

Reese, W. D. 1956. *Didymodon fuscoviridis* new to the United States. Bryologist 59: 182–183.

Reese, W. D. 1959 [1960]. *Teniola*, a new genus of mosses. Bryologist 62: 221–224.

Reese, W. D. 1961. The genus *Calymperes* in the Americas. Bryologist 64: 89–140.

Reese, W. D. 1962. *Teniolophora*, a new name for the genus *Teniola*. Bryologist 65: 67.

Reese, W. D. 1967. The discovery of *Tortula vectensis* in North America. Bryologist 70: 112–114.

Reese, W. D. 1968. New moss records from the southeastern United States. Bryologist 71: 138–139.

Reese, W. D. 1987a. *Calymperes* (Musci: Calymperaceae): world ranges, implications for patterns of historical dispersion and speciation, and comments on phylogeny. Brittonia 39: 225–237.

Reese, W. D. 1987b. World ranges, implications for patterns of historical dispersal and speciation, and comments on phylogeny of *Syrrhopodon* (Calymperaceae). Mem. New York Bot. Gard. 45: 426–445.

Reese, W. D. 1988. Rhizoid gemmae in *Weissia controversa*. Bryologist 91: 184–185.

Reese, W. D. 1989. *Scopelophila cataractae* in Texas. Bryologist 92: 413.

Reese, W. D. 1990. A third record for *Tortella flavovirens* on the northern Gulf Coast. Evansia 3: 60.

Reese, W. D. 1991. *Weissia* (*Hymenostomum*) *flavescens* new to the United States. Bryologist 94: 179–180.

Reese, W. D. & W. R. Buck. 1991. *Weissiodicranum insularum gen. et sp. nov.* Bryologist 94: 308–310.

Reese, W. D., T. Koponen & D. H. Norris. 1986. Bryophyte flora of the Huon Peninsula, Papua New Guinea. XIX. *Calymperes*, *Syrrhopodon* and *Mitthyridium* (Calymperaceae, Musci). Acta Bot. Fenn. 133: 151–202.

Reese, W. D. & B. E. Lemmon. 1965. A natural hybrid between *Weissia* and *Astomum* and notes on the nomenclature of North American species of *Astomum*. Bryologist 68: 277–283.

Reese, W. D. & H. Mohamed. 1985. A synopsis of *Calymperes* (Musci: Calymperaceae) in Malaysia and adjacent regions. Bryologist 88: 98–109.

Reese, W. D. & R. H. Zander. 1987 [1988]. Serpotortellaceae, a new family of mosses (Musci) from Madagascar and Réunion. Bryologist 90: 230–236.

Reese, W. D. & R. H. Zander. 1988a. Comparison of Calymperaceae with Pottiaceae. Bryologist 91: 18–20.

Reese, W. D. & R. H. Zander. 1988b. Typification of *Serpotortella* Dixon (Musci). Bryologist 9: 136.

Reimers, H. 1941. *Tortula brevissima* Schiffn., ein neues verderasiatisches Wustensteppenmoos im Zechstein-Kyffhauser. Notizbl. Bot. Gart. Mus. Berlin-Dahlem 15: 402–405.

Rejment-Grochowska, I. 1978. Valeur taxonomique des caractères morphologiques des spores chez les Pottiacées. In C. Suire (ed.), Congr. Internat. Bryol. Bordeaux 1977, Compt. rend., pp. 509–534. Bryophyt. Biblioth. 13. Cramer, Vaduz.

Ripetskij, R. T. 1978. Regeneration of isolated protoplasts of the moss *Pottia intermedia* (Turn.) Fürnr. Ukrajins'k Bot. Žurn. 35: 467–469. [In Ukrainian, English summary.]

Ripetskij, R. T. 1979. Apogamy in *Pottia intermedia* from isolated protoplasts. Tsitol. Genet. 12: 347–350.

Ripetskij, R. T., I. Danilkiv & E. N. Lesnyak. 1983. Genetic divergence of morphologically indistinguishable cytotypes of the moss *Pottia lanceolata*. Tsitol Genet. 17(6): 49–55. [In Russian, English summary.]

Risse, S. 1985. *Pottia intermedia* (Turn.) Fürnr. with rhizoidal tubers. J. Bryol. 13: 523–526. 1 fig. 2 tab.

Risse, S. 1987. Beobachtungen zur Ontogenie und Morphologie der Rhizoidgemmen. I. Rhizoidgemmen von *Bryoerythrophyllum ferruginascens* (Stirt.) Giac. Cryptogamie: Bryol. Lichénol. 8: 353–357.

Risse, S. 1991. *Bryoerythrophyllum ferruginascens* (Stirt.) Giac. in Deutschland, mit Hinweisen zur Bestimmung der Art. Bryol. Rundbr. 5: 1–3.

Robinson, H. 1964. New taxa and new records of bryophytes from Mexico and Central America. Bryologist 67: 446–458.

Robinson, H. 1966. The status of the moss genus *Rechingerella*. Bryologist 69: 112–113.

Robinson, H. 1968. Notes on bryophytes from the Himalayas and Assam. Bryologist 71: 82–97.

Robinson, H. 1970. A revision of the moss genus *Trichostomopsis*. Phytologia 20: 184–191.

Robinson, H. 1971a. A revision of the moss genus *Hymenostyliella*, with description of the sporophyte. Phytologia 21: 1–3.

Robinson, H. 1971b. A revised classification for the orders and families of mosses. Phytologia 21: 289–293.

Robinson, H. 1974. Scanning electron microscope studies on moss leaves and peristomes. Bryologist 74: 473–483.

Robinson, H. 1985. Comments on the cladistic approach to the phylogeny of the "bryophytes" by Mishler and Churchill. Brittonia 37: 279–281.

Robinson, H. 1986. A key to the common errors of cladistics. Taxon 35: 309–311.

Rogeon, M. A. & R. Schumacker. 1984. *Leptodontium flexifolium* (With.) Hampe et *L. gemmascens* (Mitt. ex Hunt.) Braithw. sur les toits de chaume de la haute Adour (Hautes-Pyrenées, France). Bull. Soc. Bot. Centre-Ouest, N.S. 15: 81–102.

Rohrer, J. R. 1985. A phenetic and phylogenetic analysis of the Hylocomiaceae and Rhytidiaceae. J. Hattori Bot. Lab. 59: 185–240.

Rohrer, J. R. 1988. Incongruence between gametophytic and sporophytic classifications in mosses. Taxon 37: 838–845.

Roorda van Eysinga, P. 1972. *Barbula convoluta* Hedw. var. *commutata* (Jur.) Hsn. Lindbergia 1: 250–251.

Ros, R. M. & J. Guerra. 1986. *Crossidium aberrans* Holz. & Bartr. (Musci) novedad para la flora Europea. Cryptogamie Bryol. Lichénol. 7: 71–75.

Roth, G. 1911. Die aussereuropäischen Laubmoose. Band I. Dresden.

Rumsey, F. J. 1992. The status of *Tortula freibergii* in the British Isles. J. Bryol. 17: 371–372.

Rumsey, F. J. & M. E. Newton. 1989. *Scopelophila cataractae* (Mitt.) Broth. in North Wales. J. Bryol. 15: 519–524.

Rundel, P. W., W. Stichler, R. H. Zander & H. Ziegler. 1979. Carbon and hydrogen isotope ratios of bryophytes from arid and humid regions. Oecologia (Berlin) 44: 91–94.

Rungby, S. 1957. On the variation in *Tortula subulata* Hedw. Bot. Not. 110: 232–236.

Rungby, S. 1958. *Leptodontium flexifolium* (Smith) Hampe, var. *gemmiferum* (Schpr.) *n. comb.* found in Denmark. Bot. Notis. 3:

477.

Rungby, S. 1967. *Barbula rigidula* (Hedw.) Mitt. in Denmark. Bot. Not. 120: 236.

Rushing, A. E. 1989. *Bruchia brevifolia*. Bryologist 92: 260.

Ryan, E. 1901. *Didymodon glaucus*. Rev. Bryol. 28: 39–40.

S

Sainsbury, G. O. K. 1936. A note on the anatomy of the nerve in *Tortula atrovirens*. Rev. Bryol. Lichénol. 63: 145–146.

Sainsbury, G. O. K. 1940. *Barbula rostrata* R. Br. ter., an invalid species. Bryologist 43: 113–114.

Sainsbury, G. O. K. 1948. Bryobartramiaceae, a new moss family. Bryologist 51: 9–13.

Sainsbury, G. O. K. 1950. New species of *Tortula* from subantarctic islands of New Zealand. Svensk Bot. Tidskr. 44: 72–75.

Sainsbury, G. O. K. 1955. A handbook of the New Zealand mosses. Bull. Roy. Soc. New Zealand 5: 1–490.

Sainsbury, G. O. K. 1956. A new species of *Pottia* from central Australia. Rev. Bryol. 25: 237–238.

Saito, K. 1971a. Notes on four species of the genus *Barbula*. J. Jap. Bot. 46: 55–59. [In Japanese with English summary.]

Saito, K. 1971b. Notes on Pottiaceae (1). J. Jap. Bot. 46: 139–145. [In Japanese with English summary.]

Saito, K. 1972a. Gemmae formation on rhizoids of *Eucladium verticillatum*. Misc. Bryol. Lichénol. 6: 41–42. [In Japanese.]

Saito, K. 1972b. Taxonomic and phytogeographic studies of specimens previously reported as *Molendoa sendtneriana*. J. Hattori Bot. Lab. 36: 163–170.

Saito, K. 1972c. Notes on Pottiaceae (2). J. Jap. Bot. 47: 10–19. [In Japanese with English summary.]

Saito, K. 1973a. Memoir of the Japanese Pottiaceae (I). Subfamily Pottioideae. Bull. Nat. Sci. Mus. Tokyo 16: 61–91.

Saito, K. 1973b. Pottiaceae in cities of Japan. Proc. Bryol. Soc. Japan 1: 44–45. [In Japanese.]

Saito, K. 1973c. Notes on the Pottiaceae (3). J. Jap. Bot. 48: 161–167.

Saito, K. 1973d. Taxonomic position of the genus *Neohyophila* Crum. Misc. Bryol. Lichenol. 6: 80–81. [In Japanese.]

Saito, K. 1974a. Anatomical study of the capsule in the Pottiaceae. Misc. Bryol. Lichenol. 6: 143–144. [In Japanese.]

Saito, K. 1974b. [New localities for *Barbula sublaevifolia*.] Proc. Bryol. Soc. Japan 1: 123–124.

Saito, K. 1975a. A monograph of Japanese Pottiaceae (Musci). J. Hattori Bot. Lab. 39: 373–537.

Saito, K. 1975b. Notes on Japanese Pottiaceae. Proc. Bryol. Soc. Japan 1: 123–124. [In Japanese.]

Saito, K. & T. Hirohama. 1974a. A comparative study of the spores of taxa in the Pottiaceae by use of the scanning electron microscope. J. Hattori Bot. Lab. 38: 475–488.

Saito, K. & T. Hirohama. 1974b. Observation on the spores of the species in Pottiaceae. Proc. Jap. Soc. Pl. Tax. 3(3): 11–12. [In Japanese.]

Saito, K. & T. Nakajima. 1974. New localities of *Barbula sublaevifolia*. Proc. Bryol. Soc. Japan 1: 70. [In Japanese.]

Sakurai, K. 1954. Studies of the genus *Hyophila* in Japan. Bot. Mag. 67: 787–788.

Salmon, E. S. 1903. A monograph of the genus *Streptopogon*. Ann. Bot. 17: 107–150, pl. VIII–X.

Sanderson, M. J. 1990. Flexible phylogeny reconstruction: a review of phylogenetic inference packages using parsimony. Syst. Zool. 39: 414–420.

Satake, K. 1990. A "copper moss" *Scopelophila cataractae* and copper (1). Distribution of *Scopelophila cataractae* in the world. Proc. Bryol. Soc. Japan 5: 49–53.

Satake, K. 1991. A "copper moss" *Scopelophila cataractae* and copper (2). Accumulation of copper. Proc. Bryol. Soc. Japan 5(7): 105–107.

Satake, K., K. Shibata, M. Nishikawa & K. Fuwa. 1988. Copper accumulation and location in the moss *Scopelophila cataractae*. J. Bryol. 15: 353–376.

Satake, K., M. Nishikawa & K. Shibata. 1990. A copper-rich protonemal colony of the moss *Scopelophila cataractae*. J. Bryol. 16: 109–116.

Savicz, L. 1926. *Phascum curvicollum* Ehrh. in Russland. Not. Syst. Inst. Cryptog. Horti Bot. Petropolit. 4: 3.

Savicz-Ljubitzkaja, L. I. 1955. *Trichostomum cuspidatissimum* Card. et Thér. aus dem Ostssajangebirge in Sibirien. Mitt. Thür. Bot. Gessell. 1: 199–207.

Savicz-Ljubitzkaja, L. I. 1961. Notula de *Barbula gigantea* Funck et *Trichostomo cuspidatissimo* Card. et Thér. Not. Syst. Sect. Crypt. Inst. Bot. Nom. V. L. Komarovii Ac. Sci. USSR 14: 262–267.

Savicz-Ljubitzkaja, L. I. 1964. *Barbula johansenii* Williams—species nova ad bryofloram URSS. (2 pls.) Nov. Syst. Plant. non Vasc., Acad. Sci. URSS, Inst. Bot. nom. V. L. Komarovii 1964: 286–292.

Savicz-Ljubitzkaja, L. I. 1965a. De nomenclatura *Tortulae scabrinervis* Lazar. ex Asia media. Nov. Syst. Plant. non Vasc., Acad. Sci. URSS, Inst. Bot. nom. V. L. Komarovii 1965: 251.

Savicz-Ljubitzkaja, L. I. 1965b. Supplementa nova de *Barbula perobtusa* (Broth.) Chen. Nov. Syst. Plant. non Vasc., Acad. Sci. URSS, Inst. Bot. nom. V. L. Komarovii 1965: 245–250.

Savicz-Ljubitzkaja, L. I. 1969a [1970]. De *Lydiaea Vlassovii* (Lazar.) Lazar. notula. Nov. Syst. Plant. non Vasc., Acad. Sci. URSS, Inst. Bot. nom. V. L. Komarovii 6: 245–247.

Savicz-Ljubitzkaja, L. I. & Z. N. Smirnova. 1961. On the modes of reproduction of *Sarconeurum glaciale* (Hook. f. et Wils.) Card. et Bryhn, an endemic moss of Antarctica. Rev. Bryol. Lichénol. 30: 216–222.

Savicz-Ljubitzkaja, L. I. & Z. N. Smirnova. 1963a. A contribution to the biology and geography of *Bryoerythrophyllum recurvirostre* (Hedw.) Chen. A new species in the bryoflora of the Antarctica. J. Bot. Acad. Nauk URSS 48: 350–361.

Savicz-Ljubitzkaja, L. I. & Z. N. Smirnova. 1963b. Bericht über die Gattung *Pottia* Fuern. in der Antarktis. Bot. Mat. OTD Spor. Rast. Inst. Bot. V. L. Komarov Akad. Nauk SSSR 16: 188–194 [In Russian.]

Savicz-Ljubitzkaja, L. I. & Z. N. Smirnova. 1965. De speciebus *Pottiae* Fuernr. in Antarctide [sic] inventis supplementa. Nov. Sys. Pl. non Vasc. Acad. Aci. URSS, Inst. Bot. nom. V. L. Komarovii 1965: 258–259.

Savicz-Ljubitzkaja, L. I. & Z. N. Smirnova. 1970. [Handbook of the Mosses of the U.S.S.R. The acrocarpous mosses.] Acad. Sci. USSR, Komarov Botanical Institute, Leningrad. [In Russian.]

Saxena, P. K. & R. Gill. 1986. Callus formation and plant regeneration from protonemal protoplasts of the moss *Anoectangium thomsonii* Mitt. Biol. Plantarum 28: 310–312.

Saxena, P. K. & A. Rashid. 1980. Development of gametophores from isolated protoplasts of moss *Anoectangium thomsonii* Mitt. Protoplasma 103: 401–406.

Sayre, G. 1941. A gum arabic mounting medium. Bryologist 44: 160.

Sayre, G. 1952. Key to the species of *Grimmia* in North America. Bryologist 55: 251–259.

Sayre, G., C. E. B. Bonner & W. L. Culberson. 1964. The authorities for the epithets of mosses, hepatics and lichens. Bryologist

67: 113–135.

Schatz, A. 1955. Speculations on the ecology and photosynthesis of the "copper mosses." Bryologist 58: 113–120.

Schiffner, V. 1893. *Tortula velenovskyi*, eine neue Art der Gattung *Tortula* aus Böhmen. Nov. Act. K. Leopl.-Carol. Deutsch. Akad. Naturf. 58: 481–488, pl. 17.

Schiffner, V. & J. Baumgartner. 1906. Über zwei neue Laubmoosarten aus Österreich. *Cinclidotus danubicus* Schiffn. et Baumg. *Didymodon austriacus* Schffn. et Baumg. Österr. Bot. Zeitschr. 56: 154–158.

Schlechtendal, D. de. 1836. *Gymnostomum micaceum* Schldl. species nova Brasiliensis. Linnaea 10: 443.

Schliephacke, K. 1884. *Pottia Gussfeldtii*, ein neues Laubmoos. Ber. Deutsch. Bot. Ges. 2: 461.

Schnooberger, I. 1942. Distribution of *Tortula papillosa* Wils. Bryologist 45: 24–27.

Schofield, W. B. 1966. *Crumia*, a new genus of the Pottiaceae endemic to western North America. Can. J. Bot. 44: 609–614.

Schonbeck, M. W. & J. D. Bewley. 1981 [1982]. Responses of the moss *Tortula ruralis* to desiccation treatments. I. Effects of minimum water content and rates of dehydration and rehydration. Canad. J. Bot. 59: 2698–2706.

Schonbeck, M. W. & J. D. Bewley. 1981 [1982]. Responses of the moss *Tortula ruralis* to desiccation treatments. II. Variations in desiccation tolerance. Canad. J. Bot. 59: 2707–2712.

Schornherst, R. O. 1944. *Gymnostomiella Orcuttii* Bartram in Florida. Bryologist 47: 41.

Schreber, J. C. 1770. De Phasco observationes quibus hoc genus muscorum vindicatur atque illustratur. Lipsiae.

Schumacker, R. & M. Brugués. 1991. *Scopelophila cataractae* (Mitt.) Broth. (Pottiaceae, Bryophytina), new for Spain. J. Bryol. 16: 486–488.

Schumacker, R. & P. de Zuttere. 1981. *Leptodontium flexifolium* (With.) Hampe (Pottiaceae, Musci) en Belgique. Étude chrologique, écologique et phytosociologique. Herzogia 5: 589–607.

Schumacker, R., C. C. Pedrotii, A. von Hübschmann, R. Düll & G. Schwab. 1989. *Scopelophila ligulata* (Spruce) Spruce (Pottiaceae, Bryophytina), nouveau pour l'Italie. Webbia 43: 139–144.

Scott, G. A. M. 1982. Desert bryophytes. In A. J. E. Smith (ed.), Bryophyte Ecology, pp. 105–122. Chapman and Hall, London.

Scott, G. A. M. & Ilma G. Stone. 1976. The Mosses of Southern Australia. Academic Press, New York.

Sehlmeyer, D. 1820a. Bemerkungen über *Phascum cuspidatum*. Flora 3(6): 83–85.

Sehlmeyer, D. 1820b. *Weissia fallax*, eine neue Laubmoose Art. Flora 3(41): 639–642.

Seki, T. & C. Miyagi. 1980. An ecological observation of *Gymnostomiella linginervis* Broth. in Okinawa Island. Hikobia 8: 416–423.

Seppelt, R. D. & P. M. Selkirk. 1984. Effects of submersion on morphology and the implications of induced environmental modification on the taxonomic interpretation of selected Antarctic moss species. J. Hattori Bot. Lab. 55: 273–279.

Sérgio, C. 1972a. Os Géneros *Aschisma*, *Acaulon* e *Phascum* (Musci—Pottiaceae) em Portugal. Boletim da Sociedad Broteriana 46: 457–463, est. I-III.

Sérgio, C. 1972b. Um novo musgo para a flora portuguesa *Gyroweisia luisieri sp. nov.* Bol. Soc. Port. Ciênc. Nat. 14: 81–85.

Sérgio, C. 1978. *Tortula guepinii* (B. S. G.) Limpr. dans la flore Portugaise et son intérêt phytogeographique. Bol. Soc. Broteriana, ser. 2, 52: 249–255.

Sérgio, C. 1984. Estudo taxonómico, ecológico e corológico de *Gymnostomum luisieri* (Sérgio) Sérgio ex Crundw. na península Ibérica. Anales Biol. (Univ. Murcia) 2 (Secc. Espec. 2) 1984: 357–366.

Sérgio, C. 1985. Notulae bryoflorae macaronesicae I. 2. Acerca da identidade de *Hyophila contorta* (Kunze) Jaeg. Pottiaceae da Ilha da Madiera. Portug. Acta Biol. (B) 14: 168–172.

Sérgio, C. 1992. Re-evaluation of *Acaulon piligerum* (De Not.) Limpr. (Pottiaceae) from Sardinia, as a species of a subgenus new to Europe (*Alaticosta* Stone). Cryptogamie Bryol. Lichénol. 13: 221–226.

Sérgio, C. & A. Cardoso. 1985. Notulae bryoflorae lusitanicae I. 6. Novas localidades de *Tortula solmsii* (Schgimp.) Limpr. em Portugal. Portug. Acta Biol. (B) 14: 194–196.

Sérgio, C. & C. Cassas. 1981. *Tortella flavovirens* (Bruch) Broth. var. *papillossima* C. Sérgio & Casas, C. Portug. Acta Biol. 13: 114–118.

Sérgio, C. & M. Sim-Sim. 1984. *Dialytrichia mucronata* (Brid.) Broth. au Portugal et a Madère, taxonomie, écologie, adaptation à la sécheresse. Cryptogamie Bryol. Lichénol. 5: 87–98.

Sharma, P. & R. N. Chopra. 1986. *In vitro* production and behaviour of protonemal gemmae in *Hyophila involuta* (Hook.) Jaeg. J. Hattori Bot. Lab. 60: 137–141.

Sharma, S. D. & R. N. Chopra. 1987. Effect of lead nitrate and lead acetate on growth of the moss *Semibarbula orientalis* (Web.) Wijk & Marg. grown *in vitro*. J. Pl. Physiol. 129: 243–249.

Sharma, A. P. & P. D. Sharma. 1986. Life history of *Semibarbula orientalis* (Web.) Wijk. et Marg. In Kumar, S. S., M. L. Sharma & Sunita Anand (eds.). All India Conference on Bryology, February 25–27, 1986, pp. 38–39. Abstracts of Contributed Papers. Panjab University, Chandigarh, India.

Sharma, P. D. & A. P. Sharma. 1980. Gemmae of *Semibarbula orientalis*. In Symposium on Current Researches in Plant Sciences, April 22–24, Dept. of Botany, Panjab University, Chandigarh, India.

Sharp, A. J. 1955. Factors in the distribution of *Hyophila tortula* and an extension of its known range to include Michigan. Mitt. Thür. Bot. Ges. 1: 222–224.

Sharp, A. J. 1977. The preparation of a manual of Mexican mosses. Taxon 26: 151–153.

Sharp, A. J. 1986. *Streptopogon juarezii* n. sp. and *Trematodon norrisii* n. sp. Phytologia 61: 372.

Sharp, A. J., H. Crum & P. M. Eckel, eds. 1993. Moss Flora of Mexico. Mem. New York Bot. Gard. 69. 2 vols.

Sharp, A. J. & Z. Iwatsuki. 1969. *Crumia deciduidentata*, a new species in Alaska. J. Hattori Bot. Lab. 32: 95–98.

Shaw, J. 1986. Heavy metal "tolerance" in *Scopelophila cataractae* (Mitt.) Broth. [Abstr.] Am. J. Bot. 73: 611.

Shaw, J. 1987a. Experimental taxonomy of *Weissia controversa* and *W. sharpii* (Musci: Pottiaceae). Syst. Bot. 12: 381–389.

Shaw, J. 1987b. Evolution of heavy metal tolerance in bryophytes II. An ecological and experimental investigation of the "Copper Moss," *Scopelophila cataractae* (Pottiaceae). Am. J. Bot. 74: 813–821.

Shaw, J. & L. E. Anderson. 1988. Factors affecting the distribution and abundance of the "copper moss," *Scopelophila ligulata*, in North America. Lindbergia 14: 55–58.

Shaw, J., L. E. Anderson & B. D. Mishler. 1987. Peristome development in mosses in relation to systematics and evolution. I. *Diphyscium foliosum* (Buxbaumiaceae). Mem. New York Bot. Gard. 45: 55–70.

Shaw, J. & S. C. Beer. 1989. *Scopelophila cataractae* (Mitt.) Broth. in Pennsylvania. Bryologist 92: 112–115.

Shaw, J., B. D. Mishler & L. E. Anderson. 1989. Peristome development in mosses in relation to systematics and evolution. IV. Haplolepideae: Ditrichaceae and Dicranaceae. Bryologist 92: 314–325.

Side, A. G. 1983. The occurrence of tubers on *Barbula tophacea* (Brid.) Mitt. J. Bryol. 12: 620–621.

Side, A. G. & H. L. K. Whitehouse. 1974. *Tortula amplexa* (Lesq.) Steere in Britain. J. Bryol. 8: 15–18.

Simon, T. 1987. The leaf-area index of three moss species (*Tortula ruralis*, *Ceratodon purpureus* and *Hypnum cupressiforme*). Symp. Biol. Hung. 35: 699–706.

Singh, J., B. Blackwell, R. Miller & D. Bewley. 1984. Membrane organization of the desiccation tolerant moss *Tortula ruralis* in dehydrated states. Pl. Physiol. 75: 1075–1079.

Skeist, I. 1962. Handbook of Adhesives. Reinhold: New York.

Sloover, J. L. de. 1977. Note de bryologie africaine. IX. *Andreaea*, *Racomitrium*, *Gymnostomiella*, *Thuidium*. Bull. Jard. Bot. Natl. Belgique 47: 155–181.

Sloover, J. L. de. 1980. *Acaulon muticum* (Hedw.) C. Müll., *Funaria fascicularis* (Hedw.) Lindb. et *Pottia starckeana* (Hedw.) C. Müll. à Namur (Belgique). Dumortiera 16: 23–24.

Sloover, J. L. de. 1987. Note de bryologie africaine. XIV. *Leptodontium*. Bull. Jard. Bot. Natl. Belgique 57(3/4): 425–451.

Smith, A. J. E. 1978. The Moss Flora of Britain and Ireland. Cambridge Univ. Press, Cambridge, England.

Smith, A. J. E. (ed.) 1982. Bryophyte Ecology. Chapman and Hall, London, New York.

Smith, A. J. E. 1987. On the identity of the "plant from Roscommon." J. Bryol. 14: 594–595.

Smith, A. J. E. & H. L. K. Whitehouse. 1974. The sporophyte and male plants of *Tortula stanfordensis* and the taxonomic position of this and *T. khartoumensis* Pettet and *T. rhizophylla* (Sak.) Iwats. & Saito. J. Bryol. 8: 9–14.

Smith, D. K. & K. D. McFarland. 1977. The discovery of *Tortula fragilis* Tayl. (Musci) in Tennessee. J. Tenn. Acad. Sci. 52: 5–76.

Smith, G. L. 1971. Conspectus of the genera of Polytrichaceae. Mem. New York Bot. Gard. 21(3): 1–83.

Smith Merrill, G. L. 1992. *Ozobryum ogalalense* (Pottiaceae), a new moss genus and species from the American Great Plain. Novon 2: 255–258.

Smith, R. I. L. 1985. A unique community of pioneer mosses dominated by *Pterygoneurum* cf. *ovatum* in the Antarctic. J. Bryol. 13: 509–514.

Smith, T. F. 1914. A fungus-infested moss (*Tortula muralis*). Knowledge & Sci. News London 37: 225–226., fig. 210–216.

Sögård, P. 1984. *Tortula latifolia* (Bush) Hartm. i Sverige. Moss. Vänner 22: 2–5.

Sollman, P. 1979 [1980]. *Tortula rhizophylla* (Sak.) Iwats. and Saito in Italy. Lindbergia 5: 109–110.

Sollman, P. 1983. Notes on pottiaceous mosses I. Bryologist 86: 271–272.

Sollman, P. 1984. Notes on pottiaceous mosses II. Lindbergia 10: 53–56.

Sollman, P. 1990. Some new synonyms in tropical pottiaceous mosses. Lindbergia 16: 22–24.

Sotiaux, A., P. de Zuttere, R. Schumacker, R. B. Pierrot & C. Ulrich. 1987. *Scopelophila cataractae* (Mitt.) Broth. (Pottiaceae, Musci) nouveau pour le continent européen en France, en Belgique, aux Pays-Bas et en République fédéral allemande: le genre *Scopelophila* (Mitt.) Lindb. en Europe. Cryptogamie Bryol. Lichénol. 8: 95–108.

Sowter, F. A. 1949. *Tortula subulata* Hedw. Trans. Brit. Bryol. Soc. 1: 211.

Söyrinki, N. 1987. Turrisammal (*Oxystegus tenuirostris*), lisäys Ruoveden sammalflooraan—*Oxystegus tenuirostris* (Hook. & Tayl.) A. J. E. Sm. at Ruovesi, South Häme. Lutukka 1(3): 15–16.

Springer, E. 1935. Über apogame (vegetativ entstandene) Sporogone an der bivalenten Rasse des Laubmooses *Phascum cuspidatum*. Zeitschr. Indukt. Abst. Vererb. 69: 249–262, fig. 1–25.

Squivet de Carondelet, J. 1959. Une nouvelle localité de *Trichostomum pallidisetum* K. Müll. Rev. Bryol. 28: 235.

Squivet de Carondelet, J. 1961a [1962]. *Tortula laevipila* forma *propagulifera* ratio nova et *Tortula gemifera* var. *nova* (adhuc innominata). Rev. Bryol. Lichénol. 30: 213–215.

Squivet de Carondelet, J. 1961b. *Tortula laevipila* forma *propagulifera* ratio nova et *Tortula gemmifera* var. nova (adhuc innominata). Rev. Bryol. 30; 213–215.

Stark, L. R. 1980. *Triquetrella* in North America. Bryologist 83: 363–364.

Stark, L. R. 1983. Bisexuality as an adaptation in desert mosses. Am. Midland Nat. 110: 445–448.

Stark, L. R. & R. C. Castetter. 1987. A gradient analysis of bryophyte populations in a desert mountain range. Mem. New York Bot. Gard. 45: 186–197.

Steere, W. C. 1938a. *Barbula*. In A. J. Grout, Moss Flora of North America North of Mexico. I(3): 173–185. Newfane, Vermont.

Steere, W. C. 1938b. *Didymodon*. In A. J. Grout, Moss Flora of North America North of Mexico. I(3): 185–190. Newfane, Vermont

Steere, W. C. 1939a. *Tortula*. In A. J. Grout, Moss Flora of North America North of Mexico. I(4): 228–246. Newfane, Vermont.

Steere, W. C. 1939b. *Barbula* in North America north of Mexico. Bull. Torrey Bot. Club 66: 93–119.

Steere, W. C. 1939c. *Gyroweisia tenuis* in North America. Bryologist 42: 16–23.

Steere, W. C. 1940. *Tortula* in North America north of Mexico. Bryologist 43: 12–23, 45–56, 76–86, 98–109.

Steere, W. C. 1945. *Luisierella*, a genus of mosses new to North America. Bryologist 48: 83–85.

Steere, W. C. 1947. A consideration of the concept of genus in Musci. Bryologist 50: 247–258.

Steere, W. C. 1950. On the distribution of *Aloina brevirostris* in North America. Bull. Torrey Bot. Club 77: 503–508.

Steere, W. C. 1951. *Tortula stanfordensis*, a new species from California. Bryologist 54: 119–123.

Steere, W. C. 1952. Bryophyta of Arctic America. V. The rediscovery of *Barbula Johansenii*. Bryologist 55: 259–260.

Steere, W. C. 1960. *Pterygoneurum arcticum*, a new species from northern Alaska. Bryologist 62: 215–221.

Steere, W. C. 1976. Identity of *Pterygoneurum arcticum* with *P. lamellatum*. Bryologist 79: 221–222.

Steere, W. C. 1986. *Trachyodontium*, a new genus of the Pottiaceae (Musci) from Ecuador. Bryologist 89: 17–19.

Steere, W. C. & D. E. Chapman. 1946. Mosses of El Salvador. J. Washington Acad. Sci. 36: 219–225.

Steere, W. C. & Z. Iwatsuki. 1974. The discovery of *Pterygoneurum subsessile* (Brid.) Jur. in arctic Alaska. J. Hattori Bot. Lab. 38: 463–473.

Stern, R. C. 1989. *Didymodon nicholsonii* Culm. new to France. Cyptogamie Bryol. Lichénol. 10: 171–172.

Stern, R. C. 1991. *Tortula freibergii* and *Eriopus apiculatus* in East Sussex. J. Bryol. 16: 488–489.

Stern, R. C. 1990. The "secret garden" and *Tortula brevis* at Brighton. Bull. Brit. Bryol. Soc. 55(March): 35.

Stirling, A. M. 1968. Distribution maps of bryophytes in Britain.

Gymnostomum calcareum Nees & Hornsch.; *Splachnum vasculosum* Hedw.; *Amblystegiella sprucei* (Bruch) Loeske. Trans. Brit. Bryol. Soc. 5: 595, 597, 600.

Stone, I. G. 1971. The sporophyte of *Tortula pagorum* (Milde) De Not. Trans. Brit. Bryol. Soc. 6: 270–277.

Stone, I. G. 1975. *Trachycarpidium* in Queensland, Australia. Muelleria 3(2): 122–129.

Stone, I. G. 1976a. A remarkable new moss from Queensland, Australia—*Viridivellus pulchellum* new genus and species (new family Viridivelleraceae). J. Bryol. 9: 21–31.

Stone, I. G. 1976b. *Alaticosta*, a new subgenus of *Acaulon*, in Australia. J. Bryol. 9: 213–227.

Stone, I. G. 1977a. Some morphological and anatomical features of the monotypic genus *Bryobartramia* Sainsbury (Musci). Austral. J. Bot. 25: 141–157.

Stone, I. G. 1977b. *Acaulon* subgen. *Alaticosta* validated. J. Bryol. 9: 573.

Stone, I. G. 1978. *Tortula oleaginosa*, a new moss from Australia. J. Bryol. 10: 117–124.

Stone, I. G. 1979. *Acaulon eremicola*, a new moss from the Australian arid zone. J. Bryol. 10: 467–474.

Stone, I. G. 1980a. *Phascopsis rubicundis*, a new genus and species of Pottiaceae from Australia. J. Bryol. 11: 17–31.

Stone, I. G. 1980b [1981]. *Weissia* subgenus *Astomum* in Australia and some comments on the affinities of *Viridivellus*. J. Bryol. 11: 231–243.

Stone, I. G. 1982. The combination of *Pottia willisiana* Sainsb. in *Weissia*. J. Bryol. 12: 116.

Stone, I. G. 1984. *Uleobryum curtisii sp. nov.* (Pottiaceae) from Queensland, Australia. J. Bryol. 13: 19–24.

Stone, I. G. 1985. New records of mosses in Australia. J. Bryol. 13: 475–458.

Stone, I. G. 1986. *Calymperastrum*, a new genus of Pottiaceae. J. Bryol. 14: 315–318.

Stone, I. G. 1988. *Acaulon granulosum*, a new species in the *Acaulon muticum* complex: a comparison and key to Australian species. J. Bryol. 15: 257–268.

Stone, I. G. 1989. Revision of *Phascum* and *Acaulon* in Australia. J. Bryol. 15: 745–777.

Stone, I. G. 1991. Nomenclatural changes and new moss record in Australia: including a description of the protonema of *Calomnion*. J. Bryol. 16: 261–273.

Stone, I. G. & G. A. M. Scott. 1981 [1982]. *Leptodontium paradoxum*, a new moss from Australia. J. Bryol. 11: 701–707.

Stoneburner, A. H. 1981. Variation and taxonomy of *Weissia* (Musci: Pottiaceae) in the southwestern United States. Dissert., Duke University.

Stoneburner, A. 1985 [1986]. Variation and taxonomy of *Weissia* in the southwestern United States. II. Taxonomic treatment. Bryologist 88: 293–314.

Stoneburner, A. 1986. Phyletic relationships among North American species of *Weissia* (Pottiaceae). Am. J. Bot. 73: 612. [Abstr.]

Stoneburner, A. & R. Wyatt. 1985. Variation and taxonomy of *Weissia* in the Southwestern United States. I. Biometrical Analysis. Monogr. Syst. Bot. Missouri Bot. Gard. 11: 175–185.

Størmer, P. 1968. *Tortula laevipila* (Brid.) Schwaegr. new to Norway. Saertrykk av Blyttia 26: 1–7.

Størmer, P. 1978. *Desmatodon obtusifolius* in Norway. Norweg. J. Bot. 25: 297–303.

Streimann, H. & J. Curnow. 1989. Catalogue of Mosses of Australia and its External Territories. Australian Flora and Fauna Series 10. Australian Government Publ. Serv., Canberra.

Studlar, S. M., J. D. Caponetti & A. J. Sharp. 1984. Morphology of the urban moss, *Tortula pagorum* in sterile culture. J. Hattori Bot. Soc. 56: 351–368.

Swofford, D. L. 1985. PAUP, Phylogenetic Analysis Using Parsimony, Version 2.4. Illinois Nat. Hist. Mus., Champaign, Illinois.

Swinscow, T. D. V. 1984. *Tortula solmsii* (Schimp.) Limpr. In South Devon. J. Bryol. 13: 131.

Syed, H. & A. C. Crundwell. 1973. *Barbula maxima*, nom. nov., an endemic Irish species. J. Bryol. 7: 527–529.

Synnott, D. 1983a. *Barbula icmadophila* Schimp. ex C. Muell.—a delayed addition to Irish moss flora. Irish Naturalists' J. 21(4): 177–178.

Synnott, D. 1983b. *Anoectangium warburgii*—a new Irish moss. Irish Naturalists' J. 21: 91–92.

Synnott, D. & D. W. Robinson. 1989. The moss *Trichostomopsis umbrosa* (C. Müll.) Robins. in Ireland. Irish Naturalists' J. 23: 113–114.

Synnott, D. & D. W. Robinson. 1990. The moss *Trichostomopsis umbrosa* (C. Müll.) Robinson in Ireland. Glasra 1: 15–19.

T

Takenaka, A. & K. Satake. 1991. Wave-like protonemal colonies of the moss *Scopelophila cataractae* and their consequences in light capture. J. Bryol. 16: 429–436.

Takio, S. 1989. Effect of cytokinin on morphological changes of suspension-cultured cells of the moss *Barbula unguiculata*. Pl. Cell. Rep. 7: 603–606.

Takio, S., M. Kajita, S. Takami & S. Hino. 1986. Establishment and growth characterization of suspension cultures of cells from *Barbula unguiculata*. J. Hattori Bot. Lab. 60: 407–417.

Tan, B. C., R. H. Zander & T. Taylor. 1981. *Pseudocrossidium hornschuchianum* and *P. revolutum* var. *obtusulum* in the New World. Lindbergia 7: 39–42.

Taris, H. G. M. 1964. *Leptodontium flexifolium* in Soest. Buxbaumia 18: 30–31.

Taylor, E. C. 1957. Freehand sectioning of moss leaves and stems. Bryologist 60: 17–20.

Taylor, E. C. 1961. An extension of range of *Barbula agraria*. Bryologist 64: 65–66.

Taylor, E. C., Sr. 1962. The Philibert peristome articles. An abridged translantion. Bryologist 65: 175–212.

Thériot, I. 1907. *Weisia brasiliensis* Duby. Genève Bull. Boissier (Sér. 2) 7: 277–278, 1 pl.

Thériot, I. 1923. *Gyroweisia* ou *Weisiodon*. Rev. Bryol. 50: 4–6.

Thériot, I. 1931. Mexican mosses collected by Broth Arsène Brouard III. Smithson. Misc. Coll. 85(4): 1–55.Thyssen, P. & J. Poelt. 1958. Ein neuer Fundort des Kupfermooses *Merceya ligulata* (Spr.) Schpr. in den Östalpen. Ber. Bayer. Bot. Gesell. 32: 152.

Timme, S. L. & F. D. Bowers. 1988. *Pleurochaete squarrosa* (Brid.) Lindb. new to Mississippi, and extensions of its range in Alabama. Bryologist 91: 368–369.

Toth, Z. 1987. A phytogeographic review of *Tortula* Hedw. (Pottiaceae, Musci) in Hungary. Acta Bot. Hung. 33: 249–278.

Touton, J. B. 1955. À propos de *Fissidens Warnstorfii* Fleisch. et de l'*Eucladium verticillatum* (L.) Bryol. Eur. Rev. Bryol. Lichénol. 24: 145.

Touw, A. 1963. *Hymenostomum rostellatum* (Brid.) Schimp. een nieuwe soort voor Nederland. Buxbaumia 17: 105.

Touw, A. 1992. A survey of the mosses of the Lesser Sunda Islands (Nusa Tenggara), Indonesia. J. Hattori Bot. Lab. 71: 289–366.

Townsend, C. C. 1979. A new combination in African *Tortula*. J. Bryol. 10: 576.

Townsend, C. C. 1980. [Review of:] Kramer, W., *Tortula* Hedw. sect. *Rurales* De Not. (Pottiaceae, Musci) in der östlichen Holarktis.—Verlagshous J. Cramer, Branschweig (BRD), und Verlagsgeselschaft A. R. Gantner, Vaduz (FL). 1980. Excerpta Bot., Sect. A, Taxon. Chorol. 35: 230–231.

Townsend, C. C. 1983. Notes on mosses from Ceylon and India, 7. Some species of *Trichostomum*. J. Bryol. 12: 559–566.

Townsend, C. C. 1987. *Trichostomopsis aaronis* (Lor.) Agnew & Townsend new to the U.S.S.R. J. Bryology 14: 793–794.

Townsend, C. C. & H. L. K. Whitehouse. 1979. *Desmatodon bogosicus* C. Müll. hal., an earlier name for *Tortula khartoumensis* Pettet. J. Bryol. 10: 475–478.

Tuba, Z. 1984. Changes in the photosynthetic pigment system of the drought tolerant *Tortula ruralis* during a daily desiccation. Proc. Third Meeting Bryologists from Centr. and East Eur. Univ. Karlova, Praha, 343–352.

Tuba, Z. 1985. Photosynthetic pigment responses in *Tortula ruralis* during daily desiccation. Abstr. Bot. 9, Suppl. 2: 231–239.

Tucker, E. B., J. W. Costerton & J. D. Bewley. 1975. The ultrastructure of the moss *Tortula ruralis* on recovery from desiccation. Canad. J. Bot. 53(2): 94–101.

U

Ulycna, K. L. 1977. Hybrid sporogones in *Phascum cuspidatum* Hedw. (Pottiaceae, Musci). Ukrajins'k. Bot. Žurn. 34: 155–158. [In Russian, Engish summary.]

V

Vaarama, A. 1939. *Desmatodon latifolius* (Hedw.) Bryol. Eur. in Laatokka-Karelien. Ann. Bot. Soc. Zool. Bot. Fenn. Vanamo 11(3): 44–50.

Vadam, J.-C. 1991. Présence de *Bryoerythrophyllum ferruginascens* (Stirt.) Giac. en Haute-Savoie. Cryptogamie Bryol. Lichénol. 12: 73–75.

Van der Pijl, L. 1972. Principles of Dispersal in Higher Plants. 2nd Ed. Springer-Verlag, New York.

Van der Wijk, R., W. D. Margadant & P. A. Florschütz. 1959–1969. Index Muscorum. Reg. Veg. Vols. 17, 26, 33, 48, 65.

Vaněk, R. 1952. *Pterygoneurum smardaeanum* Vaněk, n. sp. Preslia 24: 211–215.

Vashistha, B. D. & R. N. Chopra. 1984. Production of protonemal gemmae in culture by the moss *Didymodon recurvus* (Griff.) Broth. New Phytol. 97: 83–86.

Venturi, G. 1882. *Barbulae rurales*. Rev. Bryol. 9: 85–87.

Venturi, G. 1883. De la *Pottia latifolia*. Rev. Bryol. 10: 95–96.

Venturi, G. 1885. Notes sur le genre *Pottia*. *Grimmia sessitana* De Not. et *G. anceps* Boul. Rev. Bryol. 12: 51–55, 94–95.

Venturi, G. 1894. *Desmatodon gasilieni*. Rev. Bryol. 21: 75–76.

Visch, M. 1976. *Tortula papillosa*. J. Canterbury Bot. Soc. 9: 21–24.

Visotska, E. I. 1967. [A survey of the chromosome numbers in mosses of the Ukrainian SSR.] Citologija i Genetika (Kiev) [Cytology and Genetics, Acad. Sci. Ukrainian SSR] 1(4): 30–39. [In Russian, English summary.]

Vital, D. M. 1984 [1985]. The occurrence of *Gymnostomiella orcuttii* Bartr. (Pottiaceae) in Brazil. Bryologist 87: 244–245.

Vitt, D. H. 1982. Bryopsida. In S. P. Parker (ed.), Synopsis and Classification of Living Organisms, Vol. 1. New York.

Vitt, D. H. 1984. Classification of the Bryopsida. In R. M. Schuster, ed., New Manual of Bryology, Hattori Bot. Lab., Nichinan. 2: 697–759.

Vitt, D. H. & R. H. Zander. 1978 [1979]. *Crumia deciduidentata*, new to Canada from the Canadian Arctic Archipelago. Bryologist 81: 608–609.

Vohra, J. N. & B. M. Wadhwa. 1966. *Merceyopsis angulosa* Broth. et Dix.—A new record for Western Himalayas. Bull. Bot. Surv. India 8: 355–356.

W

Waclawska, Z. 1958. *Tortula Velenovskyi* Schiffn. in Poland. Fragm. Flor. Geobot. 3: 181–185.

Wadhwa, B. M. & J. N. Vohra. 1963. Name change for *Didymodon obtusifolius* Card. ex Dix. et P. Vard. Curr. Sci. 32: 482–483.

Wagner, D. H. 1981. Pohlstoffe, a good wetting agent for bryophytes. Bryologist 84: 253.

Wallace, E. C. 1972. *Tortella inflexa* (Bruch) Broth. in England. J. Bryol. 7: 153–156.

Walther, K. 1983. Bryophytina. Laubmoose. In J. Gerloff & J. Poelt, A. Engler's Syllabus der Pflanzenfamilien 5(2). Gebrüder Borntraeger, Berlin.

Warburg, E. P. 1960. *Weissia levieri* (Limpr.) Kindb., new to the British Isles. Trans. Brit. Bryol. Soc. 3: 713–714.

Warburg, E. F. & A. C. Crundwell. 1959. *Tortula virescens* De Not. new to the British Isles. Trans. Brit. Bryol. Soc. 3: 568–570.

Warburg, E. F. & A. C. Crundwell. 1965. *Tortula vectensis*, a new species from the Isle of Wight. Trans. Brit. Bryol. Soc. 4: 763–476.

Wareham, R. T. 1939a. *Pottia*. In A. J. Grout, Moss Flora of North America North of Mexico. 1(4): 197–208.

Wareham, R. T. 1939b. The moss genera *Pottia* and *Pterigoneurum* [sic] in North America North of Mexico. Abstracts of Doctoral Dissertations, No. 30, pp. 203–216. Ohio State University Press, Athens, Ohio.

Wareham, R. T. 1939c. *Pterigoneurum* [sic]. In A. J. Grout, Moss Flora of North America North of Mexico. 1(4): 208–211.

Wareham, R. T. & J. B. Whitney, Jr. 1939. The annulus of *Pottia truncata* (Hedw.) Fürnr. Bryologist 42: 106–108.

Warnstorf, C. 1909. Bemerkungen zu *Williamsiella tricolor* E. Britton = *Williamsia tricolor* Broth. Hedwigia 61: 348–350.

Warnstorf, C. 1912. Der Formenkreis der *Tortula subulata* (L.) Hedw. und deren Verhältnis zu *Tortula mucronifolia* Schwgr. Hedwigia 52: 65–80.

Warnstorf, C. 1916. *Pottia-Studien* als Verarbeiten zu einer Monographie des Genus "*Pottia*" Ehrh. sens. str. Hedwigia 58: 35–152.

Waters, D. A., M. A. Buchheim, R. A. Dewey & R. L. Chapman. 1992. Preliminary inferences of the phylogeny of bryophytes from nuclear-encoded ribosomal RNA sequences. Am. J. Bot. 79: 459–466.

Watrous, L. E. & Q. D. Wheeler. 1981. The out-group comparison method of character analysis. Syst. Zool. 30: 1–11.

Watson, W. 1913. Xerophytic adaptations of bryophytes in relation to habitat. New Phytol. 13: 149–169, 181–190.

Wearmouth, S. M., S. C. Woolven & H. L. K. Whitehouse. 1984. Some observations on the occurrence of *Tortula papillosa* Wils. on a wall. J. Bryol. 13: 9–14.

Weber, J. 1887. *Didymodon subalpinus* (De Not.) Rev. Bryol. 14: 78.

Weber, W. A. 1957. *Tortula papillosa* in New Mexico. Bryologist 60: 143–145.

Weber, W. A. 1972. The identity of *Barbula pseudopilifera* (Musci: Pottiaceae). Lindbergia 1: 214–216.

Weber, W. A. 1979. *Aloina bifrons* (Musci: Pottiaceae) new to South America. Bryologist 82: 493.

Weber, W. A. & L. D. Simone. 1977. *Tetraphis pellucida* and *T. geniculata*: scindulae as diagnostic features in bryophytes. Bryologist 80: 166–167.

Weger, M. J. A. & H. J. During. 1989. A subterranean moss greenhouse in the Californian desert. Bryologist 92: 411–412.

Werner, J. 1982. Distribution and écologie de *Tortula inermis* (Brid.) Mont. et de *Didymodon cordatus* Jur. au Grand-Duché de Luxembourg. Dumortiera 24: 15–22.

Werner, J. 1989. Zum Vorkommen von *Tortula canescens* Mont. in Moseltal (Rheinland-Pfalz). Delattinia 20: 609–610.

Werner, J. 1990. About *Weissia squarrosa* (Nees & Hornsch.) C. Müll. new to the bryophyte flora of Luxembourg and some other new or rare bryophytes of Luxembourg. Mém. Soc. Roy. Belgique 12: 37–44.

Werner, J. & J. P. Hébrard. 1986. *Tortula caninervis* (Mitt.) Broth. subsp. *spuria* (Amann) W. Kramer var. *spuria* (Pottiaceae, Musci) a Touët-sur-var (Alpes maritimes), nouveau pour la bryoflore de France. Cryptogamie Bryol. Lichénol. 7: 487–493.

Wettstein, F. von. 1928. Morphologie und Physiologie des Formwechels der Moose auf genetischer Grundlage. Biblioth. Genet. 10: 1–216.

Whitehouse, H. L. K. 1961. The occurrence of *Tortula stanfordensis* Steere in Cornwall, new to Europe. Trans. Brit. Bryol. Soc. 4: 84–94.

Whitehouse, H. L. K. 1966. The occurrence of tubers in European mosses. Trans. Brit. Bryol. Soc. 5: 103–116.

Whitehouse, H. L. K. 1972. Distribution maps of bryophytes in Britain: *Tortula stanfordensis* Steere. Trans. Brit. Bryol. Soc. 7: 94.

Whitehouse, H. L. K. 1975. How fast is *Tortula stanfordensis* spreading in Britain? Bull. Brit. Bryol. Soc. 26: 23–24.

Whitehouse, H. L. K. 1980. The production of protonemal gemmae by mosses growing in deep shade. J. Bryol. 11: 133–138.

Whitehouse, H. L. K. 1985. Nineteeth-century finds of *Anoectangium warburgii* Crundw. & M. Hill. J. Bryol. 13: 333–335.

Whitehouse, H. L. K. & A. C. Crundwell. 1991. *Gymnostomum calcareum* Nees & Hornsch. and allied plants in Europe, North Africa and the Middle East. J. Bryol. 16: 561–579.

Whitehouse, H. L. K. & A. C. Crundwell. 1992. *Gymnostomum calcareum* Nees & Hornsch. and *G. viridulum* Brid. in Europe, North Africa and the Middle East. Bull. Brit. Bryol. Soc. 35–50.

Whitehouse, H. L. K. & H. J. During. 1987. *Leptobarbula berica* (De Not.) Schimp. in Belgium and The Netherlands. Lindbergia 12: 135–138.

Whitehouse, H. L. K. & M. E. Newton. 1988. *Tortula brevis sp. nov.* and *T. stanfordensis* Steere: morphology, cytology and geographical distribution. J. Bryol. 15: 83–99.

Whittier, H. O. & B. A. Whittier. 1974. List of mosses of southeastern Polynesia. Bryologist 77: 427–446.

Wiegers, J. & B. van Geel. 1983. The bryophyte *Tortella flavovirens* (Bruch) Broth. in late glacial sediments from Usselo (The Netherlands) and its significance as a palaeo-environmental indicator. Acta Bot. Neerl. 32: 431–436.

Wilberforce, P. W. 1970. Permanent preparation of bryophyte leaves. Bull. Kent Field Club 15: 53, 1970.

Wilczek, R. & F. Demaret. 1975. *Barbula rigidula* (Hedw.) Mild. et *Cirriphyllum crassinervum* (Tayl.) Loeske et Fleisch. dans le district flandrien. Dumortiera 2: 11–12.

Wilczek, R. & F. Demaret. 1976. Trois *Barbula* propagulifères de Belgique. Dumortiera 5: 1–3.

Wiley, E. O. 1980. Phylogenetic systematics and vicariance biogeography. Syst. Bot. 5: 194–220.

Wiley, E. O. 1981. Phylogenetics: The Theory and Practice of Phylogenetic Systematics. John Wiley and Sons, New York.

Williams, C. 1966a. *Pottia intermedia* new to North America. Bryologist 49: 235–236.

Williams, C. 1966b. A natural hybrid in the genus *Weissia*. Bryologist 69: 361–365.

Williams, H. 1959a. *Phascum floerkeanum* in Ontario. Bryologist 62: 55–57.

Williams, H. 1959b. *Pterygoneurum ovatum* in Ontario. Bryologist 62: 156–158.

Williams, H. 1961. *Pottia davalliana* in Ontario and elsewhere in North America. Bryologist 64: 47–50.

Williams, R. S. 1903. Bolivian mosses. Part 1. Bull. New York Bot. Gard. 3: 104–134.

Williams, R. S. 1913. The genus *Husnotiella* Cardot. Bryologist 16: 25.

Williams, R. S. 1919. The genus *Desmatodon* in North America. Bull. Torrey Bot. Club 46: 207–220.

Williams, R. S. 1921. *Hyophila subcucullata sp. nov*. Bryologist 24: 22–24.

Willis, A. J. 1954. Mosses new to western Australia. Victorian Naturalist 71: 8–12.

Willis, A. J. 1964. Investigations on the physiological ecology of *Tortula ruraliformis*. Trans. Brit. Bryol. Soc. 4: 668–683.

Wilson, P. 1991. How to make thin free-hand sections. Evansia 7(1): 16.

Wingo, S. M. 1936. The family Tortulaceae in Tennessee. Bryologist 39: 35–39.

Winter, H. 1908. *Trichostomum viridulum* Bruch auf Rugen. Verh. Bot. Ver. Berlin 50: 160.

Wright, C. H. 1907. *Splachnobryum delicatulum* Broth. (With a note on the systematic position of the genus.) Kew Bull. (London) 1907: 136–137.

Wyatt, R. 1985. Terminology for bryophyte sexuality: toward a unified system. Taxon 34: 420–425.

Wyatt, R., I. J. Odrzykoski & A. Stoneburner. 1992. Isozyme evidence of reticulate evolution in mosses: *Plagiomnium medium* is an allopolyploid of *P. ellipticum* × *P. insigne*. Syst. Bot. 17: 532–550.

Wyatt, R. & A. Stoneburner. 1982. Range extensions for some cryptogams from granite outcrops in Alabama. Bryologist 85: 405–409.

Y

Yang, B. Y. 1965. The ontogeny of the gemmae of *Hyophila tortula*. Taiwania 11: 35–40.

Z

Zander, R. H. 1967 [1968]. The New World distribution of *Scopelophila* (= *Merceya*). Bryologist 70: 405–413.

Zander, R. H. 1968. *Barbula inaequalifolia* Tayl. new to North America. Bryologist 71: 41–44.

Zander, R. H. 1972. Revision of the genus *Leptodontium* (Musci) in the New World. Bryologist 75: 213–280.

Zander, R. H. 1975. Some pottiaceous mosses from the Aegean Islands of Greece. Ann. Mus. Goulandris 2: 83–87.

Zander, R. H. 1976. Notes on Pottiaceae in Middle America. Bryologist 79: 227–231.

Zander, R. H. 1977a. *Rhabdoweisia crenulata* and *Erythrophyllopsis andina* from Colombia. Bryologist 80: 158–160.

Zander, R. H. 1977b. *Crossidium seriatum* found in the U. S. A. Bryologist 80: 170–171.

Zander, R. H. 1977c. The tribe Pleuroweisieae (Pottiaceae, Musci) in Middle America. Bryologist 80: 233–269.

Zander, R. H. 1978a. *Melophyllum* Herz. is a synonyum of *Bryomanginia* Thér. (Ditrichaceae). Bryologist 81: 332–333.

Zander, R. H. 1978b. *Didymodon luridus* Hornsch. in Spreng. replaces *D. trifarius* (Hedw.) Roehl. [= *Saelania glaucescens*

(Hedw.) Broth.] Bryologist 81: 421–423.

Zander, R. H. 1978c. *Tortula propagulosa* of the United States is *Rhachithecium perpusillum*. Bryologist 81: 458–460.

Zander, R. H. 1978d. A propaguliferous variant of *Oxystegus tenuirostris* in Europe, India, Mexico, Cuba and Brazil. Lindbergia 4: 285–288.

Zander, R. H. 1978e. New combinations in *Didymodon* (Musci) and a key to the taxa in North America north of Mexico. Phytologia 41: 11–32.

Zander, R. H. 1978f. Synopsis of the genus *Tuerckheimia* (Pottiaceae). Misc. Bryol. Lichenol. 8: 25–28.

Zander, R. H. 1978g [1979]. A synopsis of *Bryoerythrophyllum* and *Morinia* (Pottiaceae) in the New World. Bryologist 81: 539–560.

Zander, R. H. 1978h [1979]. *Didymodon sinuosus* in Alaska and a key to species of Pottiaceae with propaguloid leaf apices. Bryologist 81: 572–575.

Zander, R. H. 1979a. Regenerated herbarium material for biosystematics and cytology. Bryologist 82: 323.

Zander, R. H. 1979b. *Molendoa hornschuchiana* new to the Americas from the Pacific coast of Alaska. Bryologist 82: 487–489.

Zander, R. H. 1979c. Lectotypification of *Tortula propagulosa* Sharp. Bryologist 82: 510.

Zander, R. H. 1979d. Patterns of sporophyte maturation dates in the Pottiaceae (Bryopsida). Bryologist 82: 538–558.

Zander, R. H. 1979e. Validation of *Tortula herzogii* Zander. Bryologist 82: 631.

Zander, R. H. 1979f. Notes on *Barbula* and *Pseudocrossidium* (Bryopsida) in North America and an annotated key to the taxa. Phytologia 44: 177–214.

Zander, R. H. 1979g. Techniques for study of Pottiaceae. Taxon 28: 643–644.

Zander, R. H. 1980a. Acid-base color reactions: the status of *Triquetrella ferruginea*, *Barbula inaequalifolia* and *B. calcarea*. Bryologist 83: 228–233.

Zander, R. H. 1980b. Spread of *Leptodontium viticulosoides* (Bryopsida) after Balsam Woolly Aphid infestation of Fraser Fir. Bull. Torrey Bot. Club 107: 7–8.

Zander, R. H. 1981a. Descriptions and illustrations of *Barbula*, *Pseudocrossidium* and *Bryoerythrophyllum* (p.p.) of Mexico. Cryptog., Bryol. Lichénol. 2: 1–22.

Zander, R. H. 1981b. [Review of:] W. Kramer. *Tortula* Hedw. sect. *Rurales* De Not. (Pottiaceae, Musci) in der östlichen Holarktis.... Bryologist 84: 144–145.

Zander, R. H. 1981c [1982]. *Didymodon* (Pottiaceae) in Mexico and California: taxonomy and nomenclature of discontinuous and nondiscontinuous taxa. Cryptog., Bryol. Lichénol. 2: 379–422.

Zander, R. H. 1981d. Evaluation of early nomenclatural types of certain moss species. American Philosophical Society, Grantees' Reports 1980 January 1, 1980–December 31, 1980: 33–34.

Zander, R. H. 1982a. *Oxystegus tenuirostris* var. *stenocarpus* (Thér.) comb. nov. includes *Trichostomum spirale* Grout. Misc. Bryol. Lichénol. 9: 72–74.

Zander, R. H. 1982b [1983]. *Leptodontium recurvifolium* (Tayl.) Lindb. is an *Oxystegus*. Lindbergia 8: 185–187.

Zander, R. H. 1982c. *Trichostomum molariforme sp. nov.* and *Atractylocarpus stenocarpus* (Wils.) in Seem.) *comb. nov.* Bryologist 85: 126–128.

Zander, R. H. 1982d. Practical species concepts. Beih. Nov. Hedw. 71: 7–11.

Zander, R. H. 1982e. Various thoughts on characters in moss taxonomy. Beih. Nov. Hedw. 71: 81–85.

Zander, R. H. 1982f. Herbarium and cultivation methods in mosses. Beih. Nov. Hedw. 71: 127–130.

Zander, R. H. 1982g. Aspects of the taxonomy of the Pottiaceae. Beih. Nov. Hedw. 71: 225–227.

Zander, R. H. 1982h. The quality of floristic and taxonomic information. Beih. Nov. Hedw. 71: 291–296.

Zander, R. H. 1983a. The genus *Streptocalypta* C. Muell. (= *Barnesia* Card.) Lindbergia 8: 161–165.

Zander, R. H. 1983b. A rapid microscopic mounting medium for delicate bryophytes. Taxon 32: 618–620.

Zander, R. H. 1983c. A reevealuation of *Neohyophila* Crum (Pottiaceae). Bryologist 86: 134–139.

Zander, R. H. 1983d. Nomenclatural changes in *Hyophila*, *Leptodontium* and *Morinia* (Pottiaceae). Bryologist 86: 156–157.

Zander, R. H. 1983e. Pottiaceae (Musci) of Cofre de Perote, Veracruz, Mexico. Am. Philos. Soc. Grantees' Rep. 1982: 12–13.

Zander, R. H. 1984. Bryophyte sexual systems: -oicous versus -oecious. Bryol. Beitr. 3: 46–51.

Zander, R. H. 1985a. Nomenclatural transfers in *Weissia* (Pottiaceae, Musci). Monogr. Syst. Bot. Missouri Bot. Gard. 11: 195–197.

Zander, R. H. 1985b. An essay on species concepts developed during revisionary studies. Bryologist 88: 215–220.

Zander, R. H. 1986a. Notes on *Bryoerythrophyllum* (Musci). Bryologist 89: 13–16.

Zander, R. H. 1986b. Two new species of Pottiaceae from Mexico. Bryologist 88: 353–356.

Zander, R. H. 1986c. Three new taxa of bryophytes from Peru. Willdenowia 16: 253–258.

Zander, R. H. 1987a. Pottiaceae of North America North of Mexico. Flora Online (Taxacom, Buffalo Museum of Science) Issue 7, Ver. 1.0, 40,853 bytes, Jan. 1987; Ver. 1.1, 40,048 bytes, Mar. 1987.

Zander, R. H. 1987b. A bibliography of the Pottiaceae (Musci). Flora Online (Taxacom, Buffalo Museum of Science) Issue 8, Ver. 1.0., 150,651 bytes.

Zander, R. H. 1988. Sporophyte reduction series in Pottiaceae (Musci). Am. J. Bot. 75(6, Part 2): 12 (Abstr.).

Zander, R. H. 1989. Seven new genera in Pottiaceae (Musci) and a lectotype for *Syntrichia*. Phytologia 65: 424–436.

Zander, R. H. 1991. [Abstract:] A phylogenetic evaluation of the genera of the Pottiaceae (Musci). Am. J. Bot. 78(6, suppl.): 11–12.

Zander, R. H. 1993a. A new combination in *Pottia* (Musci). Novon 3: 92, 1993.

Zander, R. H. 1993b. A bibliography of the Pottiaceae (Musci). Flora Online (Taxacom, Buffalo Museum of Science) Issue 8, Ver. 2.0.

Zander, R. H. & H. Crum. 1977 [1978]. *Desmatodon steereanus*, a new species of Pottiaceae from Peru. Bryologist 80: 638–640.

Zander, R. H. & A. M. Cleef. 1982 [1983]. Studies on Colombian Cryptogams. XVI. Taxonomy and ecology of *Kingiobryum paramicola* (Dicranaceae, Musci). Proc. Kon. Ned. Akad. Wetensch., Ser.C 85: 627–634.

Zander, R. H. & C. Delgadillo M. 1984 [1985]. The mosses of the Tehuacán Valley, Mexico, and notes on their distribution. Bryologist 87: 319–322.

Zander, R. H., C. Delgadillo M. & P. M. Eckel. 1980 [1981]. *Morinia saitoana*, a new species of Pottiaceae (Bryopsida) from Mexico. Bryologist 83: 508–511.

Zander, R. H. & P. M. Eckel. 1980. *Tortula cainii:* additional Ontario records and behavior in a common garden. Bryologist 83: 209–211.

Zander, R. H. & P. M. Eckel. 1982. *Hymenostylium recurvirostrum* var. *insigne* and *Barbula amplexifolia* in British Columbia,

Canada. Canad. J. Bot. 60: 1596–1600.
Zander, R. H. & P. M. Eckel. 1987. A new species of *Tuerckheimia* from Mexico. Mem. New York Bot. Gard. 45: 293–295.
Zander, R. H. & E. Hegewald. 1976. *Leptodontiella*, gen. nov. and *Leptodontium* from Peru. Bryologist 79: 16–21.
Zander, R. H. & F. J. Hermann. 1986 [1987]. *Gyroweisia monterreia*, a new moss species from Mexico. Bryologist 89: 227–229.
Zander, R. H. & W. J. Hoe. 1979. Geographic disjunction and heterophylly in *Tortella fragilis* var. *tortelloides* (= *Sarconeurum tortelloides*). Bryologist 82: 84–87.
Zander, R. H. & A. J. Sharp. 1981 [1982]. Two species of *Bryoerythrophyllum* new to Mexico. Bryologist 84: 543–547.
Zander, R. H. & W. C. Steere. 1978. *Tortula scotteri sp. nov.* from the Northwest Territories of Canada. Bryologist 81: 463–467.
Zander, R. H. & D. H. Vitt. 1979. Gametophytic distinctions of *Zygodon* (Orthotrichaceae) and *Anoectangium* and *Leptodontium* (Pottiaceae) and the status of *Anoectangium rubrigemmium* of Hawaii. Canad. J. Bot. 57: 292–296.
Zijlstra, G. 1990. On conserved types. Bryol. Times 53: 8–9.
Zuttere, P. de, A. Gohimont, R. Schumacker, A. Sotiaux & J. Werner. 1984. *Weissia rostellata* (Brid.) Lindb. (Musci), nouveau pour la Belgique, Le Grande-Duché de Luxembourg et le département du Nord (France). Dumortiera 29/30: 15–19.

GLOSSARY OF SPECIAL TERMS

This glossary includes only those terms especially appropriate to study of the Pottiaceae. See Wyatt (1985) and Magill (1990) for extensive definitions of bryological terms.

abaxial: away from the axis; the surface of the leaf facing away from the stem, i.e. the dorsal surface.

adaxial: towards the axis; the surface of the leaf facing towards the stem, i.e. the ventral surface.

anisospory: spores of two size classes occurring in a single capsule.

autoicous: antheridia, surrounded by perigonial leaves, and archegonia, surrounded by perichaetial leaves, borne in separate locations on the same plant; here usually indicating the gonioautoicous condition, perigonia borne laterally on very short branches in the axils of leaves.

bulliform: bubble-shaped, e.g. laminal cells of *Quaesticula* or *Tuerckheimia*.

carinate: keeled, with upper halves of the lamina rather appressed, diverging at a small angle at the costa and vee-shaped in transverse section.

cee: the letter "c".

cleistocarpous: a capsule lacking a differentiated operculum when mature, usually dehiscing irregularly.

clockwise: twisted (setae, peristomes) in a clockwise direction, being seen as twisted in a clockwise direction upwards towards an observer above the plant. The few taxa with clockwise twisted peristomes (species of *Timmiella* and to a lesser extent *Leptodontiella*) would have the appearance of the thread on a left-handed screw. Synonym: dextrorse.

collenchymatous: laminal cells thickened at the corners, the thickenings having the appearance of small, bright triangles.

colliculate: covered with crowded, small, rounded protrusions, used here for laminae with strongly bulging upper laminal cells, these usually smooth individually but sometimes also papillose.

correct: as used here, that most recently set right.

counterclockwise: the usual direction of twist of a peristome in the Pottiaceae is counterclockwise as seen from above with the direction of twist rising upwards. Laterally, a counterclockwise-twisted peristome would have the appearance of the thread on a standard, right-handed screw (pers. comm. D. Wagner). Synonym: sinistrorse. Jackson's (1960) Glossary of Botanic Terms (Appendix) discusses past confusion in use of terms for direction of twisting.

distal: most distant, usually from the center or body of the plant.

dorsal: as used here meaning the abaxial surface of a leaf.

emergent: a sporophyte with a portion of its capsule extending just beyond the tips of the perichaetial leaves.

eperistomate: lacking a peristome, used for capsules with a differentiated operculum but no peristome; this is the latinized form but also spelled aperistomate. See cleistocarpous.

exerted: a sporophyte with the entire capsule visible above the tips of the perichaetial leaves.

flat: In a treatment of *Grimmia* (Grimmiaceae) Sayre (1952) defined this for the leaf as when the costa is in the same plane of focus as the laminal cells at 100×, or when leaves on a microscope slide lie mostly open rather than folded ("keeled"). In practice, "flat" may also refer to leaves that are broadly channeled across the leaf but which lack a narrow groove ventrally along the costa.

gametoecia: a moss inflorescence, consisting of several leaves usually somewhat modified in size and shape terminating a stem or branch, the leaves enclosing archegonia or antheridia or both, often also paraphyses.

guide cells: large, wide-lumened, thin-walled costal cells, usually 2–4 in number, just ventral to dorsal stereid band (and sometimes an associated hydroid strand). Synonyms: eurycysts, Deuters, socii (cf. Ligrone 1985).

hyalocyst: enlarged basal laminal cells sharply differentiated from upper cells by larger size and lack of chlorophyll.

immersed: a sporophyte (capsule) entirely hidden by the perichaetial leaves.

keeled: leaves with lateral portions of the upper lamina inserted on the costa at right angles or less (being vee-shaped in transverse section) such that the leaves generally lie on their sides (at least above midleaf) when moist.

ligulate: in bryological usage, strap-shaped or narrowly elliptical, as opposed to spathulate, which usually means obovate.

lumen: the interior of a cell, of various shapes in outline. Plural: either lumina or lumens is acceptable.

lower: usually refering to the portion of a structure proximal to the stem or to the base of a plant, e.g. the differentiated basal laminal cells in the lower portion of the leaf.

mamilla (or mammilla): a smooth, broad bulge; generally the whole cell surface, when mamillose, bulges to produce collectively a colliculate laminal surface, but sometimes (not

here) simple, hollow papillae are called mamillae.
molariform: like a millstone; a papilla with a broad, rounded or flattened surface that is often tuberculated.
multifid: having many branches (or, in the case of papillae, points or salients of various shapes).
oh: the letter "o".
optical section: the shallow focal plane of a high power microscope produces an image of a section of the tissue observed. Hollow, hemispherical laminal papillae thus may appear to be oh- or cee-shaped, when in fact only a section of the hemisphere is in focus. An upright dome gives an oh-shaped optical section, a tilted dome a cee-shaped section.
p.p.: *pro parte*, in part.
prorulose: paired papillae formed by superficial projecting ends of contiguous cells, e.g. as in *Barbula indica* (see also "scindulae" fide Weber & Simone 1977).
proximal: most near, generally to the center or body of the plant.
salients: branches or points on the papillose surface of a cell.
rhexolytic: rupturing irregularly; in this case, a capsule dehiscing without the aid of an operculum.
stereid: longitudinally elongate cells with much-thickened walls and small lumina, usually evident in transverse sections of the costa and stem. Substereid cells have rather wide lumina.
trifarious: leaves distinctly arranged in three longitudinal rows, generally spiralling around the stem, e.g. as found in *Hymenostylium, Molendoa, Reimersia* and *Triquetrella.*
twist: see clockwise and counterclockwise.
upper: usually refering to the portion of a structure distal to the stem or base of a plant.
vee: the letter "v"; in certain genera, basal laminal cells are strongly differentiated from the upper laminal cells by larger size and thinner walls, and arranged so as to reach higher along the margins than medially (along the costa), thus forming a vee-shaped wedge of usually sharp differentiation.
ventral: as used here meaning the adaxial surface of a leaf.

INDEX

Species names are included when they occur with treatments other than their own genera. Subfamilies, tribes, subgenera, sections, and subsections, including synonyms, are all listed as single words. Recognized genera of the Pottiaceae are set in boldface.

CORRIGENDUM

Because it is superfluous under I.C.B.N. Art. 63, the name *Tortula atherodes* Zand., given as a nomen novum on p. 222 and used through the text, is not acceptable and therefore invalid under Art. 34.1. All the new infraspecific combinations with this binomial, none of which are "alternative names" sensu Art. 34.3, are also invalid for the same reason. *Tortula acaulon*, below, is the correct name in *Tortula*, and correct infraspecific combinations are also provided below.

New combinations:

Tortula acaulon (L. ex With.) Zand., *comb. nov.* (*Phascum acaulon* L. ex With., Syst. Arr. Brit. Pl. ed. 4., 3: 768, 1901 "*acaule*.")

Tortula acaulon var. *arcuata* (Herrnstadt & Heyn) Zand., *comb. nov.* (*Phascum cuspidatum* var. *arcuatum* Herrnstadt & Heyn, Bryologist 94: 175, 1991).

Tortula acaulon var. *affinis* (Nees & Hornsch.) Zand., *comb. nov.* (*Phascum affine* Nees & Hornsch., Bryol. Germ. 1: 74, 1823; *Phascum cuspidatum* var. *affine* (Nees & Hornsch.) Hampe).

Tortula acaulon var. *curviseta* (Dicks.) Zand., *comb. nov.* (*Phascum curvisetum* Dicks., Pl. Crypt. Brit. 4: 2, 1801; *Phascum cuspidatum* var. *curvisetum* (Dicks.) Nees & Hornsch.).

Tortula acaulon var. *diaphora* (Hag.) Zand., *comb. nov.* (*Phascum acaulon* var. *diaphorum* Hag., K. Norsk. Vid. Selsk. Skrift. 1928(3): 19, 1929; *Phascum cuspidatum* var. *diaphorum* (Hag.) C. Jens.).

Tortula acaulon var. *elata* (Brid.) Zand., *comb. nov.* (*Phascum elatum* Brid., J. Bot. (Schrader) 1800(1): 269, 1801).

Tortula acaulon var. *intertexta* (Brid.) Zand., *comb. nov.* (*Phascum intertextum* Brid., Mant. Musc. 8, 1819.)

Tortula acaulon var. *marginata* (Hernnstadt & Heyn) Zand., *comb. nov.* (*Phascum cuspidatum* var. *marginatum* Hernnstadt & Heyn, Bryologist 94: 175, 1991).

Tortula acaulon var. *mitraeformis* (Limpr.) Zand., *comb. nov.* (*Phascum cuspidatum* var. *mitraeforme* Limpr., Laubm. Deutschl. 1: 187, 1885).

Tortula acaulon var. *papillosa* (Lindb.) Zand., *comb. nov.* (*Phascum papillosum* Lindb., Oefv. K. Vet. Ak. Foerh. 21: 217, 1864; *Phascum cuspidatum* var. *papillosum* (Lindb.) Roth).

Tortula acaulon var. *pilifera* (Hedw.) Zand., *comb. nov.* (*Phascum piliferum* Scherb. ex Hedw., Spec. Musc. 20, 1801; *Phascum cuspidatum* var. *piliferum* Scherb. ex Hedw.) Hook. & Tayl.).

Tortula acaulon var. *retortifolia* (Guerra & Ros in Guerra, Jiménez, Ros & Carrión) Zand., *comb. nov.* (*Phascum cuspidatum* var. *retortifolium* Guerra & Ros in Guerra, Jiménez, Ros & Carrión, Cryptogamie, Bryol. Lichénol. 12: 390, 1991).

Tortula acaulon var. *schreberiana* (Dicks.) Zand., *comb. nov.* (*Phascum schreberianum* Dicks., Pl. Crypt. Brit. 4: 2, 1801).